Page	Amoeboid	Flagellated	Ciliated	Sporozoid	Some Pig-mented Forms s = symbionts
14		■			
39				■	
127				■	
148		■			■
190				■	
370		■			
371			■		S
656	■	■			■
690					
693	■	■			■
733		■			
744		■			■
751		■			■
754		■			
759		■			■
775	■	■			■
783	■				S
804	■				
827	■				S
860	■				S
872	■				S
952	■	■			
981	■				
994	■				S
1023	■				
1053	■				
1054	■				
1084	■				
1086	■	■			
1097	■	■			
1104	■	■			
1111		■			■
1125		■			
1135		■			■
1185		■			
1186		■			
1196		■			
731		■			■
1253		■			■
1268		■			■
1302		■			
1250		■			
1328				■	
1342					

KEY TO MAJOR GROUPS
(see page 10)

HOW THIS BOOK IS ORGANIZED
(see page 1)

AN ILLUSTRATED GUIDE TO THE PROTOZOA

SECOND EDITION

ORGANISMS TRADITIONALLY REFERRED TO AS PROTOZOA, OR NEWLY DISCOVERED GROUPS

EDITED BY

JOHN J. LEE

CITY COLLEGE OF CITY UNIVERSITY OF NEW YORK

AND

AMERICAN MUSEUM OF NATURAL HISTORY

GORDON F. LEEDALE

UNIVERSITY OF LEEDS

PHYLLIS BRADBURY

NORTH CAROLINA STATE UNIVERSITY

Volume II

(Pages 690- 1432)

SOCIETY OF PROTOZOOLOGISTS

P. O. BOX 368, LAWRENCE KANSAS 66044, U. S. A.

Cover design by Dean Jacobson
WHITWORTH COLLEGE
SPOKANE, WASHINGTON

ISBN No. 1-891276-23-9

Printed by Allen Press Inc., Lawrence, KS 66044 USA

AUTHORS

Robert A. Andersen
CCMP, Bigelow Laboratory for Ocean Sciences
West Boothbay Harbor, Maine

O. Roger Anderson
Biology Department
Lamont-Doherty Geological Observatory
Columbia University
New York, NY

Fred T. Banner
Department of Geological Sciences
University College London
London

John R. Barta
Pathobiology Department
University of Guelph
Guelph, Ontario
Canada

Demitrio Boltovskoy
Departmento de Ciencias Biólgicas
Universidad de Buenos Aires
Buenos Aires, Argentinia

Phyllis Bradbury
Department of Zoology
North Carolina State University
Raleigh, North Carolina
(Retired)

Guy Brugerolle
Laboratoire de Biologie des Protistes
Université Blaise Pascal de Clermont-Ferrand
Clermont-Ferrand, France

James P. Braselton
Dept of Environmental and Plant Biology
Ohio University
Athens, Ohio

Elizabeth U. Canning
Department of Pure and Applied Biology
Imperial College of Science, Technology and Medicine
Sillwood Park, U. K.

Tomas Cedhagen
University of Aarhus
Institute of Biological Sciences
Department of Marine Ecology
Århus, Denmark

Richard E. Clopton
Peru State College
Peru, Nebraska

Jean-Pierre Debenay
Laboratorie de Geologie
University of Angers
Angers, France

Ben L. J. Delvinquier
BP 325, 98820 Wé
Lifu Island,New Caledonia,South Pacific

John D. Dodge
Department of Botany,
Royal Holloway College
Huntersdale, U. K.
(Retired)

Michael J. Dykstra
Department of Microbiology, Pathology, and Parasitology
College of Veterinary Medicine
North Carolina State University
Raleigh, North Carolina

Walter Faber
Department of Biology
Manhattan College
New York, NY

AUTHORS

Collette Febvre
Laboratorie de Protistologie,
Station Zoologie,
Ville Franche-sur-Mer, France

Jean Febvre
Laboratorie de Protistiologie,
Station Zoologie,
Ville Franche-sur-Mer, France

Ilse Foissner
Institut für Pflauzenphysiologie
Universitüt Salzburg
Salzburg, Austria

Wilhelm Foissner
Institut für Zoologie
Universitüt Salzburg
Salzburg, Austria

Andrew J. Gooday
Institute of Oceanographic Sciences
Wormley, U. K.

John Green
Plymouth Marine Laboratory
Plymouth, U. K.
(Retired)

John R. Haynes
Department of Geology
University of Wales
Aberystywyth, U. K.

Peter Heywood
Division of Biology and Medicine
Brown University
Providence, Rhode Island

David R. A. Hill
Biology Department
Belmont University
Nashville, Tenessee

Harold W. Keller
Botanical Research Institute of Texas
Fort Worth, Texas

Michael L. Kent
Pacific Biological Station
Department of Fisheries and Oceans
Nanaimo, B.C. Canada

Paul Kugrens
Department of Biology
Colorado State University
Fort Collins, Colorado

Barry S. C. Leadbeater
School of Biological Sciences
University of Birmingham
Birmingham, U. K.

John J Lee
Biology Department
City College of City University of New York
New York, N.Y.

Robert E. Lee
Department of Biology
Colorado State University
Fort Collins, Colorado

Gordon F. Leedale
Department of Biology
University of Leeds
Leeds, U.K.
(Retired)

Denis H. Lynn
Department of Zoology
University of Guelph
Guelph, Canada

AUTHORS

Stefan Mattson
University of Arhus
Institute of Biological Sciences
Department of Marine Ecology
Århus N, Denmark

Ralph Meisterfeld
Institute of Biology II (Zoology)
RWTH- Aachen
Aachen, Germany

Michael Melkonian
Botanisches Institüt I
Universität zu Köln
Köln, Germany

Antony Michaels
Bermuda Biological Station,
Saint George, Bermuda

K. A. Mikrjukov (Deceased)
Department of Zoology and Comparative Anatomy of Invertebrates
Moscow State University
Moscow, Russia

Michael Moser
Department of Integrative Biology
University of California, Berkeley
Berkeley, California

Catherine Nigrini
161 Morris St.
Canmore, Alberta
Canada, T1W 2R7

Charles O'Kelly
CCMP, Bigelow Laboratory for Ocean Sciences
West Boothbay Harbor, Maine

David J. Patterson
School of Biological Sciences
University of Sydney
Sydney, Australia

Jan Pawlowski
Station Zoologie
Université de Genève
Geneva, Switzerland

Michael A. Peirce
MP International Consultancy
Wokingham, Berkshire
United Kingdom

Frank O. Perkins
Assistant Vice President for Research and Graduate Education
University of Hawaii at Manoa
Honollulu, Hawaii

Hans R. Preisig
Institüt fur Systematische Botanik
Universitüt Zürich
Zürich, Switzerland

Andrew Rogerson
NOVA Southeastern University
Oceanographic Center
Dania, Florida

F. L. Schuster
Department of Biology
Brooklyn College, CUNY,
Brooklyn, New York
(Retired)

F. J. Siemensma
Laboratory of Hydrobiology
Agricultural University
Wageningen, The Netherlands

AUTHORS

Alastair Simpson
Protist Research Laboratory
University of Sydney
Sydney, Australia
(now in Halifax)

Eugene B. Small
Department of Zoology
University of Maryland
College Park, Maryland

Frederick Spiegel
Department of Biological Sciences
University of Arkansas
Fayetteville, Arkansas

Neil R. Swanberg
IGBP Secretariat
Royal Swedish Academy of Sciences
Stockholm, Sweden

Kozo Takahashi
Department of Earth and Planetary Sciences
Kyushu University, Fukuoka, Japan

Ole Tendal
Zoological Museum
University of Copenhagen
Copenhagen, Denmark

Helge A. Thompsen
Botanical Institute
University of Copenhagen
Copenhagen, Denmark

Keith Vickerman
Zoology Department,
University of Glasgow
Glasgow, U. K.

Steve J. Upton
Division of Biology
Kansas State University
Manhattan, Kansas

Jiri Vavra
Department of Parasitology
Charles University
Prague, Chech Republic

Naja Vørs (Deceased)
School of Biological Sciences
University of Sydney
Sydney, Australia

John E. Whittaker
Department of Paleontology
British Museum (Nat. Hist.)
London

TABLE OF CONTENTS

*Groups listed alphabetically within each catagory

ORDER BICOSOECIDA
GRASSÉ 1926

By ØJVIND MOESTRUP

The bicosoecids form a well-defined group of heterotrophic heterokont flagellates. Recent studies on 18S rRNA (Leipe et al., 1995) have shown that *Cafeteria*, a primitive bicosoecid, is one of the most primitive heterokont protists and precedes both the oomycetes and the heterokont algae. Bicosoecids (sometimes erroneously spelled bicoecids or bikocoecids) are phagotrophic flagellates in fresh water and marine environments. Only three genera are known, comprising c. 45 described species. Most species are unicells but some form colonies. The great majority belongs to the genus *Bicosoeca*. Bicosoecids are biflagellate and the movements of the anterior hairy flagellum draws the swimming cell forward in the water. The smooth flagellum is usually directed posteriorly and may be used for attachment (Fig. 1). When attached, the movements of the hairy flagellum create a water current that draws water and food particles towards the peristome area of the cell (Figs. 1A, B). This area is supported by a broad band of microtubules, now known to be homologous with a flagellar root that serves in food uptake in other hetero- and mixotrophic heterokonts (Moestrup, 1995). In the light microscope, the peristome area may be visible as a lip-like structure. Cells of *Bicosoeca* are loricate and the cells attach to the inside bottom of the lorica with the smooth flagellum tip. *Cafeteria* and *Pseudobodo are* naked and freeswimming but may attach transiently to various surfaces, including the water film.

Key characters

1. Cells are biflagellate. The flagellar apparatus is diagnostic for the group and is described above.
2. Cells are phagotrophic and lack all traces of chloroplast, indicating that the ancestors of bicosoecids were heterotrophic rather than auto- or mixotrophic protists (comp. *Paraphysomonas*). Cells feed on bacteria or other eukaryotes.
3. Cells lack ejectile organelles. Freshwater species possess a contractile vacuole system.
4. The cell is surrounded by a lorica or naked (Figs 1-4).
5. Reproduction is by fission. Sexual reproduction is unknown.
6. Freshwater or marine species, very widely distributed, some species occur in tropical as well as temperate areas.
7. Diagnostic features used in species identification are cell size, shape, lorica shape and texture, single or colony-forming cells.

Bicosoecids were classified by Moestrup (1995) into two families, Bicosoecaceae and Cafeteriaceae, depending on the presence or absence of a lorica.

Key to the Genera

1. Cells naked, not surrounded by a lorica. 2
1´. Each cell living in a lorica.... ***Bicosoeca***

2. Smooth posterior flagellum attached to the surface directly.............. ***Cafeteria***
2´. A thin thread of mucus extends from the tip of the posterior flagellum to the substrate .... ***Pseudobodo***

Genus ***Bicosoeca*** James-Clark, 1866

Colorless cells attached by the tip of the posterior flagellum to the bottom of an organic lorica of various shapes and textures. Some lorica types are very thin and almost transparent, whilst others are thicker and variously ornamented. They may be stalked (Figs. 1A-D, 2) or unstalked (Fig. 1E), and species may be free-swimming or attached, e.g. attached to diatoms or other protists.

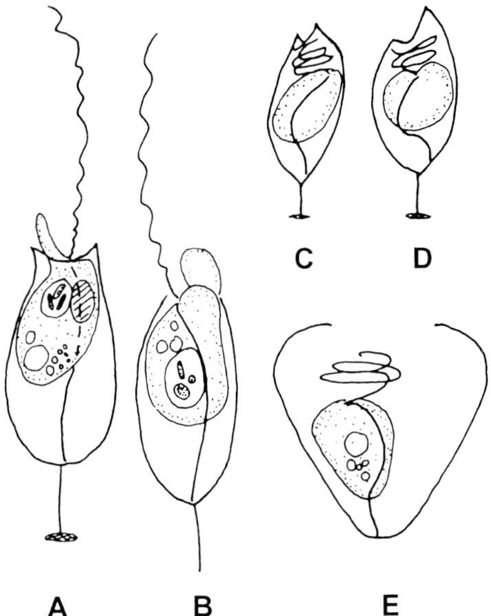

Fig. 1. *Bicosoeca* spp. A-D. *B. mignotii.* A,B. Undisturbed cells. C,D, Cells that have withdrawn into the cell with the hairy flagellum coiled up. E. *B. planctonica,* a planktonic species with stalkless lorica. x2000. (from Vørs, 1992)

Cell shape more or less rounded. The front flagellum is directed out of the lorica when the cell is undisturbed or swimming. When disturbed, the flagellum retracts and coils up rapidly on the anterior end of the cell (Figs 1C-E), and the cell withdraws to the bottom of the lorica. In some species, the daughter cells settle on the parent lorica and colonies develop. Species of *Bicosoeca* phagocytize bacteria or eukaryotic algae (Moestrup & Thomsen 1976). Nearly 40 known species, mostly in fresh water. See Preisig et al. (1991).

Bicosoeca mignotii Moestrup, Thomsen & Hibberd (Fig. 1A-E). Cells 3-4 µm long and 5-7 µm wide. Lorica elongate ovoid, transparent and delicate. Lorica chamber 10-15 µm long, 5-7 µm wide, with a 5-10 µm-long stalk that attaches to a substrate by a small disc. Anterior end of the lorica divided into two opposing beak-like parts. Lorica apparently amorphous, the stalk fibrous. Very common freshwater species, in the plankton or attached. Known from the inner Baltic, Denmark, and England but almost certainly much more widely distributed (Vørs 1992).

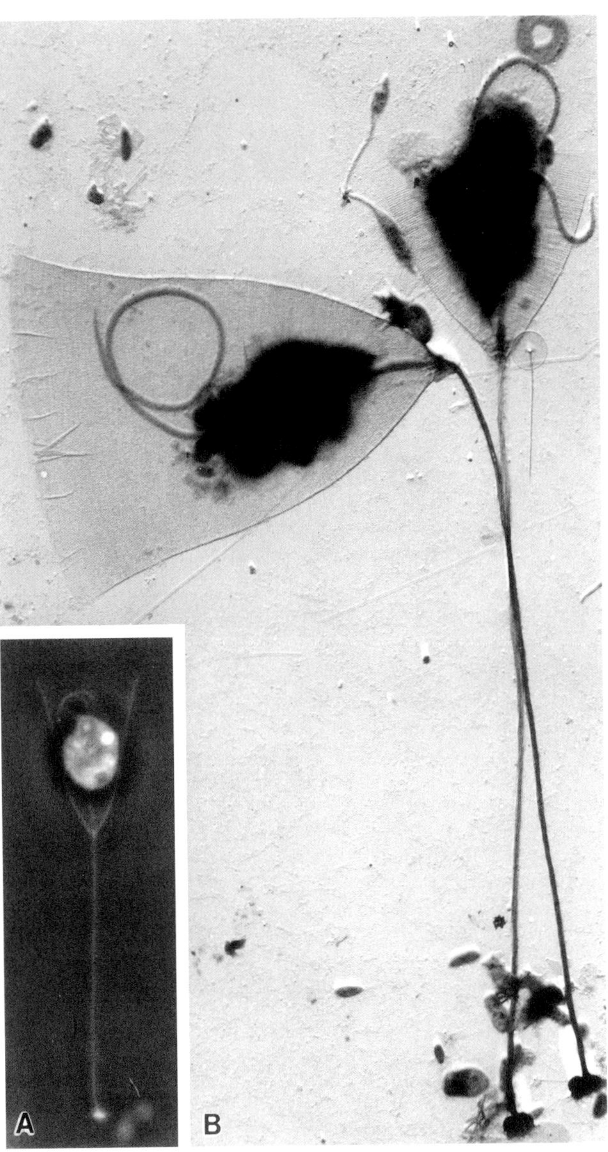

Fig. 2. *B. petiolata,* a species with very long-stalked, bell-shaped lorica. The cell on the right in Fig. 2B has only half completed formation of its lorica. A, anoptral contrast, x1000. B. shadowcast whole mount, x27,000. (Original photographs courtesy of H. A. Thomsen).

Genus ***Cafeteria***
Fenchel and Patterson, 1988

Colorless cells with the two flagella inserted subapically (Fenchel and Patterson 1988). Cells may swim with the hairy flagellum projecting forwards, or they may attach with the tip of the smooth flagellum. Four described species from temperate and tropical plankton and marine sediments. One species with a distinct furrow used in food uptake (Larsen and Patterson,1990.) See Larsen and Patterson, (1990) for a description of all four species).

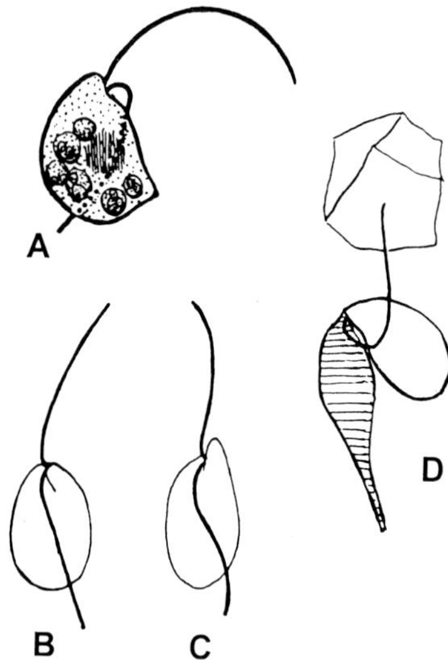

Fig. 3. *Cafeteria roenbergensis.* A. Feeding cell. B.C. Swimming cells. D. Beat envelope of anterior flagellum. x3000. (From Larsen and Patterson, 1990)

Cafeteria roenbergensis Fenchel and Patterson (Fig. 3). Cells D-shaped, cordiform or spherical, sometimes distorted by ingested food particles, mostly 3-6 µm long, compressed laterally. Larger cells occur occasionally (Larsen & Patterson 1990). Anterior flagellum 5-8 µm long, posterior flagellum c. 5 µm. The posterior flagellum passes in a shallow groove and the tip of the flagellum is often attached to a piece of debris or other surface when undisturbed. In swimming cells, the posterior flagellum trails loosely behind the cell. *Cafeteria* is easily grown in culture on a diet of bacteria. It is widely distributed and probably generally overlooked or misidentified.

Genus ***Pseudobodo*** Griessmann, 1913

As *Cafeteria*, but cells may attach to a substrate by a long thin thread of mucus which extends from the tip of the smooth flagellum (Larsen and Patterson, 1990). Swimming cells are hardly distinguishable from *Cafeteria*. Two species, both marine. In plankton and marine sediments. Distributed in both tropical and temperate areas.

Pseudobodo tremulans Griessmann (Fig. 4). Cell body rounded, 2.5-5.5 µm in diameter. Anterior flagellum 10-16 µm long, the smooth posterior flagellum 5-10 µm.

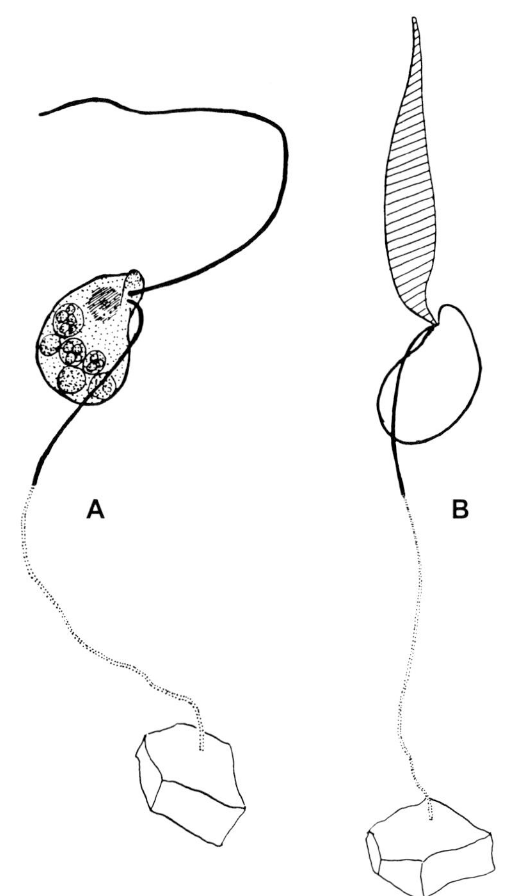

Fig. 4. *Psudobodo tremulans.* A. General aspect. B. Beat profile of hairy flagellum. x3000. (From Larsen and Patterson, 1990)

LITERATURE CITED

Fenchel, T. & Patterson, D. J. 1988. *Cafeteria roenbergensis* nov. gen., nov. sp., a heterotrophic microflagellate from marine plankton. *Mar. Microb. Food Webs*, 3:9-19.

Larsen, J. & Patterson, D. J. 1990. Some flagellates (Protista) from tropical marine sediments. *J. Nat. Hist.*, 24:801-937.

Leipe, D. D., Wainright, P. O., Gunderson, J. H., Porter, D., Patterson, D. J., Valois, F., Himmerich, S., Sogin, M. L. 1994. The stramenopiles from a molecular perspective: 16S-like rRNA sequences from *Labyrinthuloides minuta* and *Cafeteria roenbergensis*. *Phycologia*, 33:369-377.

Moestrup, Ø. 1995. Current status of chrysophyte, 'splinter groups: synuro-phytes, pedinellids, silicoflagellates, *In*: Sandgren, C. D., Smol, J. P., & Kristiansen, J. (eds)., *Chrysophyte Algae. Ecology, Phylogeny and Development,* Cambridge University Press.

Preisig, H. R., Vørs, N. & Hällfors, G. 1991. Diversity of heterotrophic heterokont flagellates, *In* Patterson D. J. & Larsen, J. (eds.), *The Biology of Freeliving Heterotrophic Flagellates, The Systematics Assoc. Spec. Vol.* 45, Clarendon Press, Oxford, pp361-399

Moestrup, Ø, and Thomsen H. A. 1976. Fine structural, studies on the genus *Bicoeca*. 1. *Bicoeca maris* with particular emphasis on the flagellar apparatus. *Protistologica*, 12: 101-120.

Vørs, N. 1992. Heterotrophic amoebae, flagellates and heliozoa from the Tvärminne area, Gulf of Finland, in 1988-1990. *Ophelia*, 36:1-109.

CHRYSOMONADA (CLASS CHRYSOPHYCEAE PASCHER, 1914)

HANS R. PREISIG and ROBERT A. ANDERSEN

This class is circumscribed here in the narrow sense, i.e. excluding organisms previously often assigned to the Chrysomonada but now referred to the Bicosoecida Grassé, 1926, Synurophyceae Andersen, 1987, Pelagophyceae Andersen & Saunders in Andersen et al., 1993, Chrysomerophyceae Cavalier-Smith in Cavalier-Smith et al., 1995, Phaeothamniophyceae Andersen & Bailey in Bailey et al., 1998, Silicoflagellata Lemmermann, 1901 (orders Dictyochales Haeckel, 1894/Pedinellales Zimmermann, Moestrup & Hällfors, 1984/Rhizochromulinales O'Kelly & Wujek, 1995), and Prymnesiophyceae Hibberd, 1976/Haptophyceae Christensen ex Silva, 1980 (=Prymnesiida/Haptomonadida). The chrysomonads comprise mainly single-celled and colonial flagellates, but amoeboid (rhizopodial), **plasmodial**, **palmelloid** (**capsoid**), coccoid, and **parenchymatous** forms also occur. In this chapter only genera with predominantly flagellate, amoeboid and plasmodial life forms are treated. About 100 such genera (including some 800 species) have been described, the 27 better known of which are surveyed in some detail below. The lesser known genera are only briefly indicated thereafter and have not been included in the "Artificial Key to Selected Genera".

Chrysomonads are mainly freshwater organisms, with only a few species reported from marine environments and a few from soil and snow habitats. They live in a wide range of freshwater environments from polar regions to the tropics, but most commonly occur in cold temperate lakes, ponds and bogs. Some species can form blooms causing taste and odor problems (see Nicholls, 1995).

Present knowledge on the Chrysomonada (Chrysophyceae) is summarized by Hibberd (1976), Pienaar (1980), Hibberd and Leedale (1985) and Kristiansen (1986, 1990). Detailed data from current research can also be found in the Proceedings of the International Chrysophyte Symposia (Kristiansen and Andersen, 1986; Kristiansen et al., 1989; Sandgren et al., 1995; Kristiansen and Cronberg, 1996). The most complete taxonomic account is that of Starmach (1985). See also Pascher (1913), Huber-Pestalozzi (1941), and Bourrelly (1957, 1968/1981). The heterotrophic taxa are reviewed in Preisig et al. (1991). A comprehensive review on ecology of planktonic chrysomonad flagellates, especially focusing on growth and perennation strategies, has been given by Sandgren (1988). Current knowledge concerning the usefulness of chrysomonads as paleoecological indicators has been summarized by Smol (1995) and Zeeb et al. (1996) (see also Sandgren, 1991). Studies on the physiology and biochemistry of chrysomonads have been done mostly using *Ochromonas danica* Pringsheim, 1955 and *Poterioochromonas malhamensis* (Pringsheim, 1952) Péterfi, 1969 (see Loeblich and Loeblich, 1978; Chen et al., 1994; Eichenberger and Gribi, 1994). For recent data from molecular biology, see Bhattacharya et al. (1992), Delaney et al. (1995), Saunders et al. (1995), Daugbjerg and Andersen (1997), Medlin et al. (1997), Saunders et al. (1997a,b), Van der Auwera and De Wachter (1997), Andersen et al. (1998a,b, 1999) and Bailey et al. (1998).

The phylogenetic relationships of the Chrysophyceae have received extensive study in the past few years. Studies based upon the 18S rRNA gene sequences show the Chrysophyceae and Synurophyceae are closely related, and the Eustigmatophyceae Hibberd & Leedale, 1971 are a sister taxon to these two classes (e.g., Bhattacharya et al., 1992; Andersen et al., 1993; Leipe et al., 1994; Saunders et al., 1995, 1997a,b; Van de Peer et al., 1996; Andersen et al., 1998a,b, 1999). Studies based upon *rbcL* gene sequences also show a close relationship between the Chrysophyceae and Synurophyceae; however, the sister taxon to the Chrysophyceae/Synurophyceae clade is the Bacillariophyceae Haeckel, 1878 in one analysis (Daugbjerg and Andersen, 1997) and is unresolved in a second analysis (Bailey et al., 1998).

A recent scheme of classification of the Chrysomonada (Chrysophyceae) has been published by Preisig (1995). Six orders were recognized, namely Bicosoecales Grassé, 1926; Chromulinales Pascher, 1910b; Hibberdiales Andersen, 1989a; Hydrurales Pascher, 1931; Sarcinochrysidales Gayral & Billard, 1977; and Chrysomeridales *nom. nud.* (O'Kelly, 1989). However, the order Bicosoecales (Bicosoecida, see chapter by Moestrup) have now been placed in a separate class in view of recent results of rRNA sequencing (Leipe et al., 1994); the Sarcinochrysidales have been transferred to the Pelagophyceae based upon gene sequence, ultrastructural, and pigment data (Saunders et al., 1997b); the Chrysomeridales have been raised to class level without explanation (Chrysomerophyceae Cavalier-Smith in Cavalier-Smith et al., 1995); thirteen coccoid, capsoid, and filamentous genera (e.g. *Phaeogloea* Chodat, 1922; *Phaeoschizochlamys* Lemmermann, 1898; *Phaeothamnion* Lagerheim, 1884; *Stichogloea* Chodat, 1897; *Tetrasporopsis* [De Toni, 1895] Lemmermann, 1899) have been placed in a new class, the Phaeothamniophyceae Andersen & Bailey in Bailey et al., 1998, based upon gene sequence, ultrastructural, and pigment data (Andersen et al., 1998b; Bailey et al., 1998).

All genera described below have been classified in the Chromulinales and the Hibberdiales. The Chromulinales (including Ochromonadales Pascher, 1910b) have been subdivided in ten families, four of which comprising flagellate forms: (1) Chromulinaceae Engler, 1897 (including Ochromonadaceae Lemmermann, 1899) (naked cells), (2) Dinobryaceae Ehrenberg, 1834 (loricate cells), (3) Paraphysomonadaceae Preisig & Hibberd, 1983 (silica-scaled cells), and (4) Chrysolepidomonadaceae Peters & Andersen, 1993a (cells covered with unique **canistrate and dendritic scales** of organic composition). Family (5) Chrysamoebaceae Poche, 1913, comprises naked amoeboid (rhizopodial) forms and family (6) Myxochrysidaceae Pascher, 1931 **plasmodial** forms. The latter family includes a single genus and species, *Myxochrysis paradoxa* Pascher, 1916b, which is known only from the original description (not described under the well-known genera below; see Pascher, 1916b). The remaining four families, (7) Chrysocapsaceae Pascher, 1912b, (8) Stichogloeaceae Lemmermann, 1899 (9) Phaeothamniaceae Hansgirg, 1886, and (10) Chrysothallaceae Huber-Pestalozzi, 1941 consist of forms which spend most of their life history as palmelloid (capsoid), coccoid, filamentous or parenchymatous stages, respectively. Many of the genera placed in the Stichogloeaceae and Phaeothamniaceae are now considered to belong to a separate class, Phaeothamniophyceae (Bailey et al., 1998).

The order Hibberdiales was established by Andersen (1989a) for the monotypic genus *Hibberdia* Andersen (q.v.) which clearly differs from other chrysomonads in the structure of the flagellar apparatus, in addition to having a unique pigment content. Subsequent studies on a species of *Chromophyton* (q.v.; Andersen, unpublished) and *Lagynion* (q.v.; O'Kelly and Wujek, 1995) revealed a similar general organization of the flagellar apparatus, suggesting that these and possibly a number of other genera should also be classified with the Hibberdiales (e.g. loricate rhizopodial genera such as *Chrysopyxis* [q.v.] and *Stylococcus*, which are normally classified with *Lagynion* in the family Stylococcaceae Lemmermann, 1899).

In the following survey the genera are arranged alphabetically and not grouped in orders and families, as details on cell ultrastructure, including the flagellar apparatus, and also molecular data are known only from very few species at present. It is likely that the classification presented above will be superseded as more detailed information becomes available.

KEY CHARACTERS

1. Most species are free-swimming flagellates or motile colonies for much of their life histories, but amoeboid (rhizopodial), plasmodial, palmelloid (capsoid), coccoid, and **parenchymatous forms** also occur.

2. Flagellate cells have two obliquely inserted **heterodynamic** flagella of unequal length; the longer (immature) flagellum is directed anteriorly, beating in a sine wave; the shorter (mature) flagellum is directed laterally when emergent (in several genera it is vestigial). The long flagellum bears tripartite tubular hairs ("**mastigonemes**"; Figs. 4, 5, 20, 25); the short flagellum is smooth and often has a photoreceptor-related swelling at the base (Fig. 17). The basal body of the short flagellum usually lies at an angle of approximately 90° with respect to the long flagellum, but narrower or wider angles also occur (e.g. ca. 30° in *Chrysonebula* [Hydrurales], 160° in *Hibberdia* [Hibberdiales]). The basal bodies are interconnected by fibrous bands. A typically cross-striated fibrous root (**rhizoplast**) is normally present; in most chrysomonads it forms a connection between the basal apparatus and the nucleus (in *Chrysosphaerella* it is directed towards the plasma membrane and attaches to a mitochondrion). There are typically four microtubular roots (R_1,R_2,R_3,R_4), but several species are known in which some of these roots are reduced or lacking. Roots R_3 and R_4 often form a loop under the shorter flagellum (no such loop occurs in Hibberdiales and Hydrurales). From the surface of some roots (i.e. R_1, R_3) cytoplasmic microtubules can be nucleated, providing cytoskeletal support to shape the cell. For more information on the flagellar apparatus, see e.g. Andersen (1989b, 1991), Preisig (1989). For details of flagellar transformation, see genus *Epipyxis* in this chapter (also Wetherbee et al., 1988).

3. Pigmented forms usually have one or two (rarely more) parietal chloroplasts, with or without pyrenoids; pyrenoids are often immersed and barely detectable with the light microscope (Fig. 4). The chloroplast is surrounded by two membranes of endoplasmic reticulum that is normally confluent with the outer membrane of the nuclear envelope (e.g. Fig. 17). The chloroplast lamellae typically have three thylakoids; a girdle lamella is usually present and the pyrenoid matrix is usually free of thylakoids. There are also several colorless genera (see Preisig et al., 1991), but these often contain a vestigial chloroplast (**leucoplast**; see Figs. 1 and 25).

4. Chloroplasts are yellow-brown or yellow-green in the pigmented forms, normally containing chlorophylls *a* and *c* (both c_1 and c_2; Andersen and Mulkey, 1983; Jeffrey, 1989). One unusual species, *Ochromonas danica*, has been reported to produce chlorophylls c_1 and c_2 (Andersen and Mulkey, 1983), but several other workers failed to detect the presence of any chlorophyll *c* pigments (Allen et al., 1960; Gibbs, 1962; Ricketts, 1965; Grevby and Sundqvist, 1992). The major carotenoid pigments are β-carotene and several xanthophylls including fucoxanthin (major xanthophyll), neoxanthin, violaxanthin, and zeaxanthin (Withers et al., 1981; Bjørnland and Liaaen-Jensen, 1989). In *Hibberdia* both fucoxanthin and antheraxanthin function as light-harvesting pigments (Alberte and Andersen, 1986).

5. An orange-red **eyespot** (**stigma**) occurs in most species; it is usually a single layer of carotenoid-containing lipid globules at the edge of a chloroplast, and this region is pressed closely against the depression in the surface of the cell into which fits the swelling on the short flagellum (Figs. 5,17). The eyespot and flagellar swelling are commonly referred to collectively as the **photoreceptor apparatus** (see Foster and Smyth, 1980; Kreimer, 1994; Walne et al., 1995).

6. The major carbohydrate storage product is a β-1,3-linked glucan, **chrysolaminaran** (**leucosin**), probably identical to that of diatoms (Quillet, 1955; Archibald et al., 1963; Smestad-Paulsen and Myklestad, 1978). It is stored as a solution, usually in one or more large vacuoles near the posterior end of the cell. Lipoidal material is also stored as generally distributed globules.

7. Most chrysomonads are naked, though some genera (*Chrysosphaerella*, *Paraphysomonas*, *Polylepidomonas*, *Spiniferomonas*) are characterized by a cell-covering of siliceous scales of complex species-specific structure (Figs. 9, 19, 20, 24). Scales are formed endogenously in **silica deposition vesicles** (**SDVs**) near the periphery of the cell, and silica is then deposited in the vesicles (see Preisig, 1986, 1994; Leadbeater and Barker, 1995). The SDVs ultimately fuse with the plasma membrane and the mature scales are extruded onto the cell surface where they are adhered in some unknown way. Unmineralized scales also occur in a small number of genera (Figs. 6 ,16).

8. Several genera are **loricate**, the loricae remaining attached at division in some species to form large colonies (see Fig. 11). Loricae are of organic composition (protein and cellulose or chitin; see Schnepf et al., 1975; Herth et al., 1977; Herth, 1979; Herth and Zugenmaier, 1979), sometimes consisting of overlapping organic scales (Fig. 12). Loricae of some chrysomonads are hyaline when first formed but later become brownish-red due to incorporation of manganese and iron compounds (Preisig, 1986; Dunlap et al., 1987).

9. Cysts (=**stomatocysts** or **statospores**) are common, usually spherical with either smooth or variously ornamented surfaces (see Figs. 11,27). They are composed of silica and have a narrow neck that is closed at maturity by a plug. At germination the plug is lost and the amoeboid protoplast emerges through the neck. The stomatocyst is a fundamental and diagnostic character of the Chrysomonada and Synurophyceae (see Sandgren, 1980a, 1980b, 1981, 1983, 1989). A species may produce morphologically indistinguishable stomatocysts both sexually and asexually, but there is great morphological variation among stomatocysts from different species (see Duff et al., 1995). Formation of stomatocysts proceeds endogenously within an internal membrane system which is considered to be homologous to the silica deposition vesicle (SDV) of diatoms (see Preisig, 1986, 1994).

10. The single nucleus in the vegetative cell is situated in the anterior end. It is typically pyriform with the narrower part extended toward the flagellar basal bodies. The nuclear envelope has usually membrane connections to the chloroplast (e.g. Fig. 17).

11. The Golgi body (dictyosome) is typically large, sigmoid, and lies against the nucleus in the anterior end of the cell, with one edge directed towards the flagellar bases (see Fig. 4). In larger cells it is clearly visible by light microscopy (referred to as the "mouthband" in older literature).

12. Cells have one or more **contractile vacuoles** (except in the relatively few marine species), usually near the anterior end of the cell (Figs. 4, 5, 17; see also Aaronson and Behrens, 1974; Wessel and Robinson, 1979; Aaronson, 1993). The discharge mechanism is poorly understood, but a small membranous channel to the plasma membrane has been observed (Andersen, 1982).

13. Many species have peripheral **muciferous bodies** that can be extruded as long threads; in some the muciferous bodies are large complex **discobolocysts**, explosively releasing discoid projectiles (Hibberd, 1970).

14. Nutrition varies from obligate phototrophy to obligate phagotrophy. Many species are mixotrophic, i.e. capable of photosynthesis as well as uptake of dissolved substances (osmotrophy) and food particles (phagotrophy). Osmotrophy is important for the uptake of sugars and amino acids (Pringsheim, 1952; Hutner et al., 1953). *Ochromonas* and *Poterioochromonas* are considered sugar flagellates (*sensu* Doflein, 1916) rather than acetate flagellates (*sensu* Pringsheim, 1935). An exogenous source of vitamins is necessary for many chrysomonads, and osmotrophic uptake of vitamins has been demonstrated in *Ochromonas* and *Poterioochromonas* (Pringsheim, 1952; Hutner et al., 1953). Alternatively, *Uroglena americana* must obtain its exogenous vitamins via phagotrophy of bacteria (Kurata, 1986; Kimura and Ishida, 1986). At least two types of phagotrophy have been reported: one involves the capture of food by flagella and the other by fine pseudopodia (Type 1 and Type 3 feeding mechanisms *sensu* Moestrup and Andersen, 1991) (Type 2 is characteristic of Pedinellids). Flagella actively draw particles to the cells which have the Type 1 mechanism (see Korshikov, 1929; Sleigh, 1964; Bird and Kalff, 1987; Wetherbee and Andersen, 1992) while organisms with Type 3 mechanisms rely passively on particles to come in contact with the pseudopodia (see Scherffel, 1901). Type 1 feeding has been reported for many genera (see Moestrup and Andersen, 1991; Holen and Boraas, 1995). Particles are captured between the flagella and held for a short period of time while the cell apparently

determines whether to accept or reject the food particle (see Wetherbee and Andersen, 1992). If rejected, the particle is pushed away by the flagella, and the capture process begins again. If accepted, the cell produces a feeding basket and the food particle is pushed into the basket by the flagella. In *Epipyxis*, the most thoroughly investigated organism, the feeding basket is formed and closed by movements of specific microtubules of the R_3 flagellar root (for details, see genus *Epipyxis* in this chapter, also Andersen and Wetherbee, 1992). Type 3 feeding is less well studied. Food particles that are captured passively by pseudopodia move along the outer surface of the pseudopods and are drawn to the cell body. The engulfment process has not been described ultrastructurally for Type 3 feeding cells. Because many chrysomonads that produce pseudopodia are reported to lack flagella, it is unclear if the basal apparatus is present without emergent flagella and engulfment is controlled by flagellar root microtubules, or if there is a distinctly different phagotrophic process. *Chrysamoeba pyrenoidifera* bears flagella that have a typical *Ochromonas*-type flagellar apparatus (O'Kelly and Wujek, 1995). Some pseudopods are supported by microtubules nucleated by the R_1 flagellar root, suggesting that it engulfs particles in a manner similar to *Epipyxis* (Andersen and Wetherbee, 1992) or *Chrysosphaerella* (Andersen, 1990). However, other pseudopod microtubules emanate from the nuclear envelope (O'Kelly and Wujek, 1995), suggesting a different type of particle uptake. The nutritional habits of chrysomonads are varied and complex, and further study is required. For review of history, cell structure, and cellular mechanisms, see Moestrup and Andersen (1991); for review of ecological implications, see Holen and Boraas (1995).

15. Vegetative reproduction: Flagellate and amoeboid cells divide longitudinally. Mitosis and cytokinesis are known in detail only from a few species. In *Ochromonas* and *Poterioochromonas* the nuclear envelope is completely dispersed at metaphase (Slankis and Gibbs, 1972; Bouck and Brown, 1973; Tippit et al., 1980), the microtubule-organizing center for the spindle poles being located on the rhizoplast. In the capsoid genus *Hydrurus* there is a more or less intact nuclear envelope with polar fenestrae during metaphase (Vesk et al., 1984).

16. Sexual reproduction has been observed, so far, in relatively few species (e.g., Fott, 1959; Lund, 1960; Kristiansen, 1963; Sandgren, 1981, 1983, 1986). Undifferentiated cells may act as gametes, fuse apically and produce a short-lived binucleate, quadriflagellate **planozygote** that subsequently undergoes encystment to become a binucleate **hypnozygote** or sexual **stomatocyst**. Autogamic processes prior to stomatocyst formation (fusion of nuclei formed by a preceding mitosis) are also known (Sandgren, 1981). Germination of sexual stomatocysts results in the formation of 1, 2, or 4 initial vegetative cells, depending on the species. The actual processes of **karyogamy** and meiosis have not yet been studied.

17. Chrysomonads usually occur in the freshwater plankton, in all bodies of water from lakes to small pools; a few species are able to grow on soil and snow. A few marine species have also been described.

ARTIFICIAL KEY TO SELECTED GENERA

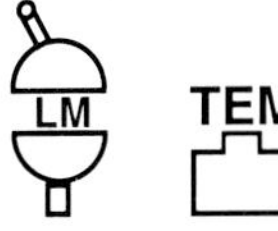

1. Vegetative cells amoeboid 29
1'. Vegetative cells flagellate (if cells have six radial chloroplasts, then see Pedinellids; if cells are covered by siliceous scales [and sometimes bristles] visible by light microscope, then also see Synurophyceae)...2

2. Colonial 20
2'. Single cells.. 3

3. Cells pigmented 6
3'. Cells colorless.. 4

4. Cells bearing siliceous scales (usually requires electron microscopic examination).................... ***Paraphysomonas***
4'. Cells naked 5

5. Cells with only one flagellum visible by light microscopy................................ ***Oikomonas***
5'. Cells with two flagella visible by light microscopy .... ***Spumella***

6. Cells without a lorica................................ 12
6'. Cells within a lorica 7

7. Lorica covering only the posterior of the cell (lorica sometimes difficult to see by light microscopy)................ ***Poterioochromonas***
7'. Lorica covering the entire cell 8

8. Lorica formed with overlapping scales (chromatic staining may be required); organism usually attached to a substratum .... ***Epipyxis***
8'. Lorica entire, not formed from overlapping scales; organism usually free-swimming ... 9

9. Cells with only one flagellum visible by light microscopy.. 11
9'. Cells with two flagella visible by light microscopy... 10

10. Cells attached to lorica by protoplasmic strand (rarely unicellular, usually forming arbusculate colonies) .... ***Dinobryon***
10'. Cells free in lorica (always unicellular) .... ***Pseudokephyrion***

11. Lorica typically urn-shaped, with narrow flagellar pore .... ***Chrysococcus***
11'. Lorica usually vase-shaped, with wide mouth-like opening..................... ***Kephyrion***

12. Cells not covered by scales (at least not with scales of marked shape)............................ 15
12'. Cells covered by scales of marked shape (usually requires electron microscopic examination)... 13

13. Scales composed of organic material (canistrate and dendritic scale types) .... ***Chrysolepidomonas***
13'. Scales composed of siliceous material...... 14 (if calcium carbonate, see Haptomonads)

14. With plate and spine-scales (plate scales not perforated)....................... ***Spiniferomonas*** (unicellular species of *Chrysosphaerella* will also key here)
14'. With plate-scales only (scales have characteristic linear perforations.......................... .... ***Polylepidomonas***

15. Cells with only one flagellum visible by light microscopy 18
15'. Cells with two flagella visible by light microscopy... 16

16. Flagella inserted on ventral surface (cells with multilayered theca; habitat marine) (see Pelagophyceae)............ ***Ankylochrysis***
16'. Flagella inserted apically or subapically (cells naked; habitat freshwater and marine) 17

17. With a vertical furrow (sulcus), running from point of flagellar insertion to posterior end of cell (short flagellum extending posteriorly in sulcus); habitat marine .... ***Sulcochrysis*** (see Pelagophyceae)
17'. No such furrow (short flagellum directed laterally; habitat freshwater and marine) .... ***Ochromonas***

18. Cells with alternating flagellate and "palmelloid" stages (the latter consisting of non-swimming flagellate cells within a colonial gel) .... ***Hibberdia***
18'. Cells without alternating flagellate and palmelloid stages....................................... 19

19. Cells typically epineustonic, sometimes sitting on small fibrillar spools (especially common in small undisturbed ponds) .... ***Chromophyton***
19'. Cells normally free-swimming, not sitting on spools.................................. ***Chromulina*** (if marine, also compare with *Pelagomonas*, see Pelagophyceae)

20. Cells colorless (radiating colonies of up to 60 cells usually attached to a substratum via thick stalks) .... ***Anthophysa***
20'. Cells pigmented... 21

21. Cells covered by both plate-scales and separate long spines....... ***Chrysosphaerella***
21'. Siliceous scales and spines absent............ 22

22. Cells covered by delicate organic scales consisting of an open framework of fibrils (requires electron microscopic examination) .... ***Lepidochrysis***
22'. Cells not covered by scales......................... 23

23. Cells without a lorica................................. 25
23'. Cells enclosed within a lorica.................... 24

24. Lorica formed with overlapping scales (chromatic staining may be required); colonies usually attached to a substratum .... ***Epipyxis***
24'. Lorica entire, not formed from overlapping scales; colonies usually free-swimming .... ***Dinobryon***

25. Cells with two flagella visible by light microscopy 27
25'. Cells with only one flagellum visible by light microscopy..26

26. Cells actively swimming within a colonial gel .... ***Saccochrysis*** (some species of *Chromulina* will also key here)
26'. Cells not swimming within a colonial gel .... ***Hibberdia***

27. Colonies not actively swimming (non-motile cells with vigorously beating flagella enclosed within gel matrix)............ ***Chrysonephele***
27'. Colonies actively swimming (flagella of all cells projecting from gel matrix)............. 28

28. Cells arranged at the periphery of gel matrix (interior of matrix ± homogeneous or with system of radiating gelatinous stalks)............ .... ***Uroglena***
28'. Cells arranged radially in gel matrix, usually attached in colony center (no system of radiating stalks as in some species of *Uroglena*) .... ***Synuropsis***

29. Cells enclosed within a lorica.................. 31
29'. Cells without a lorica 30

30. Unicells; rhizopodia arising at any point on cell surface............................ ***Chrysamoeba***
30'. Cells arranged in flattened colonies; rhizopodia arising only at distal cell poles .... ***Chrysostephanosphaera***

31. Basis of lorica extended into two thin projections forming a ring by which lorica is attached to a substratum ....***Chrysopyxis***
31'. Lorica without basal extensions....***Lagynion***

Genus ***Anthophysa*** Bory de Saint-Vincent, 1822
(The original spelling is *Anthophysis*, but *Anthophysa* is the *orthographia conservanda*).

Colorless, club-shaped cells organized into radiating colonies with up to 60 cells; colonies usually attached to a substratum via thick, dichotomously branched stalks, but colonies and cells can easily break free (especially during sample preparation) and swim freely in the water; young stalks extruded as a colorless, sticky substance; older stalks usually fairly transparent near the cells, but distally often becoming increasingly thickened and mineralized by deposits of calcium phosphate and compounds of iron and manganese; central part of the stalk occupied by one or more fine protoplasmic strands extending from posterior ends of the cells; individual cells indistinguishable from *Spumella* (q.v.), naked, with 2 unequal flagella, a leucoplast with or without an eyespot, and 1-3 contractile vacuoles near the flagellar bases; cells divide longitudinally beginning at the anterior end, and colonies also divide, leading to bifurcating stalks; nuclear division sometimes not immediately followed by cell division resulting in binucleate cells that may form globular stomatocysts, in which the 2 nuclei may fuse; sexual reproduction by fusion of whole cells (**hologamy**) never observed; commonly occurring in freshwater, especially in temporary pools, bogs and small ponds. See Belcher and Swale (1972b), Starmach (1985), Lee and Kugrens (1989).

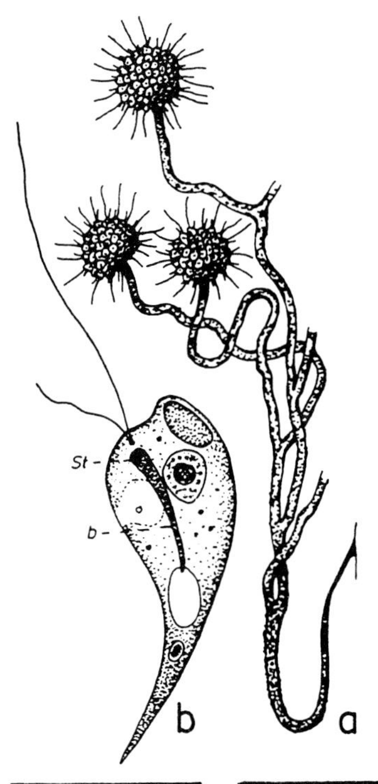

Fig. 1. *Anthophysa vegetans*. a. Colony. b. Single cell at higher magnification. b=leucoplast, st=eyespot. Scale bars: 50 µm (a) and 5 µm (b). (From Stein, 1878) and Hollande, 1942)

TYPE SPECIES: *Anthophysa vegetans* (O. F. Müller, 1786) Stein, 1878 (=*A. muelleri* Bory de Saint-Vincent, *nom. illeg.*) (Fig. 1). Colonies up to 30 µm in diameter; cells 3-10 µm long; longer flagellum approximately 1.5-2.0 times the cell length; stomatocysts approximately 10 µm in diameter, with a delicately punctate wall and a thickened pore rim.

Genus ***Chromophyton*** Woronin, 1880

Single-celled flagellates, typically found in the epineuston; cells with a long anterior and a very short posterior flagellum, a single chloroplast with a pyrenoid, no eyespot, 1-2 anterior contractile vacuoles and in at least one species with many muciferous bodies at the cell periphery; epineustic cells of some species producing and sitting on small fibrillar spools, through which thin pseudopodia of the cells may extend down to the water; epineustic cells (by some authors called pseudocysts) also completely surrounded by a sheath of dense granular material in at least one species (*C. rosanoffii*); asexual reproduction by division, in pseudocyst stage up to 8 cells being formed within the same sheath; sexual reproduction unknown; stomatocysts observed in some species; especially common in small undisturbed ponds, where a dense epineustic layer of cells may produce an intense golden sheen on the the water surface. See Petersen and Hansen (1958), Valkanov (1967), Petry (1968), Bourrelly (1968/1981), Couté (1983), Ohishi et al., (1991). Some unpublished observations show that the flagellar apparatus of *C. rosanoffii* resembles that of *Hibberdia magna* (Andersen, 1989a) rather than that of *Chromulina* or *Ochromonas,* and *Chromophyton* is therefore classified in the Hibberdiales.

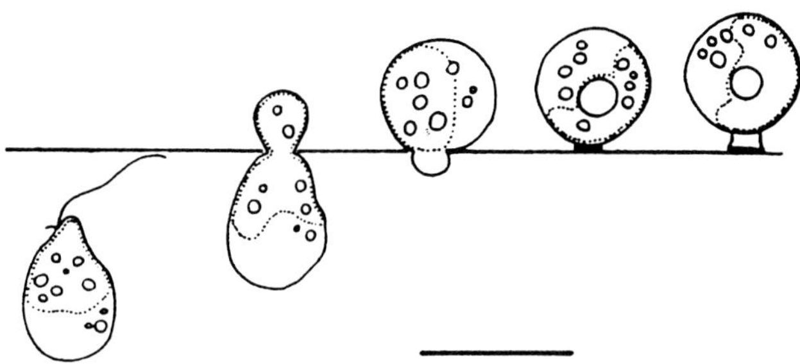

Fig. 2. *Chromophyton rosanoffii*; transition of a swarming stage (left) to a spool-forming epineustic stage (right). Scale bar=10 µm. (From Petry, 1968)

TYPE SPECIES: *Chromophyton rosanoffii* Woronin, 1880 (=*Chromulina rosanoffii* [Woronin, 1880] Bütschli, 1883-87) (Fig. 2). Swarming cells ellipsoid or pyriform, somewhat metabolic, 8 - 9 µm long and 4-6 µm wide; long flagellum approximately of same length as the cell; chloroplast with an intraplastidial pyrenoid traversed by many thylakoids; numerous peripheral muciferous bodies; pseudocysts spherical, 8 - 10 µm in diameter, sitting on a short spool; stomatocysts 4.0-7.5 µm in diameter; palmelloid stages known.

Genus ***Chromulina*** Cienkowsky, 1870 (including ***Chrysomonas*** Stein, 1878)

Usually single-celled naked flagellates with one visible apical flagellum; second very short flagellum in a pocket of the cell surface detected in species examined by electron microscopy; cells spherical, cylindrical to pyriform and often metabolic, usually free-swimming, but occasionally attaching to surfaces with posterior end, or becoming rhizopodial or palmelloid; cells with 1 - 2 chloroplasts, eyespot and pyrenoid present in some species; contractile vacuoles usually 1-2 (-6); many species with peripheral extrusomes (mucocysts, discobolocysts); nutrition phototrophic and phagotrophic; asexual reproduction by cell division in motile or palmelloid state; possible sexual fusion of palmelloid cells reported for one species (*C. pleiades* Parke, 1949); stomatocysts smooth or ornamented; many species in freshwater, some on soil and snow, and a few in brackish and marine waters. For freshwater species, see Starmach (1985); for a marine species, see Parke (1949); for cell ultrastructure, see Rouiller and Fauré-Fremiet (1958). In the type species of *Chromulina* (*C. nebulosa* Cienkowsky, 1870) as well as a few other species, the cells are not solitary, but actively swimming cells are enclosed within a colonial gel matrix as in the genus *Saccochrysis* (q.v.). *Chromulina nebulosa* should be re-examined and if it proves to be better placed with *Saccochrysis* in the same genus, the name of priority for this genus would be *Chromulina*, whereas all free-swimming, single-celled species previously assigned to *Chromulina* would have to be recombined in a separate genus which, on account of priority, would have to be called *Chrysomonas* Stein (type species: *Chrysomonas flavicans* Stein, 1878; see below). Another group of species often assigned to *Chromulina* in the past, frequently produce "spools" and aggregate in the epineuston of quiet pools. These species have been transferred to the genus *Chromophyton* (q.v.). *Chromulina placentula* Belcher & Swale, 1967, an unusual species which has distinctly flattened cells and whose cell surface is covered by minute membranous scales should be re-examined and may prove to be better placed in a different genus. Species that have cells lacking chloroplasts but otherwise closely resembling *Chromulina* are usually placed in the genus *Oikomonas* (q.v.).

REPRESENTATIVE SPECIES: *Chromulina flavicans* (Ehrenberg, 1832) Bütschli, 1883-87 (=*Monas flavicans* Ehrenberg, 1832; *Chrysomonas flavicans* [Ehrenberg, 1832] Stein, 1878; *Spirochrysis flavicans* [Ehrenberg, 1832] Conrad, 1931) (Fig. 3). Cells cylindrical to pyriform, 14-31 µm long and 6-15 µm wide, very metabolic, with a granular cell surface; one visible flagellum of ± cell length; 1-2 trough-like chloroplasts, with a red eyespot in one chloroplast; no pyrenoid; 2 anterior contractile vacuoles; phagocytosis of small chlorophytes, diatoms, etc., observed; stomatocysts spherical, 15-20 µm in diameter, ornamented with spiralling ridges; common in freshwater, apparently cosmopolitan, also known from saline waters and from soil. See Skuja (1948, 1964), for stomatocysts, see Conrad (1931).

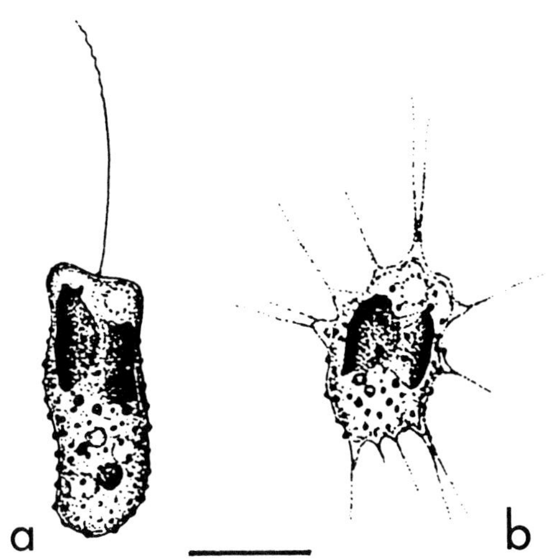

Fig. 3. *Chromulina flavicans*. a. Flagellate stage. b. Rhizopodial stage. Scale bar=10 µm. (From Skuja, 1948)

Genus ***Chrysamoeba*** Klebs, 1893

Rhizopodial state predominant; cells naked, amoeboid, solitary, or occasionally forming loose aggregates, sometimes with a very short anterior flagellum and a second reduced flagellum visible only by electron microscopy; amoeboid cells readily transform to *Chromulina*-like motile stages with a longer anterior flagellum; rhizopodia usually retracted or only a few remaining in this stage; intermediate stages common; rhizopodia granulate, slender, sometimes branched, generally all in same plane; chloroplasts 1-2; eyespot and/or pyrenoids present in some species; contractile vacuoles 1-8; phagotrophic nutrition rarely observed (see Scherffel, 1901); reproduction by division of amoeboid stage; stomatocysts known; several species in freshwater, at least one species (*C. radians*) also occurring on soil and one species (*C. nana*) marine. See Scagel and Stein (1961), Starmach (1985) and Foissner (1991). Ultrastructural studies on *Chrysamoeba radians* Klebs, 1893 and *C. pyrenoidifera* Korshikov, 1941 revealed a close relationship with chromulinalean chrysophytes (Hibberd, 1971; O'Kelly and Wujek, 1995). Rhizopodial species of which no flagellate stages are known have sometimes been placed in a separate genus, *Rhizochrysis* Pascher, 1913 (cf. Bourrelly, 1968/1981); some authors (e.g., Starmach, 1985) considered this genus to be a synonym of *Chrysamoeba*. Further studies are necessary to clarify the identity of these poorly known organisms.

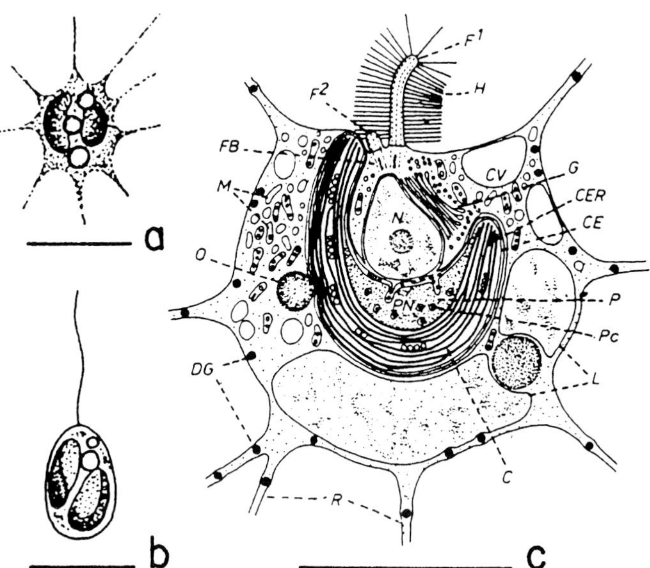

Fig. 4. *Chrysamoeba radians*. a. Amoeboid stage. b. Flagellate stage. (a, b: light microscopy) c. Cell ultrastructure. C, chloroplast; CE, chloroplast envelope; CER, chloroplast endoplasmic reticulum; CV, contractile vacuole; DG, dense globules; F1, long flagellum; F2, second flagellum; FB, flagellar basal body; G, Golgi body; H, flagellar hair; L, chrysolaminaran (leucosin) vesicles; M, mitochondria; N, nucleus; O, oil globule; P, pyrenoid; Pc, pyrenoid channels; PN, periplastidial network; R, rhizopodia. Scale bars=10 µm (a,b) and 5 µm (c). (From Klebs, 1893 and Hibberd, 1971)

TYPE SPECIES: *Chrysamoeba radians* Klebs, 1893 (Fig. 4). Rhizopodial stage with an anterior flagellum about 2 µm long, second reduced flagellum ≈ 0.4 µm long; cell 5-17 µm in diameter, with 6 - 15 mostly unbranched granulate rhizopodia 20-35 µm long; chloroplast single, U-shaped, without a girdle-lamella; no eyespot; pyrenoid lining inner face of chloroplast; nucleus in hollow of chloroplast; contractile vacuoles 2-3; chrysolaminaran vesicle often conspicuous in cell posterior; flagellate stage ovoid to elongate, 10-20 x 4-10 µm; anterior flagellum 15-18 µm long; stomatocyst spherical, 6-9 µm in diameter, smooth; known from freshwater and soil.

Genus **Chrysococcus** Klebs, 1893

Single-celled, free-swimming flagellates enclosed in spherical to pyriform to oval lorica with small flagellar pore; in some species with one or more additional small pores through which short strands of cytoplasm emerge; in some species lorica decorated with warts, spines, or other markings and/or with a short neck around flagellar pore; lorica hyaline when first formed but often becoming brownish-red in older cells due to incorporation of manganese and iron compounds; one flagellum emerging through flagellar pore; very short second flagellum detected in species studied by electron microscopy; 1 or 2 (rarely 3-5) parietal chloroplasts; eyespot and/or pyrenoid present in some species; 1 or 2 anterior contractile vacuoles; reproduction by longitudinal division within the lorica, 1 daughter cell escaping through a pore and forming a new lorica; stomatocyst formed within or outside lorica (if outside, the stomatocyst may remain attached to lorica); large genus of freshwater species, some of which are common in plankton of lakes and ponds, some also occurring in saline water. See Klebs (1893), Bourrelly (1968 /1981), Starmach (1985); for cell ultrastructure, see Belcher (1969), Belcher and Swale (1972a); for mineralization of lorica, see Heinrich et al. (1986).

Species of *Chrysococcus* are distinguished primarily by the size and shape of the lorica. "*Chrysastrella furcata*" (Dolgoff, 1922) Deflandre, 1934, a species originally described from stomatocysts and often classified in the dubious family "Chrysostomataceae," has been reassigned to the genus *Chrysococcus*, after flagellated vegetative cells with a siliceous lorica bearing 2-5 simple, branched or forked spines have been found (Nicholls, 1981). However, species with siliceous loricas should be re-examined and may prove to be better placed in a different genus.

TYPE SPECIES: *Chrysococcus rufescens* Klebs, 1893 (Fig. 5). Lorica spherical, smooth, hyaline to brown, moderately thick, with 1 or 2 openings (3 in *C. rufescens* f. *triporus* Lund, 1960), 3.5-12.0 µm in diameter; lorica compressed in *C. rufescens* var. *compressus* Skuja, 1948; cell more or less spherical, 3-7 µm in diameter; 1 emergent flagellum 1-2 1/2 times as long as the diameter of the lorica; second flagellum 0.7 µm long, with small swelling associated with eyespot; 2 chloroplasts; no pyrenoid; 1 (-2) eyespot(s) (no eyespot in *C. rufescens* var. *astigma* Huber-Pestalozzi, 1941); 1 or 2 anterior contractile vacuoles; stomatocysts smooth, ± spherical, 4-7 µm in diameter, slightly flattened on the side towards the pore; species common in freshwater, sometimes bloom-forming, also recorded from saline water (e.g. Baltic Sea).

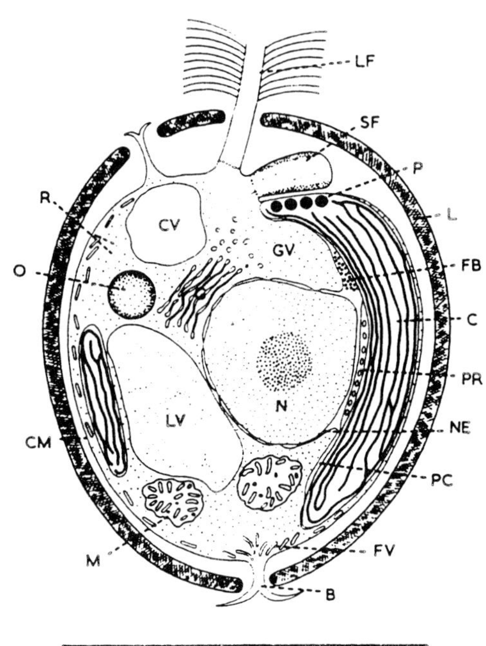

Fig. 5. *Chrysococcus rufescens*; diagrammatic section based upon electron microscopy. B, branched cytoplasmic process; C, chloroplast; CM, cell membrane; CV, contractile vacuole; FV, flattened vesicles; FB, fibrillar bundles (flagellar hairs); G, Golgi body; GV, Golgi vesicles; LF, long flagellum; LV, leucosin vacuole (chrysolaminaran); L, lorica; M, mitochondrion; N, nucleus; NE, nuclear envelope; O, oil globule; P, pigment chambers (eyespot); PC, periplastidial cisterna (ER); PR, periplastidial reticulum (ER); R, ribosomes; SF, short flagellum. Scale bar=5 µm. (From Belcher, 1969)

Chrysolepidomonas Peters & Andersen, 1993a

Single-celled flagellates, covered with characteristic canistrate and dendritic organic scales, free-swimming and with 2 unequal flagella; scales covering cell surface and at least in 2 species also both flagella (long flagellum also bearing tripartite tubular hairs); canistrate, or basket, scales with small spines, bumps or a fenestrate rim at distal surface; dendritic scales similar to canistrate scales at base, but on top of it with a tree-like outgrowth; cells spheroidal to elongate, not flattened, with a single chloroplast, containing an eyespot closely associated with the swollen region of the shorter flagellum; in one species also with a pyrenoid; 1-2 anterior contractile vacuoles; no food vacuoles and no phagotrophic feeding observed so far; scale formation performed within cisternae of the single, large Golgi body; flagellar apparatus architecture resembling that of *Ochromonas*-type flagellates; reproduction not studied in any detail so far; stomatocysts occasionally observed; 2 species in freshwater lakes and ponds, 1 species marine. See Peters and Andersen (1993a,b). Two species have earlier been placed in a different genus, *Sphaleromantis* Pascher 1910b (Harris, 1963; Manton and Harris, 1966; Pienaar, 1976), but the type species of this genus, *Sphaleromantis ochracea* (Ehrenberg, 1832) Pascher, 1910b, is apparently not scaly and also differs from species of *Chrysolepidomonas* in several other distinctive features (cells strongly flattened, with a single emergent flagellum and with two chloroplasts; see Pascher, 1910b, and Starmach, 1985). *Chrysolepidomonas* has been accommodated in a separate family, Chrysolepidomonadaceae, characterized by the unique organic scales.

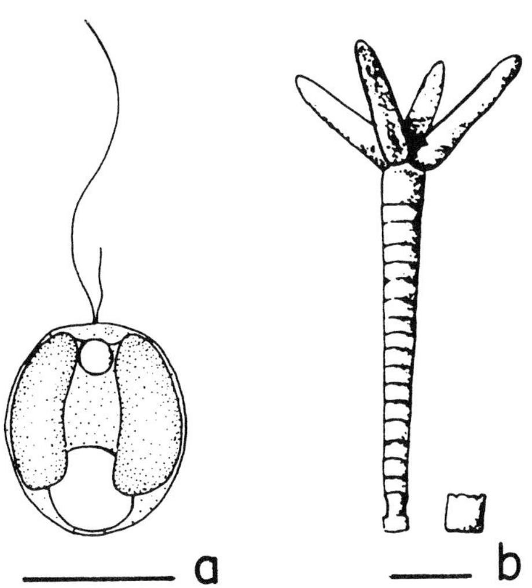

Fig. 6. *Chrysolepidomonas dendrolepidota*. a. Cell. b. Scales (left: dendritic scale, right: canistrate scale). Scale bars=5 µm (a) and 100 nm (b). (From Peters and Andersen, 1993a)

TYPE SPECIES: *Chrysolepidomonas dendrolepidota* Peters & Andersen, 1993a (Fig. 6). Cells spheroidal, 7.0-8.0 x 5.5-6.5 µm, with a single chloroplast and an eyespot; pyrenoid lacking; long flagellum about 1.25 times the cell length; short flagellum about 2-3 µm long; canistrate scales 50-70 nm in diameter and 50-80 nm in height, with 8 bump-like protuberances on the top; dendritic scales about 50-70 nm in diameter and 500 nm or more in height, with 4 terminal branch-like extensions; stomatocysts spherical, smooth, 6-8 µm in diameter.

Genus *Chrysonephele*
Pipes, Tyler & Leedale, 1989

Cells arranged peripherally in large gelatinous colonies that are free-floating or attached to macrophytes; cells naked, *Ochromonas*-like, with 2 vigorously heterodynamic flagella (cells and colonies are non-motile, however); short flagellum with a flagellar swelling (photoreceptor), lying in juxtaposition to the eyespot which is located in a lobe of the chloroplast; long flagellum also containing dense material within the flagellar membrane on the side facing the short flagellum; nutrition both phototrophic and phagotrophic; free-swimming stages, reproduction and stomatocysts not known; monotypic genus, so far only observed in an ephemeral swamp in Tasmania. See Pipes et al. (1989). For 18S rRNA gene sequence data see Saunders et al. (1997a).

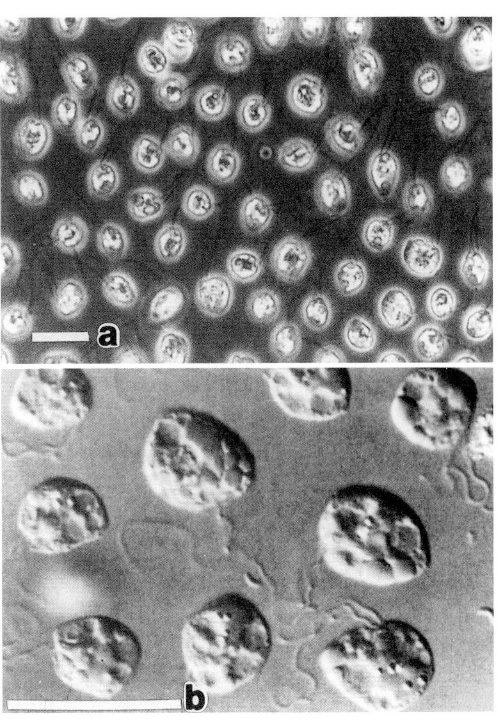

Fig. 7. *Chrysonephele palustris*. a. Cells in colony. b. Cells at higher magnification. Scale bars=10 µm. (From Pipes et al., 1989)

TYPE SPECIES: *Chrysonephele palustris* Pipes, Tyler & Leedale, 1989 (Fig. 7). Colonies saccate, delicate, more or less spherical to ovoid, up to 5 cm in diameter, usually containing many hundreds to thousands cells in a single circumferential layer of a common gelatinous matrix; cells round in face view, pear-shaped in side view, 5-9 x 4-7 µm; long flagellum 10-14 µm, short flagellum 5-10 µm; chloroplast single, bilobed, with inconspicuous eyespot; contractile vacuoles one to several, anterior to median; muciferous bodies 6-7, located at the cell periphery.

Genus ***Chrysopyxis*** Stein, 1878

Cells rhizopodial, solitary, surrounded by an ovoid, spheroidal, or flask-shaped lorica with a broadly open apical aperture; basis of lorica extended into 2 thin projections, 1 on either side, their tips being connected to form a ring, by which the lorica is attached to substratum (e.g., an algal filament); lorica and ring containing fine cellulosic fibrils (Kristiansen, 1972), older loricas often yellow or brownish owing to impregnation with iron and manganese; cell more or less spherical, attached to the lorica by a thin protoplasmic strand, at cell apex usually with a fine, sometimes branched rhizopodium extending through aperture in lorica.

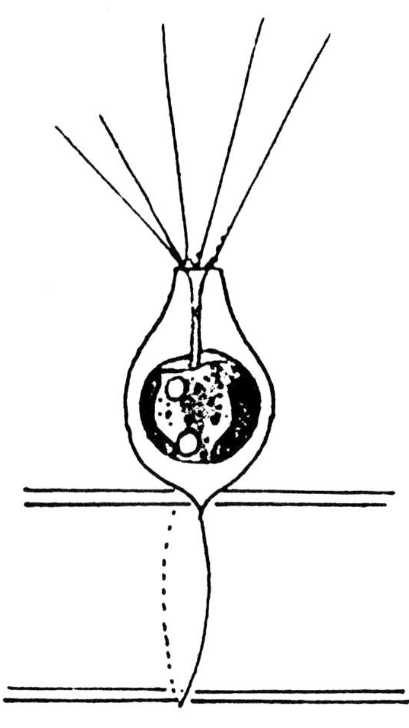

Fig. 8. *Chrysopyxis bipes*. Loricate rhizopodial cell attached to algal filament. Scale bar=10 µm. (From Penard, 1921)

Some species described as possessing a flagellum rather than a rhizopodium; chloroplasts 1-2; eyespot lacking; contractile vacuoles 1 or 2; nutrition phototrophic and phagotrophic (Scherffel, 1911); in reproduction the cell divides, one of the siblings usually remains in lorica, while the other has a single emergent flagellum and escapes from lorica; the motile cell can swim around a host algal filament several times, laying down a thin thread and then forms a lorica; stomatocysts observed in at least one species (Iwanoff, 1899); 14 species recognized by Starmach (1985), some growing commonly as epiphytes on filamentous algae in freshwater.

TYPE SPECIES: *Chrysopyxis bipes* Stein, 1878 (Fig. 8). Lorica 12-15 x 8-13 µm, ovoid or shortly flask-shaped, with a short neck and circular anterior orifice. See Penard (1921).

Genus ***Chrysosphaerella*** Lauterborn, 1896

Free-swimming; single-celled or forming globular, up to 64-celled colonies; cells spherical to oval or pyriform, in colonies mutually attached to a central mass of jelly; cell surface covered with numerous siliceous scales and one to several siliceous spines, except for the adhering posterior surfaces of colonial cells which are free of scales and spines; scales plate-like, elliptical to oval or circular, and variously patterned with ridges and hollows; spines with a hollow shaft and a bifurcate or trifurcate tip; basal part of spines elaborate, in some species bobbin- or pulley-shaped consisting of 2 discs joined by a hollow tube; in some single-celled species with a more simple base that is separated from the shaft by a septum and just above it with a circular or elliptical hole in the spine shaft; scales and spines formed internally; scale deposition vesicle not associated with the chloroplast, but may be associated with a mitochondrion; flagella 2, of unequal length; short flagellum with a swelling at the base that is closely associated with the eyespot; basal apparatus with 4 microtubular roots (R_1-R_4) and apparently a delicate system II fiber (=rhizoplast); R_3 root microtubules separating into 2 bands, and phagocytosis of bacteria taking place as the cell separates the 2 bands to create a small feeding pocket; the "f" tubule of the R_3 root forming an unusual coil on top of the nucleus; chloroplast 1 (bilobed) or 2, parietal, with an anterior eyespot; cell also containing 1 or more contractile vacuoles, a chrysolaminaran vacuole, food vacuoles and muciferous bodies; sexual reproduction not described; stomatocysts observed in some species; two colonial species, *Chrysosphaerella longispina* Lauterborn, 1896 and *C. brevispina* Korshikov, 1941, are of abundant occurrence and characteristic in clear water lakes and ponds in temperate regions of the world; blooms may cause "fishy" taste and odor problems. See Nicholls

(1980), Starmach (1985), Dürrschmidt and Croome (1985), Kristiansen and Tong (1989), Siver (1993); for stomatocysts, see Duff et al. (1995); for ultrastructure, see Preisig and Hibberd (1983) and Andersen (1990). The single-celled species have sometimes been assigned to the related genus *Spiniferomonas* (q.v.; Nicholls, 1984). Several species, originally accommodated in *Chrysosphaerella*, have later been transferred to the zooflagellate genus *Thaumatomastix* Lauterborn, 1899 (Beech and Moestrup, 1986).

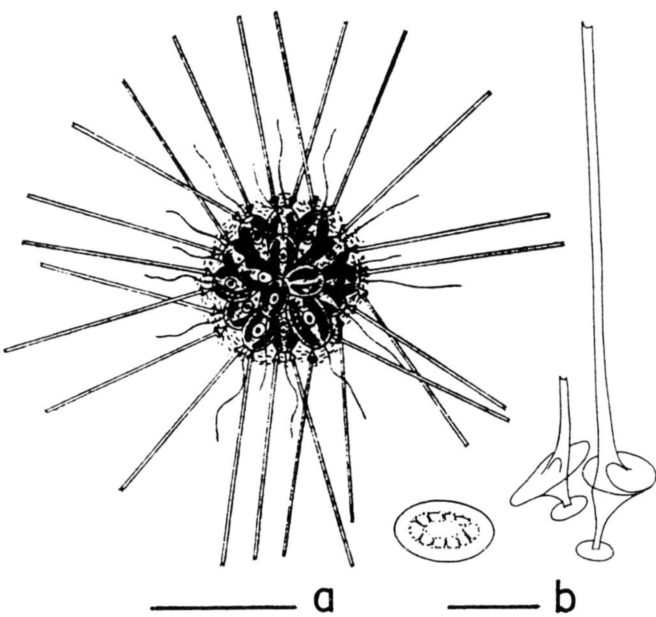

Fig. 9. *Chrysosphaerella longispina*. a. Colony. b. A plate scale and two spines of different size at high magnification (drawn from electron micrographs). Scale bars=50 µm (a) and 5 µm (b). (From Lauterborn, 1899, and Nicholls, 1984)

TYPE SPECIES: *Chrysosphaerella longispina* Lauterborn, 1896 emend. Nicholls, 1980 (Fig. 9). Colonies more or less spherical, 35-100 (-250) µm in diameter (excluding spines); cells 9-23 x 7-12 µm; typically 2-5 (-15) spines per cell; spines with a bobbin-like base and a shaft with a forked tip, spine length variable, 3-70 (-85) µm, often with 2 size categories (very short spines of 3-7 µm length and much longer spines); plate scales elliptical or oval, 3.5-6.0 x 2.0-3.0 µm, with a wide unpatterned margin and a raised central area and a series of ridges connecting the 2 areas; small, elliptical or oval unpatterned scales, 1.8-2.5 x 1.0-1.6 µm, also occurring but less common; long flagellum 1.5-3.0 times the cell length; second flagellum very short (2-3 µm); contractile vacuoles median or posterior; stomatocysts spherical, 7-15 µm in diameter, smooth and without ornamentation; species found primarily in acidic to circumneutral waters.

Genus ***Chrysostephanosphaera*** Scherffel, 1911

Usually colonial, with 2-32 non-motile naked cells arranged in a ring (sometimes more irregularly) within the periphery of a common mucilage; colony disc-shaped, flattened, the mucilage containing numerous minute globular bodies reported to be symbiotic bacteria (Geitler, 1949) but now thought to be inorganic in nature and produced endogenously (Hibberd and Leedale, 1985); cells usually with fine, straight, granular rhizopodia projecting from distal pole; chloroplasts 1 or 4; eyespot lacking, contractile vacuoles 1-2; reproduction normally by fission; a uniflagellate motile stage and stomatocysts rarely observed (Skuja, 1956); phagocytosis of bacteria reported for *C. globulifera* (Sanders et al., 1989); 2 other species rarely recorded (one is an unusual unicellular species growing in hyalocytes of *Sphagnum*; see Ellis-Adam, 1982).

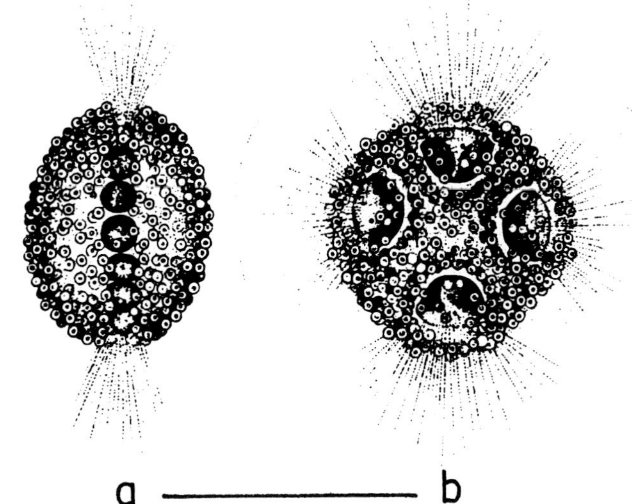

Fig. 10. *Chrysostephanosphaera globulifera*. a. 16-celled colony in equatorial view. b. 4-celled colony in face view. Scale bar=50 µm. (From Skuja, 1956)

TYPE SPECIES: *Chrysostephanosphaera globulifera* Scherffel, 1911 (Fig. 10). Colonies usually consisting of 4, 8, or 16 cells, flattened, 20-90 (rarely >100) µm in diameter in broad view; cells ellipsoid, 7-13 x 6-10 µm; rhizopodia mainly project from broader distal pole of cell towards outside of colony; chloroplast single, parietal; contractile vacuoles 1-2, usually at end of cell towards center of colony; cysts spherical to obovate, smooth; species fairly common in freshwater, especially in acid swamps and bogs. See Skuja (1956).

Genus *Dinobryon* Ehrenberg, 1834

Loricate cells, forming arbusculate colonies (rarely solitary), planktonic and free-swimming (rarely sessile); lorica cylindrical, vase- or funnel-shaped and often with a slightly broadened mouth; lorica consisting primarily or entirely of cellulose and protein, formed by successive loops of fibrils extruded during rotation of the cell; cells *Ochromonas*-like, attached to base of lorica by a thin protoplasmic strand; 2 unequal flagella; chloroplast(s) 1 (bilobed) or 2; eyespot large, associated with base of short flagellum; 1-2 contractile vacuoles, anterior, median or posterior; chrysolaminaran vacuole large, posterior; nutrition phototrophic and phagotrophic; reproduction by longitudinal cell division, after which one daughter cell swims away or often moves to mouth of parental lorica and forms a new lorica, whereas the other daughter cell occupies the parental lorica; stomatocysts formed asexually or sexually; several species wide-spread and very common in freshwater lakes and ponds; blooms sometimes causing odor problems; some species (e.g. *Dinobryon balticum* [Schütt, 1892] Lemmermann, 1900) also occurring in estuaries and coastal marine waters. See Krieger (1930), Ahlstrom (1937), Hilliard (1971), Starmach (1985); for cell ultrastructure, see Franke and Herth (1973), Kristiansen and Walne (1977), Owen et al. (1990a); for lorica ultrastructure and formation, see Herth (1979), Herth and Zugenmaier (1979); for reproduction and cyst formation, see Sheath et al. (1975), Sandgren (1980a, 1981, 1989); for nutrition, see Bird and Kalff (1987), Caron et al. (1993), McKenzie et al. (1995); for production of volatile odorous substances, see Jüttner et al. (1986).

Taxonomy based mainly upon shapes of loricas. Single-celled and sessile species should be re-examined. Species like *Dinobryon mucicolum* (Bolochonzew, 1909) Bourrelly, 1957 may be better placed in *Stylochrysalis* Stein, 1878 and species like *Dinobryon eurystoma* (Stokes, 1890) Lemmermann, 1900 may prove to belong to *Epipyxis* (q.v.) if the lorica is studied carefully.

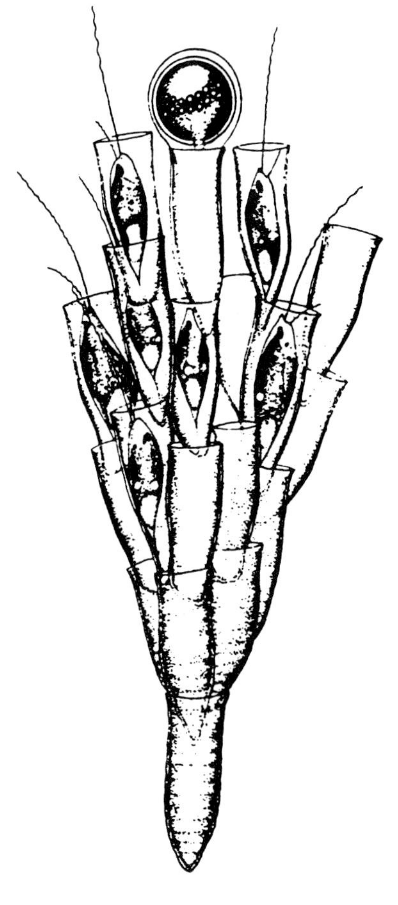

Fig. 11. *Dinobryon sertularia* colony. At mouth of one lorica with a stomatocyst forming within a thin vesicle. Scale=50 µm. (From Skuja, 1964)

TYPE SPECIES: *Dinobryon sertularia* Ehrenberg, 1834 (Fig. 11). Loricas vase-like with a distinctly broadened mouth, 25-45 µm long and 7-14 µm wide at broadest region, forming large, arbuscular colonies of up to 20-30 or more cells; individual cells fusiform, 9-12 x 4-6 µm; long flagellum more or less cell length, short flagellum 0.25-0.35 x as long; 1-2 chloroplasts, one with a prominent eyespot; stomatocysts relatively rarely observed, spherical, 10-16 µm in diameter, with a smooth or punctate wall and an inconspicuous or sometimes fairly long hooked collar; species cosmopolitan, in plankton of freshwater lakes and ponds, but also occurring in saline water (e.g. Baltic Sea).

Genus *Epipyxis* Ehrenberg, 1838
emend. Hilliard and Asmund, 1963
(including *Hyalobryon* Lauterborn, 1896)

Loricate cells; solitary, clustered or colonial; usually attached to a substratum or the water-film surface, sometimes free-swimming; lorica composed of overlapping organic scales that are often clearly visible only if stained; lorica tubular, cylindrical, fusiform or vase-like; usually hyaline, sometimes yellowish-brown; lorical scales circular, ovoid or elliptical; composed of interwoven microfibrils; upper margins of scales often bent outwards giving the lorica a denticulate appearance; cell *Ochromonas*-like, attached to base of lorica by a fine contractile protoplasmic strand containing microtubules; 2 unequal flagella; usually 1 band-shaped chloroplast, sometimes bilobed; in most species with an eyespot at apex, juxtaposed to swelling on base of short flagellum; contractile vacuoles 1-2, anterior, median or posterior; nutrition phototrophic and phagotrophic; reproduction by longitudinal cell division; stomatocysts formed inside or outside the lorica; several species very common in freshwater, often epiphytic on other algae. See Hilliard and Asmund (1963), Hilliard (1966), Starmach (1985), Ikävalko et al. (1996); for lorica fine structure, see Kristiansen (1972); for phagotrophy, see Andersen and Wetherbee (1992) and Wetherbee and Andersen (1992).

Taxonomy is based mainly on lorical shape, as well as size, shape, and arrangement of lorical scales. *Epipyxis pulchra* Hilliard & Asmund, 1963 was the first organism in which flagellar transformation was documented directly via time-lapse video sequences (Wetherbee et al., 1988). The long flagellum is the immature flagellum, and accompanying cell division, the immature flagellum retracts and transforms into the short, mature flagellum. The immature flagellum bears tubular tripartite hairs and has microtubular root R_1 associated with its basal body (the R_2 root is absent) (Andersen and Wetherbee, 1992). The mature flagellum lacks tubular hairs, contains a flagellar swelling that is closely associated with the eyespot, and has flagellar roots R_3 and R_4 associated with its basal body. Specimens of *Epipyxis pulchra*, and perhaps all species of the genus, are capable of phagocytosing bacteria. Before the cell can begin feeding, it must convert from the nonfeeding stage to the feeding stage, a process that involves a change in cell shape and reorganization of the organelles (e.g., contractile vacuoles located at the base of the flagella in nonfeeding cells and located median in feeding cells [Andersen and Wetherbee, 1992]). Once in the feeding stage, the flagella extend through the mouth scales of the lorica and establish a water current that draws particles to the cell. A bacterium is captured by the flagella and held for 1 or 2 seconds, during which the cell appears to "taste" the bacterium; after this short interval, the bacterium is either discarded or it is phagocytosed (Wetherbee and Andersen, 1992). To begin the process of phagocytosis, the cell forms a feeding basket through the dynamic separation and movement of the microtubules of the R_3 flagellar root. After the basket has formed, the flagella push the bacterium into the feeding basket, and the microtubular loop "instantaneously" contracts, closing the basket and forming a food vacuole. The bacterium is digested in the food vacuole, presumably through the fusion of primary lysosomes.

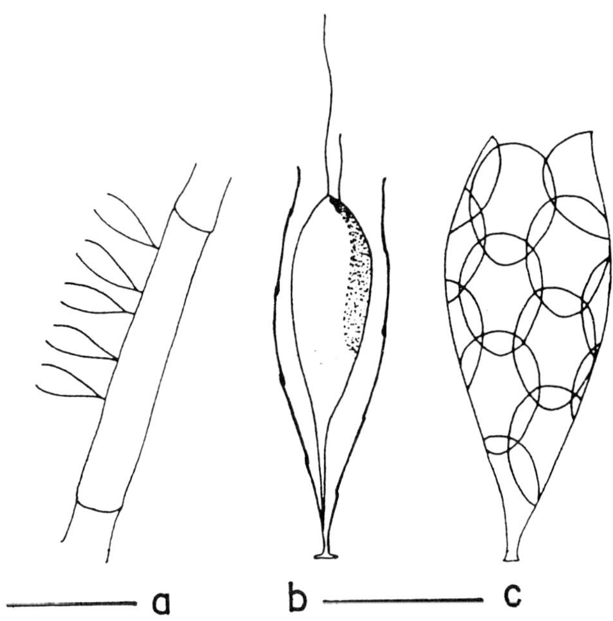

Fig. 12. *Epipyxis utriculus*. a. Group of loricas (*E. utriculus* var. *acuta*) attached to an algal filament. b. Vegetative cell enclosed in lorica at higher magnification. c. Lorica constructed of plate-like scales (drawn from an electron micrograph). Scale bars=50 µm (a) and 10 µm (b,c). (From Hilliard and Asmund, 1963)

TYPE SPECIES: *Epipyxis utriculus* (Ehrenberg, 1832) Ehrenberg, 1838 (=*Cocconema utriculus* Ehrenberg, 1832; *Dinobryon utriculus* [Ehrenberg, 1832] Klebs, 1893) (Fig. 12). Loricate cells occurring singly or as scattered clusters, usually attached to surfaces; lorica narrowly fusiform, 22-50 µm long and 7-12 µm wide at broadest region, the uppermost part cylindrical, the basal part tapered and ending in a disc-like holdfast; lorica consisting of 4-6 obliquely or spirally arranged rows of elliptical to obovoid scales, 6.5-8.4 µm in length and 4.4-5.3 µm in width (anterior row of scales discoid, 3.5-4.5 µm in diameter); vegetative cells fusiform, 9-12 x 4-6 µm; long flagellum more or less cell length, short flagellum about one-half as long; 1 chloroplast, eyespot present (in *E. utriculus f.*

astigma Scherffel, 1927 lacking); stomatocysts smooth, spherical or ovoid, 10-18 x 10-14 μm; species wide-spread in freshwater.

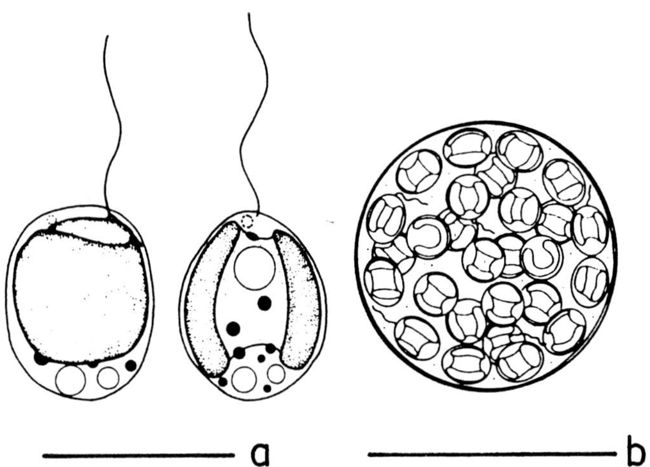

Fig. 13. *Hibberdia magna*. a. Motile stage, viewed from two different sides. b. Colonial "palmelloid" stage. Scale bars=10 μm (a) and 50 μm (b). (From Andersen, 1989a)

Genus ***Hibberdia*** Andersen, 1989a

Free-swimming, single-celled flagellate stage alternating with a "palmelloid" stage consisting of non-swimming flagellate cells within a colonial gel; motile cells with a single emergent anterior flagellum, 1-2 parietal chloroplasts, one of which containing an eyespot closely associated with a swollen region of the very short second flagellum; flagellar bases forming a wide angle (ca. 160°); basal apparatus with 3 microtubular roots (R_1, R_2 and R_4) and a rhizoplast (system II fiber); absence of R_3 root and lack of food vacuoles suggest phagocytosis does not occur; contractile vacuole single, anterior; chrysolaminaran vacuole posterior; light harvesting pigments include fucoxanthin and antheraxanthin (the latter is unique, see Alberte and Andersen, 1986); cell division both in free-swimming and colonial state; nuclear envelope mostly intact during mitosis, with polar fenestrae at metaphase; sexual reproduction and stomatocysts unknown; monotypic genus, rarely observed in nature, but intensively studied in culture (isolated from a pond in England). See Andersen (1989a); for molecular data, see Bhattacharya et al. (1992).

TYPE SPECIES: *Hibberdia magna* (Belcher, 1974) Andersen, 1989a (=*Chrysosphaera magna* Belcher, 1974) (Fig. 13). Free-swimming flagellated cells spherical to oval or ovoid, up to 10 μm long; anterior flagellum slightly longer than cell; second flagellum not visible with light microscopy (less than 1 μm long); "palmelloid" colonies spherical, 5-60 μm in diameter, adhering to each other and to the substratum in irregular clusters; each colony consisting of up to 256 closely packed flagellated cells. See Belcher (1974) and Andersen (1989a).

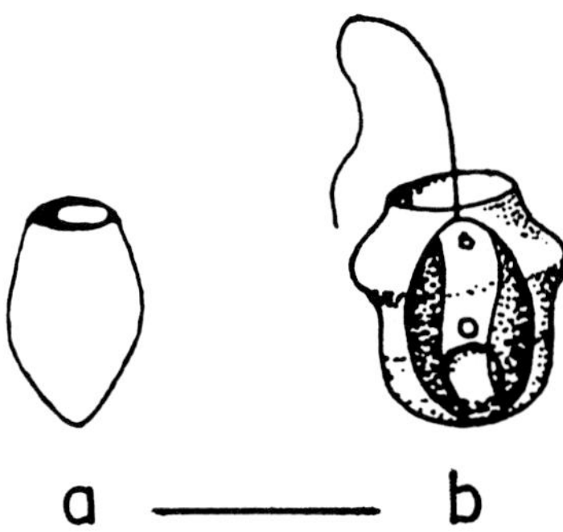

Fig. 14. *Kephyrion sitta* lorica (a). Loricate cell of *K. rubri-claustri* Conrad, 1939 (b). Scale bar=5 μm. (From Bourrelly (1968,1981) and Conrad (1939)

Genus ***Kephyrion*** Pascher, 1911

Single-celled, free-swimming flagellates enclosed in ovoid, bottle- to vase-shaped lorica that has a wide mouth opening (in *Kephyrion tubiforme* Fott, 1953 also with a basal opening); cells with 1 visible flagellum, 1 chloroplast, with or without an eyespot, and with contractile vacuoles; loricas of variable shape, thickness and ornamentation; often brown in color owing to incorporation of manganese and iron compounds; asexual reproduction by longitudinal cell division, one daughter cell swimming away and forming a new lorica, the other daughter cell remaining in the parental lorica; zygotic stomatocysts produced outside the loricas following hologamic fusion, the loricas of the 2 gametes typically remaining attached to the zygote (Kristiansen, 1963); species of *Kephyrion* commonly occur in plankton of many freshwater lakes and ponds. See Bourrelly (1957), Hilliard (1967), Starmach (1985); for mineralization of lorica, see Dunlap et al. (1987). Species ornamented with wing-like bands which encircle the loricas transversely are sometimes placed in a separate genus, *Stenokalyx* Schiller, 1926. *Pseudokephyrion* (q.v.) is the analogous biflagellate genus, distinguished from *Kephyrion* only by the presence of a second visible flagellum.

TYPE SPECIES: *Kephyrion sitta* Pascher, 1911 (Fig. 14a). Loricas pyriform, wide-mouthed, ± 5 μm long and 3 μm wide; cells sitting in the base of the lorica; 1 chloroplast; flagellum 2-3 times longer than lorica. See Pascher (1911).

Genus *Lagynion* Pascher, 1912a

Cells rhizopodial, solitary, surrounded by a flask- or bottle-shaped lorica that has a short or long neck and a single aperture through which a fine rhizopodium extends; lorica often attached to a substratum; wall sometimes dark brown owing to incorporation of iron and manganese compounds; chloroplasts 1-2; eyespot lacking; contractile vacuoles 1 or 2; reproduction by division, during which an amoeboid daughter cell leaves the lorica through the aperture; flagellate motile stages rarely observed (Meyer, 1986; O'Kelly and Wujek, 1995); cysts not known; some 25 species described. See Dop (1977), Starmach (1985).

So far only one species (*L. delicatulum* Skuja, 1964) has been studied in detail by electron microscopy; vegetative cells possess 2 basal bodies but without attached flagella; flagellar apparatus in zoospores resembles that of *Hibberdia magna* (Andersen, 1989a) rather than that of *Chromulina* or *Ochromonas*, suggesting that *Lagynion* should be classified in the Hibberdiales (O'Kelly and Wujek, 1995).

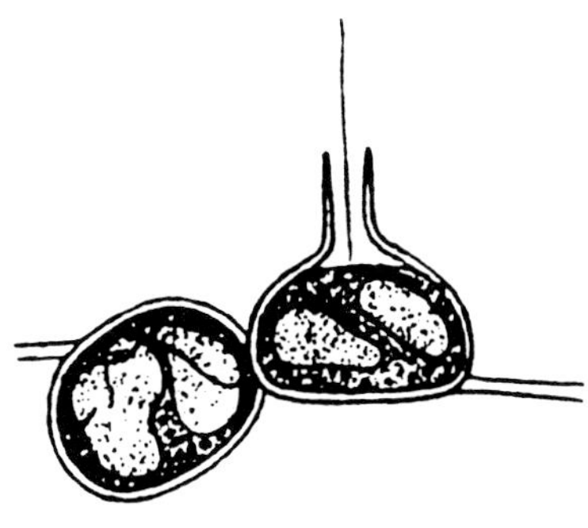

Fig. 15. *Lagynion scherffelii*. Left in face view, right in side view. Scale bar=10 µm. (From Pascher, 1913)

TYPE SPECIES: *Lagynion scherffelii* Pascher, 1912a (Fig. 15). Lorica 7-15 x 4-12 µm, ovoid to almost hemispherical, base flattened against substratum, distally extended with a short neck and apical aperture; cell with 1-2 parietal chloroplasts; 2 contractile vacuoles; species often found attached to submerged vegetation (e.g. filamentous algae) in ponds and bogs.

Genus *Lepidochrysis* Ikävalko, Kristiansen & Thomsen, 1994

Biflagellate cells, solitary or embedded in the periphery of a common gelatinous matrix and forming spherical or irregularly shaped colonies; each cell covered with delicate scales consisting of an open framework of fibrils that form a hollow structure; 2 unequal flagella; chloroplast single; eyespot and pyrenoid present; stomatocysts unknown; monospecific genus, reported from saline water (e.g. Baltic Sea, salt marsh pools), but also growing in freshwater (unpublished observation). See Clarke and Pennick (1975) and Ikävalko et al. (1994).

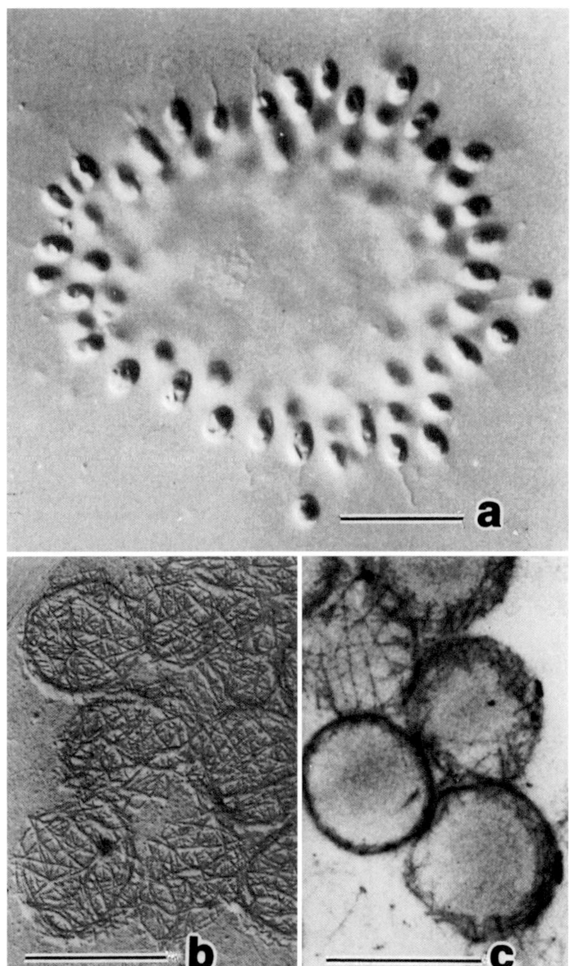

Fig. 16. *Lepidochrysis glomerifera*. a. Colony (light microscopy). b. Detached scales (shadowcast TEM preparation). c. Sections of scales showing hollow form (TEM). Scale bars=20 µm (a) and 1 µm (b,c). (From Ikävalko et al., 1994, and Clarke and Pennick, 1975)

TYPE SPECIES: *Lepidochrysis glomerifera* (Clarke & Pennick, 1975) Ikävalko, Kristiansen & Thomsen, 1994 (=*Syncrypta glomerifera* Clarke & Pennick, 1975) (Fig. 16). Colonies up to 110 µm in diameter, consisting of up to 80 cells; individual cells ± spherical to bluntly ovoid, 4.0-9.0 x 3.0-5.5 µm; scales ellipsoid, 0.7-1.4 x 0.6-1.0 µm; long flagellum 6-13 µm long; second flagellum 2-4 µm long.

Genus ***Ochromonas*** Vysotskii, 1887

Single-celled naked flagellates with 2 unequal flagella; cells spherical, cylindrical to pyriform, sometimes metabolic, usually free-swimming but occasionally attaching to surfaces with posterior end, or becoming rhizopodial or palmelloid; small and loose colonial aggregations of flagellated cells also occurring in a few species; cells with 1-2 (rarely more) chloroplasts, with or without an eyespot and/or pyrenoid; swelling at proximal end of short flagellum overlying eyespot known for some species; chloroplasts sometimes much reduced and pale or completely lost after abnormal division; contractile vacuoles 1-4; chrysolaminaran vacuole usually large, posterior; many species with peripheral extrusomes (mucocysts, discobolocysts); nutrition phototrophic, mixotrophic and heterotrophic (osmotrophic/phagotrophic); **auxotrophy** for various vitamins is common; asexual reproduction by cell division in motile or palmelloid state; during mitosis the microtubules of the spindle making direct connections with rhizoplast (cross-banded fibrous root extending from flagellar basal bodies to nucleus); nuclear envelope largely disappearing during mitosis and only small portions persisting in connection to the chloroplast endoplasmic reticulum; sexual reproduction not observed; stomatocysts of various forms known from several species; many species in freshwater, some in brackish and marine waters and on soil; some species (especially *Ochromonas danica*) are important in physiological and molecular studies. See Starmach (1985); for cell ultrastructure, see Hibberd (1970), Slankis and Gibbs (1972), Aaronson and Behrens (1973), Bouck and Brown (1973), Brown and Bouck (1973), Kahan et al. (1978), Tippit et al. (1980), Andersen (1982, 1989b); for stomatocyst formation, see Hibberd (1977); for physiology (including nutrition and toxin production), see Spiegelstein et al. (1969), Loeblich and Loeblich (1978), Preisig et al. (1991), Eichenberger and Gribi 1994; for rRNA gene sequences, see Gunderson et al. (1987), van der Auwera and de Wachter (1997) and Andersen et al. (1999); for data on mitochondrial and chloroplast genome, see Coleman et al. (1991) and Li and Cattolico (1992).

The type species, *Ochromonas triangulata* Vysotskii, 1887, is only poorly known (see Vysotskii, 1887; Starmach, 1985). *Ochromonas malhamensis* Pringsheim, 1952, a species often referred to in experimental studies, has been transferred to the genus *Poterioochromonas* (q.v.). The colorless counterpart of *Ochromonas* is *Spumella* (q.v.).

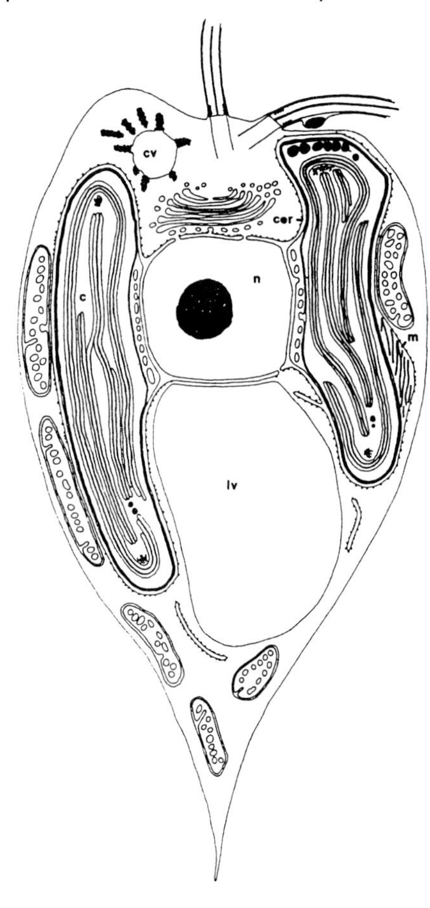

Fig. 17. *Ochromonas danica*; diagrammatic section based upon electron microscopy; c, chloroplast; cer, chloroplast endoplasmic reticulum; cv, contractile vacuole; lv, leucosin vacuole (chrysolaminaran); m, flagellar hairs (mastigonemes); n, nucleus. Scale bar=5 µm. From Gibbs (1981).

REPRESENTATIVE SPECIES: *Ochromonas danica* Pringsheim, 1955 (Fig. 17). Cells metabolic, mostly elongated, ovoid (16-18 x 8-9 µm), with a pointed, conical-shaped posterior end that is often drawn out into a thin tail (length of cell including tail up to 30 µm); less frequently ± spherical cells (9-12 x 7-9 µm) without tails also occurring; cells usually free-swimming, but occasionally attaching to surfaces; long flagellum ± cell length;

short flagellum about 1/5 to 1/6 as long; chloroplast single, bilobed, with a small anterior eyespot closely associated with swelling on short flagellum; no pyrenoid; 1 (-2) contractile vacuole(s); stomatocysts not known; freshwater species; so far rarely observed in nature, but well-known from culture studies. See Pringsheim (1955).

Genus ***Oikomonas*** Kent, 1880-82

Colorless, naked, single-celled flagellates with one visible flagellum; cells usually free-swimming, sometimes attached temporarily by a thread-like prolongation of posterior end of variable length; nucleus central or anterior; eyespot reported from one species (*O. ocellata* Scherffel, 1911); contractile vacuoles 1-3; food vacuoles often conspicuous; feeding taking place both in free-swimming and attached state; stomatocysts unknown; some 20 species described. See Preisig et al. (1991); for 18S rRNA gene data, see Cavalier-Smith et al. (1996). *Oikomonas* is considered to be the colorless counterpart of *Chromulina* (q.v.). For synonymy of *Heterochromulina* Pascher, 1912b, see Preisig et al. (1991). Despite many records of species of *Oikomonas* (especially of *O. termo* (O. F. Müller, 1773) Kent, 1880-82) in literature these organisms are poorly known; cell ultrastructure has not yet been studied in any detail.

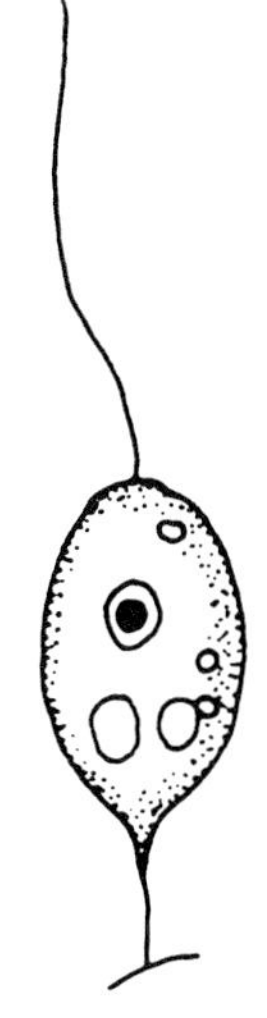

Fig. 18. *Oikomonas mutabilis*. Attached stalked cell. Scale=10 µm. (From Preisig et al., 1991)

TYPE SPECIES: *Oikomonas mutabilis* Kent, 1880-82 (Fig. 18). Cells variable in form, ovoid, spheroidal, elongate or pyriform; free-swimming or seated on a slender stalk; cell length up to 34 µm (elongate forms), spheroidal forms ±16-17 µm; flagellum up to 2 times longer than cell; contractile vacuoles 2, posterior; described from freshwater, especially among rotting vegetation, but also reported from saline soils.

Genus ***Paraphysomonas*** de Saedeleer, 1929 (incl. ***Lepidochromonas*** Kristiansen, 1980)

Colorless single-celled flagellates, either free-swimming or attached to various substrates via a thin fibrous stalk extending from cell posterior; cells *Spumella*-like, but characterized by the presence of siliceous scales on cell surface; structural details of scales used for species identification (resolved only by electron microscopy); scales variously shaped, e.g. plate-shaped, of a three-dimensional open meshwork structure (as in *P. morchella* Preisig & Hibberd, 1982b or *P. stelligera* Preisig & Hibberd, 1982b, Fig. 19c-d) or apically extended into a spine (as in *P. vestita*, Fig. 19b); majority of species with a single scale type per cell (species with 2 or 3 different scale types also occurring); scales developed in peripheral cytoplasm within flat vesicles associated with cisternae of rough endoplasmic reticulum; flagella 2, unequal; no chloroplast, but presence of an internal leucoplast demonstrated for several species; at least one species (*P. caelifrica* Preisig & Hibberd, 1982a) with a peripheral eyespot-containing leucoplast in close contact with a swelling on the short flagellum; cells also containing 1-2 contractile vacuoles, food vacuoles and vacuoles with storage product (probably chrysolaminaran); feeding via phagocytosis, with bacteria and other particles engulfed near the base of the flagella, probably using flagellar roots in a manner similar to *Chrysosphaerella* and *Epipyxis* (Andersen, 1990; Andersen and Wetherbee, 1992); asexual reproduction by division; sexual reproduction unknown; stomatocysts observed in some species; many species in freshwater, some in brackish and marine waters. See Preisig and Hibberd (1982a, 1982b) and Preisig et al. (1991); for cell ultrastructure, see Manton and Leedale (1961), Lee (1978) and Preisig and Hibberd (1983); for cell division, see Fiatte (1965); for stomatocysts, see Duff et al. (1995).

TYPE SPECIES: *Paraphysomonas vestita* (Stokes, 1885a) de Saedeleer, 1929 (=*Physomonas vestita* Stokes, 1885a) (Fig. 19a,b). Cells free-swimming or sometimes attached to a substratum by a fine stalk; cells more or less spherical, ovoid or pyriform, 4-26 µm long, with numerous scales of a single type on cell surface; scales consisting of a circular disk with an upturned rim, 0.4-4.3 µm in diameter, and with a central tapering spine, 1.0-12.5 µm long and 0.07-0.15 µm thick; long flagellum 9-54 µm; short flagellum 2-10 µm; stomatocysts rarely observed, spherical, about 8 - 11 µm in diameter, with a low collar and a smooth surface; common and widespread in freshwater, occurring in wide ranges of water temperature, conductivity and pH; also found in marine coastal waters. See Korshikov (1929), Takahashi (1978), Preisig and Hibberd (1982a).

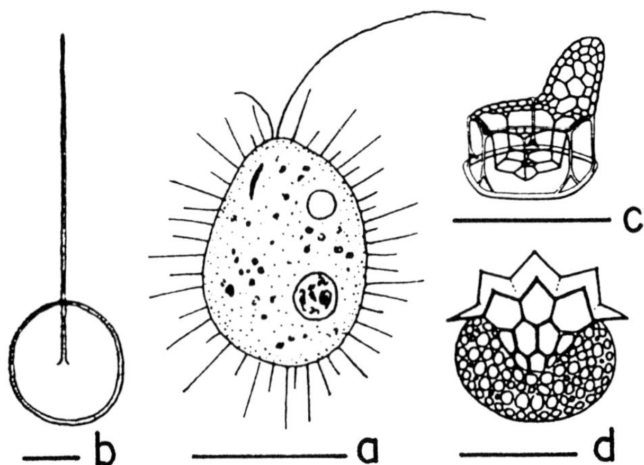

Fig. 19. *Paraphysomonas vestita*. a. Cell (light microscopic view). b. Spine scale at high magnification. c. *P. morchella*. d. *P. stelligera*, meshwork scales (drawn from electron micrographs). Scale bars=10 µm (a), 1 µm (b,d), and 0.5 µm (c). (From Heynig, 1969, and Preisig and Hibberd (1982b).

Genus ***Polylepidomonas*** Preisig & Hibberd,1983

Single-celled, free-swimming, *Ochromonas*-like, but with distinct siliceous plate-shaped scales on cell surface (scales clearly visible by electron microscopy only); flagella 2, unequal; short flagellum with a basal swelling closely appressed to cell surface underlain by anterior end of single, parietal chloroplast; eyespot lacking; cells also containing a contractile vacuole, food vacuoles and chrysolaminaran vacuole; reproduction and stomatocysts not observed; monotypic genus; occasionally observed in plankton of fresh and brackish waters. See Thomsen et al. (1981) and Preisig and Hibberd (1983). *Polylepidomonas* is the pigmented counterpart of the colorless genus *Paraphysomonas* (q.v.).

TYPE SPECIES: *Polylepidomonas vacuolata* (Thomsen in Thomsen et al., 1981) Preisig & Hibberd, 1983 (=*Paraphysomonas vacuolata* Thomsen in Thomsen et al., 1981) (Fig. 20). Cells more or less spherical in shape, 4-5 µm in diameter, covered by plate-scales of only one type; scales more or less elliptical, 1.0-1.4 x 0.6-1.1 µm, perforated by 11-42 irregularly arranged linear apertures, each 0.1-0.4 µm long and 0.05-0.08 µm wide; flagella 6-13 µm and 2-3 µm in length.

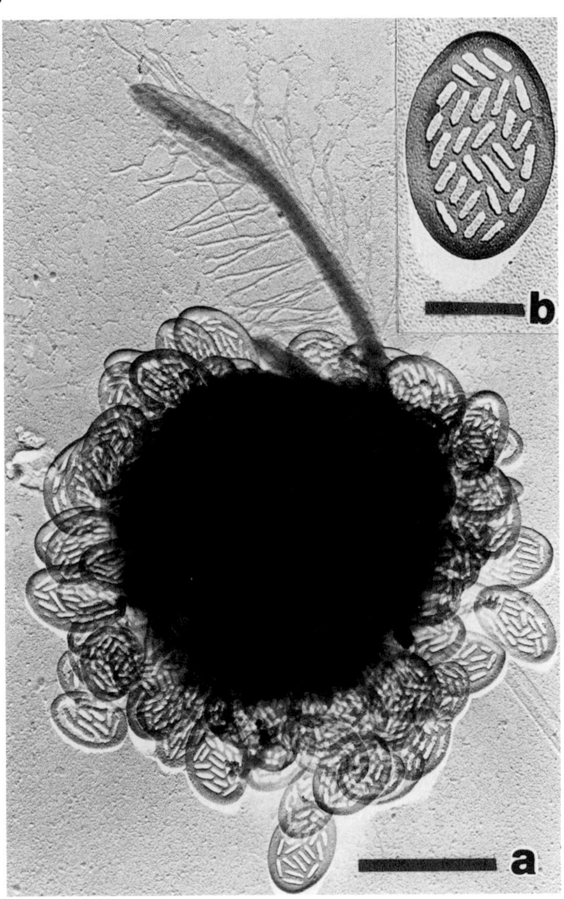

Fig. 20. *Polylepidomonas vacuolata*. a. Shadowcast preparation of cell and attached scales. b. Scale at high magnification. Scale bars=2 µm (a) and 0.5 µm (b). (From Preisig and Hibberd, 1983)

Genus ***Poterioochromonas*** Scherffel, 1901

Single-celled flagellates (or sometimes in clusters of up to 30 cells), with the posterior end of the cell sitting in a hemispherical or cone- to goblet-shaped lorica that has a short or long stalk attaching to surfaces (clustered cells joined by bases of stalks); lorica made up of interwoven microfibrils of chitin; base of lorica separated from the stalk by a plate-like septum which appears as a line when viewed in cross-section; cells *Ochromonas*-like; anterior end protruding from lorica; posterior end apparently not attached to the lorica via a protoplasmic strand and cells easily becoming free when disturbed; 2 unequal flagella, 1-3 chloroplasts (occasionally without chloroplast); eyespot lacking; 1 (-2) anterior contractile vacuole(s); nutrition phototrophic and phagotrophic; reproduction by longitudinal division; one daughter cell remaining in the parental lorica, the other escaping and forming a new lorica; stomatocysts not observed; *Poterioochromonas stipitata* (type species) by several authors considered to include *Poterioochromonas malhamensis* (Pringsheim, 1952) Péterfi, 1969 (syn. *Ochromonas malhamensis* Pringsheim, 1952), but the two have different 18S rRNA sequences (Andersen et al., 1999); *P. malhamensis* frequently used in physiological studies and also known for the production of volatile substances; common in freshwater, but often difficult to identify since lorica is inconspicuous when examined with the light microscope and cells can also readily escape from lorica and are then indistinguishable from *Ochromonas* spp. See Scherffel (1901), Pringsheim (1952; as *Ochromonas malhamensis*), Starmach (1985); for cell and lorica ultrastructure, see Péterfi (1969), Hibberd (1976), Schnepf et al. (1975, 1977), Herth et al. (1977), Wessel and Robinson (1979), Robinson and Quader (1980); for nutrition, see Sanders et al. (1990), Zhang et al. (1996); for physiological studies, see Loeblich and Loeblich (1978), Chen et al. (1994); for toxicity, see Leeper and Porter (1995); for volatile excretion products, see Jüttner and Hahne (1981).

Fig. 21. *Poterioochromonas stipitata*. Cell with attached lorica. Scale=10 µm. (From Scherffel, 1901)

TYPE SPECIES: *Poterioochromonas stipitata* Scherffel, 1901 (Fig. 21). Lorica cone-shaped, stalked, up to 20 µm long (without stalk); cell spheroidal, 7-10 µm in diameter; long flagellum about 2.5 times the cell diameter; short flagellum about 1/3 to 1/6 as long; chloroplast single, no eyespot.

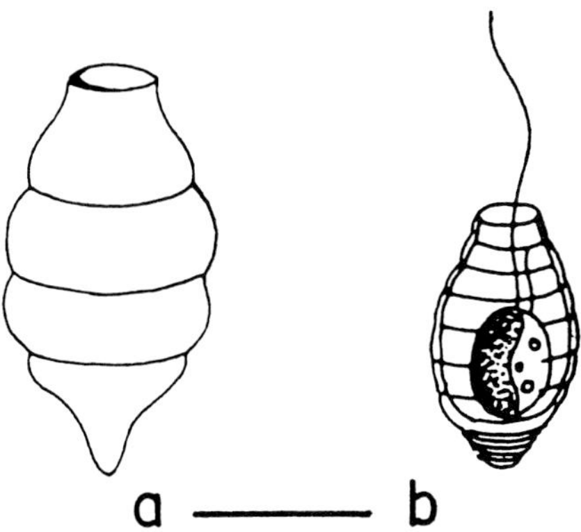

Fig. 22. *Pseudokephyrion undulatum*, lorica (a). *P. undulatissimum*, loricate cell (b). Scale bar=10 µm. (From Bourrelly, 1968/1981, and Scheffel, 1927)

Genus ***Pseudokephyrion*** Pascher, 1913

Single-celled, free-swimming flagellates enclosed in ovoid, bottle- to vase-shaped lorica that has a wide mouth opening; cells with 2 unequal flagella, 1 or 2 chloroplasts, with or without an eyespot, and with contractile vacuoles; freshwater, planktonic. See Lund (1960), Hilliard (1967), Belcher (1968), Nicholls (1977), Starmach (1985), Dunlap et al. (1987), Ikävalko et al. (1996). Except for the presence of 2 visible flagella, *Pseudokephyrion* appears to be identical to *Kephyrion* (q.v.). Species of the genus *Ollicola* Vørs, 1992 also closely resemble *Pseudokephyrion* but occur in salt water. *Ollicola vangoorili* (Conrad, 1938a) Vørs, 1992, the type species, has a chloroplast and two unequal flagella, whereas the other poorly known species combined with this genus by Vørs (previously they were assigned to *Calycomonas* Lohmann, 1908 and *Codonomonas* van Goor, 1925, respectively) were originally described to be colorless and to have a single emergent flagellum (see also Lohmann, 1908; van Goor, 1925; Lund, 1959; Preisig et al., 1991). The genus name *Dinobryopsis* Lemmermann, 1899 (type: *D. undulata* [Klebs, 1893] Lemmermann, 1899) has in fact priority over *Pseudokephyrion* (type: *P. undulatum* [Klebs, 1893] Pascher, 1913); classification of some 50 described species of *Pseudokephyrion* under *Dinobryopsis* would require many new combinations. For synonymy of *Kephyriopsis* Pascher & Ruttner in Pascher, 1913 with *Pseudokephyrion*, see Bourrelly (1968/1981).

TYPE SPECIES: *Pseudokephyrion undulatum* (Klebs, 1893) Pascher, 1913 (=*Dinobryon undulatum* Klebs, 1893; *Dinobryopsis undulata* [Klebs, 1893] Lemmermann, 1899) (Fig. 22a). Lorica ovoid to barrel-shaped, 18-25 µm by 7-12 µm, with a mouth opening of about 5 µm in diameter, and with 2-4 transverse lips encircling the lorica; cells with 2 chloroplasts, 1 eyespot, and 2 contractile vacuoles; longer flagellum approximately 1.5 times as long as the cell. See Starmach (1985).

Genus ***Saccochrysis*** Korshikov, 1941

Colonial, with flagellated cells actively swimming within a colonial gel matrix consisting of an outer, tough membranous layer and an inner, more watery gel fluid; colonial gel sometimes also containing numerous coccoid bacteria; cells with only one flagellum visible by light microscopy; a short second flagellum revealed by electron microscopy; chloroplast single, parietal, with or without a pyrenoid; eyespot lacking; 1-2 contractile vacuoles; flagellates occasionally stalked or colorless; pigmented and colorless amoebae and zoospores swimming outside the colonial gel also occurring; ingestion of solid food particles observed in colorless flagellates; dividing flagellates showing little swimming activity; after division the 2 daughter cells may remain connected for some time by a fine cytoplasmic thread at posterior end of cells before separating completely; stomatocysts rarely observed; growing in freshwater, but so far only a relatively small number of records. See Korshikov (1941), Starmach (1985), Andersen (1986), Willen (1992). Species of *Saccochrysis* resemble the type species of *Chromulina* (q.v.) and further study may reveal that they should be better placed in the latter genus.

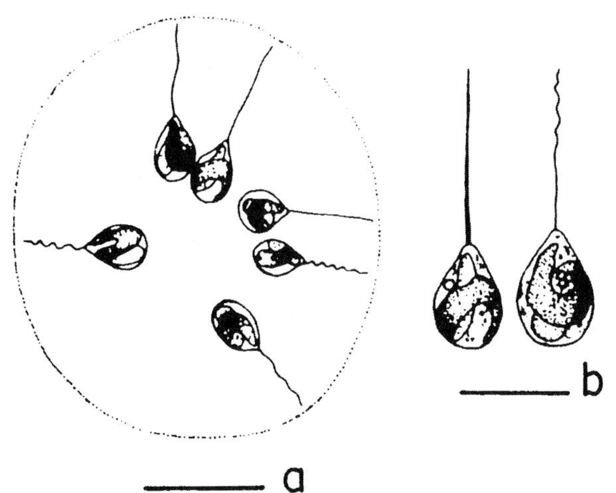

Fig. 23. *Saccochrysis piriformis*. a. Small colony. b. Single cells at higher magnification. Scale bars=20 µm (a) and 10 µm (b). (From Korshikov, 1941)

TYPE SPECIES: *Saccochrysis piriformis* Korshikov, 1941 (Fig. 23). Colonies up to 300 µm in diameter, containing up to 60 (rarely more) flagellates within a gelatinous matrix; cells pyriform or ovoid, 7-14 x 5-10 µm; long flagellum 1-2(-3) times cell length, short flagellum less than 1 µm in length; chloroplast band-shaped, slightly or greatly spiralled; no pyrenoid and eyespot; zoospores spindle-shaped, rarely spherical, 5-7 x 2-3 µm; stomatocysts smooth, laterally compressed, 7-13 µm in diameter; species apparently growing preferentially in boggy waters.

Genus ***Spiniferomonas*** Takahashi, 1973
(including ***Chromophysomonas*** Preisig & Hibberd, 1982a)

Single-celled, free-swimming, spherical to ovoid in shape, closely resembling *Ochromonas*, but characterized by the presence of siliceous scales and spines (these are clearly visible by electron microscopy only); scales plate-like, elliptical to oval or circular and variously patterned with ridges, lacunae and rods; some species with 2 or 3 different types of scales; spines tubular or triangular in cross section, with a flared funnel- or disc-shaped base and a shaft tapering to a pointed apex; terminal half of the shaft sometimes bent, flaring outwards and forming up to 4 apices; spine base simple, without double discs, septa, and holes as in *Chrysosphaerella* (q.v.); scales and spines formed within flat vesicles which are intimately associated with cisternae of rough endoplasmic reticulum in the peripheral cytoplasm; 2 flagella, of unequal length; short flagellum with a basal swelling that is closely appressed to the cell surface underlain by the eyspot (if present); chloroplast single, parietal, sometimes bilobed; in *S. bourrellyi* occasionally with a small leucoplast in place of a chloroplast; eyespot at the anterior end of the chloroplast or lacking; cells also containing 1 - 2 contractile vacuoles, food vacuoles, and a chrysolaminaran vacuole; reproduction not described; stomatocysts observed in a few species; some species widely distributed in freshwater lakes and ponds. See Takahashi (1978), Nicholls (1984, 1985), Starmach (1985), Siver (1988a, 1988b), Ito and Takahashi (1992), Nielsen (1994), Hansen (1996); for cell ultrastructure, see Preisig and Hibberd (1983); for stomatocysts, see Skogstad and Reymond (1989) and Duff et al. (1995).

Nicholls (1984) assigned all single-celled species of *Chrysosphaerella* to *Spiniferomonas* because of

matrix fairly homogeneous or containing a system of fine, radiating and branched stalks to which the cells are attached by their pointed posterior ends; individual cells *Ochromonas*-like, with 2 flagella of unequal length; chloroplasts 1-2, laminate to discoid, at least in one species containing a pyrenoid; eyespot usally 1 (rarely 2 or lacking); contractile vacuoles 1-3; numerous muciferous bodies located at the cell periphery; nutrition phototrophic and phagotrophic; cell division longitudinal; colony reproduction by constriction into two daughter colonies or by fragmentation; stomatocysts frequently observed, their ornamentation used for species identification; some species of common occurrence in the plankton of lakes and ponds, sometimes bloom-forming, one species marine. See Starmach (1985); for ultrastructure of flagellar apparatus, see Owen et al. (1990b); for stomatocysts, see Sandgren (1980b), Cronberg (1992), Duff et al. (1995); for nutrition, see Kimura and Ishida (1985, 1986); for requirements of vitamins for growth, see Kurata (1986); for production of disagreeable "fishy" odor by some species, see Yano et al. (1988); for ichthyotoxicity, see Kamiya et al. (1979). Some authors (e.g., Lemmermann, 1899; Wujek 1976) distinguished 2 separate genera, *Uroglena* and *Uroglenopsis* Lemmermann, 1899, owing to

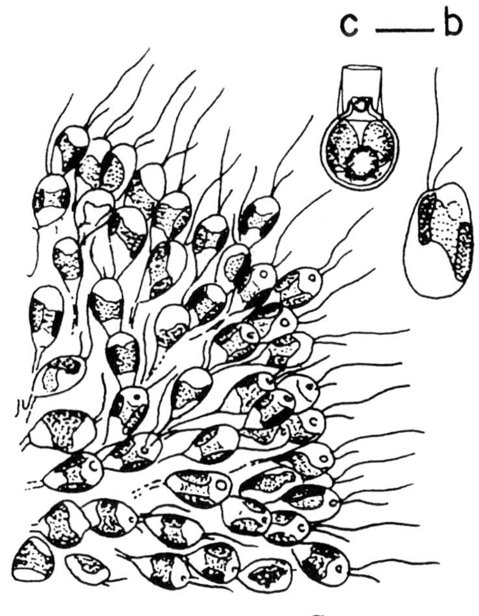

Fig. 27. *Uroglena volvox*. a. Part of a colony. b. Single cell at higher magnification. c. Stomatocyst. Scale bars=50 µm (a) and 10 µm (b,c). (From Starmach, 1985)

differences in appearance of the gelatinous stalks and minor differences in the length of the short flagellum, but this generic separation has not been generally accepted (e.g. Conrad, 1938b; Bourrelly, 1968/1981).

TYPE SPECIES: *Uroglena volvox* Ehrenberg, 1834 (Fig. 27). Colonies more or less spherical, 40-400 µm in diameter; cells ovoid or reversed pyriform, 10-20 x 8-13 µm, each sitting on a gelatinous stalk; long flagellum approximately 2-3 times the cell length; short flagellum 0.5-1.0 x the cell length; chloroplast single, laminate, spiral, with anterior eyespot; 1-2 contractile vacuoles near anterior end of cell; stomatocysts spherical, 9-16 µm in diameter, smooth (or ornamented with few to many irregular verrucae), and with a characteristic double collar above the pore; widespread species.

Lesser known genera of Chrysophyceae

Amphichrysis Korshikov, 1929. *Amphichrysis compressa* Korshikov, 1929 (Lund, 1962; Starmach, 1985)

Amphirhiza Skuja, 1948. *Amphirhiza epizootica* Skuja, 1948 (Starmach, 1985)

Angulochrysis Lackey, 1940. *Angulochrysis erratica* (Lackey, 1940)

Arthrochrysis Pascher, 1942. *Arthrochrysis leptopus* Pascher, 1942 (Starmach, 1985)

Arthropyxis Pascher, 1942. *Arthropyxis annulata* Pascher, 1942

Atraktochrysis Focke, 1957. *Atraktochrysis rotans* Focke, 1957 (Starmach, 1985)

Bitrichia Woloszynska, 1914. *Bitrichia wolhynica* Woloszynska, 1914 (Starmach, 1985)

Brehmiella Pascher, 1928. *Brehmiella chrysohydra* Pascher, 1928 (Starmach, 1985)

Calycomonas Lohmann, 1908 (see *Pseudokephyrion* Pascher, 1913)

Chrysapsis Pascher, 1910b. *Chrysapsis fenestrata* (Pascher, 1910a) Pascher, 1910b (Starmach, 1985)

Chrysarachnion Pascher, 1916a. *Chrysarachnion insidians* Pascher, 1916a (Starmach, 1985)

Chrysastridium Lauterborn, 1911 (=*Chrysidiastrum* Lauterborn in Pascher, 1913). *Chrysastridium catenatum* Lauterborn, 1911 (Starmach, 1985)

Chrysoamphipyxis Nicholls, 1987. *Chrysoamphipyxis canadensis* Nicholls, 1987

Chrysoamphitrema Scherffel, 1927. *Chrysoamphitrema brunnea* Scherffel, 1927 (Starmach, 1985)

Chrysobotriella Strand, 1928 (=***Chrysobotrys*** Conrad, 1926). *Chrysobotriella spondylomorum* (Conrad, 1926) Bourrelly, 1957 (Starmach, 1985)

Chrysocrinus Pascher, 1915. *Chrysocrinus hydra* Pascher, 1915 (Schoonoord & Ellis-Adam, 1984; Starmach, 1985)

Chrysodendron Pascher, 1927 [non *Chrysodendron* Teran & Berlandier, 1832]. *Chrysodendron ramosum* Pascher, 1927 (Starmach, 1985)

Chrysolykos Mack, 1951. *Chrysolykos planctonicus* Mack, 1951 (Starmach, 1985)

Chrysomonas Stein, 1878 (see *Chromulina* Cienkowsky, 1870)

Chrysopodocystis Billard, 1978. *Chrysopodocystis socialis* Billard, 1978

Chrysothecopsis Conrad, 1931 (incl. ***Chrysotheca*** Scherffel, 1927 and ***Stephanoporos*** Conrad & Pascher in Pascher, 1940a). *Chrysothecopsis epiphytica* (Scherffel, 1927) Conrad, 1931 (Starmach, 1985)

Chrysothylakion Pascher, 1915. *Chrysothylakion vorax* Pascher, 1915

Chrysoxys Skuja, 1948. *Chrysoxys maior* Skuja, 1948 (Starmach, 1985)

Cladonema Kent, 1880-82. *Cladonema laxum* (Kent, 1871) Kent, 1880-82 (Starmach, 1985; Preisig *et al.*, 1991)

Codonobotrys Pascher, 1942. *Codonobotrys physalis* Pascher, 1942 (Preisig et al. 1991)

Codonodendron Pascher, 1942. *Codonodendron ocellatum* Pascher, 1942 (Starmach, 1985; Preisig et al., 1991)

Codonoeca James-Clark, 1867. *Codonoeca costata* James-Clark, 1867 (Preisig et al., 1991)

Codonomonas van Goor, 1925 (see *Pseudokephyrion* Pascher, 1913)

Conradocystis Hollande, 1952. *Conradocystis dinobryonis* (Conrad, 1930) Hollande, 1952 (Starmach, 1985)

Cyclonexis Stokes, 1886a. *Cyclonexis annularis* Stokes, 1886a (Starmach, 1985)

Dendromonas Stein, 1878. *Dendromonas virgaria* (Weisse, 1845) Stein, 1878 (Starmach, 1985; Preisig et al., 1991)

Derepyxis Stokes, 1885a. *Derepyxis amphora* Stokes, 1885a (Starmach, 1985)

Didymochrysis Pascher, 1929. *Didymochrysis paradoxa* Pascher, 1929 (Starmach, 1985)

Dinobryopsis Lemmermann, 1899 (see *Pseudokephyrion* Pascher, 1913)

Eusphaerella Skuja, 1948. *Eusphaerella turfosa* Skuja, 1948 (Starmach, 1985)

Heliapsis Pascher, 1940b. *Heliapsis mutabilis* Pascher, 1940b (Starmach, 1985)

Heliochrysis Pascher, 1940a. *Heliochrysis eradians* Pascher, 1940a (Starmach, 1985)

Heterochromulina Pascher, 1912b (see *Oikomonas* Kent, 1880-82)

Heterolagynion Pascher, 1912a. *Heterolagynion oedogonii* Pascher, 1912a (Starmach, 1985)

Hyalocylix Petersen & Hansen, 1958. *Hyalocylix stipitata* Petersen & Hansen, 1958 (Starmach, 1985)

Kybotion Pascher, 1940a. *Kybotion eremita* Pascher, 1940a (Starmach, 1985)

Lepochromulina Scherffel, 1911. *Lepochromulina bursa* Scherffel, 1911 (Starmach, 1985)

Leukochrysis Pascher, 1917. *Leukochrysis pascheri* Bourrelly, 1957 (Starmach, 1985)

Leukapsis Pascher, 1940b. *Leukapsis vorax* Pascher, 1940b

Matvienkomonas Skvortzov, 1961 (see *Synuropsis* Schiller, 1929)

Microglena Ehrenberg, 1832. The type species, *M. monadina* Ehrenberg (Ehrenberg, 1832, 1838), has been referred to *Chlamydomonas* Ehrenberg, 1834 (Chlorophyceae) by Stein (1878) and though other species placed in *Microglena* are typically chrysophycean in organization, the genus should not be assigned to the Chrysophyceae (Farr et al., 1979). In addition to *M. monadina*, 6 species (and a few further questionable species) have been described and most if not all of these resemble *Chromulina* and should probably be combined with this genus. In some species of *Microglena*, unusual lens-shaped bodies of siliceous composition have been reported from the cell surface. Such bodies are not known from any species of *Chromulina*, but possibly their composition has been misinterpreted, and they merely represent mucocysts or wart-like outgrowths of the cell surface, as have been reported from *Chromulina* spp. as well as from a species of *Microglena* (*M. butcheri* Belcher, 1966) studied by electron microscopy (Couté and Preisig, 1981). All species of *Microglena* have been rarely observed, occurring in freshwater, except one which is marine (see Conrad, 1927; Droop, 1955; Belcher, 1966; Starmach, 1985).

Monochrysis Skuja, 1948. *Monochrysis maior* Skuja, 1948 (Starmach, 1985). One species, *Monochrysis lutheri* Droop 1953, has been transferred to the prymnesiophycean genus *Pavlova* Butcher, 1952 (Green, 1975). For clarification of the taxonomic position of *Monochrysis* the type species should be reexamined.

Mucosphaera Dop & van Beem in Dop (1980). *Mucosphaera geophila* Dop & van Beem in Dop (1980)

Mycochrysis Skuja, 1958. *Mycochrysis oligothiophila* Skuja, 1958 (Starmach, 1985)

Myxochrysis Pascher, 1916b. *Myxochrysis paradoxa* Pascher, 1916b (Starmach, 1985)

Ochrostylon Pascher, 1942. *Ochrostylon epiplankton* Pascher, 1942 (Starmach, 1985)

Ollicola Vørs, 1992 (see *Pseudokephyrion* Pascher, 1913)

Platychrysella Valkanov, 1970. *Platychrysella vorax* Valkanov, 1970

Platytheca Stein, 1878. *Platytheca micropora* Stein, 1878 (Starmach, 1985)

Porochrysis Pascher, 1917. *Porochrysis aspergillus* Pascher, 1917 (Starmach, 1985)

Porostylon Pascher, 1940a. *Porostylon transversale* Pascher, 1940a (Starmach, 1985)

Pseudosynura Kisselew, 1931 (see *Synuropsis* Schiller, 1929)

Pyramidochrysis Pascher, 1909. *Pyramidochrysis splendens* Pascher, 1909 (Starmach, 1985)

Rhizaster Pascher, 1915. *Rhizaster crinoides* Pascher, 1915 (Starmach, 1985)

Rhizochrysis Pascher, 1913 (see *Chrysamoeba* Klebs, 1893)

Rhizoochromonas Nicholls, 1990. *Rhizoochromonas endoloricata* Nicholls, 1990

Siderodendron Pringsheim, 1946. *Siderodendron manganiferum* Pringsheim, 1946 (Preisig et al., 1991)

Siphomonas Pringsheim, 1946. *Siphomonas fritschii* Pringsheim, 1946 (Preisig et al., 1991)

Sphaeraspis Schiller, 1954. *Sphaeraspis pascheri* Schiller, 1954 (Starmach, 1985)

Sphaerobryon Taylor, 1954. *Sphaerobryon fimbriata* Taylor, 1954 (Starmach, 1985)

Sphaleromantis Pascher, 1910b (see *Chrysolepidomonas* Peters & Andersen, 1993a)

Stephanoporos Conrad & Pascher in Pascher, 1940a (superfluous substitute name for *Chrysothecopsis* Conrad, 1931)

Stipitochrysis Korshikov, 1941. *Stipitochrysis monorhiza* Korshikov, 1941 (Starmach, 1985)

Stokesiella Lemmermann, 1910. *Stokesiella acuminata* (Stokes, 1885b) Lemmermann, 1910 (Starmach, 1985; Preisig et al., 1991)

Styloceras Reverdin, 1919. *Styloceras longissimus* Reverdin, 1919 (Starmach, 1985)

Stylochrysalis Stein, 1878. *Stylochrysalis parasita* Stein, 1878 (Starmach, 1985)

Stylococcus Chodat, 1898. *Stylococcus aureus* Chodat, 1898 (Starmach, 1985)

Syncrypta Ehrenberg, 1834 (see *Lepidochrysis* Ikävalko, Kristiansen & Thomsen, 1994 and *Synuropsis* Schiller, 1929)

Synochromonas Korshikov, 1929 (see *Synuropsis* Schiller, 1929)

Uroglenopsis Lemmermann, 1899 (see *Uroglena* Ehrenberg, 1834)

Volvochrysis Schiller, 1929 (see *Synuropsis* Schiller, 1929)

Wellheimia Pascher, 1917. *Wellheimia pfeifferi* Pascher, 1917 (Starmach, 1985)

Woronichiniella Skvortzov, 1961. *Woronichiniella pentagona* Skvortzov, 1961

LITERATURE CITED

Aaronson, S. 1993. The ultrastructure and function of the contractile vacuole. In: Berner, T. (ed.) *Ultrastructure of Microalgae*. CRC Press, Boca Raton, Florida. pp 205-219.

Aaronson, S. & Behrens, U. 1973. A note on the fine structure of the *Ochromonas danica* "tail". *Arch. Mikrobiol.*, 93:359-362.

Aaronson, S. & Behrens, U. 1974. Ultrastructure of an unusual contractile vacuole in several chrysomonad phytoflagellates. *J. Cell Sci.*, 14:1-9.

Ahlstrom, E. H. 1937. Studies on variability in the genus *Dinobryon* (Mastigophora). *Trans. Am. Microsc. Soc.*, 56:139-159.

Alberte, R. S. & Andersen, R. A. 1986. Antheraxanthin, a light-harvesting carotenoid found in a chromophyte alga. *Plant Physiol.*, 80:583-587.

Allen, M. B., Goodwin, T. W. & Phagpolngarm, S. 1960. Carotenoid distribution in certain naturally occurring algae and in some artificially induced mutants of *Chlorella pyrenoidosa*. *J. Gen. Microbiol.*, 23:93-103.

Andersen, R. A. 1982. A light and electron microscopical investigation of *Ochromonas sphaerocystis* Matvienko (Chrysophyceae): the statospore, vegetative cell and its peripheral vesicles. *Phycologia*, 21:390-398.

Andersen, R. A. 1986. Some new observations on *Saccochrysis piriformis* Korsh. emend. Andersen (Chrysophyceae). *In*: Kristiansen, J. & Andersen, R. A. (eds.), *Chrysophytes: Aspects and Problems*. Cambridge University Press, Cambridge, pp. 107-118.

Andersen, R. A. 1987. Synurophyceae classis nov., a new class of algae. *Am. J. Bot.*, 74:337-353.

Andersen, R. A. 1989a. Absolute orientation of the flagellar apparatus of *Hibberdia magna* comb. nov. (Chrysophyceae). *Nord. J. Bot.*, 8:653-669.

Andersen, R. A. 1989b. The Synurophyceae and their relationship to other golden algae. *Nova Hedwigia Beih.*, 95:1-26.

Andersen, R. A. 1990. The three-dimensional stucture of the flagellar apparatus of *Chrysosphaerella brevispina* (Chrysophyceae) as viewed by high voltage electron microscopy stereo pairs. *Phycologia*, 29:86-97.

Andersen, R. A. 1991. The cytoskeleton of chromophyte algae. *Protoplasma*, 164:143-159.

Andersen, R. A. & Mulkey, T. J. 1983. The occurrence of chlorophylls c_1 and c_2 in the Chrysophyceae. *J. Phycol.*, 19:289-294.

Andersen, R. A. & Wetherbee, R. 1992. Microtubules of the flagellar apparatus are active during prey capture in the chrysophycean alga *Epipyxis pulchra*. *Protoplasma*, 166:8-20.

Andersen, R. A., Brett, R. W., Potter, D. & Sexton, J. P. 1998a. Phylogeny of the Eustigmatophyceae based upon the 18S rRNA gene, with emphasis on *Nannochloropsis*. *Protist*, 149:61-74.

Andersen, R. A., Potter, D., Bidigare, R. R., Latasa, M., Rowan, K. & O'Kelly, C. J. 1998b. Characterization and phylogenetic position of the enigmatic golden alga *Phaeothamnion confervicola*: ultrastructure, pigment composition and partial SSU rDNA sequence. *J. Phycol.*, 34:286-298.

Andersen, R. A., Saunders, G. W., Paskind, M. P. & Sexton, J. P. 1993. Ultrastructure and 18S rRNA gene sequence for *Pelagomonas calceolata* gen. et sp. nov. and the description of a new algal class, the Pelagophyceae *classis nov*. *J. Phycol.*, 29:701-715.

Andersen, R. A., van de Peer, Y., Potter, D., Sexton, J. P., Kawachi, M. & LaJeunesse, T. 1999. Phylogenetic analysis of the SSU rRNA from members of the Chrysophyceae. *Protist*, 150:71-84.

Archibald, A. R., Cunningham, W. L., Manners, D. J. & Stark, J. R. 1963. Studies on the metabolism of

Protozoa. 10. The molecular structure of reserve polysaccharides from *Ochromonas malhamensis* and *Peranema trichophorum*. *Biochem. J.*, 88:444-451.

Bailey, J. C., Bidigare, R. R., Christensen, S. J. & Andersen, R. A. 1998. Phaeothamniophyceae *classis nova*: a new lineage of chromophytes based upon photosynthetic pigments, *rbc*L sequence analysis and ultrastructure. *Protist*, 149:245-263.

Beech, P. L. & Moestrup, Ø. 1986. Light and electron microscopical observations on the heterotrophic protist *Thaumatomastix salina* comb. nov. (syn. *Chrysosphaerella salina*) and its allies. *Nord. J. Bot.*, 6:865-877.

Belcher, J. H. 1966. *Microglena butcheri* nov. sp., a flagellate from the English Lake District. *Hydrobiologia*, 27:65-69.

Belcher, J. H. 1968. Lorica construction in *Pseudokephyrion pseudospirale* Bourrelly. *Br. Phycol. Bull.*, 3:495-499.

Belcher, J. H. 1969. A morphological study of the phytoflagellate *Chrysococcus rufescens* Klebs in culture. *Br. Phycol. J.*, 4:105-117.

Belcher, J. H. 1974. *Chrysosphaera magna* sp. nov., a new coccoid member of the Chrysophyceae. *Br. Phycol. J.*, 9:139-144.

Belcher, J. H. & Swale, E. M. F. 1967. *Chromulina placentula* sp. nov. (Chrysophyceae), a freshwater nannoplankton flagellate. *Br. Phycol. Bull.*, 3:257-267.

Belcher, J. H. & Swale, E. M. F. 1972a. Some features of the microanatomy of *Chrysococcus cordiformis* Naumann. *Br. Phycol. J.*, 7:53-59.

Belcher, J. H. & Swale, E. M. F. 1972b. The morphology and fine structure of the colourless colonial flagellate *Anthophysa vegetans* (O.F. Müller) Stein. *Br. Phycol. J.*, 7:335-346.

Belcher, J. H. & Swale, E. M. F. 1976. *Spumella elongata* (Stokes) nov. comb., a colourless flagellate from soil. *Arch. Protistenkd.*, 118:215-220.

Bhattacharya, D., Medlin, L., Wainright, P. O., Ariztia, E. V., Bibeau, C., Stickel, S. K. & Sogin, M. L. 1992. Algae containing chlorophylls *a* + *c* are paraphyletic: molecular evolutionary analysis of the Chromophyta. *Evolution*, 46:1801-1817.

Billard, C. 1978. *Chrysopodocystis socialis* gen. et sp. nov. (Chrysophyceae), une nouvelle Rhizochrysidale marine loriquée. *Bull. Soc. Bot. Fr.*, 125:307-312.

Bird, D. F. & Kalff, J. 1987. Algal phagotrophy: regulating factors and importance relative to photosynthesis in *Dinobryon* (Chrysophyceae). *Limnol. Oceanogr.*, 32:277-284.

Bjørnland, T. & Liaaen-Jensen, S. 1989. Distribution patterns of carotenoids in relation to chromophyte phylogeny and systematics. *In*: Green, J. C., Leadbeater, B. S. C. & Diver, W. L. (eds.), *The Chromophyte Algae: Problems and Perspectives. Systematics Association Special Volume*, 38:37-61. Clarendon Press, Oxford.

Bolochonzew, E. N. 1909. *Phytologie des Ladoga-Sees.* St. Petersburg.

Bory de Saint-Vincent, M. 1822. *Dictionnaire Classique d'Histoire Naturelle* 1:427, 597. Rey et Gravier, Libraires-Éditeurs, Paris.

Bouck, G. B. & Brown, D. L. 1973. Microtubule biogenesis and cell shape in *Ochromonas.* I. The distribution of cytoplasmic and mitotic microtubules. *J. Cell Biol.*, 56:340-359.

Bourrelly, P. 1957. Recherches sur les Chrysophycées. *Rev. Algol., Mém. Hors-Sér.*, 1:1-412.

Bourrelly, P. 1968/1981. *Les alges d'eau douce*, vol. 2. Boubée, Paris (réimpression revue et augmentée, 1981).

Brown, D. L. & Bouck, G. B. 1973. Microtubule biogenesis and cell shape in *Ochromonas.* II. The role of nucleating sites in shape development. *J. Cell Biol.*, 56:360-378.

Butcher, R. W. 1952. Contributions to our knowledge of the smaller marine algae. *J. Mar. Biol. Ass. U.K.*, 31:175-191.

Bütschli, 0. 1883-87. Mastigophora. *In*: *H.G. Bronn's Klassen und Ordnungen des Thierreichs*, Band 1, 2. Abteilung. C.F. Winter'sche Verlagshandlung, Leipzig und Heidelberg.

Caron, D. A., Sanders, R. W., Lim, E. L., Marrasé, C., Amaral, L. A., Whitney, S., Aoki, R. B. & Porter, K. G. 1993. Light-dependent phagotrophy in the freshwater mixotrophic chrysophyte *Dinobryon cylindricum*. *Microb. Ecol.*, 25:93-111.

Cavalier-Smith, T., Chao, E. E. & Allsopp, M. T. E. P. 1995. Ribosomal RNA evidence for chloroplast loss within Heterokonta: pedinellid relationships and a revised classification of ochristan algae. *Arch. Protistenkd.*, 145:209-220.

Cavalier-Smith, T., Chao, E. E., Thompson, C. E. & Hourihane, S. L. 1996. *Oikomonas*, a distinctive zooflagellate related to chrysomonads. *Arch. Protistenkd.*, 146:273-279.

Chen, J. L., Proteau, P. J., Roberts, M. A., Gerwick, W. H., Slate, D. L. & Lee, R. H. 1994. Structure of malhamensilipin A, an inhibitor of protein tyrosine kinase, from the cultured chrysophyte *Poterioochromonas malhamensis*. *J. Natural Products (Lloydia)*, 57:524-527.

Chodat, R. 1897. Études de biologie lacustre. Recherches sur les algues pélagiques de quelques lacs suisses et français. *Bull. Herb. Boissier*, 5:289-314.

Chodat, R. 1898. Études de biologie lacustre. *Bull. Herb. Boissier*, 6:431-476.

Chodat, R. 1922. Matériaux pour l'histoire des Algues de la Suisse, I-IX. *Bull. Soc. Bot. Genève*, 13:66-114.

Cienkowski, L. 1870. Ueber Palmellaceen und einige Flagellaten. *Arch. Mikroskop. Anat.*, 7:421-438.

Clarke, K. J. & Pennick, N. C. 1975. *Syncrypta glomerifera* sp. nov., a marine member of the Chrysophyceae bearing a new form of scale. *Brit. Phycol. J.*, 10:363-370.

Coleman, A. W., Thompson, W. F. & Goff, L. J. 1991. Identification of the mitochondrial genome in the chrysophyte alga *Ochromonas danica. J. Protozool.*, 38:129-135.

Conrad, W. 1926. Recherches sur les Flagellates de nos eaux saumâtres. *Arch. Protistenkd.*, 56:167-231.

Conrad, W. 1927. Le genre *Microglena* C. G. Ehrenberg (1838). *Arch. Protistenkd.*, 60:415-439.

Conrad, W. 1930. Flagellates nouveaux ou peu connus. I. *Arch. Protistenkd.*, 70: 657-680.

Conrad, W. 1931. Recherches sur les Flagellates de Belgique. I. *Mem. Mus. Roy. Hist. Nat. Belg.*, 47:1-65.

Conrad, W. 1938a. Notes protistologiques. III. Chrysomonadines intéressantes du nannoplankton saumâtre. *Bull. Mus. R. Hist. Nat. Belg.*, 14 (29):1-7.

Conrad, W. 1938b. Notes protistologiques. V. Observations sur *Uroglena soniaca* n. sp. et remarques sur le genre *Uroglena* Ehr. (incl. *Uroglenopsis* Lemm.). *Bull. Mus. R. Hist. Nat. Belg.*, 14 (42):1-27.

Conrad, W. 1939. Notes protistologiques. VII. Sur quelques Chrysomonadines du nannoplankton de Rouge-Cloître. *Bull. Mus. R. Hist. Nat. Belg.*, 15 (2):1-10.

Couté, A. 1983. Ultrastructure de *Chromophyton rosanoffii* Woronin emend. Couté et *Chr. vischeri* (Bourrel.) nov. comb. (Chrysophyceae, Ochromonadales, Ochromonadaceae). *Protistologia,* 19:393-416.

Couté, A. & Preisig, H. R. 1981. Sur l'ultrastructure de *Microglena butcheri* Belcher (Chrysophyceae, Ochromonadales, Synuraceae) et sur sa position systématique. *Protistologica*, 17:465-477.

Cronberg, G. 1992. *Uroglena dendracantha* n. sp. (Chrysophyceae) from Central Småland, Sweden. *Nord. J. Bot.*, 12:507-512.

Daugbjerg, N. & Andersen, R. A. 1997. A molecular phylogeny of the heterokont algae based on analyses of chloroplast encoded *rbc*L sequence data. *J. Phycol.*, 33:1031-1041.

Deflandre, G. 1934. Sur l'abus de l'emploi en paléontologie du nom de genre *Trachelomonas. Ann. Protistol.*, 4:151-165.

Delaney, T. P., Hardison, L. K. & Cattolico, R. A. 1995. Evolution of plastid genomes: inferences from discordant molecular phylogenies. *In*: Sandgren, C. D., Smol, J. P. & Kristiansen, J. (eds.), *Chrysophyte Algae: Ecology, Phylogeny and Development.* Cambridge University Press, Cambridge. pp 25-45

De Saedeleer, H. 1929. Notules systématiques. VI. *Physomonas. Ann. Protistol.*, 2:177-178.

De Toni, J. B. 1895. Sylloge Algarum. Vol.3; Fucoideae (privately published), Padova.

Doflein, F. 1916. Zuckerflagellaten. *Biol. Zentralb.*, 36:439-447.

Dolgoff, G. I. 1922. Zur Systematik von *Trachelomonas* Ehrbg. *Russ. Hydrob. Ztschr.*, 1(9-10):289-291.

Dop, A. J. 1977. A new *Lagynion* species (Chrysophyceae). *Act. Bot. Neerl.*, 26:489-491.

Dop, A. J. 1980. *Benthic Chrysophyceae from the Netherlands.* Thesis. University of Amsterdam.

Droop, M. R. 1953. On the ecology of flagellates from some brackish and fresh water rockpools of Finland. *Acta Bot. Fennica*, 51:1-52.

Droop, M. R. 1955. Some new supra-littoral protista. *J. Mar. Biol. Ass. U.K.*, 34:233-245.

Duff, K. E., Zeeb, B. A. & Smol, J. P. 1995. Atlas of chrysophycean cysts. *Develop. Hydrobiol.*, 99:1-189. Kluwer Academic Publishers, Dordrecht.

Dunlap, J. R., Walne, P. L. & Preisig, H. R. 1987. Manganese mineralization in chrysophycean loricas. *Phycologia*, 26:394-396.

Dürrschmidt, M. & Croome, R. 1985. Mallomonadaceae from Malaysia and Australia. *Nord. J. Bot.*, 5:285-298.

Ehrenberg, C. G. 1832. Über die Entwickelung und Lebensdauer der Infusionsthiere; nebst ferneren Beiträgen zu einer Vergleichung ihrer organischen Systeme. *Abh. Akad. Wiss. Berlin, Phys. Kl.*, 1831:1-154.

Ehrenberg, C. G. 1834. Dritter Beitrag zur Erkenntnis grosser Organisation in der Richtung des kleinsten Raumes. *Abh. Akad. Wiss. Berlin, Phys. Kl.*, 1833:145-336.

Ehrenberg, C. G. 1838. *Die Infusionsthierchen als vollkommene Organismen.* Voss, Leipzig.

Eichenberger, W. & Gribi, C. 1994. Diacylglyceryl-α-D-glucuronide from *Ochromonas danica* (Chrysophyceae). *J. Plant Physiol.*, 144:272-276.

Ellis-Adam, A C. 1982. *Chrysostephanosphaera hyalocytobia* spec. nov. (Chrysophyceae). *Acta Bot. Neerl.*, 31:169-174.

Engler, A. 1897. Bemerkung, betreffend der in Abteilung I:2 noch nicht berücksichtigten Chlorophyceae und Phaeophyceae. *In*: Engler, A. & Prantl, K. (eds.), *Die natürlichen Pflanzenfamilien*, 1. Teil, 2. Abt. Engelmann, Leipzig, p 570

Farr, E. R., Leussink, J. A. & Stafleu, F. A. (eds.) 1979. Index Nominum Genericorum (Plantarum), Vol. 2.

Regnum Vegetabile 101. Bohn, Scheltema and Holkema, Utrecht / W. Junk, The Hague.

Fiatte, M.-C. 1965. Observations sur la division cellulaire chez *Paraphysomonas vestita* (Stokes) De Sad., chrysomonadine décolorée. *Arch. Zool. Exp. Gén.*, 105:215-221.

Fiatte, M.-C. & Joyon, L. 1965. *Heterochromonas hovassei* (n. sp.), chrysomonadine décolorée. Arch. *Zool. Exp. Gén.*, 105:273-283.

Focke, R. 1957. "*Atraktochrysis rotans*" nov. gen., nov. spec., eine neue koloniebildende Ochromonadale. *Rev. Algol. N.S.*, 2:239-245.

Foissner, W. 1991. Diversity and ecology of soil flagellates. *In*: Patterson, D. J. & Larsen, J. (eds.), *The Biology of Free-Living Heterotrophic Flagellates. Systematics Association Special Volume*, 45:93-112. Clarendon Press, Oxford.

Foster, K. W. & Smyth, R. D. 1980. Light antennas in phototactic algae. *Microbiol. Rev.*, 44:572-630.

Fott, B. 1953. New algae and flagellata. *Preslia*, 25:143-156.

Fott, B. 1959. Zur Frage der Sexualität bei den Chrysomonaden. *Nova Hedwigia*, 1:115-129.

Franke, W. W. & Herth, W. 1973. Cell and lorica fine structure of the chrysomonad alga, *Dinobryon sertularia* Ehr. (Chrysophyceae). *Arch. Mikrobiol.*, 91:323-344.

Gayral, P. & Billard, C. 1977. Synopsis du nouvel ordre des Sarcinochrysidales (Chrysophyceae). *Taxon*, 26:241-245.

Geitler, L. 1949. Symbiosen zwischen Chrysomonaden und knospenden Bakterien-artigen Organismen sowie Beobachtungen über Organisationseigentümlichkeiten der Chrysomonaden. *Österr. Bot. Z.*, 95:300-324.

Gibbs, S. P. 1962. Chloroplast development in *Ochromonas danica*. *J. Cell Biol.*, 15:343-361.

Gibbs, S. P. 1981. The chloroplast endoplasmic reticulum: Structure, function, and evolutionary significance. *Int. Rev. Cytol.*, 72:49-99.

Grassé, P.-P. 1926. Contribution à l'étude des Flagellés parasites. *Arch. Zool. Exp. Gén.*, 65:345-602.

Green, J. C. 1975. The fine structure and taxonomy of the haptophycean flagellate *Pavlova lutheri* (Droop) comb. nov. (=*Monochrysis lutheri* Droop). *J. Mar. Biol. Ass. U.K.*, 55:785-793.

Grevby, C. & Sundqvist, C. 1992. Characterization of light-harvesting complex in *Ochromonas danica* (Chrysophyceae). *J. Plant Physiol.*, 140:414-420.

Gunderson, J. H., Elwood, H., Ingold, A., Kindle, K. & Sogin, M. L. 1987. Phylogenetic relationships between chlorophytes, chrysophytes, and oomycetes. *Proc. Natl. Acad. Sci. USA*, 84:5823-5827.

Haeckel, E. 1878. *Das Protistenreich*. Ernst Günthers Verlag, Leipzig.

Haeckel, E. 1894. *Systematische Phylogenie der Protisten und Pflanzen*. G. Reimer, Berlin.

Hansen, P. 1996. *Spiniferomonas cetrata* sp. nov. (Chrysophyceae), from an event of stomatocyst formation in the tropics. *Nova Hedwigia Beih.*, 114:71-80.

Hansgirg, A. 1886. Prodromus der Algenflora von Böhmen. Erster Theil enthaltend die Rhodophyceen, Phaeophyceen und einen Theil der Chlorophyceen. *Arch. Naturwiss. Landesdurchf. Böhmen*, 5:1-96.

Harris, K. 1963. Observations on *Sphaleromantis tetragona*. *J. Gen. Microbiol.*, 33:345-348.

Heinrich, G., Kies, L. & Schröder, W. 1986. Eisen- und Mangan-Inkrustierung in Scheiden, Gehäusen und Zellwänden einiger Bakterien, Algen und Pilze des Süsswassers. *Biochem. Physiol. Pflanzen*, 181:481-496.

Herth, W. 1979. Behaviour of the chrysoflagellate alga, *Dinobryon divergens*, during lorica formation. *Protoplasma*, 100:345-351.

Herth, W. & Zugenmaier, P. 1979. The lorica of *Dinobryon*. *J. Ultrastr. Res.*, 69:262-272.

Herth, W., Kuppel, A. & Schnepf, E. 1977. Chitinous fibrils in the lorica of the flagellate chrysophyte *Poteriochromonas stipitata* (syn. *Ochromonas malhamensis*). *J. Cell Biol.*, 73:311-321.

Heynig, H. 1969. Beobachtungen an planktischen Flagellaten. *Arch. Protistenkd.*, 111:170-191.

Hibberd, D. J. 1970. Observations on the cytology and ultrastructure of *Ochromonas tuberculatus* sp. nov. (Chrysophyceae), with special reference to the discobolocysts. *Br. Phycol. J.*, 5:119-143.

Hibberd, D. J. 1971. Observations on the cytology and ultrastructure of *Chrysamoeba radians* Klebs (Chrysophyceae). *Brit. Phycol. J.*, 6:207-23.

Hibberd, D. J. 1976. The ultrastructure and taxonomy of the Chrysophyceae and Prymnesiophyceae (Haptophyceae): a survey with some new observations on the ultrastructure of the Chrysophyceae. *Bot. J. Linn. Soc.*, 72:55-80.

Hibberd, D. J. 1977. Ultrastructure of cyst formation in *Ochromonas tuberculata* (Chrysophycae). *J. Phycol.*, 13:309-320.

Hibberd, D. J. & Leedale, G. F. 1971. A new algal class: the Eustigmatophyceae. *Taxon*, 20:523-525.

Hibberd, D. J. & Leedale, G. F. 1985. Order 4. Chrysomonadida. *In*: Lee, J. J., Hutner, S. H. & Bovee, E. C. (eds.), *An Illustrated Guide to the Protozoa*. Society of Protozoologists, Lawrence, Kansas. pp 54-70

Hilliard, D. K. 1966. New or rare chrysophytes from Lancashire County, England. *Arch. Protistenkd.*, 109:114-124.

Hilliard, D. K. 1967. Studies on Chrysophyceae from some ponds and lakes in Alaska. VII. Notes on the genera *Kephyrion*, *Kephyriopsis*, and *Pseudokephyrion*. *Nova Hedwigia*, 14:39-56.

Hilliard, D. K. 1971. Observations on the lorica structure of some *Dinobryon* species (Chrysophyceae), with comments on related genera. *Öst. Bot. Z.*, 119:25-40.

Hilliard, D. K. & Asmund, B. 1963. Studies on Chrysophyceae from some ponds and lakes in Alaska. II. Notes on the genera *Dinobryon*, *Hyalobryon* and *Epipyxis* with decriptions of new species. *Hydrobiologia*, 22:331-397.

Holen, D. A. & Boraas, M. E. 1995. Mixotrophy in chrysophytes. *In*: Sandgren, C. D., Smol, J. P. & Kristiansen, J. (eds.), *Chrysophyte Algae: Ecology, Phylogeny and Development*. Cambridge University Press, Cambridge. pp 119-140

Hollande, A. 1942. Étude cytologique et biologique de quelques flagellés libres. *Arch. Zool. Exp. Gén.*, 83:1-268.

Hollande, A. 1952. Ordre des chrysomonadines. *In*: Grassé, P.- P. (ed.), *Traité de Zoologie*. Masson, Paris, 1:471-570.

Huber-Pestalozzi, G. 1941. Das Phytoplankton des Süsswassers. Systematik und Biologie, Teil 2 (1): Chrysophyceen, farblose Flagellaten, Heterokonten. i: Thienemann, A. (ed.), *Die Binnengewässer*, Band 16, Teil 2(1). Schweizerbart, Stuttgart. pp 1-365

Hutner, S. H., Provasoli, L. & Filfus, J. 1953. Nutrition of some phagotrophic fresh-water chrysomonads. *Ann. N.Y. Acad. Sci.*, 56:852-862.

Ikävalko, J., Kristiansen, J. & Thomsen, H. A. 1994. A revision of the taxonomic position of *Syncrypta glomerifera* (Chrysophyceae), establishment of a new genus *Lepidochrysis* and observations on the occurrence of *L. glomerifera* comb. nov. in brackish water. *Nord. J. Bot.*, 14:339-344.

Ikävalko, J., Thomsen, H. A. & Carstens, M. 1996. A preliminary study of NE Greenland shallow meltwater ponds with particular emphasis on loricate and scale-covered forms (Choanoflagellida, Chrysophyceae sensu lato, Synurophyceae, Heliozoea), including the descriptions of *Epipyxis thamnoides* sp. nov. and *Pseudokephyrion poculiforme* sp. nov. (Chrysophyceae). *Arch. Protistenkd.*, 147:29-42.

Ito, H. & Takahashi, E. 1992. Chrysophytes in the southern part of Hyogo Prefecture, Japan (IV) Two new species, *Spiniferomonas hamata* and *S. nichollsii* (Chrysophyceae, Paraphysomonadaceae). *Jap. J. Phycol.*, 40:181-184.

Iwanoff, L. 1899. Beitrag zur Kenntniss der Morphologie und Systematik der Chrysomonaden. *Bull. Acad. Imp. Sci. St. Petersbourg*, 11:247-262.

James-Clark, H. 1867. On the Spongiæ Ciliatæ as Infusoria Flagellata; or, observations on the structure, animality, and relationship *of Leucosolenia botryoides*, Bowerbank. *Mem. Boston Soc. Nat. Hist.*, 1:305-340.

Jeffrey, S. W. 1989. Chlorophyll c pigments and their distribution in the chromophyte algae. *In*: Green, J. C., Leadbeater, B. S. C. & Diver, W. L. (eds.), *The Chromophyte Algae: Problems and Perspectives. Systematics Association Special Volume*, 38:13-36. Clarendon Press, Oxford.

Jüttner, F. & Hahne, B. 1981. Volatile excretion products of *Poterioochromonas malhamensis*: Identification and formation. *Z. Pflanzenphysiol.*, 103:403-412.

Jüttner, F., Höflacher, B. & Wurster, K. 1986. Seasonal analysis of volatile organic biogenic substances (VOBS) in freshwater phytoplankton populations dominated by *Dinobryon*, *Microcystis* and *Aphanizomenon*. *J. Phycol.*, 22:169-175.

Kahan, D., Oren, R., Aaronson, S. & Behrens, U. 1978. Fine structure of the cell surface and Golgi apparatus of *Ochromonas*. *J. Protozool.*, 25:30-33.

Kamiya, H., Naka, K. & Hashimoto, K. 1979. Ichthyotoxicity of a flagellate *Uroglena volvox*. *Bull. Jap. Soc. Scient. Fish.*, 45:129.

Kent, W. S. 1871. Notes on Prof. James Clark's Infusoria, with description of new species. *Month. Microsc. J.*, 6:261-265.

Kent, W. S. 1880-82. *A Manual of the Infusoria*. Bogue, London.

Kimura, B. & Ishida, Y. 1985. Photophagotrophy in *Uroglena americana*, Chrysophyceae. *Jap. J. Limnol.*, 46:315-318.

Kimura, B. & Ishida, Y. 1986. Possible phagotrophic feeding of bacteria in a freshwater red tide Chrysophyceae *Uroglena americana*. *Bull. Jap. Soc. Scient. Fish. (Nippon Suisan Gakkaishi)*, 52:697-701.

Kisselew, J. A. 1931. Zur Morphologie einiger neuer und seltener Vertreter des pflanzlichen Microplanktons. *Arch. Protistenkd.*, 73:235-250.

Klebs, G. 1893. Flagellatenstudien. II. *Z. Wiss. Zool.*, 55:353-445.

Korshikov, A. A. 1927. *Skadovskiella sphagnicola*, a new colonial chrysomonad. *Arch. Protistenkd.*, 58:450-455.

Korshikov, A. A. 1929. Studies on the Chrysomonads. I. *Arch. Protistenkd.*, 67:253-290.

Korshikov, A. A. 1941. On some new or little known flagellates. *Arch. Protistenkd.*, 95:22-44.

Kreimer, G. 1994. Cell biology of phototaxis in flagellate algae. *Int. Rev. Cytol.*, 148:229-310.

Krieger, W. 1930. Untersuchungen über Plankton - Chrysomonaden. Die Gattungen Mallomonas und *Dinobryon* in monographischer Bearbeitung. *Bot. Arch.*, 29:257-329.

Kristiansen, J. 1963. Sexual and asexual reproduction in *Kephyrion* and *Stenocalyx* (Chrysophyceae). *Bot. Tidsskr.*, 59:244-254.

Kristiansen, J. 1972. Studies on the lorica structure in Chrysophyceae. *Svensk Bot. Tidskr.*, 66:184-190.

Kristiansen, J. 1980. Chrysophyceae from some Greek Lakes. *Nova Hedwigia*, 33:167-194.

Kristiansen, J. 1986. The ultrastructural bases of chrysophyte systematics and phylogeny. *CRC Crit. Rev. Pl. Sci.*, 4:149-211.

Kristiansen, J. 1990. Phylum Chrysophyta. *In*: Margulis, L., Corliss, J. O., Melkonian, M. & Chapman, D. J. (eds.), *Handbook of Protoctista*. Jones & Bartlett, Boston. pp 438-453

Kristiansen, J. & Andersen, R. A. (eds.) 1986. *Chrysophytes: Aspects and Problems*. Cambridge University Press, Cambridge.

Kristiansen, J. & Cronberg, G. (eds.) 1996: Chrysophytes. Progress and New Horizons. *Nova Hedwigia Beih.*, 114:1-266.

Kristiansen, J., Cronberg, G. & Geissler, U. (eds.) 1989. Chrysophytes: Developments and Perspectives. *Nova Hedwigia Beih.*, 95:1-287.

Kristiansen, J. & Tong, D. 1989. *Chrysosphaerella annulata* n. sp., a new scale-bearing chrysophyte. *Nord. J. Bot.*, 9:329-332.

Kristiansen, J. & Walne P. L. 1977. Fine structure of photokinetic systems in *Dinobryon cylindricum* var. *alpinum* (Chrysophyceae). *Br. Phycol. J.*, 12:329-341.

Kurata, A. 1986. Blooms of *Uroglena americana* in relation to concentrations of B group vitamins. *In*: Kristiansen, J. & Andersen, R. A. (eds.), *Chrysophytes: Aspects and Problems*. Cambridge University Press, Cambridge. pp 185-196

Lackey, J. B. 1940. Some new flagellates from the Woods Hole area. *Am. Midl. Naturalist*, 23:463-471.

Lagerheim, G. 1884. Über *Phaeothamnion*, eine neue Gattung unter den Süsswasseralgen. *Bihang Til K. Svenska Vet.-Acad. Handlingar*, 9 (19):3-13.

Lauterborn, R. 1896. Diagnosen neuer Protozoen aus dem Gebiete des Oberrheins. *Zool. Anz.*, 19:14-18.

Lauterborn, R. 1899. Protozoen-Studien. IV. Theil. Flagellaten aus dem Gebiete des Oberrheins. *Z. Wiss. Zool.*, 65:369-391.

Lauterborn, R. 1911. Pseudopodien bei *Chrysopyxis*. *Zool. Anz.*, 38:46-51.

Leadbeater, B. S. C. & Barker, D. A. N. 1995. Biomineralization and scale production in the Chrysophyta. *In*: Sandgren, C. D., Smol, J. P. & Kristiansen, J. (eds.), *Chrysophyte Algae: Ecology, Phylogeny and Development*. Cambridge University Press, Cambridge pp 141-164

Lee, R .E. 1978. Formation of scales in *Paraphysomonas vestita* and the inhibition of growth by germanium dioxide. *J. Protozool.*, 25:163-166.

Lee, R. E. & Kugrens, P. 1989. Biomineralization of the stalks of *Anthophysa vegetans* (Chrysophyceae). *J. Phycol.*, 25:591-596.

Leeper, D. A. & Porter, K. G. 1995. Toxicity of the mixotrophic chrysophyte *Poterioochromonas malhamensis* to the cladoceran *Daphnia ambigua*. *Arch. Hydrobiol.*, 134:207-222.

Leipe, D. D., Wainright, P. O., Gunderson, J. H., Porter, D., Patterson, D. J., Valois F., Himmerich, S., Sogin, M. L. 1994. The stramenopiles from a molecular perspective: 16S-like rRNA sequences from *Labyrinthuloides minuta* and *Cafeteria roenbergensis*. *Phycologia*, 33:369-377.

Lemmermann, E. 1898. Algologische Beiträge, IV. Süsswasseralgen der Insel Wangerooge. *Abh. Natur. Ver. Bremen*, 14:501-512.

Lemmermann, E. 1899. Das Phytoplankton sächsischer Teiche. *Forschungsber. Biol. Stat. Plön*, 7:96-135.

Lemmermann, E. 1900 Beiträge zur Kenntnis der Planktonalgen. XI. Die Gattung *Dinobryon* Ehrenb. *Ber. Deutsch. Bot. Ges.*, 18:500-524.

Lemmermann, E. 1901. Silicoflagellatae. Ergebnisse einer Reise nach dem Pacific. *Ber. Deutsch. Bot. Ges.*, 19:247-271.

Lemmermann, E. 1910. Algen I. *In*: *Kryptogamenflora der Mark Brandenburg und Angrenzender Gebiete.* Gebrüder Borntraeger, Leipzig, 3:1-712.

Li, N. & Cattolico, R. A. 1992. *Ochromonas danica* (Chrysophyceae) chloroplast genome organization. *Mol. Mar. Biol. Biotechnol.*, 1:165-174.

Loeblich, A. R. III & Loeblich, L. A. 1978. Divisions Bacillariophyta, Chloromonadophyta, Chrysophyta, Eustigmatophyta, Haptophyta, Xanthophyta. *In*: Laskin, A. I. & Lechevalier, H. A. (eds.), *CRC Handbook of Microbiology*, 2nd ed. CRC Press, West Palm Beach, Florida, 2:411-423.

Lohmann, H. 1908. Untersuchungen zur Feststellung des vollständigen Gehaltes des Meeres an Plankton. *Wiss. Meeresuntersuch., Abt. Kiel*, 10:129-370.

Lund, J. W. G. 1959. Concerning *Calycomonas* Lohmann and *Codonomonas* van Goor. *Nova Hedwigia*, 1:423-429.

Lund, J. W. G. 1960. New or rare British Chrysophyceae. 3. New records and observations on sexuality and classification. *New Phytol.*, 59:349-360.

Lund, J. W. G. 1962. A genus new to Britain: *Amphichrysis* Korsh. *Br. Phycol. Bull.*, 2:116-120.

Mack, B. 1951. Morphologische und entwicklung geschichtliche Untersuchungen an Chrysophyceen. *Österr. Bot. Z.*, 98:249-279.

Manton, I. & Harris, K. 1966. Observations on the microanatomy of the brown flagellate

Sphaleromantis tetragona Skuja with special reference to the flagellar apparatus and scales. *J. Linn. Soc. (Bot.)*, 59:397-403.

Manton, I. & Leedale, G. F. 1961. Observations on the fine structure of *Paraphysomonas vestita*, with special reference to the Golgi apparatus and the origin of scales. *Phycologia,* 1:37-57.

McKenzie, C. H., Deibel, D., Paranjape, M. A. & Thompson, R. J. 1995. The marine mixotroph *Dinobryon balticum* (Chrysophyceae): phagotrophy and survival in a cold ocean. *J. Phycol.*, 31:19-24.

Medlin, L. K., Kooistra, W. H. C. F., Potter, D., Saunders, G. W. & Andersen, R. A. 1997. Phylogenetic relationships of the 'golden algae' (haptophytes, heterokont chromophytes) and their plastids. *Plant Syst. Evol. (Suppl.)*, 11:187-219.

Meyer, R. L. 1986. A proposed phylogenetic sequence for the loricate rhizopodial Chrysophyceae. *In*: Kristiansen, J. & Andersen, R. A. (eds.), *Chrysophytes: Aspects and Problems.* Cambridge University Press, Cambridge. pp 75-85

Mignot, J.-P. 1977. Étude ultrastructurale d'un flagellé du genre *Spumella* Cienk. (= *Heterochromonas* Pascher = *Monas* O.F. Müller), chrysomonadine leucoplastidiée. *Protistologica*, 13:219-231.

Moestrup, Ø. & Andersen, R. A. 1991. Organization of heterotrophic heterokonts. *In*: Patterson, D. J. & Larsen, J. (eds.), *The Biology of Free-living Heterotrophic Flagellates. Systematics Association Special Volume* 45:333-360. Clarendon Press, Oxford.

Müller, O. F. 1773. *Vermium terrestrium et fluviatilium*, vol.1 (1). Hauniae et Lipsiae (Copenhagen and Leipzig).

Müller, O. F. 1786. *Animalcula Infusoria Fluviatilia et Marina.* Hauniae (Copenhagen).

Nicholls, K. H. 1977. Two new species of *Pseudokephyrion* (Chrysophyceae) in the plankton of a lake. *J. Phycol.*, 13:410-413.

Nicholls, K. H. 1980. A reassessment of *Chrysosphaerella longispina* and *C. multispina*, and a revised key to related genera in the Synuraceae (Chrysophyceae). *Pl. Syst. Evol.*, 135:95-106.

Nicholls, K. H. 1981. *Chrysococcus furcatus* (Dolg.) comb. nov.: a new name for *Chrysastrella furcata* (Dolg.) Defl. based on the discovery of the vegetative stage. *Phycologia*, 20:16-21.

Nicholls, K. H. 1984. *Spiniferomonas septispina* and *S. enigmata*, two new algal species confusing the distinction between *Spiniferomonas* and *Chrysosphaerella* (Chrysophyceae). *Pl. Syst. Evol.*, 148:103-117.

Nicholls, K. H. 1985. The validity of the genus *Spiniferomonas* (Chrysophyceae). *Nord. J. Bot.*, 5:403-406.

Nicholls, K. H. 1987. *Chrysoamphipyxis* gen. nov.: a new genus in the Stylococcaceae (Chrysophyceae). *J. Phycol.*, 23:499-501.

Nicholls, K. H. 1990. Life history and taxonomy of *Rhizoochromonas endoloricata* gen. et sp. nov., a new freshwater chrysophyte inhabiting *Dinobryon* loricae. *J. Phycol.*, 26:558-563.

Nicholls, K. H. 1995. Chrysophyte blooms in the plankton and neuston of marine and freshwater systems. *In*: Sandgren, C. D., Smol, J. P. & Kristiansen, J. (eds.), *Chrysophyte Algae: Ecology, Phylogeny and Development.* Cambridge University Press, Cambridge. pp 181-213

Nielsen, Y. 1994. *Spiniferomonas abrupta*, sp. nov. and some rare species of Synurophyceae and Chrysophyceae, not formerly recorded from Denmark. *Nord. J. Bot.*, 14:473-480.

Ohishi, H., Yano, H., Ito, H. & Nakahara, M. 1991. Observations on a chrysophyte "Hikarimo" in a pond in Hyogo Prefecture, Japan. *Jap. J. Phycol.*, 39:37-42.

O'Kelly, C. J. 1989. The evolutionary origin of the brown algae: information from studies of motile cell ultrastructure. *In*: Green, J. C., Leadbeater, B. S. C. & Diver, W. L. (eds.), *The Chromophyte Algae: Problems and Perspectives. Systematics Association Special Volume* 38:255-278. Clarendon Press, Oxford.

O'Kelly, J. C. & Wujek, D. E. 1995. Status of the Chrysamoebales (Chrysophyceae): observations on *Chrysamoeba pyrenoidifera*, *Rhizochromulina marina* and *Lagynion delicatulum*. *In*: Sandgren, C. D., Smol, J. P. & Kristiansen, J. (eds.), *Chrysophyte Algae: Ecology, Phylogeny and Development.* Cambridge University Press, Cambridge. pp 361-372

Owen, H. A., Mattox, K. R. & Stewart, K. D. 1990a. Fine structure of the flagellar apparatus of *Dinobryon cylindricum* (Chrysophyceae). *J. Phycol.*, 26:131-141.

Owen, H. A., Stewart, K. D. & Mattox, K. R. 1990b. Fine structure of the flagellar apparatus of *Uroglena americana* (Chrysophyceae). *J. Phycol.*, 26:142-149.

Parke, M. 1949. Studies on marine flagellates. *J. Mar. Biol. Ass. U.K.*, 28:255-286.

Pascher, A. 1909. *Pyramidochrysis*, eine neue Gattung der Chrysomonaden. *Ber. Deutsch. Bot. Ges.*, 27:555-562.

Pascher, A. 1910a. Neue Chrysomonaden aus den Gattungen *Chrysococcus*, *Chromulina*, *Uroglenopsis*. *Österr. Bot. Z.*, 60:1-5.

Pascher, A. 1910b. Der Grossteich bei Hirschberg in Nord-Böhmen. *Int. Rev. Ges. Hydrobiol. Hydrogr., Monogr. Abhandl.*, 1:1-66.

Pascher, A. 1911. Über Nannoplanktonten des Süsswassers. *Ber. Deutsch. Bot. Ges.*, 29:523-533.

Pascher, A. 1912a. Eine farblose, rhizopodiale Chrysomonade. *Ber. Deutsch. Bot. Ges.*, 30:152-158.

Pascher, A. 1912b. Ueber Rhizopoden- und Palmellastadien bei Flagellaten (Chrysomonaden), nebst einer Uebersicht über die braunen Flagellaten. *Arch. Protistenkd.*, 25:153-200.

Pascher, A. 1913. Chrysomonadinae. *In*: Pascher, A. (ed.), *Die Süsswasser-Flora Deutschlands, Oesterreichs und der Schweiz.* G. Fischer, Jena, 2:7-95.

Pascher, A. 1914. Über Flagellaten und Algen. *Ber. Deutsch. Bot. Ges.*, 32:136-160.

Pascher, A. 1915. Studien über die rhizopodiale Entwicklung der Flagellaten, I. Teil. *Arch. Protistenkd.*, 36:81-117.

Pascher, A. 1916a. Rhizopodialnetze als Fangvorrichtung bei einer plasmodialen Chrysomonade. *Arch. Protistenkd.*, 37:15-30.

Pascher, A. 1916b. Fusionsplasmodien bei Flagellaten und ihre Bedeutung für die Ableitung der Rhizopoden von den Flagellaten. *Arch. Protistenkd.*, 37:31-64.

Pascher, A. 1917. Flagellaten und Rhizopoden in ihren gegenseitigen Beziehungen. *Arch. Protistenkd.*, 38:1-80.

Pascher, A. 1927. Eine Chrysomonade mit gestielten und verzweigten Kolonien. *Arch. Protistenkd.*, 57:319-330.

Pascher, A. 1928. Eine eigenartige rhizopodiale Flagellate. *Arch. Protistenkd.*, 63:227-240.

Pascher, A. 1929. Beiträge zur allgemeinen Zellehre. I. *Arch. Protistenkd.*, 68:261-304.

Pascher, A. 1931. Systematische Übersicht über die mit Flagellaten in Zusammenhang stehenden Algenreihen und Versuch einer Einreihung dieser Algenstämme in die Stämme des Pflanzenreiches. *Bot. Centralb. Beih.*, 48: 317-332.

Pascher, A. 1940a. Rhizopodiale Chrysophyceen. *Arch. Protistenkd.*, 93:331-349.

Pascher, A. 1940b. Filarplasmodiale Ausbildungen bei Algen. *Arch. Protistenkd.*, 94:295-309.

Pascher, A. 1942. Zur Klärung einiger gefärbter und farbloser Flagellaten und ihrer Einrichtungen zur Aufnahme animalischer Nahrung. *Arch. Protistenkd.*, 96:75-108.

Penard, E. 1921. Studies on some Flagellata. *Proceed. Acad. Nat. Sci. Philadelphia*, 73:105-168.

Péterfi, L. S. 1969. The fine structure of *Poterioochromonas malhamensis* (Pringsheim) comb. nov. with special reference to the lorica. *Nova Hedwigia*, 17:93-103.

Peters, M. C. & Andersen, R. A. 1993a. The fine structure and scale formation of *Chrysolepidomonas dendrolepidota* gen. et sp. nov. (Chrysolepidomonadaceae fam. nov., Chrysophyceae). *J. Phycol.*, 29:469-475.

Peters, M. C. & Andersen, R. A. 1993b. The flagellar apparatus of *Chrysolepidomonas dendrolepidota* (Chrysophyceae), a single-celled monad covered with organic scales. *J. Phycol.*, 29:476-485.

Petersen, J. B. & Hansen, J. B. 1958. On some neuston organisms I. *Bot. Tidsskr.*, 54:93-110.

Petry, K. 1968. Entwicklungsgeschichtliche Untersuchungen an *Chromophyton rosanoffii* und einigen Chlorophyceen. *Österr. Bot. Z.*, 115:447-481.

Pienaar, R. N. 1976. The microanatomy of *Sphaleromantis marina* sp. nov. (Chrysophyceae). *Br. Phycol J.*, 11:83-92.

Pienaar, R. N. 1980. Chrysophytes. *In*: Cox E. R (ed.), *Phytoflagellates.* Elsevier/North Holland, New York. pp 213-242

Pipes, L. D., Tyler, P. A. & Leedale, G. F. 1989. *Chrysonephele palustris* gen. et sp. nov. (Chrysophyceae), a new colonial chrysophyte from Tasmania. *Nova Hedwigia Beih.*, 95:81-97.

Poche, F. 1913. Das System der Protozoa. *Arch. Protistenkd.*, 30:125-321.

Preisig, H. R. 1986. Biomineralization in the Chrysophyceae. *In*: Leadbeater, B. S. C. & Riding, R. (eds.), *Biomineralization in Lower Plants and Animals.* Systematics Association Special Volume, 30:327-344. Clarendon Press, Oxford.

Preisig, H. R. 1989. The flagellar base ultrastructure and phylogeny of chromophytes. *In*: Green, J. C., Leadbeater, B. S. C. & Diver, W. L. (eds.), *The Chromophyte Algae: Problems and Perspectives.* Systematics Association Special Volume, 38:167-187. Clarendon Press, Oxford.

Preisig, H. R. 1994. Siliceous structures and silicification in flagellated protists. *Protoplasma*, 181:29-42.

Preisig, H. R. 1995. A modern concept of chrysophyte classification. *In*: Sandgren, C. D., Smol, J. P. & Kristiansen, J. (eds.), *Chrysophyte Algae: Ecology, Phylogeny and Development.* Cambridge University Press, Cambridge. pp 46-74

Preisig, H. R. & Hibberd, D. J. 1982a. Ultrastructure and taxonomy of *Paraphysomonas* (Chrysophyceae) and related genera (1). *Nord. J. Bot.*, 2:397-420.

Preisig, H. R. & Hibberd, D. J. 1982b. Ultrastructure and taxonomy of *Paraphysomonas* (Chrysophyceae) and related genera (2). *Nord. J. Bot.*, 2:601-638.

Preisig, H. R. & Hibberd, D. J. 1983. Ultrastructure and taxonomy of *Paraphysomonas* (Chrysophyceae) and related genera (3). *Nord. J. Bot.*, 3:695-723.

Preisig, H. R., Vørs, N. & Hällfors, G. 1991. Diversity of heterotrophic heterokont flagellates. *In*: Patterson, D. J. & Larsen, J. (eds.), *The Biology of*

Free-Living Heterotrophic Flagellates. Systematics Association Special Volume, 45:361-399. Clarendon Press, Oxford.

Pringsheim, E. G. 1935. Über Azetatflagellaten. *Naturwissenschaften*, 23:110-114.

Pringsheim, E.G. 1946. On iron flagellates. *Phil. Trans. R. Soc., Ser. B*, 232:311-342.

Pringsheim, E. G. 1952. On the nutrition of *Ochromonas. Q. J. Microsc. Sci.*, 93:71-96.

Pringsheim, E. G. 1955. Über *Ochromonas danica* n. sp. und andere Arten der Gattung. *Arch. Mikrobiol.*, 23:181-192.

Quillet, M. 1955. Sur la nature chimique de la leucosine, polysaccharide de réserve caractéristique des Chrysophycees, extraite d'*Hydrurus foetidus. C. R. Heb. Séanc. Acad. Sci., Paris*, 240:1001-1003.

Reverdin, L. 1919. Étude phytoplanctonique, expérimentale et descriptive des eaux du Lac de Genève. *Arch. Sci. Phys. Nat., Ser.* 5, 1:302-345 and 403-451.

Ricketts, T. R. 1965. Chlorophyll *c* in some members of the Chrysophyceae. *Phytochemistry,* 4:725-730.

Robinson, D. G. & Quader, H. 1980. Topographical features of the membrane of *Poterioochromonas malhamensis* after colchicine and osmotic treatment. *Planta*, 148:84-88.

Rouiller, C. & Fauré-Fremiet, E. 1958. Structure fine d'un flagellé chrysomonadien: *Chromulina psammobia. Exp. Cell Res.*, 14:47-67.

Sanders, R. W., Porter, K. G., Bennett, S. J. & DeBiase, A. E. 1989. Seasonal patterns of bacterivory by flagellates, ciliates, rotifers, and cladocerans in a freshwater planctonic community. *Limnol. Oceanogr.*, 34:673-687.

Sanders, R. W., Porter, K. G. & Caron, D. A. 1990. Relationship between phototrophy and phagotrophy in the mixotrophic chrysophyte *Poterioochromonas malhamensis. Microb. Ecol.*, 19:97-109.

Sandgren, C. D. 1980a. An ultrastructural investigation of resting cyst formation in *Dinobryon cylindricum* Imhof (Chrysophyceae, Chrysophycota). *Protistologica*, 16:259-276.

Sandgren, C. D. 1980b. Resting cyst formation in selected chrysophyte flagellates: an ultrastructural survey including a proposal for the phylogenetic significance of interspecific variations in the encystment process. *Protistologica*, 16:289-303.

Sandgren, C. D. 1981. Characteristics of sexual and asexual resting cyst (statospore) formation *in Dinobryon cylindricum* Imhof (Chrysophyta). *J. Phycol.*, 17:199-210.

Sandgren, C. D. 1983. Survival strategies of chrysophycean flagellates: reproduction and the formation of resistant resting cysts. *In*: Fryxell, G. A. (ed.), *Survival Strategies of the Algae.* Cambridge University Press, Cambridge. pp 23-48

Sandgren, C. D. 1986. Effects of environmental temperature on the vegetative growth and sexual life history of *Dinobryon cylindricum* Imhof. *In*: Kristiansen, J. & Andersen, R. A. (eds.), *Chrysophytes: Aspects and Problems.* Cambridge University Press, Cambridge. pp 207-225

Sandgren, C. D. 1988. The ecology of chrysophyte flagellates: their growth and perennation strategies as freshwater phytoplankton. *In*: Sandgren, C. D. (ed.), *Growth and Reproductive Strategies of Freshwater Phytoplankton.* Cambridge University Press, Cambridge. pp 9-104

Sandgren, C. D. 1989. SEM investigations of statospore (stomatocyst) development in diverse members of the Chrysophyceae and Synurophyceae. *Nova Hedwigia Beih.*, 95:45-69.

Sandgren, C. D. (ed.) 1991. Applications of chrysophyte stomatocysts in paleolimnology. *J. Paleolimnol.*, 5:1-113.

Sandgren, C. D., Smol, J. P. & Kristiansen, J. (eds.) 1995. *Chrysophyte algae: Ecology, Phylogeny and Development.* Cambridge University Press, Cambridge.

Saunders, G. W., Potter, D., Paskind, M. P. & Andersen, R. A. 1995. Cladistic analyses of combined traditional and molecular data sets reveal an algal lineage. *Proc. Natl. Acad. Sci. USA*, 92:244-248.

Saunders, G. W., Hill, D. R. A. & Tyler, P. A. 1997a. Phylogenetic affinities of *Chrysonephele palustris* (Chrysophyceae) based on inferred nuclear small-subunit ribosomal RNA sequence. *J. Phycol.*, 33:132-134.

Saunders, G. W., Potter, D. & Andersen, R. A. 1997b. Phylogenetic affinities of the Sarcinochrysidales and Chrysomeridales (Heterokonta) based on analyses of molecular and combined data. *J. Phycol.*, 33:310-318.

Scagel, R. F. & Stein, J. R. 1961. Marine nannoplankton from a British Columbia fjord. *Can. J. Bot.*, 39:1205-1213.

Scherffel, A. 1901. Kleiner Beitrag zur Phylogenie einiger Gruppen niederer Organismen. *Bot. Zeit.*, 59:143-158.

Scherffel, A. 1911. Beitrag zur Kenntnis der Chrysomonadineen. *Arch. Protistenkd.*, 22:299-344.

Scherffel, A. 1927. Beitrag zur Kenntnis der Chrysomonadineen. II. *Arch. Protistenkd.*, 57:331-361.

Schiller, J. 1926. Der thermische Einfluss und die Wirkung des Eises auf die planktischen Herbstvegetationen in den Altwässern der Donau bei Wien. *Arch. Protistenkd.*, 56:1-62.

Schiller, J. 1929. Neue Chryso- und Cryptomonaden aus Altwässern der Donau bei Wien. *Arch. Protistenkd.*, 66:436-458.

Schiller, J. 1954. Neue Mikrophyten aus künstlichen betonierten Wasserbehältern. 2. Mitteilung über neue Cyanosen. *Arch. Protistenkd.*, 100:116-126.

Schnepf, E., Deichgräber, G., Röderer, G. & Herth, W. 1977. The flagellar root apparatus, the microtubular system and associated organelles in the chrysophycean flagellate, *Poterioochromonas malhamensis* Peterfi (syn. *Poteriochromonas stipitata* Scherffel and *Ochromonas malhamensis* Pringsheim). *Protoplasma*, 92:87-107.

Schnepf, E., Röderer, G. & Herth, W. 1975. The formation of the fibrils in the lorica of *Poteriochromonas stipitata*: tip growth, kinetics, site, orientation. *Planta (Berl.)*, 125:45-62.

Schoenichen, W. 1925. *Einfachste Lebensformen des Tier- und Pflanzenreiches*, 5. Aufl., Band 1. *Spaltpflanzen, Geissellinge, Algen, Pilze*. Hugo Bermühler Verlag, BerlinLichterfelde.

Schoonoord, M. P. & Ellis-Adam, A. C. 1984. *Chrysocrinus honorarius* spec. nov. (Stylococcaceae, Chrysophyceae); some stages of its lorical development and its distribution on the substrate. *Acta Bot. Neerl.*, 33:399-418.

Schütt, F. 1892. Das Pflanzenleben der Hochsee (2. Anhang zu Kapitel VIII). *In*: Hensen, V. (ed.), *Ergebnisse der Plankton-Expedition der Humboldt-Stiftung* 1A:243-314. Verlag Lipsius & Tischer, Kiel.

Sheath, R. G., Hellebust, J. A. & Sawa, T. 1975. The statospore of *Dinobryon divergens* Imhof: formation and germination in a subarctic lake. *J. Phycol.*, 11:131-138.

Silva, P. C. 1980. Names of classes and families of living algae. *Regnum Vegetabile*, 103:1-156.

Siver, P. A. 1988a. The distribution and ecology of *Spiniferomonas* (Chrysophyceae) in Connecticut (USA). *Nord. J. Bot.*, 8:205-212.

Siver, P. A. 1988b. *Spiniferomonas triangularis* sp. nov., a new silica-scaled freshwater flagellate (Chrysophyceae, Paraphysomonadaceae). *Br. Phycol. J.*, 23:379-383.

Siver, P.A. 1993. Morphological and ecological characteristics of *Chrysosphaerella longispina* and *C. brevispina* (Chrysophyceae). *Nord. J. Bot.*, 13:343-351.

Skogstad, A. & Reymond, O. L. 1989. An ultrastructural study of vegetative cells, encystment, and mature statospores in *Spiniferomonas bourrellyi* (Chrysophyceae). *Nova Hedwigia Beih.*, 95:71-79.

Skuja, H. 1948. Taxonomie des Phytoplanktons einiger Seen in Uppland, Schweden. *Symb. Bot. Upsal.*, 9 (3):1-399.

Skuja, H. 1956. Taxonomische und biologische Studien über das Phytoplankton schwedischer Binnengewässer. *Nov. Act. Reg. Soc. Sci. Upsal., Ser.* 4/16 (3):1-404.

Skuja, H. 1958. *Mycochrysis* nov. gen., Vertreterin eines neuen Typus der Koloniebildung bei den gefärbten Chrysomonaden. *Svensk Bot. Tidskr.*, 52:23-36.

Skuja, H. 1964. Grundzüge der Algenflora und Algenvegetation der Fjeldgegenden um Abisko in Schwedisch-Lappland. *Nov. Act. Reg. Soc. Sci. Upsal., Ser.* 4/18 (3):1-465.

Skvortzov, B. V. 1961. Harbin Chrysophyta, China Boreali-Orientalis. *Bull. Herb. N.-E. Forest. Acad.*, 12 (3):1-70.

Slankis, T. & Gibbs, S. P. 1972. The fine structure of mitosis and cell division in the chrysophycean alga *Ochromonas danica*. *J. Phycol.*, 8:243-256.

Sleigh, M. A. 1964. Flagellar movement of the sessile flagellates *Actinomonas*, *Codonosiga*, *Monas*, and *Poteriodendron*. *Q. J. Microscop. Sci.*, 105:405-414.

Smestad-Paulsen, B. & Myklestad, S. 1978. Structural studies of the reserve glucan produced by the marine diatom *Skeletonema costatum* (Grev.) Cleve. *Carbohydr. Res.*, 62:386-388.

Smol, J. P. 1995. Application of chrysophytes to problems in paleoecology. *In*: Sandgren, C. D., Smol, J. P. & Kristiansen, J. (eds.), *Chrysophyte Algae: Ecology, Phylogeny and Development*. Cambridge University Press, Cambridge. pp 303-329

Spiegelstein, M., Reich, K. & Bergmann, F. 1969. The toxic principles of *Ochromonas* and related Chrysomonadina. *Verh. Internat. Verein. Limnol.*, 17:778-783.

Starmach, K. 1985. Chrysophyceae und Haptophyceae. *In*: Ettl, H., Gerloff, J., Heynig, H. & Mollenhauer, D. (eds.), *Süsswasserflora von Mitteleuropa*. G. Fischer, Stuttgart, 1:1-515.

Stein, F. von 1878. *Der Organismus der Infusionsthiere*, Abt. 3 (1). Engelmann, Leipzig.

Stokes, A. C. 1885a. Notes on some apparently undescribed forms of freshwater infusoria. No 2. *Am. J. Sci.*, 29:313-328.

Stokes, A. C. 1885b. Notices of new fresh-water infusoria. III. *Am. Month. Microsc. J.*, 6:121-127.

Stokes, A. C. 1886a. Notices of new fresh-water infusoria. *Proc. Amer. Philos. Soc.*, 23:562-568.

Stokes, A. C. 1886b. Notices of new fresh-water infusoria. V. *Am. Month. Microsc. J.*, 7:81-86.

Stokes, A. C. 1890. Notices of new fresh-water infusoria. *Proc. Am. Phil. Soc.*, 28:74-80.

Strand, E. 1928. Miscellanea nomenclatorica zoologica et palaeontologica. *Arch. Naturgesch.*, 92A (8):30-75.

Takahashi, E. 1973. Studies on genera *Mallomonas* and *Synura*, and other plankton in freshwater with the electron microscope. VII. New genus *Spiniferomonas* of the Synuraceae (Chrysophyceae). *Bot. Mag. Tokyo*, 86:75-88.

Takahashi, E. 1978. *Electron microscopical studies of the Synuraceae (Chrysophyceae) in Japan: taxonomy and ecology.* Tokai University Press, Tokyo.

Taylor, F. J. 1954. A new chrysophycean flagellate: *Sphaerobryon fimbriata* gen. et sp. nov. *Hydrobiologia*, 6:369-371.

Thomsen, H. A., Zimmermann, B., Moestrup, Ø. & Kristiansen, J. 1981. Some new freshwater species of *Paraphysomonas* (Chrysophyceae). *Nord. J. Bot.*, 1:559-581.

Tippit, D. H., Pillus, L. & Pickett-Heaps, J. D. 1980. Organization of spindle microtubules in *Ochromonas danica*. *J. Cell Biol.*, 87:531-545.

Valkanov, A. 1967. Untersuchungen *über Chromophyton rosanoffii* Woronin. *Bull. Inst. Zool. Mus.*, 23:97-108.

Valkanov, A. 1970. Beitrag zur Kenntnis der Protozoen des Schwarzen Meeres. *Zool. Anz.*, 184:241-290.

Van de Peer, Y., Van der Auwera, G. & de Wachter, R. 1996. The evolution of Stramenopiles and Alveolates as derived by "substitution rate calibration" of small ribosomal subunit RNA. *J. Mol. Evol.*, 42:201-210.

Van der Auwera, C. & de Wachter, R. 1997. Complete large subunit ribosomal RNA sequences from the heterokont algae *Ochromonas danica*, *Nannochloropsis salina*, and *Tribonema aequale*, and phylogenetic analysis. *J. Mol. Evol.*, 45:84-90.

van Goor, A. C. J. 1925. Über einige bemerkenswerte Flagellaten der holländischen Gewässer. *Recueil Trav. Bot. Néerl.*, 22:315-319.

Vesk, M., Hoffman, L. R. & Pickett-Heaps, J. D. 1984. Mitosis and cell division in *Hydrurus foetidus* (Chrysophyceae). *J. Phycol.*, 20:461-470.

Vørs, N. 1992. Heterotrophic amoebae, flagellates and heliozoa from the Tvärminne area, Gulf of Finland, in 1988-1990. *Ophelia*, 36:1-109.

Vysotskii, A. V. 1887. Mastigophora i Rhizopoda, najdennyja v Vejsovom i Repnom ozerach. *Trudy Ob__. Isp. Prir. Imp. Khar'kovsk. Univ.*, 21:119-140.

Walne, P. L., Passarelli, V., Lenzi, P., Barsanti, L. & Gualtieri, P. 1995. Isolation of the flagellar swelling and identification of retinal in the phototactic flagellate, *Ochromonas danica* (Chrysophyceae). *J. Euk. Microbiol.*, 42:7-11.

Weisse, J. F. 1845. Beschreibung einiger neuer Infusorien, welche in stehenden Wässern bei St. Petersburg vorkommen. *Bull. Acad. Imp. Sci. Saint-Pétersbourg, Cl. Phys. Math.*, 4:138-144.

Wessel, D. & Robinson, D. G. 1979. Studies on the contractile vacuole of *Poterioochromonas malhamensis* Peterfi. I. The structure of the alveolate vesicles. *Eur. J. Cell Biol.*, 19:60-66.

Wetherbee, R. & Andersen, R. A. 1992. Flagella of a chrysophycean alga play an active role in prey capture and selection. Direct observations on *Epipyxis pulchra* using image enhanced video microscopy. *Protoplasma*, 166:1-7.

Wetherbee, R., Platt, S. J., Beech, P. L. & Pickett-Heaps, J. D. 1988. Flagellar transformation in the heterokont *Epipyxis pulchra* (Chrysophyceae): Direct observations using image-enhanced light microscopy. *Protoplasma*, 145:47-54.

Willen, T. 1992. The meticulous phycological investigation of Lake Rudträsket, Central Sweden, 1947-1949, by Heinrichs Skuja. *Nord. J. Bot.*, 12:589-616.

Withers, N. W., Fiksdahl, A., Tuttle, R. C. & Liaaen-Jensen, S. 1981. Carotenoids of the Chrysophyceae. *Comp. Biochem. Physiol.*, 68 B:345-349.

Woloszynska, J. 1914. Zapiski algologiczne. *Spraw. Towarz. Nauk. Warszawsk., Wydz.*, 3 (7):1-4.

Woronin, A. 1880. *Chromophyton rosanoffii. Bot. Z.*, 38:625-631 and 641-648.

Wujek, D. E. 1976. Ultrastructure of flagellated chrysophytes. II. *Uroglena* and *Uroglenopsis. Cytologia*, 41:665-670.

Yano, H., Nakahara, M. & Ito, H. 1988. Water blooms of *Uroglena americana* and the identification of odorous compounds. *Wat. Sci. Tech.*, 20 (8/9):75-80.

Zeeb, B., Duff, K. E. & Smol, J. P. 1996. Recent advances in the use of chrysophyte stomatocysts in paleoecological studies. *Nova Hedwigia Beih.*, 114:247-252.

Zhang, X., Watanabe, M. M. & Inouye, I. 1996. Light and electron microscopy of grazing *by Poterioochromonas malhamensis* (Chrysophyceae) on a range of phytoplankton taxa. *J. Phycol.*, 32:37-46.

Zimmermann, B., Moestrup, Ø. & Hällfors, G. 1984. Chrysophyte or heliozoon: ultrastructural studies on a cultured species of *Pseudopedinella* (Pedinellales ord. nov.), with comments on species taxonomy. *Protistologica*, 20:591-612.

CLASS PEDINOPHYCEAE
MOESTRUP 1991

By Øjvind Moestrup

Pedinomonads are very small green flagellates that differ from other green algae in details of the flagellar apparatus and in mitosis and cell division. The genus *Pedinomonas is* well known, while two additional genera of the class, *Resultor* and *Marsupiomonas,* were described recently. Only c. 10 species, distributed in fresh water and marine environments. The class was formally erected by Moestrup (1991), see also Jones et al. (1994).

Key characters

1. All species uniflagellate, but a second basal body is present in the cell. At cell division the latter grows into a new flagellum and the cell forms two new basal bodies. The emergent flagellum is therefore no. 1 according to the flagellar numbering system for flagella introduced by Moestrup and Hori (1989) and Heimann et al. (1989). The flagellum inserts laterally and lacks scales. Thin hairs have been described on the flagella of *Pedinomonas* and *Marsupiomonas.* The flagellar basal bodies short, resembling those of more advanced green algae. They are inserted at opposite polarity to each other, shifted in a counterclockwise direction (resembling the green algal class Ulvophyceae) but not overlapping (resembling the Chlorophyceae).
2. Cells contain a single parietal chloroplast with a pyrenoid more or less directly opposite the flagellar insertion and an eyespot on the pyrenoid-containing part of the chloroplast. A contractile vacuole is present in the front end of freshwater species. Pigments include chlorophylls *a* and *b*, lycopene, carotenes, zeaxanthin, lutein, lutein 5,6-epoxide, violaxanthin, luteoxanthin, neoxanthin, and neochrome (Ricketts 1967).
3. The cell is without scales or cell wall. In *Marsupiomonas* the cell is located in a lorica-like close-fitting bowl, which opens anteriorly.
4. The cells are uninuclear.
5. Ejectile organelles are absent.
6. Starch is stored in the chloroplast.
7. Mitosis and cell division show very primitive traits, compared to other green algae: mitosis is closed throughout, without polar gaps. The mitotic spindle is therefore entirely internal. Cytokinesis takes place without a phragmoplast or phycoplast.
8. Cell division is by longitudinal fission. Sexual reproduction is unknown.
9. The monotypic genera *Resultor* and *Marsupiomonas* are members of the marine nanoplankton, while species of *Pedinomonas are* from fresh water. Cells of *Pedinomonas tenuis* from salt lakes in Middle Asia are surrounded by a lorica and almost certainly belong in *Marsupiomonas.* The generic affinity of *Pedinomonas noctilucae*, a symbiont of green *Noctiluca*, is uncertain.
10. Diagnostic features used for species identification are cell shape, size, etc. Electron microscopy is of little use.

With a single order Pedinomonadales Moestrup, 1991, and a single family Pedinomonadaceae Korshikov, 1938.

Key to the genera

1. Cell surrounded by close-fitting bowl-shaped lorica with large aperture anteriorly . ... ***Marsupiomonas***

1´. No lorica, cells naked 2

2. Cells often non-motile with the flagellum curved around the cell. Cells jump by stretching the flagellum to full length....................... ***Resultor***

2´. No jumping movements.............. ***Pedinomonas***

Genus *Pedinomonas* Korshikov, 1923

Unicellular naked flagellate. Cells swim with the flagellum undulating behind the cell. Very thin hairs are present on the flagellum. With 0, 1, or 2 pyrenoids. Palmella stages known. Asexual cysts smooth walled or with warts.

Free-living in fresh water or symbiotic in *Noctiluca* or the radiolarian *Thalassolampe.* See Ettl & Manton (1964), Pickett-Heaps and Ott (1974).

Pedinomonas minor Korshikov (Fig. 1). Cells flattened, 2.5-7 µm long, 1.8-4.4 µm wide, and 1.2-2 µm thick. With single pyrenoid. Widely distributed.

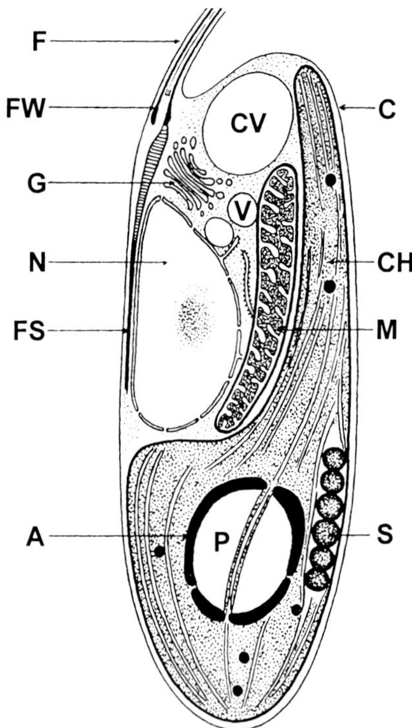

Fig. 1. *Pedinomonas minor.* A, starch grains; C, plasmalemma; CH, chloroplast; CV, contractile vacuole; F, emergent flagellum; FS, rhizoplast; FW, basal body; G, Golgi body; M, mitochondrion; N, nucleus; P, pyrenoid; S, eyespot; V, vacuole. Not to scale. (From Ettl, 1972).

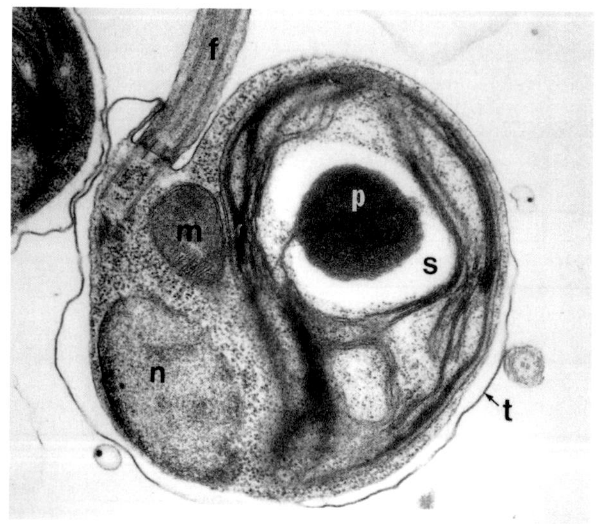

Fig. 2. *Marsupiomonas pelliculata.* Longitudinal section. f, flagellum; m, mitochondrion; n, nucleus; p, pyrenoid; s, starch grain; t, lorica. x23,500. (From Jones et al., 1994).

Genus ***Marsupiomonas***
Jones, Leadbeater & Green, 1994

Cells surrounded by close-fitting lorica, except at the anterior end, otherwise similar to *Pedinomonas.* The flagellum emerges tangentially from the cell near one side of the lorica aperture. See Jones et al. (1994).

Marsupiomonas pelliculata Jones, Leadbeater & Green (Fig. 2). Cells compressed, c. 3 µm long, 2.3 µm wide and 1.8 µm thick. Isolated from a salt marsh near Plymouth, England.

Genus ***Resultor*** Moestrup, 1991

Tiny flagellated cell, usually with the single emergent flagellum curved around the cell. The flagellum occasionally stretches out rapidly to full length but immediately recoils around the cell. Cells may swim for a short distance with the flagellum undulating behind the cell. Otherwise very similar to *Pedinomonas.* See Moestrup (1991).

Resultor minutissima (Throndsen) Moestrup (Fig. 3). Cells compressed, c. 3 µm long, 2 µm wide, and 1 µm thick. Widely distributed marine nanoplankton flagellate.

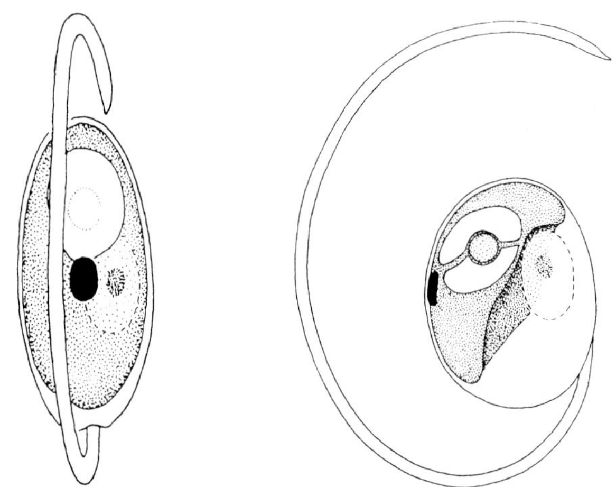

Fig. 3. *Resultor micron*, cells in two different views. x10,000. (From Moestrup, 1991).

LITERATURE CITED

Ettl, H. & Manton, I. 1964. Die feinere Struktur von *Pedinomonas minor* Korschikoff. *Nova Hedwigia*, 8: 21-444.

Heiman, K., Reize, I. B. & Melkonian, M. 1989. The flagellar developmental cycle in algae: flagellar transformation in *Cyanophora paradoxa* (Glaucocystophyceae). *Protoplasma*, 148:106-110.

Jones, H. L. J., Leadbeater, B. S. C. & Green J. C. 1994. An ultrastructural study of *Marsupiomonas pelliculata* gen. et sp. nov., a new member of the Pedinophyceae. *Eur. J. Phycol.*, 29:171-181.

Moestrup, Ø. 1991. Further studies of presumedly primitive green algae, including the description of Pedinophyceae class nov. and *Resultor* gen. nov. *J. Phycol.*, 27:119-133.

Moestrup, Ø. & Hori, T. 1989. Ultrastructure of the flagellar apparatus in *Pyramimonas octopus* (Prasinophyceae). II. Flagellar roots, connecting fibres, and numbering of individual flagella in green algae. *Protoplasma*, 148:41-56.

Pickett-Heaps, J. D. & Ott, D. W.. 1974. Ultrastructural morphology and cell division in *Pedinomonas. Cytobios*, 11:41-58.

Ricketts, T. R. 1967. The pigments of the phytoflagellates, *Pedinomonas minor* and *Pedinomonas tuberculata. Phytochemistry*, 6: 19-24.

CLASS **PELAGOPHYCEAE**
ANDERSEN & SAUNDERS 1993

ROBERT A. ANDERSEN and HANS R. PREISIG

The Pelagophyceae are a group of flagellated, coccoid, palmelloid, sarcinoid and filamentous organisms. Some of these were once placed in the Chrysophyceae (See chapter on Chrysomonada) but they differ from the chrysomonads in a significant number of features (Andersen et al., 1993, Saunders et al., 1997; Bailey and Andersen, 1999). The pelagophytes are marine and usually microscopic although a few reach macroscopic size. Eleven genera are presently known, but because the class was recognized only recently, it is likely that additional genera will be added. Cavalier-Smith et al. (1995) suggested that the Sarcinochrysidales be placed in a separate class, but more recently based upon additional molecular data, one class (Pelagophyceae) is recognized with two orders, the Pelagomonadales Andersen & Saunders 1993 and the Sarcinochrysidales Gayral & Billard, 1977 (Saunders et al., 1997). *Aureococcus, Pelagococcus*, and *Pelagomonas* are placed in the Pelagomonadales, and the remaining genera are placed in the Sarcinochrysidales.

The phylogenetic relationships of the Pelagophyceae, based upon the 18S rRNA and *rbc*L genes, have been addressed in several recent papers. They have a close relationship with the silicoflagellates, pedinellids, and the marine photosynthetic amoeba *Rhizochromulina* Hibberd & Chrétiennot-Dinet, 1979 (see Chapter Silicoflagellata; Saunders et al.,1995, Daugbjerg and Andersen 1997, DeYoe et al., 1997, Potter et al., 1997, Saunders et al., 1997, Andersen et al., 1998, Bailey and Andersen, 1999). These organisms, in turn, may have a sister relationship with the diatoms, possibly comprising a group of chromophytes (= heterokont algae) with a 'reduced flagellar apparatus' and with similar photosynthetic pigments (Saunders et al., 1995). However, phylogenetic analyses based upon molecular data have not, to date, provided a clear phylogenetic relationship for the various classes of chromophytes.

KEY CHARACTERS

1. Vegetative cells are free-swimming flagellates, planktonic, coccoid, single cells or attached filamentous, palmelloid, and sarcinoid masses. With the exception of zoospore formation, alternate life history stages are unknown.

2. Some genera are free-swimming, single-cell flagellates (*Ankylochrysis*, *Pelagomonas*, *Sulcochrysis*). Zoospores are produced by some nonflagellated genera (*Chrysocystis, Chrysonephos, Chrysoreinhardia, Nematochrysopsis, Sarcinochrysis*) while other genera have no known flagellated stage (*Aureococcus, Aureoumbra, Pelagococcus*). The flagellar basal bodies of all swimming cells are positioned on or very close to the nuclear envelope, and no evidence has been found for the presence of a rhizoplast. There is a transitional helix (typically two gyres) below the major transitional plate. Flagellar hairs are present on the immature (longer, anteriorly directed) flagellum, and the hairs are typically tripartite without lateral filaments although *Pelagomonas* appears to have bipartite hairs (Andersen et al.,

1993). Members of the Sarcinochrysidales have two emergent flagella but *Pelagomonas* (Pelagomonadales) has only one flagellum and lacks any visible remnant of the second, mature (shorter, posteriorly directed) flagellum. The insertion of flagella in the cell is variable; the flagella of *Ankylochrysis* are inserted almost apically; those for *Sarcinochrysis* are inserted about 1/3 the way down from the anterior end, while those of *Pelagomonas* are inserted at the extreme posterior end of the cell. Eyespots are not present and a photoreceptor apparatus, if present, is probably formed by the association of a swollen region on the mature (posterior) flagellum and a plasma membrane/chloroplast complex. The flagellar apparatus of *Pelagomonas*, so highly reduced in other ways, completely lacks microtubular roots (Andersen et al., 1993). Conversely, flagellated cells of sarcinochrysidalean species have well-developed microtubular roots. The R_1 root originates on the right side of the immature flagellum and rises slightly while gently curving anti-clockwise (viewed from the anterior). In *Ankylochrysis* and *Chrysonephos*, the R_1 root nucleates a special band of cytoskeletal microtubules reminiscent of the bypassing root of brown algal zoospores. The R_2 root is absent in *Ankylochrysis* but is present in other sarcinochrysidalean taxa. It arises on the left side of the immature flagellum and extends anteriorly and dorsally. The R_3 root originates on the right side of the mature flagellum and extends posteriorly. The R_3 root consists of a tightly pressed band of microtubules, some of which extend a considerable distance towards the cell posterior. However, in *Sulcochrysis* the R_3 root reaches the posterior and then curves back up the dorsal side of the cell and ends by coiling several times (Honda et al., 1995). The R_4 root arises on the left side of the mature flagellum and extends more or less parallel to that flagellum and ends by curving anti-clockwise under the flagellum. The single basal body of *Pelagomonas* is highly unusual amongst all eukaryote cells (including centrioles in animal cells) because the other associated second basal body is completely lacking. Flagellar transformation has been described to document this unusual situation in *Pelagomonas* (Heimann et al., 1995). Prior to cell division in *Pelagomonas*, two new basal bodies/flagella appear *de novo* as the parental flagellum retracts (or abscises). The parental basal body segregates with one of the new basal body/flagellum during cytokinesis so that briefly one of the daughter cells has a basal body/flagellum and a barren parental basal body. Within a short time period, the parental basal body transforms or disintegrates so that no trace remains.

3. All members are photosynthetic and have one to four (in *Chrysocystis* up to eight) parietal chloroplasts. Distinctly stalked pyrenoids are present in *Aureoumbra, Chrysocystis, Chrysonephos, Chrysoreinhardia,* and *Sarcinochrysis*; bulging pyrenoids are found in *Ankylochrysis* and *Sulcochrysis*, an immersed pyrenoid is present in *Aureococcus* and pyrenoids are absent in *Pelagococcus* and *Pelagomonas*. The chloroplast is surrounded by two additional membranes, the chloroplast endoplasmic reticulum, and these two membranes are confluent with the outer membrane of the nuclear envelope.

4. Pelagophytes are yellowish-brown in color. In addition to chlorophyll c_1 and c_2 (Jeffrey 1989), pelagophytes contain fucoxanthin, 19′-butanoxylofucoxanthin, diatoxanthin, diadinoxanthin and β-carotene (Bidigare, 1989; Bjørnland and Liaaen-Jensen ,1989).

5. The storage product(s) of the Pelagophyceae are unknown, but presumably the major carbohydrate storage product is a β-1,3-linked glucan similar or identical to chrysolaminaran. Sterols have been described for several genera (*Chrysoreinhardia* (as *Chrysoderma*), *Nematochrysopsis,* and *Sarcinochrysis* possibly). Sterols include 24-propylidenecholesterol, which appears to be unique to the Pelagophyceae and is therefore a chemotaxonomic marker (Billard et al., 1990, Giner and Djerassi, 1991, Giner and Boyer, 1998). *Aureococcus anophagefferens* contains, as free sterols, 48% 24-methylenecholesterol, 32% (E)-24-propylidenecholesterol and 12% (Z)-24-propylidenecholesterol (Giner and Boyer, 1998).

6. Cell walls are the most common cell covering in the Pelagophyceae, but the flagellate genera *Ankylochrysis* and *Pelagomonas* have a thin organic theca surrounding the cells. Cells of *Aureococcus, Sulcochrysis*, and cultured (but not wild) cells of *Aureoumbra* are naked. The

chemical composition of the thecae and cell walls is unknown. Cysts are unknown.

7. Bloom formation is common for *Aureococcus* and *Aureoumbra*, but thus far these blooms are limited to coastal areas of the United States. *Aureococcus anophagefferens* has formed blooms in Narragansett Bay, Rhode Island, various bays on Long Island, New York, and Barnegat Bay, New Jersey (Cosper et al., 1987, 1989, 1993; Sieburth et al., 1988; pers. observ.). The reason(s) for its bloom formation are not known, but hypotheses include iron availablity, nitrogen concentrations, ground water flows, general eutrophication, elevated salinities, and disruption of grazer populations (Boyer et al., 1998; Bricelj and Lonsdale, 1997; Buskey et al., 1997; Gobler and Cosper, 1996; Lomas et al., 1996; LaRoche et al., 1997; Milligan and Cosper, 1997). It has also been suggested that *A. anophagefferens* either originated from open ocean water or has recently speciated in coastal waters (Yentsch et al., 1989). Viruses apparently play a role in the disruption of blooms (Gobler et al., 1997; Garry et al., 1998; Gastrich et al., 1998). Antibodies to detect *Aureococcus* and *Aureoumbra* have been developed (Anderson et al., 1989; Lopez-Barreiro et al., 1998). *Aureoumbra lagunensis* formed a bloom along the Texas coastline, especially the Laguna Madre (Stockwell et al., 1993; DeYoe and Suttle, 1994; DeYoe et al., 1997; Buskey et al., 1998). A bloom began in January 1990 and persisted continuously until October 1998, the longest algal bloom in recorded history (Street et al., 1997; Buskey et al., 1998). *Aureoumbra* is unable to use nitrate, and high levels of ammonia have been hypothesized as the reason for bloom formation and longevity (DeYoe and Suttle, 1994; DeYoe et al., 1997). Others have attributed bloom longevity to long turnover time and hypersalinity (Buskey et al., 1998).

ARTIFICIAL KEY TO GENERA

1. Cells flagellate.. 2
1.' Cells coccoid, palmelloid, sarcinoid, or filamentous.. 4

2. Uniflagellate.................................. ***Pelagomonas***
2'. Biflagellate.. 3

3. Cells covered with thin theca.***Ankylochrysis***
3'. Cells naked.................................. ***Sulcochrysis***

4. Cells coccoid.. 5
4'. Cells palmelloid, sarcinoid, or filamentous....7

5. Cells with stalked pyrenoid....... ***Aureoumbra***
5'. Cells without stalked pyrenoid 6

6. Cells with cell walls, pelagic in open ocean .... ***Pelagococcus***
6.' Cells without cell walls, coastal, often forming blooms.................................. ***Aureococcus***

7. Palmelloid.. 8
7'. Sarcinoid or filamentous.................................. 9

8. Colonies up to 30-50 mm long and 5 mm wide .... ***Chrysocystis***
8'. Colonies much smaller, not over few mm in diameter.......................... ***Chrysoreinhardia***

9. Sarcinoid.................................. ***Sarcinochrysis***
9'. Filamentous.. 10

10. Dichotomously branching filaments................ .... ***Chrysonephos***
10'. Unbranched filaments..................................11

11. Filaments long and well developed................... .... ***Nematochrysopsis***
11'. Filaments short, rudimentary .... ***Sarcinochrysis marina*** .... var. ***filamentosa***

Genus Ankylochrysis Billard in Honda & Inouye, 1995

Single-celled, free-swimming flagellates with a multilayered external cell covering (theca); cells reniform, dorsally often convex, ventrally with a furrow subtending the flagellar insertion and extending obliquely anteriorly; 2 unequal flagella, inserted side by side on ventral surface; single chloroplast, with a pyrenoid transected by a pair of thylakoids; eyespot lacking; marine, monospecific genus. See Billard, 1984; Honda and Inouye, 1995.

TYPE SPECIES: *Ankylochrysis lutea* (van der Veer) Billard in Honda et Inouye, 1995 (=*Ankylonoton luteum* van der Veer) (Fig. 1). Cell bean-shaped, 4.0-10.5 x 3.0-6.0 µm; anterior flagellum 6-9 µm long, posterior flagellum 5-8 µm long, euryhaline species. See van der Veer, 1970; Billard, 1984; Honda & Inouye, 1995.

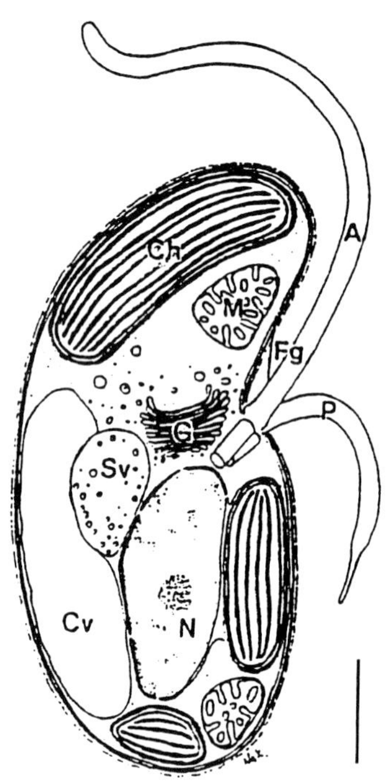

Fig. 1. *Ankylochrysis lutea* cell viewed from right side. A, anterior flagellum; Ch, chloroplast, Cv, chrysolaminaran vacuole; Fg, flagellar groove; G, Golgi body; M, mitochondrion; N, nucleus; P, posterior flagellum; Sv, vacuole with scale-like bodies. Scale=2 µm. (From Honda and Inouye, 1995)

Genus ***Aureococcus*** Hargraves & Sieburth,1988 in Sieburth et al., 1988

Solitary coccoid cells, spherical or subspherical; without a cell wall; single parietal chloroplast with an emersed pyrenoid; flagellate cells and cysts unknown; marine planktonic organism forming dense blooms in coastal bays; monospecific genus. See Sieburth et al., 1988.

TYPE SPECIES: *Aureococcus anophagefferens* Hargraves & Sieburth, 1988 in Sieburth et al., 1988 (Fig. 2). Cells 1.5–4 µm in diameter.

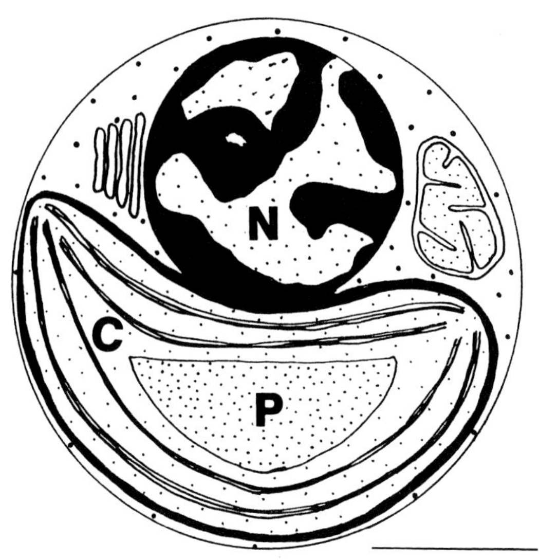

Fig. 2. *Aureococcus anophagefferens*. C, chloroplast; N, nucleus; P, pyrenoid. Scale=1 µm.

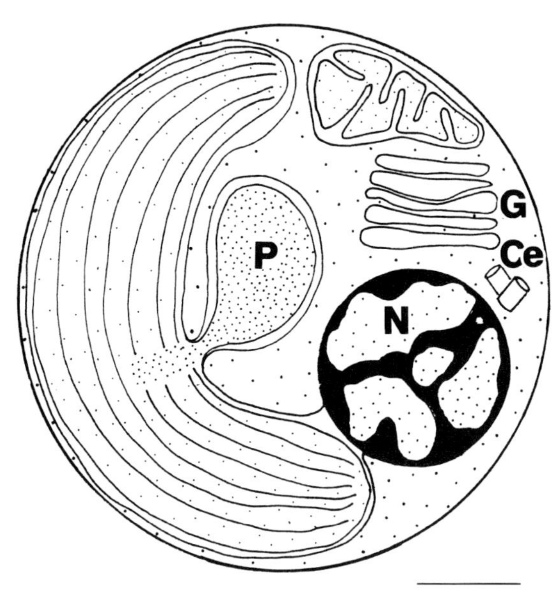

Fig. 3. *Aureoumbra lagunensis*. Ce, centriole; C, chloroplast; G, Golgi body; N, nucleus; P, stalked pyrenoid;. Scale=1 µm.

Genus ***Aureoumbra*** Stockwell, DeYoe, Hargraves & Johnson, 1997 in DeYoe et al., 1997

Solitary coccoid cells, spherical or subspherical; with or without a cell wall; single parietal chloroplast with a stalked pyrenoid; flagellate cells and cysts unknown; marine planktonic organism forming dense blooms in coastal embayments; monospecific genus. (See DeYoe et al., 1995, 1997; Villareal et al., (1998).

TYPE SPECIES: *Aureoumbra lagunensis* Stockwell, DeYoe, Hargraves & Johnson, 1997 in DeYoe et al., 1997. (Fig. 3). Cells 2.5-5 µm in diameter.

Genus ***Chrysocystis*** Lobban, Honda & Chihara in Lobban , 1995

Solid gelatinous colonies with cells located around the periphery; colonies macroscopic, appearing pale yellow in color; zoospores formed by 2 successive divisions of vegetative cells; zoospores pyriform or ovoid with one to several chloroplasts and laterally inserted flagella; longer flagellum directed anteriorly, shorter flagellum directed posteriorly; eyespots absent; asexual reproduction by colony fragmentation and zoospore formation; sexual reproduction unknown; marine, monospecific genus. See Lobban et al., 1995.

TYPE SPECIES: *Chrysocystis fragilis* Lobban, Honda & Chihara in Lobban et al., 1995 (Fig. 4). Colony fragile, cylindrical in shape, 30-50 mm long and approximately 5 mm wide; cells spherical, 8-14 µm in diameter; cells covered with a thin cell wall; numerous muciferous bodies located near the periphery of the cells; 4 to 8 parietal chloroplasts, elliptical to saucer-shaped; chloroplasts with 1 or 2 projecting pyrenoids; zoospores 4-5 µm long, 2-3 µm wide, with a single posterior chloroplast; anterior flagellum 9-13 µm long, posterior flagellum 5-6 µm long.

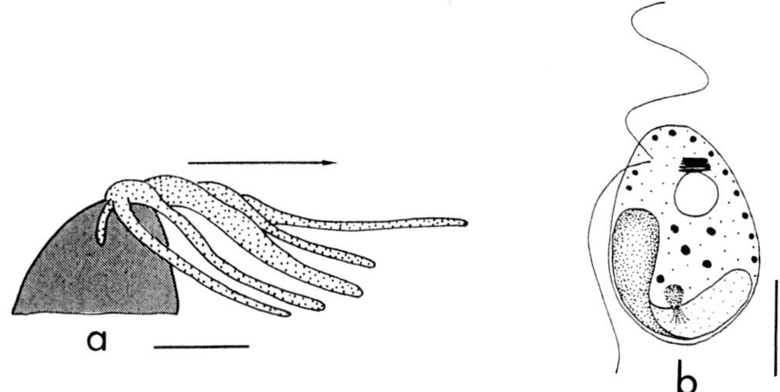

Fig. 4. *Chrysocystis fragilis*. a. Colony attached to a rock, extended by water current. Scale=10 mm. b. zoospore. Scale=5 µm.

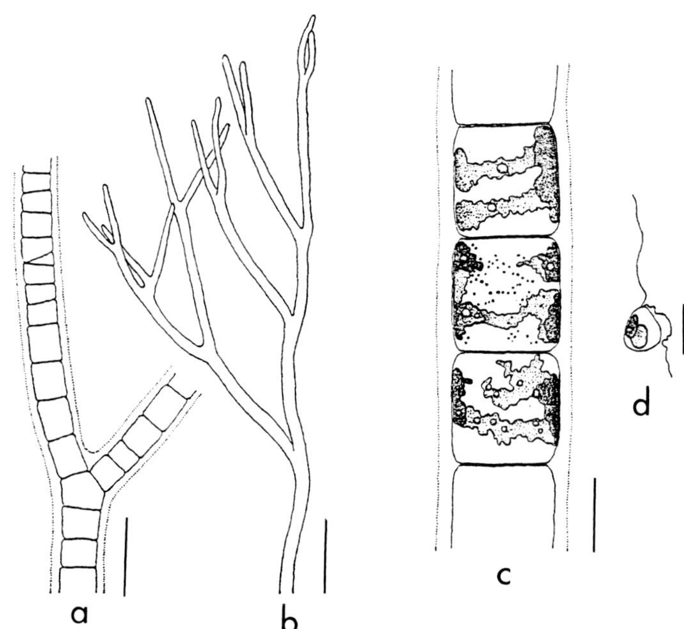

Fig. 5. *Chrysonephos lewisii*. a. Segment from a typical filament. b. lower magnification showing branching of filament. c. Enlarged cells showing chloroplast structure. d. Zoospore. Scale bars=50 µm (a), 200 µm (b), and 10 µm (c, d). (From Taylor, 1951)

Genus ***Chrysonephos*** Taylor, 1952

Filamentous, with falsely dichotomous branching; mostly uniseriate, delicate, and forming golden brown tufts; cell walls indistinct, but more apparent in basal cells than in apical cells; zoospores released by dissolution of cell walls; zoospores with longer anteriorly directed flagellum and shorter posteriorly directed flagellum; flagella inserted laterally; eyespot absent; marine, monospecific genus. See Taylor (1951, 1952), Honda (1995), Lobban et al. (1995), Sartoni et al. (1995), Boddi et al., (1999).

TYPE SPECIES: *Chrysonephos lewisii* (Taylor) Taylor, 1952 (=*Chrysophaeum lewisii* Taylor, 1951) (Fig. 5). Filaments 1-3 cm long; cells near the base 30-38 µm in diameter, cells near the tips tapering to 4.5-8 µm; cell length typically 1.5-2 times the diameter; 2-4 parietal chloroplasts with erose margins; stalk-like pyrenoids; zoospores pyriform, 7-12 µm in diameter, 8-9 µm long.

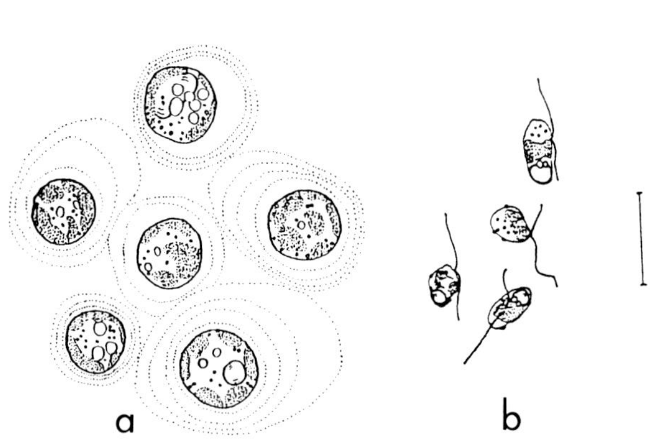

Fig. 6. *Chrysoreinhardia feldmannii*. a. Vegetative cells independently surrounded by gelatinous material. b. Zoospores. Scale bar=10 μm. (From Billard, 1984)

Genus ***Chrysoreinhardia*** Billard in Hoffman et al. (2000)

Vegetative cells palmelloid, lacking cell walls but surrounded by gelatinous matrix; chloroplast(s) deeply divided with several lobes; pyrenoids stalked; zoospores biflagellated, with long anteriorly directed flagellum bearing tripartite tubular hairs (without lateral filaments) and shorter posteriorly directed smooth flagellum; eyespot absent. *Chrysoreinhardia algicola* (Reinhard, 1885) Billard in Hoffman et al. (2000), the type species, is only known from its incomplete original description. The basionym *Pulvinaria algicola* Reinhard is illegitimate because the homonym *Pulvinaria* Bonorden, 1851, a fungal name, predates the algal name.

REPRESENTATIVE SPECIES: *Chrysoreinhardia feldmannii* (Bourrelly & Magne) Billard and Fresnel in Hoffmann et al. (2000) (=*Chrysobotrys feldmannii* Bourrelly & Magne, 1953; (=*Chrysosphaera feldmannii* [Bourrelly & Magne] Bourrelly, 1957; *Pulvinaria feldmannii* [Bourrelly & Magne] Billard & Fresnel, 1980) (Fig. 6). Cells 4.5 – 7 (11) μm in diameter; common in the littoral zone of rocky temperate marine coastlines. See Billard and Fresnel (1980), Billard (1984), Hoffman et al. (2000).

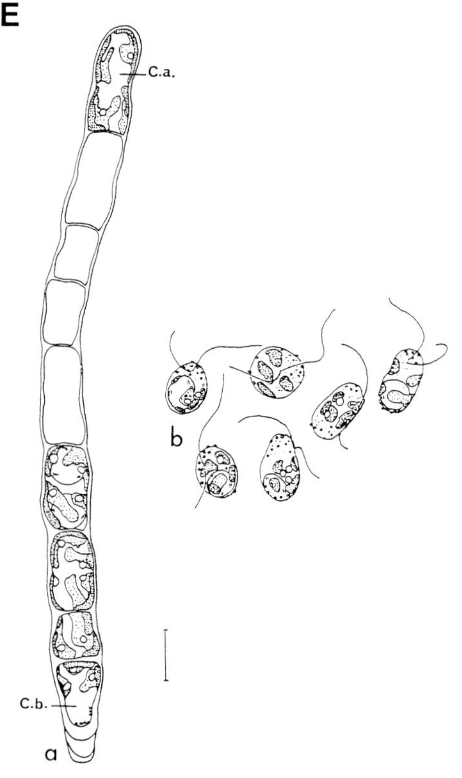

Fig. 7. *Nematochrysopsis roscoffensis*. a. Typical filament with four chloroplasts per cell. C.b., basal cell; C.a., apical cell. b. Zoospores. Scale bar=2 μm. (From Billard, 1984)

Genus ***Nematochrysopsis*** Chadefaud, 1947

Uniseriate filaments attached to the substrate with a gelatinous holdfast; cell wall thick and appearing laminate in some electron micrographs; terminus of the filament sometimes broken and erroneously giving the appearance of a *Tribonema*-like (Xanthophyceae) H-shaped cell wall structure; typically with 4 yellow-brown parietal chloroplasts; stalked pyrenoids present; zoospores with a longer anteriorly directed flagellum bearing tripartite tubular hairs that lack lateral filaments and a shorter posteriorly directed flagellum lacking tripartite hairs; palmelloid stage was reported in original description; marine. See Chadefaud (1947); Gayral and Lepailleur (1971); Billard (1984).

TYPE SPECIES: *Nematochrysopsis marina* (Feldmann, 1941) Billard in Hoffmann et al, 2000 (=*Tribonema marinum* Feldmann, 1941; *Nematochrysopsis roscoffensis* Chadefaud, 1947; *Nematochrysopsis roscoffensis* f. *giganteus* Billard, 1988). Filaments 8-15 (17) μm diameter; cells 15-30 μm long. See Feldmann (1941), Chadefaud (1947), Gayral and

Lepailleur (1971), Billard (1984, 1988), Hoffmann et al. (2000).

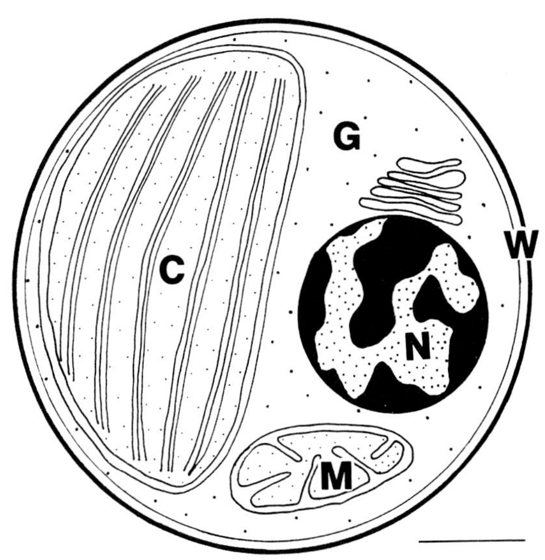

Fig. 8. *Pelagococcus subviridis*. C, chloroplast; G, Golgi body; M, mitochondrion; N, nucleus; W, cell wall. Scale=500 nm.

Genus ***Pelagococcus*** Norris, 1977 in Lewin et al., 1977

Solitary coccoid cells, spherical or subspherical; with tri-layered cell wall; single parietal chloroplast without pyrenoid; flagellate and cyst stages unknown; marine, monospecific genus. See Lewin et al. (1977), Vesk and Jeffrey (1987), Andersen et al. (1996).

TYPE SPECIES: *Pelagococcus subviridis* Norris in Lewin et al., 1977 (Fig. 8). Cells 2-4 (7) µm in diameter; planktonic in open ocean waters.

Genus ***Pelagomonas*** Andersen & Saunders, 1993 in Andersen et al., 1993

Single-celled, free-swimming flagellates; with thin cell covering (theca); single flagellum arising from basal body inserted in the posterior of the cell, with flagellum extending anteriorly along a cellular groove and extending forward beyond the anterior of the cell; no second flagellum and no second basal body present; microtubular flagellar roots absent; transitional helix below the major transitional plate; marine, monospecific genus. See Andersen et al. (1993), Heimann et al. (1995).

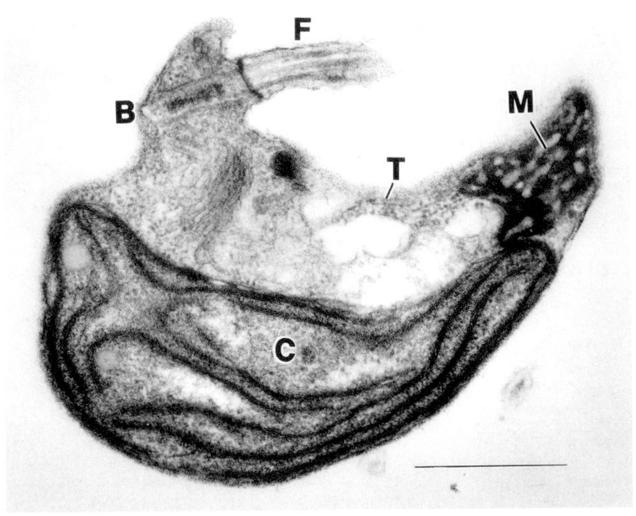

Fig. 9. *Pelagomonas calceolata*. B, single basal body; C, chloroplast; F, flagellum; G, Golgi body; M, mitochondrion; T, theca. Scale=500 nm.

TYPE SPECIES: *Pelagomonas calceolata* Andersen & Saunders, 1993 in Andersen et al., 1993 (Fig. 9). Cells slipper-shaped, 1.5-2 µm long and 1 - 1.5 µm wide.

Genus ***Sarcinochrysis*** Geitler, 1930

Small sarcinoid packets of cells; cells spheroid to elliptical, united by thin gelatinous matrix; distinct cell wall lacking; 2 parietal chloroplasts with stalked pyrenoid; zoospores spheroid to elliptical with 2 flagella; longer anteriorly directed flagellum with tripartite tubular hairs (lacking lateral filaments), shorter posteriorly directed flagellum without tripartite hairs; eyespot absent; marine, coastal, in temperate waters, growing attached to substrates. See Geitler (1930), Billard (1984).

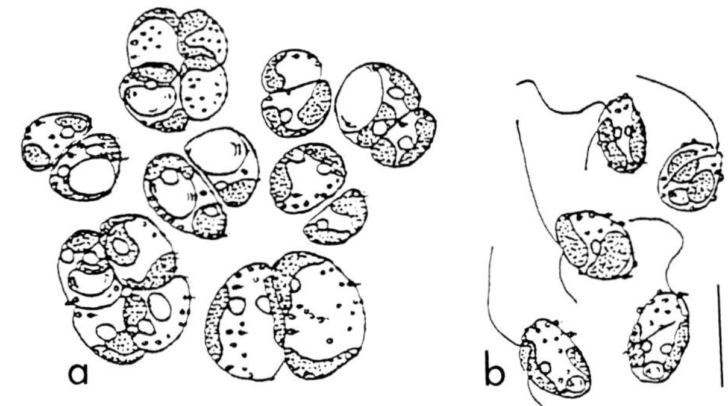

Fig. 10. *Sarcinochrysis marina*. a. Vegetative cells. b. Zoospores. Scale bar=10µm. (From Billard, 1984)

TYPE SPECIES: *Sarcinochrysis marina* Geitler, 1930 (Fig. 10). Cells 5-9 µm, zoospores 4 - 5 µm wide and 5-6 µm long. For *S. marina* var. *filamentosa*, see Billard (1984, 1988).

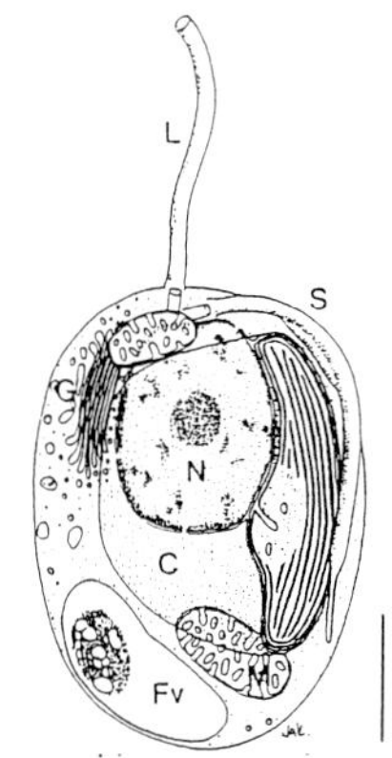

Fig. 11. *Sulcochrysis biplastida* viewed from ventral surface. C, chloroplast; Fv, food vacuole; G, Golgi body; L, long flagellum; M, mitochondrion; N, nucleus; S, short flagellum. Scale bar=2 µm. (From Honda et al., 1995)

Lesser known members of the Pelagophyceae

The genus *Sulcochrysis* may be a member of the Pelagophyceae, but it also shows some morphological resemblance with Silicoflagellata. Molecular data may show to which group it is most closely related.

Genus ***Sulcochrysis***
Honda, Kawachi & Inouye, 1995

Single-celled, naked, free-swimming flagellates; with 2 unequal flagella emerging from anterior end of cell; longer flagellum extending anteriorly, bearing tripartite tubular flagellar hairs (without lateral filaments); shorter flagellum naked, extending posteriorly, lying in a vertical furrow (sulcus) that runs from base of short flagellum to posterior end of cell; with a flagellar swelling near base of short flagellum; transitional helix below the flagellar transition plate; basal bodies situated in anterior depression of the nucleus, rhizoplast absent; two chloroplasts, discoid and yellow-brown, each with a single projecting pyrenoid at central part of chloroplast; eyespot lacking; nucleus anterior in cell, slightly depressed at anterior end; cells mixotrophic, with large food vacuole at posterior end of cell; asexual reproduction by longitudinal division; sexual reproduction and cysts unknown; marine, monospecific genus. See Honda et al., 1995.

TYPE SPECIES: *Sulcochrysis biplastida* Honda, Kawachi & Inouye 1995 (Fig. 11). Cells spherical to ovoid, 6.0-8.0 x 3.5-5.5 µm; anterior flagellum 12-20 µm long, posterior flagellum 4-6 µm long.

There are no other described genera that we can place in the Pelagophyceae. *Chrysophaeum* Lewis & Bryan, 1941 (type species: *Chrysophaeum taylori* Lewis & Bryan, 1941) has been listed under the Sarcinochrysidales by Lobban et al. (1995), but from the incomplete data known to date, it is not clear whether it is really related to this group. Also, there are two undescribed organisms that belong in the class based upon phylogenetic analyses of gene sequences: culture isolates CCMP1395 and CCMP1410 from the Provasoli-Guillard National Center for Culture of Marine Phytoplankton (Andersen et al., 1997) (see Daugbjerg & Andersen, 1997; Potter et al., 1997), Saunders et al., 1997).

LITERATURE CITED

Andersen, R. A., Saunders, G. W., Paskind, M. P. & Sexton, J. P. 1993. Ultrastructure and 18S rRNA gene sequence for *Pelagomonas calceolata* gen. et sp. nov. and the description of a new algal class, the Pelagophyceae classis nov. *J. Phycol.*, 29:701-715.

Andersen, R. A., Bidigare, R. R., Keller, M. D. & Latasa, M. 1996. A comparison of HPLC pigment signatures and electron microscopic observations for oligotrophic waters of the North Atlantic and Pacific Oceans. *Deep-Sea Res., II*, 43:517-537.

Andersen, R. A., Morton, S. L. & Sexton, J. P. 1997. CCMP-Provasoli-Guillard National Center for Culture of Marine Phytoplankton—List of Strains. *J. Phycol.*, 33 (Suppl.):1-75.

Andersen, R. A., Potter, D., Bidigare, R. R., Latasa, M., Rowan, K. & O'Kelly, C. J. 1998. Characterization and phylogenetic position of the enigmatic golden alga *Phaeothamnion confervicola*: Ultrastructure, pigment composition and partial SSU rDNA sequence. *J. Phycol.*, 34:286-298.

Anderson, D. M., Kulis, D. M. & Cosper, E. M. 1989. Immunofluorescent detection of the brown tide organism, *Aureococcus anophagefferens*. *In*: Cosper, E. M., Bricelj, V. M. & Carpenter, E. J. (Eds.), *Novel Phytoplankton Blooms, Causes and Impacts of Recurrent Brown Tides and other*

Unusual Blooms. Springer-Verlag, Berlin. pp 213-28

Bailey, J. C. & Andersen, R. A. 1999. Analysis of clonal cultures of the brown tide algae *Aureococcus* and *Aureoumbra* (Pelagophyceae) using 18S rRNA, *rbc*L and RUBISCO spacer sequences. *J. Phycol.*, 35:570-574.

Bailey, J. C., Bidigare, R. R., Christensen, S. J. & Andersen, R. A. 1998. Phaeothamniophyceae classis nova: a new lineage of chromophytes based upon photosynthetic pigments, *rbc*L sequence analysis and ultrastructure. *Protist*, 149:245-263.

Bidigare, R. R. 1989. Photosynthetic pigment composition of the brown-tide alga: unique chlorophyll and carotenoid derivatives. *In*: Cosper, E. M., Bricelj, V. M. & Carpenter, E. J. (eds.), *Novel Phytoplankton Blooms: Causes and Impacts of Recurrent Brown Tides and Other Unusual Blooms.* Springer-Verlag, Berlin. pp 57-75

Billard, C. 1984. Recherches sur les Chrysophyceae marines de l'ordre des Sarcinochrysidales. Biologie, systématique, phylogénie. Ph.D. thesis, l'Université de Caen. 224 pp

Billard, C. 1988. Les Sarcinochrysidales des côtes françaises. *Bull. Soc. Linn. Normandie*, 110-111:61-69.

Billard, C. & Fresnel, J. 1980. Nouvelles observations sur le *Pulvinaria feldmannii* (Bourrelly et Magne) comb. nov. (Chrysophycée, Sarcinochrysidale) formant une ceinture sur substrat meuble. *Cryptogamie: Algologie* , 1:281-292.

Billard, C., Dauguet, J.-C., Maume, D. & Bert, M. 1990. Sterols and chemotaxonomy of marine Chrysophyceae. *Bot. Mar.*, 33:225-228.

Bjørnland, T. & Liaaen-Jensen, S. 1989. Distribution patterns of carotenoids in relation to chromophyte phylogeny and systematics. *In*: Green J. C., Leadbeater, B. S. C. & Diver, W. L. (eds.), *The Chromophyte Algae, Problems and Perspectives.* Systematics Association Special Volume No. 38, Clarendon Press, Oxford. pp 37-60

Boddi, S., Bigazzi, M. & Sartoni, G. 1999. Ultrastructure of vegetative and motile cells and zoosporogenesis in *Chrysonephos lewisi* (Taylor) Taylor (Sarcinochrysidales, Pelagophyceae) in relation to taxonomy. *Eur. J. Phycol.*, 34:297-306.

Bourrelly, P. 1957. Recherches sur les Chrysophycées. *Rev. Algol. Mém. Hors-Sér.*, 1:1-412.

Bourrelly, P. & Magne, F. 1953. Deux nouvelles espèces de Chrysophycées marines. *Rev. gén. Bot.*, 60:684-687.

Boyer, G. L., Szmyr, D. B. & Alexander, J. A. 1998. Iron and nitrogen nutrition in the brown tide organism *Aureococcus anophagefferens. Can. Tech. Rep. Fish. Aquat. Sci.*, 2261:11-13.

Bricelj, V. M. & Lonsdale, D. J. 1997. *Aureococcus anophagefferens*: causes and ecological consequences of brown tides in U. S. mid-Atlantic coastal waters. *Limnol. Oceanogr.*, 42:1023-1038.

Buskey, E. J., Wysor, B. & Hyatt, C. 1998. The role of hypersalinity in the persistence of the Texas 'brown tide' in the Laguna Madre. *J. Plankton Res.*, 20:1553-1565.

Buskey, E. J., Montagna, P. A., Amos, A. F. & Witledge, T. E. 1997. Disruption of grazer populations as a contributing factor to the initiation of the Texas brown tide algal bloom. *Limnol. Oceanogr.*, 42:1215-1222.

Cavalier-Smith, T., Chao, E. E. & Allsopp, M. T. E. P. 1995. Ribosomal RNA evidence for chloroplast loss within Heterokonta: pedinellid relationships and a revised classification of ochristan algae. *Arch. Protistenkd.*, 145:209-220.

Chadefaud, M. 1947. Une nouvelle Chrysophyée marine filamenteuse: *Nematochrysopsis roscoffensis* n. g. n. sp. *Bull. Soc. Bot. France.*, 94:239-243.

Cosper, E. M., Garry, R. T., Milligan, A. J. & Doall, M. H. 1993. Iron, selenium and citric acid are critical to the growth of the "brown tide" microalga, *Aureococcus anophagefferens. In*: Smayda, T. J. & Shimizu, Y. (eds.), *Toxic Phytoplankton Blooms in the Sea.* Elsevier, Amsterdam. pp 667-673

Cosper, E. M., Dennison, W., Milligan, A., Carpenter, E. J., Lee, C., Holzapfel, J. & Milanese, L. 1989. An examination of the environmental factors important to initiating and sustaining "brown tide" blooms. *In* Cosper, E. M., Bricelj, V. M. & Carpenter, E. J. (eds.), *Novel Phytoplankton Blooms: Causes and Impacts of Recurrent Brown Tides and Other Unusual Algal Blooms.* Springer-Verlag, Berlin. pp 317-340

Cosper, E. M., Dennison, W., Carpenter, E. J., Bricelj, V. M., Mitchell, J. G., Kuenstner, S. H., Colflesh, D. & Dewey, M. 1987. Recurrent and persistent brown tide blooms perturb coastal marine ecosystems.. *Estuaries, 10:287-290*

Daugbjerg, N. & Andersen, R. A. 1997. A molecular phylogeny of the heterokont algae based on analyses of chloroplast-encoded *rbc*L sequence data. *J. Phycol.*, 33:1031-1041.

DeYoe, H. R. & Suttle, C. A. 1994. The inability of the Texas "brown tide" alga to use nitrate and the role of nitrogen in the initiation of a persistent bloom of this organism. *J. Phycol.*, 30:800-806.

DeYoe, H. R., Chan, A. M. & Suttle, C. A. 1995. Phylogeny of *Aureococcus anophagefferens* and a morphologically similar bloom-forming alga from Texas as determined by 18S ribosomal RNA sequence analysis. *J. Phycol.*, 31:413-418.

DeYoe, H. R., Stockwell, D. A., Bidigare, R. R., Latasa, M., Johnson, P. W., Hargraves, P. E. & Suttle, C. A. 1997. Description and characterization of the algal species *Aureoumbra lagunensis* gen. et sp. nov. and referral of *Aureoumbra* and *Aureococcus* to the Pelagophyceae. *J. Phycol.*, 33:1042-1048.

Feldmann, J. 1941. Une nouvelle Xanthophycée marine: *Tribonema marinum* nov. sp. Bull. Soc. Hist. Nat. Afrique Nord, 32:56-61.

Garry, R. T., Hearing, P. & Cosper, E. M. 1998. Characterization of a lytic virus infectious to the bloom-forming microalga *Aureococcus anophagefferens* (Pelagophyceae). *J. Phycol.*, 34:616-621.

Gastrich, M. D., Anderson, O. R., Benmayor, S. S. & Cosper, E. M. 1998. Ultrastructural analysis of viral infection in the brown-tide alga, *Aureococcus anophagefferens* (Pelagophyceae). *Phycologia* , 37:300-306.

Gayral, P. & Billard, C. 1977. Synopsis du nouvel ordre des Sarcinochrysidales (Chrysophyceae). *Taxon* , 26:241-245.

Gayral, P. & Lepailleur, H. 1971. Etude de deux Chrysophycées filamenteuses: *Nematochrysopsis roscoffensis* Chadefaud, *Nematochrysis hieroglyphica* Waern. *Rev. Gén. Bot.*, 78:61-74

Geitler, L. 1930. Ein grünes Filarplasmodium und andere neue Protisten. *Arch. Protistenkd.*, 69:615-636.

Giner, J.-L. & Boyer, G. L. 1998. Sterols of the brown tide alga *Aureococcus anophagefferens*. *Phytochemistry*, 48:475-477.

Giner, J.-L. & Djerassi, C. 1991. Biosynthetic studies of marine lipids. 31. Evidence for a protonated cyclopropyl intermediate in the biosynthesis of 24-propylidenecholesterol. *J. Am. Chem. Soc.*, 113:1386-1393.

Gobler, C. J. & Cosper, E. M. 1996. Stimulation of "brown tide" blooms by iron. *In*: Yasumoto, T., Oshima, Y. & Fukuyo, Y. (eds.), *Harmful and Toxic Algal Blooms*. IOC UNESCO, Paris. pp 321-324

Gobler, C. J., Hutchins, D. A., Fisher, N. S., Cosper, E. M. & Sañudo-Wilhelmy, S. A. 1997. Release and bioavailability of C, N, P, Se and Fe following viral lysis of a marine chrysophyte. *Limnol. Oceanogr.*, 42:1492-1504.

Heimann, K., Andersen, R. A. & Wetherbee, R. 1995. The flagellar development cycle of the uniflagellate *Pelagomonas calceolata* (Pelagophyceae). *J. Phycol.*, 31:577-583.

Hibberd, D. J. & Chrétiennot-Dinet, M.-J. 1979. The ultrastructure and taxonomy of *Rhizochromulina marina* gen. et sp. nov., an amoeboid marine chrysophyte. *J. Mar. Biol. Assoc. U.K.*, 59:179-193.

Hoffmann, L., Billard, C., Janssens, M., Leruth, M. & Demoulin, V. 2000. Mass development of marine benthic Sarcinochrysidales (Chrysophyceae s. l.) in Corsica. *Botanica Marina*, 43:223-231.

Honda, D. 1995. The taxonomy and phylogeny of the marine Chrysophyceae with special reference to the Sarcinochrysidales. Ph.D. Thesis, University of Tsukuba, Tsukuba, Japan. 84 pp + tables and figures.

Honda, D. & Inouye, I. 1995. Ultrastructure and reconstruction of the flagellar apparatus architecture in *Ankylochrysis lutea* (Chrysophyceae, Sarcinochrysidales). *Phycologia*, 34:215-227.

Honda, D., Kawachi, M. & Inouye, I. 1995. *Sulcochrysis biplastida* gen. et sp. nov.: cell structure and absolute configuration of the flagellar apparatus of an enigmatic chromophyte alga. *Phycol. Res.*, 43:1-16.

Jeffrey, S. W. 1989. Chlorophyll *c* pigments and their distribution in the chromophyte algae. *In*: Green, J. C., Leadbeater, B. S. C. & Diver, W. L. (eds), *The Chromophyte Algae, Problems and Perspectives*. Systematics Association Special Volume No. 38, Clarendon Press, Oxford. pp 13-36

LaRoche, J., Nuzzi, R., Waters, R., Wyman, K., Falkowski, P. G. & Wallace, D. W. R. 1997. Brown tide blooms in Long Island's coastal waters linked to interannual variability in groundwater flow. *Global Change Biology*, 3:397-410.

Lewin, J., Norris, R. E., Jeffrey, S. W. & Pearson, B. E. 1977. An aberrant chrysophycean alga *Pelagococcus subviridis* gen. et sp. nov. from the North Pacific Ocean. *J. Phycol.*, 13:259-266.

Lewis, I. F. & Bryan, H. F. 1941. A new protophyte from the Dry Tortugas. *Amer. J. Bot.*, 28:343-348.

Lobban, C. S., Honda, D., Chihara, M. & Schefter, M. 1995. *Chrysocystis fragilis* gen. nov., sp. nov. (Chrysophyceae, Sarcinochrysidales), with notes on other macroscopic Chrysophytes (golden algae) on Guam reefs. *Micronesica*, 28:91-102.

Lomas, M. W., Glibert, P. M., Berg, G. M. & Burford, M. 1996. Characterization of nitrogen uptake by natural populations of *Aureococcus anophagefferens* (Chrysophyceae) as a function of incubation duration, substrate concentration, light, and temperature. *J. Phycol.*, 32:907-916.

Lopez-Barreiro, T., Villareal, T. A. & Morton, S. L. 1998. Development of an antibody against the Texas brown tide (*Aureoumbra lagunensis*). *In*:

Reguera, B., Blanco, J., Fernandez, M. L., & Wyatt, T. (eds.), *Harmful Algae.* Xunta de Galicia and Intergovernmental Oceanographic Commission of UNESCO, Paris. pp 263-265

Milligan, A. J. & Cosper, E. M. 1997. Growth and photosynthesis of the 'brown tide' microalga *Aureococcus anophagefferens* in subsaturating constant and fluctuating irradiance. *Mar. Ecol. Prog. Ser.*, 153:67-75.

Potter, D., LaJeunesse, T. C., Saunders, G. W. & Andersen, R. A. 1997. Convergent evolution masks extensive biodiversity among marine coccoid picoplankton. *Biodiversity and Conservation* , 6:99-107.

Reinhard, L. 1885. Recherches algologiques. I. Matériaux sur la morphologie et la systématique des algues de la Mer Noire. Zapiski Novorossijskago Obshchestva Estestvoispytatelij, Odessa. (Mémoires de la Société des Naturalistes de la Nouvelle Russie), 9(2):201-502.

Sartoni, G., Boddi, S. & Hass, J. 1995. *Chrysonephos lewisii* (Sarcinochrysidales, Chrysophyceae), a new record for the Mediterranean algal flora. *Bot. Mar.*, 38:121-125.

Saunders, G. W., Potter, D., Paskind, M. P. & Andersen, R. A. 1995. Cladistic analyses of combined traditional and molecular data sets reveal an algal lineage. *Proc. Natl. Acad. Sci. USA*, 92:244-248.

Saunders, G. W., Potter, D. & Andersen, R. A. 1997. Phylogenetic affinities of the Sarcinochrysidales and Chrysomeridales (Heterokonta) based on analyses of molecular and combined data. *J. Phycol.*, 33:310-318.

Sieburth, J. McN., Johnson, P. W. & Hargraves, P. E. 1988. Ultrastructure and ecology of *Aureococcus anophagefferens* gen. et sp. nov. (Chrysophyceae): the dominant picoplankton during a bloom in Narragansett Bay, Rhode Island, summer 1985. *J. Phycol.*, 24:416-425.

Stockwell, D. A., Buskey, E. J. & Whitledge, T. E. 1993. Studies on conditions conducive to the development and maintenance of a persistent "brown tide" in Laguna Madre, Texas. *In* Smayda, T. J. & Shimizu, Y. (eds.), *Toxic Phytoplankton Blooms in the Sea.* Elsevier Science Publishers, Amsterdam. pp 693-698

Street, G. T., Montagna, P. A. & Parker, P. L. 1997. Incorporation of brown tide into an estuarine food web. *Mar. Ecol. Prog. Ser.*, 152:67-78.

Taylor, W. R. 1951. Structure and reproduction of *Chrysophaeum lewisii. Hydrobiologia*, 3:122-130.

Taylor, W. R. 1952. The algal genus *Chrysophaeum. Bull. Torrey Bot. Club*, 79:79.

van der Veer, J. 1970. *Ankylonoton luteum* (Chrysophyta), a new species from the Tamar Estuary, Cornwall. *Acta Bot. Neerl.*, 19:616-636.

Vesk, M. & Jeffrey, S. W. 1987. Ultrastructure and pigments of two strains of the picoplanktonic alga *Pelagococcus subviridis* (Chrysophyceae). *J. Phycol.*, 23:322-336.

Villareal, T. A., Mansfield, A. & Buskey, E. J. 1998. Growth and chemical composition of the Texas brown tide-forming pelagophyte *Aureoumbra lagunensis. In*: Reguera, B., Blanco, J., Fernandez, M. L. & Wyatt, T. (eds.), *Harmful Algae.* Xunta de Galicia and Intergovernmental Oceanographic Commission of UNESCO, Paris. pp 359-362

Yentsch, C. S., Phinney, D. A. & Shapiro, L. P. 1989. Absorption and fluorescent characteristics of the brown tide chrysophyte. Its role on light reduction in coastal marine environments. *In*: Cosper, E. M., Bricelj, V. M. & Carpenter, E. J. (eds.), *Novel Phytoplankton Blooms, Causes and Impacts of Recurrent Brown Tides and other Unusual Blooms.* Springer-Verlag, Berlin. pp 77-83

ORDER RAPHIDOMONADIDA
HEYWOOD & LEEDALE, 1983
by
PETER HEYWOOD and GORDON F. LEEDALE

Raphidomonads are biflagellated motile or palmelloid unicells lacking cell walls. Cells are ovoid to almost spherical, occasionally pyriform, occasionally lanceolate. They are often flattened dorsoventrally with a furrow on the ventral surface. Chloroplasts are few (5) to many in number and contain chlorophylls *a* and *c*. An extensive Golgi apparatus often forms a ring over the anterior surface of the nucleus. Some species contain **extrusomes** (trichocysts or mucocysts). Interphase nuclei and mitotic chromosomes are comparatively large. Meiosis and fertilization have not been described. Cysts are known for some genera. No fossils have been reported.

Freshwater genera (*Gonyostomum, Merotricha*, and *Vacuolaria*) and marine genera (*Chattonella*, *Fibrocapsa*, *Heterosigma,* and *Olisthodiscus*) are recognized. Raphidomonads may be locally abundant, giving rise to blooms. The order Raphidomonadida was often termed the Chloromonadophyceae (or some variant of this name) in algal literature; in classifications of the Protozoa the group was usually the Chloromonadida. These names were inappropriate, since the genus *Chloromonas* is not included within the class or order. We proposed (Heywood and Leedale, 1985) that the order be named Raphidomonadida based upon the validation of the algal class name Raphidophyceae by Silva (1980). The group has been reviewed by Fott (1968) and by Heywood (1980, 1990). A useful taxonomic guide to the marine species has been provided by Fukuyo et al. (1990).

KEY CHARACTERS

1. Two flagella of unequal length arise from a shallow pit at or near the cell apex. The anteriorly directed flagellum is approximately as long as the cell and bears stiff, tubular hairs (Heywood, 1980) resembling mastigonemes of other "heterokont" algae (see 19 below). The posteriorly directed flagellum is up to twice the length of the cell and lacks appendages (Fott, 1968). Most raphidomonads contain an extensive flagellar root system, one part of which is a layered structure that appears to be characteristic of the order (Vesk and Moestrup, 1987).

2. During locomotion the flagellum that bears tubular mastigonemes projects anteriorly and beats rapidly. The other flagellum trails posteriorly and moves only occasionally.

3. Chloroplast-containing and color-less forms have been described (Fott, 1968); the latter are omitted here since evidence is lacking to warrant their inclusion in the order; some recent observations argue against such inclusion. Pigmented genera have many discoid or plano-convex chloroplasts up to 3 x 5 µm in size in the peripheral layer (termed "ectoplasm" by some authors). Pyrenoids are absent in *Gonyostomum, Merotricha,* and *Vacuolaria,* but are present in some of the marine species. Chloroplast lamellae consist of three thylakoids between which frequent interconnections occur; a girdle lamella is is present in most species, but absent in some species (Heywood, 1977; Mignot, 1976).

4. Chloroplasts may vary in color from bright green to yellow-green to yellow-brown. Chlorophylls *a* and *c* are present (Guillard and Lorenzen, 1972). In *Gonyostomum semen* and *Vacuolaria virescens* (freshwater species) the dominant carotenoids are diadinoxanthin, dinoxanthin, ß,ß-carotene, and hetero-xanthin, while in *Chattonella japonica* and *Fibrocapsa japonica* (marine species) the dominant carotenoids are fucoxanthin, fucoxanthinol, ß,ß-carotene, and viola-xanthin (Fiksdahl et al., 1984).

5. The chloroplast-containing raphidomonads are photoautotrophic; possibly some species also utilize heterotrophy.

6. Reserve food material appears to be oil, present as osmiophilic spheres in the cytoplasmic matrix and in the chloroplast stroma. Lipid-containing structures in *Heterosigma akashiwo* show diurnal changes (Wada et al., 1987). The major sterol of *Heterosigma akashiwo* and *Chattonella antiqua* is 24-ethylcholesterol (Nichols et al., 1987).

7. No eyespots are present.

8. No flagellar swellings have been observed.

9. All species are monads. Cells are ovoid to almost spherical, occasionally pyriform, occasionally lanceolate. They are often flattened dorsoventrally with a furrow on the ventral surface. Some species (e.g. *Vacuolaria virescens*, q.v.) may be variable in shape (Fott, 1968; Spencer, 1971). The variable size and morphology of *Chattonella,* which made species recognition problematic, has led to the use of monoclonal antibodies for identification (Hiroishi et al., 1988).

10. A cell wall is absent, but some species secrete mucus and form palmellae (Heywood, 1980).

11. A contractile vacuole occurs in freshwater genera (*Gonyostomum, Merotricha*, and *Vacuolaria*), but is absent in the marine genera. The contractile vacuole may reach 1 0 µm diameter; it lies between the nucleus and the region of flagellar insertion. In *Vacuolaria virescens* a large Golgi body occurs between the anterior surface of the nucleus and the contractile vacuole; it is thought to be involved in osmoregulation (Heywood, 1978b).

12. The nucleus is large (to 20 µm long) and contains one or more nucleoli; interphase chromatin occurs as distinct fine threads (Heywood, 1980; Hovasse, 1945). Chromosome numbers are large: 97 ± 2 in *Vacuolaria virescens* and 65-75 in *Gonyostomum semen* (Heywood, 1980); chromosome length at metaphase varies from 1 to 12 µm. In *Vacuolaria virescens* kinetochores are present, attached to spindle microtubules; the latter form around the flagellar basal bodies and enter the nucleus at prophase through polar gaps in the nuclear envelope (Heywood, 1978a). Condensation of chromosomes, dispersion of nuclei, and formation of a metaphase plate of chromosomes resemble similar events in the "classical mitosis" of other organisms. Peculiar features of mitosis in *Vacuolaria virescens* are that Golgi bodies and contractile vacuoles occur near the poles of the mitotic nucleus and the original nuclear envelope remains intact until telophase, by which time a new nuclear envelope has begun to form around the daughter nuclei (Heywood, 1978a).

13. Asexual reproduction is by binary fission. Possible sexual reproduction has been reported for *Chattonella marina* (Subrahmanyan, 1954). Microfluoro-metric studies of *Chattonella antiqua* and *Chattonella marina* suggest a diplontic life cycle in which meiosis precedes encystment (Yamaguchi and Imai, 1994).

14. Some species form cysts, probably in response to nutrient depletion (Nakamura and Umemori, 1991). In *Chlorella antiqua,* silicon predominates in the cyst wall with magnesium and aluminum also being present (Meksumpun et al., 1994).

15. Extrusomes (trichocysts or mucocysts) occur in most (but not all) species. The rod-shaped trichocysts of *Gonyostomum* can discharge mucilaginous threads for distances in excess of 100 µm. The spherical mucocysts of *Vacuolaria* secrete mucilage around the cell.

16. Freshwater species frequently occur in the plankton, among aquatic plants or immediately above the benthic muds of bogs, ponds, and lakes, the pH of which ranges from 4.0 to 7.0. *Chattonella*, *Fibrocapsa, Heterosigma,* and *Olisthodiscus* occur in marine environments.

17. Diel vertical migrations between lower layers at night where inorganic nutrients are more abundant and euphotic zones during the day, have been observed in both freshwater (Cronberg et al., 1988) and marine (Watanabe et al., 1988) species. In *Heterosigma akashiwo*, phosphate taken up at night in deeper water was accumulated as polyphosphate and subsequently used for photophosphorylation in the photic layer (Watanabe, 1988).

8. Raphidomonads may occur in large numbers causing mortality to marine organisms (Subrahmanyan, 1954) or health

problems to bathers (Cronberg et al., 1988). Red tides of raphidomonads may be controlled by viruses (Nagasaki et al., 1994) or by heterotroph predation (Nakamura et al., 1992).

19. Ultrastructural and biochemical evidence suggests that the closest relative of the raphidomonads are members of the algal division Heterokontophyta *sensu* Leedale (see van den Hoek, 1978) which also have heterokont flagella, tubular mastigonemes, chlorophylls *a* and *c*, three-thylakoid lamellae in the chloroplasts, and other common features. In this book those organisms are represented by the Chrysomonadida (q.v.) (and Heterochlorida), but they also include several major algal groups such as the diatoms and brown seaweeds.

ARTIFICIAL KEY TO SELECTED GENERA

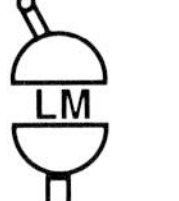

1. Freshwater forms, contractile vacuole present.. 2
1′. Brackish or marine forms, contractile vacuole absent................................... 4

2. Cells with rod-shaped trichocysts...... 3
2′. Cells with spherical mucocysts but lacking rod-shaped trichocyst .... ***Vacuolaria***

3. Trichocysts mainly in swollen anterior region of cells; flagella arise on lateral surface immediately below this region............................. ***Merotricha***
3′. Trichocysts around cell periphery (but may be locally dense at cell anterioand/or posterior); flagella arise near cell apex .... ***Gonyostomum***

4. Flagella arise at or near cell apex 5
4′. Flagella arise sub-apically to laterally on ventral surface............................. 6

5. Prominent rod-shaped trichocysts predominantly present at cell posterior .... ***Fibrocapsa***
5′. Mucocysts if present and conspicuous are not predominantly grouped at the cell posterior .... ***Chattonella***

6. Chloroplasts are yellowish-brown to brown; during swimming, cell rotates helically .... ***Heterosigma***
6′. Chloroplasts are yellowish-green to green; during swimming, cell moves without rotation .... ***Olisthodiscus***

Genus ***Chattonella*** Biecheler, 1936

Cell shape is variable both within and between species; cells are frequently obovate, but vary from globose to lanceolate, with the latter often possessing a distinct caudus; 2 flagella, about the same length as the cell; many chloroplasts, often present as a single layer below the plasmalemma, give the cell a yellow-brown color; pyrenoids present in some species; some species form cysts; mucocysts present in some species; contractile vacuole absent; saline, periodically abundant, causes red tides; synonymous with *Hornellia* (Subrahmanyan, 1954). See Biecheler (1936); Fukuyo et al. (1990); Mignot (1976).
TYPE SPECIES: *Chattonella subsalsa* Biecheler, 1936 (Fig. 6). Cell 30-50 x 15-25 µm; brackish and marine habitats, often rich in organic matter (Mignot, 1976); may reach high densities; 6 other species have been recognized by Fukuyo et al, (1990): *C. antiqua, C. globosa, C. marina* (synonymous with *Hornellia marina* Subrahmanyan), *C. minima, C. ovata,* and *C. verruculosa.*

Genus ***Fibrocapsa*** Toriumi & Takano, 1973

Cells ovate to obovate, slightly flattened; flagella arise from an anterior depression; anterior flagellum is approximately the same length as the cell, but the trailing flagellum is slightly longer; trichocysts predominantly grouped at posterior end; chloroplasts

forming a single layer below the plasmalemma are yellowish-green or brown. See Hara and Chihara (1985), Toriumi and Takano (1973).
TYPE SPECIES: *Fibrocapsa japonica* Toriumi & Takano, 1973 (Fig. 2). Cell 20-30 x15-17 µm; many rod- shaped trichocysts often grouped at posterior end can eject threads up to 300 µm in length; each chloroplast possesses a pyrenoid ; chloroplasts may be so tightly abutted at the cell periphery that they appear as one large reticulate chloroplast.

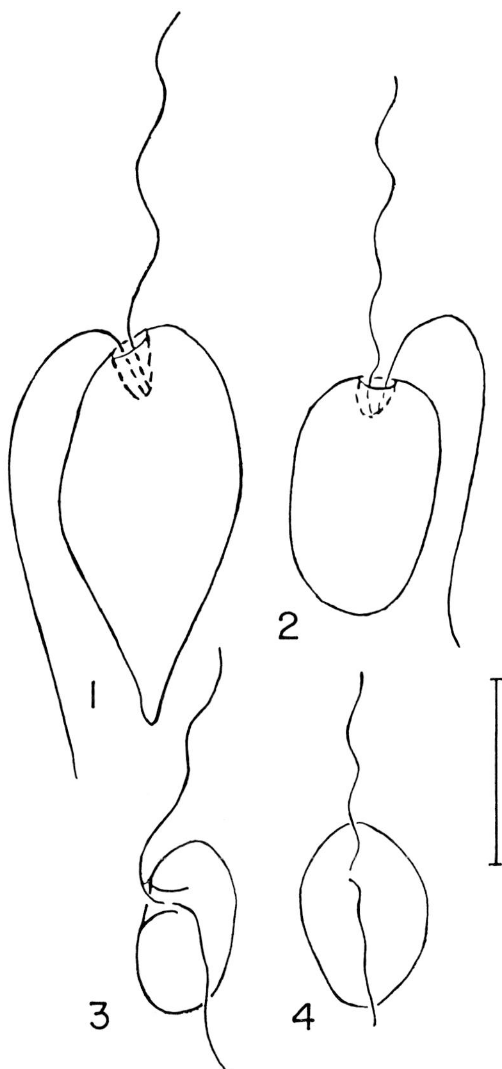

Figs. 1-4. Diagram of marine raphidomonads to allow comparison of cell size, shape, and insertion of flagella; all at same magnification; scale bar=10 µm. Fig. 1. *Chattonella subsalsa* drawn from lateral surface to illustrate insertion of the flagella and structure of the caudus. Figs. 2-4 are drawn in ventral view to illustrate cell shape and insertion of flagella; Fig. 2. *Fibrocapsa japonica.* Fig. 3. *Heterosigma akashiwo.* Fig. 4. *Olisthodiscus luteus.* Figs. 1 and 3 are based on the illustrations in Table 3 of Hara and Chihara (1987); Fig. 2 is based on Fig. 1 of Hara and Chihara (1985); Fig. 4 is based on Fig. 1 of Hara et al. (1985).

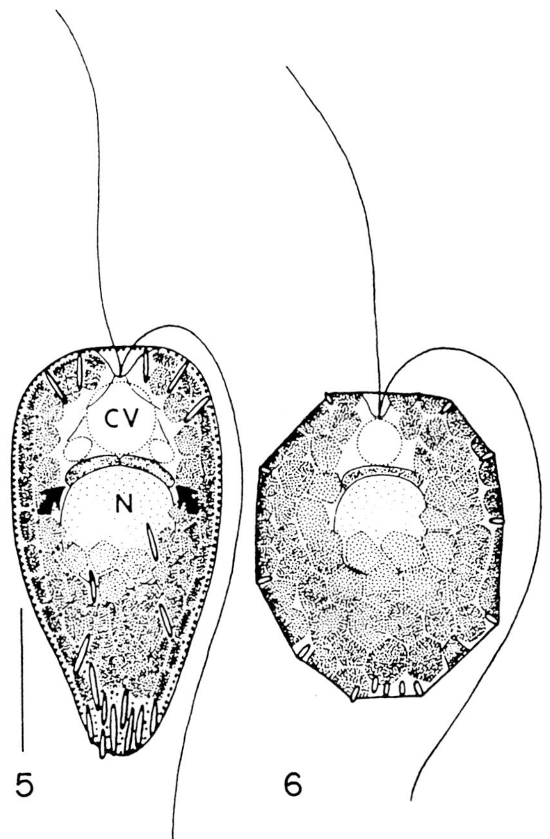

Fig. 5. *Gonyostomum semen*; ventral view, but trailing flagellum (usually over ventral surface) shown at side of cell; chloroplasts stippled; Golgi apparatus and vesicles arrowed between nucleus (N) and contractile vacuole (CV); group of trichocysts shown at posterior end of cell. x1000. Fig. 6. *Gonyostomum latum*, ventral view; organelles as Fig. 5. x1,000. Figs. 5-8 from Fott (1968) by permission of E. Schweizerbart'sche Verlagsbuchhandlung

Genus *Gonyostomum* Diesing

Cells obovate-lanceolate, obovate or almost spherical, flattened dorsoventrally with a furrow on the ventral surface, caudus short; rod-shaped trichocysts around cell periphery or concentrated at the 2 ends of the cell; flagella usually the same length as the cell, but the trailing flagellum may reach twice that length; many chloroplasts in a single layer in the cell periphery; frequently among aquatic plants in acidic bogs or pools. See Hovasse (1945).

TYPE SPECIES: *Gonyostomum semen* Diesing, (Fig. 5). Cell 40-90 µm long; many rod-shaped trichocysts often grouped at ends of cell; grown on defined medium at pH 5.5-5.8 (Heywood, 1980); division stages observed during the dark period in laboratory cultures (Heywood, 1980); during unfavorable conditions may be present as cysts (Cronberg et al., 1988); frequently collected, usually from acidic bogs, pools, and lakes; diurnal movement occurs in natural populations (Cronberg et al., 1988), cells being near the surface in early hours of daylight; 3 other species have been recognized (Fott, 1968): *G intermedium*, *G. latum* (Fig. 2) and *G. ovatum*

Genus *Heterosigma* Hada, 1968

Cell shape variable, from approximately spherical to ovoid, somewhat compressed dorsoventrally; flagella arise from separate pits in a furrow running from near the apex along the ventral side; both flagella approximately the same length as the cell; the beat of the anterior flagellum causes cell to rotate helically while swimming; discoid chloroplasts, 10-30 per cell, in peripheral layer, yellowish-brown to brown in color; mucocysts present. See Fukuyo et al. (1990); Hara and Chihara (1987); Vesk and Moestrup (1987).

TYPE SPECIES: *Heterosigma akashiwo* (Hada) Hada, 1968 (Fig. 3). Cell 10-22 x 8-15 µm; only species of the genus; was maintained in some culture collections as *Olisthodiscus luteus*; other synonyms: *Heterosigma inlandica, Chattonella inlandica, Chattonella luteus.*

Genus *Merotricha* Mereschkowsky, 1879

Cells appear elliptical in front view; anterior end of cell enlarged into a distinctive convex structure containing a group of trichocysts; flagella inserted laterally below this region; bright green discoid chloroplasts occur in cell periphery except for the cell anterior where the trichocysts occur.

TYPE SPECIES: *Merotricha bacillata* Mereschowsky, 1879 (Fig. 7). Cell 40-50 x 18-25 µm; only species of the genus; trailing flagellum overlooked in original description but subsequently described by Skuja (1934) as *M. capitata,* which is almost certainly synonymous with *M. bacillata* (Fott, 1968); no accounts of laboratory culture, biochemistry, or ultrastructure; from acidic bogs and ponds (Skuja, 1934; Graffius, 1966).

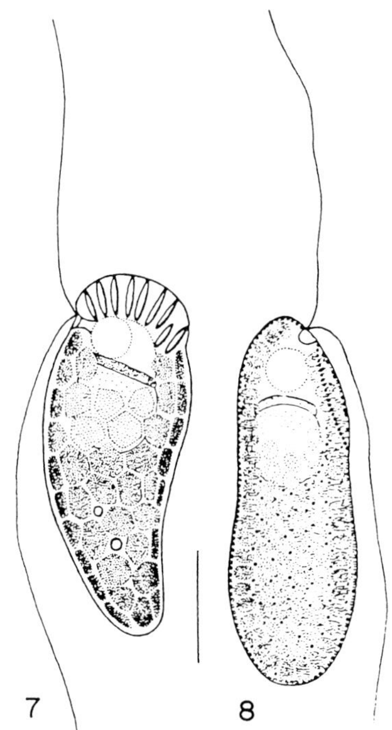

Fig. 7. *Merotricha bacillata*, side view; organelles as Fig. 1; distinctive anterior region contains battery of trichocysts. x1000. Fig. 8. *Vacuolaria virescens*, side view; organelles as Fig. 1; peripheral cytoplasm contains small, spherical mucocysts between the chloroplasts. x1000.

Genus ***Olisthodiscus*** Carter, 1937

Cells discoid in ventral view; pronouncedly flattened dorsoventrally with dorsal surface concave and ventral surface convex, giving a structure resembling a saddle or a turtle's carapace; flagella arise from a shallow groove on the ventral surface; anterior flagellum somewhat longer than the cell, trailing flagellum is shorter, being of the same length as the cell or less; the beat of the anterior flagellum causes cell to move forward smoothly, without rotation; discoid chloroplasts, 5-13 in number with 6 being common, in peripheral layer, yellowish-green in color; mucocysts absent. See Fukuyo et al. (1990); Hara et al. (1985).

TYPE SPECIES: *Olisthodiscus luteus* Carter, 1937 (Fig. 4). Cell 15-25 long x 10-16 wide by 5-7 µm thick; only species of the genus; has been confused with *Heterosigma akashiwo*, which has similar size, shape, and environment, but the chloroplasts of *Olisthodiscus* are greener, and it swims without rotating.

Genus ***Vacuolaria*** Cienkowsky, 1870

Cell shape variable both between and within species, from almost spherical to oblong-cylindrical; cell dimensions also variable; flagella approximately the same length as the cell; many chloroplasts, not restricted to a single layer; spherical mucocysts occur at cell periphery. See Spencer (1971), Heywood (1977, 1978a,b).

TYPE SPECIES: *Vacuolaria virescens* Cienkowsky, 1870 (Figs. 4,5). Cell 30-85 x 18-25 µm; var. *minuta* only 19-23 x 17-19 µm (Fott, 1970); cell shapes vary in single natural collection (Spencer, 1971) or in single laboratory culture; spherical mucocysts occur near cell periphery and are involved in formation of palmelloid stages; species maintained in culture (Guillard and Lorenzen, 1972; Heywood, 1990) and several ultrastructural investigations made (Schnepf and Koch, 1966; Mignot, 1967; Heywood, 1977, 1978a,b); some aspects of chloroplast biochemistry also studied (Guillard and Lorenzen, 1972); occurs in ponds, bogs, and lakes usually rich in vegetation (Poisson and Hollande, 1943; Graffius, 1966; Spencer, 1971) and usually of pH 4.0-6.5, but also reported from slightly alkaline environments (Graffius, 1966); related species, *V. viridis* , pear-shaped with broad cell anterior (Fott, 1968, 1970) though Spencer (1971) thinks this is synonymous with *V. virescens* ; Fott (1968, 1970) described. *V. penardii* , a species in which chloroplasts are absent from the cytoplasm immediately below the cell membrane.

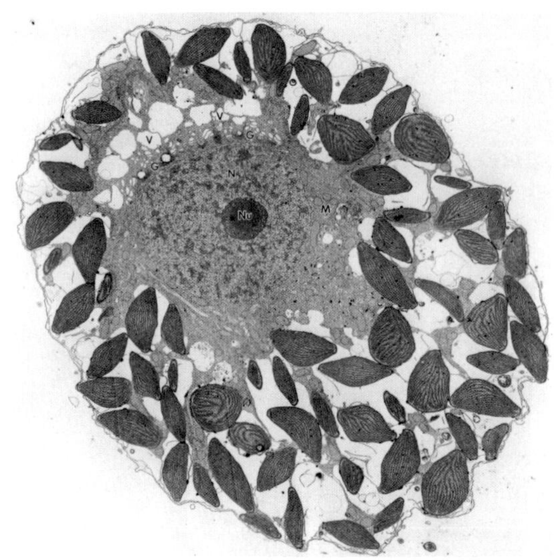

Fig. 9. *Vacuolaria virescens*; EM longitudinal section showing many chloroplasts in cell periphery, not in a single layer; nucleus (N) with nucleolus (Nu) lies in anterior half of cell surrounded by characteristic zone of cytoplasm including mitochondria (M) and Golgi bodies (G); Golgi vesicles (V) empty into the contractile vacuole. x3000. (from Heywood [1977] by permission of *J. Phycol.*)

LITERATURE CITED

Biecheler, B. 1936. Sur une chloromonadine nouvelle d'eau saumâtre *Chattonella subsalsa* n. gen., n. sp. *Arch. Zool. Exp. Gén*., 78:79-83.

Cronberg, G., Lindmark, G. & Björk, S. 1988. Mass development of the flagellate *Gonostomum semen* (Raphidophyta) in Swedish forest lakes—an effect of acidification? *Hydrobiologia,* 161:217-236.

Fiksdahl, A., Withers, N., Guillard, R. R. L. & Liaaen-Jensen, S. 1984. Carotenoids of the

Raphidophyceae—A chemosystematic contribution. *Comp. Biochem. Physiol.*, 78B:265-271.

Fott, B. 1968. VIII. Klasse: Chloromonadophyceae *In*: Huber-Pestalozzi, G. G. (ed.), *Das Phytoplankton des Susswassers*. 2nd ed. Stuttgart, 3:79-93.

Fott, B. 1970. Taxonomische Übertragungen und Namensänderungen unter den Algen. *III. Chloromonadophyceae*. *Preslia*, 42:16-20.

Fukuyo, Y., Takano, H., Chihara, M. & Matsuoka, K. 1990. *Red Tide Organisms in Japan—An Illustrated Taxonomic Guide*. Uchida Rokakuho. pp 332-349

Graffius, J. H. 1966. Additions to our knowledge of Michigan Pyrrhophyta and Chloromonadophyta. *Trans. Am. Microsc. Soc.*, 85:260-270.

Guillard, R. R. L. & Lorenzen, C. J. 1972. Yellow-green algae with chlorophyllilde *c*. *J. Phycol.*, 8:10-14.

Hara, Y. & Chihara, M. 1985. Ultrastructure and taxonomy of *Fibrocapsa japonica* (Class Raphidophyceae). *Arch Protistenkd.*, 130:133-141.

Hara, Y. & Chihara, M. 1987. Morphology, ultrastructure and taxonomy of the raphidophycean alga *Heterosigma akashiwo*. *Bot. Mag. Tokyo,* 100:151-163.

Hara, Y., Inouye, I. & Chihara, M. 1985. Morphology and ultrastructure of *Olisthodiscus luteus* (Raphidophyceae) with special reference to the taxonomy. *Bot. Mag. Tokyo,* 98:251-262.

Heywood, P. 1977. Chloroplast structure in the chloromonadophycean alga *Vacuolaria virescens*. *J. Phycol.*, 13:68-72.

Heywood, P. 1978a. Ultrastructure of mitosis in the chloromonadophycean alga *Vacuolaria virescens*. *J. Cell Sci.,* 31:37-51.

Heywood, P. 1978b. Osmoregulation in the alga *Vacuolaria virescens*. Structure of the contractile vacuole and the nature of its association with the Golgi apparatus. *J. Cell Sci.*, 31:213-224.

Heywood, P. 1980. Chloromonads. *In*: Cox, E. R. (ed.), *Phytoflagellates*. Elsevier/North Holland, New York. pp 351-379

Heywood, P. 1990. Raphidophyceae. *In*: Margulis, L., Corliss, J. O., Melkonian, M. & Chapman, D. J. (eds.), *Handbook of Protoctists*. Jones & Bartlett Publishers, Boston, Massachusetts. pp 318-325

Heywood, P. & Leedale, G. F. 1985. Raphidomonadida. *In:* Lee, J. J., Hutner, S. H., & Bovee, E. C. (eds.), *Illustrated Guide to the Protozoa*. Society of Protozoologists, Lawrence, Kansas. pp 70-74

Hiroishi, S., Uchida, A., Nagasaki, J. & Ishida, Y. 1988. A new method for identification of inter- and intra-species of the red tide algae *Chattonella antiqua* and *Chattonella marina* (Raphidophyceae) by means of monoclonal antibodies. *J. Phycol.,* 24:442-444.

Hoek, C., van den 1978. *Algen*. Thieme, Stuttgart.

Hovasse, R. 1945. Contribution à l'étude des Chloromonadines: *Gonyostomum semen* Diesing. *Arch. Zool. Exp. Gén.,* 84:239-269.

Meksumpun, S., Montani, S. & Uematsu, M. 1994. Elemental components of cyst walls of three marine phytoflagellates, *Chattonella antiqua* (Raphidophyceae), *Alexandrium catenella* and *Scrippsiella trochoidea* (Dinophyceae). *Phycologia,* 33:275-280.

Mignot, J. P. 1967. Structure et ultrastructure de quelques Chloromonadines. *Protistologica* 3:5-23.

Mignot, J. P. 1976. Compléments à l'étude de Chloromonadines. Ultrastructure de *Chattonella subsalsa* Biecheler flagellé d'eau saumâtre. *Protistologica*, 12:279-93.

Nagasaki, K., Ando, M., Itakura, S., Imai, I. & Ishida, Y. 1994. Viral mortality in the final state of *Heterosigma akashiwo* (Raphidophyceae) red tide. *J. Plank. Res.*, 16:1595-1599.

Nakamura, Y. & Umemori, T. 1991. Encystment of the red tide flagellate *Chattonella antiqua* (Raphidophyceae): cyst yield in batch cultures and cyst flux in the field. *Mar. Ecol. Progr. Ser.*, 78:273-284.

Nakamura, Y., Yamzaki, Y. & Hiromi, J. 1992. Growth and grazing of a heterotrophic dinoflagellate, *Gyrodinium dominans,* feeding on a red tide flagellate, *Chattonella antiqua. Mar. Ecol. Progr. Ser.*, 82:275-279.

Nichols, P. D., Volkman, J. K., Hallegraeff, G. M. & Blackburn, S. I. 1987. Sterols and fatty acids of the red tide flagellates *Heterosigma akashiwo* and *Chattonella antiqua* (Raphidophyceae). *Phytochemistry,* 26:2537-2541.

Poisson, R. & Hollande, A. 1943. Considérations sur la cytologie, la mitose et les affinités des Chloromonadines. Étude de *Vacuolaria virescens* Cienk. *Ann. Sci. Nat. (Zool.)*, Ser. II, 5:147-160.

Schnepf, E. & Koch, W. 1966. Über die Entstehung der pulsierenden Vacuolen von *Vacuolaria virescens* (Chloromonadophyceae) aus dem Golgi-apparat. *Arch. Mikrobiol.*, 54:229-36.

Silva, P. C. 1980. Names of classes and families of living algae. *Regnum Veg.*, 103:1-156.

Skuja, H. 1934. Beitrag zur Algenflora Lettlands. I. *Acta Horti. Bot. Univ. Latv.*, 7:25-86.

Spencer, L. B. 1971. A study of *Vacuolaria virescens* Cienkowski. *J. Phycol.*, 7:274-279.

Subrahmanyan, R. 1954. On the life-history and ecology of *Hornellia marina* gen. et sp. nov., causing green discolouration of the sea and mortality among marine organisms off the Malabar Coast. *Indian J. Fish*, 1:182-203.

Toriumi, S. & Takano, H. 1973. *Fibrocapsa*, a new genus in Chloromonadophyceae from Atsumi Bay, Japan. *Bull Tokai Reg. Fish. Res. Lab.*, 76:25-35.

Vesk, M. & Moestrup, Ø. 1987. The flagellar root system in *Heterosigma akashiwo* (Raphidophyceae). *Protoplasma*, 137:15-28.

Wada, M., Hara, Y., Kato, M., Yamada, M. & Fujii, T. 1987. Diurnal appearance, fine structure, and chemical composition of fatty particles in *Heterosigma akashiwo* (Raphidophyceae). *Protoplasma*, 137:134-139.

Watanabe, M., Kohata, K. & Kunugi, M. 1988. Phosphate accumulation and metabolism by *Heterosigma akashiwo* (Raphidophyceae) during diel vertical migration in a stratified microcosm. *J. Phycol.*, 24:22-28.

Yamaguchi, M. & Imai, I. 1994. A microfluorometric analysis of nuclear DNA at different stages in the life history of *Chattonella antiqua* and *Chattonella marina* (Raphidophyceae). *Phycologia.*, 33:163-170.

RESIDUAL HETEROTROPHIC STRAMENOPILES

by DAVID J. PATTERSON

The stramenopiles are a large group of protists which are defined by reference to a synapomorphy—tripartite, tubular hairs (Patterson, 1989). The composition of the group was indicated elsewhere (Patterson, 1994) and is upgraded to include genera described subsequent to 1994 as follows:

- Stramenopiles
  - Bicosoecales
  - Slabyrinthulids
    - *Diplophrys*
    - Labyrinthulids
    - Thraustochytrids
  - Sloomycetes
    - Oomycetes
    - Hyphochytridiomycetes
  - Slopalines
    - Proteromonadidae
    - Opalinidae
  - Stramenochromes
    - Core chrysophytes
      - Chromulinales
      - Hibberdiales
      - Hydrurales
      - Pelagophytes
      - Chrysomeridales
    - Axodines
      - Dictyochales
      - Pedinellales
      - Rhizochromulinales
    - Synurids
    - Xanthophytes
    - Raphidophytes
    - Eustigmatophytes
    - Brown algae
    - Diatoms
    - Parmales
    - *Commation*
    - *Developayella*
    - *Ollicola*
    - *Pendulomonas*
    - *Reticulosphaera*
    - *Siluania*

The group contains organisms with a diverse array of organizational types from the fungus-like oomycetes to the multicellular, massive brown algae. It includes autotrophs such as diatoms and chrysophytes as well as mixotrophs and heterotrophs. *Reticulosphaera* is an amoeboid autotroph, but there is dispute as to whether it is a member of this taxon. The group has some compositional overlap with the chrysophytes, chromophytes, heterokonts, or chromists — but not to a level that might regarded as sufficient to regard the stramenopiles as being identical with any of these (despite assertions by some to the contrary). The principal difference between this taxon and the others named is its definition by a single synapomorphy. The classification given above seeks to ensure that all taxa are monophyletic. Consequently, it differs from alternative groupings (heterokonts, chrysophytes, chromists) not only in concept, but from alternative groupings (heterokonts, chrysophytes, chromists) not only in concept, but also in composition, hierarchical structure, and our present inability to rank its members.

Many stramenopiles have plastids, such that the grouping that contains the core chrysophytes has intuitively been regarded as algal (e.g. Cavalier-Smith, 1989). Comparisons of sequences and explicit analyses (Leipe et al., 1994, 1996; Williams, 1991), however, suggest that the ancestral stramenopiles were heterotrophs and that definitions that are dependent on plastids will only define a subset of this group.

In this book, various subsets of the stramenopiles have been included in other chapters; however, as criteria of monophyly have not been applied, the coverage has been incomplete. This chapter therefore serves to include organisms which are stramenopiles but have not been included elsewhere.

KEY CHARACTERS

Stramenopiles are identified by having tripartite hairs which are usually attached to the anterior flagellum. The presence of these hairs can be inferred from the flow of fluid which flows from tip to base of beating flagella. Some stramenopiles (e.g. opalines) have secondarily lost the hairs, and their membership of the group must then be determined by characters derived within the group (e.g. a double gyre in the transitional region of the flagellum).

Genera and species are distinguished on the number and disposition of the flagella, motility, and cell shape. Within this group, one subset is distinguished because the cell attaches to the substrate by the tip of the flagellum.

KEY TO GENERA

1. Taxa covered elsewhere in this text. See appropriate chapters
1'. Taxa not covered elsewhere in this text.. 2

2. Cells with two or no flagella.......................... 3
2'. Cells always with a single flagellum ***Siluania***

3. Cells attaching to substrate via the distal tip of recurrent flagellum 4
3'. Cells not attaching by tip of recurrent flagellum but may attach using some of the distal length of the flagellum 6

4. Cell within lorica .... ***Bicosoeca***
4'. Cells aloricate 5

5. Recurrent flagellum long; cell with apical collar; cell attaching indirectly by a strand of mucus .... ***Pseudobodo***
5'. Recurrent flagellum short, no mucus strand, no collar.. ***Cafeteria***

6. Cell flattened and gliding, flagella normally absent.. ***Commation***
6'. Cell with active flagella 7

7. Attached cells attach by distal section of recurrent flagellum................ ***Pendulomonas***
7'. Both flagella active in attached cells, cells attached by cytoplasmic strands .... ***Developayella***

ORDER BICOSOECIDA

Cells bearing two flagella that insert anterio-laterally; one flagellum attaching to the substrate or lorica either directly or indirectly (via a thread of mucus). Anterior flagellum creates currents of water from which particles are ingested at a discrete ingestion area. Freshwater and marine. Three genera in two families: Bicosoecidae for the loricate taxa and the Cafeteriidae for the aloricate taxa.

Genus ***Bicosoeca*** James-Clark, 1867

Biflagellated cells enclosed within an organic lorica, usually vase-shaped. Solitary or colonial; if colonial, may be arborescent or ball-shaped (Fig. 1). Lorica may attach directly to substrate or have stalk. Anterior margin of cell raised on one side to form a collar. This forms part of the ingestion area. Cells may retract into their lorica by action of the recurrent flagellum and anterior flagellum coils up. Sometimes incorrectly spelled *Bicoeca*. About 40 species, freshwater and marine. Reference: Preisig et al., 1990.
TYPE SPECIES: *Bicosoeca lacustris* James-Clark, 1867.

Genus ***Cafeteria*** Fenchel & Patterson, 1988

Genus of marine flagellates, attached by the tip of the shorter recurrent flagellum (Fig.

2.) Food ingested in ventral depression or posteriorly. Very common in marine habitats. Several species. *Acronema* (Teal et al., 1998) cannot be clearly distinguished from this genus. Cosmopolitan type species. References: Preisig et al., 1990; Teal et al., 1998.
TYPE SPECIES: *Cafeteria roenbergensis* Fenchel & Patterson, 1988

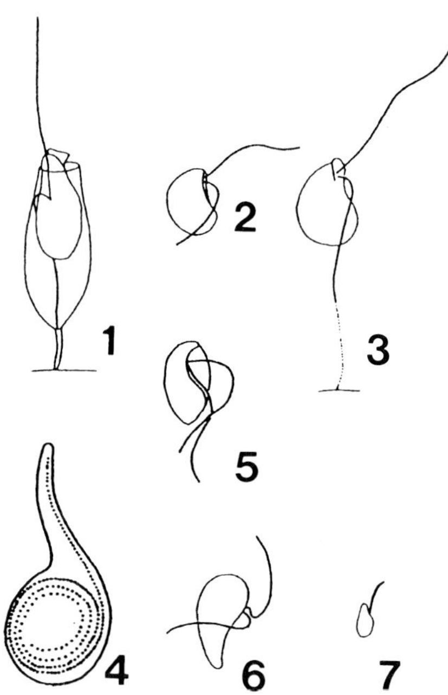

Figs. 1-7. 1. *Bicosoeca lacustris.* (after Mignot, 1974) 2. *Cafeteria roenbergensis*, original. 3. *Pseudobodo tremulans*, original. 4. *Commation esposianum.* (after Thomsen & Larsen, 1993) 5. *Developayella elegans.* (after Tong, 1995) 6. *Pendulomonas adriperi.* (after Tong, 1997) 7. *Siluania monomastiga* (original). Scale=10 μm.

Genus ***Pseudobodo*** Griessmann, 1913

Genus of marine flagellates, attaching usually by a thin strand of mucus arising from the tip of the shorter recurrent flagellum. With a raised collar around part of the antero-lateral margin of the cell (Fig. 3). Common in marine habitats. Probably two species, type species: *P. tremulans* Griessmann, 1913 is cosmopolitan. Reference: Preisig et al., 1990.

Other genera

Genus ***Commation*** Thomsen & Larsen, 1993

Gliding circular or ovoid cell, usually flattened, with proboscis, the single (hairy) flagellum occasionally emerging at the base of the proboscis (Fig. 4). Move by gliding involving the secretion of mucus. Phagotrophic and marine, described from Antarctic waters. Two species, type species *C. esposianum* Thomsen & Larsen, 1993. Reference: Thomsen and Larsen, 1993.

Genus ***Developayella*** Tong, 1995

Free swimming naked stramenopiles with 2 flagella which insert into a considerable depression on the right anterior ventral face of the cell (Fig. 5.). One flagellum is held in an arc in front of the swimming cells and bears hairs; the other is smooth and attached along the base of the depression. Stationary cells attach by means of cytoplasmic strands and when stationary, both flagella are directed posteriorly, the anterior flagellum sweeps up and down slowly, the posterior flagellum beats rapidly. Marine (estuarine). Monotypic, type species: *D. elegans* Tong, 1995. Reference: Tong, 1995.

Genus ***Pendulomonas*** Tong, 1997

Comma-shaped cells with 2 equal flagella. May swim or may attach to the substrate using the distal third of the recurrent flagellum (Fig. 6). Anterior flagellum used to generate feeding current, but posterior part of the cell may be extended to form a feeding cup. Marine, described from south of England. One and type species, *P. adriperis* Tong, 1997. Reference: Tong, 1997

Genus ***Siluania***
Karpov, Kersanach & Williams, 1998

Freshwater stramenopile with one short anterior flagellum with tripartite hairs. Small. With small apical cytostome (Fig. 7). One and type species, *S. monomastiga* Karpov, Kersanach & Williams, 1998. Reference: Karpov et al., 1998.

ACKNOWLEDGEMENTS:
Support from the Australian Research Council, the Australian Biological Resources Study, and the University of Sydney is gratefully recognized.

References

Cavalier-Smith, T. 1989. The Kingdom Chromista. *In*: Green, J. C., Leadbeater, B. S. C. & Diver, W. L. (eds.). *The Chromophyte Algae: Problems and Perspectives.* Clarendon Press, Oxford. pp 381-408

Karpov. S. A., Kersanach, R. & Williams, D. M. 1998. Ultrastructure and 18S rRNA gene sequence of a small heterotrophic flagellate *Siluania monomastiga* gen. et sp. nov. (Bicosoecida). *Europ. J. Protistol.*, 34:415-425.

Leipe, D. D., Tong, S. M., Goggin, C. L., Slemenda, S. B., Pieneazek, N. J. & Sogin, M. L. 1996. 16S-like rDNA sequences from *Developayella elegans*, *Labyrinthuloides haliotidis*, and *Protoeromonas lacertae* confirm that the stramenopiles are a primarily heterotrophic group. *Europ. J. Protistol.*, 32:449-458.

Leipe, D. D., Wainright, P. O., Gunderson, J. H., Poter, D., Patterson, D. J., Valois, F., Himmerich, S. & Sogin, M. L. 1994. The stramenopiles from a molecular perspective: 16S-like rRNA sequences from *Labyrinthuloides minuta* and *Cafeteria roenbergensis*. *Phycologica*, 33:369-377.

Patterson, D. J.1989. Stramenopiles: chromophytes from a protistological perspective. *In*: Green, J. C., Leadbeater, B. S. C. & Diver, W. L. 1989. *The Chromophyte Algae: Problems and Perspectives*. Clarendon Press, Oxford. pp 357-379

Preisig, H. R., Vørs, N. & Hällfors, G. 1989. Diversity of heterotrophic heteokont flagellates. In: Patterson, D. J. & Larsen, J. (eds.), *The Biology of Free-Living Heterotrophic Flagellates*. Clarendon Press, Oxford. pp 361-399

Teal, T. M., Guillemette, T., Chapman, T. & Margulis, L. 1998. *Acronema sippiwissettensis* gen. nov. et sp. nov., microbial mat bicosoecid (Bicosoecales= Bicosoecida). *Europ. J. Protistol.*, 34:402-414.

Thomsen, H. A & Larsen, J. 1993. The ultrastructure of *Commation* gen. nov. (Stramenopiles incertae sedis), a genus of heterotrophic nanoplanktonic flagellates from Antarctic waters. *Europ. J. Protistol.*, 29:462-477.

Tong, S. M. 1995. *Developayella elegans* nov. gen., nov. spec., a new type of heterotrophic flagellate from marine plankton. *Europ. J. Protistol.*, 31:24-31.

Tong, S. M. 1997. Heterotrophic flagellates and other protists from Southampton Water, U.K. *Ophelia*, 47:71-131

Vørs, N. 1992. Heterotrophic amoebae, flagellates and heliozoa from the Tvärminne area, Gulf of Finland, in 1988-1990. *Ophelia*, 36:1-109.

Williams, D. M. 1991. Phylogenetic relationships among the chromista: a review and preliminary analysis. *Cladistics*, 7:141-156.

ORDER SLOPALINIDA

Patterson, 1989

BY

B. L. J. DELVINQUIER and DAVID J. PATTERSON

Slopalines embrace seven genera of endocommensal flagellates occurring mostly in the posterior regions of the intestines of cold-blooded vertebrates. Most opalines have been described from Anura. The order comprises two families, the Opalinidae and the Proteromonadidae. The Opalinidae consist of five genera and about 400 species which have been described in amphibia, reptiles, freshwater and marine fish, and in a few invertebrates. The group has been most recently reviewed by Delvinquier and Patterson (1993).

The Proteromonadidae consist of two genera. There is only a few species of *Karotomorpha* (from amphibia). There are several species of *Proteromonas* occurring mainly in reptiles and more rarely in amphibia and mammals. The opalines have been more extensively studied than proteromonads (Kulda and Nohyková, 1978; Metcalf, 1923, 1940; Wessenberg, 1978; Delvinquier and Patterson, 1993). The relationships among genera of slopalines is illustrated in Fig. 1.

The surface of opalines is covered with rows (kineties) of flagella. As a result, they were first considered to be ciliates, and Metcalf (1923, 1940) referred to them as Protociliates; however, they lack a number of the pecularities of the ciliates such as dimorphic nuclei and were transferred from the ciliates to the Sarcomastigophora by Corliss and Balamuth (1963). With the recognition of the polyphyly of the Sarcomastigophora, they have been segregated as an independant group. Opalines and proteromonads are placed together because they have a similar and distinctive structure of the flagellum and because in both families the cell surface is underlain by a cortex of evenly spaced microtubules. In the opalines and *Karotomorpha*, these take the form of ribbons. In *Proteromonas*, the microtubules are single. On the basis of these ultrastructural similarities proteromonads and opalines were combined to form the slopalines (see Delvinquier and Patterson, 1993).

Proteromonas has tripartite hairs attached to the posterior end of the body which together with double-gyred transitional helices in the flagellar transitional region indicate an affinity with the chrysophyte algae and other stramenopiles (Delvinquier and Patterson, 1993).

PROTEROMONAS
KAROTOMORPHA
PROTOZELLERIELLA
ZELLERIELLA
PROTOOPALINA
CEPEDEA
OPALINA

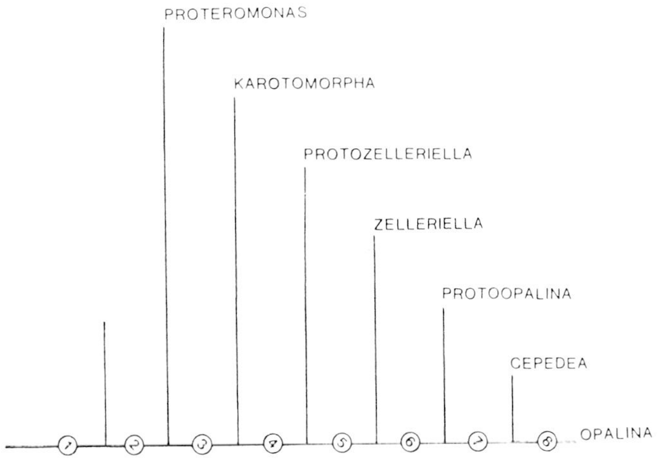

Fig. 1. Proposed relationships within the slopalines. Numbered items indicate proposed evolutionary steps. (1) Origins of stramenopiles from biflagellated tubulocristate protists by appearance of tripartite hairs and derivation of a subset with a double-gyred transitional helix. (2) Tripartite hairs reduced to body surface; evenly spaced microtubules underlie cell surface. (3) Complete loss of tripartite hairs, cortical microtubules transform into ribbons, and number of flagella increased to four. (4) Kineties begin to develop; cells become binucleated, enlarge, and become flattened. (5) Kineties become continuous; falx distinctive and marginal. (6) Falx becomes axial. (7) Nuclei proliferate and become smaller. (8) Return to marginal falx.

Opalines are thought to have evolved in Anura from Gondwanaland prior to the break-up of the continent in late Jurassic (Delvinquier and Patterson, 1993). They now have a world-wide distribution.

The opalines have a life cycle that is linked to that of the host in those species that have been studied. Our understanding is based on opalines that live in Anura. Opalines grow and divide along and across the kineties while in their host. Cysts are produced during the sexual activity of the host and are released into water. Tadpoles are infected by feeding on the cysts. Excysting opalines are destined to produce gametes. Conjugation occurs between dissimilar gametes (Wessenberg, 1961) (Fig. 2).

Ω

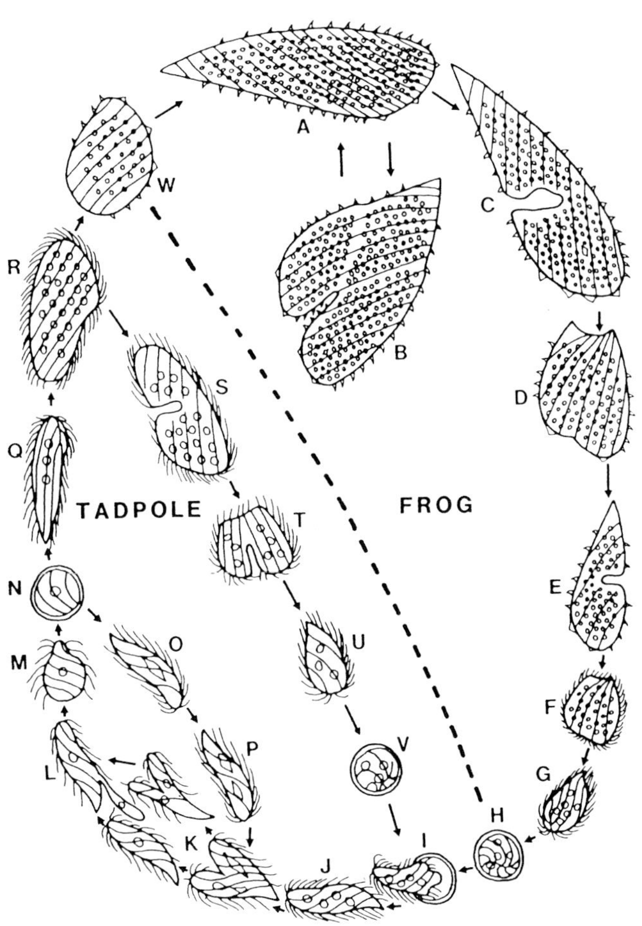

Fig. 2. Life cycle of *Opalina* proposed by Wessenberg. A,B. Growth and division of trophonts. C-G. Palintomy (divisions) leading to the production of small cells which form cysts (H). Cysts are shed from the frog and excyst (I) upon ingestion by tadpoles. The cells which excyst (K) will produce anisogametes (K), which fuse (L) to produce zygocysts (N). The sexual cycle may be repeated (O,P), or, alternatively, cells may grow and divide with or without cyst formation.

Proteromonads are small protists, 10-25 μm in length, mononucleated with apically inserting flagella. *Proteromonas* is biflagellated with a smooth cortex and material adhering to its posterior end. *Karotomorpha* is quadriflagellated with a pleated or striated cortex. The two genera can be distinguished on the basis of the number of flagella.

Opalines are usually greater than 100 µm in length and are binucleated or multinucleated. Nuclei are monomorphic. All opalines have flagella arranged as kineties. The kineties arise anteriorly at the **falx**, which is a morphogenetic center.

Opaline genera are distinguished by the number of nuclei and by the orientation of the falx relative to the antero-posterior axis of the cell (Fig. 3). *Protozelleriella*, *Zelleriella,* and *Protoopalina* typically have two nuclei with a diameter greater than 10 µm, but a few exceptions are reported. *Opalina* and *Cepedea* are multinucleated with nuclei typically less than 10 µm in diameter. A long and thin falx extends along the margin and mostly lies transverse to the antero-posterior axis of the cell in *Protozelleriella*, *Zelleriella,* and *Opalina*. This is referred to as a marginal falx. Alternatively, a short and broad falx lies mostly along the antero-posterior axis of the cell in *Protoopalina* and *Cepedea* and is referred to as axial.

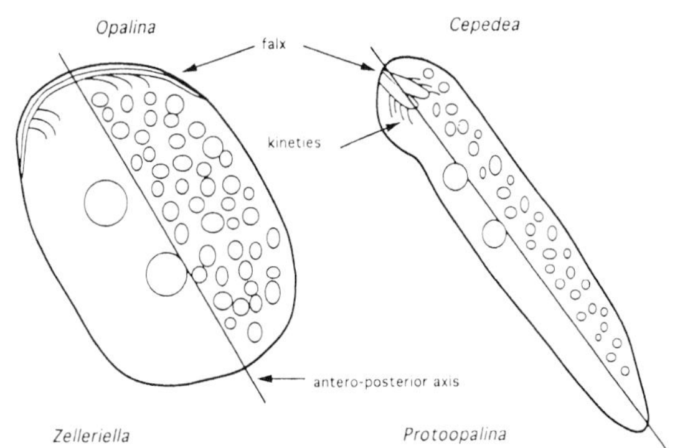

Fig. 3. Orientation of falx: marginal (left) and axial (right) for the four major genera.

Standard staining techniques to be used for slopalines include Bodian's protargol, silver nitrate impregnation, or Fernandez-Galiano's ammoniacal silver impregnation.

Opaline distribution in hosts is reviewed in Delvinquier and Patterson (1993).

KEY TO THE FAMILIES AND GENERA

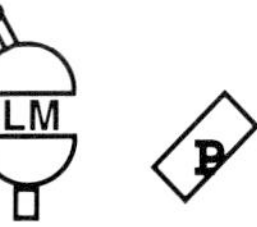

1. Mononucleated, flagella not arranged as kineties .... Family **Proteromonadidae**............. 2
1'. Multinucleated, flagella arranged as kineties.. ....Family **Opalinidae** 3

2. Two anterior flagella.............. ***Proteromonas***
2'. Four anterior flagella............ ***Karotomorpha***

3. Binucleated.. 4
3' Multinucleated... 6

4. With long, thin, marginal falx extending along the margin and mostly transverse to the antero-posterior axis of the cell.................. 5
4' With short, broad axial falx running mostly in the direction of the antero-posterior axis of the cell.................................. ***Protoopalina****

5. Peripheral hyaline margin devoid of flagella .... ***Protozelleriella***
5'. Body evenly covered with kineties .... ***Zelleriella***

6. With long, thin, marginal falx transverse to the anteroposterior axis of the cell; body evenly covered with kineties............ ***Opalina***
6'. With short, broad axial falx parallel to the anteroposterior axis of the cell ....***Cepedea***

* A few species of *Protoopalina* have been described with four, six, or eight nuclei. All of these species have large and aligned nuclei by which they can be distinguished (Delvinquier and Patterson, 1993).

FAMILY PROTEROMONADIDAE Grassé, 1952

Genus ***Proteromonas*** Kunstler, 1883

Endocommensal flagellates up to about 25 µm in length, with 2 apical flagella. Surface not markedly ridged, but posterior end may carry fine hairs. A large dictyosomal complex is located near the anterior of the cell. Complex life cycle involves a cyst in which multiple divisions occur. Records mainly from the guts of lizards but also reported from caudate amphibia and mammals. Four species, e.g. *Proteromonas lacertaeviridis*.

Fig. 4. *Proteromonas lacertaeviridis*.

Genus ***Karotomorpha*** Travis, 1934

Small (12-16 µm) flagellates with 4 long apical flagella. Single nucleus, body surface ridged. Massive dictyosomal complex (parabasal apparatus) is located in the anterior part of the cell. Reported from the guts of Australian, North American, and European amphibia. Several species.
TYPE SPECIES: *Karotomorpha bufonis* (Alexeieff, 1916)
TYPE HOST: *Triton*, *Rana*, *Amblystoma*

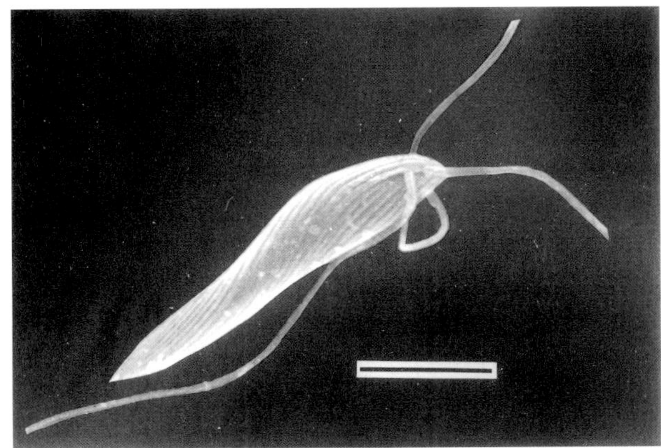

Fig. 5. *Karatomorpha bufonis*. Scale bar=5 µm.

FAMILY OPALINIDAE Claus, 1874

Genus ***Protozelleriella***
Delvinquier, Markus & Passmore, 1991

Binucleated, with a long, thin, marginal falx almost perpendicular to the antero-posterior axis of the cell, with a peripheral, hyaline margin that is devoid of flagella and with a crenulate posterior margin. One report only from bufonid toad in South Africa.
TYPE SPECIES: *Protozelleriella devilliersi* Delvinquier, Markus & Passmore, 1991
TYPE HOST: *Capensibufo rosei*

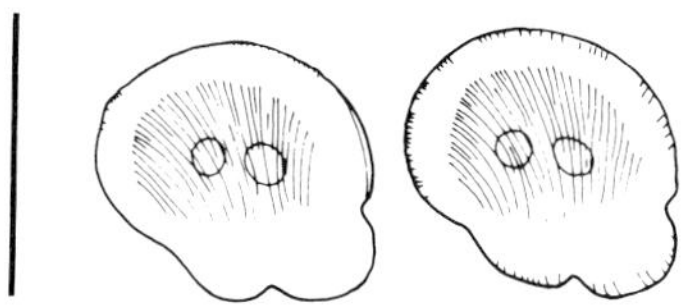

Fig. 6. *Protozelleriella devilliersi*. Scale bar=100 µm.

Genus ***Zelleriella*** Metcalf, 1920

Binucleated, with a long, thin, marginal falx, transverse to the antero-posterior axis of the cell. Kineties cover the body evenly. Mostly Southern Hemisphere distribution. Mostly in anuran amphibia but occasional records for caudate and caecilian amphibia, reptiles, and fish. *Zelleriella antilliensis, Z. falcata, Z. devincki.*
TYPE SPECIES: *Zelleriella antillliensis* (Metcalf, 1914) Metcalf, 1923
TYPE HOST: *Bufo marinus* (as *Bufo aqua*)

Genus ***Protoopalina*** Metcalf, 1918

Binucleated, with a short, broad, axial falx running parallel to the antero-posterior axis of the cell. Kineties cover the body evenly. A few species with 4, 6, or 8 nuclei have been reported, mostly in Asia. Predominantly Old World. Reported mostly from anuran amphibia, but also from caudate amphibia, reptiles, and fish. Examples: *Protoopalina intestinalis, P. mitotica, P. tenuis, P. saturnalis.*
TYPE SPECIES: *Protoopalina intestinalis* (Stein, 1856)
TYPE HOST: *Bombina bombina*

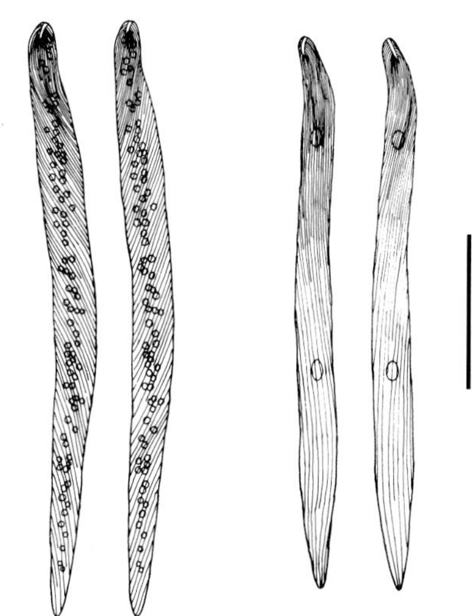

Fig. 7 (left). *Zelleriella falcata.* Fig. 8 (right). *Protoopalina tenuis.* Scale bar=100 µm.

Genus ***Cepedea*** Metcalf, 1920

Multinucleated, with a short, broad, axial falx parallel to the antero-posterior axis of the cell. Kineties cover the body evenly. More common in Old World than in New World and unreported in Australia. Mostly in anuran amphibia, but also in fish and one report in a freshwater snail. *Cepedea dimidiata, C. affinis, C. ciliata.*
TYPE SPECIES: *Cepedea dimidiata* (Stein, 1860) Metcalf, 1923

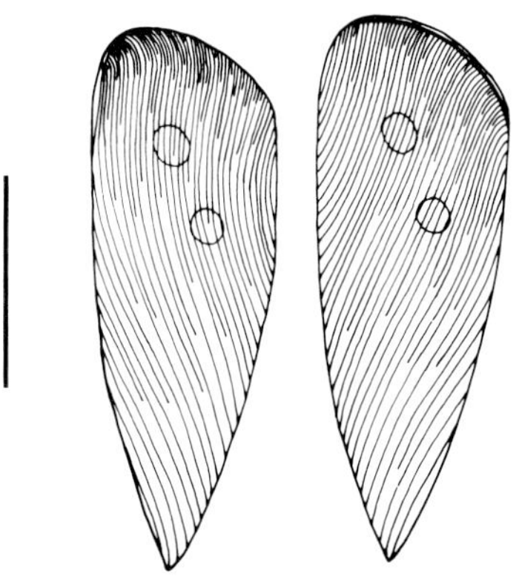

Fig. 9. *Cepedea affinis.* Scale bar=100 µm.

Genus ***Opalina*** Purkinje & Valentin, 1835

Multinucleated, with a long, thin, marginal falx transverse to the antero-posterior axis of the cell. Kineties cover the body evenly. Body shape either elongate or broad. May have a suture line running dorso-ventrally at the back of the cell. Mostly in Asia and in North America and unreported in Australia. Mostly in Anura but occasional records in caudate amphibia, reptiles, and one freshwater mussel. Many species, e.g. *Opalina ranarum, O. obtrigonoidea, O. duquesnei, O. elongata*.
TYPE SPECIES: *Opalina ranarum* (Ehrenberg, 1831) Purkinje & Valenin, 1835
TYPE HOST: Frogs

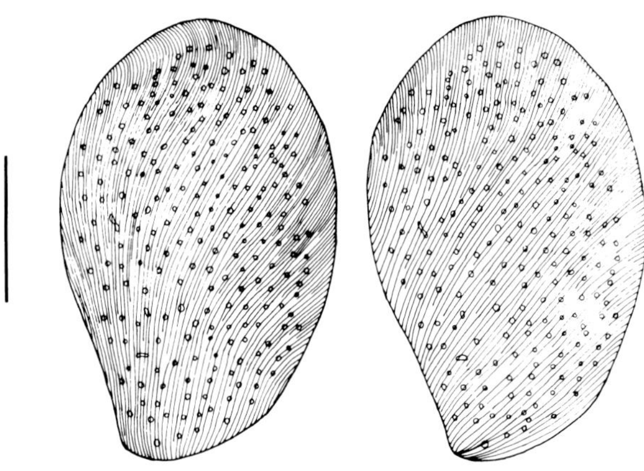

Fig. 10. *Opalina duquesnei.* Scale bar=100 µm.

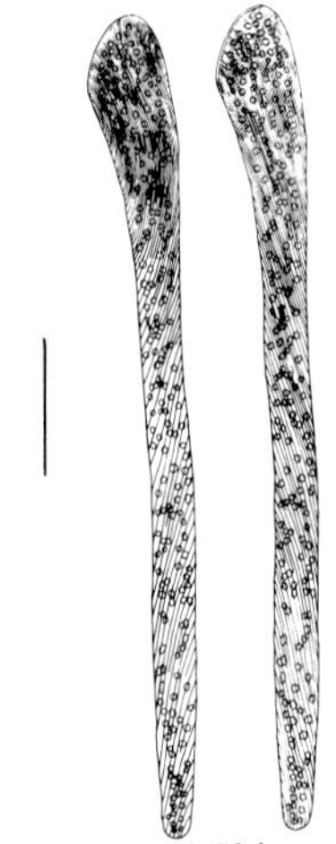

Fig. 11. *Opalina elongata.* Scale bar=100 µm.

Bezzenbergeria and *Hegneriella* (Opalinidae) are not regarded as valid genera (Delvinquier and Patterson, 1993).

LITERATURE CITED

Corliss, J. O. & Balamuth, W. 1963. Consideration of the opalinids as a new superclass in the Subphylum Sarcomastigophora. *J. Protozool.*, 10 (Suppl.):26.

Delvinquier, B. L. J. & Patterson, D. J. 1993. The Opalines. *In*: Kreier, J. P. & Baker, J. R. (eds.),

Parasitic Protozoa. 2nd ed., Vol. 3. Academic Press, San Diego. pp 247-325

Grassé, P.-P. 1929. Sur la cytologie du flagelle parasite *Proteromonas lacetaeviridis* Grassé. *C. R. Assoc. Anat. Bordeaux*, 18:267-275.

Kulda, J. & Nohynková, E. 1978. Flagellates of the human intestine and of intestines of other species. *In*: Kreier, J. P. (ed.), *Parasitic Protozoa.* Academic Press, New York. pp 1-138

Metcalf, M. M. 1923. The opalinid ciliate infusorians. *Bull. U. S. Natl. Mus.*, 120:1-484.

Metcalf, M. M. 1940. Further studies on the opalinid ciliate infusorians and their hosts. *Proc. U. S. Natl. Mus.*, 87:465-634. Wessenberg, H. 1961. Studies on the life cycle and morphogenesis of *Opalina. U. C. Publ. Zool.*, 61:315-369.

Wessenberg, H. 1978. Opalinata. *In*: Kreier, J. P. (ed.), *Parasitic Protozoa.* Vol. 2. Academic Press, New York and London. pp 551-581

CLASS SYNUROPHYCEAE Andersen, 1987

ROBERT A. ANDERSEN and HANS R. PREISIG

The Synurophyceae are separated from the Class Chrysophyceae (See Chapter Chrysomonada; Andersen, 1987; Andersen, 1989; Moestrup, 1995; Preisig, 1995). A common and obvious feature of all common genera of Synurophyceae is the presence of bilaterally symmetrical silica scales. The scales remain as fossils in lake sediments, and numerous paleolimnological studies have been based upon the distribution of these fossil scales (e.g. Battarbee et al., 1980; Smol, 1980; Smol et al., 1984a,b; Smol, 1995). The Synurophyceae have a number of significant cellular features that are not found in the Chrysophyceae. These include a unique flagellar apparatus, chlorophyll c_1 only—rather than chlorophylls c_1 and c_2 combined—chloroplasts that usually lack membrane connections to the nucleus, as well as the production of bilaterally symmetrical scales (Andersen, 1987; Andersen, 1989). In the older literature these organisms were classified in the Chrysophyceae, and many papers still refer to these two groups informally as "chrysophytes". The history of these two classes is described in Kristiansen (1995). The Synurophyceae are predominantly freshwater flagellates, occurring in a wide range of habitats from polar regions to the tropics (Kristiansen, 1981, 1986a,b; Cronberg, 1989; Kristiansen et al., 1995). There are only a few reports of Synurophyceae from marine environments, e.g. *Mallomonas epithalattia* Droop, 1955 and *M. subsalina* Conrad & Kufferath, 1954 (Parke and Dixon, 1976) and *Synura uvella* Ehrenberg, 1834 (McAlice, 1975). Although representatives can be found in all types of freshwater, they are most commonly encountered in acidic or circumneutral waters in cold temperate bogs, ponds, and lakes (Kristiansen, 1986b; Asmund and Kristiansen, 1986; Siver 1991; see also papers in Kristiansen and Andersen, 1986; Kristiansen et al., 1989; Sandgren et al., 1995; Kristiansen and Cronberg 1996). Almost two hundred species have been described (e.g. Starmach, 1985; Asmund and Kristiansen, 1986), but there are numerous synonymies caused by independent descriptions using light and electron microscopy.

Very few have studied the biochemistry, cell biology, molecular biology and physiology of the Synurophyceae. The major carbohydrate storage product has not been determined for any member of the Synurophyceae but it is often assumed to be leucosin (Klebs, 1893; Archibald et al., 1963), chrysolaminaran (Quillet, 1955; Beattie et al., 1961; Smestad-Paulsen and Myklestad, 1978) or similar relatively small (dp < 40) ß-1,3 linked glucan. The chloroplast genome of *Synura petersenii* Korshikov, 1929 is only 91.5 kb (Wee et al., 1993), one of the smallest chloroplast genomes known (see Coleman and Goff, 1991). Another member of the class, *Chrysodidymus synuroideus* Prowse, 1962, has a chloroplast genome size of 102 kb (Graham et al., 1993), and these data suggest the chloroplast genome size of the Synurophyceae is quite small compared to other algae and plants. The organization of the plastid genome in *Synura* is scattered rather than as a ring (Coleman, 1985). The mitochondrial genome of *Chrysodidymus synuroideus* is 33 kb (Graham et al., 1993) and the putative mitochondrial genome of *Synura petersenii* is 34.5 kb (Wee et al., 1993); these mitochondrial genomes are comparable in size to those of other algal groups (see Coleman and Goff, 1991). Other molecular biology details are summarized by Wee et al. (1996b).

The Synurophyceae are closely related to the Chrysophyceae based upon phylogenetic analyses using the 18S rRNA gene (Ariztia et al., 1991; Bhattacharya et al., 1992; Leipe et al., 1994; Saunders et al., 1995, 1997; Lavau et al., 1997; Bailey et al., 1998; Andersen et al., 1999) and they are placed together in the Division Chrysophyta. The relationship is also supported by the occurrence of the unique resting cyst (**stomatocyst, statospore**) that is produced only by members of these two classes.

Numerous floristic and ecological studies of Synurophyceae have been published (see Takahashi, 1978; Starmach, 1985; Asmund and Kristiansen, 1986; Sandgren, 1988; Siver, 1991; Smol, 1995). Synurophyceae have been reported from almost all types of freshwater habitats, but they are more frequently collected in relatively cooler waters from temperate regions. Many studies have attempted to link the distribution of specific species with physical and chemical factors, especially temperature, pH, and conductivity (e.g. Kristiansen, 1986a; Siver and Hamer, 1992; Siver, 1995). Generally, Synurophyceae are collected from waters with pH <7.5; however, there are reports of a number of species found predominantly in alkaline waters (Kristiansen, 1985, 1986a,b, 1988a; Wee and Gabel, 1989). Alternatively, Cumming et al. (1992) suggest aluminum concentration in waters may be an important factor in determining species occurrence.

Some species are considered indicator species because they occur in waters with narrowly defined physical and chemical parameters. It is possible to make paleolimnological reconstructions of lake histories based upon these indicator species because their silica scales and siliceous cysts are often preserved as fossils in lake sediments. The paleolimnological reconstructions are used to hypothesize climatic change, productivity, and pH shifts, especially with regard to lake acidification (see Smol, 1995).

There are relatively few genera in the Synurophyceae, and these are classified in the single order Synurales Andersen, 1987 and in the two families Synuraceae Lemmermann, 1899 and Mallomonadaceae Diesing, 1866. Members of the Synuraceae are colonial. The common genera in this family are *Synura* Ehrenberg, 1834, *Chrysodidymus* Prowse, 1962, and *Tessellaria* Playfair, 1918. *Chlorodesmos* Phillips, 1884 is now included in the genus *Synura* (Calado and Rino, 1994), and so is *Catenochrysis* Perman, in Perman and Vinniková, 1955, a name which has sometimes been used in the literature (Bourrelly, 1981; Starmach, 1985), but is a superfluous substitute name for *Chlorodesmos* Phillips, 1884. *Pseudosyncrypta* Kisselew, 1931 is a less known taxon, but its two nearly equal flagella, posterior contractile vacuoles and lack of an eyespot suggest that it belongs in the Synuraceae (see Kisselew, 1931). The genus *Syncrypta* Ehrenberg, 1834 has been problematic since its description (Ehrenberg, 1834, 1838). It is possible that *Syncrypta volvox* Ehrenberg, 1834 is identical to *Synura sphagnicola* (Korshikov, 1927) Korshikov, 1929, but it is probably impossible to resolve this problem. One species, *Syncrypta glomerifera* Clarke & Pennick, 1975, is now classified as *Lepidochrysis glomerifera* (Clarke & Pennick, 1975) Ikävalko, Kristiansen & Thomsen, 1994 (Chrysophyceae) (Ikävalko et al., 1994). The remaining species of *Syncrypta* (see Bourrelly, 1957; Starmach, 1985) should also be treated as members of the Chrysophyceae, and probably should be placed with species of the genus *Synuropsis* Schiller, 1929 (see Chapter Chrysomonada).

Members of the Mallomonadaceae are single-celled flagellates. The most common genus is *Mallomonas* Perty, 1852 (including *Mallomonopsis* Matvienko, 1941), and it is a large genus with subgeneric Sections and Series (Asmund and Kristiansen, 1986; Siver, 1988; Preisig, 1989). The genus *Conradiella* Pascher, 1925 is less well known and has been described using light microscopy only (Conrad, 1914, 1926; Pascher, 1925). Unlike the scale-bearing genera of the Synurophyceae, cells of *Conradiella* have a stack of several siliceous rings that surround the cell. Based upon treatment with various acids, these rings have been determined to be composed of silica (Conrad, 1926). Perhaps surprisingly, *Conradiella* has not been observed in any of the numerous electron microscopical studies of the Synurophyceae. The genus *Microglena* Ehrenberg, 1832 has been described to have siliceous granules that are located around the periphery of the cell (Conrad, 1927a), and based upon these putative siliceous structures, *Microglena* has been aligned with the silica-scaled genera (Bourrelly, 1957, 1968, 1981; Starmach, 1985). It should be noted, however, that the type species, *M. monadina* Ehrenberg, 1832, has been referred to *Chlamydomonas* (Chlorophyceae) by Stein (1878), and therefore the genus cannot be assigned to either the Synurophyceae or Chrysophyceae. Ultrastructural studies of *M. punctifera* (O. F. Müller 1786) Ehrenberg, 1832 (see Wujek, 1967) and *M. butcheri* Belcher, 1966 (Couté and Preisig, 1981) did not demonstrate the presence of siliceous structures. Based upon ultrastructural observations, the latter species appears to be closely related to *Chromulina* (see Chapter Chrysomonada). *Microglena cordiformis* Conrad, 1927a has two chloroplasts and no eyespot which might suggest affinities to the Synurophyceae, but it also has anterior contractile vacuoles and this is not a feature found in well-studied Synurophyceae. Other species of *Microglena* have been described to have a single chloroplast with an eyespot and these seem unlikely to be members of the Synurophyceae.

Key Characters

1. All species are free-swimming flagellate cells, either single-celled or colonial, during some or all of their life history. Two subequal flagella are present on colonial forms (*Chrysodidymus, Pseudosyncrypta, Synura, Tessellaria*) and some species of *Mallomonas* (formerly classified as *Mallomonopsis*). Two very unequal flagella, or a single flagellum with a barren basal body, are present on most species of *Mallomonas* as well as on the enigmatic *Conradiella* (Bourrelly, 1957, 1981; Starmach, 1985). The longer flagellum bears tripartite tubular hairs and beats in a sine wave. This flagellum is the immature flagellum with respect to flagellar transformation (see Wetherbee et al., 1988). The shorter flagellum, when present and visible, is less active. In *Mallomonas guttata* Wujek, 1984, unusual appendages have been observed at the distal end of the shorter flagellum (Zhang et al., 1990b). Small organic scales may be present on the flagella (Hibberd, 1973; Zimmermann, 1977).

The ultrastructure of the flagellar apparatus of Synurophyceae is unique amongst protists (Andersen, 1985, 1991). The absolute orientation of the flagellar apparatus, i.e., the arrangement of the flagellar apparatus with respect to other cellular organelles, has more than one pattern (two or four); for all other protistan flagellates only one absolute orientation is known. The two basal bodies are arranged in a nearly parallel fashion, and they are connected with two or more striated fibers. The basal apparatus is attached to the nucleus with one or two large rhizoplasts (=system II fibers). Cells have either R_1 microtubular roots only, or R_1 and R_3 roots; the R_2 and R_4 microtubular roots that are typical for members of the Chrysophyceae are absent. The distinctive R_1 root is attached to the rhizoplast, and it extends anteriorly where it forms a loop around the flagella. Numerous cytoskeletal microtubules extend from this root, providing cytoskeletal support to shape the cell. Specialized cytoskeletal microtubules also extend down from the R_1 root along the chloroplasts, maintaining the chloroplasts in a fixed position. Additional specialized cytoskeletal microtubules extend down along the outside of one chloroplast and are involved in silica scale formation (Schnepf and Deichgräber, 1969; Mignot and Brugerolle, 1982; Andersen, 1985, 1989; Beech and Wetherbee, 1990b; Pipes et al., 1991; Pipes and Leedale, 1992; Graham et al., 1993).

2. All examined species have chlorophyll *a* and c_1 but lack chlorophyll c_2 (Andersen and Mulkey, 1983; Jeffrey, 1989). The lack of chlorophyll c_2 distinguishes the Synurophyceae from the Chrysophyceae and other heterokont chromophytic phytoflagellates. In place of chlorophyll c_2, however, a new form of chlorophyll *c* has been found in *Synura petersenii* (Fawley, 1989). The chemical structure of this new pigment has not been determined and the extent of its occurrence in the Synurophyceae is not known. The major carotenoid pigments are fucoxanthin and violaxanthin (Bidigare and Andersen, unpubl. observ.), similar to the carotenoids of the Chrysophyceae, Eustigmatophyceae, Phaeophyceae and marine Raphidophyceae (Bjørnland and

Liaaen-Jensen, 1989). Two dubious colorless species of *Mallomonas* have also been described in the literature (see Preisig et al., 1991).

3. Most species are reported to have two chloroplasts. However, Harris (1953) reported the presence of a single chloroplast with two very deeply divided lobes for 25 species of *Mallomonas*. Thus, the number of plastids in the Synurophyceae should be carefully re-examined. The chloroplast ultrastructure consists of a chloroplast (with two membranes) surrounded by two membranes of endoplasmic reticulum. Typically, the endoplasmic reticulum membranes are either not confluent with the outer membrane of the nuclear envelope or are only weakly connected to the outer membrane of the nuclear envelope (Wujek and Kristiansen, 1978; Andersen, 1987, 1989; Zhang et al., 1990b; Pipes et al., 1991; Graham et al., 1993). The central chloroplast lamellae, usually 5-10 per plastid, are each formed from three adpressed thylakoids. The central lamellae are sheet-like in structure, and they are surrounded by a single girdle lamella that is bag-like in structure. Pyrenoids are apparently rare in the Synurophyceae, but they have been reported for *Synura sphagnicola* (Hibberd, 1978).

4. Members of the Synurophyceae do not possess a red eyespot in the anterior margin of the chloroplast, although representatives of *Synura* and *Mallomonas* are known to be phototactic (K. Foster, pers. comm.). A swelling occurs on one or both of the flagella in Synurophyceae and this has been suggested to be the location of phototactic reception (Andersen, 1985, 1989). Some species of *Synura* (e.g. *S. curtispina* [Petersen & Hansen, 1956] Asmund, 1968 and *S. sphagnicola*) occasionally have red droplets in the anterior end of the cells, but these droplets are not associated with the chloroplast (Hibberd, 1978; Andersen, unpubl. observ.). *Chrysodidymus synuroideus* also has red droplets but these are located in the cell posterior (Graham et al., 1993).

5. The presence of more or less bilaterally symmetrical silica scales is a characteristic of Synurophyceae; for a perspective of silica scales produced by the Chrysophyceae, see Preisig and Hibberd (1982a,b, 1983). Silica scale morphology is used as the primary, and often only, character for identification of taxa. Recently, gene sequences have been used to test the validity of this scale-based taxonomy, and preliminary results suggest that scale morphology is a good taxonomic character (Lavau et al., 1997). The scales are arranged in a highly organized fashion around each cell in most taxa; however, *Tessellaria* and *Synura lapponica* Skuja, 1956 have scales that do not surround individual cells, and in some cases the scales are not around individual cells of *Synura sphagnicola*; in these cases, the scales are arranged loosely in an unorganized fashion around the colony (Skuja, 1956; Pipes et al., 1991, unpubl. observ.). Laboratory experiments have shown that scaleless cells can be produced under conditions of extreme silicon starvation (Sandgren and Barlow, 1989; Leadbeater and Barker, 1995; Sandgren et al., 1996). Scale formation, or scale biogenesis, has been described in some detail (see Schnepf and Deichgräber, 1969; Mignot and Brugerolle, 1982; Brugerolle and Bricheux 1984; Wetherbee et al., 1989, 1992, 1995; Preisig, 1986, 1994; Leadbeater and Barker, 1995; Ludwig et al., 1996; Wee, 1997); for specific information, see generic descriptions of *Mallomonas, Synura,* and *Tessellaria* below. The deployment of scales around the cell has also been described. In *Mallomonas adamas* Harris & Bradley, 1960, cells immediately following cell division have 30 scales (5 rows of 6 scales); scales are formed and secreted throughout interphase, and just before division, cells have 60 scales (5 rows of 12 scales). Scales are added singly, not in rows, and specific scale types are excreted and deployed to their specific locations (Lavau and Wetherbee, 1994). In species such as *Mallomonas splendens* (G. S. West, 1909) Playfair, 1921 emend. Croome, Dürrschmidt & Tyler, 1985, scales have attached bristles. The bristles are excreted base-first, but the bristle is attached to the cell via a microtubule-enforced cytoplasmic strand, and once the bristle is completely excreted, the base of the bristle is pulled into place (Beech et al., 1990). Ecologically, scales and bristles are believed to play a role as antipredatory devices against smaller predators (e.g. ciliates), but they seem to be ineffective against larger predators (Sandgren and Walton, 1995).

6. Synurophyceae produce endogenous siliceous cysts (=statospores, stomatocysts) like those of

the Chrysophyceae. A silica deposition vesicle forms near the periphery of the cell, and silica is then deposited in the vesicle. Initially, a smooth-surfaced cyst is formed and if morphogenesis is interrupted, this may be the final cyst morphology (Sandgren, 1980; Leadbeater and Barker, 1995). However, some species have more ornate cysts, and the ornamentation is added after the initial smooth-surfaced cyst shape is formed. Sandgren (1980, 1989) suggests that the morphogenesis of Synurophyceae cysts differs from that of Chrysophyceae. Scanning electron microscopy of cysts from field samples has revealed many morphological details, but unfortunately, few careful taxonomic studies have been made, and field sample morphotypes often cannot be assigned to specific species (see Sandgren, 1991; Duff et al., 1994).

7. Vegetative cells divide by mitosis and cytokinesis. Mitosis begins as flagellar transformation proceeds, and the two sets of basal bodies migrate apart (Beech and Wetherbee, 1990b). The rhizoplast acts as a **microtubular organizing center (MTOC)** for the spindle microtubules. The nuclear envelope remains partially intact during mitosis in *Synura* and *Chrysodidymus* (Andersen, 1989; Graham et al., 1993) but breaks down completely in *Mallomonas* (Beech and Wetherbee, 1990b). Cytokinesis begins as a cleavage furrow forms at the anterior end of the cell. Cleavage proceeds posteriorly and the two daughter cells are formed.

8. Sexual reproduction is poorly known. It is suggested that two cells, acting as isogamous gametes, meet at their posterior ends and fuse in *Mallomonas* (Wawrik, 1960, 1970, 1972; Kristiansen, 1961), whereas in *Synura* isogametes fuse at their anterior ends (Sandgren and Flanagin, 1986). Sandgren and Flanagin (1986) suggest *Synura* is heterothallic because mixtures of different clones produce cyst frequencies of 1 - 20% while individual clones produce cysts at 0.001 to 0.1%. Conversely, Wawrik (1972) suggests that *Mallomonas* is homothallic. Despite these studies, definitive evidence for ploidy change is absent and meiosis has never been demonstrated.

9. Palmelloid stages have been reported for both *Synura* and *Mallomonas* (Pascher, 1912; Conrad, 1914; Andersen, 1989). In *Synura*, palmelloid cells grow and divide, producing silica scales around each cell. Both the cells and scales are embedded in a gelatinous matrix (Andersen, 1989). Palmelloid colonies of *Synura* occur in culture when cells are grown on agar, and they can be collected from the sediments of lakes and ponds, although they are rarely reported. Amoeboid stages have also been reported for both *Mallomonas* and *Synura* (Pascher, 1912; Conrad, 1914), but they have not been observed in culture studies.

10. The Synurophyceae are not usually associated with blooms because they are more common in oligotrophic waters and during cooler periods of the year. However, *Synura* can form extensive blooms under the ice and at the time of ice melt, and a few species of *Mallomonas* (*M. acaroides* Perty, 1852; *M. caudata* Iwanoff, 1899 emend. Krieger, 1930; *M. crassisquama* [Asmund, 1959] Fott, 1962) will bloom in the summer (Thomasson, 1970; Kristiansen, 1971; Nicholls and Gerrath, 1985; Nicholls, 1995). The possible causes for synurophyte blooms are reviewed by Nicholls (1995). The Synurophyceae, especially *Synura,* have been implicated as a source of taste and odor for many years (Guseva, 1935; Taft, 1965; Collins and Kalnins, 1972; Nicholls and Gerrath, 1985; Nicholls, 1995). Evidence suggests that *Synura petersenii* is the most common source and that it imparts a fishy, melon, or cucumber odor to drinking water (Collins and Kalnins, 1972; Nicholls and Gerrath, 1985; Hayes and Burch, 1989; Wee et al., 1994; Nicholls, 1995). Several compounds are known or suspected to cause taste and odor, including oct-1-en-3-ol; pentanone-3; penten-3-one; ocatanone-3; (E,E)-2,4-heptadienal; (E,Z)-2,4-heptadienal; *trans,cis*-2,4-decadienal; *trans*-2, *cis*-6-nonadienal; ß-cyclocitral; etc. (see Hayes and Burch, 1989; Wee et al., 1994; Nicholls, 1995).

11. Contractile vacuoles probably occur in all Synurophyceae, and except for their position in the cell, they resemble those of chrysophytes. In most species, with the possible exception of some species of *Conradiella* and *Mallomonas* (e.g., *M. bacterium* Conrad, 1927b; *M. coronifera* Matvienko, 1941; *M. mesolepis* Skuja, 1934), the contractile vacuoles are posterior to the

nucleus and chloroplasts, whereas in the Chrysophyceae the contractile vacuoles are usually located near the flagellar bases at the cell anterior. The periodicity of contraction appears to vary somewhat by species (e.g. see Skuja, 1956) but to what degree this is due to environmental factors is unknown. Presumably, the ultrastructure of the contractile vacuole is similar to that described for *Ochromonas* Wyssotzki, 1887 (Chrysophyceae) (Aaronson and Behrens, 1974; Andersen, 1982).

12. Nutritionally, the Synurophyceae are phototrophic, but it seems likely (although not established) that cells will take up dissolved organic matter by osmotrophy. There is no clear evidence for phagotrophy in the group, though particle ingestion has been reported by Bird and Kalff (1987) for *Catenochrysis* (=*Synura*), and Conrad (1914) described a species of *Mallomonas* in which he observed cells of *Chlorella* Beijerinck, 1890 that apparently had been taken up in food vacuoles. Phagotrophy, if it exists in the Synurophyceae, must occur by a means different than for the Chrysophyceae because the flagellar microtubular root system is fundamentally different. Some apparently colorless species of *Mallomonas* have also been described (see Preisig et al., 1991).

ARTIFICIAL KEY TO GENERA

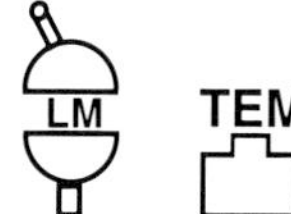

1. Colonial organisms 2
1'. Single-celled organisms 4

2. Silica scales surrounding the colony .... ***Tessellaria*** and ***Synura lapponica***
2'. Silica scales surrounding individual cells.... 3

3. Colonies consisting of two cells........................ .... ***Chrysodidymus***
3'. Colonies consisting of more than two cells....... .... ***Synura***

4. Cells surrounded by silica scales...................... .... ***Mallomonas***
4'. Cells surrounded by siliceous rings................ .... ***Conradiella***

Genus ***Chrysodidymus*** Prowse, 1962

Colonial flagellates typically consisting of only 2 cells, with both cells elongate in shape and slightly inflated at the posterior ends; cells surrounded by silica scales; scales near the cell anterior bearing spines, scales near the cell posterior lacking spines; 2 subequal and heterodynamic flagella; the longer flagellum bearing tripartite flagellar hairs and flagellar scales; the shorter flagellum lacking tripartite hairs and apparently lacking flagellar scales; shorter flagellum containing a green autofluorescing substance near the flagellar base; ultrastructurally, a lamellate structure occurs on the inner surface of a chloroplast; during mitosis, the nuclear envelope is intact along the posterior region of the nucleus but breaks down in other regions; freshwater, rare in acid bogs and pools; one currently recognized species; cysts unknown. *C. gracilis* Prowse, 1962 has been considered a synonym of *C. synuroideus*; however, the basis of the synonymy has been cell shape. The original description of *C. gracilis* reports the cell size to be 22-24 µm long, 8.5-9.0 µm wide at the anterior end and 6.0-6.5 µm wide at the posterior end, and because no reports for *C. synuroideus* are that large, this synonymy (Wujek and Wee, 1983) may be unjustified. For taxonomy, see Prowse (1962), Wujek and Wee (1983), Nicholls and Gerrath (1985); for ultrastructure and molecular biology, see Graham et al. (1993).

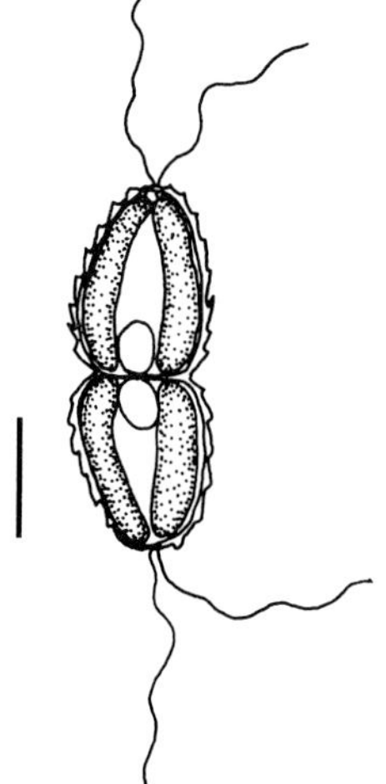

Fig. 1. *Chrysodidymus synuroideus*, two-celled colony based upon light microscopy. Scale bar=10 µm. (From Prowse, 1962)

TYPE SPECIES: *Chrysodidymus synuroideus* Prowse, 1962. Synonym: *Chrysodidymus gracilis* Prowse. Cells 14-15 µm long, 10-11 µm wide at posterior end and 7-8 µm wide at the anterior end (Prowse, 1962), or smaller (see Graham et al., 1993). Cells more globular immediately following cell division, but then quickly becoming the typical elongate shape (Fig. 1). Found in acidic (bog) waters.

Genus **Conradiella** Pascher, 1925

Single-cell flagellates with one flagellum visible by light microscopy; cells covered with armor consisting of hoop-like siliceous rings; some species with bristles or spines; 2 parietal chloroplasts; contractile vacuoles in anterior (?) end of the cells; chrysolaminaran vacuole in posterior end of the cell; cysts known for some species; a somewhat dubious genus observed only by light microscopy and therefore possibly representing certain species of *Mallomonas*; found in the freshwater plankton (Fig. 2). See Pascher (1925), Conrad (1926), Bourrelly (1957), Starmach (1985), Kristiansen (1988b).

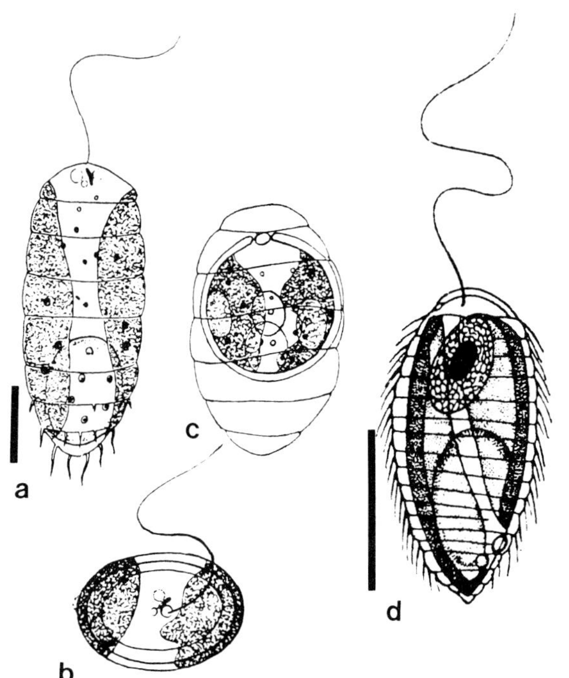

Fig. 2. *Conradiella.* Fig. 2a-c. *Conradiella pascheri* Conrad, 1926 based upon light microscopy. a. Lateral view. b. Anterior view. c. With cyst. Scale bar=10 µm. (From Fott, 1959) Fig. 2d. *Conradiella calva* based upon light microscopy; lateral view. Scale bar=10 µm. (From Pascher, 1925)

TYPE SPECIES: *Conradiella calva* (Massart in Conrad, 1914) Pascher, 1925 (Fig. 2d) (=*Mallomonas calva* Massart in Conrad, 1914). Cells pyriform, 21 µm long and 10 µm wide, approximately 15 rings, flagellum approximately 1.5 times the cell length, 2 parietal chloroplasts, cysts not reported. See Conrad (1914), Pascher (1925), Starmach (1985).

Genus **Mallomonas** Perty, 1852

Single-cell flagellates, varying in size (6 µm to > 60 µm long) and shape (globose to elongate); cells covered by silica scale armor; silica scales variable and taxon specific; bristles attached to silica scales in many species, and spines on scales of a few species. Typically with one visible flagellum when observed by light microscopy although some species (formerly as *Mallomonopsis*) with 2 visible flagella; second flagellum, if present, visible by electron microscopy; swelling on second flagellum thought to be functional in phototaxis; some species (e.g. *M. splendens* [G. S. West, 1909] Playfair, 1921) with only one flagellum. Long flagellum bearing tripartite flagellar hairs and at least sometimes bearing organic flagellar scales; 2 parietal chloroplasts (see Harris, 1953, however, who reports one deeply bilobed chloroplast in 25 species); no eyespot; contractile vacuoles usually in the posterior region of the cell; chrysolaminaran vacuole posterior to the nucleus; cell division by longitudinal cleavage from anterior to posterior of cell; endogenously formed siliceous cysts known for many species; amoeboid and palmelloid stages reported; sexual reproduction apparently present and occurring by lateral or posterior fusion of isogamous (vegetative?) cells; large genus with over 150 described taxa; many taxa described by light microscopy have been ignored, and many new taxa described more recently by electron microscopy may be synonymous with earlier species; genus subdivided into some 18 Sections, with up to 6 Series within Sections; further study may show that at least some Sections represent distinct genera; predominately freshwater (rarely marine); widespread in distribution in temporary pools, bogs, ponds, and lakes of all sizes. For taxonomy, see Perty (1852), Conrad (1927b, 1933), Harris and Bradley (1960), Takahashi (1978), Starmach (1985), Asmund and

Kristiansen (1986), Siver (1991); for pigment analysis see Andersen and Mulkey (1983); for general ultrastructure see Belcher (1969), Hibberd (1976), Zimmermann (1977), Wujek and Kristiansen (1978), Mignot and Brugerolle (1982), Beech and Wetherbee (1990a, 1990b), Beech et al. (1990), Zhang et al. (1990a,b); for scale and bristle formation and deployment see Brugerolle and Bricheux (1984), Lavau and Wetherbee (1994), Wetherbee et al. (1989, 1992, 1995), Miller et al. (1996); for cyst stucture and development see Wawrik (1960, 1972), Kristiansen (1961), Sandgren (1980, 1989); for paleolimnological applications see Smol (1980, 1995); for molecular biology, see Ariztia et al. (1991), Bhattacharya et al. (1992), Lavau et al. (1997); for parasites see Kristiansen (1989, 1991, 1993). (see Figs. 3, 4).

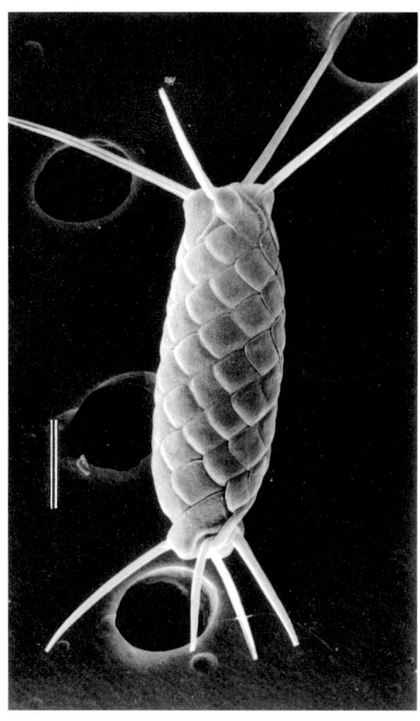

Fig. 3. *Mallomonas splendens* (G. S. West, 1909) Playfair 1921. Single cell, SEM. Scale bar=10 µm. (From S. Lavau, with permission)

TYPE SPECIES: *Mallomonas acaroides* Perty, 1852 emend. Iwanoff, 1899 (=*M. pediculus* Teiling, 1946; *M. acaroides* var. *striatula* Asmund, 1959; *M. acaroides* var. *galeata* Harris & Bradley, 1960; *M. acaroides* var. *echinospora* [Nygaard, 1947] Fott, 1962). Cells broadly ovoid or broadly ellipsoidal, posterior wider than anterior, 13-38 (45) µm long and (7)10-18 (28) µm wide; flagellum approximately as long as the cell; cysts spheroidal with a collar, 20-22 µm in diameter, and with regularly spaced short, blunt spines; scales domed, with V-rib and struts; bristles dimorphous, at least in some populations (serrated bristle and helmet bristle) (Fig. 4b). Widely distributed in meso- to eutrophic waters that are alkaline. See Perty (1852), Iwanoff (1899), Asmund and Kristiansen (1986), Nicholls (1987), Siver (1991).

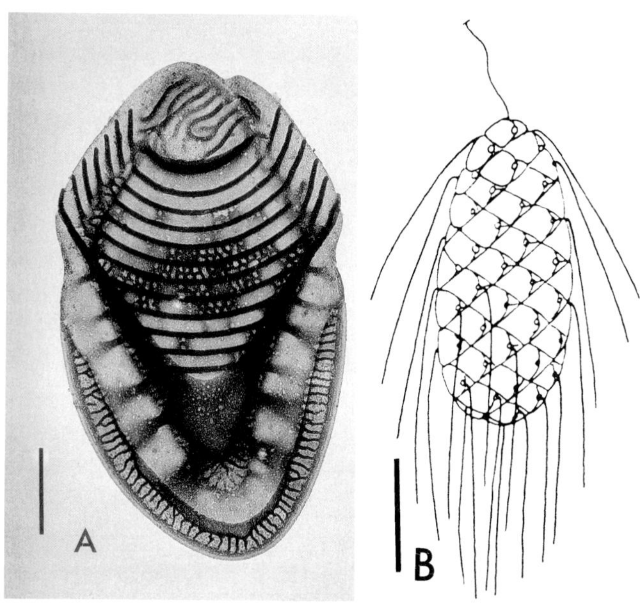

Fig. 4a. *Mallomonas striata* Asmund, 1959, typical scale. TEM. Scale bar=1 µm. Fig. 4b. *Mallomonas acaroides*, illustration based upon light microscopy. Scale bar=10 µm. (From Starmach, 1985)

Genus ***Synura*** Ehrenberg, 1834

Colonial flagellates with radiating pyriform cells; cells with elongate posteriors that are attached at the colony center; 2 subequal heterodynamic flagella inserted in a parallel arrangement at the anterior of the cells; longer flagellum bearing tripartite flagellar hairs and both flagella often covered with tiny organic scales; 2 chloroplasts, typically elongate and parietal (not elongate in *Synura lapponica,* axial orientation in *Synura sphagnicola*); cells typically covered with an armor of silica scales (scales often surrounding the colony rather than cells in *Synura lapponica*); silica scales variously ornamented and species specific in morphology; silica scales at the cell anterior often bearing well-developed spines, and in the case of *Synura spinosa* Korshikov, 1929 there are special elongate scales at the anterior as well; silica scales typically with a rimmed margin around half to three fourths of the scale margin; silica scales absent under conditions of extreme silicon starvation; contractile vacuoles situated below the nucleus in the posterior half of the cell; a large chrysolaminaran vacuole situated in the posterior of cells; colonies rotating or tumbling when swimming; cell division by longitudinal cleavage from anterior to posterior of cells; colonies dividing into 2 approximately equal daughter colonies as colonies reach maximum size; palmelloid non-motile colonies that occur on sediments known for some species, acting as an alternative life stage; endogenously formed siliceous cysts formed by most species; freshwater (rarely marine) and common in the plankton during certain periods of the year; approximately 15 recognized species; genus divided into Section *Synura* Ehrenberg, 1834, Section *Petersenia* Petersen & Hansen, 1956 and Section *Lapponica* Balonov & Kuzmin, 1974; for taxonomic descriptions, see Ehrenberg (1834, 1838), Korshikov (1929), Conrad (1946), Petersen and Hansen (1956, 1958), Bourrelly (1957, 1981), Nicholls and Gerrath (1985), Takahashi (1978), Starmach (1985), Siver (1987); for ultrastructural descriptions, see Schnepf and Deichgräber (1969), Hibberd (1973, 1976, 1978), Mignot and Brugerolle (1982), Brugerolle and Bricheux (1984), Andersen (1985, 1987, 1989), Leadbeater (1986, 1990), Sandgren and Barlow (1989), Leadbeater and Barker (1995), Sandgren et al. (1996); for a culture study see Wee et al. (1991); for sexual reproduction see Wawrik (1970), Sandgren and Flanagin (1986); for cyst formation see Sandgren (1989); for molecular studies, see Bhattacharya et al. (1992), Wee et al. (1993), Wee et al. (1996a,b), Lavau et al. (1997).

TYPE SPECIES: *Synura uvella* Ehrenberg, 1834 (Fig. 5A,B). Motile colonies up to 100-150 μm in diameter, spherical to elongate; cells pyriform, 8-15 μm in diameter and 20-38 μm long. Scales at the anterior of the cell with a well developed spine, a honey-comb type pattern, and a rimmed margin along the posterior half of the scale; scales at the posterior of the cell with a small spine or without a spine; scales approximately 4-5 μm in size. Silica cyst pyriform, 18-22 μm long, 12-15 μm wide at the posterior end. This species is widely reported in the older literature (before 1960), but many of these records are not documented with scale descriptions and may represent other species of the genus. A palmelloid vegetative state is common for this species (Andersen, 1989). For taxonomy, see Ehrenberg (1834, 1838), Korshikov (1929), Conrad (1946), Bourrelly (1957, 1968, 1981), Petersen and Hansen (1956, 1958), Takahashi (1978) Starmach (1985); for ultrastructure, see Andersen (1985); for pigment analysis, see Andersen and Mulkey (1983).

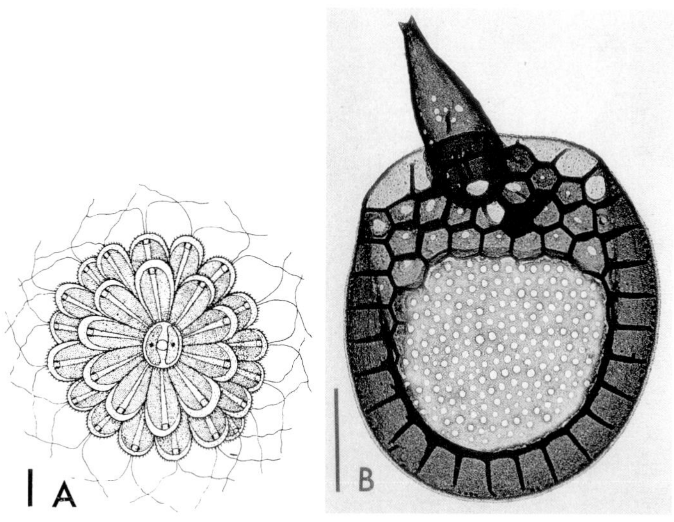

Fig. 5. *Synura uvella*. A. Colony based upon light microscopy. Scale bar=10 μm. (From Fott, 1959) (originally from Stein, 1878) B. Typical spine scale. TEM. Scale bar=1 μm.

Genus ***Tessellaria*** Playfair, 1918

Colonial flagellates with many tightly packed cells which are attached in the colony center, colony more or less spherical; colony surrounded by diffuse gelatinous matrix with plate and spine scales embedded in the gel; 2 subequal flagella extend through the scale layers and beat heterodynamically beyond the gel matrix; 2 chloroplasts in the anterior of the cell; contractile vacuoles in the posterior of the cell; cysts unknown (Playfair, 1915, 1918; Tyler *et al.*, 1989). The genus is monotypic. It is likely that *Synura lapponica* Skuja, 1956 belongs in this genus because its colonies are also surrounded by layers of scales and its plate scale morphology is very similar (Skuja, 1956; Petersen and Hansen, 1958; Siver, 1987).

TYPE SPECIES: *Tessellaria volvocina* (Playfair, 1915) Playfair, 1918 (Fig. 6A-C). Colonies with a few to more than 100 cells, being up to 200 µm in diameter; cells approximately 10 µm in diameter; plate scales 3-11 µm by 2.2-5.5 µm in size, with a perforated pattern on the upper surface and sometimes with a bubble-like projection in the center, and with peripheral upturned rim; spine scales 3.5-5.5 µm long and 0.3-0.5 µm in diameter, formed as a rolled sheet with a flattened base that is directed at an approximately 90° angle. Scales apparently embedded within a diffuse mucilage with no adhesion of scales to produce a highly organized scale case. Cells with 2 chloroplasts lying more or less parallel to the anterior cell surface, no membrane connection between the chloroplast ER and the nuclear envelope, with ring-like plastid genome, and with 10-15 typical lamellae formed from 3 adpressed thylakoids plus an encircling girdle lamella. Two flagella, inserted in a small depression at the anterior end of the cell, and subequal in length (20-25 µm long); basal bodies nearly parallel in orientation, connected to the nucleus via a rhizoplast, and with typical synurophyte striated bands connecting them. Flagellar root R_1, consisting of 3 microtubules, extending up from the rhizoplast and forming a loop around the periphery of the flagellar depression, cytoskeletal microtubules extending from the R_1 root to the chloroplasts and towards the posterior of the cell and towards the colony center. Additional flagellar roots absent. Transitional helix (8.5 gyres) is located just above the transitional plate. Flagellar swellings with amorphous material located just above the transitional helices. Silica scale biogenesis by progressive silicification of a silica deposition vesicle that forms by the fusion of smaller vesicles which apparently originate from the Golgi body; the silica deposition vesicle lies away from the chloroplast just below the plasma membrane at the anterior end of the cell, closely associated with microtubules from the R_1 root; mitochondrial profiles, nucleus, Golgi body, and chrysolaminaran vacuole present and typical. Known only from lakes, ponds, and wet meadows in Australia. For ultrastructural details, see Pipes et al. (1991), Pipes and Leedale (1992); for taxonomic and ecological details, see Playfair (1915, 1918), Tyler (1989).

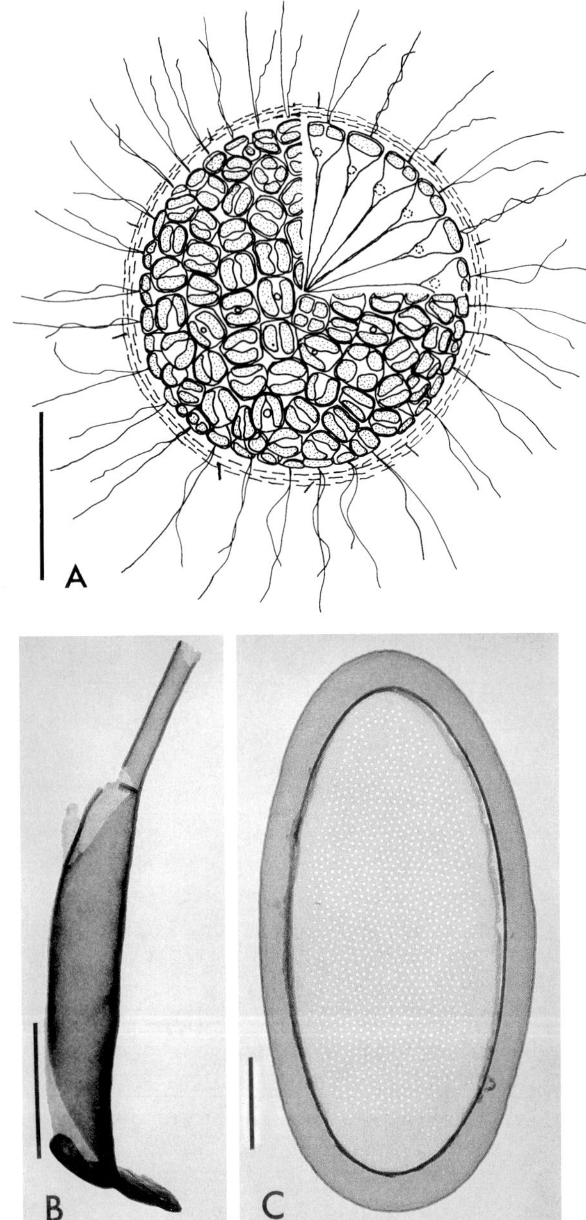

Fig. 6. *Tessellaria volvocina*. A. Colony illustrating the tightly packed, or tessellated, arrangement of the cells, and single cells showing flagella, chloroplasts, and contractile vacuoles. Scale bar=50 µm. B. Spine scale. C. Oval body scale. Scale bars of B and C = 1 µm.

Lesser Known Genus

Genus ***Pseudosyncrypta*** Kisselew, 1931.

TYPE SPECIES *Pseudosyncrypta volvox* Kisselew, 1931 (Bourrelly, 1957; Starmach, 1985).

Literature Cited

Aaronson, S. & Behrens, U. 1974. Ultrastructure of an unusual contractile vacuole in several chrysomonad phytoflagellates. *J. Cell Sci.*, 14:1-9.

Andersen, R. A. 1982. A light and electron microscopical investigation of *Ochromonas sphaerocystis* Matvienko (Chrysophyceae): the statospore, vegetative cell and its peripheral vesicles. *Phycologia*, 21:390-398.

Andersen, R. A. 1985. The flagellar apparatus of the golden alga *Synura uvella*: four absolute orientations. *Protoplasma,* 128:94-106.

Andersen, R. A. 1987. Synurophyceae *classis nov.*, a new class of algae. *Amer. J. Bot.*, 74:337-353.

Andersen, R. A. 1989. The Synurophyceae and their relationship to other golden algae. *Beih. Nova Hedwigia*, 95:1-26.

Andersen, R. A. 1991. The cytoskeleton of chromophyte algae. *Protoplasma*, 164:143-159.

Andersen, R. A. & Mulkey, T. J. 1983. The occurrence of chlorophylls c_1 and c_2 in the Chrysophyceae. *J. Phycol.*, 19:289-294.

Andersen, R. A., van de Peer, Y., Potter, D., Sexton, J. P., Kawachi, M. & La Jeunesse, T. 1999. Phylogenetic analysis of the SSU rRNA from members of the Chrysophyceae. *Protist*, 150:71-84.

Archibald, A. R., Cunningham, W. L, Manners, D. J. & Stark, J. R. 1963. Studies on the metabolism of Protozoa. 10. The molecular structure of reserve polysaccharides from *Ochromonas malhamensis* and *Peranema trichophorum*. *Biochem. J.*, 88:444-451.

Ariztia, E.,V., Andersen, R.,A. & Sogin, M. L. 1991. A new phylogeny for chromophyte algae using 16S-like rRNA sequences from *Mallomonas papillosa* (Synurophyceae) and *Tribonema aequale* (Xanthophyceae). *J. Phycol.*, 27:428-436.

Asmund, B. 1959. Electron microscope observations on *Mallomonas* species and remarks on their occurrence in some Danish ponds and lakes III. *Dan. Bot. Ark.,* 18:1-50.

Asmund, B. & Kristiansen, J. 1986. The genus *Mallomonas* (Chrysophyceae). *Opera Bot.,* 85:1-128.

Asmund, B. 1968. Studies on Chrysophyceae from some ponds and lakes in Alaska. VI. Occurrence of *Synura* species. *Hydrobiologia,* 31:497-515.

Bailey, J. C., Bidigare, R. R., Christensen, S. J. & Andersen, R. A. 1998. Phaeothamniophyceae *classis nova:* a new lineage of chromophytes based upon photosynthetic pigments, rbcL sequence analysis and ultrastructure. *Protist*, 149:249-263.

Balonov, I. M. & Kuzmin, G. V. 1974. Species of the genus *Synura* (Ehr.) (Chrysophyta) in water reservoirs of the Volga Cascade. *Bot. Zh.*, 59:1675-1686.

Battarbee, R. W., Cronberg, G. & Lowry, S. 1980. Observations on the occurrence of scales and bristles of *Mallomonas* spp. (Chrysophyceae) in the micro-laminated sediments of a small lake in Finnish North Karelia. *Hydrobiologia*, 71:225-232.

Beattie, A., Hirst, E. L. & Percival, E. 1961. Comparative structural investigations on leucosin (chrysolaminarin) separated from diatoms and laminarin from the brown algae. *Biochem. J.* 79:531-537.

Beech, P. L. & Wetherbee, R. 1990a. Direct observations on flagellar transformation in *Mallomonas splendens* (Synurophyceae). *J. Phycol.*, 26:90-95.

Beech, P. L. & Wetherbee, R. 1990b. The flagellar apparatus of *Mallomonas splendens* (Synurophyceae) at interphase and its development during the cell cycle. *J. Phycol.*, 26:95-111.

Beech, P. L., Wetherbee, R. & Pickett-Heaps, J. D. 1990. Secretion and deployment of bristles in *Mallomonas splendens* (Synurophyceae). *J. Phycol.*, 26:112-122.

Beijerinck, M. W. 1890. Culturversuche mit Zoochlorellen, Lichengonidien und anderen niederen Algen. *Bot. Zeitung (Berlin),* 48:725-785.

Belcher, J. H. 1966. *Microglena butcheri* nov. sp., a flagellate from the English Lake District. *Hydrobiologia,* 27:65-69.

Belcher, J. H. 1969. Some remarks upon *Mallomonas papillosa* Harris & Bradley and *M. calceolus* Bradley. *Nova Hedwigia*, 18:257-270.

Bhattacharya, D., Medlin, L., Wainright, P. O., Ariztia, E. V., Bibeau, C., Stickel, S. K. & Sogin, M. L. 1992. Algae containing chlorophylls *a* and *c* are paraphyletic: molecular evolutionary analysis of the Chromophyta. *Evolution*, 46:1801-1817.

Bird, D. F. & Kalff, J. 1987. Algal phagotrophy: regulating factors and importance relative to photosynthesis in *Dinobryon* (Chrysophyceae). *Limnol. Oceanogr.*, 32:277-284.

Bjørnland, T. & Liaaen-Jensen, S. 1989. Distribution patterns of carotenoids in relation to chromophyte phylogeny and systematics. *In*: Green, J. C., Leadbeater, B. S. C. & Diver, W. L. (eds.), *The Chromophyte Algae: Problems and Perspectives*. Clarendon Press, Oxford. pp 37-61

Bourrelly, P. 1957. Recherches sur les Chrysophycées: morphologie, phylogénie, systématique. *Rev. Algol., mémoire hors-séries*, 1:1-412.

Bourrelly, P. 1968. *Les Algues d'eau douce. II. Algues jaunes et brunes.* Boubée, Paris.

Bourrelly, P. 1981. *Les Algues d'eau douce. II. Algues jaunes et brunes, réimpression revue et augmentée.* Boubée, Paris.

Brugerolle, G. & Bricheux, G. 1984. Actin microfilaments are involved in scale formation of the chrysomonad cell *Synura*. *Protoplasma*, 123:203-212.

Calado, A. J. & Rino, J. A. 1994. *Chlorodesmos hispidus*, a morphological expression of *Synura spinosa* (Synurophyceae). *Nord. J. Bot.*, 14:235-239.

Cienkowski, L. 1870. Ueber Palmellaceen und einige Flagellaten. *Arch. Mikroskop. Anat.*, 7:421-438.

Clarke, J. K. & Pennick, N. C. 1975. *Syncrypta glornerifera* sp. nov., a marine member of the Chrysophyceae bearing a new form of scale. *Br. Phycol. J.*, 10:363-370.

Coleman, A. W. 1985. Diversity of plastid DNA configuration among classes of eukaryotic algae. *J. Phycol.*, 21:1-6.

Coleman, A. W. & Goff, L. J. 1991. DNA analysis of eukaryotic algal species. *J. Phycol.*, 27:463-473.

Collins, R. P. & Kalnins, K. 1972. Analysis of the free amino acids in *Synura petersenii*. *Phyton*, 29:89-94.

Conrad, W. 1914. Contribution à l'étude des Flagellates. *Arch. Protistenkd.*, 34:79-94.

Conrad, W. 1926. Recherches sur les Flagellates de nos eaux saumâtres. 2e Partie: Chrysomonadines. *Arch. Protistenkd.*, 56:167-231.

Conrad, W. 1927a. Le genre *Microglena* C. G. Ehrenberg (1838). *Arch. Protistenkd.*, 60:415-439.

Conrad, W. 1927b. Essai d'une monographie des genres *Mallomonas* Perty (1852) et *Pseudomallomonas* Chodat (1920). *Arch. Protistenkd.*, 59:423-505.

Conrad, W. 1933. Revision du genre *Mallomonas* Perty incl. *Pseudomallomonas* Chodat. *Mem. Mus. Hist. Nat. Belgique*, 56:1-82.

Conrad, W. 1946. Notes protistologiques XXXI. Matériaux pour la morphologie des *Synura* Ehrenberg. *Bull. Mus. R. Hist. Nat. Belg.*, 22 (11):1-12.

Conrad, W. & Kufferath, H. 1954. Recherches sur les eaux saumâtres des environs de Lilloo. II. *Mém. Inst. R. Sci. Nat. Belg.*, 127:1-346.

Couté, A. & Preisig, H. R. 1981. Sur l'ultrastructure de *Microglena butcheri* Belcher (Chrysophyceae, Ochromonadales, Synuraceae) et sur sa position systématique. *Protistologica*, 17:465-477.

Cronberg, C. 1989. Scaled chrysophytes from the tropics. *Beih. Nova Hedwigia*, 95:191-132.

Croome, R. L., Dürrschmidt, M. & Tyler, P. 1985. A light and electron microscopical investigation of *Mallomonas splendens* (G. S. West) Playfair (Mallomonadaceae, Chrysophyceae). *Nova Hedwigia*, 41:463-470.

Cumming, B. F., Smol, J. P. & Birks, H. J. B. 1992. Scaled chrysophytes (Chrysophyceae and Synurophyceae) from Adirondack drainage lakes and their relationship to environmental variables. *J. Phycol.*, 28:162-178.

Diesing, K. M. 1866. Revision der Prothelminthen. Abtheilung: Mastigophoren. *Sitzungsber. Kaiserl. Akad. Wiss., Math. Naturwiss. Cl.,* Abt. 1, 52:287-401.

Droop, M. R. 1955. Some new supra-littoral protista. *J. Mar. Biol. Assoc. U.K.*, 34:233-245.

Duff, K. E., Zeeb, B. A. & Smol, J. P. 1994. *Atlas of Chrysophycean Cysts*. Kluwer Acad. Publ., Dordrecht. 189 pp

Ehrenberg, C. G. 1832. Über die Entwickelung und Lebensdauer der Infusionsthiere; nebst ferneren Beiträgen zu einer Vergleichung ihrer organischen Systeme. *Abhandl. Königl. Akad. Wiss. Berlin* (1831), 1-154.

Ehrenberg, C. G. 1834. Dritter Beitrag zur Erkenntnis grosser Organisation in der Richtung des kleinsten Raumes. *Abhandl. Königl. Akad. Wiss. Berlin* (1833):145-336.

Ehrenberg, C. G. 1838. *Die Infusionsthierchen als vollkommene Organismen.* Voss, Leipzig.

Fawley, M. W. 1989. A new form of chlorophyll *c* involved in light harvesting. *Plant Physiol.*, 91:727-732.

Fott, B. 1959. *Algenkunde*. Gustav Fischer Verlag, Jena. 482 pp

Fott, B. 1962. Taxonomy of *Mallomonas* based upon electron micrographs of scales. *Preslia*, 34:69-84.

Graham, L .E., Graham, J. M. & Wujek, D. E. 1993. Ultrastructure of *Chrysodidymus synuroideus* (Synurophyceae). *J. Phycol.*, 29:330-341.

Guseva, K. A. 1935. The conditions of mass development and the physiology of nutrition of *Synura*. *Mikrobiologiya*, 4:24-43. (in Russian)

Harris, K. 1953. A contribution to our knowledge of *Mallomonas*. *Bot. J. Linn. Soc.*, 55:88-102.

Harris, K. & Bradley, D. E. 1960. A taxonomic study of *Mallomonas*. *J. Gen. Microbiol.*, 22:750-777.

Hayes, K. P. & Burch, M. D. 1989. Odorous compounds associated with algal blooms in south Australian waters. *Wat. Res.*, 23:115-121.

Hibberd, D. J. 1973. Observations on the ultrastructure of flagellar scales in the genus *Synura* (Chrysophyceae). *Arch. Mikrobiol.*, 89:291-304.

Hibberd, D. J. 1976. The ultrastructure and taxonomy of the Chrysophyceae and

Prymnesiophyceae (Haptophyceae): a survey with some new observations on the ultrastructure of the Chrysophyceae. *Bot. J. Linn. Soc.*, 72:55-80.

Hibberd, D. J. 1978. The fine structure of *Synura sphagnicola* (Korsh.) Korsh. (Chrysophyceae). *Br. Phycol. J.*, 13:403-412.

Ikävalko, J., Kristiansen, J. & Thomsen, H. A. 1994. A revision of the taxonomic position of *Syncrypta glomerifera* (Chrysophyceae), establishment of a new genus *Lepidochrysis* and observations on the occurrence of *L. glomerifera* comb. nov. in brackish water. *Nord. J. Bot.*, 14:339-344.

Iwanoff, L. 1899. Beiträge zur Kenntniss der Morphologie und Systematik der Chrysomonaden. *Bull. Acad. Imp. Sci. St. Petersbourg,* 11:247-262.

Jeffrey, S. W. 1989. Chlorophyll *c* pigments and their distribution in the chromophyte algae. *In*: Green, J. C., Leadbeater, B. S. C. & Diver, W. L. (eds.), *The Chromophyte Algae: Problems and Perspectives.* Clarendon Press, Oxford. pp 13-36

Kisselew, J. A. 1931. Zur Morphologie einiger neuer und seltener Vertreter . des pflanzlichen Microplanktons. *Arch. Protistenkd.*, 73:235-250.

Klebs, G. 1893. Flagellatenstudien Teil II. *Z. Zool.*, 55:353-445.

Korshikov, A. A. 1927. *Skadovskiella sphagnicola,* a new colonial Chrysomonad. *Arch. Protistenkd.*, 58:450-455.

Korshikov, A. A. 1929. Studies on the Chrysomonads. I. *Arch. Protistenkd.*, 67:253-290.

Krieger, W. 1930. Untersuchungen über Plankton-Chrysomonaden. *Bot. Arch.*, 29:258-329.

Kristiansen, J. 1961. Sexual reproduction in *Mallomonas caudata. Bot. Tidsskr.*, 57:306-309.

Kristiansen, J. 1971. A *Mallomonas* bloom in a Bulgarian mountain lake. *Nova Hedwigia*, 21:877-882.

Kristiansen, J. 1981. Distribution problems in the Synuraceae (Chrysophyceae). *Verh. Internat. Verein. Limnol.*, 21:1444-1448.

Kristiansen, J. 1985. Occurrence of scale-bearing Chrysophyceae in a eutrophic Danish lake. *Verh. Internat. Ver. Limnol.*, 22:2826-2829.

Kristiansen, J. 1986a. Identification, ecology, and distribution of silica-scale-bearing Chrysophyceae, a critical approach. *In*: Kristiansen, J. & Andersen, R. A. (eds.), *Chrysophytes, Aspects and Problems*. Cambridge Univ. Press, Cambridge. pp 229-239

Kristiansen, J. 1986b. Silica-scale bearing chrysophytes as environmental indicators. *Br. Phycol. J.*, 21:425-436.

Kristiansen, J. 1988a. Seasonal occurrence of silica-scaled chrysophytes under eutrophic conditions. *Hydrobiologia*, 161:171-184.

Kristiansen, J. 1988b. The problem of "enigmatic chrysophytes". *Arch. Protistenkd.*, 135:9-15.

Kristiansen, J. 1989. Occurrence and seasonal cycle of *Mallomonas teilingii* (Synurophyceae), with special reference to the effects of parasitism. *Beih. Nova Hedwigia*, 95:179-189.

Kristiansen, J. 1991. Possible chloroplast retention by a colorless parasite of the chrysophyte *Mallomonas teilingii. Endocytobiosis & Cell Res.*, 8:83-87.

Kristiansen, J. 1993. The "*tridentata* parasite" of *Mallomonas teilingii* (Synurophyceae)—a new dinophyte?—or what? *Arch. Protistenkd.*, 143:195-214.

Kristiansen, J. 1995. History of chrysophyte research: origin and development of concepts and ideas. *In*: Sandgren, C. D., Smol, J. P. & Kristiansen, J. (eds.), *Chrysophyte Algae: Ecology, Phylogeny and Development.* Cambridge Univ. Press, Cambridge. pp 1-22

Kristiansen, J. & Andersen, R. A. (eds.) 1986. Chrysophytes: Aspects and Problems. Cambridge Univ. Press, Cambridge. 337 pp

Kristiansen, J. & Cronberg, G. (eds.) 1996. Chrysophytes: New Horizons. *Beih. Nova Hedwigia*, 114:1-266.

Kristiansen, J., Cronberg, G. & Geissler, U. (eds.) 1989. Chrysophytes: Developments and Perspectives. *Beih. Nova Hedwigia*, 95:1-287.

Kristiansen, J., Wilken, L. R. & Jürgensen, T. 1995. A bloom of *Mallomonas acaroides*, a silica-scaled chrysophyte, in the crater pond of a pingo, Northwest Greenland. *Polar Biology*, 15:319-324.

Lavau, S., Saunders, G. W. & Wetherbee, R. 1997. A phylogenetic analysis of the Synurophyceae using molecular data and scale case morphology. J. Phycol., 33:135-151.

Lavau, S. & Wetherbee, R. 1994. Structure and development of the scale case of *Mallomonas adamas* (Synurophyceae). *Protoplasma*, 181:259-268.

Leadbeater, B. S. C. 1986. Scale-case construction in *Synura petersenii* Korsh. (Chysophyceae). *In*: Kristiansen, J. & Andersen, R. A. (eds.), *Chrysophytes: Aspects and Problems.* Cambridge Univ. Press, Cambridge. pp 121-131

Leadbeater, B. S. C. 1990. Ultrastructure and assembly of the scale case in *Synura* (Synurophyceae Andersen). *Br. Phycol. J.*, 25:117-132.

Leadbeater, B. S. C. & Barker, D. A. N. 1995. Biomineralization and scale production in the Chrysophyta. *In*: Sandgren, C. D., Smol, J. P. &

Kristiansen, J. (eds.), *Chrysophyte Algae: Ecology, Phylogeny and Development.* Cambridge Univ. Press, Cambridge. pp 141-164

Leipe, D. D., Wainright, P. O., Gunderson, J. H., Porter, D., Patterson, D. J., Valois, F., Himmerich, S. & Sogin, M. L. 1994. The Stramenopiles from a molecular perspective: 16S-like rRNA sequences from *Labyrinthuloides minuta* and *Cafeteria roenbergensis*. *Phycologia*, 33:369-377.

Lemmermann, E. 1899. Das Phytoplankton sächsischer Teiche. *Forschungsber. Biol. Stat. Plön,* 7:96-140.

Ludwig, M., Lind, J. L., Miller, E. A. & Wetherbee, R. 1996. High molecular mass glycoproteins associated with the siliceous scales and bristles of *Mallomonas splendens* (Synurophyceae) may be involved in cell surface development and maintenance. *Planta*, 199:219-228.

Matvienko, O. M. 1941. Do systematiki rodi *Mallomonas. Trudi Inst. Bot. Charkow*, 4:41-47.

McAlice, B. J. 1975. Preliminary checklist of planktonic microalgae from the Gulf of Maine. Maine Sea Grant Information Leaflet #9.

Mignot, J. P. & Brugerolle, G. 1982. Scale formation in chrysomonad flagellates. *J. Ultrastruct. Res.*, 81:13-26.

Miller, E., Lind, J. L. & Wetherbee, R. 1996. The scale-associated molecules of *Mallomonas* spp. (Synurophyceae) are immunologically distinct. *In*: Kristiansen, J. & Cronberg, G. (eds.), *Chrysophytes: Progress and New Horizons. Beih. Nova Hedwigia*, 114: 45-55.

Moestrup, Ø. 1995. Current status of chrysophyte 'splinter groups': Synurophyceae, pedinellids, silicoflagellates. *In*: Sandgren, C. D., Smol, J. P. & Kristiansen, J. (eds.), *Chrysophyte Algae: Ecology, Phylogeny and Development.* Cambridge Univ. Press, Cambridge. pp 75-91

Müller, O. F. 1786. *Animalcula Infusoria Fluviatilia et Marina.* Hauniae (Copenhagen).

Nicholls, K. H. 1987. The distinction between *Mallomonas acaroides* var. *acaroides* and *Mallomonas acaroides* var. *muskokana* var. nov. (Chrysophyceae). *Can. J. Bot.*, 65:1779-1784.

Nicholls, K. H. 1995. Chrysophyte blooms in the plankton and neuston of marine and freshwater systems. *In*: Sandgren, C. D., Smol, J. P. & Kristiansen, J. (eds.), *Chrysophyte Algae: Ecology, Phylogeny and Development.* Cambridge Univ. Press, Cambridge. pp 181-213

Nicholls, K. H. & Gerrath, J. F. 1985. The taxonomy of *Synura* (Chrysophyceae) in Ontario with special reference to taste and odour in water supplies. *Can. J. Bot.*, 63:1482-1493.

Nygaard, G. 1949. Hydrobiological studies on some Danish ponds and lakes, part II. *Biol. Skr., Kongel. Danske Vidensk. Selsk.*, 7(1):1-293.

Parke, M. & Dixon, P. S. 1976. Check-list of British marine algae—third revision. *J. Mar. Biol. Assoc. U.K.*, 56:527-594.

Pascher, A. 1912. Über Rhizopoden und Palmellastadien bei Flagellaten (Chrysomonaden), nebst einer Übersicht über die braunen Flagellaten. *Arch. Protistenkd.*, 25:153-200.

Pascher, A. 1914. Über Flagellaten und Algen. *Ber. Dtsch. Bot. Ges.*, 32:136-160.

Pascher, A. 1925. Neue oder wenig bekannte Flagellaten. XVI. *Arch. Protistenkd.*, 52:565-584.

Perman, J. & Vinniková, A. 1955. Tri zimní planktonní chrysomonády. *Preslia*, 27:272-279.

Perty, M. 1852. *Zur Kenntniss kleinster Lebensformen nach Bau, Funktionen, Systematik, mit Specialverzeichniss der in der Schweiz beobachteten.* Jent & Reinert, Bern. 228 pp

Petersen, J.-B. & Hansen, J. B. 1956. On the scales of some *Synura* species. *Biol. Medd. Dan. Vid. Selsk.*, 23(2):3-27.

Petersen, J.-B. & Hansen, J. B. 1958. On the scales of some *Synura* species. II. *Biol. Medd. Dan. Vid. Selsk.*, 23(7):3-13.

Pipes, L. D. & Leedale, G. F. 1992. Scale formation in *Tessellaria volvocina* (Synurophyceae). *Br. Phycol. J.*, 27:1-29.

Pipes, L. D., Leedale, G. F. & Tyler, P. A. 1991. The ultrastructure of *Tessellaria volvocina* (Synurophyceae). *Br. Phycol. J.*, 26:259-278.

Phillips, F. W. 1884. On *Chlorodesmos hispida*, a new flagellate animalcule. *Trans. Hertfordshire Nat. Hist. Soc. Field Club*, 2:92-94.

Playfair, G. I. 1915. Freshwater algae of the Lismore district: with an appendix on the algal fungi and Schizomycetes. *Proc. Linn. Soc. New South Wales*, 40:310-362.

Playfair, G. I. 1918. New and rare freshwater algae. *Proc. Linn. Soc. New South Wales*, 43:497-543.

Playfair, G. I. 1921. Australian freshwater flagellates. *Proc. Linn. Soc. New South Wales*, 46:99-146.

Preisig, H. R. 1986. Biomineralization in the Chrysophyceae. *In*: Leadbeater, B. S. C. & Riding, R. (eds.), *Biomineralization in Lower Plants and Animals.* Systematics Assoc. Spec. Vol. 30. Clarendon Press, Oxford. pp 327-344

Preisig, H. R. 1989. *Mallomonas alphaphora* (Chrysophyceae), a new species from Western Australia. *Plant Syst. Evol.*, 164:209-214.

Preisig, H. R. 1994. Siliceous structures and silicification in flagellated protists. *Protoplasma*, 181:29-42.

Preisig, H. R. 1995. A modern concept of chrysophyte classification. *In*: Sandgren, C. D., Smol, J. P. & Kristiansen, J. (eds.), *Chrysophyte*

Algae: Ecology, Phylogeny and Development. Cambridge Univ. Press, Cambridge. pp 46-74

Preisig, H. R. & Hibberd, D. J. 1982a. Ultrastructure and taxonomy of *Paraphysomonas* (Chrysophyceae) and related genera 1. *Nord. J. Bot.*, 2:397-420.

Preisig, H. R. & Hibberd, D. J. 1982b. Ultrastructure and taxonomy of *Paraphysomonas* (Chrysophyceae) and related genera 2. *Nord. J. Bot.*, 2:601-638.

Preisig, H. R. & Hibberd, D. J. 1983. Ultrastructure and taxonomy of *Paraphysomonas* (Chrysophyceae) and related genera 3. *Nord. J. Bot.*, 3:695-723.

Preisig, H. R., Vørs, N. & Hällfors, G. 1991. Diversity of heterotrophic heterokont flagellates. *In*: Patterson, D. J. & Larsen, J. (eds.), *The Biology of Free-Living Heterotrophic Flagellates.* Systematics Assoc. Special Vol. 45. Clarendon Press, Oxford. pp 361-399

Prowse, G. A. 1962. Further Malayan freshwater flagellata. *The Gard.' Bull., Singapore,* 19:105-146.

Quillet, M. 1955. Sur la nature chimique de la leucosine, polysaccharide de réserve caractéristique des Chrysophycées, extraite d'*Hydrurus foetidus. C. R. Heb. Séanc. Acad. Sci., Paris*, 240:1001-1003.

Sandgren, C. D. 1980. Resting cyst formation in selected chrysophyte flagellates: an ultrastructural survey including a proposal for the phylogenetic significance of interspecific variations in the encystment process. *Protistologica*, 16:289-303.

Sandgren, C. D. 1988. The ecology of chrysophyte flagellates: their growth and perennation strategies as freshwater phytoplankton. *In*: Sandgren, C. D. (ed.), *Growth and Reproductive Stategies of Freshwater Phytoplankton.* Cambridge Univ. Press, Cambridge. pp 9-104

Sandgren, C. D. 1989. SEM investigations of statospore (stomatocyst) development in diverse members of the Chrysophyceae and Synurophyceae. *Beih. Nova Hedwigia*, 95:45-69.

Sandgren, C. D. (ed.) 1991. Applications of chrysophyte stomatocysts in paleolimnology. *J. Paleolimnol.*, 5:1-113.

Sandgren, C. D. & Barlow, S. B. 1989. Siliceous scale production in chrysophyte algae. II. SEM observations regarding the effects of metabolic inhibitors on scale regeneration in laboratory population of scale-free *Synura petersenii* cells. *Beih. Nova Hedwigia*, 95:27-44.

Sandgren, C. D. & Flanagin, J. 1986. Heterothallic sexuality and density dependent encystment in the chrysophycean alga *Synura petersenii* Korsh. *J. Phycol.*, 22:206-216.

Sandgren, C. D., Hall, S. A. & Barlow, S. B. 1996. Siliceous scale production in chrysophyte and synurophyte algae. I. Effects of silica-limited growth on cell silica content, scale morphology, and the construction of the scale layer of *Synura petersenii. J. Phycol.*, 32:675-692.

Sandgren, C. D., Smol, J. P. & Kristiansen, J. (eds.) 1995. *Chrysophyte Algae: Ecology, Phylogeny and Development.* Cambridge Univ. Press, Cambridge. 399 pp

Sandgren, C. D. & Walton, W. E. 1995. The influence of zooplankton herbivory on the biogeographiy of chrysophyte algae. *In*: Sandgren, C. D., Smol, J. P. & Kristiansen, J. (eds.), *Chrysophyte Algae: Ecology, Phylogeny and Development.* Cambridge Univ. Press, Cambridge. pp 269-302

Saunders, G. W., Potter, D., Paskind, M. P. & Andersen, R. A. 1995. Cladistic analyses of combined traditional and molecular data sets reveal an algal lineage. *Proc. Natl. Acad. Sci. USA*, 92:244-248.

Saunders, G. W., Potter, D. & Andersen, R. A. 1997. Phylogenetic affinities of the Sarcinochrysidales and Chrysomeridales based on analyses of molecular and combined data. *J. Phycol.*, 33:310-318.

Schiller, J. 1929. Neue Chryso- und Cryptomonaden aus Altwässern der Donau bei Wien. *Arch. Protistenkd.*, 66:436-458.

Schnepf, E. & Deichgräber, G. 1969. Über die Feinstruktur von *Synura petersenii* unter besonderer Berücksichtigung der Morphogenese ihrer Kieselschuppen. *Protoplasma*, 68:85-106.

Siver, P. A. 1987. The distribution and variation of *Synura* species (Chrysophyceae) in Connecticut, USA. *Nord. J. Bot.*, 7:107-116.

Siver, P. A. 1988. *Mallomonas retrorsa*, a new species of silica-scaled Chrysophyceae with backwards oriented scales. *Nord. J. Bot.*, 8:319-323.

Siver, P. A. 1991. *The Biology of Mallomonas. Morphology, Taxonomy and Ecology.* Kluwer Acad. Publ., Dordrecht. 230 pp

Siver, P. A. 1995. The distribution of chrysophytes along environmental gradients: their use as biological indicators. *In*: Sandgren, C. D., Smol, J. P. & Kristiansen, J. (eds.), *Chrysophyte Algae: Ecology, Phylogeny and Development.* Cambridge Univ. Press, Cambridge. pp 232-268

Siver, P. A. & Hamer, J. S. 1992. Seasonal periodicity of Chrysophyceae and Synurophyceae in a small New England lake: implications for paleolimnological research. *J. Phycol.*, 28:186-198.

Skuja, H. 1934. Beitrag zur Algenflora Lettlands, I. *Acta Horti Bot. Univ. Latv.*, 7:241-302.

Skuja, H. 1956. Taxonomische und biologische Studien über das Phytoplankton schwedischer Binnengewässer. *Nov. Act. Reg. Soc. Sci. Upsal.*, Ser 4/16 (3):1-404.

Smestad-Paulsen, B. & Myklestad, S. 1978. Structural studies of the reserve glucan produced by the marine diatom *Skeletonema costatum* (Grev.) Cleve. *Carbohydr. Res.*, 62:386-388.

Smol, J. P. 1980. .Fossil synuracean (Chrysophyceae) scales in lake sediments: a new group of paleoindicators. *Can. J. Bot.*, 58:458-465.

Smol, J. P. 1995. Application of chrysophytes to problems in paleoecology. *In*: Sandgren, C. D., Smol, J. P. & Kristiansen, J. (eds.), *Chrysophyte Algae: Ecology, Phylogeny and Development.* Cambridge Univ. Press, Cambridge. pp 303-329

Smol, J. P., Charles, D. F. & Whitehead, D. R. 1984a. Mallomonadacean microfossils provide evidence of recent lake acidification. *Nature*, 307 (5952):628-630.

Smol, J. P., Charles, D. F. & Whitehead, D. R. 1984b. Mallomonadacean (Chrysophyceae) assemblages and their relationships with limnological characteristics in 38 Adirondack (New York) lakes. *Can. J. Bot.*, 62:911-923.

Starmach, K. 1985. Chrysophyceae und Haptophyceae. *In*: Ettl, H., Gerloff, J., Heynig, H. & Mollenhauer, D. (eds.), *Süsswasserflora von Mitteleuropa*, Band 1. Gustav Fischer Verlag, Stuttgart. 515 pp

Stein, F. 1878. *Der Organismus der Infusionsthiere.* III. 1. Hälfte. Engelmann, Leipzig. 154 pp

Taft, C. E. 1965. *Water and Algae: World Problems.* Educational Publ., Chicago. 236 pp

Takahashi, E. 1978. *Electron Microscopical Studies of the Synuraceae (Chrysophyceae) in Japan. Taxonomy and Ecology.* Tokai Univ. Press, Tokyo. 194 pp

Teiling, E. 1946. Zur Phytoplanktonflora Schwedens. *Bot. Notiser*, (1946):61-88.

Thomasson, K. 1970. A *Mallomonas* population. *Sven. Bot. Tidskr.*, 64:303-311.

Tyler, P. A., Pipes, L. D., Croome, R. L. & Leedale, G. F. 1989. *Tessellaria volvocina* rediscovered. *Br. Phycol. J.*, 24:329-337.

Wawrik, F. 1960. Sexualität bei *Mallomonas fastigata* var. *kriegerii. Arch. Protistenkd.*, 104:541-544.

Wawrik, F. 1970. Isogamie bei *Synura petersenii* Korschikov. *Arch. Protistenkd.*, 112:259-261.

Wawrik, F. 1972. Isogame Hologamie in der Gattung *Mallomonas* Perty. *Nova Hedwigia*, 23:353-362.

Wee, J. L. 1997. Scale biogenesis in synurophycean protists. Crit. Rev. Plant Sci., 16:497-534.

Wee, J. L., Chesnick, J. & Cattolico, R. A. 1993. Partial characterization of the chloroplast genome from the chromophytic alga *Synura petersenii* (Synurophyceae). *J. Phycol.*, 29:96-99.

Wee, J. L. & Gabel, M. 1989. Occurrences of silica-scaled chromophyte algae in a region of predominantly alkaline habitats. *Amer. Midl. Nat.*, 121:32-40.

Wee, J. L., Harris, S. A., Smith, J. P., Dionigi, C. P. & Millie, D. F. 1994. Production of the taste/odor-causing compound *trans-2,cis-6-nonadienal*, within the Synurophyceae. *J. Appl. Phycol.*, 6:365-369.

Wee, J. L., Hinchey, J. M., Nguyen, K. X., Kores, P. & Hurley, D. L. 1996a. Investigating the comparative biology of the heterokonts with nucleic acids. *J. Euk. Microbiol.*, 43:106-112.

Wee, J. L., Hinchey, J. M., Nguyen, K. X., Kores, P. J. & Hurley, D. L. 1996b. Assessment of the phylogenetic utility of *psb*A and ITS/5.8S nucleotide sequences for the Synurophyceae. *In*: Kristiansen, J. & Cronberg, G. (eds.), *Chrysophytes: Progress and New Horizons. Beih. Nova Hedwigia*, 114: 29-43.

Wee, J. L., Millie, D. F. & Walton, S. P. 1991. A statistical characterization of growth among clones of *Synura petersenii* (Synurophyceae). *J. Phycol.*, 27:570-575.

West, G. S. 1909. The algae of the Yan Yean Reservoir, Victoria. J. *Linn. Soc. Bot.*, 39:1-88.

Wetherbee, R., Koutoulis, A. & Andersen, R. A. 1992. The microarchitecture of the chrysophycean cytoskeleton. *In*: Menzel, D. (ed.), *The Cytoskeleton of Algae.* CRC Press, Boca Raton. pp 1-17

Wetherbee, R., Koutoulis, A. & Beech, P. L. 1989. The role of the cytoskeleton during the assembly, secretion and deployment of scales and spines. *In*: Coleman, A. W., Goff, L. J. & Stein-Taylor, J. R. (eds.), *Algae as Experimental Systems.* A. R. Liss, New York. pp 93-108

Wetherbee, R., Ludwig, M. & Koutoulis, A. 1995. Immunological and ultrastructural studies of scale development and deployment in *Mallomonas* and *Apedinella*. *In*: Sandgren, C. D., Smol, J. P. & Kristiansen, J. (eds.), *Chrysophyte Algae: Ecology, Phylogeny and Development.* Cambridge Univ. Press, Cambridge. pp 165-178

Wetherbee, R., Platt, S. J., Beech, P. L. & Pickett-Heaps, J. D. 1988. Flagellar transformation in the heterokont *Epipyxis pulchra* (Chrysophyceae): direct observations using image-enhanced light microscopy. *Protoplasma*, 145:47-54.

Wujek, D. E. 1967. *Microglena punctifera* (O.F.M.) Ehrenberg in the United States. *Trans. Am. Microsc. Soc.*, 86:340-341.

Wujek, D. E. 1984. Chrysophyceae (Mallomonadaceae) from Florida. *Florida Sci.*, 47:161-170.

Wujek, D. E. & Kristiansen, J. 1978. Observations on the bristle- and scale-production in *Mallomonas caudata* (Chrysophyceae). *Arch. Protistenkd.*, 120:213-221.

Wujek, D. E. & Wee, J. L. 1983. *Chrysodidymus* in the United States. *Trans. Am. Microsc. Soc.*, 102:77-80.

Wyssotzki, A. 1887. Les mastigophores et rhizopodes trouvés dans les lacs Weissowo et Repnoie. *Trudy Obshchestvo Ispytatelei Prirody pri Imperatorskom Kharkovskom Universitete, Kharkov,* 21:119-140.

Zhang, X., Inouye, I. & Chihara, M. 1990a. An unusual short flagellum in *Mallomonas guttata* (Synurophyceae, Chrysophyta). *Bot. Mag. Tokyo*, 103 (1069):97-101.

Zhang, X., Inouye, I. & Chihara, M. 1990b. Taxonomy and ultrastructure of a freshwater scaly flagellate *Mallomonas tonsurata* var. *etortisetifera* var. nov. (Synurophyceae, Chromophyta). *Phycologia*, 29:65-73.

Zimmermann, B. 1977. Flagellar and body scales in the chrysophyte *Mallomonas multiunca* Asmund. *Br. Phycol. J.*, 12:287-290.

CLASS SILICOFLAGELLATA
LEMMERMANN, 1901

By ØJVIND MOESTRUP and CHARLES J. O'KELLY

The Silicoflagellata (known in botanical literature as the Dictyochophyceae Silva, 1982) comprises three small (c. 30 extant species) but distinctive groups of free-living protists sometimes counted among the "chrysomonads" ("chrysophytes"): the silicoflagellates *sensu stricto,* the pedinellids, and the rhizochromulinids. These organisms, according to recent research, form a clade that is most closely related to pelagomonads and diatoms (Saunders et al., 1995). In some classifications (Cavalier-Smith 1993, Saunders et al., 1995) silicoflagellates, pedinellids, and pelagomonads are each afforded class status.

Silicoflagellata are small to medium size unicellular protists (range 5-50, exceptionally to 500 µm), usually occurring either as flagellates or as axopodial amoebae. As in most other members of the heterokont/stramenopile clade; the youngest (anterior or only) flagellum on swimming cells bears tripartite tubular hairs. The defining characteristic is the presence of cytoplasmic microtubules that arise from differentiated pads (presumed nucleating sites) on the nuclear envelope (Fig. 4). These microtubules extend into the axopodia, although they may be present even when axopodia are not expressed. The axopodial microtubules are either not united into axonemes or they form axonemes consisting of three microtubules; this feature distinguishes Silicoflagellata from the actinophryid amoebae, in which the axopodial axonemes contain numerous, highly organized microtubules.

No root microtubules are associated with the kinetid in flagellate cells, a feature that links Silicoflagellata with pelagomonads and separates them from other heterokont/stramenopile taxa. Also lacking is the "transitional helix" typically found distal to the transition zone between kinetosomes and flagella in other heterokonts, but two rings (or a helix of two gyres) may be present proximal to the transition zone. This feature also links Silicoflagellata and pelagomonads.

Most species of Silicoflagellata are photosynthetic, containing chloroplasts with thylakoid membranes in stacks of three and the "chrysomonad" pigments (chlorophylls *a* and *c*, fucoxanthin), but a few pedinellids are known that lack chloroplasts.

Silicoflagellates *sensu stricto* form complex siliceous external skeletons, which fossilize well. From these palynomorphs, a silicoflagellate fossil record including several form genera and extending back to the Cretaceous is known. Other Silicoflagellata are naked or have a covering of organic scales. For these taxa, no fossil record is known.

Silicoflagellata are common in marine planktonic and benthic habitats. Blooms are usually benign, but a few have been associated with fish kills although toxins have not been identified. Some species of pedinellids are known from fresh water.

Species of Silicoflagellata may have complex asexual life histories including several different trophic stages and cysts, but sexual reproduction has yet to be observed. Accounts of the structure, reproduction and diversity of Silicoflagellata have been given by Moestrup and Thomsen (1990), Moestrup (1995) and O'Kelly and Wujek (1995).

Key to the Orders

1. Siliceous exoskeleton present................ .... Order **Dictyochales**
1'. Siliceous exoskeleton absent, cells radially symmetrical, axopodial microtubules in groups of three....................... **Pedinellales**
1''. Siliceous exoskeleton absent, cells bilaterally or not symmetrical, no fixed number of axopodial microtubules .... **Rhizochromulinida**

ORDER DICTYOCHALES HAECKEL, 1894
(Silicoflagellata Borgert 1890)

by ØJVIND MOESTRUP

The silicoflagellates *sensu stricto* are very common and important members of the marine plankton in many seas. When occurring in bloom conditions, they may cause mortality of fish, due to mechanical irritation of the fish gills by the spines of the skeleton (Fig. 1A) (Erardle Denn and Ryckaert, 1990). Numerous fossil representatives are known from the Cretaceous onwards, peaking with more than a hundred species and varieties in the Miocene (Tappan, 1980), but the number of extant species is probably limited to three (Moestrup and Thomsen, 1990). These are readily identified as silicoflagellates because of the characteristic siliceous skeleton that surrounds the cell as a lateral bowl (Figs. 2A,B). Naked stages can be confused with raphidophytes but differ in the presence of only one long flagellum (Fig. 1B) (van Valkenburg and Norris 1970; Moestrup and Thomsen 1990). Classification of the silicoflagellates is blessed with many names. Apparently the oldest class name is Silicoflagellata Lemmermann 1901 and the oldest order name Silicoflagellata Borgert 1890, p. 231. Borgert coined the term silicoflagellates.

The extant species are classified into a single family, for which the oldest name appears to be Dictyochida Borgert, 1891, p. 661 (=Dictyochaceae in the botanical nomenclature, Dictyochidae in the zoological nomenclature). They belong to a single genus *Dictyocha.*

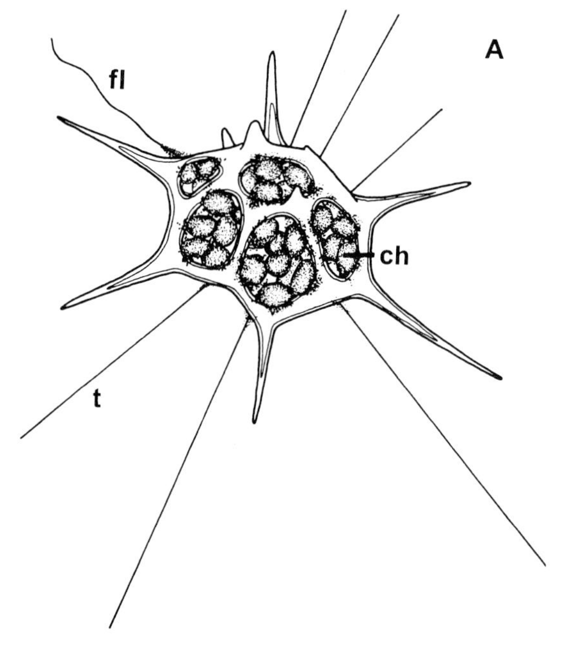

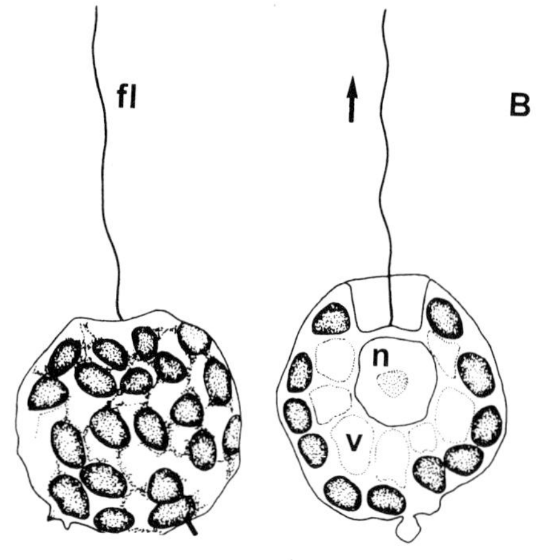

Fig. 1. *Dictyocha (=Distephanus) speculum.* A. Skeleton-bearing stage; B. Naked stage. ch, chloroplast; fl, long flagellum; n, nucleus; t, tentacle; v, vacuole. Arrow indicates direction of swimming. Not to scale. (From Moestrup and Thomsen, 1990)

Key characters

1. Only a single (hairy) flagellum is present, but a second, barren, basal body is located within the cell. Cells of the skeleton-lacking stage have two flagella, a long hairy flagellum and a very short stubby one. The flagella insert in an apical pit (Fig. 1B). The hairy flagellum

and a very short stubby one. The flagella insert in an apical pit (Fig. 1B). The hairy flagellum membrane extends into a wing-like structure supported by a cross-banded paraxonemal rod, as in the order Pedinellales. Microtubular flagellar roots are unknown and a distinct rhizoplast has not been identified (Moestrup and Thomsen, 1990).

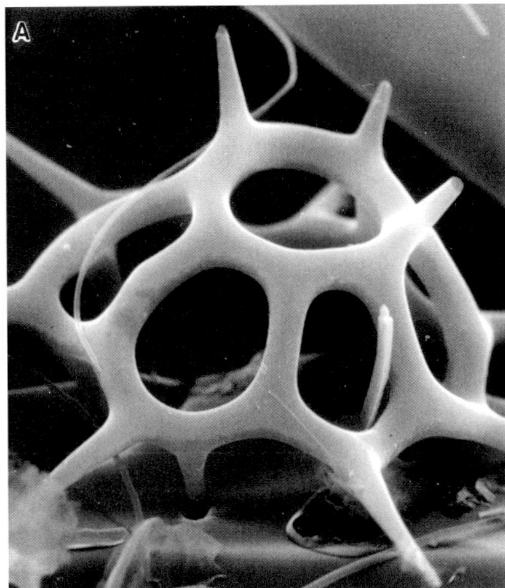

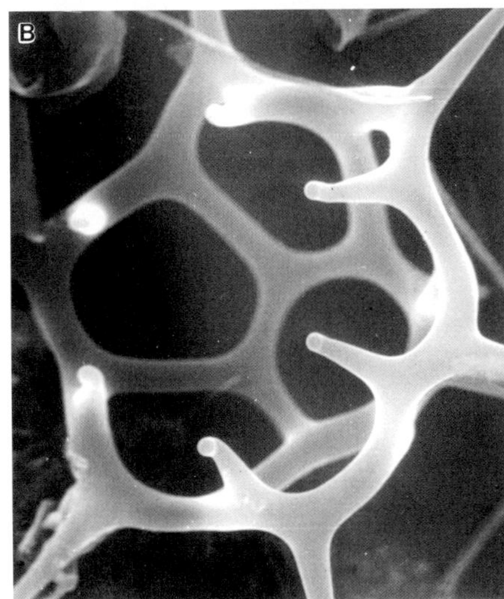

Fig. 2. *Dictyocha speculum.* Scanning electron micrographs of the siliceous skeleton, in B showing the cavity that houses the living cell. x3600. (From Moestrup and Thomsen, 1990).

2. Many chloroplasts, located peripherally (Fig. 1), each containing a single embedded pyrenoid.
3. Numerous Golgi bodies located as an anterior collar around the centrally located nucleus.
4. No cell wall, but cell surrounded by external silicified skeleton (Figs 1A, 2A, B) or naked (Fig. 1 B) (see below).
5. No eyespot or ejectile organelles.
6. The naked stage of the life cycle with an extensive layer of vacuoles between the outer layer of chloroplasts and the central part (Fig. 1B) which comprises the nucleus and the Golgi apparatus.
7. The skeleton-bearing stage with an internal canal system surrounding the central nucleus and Golgi apparatus-containing region. The canal system is connected by radial canals to the exterior.
8. Tentacles (Fig. 1A) supported by variable number of microtubules that extend from the nuclear envelope. Tentacles apparently emerging from any part of the cell.

9. The life cycle not yet fully understood, but comprising at least three main stages: the typical skeletonbearing stage, the naked stage, and a multinucleate stage which may reach 500 μm in diameter (Moestrup and Thomsen, 1990; Henriksen et al., 1993). All stages are present in marine plankton. A multinucleate, amoeboid stage has been seen in culture.
10. Classification based mainly on morphology of the siliceous skeleton. This can be highly variable (see e.g. Boney, 1973; van Valkenburg and Norris, 1970) and in large populations, specimens which cannot be referred to species level may occur.

Genus ***Dictyocha,*** Ehrenberg, 1837

The extant species may all be referred to this genus. Only three species are well known. They have been classified into three different genera, but this appears to be unwarranted. For distribution of the species in the seas in relation to temperature and salinity, see Henriksen et al. (1993).

Dictyocha speculum Ehrenberg, 1837. The skeleton comprises two well-separated rings, interconnected by six bars. The outer ring hexagonal, the inner almost circular Six spines extend in a radial direction from the outer ring, two opposite spines usually slightly longer than the others. The cell is located in a cavity formed by the rings. The diameter of the outer ring is c. 20 μm and in Danish material the distance between the tips of the longest spines was 37.5-56 μm (Moestrup and Thomsen, 1990). The naked stage is more or less spherical,

15-20 μm in diameter, occasionally smaller (to 10 μm in culture), the front end slightly flattened with a distinct pit. Flagellum 20-30 μm long. Cells contain 30-50 chloroplasts. The multinucleate stage with numerous. organelles; of all types, more or less spherical, 45-500 μm in diameter. Widely distributed in cold and temperate seas, at temperatures up to 18-20°C (Henriksen et al., 1993).

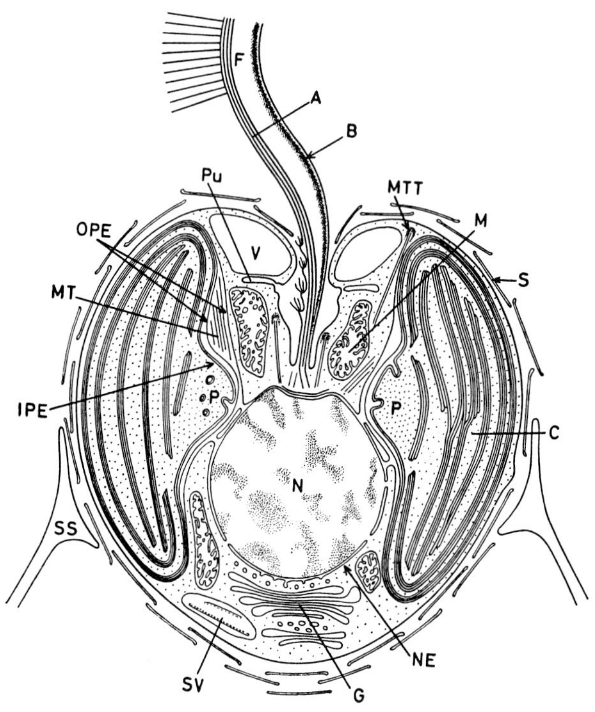

Fig. 3. *Apedinella radians (=A. spinifera).* A, axoneme of flagellum; B, paraxonemal rod supporting the flagellar wing; C, chloroplast; F, flagellum; G, Golgi body; IPE, OPE, chloroplast envelope; MT, immature flagellar hairs, MTT, microtubules; N, nucleus; NE, nuclear envelope; Pu, pusule-like cavity; S, plate scales; SS, spine scales; V, vacuole. Not to scale. (From Throndsen, 1971).

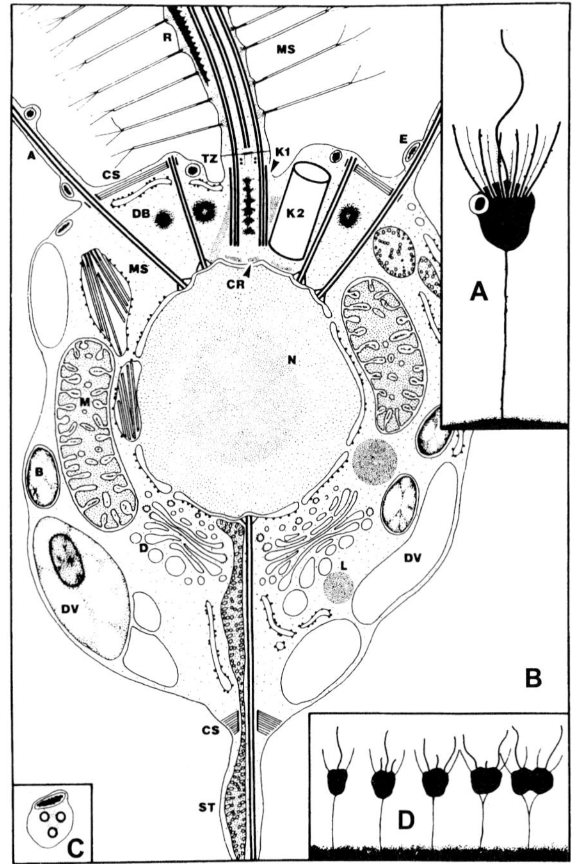

Fig. 4. *Pteridomonas danica.* A. Cell with food vacuole. B, Details of cell structure. C. Tentacle with triplet of microtubules. D. Cell division. Abbreviations in Fig. 4B: A, tentacle; B, endosymbiotic bacteria; CR, dense material on the nuclear envelope; CS, cytoskeletal elements; D, Golgi body; DB, dense bodies; DV, digestion vesicles; E, extrusomes; KI, K2, basal bodies; L, lipid; M, mitochondrion; MS, tripartite flagellar hairs; N, nucleus; R, paraxonemal rod; TZ, flagellar transition zone; ST, stalk. Fig. 4A=18,000x. (After Patterson and Fenchel, 1985).

ORDER PEDINELLALES

Zimmermann, Moestrup and Hällfors 1984

By ØJVIND MOESTRUP

The pedinellid flagellates comprise more or less spherical cells with a single anterior (hairy) flagellum. The flagellum beats in a planar sine wave and bears a more or less well-developed wing supported internally by a paraxonernal rod (Figs. 3,4). Tentacles emanate from the cell surface, supported by triplets of microtubules attached proximally to the nuclear envelope

(Figs. 4A-C). Cells with three or six chloroplasts and auto- or mixotrophic(*Pedinella*) or without chloroplasts and heterotrophic. Some species free swimming, often with a trailing stalk, others attached. The body in some species clad in a cover of organic scales (Fig. 3), some of which may be spine-shaped. In fresh water and marine plankton. Synonym: order Ciliophryida Febre-Chevalier 1985.

With a single family Actinomonadidae Kent, 1880.

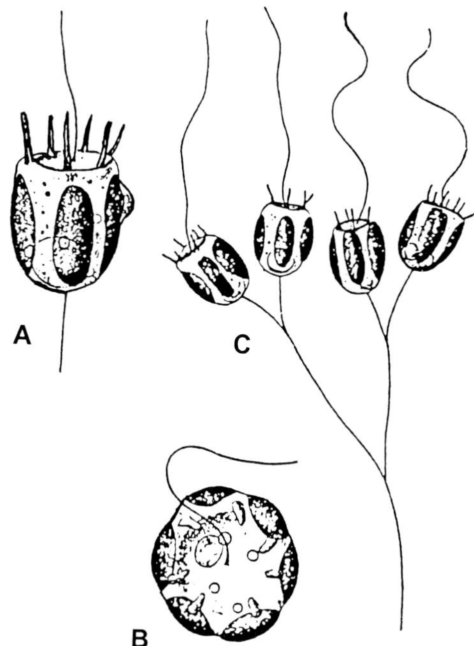

Fig. 5. *Pedinella hexacostata,* a colony- forming, mixotrophic pedinellid. A,B=2000x. C: x1000. (From Conrad, 1926).

Key characters

1. One emergent flagellum, but a second, barren, basal body is present in the cell (Fig. 4B). Flagellum with more or less distinct wing, supported by paraxial rod (Figs. 3,4). Two rows of tripartite flagellar hairs. Small ring-shaped flagellar scales found in two genera. No microtubular or striated flagellar roots are known.
2. Three or six lateral chloroplasts in one tier, or without chloroplasts. Pyrenoids present or absent.
3. No eyespot.
4. With stalk of very variable length. Cells free swimming (trailing stalk) or attached.
5. Contractile vacuoles in freshwater species.
6. Cell body covered with organic scales in some genera.
7. Ejectile organelles unknown.
8. Division by fission.
9. Organic cysts known, sexuality unknown.
10. Species are distinguished by cell size and shape, etc. As these are often highly variable characters, species taxonomy is unsatisfactory and awaits alternative characters. A few species are recognized by the morphology of the scales.

Key to the genera

1. Cells with three or six chloroplasts............. 2
1´. Cells without chloroplasts.......................... 4

2. Cells attached by means of a stalk, pyrenoids absent .... ***Pedinella***
2´. Cells with trailing stalk or stalkless, pyrenoids present.. 3

3. Cells with long to very long trailing stalk............................... ***Pseudopedinella***
3´. Cells with six posteriorly-directed spines, held together in swimming cells.. ***Apedinella***

4. Cells with long or very long trailing stalk.................................... ***Actinomonas***
4´. Cells attached with shorter or longer stalk 5

5. With very short stalk, hardly visible in the light microscope, flagellum in a figure-of-eight, numerous tentacles from all sides... ***Ciliophrys***
5´. With long stalk .... ***Pteridomonas***

Genus ***Pedinella*** Vysotskii, 1888

Cells with an anterior ring of c. 12 tentacles surrounding the flagellum. Cells naked. Six chloroplasts without pyrenoids. Contractile vacuoles anteriorly. With posterior stalk. Cells mixotrophic, mainly ingesting bacteria, which are caught near the ring of tentacles. Uptake process unknown, but the tentacles contain small vacuoles with dense contents that may be involved. One species.

Pedinella hexacostata Vysotskii (Fig. 5). Cells apple-shaped, 7-8 µm long (range 6-11 µm) and 9-10 (8-11) µm wide. Flagellum 17-28 µm long. Cells contain numerous contractile vacuoles. In fresh water and inland saline waters. Rarely sighted, the most recent detailed description by Swale (1969).

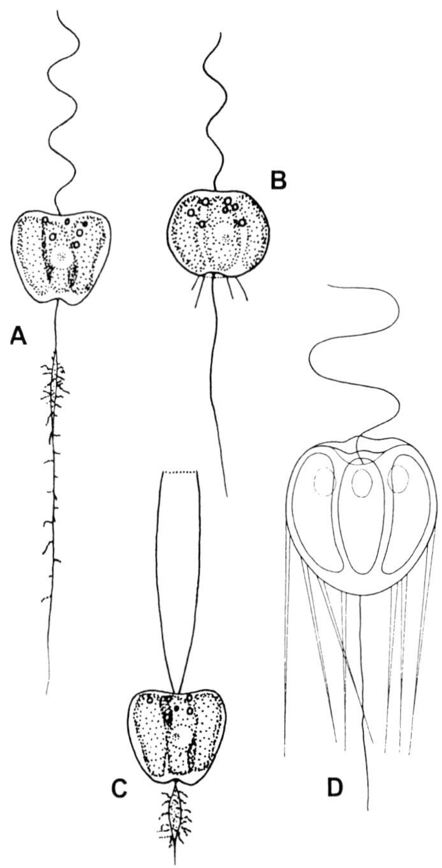

Fig. 6. Autotrophic pedinellids. A-C, *Pseudopedinella elastica,* with naked cells. D, *Apedinella radians,* with spines. A-C=1200x, (From Zimmermann et al., 1984) D=2700x. (From Throndsen, 1969)

Genus ***Pseudopedinella***, N. Carter, 1937

As *Pedinella,* but cells free-swimming with trailing stalk that may reach over 100 µm in length. Three or six chloroplasts, each with a pyrenoid. Ultrastructural descriptions by Zimmermann et al. (1984) and Thomsen (1988). Common in freshwater and marine plankton. Half a dozen species but taxonomy very uncertain due to variability.

Pseudopedinella elastica Skuja (Fig. 6A-C). Cells 9-15 µm in diameter. With six chloroplasts in a tier along the cell periphery. Trailing stalk up to c. 140 µm long. Widely distributed in brackish and marine plankton.

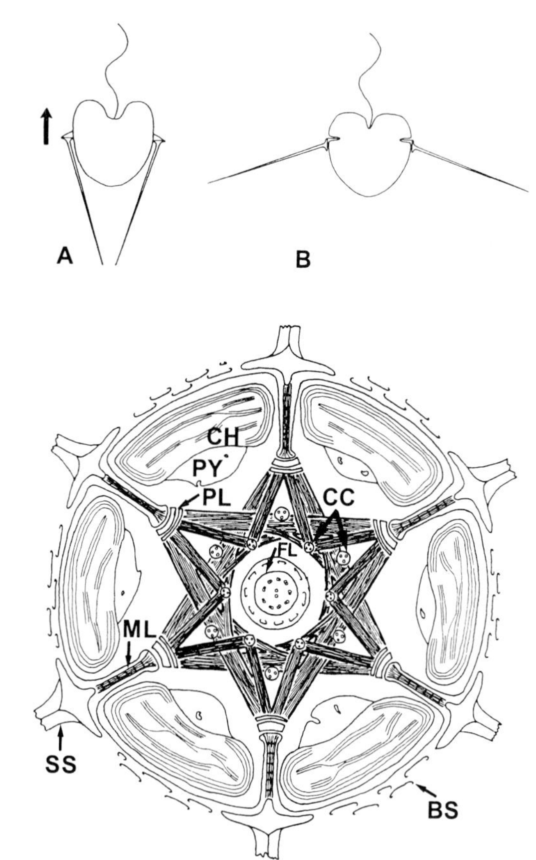

Fig. 7. *Apedinella radians.* A,B. Movement of scales. C. Transverse section of the cell near the base of the flagellum, showing the very complex system of fibers that interconnects the bases of the six spines and serve in motility. BS, flat body scales; CC, triplets of microtubules; CH, chloroplast; FL, flagellum; ML microligament; PL, plaque; PY, pyrenoid; SS, spine scales. Not to scale. (From Koutoulis et al., 1988)

Genus ***Apedinella*** Throndsen, 1971

Cells free-swimming, without or very rarely with a trailing stalk. Tentacles rarely present externally, but triplets of microtubules present in the cell. Some tentacles associate with the spines but are not usually visible in the light microscope. Cell covered with plate scales and 6 spine scales. The latter insert laterally and are directed backwards during swimming. They may serve as brakes by being swung to a position perpendicular to the direction of swimming. Movement of spines mediated by a complex system of fibers in the cell (Koutoulis et al., 1988). Six chloroplasts, each with a pyrenoid. One species.

Apedinella radians (Throndsen) Campbell (Figs. 3,6D,7). Cells 7-10 μm in diameter. Flagellum 1-2 x cell length. Very common in marine plankton and widely distributed.

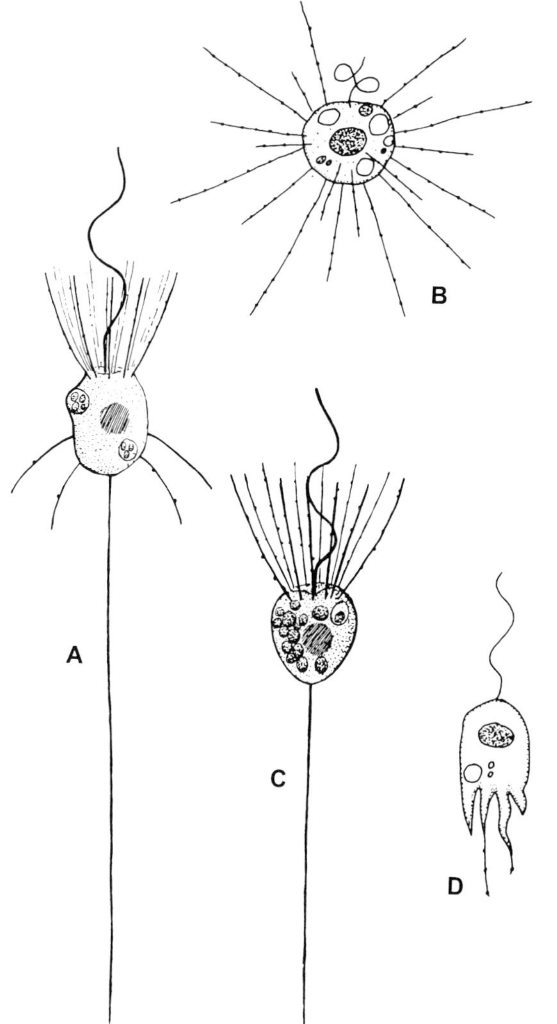

Fig. 8. Heterotrophic pedinellids. A. *Actinomonas mirabilis.* B,D, *Ciliophrys infusionum,* in B a sessile cell, in D a swimming cell. C. *Pteridomonas danica.* A,C:=3000x, (From Larsen and Patterson, 1990) B, D=1000x, (After Preisig et al., 1991)

Genus ***Actinomonas*** Kent, 1880

Cells colorless, with tentacles emanating from the cell in all directions but particularly numerous at the front end. Tentacles of the front end interconnected by complex system of fibers (Larsen, 1985). With well-developed flagellar wing supported by paraxonemal rod. Heterotrophic. Very common in marine plankton. One well-described species.

Actinomonas mirabillis Kent, 1880 (Fig. 8A). Cells 4-8 μm in diameter. With trailing stalk up to c. 120 μm long. Sometimes more than one trailing stalk present. With annular scales on the flagellum. Ultrastructural study by Larsen (1985, as *Actinomonas pusilla*). *Parapedinella reticulata* S. M. Pedersen and Thomsen somewhat resembles *Actinomonas.* It has numerous tentacles emanating from the cell surface but the cell, which is 7-15 μm in diameter, is covered with plate-like organic scales (Pedersen et al. 1986). Annular scales are present on the flagellar surface. Known from Australia and Denmark.

Genus ***Pteridomonas*** Penard, 1890

Cells very similar to *Actinomonas*, but tentacles are mainly confined to an anterior ring around the flagellum, and the cells are usually attached to a substratum. They also differ in the presence of a very reduced flagellar wing and paraxonemal rod and in the presence of 2 ring-like structures below the transverse septum of the flagellar transition region (Patterson and Fenchel 1985). Such rings are also present in *Apedinella* but not in any other pedinellid. Heterotrophic, feeding on bacteria and on other protists. Three species in marine, freshwater, and brackish environments.

Pteridomonas danica Patterson and Fenchel (Figs. 4A-D,8C). Cells 3.5-5.5 μm in diameter. Attached with a narrow contractile stalk of variable length, mostly 15-25 long. With c. 12 tentacles around the flagellum base. Cells may detach from the substrate and swim freely, often with the tentacles resorbed and occasionally with the stalk resorbed. Cells caught in debris may deform and move by amoeboid locomotions (Patterson and Fenchel 1985). Common but often confused with *Actinomonas.*

Genus ***Ciliophrys*** Cienkowsky, 1875

Colorless cells occurring in 2 different stages. Sessile (feeding) cells with granular 'tentacles over the entire cell surface. Flagellum inactive or moving very slowly in a figure-of-eight posture. Attached to a substrate with a very short stalk. The cell may change within a few minutes to a swimming condition, in which the tentacles are more or less retracted and the flagellum moves in a typical pedinellid fashion. The cell feeds on bacteria or very small protists which are caught on the tentacles, transported, to the cell body and ingested. Ultrastructural work by Davidson (1982), see also Larsen and Patterson (1990). Two very similar species, one marine, the other from fresh water.

Ciflophrys infusionum Cienkowsky, 1875 (Fig. 8B, D). Cells 5.5-11 µm in diameter with tentacles extending another 10-15 µm. Very widely distributed, attached to detritus or other objects.

LITERATURE CITED

Boney, A. D. 1973. Observations on the silicoflagellate *Dictyocha speculum* Ehrenb. from the Firth of Clyde. *J. Mar. Biol. Ass. U.K.*, 53:263-268.

Cavalier-Smith, T. 1993. Kingdom Protozoa and its 18 phyla. *Microbiol. Rev.* 67:953-994.

Davidson, L. A. 1982. Ultrastructure, behavior, and algal flagellate affinities of the Helioflagellate *Ciliophrys marina*, and the classification of the Helioflagellates (Protista, Actinopoda, Heliozoea). *J. Protozool.*, 29:19-29.

Erard-le Denn, E. & Ryckaert, M. 1990. Trout mortality associated to *Distephanus speculum*. *In*: Granéli, E., Sundström, B. Edler, L. and Anderson, D. M. (eds.), *Toxic Marine Phytoplankton.* Elsevier, New York. p. 137

Henriksen, P., Knipschildt, F., Moestrup, Ø. & Thomsen H. A. 1993. Autecology, life history and toxicology of the silicoflagellate *Dictyocha speculum* (Silicoflagellata, Dictyochophyceae). *Phycologia*, 32:29-39.

Koutoulis, A., McFadden, G. I. & Wetherbee, R. 1988. Spine-scale orientation in *Apedinella radians* (Pedinellales, Chrysophyceae): the microarchitecture and immunocytochemistry of the associated cytoskeleton. *Protoplasma*, 147:25-41.

Larsen, J. 1985. Ultrastructure and taxonomy of *Actinomonas pusilla*, a heterotrophic member of the Pedinellales (Chrysophyceae). *Br. Phycol. J.*, 20:341-355.

Moestrup, Ø. 1995. Current status of chrysophyte 'splinter groups': synurophytes, pedinellids, silicoflagellates. *In*: Sandgren, C. D., Smol, J. P. & Kristiansen, J. (eds.), *Chrysophyte Algae: Ecology, Phylogeny and Development.* Cambridge University Press, Cambridge. pp 75-91

Moestrup, Ø. & Thomsen, H. A. 1990. *Dictyocha speculum* (Silicoflagellata, Dictyochophycaeae), studies on armoured and unarmoured stages. *Biol. Skr. Dan. Vid. Selsk.*, 37:1-57.

O'Kelly, C. J. & Wujek, D. E. 1995. Status of the Chrysamoebales (Chrysophyceae): observations on *Chrysamoeba pyrenoidifera, Rhizochromulina marina* and *Lagynion delicatulum*, *In*: Sandgreen, C. D., Smol, J. P. & Kristiansen J. (eds.), *Chrysophyte Algae: Ecology, Phylogeny and Development.* Cambridge University Press, Cambridge, pp 361-372

Patterson, D. J. & Fenchel, T. 1985. Insights into the evolution of heliozoa (Protozoa, Sarcodina) as provided by ultrastructural studies on a new species of flagellate from the genus *Pteridomonas. Biol. J. Linn. Soc.*, 34:381-403.

Pedersen, S. M., Beech, P. L. & Thomsen, H. A. 1986. *Parapedinella reticulata* gen. et sp. nov. (Chrysophyceae) from Danish waters. *Nord. J. Bot.*, 6:507-513.

Saunders, G. W., Potter, D., Paskind, M. P & Andersen, R. A. 1995. Cladistic analyses of combined traditional and molecular data sets reveal a new algal lineage. *Proc. Natl Acad. Sci. USA*, 92:244-248.

Swale, E. M. F. 1969. A study of the nanoplankton flagellate *Pedinella hexacostata* Vysotskii by light and electron microscopy. *Br. Phycol. J.*, 4:65-86.

Tappan , H. 1980. *The Paleobiology of Plant Protists.* WH Freeman and Co., San Francisco.

Thomsen, H. A. 1988. Ultrastructural studies of the flagellate and cyst stages of *Pseudopedinella tricostata* (Pedinellales, Chrysophyceae). *Br. Phycol. J.*, 23:1-16.

van Valkenburg, S. D. & Norris RE. 1970. The growth and morphology of the silicoflagellate *Dictyocha fibula* Ehrenberg in culture. *J. Phycol.* 6:48-54.

Zimmermann, B., Moestrup, Ø. & Hällfors, G. 1984. Chrysophyte or heliozoan: ultrastructural studies on a cultured species of *Pseudopedinella* (Pedinellales ord. nov.), with comments on species taxonomy. *Protistologica*, 20:591-612.

ACANTHARIA Haeckel, 1881

By Colette Febvre, Jean Febvre, and Anthony Michaels

The Acantharia Haeckel, 1881 are usually classified with the Radiolaria and have recently been placed in a single phylum—Radiozoa Cavalier-Smith, 1987 (Corliss, 1994). The Acantharia share a common feature with the Radiolaria and the related Heliozoa Haeckel, 1866, i.e. the presence of **axopodia**. These are long radiating, unbranched processes stiffened by patterned arrays of microtubules (Febvre-Chevalier and Febvre, 1993). For this reason, all these taxa are often placed in the phylum Actinopoda (Lee et al., 1985; Puytorac et al., 1987, Margulis et al., 1990, 1993, Mehlhorn and Rüthmann, 1992). However, as this character may have been acquired on multiple occasions, these groupings may create artificial taxa.

The Acantharia are delicate, exclusively marine protozoa. They are free living, microphagic, and contribute with other unicellular eukaryotes to the microzooplankton. They are generally spherical, oblong, or sometimes flattened, ranging in size from 0.05-5 mm in diameter. The most important features to identify a protozoan as a member of the Acantharia are as follows: 1) The characteristic skeleton consists of usually 20 radial spicules, composed of strontium sulfate, joined at the center of the cell and emerging from the cell surface in a regular pattern called **Müller's law** (Müller, 1859). 2) The cell body is divided into two parts, a centrally located endoplasm surrounded by the ectoplasm. There are two extracellular lattices, the capsular wall which divides the endoplasm from the ectoplasm and the periplasmic cortex which bounds the outer edge of the ectoplasm. Note: only the latter one is distinctive for the acantharia. 3) The presence of contractile bundles, the **myonemes**, which occupy a peripheral position in the cell and are grouped as stretchers around the tip of the spicules.

Reproduction and the life cycle are incompletely elucidated. The vegetative form is called a **trophont**. There is no evidence for asexual reproduction although it was previously reported in one genus (Schewiakoff, 1926). Sexual reproduction takes place in a **gamont** which maintains the aspect of the trophont or in a cyst (**gamontocyst**). Encystment occurs in some species and requires a complete reorganization of the mineral skeleton and a series of mitotic and meiotic fissions. Thousands of biflagellated **isogametes** are formed in the gamonts and gamontocysts and shed within minutes. Further steps to the formation of the young acantharian remain unknown.

There is no direct evidence concerning the origin of the Acantharia. Since their skeleton dissolves rapidly in sea water after death, they do not fossilize. It seems likely that they emerged in the late Precambrian or Cambrian together with many protozoa such as the radiolaria and the Foraminifera. The Acantharia are usually placed in the Class Acantharea which comprises around 50 genera and 140 species grouped into 20 families and 4 orders: Holocanthida, Symphyacanthida, Chaunacanthida, and Arthracanthida. The Arthracanthida are distributed among two suborders.

The first work on Acantharia was published by Müller (1855), who observed living Mediterranean specimens and named them "Acanthometren". Most of the earliest treatises described Acantharia (*alias* Acanthometren) from plankton samples fixed "en masse" during expeditions (the RV Challenger, the Humboldt, and the Südpolar expeditions). Therefore, the morphology of the skeleton rather than the structure of the cell body was taken into account for the systematics (Hertwig, 1879; Haeckel, 1888; Popovsky, 1904; Mielk, 1906). Despite the prominent distinct characters, the Acantharia were often classified among the Radiolaria. The primary reference for the systematics is the monograph by Schewiakoff (1926). This book contains a set of remarkable hand-drawn plates and a clear and expansive description of acantharian taxonomy based on the accurate *in vivo* observation of 80 species among the 130 described by Müller (1858) and Haeckel (1888). Schewiakoff's system takes the character of the cell body into account, whereas his predecessors used almost exclusively the skeletal morphology as taxonomic criteria. Although his work gives useful information on the ecology of the mediterranean species, it contains several descriptions of the biology of Acantharia which are now thought to be incorrect, especially myoneme formation, nuclear events, and the life cycle. More recently, general descriptions of morphology and systematics were compiled from Schewiakoff (1926) by Tregouboff (1953), Rechetniak (1981), and Cachon and Cachon (1985). Present knowledge on the fine structure of these protozoa, their life cycle, nuclear fission, endosymbiosis, and myoneme behavior are summarized in Febvre (1990). Few modern ecological data are available including some assessments of their abundance and distribution (Beers et al., 1975, 1977, 1982; Bishop et al., 1977 1978, 1980; Massera-Bottazzi et al., 1971; Massera-Bottazzi and Andreoli, 1974, 1982a, 1982b; Michaels, 1988a, 1988b, 1991; Michaels et al., 1995). The role of acantharia in the cycling of strontium and biogenic elements is only now becoming appreciated (Bernstein et al., 1987, 1992). They may be important in the cycling of carbon through both the photosynthesis of their symbionts (Michaels, 1991, Caron et al., 1995) and through their contributions to the rain of sinking particles (Caron and Swanberg, 1990, Michaels. 1991, Michaels et al., 1995).

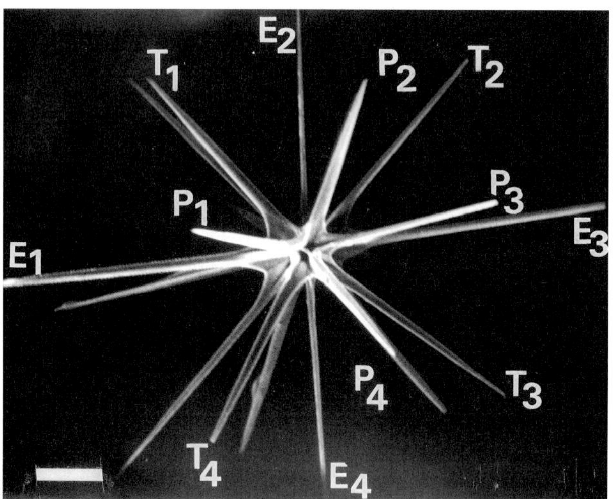

Fig. 1. Radial organization of the spicules of the arthracanthidan, *Acanthostaurus purpurascens*, according to Müller's law. Four polar spicules labelled P1-P4; tropical spicules labelled T1-T4; equatorial spicules labelled E1-E4. Remaining spicules belong to the other hemisphere. SEM. Scale bar=75 µm.

KEY CHARACTERS

Characters of the skeleton. The skeleton consists of 10 diametral or 20 radial spicules. The distal ends of the spicules arise at precise points on the cell surface, which Müller described in an analogy to the surface of the earth (Müller, 1859). Two quartets of polar tips of the spicules emerge at 60°N and 60°S, two quartets of tropical tips of the spicules emerge at 30°N and 30°S and one quartet of equatorial spicules arises at 0°. Each quartet is turned at 45° with respect to its neighbors so that the polar and equatorial tips arise at longitudes 0°, 90°, 180° and 270°, while the tropical tips arise at 45°, 135°, 225°, and 315° (Fig. 1, 2).

Another distinctive character of these protists is the mineral composition of their skeleton which was recognized very early as strontium sulfate by Bütschli (1906). This was confirmed later by crystallography and X-ray analysis. The crystaline structure of the spicule was found to be a rhombic monocrystal of celestite ($SrS0_4$) (Massera-Bottazzi and Vinci, 1965), and strontium sulfate was characterized in the Arthracanthida *Acanthometra pellucida* (Hollande and Cachon-Enjumet, 1963

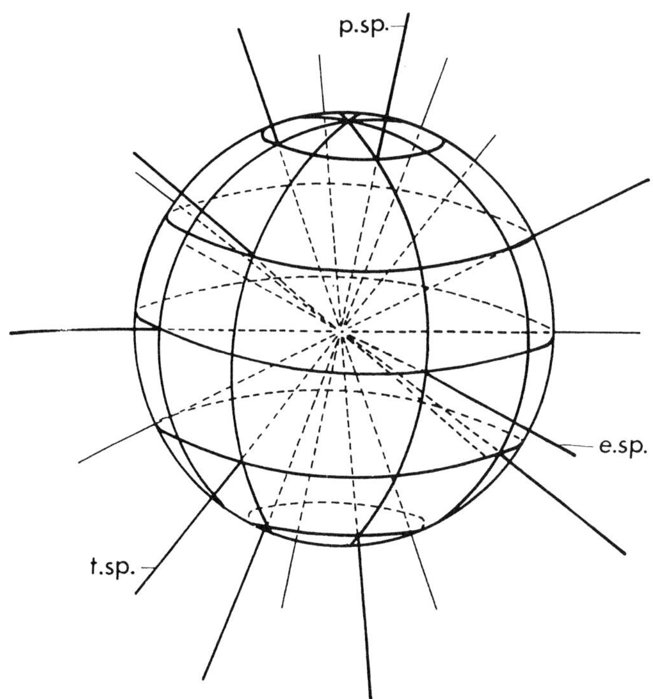

Fig. 2. Müllerian orientation of spines.

Cell body organization. The cell body consists of a central endoplasm and a peripheral ectoplasm (Fig.3). In most species the endoplasm is colored brown, red, or black by pigments and contains the nuclei. In three Orders (Symphyacanthida, Holacanthida, Chaunacanthida), the endoplasm is dense in the cell center and becomes progressively clearer towards the periphery. Moreover, it is not evidently separated from the ectoplasm. In contrast, in the fourth Order (Arthracanthida), the endoplasm is easily distinguished from the ectoplasm since it is limited by a thick capsular wall (Fig. 3). The endoplasm includes nuclei and most cell organelles, such as large mitochondria with tubular cristae, ribosomes, areas of rough endoplasmic reticulum, dictyosomes, membrane-bounded organelles such as peroxisomes and **extrusomes** resembling the **kinetocysts** found in Heliozoa (Fig. 4). In some Symphyacanthida (*Haliommatidium* and *Dicranophora*), the endoplasm contains during the vegetative life a single large nucleus probably polyploid, while in all other Acantharia there are numerous small round or oblong nuclei (Fig. 4). In many species, the **trophont** contains hundreds to thousands of nuclei immediately prior to gametogenesis.

In many species, the endoplasm contains symbiotic algae such as **zooxanthellae** (Fig. 5, 6). In natural assemblages of acantharia, approximately half of the individuals have symbionts and the smallest and largest size classes have the lowest proportions of symbionts (Michaels 1988a, 1988b, 1991). Symbionts are found in all Arthracanthida (Febvre and Febvre-Chevalier, 1979; Febvre, 1990) at some part of the life cycle. Within a species, young trophonts and gamonts lack symbionts

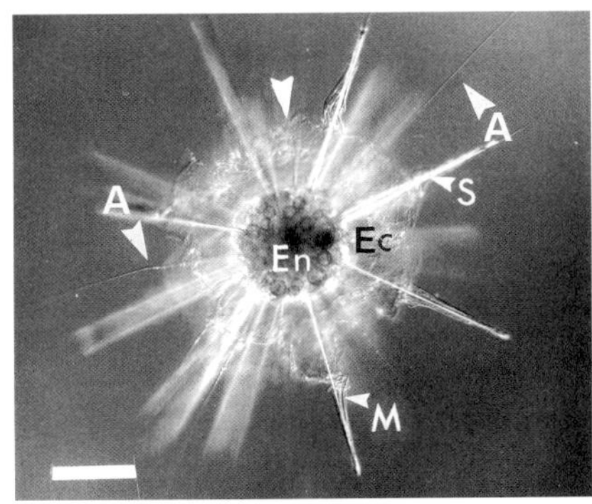

Fig 3. *Phyllostaurus siculus* showing endoplasm (En) including round zooxanthellae and ectoplasm (Ec) sustained by radial spicules (S), axopodia (A), myonemes (M), periplasmic cortex (white arrowhead), capsular wall (CW). LM. Scale bar=60 µm.

and symbiont number appears to increase with size in the trophont stage. Symbionts appear to be shed or consumed immediately prior to gametogenesis. Acantharia appear to have a diversity of symbiont types, including dinoflagellates and haptophytes, and some species appear to have more than one symbiont within the same host. A dinoflagellate of the genus *Amoebophrya* often parasitizes the endoplasm of the Arthracanthida *Acanthometra* (Borgert, 1897; Cachon and Cachon, 1969).

In the Holacanthida, Chaunacanthida, and Symphyacanthida, which encyst just before gametogenesis, the endoplasm is progressively enriched in tiny oval plaques (Fig. 7). These plaques are called **lithosomes**. They are birefringent and generated by the Golgi. They are transported towards the cell surface and arranged to form the cyst wall. During this morphogenetic change the external networks and the **myonemes** are shed.

The ectoplasm consists of anastomosed cytoplasmic strands and radial axopodial arrays. This cytoplasmic network limits large lacunae and complicated channels filled with sea water. This system extends at the surface of the acantharian in a very flexible, sensitive and highly dynamic reticulopodial network in which prey organisms such as diatoms, silicoflagellates, coccolithophorids, and tintinnid ciliates are trapped.

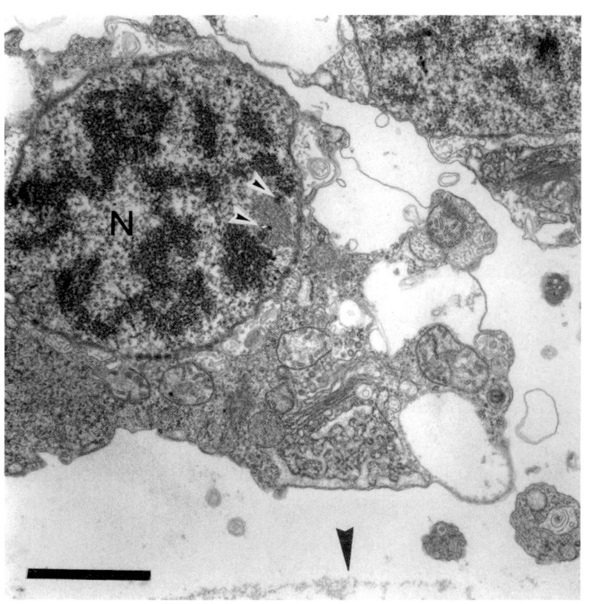

Fig. 4. The endoplasm of *Lithoptera mülleri* is limited by the capsular wall (black arrowhead). It consists of numerous islets separated by large channels filled with sea water. It contains round nuclei (N), which are here dividing; note the chromosomes attached to the spindle by kinetochores (white arrowheads). EM. Scale bar=1 µm.

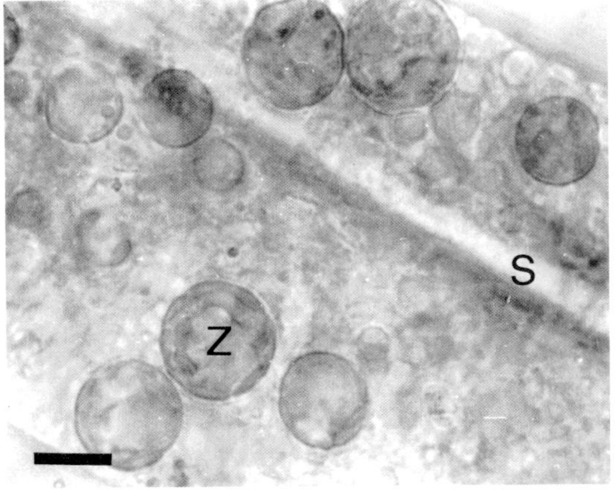

Fig. 5. Zooxanthellae (Z) in the endoplasm of *Amphilonche elongata*. The main spicule (S) is visible as a clear band in the endoplasm. LM. Scale bar=10 µm

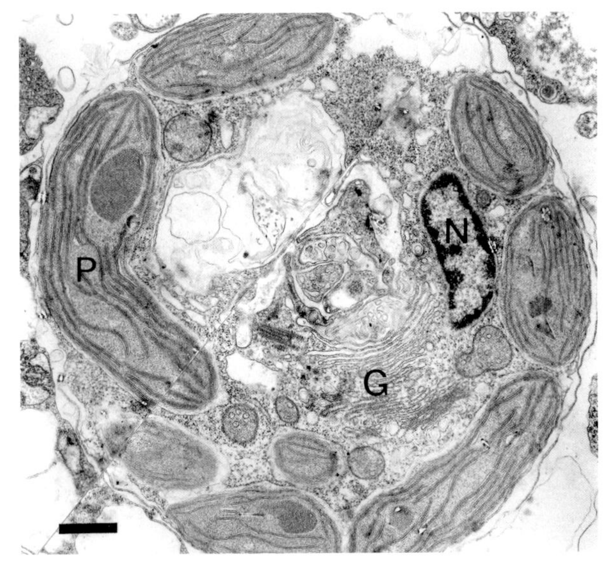

Fig. 6. Thin section through a zooxanthella in the endoplasm of *Lithoptera mulleri*. Plastid (P), Nucleus (N), Golgi (G). EM. Scale bar=1 µm.

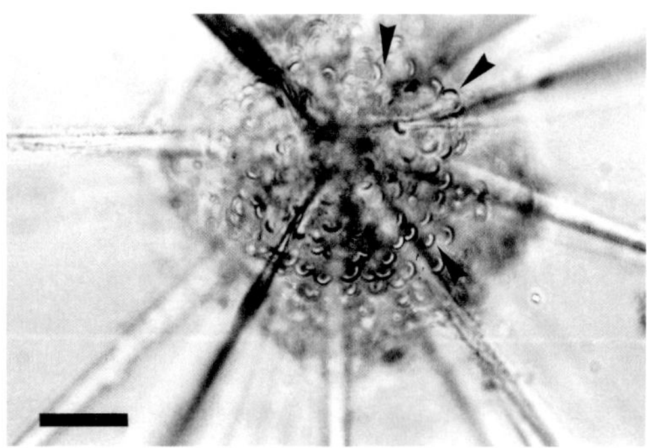

Fig. 7. Lithosomes (arrowheads) in the Chaunacanthida, *Stauracon pallida*. EM. Scale bar=25 µm. (Courtesy of J. and M Cachon)

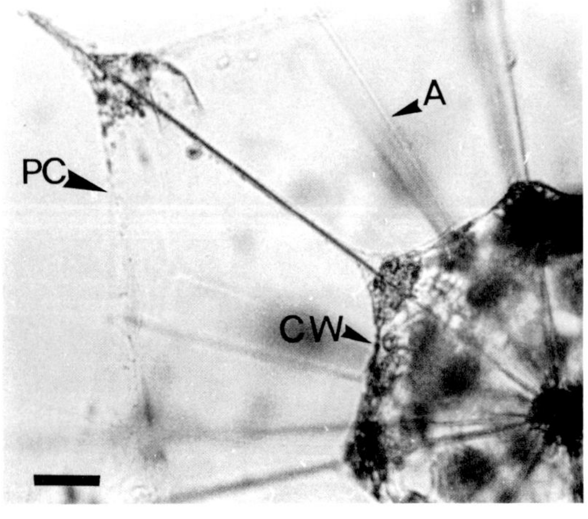

Fig. 8. *Acanthometra pellucida*. The capsular wall (CW) limits the endoplasm while the periplasmic cortex (PC) limits the ectoplasm. Axopod (A). EM. Scale bar=50 µm.

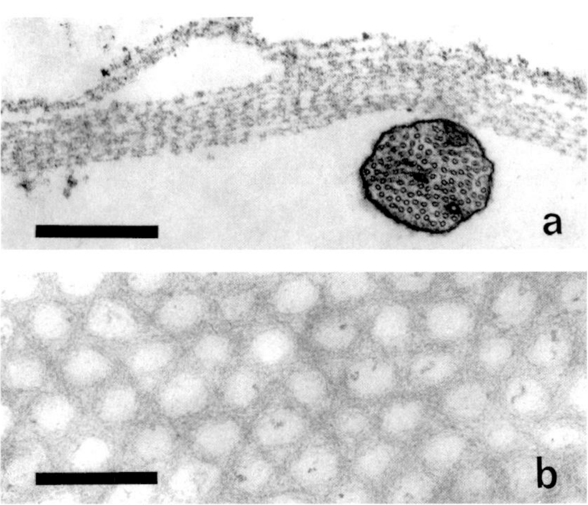

Fig. 9a,b. Cross and tangential sections through the capsular wall of *Heteracon mulleri* and *Haliommatidium mulleri*, respectively. EM. Scale bar=0.5 µm.

Extracellular networks. The cytoplasm is covered with two kinds of extracellular lattices, the capsular wall and the periplasmic cortex (Fig. 8). The **capsular wall**, previously called **capsular membrane** (Schewiakoff, 1926) consists of a thin or thick inner fibrillar layer resembling a meshwork (Fig. 9a,b). It overlays the cell membrane and the external margin of the endoplasm. In the Arthracanthida, it is thick, located between the dense endoplasm and the clear ectoplasm. Cytoplasmic protrusions and **axopodia** pass through temporary holes in this extra-cellular lattice (Figs.10, 12).

In the Holacanthida, Symphyacanthida, and Chaunacanthida, the capsular wall is not obvious and consists of a thin **fibrillar layer** (Fig. 9ab) called **"inner Gallert Mantel"** by Schewiakoff (1926). Although thinner and more superficial, this inner coat shows the same aspect as in the Arthracanthida. Therefore, we use the common term **capsular wall** to designate this feature whatever the order.

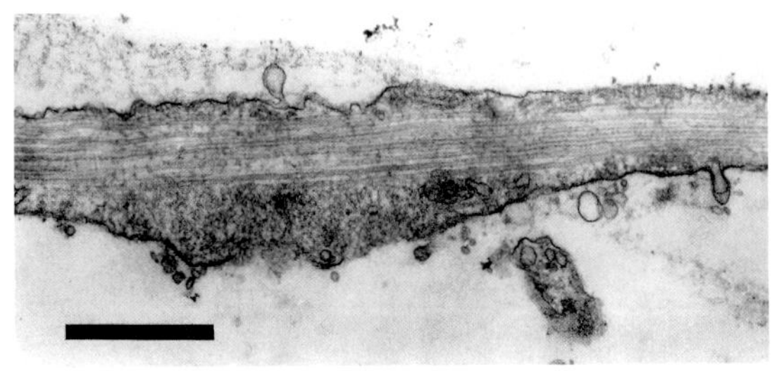

Fig. 10. Axopod passing across the capsular wall in *Heteracon mulleri*. EM. Scale bar=0.5 µm.

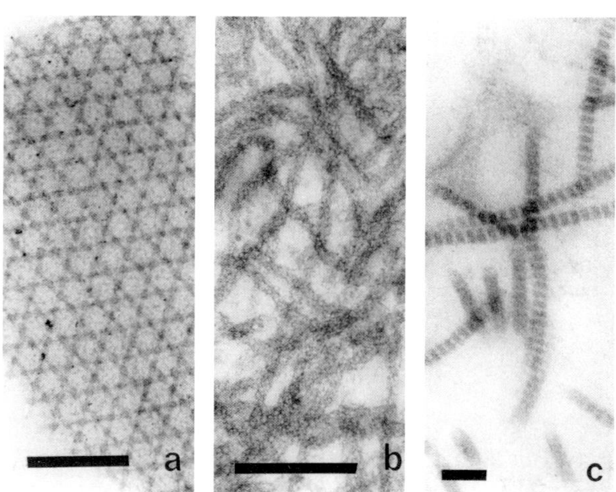

Fig. 11a-c. Tangential sections through the periplasmic cortex. EM. (a) *Haliommatidium mulleri*. Scale bar=0.5 μm. (b) *Acantholithium stellatum*. Scale bar=0.5 μm. (c) *Astrolithium bulbiferum*. Scale bar=0.1 μm.

The **periplasmic cortex** is the most exernal layer (Fig. 8). It consists of 20 polygonal pieces joined together with each other by elastic junctions. Each of these 20 pieces is centered around the spicule. The fine structure of the cortex consists of a lattice of thin fibrils arranged in complicated specific patterns (Fig. 11 a-c). Around each spicule, the capsular wall turns up and binds to the periplasmic cortex so as to limit a narrow space filled with ectoplasm (Fig. 12).

Axopodia. The axopodia are thin, slender processes radiating from the cell surface. These extensions are sensitive and can retract upon chemical (colchicine, podophyllotoxine, and diverse microtubule poisons) or physical stimulation (cold, mechanical shocks), then regrow progressively and slowly. They are stiffened by a central axis (axoneme) consisting of patterned arrays of microtubules. A thin sheet of cytoplasm covered with the cell membrane surrounds the axonemes (Figs. 10, 12, 13). It contains diverse organelles such as extrusomes, mitochondria, coated vesicles, and Ca^{2+} sequestering vesicles which move in a saltatory motion.

Two main patterns of microtubules have been found: dodecagonal (Holacanthida)(Fig. 14a); hexagonal (Arthracanthida, and Chaunacanthida) (Fig. 14b). The axonemes (Mt-rods) originate from tiny **microtubule-organizing centers (MTOCs)** located in the central endoplasm and consisting of dense plaques located at the bases of each spicule. These plaques are closely fitted to the membrane of a large perispicular vacuole (Fig. 15).

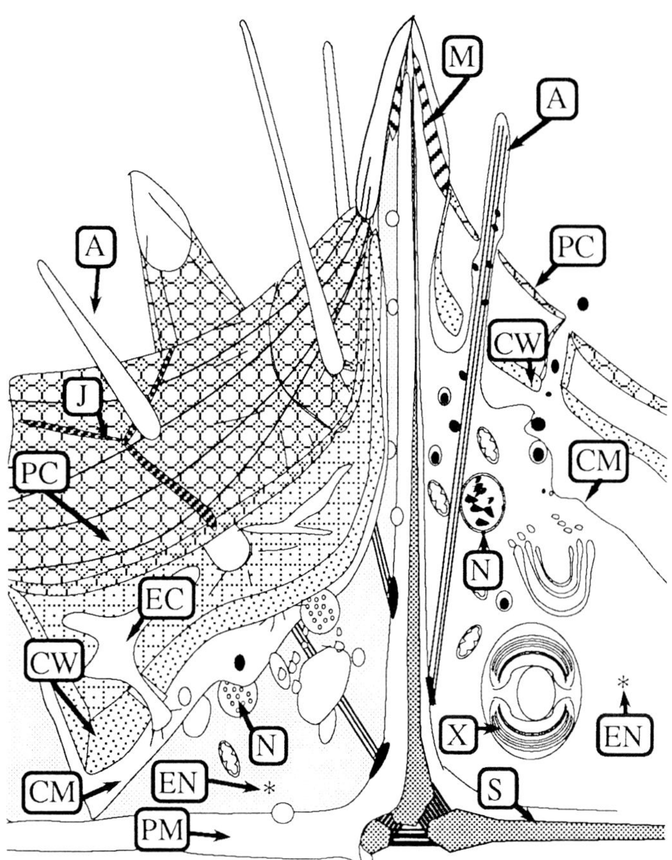

Fig. 12. Interpretative drawing of the tridimensional left part and the fine organization (right part) around a spicule (S), endoplasm (En), capsular wall (CW), ectoplasm (Ec), periplasmic cortex (PC), elastic junctions (J), axopod (A), myoneme (M), cell membrane (CM), perispicular membrane (PM), nuclei (N), xanthella (X).

Fig. 13. *Acanthochiasma rubescens*. Numerous axopodia arise from the ectoplasm. The endoplasm is opaque in the cell center while clearer at the periphery. The ectoplasm surrounds the spicule. LM. Scale bar=100 μm.

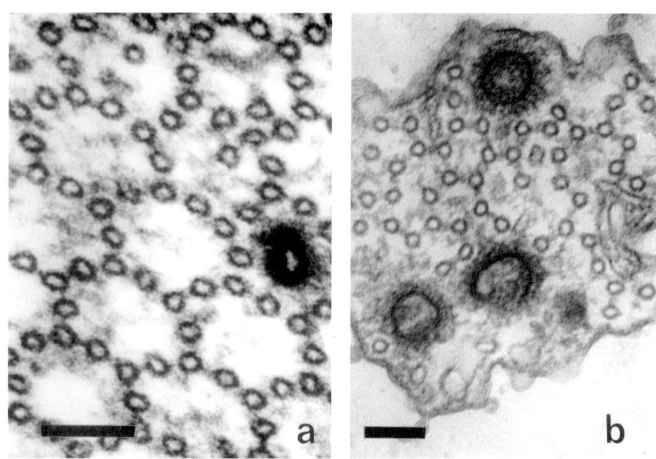

Fig. 14a,b. Cross-section through axopods. EM. (a) *Acanthochiasma fusiforme.* Scale bar=0.1 µm. (b) *Lithoptera mulleri.* Scale bar=0.1 µm.

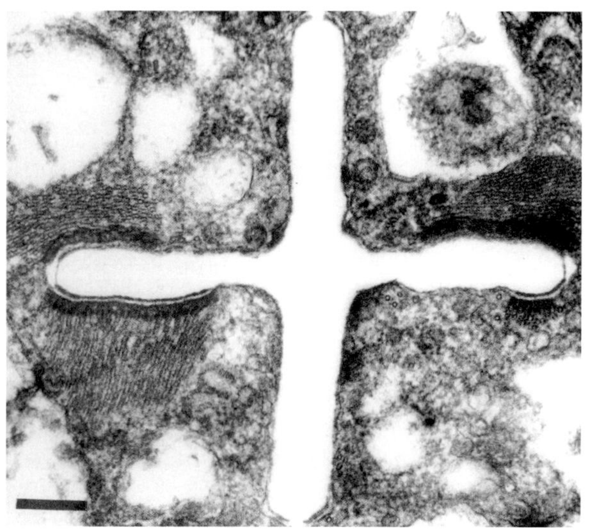

Fig. 15. Cross-section through the base of a spicule in *Gigartacon mulleri* showing the dense MTOCs and oblique sections of microtubules. Scale bar=0.5 µm.

Myonemes. The cytoplasm around each spicule is limited by the cell membrane. This narrow space contains 2-60 myonemes (Figs. 16, 17). These are birefringent fibrillar bundles anchored by one end to the spicules and by the other end to the periplasmic cortex, itself linked to the capsular wall. Depending on the species, their number varies between 40 to 1200 i.e 2-60 myonemes per spicule. Their size scale ranges in 5-90 µm long. They are either flat, triangular, or as long ribbons or short cylinders. The myonemes are motor organelles capable of three kinds of movements: slow undulating movement which produces no obvious change in the position of the periplasmic cortex (Fig. 16 a-b); rapid contraction which suddenly pulls the cortex to the tip of the spicule (Fig. 16c); and slow subsequent relaxation. Rapid and simultaneous contraction of all the myonemes produces subsequent inflation of the cortex. Therefore, the volume of the ectoplasmic region is suddenly increased. Such a coordinated activity of the myonemes occurs every 10-20 minutes and may play a part in buoyancy regulation (Febvre-Chevalier and Febvre, 1994). The myoneme consists of a bundle of nonactin filaments (2-4 nm in diameter), twisted by pairs in elementary microstrands (Fig. 17). It shows a series of clear zones and thin dark bands whose length depends on the degree of contraction (Fig. 18). The myoneme is

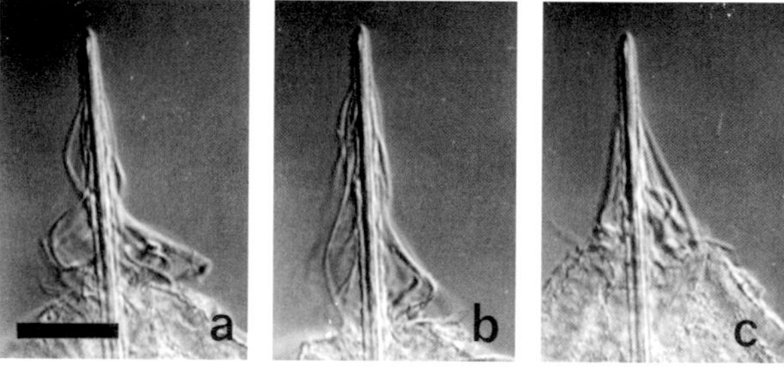

Fig. 16a-c. Three steps of myoneme movement in *Heteracon biformis*. 16a-b. Slow undulating movement. 16c. Rapid contraction. Note that the position of the ectoplasm (arrowhead) is higher after rapid shortening than before. LM. Scale bar=30 µm.

made of a Ca^{2+}-binding protein which does not require ATP for driving contraction (Febvre and Febvre-Chevalier, 1982, 1989a,b, 1990). It shares these physiological characteristics with the myonemes of other protists such as ciliates and flagellates.

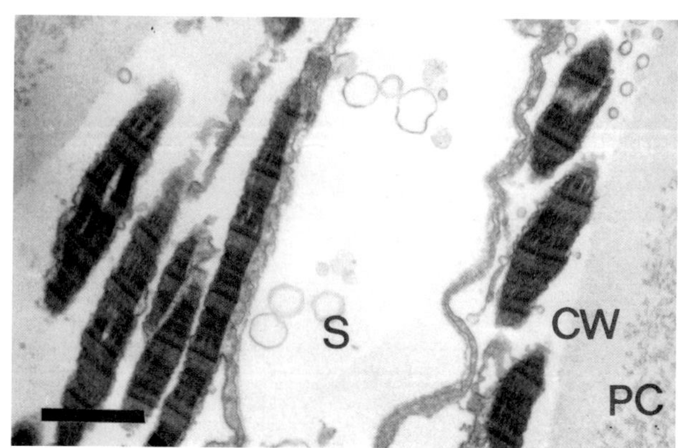

Fig. 17. Longitudinal section through the myonemes around a spicule (S), capsular wall (CW), cortex (PC). EM. Scale bar=1 µm.

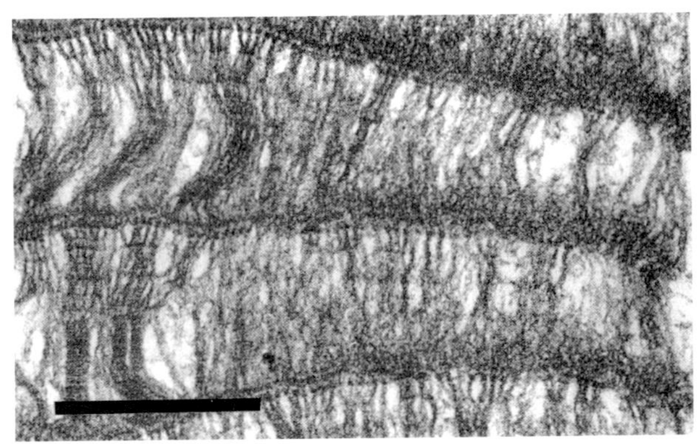

Fig. 18. Thin longitudinal section through a myoneme. Note the variable spacing of the clear bands indicating that the myoneme was contracting. EM. Scale bar=0.5 µm.

Ecology and Distribution

Acantharia are common in the surface waters of tropical and subtropical oceans worldwide and occur in lower numbers in the surface waters of temperate and polar seas (summarized in Caron and Swanberg, 1990). They tend to be less common in very nearshore environments and coastal upwelling zones. They are generally restricted to the illuminated upper few 100 meters of the ocean although some individuals have been collected in sediment traps many thousands of meters below the surface (Bernstein et al., 1987).

Early assessments of the abundance and distribution used plankton nets with 160-350 µm mesh sizes. Abundances were typically from 1-500 m^{-3} (e.g. Massera-Bottazzi et al., 1971; Massera-Bottazzi and Andreoli, 1974, 1982a, 1982b). More recent estimates using pump systems or the filtration of water samples yield abundance estimates of 1000-50,000 m^{-3} (Beers et al., 1975, 1977, 1982; Bishop et al., 1977 1978, 1980; Michaels, 1988a, 1988b, 1991; Michaels et al., 1995). The earlier estimates from nets are lower because of a combination of the passage of small cells through the coarse mesh, the destruction of cells in the mesh, dissolution of the skeletons, and intrinsic underestimates of microplankton abundances by biases in plankton net collections (Michaels, 1988a, 1988b).

The dissolution of the strontium sulfate skeleton in preserved plankton samples also contributes to the lack of recognition of this taxa by biological oceanographers. Although acantharia are often the most abundant of the shelled sarcodines in the upper ocean (Caron and Swanberg, 1990; Michaels, 1988a, 1988b, 1991; Michaels et al., 1995), their skeletons dissolve in most common preservatives (Beers and Stewart, 1971), and thus they tend to be missed in most assessments of microplankton abundance. The addition of 72 mg Sr l^{-1} to a formalin-preserved sample serves to reduce this dissolution. Alternative methods of preservation (e.g drying on filters) also allow for quantitative assessments of abundance, but can obscure other features of the cells.

Acantharia abundances tend to be highest near the surface and to decrease markedly with depth. They have active buoyancy control, and on calm days larger acantharia can aggregate in the upper few meters of the water column or be concentrated in Langmuir cells (windrows). This surface affinity is likely related to the photosynthetic physiology of their symbiotic algae. During periods of strong vertical mixing, their abundances become uniform with depth, but they reassert the surface tendencies within a few days of restratification.

There are few definitive time-series studies of acantharia abundance, and in those studies, there are no strong seasonal patterns in abundance. Acantharia tended to be more abundant in the summer and fall at the VERTEX time-series station (33°N, 139°W) although acantharian biomass did not show the same pattern (Michaels 1991). Integrated abundances were variable at the site of the Bermuda Atlantic Time-series Study (32°N, 64°W) in the Sargasso Sea (Michaels et al., 1995). Samples collected with plankton nets at a variety of locations throughout the Atlantic suggest higher abundances in the spring and lower abundances in the summer although the data were not collected as an explicit time-series (Massera Bottazzi and Andreoli, 1982a). In coastal waters, seasonal patterns are sometimes observed if "oceanic" waters displace eutrophic waters near the coast for a defined portion of the year. In the oligotrophic tropical and subtropical seas, acantharia are a regular, common, and conspicuous component of the microplankton year-round.

Acantharia consume a wide variety of prey. Reports from microscopical observations of feeding vacuoles describe the remains of tintinnids and other types of ciliates, diatoms, dinoflagellates, copepod nauplii, copepodites, and adults, pelagic molluscs, and other sarcodine taxa (Caron and Swanberg, 1990). Examinations of cells under epifluorescent illumination (blue exitation, red and orange emmision) reveal signs of microalgal prey including picoplankton-sized (<2 µm) cyanobacteria

(Michaels, personal observations). Small bacteria can also be observed within food vacuoles with the appropriate staining techniques; however, it is unclear if they are actively consumed or accidently incorporated with larger food. There are no published quantitative estimates of grazing rates.

Symbiotic algae are an important part of the nutrition of acantharia. These symbionts have extremely high saturation irradiances and can continue to fix carbon, even near the surface. This carbon is thought to contribute to the nutrition of the acantharian host and can sometimes meet or exceed the inferred daily metabolic requirements of the host for carbon (Michaels, 1988a, 1988b, 1991; Caron et al., 1995).

Little is known about the host-symbiont nitrogen dynamics. In the oligotrophic settings in which acantharia are most common, it is likely that nitrogen and phosphorus are supplied to the association by the feeding activity of the acantharian as the dissolved nutrients are present at very low concentrations; however, the symbionts may play a role in the retention of these nutrients either by recycling through the host or by meeting the host energy demands and thus allowing a more efficient use of particulate nutrient material.

Symbiont abundances within a host vary from a few large dinoflagellates in some species to many thousands of smaller symbionts in others. Most acantharia have symbionts at some stage in their life cycle although in heterogeneous natural populations, generally about half of the individual acantharia are aposymbiotic (Michaels 1988, 1991). These aposymbiotic individuals are a mix of juveniles, near-reproductive adults (which lose the symbionts) and aposymbiotic species. A significant proportion of the capsule biovolume can be filled with the symbionts. Symbiont reproduction is likely under host control as the slow symbiont doubling times inferred by cell division (very few of the cells are seen in mitosis at any one time) are much lower than the rapid growth that could be inferred from the carbon-specific rates of symbiont production (9.4 ng C/ng symbiont C/day, Michaels, 1991). Most of the symbiont-fixed carbon likely is translocated to the host.

Acantharia and their reproductive cysts are a common constituent in sediment trap samples collected in oceanic regimes (Antia et al., 1993; Bernstein et al., 1987, 1992; Michaels, 1991; Michaels et al., 1995; Spindler and Beyer, 1990). They can account for up to 5-10% of the particulate organic carbon in sediment traps placed just below the euphotic zone of oligotrophic seas (Michaels, 1991; Michaels et al., 1995). This export is surprising as acantharia rarely account for even one percent of the standing stock of particulate carbon.

A number of features of acantharian biology and life history account for their disproportionate role in carbon export. They are large and have a skeleton with a high specific gravity; this leads to rapid sinking rates. They have an encystment stage as part of the reproductive cycle in many species, and these cysts seem designed for rapid sinking. Symbiont production may allow the host to have a faster growth rate than other cells of the same size and thus compensate for losses due to sinking, grazing, and reproduction. Although there are few data on predation, if eaten, it is likely by a large animal which would in turn make a large, fast-sinkng fecal pellet. Thus a combination of their unique biology and place in the trophic web leads to their disproportionate representation in sinking materials.

As part of the group of particles that mediate vertical export, they also may play a role in the export of strontium and some trace elements. Surface waters are frequently slightly depleted in dissolved strontium, and the sinking of acantharia skeletons likely accounts for most of this disequilibrium (Brass and Turekian, 1974; Bernstein et al., 1987). Lead, barium, and other trace metals are incorporated into the celestite matrix during growth. Because of the relatively high distribution coefficients for metals in celestite (Brass, 1980; Michaels and Coale, unpublished data), the skeletons of acantharia may account for a disproportionate amount of the trace metal flux for certain metals in environments with a large acantharian flux (Bernstein et al., 1987, 1992; Brass, 1980; Michaels and Coale, unpublished data).

KEY TO THE ORDERS

1. Ten diametral spicules of equal or unequal length, crossing or fitting together in the cell center.. **Holacanthida**

1'. 20 radial spicules 2

2. Bases of the spicules indissociable in H_2SO_4......... .... **Symphyacanthida**

2'. Bases of the spicules dissociable with H_2SO_4 3

3. Bases of the spicules oblong, resembling grape-pips or ninepins. Spicules cruciform in cross-section, generally serrate on two opposite edges, smooth on the two others......... **Chaunacanthida**

3'. Bases of the spicules pyramidal with 4-6 facets with or without basal extensions; juxtaposed or linked so as to form a more or less cohesive system. .... **Arthracanthida**

ORDER HOLACANTHIDA Schewiakoff, 1926

Ten diametral spicules of equal or unequal length, crossing or fitting together in the cell center, dissociated after treatment with sulfuric acid. Endoplasm with pigments and numerous nuclei. Ectoplasm clearer and less granular than the endoplasm, resemblng a gelatinous sheet. Limit between endoplasm and ectoplasm generally ill-defined. Capsular wall near the cortex, poorly visible in light microscopy (LM). Cell surface irregular or limited by a thin periplasmic cortex. Cytoplasm crossed by numerous thin sensitive axonemes. Flat myonemes. Gametogenesis within a cyst

KEY TO FAMILIES

1. Long, equal, generally cylindrical spicules crossing in the cell center. Endoplasm with pigments, clear at the periphery, not obviously limited by a capsular wall. Ectoplasm capable of inflating far from the endoplasm if undisturbed for some hours. **Gamontocyst** round or oval, limited by small plaques; showing remnants of spicules forming a small bundle in the center of the cyst.. 2

1'. Spicules equal or unequal, twisted at half length, more or less intimately fitted or cemented with one another in the cell center of the adult vegetative forms...... **Acanthocollidae**

2. Ectoplasmic rim irregular. Cortex and myonemes very thin, poorly visible. Numerous granule-studded axopodia. **Acanthochiasmidae**

2'. Ectoplasm very narrow, limited by an elastic periplasmic cortex. Capsular wall very thin, closely fitted in some areas to the periplasmic cortex. Two flat myonemes per spicule, totally included under the periplasmic cortex; 10-15 large spherical granules scattered around each spicule just under the periplasmic cortex .... **Acanthoplegmidae**

FAMILY ACANTHOCHIASMIDAE Haeckel, 1862

Long straight spicules of equal length, simply crossed in the cell center. Spicules generally cylindrical, smooth. Endoplasm consisting of two concentric zones; ectoplasm vacuolar. Myonemes very thin. Gametogenesis in oblong or spherical cysts. A single genus, *Acanthochiasma*, with six species: *A. rubescens*, *A. fusiforme*, *A. serrulata*, *A. planum*, *A. quadrangulatum*, *A. hertwigi.* Among these species, only two are common: *A. rubescens* and *A. fusiforme* .
Localization: Atlantic (Sargasso sea, South-equatorial stream), Indian Ocean, Mediterranean, China sea.

Genus *Acanthochiasma* Haeckel, 1862

In this species, the spicules simply cross in the cell center. The cell body is bulky, with numerous sensitive axopodia radiating in all the directions. The endoplasm consists of a central reddish-black sphere surrounded by a clear vesicular sheet including most nuclei. The ectoplasm resembles a transparent jelly striated by anastomosed cytoplasmic strands and axopodial axonemes. As in *A. rubescens*, the periplasmic cortex is more visible when the cell is completely inflated after a long period of calm (Fig. 13). Gametogenesis occurs in an oblong cyst limited by very thin birefringent plaques. Remnants of spicules form a bundle in the cyst center. The myonemes are extremely thin and appear in EM as narrow flat bands.

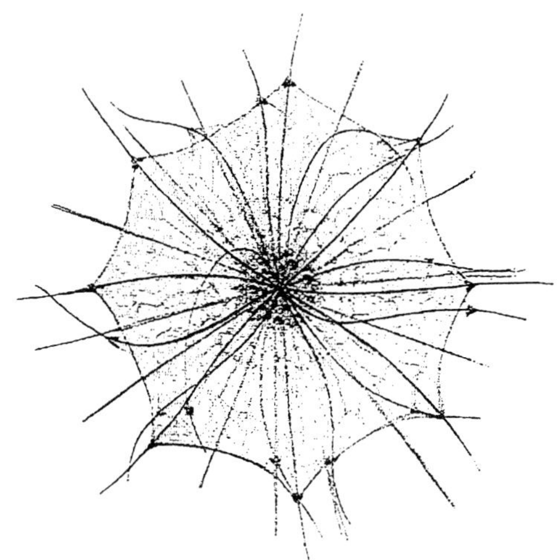

Fig. 19. *Acanthoplegma krohni,* from Schewiakoff (1926).

FAMILY ACANTHOPLEGMIDAE Schewiakoff, 1926

The 10 equatorial spicules simply cross in the cell center. They are long, equal, cylindrical, straight and very thin. The endoplasm contains brown or yellowish inclusions and numerous nuclei. It becomes clearer towards the perisphery. The capsular wall is very thin, and consists of a thin network. The ectoplasm is limited by a periplasmic cortex consisting of a hexagonal elastic lattice. Two flat large triangular myonemes are anchored on each spicule and completely covered by the periplasmic cortex. Large spherical granules are scattered around each spicule just under the periplasmic cortex, so that the ectoplasm seems to bear small umbrellas around the 20 spicules (Fig. 19). These spheres have been confounded with zooxanthellae by Schewiakoff (1926). As in the other members of the family Holacanthidae, gametogenesis takes place in an oval cyst after complete remodeling of the cell. A single genus, *Acanthoplegma*, with only one species, *A. krohni*, present in all the oceans though not abundant.

In 1981, Rechetniak created a new order for the genus *Acanthoplegma*; however, the characters of both the skeleton and the cell body appear in agreement with the definition of the Order Holacanthida so the order does not need modification.

FAMILY ACANTHOCOLLIDAE Schewiakoff, 1926

This new family is characterized by 10 cylindrical spicules which are equal or unequal, twisted at half length and fitting more or less tightly or cemented in the center. The endoplasm includes brown pigments, some nuclei, and organelles. The ectoplasm is clear, transparent with anastomosed strands. The two extracellular networks consist of a thin flexous capsular wall overlaid by a superficial periplasmic cortex. A few thin axopodia emerge from the cell body. The myonemes (2-6 per spicule) are triangular and very short. Gametogenesis occurs within oblong cysts.

Three genera: *Acanthospira*, *Acanthocyrtha*, and *Acanthocolla*., which were placed by Schewiakoff in the Family Acanthoplegmidae; however, the caracteristics of the skeletons are so distinct that it appears necessary to erect this new family for this genus.

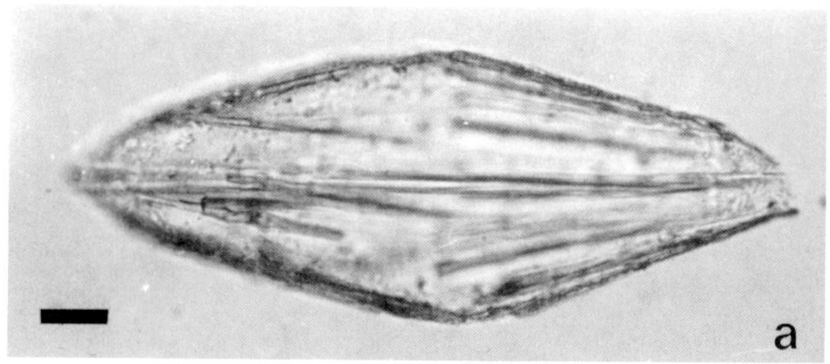

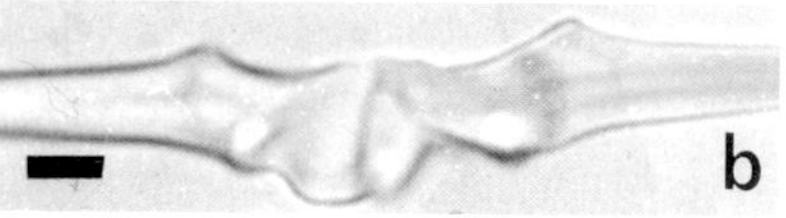

Fig. 20a,b. (a) Cyst of *Acanthocyrtha haeckeli* with a bundle of remnant spicules within the cyst. LM. Scale bar=15 µm. (b) Detail of the central part of a spicule remaining in the cyst cytoplasm after encystment. LM. Scale bar=2.5 µm.

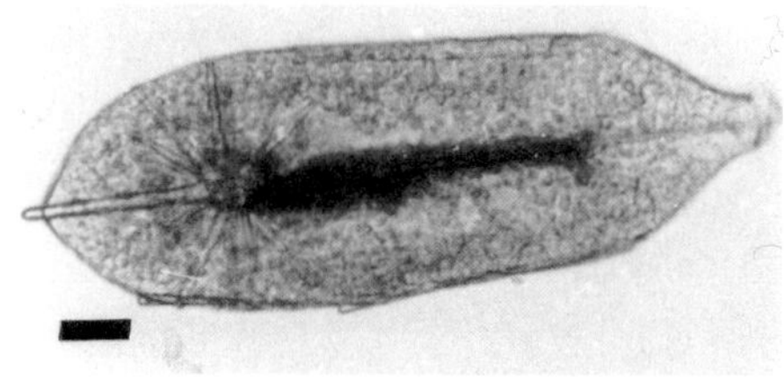

Fig. 21. Cyst of *Acanthocolla cruciata*. Scale bar=20 µm.

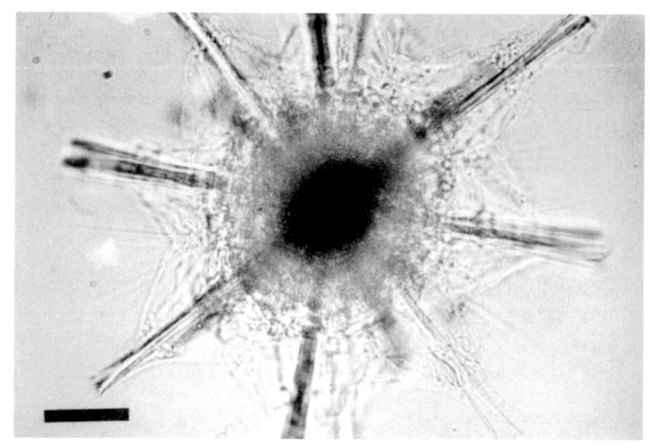

Fig. 22. *Acantholithium dicopum*. LM. Scale bar=50 µm.

Fig. 23. Spicule of *Astrolonche serrata* showing thorns at the proximal part (arrowheads). LM. Scale bar=3 µm.

KEY TO GENERA

1. Spicules of equal length.................................... 2
1'. Spicules of unequal length and similar shape 3

2. Spicules simply twisted crossing in the cell center. Two flat myonemes. Cyst long and cylindrical.................................. ***Acanthospira***
2'. Spicules of approximately equal length and same shape twisted and fitted in the cell center; 5 - 6 flat myonemes per spicule. Cyst biconical......... .... ***Acanthocyrtha***

3. Two equatorial spicules much thicker than the eight others. Central part twisted, lateral part cylindrical, flattened at both distal ends. Cell body flattened, square or rectangular-shaped. Four myonemes. The two diametral spicules still visible in the cyst center .... ***Acanthocolla***

ORDER SYMPHYACANTHIDA
Schewiakoff, 1926

Bases of the spicules indissociable in H_2SO_4, tightly fused into a tiny sphere or a star-like mass. Cell body round or oblong. Endoplasm consisting of two sheets, the inner dense, containing nuclei, brown, yellow, or reddish pigments, the outer with anastomosed cytoplasmic strands and axopodial axonemes. Capsular wall visible in LM, consisting of a thin meshwork if seen by electron microscopy. Ectoplasm transparent. Periplasmic cortex obvious in LM, consisting of an elastic lattice with numerous pores through which axopodia arise. Myonemes flat. Gametogenesis in a **gamontocyst**.

KEY TO FAMILIES

1. Spicules of equal length and same shape 2
1'. Two spicules longer and thicker than the others .... **Amphilithidae**

2. Spicules fitting with each other in a spherical or star-like structure. Capsular wall peripheral. Ectoplasmic cortex made of elastic threads.. **Astrolithidae**
2'. Spicules very thin, long, all equal, smooth or bearing apophyses or latticed plaques connected with each other to form a perforated shell........... .... **Pseudolithidae**

FAMILY AMPHILITHIDAE Haeckel, 1887
emend. Schewiakoff, 1926

The skeleton is made of 20 spicules, which are of different size and shape, two of them being longer and thicker than the 18 others. All of them are joined in the cell center. The cell body is cylindrical or oval, and the endoplasm contains numerous nuclei and zooxanthellae. The capsular wall is thin and peripheral. The ectoplasm is covered with a periplasmic cortex. It is attached to each spicule by 6-24 myonemes.

KEY TO GENERA

1. Two equatorial spicules cylindrical (conventionnally referred to as anterior and posterior), much longer and thinner than the 18 others. The anterior 20 times shorter than the posterior. The other spicules 5-20 times shorter than the posterior. Endoplasm with numerous zooxanthellae....................................... ***Amphilithium***
1'. Two equatorial spicules 1.5-2 times longer than the 18 others, both of about the same length. Endoplasm cylindrical with zooxanthellae and numerous nuclei; 6-8 short myonemes .... ***Amphibelone***

FAMILY ASTROLITHIDAE Haeckel, 1887
emend. Schewiakoff, 1926

The skeleton consists of 20 spicules of equal size and same shape linked at their base into a single homogeneous, either spherical or star-like mass. Endoplasm and ectoplasm separated by a thin peripheral capsular wall. The periplasmic cortex consists of a network of elastic threads between which narrow channels leave the way open to the internal space. These splits often contain electron-dense material which may correspond to defecation vesicles or degraded organelles; 6-12 short, flat, or 16-32 thread-like myonemes.

KEY TO GENERA AND SPECIES

1. Spicules cylindrical, of equal length, sharp-pointed and intimately united at their base, forming a spherical central mass. Endoplasm golden-yellow in the center, granular at the periphery.................................... ***Heliolithium***

1'. Spicules rhombic or cylindrical only on a part of their length.. 2

2. Spicules rhombic:................... ***Acantholithium***

2'. Spicules cylindrical bulbous at their base .... ***Astrolithium***

2''. Spicules cylindrical, with 2-6 rows of sharp thorns at their base (Fig. 23). Numerous axopodia .... ***Astrolonche***

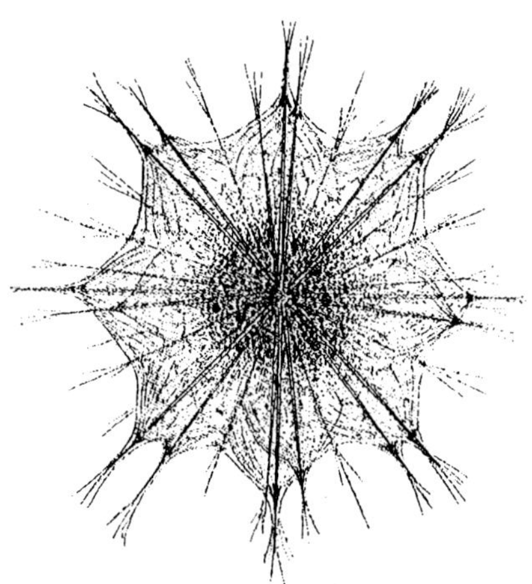

Fig. 24. *Heliolithium aureum*, from Schewiakoff (1926).

Genus ***Heliolithium*** Schewiakoff, 1926

The twenty radial spicules are cylindrical, all of equal size and length. They are fused at their proximal end into a small spherule. The tips of the spicules are sharp-pointed like needles. The bases are closely connected forming a central skeletal sphere. The whole skeleton is included in a vacuole whose membrane is closely fitted to the spicules. The cytoplasm is divided into 2 regions: a central golden-yellow endoplasm, slightly granular at the periphery, containing numerous tiny spherical nuclei, lipid inclusions, mitochondria, and large **extrusomes** (**kinetocysts**). The ectoplasm is clear, alveolar with cytoplasmic strands and axopodial axonemes. The axopodia emerge from holes in the periplasmic cortex. Each passage is localized between neighboring spicules. The capsular wall lies between endo- and ectoplasm near the periphery of the cell. It consists of a fleecy meshwork. The periplasmic cortex is made of a multi-layered lattice of elastic strands. There are 24-32 myonemes per spicule. Species frequently found, but not very numerous.

FAMILY PSEUDOLITHIDAE Schewiakoff, 1926

The twenty spicules are all of equal length and same shape, very thin, with or without **apophyses** or plaques. The endoplasm is yellow-brownish in the center and clearer at the periphery and includes a single large nucleus and **lithosomes**. The ectoplasm is very clear, transparent with thin cytoplasmic strands forming a wide-meshed pseudopodial network. A thin capsular wall limits the endoplasm near the cell center. The periplasmic cortex is stretched between the tips of the spicules. There are numerous axopodia and 16 flat myonemes per spicule.

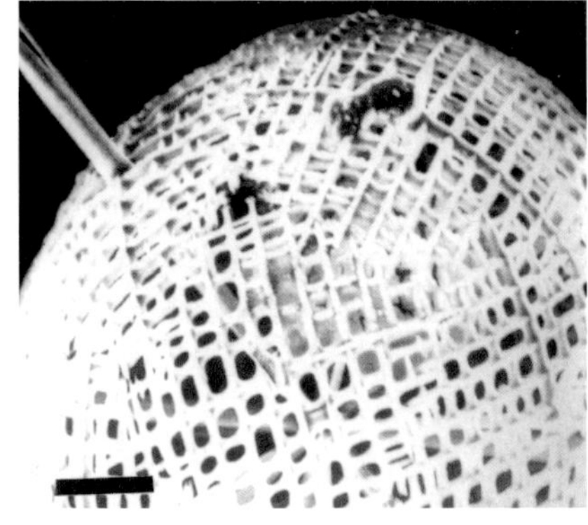

Fig. 25. Cyst wall of *Haliommatidium mülleri*. SEM. Scale bar=25 µm.

KEY TO GENERA

1. Spicules smooth, cylindrical, or flattened; with sharp edges linked into a small sphere in the cell center. Cell body golden yellow or brown............. .... ***Pseudolithium***

1'. Spicules with four apophyses perpendicular to the axis of the spicules in the proximal third of the length..2

2. Apophyses short and truncate .. ***Dicranophora***

2'. Apophyses long, bearing secondary and tertiary apophyses all orthogonal between them. These apophyses form slightly concave grids which join with each other just before gametogenesis

to form a cyst (Fig. 25). Biflagellated gametes are formed within this gamontocyst, then shed through this perforated shell.............................
.... *Haliommatidium*

ORDER CHAUNACANTHIDA
Schewiakoff, 1926

Bases of the spicules oblong resembling grapepips or ninepins. Spicules cruciform in cross section, generally serrate on two opposite edges, smooth on the two others. Spicules included at their base in a contractile matrix. Endoplasm black or reddish in the cell center, clearer at the periphery. Capsular wall very conspicuous in some genera. Ectoplasm transparent. Cortex showing a square pattern of elastic junctions. Myonemes ribbon-shaped, very long and mobile. Morphogenetic remodeling before gametogenesis leading to a special stage called *Litholophus*. Gamontocyst oblong with small adjacent polygonal plaques forming a wall and two kinds of pores, small pores at the angles of the plaques and large apertures at one pole.

KEY TO FAMILIES

1. Twenty spicules of equal length and same shape with their proximal ends conical 2
1'. Spicules of unequal length. Ectoplasm with elastic junctions; 8-12 myonemes per spicule....
.... **Stauraconidae**

2. Ectoplasm homogeneous, clear. Periplasmic cortex without elastic junctions; 4-6 myonemes per spicule.................................... **Conaconidae**
2'. Ectoplasm granular. Periplasmic cortex with elastic junctions; large size; spicules 0.5-2 m in diameter); 6-8 myonemes per spicule
.... **Gigartaconidae**

FAMILY STAURACONIDAE Schewiakoff, 1926

One, two, or four spicules are longer than the others. The cell center is black or reddish. The peripheral endoplasm and the ectoplasm are clear. The ectoplasm shows numerous elastic junctions. The myonemes are very long, mobile ribbons. Encystment is preceded by *Litholophus* stage (Fig. 26).

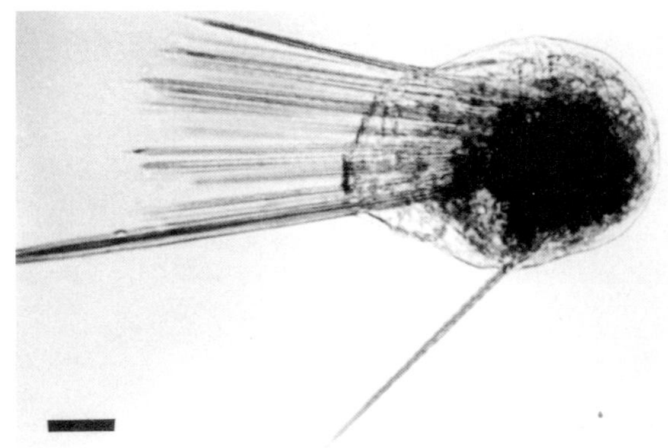

Fig. 26. *Litholophus* stage of *Heteracon mülleri*. LM. Scale bar=12 μm.

KEY TO GENERA

1. One smooth, spatula-shaped spicule longer than the 19 others. Bases of the spicules resembling a ninepin; included in a very contractile central mass. Cell center black. Endoplasm containing the nuclei. Ectoplasm clear, surrounded by a thin periplasmic cortex. Eight ribbon-like and very mobile myonemes. *Litholophus* and cyst (Fig. 26). Size 0.08-0.1 mm. Spicules 0.2-0.4mm.. ***Heteracon***
1'. Two spicules three times longer than the others, all of them serrate along two opposite sharp edges, the other edges smooth. Cell body brown or reddish. Cytoplasm granulous; 6-8 myonemes. .... ***Amphiacon***
1". Four equatorial spicules longer than the 16 others (Fig. 27). Endoplasm granular with nuclei. Separated from the ectoplasm by an obvious capsular wall. Ectoplasm clear, limited by a periplasmic cortex bearing a rectangular lattice of elastic junctions; 6-8 ribbon-like, long and very dynamic myonemes per spicule.
.... ***Stauracon***

Genus ***Stauracon*** Schewiakoff, 1926

This very large species bears 4 diametral spicules much longer than the others (Fig. 27). The endoplasm is black while clear at the periphery. It is limited by a thick capsular wall, which is very obvious in LM. The ectoplasm is transparent, limited by the periplasmic cortex. This bears elastic junctions forming a very conspicuous rectangular pattern around each spicule. The myonemes are flat, long, and very mobile, capable of undulating slowly or contracting very rapidly and efficiently. Gametogenesis occurs in a gamontocyst.

Cyst formation is preceded by a complete remodeling of the shape leading to a *Litholophus*. Contrary to *Heteracon*, which resembles a closed umbrella (Fig. 26), the spicules remain here "half-closed".

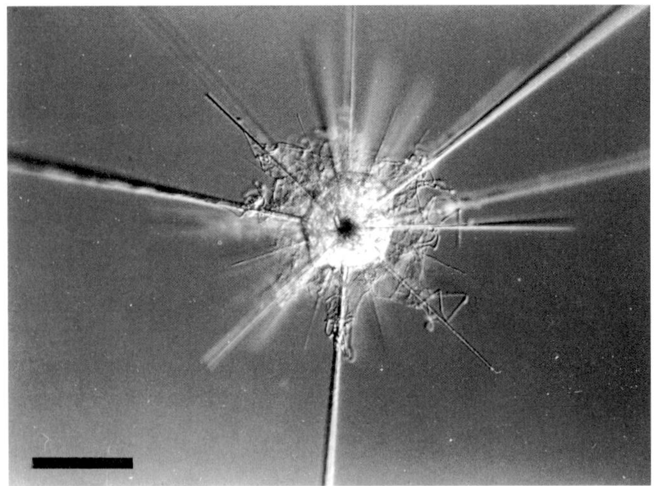

Fig. 27. *Stauracon pallidus*. Note the four equatorial spicules longer than the others. LM. Scale bar=100 µm.

FAMILY CONACONIDAE Schewiakoff, 1926

All the spicules are of equal length (0.1-0.2mm long) with two opposite sharp and serrate edges, the others smooth. The bases of the spicules are conical, quadrangular in cross section, loosely linked in the cell center and embedded in a central mass of contractile material. Cell body spherical. Endoplasm clear containing the nuclei, some zooxanthellae (Dinoflagellata). Tiny, polygonal, and birefringent plaques formed just before encystment. Very thin capsular wall. Ectoplasm transparent. One genus, *Conacon*, with one common species: *C. foliaceus*

FAMILY GIGARTACONIDAE Schewiakoff, 1926

The spicules are long, crisp, of equal size (1-2 mm long). They are flattened and serrate along two opposite edges, smooth on the two other edges. The bases of the spicules resemble grape pips embedded in a contractile matter. The periplasmic cortex shows elastic junctions. One genus, *Gigartacon*, with five species: *G. fragilis*, *G. denticulatus*, *G. mulleri*, *G. abcisus*, *G. cordiformis*, *G. bifidus*.

ORDER ARTHRACANTHIDA
Schewiakoff, 1926

Bases of the spicules pyramidal with 4-6 facets with or without basal extensions juxtaposed or linked so as to form a more or less cohesive system. Endoplasm with numerous nuclei, pigments, inclusions, and zooxanthellae (Haptophyta). Thick capsular wall. Ectoplasm limited by a periplasmic cortex. Myonemes cylindrical, generally numerous. A few axopodia emerging between the spicules. This order is divided into two suborders on the basis of the shape and arrangement of the spicules in their basal region. The endoplasm always contains numerous zooxanthellae. Contrary to the three other Orders of Acantharia, gametogenesis occurs in a gamont which keeps the aspect of the trophont. The whole endoplasm is converted into biflagellated gametes, which are shed across the capsular wall at maturity.

KEY TO SUBORDERS

1. Proximal end of each spicule looking like a 5 - 6-edged pyramid...................... **Sphaenacantha**
1'. Proximal end of each spicule bearing 4 - 6 lateral extensions resembling leaves or blades. Spicules assembled in the cell center into a star-like building with the bases limiting 22 pyramidal cavities ....**Phyllacantha**

SUBORDER SPHAENACANTHA
Schewiakoff, 1926

In this suborder the base of the spicule looks like a small pyramid without lateral extensions on the ridges of the pyramid (Fig. 29).

KEY TO FAMILIES

1. Spicules of equal or unequal length without apophyses. Thin capsular wall. Periplasmic cortex with elastic junctions; 16-40 myonemes per spicule........................... **Acanthometridae**
1'. Spicules bearing lateral hypophyses 2

2. Spicules all of the same length.......................... 3
2'. Two to six spicules longer than the others which bear lateral apophyses or armored plaques....... 4

3. Spicules all of the same shape, bearing a single lattice of concave plaques which are joined with each other forming a single intracytoplasmic

shell. Endoplasm yellow, red, or green. Thin capsular wall. Periplasmic cortex; 6-16 myonemes per spicule.................. **Dorataspidae**

3'. Spicules bearing 2-3 sets of plaques and/or apophyses forming two or three concentric latticed shells; 6-8 myonemes per spicule.......... .... **Phractopeltidae**

4. Two spicules longer and stronger than the others, bearing longitudinal cornets and a shell with two small, oval perforations around each small spicule. Acantharian resembling a butterfly-bow .... **Diploconidae**

4'. Four spicules longer and thicker than the others, bearing on the apical third of their length flat, triangular, latticed plaques. Each plaque consists of 1-5 lateral apophyses cross-linked by 3-5 orthogonal bars delineating a square-based lattice. Species flattened in the plane of the longest spicules. Endoplasm including numerous zooxanthellae. Six-eight short cylindrical myonemes....... **Lithopteridae**

4". Two to six spicules longer than the others, all of them bearing lateral thick armored plaques or apophyses. Cell body spherical or oblong. Numerous axopodia. Capsular wall overlaying the armored shell, ectoplasm around the shell, covered with a periplasmic cortex; 8-12 short myonemes.................................... **Hexalaspidae**

FAMILY ACANTHOMETRIDAE Haeckel, 1887 emend. Schewiakoff, 1926

Skeleton consisting of 20 spicules of the same length and the same shape or of two spicules longer than the others. Cell body polygonal or oblong. Endoplasm, including zooxanthellae, limited by a capsular wall. Ectoplasm clear, limited by a periplasmic cortex. Cylindrical myonemes.

KEY TO GENERA

1. Thin spicules equal or subequal, all of the same shape. Cell body a polygon.......... ***Acanthometra***

1'. Two or four equatorial spicules longer and thicker than the others. 2

2. Two equatorial spicules longer than the others. Cell body lengthened. Endoplasm, including zooxanthellae (Fig. 5), limited by a capsular wall. Ectoplasm clear limited by a periplasmic cortex. Numerous axopodia; 16-24 myonemes per spicule................................... ***Amphilonche***

2'. Four equatorial spicules longer than the others.. ***Tetralonche***

Genus ***Acanthometra*** J. Müller, 1856

The twenty spicules are of equal size and same shape. They are linked in the cell center by the edges of their 5-edged pyramidal bases. The cell body is polygonal with a yellow-green endoplasm including numerous zooxanthellae and nuclei. The endoplasm is limited by a thin capsular wall. The ectoplasm is clear with delicate cytoplasmic strands. The periplasmic cortex consists of 20 polygonal pieces interconnected by elastic junctions; 25-40 thin cylindrical myonemes (Fig. 8).

FAMILY DORATASPIDAE Haeckel, 1887

The twenty spicules are of the same shape and generally the same length. They bear 2-4 cross-shaped ramified processes or 5-6 angular porous plaques which link with one another to form a latticed or armored shell. The conical bases of the spicules are linked or closely associated forming a mass in the cell center. The cell body is spherical with a yellow or red endoplasm, surrounded by a capsular wall. The ectoplasm contains numerous zooxanthellae. The periplasmic cortex covers the ectoplasm and is linked to the spicules by 6-16 myonemes.

Eleven genera, thirty-five species. Five genera are cosmopolitan, the others found in the Pacific Ocean.

KEY TO GENERA

1. Spicules with 2-4 cross-shaped apophyses linked with their neighbors forming a latticed shell... 2

1'. Spicules with 5-6 polygonal, porous, or armored plaques forming a spherical or oblong shell 5

2. Cylindrical spicules bearing two apophyses on their proximal third. These are either unramified or sub-branched as a fork and linked with the apophyses of the neighboring spicules so as to delineate a large meshed shell. At the junction points, 1-2 harpoon-shaped spines pointed towards the exterior. Endoplasm

yellowish. Eight cylindrical myonemes per spicule.. ***Pleuraspis***

2'. Spicules with four opposite processes which can be ramified, forming a latticed plaque. 3

3. Secondary or tertiary apophyses linked with the secondary or tertiary apophyses of the neighboring spicules, forming a small meshed latticed shell. Cylindrical spicules. ***Stauraspis***

3'. Apophyses linked partly with each other, partly with the apophyses of the neighboring spicules forming a latticed shell 4

4. Each spicule bearing perforated plaques with four cross-shaped pores (aspinal pores). These plaques link with the neighboring plaques by processes delineating 8-12 large pores (sutural pores) (Fig. 28, 29a,b). Endoplasm yellowish with 8-16 myoneme per spicule.... ***Lychnaspis***

4'. Each plaque consists of eight aspinal pores surrounded by 1-2 cycles of smaller "coronal pores" and one circle of small sutural pores. Central capsule generally pink; 12-16 myonemes per spicule............................... ***Icosaspis***

5. Shield plaques generally linked with each other along a polygonal line. External surface of the plaques smooth.. 6

5'. Shields without sutural line. External surface of the shell bearing caveolae and a network of combs and perpendicular spines....................... 8

6. In each shield two aspinal pores surrounded by one circle of sutural pores localized along the sutural line. Endoplasm uncolored or red. 6 - 1 2 myonemes per spicule..................... ***Dorataspis*** Six rare species.

6'. In each shield two aspinal pores surrounded by one or several circles of coronal pores............. 7

7. Coronal pores forming circles between aspinal and sutural pores .... ***Coscinaspis***

7'. No sutural pores; junctions between adjacent shields conspicuous. Numerous coronal pores surrounding the aspinal pores......... ***Craniaspis***

8. All the shields identical..................................... 9

8'. Six shields on two equatorial spicules and four polar spicules bearing large aspinal furrows, each with six aspinal pores. The 14 other shields with caveolae, each with two pores. Each aspinal furrow surrounded by 12-16 smaller sutural furrows bearing pores ..... ***Siphonaspis***

9. Each aspinal caveola surrounded by aspinal pores and by 10-12 sutural furrows...................... 10

9' Outside of the sutural furrows coronal furrows all with pores and spines forming combs............. .... ***Aconthaspis***

10. Sutural furrows with an alternation of parts without pores and parts with blind pores. Endoplasm dark-red................ ***Hystrichaspis***

10'. All the sutural furrows lacking pores. Endoplasm colorless; spherical or ellipsoidal; 8 myonemes per spicule...................... ***Dictyaspis***

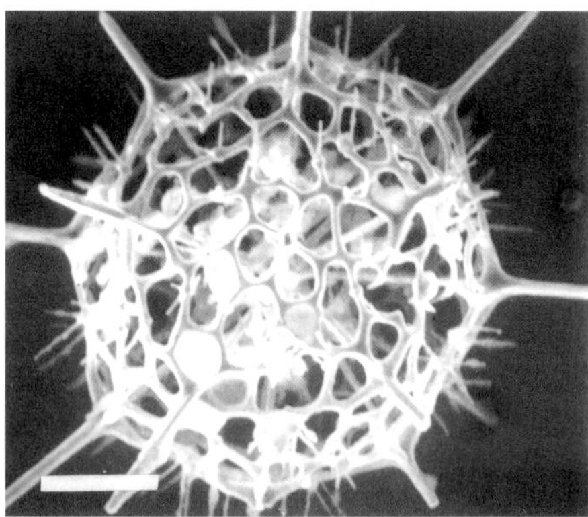

Fig. 28. *Lychnaspis giltschi.* SEM. Scale bar=30 µm.

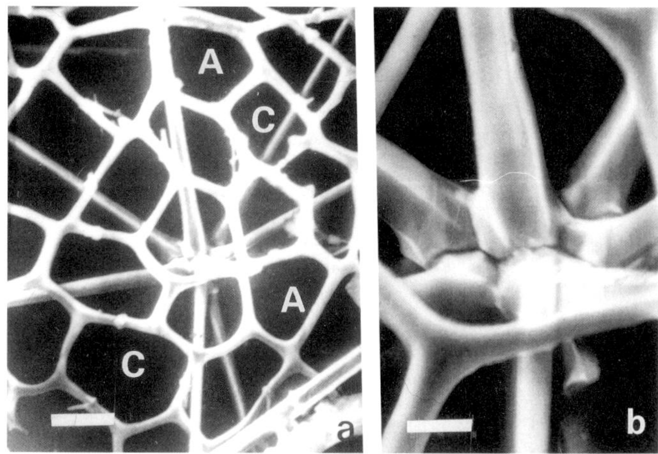

Fig. 29a,b. (a) Details of the shell of *L. giltschi.* A, aspinal pore; C, coronal pores. Scale bar=10 µm. (b) Bases of the spicules of *L. giltschi.* SEM. Scale bar=2 µm

Genus ***Pleuraspis*** Haeckel, 1887

In this cosmopolitan and beautiful species, the shell has large pores delineated by lateral strong cylindrical apophyses extending half the length of the spicules (Fig. 31). In this species, the lateral apophyses fork at their tips. A single or two harpoon-shaped processes extend from this latticed shell. Outside of the shell, the spicules are cylindrical and all of the same length. The endoplasm colored yellow-green by the numerous zooxanthellae is visible through this strong latticed shell and the thin transparent capsular wall. It contains numerous small nuclei and lipid inclusions. The ectoplasm extends as a clear sheet overlying the capsular wall and containing anastomosed strands. The axopodia radiate from the thin ectoplasmic cortex attached to the spicules by 8 myonemes.

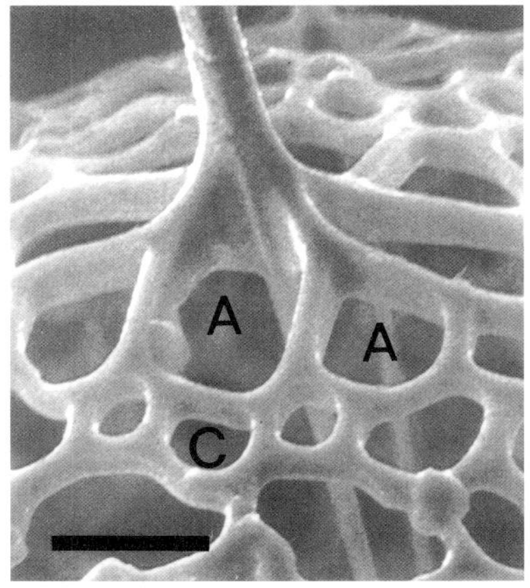

Fig. 30. *Icosapsis elegans*. Aspinal pores (A), coronal pores (C). SEM. Scale bar=15 μm.

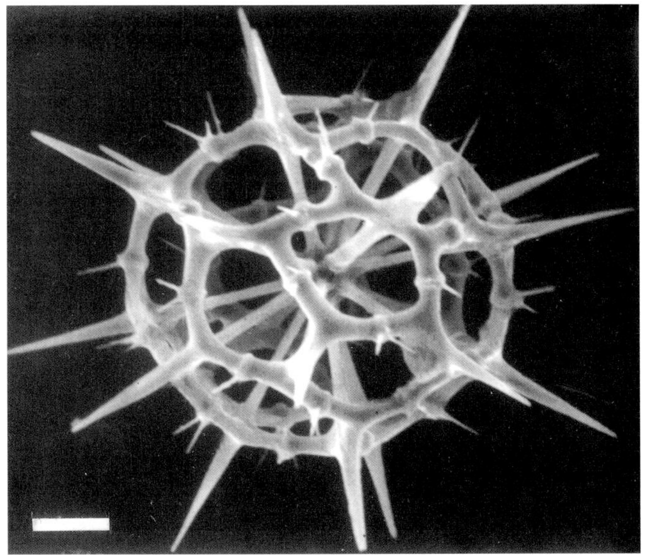

Fig. 31. Skeleton of *Pleuraspis costata*. SEM. Scale bar=25 μm.

FAMILY PHRACTOPELTIDAE Haeckel, 1887

The skeleton is composed of 20 radial spicules joined at their proximal ends, bearing two concentric latticed shells, each formed of 20 latticed plaques. Distally, the spicules can bear branched apophyses which never form a third shell. Endoplasm containing pigments and zooxanthellae. Genus *Phractopelta*. Five species: *P. dorataspis, P. cruciata, P histrix, P. stauropora, P. tessaraspis*. The last species is commonly found in the Sargasso sea.

FAMILY DIPLOCONIDAE Haeckel, 1887

The skeleton consists of 20 radial spicules of different shape and different length, among which two equatorial spicules are longer and stronger than the 18 others. Each of these two main spicules bear a cornet. The spicules pierce a thick ellipsoid armored shell. Around each spicule on the upper part of the shell, there are two oval aspinal pores surrounded by 5-6 sutural pores. The cornets are striated longitudinally and serrate at their edge. The endoplasm contains nuclei, lipid inclusions, and zooxanthellae. The axopods pass across the sutural pores. The periplasmic cortex is suspended to the apex of the small spicules by 6-12 myonemes. The single genus, *Diploconus*, has five rare species: *D. fasces*, *D. cylindrus*, *D. saturnus*, *D. cyathiscus*, *D. hexaphilus*.

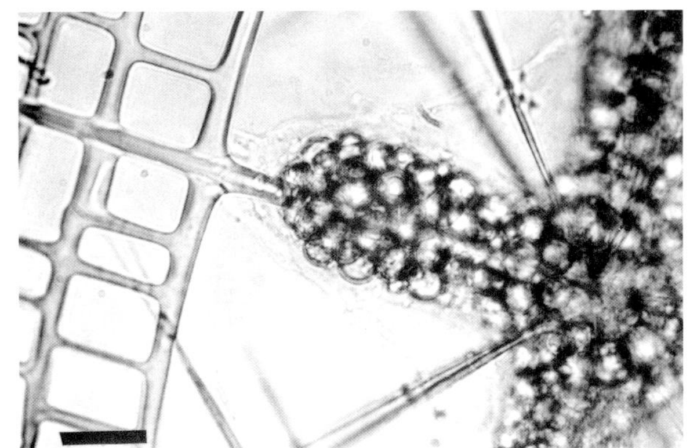

Fig. 32. Sector of *Lithoptera mulleri* showing one main spicule, the endoplasm with zooxanthellae. LM. Scale bar=40 μm.

FAMILY LITHOPTERIDAE Haeckel, 1887
emend. Popofsky, 1904

This family is characterized by four spicules longer and thicker than the others, bearing on the apical third of their length flat, triangular, latticed plaques. ,Each plaque consists of 1-5 lateral apophyses cross-linked by 3-5 orthogonal bars delineating a square-based lattice. The two species of this family are flattened in the plane of the largest spicules. The endoplasm includes numerous zooxanthellae belonging to the Haptophyta. It includes also small nuclei (Fig. 32) and is limited by a thick fibrillar meshwork through which axopodia pass. An active traffic of tiny kinetocysts takes place along the axopodia. The short myonemes are cylindrical and not very mobile. The family has a single genus, *Lithoptera*, with two rare species: *L. fenestrata*, *L. mulleri*.

FAMILY HEXALASPIDAE Haeckel, 1887

Among the 20 spicules, 2-6 have a different shape. They are longer and stronger than the others. The spicules are assembled into a central mass resembling a star. They bear at half-length armored thick plates with numerous spines at their external surface and pores edged with serrate rings. These plaques also bear sheets surrounding all the spicules or only the major ones. The endoplasm is included in the armored shell. The capsular wall lies outside of this shell. There are 8-12 myonemes. the family has three genera: *Coleaspis Hexalaspis*:, and *Hexaconus*. The genus *Coleaspis* has three infrequently observed species, *C. coronata, C. vaginata*, *C. obscura.* There are two very rare species in the genus *Hexalaspis—H. heliodiscus*, *H. sexalata*—and six rare species in the genus *Hexaconus—H. ciliatus, H. serratus*, *H. hexaconus*, *H. unfundibulum*, *H. nivalis*, *H. tripanon.*

SUBORDER PHYLLACANTHA Schewiakoff, 1926

In this suborder the pyramidal base of the spicules bear 4-6 lateral leaves or blades on each ridge (Fig. 33)

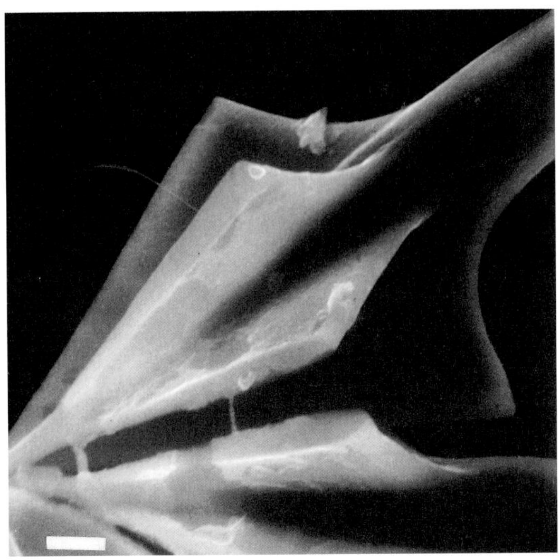

Fig. 33. Basis of the spicule in the Phyllacanthan, *Acanthostaurus purpurascens.* Scale bar=5 μm.

KEY TO FAMILIES

1. Spicules generally cylindrical, without apophyses, all equal and of the same shape or 2 - 4 spicules longer and stronger than the others. Endoplasm spherical, ellipsoidal, or cushion-shaped, limited by a thick capsular wall; 16-80 cylindrical myonemes............ **Phyllostauridae**

1'. Spicules bearing apophyses............................... 2

2. Four apophyses looking like either simple or branched lateral wings or thorns, or bearing one or several secondary orthogonal apophyses which can form a latticed plaque. Endoplasm spherical or polygonal, limited by a thick capsular wall....................................... **Stauracanthidae**

2'. Latticed plaques formed by the slightly curved apophyses joining together into a latticed sphere. Endoplasm spherical.............. **Dictyacanthidae**

FAMILY PHYLLOSTAURIDAE Schewiakoff, 1926

The skeleton consists of 20 spicules without apophyses which are either of the same length and similar shape or dfferent, 2-4 of them being longer and stronger than the others. The base of each spicule bears four large lateral leaf-shaped extensions, cruciform in cross section. The cell body, surrounded by a thick capsular wall, is spherical, oblong, or quadrangular looking like a cushion. It consists of endoplasm containing various inclusions and peripheral ectoplasm. The axopodia arise from the ectoplasmic cortex between the spicules. The ectoplasmic cortex is formed of polygonal pieces connected by elastic junctions.

KEY TO GENERA

1. Spicules of the same length and the same shape. One or two spicules can be longer than the others. .... ***Phyllostaurus***

1'. Two to four spicules longer than the others and of different shape.. 2

2. Two spicules longer than the others and of different shape. Endoplasm oblong; 16-20 myonemes per spicule................ ***Amphistaurus***

2'. Four spicules longer and stronger than the 16 others, which are identical or different 3

3. Four main spicules all identical, the 16 others of different shape, though all similar, or the polar spicules a little shorter and flatter than the tropical spicules. Endoplasm quadrangular, red; 16-30 myonemes................... ***Acanthostaurus***

3'. Main spicules different from one another.......... 4

4. Two main opposite spicules identical two by two. Endoplasm oblong up to rhomboidal.. ***Lonchostaurus***
4'. Rhomboidal acantharian. Only two main spicules along the transversal axis are identical with one another; the two other spicules lying along the longitudinal axis are different. The tropical and polar spicules are also different... ***Zygostaurus***

FAMILY STAURACANTHIDAE Schewiakoff, 1926

The skeleton is made of 20 spicules of equal length and generally same shape, with four longitudinal wings so that the spicule is cruciform in cross section. The wings bear simple or branched apophyses which can link with four or more apophyses of the neighboring spicules, forming a latticed shell. The bases of the spicules bear four lateral blade-shaped extensions so as to delineate deep cavities between the spicules. The endoplasm is spherical to polygonal, surrounded by a thick capsular wall. The periplasmic cortex is thick with elastic junctions. It is linked to each spicule by 40-60 myonemes resembling long cylindrical threads

KEY TO GENERA

1. Spicules with apophyses.................................... 2
1'. Spicules bearing rectangular latticed plaques orthogonal to the direction of the spicule. Each plaque consists of a series of orthogonal apophyses.................................... ***Phatnacantha***

2. Apophyses unbranched..3
2'. Apophyses branched. Secondary and tertiary apophyses at right angles, crossing but never forming latticed plaques. .... ***Stauracantha***

3. Each spicule bearing four crosswise apophyses which can resemble thorns, stumps, or wings..***Xiphacantha***
3'. On each spicule 2-6 groups of apophyses never crosswise .... ***Pristacantha***

Genus ***Xiphacantha*** Haeckel 1887

The skeleton of this large genus (1.3-1.6 mm in diameter) consists of 20 long and sharp radial spicules, all of the same shape and the same size (0.008-0.012 mm wide). At about 0.1 mm from the base, the spicule bears 4 large crosswise lamellar wings measuring 20-70 µm long (Fig. 35, arrows). The edges and the surface of the wings bear fine serrations. The basal ends of the spicules are pyramidal, bearing 6 lateral blades limiting pyramidal cavities. The cell body is polygonal with a central endoplasm surrounded by a thick capsular wall and a large ectoplasm limited by a delicate ectoplasmic cortex. The endoplasm contains numerous nuclei, zooxanthellae, and diverse inclusions. It is opaque yellow-brown or green-brown depending on the relative abundance of pigments and zooxanthellae. The myonemes are long and very numerous cylindrical threads. This species is cosmopolitan and frequent in the spring and the autumn in the Mediterranean.

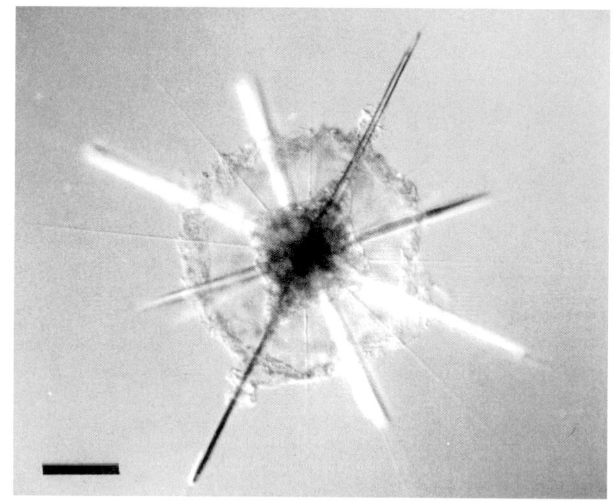

Fig. 34. *Acanthostaurus purpurascens.* LM. Scale bar=100 µm.

Fig. 35. *Xiphacantha alata.* SEM. Scale bar=230 µm.

FAMILY DICTYACANTHIDAE Schewiakoff, 1926

The skeleton consists of 20 spicules of the same length and the same shape, bearing latticed plaques which link with each other so as to form a latticed shell. The cell body is spherical, surrounded by a thick capsular wall, and contains several nuclei, pigments, and zooxanthellae. The ectoplasm is limited by a thick periplasmic cortex showing radial and concentric elastic junctions. This outer layer is suspended to the spicules by means of 50 thread-like myonemes. One genus, *Dictyacantha*, with two rare or very rare species: *D. tetragona* and *D. tabulata* .

LITERATURE CITED

Antia, A. N., Bauerfeind, E., Von Bodunger, B., Zeller, A. 1990. Abundance, encystment and sedimentation of Acantharia during Autumn, 1990 in the East Greenland Sea. *J. Plankt. Res.*, 15:99-114.

Beers, J. R., Reid, F. M. H. & Stewart, G. L. 1975. Microplankton of the North Pacific Central Gyre. Population structure and abundance, June 1973. *Int. Revue ges. Hydrobiol.,* 60:607-638.

Beers, J. R., Reid, F. M. H. & Stewart, G. L. 1982. Seasonal abundance of the microplankton population in the North Pacific central gyre. *Deep-Sea Res.,* 29:227-245.

Beers, J. R. and Stewart, G. L. 1970. The preservations of acantharians in fixed plankton samples. *Limnol. Oceanogr.* 15:825-827.

Bernstein, R. E., Byrne, R. H., Betzer P. R.& Greco A. M. 1992. Morphologies and transformations of celestite in seawater - the role of acantharians in strontium and barium geochemistry. *Geochim. Cosmochim. Acta.*, 56:3273-3279

Bernstein, R. E., Betzer P. R, Feely, R. A., Byrne, R. H., Lamb, M. F. & Michaels, A. F. 1987. Acantharian fluxes and strontium to chlorinity ratios in the North Pacific Ocean. *Science*, 237:1490-1494.

Bishop, J. K. B., Ketten, D. R. & Edmond, J. M. 1978. The chemistry, biology and vertical flux of particulate matter from the upper 400 m of the Cape Basin in the southeast Atlantic Ocean. *Deep-Sea Res.,* 25:1121.

Bishop, J. K. B., Collier, R. W., Ketten, D. R. & Edmond, J. M. 1980. The chemistry, biology and vertical flux of particulate matter from the upper 400 m of the Panama Basin. *Deep-Sea Res.,* 27:615.

Bishop, J. K. B., Edmond, J. M., Ketten, D. R., Bacon, M. P. & Silker, W. B. 1977. The chemistry, biology and vertical flux of particulate matter from the upper 400 m of the equatorial Atlantic Ocean. *Deep-Sea Res.,* 24:511-548.

Borgert, A. 1897. Beiträge zur Kenntnis des in *Sticholonche zanclea* und Acanthometriden-Arten vorkommenden Parasiten. *Z. wiss. Zool.*, 63:141-186.

Brass, G. W. 1980. Trace elements in acantharian skeletons. *Limnol. Oceanogr.*, 25:146-149.

Brass, G. W. & Turekian, K. K. 1974. Strontium distribution in GEOSECS oceanic profiles. *Earth. Planet. Sci. Letters.*, 23:141-148.

Bütschli, O. 1906. Ueber die chemische Natur der Skeletsubstanz der Acantharia. *Zool. Anz.*, 30:784-789.

Cachon, J. & Cachon, M. 1969. Ultrastructures des Amoebophryidae (Péridiniens Duboscquodinida). I. Manifestations des rapports entre l'hôte et le parasite. *Protistologica*, 5:535-547.

Cachon, J. & Cachon, M. 1985. Superclass Actinopoda, Calkins 1902. Class Acantharea. *In*: Lee, J. J., Hutner, S. H. & Bovee, E. C. (eds.), *An Illustrated Guide to the Protozoa.* Society of Protozoologists. Lawrence, Kansas. pp 274-283

Caron, D. A., Michaels, A. F., Swanberg, N. R. & Howse, F. A. 1995. Primary productivity by symbiont-bearing planktonic sarcodines (Acantharia, Radiolaria and Foraminifera) in surface waters near Bermuda. *J. Plankton Res.*

Caron, D. A. & Swanberg, N. R. 1990. The ecology of planktonic sarcodines. *Rev. Aquat. Sci.,* 3:147-180.

Corliss, J. O. 1994. An interim utilitarian ("user-friendly) hierarchical classification and characterization of the protists. *Acta Protozool.,* 33:1-51.

Febvre, J. 1990. Phylum Actinopoda. Class Acantharia. *In*: Margulis, L., Corliss, O., Melkonian, M. & Chapman, D. J. (eds.), *Handbook of Protoctista*. Jones and Bartlett, Boston. pp 363-379

Febvre, J. & Febvre-Chevalier, C. 1979. Ultrustructural study of Zooxanthellae of 3 species of Acantharia with details of their taxonomic position in the Prymnesiales (Prymnesiophyceae, Hibberd, 1976). *J. Mar. Biol. Assoc. U. K.*, 59:215-226.

Febvre, J. & Febvre-Chevalier. C. 1982. Motility processes in Acantharia (Protozoa). I. Cinematographic and cytological study of the myonemes. Evidence for a helix-coiled mechanism of the constituent filaments. *Biol. Cell*, 44:283-304.

Febvre, J. & Febvre-Chevalier, C. 1989a. Motility processes in Acantharia II. A Ca^{2+}-dependent system of contractile 2-4 nm filaments isolated from demembranated myonemes. *Biol. Cell*, 67:243-249.

Febvre, J. & Febvre-Chevalier, C. 1989b. Motility processes in Acantharia III. Calcium regulation of the contraction-relaxation cycles of *in vivo* myonemes. *Biol. Cell*, 67:251-261.

Febvre, J. & Febvre-Chevalier, C. & Sato, H. 1990. A study in polarizing microscopy of a contractile system of nanofilaments: the myoneme of acantharians. *Biol. Cell.*, 69:41-51.

Febvre-Chevalier, C. & Febvre, J. 1993. Structural and physiological basis of axopodial dynamics. *Acta Protozool.*, 32:211-227.

Febvre-Chevalier, C. & Febvre, J. 1994. Buoyancy and swimming in marine planktonic protists. *In*: Maddock, L., Bone, Q. & Rayner, J. L. V. (eds.), *Mechanics and Physiology of Animal Swimming.* Cambridge University Press.

Haeckel, E. 1888. Die Radiolarien. Eine Monographie. 3ter Teil, Die Acantharien oder Actinopyleen-Radiolarian. Berlin.

Hertwig, R. 1879. Der Organismus der Radiolarien. Jena. pp 1-149

Hollande, A. & Cachon-Enjumet, M. 1963. Sur la constitution chimique des spicules d'Acanthaires. *Bull.'Inst. Océanogr. (Monaco),* 60:1-4.

Hollande, A., Cachon, J. & Cachon-Enjumet, M. 1965. Les modalités de l'enkystement présporogénétique chez les Acanthaires. *Protistologica*, 1:91-104.

Lee, J. J., Hutner, S. H. & Bovee, E. C. (eds.) 1985. *An Illustrated Guide to the Protozoa*. Society of Protozoologists. Lawrence, Kansas. 1-628 pp

Margulis, L., Corliss, J.,O., Melkonian, M. & Chapman, D. J. 1990. *Handbook of Protoctista*. Jones and Bartlett, Boston. 1-288.

Margulis, L., McKhann, H. I. & Olendzenski, L. 1993. *Illustrated glossary of Protoctista*. Jones and Bartlett, Boston. 1-288.

Massera Bottazzi, E. & Andreoli, M. G. 1974. Distribution of Acantharia in the North Atlantic. *Arch. Oceanogr. Limnol.,* 18:115-145.

Massera Bottazzi, E. & Andreoli, M. G. 1982a. Distribution of adult and juvenile Acantharia (Protozoa Sarcodina) in the Atlantic Ocean. *J. Plankton Res.,* 4:757-777.

Massera Bottazzi, E. & Andreoli, M. G. 1982b. Distribution of the Acantharia in the Western Sargasso Sea in correspondence with "thermal fronts". *J. Protozool.,* 29:162-169.

Massera-Bottazzi, E. & Vinci, A. 1965. Verifica della legge di Müller e osservazioni sulla struttura mineralogica delle spicole degli Acanthari (Protozoa). *Ateneo Parmense, Acta Naturalia*. 3-12.

Mehlhorn, H. & Rüthmann, A. 1992. *Allgemeine Protozoologie.* G. Fischer Verlag, Jena. 326 pp

Mielk, W. 1906. Untersuchungen an Acanthometriden des Pacifischen Oceans. *Zool. Anz.*, 30:754-763.

Michaels, A. F. 1988a. Vertical distribution and abundance of Acantharia and their symbionts. *Mar. Biol.,* 97:559-569.

Michaels, A. F. 1988b. Acantharia in the carbon and nitrogen cycles of the Pacific Ocean. Ph.D. Dissertation, University of California, Santa Cruz.

Michaels, A. F. 1991. Acantharian abundance and symbiont productivity at the VERTEX seasonal station. *J. Plankton Res.,* 13:399-418.

Michaels, A. F., Caron, D. A., Swanberg, N. R., Howse, F. A. & Michaels, C. M. 1995. Planktonic sarcodines (Acantharia, Radiolaria and Foraminifera) in surface waters near Bermuda: abundance, biomass and vertical flux. *J. Plankton Res.*

Müller, J. 1855. Über Sphaerozoum und Thalassicola. *Monatsberichte der Königlischer Preussischer Akademie der Wissenschaft zu Berlin.* Berlin, 229-253.

Müller, J. 1858. Thalassicolen, Polycysten und Acanthometren des Mittelmeeres. *Abhandlungen der Königlisher Akademie der Wissenschaften zu Berlin.* 1-62.

Müller, J. 1859. Einige neue Polycysten und Acanthometren des Mittelmeers. *Physic abhandlungen der KöniglisherAkademie der Wissenschaften aus der Jahre 1858.* Berlin. 1-62.

Popovsky, A. 1904. Die Acantharia der Plankton Expedition. 1. Acanthometra. Ergebniss der Plankton Expedition der Humboldt-Stiftung. 3:1-158.

Puytorac, P. de, Grain, J. & Mignot, J-P. 1987. *Précis de Protistologie*. Société Nouvelle des Editions Boubée, Paris. 1-551.

Rechetniak, V. V. 1981. Acantharia. *In. OAYHA. CCCP.* Leningrad, EHHTRA CKOE OT E EH E, 3-210.

Schewiakoff, W. 1926. Dia Acantharia des Golfes von Neapel. *In*: :Fauna und Flora des Golfes von Neapel. *Monographie.* Bardi, Roma, Friendlader und Sohn, Berlin. 1-753. Atlas with 46 plates.

Spindler, M & Beyer, K. 1990. Distribution, abundance and diversity of Antarctic Acantharian cysts. *Mar. Micropaleon.*, 15:209-218.

Tregouboff, G. 1953. Classe des Acanthaires. *In*: Grassé, P.-P. (ed.), *Traité de Zoologie*. Paris Masson. pp 271-320

AMOEBAE OF UNCERTAIN AFFINITIES

by
D. J. Patterson, A. G. B. Simpson, and
A. Rogerson

The amoebae of traditional literature (e.g. Levine et al., 1980) are a polyphyletic group of organisms (see Introduction to chapter on ramicristate amoebae). Ultrastructural and molecular evidence (Bhattacharya et al., 1995; Hinkle and Sogin, 1993; Patterson, 1994) has shown that the amoeboid body form has arisen on numerous occasions mostly from the flagellated protists. There are probably no primitively aflagellated amoeboid organisms. The retention of a single taxonomic group (e.g. Sarcodina) for the amoebae is not satisfactory as it suggests that a polyphyletic cluster of organisms has much more than just the body form in common. Many of the organisms previously assigned to the 'Sarcodina' are now being distributed to well-defined lineages (e.g. Acantharea, Polycystinea, Phaeodarea, Actinophryida, Centrohelida, Desmothoracida, Vampyrellidae, nucleariid filose amoebae (Cristidiscoidea), Pelobiontida, and Heterolobosea). Other members (e.g. *Gymnophrys* and *Microcometes*) have been shown to have flagella in the trophic phase of their life cycle and these have been exported to relevant sections of this volume. For other areas, such as the Foraminifera, one or more characters which unambiguously define a monophyletic taxon have yet to emerge. Traditionally, amoebae with shells have been linked either to the Foraminifera within the Granuloreticulosea, or with the shelled lobose or filose amoebae. A number of these taxa have been examined ultrastructurally and found to have different complements of organelles when compared with their taxonomic bed-fellows. Despite the improvements in the taxonomy of amoebae, many inhabit one of the last uncharted taxonomic territories of protozoology, and high-level assignment remains a matter of dispute. The taxa of uncertain affinities, together with some of the smaller yet well circumscribed groups, form the basis of this chapter. There is no evidence that these organisms are closely related—except where specified. The only character which the taxa included here hold in common is that they are unicellular or **syncytial** amoeboid eukaryotes which lack flagella or cilia in the trophic phase of the life cycle.

This chapter seeks to summarize available information on as many nominal genera as possible. Many of these are listed in Patterson (1994). Many genera have unclear identities, and some original descriptions fail to distinguish new taxa from previously described taxa. There are particular problems with the parasitic amoebae with lobose pseudopodia and with the large **reticulate** amoebae such as *Labyrinthomyxa, Myxastrum, Pontomyxa.* We have taken no responsibility for synonymizing these names. Except by reference to characteristics which we do not believe to be taxonomically meaningful, we cannot distinguish between some of these taxa, and the key below reflects this situation.

Many species covered here are described from only a few cells at one stage of the life cycle and are in need of further attention before their status can be confirmed. The genera *Malamoeba* and *Malpighamoeba*—parasites of insects and with unclear inter-relationships—are **ramicristate** (Braun et al., 1988, Harry and Finlayson, 1976; Schwantes and Eichelberg, 1984) and are not included here. The reader is also directed to the entries in this volume on pelobionts, heterolobosea, ramicristate amoebae, Xenophyophorea, Granuloreticulosea, shelled lobose amoebae, shelled filose amoebae, and flagellates of uncertain affinities for other amoeboid organisms.

Key characteristics

The organisms placed in this group are distinguished on the basis of size, locomotive morphology, and presence and nature of a lorica or shell, pseudopodial form, life cycle characterictics, and occasionally on ultrastructural attributes. The most widely used character is the form of the pseudopodia. This character is discussed in the introduction to the ramicristate amoebae. We distinguish those taxa with broad pseudopodia from those with thin pseudopodia. The amoebae with the latter form of pseudopodia are particularly poorly known. Fine pseudopodia may or may not have granules (**extrusomes**); they may or may not taper, may or may not branch and, if branching, may or may not anastomose.

KEY TO GENERA

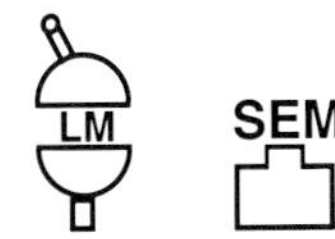

1. Amoebae parasitic... 2
1'. Amoebae ectocommensals or free-living......... 8

2. From brains of primates (see also Chapters on Heterolobosea and Ramicristate Amoebae) .... ***Balamuthia***
2'. From other sites or organisms........................3

3. From cloaca of amphibia.....***Hyalodaktylethra***
3'. From elsewhere... 4

4. No nucleolus (nuclear **endosome**), reported from guts of termites and wood-eating roaches.. ***Endamoeba***
4'. With nucleoli... 5

5. With **lobose** pseudopodia, nucleus with irregular nucleolus. Reported from a variety of invertebrates and vertebrates... ***Endolimax***
5'. With rounded nucleolus................................ 6

6. Amitochondriate, smooth rounded cysts, mostly reported from vertebrates including man .... ***Entamoeba***
6'. Cysts ovoid or irregular, in vertebrates .... ***Iodamoeba***
6''. From other sites... 7

7. Most of the following genera have been reported on one or a few occasions, and are listed by host.

From chaetognaths, eruptive.......... ***Janickina***
From *Daphnia* haemocoel............ ***Pansporella***
From *Branchipus* (Crustacea) ***Branchipicola***
From sponges.............................. ***Topsentella***
From iguanas.......................... ***Martineziella***
From lizard guts...................... ***Hartmannina***
From rat flea............................ ***Malpighiella***
From gnat guts............................... ***Dobellina***
From *Hydrophilus*, binucleated amoebae with cysts... ***Liegeoisia***
From orthopteran and hymenopteran malpighian tubules......................... ***Malamoeba***
.... ***Malpighamoeba***

Other parasitic amoebae may be encountered in the chapters on ramicristate amoebae and on the Heterolobosea.

8. Amoebae enclosed within a test....................... 9
8'. Naked amoebae or amoebae with adhering material 17

9. Single aperture giving rise to pseudopodia 11
9'. Pseudopodia arising from more than one aperture.. 10

10. Pseudopodia arising in two tufts................ .... ***Diplophrys***
10'. Pseudopodia arising from a number of sites .... ***Microcometes***

11. Amoebae under 25 µm................................. 12
11'. Amoebae over 25 µm................................. 15

12. Aperture with neck and basal septum......... 13
12'. Aperture uncollared.................................... 14

13. Aperture with septum with central pore......... .... ***Belaria***
13'. Aperture with septum having lateral opening .... ***Microgromia***

14. Lorica flask-shaped .... ***Apogromia***
14'. Lorica irregular.. ....***Kibisidytes/Heterogromia***

15. Test with aperture extending internally as a tube....................................... ***Lagenidiopsis***
15'. Test without tube.. 16

16. Test opaque brown, egg-shaped; may be lobed, pseudopodia not reticulate.................. ***Gromia***
16'. Test composed of fine particles; pseudopodia may form reticulum................. ***Pleurophrys***

17. With **sorocarps** (see also chapters on acrasids, dictyostelids, myxomycetes) 18
17'. Without sorocarps..................................... 20

17. With **sorocarps** (see also chapters on acrasids, dictyostelids, myxomycetes) 18
17'. Without sorocarps...................................... 20

18. Sorocarp shaped like a volcano...... ***Fonticula***
18'. Sorocarp branched...................................... 19

19. Sorocarps delicate................. ***Copromyxella***
19'. Sorocarps not delicate..................***Copromyxa***

20. With fine or radiating pseudopodia............. 22
20'. With broad pseudopodia.............................. 21

21. Broad pseudopodial front is granular.............. .... ***Hyalodiscus***
21'. Pseudopodia not granular........... ***Stygamoeba***

(Other amoebae with broad pseudopodia are included in chapters on ramicristate amoebae, Heterolobosea, and pelobionts)

22. Pseudopodia unbranched................................ 23
22'. Pseudopodia branched or reticulate............ 33

23. With adherent material................................ 24
23'. Without adherent material.......................... 30

24. Body with spines.. 25
24'. Body without spines.................................. 26

25. Spines organic, with broad abutting bases, no further adhering material...... ***Belonocystis***
25'. Spines siliceous, with spines and scales .... ***Rabdiophrys***

26. Cell surrounded by mucus............. ***Nuclearia***
26'. Cell surrounded by adherent particles....... 27

27. Adhering elements are similar.................... 28
27'. Adhering material is irregular................... 29

28. Adherent material in the form of hollow spherical or ovoidal **perles**.................... .... ***Pompholyxophrys***
28'. Adherent material in the form of flattened plates....................................... ***Pinaciophora***
28''. Adhering material concave scales................. .... ***Clathrella***

29. Pseudopodia branched................ ***Elaeorhanis***
29'. Pseudopodia not branched............. ***Lithocolla***

30. Pseudopodia arise from opposite poles as two tufts .... ***Diplophrys***
30'. Multiple appendages arising from all over the cell surface... 34

31. Medium-sized cells with long appendages which may move slowly................ ***Artodiscus***
31'. Arms not motile... 32

32. Small cells, with short stiff arms.***Ministeria***
32'. Pseudopodia emerging in numerous tufts......... .... ***Actinocoma***

33. Small cells with a small number of fine branched pseudopodia ....***Gymnophrys***
33'. Medium to large amoebae with pseudopodia which may branch and fuse.......................... 34

34. Amoebae which feed on algae and other organisms by cutting holes into surrounding walls; usually colored yellow/orange.......... 35
34'. Other.. 39

35. With irregular cysts........................ ***Gobiella***
35'. With rounded cysts...................................... 36

36. Amoebae with one or few nuclei and granulated or warty surface; pseudopodia emerging from the edge of a lamellopodial structure.................................... ***Hyalodiscus***
36'. Amoebae without granulated surface and usually multinucleated................................. 37

37. Branching body form, typically from soils, colorless or lightly colored............ ***Arachnula***
37'. Body form usually not branching, often orange or red, typically from freshwater habitats and consuming algae........................ 38

38. Cysts formed external to algal cells.............. .... ***Vampyrella***
38'. Cysts formed inside algal cells which have been attacked; known to consume cytoplasm of *Oedogonium*................................. ***Lateromyxa***

39. Consumes nematodes.................. ***Leptophrys***
39'. Consumes other organisms......................... 40

40. Amoebae without extensive plasmodial appearance.. 41
40'. Amoebae able to develop into **plasmodia** or meroplasmodia (cell masses interconnected by strands of cytoplasm)...................................46

41. With flagellated stage in life cycle................... .... ***Protomonas***
41'. Without flagellated stage............................ 42

42. Rounded bodies... 43
42'. Flattened cell bodies................................... 45

43. Cells may occur in a loose cluster .... ***Myxodictyum***
43'. Cells not reported to form clusters.......... 44

44. Typically uninucleated .... ***Vampyrellidium***
44'. Two to four large nuclei; eats algae.............. .... ***Asterocaelum***

45. Pseudopodia do not appear stiffened............. .... **Arachnula**
45'. Pseudopodia apparently stiffened internally .... ***Biomyxa/Penardia***

46. Rounded body mass with radiating pseudopodia .... ***Myxodictyum/Dictyomyxa***
46'. Body mass more extended........................... 47

Note: Some large amoebae such as *Stereomyxa* and *Corallomyxa* have been shown to have ultrastructural features resembling *Acanthamoeba* and have been included in the chapter on ramicristate amoebae.

47. With lidded spores .... ***Cichkovia***
47'. Without lidded spores............................... 48

48. Associated with ascidia................ ***Pontomyxa***
48'. Not associated with ascidia......................... 49

49. Organisms **meroplasmodial** with cytoplasm in discrete masses interconnected by a system of pseudopodia or cytoplasmic strands 50
49'. Organisms plasmodial—i.e. with a single usually irregular mass of cytoplasm from which pseudopodia emerge.............................. 58

50. Interconnecting pseudopodia both thick and thin.. ***Synamoeba***
50'. Interconnecting pseudopodia generally fine.51

51. With a photosynthetic compartment .... ***Chlorarachnion***
51'. No photosynthetic compartment................. 52

52. Freshwater.......................... ***Leukarachnion***
52'. Marine.. 53

53. Without flagellated stage in life cycle........ 54
53'. With flagellated stage in life cycle.............. 56

54. Not known to form plasmodia (i. e. occurs only as isolated amoebae or as meroplasmodia)......................... ***Gymnophrydium***
54'. Plasmodia may be formed........................... 55

55. With surrounding layer of sand particles, sponge spicules, etc....................... ***Urbanella***
55'. Not.. ***Cinetidomyxa***

56. With delicate lorica................ ***Leucodictyon***
56'. Without lorica... 57

57. Not known to form plasmodia.***Reticulamoeba***
57'. Forms plasmodia................... ***Thalassomyxa***

58. Described as white............. ***Megamoebamyxa***
58' Color yellow/orange or not defined. The following taxa are not easily distinguished:
Terrestrial............................ ***Theratromyxa***
Freshwater.............. ***Enteromyxa, Penardia***
Marine.................... ***Aletium, Cinetidomyxa, ... Protogenes, Protomyxa, Rhizoplasma***

Genus *Actinocoma* Penard, 1903

Uninucleate amoeba, branching but non-anastomosing pseudopodia radiating from symmetrical body mass, initially arising as broad trunks (Fig. 1). Body mass about 20 µm wide, pseudopodia with granules. Freshwater, rare. Ref. Penard (1904).

Genus ***Aletium*** Trinchese, 1881

Large (up to 5 mm) marine plasmodial organism usually with bipolar appearance, associated with green algae (Fig. 2). Golden. One species. Ref. Schepotieff (1912a).
TYPE SPECIES: *Aletium pyriforme* Trichese, 1881

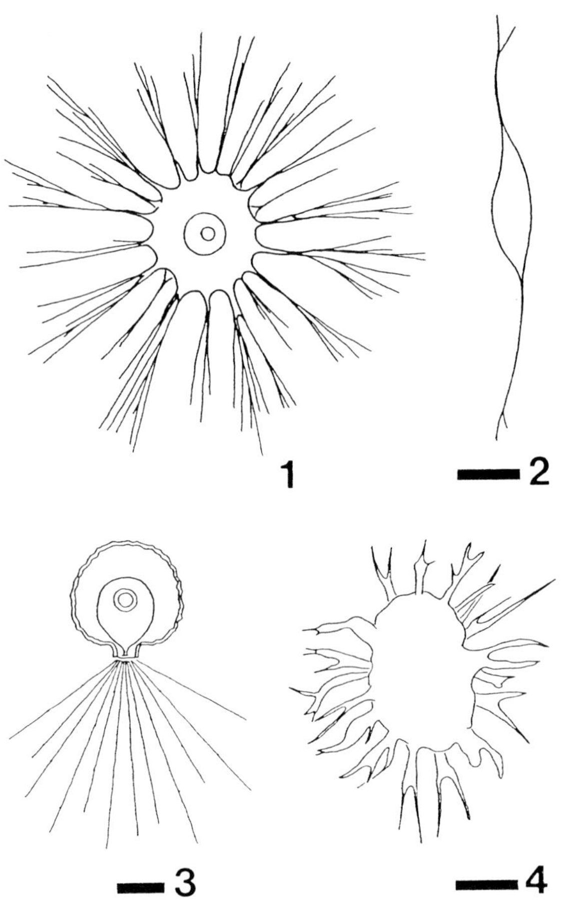

Fig. 1. *Actinocoma ramulosa* after Penard, 1903. Fig. 2. *Aletium pyriforme* after Schepotieff, 1912. Scale bar=1 mm. Fig. 3. *Apogromia pagei* after Ertl, 1984. Scale bar=20 µm. Fig. 4. *Asterocaelum algophilum* after Page, 1991. Scale bar=20 µm.

Genus ***Apogromia*** de Saedeleer, 1934

Species in a tiny flask-shaped thin organic test, with single terminal un-collared aperture giving rise to fine branched granular and occasionally anastomosing pseudopodia (Fig. 3). Several species, mostly freshwater. We are unable to distinguish the genus from *Paraleiberkuhni* de Saedeleer or *Heterogromia* de Saedeleer. Ref. de Saedeleer (1934); Ertl (1984).
TYPE SPECIES: *Apogromia mucicola* (Archer, 1877)

Genus ***Artodiscus*** Penard, 1890

See chapter on Flagellates of Uncertain Affinities.

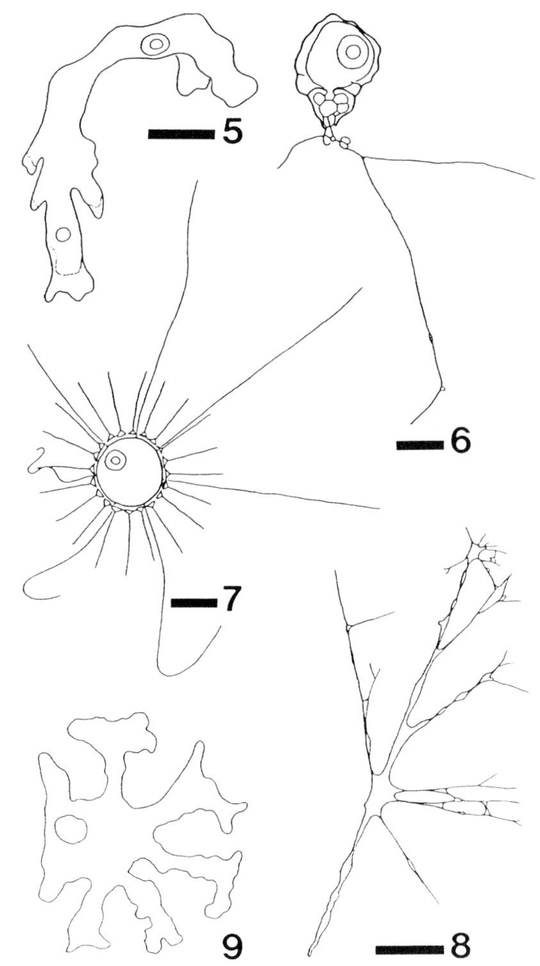

Fig. 5. *Balamuthia mandrillaris* after Visvesvara et al., 1993. Scale bar=5 µm. Fig. 6. *Belaria bicorpor* after de Saedeleer, 1934. Scale bar=5 µm. Fig. 7. *Belonocystis tubistella* after Rainer, 1968. Scale bar=10 µm. Fig. 8. *Biomyxa vagans* after Leidy, 1879. Scale bar=10 µm. Fig. 9. *Branchipocola nucleomira* after Chatton, 1953.

Genus ***Asterocaelum*** Canter, 1973

Genus of medium sized amoebae (usually under 50 µm, but reportedly up to 300 µm) described to date from freshwater. Body usually compact but sometimes extending, pseudopodia radiating in most directions, but tapering from base to tip and without visible extrusomes (Fig. 4). Ingest algae whole, and may form spined cysts after food is taken up. From freshwater, Europe (Lough Neagh and Lough Kinross) and North America. Ref. Canter (1973, 1980).

Genus ***Balamuthia***
Visvesvara, Schuster & Martinez, 1993

Locomotive amoebae 12 to 60 µm in length, moving with leading lobose pseudopodium or by spider-like activity of the many radiating unbranched pseudopodia. Usually uninucleate with a vesicular nucleus about 5 µm diameter (Fig. 5). Forms cysts. Mature cysts with tripartite wall consisting of an outer loose ectocyst, an inner endocyst and a middle mesocyst. Pathogenic amoeba that causes amoebic encephalitis in animals. Isolated from the brain. Ref. Visvesvara et al. (1993).
TYPE SPECIES: *Balamuthia mandrillaris* Visvesvara, Schuster & Martinez, 1993
TYPE HOST: mandrill baboon

Genus ***Belaria*** de Saedeleer, 1934

Amoeba within a test and measuring 10-20 µm linear dimension, with conical neck having a basal septum with central pore, one terminal aperture from neck from which relatively long, granular thin pseudopodia arise (Fig. 6). Consumes bacteria and reported (once only?) from freshwater. Has similarities to *Apogromia* and *Microgromia* to which it may be related. Ref. de Saedeleer (1934).
TYPE SPECIES: *Belaria bicorpor* de Saedeleer, 1934

Genus ***Belonocystis*** Rainer, 1969

Amoeba with thin pseudopodia emerging from a **case** (periplast) comprised of spines with a broad open filamentous distal region (Fig. 7). Body measures approximately 10-25 µm. One species reported to date, from freshwater. Ref. Rainer (1968).
TYPE SPECIES: *Belonocystis tubistella* Rainer, 1968

Genus ***Biomyxa*** Leidy, 1879

Amoeboid organism with stiffened and branching pseudopodia (Fig. 8). Cytoplasm most usually forming a single large mass. Pseudopodia with granules, extending and moving actively. Up to 500 µm in size. Two species, one originally described from freshwater the other coprophilic, but also reported from marine sites. Some reports are of organisms similar to other amoebae (e.g. *Arachnula, Gymnophrydium*) and the identity of this genus needs clarification. Held by some to be related to the Granuloreticulosea. Refs. Leidy (1879); Anderson and Hoeffler (1979).
TYPE SPECIES: *Biomyxa vagans* Leidy, 1879.

Genus ***Branchipocola*** Breindl, 1925

Star-shaped organism (up to 150 µm) parasitic in the intestine of *Branchipus*. Encystment accompanies death of host, up to 8 individuals may be contained, each within an envelope, with each cyst (Fig. 9). Possibly related to *Pansporella*. Refs. Chatton, (1953).

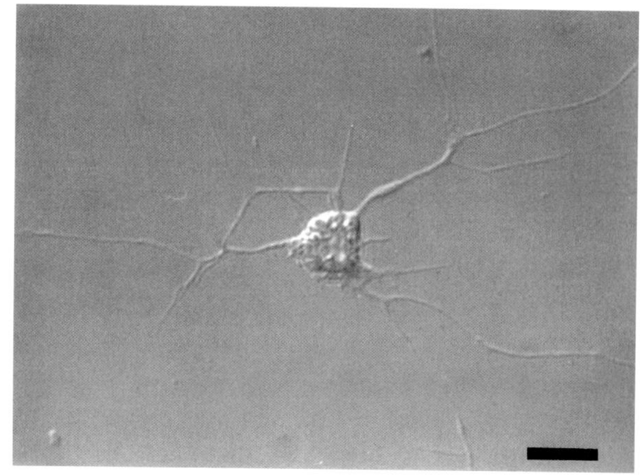

Fig. 10. *Chlorarachnion reptans* courtesy of Dr G. McFadden. Scale bar=10 µm.

Genus ***Chlorarachnion*** Geitler, 1930

Large reticulate amoeboid organism with uni-flagellated dispersal stage (Fig. 10). The vegetative stage is comprised of nucleated regions (amoebae) up to 20 µm in diameter, interconnected by thinner pseudopodia. With green (chlorophylls *a* and *b*) photosynthetic compartments—endosymbiotic eukaryotic algae.

Marine, originally described from the Canary islands. Molecular studies suggest affinities with testate amoebae. One described species. Refs. Geitler (1930); Hibberd and Norris (1984).
TYPE SPECIES: *Chlorarachnion reptans* Geitler, 1930

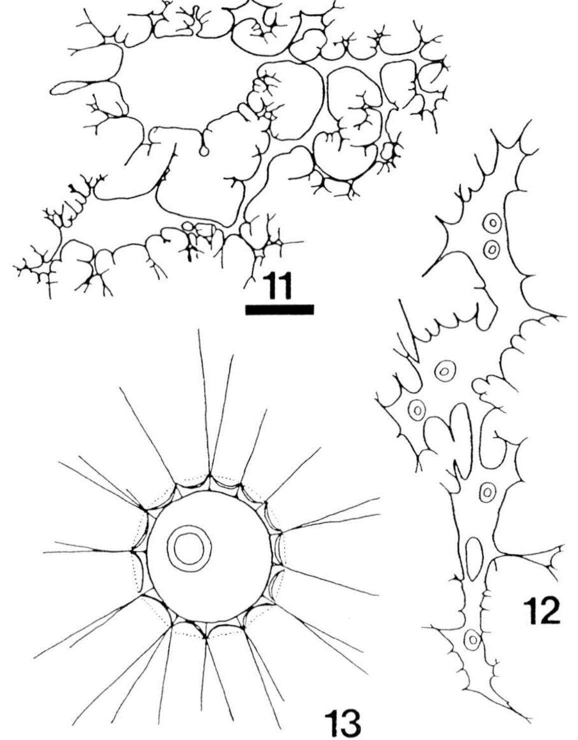

Fig.11. *Chichkovia reticulata* after Valkanov, 1931. Scale bar=20 µm. Fig. 12. *Cinetidomyxa nucleoflagellata* after Trégouboff, 1953. Fig. 13 *Clathrella foreli* after Penard, 1904.

Cichkovia Valkanov, 1931

Large reticulate amoeba with local condensations of cytoplasm Fig. 11); with unusually lidded or capped spores. Reported on one occasion from Northern Europe. Ref. Valkanov (1931).

Cinetidomyxa Chatton & Lwoff, 1924

Plasmodial amoeba from marine habitats forming a network of branching and anatomosing pseudopodial strands. With variations in life cycle, including a compact form, cysts, and small distributive cells (Fig. 12). Nuclei with associated cytoskeletal (= flagellar?) structures. Superficial similarity with large centramoebid amoebae such as *Stereomyxa*, as well as with other large marine amoebae such as *Leukarachnion.* This area is in need of further attention. Two species described to date. Ref. Cachon and Cachon-Enjume (1964).
TYPE SPECIES: Cinetidomyxa nucleoflagellata Chatton & Lwoff, 1924

Clathrella Penard, 1903

Medium-sized (about 50 µm) amoeboid organism living in a tight fitting, unstalked, siliceous lorica from which branched and granulated pseudopodia arise (Fig. 13). With one nucleus. Reported from a freshwater site in Northern Europe, consuming diatoms and other detritus. Ref. Penard (1903).

FAMILY COPROMYXIDAE Olive & Stoianovitch in Olive, 1975

Cellular slime molds, amoebae described as uninucleate and lobose, with broad **hyaloplasmic extensions** anteriorly and possibly with filose extensions posteriorly. Cells aggregate to form a non-streaming (=non-moving?) structure prior to forming cysts (=spores). The aggregate structure (**sorocarp**) may be branched and erect up to several millimeters long. Single amoebae emerge from cysts. Mitochondria with tubular cristae. Normally classified with the acrasid slime molds but from which they differ (inter alia) by lack of streaming **pseudoplasmodium**, sorocarp formation, and mitochondrial organization. Ref. Blanton (1989).

Genus ***Copromyxa*** Zopf, 1885

Copromyxid with characters of the family, sorocarps tree-like (Figs. 14,15). Probably monospecific; described from Europe and America from cow dung. Ref. Spiegel and Olive (1978).
TYPE SPECIES: *Copromyxa protea* Zopf, 1885

Genus ***Copromyxella***
Raper, Worley & Kurzynski, 1978

Copromyxid said to differ from *Copromyxa* by delicate sorocarps, smallish amoebae frequently without contractile vacuoles, smaller and generally globose sorocysts with thin walls, and grown with relative ease on dung-free substrates Figs. 17,18). With several species. Ref. Raper et al. (1978)

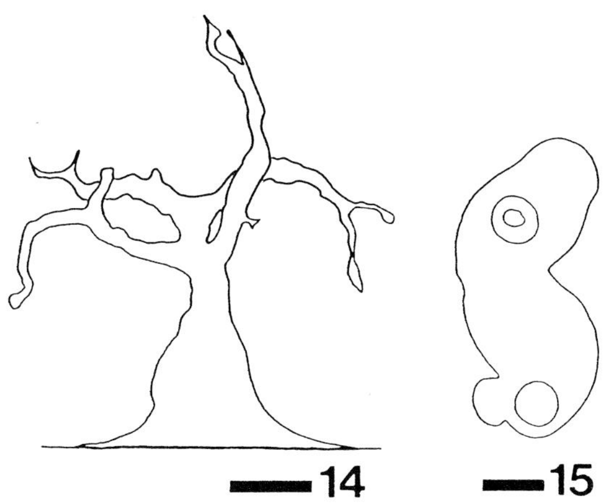

Figs. 14,15. *Copromyxa arborescens* after Nesom and Olive (1972). 14. Sorocarp. Scale bar=200 µm. 15. Amoeboid form. Scale bar=5 µm.

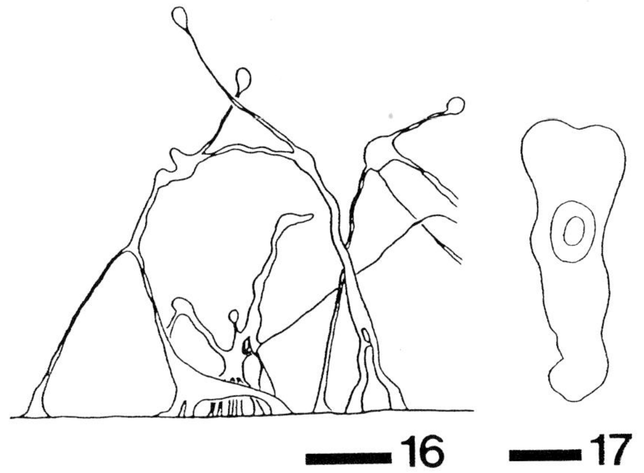

Figs. 16,17. *Copromyxella silvatica* after Raper et al. (1971). 16. Sorocarp. Scale bar=200 µm. 17. Amoeboid form. Scale bar=5 µm.

Genus ***Dictyomyxa*** Monticelli, 1897

Amoeboid organism, with massive central body up to about 1 mm in size, from which extend fine pseudopodia (Fig. 18). Body orange, pseudopodia colorless. Marine (Mediterranean). Monospecific. Ref. Rhumbler (1904).
TYPE SPECIES: *Dictyomyxa kaiseri* Monticelli, 1897

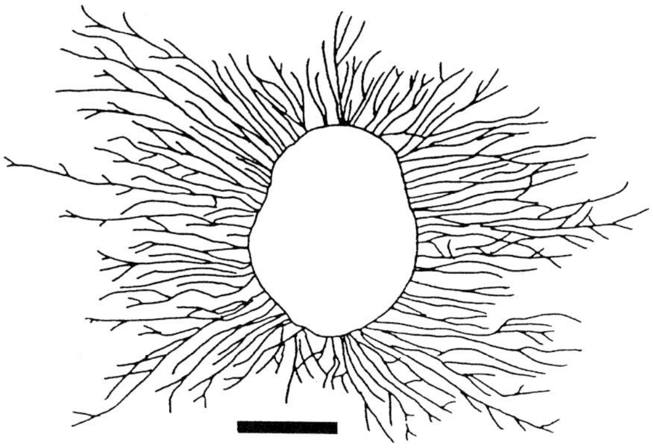

Fig. 18. *Dictyomyxa kaiseri* after Rhumbler (1903). Scale bar=1 mm.

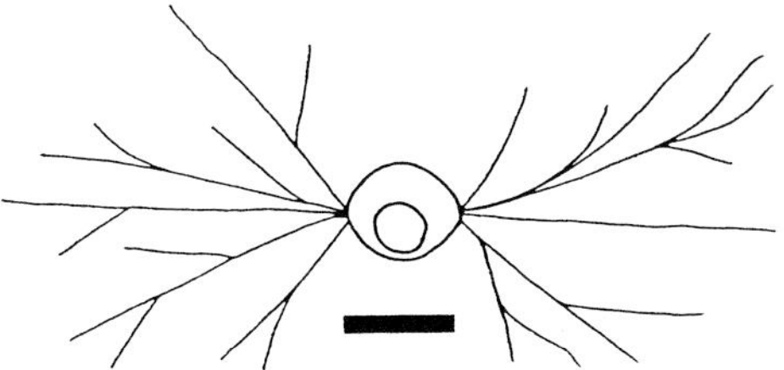

Fig. 19. *Diplophrys archeri* after Vørs (1992). Scale bar=5 µm.

Genus ***Diplophrys*** Barker, 1867

Small (under 10 µm), round-bodied organisms with 2 tufts of fine pseudopodia emerging from either pole (Fig. 19). With one large orange inclusion (lipid?). May form aggregates. Ultrastructural studies reveal surface coating of rounded scales similar to those of thraustochytrids and thereby affinities with other stramenopiles. Several nominal species, but probably the same, from freshwater and marine habitats. Ref. Patterson (1989).

Genus ***Dobellina*** Bishop & Tate, 1939

Parasitic amoeba, small to medium (3.5 to 25 µm), uninucleate to multinucleate, reported from the intestines of gnats, producing cysts with 2 or more nuclei (Figs. 20,21). Ref. Kudo (1971).

Genus ***Elaeorhanis*** Greeff, 1873

Amoeboid organism measuring up to 20 µm in diameter with filose pseudopodia, sometimes branching, but neither stiffened nor with **extrusomes** (Fig. 22). Cytoplasm enclosed within an aggregation of small pieces of adhering detritus—collectively with a reported diameter up to 59 µm. Frequently with orange droplet. Often previously placed with heliozoa. Ref. Rainer (1968).

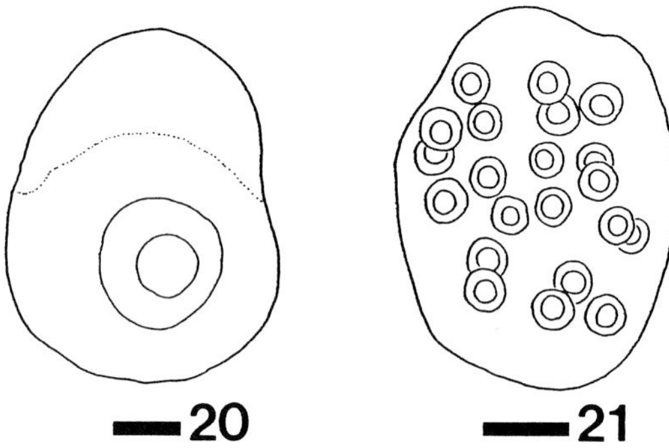

Figs. 20,21. *Dobellina mesnili* after Kudo (1971). 20. Uninucleate stage. Scale bar=2 µm. 21. Multinucleate stage. Scale bar=5 µm.

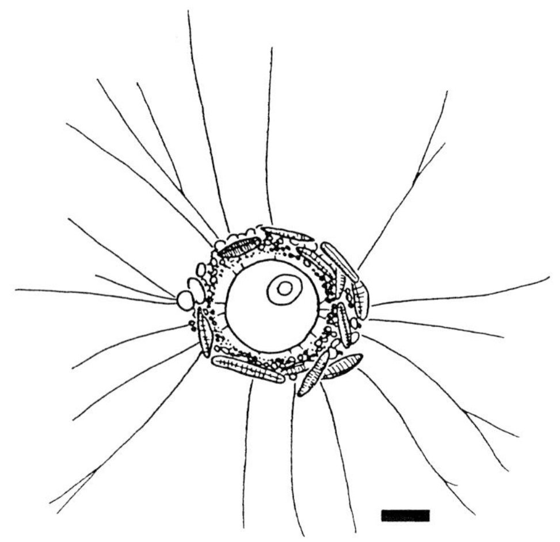

Fig. 22. *Elaeorhanis cincta* after Rainer (1968). Scale bar=10 µm.

Genus ***Endamoeba*** Leidy, 1879

Locomotive cell usually between 10 and 165 µm with broad, eruptive lobate pseudopodia. Ectoplasm clear, endoplasm granular with internal ridges (Fig. 23). Nucleus oval or spherical, with a granular periphery and a clear center. No nulceolus (endosome). Cytoplasm without crystals or contractile vacuoles. Cysts spherical with a thick membrane and often 60 or more nuclei. Despite early nomenclatural confusion, the genus is now considered to be restricted to intestines of cockroaches, termites, and wood-eating roaches. Ref. Griffin (1978).

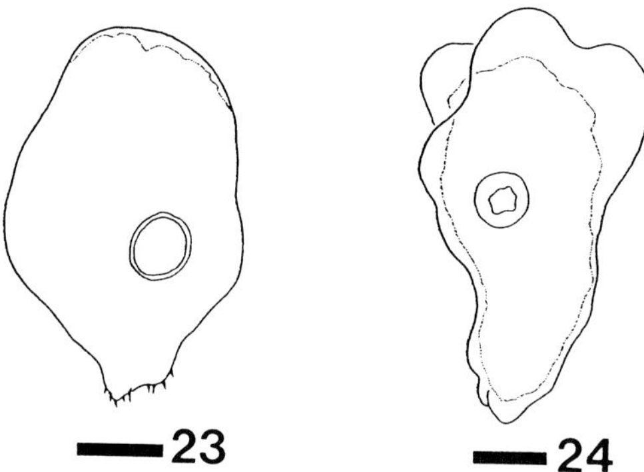

Fig. 23. *Endamoeba blattae* after Kudo (1971). Scale bar=20 µm. Fig. 24. *Endolimax nana* after Kudo (1971). Scale bar=5 µm.

Genus ***Endolimax*** Kuenen & Swellengrebel, 1913

Locomotive cell elongate when in active monopodial movement (Fig. 24). Pseudopodia clear and eruptive. Uroid temporary and bulbous. Clear anterior hyaline region with a granular cytoplasm without contractile vacuoles or crystals. Nucleus spherical with a few chromatin granules and a large irregularly shaped endosome. Cysts round or oval, with 1 to 8 nuclei with **glycogen bodies** in immature cysts. Habitat: Commensal, intestine of man, apes, guinea-pigs, cockroaches, turtles, frogs, and fowls. Ref. Levine (1973).

FAMILY ENTAMOEBIDAE Chatton, 1925

Amitochondriate parasitic amoebae with pseudopodia as clear semi-eruptive bulges from anterior of cell. Central nucleolus usually less than half the diameter of the nucleus. The ability to form cysts is very common.

Genus ***Entamoeba***
Cassigrandi & Barbagallo, 1895

Pseudopodia as semi-eruptive anterior bulges with clear hyaline caps. Endoplasm granular with

vesicles but without crystals or contractile vacuole. Nucleus ring-like, with peripheral granules and a central endosome. Cysts round and smooth. Mature cysts of some species multinucleate. Many varieties reported. Habitat: parasite of the intestinal mucosa of man, apes, monkeys, dogs, cats, pigs, rats, cattle, and other animals; some species commensal in mouth; can migrate to other tissues via the bloodstream; causes mild to severe diarrhoea, on occasion lethal. Ref. Albach and Booden, 1978.
TYPE SPECIES: *Entamoeba coli* Cassigrandi & Barbagallo, 1895

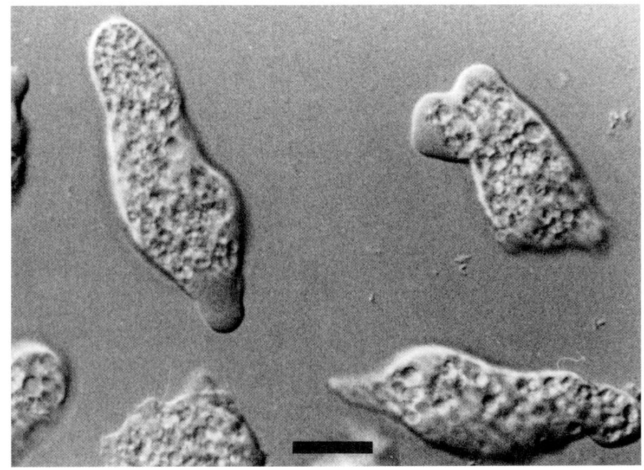

Fig. 25. *Entamoeba invadens.* Scale bar=5 μm.

Fig. 26. *Enteromyxa paludosa* after Canter (1966). Scale bar=50 μm.

Genus *Enteromyxa* Cienkowski in Zopf, 1885

Amoeboid organism existing in a small (less than 20 μm) uninucleate form, or large (up to 1 mm), multinucleate form (Fig. 26). The latter usually has one dimension limited to 40 μm. Outer cytoplasm is hyaline, from which conical or thin pseudopodia which are not stiffened nor anastomose arise. Body may be surrounded by mucilage. Cysts are formed, with large organisms breaking into small ectocyst elements within which a number of endocysts may form. Feeds on cyanobacteria by ingestion; from freshwater habitats and described only from N. Europe. Ref. Canter (1966).

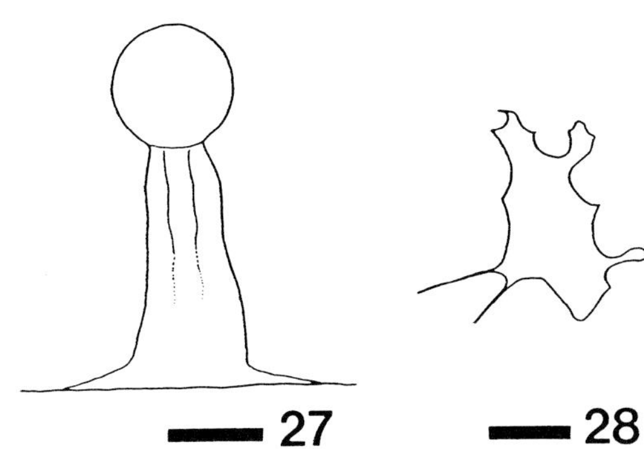

Figs. 27,28. *Fonticula alba* after Worley et al. (1979). 27. Sorocarp. Scale bar=100 μm. 28. Amoeboid form. Scale bar=5 μm.

Genus *Fonticula* Worley, Raper & Hohl, 1979

Amoeba with aggregating cellular and encysting stage resembling slime molds (Figs. 27,28). Sorocarp volcaniform. Amoebae with filose pseudopodia and discoid cristae—a combination of characteristics which makes assignment to either acrasids or dictyostelids uncertain. The proposition of *Fonticula* being a bridge between these 2 categories of slime molds is inconsistent with distant relatedness of these 2 types of protist and was excluded from the Heterolobosea by Page and Blanton (1985). Type locality: dog dung, Kansas. Ref. Blanton (1989); Deasey and Olive (1981); Worley et al. (1979).

Genus *Gromia* Dujardin, 1835

Shelled amoeba up to 5 mm, with brown opaque organic shell with single aperture from which extend branching pseudopodia which are not granular, may fuse again but do not form a reticulum (Fig. 29). Shell egg-shaped or variously lobed. Cytoplasm multinucleated. Produces flagellated gametes/swarmers. Marine, reported world-wide, affinities with foraminifera and shelled filose amoebae previously proposed, but not currently substantiated. Ref. Hedley and Wakefield (1969).

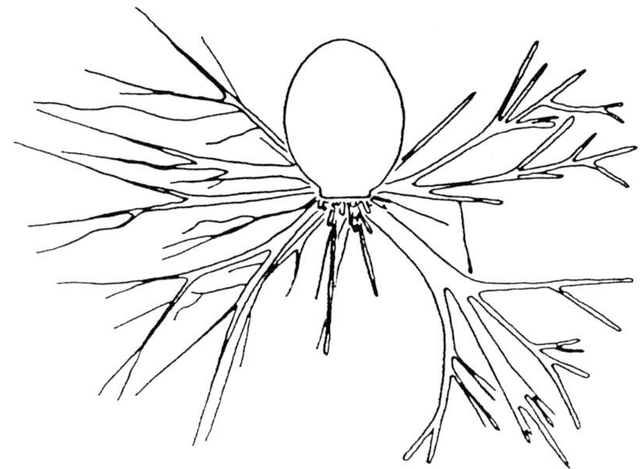

Fig. 29. *Gromia oviformis* after Doflein (1916).

Genus ***Gymnophrydium*** Dangeard, 1910

Freshwater and marine amoeboid and **plasmodial** organism measuring from 10 to 500 µm, pseudopodia thin (filopodia) and forming a network linking larger aggregates of cytoplasm—a kind of organization referred to as **meroplasmodial** (Fig. 30). Small amoeboid forms are produced by fission or budding. Considerable similarity with *Leucodictyon* and *Reticulamoeba*, but no cysts or flagellated stages reported, and no ultrastructural work. Two species, relationships with other branching filose plasmodial protists uncertain. Reported from freshwater and marine environments. Refs. Dangeard (1968); Kumar (1980)

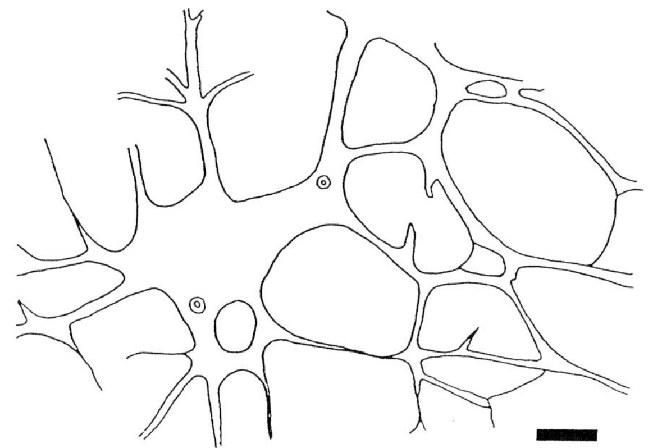

Fig. 30. *Gymnophrydium marinum* after Dangeard (1968). Scale bar=20 µm.

Genus ***Gymnophrys*** Cienkowski, 1876

See Chapter on Flagellates of Uncertain Affinities.

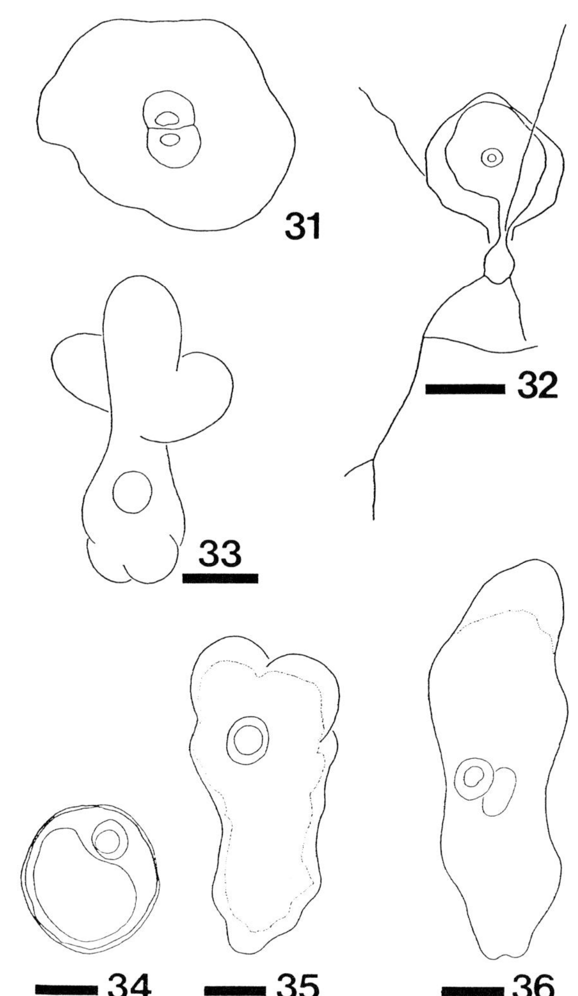

Fig. 31. *Hartmannina diploidea* after Chatton (1953). Fig. 32. *Heterogromia intermedia* after de Saedeleer (1934). Scale bar=5 µm. Fig. 33. *Hyalodaktylethra renacuajo* after Frenzel (1897). Scale bar=10 µm. Figs. 34,35. *Iodamoeba buetschlii* after Kudo (1971). Scale bars=5 µm. 34. Cyst. 35. Active. Fig. 36. *Janickina chaetognathi* after Janicki (1928). Scale bar=5 µm.

Genus ***Hartmannina*** Chatton, 1953

Parasitic amoeba from gut of lizards in Europe, up to 30 µm, typically with 2 adhering nuclei (Fig. 31). Believed to exhibit sexuality. Also forms cysts. Ref. Chatton (1953).

Genus ***Heterogromia*** de Saedeleer, 1934

Small irregularly shaped amoeba occupying a test (about 10 µm linear dimension), with a short terminal neck, aperture without collar giving rise to thin granular pseudopodia which may branch near tips (Fig. 32). Eats bacteria and reported from freshwater, but probably not distinguishable from *Kibisidytes* from marine habitats and published in the same year. Ref. de Saedeleer (1934).

Genus ***Hyalodaktylethra*** Delvinquier & Freeland, 1988

Medium-sized amoeba measuring about 45 µm in diameter, encountered in the cloaca of *Bufo marinus* (Fig. 33). **Limax amoeba** with extensive hyaline cap, broad lobose pseudopodia, and a spherical uroid devoid of villi. Monospecific. Ref. Delvinquier and Freeland (1988).
TYPE SPECIES: *Hyalodaktylethra renacuajo* (Frenzel, 1897)

Genus ***Iodamoeba*** Dobell, 1919

Rounded cell approximately 6 to 25 µm, locomotive amoeba elongate, 10 to 40 µm. With clear, slowly eruptive pseudopodia (Figs. 34,35). Indistinct hyaloplasm grades into a granular cytoplasm. No uroid, contractile vacuoles, or cytoplasmic crystals. Spherical nucleus with a prominent central nucleolus. Uninucleate cysts oval or angular. *Pseudolimax* is believed to be the same genus. Habitat: commensal, intestine of man, apes, monkeys, and pigs. Refs. Chatton (1953); Dobell (1919).

Genus ***Janickina*** Chatton, 1953

Active locomotive cell elongate, usually 75 to 100 µm in length, with an eruptive motion. Nucleus ovate with an irregular nucleolus (endosome) (Fig. 36). Adjacent to the nucleus is a paranuclear **'amphosome'**, ultrastructurally identical to the **parasomes** of *Neoparamoeba* and *Paramoeba*. In *Janickina*, this DNA-rich inclusion is approximately 30 µm and divides synchronously with the nucleus. Cytoplasm without crystals or contractile vacuoles but with numerous vesicles and granules. No cysts reported. Parasite of the chaetognaths, *Spadella* spp. and *Sagitta* spp. Refs. Hollande (1980); Janicki (1932).

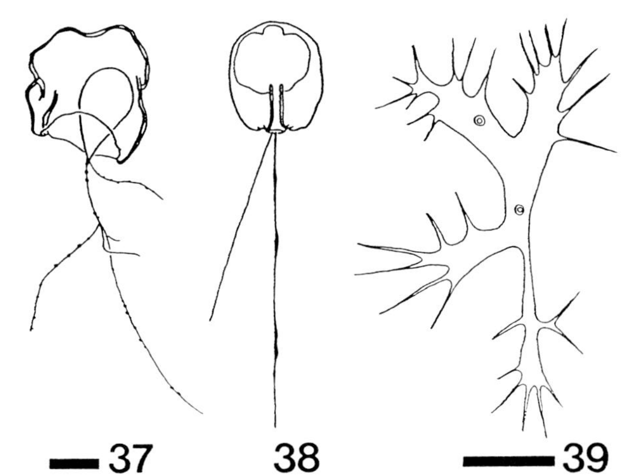

Fig. 37. *Kibisidytes marinus* after Jepps (1934). Scale bar=5 µm. Fig. 38. *Lagediniopsis* sp. after Sudzuki (1979). Fig. 39. *Leptophrys elegans* after Page (1991). Scale bar=50 µm.

Genus ***Kibisidytes*** Jepps, 1934

Small marine amoeba housed in an organic shell that may become brown and usually with one opening giving rise to delicate, branched, and granulate pseudopodia (Fig. 37). Divides within the shell. The original description is poor and it is hard to distinguish between this and *Heterogromia* or *Microcometes* (the flagella of which are very hard to see). Ref. Jepps (1934).
TYPE SPECIES: *Kibisidytes marinus* Jepps, 1934

Genus ***Lagenidiopsis*** Golemansky, 1974

Small to medium-sized (20-90 µm), shelled amoebae, shell organic but may have adhering material (Fig. 38). Aperture extends into the center of the shell as a tube. Aperture gives rise to thin filose pseudopodium/a. The test has some similarities with *Lieberkuehnia* and other taxa of uncertain affinities (Rhumbler, 1904), but no detailed studies have been carried out. Northern Europe, marine sediments. Ref. Golemansky (1974); Sudzuki (1979).

Genus ***Leptophrys*** Hertwig & Lesser, 1874

Amoeba with filose pseudopodia and consuming algae by ingestion (Fig. 39). Previously classified with the vampyrellids (e.g. Röpstorf et al., 1994) but separated here because they do not feed in the same characteristic fashion (i.e. by penetration of the cell wall) and because the dividing nucleus is characterized by a fragmenting nuclear envelope. Catholic diet, including algae and nematodes. In the absence of extensive ultrastructural studies, the present position of this genus remains uncertain, Dobell (1913) regarding this as being synonymous with *Arachnula*. Refs Dobell (1913); Röpstorf et al. (1994).

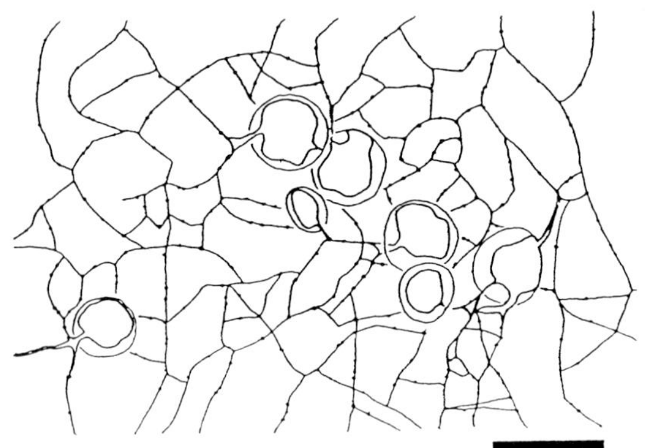

Fig. 40. *Leucodictyon marinum* after Grell (1991). Scale bar=20 μm.

TYPE SPECIES: *Leucodictyon marinum* Grell, 1991

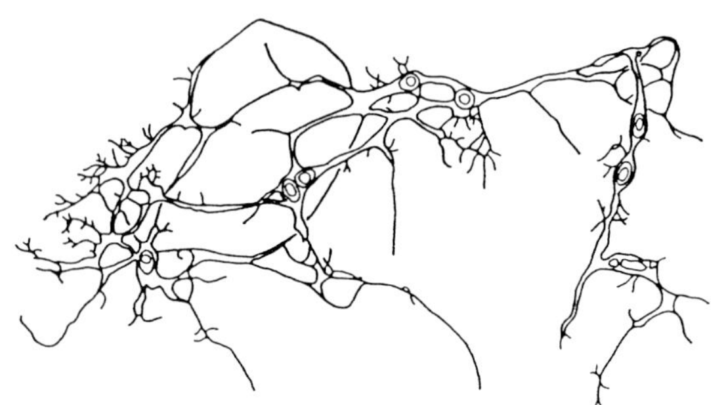

Fig. 41. *Leukarachnion batrachospermi* after Geitler (1942).

Genus ***Leucodictyon*** Grell, 1991

Large marine meroplasmodial organism, with discrete masses of cell bodies giving rise to a reticulum of fine pseudopodia with extrusomes (Grell 1991; Grell and Schüller, 1991) (Fig. 41). Body masses aggregate while feeding. Cytoplasm of cell bodies contained within fine organic lorica. **Extrusomes** resemble those of some small filopodial amoebae (Mikrjukov and Mylnikov, 1995), mitochondrial cristae said to be tubular. Some nuclei with rod-like fibrous inclusions. Has a distributive stage with 2 flagella of dissimilar lengths. These stages may swim or glide. One species described from marine sites in Japan. Refs. Grell (1991); Grell and Schüller (1991).

Genus ***Leukarachnion*** Geitler, 1942

Plasmodial, marine, multinucleate amoebae able to aggregate into anastomosing plasmodia several millimeters in size (Fig. 40). Forms cysts. Described from Austria in association with the rhodophyte *Batrachospermum*. Resembles a number of other large marine amoebae, such as *Cinetidomyxa*, *Thalassomyxa*, and *Stereomyxa*. Refs. Geitler (1942, 1959).

Genus ***Liegeoisia*** Chatton, 1953

Binucleated amoebae found in the malpighian tubules of *Hydrophilus* (Fig. 42). Cell about 25 μm. Spores not known, but has cysts. Said by some to be related to *Pansporella*. Ref. Chatton (1953).

Genus ***Lithocolla*** Schulze, 1874

Amoeba with thin pseudopodia, rounded, small to medium (10-50 μm) body, but with various irregular particles adhering to the outer surface (Fig. 43). Most similar in appearance to the nucleariid filose amoebae which also have adhering structures. From marine, brackish water, and freshwater. Cysts and non-amoeboid stages in life cycle are not reported. Monospecific. Ref. Page and Siemensma (1991).

TYPE SPECIES: *Lithocolla globosa* Schulze, 1874

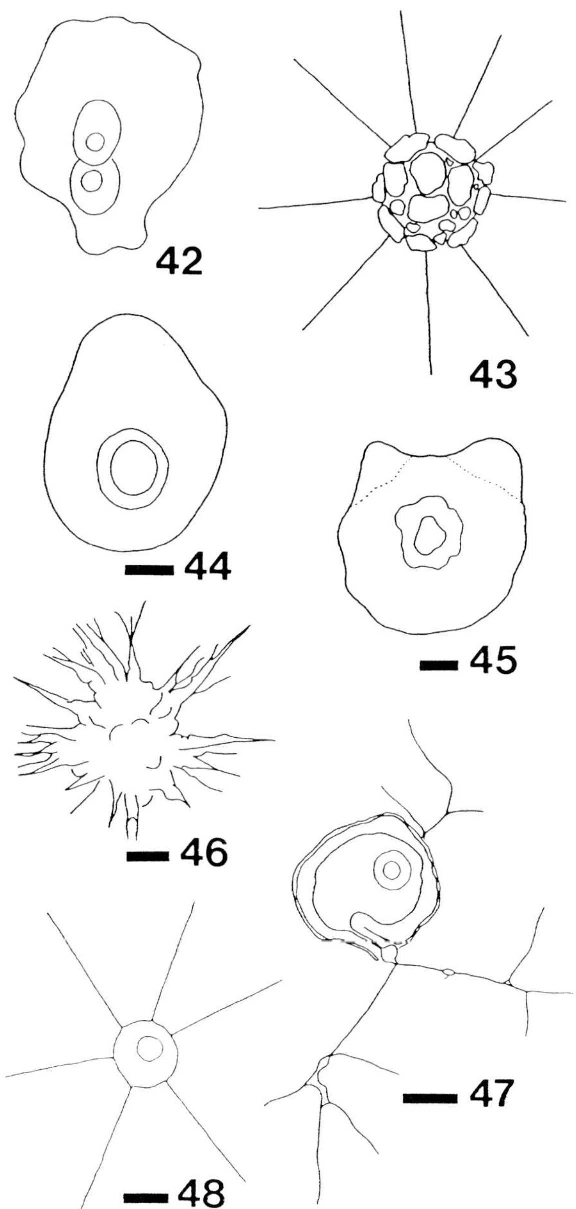

Fig. 42. *Liegeoisia hydrophili* after Chatton (1953). Fig. 43. *Lithocola globosa* after Siemensma (1991). Fig. 44. *Malpighiella refringens* after Doflein and Riechenow (1961). Scale bar=2 µm. Fig. 45. *Martinezia baezi* after Kudo (1971). Scale bar=2 µm. Fig. 46. *Megamoebamyxa argillobia* after Nyholm (1950). Scale bar=2 mm. Fig. 47. *Microgromia longisaepinen* after de Saedeleer (1934). Scale bar=5 µm. Fig. 48. *Ministeria marisola* after Patterson et al. (1993). Scale bar=2 µm.

Genus ***Malpighiella*** Minchin, 1910

Small amoebae (about 10 µm), first described from the malpighian tubules of the rat flea, moves slowly, feeding is osmotrophic, with spherical cysts with 2 or 4 nuclei (Fig. 44). Said by some to be related to the hartmannellids. Ref. Chatton (1953).

Genus ***Martineziella*** Hegner & Hewitt, 1941

Amoeba parasitic in the intestines of iguanas (Fig. 45). Cell body up to 20 µm. Nucleus up to 4 µm. Cysts not known. Ref. Kudo (1971).

Genus ***Megamoebomyxa*** Nyholm, 1950

Large marine organism measuring up to 25 mm, reported once in the literature (Nyholm, 1950) from organically enriched sediments in N. Europe (Fig. 46). Takes the form of large masses of cytoplasm which may range in form from linear and with bulbous termini to star-shaped. Opaque white, with filopodia which possibly anastomose, and with detrital particles adhering to outer surface. Monospecific. Affinities unknown. Said to have 3 nuclei. Ref. Nyholm (1950).
TYPE SPECIES: *Megamoebomyxa argillobia* Nyholm, 1950

Genus ***Microcometes*** Cienkowski, 1876

See chapter on Flagellates of Uncertain Affinities.

Genus ***Microgromia*** Hertwig, 1874

Freshwater amoeba; cell bodies are small and located within organic lorica with one aperture from which emerge filose pseudopodia which are stiffened, have extrusomes, and are able to branch and anastomose (Fig. 47). Lorica with neck that has a basal septum with lateral opening. Cells purportedly able to leave lorica. Cell bodies with one nucleus and contractile vacuole. Pseudopodial system of many cells able to form an anastomosing reticulum, and cells may form aggregates (under which they have been referred to as *Cystophrys*). Also able to produce co-operative pseudopodia. May produce swarmers with 2 homodynamic flagella. Several species, with some similarities to *Leucodictyon* (see *Leucodictyon*). Refs. de Saedeleer (1934); Hertwig (1874).

Genus ***Ministeria*** Patterson et al., 1993

Small marine organism measuring usually less than 5 µm in diameter, with equi-spaced smooth arms radiating from a spherical body (Fig. 48). Two species described to date, one is able to vibrate and this suggests that it may have a flagellum. Mitochondrial cristae are flattened. From Atlantic and Pacific locations. Ref. Patterson et al. (1993).

Fig. 49. *Myxodictyum sociale* after Rhumbler (1903).

Genus ***Myxodictyum*** Haeckel 1868

Amoeboid organism, rounded central body mass, gives rise to radiating fine pseudopodia, body about 30-40 µm, reported as occurring in polygonal colonies (clusters). Marine. Ref. Rhumbler (1904).

FAMILY NUCLEARIIDAE

Amoebae with filose pseudopodia, previously assigned to the Aconchulinida or Athalamida but segregated from other amoebae with thin pseudopodia on the basis of ultrastructural characteristics, principally the presence of discoidal or flattened mitochondrial cristae (Page and Siemensma, 1991; Patterson et al., 1987). Referred to as the nucleariid filose amoebae or as the Cristidiscoidea. The amoebae consume bacteria, detritus, algae (some penetrating the cells of healthy algae). The pseudopodia lack microtubules but are probably supported by actin microfilaments. The pseudopodia tend to project more from one face of the cell than others. With or without perforated siliceous structures adhering to the body surface. Mostly reported from freshwater locations. Widespread in distribution but not necessarily common.

Genus ***Nuclearia*** Cienkowski, 1865

Small to medium-sized amoebae, either naked or with mucus sheath, either flattened with pseudopodia extending predominantly from one margin and/or adopting a rounded body form with radiating pseudopodia, in which form they may be confused with heliozoa but can be distinguished by the absence of **extrusomes** (Fig. 50). May form organic walled (mucus-sheath) cysts. Typically without color (cf. vampyrellids with which they have previously been classified). With about 9 species. Genus revised by Patterson (1984) to accommodate species previously assigned to *Astrodisculus, Nuclearella, Nuclearina*, *Heliosphaerium*, and *Nucleosphaerium.* Refs. Patterson (1983, 1984).

Genus ***Pinaciophora*** Greeff, 1873

Filose amoebae with flattened hollow scales with delicately perforated texture (Fig. 51). Central perforations larger and visible with light microscope, usually as a hexagonal pattern. Normally consume detritus and algae. Two species recognized and reviewed by Page and Siemensma (1991). *Potamodiscus* Gerloff is now regarded as being the same as *Pinaciophora.* Ref. Thomsen (1978).

Genus ***Pompholyxophrys*** Archer, 1869

Medium-sized amoebae, typically with a spherical or near spherical body coated with numerous delicate and perforated siliceous **perles** (Fig. 52). Pseudopodia emerge through the layer of perles. Cells feed on algae and detritus and have been reported from freshwater habitats. This is the same as *Hyalolampe* Greeff, 1869. With 7 species. Refs. Page and Siemensma (1991); Patterson (1985); Rainer (1968).

Genus ***Rabdiophrys*** Rainer, 1968

Filose amoeba with hollow siliceous structures forming scales and spines (Fig. 53). Six species recognized and reviewed by Page and Siemensma, 1991. Mostly from freshwater habitats but some from brackish water or marine sites. Ref. Page and Siemensma (1991).

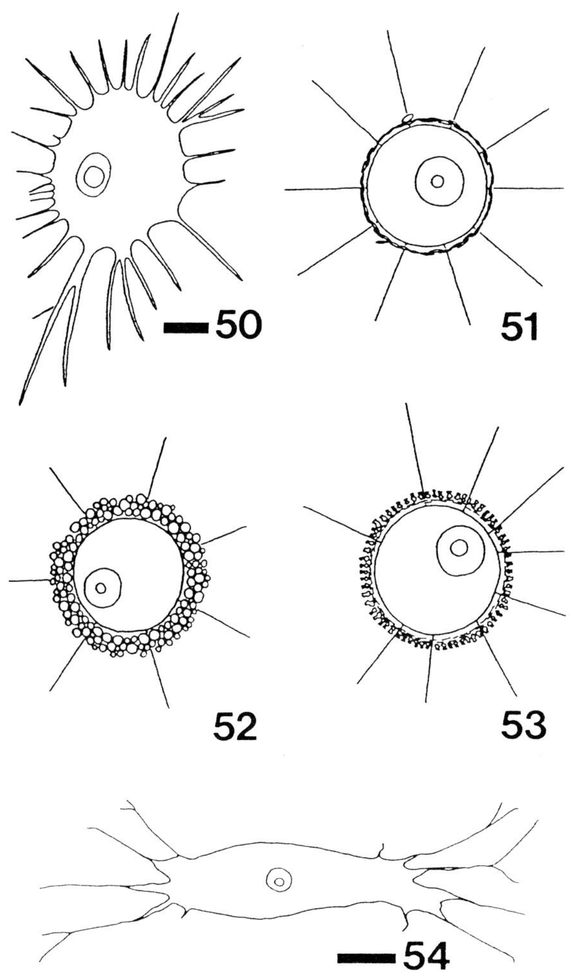

Fig. 50. *Nuclearia moebiusi* after Patterson (1984). Scale bar=5 µm. Fig. 51. *Pinaciophora fluviatilis* after Siemensma (1991). Fig. 52. *Pompholyxophrys punicea* after Siemensma (1991). Fig. 53. *Rabdiophrys triangula* after Siemensma (1991). Fig. 54. *Vampyrellidium perforans* after Surek and Melkonian (1980). Scale bar=10 µm.

Genus ***Vampyrellidium*** Zopf, 1885

Small to medium (10-40 µm), naked amoebae, typically flattened with pseudopodia mostly emerging from an anterior margin, but may adopt rounded form; consuming the contents of organic-walled prokaryotic or eukaryotic algae following wall penetration (Fig. 54). Previously placed with vampyrellid amoebae but distinguished by ultrastructural characters such as mitochondrial cristae. Also with a striated band located near the nucleus and giving rise to microtubules. Two species. Refs. Surek and Melkonian (1980); Patterson et al. (1987).

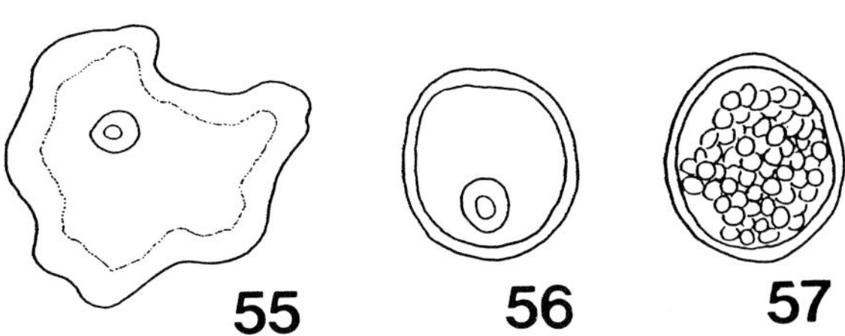

Figs. 55-57. *Pansporella perplexa* after Bovee (1985). 55. Active. 56. Cyst, early stage. 57. Late cyst.

Genus ***Pansporella*** Chatton, 1925

Rounded cell from 10 to 100 µm, locomotive cell from 100 to 125 mm with eruptive, lobate pseudopodia (Figs. 55-57). Cytoplasm finely granular without crystals or contractile vacuoles. Nucleus spherical, from 1 to 15 mm diameter. Cysts round and smooth with many spores often around 8.5 mm diameter. Habitat: cyst in haemocoel of the water flea *Daphnia*. Trophic stage free-living in freshwater. Ref. Chatton (1925).

Genus ***Penardia*** Cash, 1904

Medium-sized amoeba with branching body up to 100 µm long from which extend fine pseudopodia which branch and anastomose (Fig. 58). Distinctions between this and some other genera, such as *Arachnula*, *Theratomyxa,* etc. remain unclear and these medium-sized amoebae with fine branching pseudopodia are in need of further attention. Ultrastructure unknown; one ultrastructural study has been carried out on "*P. cometa*" (Mikrjukov and Mylnikov, 1995), but we do not regard this as being of a species of *Penardia* (see *Gymnophrys*). Page and Siemensma (1991) recognize one species, *P. mutabilis*, and we concur. From bogs, Europe. *Penardia* van Oye, 1923 (van Oye, 1923) is a homonym for a species which we assign to *Artodiscus*. Ref. Cash (1904); Mikrjukov and Mylnikov (1995); Page and Siemensma (1991).

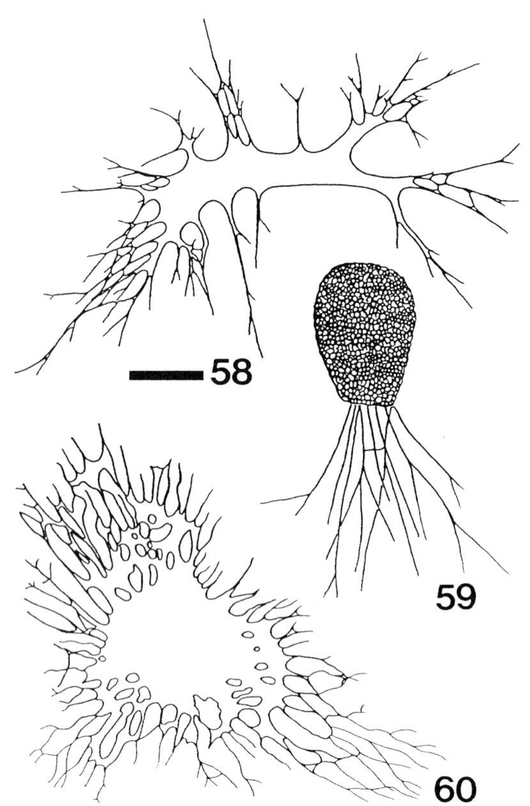

Fig. 58. *Penardia mutabilis* after Cash (1904). Scale bar=50 µm. Fig. 59. *Pleurophrys sphaerica* after de Saedeleer (1934). Fig. 60. *Pontomyxa flava* after Deflandre (1953).

Genus ***Pleurophrys*** Claparède & Lachmann, 1859

Marine shelled amoeba, about 70 µm long. Shell composed of fine particles, ovate, but with truncated narrow end occupied by the single aperture from which arise fine pseudopodia which may branch and fuse (Fig. 59). Ref. de Saedeleer (1934).

Genus ***Pontomyxa*** Topsent, 1892

Large multinucleate amoeba up to 6 cm when extended (Fig. 60). Bidirectional particle movement occurs in thin anastomosing pseudopodia. Type species is yellow color and was reported as an ectocommensal in tunicates. Two species. Suggested by some to be related to vampyrellids. Ref. Topsent (1892).

TYPE SPECIES: *Pontomyxa flava* Topsent, 1892

Genus ***Protogenes*** Haeckel, 1865

Medium to large (from 100 µm up to 4 mm) amoeboid organism with central mass of cytoplasm giving rise to radiating pseudopodia (Fig. 61). Reported from marine locations, Northern Europe. Two nominal species (but see *Vampyrelloides*). Reference: Rhumbler, 1904.

Protomonas Haeckel, 1866

Small amoeba with fine pseudopodia but may form pseudoplasmodia. May produce cysts (Fig. 62). Biflagellated dispersal stages arise from the cysts and transform subsequently into heliozoon-like amoebae or more normal amoeboid forms. From fresh water. Ref. Kudo, (1971).

TYPE SPECIES: *Protomonas primordialis* Haeckel, 1866

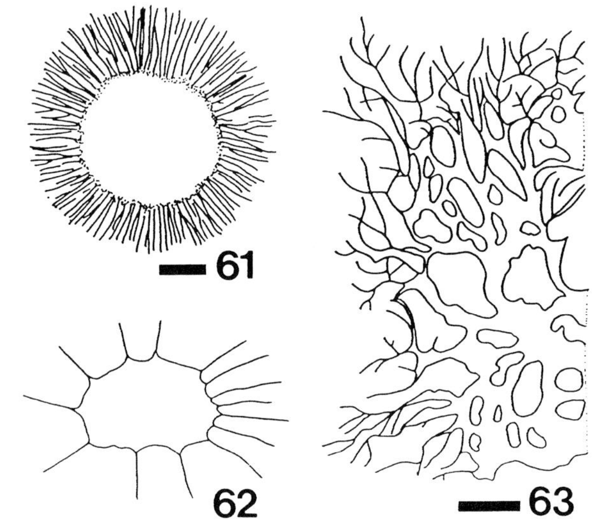

Fig. 61. *Protogenes primordialis* after Rhumbler (1903). Scale bar=200 µm. Fig. 62. *Protomonas amyli* after Kudo (1981). Fig. 63. *Protomyxa aurantiaca* after Rhumbler (1903). Scale bar=50 µm.

Genus ***Protomyxa*** Haeckel, 1868

Amoeba forming a large plasmodium with fine radiating and anastomosing pseudopodia—with complex life cycle, including cysts, and dispersal of small heliozoa-like amoeboid cells (Fig. 63). Colored yellow or orange. Marine (Canary Islands). Ref. Rhumbler (1904).

Genus ***Reticulamoeba*** Grell, 1994

Marine amoeba with delicate, radiating, branching and anastomosing pseudopodia linking smaller masses of cytoplasm (Fig. 64). Feeds on diatoms and produced biflagellated swarmers. Very similar to *Leucodictyon* which differs because cell bodies are enclosed in loricas and because cells tend to be more aggregated, especially when feeding. Two species, marine. Refs. Grell (1994a, 1995).

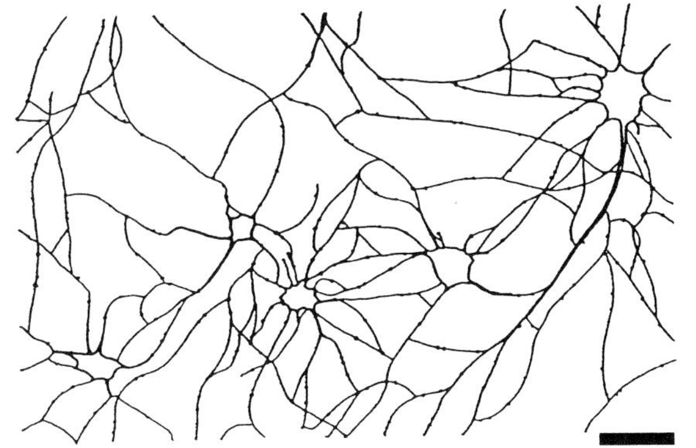

Fig. 64. *Reticulamoeba gemmipara* after Grell (1994). Scale bar=20 µm.

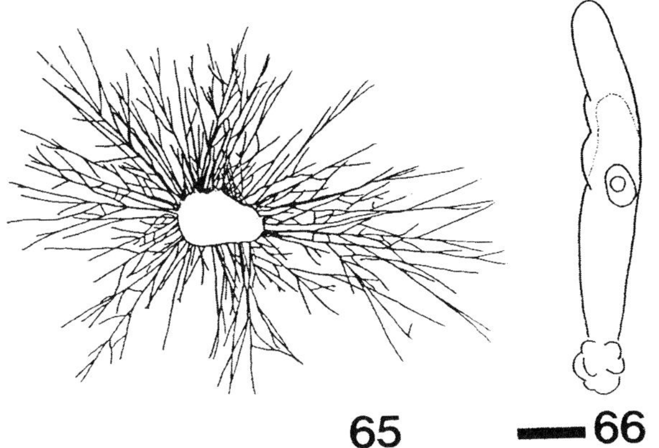

Fig. 65. *Rhizoplasma kaiseri* after Doflein (1916). Fig. 66. *Stygamoeba regulata* after Smirnov (1995). Scale bar=5 µm.

Genus ***Rhizoplasma*** Verworn, 1896

Large amoeba with a central body mass up to 1 cm across and with a pseudopodial network extending to 3 cm, cytoplasm may stream in both directions (Fig. 65). Orange color. Marine (Red Sea). Refs. Rhumbler (1904); Schepotieff (1912b).

Genus ***Stygamoeba*** Sawyer, 1975

Small amoeba (about 20 µm), thin, elongate, and sometimes branching (Fig. 66). Locomotion steady, monopodial. Anterior hyaloplasm may form thin lateral hyaline extensions in slow moving amoebae. When not active, amoebae can adopt a broad, leaf-like morphology with peripheral branches. Floating form with radiating pseudopodia. Uninucleate. No flagellate stages. Two marine species. This genus was previously classified in the Stereomyxidae, but Smirnov (1996) has shown that it is not a ramicristate amoeba. Ref. Smirnov (1996).

TYPE SPECIES: *Stygamoeba polymorpha* Sawyer, 1975

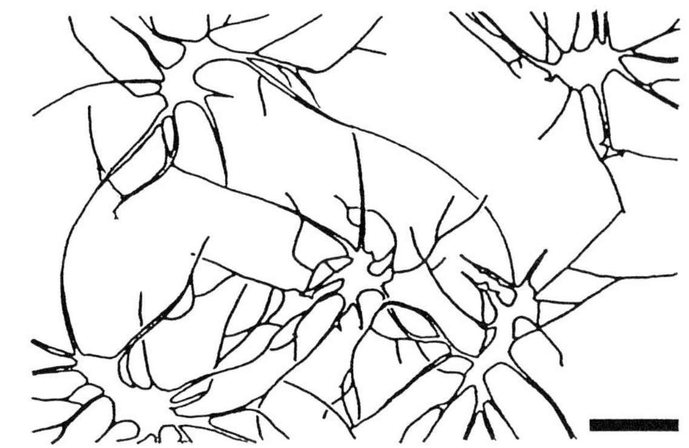

Fig. 67. *Synamoeba arenaria* after Grell (1994b). Scale bar=50 µm.

Genus ***Synamoeba*** Grell, 1994

Marine meroplasmodial organism in that it is comprised of a number of interlinked masses of cytoplasm (Fig. 67). The pseudopodia linking the organisms may be delicate (thread-like) or more substantial tracts. No flagellated or encysted forms have been described. The organism has been described once from Tenerife, with few details (e.g. no size is given). With similarities to organisms such as *Gymnophrydium* from which it was not distinguished. Ref. Grell (1994b).

TYPE SPECIES: *Synamoeba arenaria* Grell, 1994

Genus ***Thalassomyxa*** Grell, 1985

An amoebae with thin stiff pseudopodia, body form ranges from a medium-sized discrete mass of cytoplasm to a large marine plasmodium forming a network of broad pseudopodia with filose extensions (Fig. 68). Alternates between actively feeding motile stage and resting stage which may undergo multiple divisions to produce small distributive 'buds'. Several species. Similar superficially to several other genera of large plasmodial amoebae. Ref. Grell (1985, 1992).

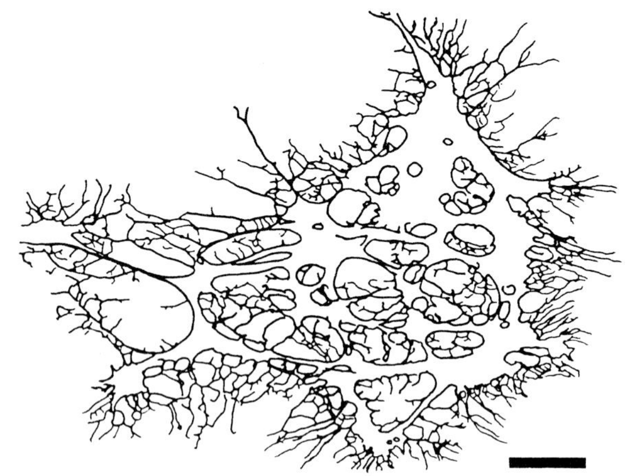

Fig. 68. *Thalassomyxa australis* Grell, 1985. Scale bar=100 µm.

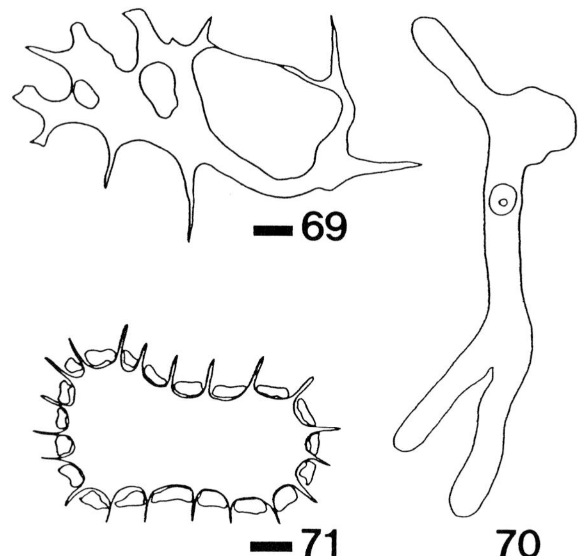

Fig. 69. *Theratromyxa weberi* after Zwillenberg (1953). Scale bar=20 µm. Fig. 70. *Topsentella fallax* after Chatton (1953). Fig. 71. *Urbanella napoletana* after Schepotieff (1912). Scale bar=100 µm.

Genus ***Theratromyxa*** Zwillenberg, 1953

Medium to large (40-300 µm), amoeboid and multinucleated, plasmodial organism described from soils, moves by creeping with cytoplasm described as 'flowing'; consumes nematode larvae which adhere to the cell and may also consume fungi (Old and Oros, 1980); life cycle may involve cysts, and cell can divide into many smaller amoebae (Fig. 69). Cytoplasmic reticulum usually of broad tracts of cytoplasm but may produce thin pseudopodia although detailed information is not available. Considerable similarity with other large reticulate amoebae, several of which have also been reported from soils (e.g. *Arachnula* and *Leptophrys*). Refs. Old and Oros (1980); Zwillenberg (1953).

Genus ***Topsentella*** Duboscq & Tuzet, 1936

Small, worm-shaped amoebae (up to 15 µm) (Fig. 70). Parasite of sponges. Affinities uncertain. Ref. Chatton (1953).

Genus ***Urbanella*** Schepotieff, 1912

Medium to large, marine, gold or red organism up to 3 mm in size, body typically enclosed in a sheath which incorporates sand grains and sponge spicules; produces finger-like pseudopodia (Fig. 71). May found large masses with bulbous outline. Said to have a complex life cycle. Monospecific. Ref. Schepotieff (1912a).
TYPE SPECIES: *Urbanella napoletana* Schepotieff, 1912

FAMILY VAMPYRELLIDAE Zopf, 1885

Medium to large (30 to >1000 µm), naked amoebae with numerous tapering filose pseudopodia; rarely anastomosing. Many consume algae or fungi by penetration through the cell wall (Fig. 72, Röpstorf et al., 1994), but this may not be the exclusive mode of feeding of some species (Dobell, 1913). With a cyst stage in the life cycle and cell division usually occurring in cyst or at excystation. Usually with orange color. Previously classified with the nucleariid filose amoebae (as the Aconchulinidae). Separated on the basis of fine structure—principally the vesiculate nature of the mitochondrial cristae, but also osmiophilic granules and arrays of ribosomes. Affinities unclear, but some similarities with protostelids. Five genera recognized here, *Leptophrys* excluded because of differences in feeding behavior and lack of confirmatory ultrastructural evidence.

Fig. 72. Typical vampyrellid (*Vampyrella*) life cycle after Röpstorf et al. (1994).

Genus ***Arachnula*** Cienkowski, 1876

Typically multinucleate, branching organism, size ranges from 50 µm linear dimension to organisms covering several square millimeters, typically found in soils where it may penetrate fungal mycelia and spores by cutting a hole in the wall and then consume cytoplasm (Fig. 73). May ingest other protists. Usually colorless but may have light pink or orange color. Cysts usually produced after feeding. Reported from Europe and Gondwanaland. Refs. Dobell (1913); Old and Darbyshire (1980).

Genus ***Gobiella*** Cienkowski, 1881

Medium to large amoebae consuming algal cells penetrating by removing a hole in the wall (mentioned in passing in Hülsmnann, 1993) (Fig. 74). Distinguished from *Vampyrella* by having irregular cysts. Ref. Röpstorf et al. (1994).

Genus ***Hyalodiscus*** Hertwig & Lesser, 1874

Medium-sized (up to 150 µm or so) amoebae, from soils, with cytoplasm located usually in a hump and with a leading flattened **lamellipodium** from which fine pseudopodia emerge (Fig. 75). Distinctive warty surface. Consuming algae, eggs of rotifers, and cysts of ciliates. Ref. Röpstorf et al. (1994).

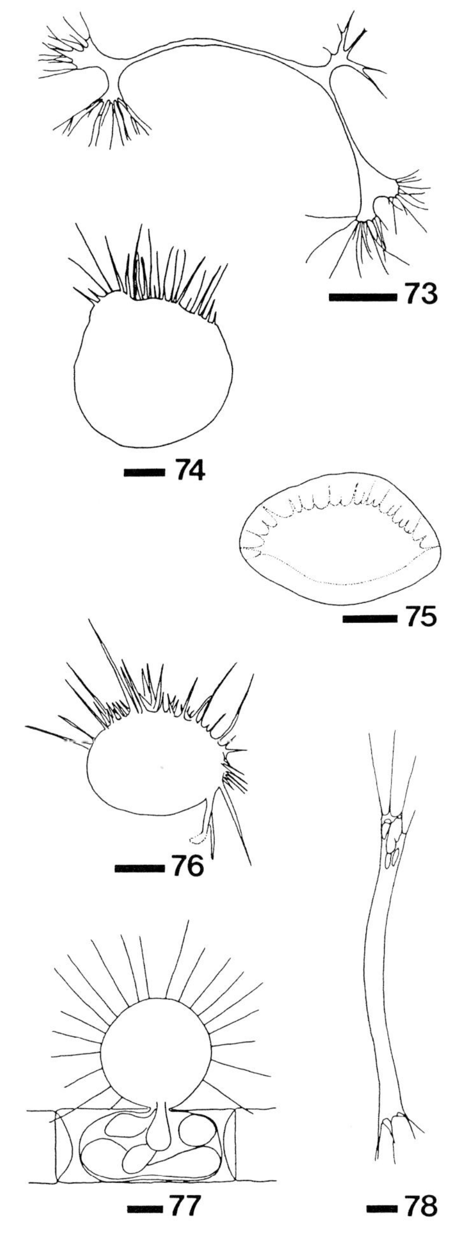

Fig. 73. *Arachnula impatiens* after Page (1991). Scale bar=100 µm. Fig. 74. *Gobiella borealis* after Röpstorf et al. (1994). Scale bar=20 µm. Fig. 75. *Hyalodiscus rubicundus* after Page (1991). Scale bar=20 µm. Fig. 76. *Lateromyxa gallica* after Hülsmann (1993). Scale bar=20 µm. Fig. 77. *Vampyrella lateritia* after Page (1991). Scale bar=20 µm. Fig. 78. *Vampyrelloides roseus* after Schepotieff (1912). Scale bar=200 µm.

Genus *Lateromyxa* Hülsmann, 1993

Amoebae from 25-800 µm in diameter, with one to multiple nuclei; pseudopodia usually tapering and rising from anterior margins of moving cells (Figs. 76,79). Move from cell to cell of *Oeodogonium* through lateral walls, forming 'pseudocyst' in the process. Monospecific. Ref. Hülsmann (1993); Röpstorf et al. (1993).
TYPE SPECIES: *Lateromyxa gallica* Hülsmann, 1993

Genus *Vampyrella* Leidy, 1879

Usually orange-colored cells with numerous fine pseudopodia. They are typically encountered in freshwater ponds where they consume the cytoplasm of algae after cutting a hole in the cell wall and entering the algal cell (Fig. 77). Hülsmann (1993) gives the type as *Vampyrella lateritia* (Cienkowski, 1865) Leidy, 1879; and later as *V. lateritia* (Fresenius, 1865) Leidy 1879. Ref. Hülsmann (1993).

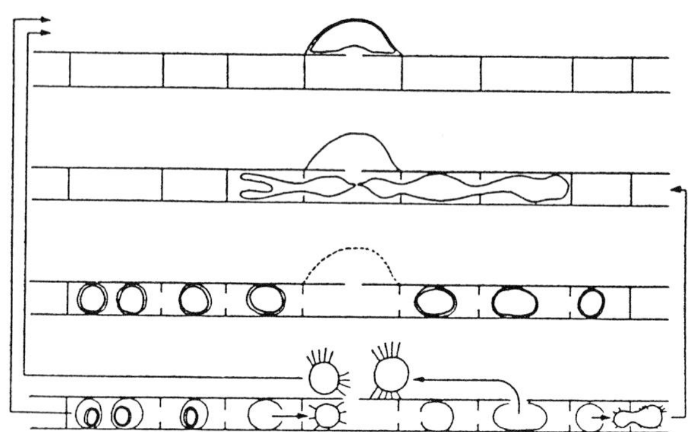

Fig. 79. Lifecycle of *Lateromyxa* after Hülsmann (1993).

Genus *Vampyrelloides* Schepotieff, 1912

The only genus of vampyrellids encountered exclusively in marine conditions where it is associated with algae (Fig. 78). Can be quite large, multiple tracts of fine pseudopodia. Encystment involves cells about 0.5 mm in size breaking up to give cysts about 50 µm in diameter. Ref. Schepotieff (1912b) as *Protogenes roseus*.

LITERATURE CITED

Albach, R. A. & Booden, T. 1978. Amoebae. *In*: Kreier, J. P. (ed.), *Parasitic Protozoa*, Vol. 2. Academic Press, New York. pp 455-506

Anderson, O. R. & Hoeffler, W. K. 1979. Fine structure of a marine proteomyxid and cytochemical changes during encystment. *J. Ultrastruct. Res.*, 66:276-287.

Bhattacharya, D., Helmchen, T. & Melkonian, M. 1995. Molecular evolutionary analyses of nuclear-encoded small subunit ribosomal RNA identify an independent rhizopod lineage containing the Euglyphina and Chlorarachniophyta. *J. Euk. Microbiol.*, 42:65-69.

Blanton, R. L. 1989. Phylum Acrasea. *In*: Margulis, L., Corliss, J. O., Melkonian, M. & Chapman, D. J. *Handbook of Protoctista*. Jones and Bartlett, Boston. pp 75-87

Braun, L., Ewen, A. B. & Gillott, C. 1988. The life cycle and ultrastructure of *Malamoeba locustae* (King and Taylor) (Amoebidae) in the migratory grasshopper *Melanoplus sanguinipes* (F.) (Acrididae). *Can. Ent.*, 120:759-772.

Cachon, J., & Cachon-Enjumet, M. 1964. Cytologie et cycle évolutif de *Cinetidomyxa chattoni* nov. sp., Héliozoaire Protéomyxée. *Arch. Zool. Gen. Exp.*, **104**: 47-60

Canter, H. M. 1966. On the protozoan *Enteromyxa paludosa* Cienkowski. *J. Linn. Soc. (Zool.)*, 46:143-154.

Canter, H. M. 1973. A new primitive protozoan devouring centric diatoms in the plankton. *Zool. J. Linn. Soc.*, 52:63-83.

Canter, H. M. 1980. Observations on the amoeboid protozoan *Asterocaelum* (Proteomyxida) which ingests algae. *Protistologica*, 16:475-483.

Cash, J. 1904. On some new and little-known British freshwater Rhizopoda. *J. Linn. Soc. (Zool).*, 24:218-225.

Chatton, E. 1925. *Pansporella perplexa*. Réflexions sur la biologie et la phylogénie des protozoaires. *Ann. Sci. Nat. Zool. Ser.*, 10:8, 5-84.

Chatton, E. 1953. Classe des Lobosa Leidy, 1879. Ordre des amoebienes nus ou Amoebaea. In Grassé, P.-P. *Traité de Zoologie*. Tome I, fasc. II. Masson et Cie, Paris. pp 5-91

Chatton, E. & Lwoff, A. 1924. Un proteomyxée, *Cinetidomyxa nucleoflagellata* nov. gen. nov. sp., à cintétide intracytoplasmiques, et sa multiplication. *C. R. Sc. Soc. Biol.*, 91:584-587

Dangeard, P. 1968. Sue une protéomyxée (protiste plasmodial) développée dans des cultures d'algues marine et sur son rattachement au genre

Gymnophryidum de P. A. Dangeard. *C. R. Acad. Sci., Paris*, 266:1279-1282.

Deasey, M. C. & Olive, L. S. 1981. Role of Golgi apparatus in sorogenesis by the cellular slime mold *Fonticula alba*. *Science*, 213:561-563.

Delvinquier, B. L. J. & Freeland, W. J. 1988. Protozoan parasites of the cane toad, *Bufo marinus*, in Australia. *Aust. J. Zool.*, 36:301-316.

Dobell, C. 1913. Observations on the life-history of Cienkowski's "*Arachnula*". *Arch. Protistenkd.*, 31:317-353

Dobell, C. 1919. *The Amoebae Living in Man.* Bale, Sons and Danielsson, London.

Ertl, M. 1984. *Apogromia pagei*, n. sp., a new shelled rhizopod from fresh water plankton. *Arch. Protistenkd.*, 128:335-339.

Geitler, L. 1942 Ein neuer filarplasmodialer Organismus, *Leukarachnion batrachospermi*, und seine Lebensweise. *Biol. Zbl.*, 62:541-549.

Geitler, L. 1930. Ein grunes Filarplasmodium und andere neue Protisten. *Arch Protistenkd.*, 69:615-636.

Geitler, L. 1959. Über einen Leukarachnion-ähnliche, filarplasmodialen Rhizopoden. *Arch. Protistenkd.*, 103:573-580

Golemansky, V. 1974. *Lagenidiopsis valkanovi* gen. n., sp. n.—un nouveau thécamoebien (Rhizopoda: Testacea) du psammal supralittoral des mers. *Acta Protozool.*, 13:1-4.

Grell, K. G. 1985. Der Formwechsel des plasmodialen Rhizopoden *Thalassomyxa australis* n. g., n. sp. *Protistologica*, 21:215-233.

Grell, K. G. 1991. *Leucodictyon marinum*, n. gen., n. sp., a plasmodial protist with zoospore formation from the Japanese coast. *Arch. Protistenkd.*, 140:1-21.

Grell, K. G. 1992. A species of *Thalassomyxa* from the North coast of Jamaica. *Arch. Protistenkd.*, 142:15-33.

Grell, K. G. 1994a. *Reticulamoeba gemmipara* n. gen., n. sp., an "amoebo-flagellate" with reticulopodia and zoosporogenesis. *Arch. Protistenkd.*, 144:55-61.

Grell, K. G. 1994b. The feeding community of *Synamoeba arenaria* n. gen., n. sp. *Arch. Protistenkd.*, 144:143-146.

Grell, K. G. 1995. *Reticulamoeba minor* n. sp. and its reticulopodia. Arch. Protistenkd., 145: -9.

Grell, K. G. & Schüller, S. 1991. The ultrastructure of the plasmodial protist Leucodictyon *marinum Grell. Eur. J. Protistol.*, 27:168-177.

Griffin, J. L. 1978. Pathogenic free-living amoebae. *In*: Kreier, J. P. (ed.), *Parasitic Protozoa*. Vol. 2. Academic Press. pp 507-549

Harry, O. G. & Finlayson, L. H. 1976. The life-cycle, ultrastructure and mode of feeding of the locust amoeba *Malpighamoeba locustae*. *Parasitology*, 72:127-135.

Hedley, R. H. & Wakefield, J. S. 1969. Fine structure of *Gromia oviformis* (Rhizopodea: Protozoa). *Bull. Br. Mus (Nat. Hist.)*, 18:69-89.

Hertwig, R. 1874. Ueber *Mikrogromia socialis*, eine Colonie bildende Monothalamie des süssen Wassers. *Arch. Mikr. Anat.*, 1873 (suppl.):1-34.

Hibberd, D. J. & Norris, R. A. 1984. Cytology and ultrastructure of *Chlorarachnion reptans* (Chlorarachniophyta *divisio nova*, Chlorarachniophyceae *classis nova*). *J. Phycol.*, 20:310-330.

Hinkle, G. & Sogin, M. L. 1993. The evolution of the Vahlkamfiidae as deduced from 16S-like ribosomal RNA analysis. *J. Euk. Microbiol.*, 40:599-603.

Hollande, A. 1980. Identification du parasome (Nebenkern) de *Janickina pigmentifera* à un symbionte (*Perkinsiella* amoebae nov. gen.-nov. sp.) apparenté aux flagellé kinetoplastidies. *Protistologica*, 16: 613-625.

Hülsmann, N. 1993. *Lateromyxa gallica* n. gen., n. sp. (Vampyrellidae): a filopodial amoeboid protist with a novel life cycle and conspicuous ultrastructural characters. *J. Euk. Microbiol.*, 40:141-149.

Jankicki, C. 1932. Studium aus der genus *Paramoeba* Schaudinn. *Z. Wiss. Zool.*, 103:449-518.

Jepps, M. W. 1934. On *Kibisidytes marinus* n. gen., n. sp., and some other rhizopod protozoa found on surface films. *Q. J. Microsc. Sci.*, 77:121-127.

Kudo, R. R. 1971. *Protozoology*. 5th edition. Thomas, Springfield

Kumar, S. R. 1980. Morphology and taxonomy of the protist *Gymnophrydium marinum* Dangeard from the North Sea. *Bot. Mar.*, 23:353-360.

Leidy, J. 1879. Freshwater Rhizopods of North America. Rep. U. S. Geological Survey of the Territories 12, Washington, USA.

Levine, N. D. 1973. *Protozoan Parasites of Man and Domestic Animals,* 2nd ed. Burgess, Minneapolis.

Levine, N. D., Corliss, J. O., Cox, F.E.G., Deroux, G., Grain, J., Honigberg, B. M., Leedale, G. F., Loeblich II, A. R., Lom, J., Lynn, D. H., Merinfeld, E. G., Page, F. C., Poljansky, G., Sprague, V., Vavra, J., and Wallace, F. G. 1980. A newly revised classification of the protozoa. *J. Protozool.*, 27:37-58.

Mikrjukov, K. J. A. & Mylnikov, A. P. 1995. Fine structure of an unusual rhizopod, *Penardia cometa*, containing extrusomes and kinetosomes. *Eur. J. Protistol.*, 31:90-96.

Mikrjukov, K. A. & Mylnikov, A. P. 1996. New information of structure and life cycle of Athalamida amoebae (Protista Athalamida). *Zool. Zh.*, 75:1283-1293. (in Russian)

Nyholm, K-G. 1950. A marine nude Rhizopod type *Megamoebomyxa argillobia*. *Zool. Bidr. Upps.*, 29:93-102.

Old, K. M. & Darbyshire, J. F. 1980. *Arachnula impatiens* Cienk., a mycophagous giant amoeba from soil. *Protistologica*, 16:277-287.

Old, K. M. & Oros, J. M. 1980. Mycophagous amoebae in Australian forest soils. *Soil. Biol. Biochem.*, 12:169-175.

Page, F. C. & Siemensma, F. J. 1991. *Nackte Rhizopoda und Heliozoa*. Fischer Verlag, Stuttgart.

Patterson, D. J. 1983. On the organization of the naked filose amoeba, *Nuclearia moebiusi* Frenzel, 1897 (Sarcodina, Filosea) and its implications. *J. Protozool.* 30:301-307.

Patterson, D. J. 1984. The genus *Nuclearia* (Sarcodina: Filosea) species composition and characteristics of the taxa. *Arch. Protistenkd.*, 128:127-139.

Patterson, D. J. 1985. On the organization and affinities of the amoeba, *Pompholyxophrys punicea* Archer, based on ultrastructural examination of individual cells from wild material. *J. Protozool.*, 32:241-246.

Patterson, D. J. 1989. Stramenopiles: chromophytes from a protistan perspective. *In*: Green, J. C., Leadbeater, B. S. C. & Diver, W. L. (eds.), *The Chromophyte Algae: Problems and Perspectives*. Clarendon Press, Oxford. pp 357-379

Patterson, D. J. 1994. Protozoa, Evolution and Classification. Prog. Protozool., Proc. IX Int. Cong. Protozool. Fischer Verlag, Stuttgart. pp 1 - 14

Patterson, D. J., Nygaard, K., Steinberg, G. & Turley, C. M. 1992. Heterotrophic flagellates and other protists associated with detritus in the mid North Atlantic. *J. Mar. Biol. Assoc., UK.*, 73:67-95.

Patterson, D. J., Surek, B. & Melkonian, M. 1987. The ultrastructure of *Vampyrellidium perforans* Surek & Melkonian and its taxonomic position among the naked filose amobae. *J. Protozool.*, 34:63-67

Patterson, D. J. & Zölffel, M. 1991. Heterotrophic flagellates of uncertain taxonomic position. *In*: Patterson, D. J. & Larsen, J. (eds.), *The Biology of Free-Living Heterotrophic Flagellates*. Clarendon Press, Oxford. pp 427-475

Penard, E. 1903. Sur quelques protistes voisin des héliozoaires ou des flagellates. *Arch. Protistenkd.*, 2:283-304.

Penard, E. 1904. *Les Héliozoaires d'Eau Douce*. Kündig, Geneva.

Rainer, H. 1968. Urtiere, Protozoa Wurzelfüßler, Rhizopoda Sonnentierchen, Heliozoa. Part 56 in *Die Tierwelt Deutschlands*. F. Dahl (ed.). Fischer Verlag, Jena.

Rhumbler, L. 1904. Systematische Zusammenstellung der recenten Reticulosa. *Arch. Protistenkd.*, 3:181-294.

Raper, K. B., Worley, A. C. & Kurzynski, T. A. 1978. *Copromyxella*: a new genus of Acrasidae. *Amer. J. Bot.*, 65:1011-1026.

Röpstorf, P., Hülsmann, N. & Hausmann, K. 1993. Karyological investigations on the vampyrellid filose amoeba *Lateromyxa gallica* Hülsmann 1993. *Eur. J. Protistol.*, 29:302-310.

Röpstorf, P., Hülsmann, N. & Hausmann, K. 1994. Comparative fine structural investigations of interphase and mitotic nuclei of vampyrellid filose amoebae. *J. Euk. Microbiol.*, 41:18-30.

Saedeleer, H. de 1934. Beitrag zur Kenntnis der Rhizopoden: morphologische und systematische Untersuchungen und ein Klassifikationsversuch. *Mem. Mus. Roy. Hist. Nat. Belg.*, 60:1-112.

Schepotieff, A. 1912a. Untersuchungen über niedere Organismen. III. Monerenstudien. *Zool. Jahrb., Abt. F. Anat.*, 32:367-400.

Schepotieff, A. 1912b. Rhizopodienstudien. *Zool. Jahrb. (Suppl.)*, 15: 219-242.

Schwantes, U. & Eichelberg, D. 1984. Elektronenmikroskopische Untersuchungen zum Parasitenbefall der Malpighischen Gefässe von *Apis mellifera* durch *Malpighamoeba mellificae* (Rhizopoda). *Apidologie*, 15:435-450.

Smirnov, A. V. 1996. *Stygamoeba regulata* n. sp. (Rhizopod) - a marine amoeba with an unusual combination of light-microscopical and ultrastructural features. *Arch. Protistenkd.*, 146:299-302.

Spiegel, F. W. & Olive, L. S. 1978. New evidence for the validity of *Copromyxa protea*. *Mycologia*, 70:843-847.

Sudzuki, M. 1979. Marine testacea of Japan. *Sesoko Mar. Sci. Lab., Tech. Rep.*, 6:51-61.

Surek, B. & Melkonian, M. 1980. The filose amoeba *Vampyrellidium perforans* nov. sp. (Vampyrellidae, Aconchulinida): axenic culture, feeding behaviours and host range specificity. *Arch. Protistenkd.*, 123:166-191.

Thomsen, H. A. 1978. On the identity between the heliozoan *Pinaciophora fluviatilis* and *Potamodiscus kalbei*; with the description of eight new *Pinaciophora* species. *Protistologica*, 14:359-373.

Topsent, E. 1892. Description of *Pontomyxa flava*, Rhizopode marin etc. *Arch. Zool. Exp.*, 3rd series 1:385.

Valkanov, A. 1931. Über eine neue Vampyrellidae—*Cichkovia reticulata*. *Annuaire Univ. Sofia II Fac Phys. Math.*, Livre 3 (Sci. Nat.), 27:199-214.

van Oye, P. 1923. Deux rhizopodes nouveau du Congo Belge. *Rev. Zool. Afr.*, 11:B1-B9.

Visvesvara, G. S., Schuster, F. L. & Martinez, A. J. 1993. *Balamuthia mandrillaris*, n. g., n. sp., agent of amebic meningoencephalitis in humans and other animals. *J. Euk. Microbiol.*, 40:504-514.

Worley, A. C., Raper, K. B. & Hohl, M. 1979. *Fonticula alba*; a new cellular slime mould (Acrasiomycetes). *Mycologia*, 71:746-760.

Zopf, W. 1885. Die Pilzthiere oder Schleimpilze. *Encykl. Naturwiss, Breslau.*

Zwillenberg, L. O. 1953. *Theratromyxa weberi*, a new proteomyxean organism from soil. *J. Microbiol. Serol.*, 19:101-116.

ORDER ARCELLINIDA KENT, 1880

By RALF MEISTERFELD

The Arcellinida are the largest group of the testacea (shelled amoebae) containing about three quarters of the known species of these amoebae. All Arcellinida have a test or tectum. This key character that binds the Arcellinida together may not be homologous, but at moment the ultrastructural and molecular data do not clarify the relations within this group or with other sarcodina. They are distinguished by their tests, tectum, or other external membrane with a single **aperture** and composed of either organic material or inorganic material in an organic cement.

Shell Morphology. The architecture and composition of the shell has been central in the taxonomy of all shelled amoebae whether lobose or filose. Two principal types are common: proteinaceous and agglutinated. Those with proteinaceous tests can be divided into:

a) families which have a more or less flexible membrane that encloses the cytoplasm (Microcoryciidae)
b) those with a rigid sheet of fibrous material (Hyalospheniidae)
c) those with a test constructed of regularly arranged hollow building units to form an **areolate surface** (Microchlamyiidae, Arcellidae) (Fig. 1).

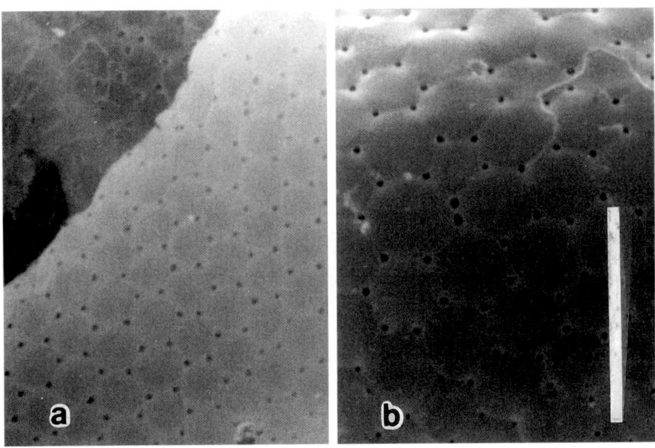

Fig. 1. Shell surface of *Arcella hemisphaerica* (a) and (b) *A. gibbosa* with typical building units (areoles). Scale bar=5 µm.

Agglutinated shells have either a cement matrix of often perforated building units or a sheet-like cement in which foreign material (**xenosomes**) is incorporated (Difflugina) (Fig. 2). The morphology of this cement is of growing importance for the classification of testate amoebae (Ogden and Ellison, 1988; Ogden, 1990). Few genera (Lesquereusiidae) have **siliceous** tests, composed of endogenous rods, nails, or rectangular plates which are produced by the organisms themselves (**idiosomes**). **Calcareous** shells are characteristic for the Paraquadrulidae and the genus *Cryptodifflugia*. The former has rectangular calcite plates bound by an internal sheet of cement, while in the latter a thick layer of calcium phosphate is deposited within an organic template.

Genera and species are mainly distinguished using differences in shell shape and dimensions. The Arcellinida display a great diversity of shell shapes. Especially in drier habitats such as mosses and soil the apertural apparatus has evolved to a slit-like structure often covered by the anterior lip (*Plagiopyxis*) or a **visor** (*Planhoogenraadia*). The **pseudostome** (aperture) can be protected by teeth or located at the end of a **tubus** (tubular section of a test leading inwards from aperture). Bonnet (1975b) has classified the different types of shell shape and apertural morphology and used these types to suggest phylogenetic relationships.

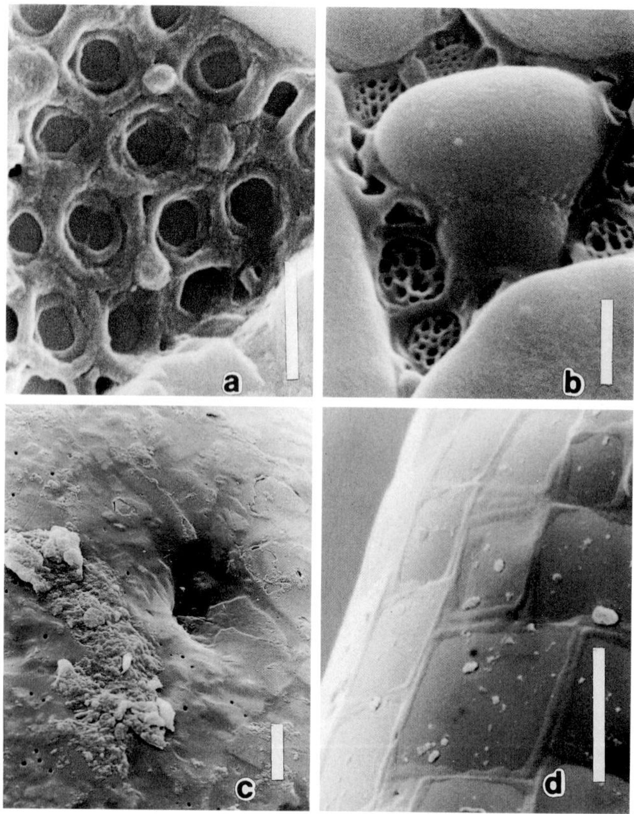

Fig. 2. Different organic cement types in Arcellinida. Structured cement: a) *Difflugia elegans*; b) *Netzelia wailesi*. Scale bar=1 µm. Sheet-like cement: c) *Cyclopyxis puteus*; d) *Quadrulella symmetrica*. Scale bar=5 µm. *Netzelia* and *Quadrulella* build siliceous tests with charcteristic idiosomes

Cell Morphology. For many species and genera of testate amoebae, few if any details are known about the cytoplasm because they have thick, opaque shells which prevent or hinder study by light or transmission electron microscopy. Those which have been studied are *Microchlamys* (Golemansky et al., 1987); *Arcella* (Netzel, 1975a, Raikov et al., 1989), *Centropyxis* (Netzel, 1975b), *Cryptodifflugia* (Griffin, 1972), *Difflugia* (Ogden and Meisterfeld, 1991; Ogden, 1991a), *Heleopera* (Ogden, 1991b), *Hyalosphenia* (Charret, 1964; Joyon et al., 1962), *Lesquereusia* (Harrison et al., 1976), *Nebela* (Bonnet et al., 1981a), *Netzelia* (Netzel, 1983; Anderson, 1987, 1990), *Phryganella* (Ogden and Pitta, 1990). Common to

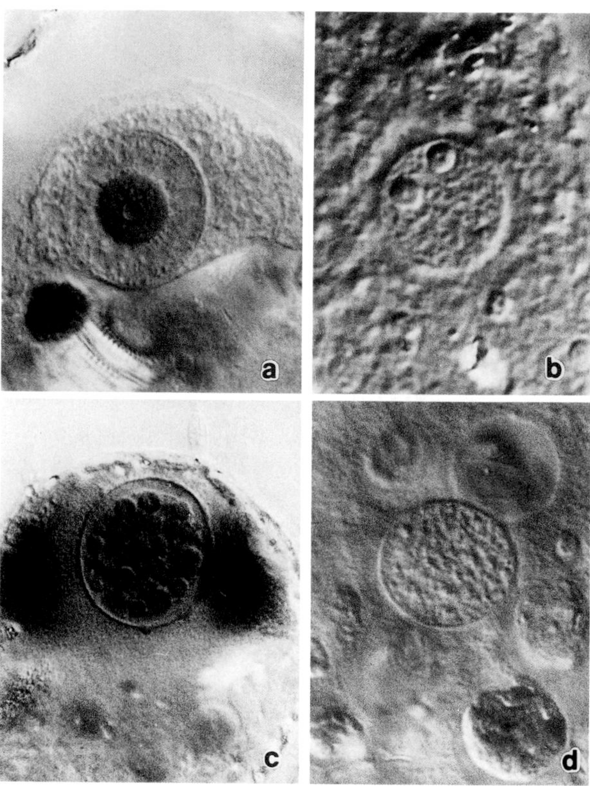

Fig. 3. a) **Vesicular** nucleus of *Difflugia hydrostatica*, diameter 22 µm. b) **Ovular** nuclei of *Archerella flavum*, diameter 11.5 µm, c) of *Netzelia wailesi*, diameter 25 µm, and d) *Apodera vas*, diameter 16 µm.

the species studied is a zonation of the cytoplasm into an anterior part close to the aperture with numerous food and digestive vacuoles, and a posterior more dense zone which stains with basic dyes ("**chromidium**") and usually contains one, or rarely two or more nuclei (genus *Arcella*). This region is a dense rough ER with one or more golgi complexes within or more often at the periphery. The dictyosomes are involved in the secretion of organic building units and cement and perhaps in the synthesis of siliceous idiosomes (e.g. Harrison et al., 1976, 1981). Contractile vacuoles are located laterally to this region. Mitochondria are usually ovoid or spherical and have branched tubular cristae. They are concentrated in the posterior part of the cell. The nucleus has two unit membranes, nuclear pores, and inner and outer additions to the membrane have been described (Raikov et al., 1989). According to the classification of Raikov (1982) two principal types of nuclei are found in testate amoebae: 1.

vesicular nuclei with one, often central nucleolus, sometimes with a few additional very small nucleoli; 2. **ovular nuclei** with several to many small nucleoli (Fig. 3). Small vesicles of cement are usually scattered throughout the cytoplasm in genera with a test constructed of hollow, organic building units (Netzel, 1975a; Mignot and Raikov, 1990).

Pseudopodia. Many species and even generic diagnoses are based on empty tests alone. Especially in soil-dwelling species, pseudopodia are difficult to observe. Arcellinida have either **endolobopodia**, which are granular or completely hyaline or **ectolobopodia** that can anastomose (**reticulo-lobopodia**). Bonnet (1961, 1963), who has studied the pseudopodia of 17 soil species has found transitions between these two types in *Plagiopyxis.* This brings into question distinctions between taxa based on the differences between pseudopodia. Pseudopodial characters in this group change with the kind of activity.

Reproduction. Arcellinida reproduce by binary fission. The first stage is the construction of an identical daughter shell. Depending on the type of shell, organic building units or mineral particles and cement are arranged around a cytoplasmic extension functioning as a template. On completion of the new test, fission takes place usually as a closed orthomitosis. There have been few studies on the ultrastructure of mitosis in the group (Raikov and Mignot, 1991) or studies of the morphogenesis of the shell (*Arcella*: Netzel, 1975a; Netzel and Grunewald, 1977; Mignot and Raikov, 1990. *Centropyxis*: Netzel, 1972. *Heleopera*: Ogden, 1989; *Netzelia*: Netzel, 1983. *Phryganella*: Ogden and Pitta, 1989).

Although binary fission is the normal form of reproduction, the Arcellinida can no longer be considered asexual. Mignot and Raikov (1992) have studied the details of meiosis and autogamy in the cysts of *Arcella.* Nuclear division and fusion within cysts have been studied in *Paraquadrula* (Lüftenegger and Foissner, 1991) and *Heleopera* (R. M. pers. observ.).

Cysts. Under unfavorable environmental conditions most testacea produce cysts. A few aquatic genera (e.g. many *Difflugia*) are believed not to form cysts. Resting cysts are formed within the test and have a thick organic wall. Soil and moss species also have a special kind of short-term cyst, the **precyst**. It is characterized by a relatively thin membrane enclosing the cell and a diaphragm, which closes the aperture. Often this diaphragm is externally enforced by a detrital plug. Some Microcoryciidae have external cysts.

Ecology. Testate amoebae are found world-wide, but they are neither cosmopolitan nor ubiquitous. Several species, mainly from the Nebelidae, Distomatopyxidae, and Lamtopyxidae have a restricted geographical distribution, which is certainly not a result of uneven sampling effort. Arcellinida are common in all freshwater habitats, inclusive of sediments, "Aufwuchs," plankton (Schönborn, 1962a), and mosses. Especially rich gatherings can be made in wet or moist sphagnum (e.g. Grospietsch, 1954; Meisterfeld, 1977, 1978; Schönborn, 1962b). Bonnet, Chardez, Coûteaux and Schönborn (review by Foissner, 1987) have shown that soils are characterized by a characteristic and diverse biota of Arcellinida and Euglyphida.

Their empty tests remain intact after the death of the amoebae and can be identified to species in most cases. In contrast to most other freshwater and soil protozoa, these remains give an accurate estimate of production rates and standing stocks (Lousier, 1974, 1984a-c, 1985). Testate amoebae are an important component in aquatic and terrestrial ecosystems. In soils they are as abundant as 10^6 to $10^8/m^{-2}$ and annually produce biomass of 1 to more than 200 g (Schönborn, 1977, 1982; Foissner and Adam, 1981; Lousier and Parkinson, 1984; Meisterfeld, 1986, 1989). In many forest soils they are as important as earthworms! They are consumers of bacteria, fungi, and different algal groups. Testate amoebae may feed selectively. Although some species can perforate fungal mycelia, they usually prefer smaller microbiota (Meisterfeld et al., 1992). Larger Heleoperidae and Nebelidae are predators of small Euglyphida (e.g. Ogden, 1989).

Availability of water determines the distribution of testate amoebae in mosses (Meisterfeld, 1977; 1979b, Tolonen et al.,1992) and soils as well as the population turnover (Lousier, 1974; Coûteaux, 1976a,b). Bonnet (e.g. 1964, 1985, 1988, 1990) used multivariate statistics to analyze the distribution of shelled amoebae in different soil types and found significant correlations between soil or humus type and testacean communities. Similar relations have been found in lake sediments

(Schönborn et al., 1966). This "bioindicative potential" has been used to monitor changes in the soil biota as effects of organic and mineral fertilizers (Chardez et al., 1972; Aescht and Foissner, 1992), liming (Wanner, 1991), or organic and conventional farming (Foissner et al., 1987).

Although tests fossilize and have been found in deposits as old as Mississippian (Loeblich and Tappan, 1964) or Triassic (Poinar et al., 1993), most records are from Pleistocene and younger deposits. The well defined niches of several species and habitat-specific communities have made testate amoebae useful in palaeolimnological studies. Analogous to pollen analysis, subfossil testacea assemblages have been used to describe changes of water tables during bog development and in investigations of the ageing of lakes (reviewed by Tolonen, 1986).

Phylogeny and classification. The evolutionary origins of the Arcellinida are uncertain. Page (1987) and Schönborn (1967, 1989) suggest that the Cochliopodiidae are the ancestors of the Arcellinida. As a result of his "Topophenetic Analysis," Schönborn (1989) proposes lines to the Centropyxidae, Arcellidae, and Microcoryciidae. Although the Arcellidae are relatively similar, the Centropyxidae form the 'stem' group of most other families of lobose testate amoebae. The Phryganellina present difficulties because of their distinctive reticulolobopodia. Griffin (1972), Hedley et al. (1977), and Ogden and Pitta (1990) have shown that the ultrastructure of these pseudopodia with their system of pointed pseudopodia and a connecting network of thin pseudopodial strands is quite different from normal lobopodia. Ogden and Pitta (1990) consider the pseudopodia of *Phryganella* to be filopodia! Molecular data will help to solve these problems and to reconstruct the true phylogeny of this group.

Considering these uncertainties, the classification of the Arcellinida used here remains provisional and follows conventional lines. Three suborders are recognized: the Arcellina with three, the Difflugina with twelve, and the Phryganellina with two families.

Key characters. Determination of testate amoebae is based mainly on characters of the test. Features of the cell are of minor importance. The nuclear morphology, whether ovular or vesicular, can be very useful. Unfortunately, many species are only known as empty tests.

Observation and identification. Arcellinida are common in almost all aquatic and terrestrial habitats. High numbers are often found in wet mosses like sphagnum. Empty tests of soil species are best separated by placing an air-dried soil sample in a cylinder of water and allowing the air-filled tests to float to the surface. Bonnet and Thomas (1958) designed an isolation apparatus which was later modified by Decloitre (1960) and Chardez and Krizelj (1970).

Embedding in resin (e.g. Euparal Chroma, refractive index 1.535) is recommended to clear the opaque shells and to inspect the internal morphology of the shells of Centropyxidae, Plagiopyxidae, and Distomatopyxidae. For light microscopy, fixation in Bouin's fixative or sublimate and staining with Meyer's acid haemalum, bromophenol blue, or borax carmine have been used successfully. Scanning electron microscopy (SEM) is the best means to document test morphology and to study the organic cement matrix. Standard preparative procedures (Lee and Soldo, 1992) are sufficient. A promising approach for the identification of clones of lobose as well filose testate amoebae is RAPD-PCR (Wanner et al., 1997).

Many aquatic and soil species can be cultured in soil extract or Prescott and James' media. The media should not be organically enriched. Yeast, *Chlorella vulgaris*, *Chlorogonium elongatum*, and *Enterobacter aerogenes* are useful food species. Additionally testate amoebae with agglutinate shells need small mineral particles as test-building material.

Literature. No modern monographs are available. Penard (1902), Cash and Hopkinson (1905, 1909), Cash et al. (1915), Wailes and Hopkinson (1919) are useful. Keys to many aquatic and planktonic species can be found in Grospietsch (1972a,b) and Harnisch (1958), but the descriptions of most soil species are too recent to be included in these publications. Bonnet and Thomas (1960) describe about a hundred taxa. Schönborn (1966) contains information on the general biology and ecology. Ogden and Hedley (1980) illustrate 95 mainly aquatic and moss species by SEM micrographs. Decloitre (1976a, 1977a,b, 1978, 1979a,b, 1982, 1986) has published a series of compilations although without any attempt to revise the numerous insufficiently described taxa. These papers may help to provide access to the scattered literature.

KEY TO SUBORDERS OF THE ORDER

1. Test membraneous or chitinoid, pliable or rigid; no scales or plates; may have attached debris; pseudopodia digitate, finely granular.................. .... **Arcellina**

1'. Test rigid; usually with embedded and/or attached mineral particles, plates, scales, siliceous granules in organic cement, sometimes chitinoid; pseudopodia digitate, finely granular................................ **Difflugina**

1". Test with attached or embedded siliceous material, sometimes completely chitinoid; pseudopodia conical, clear (ectoplasmic), may anastomose............................... **Phryganellina**

SUBORDER ARCELLINA Haeckel, 1894

With characters as in the key, above.

KEY TO THE FAMILIES IN THE SUBORDER

1. Tectum or test flexible...................................... 2
1.'Test rigid areolar, pseudostome distinct, round... .... **Arcellidae**

2. Test flexible, areolate, cytoplasm in a separate membrane sac.................. **Microchlamyiidae**
2'. Test flexible to semi-rigid, no areoles; pseudostome distinct, variable....... **Microcoryciidae**

FAMILY MICROCORYCIIDAE de Saedeleer, 1934

Test proteinaceous, flexible to semi-rigid, not areolate; pseudostome ventral, distinct, variable.

None of the genera in this family has been studied with modern methods; therefore this group may be artificial, and affinities to other families like Microchlamyiidae remain unclear.

KEY TO THE GENERA OF THE FAMILY

1. External layer of test gelatinous; usually one nucleus; plasma violet; pseudostome invaginated .... ***Amphizonella***
1'. Test not gelatinous... 2

2. Pseudostome, like a slit.................................... 5
2'. Pseudostome variable or round sometimes wide open..3

3. Dorsal face with mineral particles or attached debris..4
3'. No foreign material, test smooth......................... .... ***Penardochlamys***

4. Test hemispherical, bilayered; flexible, wide, ventral pseudostome .... ***Diplochlamys***
4'. Dorsal test ±rigid; ventral like a flexible skirt.. .... ***Microcorycia***

5. Test pyriform or discoid, violet........ ***Zonomyxa***
5'. Test laterally compressed, in oral view ovoid..... .... ***Parmulina***

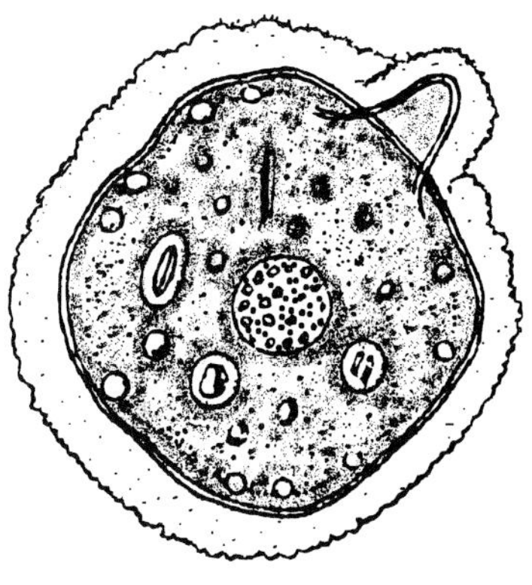

Fig. 4. *Amphizonella violacea*. Scale bar=50 µm. (After Penard, 1906)

Genus ***Amphizonella*** Greeff, 1866

Test ±round; bilayered pellicle, outer layer gelatinous 8-12 µm with fine denticles, inner layer thin, chitinoid, undulates with internal movements, sac-like; pseudostome invaginated,

variable (Fig. 4). Size 125-200 µm. Pseudopods cylindroid, finely granular, rounded ends. Movements sluggish. Endoplasm clear, violet; yellow granules in purple vesicles (digestion products of cyanobacteria?). One ovular nucleus. Contractile vacuoles 20 to 30 µm. Crystals none. Cysts reported. Feeding: algivorous. Habitat: sphagnum mosses. Monospecific. Refs. Penard (1906), Thomas (1957).
TYPE SPECIES: *Amphizonella violacea* Greeff, 1866

Genus ***Diplochlamys*** Greeff, 1888

Test small, round, greyish-yellow, bilayered; inner hyaline sac enclosing the cell with flexible aperture, outer layer consists of loosely arranged debris (Fig. 5). Size 40 to 60 µm. Pseudopods clear, 10 to 20 µm long, tapered, nearly pointed. Endoplasm granular; one vesicular nucleus. Feeding: herbivorous. Habitat: mosses on trees, soil. Five species. Ref. Penard (1909).
TYPE SPECIES: *Diplochlamys leidyi* Greeff, 1888

Fig. 5. *Diplochlamys timida.* Scale bar= 50 µm. (After Penard, 1909)

Genus ***Microcorycia*** Cockerell, 1911

Test composed of flexible membrane, aboral part like a dome, sometimes with concentric ridges or small horns at the dorsal side, covered with small debris or mineral particles. Test bowl-shaped, brown, extends as membranous border at rim, pursed shut around body (Fig. 6). Most species smaller than 100 µm. Pseudopods lobate, through aperture. Ectoplasm pale, at tips of pseudopods. Endoplasm granular, yellowish to reddish, some species with glycogen spheres (2 to 5 µm). One ovular nucleus. Cyst external (Hallas, 1975). Feeding: herbivorous. Habitat: dry mosses. Eleven species. Refs: Penard (1902), Chardez (1984).

TYPE SPECIES: *Microcorycia flava* (Greeff, 1866, emend Penard, 1902)

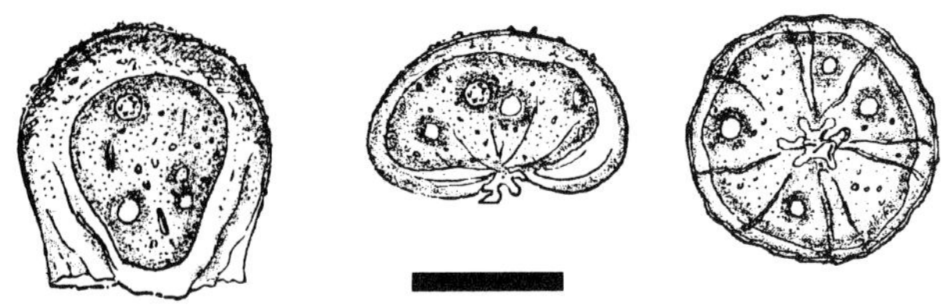

Fig. 6. *Microcorycia flava.* Scale bar=50 µm. (After Penard, 1902)

Genus ***Penardochlamys*** Deflandre, 1953

Test round, wavy border, chitinoid, flexible, punctate; pseudostome round or wavy, rarely visible, slightly invaginated, 10 µm (Fig. 7). Size 60 to 70 µm; ovoid laterally. Pseudopods cylindroid, finely granular, round ends. Endoplasm granular. Two vesicular nuclei, 1 - 2 contractile vacuoles. Feeding: herbivorous. Habitat: freshwater, on plants. Monospecific. Synonym: *Pseudochlamys arcelloides*. Refs.: Penard (1904), Deflandre (1953).
TYPE SPECIES: *Penardochlamys arcelloides* (Penard, 1904)

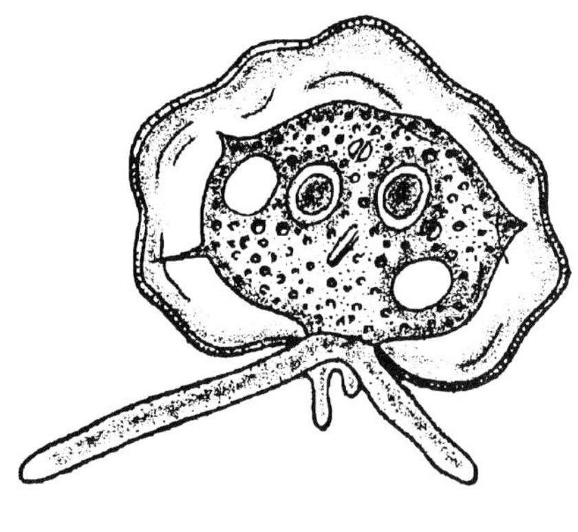

Fig. 7. *Penardochlamys arcelloides.* Scale bar=20 µm. (After Penard, 1904)

Genus ***Zonomyxa*** Nüsslin, 1882

Test discoid (resting form), 140 to 160 µm; locomotive form pyriform, 220 to 250 µm long (Fig. 8). Test chitinoid pellicle, without mucilaginous envelope, flexible, follows the

movements of the cell. Surface with small temporal perforations through which thin plasma threads emerge. Pseudopods single, clear, conical, from slit-like pseudostome. Endoplasm granular, violet-tinted. Up to 32 vesicular nuclei. Crystalloid bodies, 3 to 5 µm. Feeding: herbivorous. Habitat: sphagnum mosses. Monospecific. Ref. Penard (1906).
TYPE SPECIES: *Zonomyxa violacea* Nüsslin, 1882

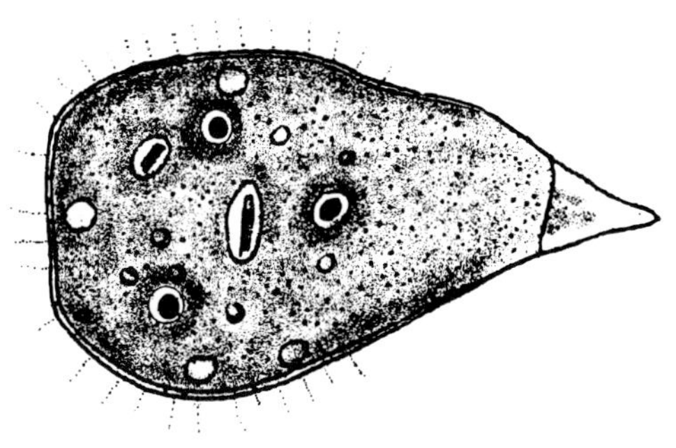

Fig. 8. *Zonomyxa violacea*. Scale bar=50 µm. (After Penard, 1906)

Genus ***Parmulina*** Penard, 1902

Test in lateral view hemispherical, compressed laterally, in oral view ovoid; slit-like pseudostome when folded. Test clear, flexible, chitinoid, opaque, attached debris (Fig. 9). Pseudopods rare, bluntly lobate. Endoplasm granular. One vesicular nucleus. Feeding: herbivorous. Habitat: mosses on trees. Four species. Ref. Penard (1909).
TYPE SPECIES: *Parmulina cyathus* Penard, 1902

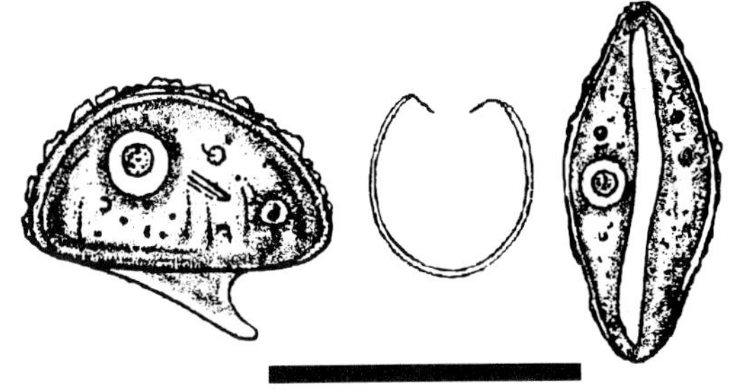

Fig. 9. *Parmulina cyathus*. Scale bar=50 µm. (After Penard, 1909)

FAMILY MICROCHLAMYIIDAE Ogden, 1985

Test proteinaceous, flexible or rigid, finely **areolate**; cytoplasm enclosed in a separate membrane sac with single aperture, attached to main test; nucleus central.

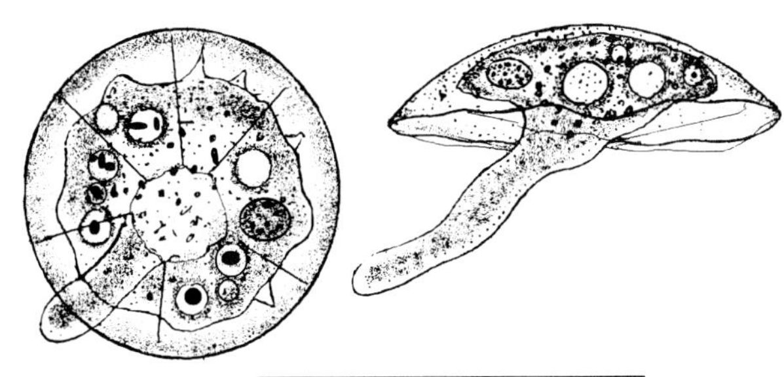

Fig. 10. *Microchlamys patella*. Scale bar=50 µm. (After Penard, 1902)

Genus ***Microchlamys*** Cockerell, 1911

Test clear, yellow to brown, finely areolated (~0.45 µm), chitinoid, flexible, folded; cell enclosed within a membranous sac which is fixed to the shell at intervals, lost in empty shells. The sac has a central pseudostome (Fig. 10). Pseudopods cylindroid, finely granular. Endoplasm granular; cell moves inside the shell eruptively. One vesicular nucleus. Crystals 2 µm or less. Feeding: herbivorous. Habitat: sphagnum mosses, soil, freshwater. Two species. Refs: Penard (1902), de Saedeleer (1934), Ogden (1985), Golemansky et al. (1987).
TYPE SPECIES: *Microchlamys patella* (Claparéde & Lachmann, 1859) Synonym=*Pseudochlamys*.

FAMILY ARCELLIDAE Ehrenberg, 1843

Test rigid, transparent, colorless, yellow, or brownish, composed of organic building units, areolar, smooth; aperture round.

KEY TO GENERA OF THE FAMILY

1. Aperture smaller than half the shell diameter.. 2
1'. Aperture almost as large as the shell diameter, one vesicular nucleus.................... ***Pyxidicula***

2. Shell of variable shape, bi- or multinucleate, common.. ***Arcella***
2'. Test hemispherical, one ovular nucleus............... .... ***Antarcella***

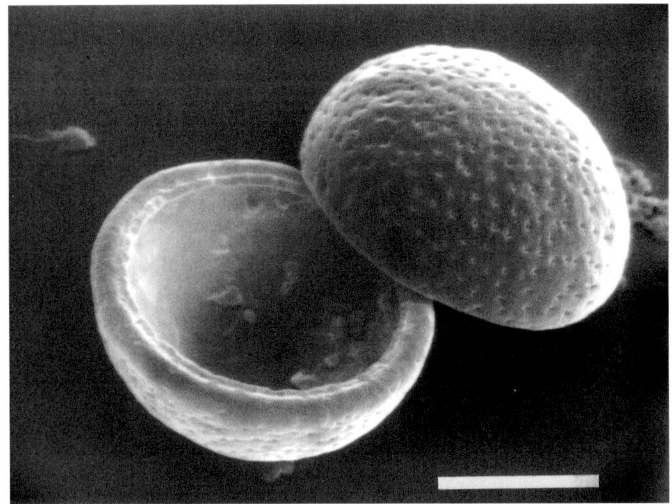

Fig. 11. *Pyxidicula operculata.* Scale bar=10 μm.

Genus ***Pyxidicula*** Ehrenberg, 1838

Shell hemispherical, bowl-shaped; the aperture is almost as wide as the diameter of the shell, apertural rim usually recurved outside (Fig. 11). Shell wall areolate, composed of organic, hollow building units, which in older, brown specimens become filled with inorganic material like manganese. One vesicular nucleus. During mitosis the nuclear membrane and the nucleolus remain intact with an intranuclear spindle apparatus (closed orthomitosis). One contractile vacuole. Feeding on algae and bacteria. Habitat: freshwater plants. Eight species. Refs. Penard (1902), Doflein (1916), Ogden (1987a).
TYPE SPECIES: *Pyxidicula operculata* (Agardh, 1827) Ehrenberg, 1838.

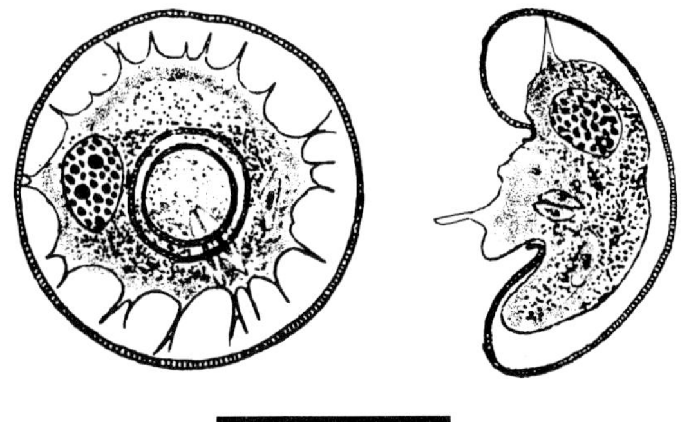

Fig. 12. *Antarcella atava.* Scale bar=50 μm. (After Collin, 1914)

Genus ***Antarcella*** Deflandre 1928, emend 1953

Test in lateral view hemispherical; aperture circular, invaginated. In contrast to *Arcella*, *Antarcella* has one ovular nucleus (Fig. 12). Contractile vacuole 10 to 15 μm. Feeding: herbivorous. Habitat: sphagnum mosses. Two species. Refs: Collin (1914), Deflandre (1928a).
TYPE SPECIES: *Antarcella atava* (Collin, 1914)

NOTE: To distinguish *Antarcella* species from *Arcella* spp., staining of the nucleus is recommended.

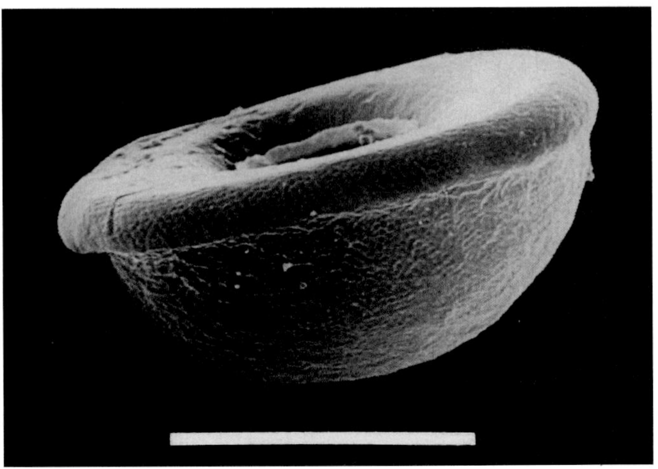

Fig. 13. *Arcella vulgaris.* Scale bar=50 μm.

Genus ***Arcella*** Ehrenberg, 1832

Species with a more or less circular shell, aperture central, invaginated, in many species surrounded by a tubus or a circle of pores (Figs. 13–15). Test completely organic, composed of box-like building units arranged in a single layer and cemented together, resulting in an areolar surface. Size of the building units varies between species. Young shells are colorless, older become brown due to iron and manganese storage into the building units. Most species are binucleate, but several species have more; *A. megastoma* may have up to 200. These nuclei are always vesicular. Several contractile vacuoles. Cell does not fill the test, fixed with small epipodes at the shell wall. Cyst round, within test. About 50 species and many more varieties and forms. Deflandre (1928a) in his monograph has subdivided this genus in 4 artificial sections and 4 derived groups on the basis

of the height/diameter ratio, the presence of angular facets, or a flat circular rim.
TYPE SPECIES: *Arcella vulgaris* Ehrenberg, 1832.

Arcella species can be found in all freshwater biotopes, wet and dry mosses, and with a few species in soils. Feeding mainly herbivorous. Refs: Deflandre (1928a), Decloitre (1976a, 1979b, 1982, 1986), Netzel and Grunewald (1977), Ogden and Hedley (1980).

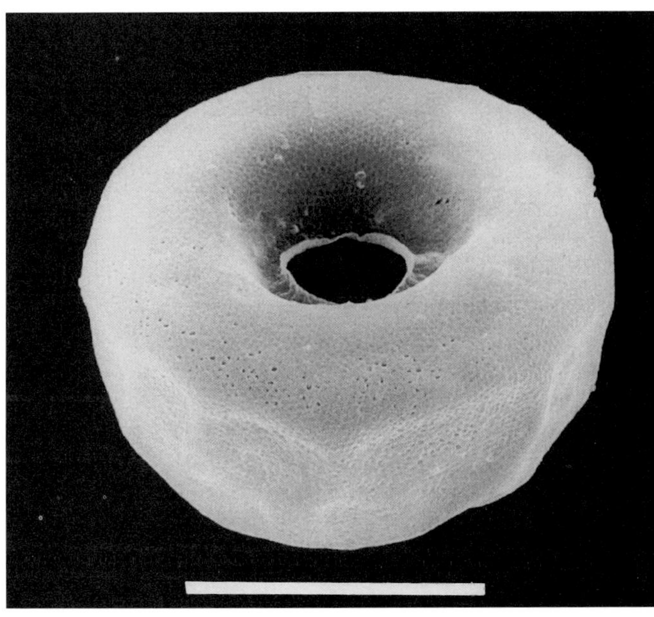

Fig. 14. *Arcella gibbosa*. Scale bar=50 μm.

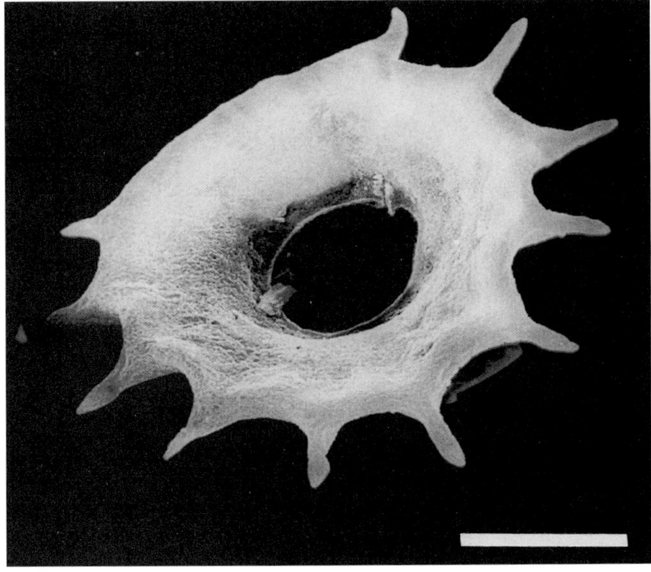

Fig. 15. *Arcella dentata*. Scale bar=40 μm.

1. **Section Vulgares** with a h/d ratio 0.35 - 0.95 (Figs. 13, 14).
2. **Section Carinatae** with a keel or ring of spines (Fig. 15).
3. **Section Aplanatae** with a h/d ratio < 0.3.
4. **Section Altae** with a h/d ratio > 1.0.

SUBORDER DIFFLUGINA

Test either composed of mineral grains, diatoms, or collected scales or plates, or composed of endogenously formed siliceous or calcite elements of various shapes (**idiosomes**) both held together by an ± abundant organic cement, or completely chitinoid. Pseudopods digitate, granular.

KEY TO THE FAMILIES OF THE SUBORDER

1. Pseudostome terminal or central at ventral face if test is circular or oval in oral view............ 2
1'. Pseudostome sub-terminal or at the end of an asymmetrical neck 8

2. Ventral face flat, aperture more or less invaginated.. 10
2'. Aperture terminal 3

3. Aperture slit-like or lenticular.. **Heleoperidae**
3'. Aperture round, oval or lobed 4

4. Test completely organic......... **Hyalospheniidae**
4'. Test with mineral grains, diatoms, or endogenously formed elements (idiosomes) such as rectangular, worm-like, or nail-shaped plates 5

5. a) Test composed mainly of mineral grains or diatoms, often opaque greyish or b) test hexagonal or polygonal in cross-section, mainly organic with adherent mineral particles .... **Difflugiidae**
5'. Test composed of round, oval, nail-shaped, sausage-like, or rectangular plates, collected or endogenously formed; shell sometimes mixed with mineral grains, usually transparent, sometimes compressed 6

6. Test completely or partly composed of endogenously formed elements 7

6'. With collected or predated round or oval siliceous plates, fragments of diatoms or mineral grains................................. **Nebelidae**
6". Test is dark violet..................... **Heleoperidae**

7. Siliceous rectangular plates, nails, or sausage-like beads sometimes mixed with mineral particles .... Family **Lesquereusiidae**
7'. Calcite plates.. (These species are all small, use microscope with polarizer).**Paraquadrulidae**

8. Test with an asymmetrical neck.......................... .... **Lesquereusiidae**
8'. Test with sub-terminal invaginated or slit-like aperture... 9

9. Aperture oval, round, or semicircular always invaginated.............................. **Centropyxidae**

9'. Aperture a more or less invaginated slit, overhung by the anterior lip, in ventral view often difficult to observe .... **Plagiopyxidae**

10. Aperture circular, oval, triangular, three-lobed, or crescent-shaped..... **Trigonopyxidae**
10'. Apertural apparatus with an internal opening at the end of an invaginated tubus and an external aperture with either teeth-like structures or two crescent shaped openings ..11

11. External aperture with two crescent-shaped openings............................ **Distomatopyxidae**
11'. External aperture bordered by large teeth........ .... **Lamptopyxidae**

FAMILY DIFFLUGIIDAE Wallich, 1864

Pseudostome terminal, of variable shape.

KEY TO GENERA OF THE FAMILY AND GENERA *INCERTAE SEDIS*

LM

1. Test round or oval in cross-section.................. 2
1'. Test polygonal in cross-section........................ 9

2. Test with 3-4 lateral bulges near pseudostome.. .... ***Maghrebia***
2'. No lateral bulges.. 3

3. Test partitioned in two parts by a diaphragm between neck and main body, often visible as constriction... 6
3'. No such diaphragm.. 4

4. Aperture surrounded by a distinct lobed or denticulate collar and an internal circular opening or complex, consisting of a central pore and irregularly elongate oval pores radiating around the central pore.................................. 8
4'. Aperture round, oval, lobed or denticulate, no distinction between internal and external opening.. 5

5. Test composed mainly of angular mineral particles or diatoms; aperture relatively large, round, oval or lobed; many different shell shapes, very common......................... ***Difflugia***
5'. Test elongate oval, brownish, chitinoid with attached mineral particles; with small aperture .... ***Schwabia***
5". Test ovoid, aperture at the broader end, brownish, smooth chitinoid surface.................. .... ***Pseudawerintzewia***

6. Internal diaphragm with one opening.................... .... ***Lagenodifflugia***
6'. Internal diaphragm with two openings............. 7

7. Diaphragm composed of small mineral grains, two circular openings.................... ***Zivkovicia***
7'. Two openings formed by a mainly organic bridge with few attached mineral particles which connects both broad sides................ ***Pontigulasia***

8. Test with incurved collar, three to six lobes .... ***Cucurbitella***
8'. Aperture denticulate with more than 12 teeth..... .... ***Protocucurbitella*** (see ***Difflugia***)
8". Aperture complex with central opening and numerous elongate oval pores radiating around the central pore....................... ***Suiadifflugia***

9. Test chitinoid or with few attached mineral particles, usually hexagonal in cross-section, no collar.................................... ***Sexangularia***
9'. Test agglutinated of mineral grains; cross-section triangular to pentagonal, short collar..... .... ***Pentagonia***

Genus ***Difflugia*** Leclerc, 1815

Species with an agglutinated shell, with terminal, round, oval, lobed, or teethed (but never slit-like) aperture, some with necklace but never with internal diaphragm (Figs. 16-23). Test always composed of mineral particles or diatoms in a structured or sheet-like organic cement. Many *Difflugia* select and arrange the building material according to size and shape to construct a species-specific shell. The nucleus is mostly ovular, in some species vesicular. Several freshwater species have green symbionts.

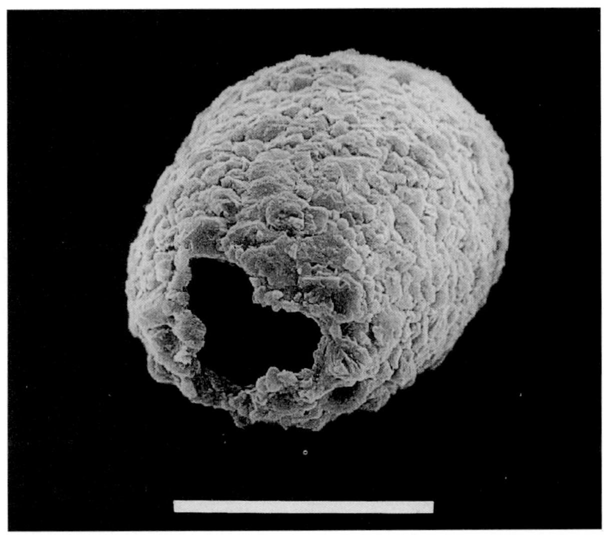

Fig. 16. *Difflugia gramen*. Scale bar=100 µm.

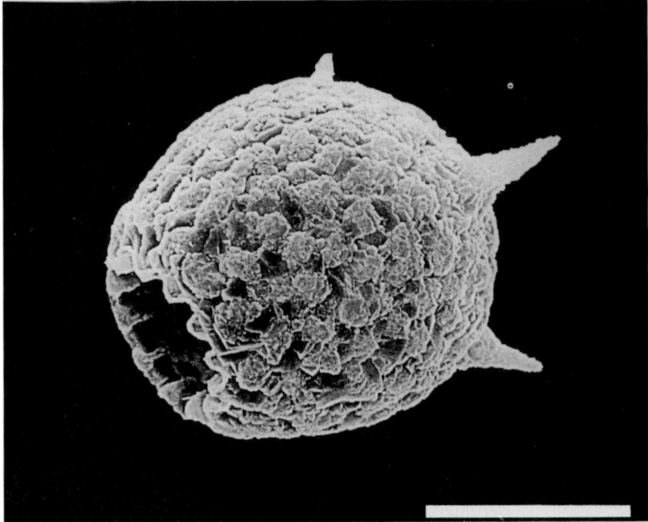

Fig. 17. *Difflugia corona*. Scale bar=100 µm.

The taxonomy of this genus is based mainly on differences in shape and size of their agglutinated shells. As the test is often opaque, cytoplasmic characters are rarely used. Small differences in shell size, shape, or composition have been sufficient for many authors to describe more than 300 species and about 200 subspecies, varieties, or forms with little regard to the value of the characters used, the previous literature, or the rules of nomenclature. Many of these descriptions are inadequate by modern standards and therefore the determination to species level is extremely difficult, even for the specialist. A promising specific character is the patterning of the organic cement (Ogden, 1983a, 1990; Ogden and Zivkovic, 1983). Gauthier-Lièvre and Thomas (1958) in their survey of the African species have divided the genus *Difflugia* into ten artificial groups based on the morphology of the test: lobed (Figs. 16,17), collared, compressed (Fig. 18), urceolate (Fig. 19), globose, ovoid-globose (Fig. 20), elongate, acute angled (Fig. 21), horned (Fig. 22), and pyriform (Fig. 23). Like the shell morphology the habitats of the numerous *Difflugia* species are very diverse. Many species are common in freshwater sediments or between water plants; others like *D. hydrostatica* are planktonic; *D. lucida* lives in dry mosses and soil. Feeding on mainly algae and fungi. Refs: Jennings (1916), Gauthier-Lièvre and Thomas (1958), Ogden (1979b; 1980; 1983a), Ogden and Hedley (1980), Ogden and Meisterfeld (1989; 1991), Ogden and Zivkovic (1983).

The type species *D. proteiformis* Lamarck, 1816 has not been studied with modern methods, and its nature is questionable (Ogden and Ellison, 1988).

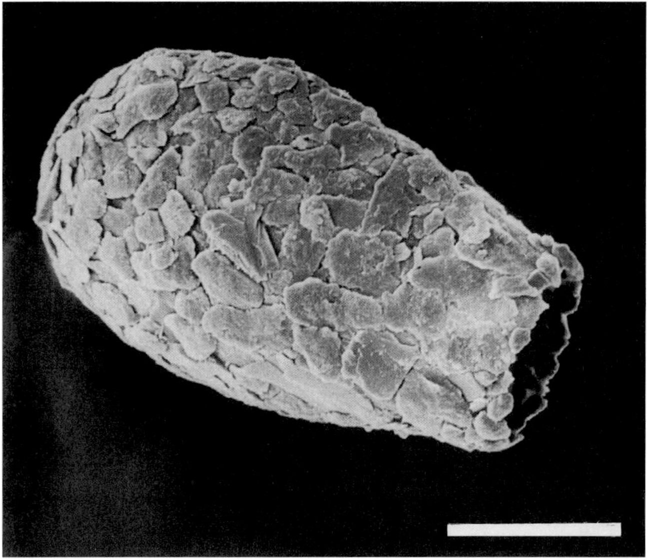

Fig. 18. *Difflugia lucida*. Scale bar=20 µm.

NOTE: Gauthier-Lièvre and Thomas (1960) have described the following:

1. The subgenus ***Pseudocucurbitella*** which has affinities to the lobed *Difflugia spp.* (Fig. 16) as well as to the genus *Cucurbitella* (Fig. 30).

2. The test of the genus ***Protocucurbitella*** resembles that of *Difflugia corona* but differs by having a diaphragm around the peristome. Feeding: herbivorous. Habitat: freshwater plants. Five species, which differ mainly in size and length of horns. This genus is not generally accepted; see Dekhtyar (1993).

TYPE SPECIES: *Protocucurbitella coroniformis*

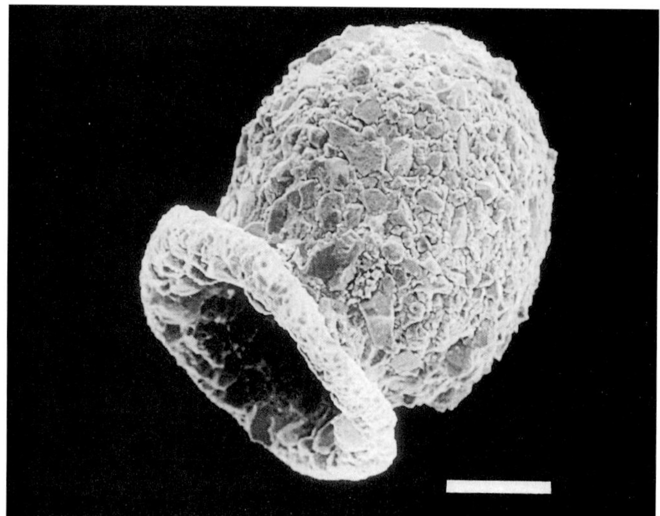

Fig. 19. *Difflugia urceolata.* Scale bar=50 µm.

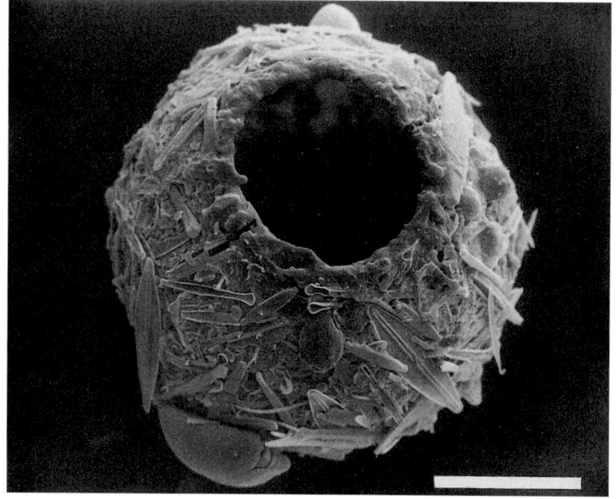

Fig. 20. *Difflugia globulosa.* Scale bar=50 µm.

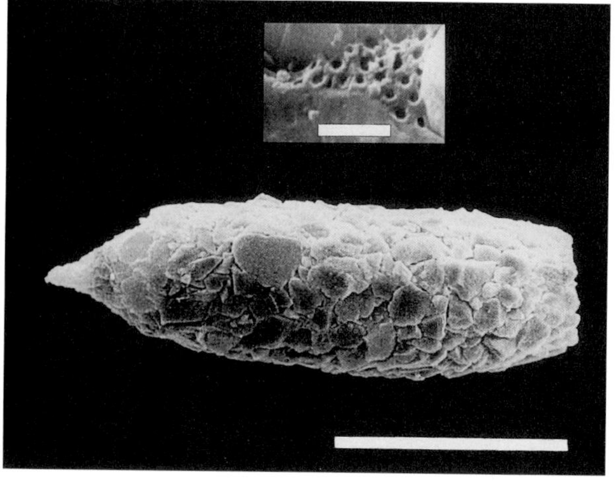

Fig. 21. *Difflugia smilion.* Inset with organic cement. Scale bar=100 µm (inset 2 µm).

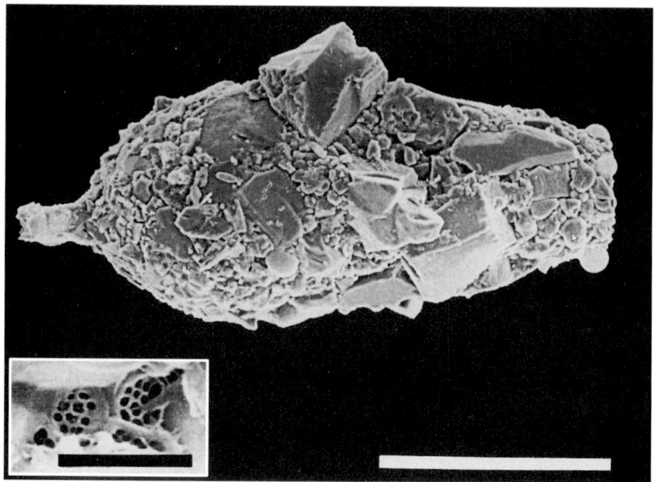

Fig. 22. *Difflugia acuminata.* Inset shows the organic cement. Scale bar=100 µm (inset 2 µm).

Fig. 23. *Difflugia bacillifera.* Characteristic species of wet *Sphagnum.* Scale bar=50 µm.

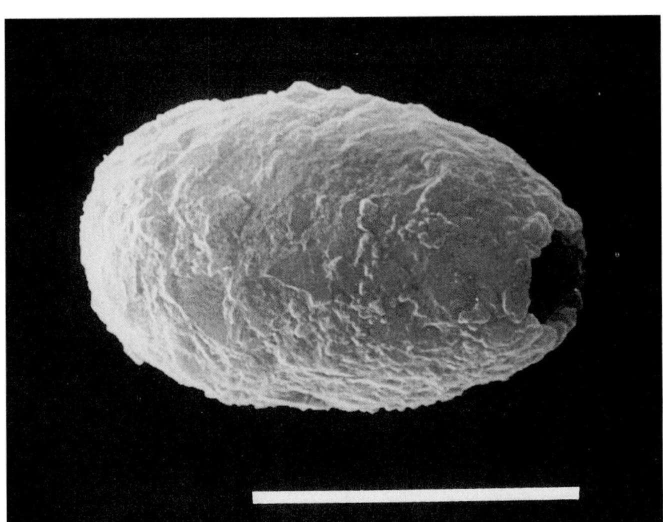

Fig. 24. *Schwabia terricola*. Scale bar=50 µm.

Genus ***Schwabia*** Jung, 1942

Shell ovoid, circular in cross-section, chitinous with small mineral particles which produce a smooth surface, aperture terminal (Fig. 24). Four species. Habitat: either freshwater or alkaline forest soils (*S. terricola*). Refs.: Bonnet and Thomas (1960), Chardez (1966).
TYPE SPECIES: *Schwabia irregularis* Jung, 1942.

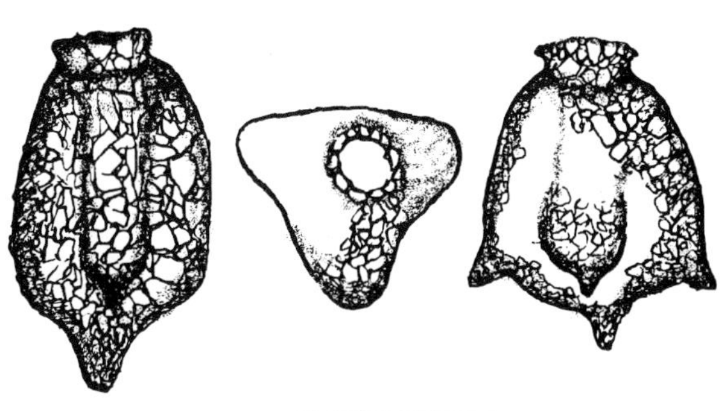

Fig. 25. *Pentagonia maroccana*. Scale bar=50 µm. (After Gauthier-Lièvre and Thomas, 1958)

Genus ***Pentagonia***
Gauthier-Lièvre & Thomas, 1958

Shape pyriform, in cross section polygonal with 3-5 lateral swellings sometimes terminating in a horn, short (5-8 µm) collar, aperture circular, fundus with horn. Agglutinate shell composed of mineral particles as in *Difflugia*. Habitat: freshwater, sediments. Monospecific. (Fig. 25). Ref. Gauthier-Lièvre and Thomas (1958).
TYPE SPECIES: *Pentagonia maroccana* Gauthier-Lièvre & Thomas, 1958

Genus ***Maghrebia***
Gauthier-Lièvre & Thomas, 1958

Test cylindroid, round base, constricted at short neck with 4 ridges, like water-jar with handles; chitinoid; embedded siliceous particles (Fig. 26). Pseudostome round. Feeding: herbivorous. Habitat: warm freshwater, on plants, Africa. Monospecific. Ref. Gauthier-Lièvre and Thomas (1958).
TYPE SPECIES: *Maghrebia spatulata* Gauthier-Lièvre & Thomas, 1958.

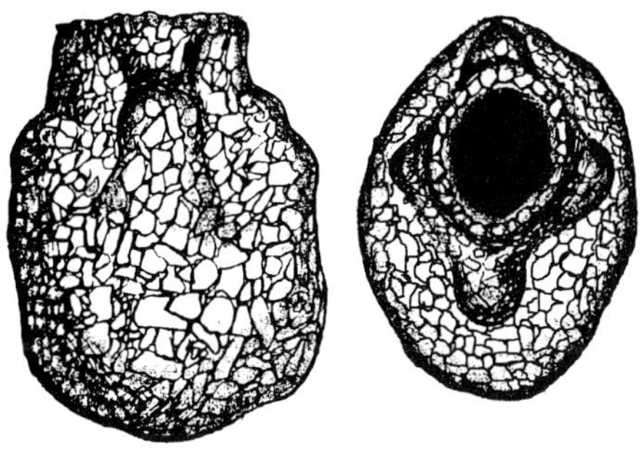

Fig. 26. *Maghrebia spatulata*. Scale bar=100 µm. (After Gauthier-Lièvre and Thomas, 1958)

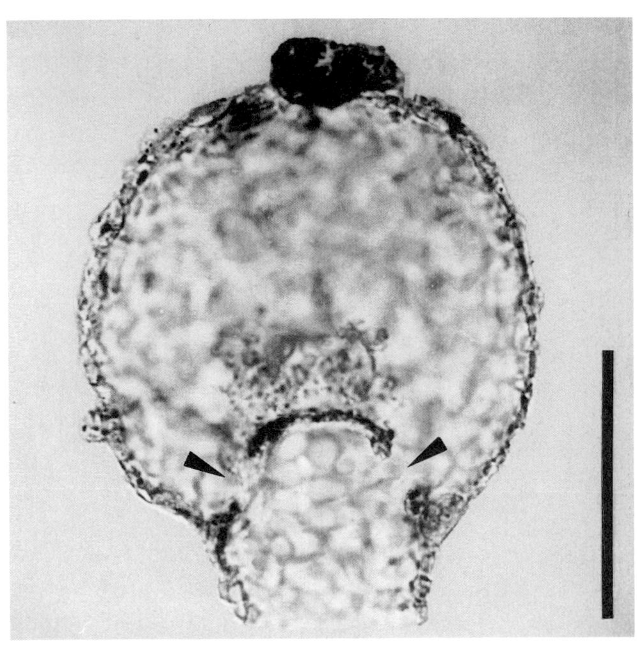

Fig. 27. *Zivkovicia compressa*. The arrows mark the two internal openings. Note: specimen mounted in Euparal. Scale bar=100 µm.

Genus ***Zivkovicia*** Ogden, 1987

Shell pyriform, mostly with a distinct constriction of the neck which sometimes can be obscured by larger mineral particles, circular or compressed in transverse-section, aperture terminal, internal mineral diaphragm with 2 circular openings (Fig. 27). Test built of agglutinated mineral particles bound by a structured organic cement. From freshwater sediments. Recent descriptions: Ogden (1983b, 1987b).
TYPE SPECIES: *Zivkovicia compressa* (Carter, 1864)

Genus ***Lagenodifflugia*** Medioli & Scott, 1983

Shell pyriform, often with constriction of the neck, cross-section circular or slightly compressed, composed of mineral particles bound by a structured organic cement network with perforated meshes; pseudostome terminal, circular (Fig. 28). The test is partitioned into 2 regions by a diaphragm-like part of the shell wall with a single central opening. Ovular nucleus. Habitat: freshwater plants, sphagnum, and sediments. Four species. Refs. Ogden (1983b, 1987b).
TYPE SPECIES: *Lagenodifflugia vas* (Leidy, 1874)
Synonyms: *Difflugia vas* Leidy, 1874, *Pontigulasia spectabilis* Penard, 1902 (in part), *Pontigulasia vas* (Leidy, 1874) Schouteden, 1906.

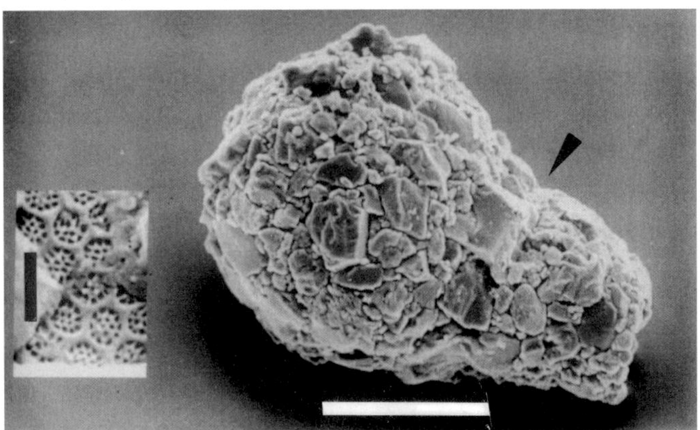

Fig. 28. *Lagenodifflugia vas.* Inset shows the organic cement. Scale bar=50 µm (inset 1 µm).

Genus ***Pontigulasia*** Rhumbler, 1896

Shell pyriform, in transverse-section either circular or compressed, sometimes with a constriction between the main body and the neck, aperture terminal, circular. In the region of the constriction the shell is divided internally into 2 parts in this genus, in contrast to *Zivkovicia*, by a narrow mainly organic bridge with few attached mineral particles, which connects both broad sides (Fig. 29). The test is composed mainly of agglutinate mineral particles with some diatom frustules. In embedded specimens the internal bridge can be observed as a dark structure in lateral view. Feeding: herbivorous. Habitat: freshwater plants, sphagnum, and sediment. Five species. Refs. Ogden (1983b, 1987b).
TYPE SPECIES: *Pontigulasia rhumbleri* Hopkinson, 1919

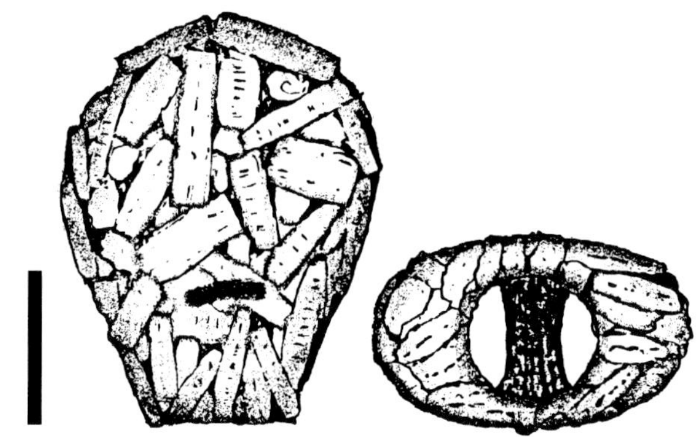

Fig. 29. *Pontigulasia rhumbleri.* Scale bar=50 µm. (After Wailes and Hopkinson, 1919)

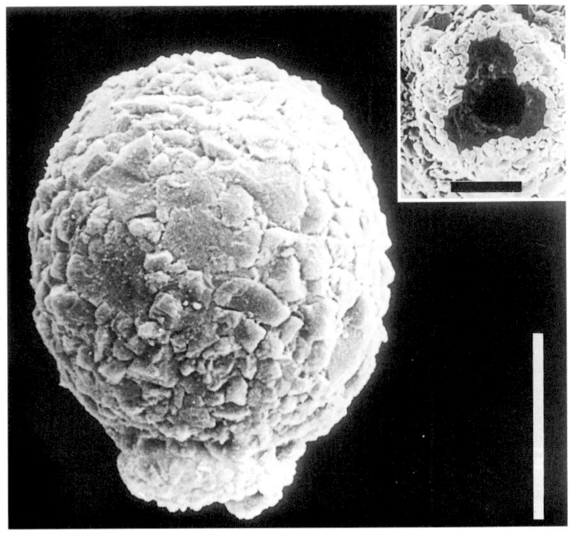

Fig. 30. *Cucurbitella mespiliformis.* Scale bar=50 µm (inset 20 µm).

Genus ***Cucurbitella*** Penard, 1902

Shell ovoid, with distinct apertural collar, dark grey or opaque in color, outline regular, body constructed of small to medium, angular mineral

grains. Organic cement seldom visible as surface structure (Fig. 30). Aperture with 3 to 12 lobes, composed of small mineral grains. Internal aperture a circular or lobed opening in diaphragm on a level with main body wall. Nucleus vesicular, cytoplasm of some species with zoochlorellae. Feeding: herbivorous. Habitat: freshwater plants and sediments. Thirteen species. Refs. Penard (1902), Gauthier-Lièvre and Thomas (1960), Ogden and Meisterfeld (1989).
TYPE SPECIES: *Cucurbitella mespiliformis* Penard, 1902.

Genus ***Suiadifflugia*** Green, 1975

Test sub-spherical, (130–200 µm), the aperture is complex, consisting of a central pore and numerous elongate oval pores radiating around the central pore like the petals of a flower (Fig. 31). Habitat: freshwater plants and sediments. Distribution: Mato Grosso (Brazil). Monospecific. Ref. Green (1975), Velho and Lansac-Tôha (1996).
TYPE SPECIES: *Suiadifflugia multipora* Green, 1975

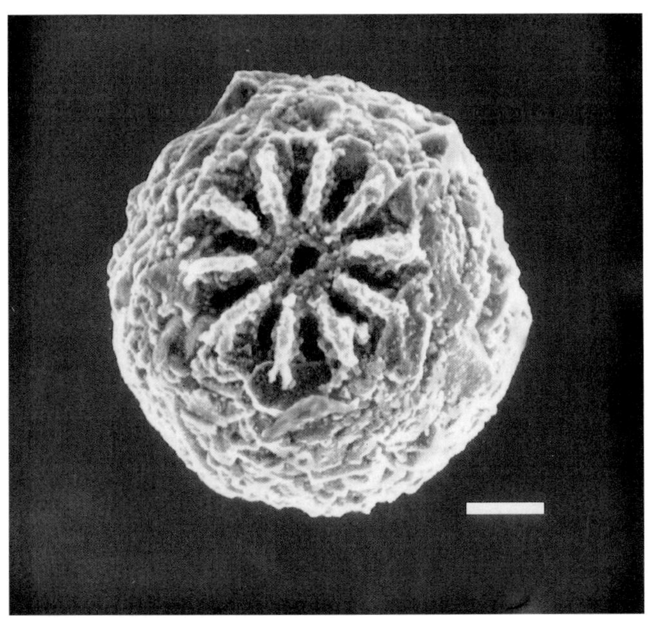

Fig. 31. *Suiadifflugia multipora.* Scale bar=20 µm. (From Green, 1975)

Genus ***Sexangularia*** Awerintzew, 1906

Organic test, sometimes with attached mineral particles, cross-section polygonal mostly hexagonal (Fig. 32). Nucleus vesicular. Habitat: freshwater plants, sphagnum. Three small species (10-70 µm). Refs. Penard (1904), Deflandre (1931).
TYPE SPECIES: *Sexangularia minutissima* (Penard, 1904) Awerintzew, 1906

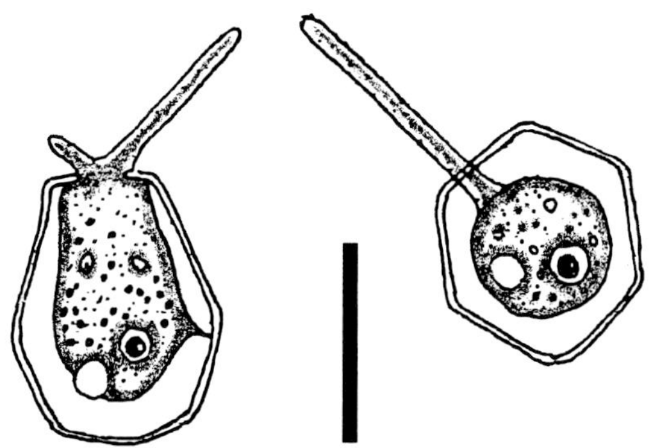

Fig. 32. *Sexangularia minutissima.* Scale bar=10 µm. (After Deflandre, 1953)

FAMILY CENTROPYXIDAE Jung, 1942

Pseudostome eccentric; often invaginated.

KEY TO THE GENERA OF THE FAMILY

1. Marine species........................ ***Centropyxiella***
1'. Freshwater, moss or soil species..................... 2

2. Aperture not invaginated......... ***Proplagiopyxis***
2'. Aperture invaginated...................................... 3

3. Ventral side flat............................ ***Centropyxis***
3'. Ventral side convex.............................. ***Oopyxis***

NOTE: Traditionally *Oopyxis* is included in this key although the affinities to the Centropyxidae are not clear and the genus is *incertae sedis.*

Genus ***Centropyxis*** Stein, 1857

More than 130 species and many varieties have been described. Many of these descriptions are incomplete. In this genus we find 2 different shell types (Figs. 33–34):

1. Test: bilaterally symmetrical, rounded, flattened more at front than at rear; ventral face flat; often spines at sides and rear. Pseudostome ventral, anterior, roundish, dorsal and ventral lips recurved, invaginated,

often bridged from ventral margin to dorsal face of the test. Shell can be organic or constructed with mineral particles or diatoms. Shape, number of spines, and size extremely variable. These species prefer freshwater habitats. A typical representative is *C. aculeata* (Fig. 33).

2. Shell circular or elongated in ventral view, in lateral view oral region slightly flattened, aperture subterminal, only ventral lip incurved, no bridges from ventral to dorsal face. Although some are common in freshwater habitats and sphagnum, most species of this group are found in drier mosses and humus. A typical species is *C. sphagnicola* (Fig. 34).

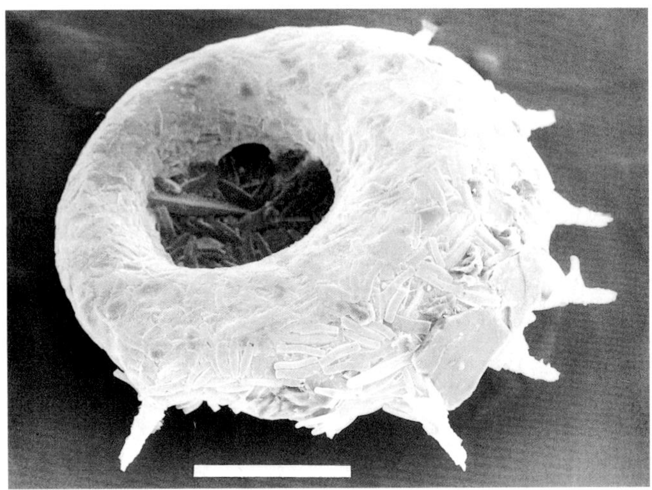

Fig. 33. *Centropyxis aculeata.* Scale bar=50 µm.

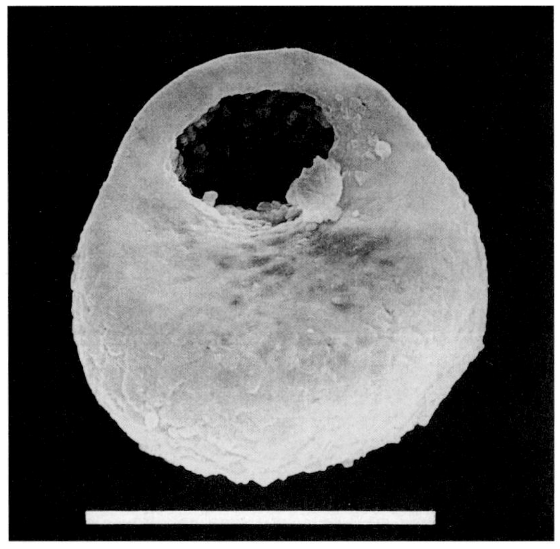

Fig. 34. *Centropyxis sphagnicola.* Scale bar=50 µm.

All *Centropyxis* species have an ovular nucleus. Feeding: all types of microbiota. Refs. Deflandre (1929), Bonnet (1989a), Bonnet and Thomas (1960), Decloitre (1978, 1979a,b, 1982, 1986), Lüftenegger et al. (1988), Ogden and Hedley (1980).
TYPE SPECIES: *Centropyxis aculeata* (Ehrenberg, 1838)

NOTES: Laminger (1971) has erected a subgenus *Toquepyxis* to allow for the divergent morphology of *C. lapponica* Grospietsch, 1954; later Aescht and Foissner (1989) changed the rank to genus. Zivkovic (1975) described the genus *Collaripyxidia* with a recurved, dentate apertural collar, which is similar to *Centropyxis marsupiformis.* Neither taxon is accepted and they can only be evaluated in the context of a revision of the whole genus *Centropyxis.*

Genus ***Centropyxiella*** Valkanov, 1970

Test oval. Pseudostome: antero-ventral, 10 to 20 µm across with flared rim (Fig. 35). Shell surface smooth, composed of overlapping flattish mineral particles, held together by organic cement. Ecology: marine interstitial, common. Eight species. Refs. Chardez (1977), for SEM, biometry, see Golemansky and Ogden (1980).
TYPE SPECIES: *Centropyxiella arenaria* Valkanov, 1970

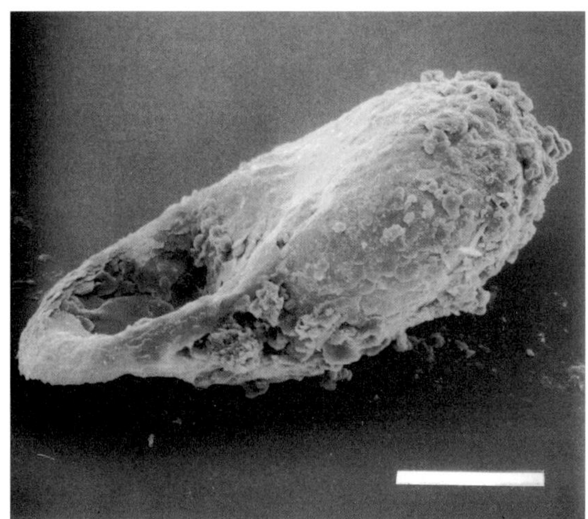

Fig. 35. *Centropyxiella arenaria.* Scale bar=10 µm.

Genus ***Proplagiopyxis*** Schönborn, 1964

Shell intermediate between *Centropyxis* and *Plagiopyxis* but with very little slant towards aperture (Fig. 36). Pseudostome eccentric, circular, no invagination. Shell brown, half spherical in shape,

slight slope to the aperture in side view, xenosomes lacking or rare. Differs from *Centropyxis* (e.g. *C. plagiostoma*) by lack of invagination and border of mineral particles around aperture. Habitat: soil. Monospecific. Ref. Schönborn (1964).
TYPE SPECIES: *Proplagiopyxis nuda* Schönborn, 1964

This genus is placed in the Centropyxidae and not the Plagiopyxidae due to its circular aperture.

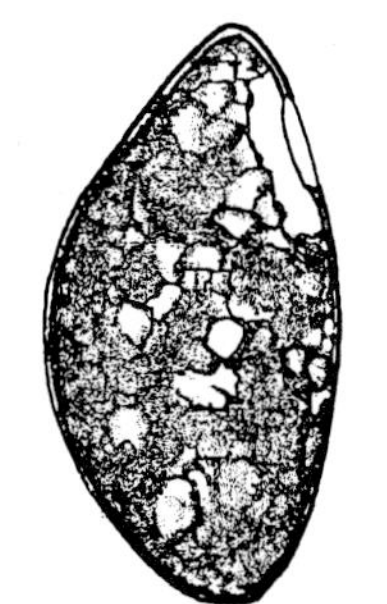

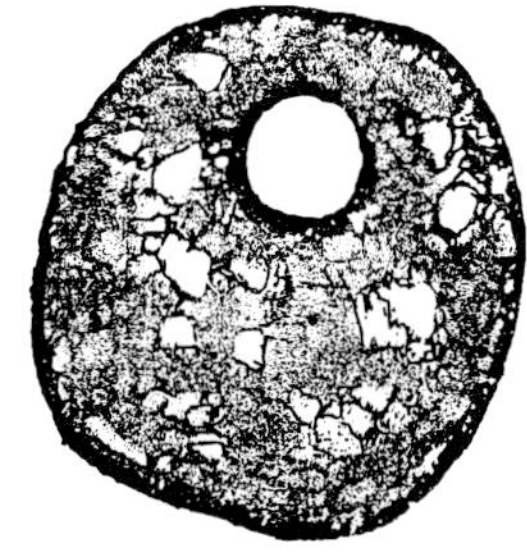

Fig. 36. *Proplagiopyxis nuda.* Scale bar=50 μm. (After Schönborn, 1964)

FAMILY TRIGONOPYXIDAE Loeblich & Tappan, 1964

Shell circular or elliptical, aperture central, often invaginated. Test composed of mineral particles in sheet-like cement.

KEY TO THE GENERA OF THE FAMILY

1. Test in oral view circular................................. 2
1'. Test elliptical.. 5

2. Aperture triangular, three-lobed, or irregular; always with thickened organic rim.................... .... ***Trigonopyxis***
2'. Aperture circular, irregular, crescent-shaped, or with ≥ 5 lobes; never with thickened organic rim... 3

3. Aperture crescent-shaped........... ***Cornuapyxis***
3'. Aperture not crescent-shaped........................... 4

4. Aperture invaginated...................... ***Cyclopyxis***
4'. Aperture not invaginated................ ***Geopyxella***

5. Aperture elliptical...................... ***Ellipsopyxis***
5'. Aperture circular.................... ***Ellipsopyxella***

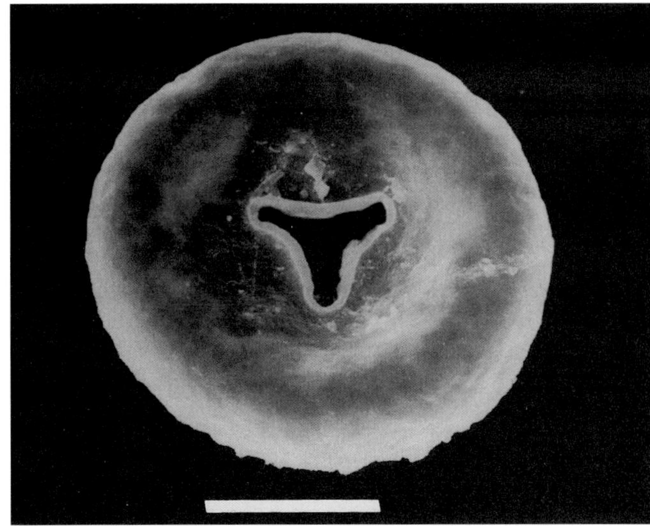

Fig. 37. *Trigonopyxis arcula.* Scale bar=50 μm.

Genus ***Trigonopyxis*** Penard, 1912

Shell brownish, hemispherical; aperture central, invaginated, frequently triangular but more often irregular, one species with three apertural lobes, always surrounded by a ring of organic cement, with a second internal membranous wall (Fig. 37). Nucleus ovular. Feeding: bacteria and fungi. Habitat: common in acid forest soil, litter, mosses (e.g. sphagnum). Five species. Refs. Penard (1912), Rauenbusch (1987), Lüftenegger et al. (1988).
TYPE SPECIES: *Trigonopyxis arcula* (Leidy, 1879)

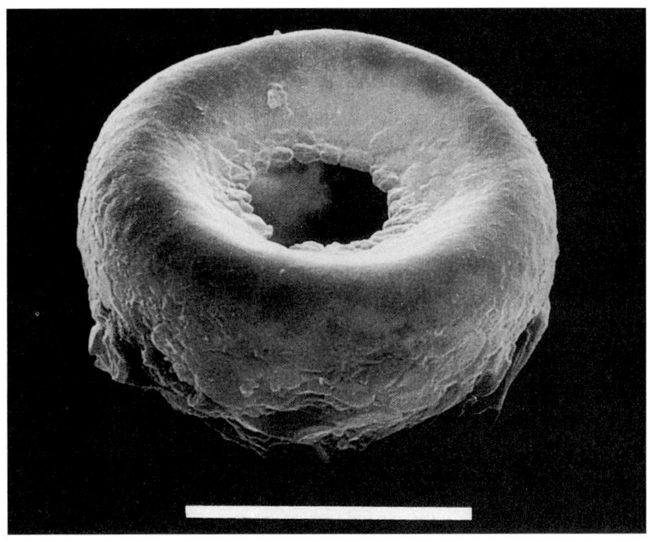

Fig. 38. *Cyclopyxis kahli.* Scale bar=50 μm.

Genus ***Cyclopyxis*** Deflandre, 1929

In oral view circular, aperture central, invaginated sometimes with internal tube (e.g. *C. puteus* Thomas, 1960 [Fig. 39]), in most species circular, few species with irregular or lobed pseudostome, margin never with thick organic lip but often with small mineral particles (Fig. 38). Ventral surface usually smooth, dorsal rough with larger sand grains. More than 30 species and twice as many varieties. Habitat: sphagnum mosses, soil. Refs. Deflandre (1929), Decloitre (1977a, 1979b, 1982, 1986), Ogden and Hedley (1980), Ogden (1988).
TYPE SPECIES: *Cyclopyxis arcelloides* (Penard, 1902)

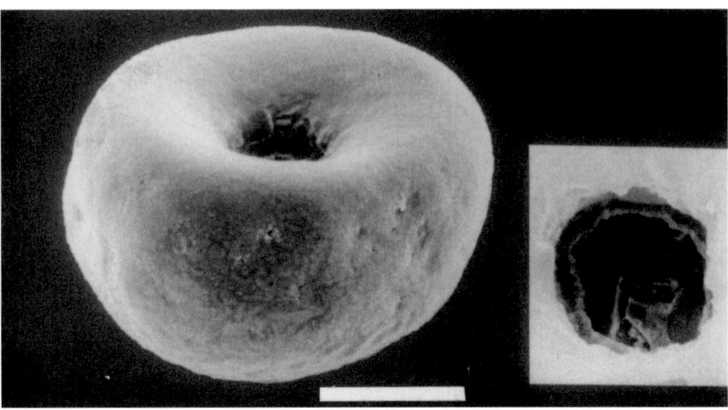

Fig. 39. *Cyclopyxis puteus.* Scale bar=50 µm.

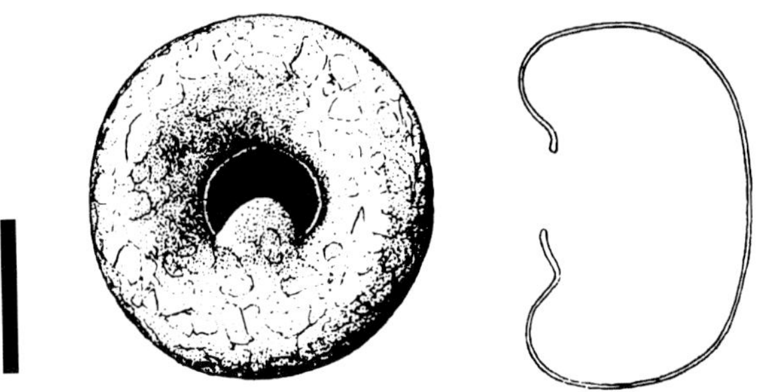

Fig. 40. *Cornuapyxis lunaristoma.* Scale bar=50 µm. (After Coûteaux and Chardez, 1981)

Genus ***Cornuapyxis*** Coûteaux & Chardez, 1981

Form similar to *Cyclopyxis* but aperture crescent-shaped (Fig. 40). Shell hemispherical in lateral view, ventral sole weakly invaginated, smooth, dorsal surface covered with small mineral grains. From mosses and soil, tropical rain forests, French Guyana. Monospecific. Ref. Coûteaux and Chardez (1981).
TYPE SPECIES: *Cornuapyxis lunaristoma* Coûteaux & Chardez, 1981.

Genus ***Geopyxella*** Bonnet & Thomas, 1955

Test thick membrane, attached thin siliceous particles; subglobular or hemispheroid (Fig. 41). Pseudostome ventral, central, round, in contrast to *Cyclopyxis*, not invaginated. Ventral surface smooth. Nucleus with one large central caryosome and several smaller peripheral nucleoli. Two species and several subspecies. Due to its narrow niche the type species is a good indicator species for calcareous rendziniform soils. Similar species: *G. aquatica*. Refs. Bonnet and Thomas (1960), Bonnet (1974a), Chardez (1967b).
TYPE SPECIES: *Geopyxella sylvicola* Bonnet & Thomas, 1955

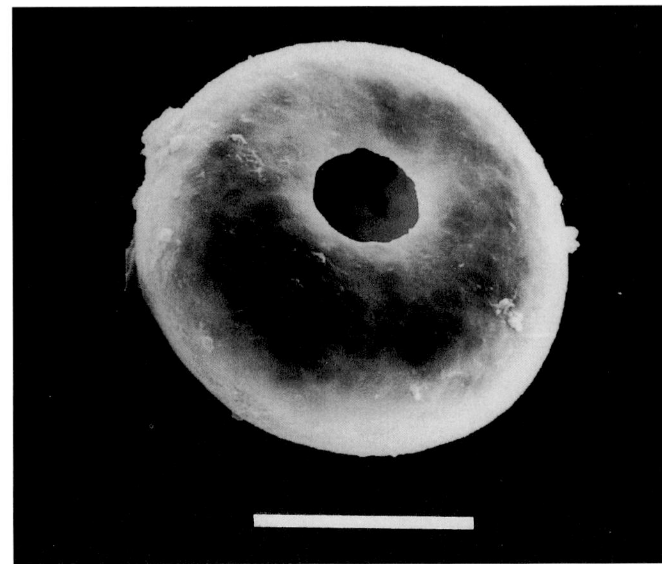

Fig. 41. *Geopyxella sylvicola.* Scale bar=50 µm.

Genus ***Ellipsopyxis*** Bonnet, 1965

Shell as in *Cyclopyxis* but bilaterally symmetrical, with elliptical pseudostome with no or only slight invagination (Fig. 42). Test yellowish or clear maroon, ventral surface smooth due to organic cement, dorsal and lateral surfaces encrusted with small mineral particles. In dorsal view elliptic, but in lateral view on to the large axis semi-elliptic. Mainly from rich organic tropical forest soils. Distribution: Africa, Asia, Central- and South America. Four species. Ref. Bonnet (1965).
TYPE SPECIES: *Ellipsopyxis pauliani* Bonnet, 1965.

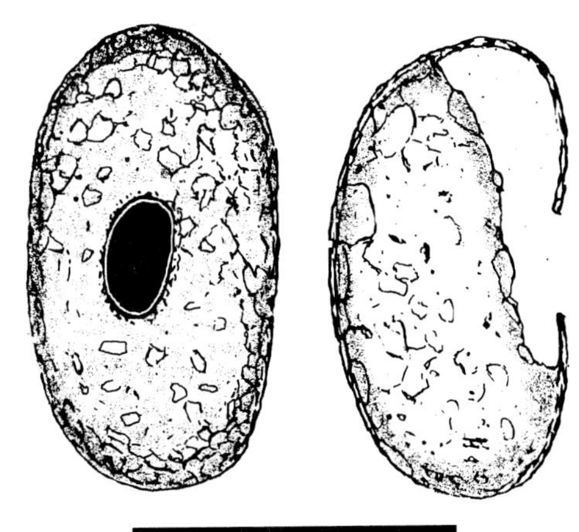

Fig. 42. *Ellipsopyxis pauliani.* Scale bar=50 µm. (After Bonnet, 1965)

Genus ***Ellipsopyxella*** Bonnet, 1975

Test resembles *Geopyxella* but with biaxial symmetry (Fig. 43). In contrast to *Ellipsopyxis* the aperture is circular, not or only very weakly invaginated. During resting phases the pseudostome is closed by a mucous diaphragm. The test is composed of relatively flat mineral particles (crystalline quartz) giving a smooth surface, ventral sole thicker than the rest of the shell, pseudostome without organic lip. Found in grassland soils. Geographic distribution: Ivory-Coast, French Guyana. Monospecific. Refs. Bonnet (1975a), Coûteaux (1979).
TYPE SPECIES: *Ellipsopyxella regularis* Bonnet, 1975.

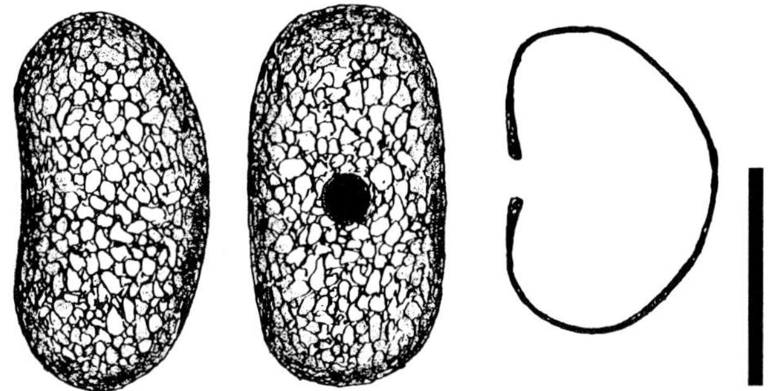

Fig. 43. *Ellipsopyxella regularis.* Scale bar=50 µm. (After Bonnet, 1975a)

FAMILY LAMTOPYXIDAE Bonnet, 1974

Test with flat sole, bilaterally symmetrical due to its elliptical internal opening at the end of a deeply invaginated tube; the external aperture is bordered by large teeth. This family is believed to be intermediate between *Cyclopyxis* and *Distomatopyxis* (Bonnet, 1979a).

Genus ***Lamtopyxis*** Bonnet, 1974

In dorsal and lateral view like *Cyclopyxis puteus* but external opening with 3 to 5 teeth (Fig. 44). At its base the apertural tube is reinforced by a more or less quadratic organic frame; the internal opening is bordered by a collar. The test is composed of flat mineral particles which are held together by an unstructured organic cement giving the shell a smooth surface as in *Cyclopyxis.* All 5 species were described from tropical forest soils. Ref. Bonnet (1974c).
TYPE SPECIES: *Lamtopyxis callistoma* Bonnet, 1974.

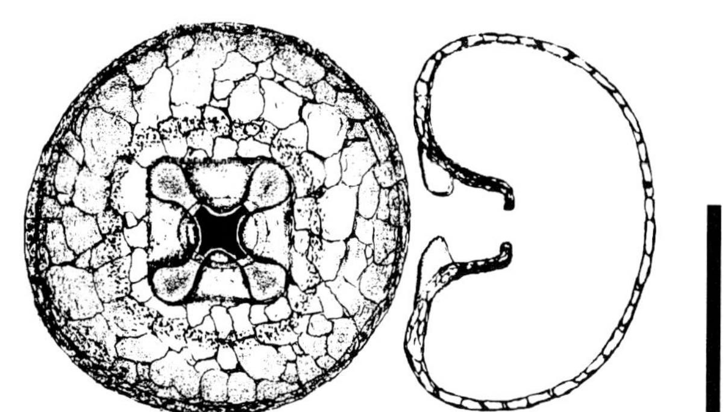

Fig. 44. *Lamtopyxis callistoma.* Scale bar=100 µm. (After Bonnet, 1974c)

FAMILY DISTOMATOPYXIDAE Bonnet, 1970

Test hemispherical like *Cyclopyxis.* Pseudostome central, elliptical, invaginated, and lying at the bottom of a tubular vestibule. The vestibule is partly covered by a diaphragm which is fixed by two bridges forming two crescent-shaped openings at opposite sides.

Genus ***Distomatopyxis*** Bonnet, 1964

Shell morphology with characters of the family. Test composed of flat mineral particles in an organic cement, surface smooth (Fig. 45). Habitat:

organic forest soils, North- and Central America, East Asia (Korea, Japan, Philippines), and Spain (Asturias). Two species. Refs. Bonnet and Vandel (1970), Bonnet (1979a), Bonnet and Gomez-Sanchez (1984).
TYPE SPECIES: *Distomatopyxis couillardi* Bonnet, 1964

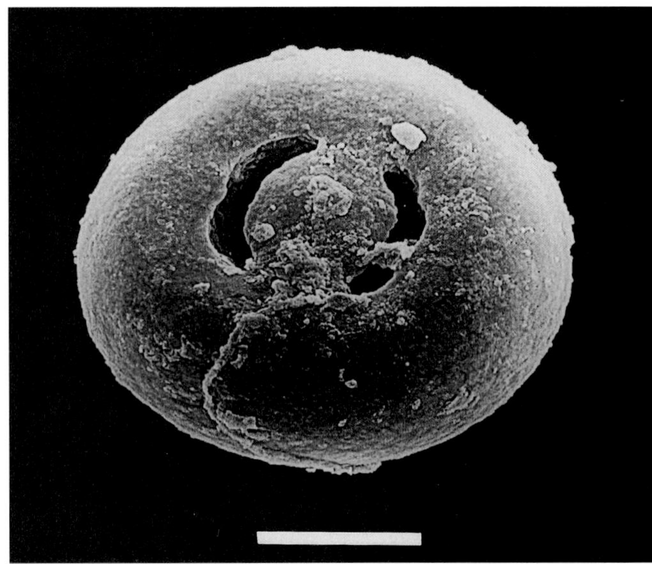

Fig. 45. *Distomatopyxis couillardi.* Scale bar=50 µm.

FAMILY PLAGIOPYXIDAE Bonnet & Thomas, 1960

Ventral side flat. Pseudostome antero-ventral, a more or less invaginated overhung slit.

KEY TO GENERA OF THE FAMILY.

1. Anterior lip of pseudostome a distinct, overhanging hood...6
1'. Aperture more or less slit-like, no overhanging hood...2

2. Anterior lip of pseudostome with pores.. **Bullinularia**
2'. No such pores...3

3. Anterior (dorsal) lip curves below convex ventral face............................ **Geoplagiopyxis**
3'. Posterior lip curves below anterior lip, or pseudostome an open slit....................................4

4. Aperture covered by the anterior lip................ 5
4'. Aperture open slit.............. **Protoplagiopyxis**

5. Posterior (ventral) lip with extension towards anterior lip (Fig. 49)......... **Paracentropyxis**
5'. No extension of the posterior lip.. **Plagiopyxis**

6. Ventral face rounded................... **Hoogenraadia**
6'. Ventral face flattened.......... **Planhoogenraadia**

Genus ***Protoplagiopyxis*** Bonnet, 1962

Shell chitinous with covering of extraneous mineral particles (xenosomes), morphology intermediate between those forms of *Centropyxis* with visor and *Plagiopyxis*; aperture a straight or crescent-shaped slit of 'plagiostome' type with tendency to 'cryptostome' (pseudostome located back into the test), no vestibule, but the apertural region is well separated by the fairly abrupt dorso-ventral depression in the anterior part. Ventral sole flattened very clearly and little or not inclined (Fig. 46). Three species. Habitat: soil. Ref. Bonnet (1962).
TYPE SPECIES: *Protoplagiopyxis delamarei* Bonnet, 1962

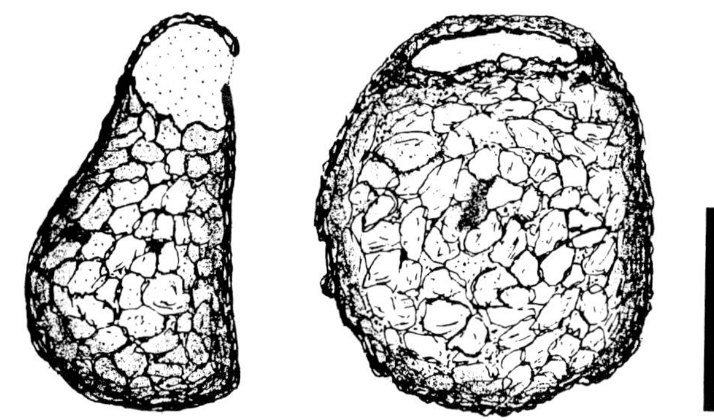

Fig. 46. *Protoplagiopyxis delamarei.* Scale bar=50 µm. (After Bonnet, 1962)

Genus ***Plagiopyxis*** Penard, 1910

Shell in ventral view circular or oval, in lateral view with a flat ventral face; aperture a subterminal, elongate slit, perpendicular to long axis of the shell; dorsal apertural lip somewhat incurved usually hiding the opening (Fig. 47). Twenty-two species. Almost all species are edaphobiont (live exclusively in soil). The most widely distributed and common species is *Plagiopyxis declivis* Bonnet & Thomas, 1955. Shell hemispherical, ventral side covered with flat mineral particles, dorsal side with rough

xenosomes; dorsal (anterior) lip with sharp, sometimes irregular margin overhanging, ventral (posterior) lip projects straight inside shell as elongation of ventral sole. All species have an ovular nucleus. Habitat: soil. Refs. Bonnet and Thomas (1960), Thomas (1958), Bonnet (1984a, 1988).
TYPE SPECIES: *Plagiopyxis callida* Penard, 1910.

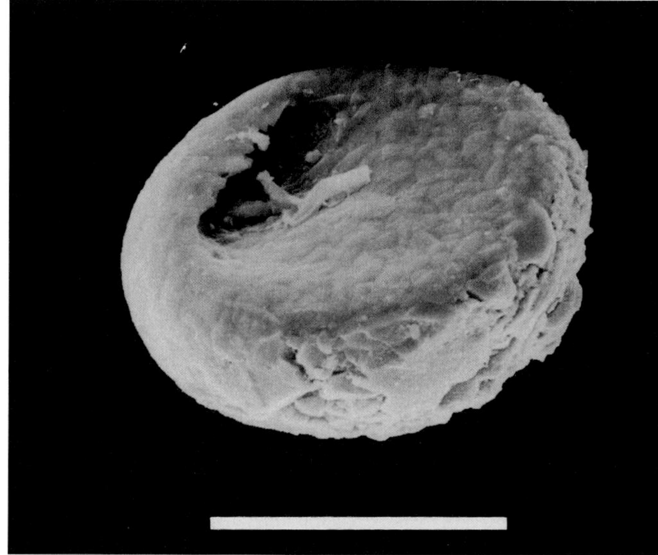

Fig. 47. *Plagiopyxis declivis.* Scale bar=50 µm.

Genus ***Geoplagiopyxis*** Chardez, 1961

Test plagiostome, close to *Plagiopyxis*, but differs from it by the dorsal face curving below the convex ventral face (Fig. 48). Shell hemispherical, wider than long, ventral surface slightly vaulted, aperture not easily visible as an irregular slit. Test composed of amorphous plates without larger mineral particles. Habitat: soil. Two species. Ref. Chardez (1961).
TYPE SPECIES: *Geoplagiopyxis declivus* Chardez, 1961

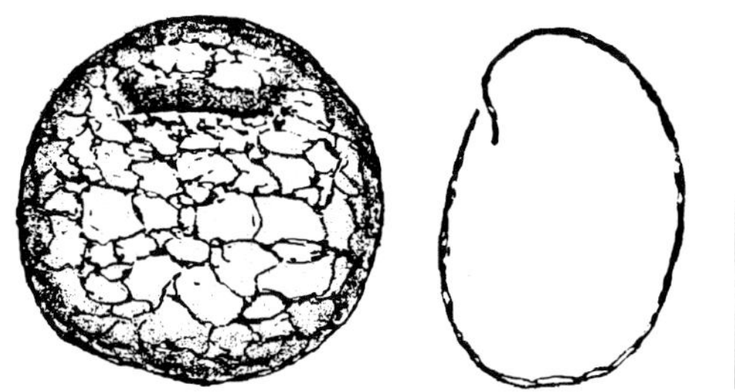

Fig. 48. *Geoplagiopyxis declivus.* Scale bar=50 µm. (After Chardez, 1961)

Genus ***Paracentropyxis*** Bonnet, 1960

Shell in dorsal view circular or slightly elliptical; at the dorsal face a groove separates the remainder of the test from a small visor in form of a tapered crescent (Fig. 49). Ventral face concave; at median level the visor is rounded, forming a buccal cavity, ventral face and the side walls of this **vestibulum** are connected to the dorsal part of the shell; with a small internal opening as in *Centropyxis sylvatica*. The slit-like pseudostome is hidden by the dorsal lip and an extension of the ventral shell wall (**cryptostomy**). Shell transparent and composed of exogenous mineral particles, which are embedded in a hyaline cement. From tropical soils. Monospecific. Ref. Bonnet (1960a).
TYPE SPECIES: *Paracentropyxis mimetica* Bonnet, 1960

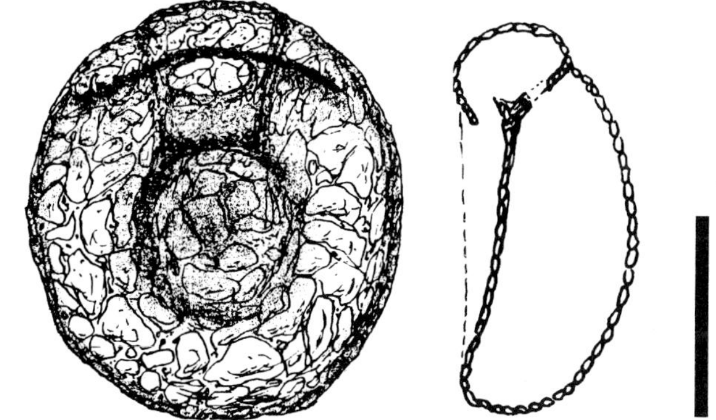

Fig. 49. *Paracentropyxis mimetica.* Scale bar=20 µm. (After Bonnet, 1960a)

Genus ***Bullinularia*** Deflandre, 1953

Test dark brown and opaque, oval in oral view, attached siliceous particles; ventral surface flattened, smooth; hemispheroid laterally (Fig. 50). Pseudostome: a long curving slit, ventral lip prolonged, dorsal lip with pores. Vesicular nucleus! Feeding: herbivorous. Habitat: dry sphagnum, soils. Ecological data (*B. indica*): Tolonen et al. (1992). Seven species which differ mainly in their dimensions. Refs. Penard (1912), Bonnet and Thomas (1960), Rauenbusch (1987).
TYPE SPECIES: *Bullinularia indica* (Penard, 1907).

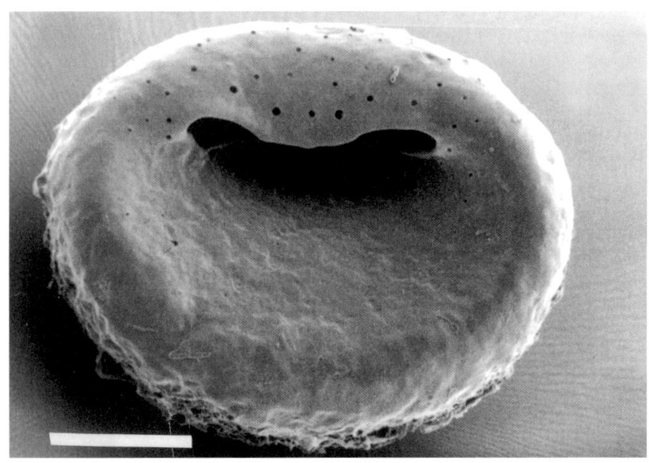

Fig. 50. *Bullinularia indica.* Scale bar=50 μm.

Genus ***Hoogenraadia*** Gauthier-Lièvre & Thomas, 1958

Test ovoid-globular, overhanging anterior visor often with 2 lateral grooves, incurved ventral lip. Ventral face rounded (Fig. 51). Pseudostome more or less round, partly covered by the visor. Habitat: acid bogs and swamps or from soils, tropical Africa, South America. Five species, all from the tropics. Ref. Gauthier-Lièvre and Thomas (1958).
TYPE SPECIES: *Hoogenraadia africana* Gauthier-Lièvre & Thomas, 1958

NOTE: Stepanek (1963) has erected the similar perhaps synonymous genus *Gillardella*, which has a much smaller visor.

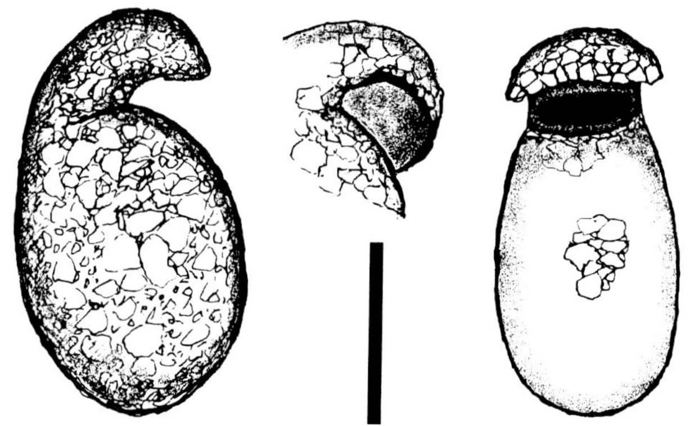

Fig. 51. *Hoogenraadia africana.* Scale bar=50 μm. (After Gauthier-Lièvre and Thomas, 1958)

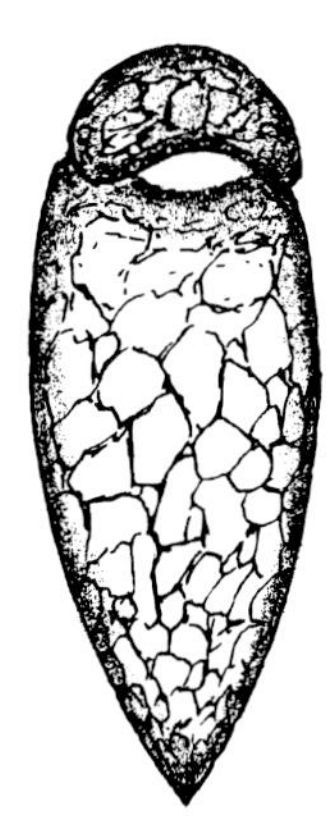

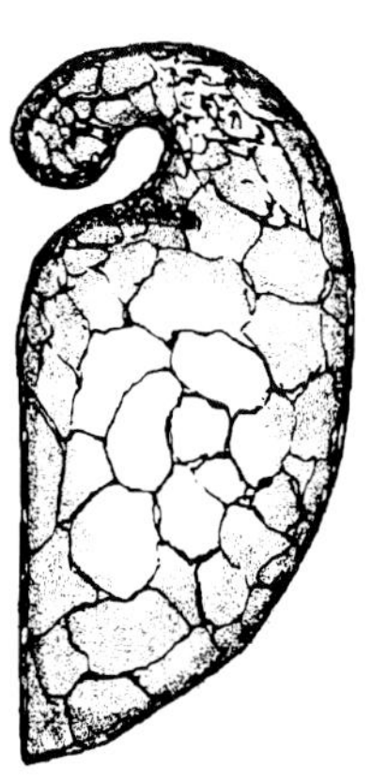

Fig. 52. *Planhoogenraadia acuta.* Scale bar=50 μm. (After Bonnet, 1977)

Genus ***Planhoogenraadia*** Bonnet, 1977

Test bilaterally symmetrical; aperture more or less covered by a visor-like extension of the dorsal surface, 2 lateral grooves at the junction of visor and belly (Fig. 52). In contrast to *Hoogenraadia* ventral side always flat; this is interpreted as an adaptation to the living in soil as in *Plagiopyxis* or certain *Centropyxis*. Pseudostome usually semi-circular or elongated oval, located back into the test. Habitat: forest soils often on calcareous bedrock. Eight species. Refs: Bonnet (1977, 1984b).
TYPE SPECIES: *Planhoogenraadia acuta* Bonnet, 1977

FAMILY PARAQUADRULIDAE Deflandre, 1953

The shells of members of this family are composed of endogenously formed, rectangular calcite plates embedded in a sheet-like organic cement. The presence of calcite should be verified by polarizing microscopy, with hydrochloric acid, or X-ray-microanalysis.

KEY TO THE GENERA OF THE FAMILY

1. Test oval or round .... ***Paraquadrula***
1'. Test flask-shaped.................... ***Lamtoquadrula***

Genus ***Paraquadrula*** Deflandre, 1932

Test transparent, roundish oval in broad view; constructed of secreted quadratic or rectangular,

calcite plates in rows, sheet-like cement (Fig. 53). Vesicular nucleus. Feeding: herbivorous. Habitat: calcareous substrates like freshwater, mosses, and soil. Six species. Refs. Deflandre (1932), Decloitre (1961), Bonnet (1989b), Lüftenegger and Foissner (1991).
TYPE SPECIES: *Paraquadrula irregularis* (Archer, 1877)

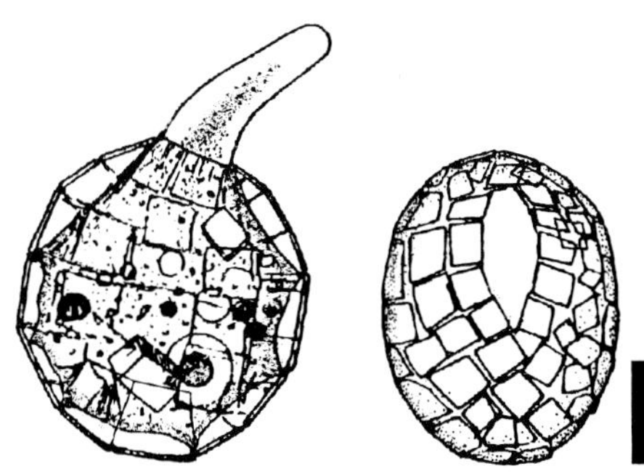

Fig. 53. *Paraquadrula irregularis.* Scale bar=50 µm. (After Deflandre, 1953)

Genus ***Lamtoquadrula*** Bonnet, 1974

Shell flask-shaped, neck bent, test and aperture in cross-section circular, covered with rectangular calcite plates (idiosomes) as in *Paraquadrula* (Fig. 54). Aperture with a thin transparent organic lip, only visible with phase or interference contrast. Habitat: sometimes abundant in humus on calcium-rich amphibolite rocks and tropical gallery forests. Geographical distribution: Ivory Coast. Monospecific. Ref. Bonnet (1974b).
TYPE SPECIES: *Lamtoquadrula deflandrei* Bonnet, 1974

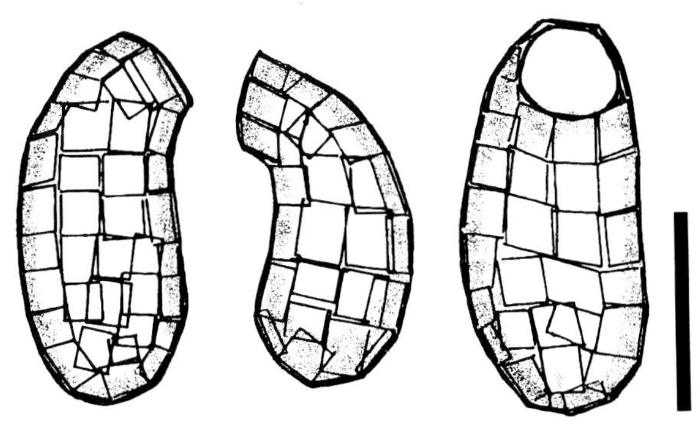

Fig. 54. *Lamtoquadrula deflandrei.* Scale bar=20 µm. (After Bonnet, 1974b)

FAMILY LESQUEREUSIIDAE Jung, 1942

This family contains genera which build shells of endogenous rod-like, nail-shaped, or rectangular siliceous plates to which mineral particles may be added. Whether it is a monophyletic unit is questionable. *Lesquereusia* shows affinities in shell shape and cement structure with *Lagenodifflugia*, as does *Netzelia* with certain *Difflugia ssp.* The rectangular shell plates of *Quadrulella, Microquadrula*, and *Pomoriella* resemble those of the Paraquadrulidae, although the latter are calcite and have an unstructured cement!

KEY TO THE GENERA OF THE FAMILY

1. Test with attached neck, usually with rod-like idiosomes or a mixture with foreign material. .... ***Lesquereusia***
1'. Aperture terminal or bent.............................. 2

2. Aperture lobed.................................... ***Netzelia***
2'. Aperture round or elliptical, body plates rectangular...3

3. Marine species.............................. ***Pomoriella***
3'. Species found in freshwater, mosses, or soil... 4

4. Aperture terminal, shell > 45 µm..***Quadrulella***
4'. Aperture bent, round, small species < 45 µm...... .... ***Microquadrula***

Genus ***Lesquereusia*** Schlumberger, 1845

Test with asymmetrical neck, more or less attached to the body; composed of endogenous siliceous rods or in other species with collected mineral particles, always structured mesh-like cement (Fig. 55). Ovular nucleus. Feeding: herbivorous (diatoms, green algae). Habitat: aquatic mosses or sediments. Species differ mainly in shape of the siliceous rods, the amount of mineral particles used, and the orientation of the neck. Sixteen species. Refs. Thomas and Gauthier-Lièvre (1959a, 1961), Ogden and Hedley (1980); for mitosis, culture, see Stump (1959), Freeman (1974), for EM and cytochemistry, see Harrison et al. (1975, 1976).
TYPE SPECIES: *Lesquereusia spiralis* (Ehrenberg, 1840)

NOTE: This genus was erected by Schlumberger (1845) and obviously named after the Swiss naturalist Lesquereux. By *lapsus* it was first published as *Lecqereusia*. This was emended by Hopkinson (in Cash and Hopkinson, 1909).

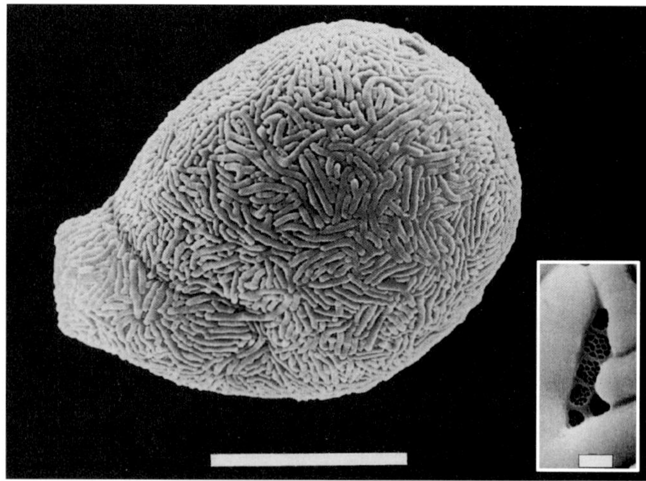

Fig. 55. *Lesquereusia spiralis*. Inset shows the organic cement. Scale bar=50 µm (inset 1 µm).

Fig. 56. *Netzelia wailesi*. Apertural view. Scale bar=50 µm.

Genus ***Netzelia*** Ogden, 1979

Shell ovoid, circular in cross-section, pseudostome lobed with a thick organic rim or a necklace made of small idiosomes (Fig. 56). Species of *Netzelia* differ from lobed species of *Difflugia* in their ability to build their test completely from endogenous siliceous elements (idiosomes) although they can use small sand grains, diatoms, or undigested algal cell walls as supplementary building material. These xenosomes are always smoothed and modified by the deposition of silica (Anderson, 1987). The idiosomes often have a nail-like shape. All particles are held in position by perforated cement units and are arranged in a single layer. Outline usually regular, but one species (*N. tuberculata*) often has protuberances to give a mulberry-like appearance. Habitat: aquatic vegetation and sphagnum. Three species. Refs. Anderson (1989, 1990), Netzel (1976), Ogden (1979a), Ogden and Meisterfeld (1989).
TYPE SPECIES *Netzelia oviformis* (Cash, 1909)

Genus ***Quadrulella*** Cockerell, 1909

Test pyriform or with distinct neck, some species with lateral ridge, mostly oval in cross-section, composed of endogenous, siliceous, quadrangular, non-overlapping plates. Test colorless, transparent (Fig. 57). Shape very similar to certain *Nebela spp.* Nucleus always ovular. Feeding: herbivorous. Habitat: sphagnum mosses, soil. One dozen species. Refs. Deflandre (1936), Chardez (1967a), Ogden and Hedley (1980).
TYPE SPECIES: *Quadrulella symmetrica* (Wallich, 1863)

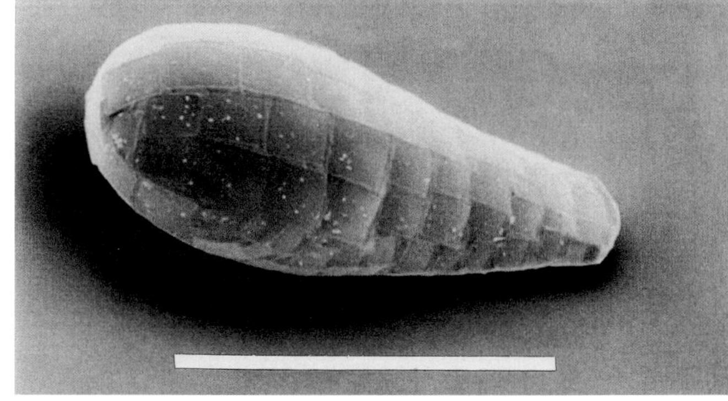

Fig. 57. *Quadrulella symmetrica*. Scale bar=50 µm.

Genus ***Microquadrula*** Golemansky, 1968

Shell ovoid, circular in cross-section, aperture round, obliquely truncated, covered with square or rectangular siliceous plates (Fig. 58). Transparent, plates imbedded in an organic, hyaline cement. Habitat: humid moss on rocks, rare, Cuba. Monospecific. Ref. Golemansky (1968).

TYPE SPECIES: *Microquadrula musciphila* Golemansky, 1968

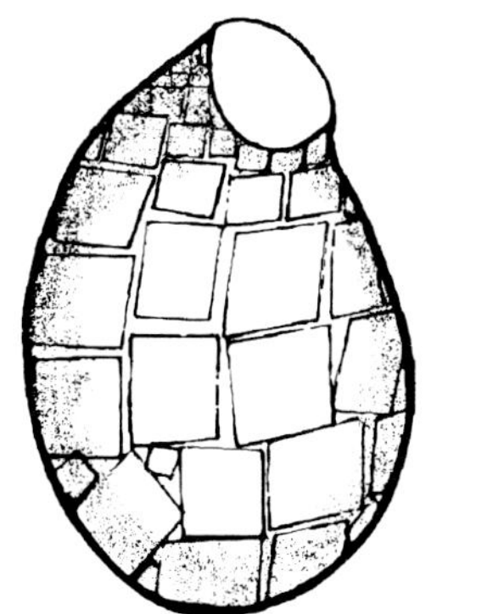

Fig. 58. *Microquadrula musciphila.* Scale bar=10 μm. (After Golemansky, 1968)

Genus ***Pomoriella*** Golemansky, 1970

Test flask-shaped, neck bent; with siliceous, non-overlapping, intrinsic plates (Fig. 59). Pseudostome round, terminal. Size 40 to 50 μm; vesicular nucleus. Habitat: marine; sandy beaches. *Pomoriella* differs from *Microquadrula* by its distinctly bent neck, size, and ecology. Monospecific. Refs. Golemansky (1970b), Golemansky and Ogden (1980).
TYPE SPECIES: *Pomoriella valkanovi* Golemansky, 1970

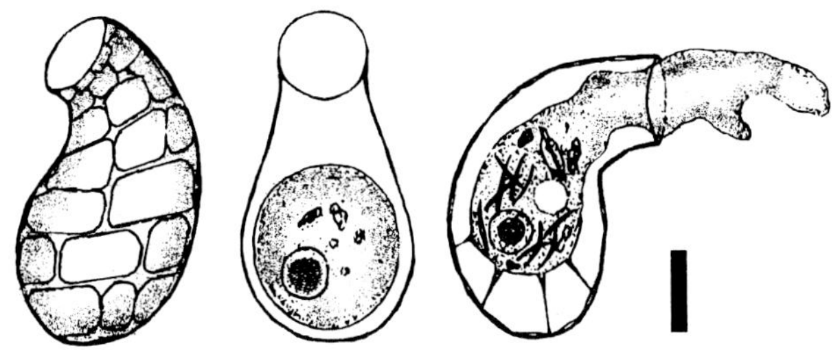

Fig. 59. *Pomoriella valkanovi.* Scale bar=10 μm. (After Golemansky, 1970b)

FAMILY HYALOSPHENIIDAE Schultze, 1877

Test rigid, chitinoid, clear, completely organic, non-areolar; oval to flask-shaped, pseudostome terminal.

KEY TO GENERA OF THE FAMILY

1. Shell compressed........................ ***Hyalosphenia***
1'. Cross-section circular.............. ***Leptochlamys***

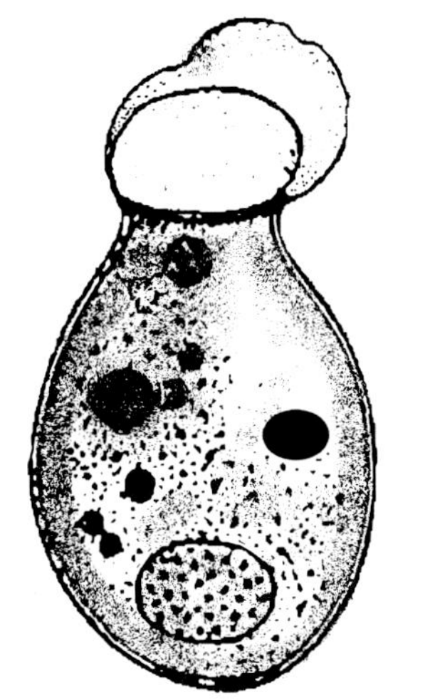

Fig. 60. *Leptochlamys ampullacea.* Scale bar=10 μm. (After West, 1901)

Genus ***Leptochlamys*** West, 1901

Test ovoid, cross-section circular, truncated; clear (Fig. 60). Pseudostome terminal, round, with short collar. Size: 48 to 55 μm. Nucleus ovular, oval, nucleoli concentrated below nuclear membrane. Feeding: herbivorous. Habitat: freshwater (rare!). Monospecific. Ref. Cash and Hopkinson (1909).
TYPE SPECIES: *Leptochlamys ampullacea* West, 1901

Genus ***Hyalosphenia*** Stein, 1859

Species with compressed, chitinoid transparent shell with no foreign material; aperture often with a thickened lip (Fig. 61). About 20 species. Two principal shell types, wedge-shaped or flask-shaped. The most common species of the first group is *H. papilio* Leidy,1879, which has zoochlorellae. It is a dominant species in the green parts of wet sphagnum, and therefore an excellent indicator in palaeoecological studies. *Hyalosphenia elegans* (Fig. 62) is a typical flask-shaped species; it is a herbivore and common in mosses and acid soils. For ecological data, see Schönborn (1962b),

Meisterfeld (1977), Bonnet (1989b). Refs. Grospietsch (1965), Charret (1964), Ogden and Hedley (1980).
TYPE SPECIES: *Hyalosphenia ligata* (Tatem, 1870)
Synonym: *H. cuneata* Schulze, 1875

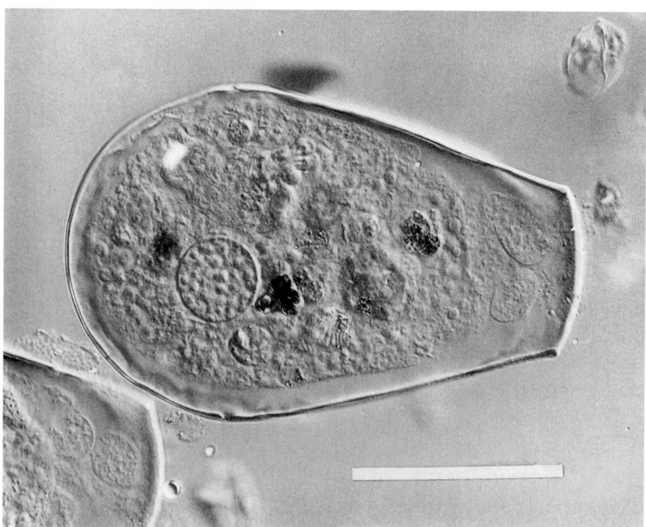

Fig. 61. *Hyalosphenia papilio.* Scale bar=50 µm.

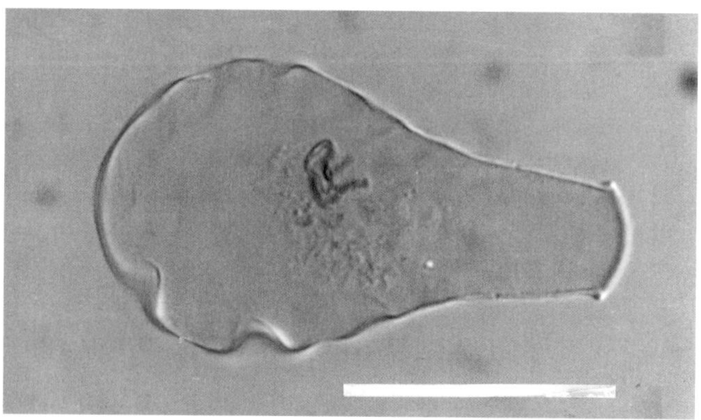

Fig. 62. *Hyalosphenia elegans.* Scale bar=50 µm.

NOTE: Stepanek (1967) proposed the genus *Pseudohyalosphenia* for individuals which resemble *Hyalosphenia papilio* in shape but have a shell constructed from siliceous plates. Unfortunately, his description is not detailed. enough! Chardez (1967c) recognizes several other genera in this family, which are included here in the Heleoperidae and Nebelidae.

FAMILY HELEOPERIDAE Jung, 1942

KEY TO THE GENERA OF THE FAMILY

1. Aperture lenticular or slit-like.......***Heleopera***
1'. Aperture oval, shell deep violet.***Awerintzewia***

Genus ***Heleopera*** Leidy, 1879

Shells of the genus *Heleopera* are always laterally compressed; aperture terminal, lenticular or slit-like, with thin organic rim, in contrast to *Nebela* with acute notches at edges (Fig. 63). Besides colorless or yellow, several red or purple species are common. Test composed of collected euglyphid body plates, mineral particles, or diatoms. These elements are often coated and reinforced with siliceous material. In contrast to *Nebela* numerous finger-like pseudopodia. All species have an ovular nucleus. More than a dozen species. Feeding: *H. sphagni* has zoochlorellae and is abundant in the green horizon of wet sphagnum; it uses mainly its symbionts; most other species are predators of small euglyphids. Habitats: lake sediments, mosses, and soil. Refs. Brabet et al. 1982, Ogden and Hedley (1980), Bonnet (1989b), Ogden (1989, 1991b).
TYPE SPECIES: *Heleopera sphagni* (Leidy, 1874)

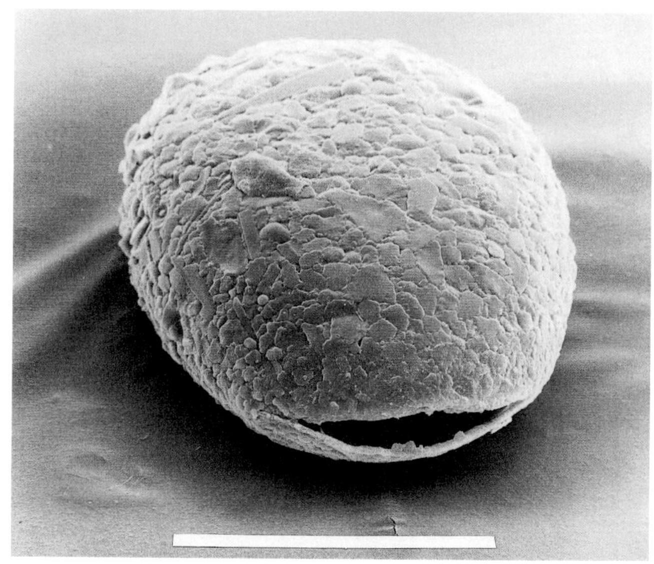

Fig. 63. *Heleopera sphagni.* Scale bar=50 µm.

Genus ***Awerintzewia*** Schouteden, 1906

Like *Heleopera p. amethystea*, dark violet color, but aperture oval and pseudostomal border internally thickened (Fig. 64). Size: 135 to 180 µm long. Pseudostome: 40 by 12 µm. Test oval, with irregular, embedded siliceous particles. Nucleus ovular. Feeding: herbivorous. Habitat: sphagnum, aquatic vegetation but also forest soils. Monospecific. Ref. Penard (1902).
TYPE SPECIES: *Awerintzewia cyclostoma* (Penard, 1902)

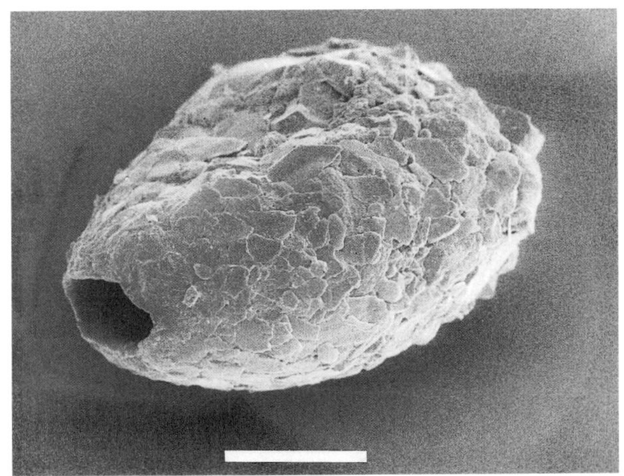

Fig. 64. *Awerintzewia cyclostoma.* Scale bar=50 µm.

FAMILY NEBELIDAE Taranek, 1882

Shell composed of shell plates of small euglyphids (e.g. *Euglypha, Trinema, Tracheleuglypha*) or plates of *Quadrulella* and diatom fragments. The assumption in the older literature that the Nebelidae can synthesize their own siliceous ideosomes has not been proven. There is some evidence from scanning electron microscopy that members of the genus *Argynnia* can coat foreign building material. Usually Nebelidae fix the building material in a sheet-like organic cement matrix, which is normally visible between the plates; only *Argynnia* and *Physochila* use distinct cement units. When building materials are limited, the test of some species are completely organic (e.g. *N. tincta*, *A. cockayni*).

Classification: In contrast to the traditional view, the family Nebelidae is here restricted to taxa which formerly were included in the genus *Nebela.* Deflandre (1936) proposed several groups to accommodate the large morphological diversity in this genus. Jung (1942b) split *Nebela* into 11 genera, unfortunately without type designations. Only the monotypic genera are valid; all others are *nomina nuda*. With the exception of *Physochila* none of the genera have been recognized by subsequent workers. Loeblich and Tappan (1961) validated *Apodera* and *Certesella.* Here, genera with structured cement and distinct shell morphology (*Argynnia*, *Physochila*) and those taxa with sheet-like cement, sufficient morphological particularity, and restricted geographic distributions are recognized as distinct genera, the remaining species being assigned to *Nebela.* Although usually grouped with the Nebelidae, *Schoenbornia* is more isolated—and should probably be thought of as *sedis mutabilis*. It is the only genus of this family with a vesicular nucleus. Although *incertae sedis*, *Geamphorella*, *Jungia*, and *Pseudonebela* are included in the key of this family!

KEY TO GENERA IN THE FAMILY

1. Pseudostome with organic rim or lip................ 3
1'. Margin of the pseudostome bordered by siliceous plates, or quartz grains, no organic rim......... 2

2. Aperture with recurved collar....... ***Physochila***
2'. Aperture oval with siliceous plates, no collar..... .... ***Argynnia***
2''. Aperture circular, quartz grains forming a short collar.. ***Jungia***

3. In broad view two invaginated frontal pores on each side, connected by an internal tube.......... 9
3'. If pores, then on the narrow side of the shell... 4

4. Shell pyriform, oval, elongate, or with a lateral indentation at each side.................................... 5
4'. With distinctly separated, sometimes swollen neck...8

5. Shell with flat lateral ridge of variable length .... ***Nebela*** ...**Subgroup Carinatoides**
5'. No lateral ridge... 6

6. No lateral indentation.. 7
6'. Lateral indentation at each side, shell hyaline..... .... ***Alocodera***

7. Test compressed, medium to large species, ovular nucleus....................................... ***Nebela*** .... **Subgroup *Collaroides***

7'. Test circular in cross section, small (< 45 µm), vesicular nucleus............. ***Schoenbornia***

8. Aperture with organic, denticulated lip, appears lobed.. ***Pseudonebela***
8'. Lip not denticulated.. 10

9. Two groups of small internal teeth near the aperture.. ***Certesella***
9'. No groups of small internal teeth......... ***Porosia***

10. Neck short ≈ 5 µm.................. ***Geamphorella***
10'. Neck longer.. 11

11. Neck not deeply constricted.................. ***Nebela***
......................................**Subgroup *Lagenoides***
11'. Neck deeply constricted at the junction with the body.. ***Apodera***

Genus ***Nebela*** Leidy, 1874

Test ovate, pyriform, elongate or with a long neck, always compressed, sometimes with lateral pores. Yellowish, transparent often with predated

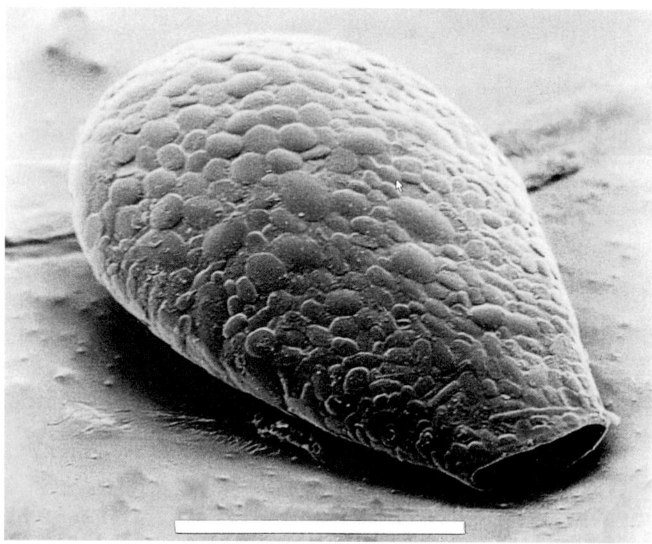

Fig. 65. *Nebela collaris.* Scale bar=50 µm.

siliceous plates or diatom frustules in an unstructured organic cement (Figs. 65-67). Ovular nucleus. Most *Nebela* sp. are predators of small Euglyphida. Habitat: common in mosses (sphagnum) and soil. Loeblich and Tappan (1964) have designated *Nebela numata* Leidy, 1874 as valid type species, which unfortunately is a junior subjective synonym of *Nebela collaris* (Ehrenberg, 1848). Here this large genus is subdivided into three groups:

1. ***Collaroides*** with an ovate or elongate shell (*N. collaris* Fig. 65).
2. ***Lagenoides*** with a long neck (*N. lageniformis* Fig. 66).
3. ***Carinatoides*** with a lateral ridge (*N. carinata* Fig. 67).

Refs. Deflandre (1936), Jung (1942b), Ogden and Hedley (1980), Rauenbusch (1987). For ecological data, see Bonnet (1990), Tolonen et al. (1992).

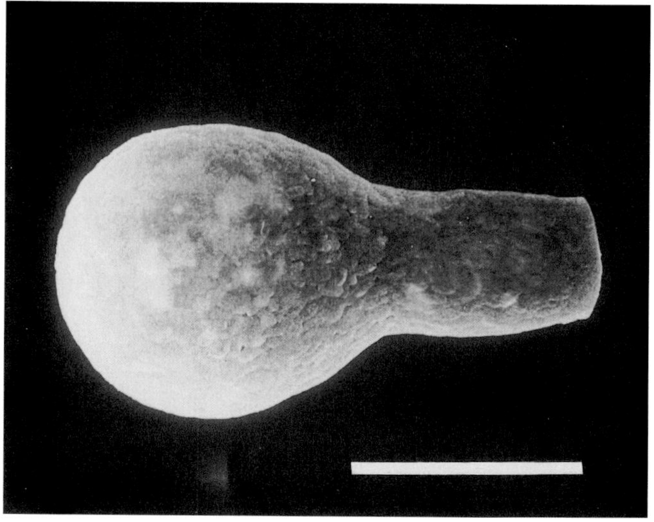

Fig. 66. *Nebela lageniformis.* Scale bar 50 µm.

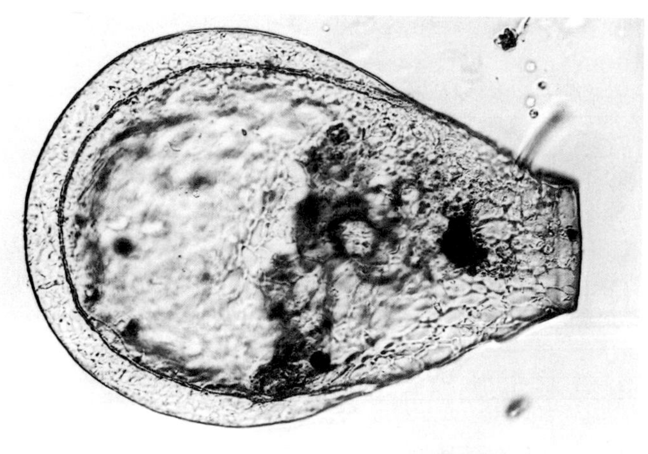

Fig. 67. *Nebela carinata.* Scale bar=50 µm.

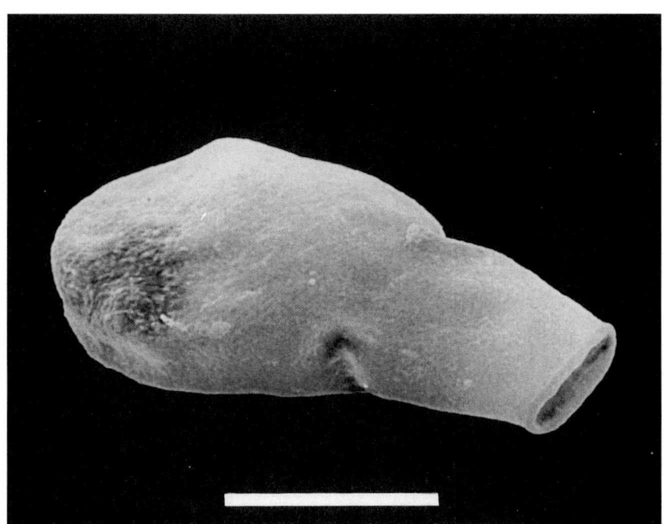

Fig .68. *Alocodera cockayni.* Scale bar=50 µm.

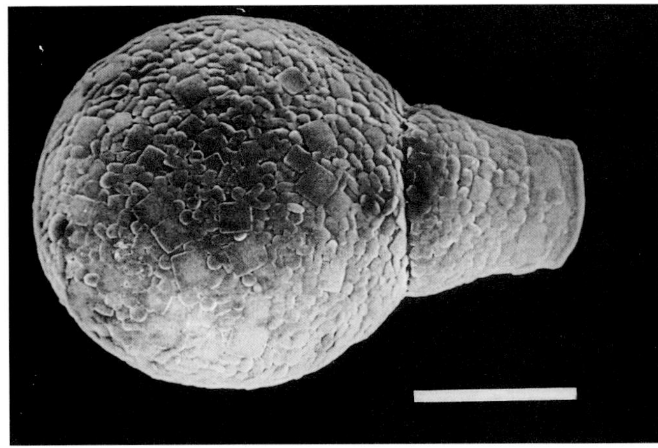

Fig. 69. *Apodera vas.* Scale bar=50 µm.

Genus ***Alocodera*** Jung, 1942

Shell pyriform, compressed, neck separated by a lateral indentation from the posterior part of the test and sometimes with 2 difficult to observe lateral pores; aperture arched in narrow view, thickened rim (Fig. 68). Shell very transparent, yellowish, smooth surface, small xenosomes in organic cement. Habitat: sphagnum, rare, only known from the southern hemisphere. Monospecific. Ref.: Deflandre (1936), Jung (1942a).
TYPE SPECIES: *Alocodera cockayni* (Penard, 1910)

Genus ***Apodera*** Loeblich & Tappan, 1961

Shell subspherical or ellipsoidal, compressed; constriction between body and swollen neck which tapers from its junction with the body towards the aperture (Fig. 69). Aperture oval, slightly arched. Feeds on a wide range of foods including algae. Habitat: mosses, organic soils. *Apodera vas* is one of the most common nebelids in the southern hemisphere but so far not found in Europe or North America. This genus was proposed by Jung (1942b) without type designation. This was corrected by Loeblich and Tappan (1961). Three species. Ref. Deflandre (1936), Jung (1942a).
TYPE SPECIES: *Apodera vas* Certes, 1889

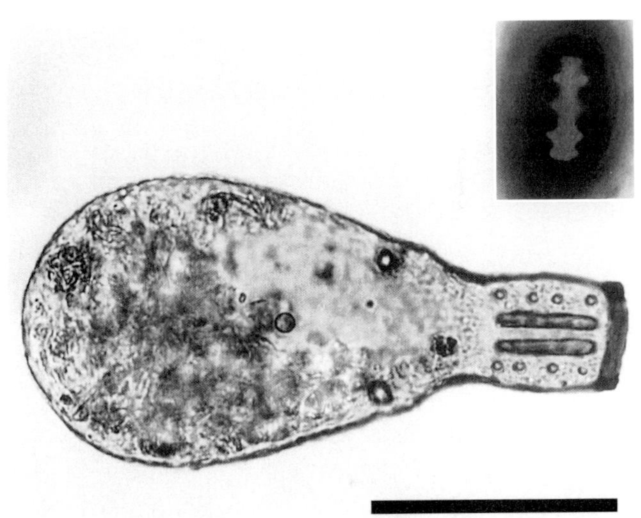

Fig. 70. *Certesella certesi.* Scale bar=50 µm. (Neck with two internal thickenings at both sides, not a tube as in the original diagnosis.)

Genus ***Certesella*** Loeblich & Tappan, 1961

Shell flask-shaped, compressed, neck with parallel sides, organic apertural rim (Fig. 70). In broad view on each side 2 pores in depressions connected by an internal tube as in *Porosia bigibbosa*. Neck with 1 or 2 internal organic thickenings and rows of teeth at both sides. Scanning electron microscopy reveals surface to be smooth. Shell composed of collected euglyphid idiosomes in an unstructured cement, transparent. Habitat: sphagnum. This genus is restricted to the southern hemisphere. It was proposed by Jung (1942b) as *Penardiella* without type designation (Loeblich and

Tappan, 1961). Three species. Ref. Deflandre (1936).
TYPE SPECIES: *Certesella martiali* (Certes, 1889)

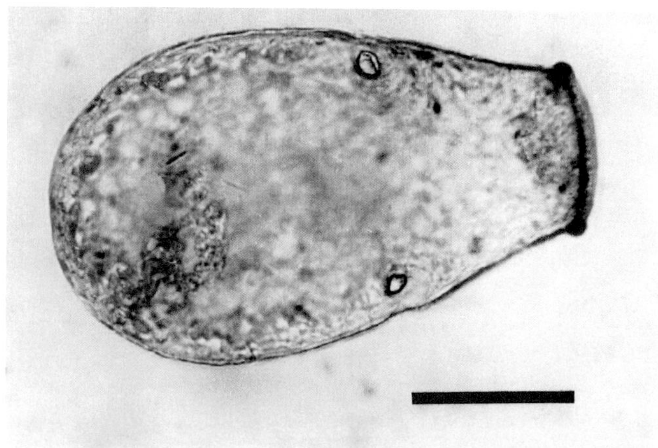

Fig. 71. *Porosia bigibbosa.* Scale bar=50 µm.

Genus ***Porosia*** Jung, 1942

This genus is closely related to *Certesella* but lacks the punctate neck (Fig. 71). Test pyriform, base rounded, in broad view 2 invaginated pores connected by internal tubes, in narrow view a small lateral pore just anterior to the larger pores. Shell composed of euglyphid shell plates embedded in unstructured cement. Pseudostome arched, thickened organic rim. Feeding: herbivorous and carnivorous? Habitat: sphagnum, soil, rare. Monospecific. Ecological data: Bonnet (1990). Refs. Penard (1890), Deflandre (1936), Ogden and Hedley (1980).
TYPE SPECIES: *Porosia bigibbosa* Jung, 1942

Genus ***Argynnia*** Vucetich, 1974

Shell compressed; aperture surrounded by siliceous plates, giving it a rough outline. Shell greyish, composed of various euglyphid plates, mixed with diatom fragments and mineral particles, porous cement (Fig. 72). The structure of the test is intermediate between *Nebela* and *Difflugia*. Nucleus ovular. This genus was proposed by Jung (1942b) without type designation and validated by Vucetich (1974) including only 3 species from the southern hemisphere. Here *Argynnia* is used in a less restricted sense to allow for common and better known species with similar test structure, such as *Argynnia dentistoma* (Penard, 1890) (synonym: *N. dentistoma*). Habitat: *A. dentistoma* is common in sphagnum or in acid humus; other species live in freshwater sediments. Used in this broader sense *Argynnia* has about fifteen species. Refs: Deflandre (1936), Ogden and Hedley (1980).

TYPE SPECIES: *Argynnia schwabei*

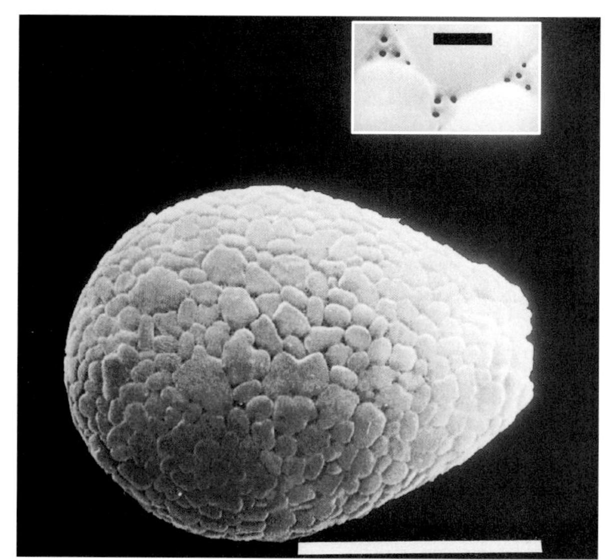

Fig. 72. *Argynnia dentistoma.* Scale bar=50 µm (inset 2 µm).

Fig. 73. *Physochila griseola.* Scale bar=50 µm.

Genus ***Physochila*** Jung, 1942

Test grey, pyriform, slightly compressed, surface with indentations and covered with siliceous exogenous plates (Fig. 73). Pseudostome more or less circular, terminal, with collar recurved posteriorly. Nucleus-ovular. Feeding: herbivorous. Habitat: sphagnum mosses. Old name: *Nebela griseola.* This genus was proposed by Jung (1942b) without type designation. Nine species. Refs. Deflandre (1936), Decloitre (1977b), Ogden and Hedley (1980).
TYPE SPECIES: *Physochila griseola* (Wailes & Penard, 1911)

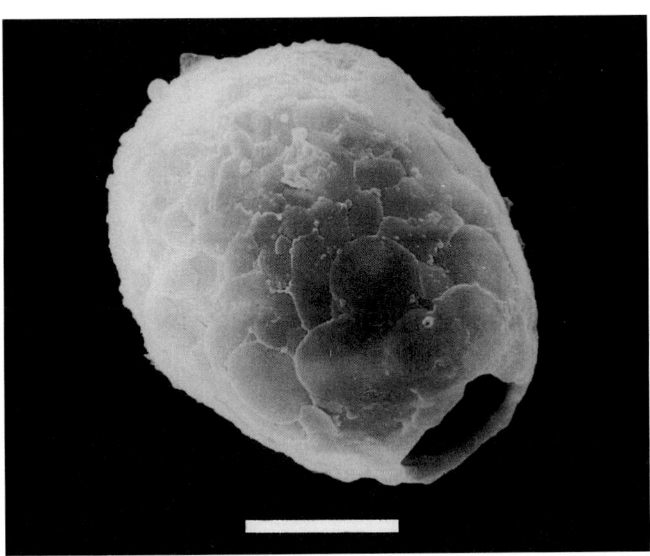

Fig. 74. *Schoenbornia humicola.* Scale bar=10 µm.

Genus ***Schoenbornia*** Decloitre, 1964

Shell ovoid, circular in cross section, hyaline, transparent, aperture circular (Fig. 74). Test composed of collected idiosomes of small euglyphids, angular quartz, and amorphous siliceous elements. Schönborn et al. (1987) suppose that the latter are produced by the amoeba, but this remains unproved. Large nucleus with central nucleolus, up to 3 contractile vacuoles, usually only 1 extended lobopodium. Resting stages spherical cysts with thick membrane separate from the shell wall and precysts with plasma retracted in the posterior part of the shell and covered with a thin membrane. Both stages have their aperture sealed with a diaphragm often hidden by a detrital plug. Feeding: collects detritus with bacteria and stores it around the aperture. Ecology: acid moder and raw humus, known from Europe, Africa, and Asia. Two species. Ref. Coûteaux (1978), Schönborn et al. (1987).
TYPE SPECIES: *Schoenbornia humicola* (Schönborn, 1964). *Heleoporella* Coûteaux, 1978 is probably a synonym of this genus.

SUBORDER PHRYGANELLINA Bovee, 1985

Pseudopods are clear ectoplasmatic, conical, pointed, sometimes branched, may anastomose. *Cryptodifflugia* has an open orthomitosis! (Page, 1966). Whether this is a widespread character of the suborder is unknown.

NOTE: The status of this suborder, which is based on pseudopodial characters alone, is unclear.

KEY TO THE GENERA OF THE SUBORDER

1. Test composed of mineral particles in organic matrix.. ***Phryganella***
1'. Test hyaline; if mineral particles present, then superficially attached...................................... 2

2. Aperture terminal.................. ***Cryptodifflugia***
2'. Aperture eccentric, test brownish ..***Wailesella***

FAMILY PHRYGANELLIDAE Jung, 1942

One genus with agglutinated shells.

Genus ***Phryganella*** Penard, 1902

Shell in lateral view hemispherical or higher; aperture large often ≈ two thirds of the shell diameter, not or only slightly invaginated. Test composed of mineral particles of variable size embedded in an organic matrix, which in older specimens can become dark brown due to manganese and iron deposition (Fig. 75). Around the aperture and the ventral face these particles are small, giving it a regular and smooth outline while at the aboral extremity larger grains are incorporated. When *P. acropodia* is cultured without a supply of mineral particles, it constructs completely organic tests. Ovular nucleus. Cyst internal with aperture closed by an organic diaphragm. Feeding: mainly bacteria, fungi, and algae (aquatic species). About 6 species and several subspecies. One of the most common species in sphagnum and soils is *Phryganella acropodia* (Hertwig & Lesser, 1874);

for population dynamics see Lousier (1984b). Refs. Chardez (1969), Lüftenegger et al. (1988), Ogden and Pitta (1989, 1990).
TYPE SPECIES: *Phryganella nidulus* Penard, 1902

Note: There is much confusion over the systematics of this genus due mainly to reports of different pseudopodia. Probably *P. acropodia* is a collective species. Although most species do not have an invaginated aperture, or an only slightly invaginated one with a smooth rim, there is a risk that empty tests are confounded with *Cyclopyxis* species.

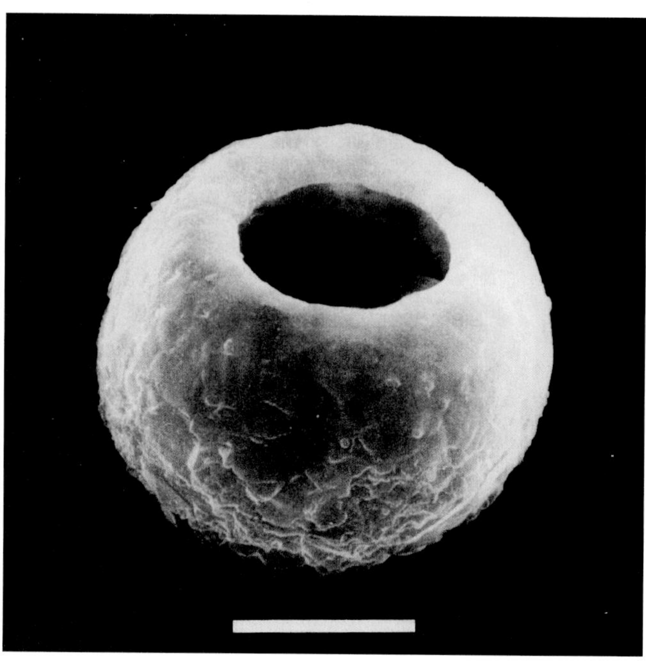

Fig. 75. *Phryganella acropodia.* Scale bar=25 μm.

FAMILY CRYPTODIFFLUGIIDAE Jung, 1942

Two genera with hyaline shells. Surface organic or with attached mineral particles. Aperture terminal or eccentric.

Genus ***Cryptodifflugia*** Penard, 1890

About 20 usually small species, with circular or oval cross-section, with adhering foreign particles or smooth surface (Fig. 76). Test smooth, clear, flask- or egg-shaped, circular in cross-section, aperture terminal. Shell wall in *C. oviformis* with 2 distinct layers, outer surface thin organic, inner layer thick from calcified material (Hedley et al., 1977). Whether this is a common character of all *Cryptodifflugia* spp. has to be proved. Nucleus vesicular. Cyst in test, by mucous plug. Food: bacteria, yeast. Habitat: freshwater, mosses, soil. Deflandre (1953) has placed species with circular cross-section in the genus *Difflugiella* and all compressed forms in *Cryptodifflugia.* This distinction is not made here. Following Page (1966) *Difflugiella* is considered to be a synonym of *Cryptodifflugia.* Refs: Grospietsch (1964), Schönborn (1965a), Page (1966). For ultrastructure see Griffin (1972), Hedley et al. (1977).
TYPE SPECIES: *Cryptodifflugia oviformis* Penard, 1890.

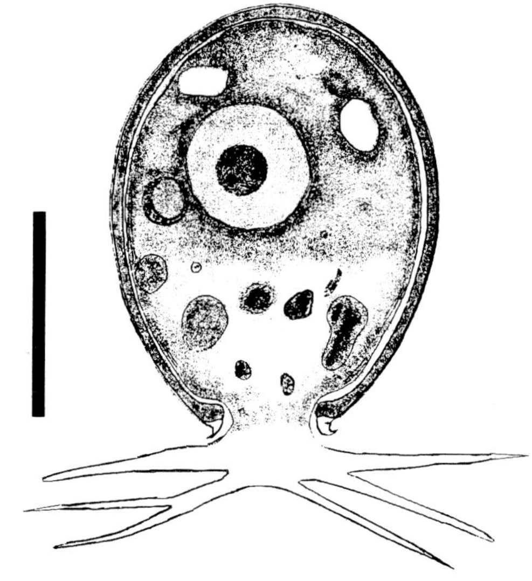

Fig. 76. *Cryptodifflugia oviformis.* Scale bar=10 μm. (After Hedley et al., 1977)

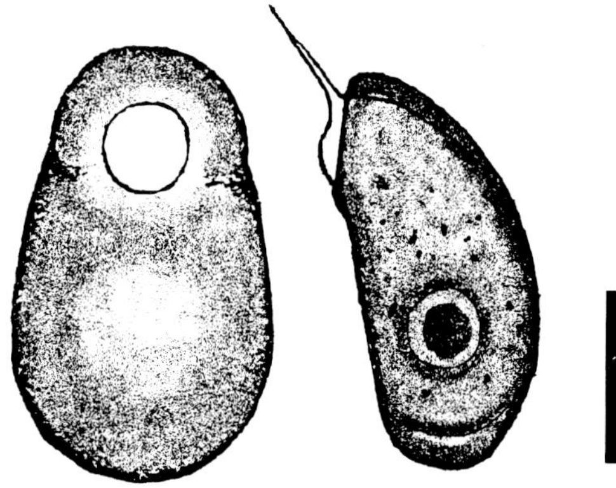

Fig. 77. *Wailesella eboracensis.* Scale=10 μm. (After Deflandre, 1928b)

Genus ***Wailesella*** Deflandre, 1928

Shell ovate, brown or yellow-brown, chitinoid. Size: 20 to 28 μm long; 13 · to 17 μm wide. Pseudostome a simple, round (5 to 6 μm) perforation, subterminal; near one end, forms an angle of app. 45° with long axis. Pseudopods clear, conical. Endoplasm: granular. Nucleus spherical, vesicular, 4 μm; endosome 2 μm. Contractile vacuole 3 to 4 μm. Feeding: herbivorous. Habitat: dryer sphagnum mosses, litter, and peaty soils. Monospecific. (Fig. 77). Refs. Wailes and Penard (1911), Deflandre (1928b).
TYPE SPECIES: *Wailesella eboracensis* (Wailes & Penard, 1911)

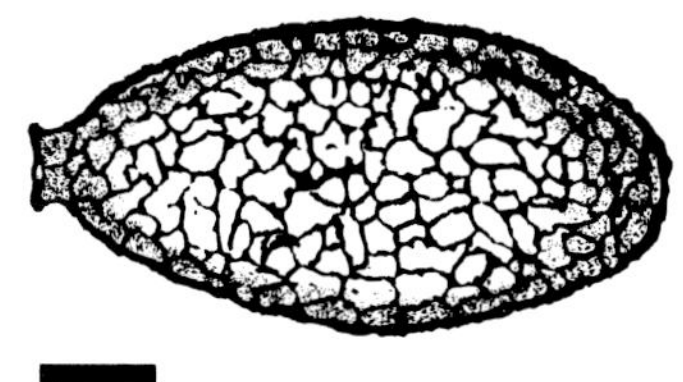

Fig. 78. *Geamphorella lucida.* Scale bar=10 μm. (After Bonnet, 1959)

INCERTAE SEDIS

Genus ***Geamphorella*** Bonnet, 1959

Shell transparent, ovoid, circular in cross section, tinted clear yellow or greyish white, with a chitinoid collar, slightly widened at the circular aperture, covered with amorphous silica elements (Fig. 78). Length 55 μm, breadth 25 μm, length of the neck 5 μm, diameter 7 μm. Habitat: humic calcareous soils, rare. Monospecific. Ref. Bonnet (1959).
TYPE SPECIES: *Geamphorella lucida* Bonnet, 1959
NOTE: Decloitre (1964) described *Pseudogeamphorella* with a compressed shell. The description is too vague to be included here.

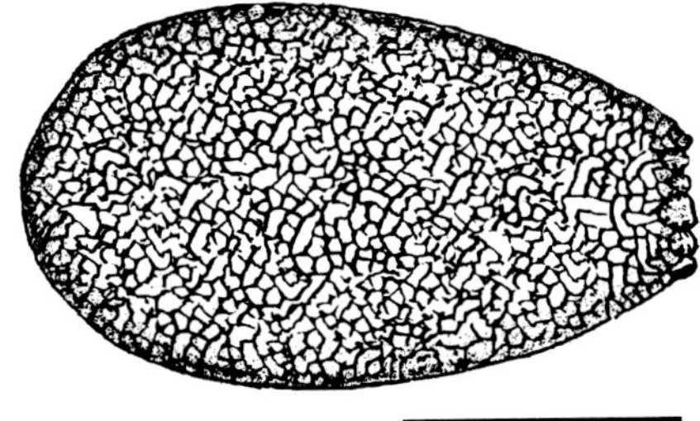

Fig. 79. *Jungia sundanensis.* Scale bar=50 μm.

Genus ***Jungia*** Loeblich & Tappan 1961

Test sack-like, cross section and aperture circular, opening with a rim of sand grains forming a short collar. Shell made of irregular plates as in *Nebela.* Habitat: mosses. Known from Indonesia, Congo, and Venezuela. Four species. (Fig. 79). Ref. Chardez (1966).
TYPE SPECIES: *Jungia sundanensis* van Oye, 1949

Genus ***Oopyxis*** Jung, 1942

Test brown, ovoid, in cross-section round or oval, ventral face never flat. Pseudostome subterminal. Habitat: freshwater mosses. Four species. (Fig. 80). Ref. Jung (1942a).
TYPE SPECIES: *Oopyxis cophostoma* Jung, 1942

Genus **Pseudawerintzewia** Bonnet, 1959

Test oviform; cross-section circular, brown, attached quartz particles (Fig. 81). Pseudostome elliptical or round, in flattened center of the broader end, thickened lip, no collar. Feeding herbivorous? Ecology: *P. calcicola* and *P. orbistoma* usually associated with calcareous soils (Bonnet, 1989b). Three species; *P. orbistoma* is similar to the type species but with a circular pseudostome. Refs. Bonnet (1959, 1974a), Schönborn et al. (1983).
TYPE SPECIES: *Pseudawerintzewia calcicola* Bonnet, 1959

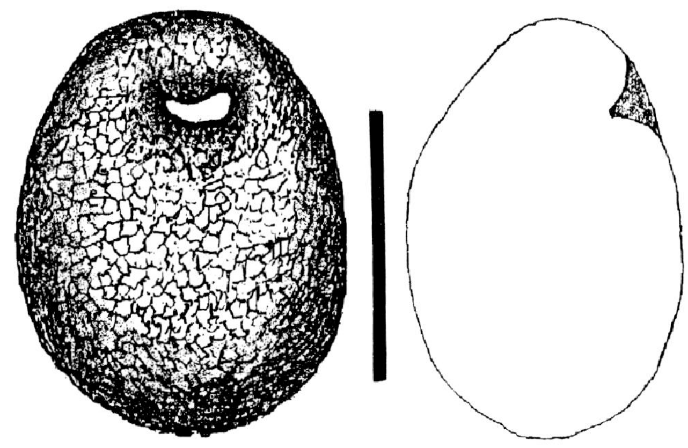

Fig. 80. *Oopyxis cophostoma.* Scale bar=50 μm. (After Jung, 1942a)

Genus ***Pseudonebela*** Gauthier-Lièvre, 1953

Test lageniform, in cross-section circular. Aperture surrounded by an organic lip with 3 - 5

denticulations that give the opening a lobed appearance. Shell hyaline, built of various platelets in an organic sheet-type(?) cement. Habitat: aquatic vegetation in tropical swamps (Africa). Monospecific. (Fig. 82). Ref.: Gauthier-Lièvre (1953).
TYPE SPECIES: *Pseudonebela africana* Gauthier-Lièvre, 1953.

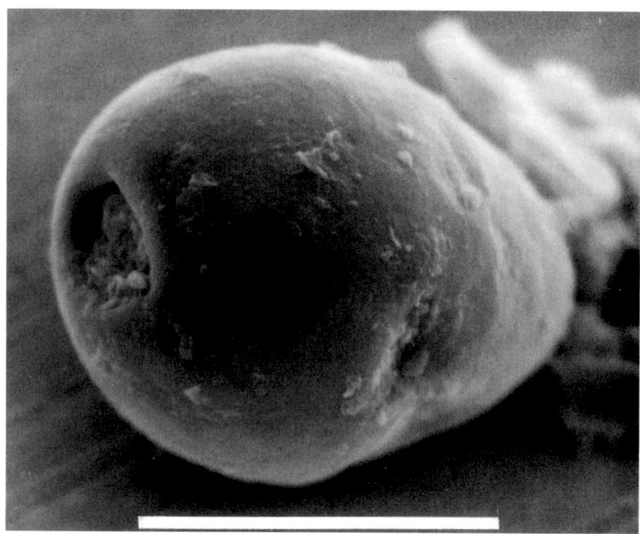

Fig. 81. *Pseudawerintzewia calcicola.* Scale bar=50 µm.

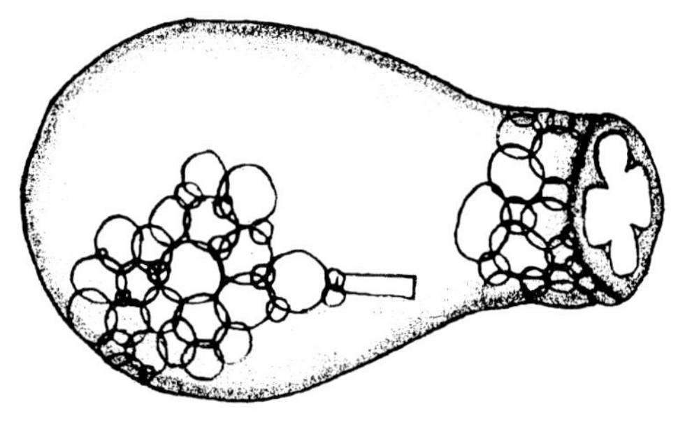

Fig. 82. *Pseudonebella africana.* Scale bar=50 µm.

LITERATURE CITED: See Testate Filose Amoebae, p 1065.

PHYLUM HELIOZOA

by K. A. Mikrjukov, F. J. Siemensma, and D. J. Patterson

The heliozoa are a polyphyletic assemblage of protists distinguished by having stiff arms which radiate from a rounded body. Unlike radiolaria, another polyphyletic group of organisms which have a similar body shape, heliozoa do not have an internal inorganic skeleton or a central capsule. The arms (**axopodia**) are supported internally by **axonemes**, geometric arrays of microtubules. The cytoplasm which covers the axonemes contains extrusible bodies (**extrusomes**) which are believed to be involved in the capture of prey.

The microtubules of the axonemes terminate on the surface of the nuclei or on other structures. These structures are typically located in the center of the cell. If they appear by electron microscopy as an amorphous mass of fibro-granular electron-dense material, they are referred to as '**axoplasts**'. If the nucleating structure is a tripartite disc with adhering, electron-dense hemispheres, it is referred to as a '**centroplast**'. The arms may extend or retract, sometimes quickly, by polymerization and depolymerization of the microtubules.

The heliozoa are mostly found in benthos of freshwater and, to a lesser extent, in marine or soil ecosystems or in the water column. They are passive predators, with prey (flagellates, ciliates, algae, chytrid zoospores, and small metazoa) being trapped on the arms by mechanisms not fully understood but involving the extrusomes in at least some species. The heliozoan body form is well adapted to diffusion feeding. This body form has arisen in many lines of eukaryotic evolution, making the heliozoa an ecological group. We refer to the 'heliozoa' only in a colloquial sense.

Heliozoa range in size from under 10 µm to over 500 µm. They are mostly naked, but may be coated with a periplast of organic or inorganic scales and/or spicules or may live in a perforated test. Cysts are formed by many species. Heliozoa are usually uninucleate, but multinucleated genera are known. Sexual activity has not been widely reported. Reproduction is by binary fission, but budding, autogamy, production of flagellated zoospores or even gametes have been reported for some heliozoa. Although heliozoa mostly have a spherical symmetry, some are supported on stalks. Heliozoa as a group are cosmopolitan, as are—to the best of our knowledge—most species. The general diversity and biology have been reviewed by Rainer (1968), Siemensma (1991), and Febvre-Chevalier (1985).

The heliozoa have been traditionally considered a discrete group, but the polyphyletic nature of the heliozoa is indicated by the different patterns of ultrastructural organization (Smith and Patterson, 1986). There are four well circumscribed types of heliozoa (actinophryids, centrohelids, desmothoracids, and gymnosphaerids). A number of heliozoa fall outside these groups (included here as "Other Heliozoa"). The heliozoa of earlier workers have included a variety of helioflagellates (such as dimorphids and pedinellid stramenopiles) and some nucleariid amoebae bearing siliceous scales. These are not included in this chapter.

By light microscopy, the different types can be distinguished by the shape of the actinopods, location and nature of the microtubule-organizing center, and material adhering to the outside of the cell. The largest group are the centrohelids; absolute (species level) identification requires electron microscopy for details of scale organization.

The following key provides the basis for assigning organisms to groups on the basis of their appearance by light microscopy.

KEY TO TYPES OF HELIOZOA

1. The axopodia are flattened and motile, marine.. taxopodids (see section on "Other Heliozoa")
1'. The axopodia are not motile.......................... 2

2. Heliozoon located within a perforated capsule. .... Family **Clathrulinidae**
2'. Not within a perforated lorica....................... 3

3. Heliozoa with stalk.. 4
3'. Heliozoa without stalk.................................. 8

4. Stalk mucoid... 5
4'. Stalk cytoplasmic.. 6

5. Stalk material extends around the cell body and encloses it as a capsule; both stalk and cell body may be supported by siliceous scales. .... (part).. Order **Centrohelida**
5'. Mucous capsule absent; stalk material may cover basal part of the cell body..... ***Servetia*** .... (see section on "Other Heliozoa")

6. Stalk contractile, body naked.......................... .... (part).....Family **Gymnosphaeridae**
6'. Cell with mucus and/or spicules, stalk not contractile... 7

7. Base of stalk small.................... ***Actinolophus*** (see section on 'Other Heliozoa')
7'. Stalk base expanded and amoeboid .... ***Wagnerella*** .... (see section on "Other Heliozoa")

8. Axonemes terminate on central non-nuclear body.. 9
8'. Axonemes terminate on a central nucleus or there are numerous peripheral nuclei; arms tapering.................... Family **Actinophryidae**

9. Large marine heliozoa (greater than 250 μm diameter), naked or with a coat of tangential siliceous spicules; axonemes terminate on axoplast................. Family **Gymnosphaeridae**
9'. Heliozoa less than 100 μm in diameter, naked, with mucous, organic, or siliceous secretions; axonemes terminate not on nucleus but on centroplast...................... Order **Centrohelida**

ORDER CENTROHELIDA Kühn 1926

Heliozoa with axonemes arising from **centroplast** comprised of a tripartite disc sandwiched between two hemispheres of dense material. The nucleus is located eccentrically. The pattern of microtubules in the axoneme is one of hexagons and triangles (Bardele, 1977). Mitochondrial cristae are lamellate; the kinetocysts are structurally complex ball-and-cone-shaped structures which can kill motile prey (Bardele, 1976). Members of two genera have a mucous stalk. Division in the order usually by binary fission, but there are records of budding and one account of the production of biflagellated forms. Many species form cysts. This order contains the most (85) species of heliozoa (more then 70 %). The species level taxonomy has been recently reviewed (Mikrjukov, 1996a, 1996b; 1997). Most are described from freshwater habitats, but they also occur in marine and terrestrial sites.

The ultrastructure has been well studied (e.g. Bardele, 1975; Siemensma, 1981) and, as the group is defined here, is structurally very conservative. One species has a naked cell surface; members of one genus have a mucous cell envelope; one genus has tangential organic spicules; and one genus has radial organic spicules. The remainder have a periplast of tangential and/or radial siliceous rods, plate-scales, or tube-like, trumpet-like, or spine-like radial spicules. There are 14 genera in three families.

KEY TO THE GENERA OF THE CENTROHELIDA

1. Without siliceous elements in periplasts....... 2
1'. Periplasts with siliceous elements............... 5

2. Cell surface is naked....................... ***Oxnerella***
2'. Cell surface is covered by a mucous coat and may bear tangential or radial organic spicules.. 3

3. Organic spicules are absent.... ***Chlamydaster***
3'. Organic spicules are present......................... 4

4. Spicules are tangential............ ***Phaerastrum***
4'. Spicules are radial.................... ***Heterophrys***

5. Periplast comprised of tangential elements only... 6
5'. Periplast with tangential plate scales and radial elements.. 9

6. Periplast comprised of tangential rod-like spicules .... ***Parasphaerastrum***
6'. Periplast comprised of plate-scales which may be elliptical or spindle-shaped.............. 7

7. All scales with a broad marginal rim............ 8
7'. Two types of plate scales, both lacking broad marginal rim............. ***Pseudoraphidiophrys***

8. Scales with smooth or reticulate texture .... ***Polyplacocystis***
8'. Plate scales with radial slits .... ***Raphidiophrys***

9. Radial elements are tube-shaped, trumpet-shaped or funnel-shaped.............................. 10
9'. Radial elements are spine-like (spicules); plate scales do not have a broad marginal rim.. 11

10. With one type of funnel-shaped radial elements; plate scales without broad marginal rim............................ ***Pseudoraphidocystis***
10'. Two types of radial elements: funnel-shaped and trumpet-shaped or tube-like; plate scales with broad marginal rim ....***Raphidocystis***

11. Spicules are bilaterally symmetrical and consist of a shaft and a base formed as a continuation of a membranous flange to the shaft... 12
11'. Spicules consist of a shaft and a discrete base plate.. 13

12. The membranous base continues as two lateral wings extending along the shaft, and a further basal wing forms the base of the spicule extending perpendicularly to the shaft and lateral wings.............................. ***Echinocystis***
12'. Without well-developed basal wing; spicules often with stalk........................... ***Pterocystis***

13. Spicule base-plates are heart-shaped; the shafts attach centrically.......... ***Choanocystis***
13'. Base plates of spicule are circular; shafts attach centrally..................... ***Acanthocystis***

FAMILY HETEROPHRYIDAE Poche,1913

Centrohelid heliozoa with a mucous coat with or without tangential or radial organic spicules. Not with siliceous elements in the periplast. Four genera.

Genus ***Oxnerella*** Dobell, 1917

Centroheliozoa with a naked cell surface. Nucleus with a large endosome. Marine; monotypic. Reference: Dobell (1917).
TYPE SPECIES: *Oxnerella maritima* Dobell, 1917.

Genus ***Chlamydaster*** Rainer, 1968

Centroheliozoa covered with a mucous coat only. From fresh and brackish waters; 2 species. Reference: Mikrjukov (1995a).
TYPE SPECIES: *Clamydaster sterni* Rainer, 1968.

Genus ***Sphaerastrum*** Greeff, 1875

Centroheliozoa covered with a mucous coat incorporating tangential organic spicules. Monotypic, freshwater, often forms colonies of 6—12 specimens connected by cytoplasmic bridges. The taxonomy has been reviewed by Mikrjukov (1996a).

TYPE SPECIES: *Sphaerastrum fockii* (Archer, 1869) West, 1901

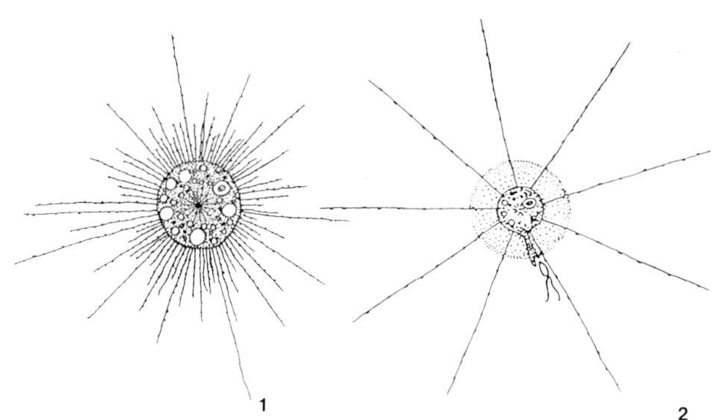

Fig. 1. *Oxnerella maritima* (after Dobell, 1917). Fig. 2. *Chlamydaster sterni* (after Rainer, 1968).

Genus ***Heterophrys*** Archer, 1869

Centroheliozoa with a mucous coat incorporating radial organic spicules. Often with algal symbionts. References: taxonomy (Mikrjukov, 1996a), ultrastructure (Bardele, 1975, 1977). TYPE SPECIES: *Heterophrys myriapoda* Archer, 1869

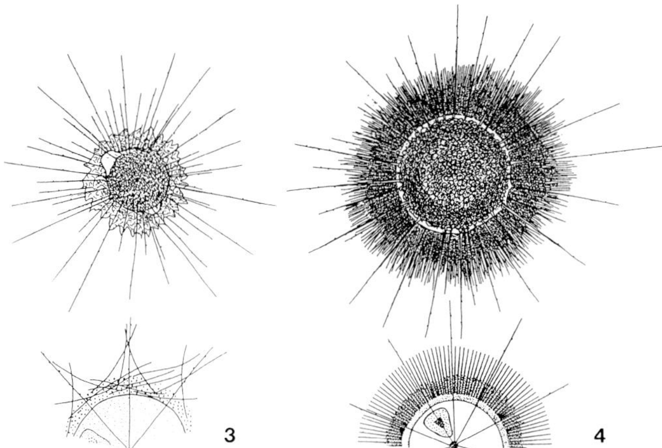

Fig. 3. *Sphaerastrum fockii* (after Archer, 1869) and a fragment of its periplast. Fig. 4. *Heterophrys myriapoda* (after Archer) and a fragment of its periplast.

FAMILY RAPHIDIOPHRYIDAE Mikrjukov, 1996

Centrohelid heliozoa with a periplast of siliceous elements, mostly tangential spicules or scales, radial elements if present occupy an outer position and are radially symmetrical with a tube-like, trumpet-like, or funnel-like shape. Four genera.

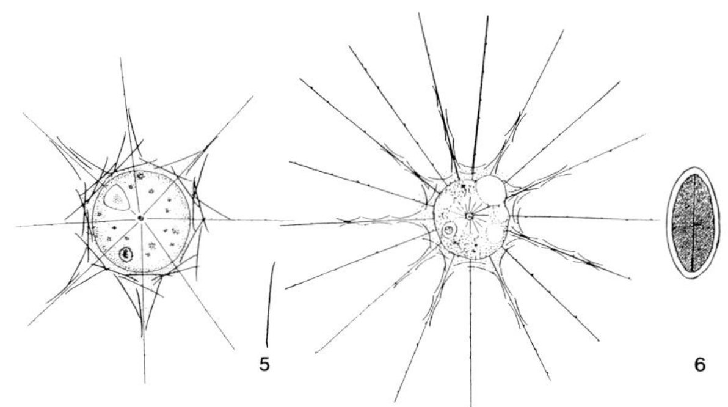

Fig. 5. *Parasphaerastrum marina* and a typical scale. Fig. 6. *Polyplacocystis symmetrica* (after Siemensma, 1991) and a typical scale.

Genus ***Parasphaerastrum*** Mikrjukov, 1996

Marine centroheliozoa with a periplast of only tangential, solid, siliceous, straight, or curved rods. Radial elements are absent. Monotypic. Reference: Mikrjukov (1996a).
TYPE SPECIES: *Parasphaerastrum marina* (Ostenfeld, 1904) Mikrjukov, 1996

Genus ***Polyplacocystis*** Mikrjukov, 1996

Centroheliozoa with a periplast comprising one or several types of tangential, spindle-shaped or flat, siliceous scales, which have a reticular or smooth upper surface and a hollow marginal rim. There are no internal septae. Radial elements absent. Five species from marine and freshwater habitats. References: Dürrschmidt and Patterson (1987b), Siemensma and Roijackers (1988b); Mikrjukov (1996b) for taxonomy.
TYPE SPECIES: *Polyplacocystis symmetrica* (Penard, 1904) Mikrjukov, 1996

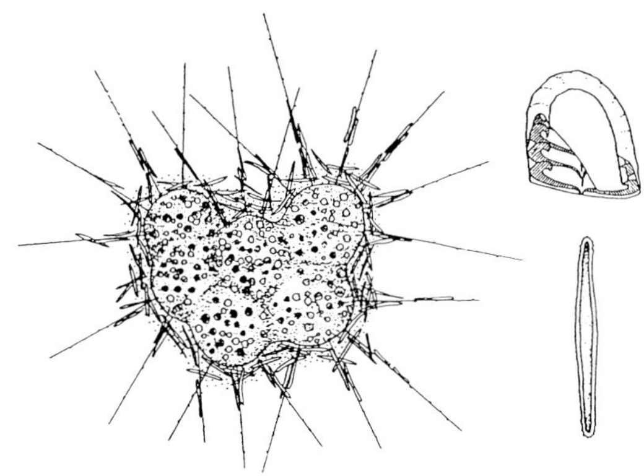

Fig. 7. A colonial form of *Raphidiophrys viridis*, its scale, and a scheme of the scale of *Raphidiophrys* (after Siemensma 1991).

Genus *Raphidiophrys* Archer, 1867

Centroheliozoa with flat, tangential, siliceous scales consisting of 2 parallel plates connected by internal radial septa (seen in scanning electron micrographs as radial striations or ribs). The scale has a hollow marginal rim. Radial elements are absent. May form clusters of cells. Reference: Siemensma and Roijackers (1988b).
TYPE SPECIES: *Raphidiophrys viridis* Archer, 1867

Genus *Raphidocystis* Penard, 1904

Centroheliozoa with a periplast comprising 3 types of scales: straight tubular or trumpet-like radial scales, smaller funnel-shaped radial scales, and tangential, elliptical plate-scales. The plate scales have a broad, hollow marginal rim, a smooth or reticular central area, and a well developed or vestigial medial rib as in *Polyplacocystis.* References: Mikrjukov (1996b), Nicholls and Dürrschmidt (1985).
TYPE SPECIES: *Raphidocystis lemani* (Penard, 1891) Penard, 1904.

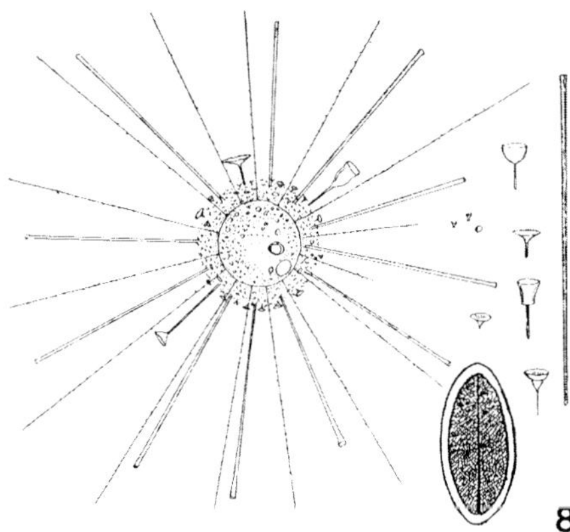

Fig. 8. *Raphidocystis lemani* and elements of its periplast (after Penard, 1891).

FAMILY ACANTHOCYSTIDAE Claus, 1874

Centrohelid heliozoa with a periplast of external and internal layers of siliceous elements. The internal elements are tangential plate-scales lacking a hollow marginal rim, while the external elements may be tangential plates with a well developed central sternum and sometimes with a concave base, or funnel-like structures, or radial spicules with a well developed shaft. Six genera.

Genus *Pseudoraphidocystis* Mikrjukov, 1997

Centroheliozoa with a periplast comprising tangential plate-scales lacking a hollow marginal rim, and radial funnel-like structures lacking an axial sternum/shaft. Three species. References: Siemensma (1991), Mikrjukov (1997).
TYPE SPECIES: *Pseudorapidocystis glutinosa* (Penard, 1904) Mikrjukov, 1997

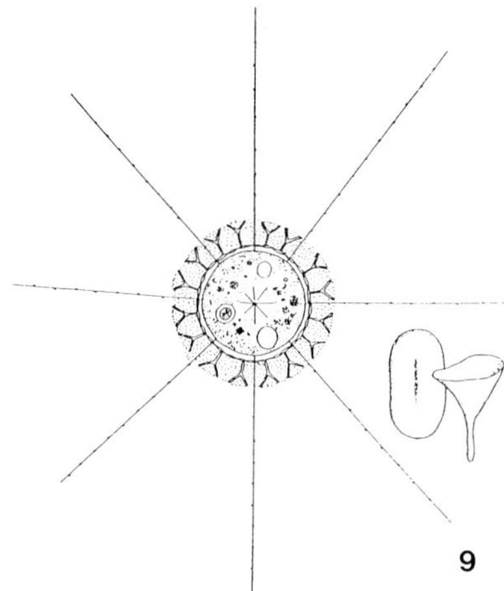

Fig. 9. *Pseudoraphidocystis glutinosa* and elements of its periplast (after Siemensma).

Genus *Pseudoraphidiophrys* Mikrjukov, 1997

Centroheliozoa with a periplast comprising 2 types of tangential scales: the internal layer plate-scales without a hollow marginal rim, and an external layer of plates with a well developed sternum and radiating ribs. The base of the sternum may be expanded and concave to create a velum-like extension of the scale. No records from light microscopy. Four species, all freshwater. References: Dürrschmidt (1985), Croome (1987); Mikrjukov (1997).
TYPE SPECIES: *Pseudoraphidiophrys formosa* (Dürrschmidt, 1985) Mikrjukov, 1997

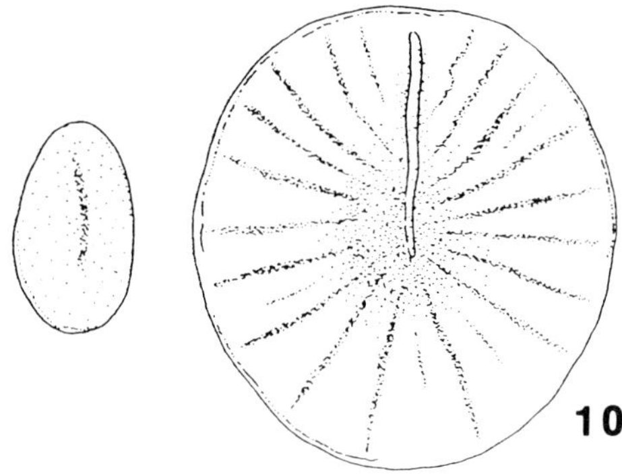

Fig. 10. Two types of scales of *Pseudoraphidiophrys formosa* (after Dürrschmidt, 1987).

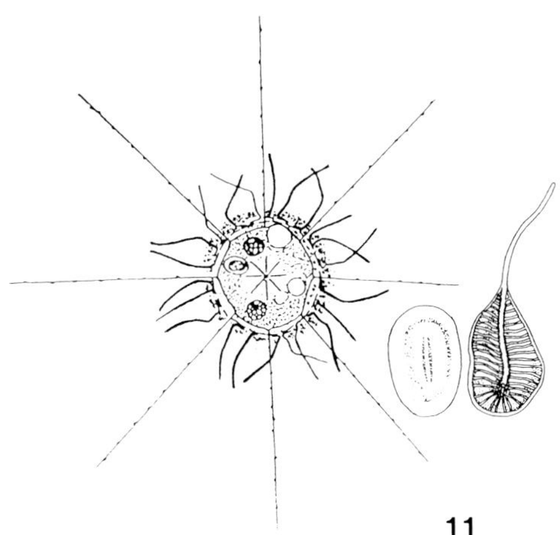
11

Fig. 11. *Pterocystis pteracantha* and elements of its periplast (after Siemensma, 1991).

Genus ***Pterocystis*** Siemensma & Roijackers, 1988

Centroheliozoa with a periplast with an internal layer of tangential plate-scales and external layer of radial spicules. These spicules consist of a shaft and a thin, membranous basal region which may extend along the shaft and may form a stalk-like protrusion at the base of the spicule. The membranous region is not clearly differentiated into lateral and basal regions as in *Echinocystis*. Ten species, all from freshwater. Reference: Mikrjukov (1997).
TYPE SPECIES: *Pterocystis pteracantha* (Siemensma, 1981) Siemensma & Roijackers, 1988

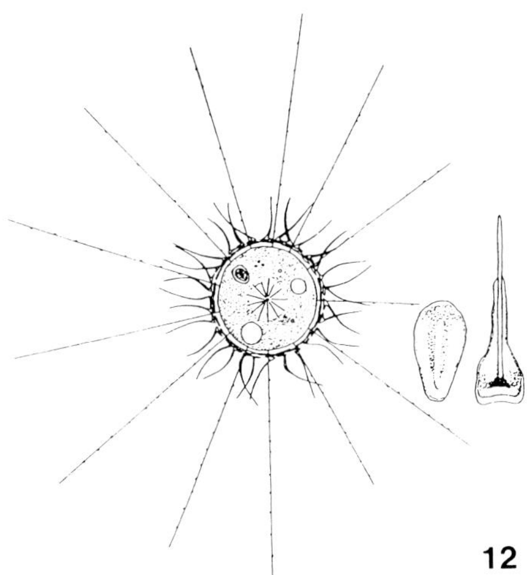
12

Fig. 12. *Echinocystis erinaceoides* and elements of its periplast (after Siemensma, 1991).

Genus ***Echinocystis*** Mikrjukov, 1997

Centroheliozoa with a periplast of internal tangential plate-scales and external radial spicules. The radial spicules have a shaft with 2 lateral wings extending along it, and a basal wing oriented perpendicular to the lateral ones. Eight freshwater, rarely brackish, species. References: Mikrjukov (1997), Siemensma (1991), Vørs (1992).
TYPE SPECIES: *Echinocystis erinaceoides* (Petersen & Hansen, 1960) Mikrjukov, 1997

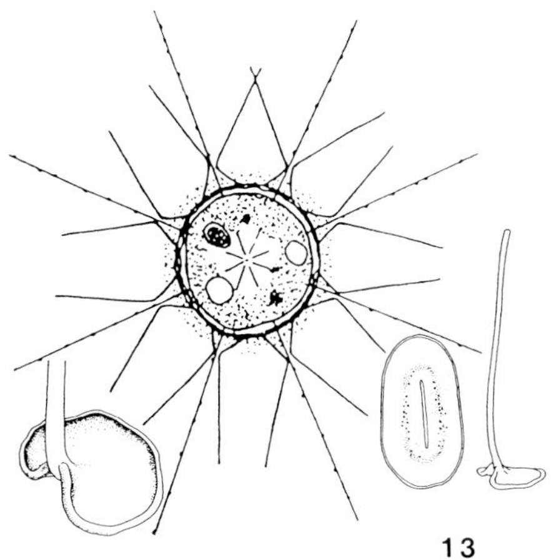
13

Fig. 13. *Choanocystis lepidula* and elements of its periplast (after Siemensma, 1991).

Genus ***Choanocystis*** Penard, 1904

Centroheliozoa with a periplast with internal tangential plate-scales and external radial spicules. The radial spicules have a cylindrical shaft, which inserts eccentrically at the apex of an invagination of a heart-shaped base-plate. Thirteen species, mostly freshwater. References: Mikrjukov (1995b), Siemensma and Roijackers (1988a).
TYPE SPECIES: *Choanocystis lepidula* Penard, 1904

Genus ***Acanthocystis*** Carter, 1863

Centroheliozoa with a periplast with tangential plate-scales and external radial spicules. The radial spicules have a cylindrical shaft attached to the central part of a radially symmetric base-plate. Fourteen species, mostly freshwater, the type species *A. turfacea* Carter, 1863 has been recorded in both marine, brackish, and fresh waters. Reference: Siemensma (1991).
TYPE SPECIES: *Acanthocystis turfacea* Carter, 1863

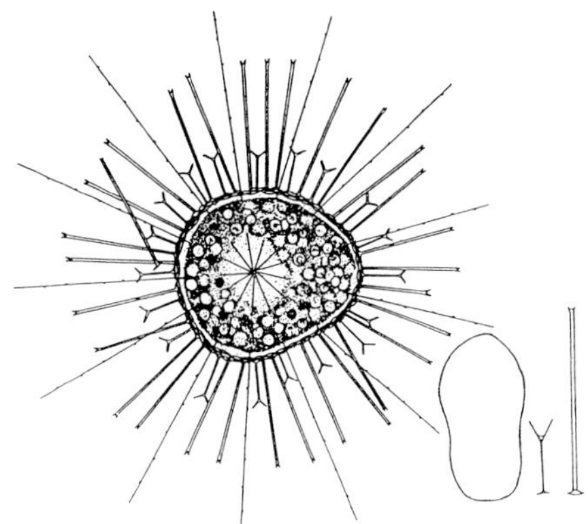

Fig. 14. *Acanthocystis turfacea* and its periplast elements (after Siemensma, 1991).

FAMILY ACTINOPHRYIDAE Claus, 1874

Heliozoa with axonemes terminating at or near nuclei; nucleus may be large and central (one genus) or small and numerous and located towards the periphery of the cell. Axopodia taper from base to tip. Microtubules within the axonemes in double interlocking coils. Mitochondrial cristae are tubular. With 2 types of simple extrusomes which cause prey to adhere to the heliozoon so it can be ingested (Patterson and Hausmann, 1981). Cells often fuse around a large accumulation of food. Reproduction mainly by binary fission, but autogamy (within one diploid cell giving rise to two haploid cells which fuse again) occurs in a cyst. Trophic cells are naked; cysts are surrounded by a layer of siliceous elements. Common and widespread in freshwater, also in soil, occasionally encountered in brackish and marine environments.
Affinities unclear, but may lie with the stramenopiles (Patterson, 1986). Two genera. References: Mignot (1979, 1980a, 1980b, 1984) and Patterson (1979).

KEY TO THE GENERA OF ACTINOPHRYIDAE

1. With a central nucleus............... ***Actinophrys***
1'. Cell with several peripheral nuclei................. .... ***Actinosphaerium***

Genus ***Actinophrys*** Ehrenberg, 1830

Spherically symmetrical heliozoa with a large central nucleus. May have a layer of peripheral vacuoles after feeding. Axonemes terminate on the nucleus. Several species; type-species freshwater. References: Mignot (1979, 1980a, 1980b), Patterson (1979).
TYPE SPECIES: *Actinophrys sol* Ehrenberg, 1830.

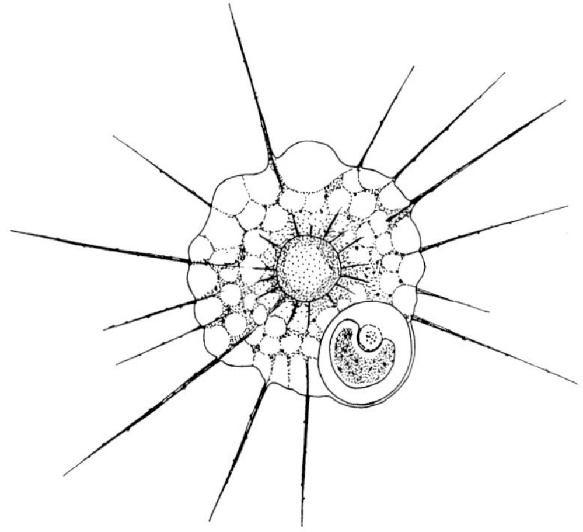

Fig. 15. *Actinophrys sol* (after Grenacher, 1869).

Genus ***Actinosphaerium*** Stein, 1857

Spherical naked heliozoa with numerous small nuclei. Cytoplasm often has a discrete peripheral layer of large vacuoles, and the nuclei occur at the inner boundary of this layer. Axonemes terminate on the nuclei or an electron-dense material near the nuclei. We regard *Camptonema* of Schaudinn and *Echinosphaerium* of Hovasse as synonyms. With several species. References: Anderson and Beams (1960).
TYPE SPECIES: *Actinosphaerium eichhornii* Ehrenberg, 1840

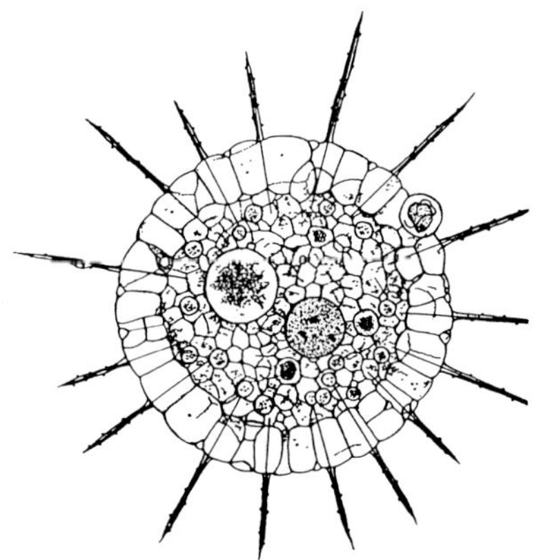

Fig. 16. *Actinosphaerium eichhornii* (after Siemensma, 1991).

FAMILY CLATHRULINIDAE Claus, 1874

Heliozoa (colloquially called the desmothoracids) with body located in a perforated, extracellular and usually stalked lorica. Cells with single central nucleus. Long, sometimes forked axopodia extend through the pores of the capsule. Axonemes terminate on the nuclear envelope. Microtubular cristae are tubular. Extrusomes are complex and resemble those of some cercomonads. There is a complex life cycle with the trophic heliozoon stage dividing by binary fission. One daughter may become a free-swimming form with one (in one species) or two flagella, which settles to form an amoeba, subsequently developing stiff pseudopodia and eventually secreting the capsule and stalk. Products of division may encyst within the capsule. Stalked capsules may join end-to-end to form colonies. With three genera and 11 species. References: Bardele (1972), Brugerolle (1985), Siemensma (1991).

KEY TO THE GENERA OF CLATHRULINIDAE:

1. Capsule is lattice-like, with large openings.... .... ***Clathrulina***
1'. Capsule is non-lattice-like, penetrated by pores or small openings................................. 2

2. Capsule tapers into a short conical stalk.......... .... ***Cienkowskya***
2'. Capsule is sharply differentiated from a long tube-like stalk .... ***Hedriocystis***

Genus ***Clathrulina*** Cienkowski, 1867

The cell body is loosely enclosed in a capsule with usually wide perforations through which protrude branched or unbranched axopodia. In 2 species the lorica is supported by a long, hollow stalk. Two stalked and 2 stalkless freshwater species. Reference: Bardele (1972).
TYPE SPECIES: *Clathrulina elegans* Cienkowski, 1867

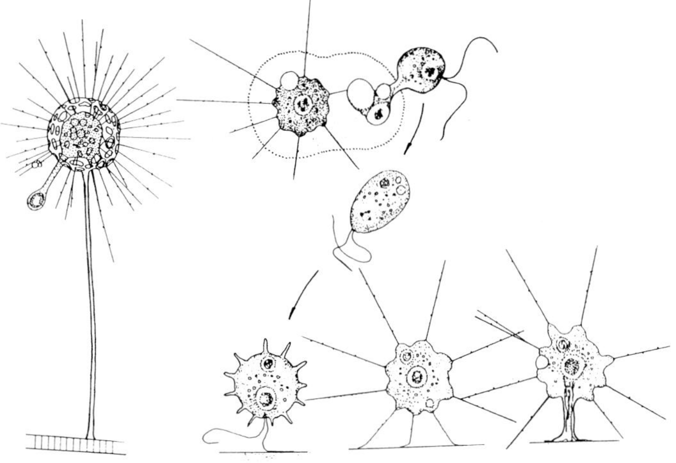

Fig. 17. *Clathrulina elegans* (after Cienkowski, 1867) and stages of its life cycle (after Siemensma, 1991).

Genus ***Hedriocystis*** Hertwig & Lesser, 1874

Cell body enclosed within an extracellular capsule which is either homogeneous (3 species) or composed of regular polygonal facets (2 species); axopodia emerge from small perforations. Three species with a long hollow stalk, one species without. Reference: Brugerolle (1985).
TYPE SPECIES: *Hedriocystis pellucida* Hertwig & Lesser, 1874

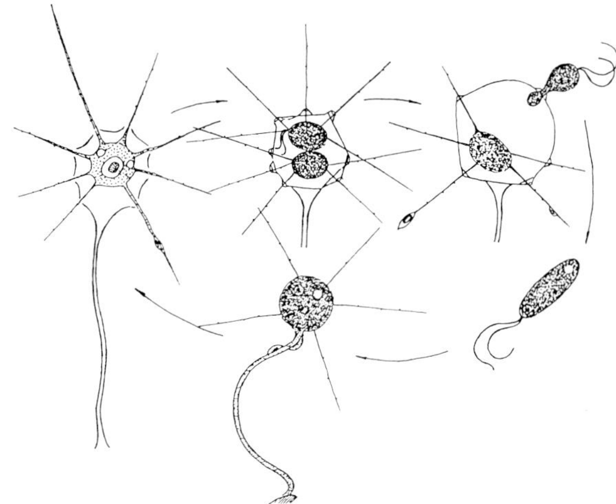

Fig. 18. *Hedriocystis pellucida* (after Hertwig and Lesser, 1874) and stages of its life cycle (after Hoogenraad, 1927).

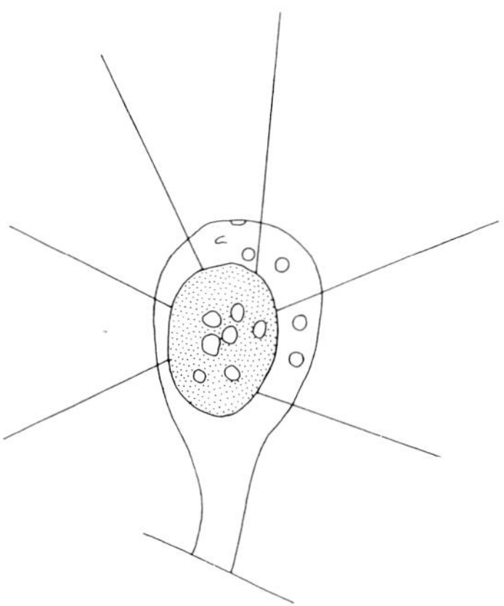

Fig. 19. *Cienkowskya mereschkovskii* (after Cienkowski, 1881).

Genus ***Cienkowskya*** Schaudinn, 1896

The cell body is enclosed in an irregular homogeneous capsule; axopodia emerge through small pores. The capsule tapers into a short, conical stalk. The type species, *C. mereschkovskii*, was described from a marine habitat, but has not been recorded subsequently. A second, freshwater species, *C. brachiopus,*

produces zoospores with one flagellum. Reference: De Saedeleer (1930).
TYPE SPECIES: *Cienkowskya mereschkovskii* (Cienkowski, 1881) Schaudinn, 1896

FAMILY GYMNOSPHAERIDAE Poche, 1913

Heliozoa with axonemes arising from a central homogeneous axoplast and the microtubules in the axonemes arranged in irregular hexagonal prisms. One eccentric nucleus or multiple nuclei; mitochondrial cristae are tubular; different types of extrusomes. Usually attached to the substrate by a cytoplasmic stalk; with the cell body differentiated into a 'head' with radiating axopodia and an extensive multinucleate amoeboid base. The stalk of *Actinocoryne* is contractile. May occasionally be free-swimming. Cell surface may be covered by tangential siliceous (?) spicules, mucus, or be naked. Life cycle is complex but not fully resolved. *Actinocoryne* may produce multiple buds from the 'head' which form small, temporarily floating, spherical, amoeboid cells. Motile cells with two flagella are produced by *Gymnosphaera* and *Actinocoryne* (possibly with an annual cycle). Three monotypic marine genera. References: Febvre-Chevalier (1980, 1990), Jones (1976).

KEY TO THE GENERA OF GYMNOSPHAERIDAE:

1. Cell surface naked or covered with a thin mucus coat... 2
1'. Cell surface with mucus and tangential siliceous (?) spicules, uninucleate, attached or floating................................ ***Hedraiophrys***

2. Unattached, multinucleate, with cytoplasm differentiated into ecto- and endoplasm .... ***Gymnosphaera***
2'. Attached, with a contractile cytoplasmic stalk and an extensive multinucleate amoeboid base; cytoplasm uniform.................... ***Actinocoryne***

Genus ***Hedraiophrys*** Febvre-Chevalier, 1973

Cell usually a truncated cone attached to a substrate, with a vesicular 'ectoplasm' and a denser, granular 'endoplasm'. The cell is covered with microfibrillar mucus and numerous tangential siliceous spicules which are as long or longer then the body diameter. There is a large, central axoplast, and the nucleus lies near the edge of the cell. May detach and float as spherical cells for a short period. With one marine species, *H. hovassei*, which contains numerous algal and bacterial symbionts. Reference: Febvre-Chevalier (1973).
TYPE SPECIES: *Hedraiophrys hovassei* Febvre-Chevalier, 1973

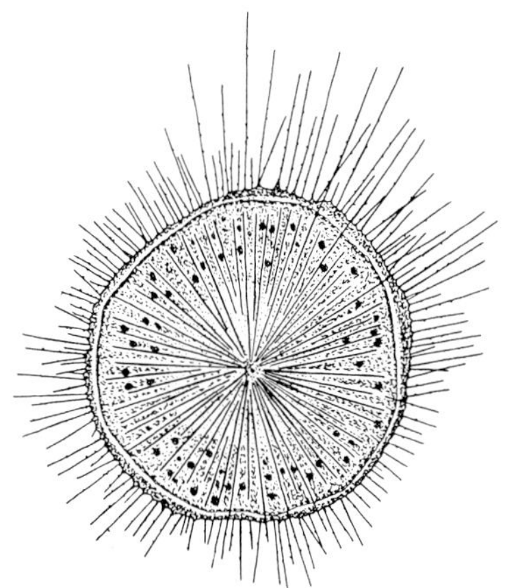

Fig. 20. *Hedraiophrys hovassei* (after Febvre-Chevalier, 1973).

Genus ***Actinocoryne*** Febvre-Chevalier, 1980

With an ovoid cell body ('head') from which radiate axopodia. The head attaches to the substrate by a contractile stalk which arises from an extensive amoeboid and multinucleated base. Surface naked. The axoplast occupies the center of the head. Monotypic; marine. Reference: Febvre-Chevalier (1980).
TYPE SPECIES: *Actinocoryne contractilis* Febvre-Chevalier, 1980

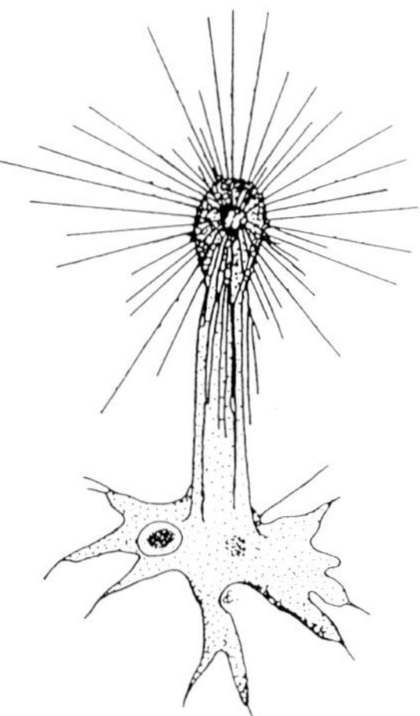

Fig. 21. *Actinocoryne contractilis* (after Febvre-Chevalier et al., 1986).

Genus *Gymnosphaera* Sassaki, 1894

Unstalked heliozoa with an outer vacuolated 'ectoplasm' and a more central granular, denser 'endoplasm'. The surface is covered by mucus. With a central homogeneous axoplast. With many small nuclei distributed throughout the endoplasm. Monotypic; marine. References: Jones (1976), Febvre-Chevalier (1975).
TYPE SPECIES: *Gymnosphaera albida* Sassaki, 1894.

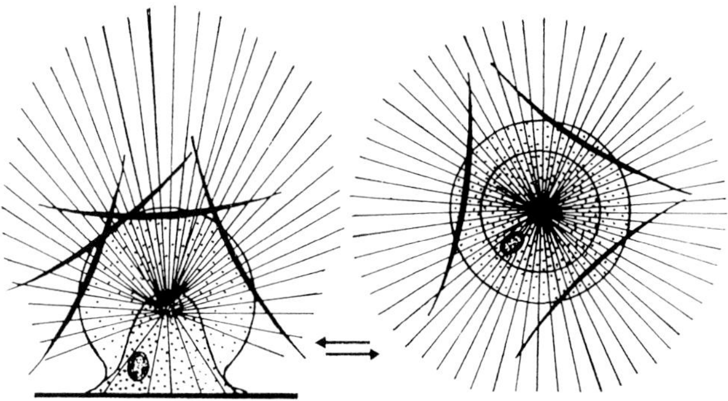

Fig. 22. *Gymnosphaera albida* (after Sassaki, 1894).

OTHER HELIOZOA

A number of taxa of uncertain affinities have a heliozoan-like appearance and have been Included as heliozoa in earlier treatments (Rainer, 1968; Fcbvrc-Chevalier, 1985). We consider these organisms here, assuming that later work will establish their affinities more clearly.

Genus *Sticholonche* Hertwig, 1877

Large, marine, pelagic protists with 50-60 rows of motile oar-like pseudopodia inserting into depressions on the surface of a large, single, central nucleus. Body heart-shaped, bilaterally symmetrical, dorsoventrally compressed. The outer cytoplasm contains numerous small, curved, siliceous, tangential spicules and has 14 rosettes of long, external, radial spicules. The microtubules of the axonemes are arranged in irregular hexagons. Often treated as an order—the Taxopodida Fol, 1883. Reference: Cachon and Cachon (1977).
TYPE SPECIES: *Sticholonche zanclea* Hertwig, 1877.

Genus *Wagnerella* Mereschkowsky, 1878

Stalked heliozoa with a spherical 'head' supported by a cylindrical, non-contractile cytoplasmic stalk with an enlarged base. The nucleus is at the base of the stalk. The center of the head is occupied by an axoplast comprised of a central element surrounded by granular material. The head is covered by mucilage with associated small, tangential, siliceous (?) spicules. Reproduction by multiple budding which produces amoeboid cells from the head like *Actinocoryne* and also from the base. A complex life cycle described by Zuelzer (1909) needs verification. Monotypic; the single marine species, *W. borealis,* may possibly be a gymnosphaerid. Reference: Zuelzer (1909).
TYPE SPECIES: *Wagnerella borealis* Mereschkowsky, 1878

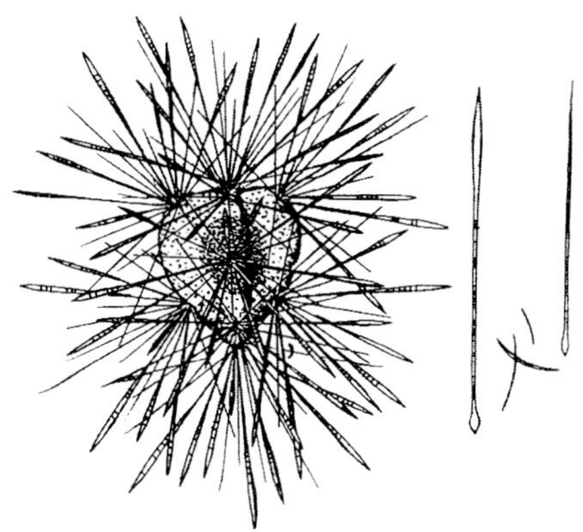

Fig. 23. *Sticholonche zanclea* (after Hollande and Enjumet, 1954).

Genus *Actinolophus* Schulze, 1874

Stalked heliozoa with a pyriform 'head' at the top of a cylindrical, non-contractile, cytoplasmic stalk, which seems to be hollow. The center of the 'head' includes a homogeneous, pear-shaped axoplast; the nucleus is eccentric but in the 'head'. The axonemes support axopodia and also extend into the stalk. The 'head' is covered by a gelatinous layer. Monotypic, marine. Reference: Villeneuve (1937).
TYPE SPECIES: *Actinolophus pedunculatus* Schulze, 1874

Genus *Servetia* Poche, 1913

Stalked heliozoon with a spherical cell body at the top of a long, hollow, non-cytoplasmic, tube-like stalk, which may form a broadened base. No capsule nor extracellular gelatinous matrix. Numerous unbranched pseudopodia extend in all directions. Reproduction is by division into 2—4 identical heliozoon cells. Monotypic, marine. Reference: Cienkowski (1881).
TYPE SPECIES: *Servetia borealis* (Mereschkowsky, 1879) Poche, 1913.

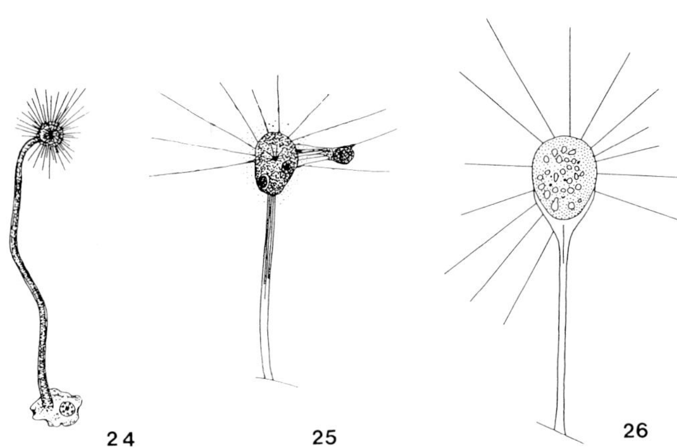

Fig. 24. *Wagnerella borealis* (after Zuelzer, 1909). Fig. 25. *Actinolophus pedunculatus* (after Schulze, 1874). Fig. 26. *Servetia borealis* (after Cienkowski, 1881).

LITERATURE CITED

Anderson, E. & Beams, H. W. 1960 The fine structure of the heliozoan, *Actinosphaerium nucleofilum*. *J. Protozool.*, 7:190-199.

Archer, W. 1869. On some freshwater Rhizopoda, new or little-known. *Q. J. Microsc. Sci.*, 9:250-271.

Bardele, C. F. 1972. Cell cycle, morphogenesis and ultrastructure in pseudoheliozoan *Clathrulina elegans*. *Z. Zellforsch.*, 130:219-242.

Bardele, C. F. 1975. The fine structure of the centrohelidian heliozoan *Heterophrys marina*. *Cell Tissue Res.*, 161:85-102.

Bardele, C. F. 1976. Particle movement in heliozoan axopods associated with lateral displacement of highly ordered membrane domains. *Z. Naturforsch.*, 31c:190-194.

Bardele, C. F. 1977. Comparative study of axopodial microtubule patterns and possible mechanisms of pattern control in the centrohelidian Heliozoa *Acanthocystis*, *Raphidiophrys* and *Heterophrys*. *J. Cell Sci.*, 25:205-232.

Brugerolle, G. 1985. Ultrastructure d'*Hedriocystis pellucida* (Heliozoa Desmothoracida) et de sa forme migratice flagellée. *Protistologica*, 21:259-265.

Cachon, J. & Cachon, M. 1977. *Sticholonche zanclea* Hertwig: a reinterpretation of its phylogenetic position based upon new observations on its ultrastructure. *Arch. Protistenkd.*, 120:148-168.

Cienkowski, L. 1867. Über *Clathrulina*, eine neue Actinophryengattung. *Arch. Mikr. Anat.*, 3:311-316.

Cienkowski, L. 1881. An account of the White Sea excursion in 1880. *Proc. St. Petersb. Imp. Soc. Nat.*, 12:130-171.

Croome, R. 1987. Observations of the genera *Acanthocystis, Raphidiophrys, Clathrulina* and *Pompholyxophrys* (Protozoa, Sarcodina) from australian freshwaters. *Arch. Protistenkd.*, 133:237-243.

De Saedeleer, H. 1930. Structure, nutrition, reproduction de *Monomastigocystis brachipous* n. g., n. sp. *Ann. Protistol.*, 3:1-11.

Dobell, C. 1917. On *Oxnerella maritima*, nov.gen., nov.spec., a new heliozoan, and its method of division; with some remarks on the centroplast of Heliozoa. *Q. J. Microsc. Sci.*, 62:515-538.

Dürrschmidt, M. 1985. Electron microscopic observations on scales of species of the genus *Acanthocystis* (Centrohelidia, Heliozoa) from Chile. *Arch. Protistenkd.*, 129:55-87.

Dürrschmidt, M. 1987. An electron microscopic study on fresh water Heliozoa (genus *Acanthocystis*, Centrohelidia) from Chile, New Zealand, Malaysia and Sri-Lanka. III. *Arch. Protistenkd.*, 133:49-80.

Dürrschmidt, M. & Patterson, D. J. 1987a. A light and electron microscopical study of a new species of centroheliozoon, *Chlamydaster fimbriatus. Tissue Cell,* 19:365-376.

Dürrschmidt, M. & Patterson, D. J. 1987b. On the organization of the heliozoa *Raphidiophrys ambigua* Penard and *R. pallida* Schulze. *Ann. Sci. Nat. Zool. Biol. Anim.*, 13th ser., 8:135-155.

Febvre-Chevalier, C. 1973. *Hedraiophrys hovassei*, morphologie, biologie et cytologie. *Protistologica*, 9:503-520.

Febvre-Chevalier, C. 1975. Etude cytologique de *Gymnosphaera albida* Sassaki,1894 (Heliozoaire Centrohelidie). *Protistologica*, 11:331-344.

Febvre-Chevalier, C. 1980. Behaviour and cytology of *Actinocoryne contractilis*, nov. gen., nov. sp., a new stalked heliozoan (Centrohelidia): comparison with the other related genera. *J. Mar. Biol. Assoc. U.K.*, 60:909-928.

Febvre-Chevalier, C. 1985. Class Heliozoea Haeckel 1866. *In*: Lee, J. J., Hutner, S. H. & Bovee, E. C. (eds.), *An Illustrated Guide to the Protozoa.* Society of Protozoologists, Lawrence, Kansas. pp 302-317

Febvre-Chevalier, C. 1990. Phylum Actinopoda Class Heliozoa. *In*: Margulis, L., Corliss, J. O. , Melkonian, M. & Chapman, D. J. (eds.), *Handbook of Protoctista*. Jones and Barlett Publishers, Boston. pp 347-362

Grenacher, H. 1869. Über *Actinophrys sol. Verhand. Phys. Med. Gesell. Wurzburg, N. F.,* 1:166-178.

Hertwig, R. & Lesser, E. 1874. Über Rhizopoden und denselben nahestehende Organismen. III. Teil Heliozoa. *Arch Mikr. Anat.*, 10:147-236.

Hollande, A. & Enjumet, M. 1954. Morphologie et affinités Radiolaire *Sticholonche zanclea* Hertwig. *Ann. Sci. Nat. Zool. 11e ser.*, 16:337-343.

Hoogenraad, H. R. 1927. Beobachtungen über den Bau, die Lebensweise und die Entwicklung von *Hedriocystis pellucida. Arch. Protistenkd.*, 58:321-343.

Jones, W. C. 1976. The ultrastructure of *Gymnosphaera albida* Sassaki, a marine axopodiate protozoon. *Phil. Trans. R. Soc. London* (B), 275:349-384.

Mignot, J.-P. 1979. Etude ultrastructurale de la pédogamie chez *Actinophrys sol* (Heliozoaire). I. La division progamique. *Protistologica*, 15:387-406.

Mignot, J.-P. 1980a. Etude ultrastructurale de la pédogamie chez *Actinophrys sol* (Heliozoaire). II. Les divisions de maturation. *Protistologica*, 16:205-226.

Mignot, J.-P. 1980b. Etude ultrastructurale de la pédogamie chez *Actinophrys sol* (Heliozoaire). III. Gamètogenèse, fécondation, enkystement. *Protistologica*, 16:533-547.

Mignot, J.-P. 1984. Etude ultrastructurale de la mitose végétative chez l'Heliozoaire *Actinophrys sol. Protistologica*, 20:247-264.

Mikrjukov, K. A. 1995a. Formation of extrusive organelles — kinetocysts in the heliozoon, *Clamydaster sterni. Cytology* (Leningrad), 37:153-158.

Mikrjukov, K. A. 1995b. Revision of species composition of the genus *Choanocystis* (Centroheliozoa, Sarcodina) and its members in Eastern Europe. *Zool. Zhurn.* (Moscow), 74:3-17.

Mikrjukov, K. A. 1996a. Revision of genera and species composition in lower Centroheliozoa. I. Family Heterophryidae Poche. *Arch. Protistenkd.*, 147:107-113.

Mikrjukov, K. A. 1996b. Revision of genera and species composition in lower Centroheliozoa. II. Family Raphidiophryidae n. fam. *Arch. Protistenkd.*, 147:205-212.

Mikrjukov, K. A. 1997. Revision of genera and species composition of the family Acanthocystidae (Centroheliozoa; Sarcodina). *Zool. Zhurn.* (Moscow), 76:1-13.

Nicholls, K. H. & Dürrschmidt, M. 1985. Scale structure and taxonomy of some species of *Raphidocystis, Raphidiophrys* and *Pompholyxophrys* (Heliozoea) including description of six new taxa. Can. J. Zool., 63:1944-1961.

Patterson, D. J. 1979. On the organization and classification of the protozoon, *Actinophrys sol* Ehrenberg, 1830. *Microbios*, 26:165-208.

Patterson, D. J. 1986. The actinophryid heliozoa (Sarcodina, Actinopoda) as chromophytes. *In*: Kristiansen, J. & Andersen, R. A. (eds.), *Chrysophytes: Aspects and Problems.* Cambrige University Press, pp.49-67.

Patterson, D. J. & Hausmann, K. 1981. Feeding by *Actinophrys sol* (Protista, Heliozoa): I. Light microscopy. *Microbios*, 31:39—55.

Penard, E., 1891. Contributions à l'étude des Rhizopodes du Leman. *Arch. Sci. Nat.*, 26:134-156.

Rainer, H. 1968. Urtiere, Protozoa; Würzelfüßler, Rhizopoda; Sonnentierchen, Heliozoa. *In*: Dahl, F. (ed.), *Die Tierwelt Deutschlands.* Bd. 56. Gustav Fischer Verlag, Jena.

Sassaki, C. 1894. Untersuchungen über *Gymnosphaera albida*, eine neue marine Heliozoe. *Jenaische Z. Naturw.*, 28:45-52.

Schulze, F. E. 1874. Rhizopodenstudien. *Arch Mikr. Anat.,* 10:377-400.

Siemensma, F. J. 1981. De Nederlandse Zonnendiertjes (Actinopoda, Heliozoa). *Wet. Meded. K.N.N.V.*, Vol. 149. 118 pp

Siemensma, F. J. 1991. Heliozoea. *In*: Page, F. C. & Siemensma, F. J. (eds.), *Nackte Rhizopoda und Heliozoea. Protozoenfauna*, Vol. 2, Fischer Verlag, Stuttgart. pp 171—290

Siemensma, F. J. & Roijackers, R. M. M. 1988a. A study of little-known acanthocystid heliozoeans, and a proposed revision of the genus *Acanthocystis* (Actinopoda, Heliozoea). *Arch. Protistenkd.*, 135:197-212.

Siemensma, F. J. & Roijackers, R. M. M. 1988b. The genus *Raphidiophrys* (Actinopoda, Heliozoa): scale morphology and species distinction. *Arch. Protistenkd.*, 136:237-248.

Smith, R. M. & Patterson, D. J. 1986. An analysis of heliozoan interrelationships: an example of the potentials and limitations of ultrastructural approaches to the study of protistan phylogeny. *Proc. R. Soc. Lond.* B, 227:325-366.

Villeneuve, F. 1937. Sur la structure de *Cienkowskya mereschkowskyi* Cienk. et d'*Actinolophus pedunculatus* Schulz., Heliozoaires des eaux saumâtres de Sête. *Arch. Zool. Exp. Gen.*, 78:243-250.

Vørs, N. 1992. Heterotrophic amoebae, flagellates and heliozoa from the Tvarminne area, Gulf of Finland, in 1988-1990. *Ophelia*, 36:1-109.

Zuelzer, M. 1909. Bau und Entwicklung von *Wagnerella borealis* Mereschk. *Arch. Protistenkd.*, 17:135-202.

PHYLUM GRANULORETICULOSA
Lee, 1990

By JOHN J. LEE, JAN PAWLOWSKI, JEAN-PIERRE DEBENAY, JOHN WHITTAKER, FRED BANNER, ANDREW J. GOODAY, OLE TENDAL, JOHN HAYNES, and WALTER W. FABER

Members of this phylum are characterized by the possession of **granular reticulopods**, with distinctive bidirectional streaming, that are not reinforced with geometrically arrayed microtubules and that form **anastomosing networks** and occasional filopods (Figs. 1 and 2). The Phylum is divided into two classes on the basis of the possession, or not, of a test with one or more permanent openings.

In actual practice, the pseudopods of very few Foraminifera, the largest class, have been studied to see if their pseudopodia meet the criteria of the phylum. We presume they do, and all are included based on our knowledge of other criteria (test and life cycles). Unlike the first edition, in this edition of the book we have treated the members of the Class Athalamea differently. We do not know if the pseudopods of most of the athalamate organisms meet the criteria which are used to presently define the phylum. If the pseudopods differ, we should in the future find molecular or other signatures which will help define the phylum better.

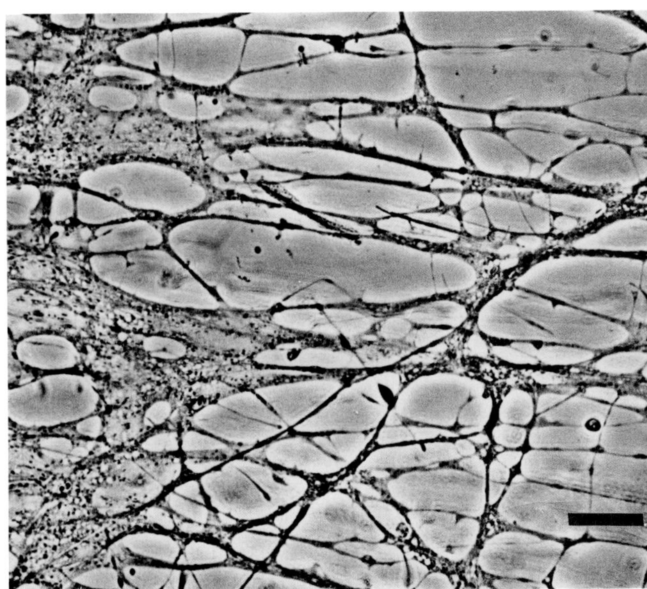

Fig. 1. Phase constrast micrograph of *Astrammina rara*. Scale=10 μm. (by Samuel Bowser and Jeffrey Travis)

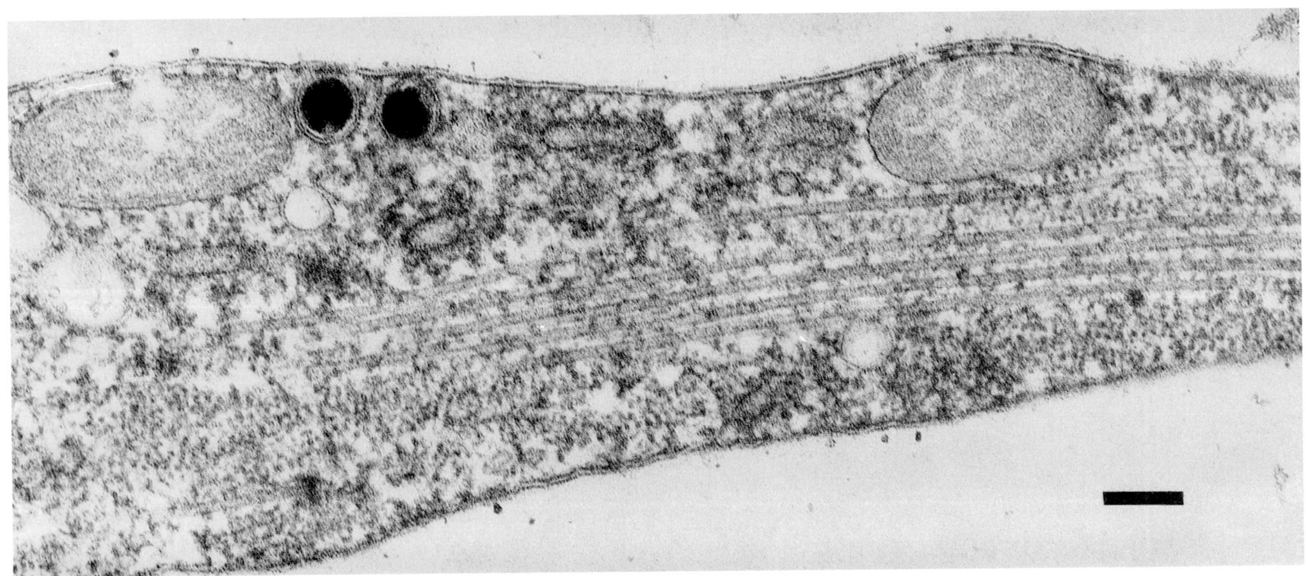

Fig. 2. TEM of pseudopods of *Allogromia laticollaris*.

CLASS ATHALAMEA
Haeckel, 1862

Ameboid organisms lacking a test or shell, though some forms might be covered by a thin lorica. Pseudopods could arise anywhere over the surface of the body, and could be branched to a greater or lesser extent in different representatives of the group, with or without anastomosing connections in the pseudopodial network. Bi-directional streaming of granules in pseudopodial extensions must be the rule in all representatives. Organisms that have not been examined by modern techniques, nor have been seen in recent years, whether they do have granular reticulopodial bidirectional streaming, have been removed from this class and placed with the amoebae of uncertain affinities. One genus remains: *Reticulomyxa*. This organism has been studied extensively using modern techniques and meets the criteria used to define the phylum.

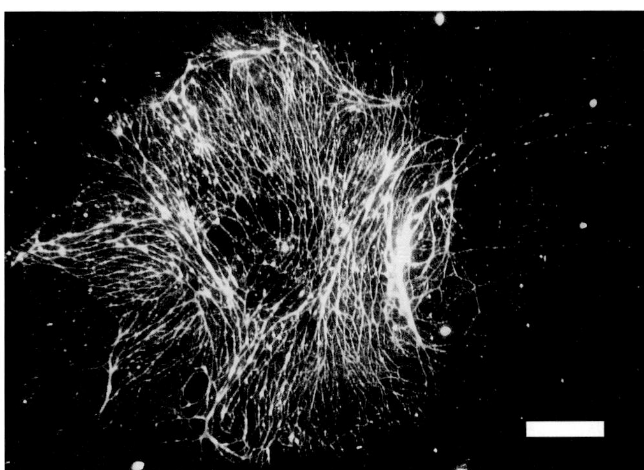

Fig. 3. Dark-field photograph of a small *Reticulomyxa filosa* organism. Scale=400 μm. (by Samuel Bowser)

FAMILY RETICULOMYXIDAE

Amebae with green or brown pigmentation or colorless. Found in freshwater or in litter.

Genus **Recticulomyxa** Nauss, 1949

Reticulomyxa filosa Nauss, 1949 (Fig.s. 3-6A). Body white in appearance. Body size up to 6 mm; with pseudopodial network, diameter extends to 25 cm. Multinucleate, with intranuclear mitoses following feeding. Multidirectional streaming observed in plasmodial veins. This organism has been used in studies of ameboid movement, and we know more about its reticulopodia than most other members of the Granuloreticulosa. Refs. Hülsmann (1984); Nauss (1949).

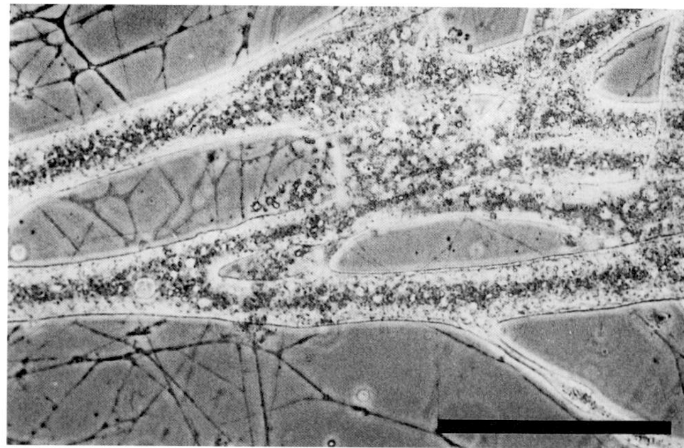

Fig. 4 Phase contrast micrograph showing the enlarged pseudopodial network. Scale 20μm. (by Norbert Hülsmann)

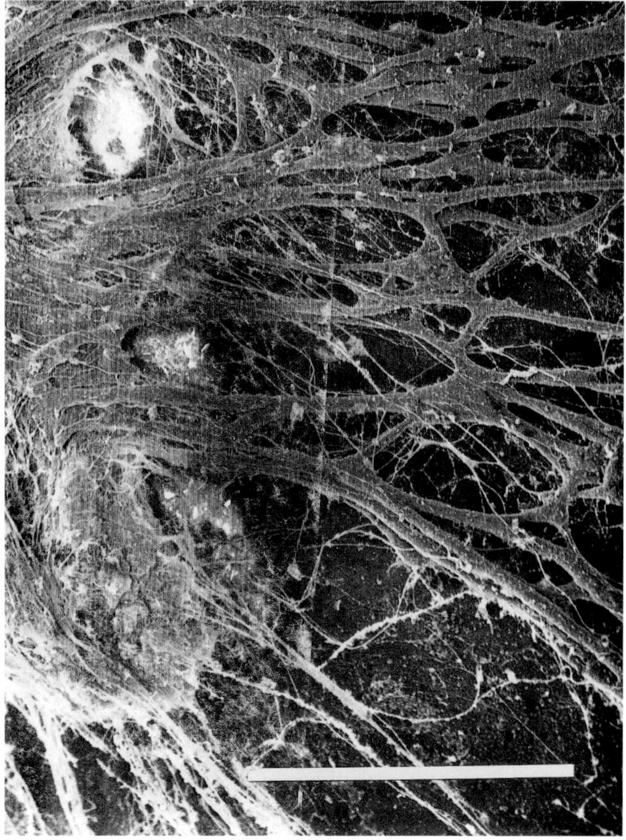

Fig. 5. Critical-point dried specimen of *Reticulomyxa filosa* as viewed under the scanning electron microscope. Scale 100=μm. (by Norbet Hülsmann)

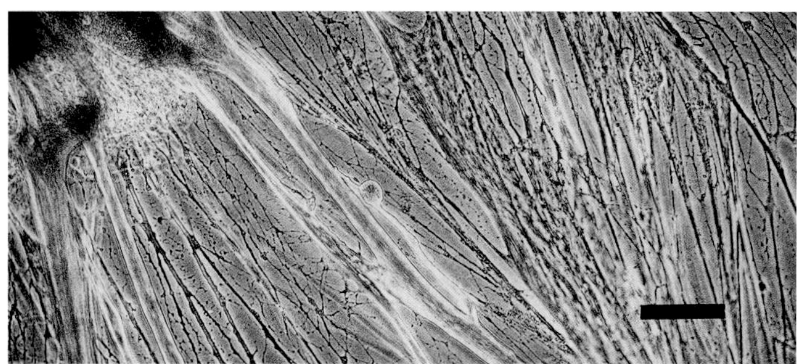

Fig. 6A. Phase-contrast micrograph of the details of granular reticulopods. Scale=10 µm. (by Samual Bowser)

LITERATURE CITED

Hülsmann, N. 1984. Biology of the genus *Reticulomyxa* (Rhizopoda). *J. Protozool.*, 31:55A-56A (Abstr.)

Nauss, R. N. 1949. *Reticulomyxa filosa* gen. et sp. nov., a new primitive plasmodium. *Bull. Torrey Bot. Club*, 76:161-173.

MOLECULAR PERSPECTIVE ON SYSTEMATICS

Recent development of molecular systematics based on analysis of ribosomal DNA sequences sheds new light on the origin of foraminifera and their higher-level relationships.

In spite of the difficulty in obtaining pure foraminiferal DNA that led to conflicting results of the first molecular studies of foraminifera (Pawlowski et al. 1994.; Wray et al. 1995), the position of foraminifera at the universal ribosomal tree of life is now relatively well established. Analysis of both LSU and SSU rDNA sequences places foraminifera in the lower part of the eukaryotic tree (Fig. 6B). They branch close to the slime molds (*Dictyostelium, Physarum*) and *Entamoeba* on the LSU rDNA tree (Pawlowski et al., 1994) and next to the Euglenozoa on the SSU rDNA tree (Pawlowski et al., 1996). The position of foraminifera on the ribosomal trees suggests their relatively early origin in evolution of eukaryotes, however, it cannot be excluded that this position is biased due to unusually rapid rates of rRNA evolution that produce an artificial grouping of foraminifera with early evolved protists lineages (Sogin, 1997).

Analysis of more than hundred partial SSU rDNA sequences obtained from representatives of all major taxonomic groups shows clearly a monophyletic origin of the foraminifera. The higher-level relationships among foraminifera; however, seem to be much more complex than suggested by morphology-based systematics. In opposition to the classical view of foraminiferal evolution, from the primitive membraneous-walled group to more evolved agglutinated and calcareous groups, the molecular data suggest rather early divergence of calcareous Miliolina and Spirillinina (Fig. 6B). The analysis of rDNA sequences does not confirm also the taxonomic separation of unilocular tectinous Allogromiina and unilocular agglutinated Astrorhizina, showing very close relationships between the representatives of both groups. Moreover, the naked, freshwater reticulopodia-bearing athalamiid, *Reticulomyxa filosa*, branch closely to some Allogromiids, invalidating the taxonomic separation between naked and testate Granuloreticulopodia. The molecular data indicate also close relationships between Textulariina, Rotaliina and Lageniina, as well as the polyphyletic origin of planktonic Globigerinina (de Vargas et al., 1997).

LITERATURE CITED

Pawlowski, J., Bolivar, I., Guiard-Maffia, J. & Gouy, M. 1994. Phylogenetic position of foraminifera inferred from LSU rRNA gene sequences. *Mol. Biol. Evol.*, 11:929-938.

Pawlowski, J., Bolivar, I., Fahrni, J., Cavalier-Smith, T. & Gouy, M. 1996. Early origin of foraminifera suggested by SSU rRNA gene sequences. *Mol. Biol. Evol.*, 13:445-450.

Sogin, M. L. 1997. History assignment: when was the mitochondrion founded? *Current Opinion in Genetics & Development*, 7:792-799.

Vargas, C. de, Zaninetti, L., Hilbrecht, H. & Pawlowski, J. 1997. Phylogeny and rates of molecular evolution of planktonic foraminifera: SSU rDNA sequences compared to the fossil record. *J. Mol. Evol.*, 45:285-294.

Wray, C. G., Langer, M. R., De Salle, R., Lee, J. J. & Lipps, J. H. 1995. Origin of the foraminifera. *Proc. Nat. Acad. Sci. USA*, 92:141-145.

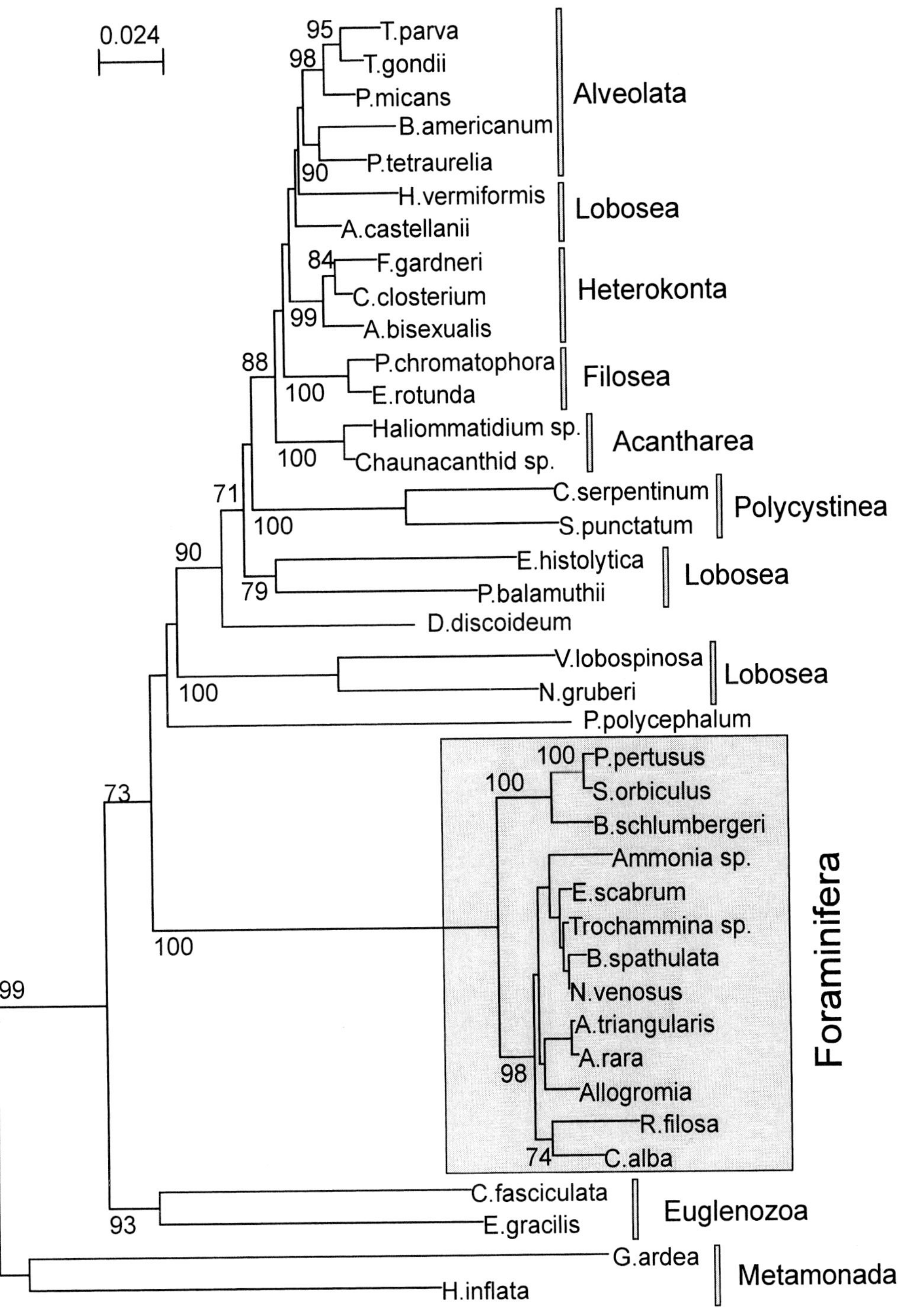

Fig. 6B. SSU rDNA tree of eukaryotes showing the phylogenetic position of the foraminifera. The tree was inferred by neighbor-joining method. Horizontal distances are proportional to inferred evolutionary distances according to a scale given in substitutions per site. Bootstrap percentage values greater than 50% (out of 1000 replicates) are given next to each internal branch.according to a scale given in substitutions per site. Bootstrap percentage values greater than 50% (out of 1000 replicates) are given next to each internal branch.

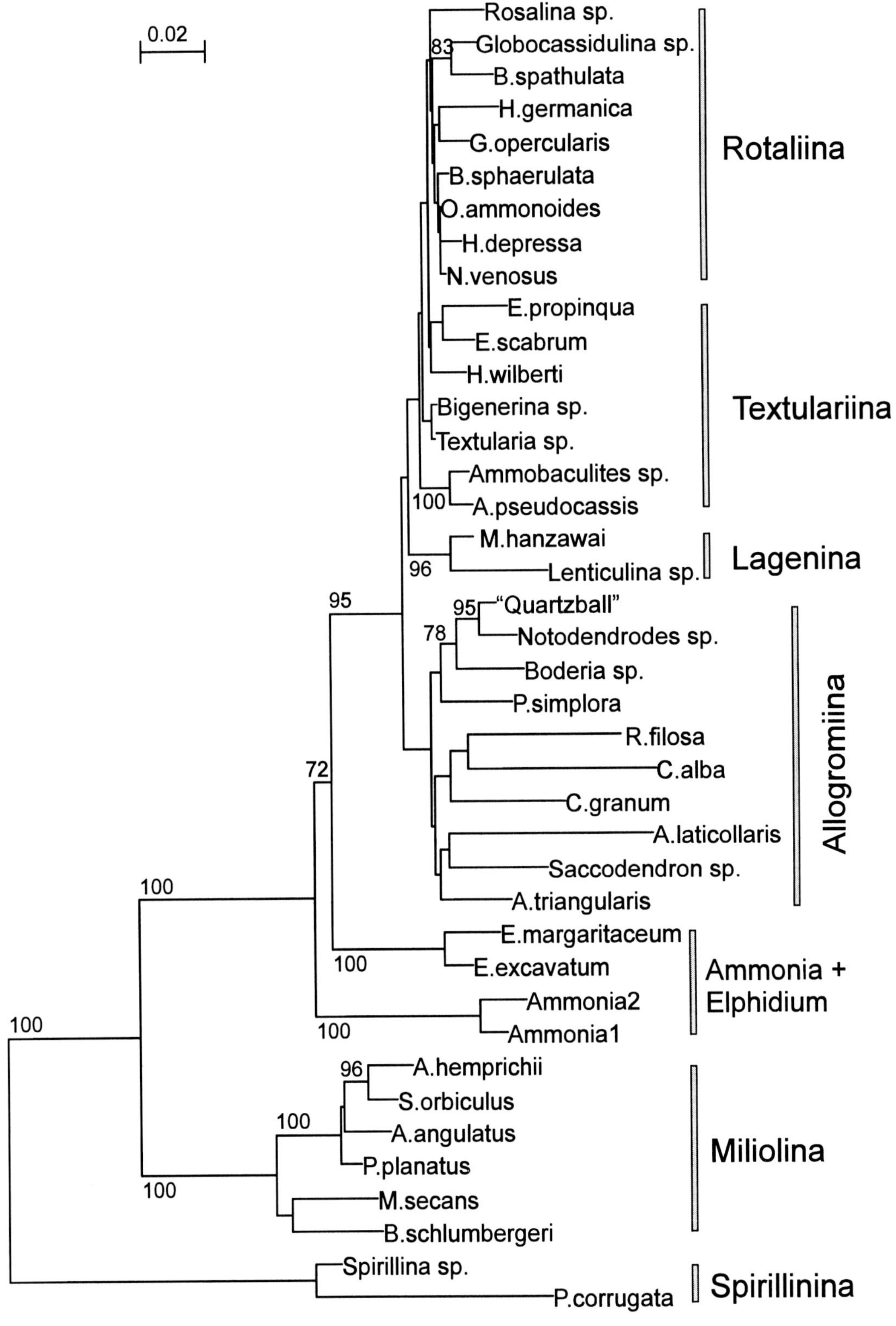

Fig. 6C. Phylogeny of benthic foraminifera inferred by neighbor joining method from 40 partial sequences of SSU rDNA. Bootstrap percentage values greater than 50% (out of 1000 replicates) are indicated along branches.

CLASS FORAMINIFERA Lee, 1990

By JOHN J. LEE, JAN PAWLOWSKI, JEAN-PIERRE DEBENAY, JOHN WHITTAKER, FRED BANNER, ANDREW J. GOODAY, OLE TENDAL, JOHN HAYNES and WALTER W. FABER

Foraminifera are among the more abundant and most conspicuous protozoa in most marine and brackish water habitats. Many species have durable shells (tests) which are an important component of marine sediments above the lysocline and fossilize well. Some species, however, have soft, delicate, organic, or agglutinated tests and therefore, little fossilization potential. The best known foraminiferan-containing sedimentary formations are the white cliffs of Dover and the limestone used to build the Egyptian pyramids. Because they fossilize well and have been abundant in the seas since the Cambrian, there are many named species (about 40,000 species) of foraminifera, of which about 4000 are alive today. Some species are found in the plankton, but the majority are benthic. The distribution of modern foraminifera has been studied extensively and has been found useful in paleoecological interpretations (reviewed in Murray, 1991). The fossils of foraminifera have been very useful biostratigraphic markers in petroleum exploration.

Foraminifera are large protozoa (range 60 µm-12 cm) which have life-spans often proportional to their sizes (days-years). A few of the **monothalamic** species can reproduce by binary fission, budding, or **cytotomy**. Many species are known to have a variety of complex life cycles, involving both sexual and asexual reproduction (Fig.s. 7-9), but the life cycles of most species are unknown (reviewed by Lee et al. 1991). Interestingly, there may be morphological differences between the sexual (**gamont**) and asexual (**agamont, schizont**) phases of the life cycles which might mislead a beginner to think that he/she was dealing with two separate species. The materials used in the construction of the test (e.g. organic, agglutinated, various types of mineralized calcareous), the geometry of chambers (in **multichambered** species) and their construction, the form of the aperture(s) are some of the important characters used in foraminiferan identification and classification. The Catalogue of Foraminifera (Ellis and Messina 1944 and following more than 94 vols.) lists almost all the species ever described. Two excellent briefer, more comprehensive treatments of their structure and taxonomy are those of Haynes (1981) and Loeblich and Tappan (1988). Quite a number of recent families of larger and planktonic foraminifera (12) are the hosts for endosymbiotic algae. One superfamily of larger foraminifera, Soriticea, has four families each of which hosts different types of algae (unicellular red, chlorophytes, dinoflagellates, and diatoms). Three additional families of larger foraminifera are the hosts to diatoms. Planktonic foraminifera are hosts for chrysophytes and dinoflagellates.

Our classification is based on that of Loeblich and Tappan (1988) for most of the groups, excepting planktonic foraminifera (F.B. and J.W.). For convenience, only the major taxonomic groups are distinguished, in spite of recent tendencies by specialists to multiply the higher taxonomic categories. In our opinion, these classifications, based mostly on the structural details of the wall (Fig.s. 10-15), need to be confirmed by studies of living foraminifera, in particular by analysis of molecular and biochemical characters (e.g. comparison of rRNA gene sequences or proteins).

KEY TO THE ORDERS

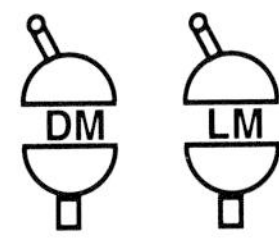

1. Test membranous or agglutinated 2
1'. Test of secreted calcite or silica.................... 4

2. Test flexible, membranous or proteinaceous .... **Allogromiida**
2'. Test of agglutinated material 3

3. Test unilocular or with second tubular chamber .... **Astrorhizida**
3'. Test multilocular..................... **Textulariida**

4. .Test of secreted silica .... **Silicoloculinida***

4'. Test of secreted calcite................................... 5

5. Test of secreted calcitic spicules........ ....,,,....... **Carterinida**
5'. Test of aragonitic or calcitic crystals......... 6

6. Test aragonitic.. 7
6'. Test calcitic... 8

7. Test aragonitic, non-septate (fossil only)...... .... **Involutinida***
7'. Test aragonitic, septate .... **Robertinida**

8. Test of microgranular calcite (Fossil only) .... **Fusulinida***
8'. Test of optically radial calcite...................... 9

9. Test porcellaneous of imperforate calcite .... **Miliolida**
9'. Test hyaline of perforate calcite................ 10

.... **Spirillinida**
10'. Test of optically radiate crystals.............11

11. Test monolamellar....................... **Lagenida**
9. Test porcellaneous of imperforate calcite .... **Miliolida**
9'. Test hyaline of perforate calcite............... 10

10. Test of optically single crystal of calcite... .... **Spirillinida**
10'. Test of optically radiate crystals............ 11

11. Test monolamellar....................... **Lagenida**
11'. Test bilamellar... 12

12. Benthonic in habit........................ **Rotaliida**
12'. Planktonic in habit .... **Globigerinida**

(* not mentioned further in this chapter)

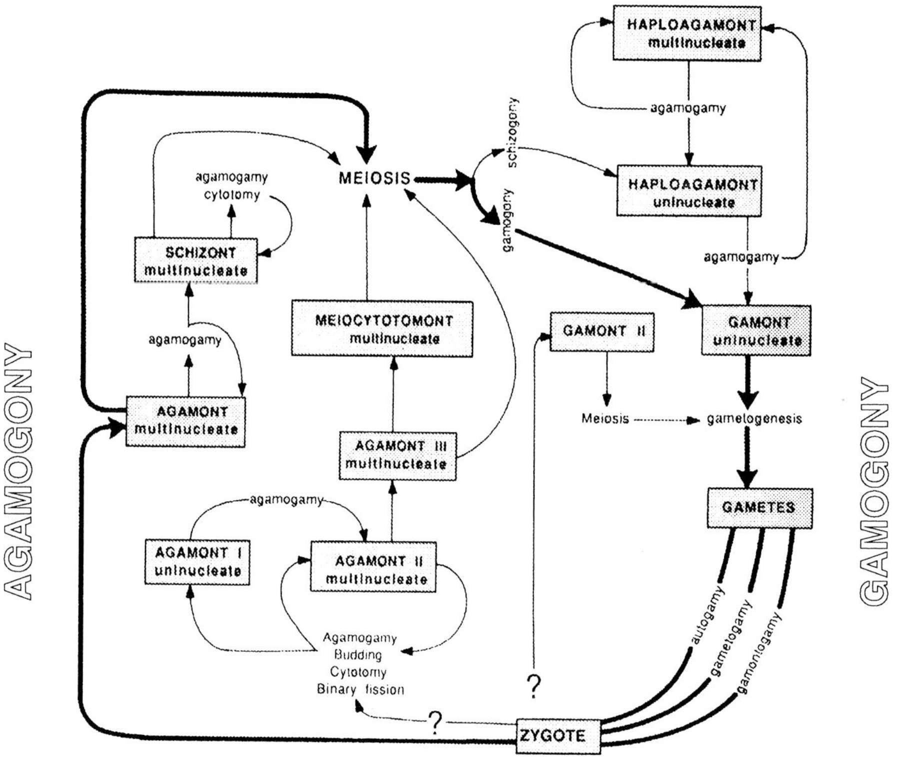

Fig. 7. A diagram of the life-cycle options in foraminifera. The steps in the **classical** and **paraclassical** life-cycles are indicated by the bold arrows. In the **classical** and paraclassical cycles, there is a regular alternation between a haploid, uninucleate, **megalospheric gamont** and a diploid, multinucleate, **microspheric agamont**. In the classical life-cycle, sexual reproduction is **gametogamic**. In the paraclassical cycles, sexual processes are either **gamontogamic** or **autogamic.**

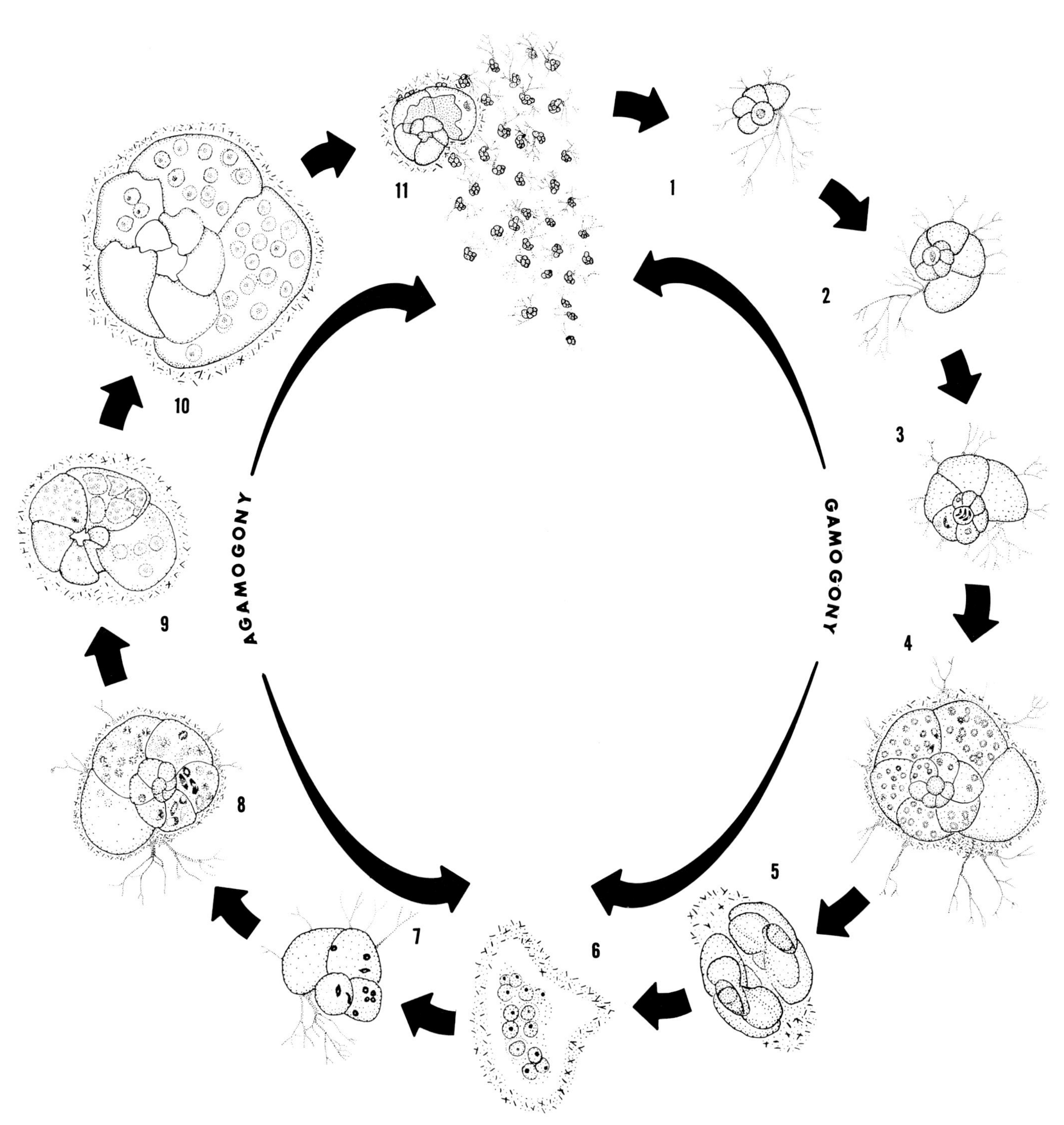

Fig. 8. Life cycle of *Rosalina leei*. **Gamogony**, the sexual generation, 1-6: 1, gamont, enlarged; 2-4, growth of gamont with mitoses of nuclei; 5, pairing of gamonts with fertilization; 6, zygotes produced by fertilization. **Agamogony**, the asexual generation, 7-11: 7, growth of agamont; 8, meiosis in agamont, 9-10, gamonts developing in agamont; 11, emergence of gamonts from agamont.

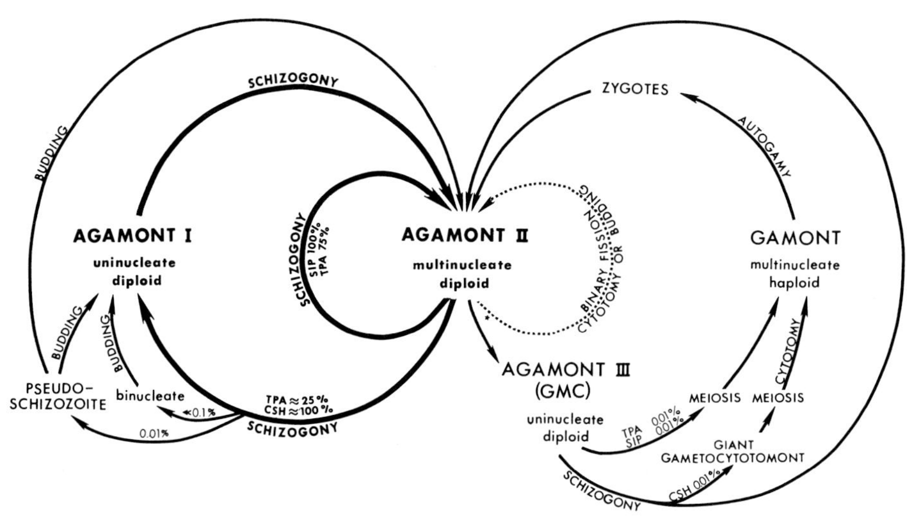

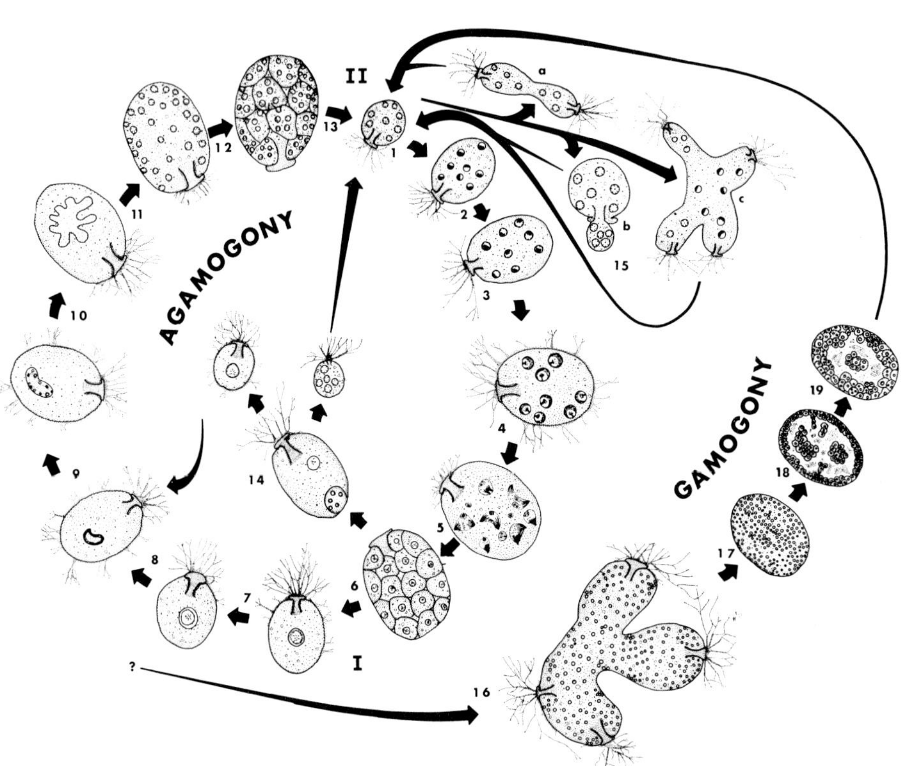

Fig. 9A. Nonclassical life-cycle of three strains of *Allogromia laticollaris*. 9B. Life-cycle of *A. laticollaris*. Agamont phase, 1-15: 1, juvenile agamont II, early G; 2, young agamont II, mid G (RNA synthesis also occurs in this phase); 3, growing agamont II, late G; 4, mature agamont II, chromosomes in "mushroom-like conFig.uration," RNA accumulate at the periphery; 5, karyokinesis; 6, cytokinesis (schizogony); 7, young agamont I, S phase; 8, maturing agamont I, G2 phase; 9, mature agamont I, nucleus differentiating, RNA granules at the periphery of the nucleus; 10, mature agamont I, early "amoeba-form" nucleus; 11, agamont I, amoeba-like nucleus; 12, agamont I, post *Zerfall*; 13, agamont I (schizogony); 14, agamont I, relatively uncommon life-cycle alternate pathway in which budding gives rise to an agamont II; 15, agamonts II, relatively uncommon alternate life-cycle pathways including (a) binary fission, (b) budding, and (c) cytotomy. Gamont phase 16-19: 16, giant gametocytotomont; 17, multinucleate gamont prior to the formation of gametes; 18, gamont filled with gametes; 19, gamont with some gametes and zygotes.

Fig. 10. Test characteristics used describing and classifying foraminifera. a. Agglutinated **biserial test**. b. Mixed growth; biserial to **monoserial**. c. Higher magnification of b showing mixed sizes of agglutinated particles. d. Uniserial ribbed test with simple terminal **aperture**. e. High magnification of volcanic ash particles agglutinated to the surface of a deep-sea foraminifer. f. Crystalline surface structure of a **porcelaneous** foriminifer. g,h. Winding, **milioline coiling** mode: g, surface striate; h, smooth surface. Scale=200 µm for (a); scale=150 µm for (b); scale=5 µm for (c); scale=10 µm for (d,e); scale=2µm for (f); scale=500 µm for (g); and scale=100 µm for (h).

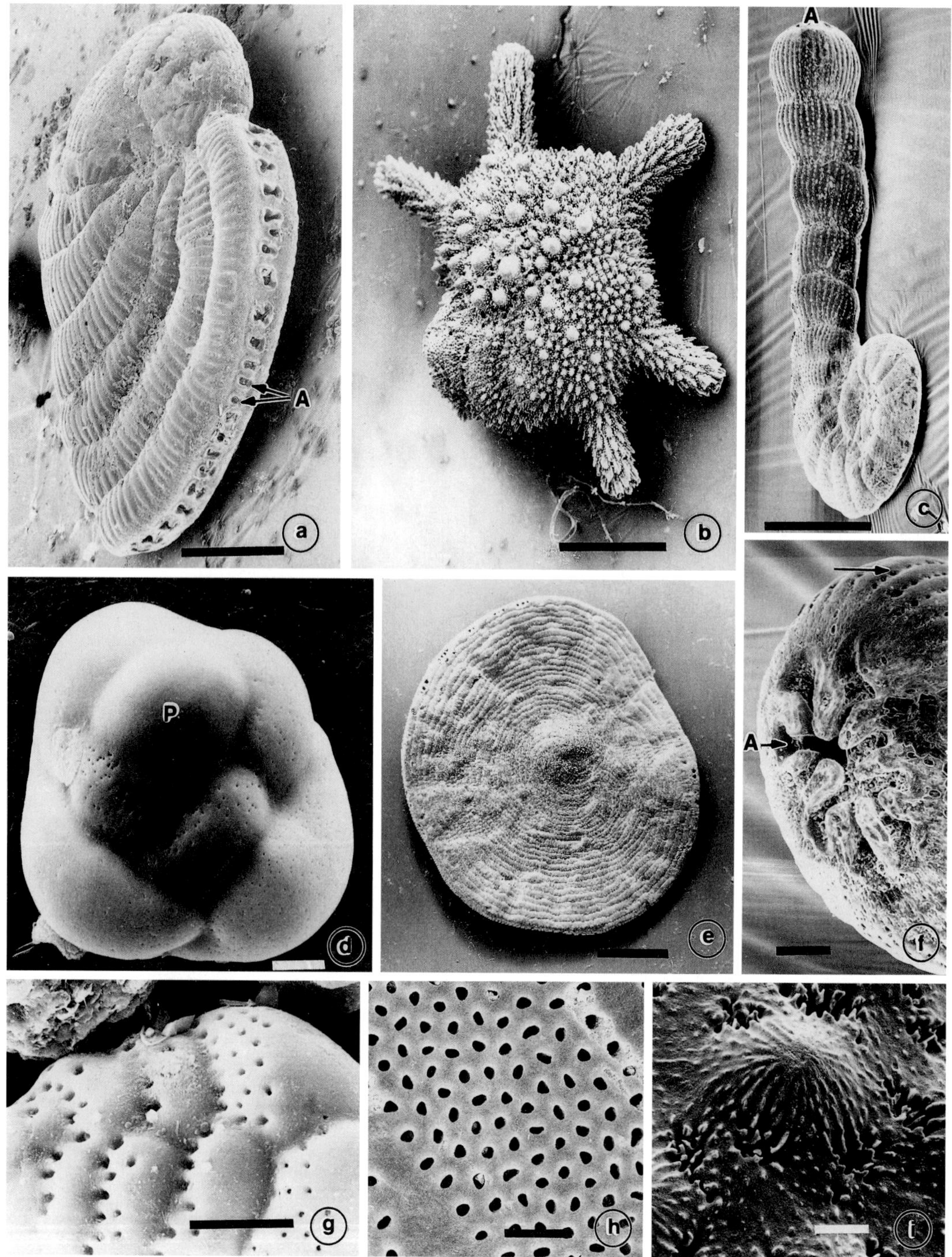

Fig. 11. Test characteristics used describing and classifying foraminifera. a. **Flabelliform** (involute planispiral to uncoiled) miliolid with multiple **marginal apertures** (A) on terminal chamber. Each chamber is subdivided into **chamberlets**. b. Biconvex **trochospiral** form with peripheral spines. c. Uncoiled, **involuted**, **planispiral**, ribbed, miliolid with terminal **dendritic aperture** (f). d. Low trochospiral test; (P) **proloculum**. e. **Annular discoid** foraminifer; (J) **juvenarium**. f. **Dendritic aperture**; arrow points to **pits** between **ribs**. g. **Pores** distributed near the sutures on the surface of a biserial foraminifer. h. More uniformly distributed pores on another foraminifer. i. Higher magnification of foraminifer in b showing pore distribution around **blisters** on surface. Scale=500 µm for a; scale=400 µm for b,c; scale=10 µm for d,g,i; scale=800 µm for e; and scale=20 µm fo f,h.

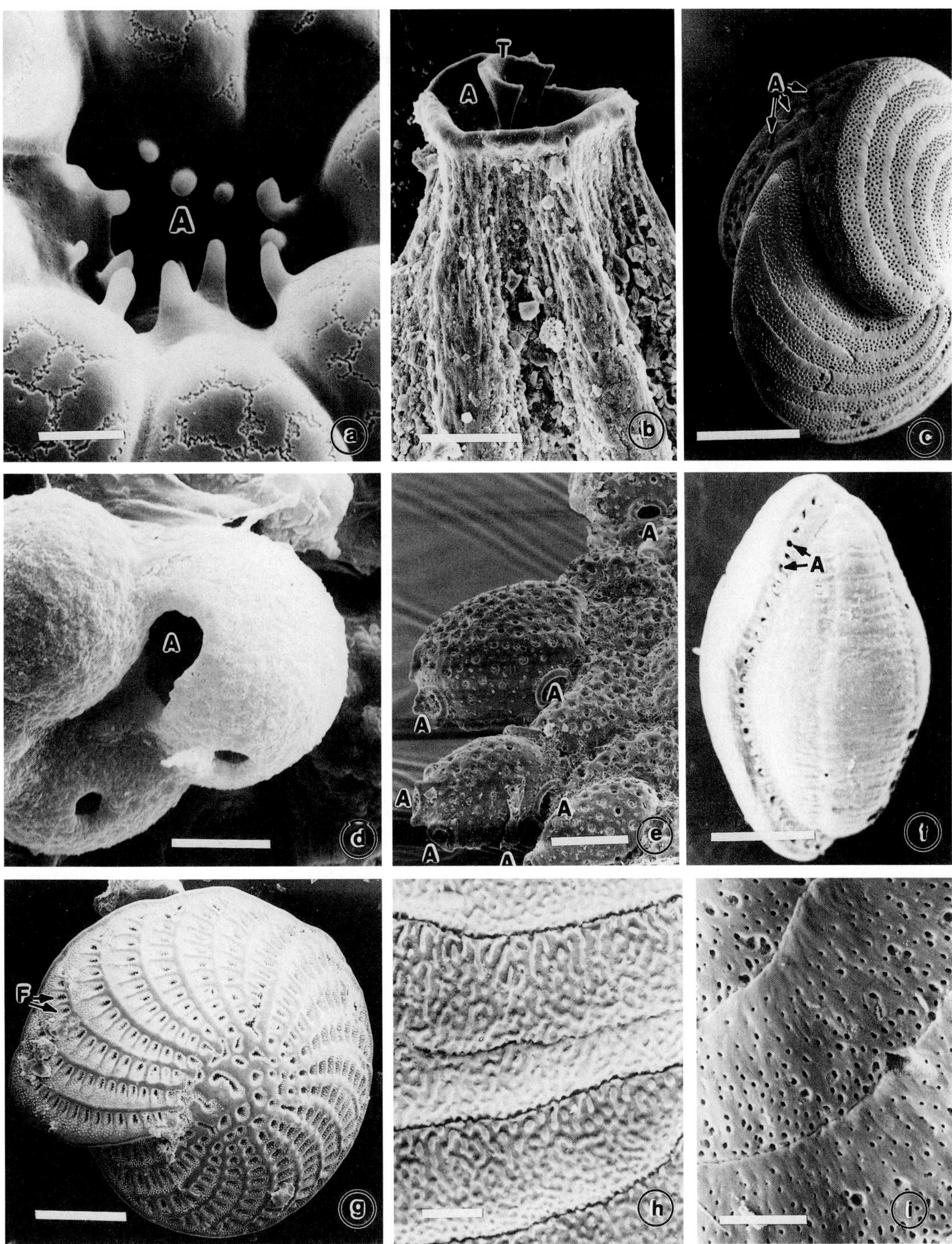

Fig. 12. Test characteristics used describing and classifying foraminifera. a. **Umbilical aperture** (A), surrounded by **denticles**, of a trochospiral foraminifer. b. Aperture (A) of a miliolid showing a spoon-shaped **tooth** (T). c. Multiple apertures (A) of an **involute** archasine larger foraminifer. d. **Umbilico-lateral aperture** (A). e. Multiple terminal and **septal apertures** (A). f. **Planispiral-fusiform** (roller-shaped) foraminifer with multiple apertures (A). g. Planispiral foraminifer with an **umbilicus** with a **canaliculate boss** and sutural fissures crossed by **septal bridges** arching around **fossettes** (F) which lead into an internal canal system. h,i. Surfaces of annular miliolid chambers showing **rugose** and pitted textures. Scale=10 μm for a,b; Scale=400 μm for c,f; scale=20 μm for d,e,h,i; and scale=250 μm for g.

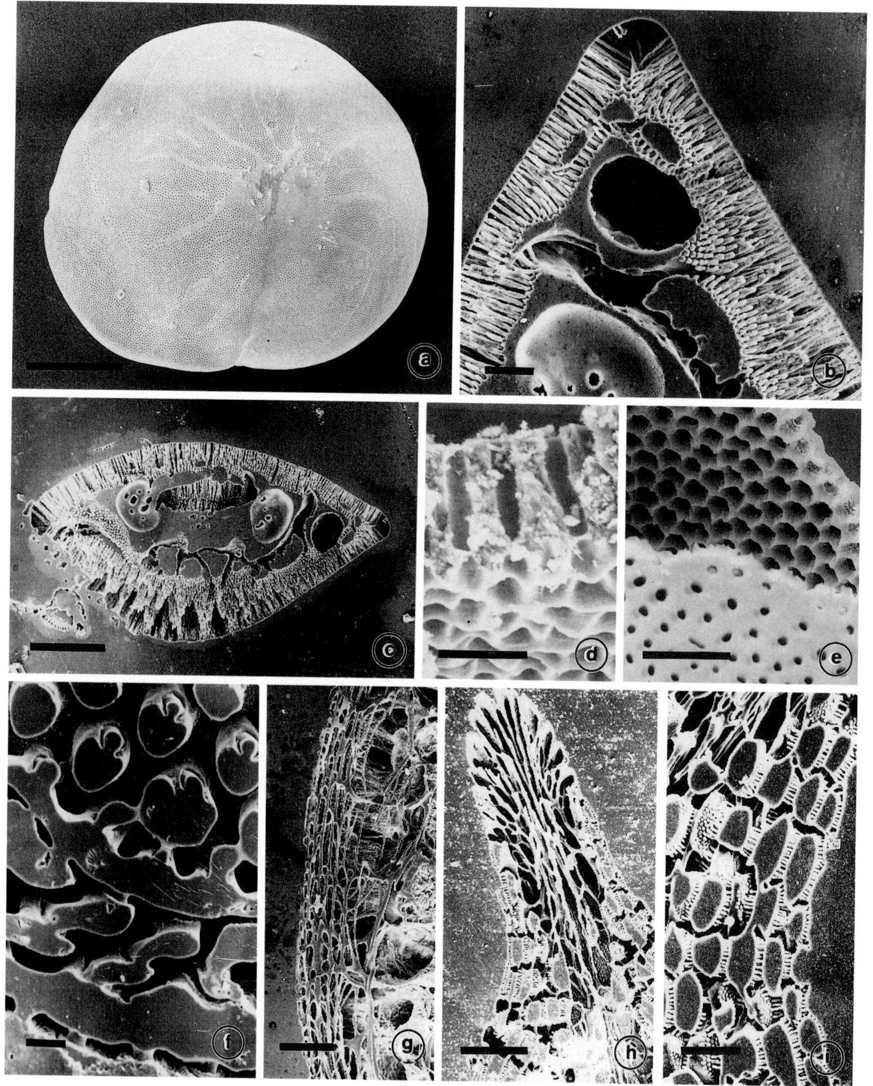

Fig. 13. All Figures show some of the complexities of the foraminiferal test and cytoplasmic spaces. The aim is to show some of the ways in which the test compartmentalizes internal cytoplasm and the canal and **stolon systems** which serve to circulate materials between **chambers** and chamberlets and with the external environment. Studies on this topic have been greatly aided by the embedding technique developed by Hottinger (1979). Foraminifera are embedded in epoxy resins used for TEM. The resin containing the specimens is then mounted on a stub and ground on a fine polishing table to a depth in the specimen desired by the investigator. Then mild acid (1N HCl) is used to dissolve the test. The result is a cast of the spaces within the test occupied by the external seawater and spaces and organelles within the organism. Fig.s. b,c,f-i are examples of such casts. Fig.s. a-e are different anatomical views of the larger foraminifer, *Amphistegina lobifera*. a. The dorsal surface of the **biconvex** organism is covered with fine pores. c.The solid grey material in the center of the cell is occupied by the cytoplasm. The fiber-like fringe at the periphery of the cell is formed by individual pore-lining tubes, seen in enlargement in b. Seawater enters the pores and is separated only by a thin organic lining and the host cell's membrane from diatom endosymbionts. d,e. Pieces of a fractured test which shows the pores and cup-shaped **internal pore rims**. e. A different perspective of the pore rims than shown in d. An individual, frustule-less, spherical diatom is normally located in the cytoplasm of the host in each pore rim. f. Part of a cast of *Amphisorus hemprichii* showing chamberlets, stolon system, and apertures. g. The **marginal canal system** in part of a cast of *Heterostegina depressa*. h,i. From a cast of *Calcarina gaudichaudii* which show the canal system within the spines of these 'star sands' and the connections of the canals to the chamberlets. Scale=200 μm for a,c; scale=50 μm for b,f,i; scale=20 μm for d,e; and scale=100 μm for g,h.

ORDER ALLOGROMIIDA
Loeblich & Tappan, 1961

KEY TO THE FAMILIES

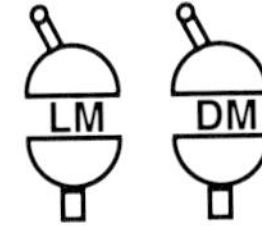

1. Test membraneous, flexible; where known, gametes biflagellate...................... **Lagynidae**
1'. Test proteinaceous, firm; where known, gametes amoeboid.. 2

2. Test free, unilocular..................................... 3
2'. Test free or attached, multilocular .... **Hospitellidae***

3. Test globular, ovate to cylindrical, aperture single, terminal.................... **Allogromiidae**
3'. Test tubular and elongate, sometimes thread-like; aperture at both ends of the test................................. **Shepheardellidae**

*-Family not illustrated in this volume

FAMILY LAGYNIDAE Schultze, 1854

Test small, unilocular with single or multiple apertures; wall membraneous, may have ferrugin-ous surface crust; numerous fresh water and marine genera.

KEY TO REPRESENTATIVE GENERA

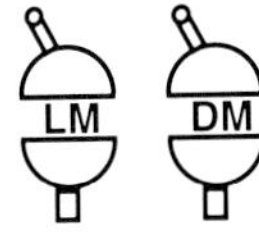

1. Test spherical to hemispherical with loosely attached foreign particles........... ***Myxotheca***
1'. Test without foreign particles...................... 2

2. Test elongate, tubular, convoluted, divided distally into numerous radiating, irregularly convoluted branches; inhabiting empty tests of another foraminifer................ ***Ophiotuba***
2'. Test free... 3

3. Test >1 mm in length, conical to angular outline, sometimes flattened; apertures located at angles of test .... ***Boderia***
3'. Test <0.1 mm in length, elongate, ovate or slightly compressed and narrowing towards terminal aperture............................ ***Lagynia***

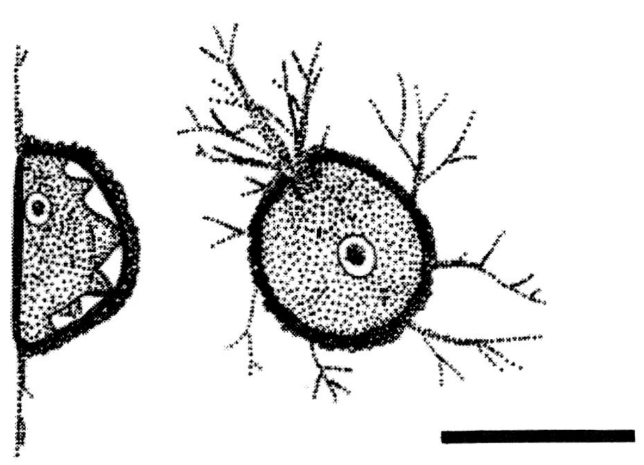

Fig. 14. *Myxotheca arenilega*. Scale=100 μm.

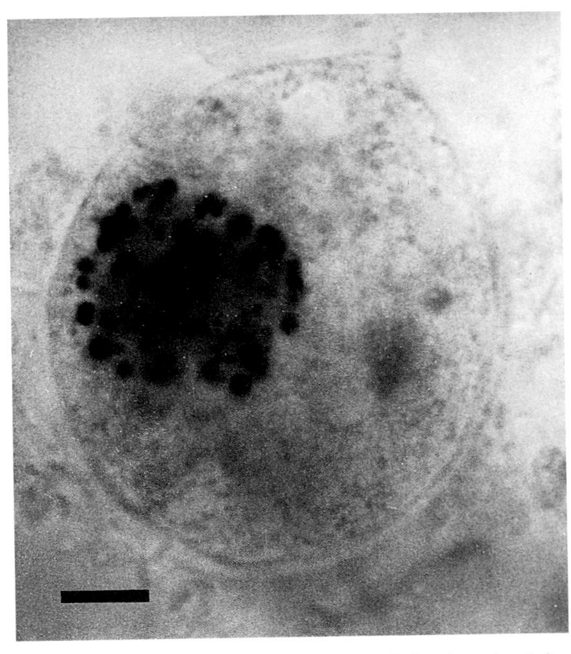

Fig. 15. Recently released agamont I (uninucleate) of *Allogromia laticolaris*. Feulgen stained and photographed with the aid of a green filter to increase contrast. Scale=100 μm.

FAMILY ALLOGROMIIDAE Rhumbler, 1904

Test globular to ovate with single, terminal aperture; may have entosolenian tube projecting into interior; wall proteinaceous, sometimes incorporating agglutinated particles.

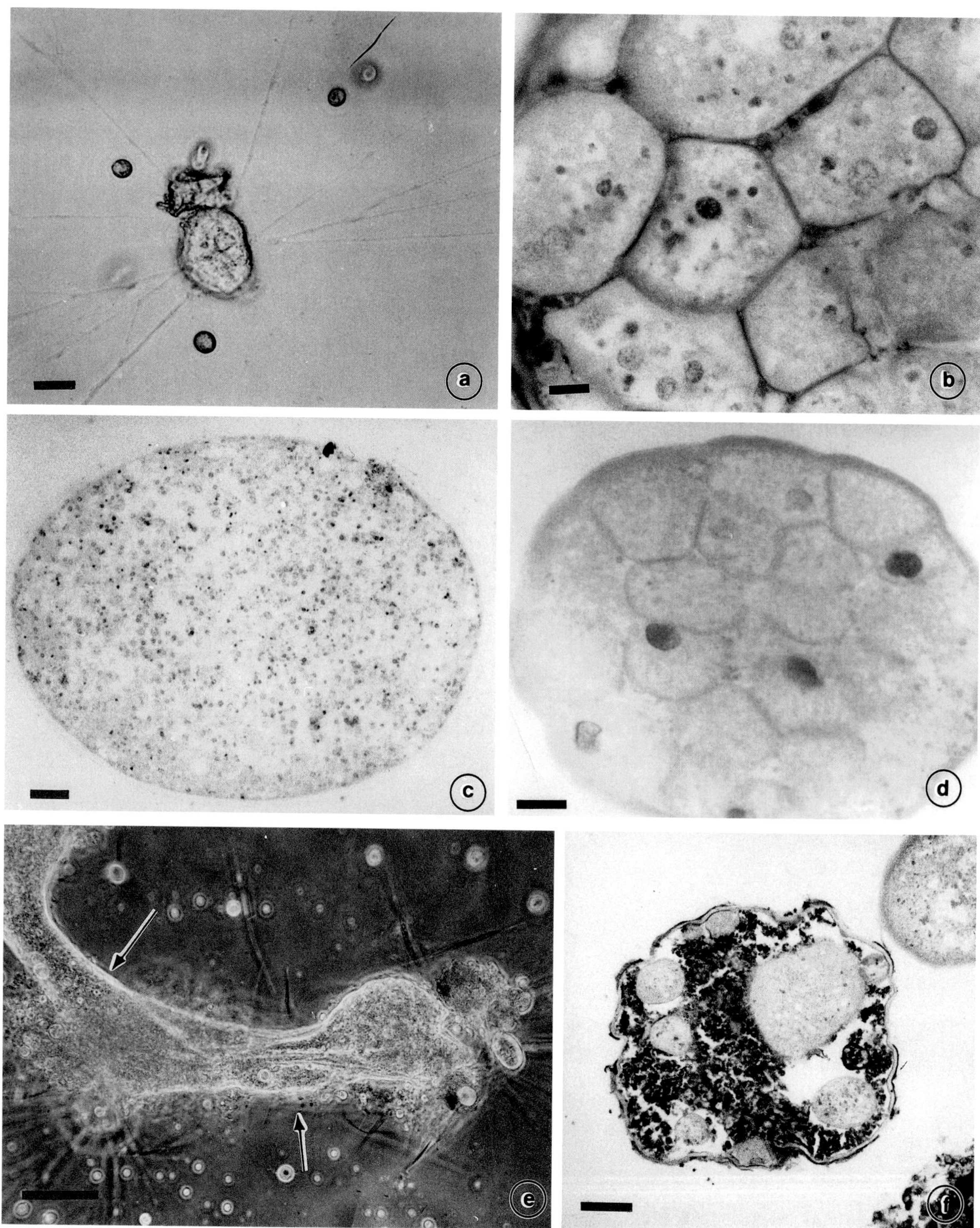

Fig. 16. With the exception of a, all are Figures of different stages of the life cycle of *Allogromia laticollaris.* The nuclei in Fig.s. b-d,f are Feulgen stained and photographed through a green filter to enhance contrast. a. Low-power light micrograph of *Allogromia* sp. (NF) which has a very translucent test allowing one to observe the nucleus. About 1/2 the pseudopodal net is shown. Scale=100 µm. b. Portion of agamont II showing multinucleate schizoites just before release. Scale=50 µm. c. A gamont filled with gametic nucleii. Cytokinesis has not yet been completed in all parts of the organism. Scale=100 µm. d. Agamont I just before the release of uninucleate schizozoites. Scale=100 µm. e. Portions of a cell undergoing cytotomy. Phase contrast. Scale=100 µm. f. An agamont II showing very uneven-sized multinucleate schizozoites just before release. Scale=20 µm

KEY TO REPRESENTATIVE GENERA

1. Test with spherical, ovoid, or flask-like proloculus and crescentic second chamber extending part of way around circumference of proloculus................................. ***Periptygma***
1'. Test unilocular... 2

2. Test with adhering fine detrital particles and single wide aperture................ ***Gloiogullmia***
2'. Test without adherent particles...................... 3

3. Test oval to spherical.................... ***Allogromia***
3'. Test elongate, subcylindrical..................... .... ***Cylindrogullmia***

FAMILY SHEPHEARDELLIDAE
Loeblich & Tappan, 1984

Test free, tubular, and elongate to thread-like with apertures at both ends; wall perspicuous, firm (ex. *Nemogullmia, Tinogullmia, Shepheadella* [Fig. 17]).

Fig. 17. *Shepheardella taeniformis*. Scale=1 mm.

ORDER ASTRORHIZIDA Brady, 1881

Test unilocular, globular, tubular, or branching, may have globular proloculus and tubular second chamber or consist of loosely attached subglobular chambers; wall agglutinated.

KEY TO THE SUPERFAMILIES

1. Test unilocular or with loosely attached chambers 2
1'. Test consisting of proloculus followed by tubular second chamber................................ 3

2. Test globular or tubular.......... **Astrorhizacea**
2'. Test consisting of a complex of fine tubules..... .... **Komokiacea**

3. Second chamber rectilinear.......................... .... **Hippocrepinacea***
3'. Second chamber spirally enrolled.................. .... **Ammodiscacea***

*Superfamily not illustrated in this volume.

SUPERFAMILY ASTRORHIZACEA Brady, 1881

Agglutinated test usually free but sometimes attached and irregular; globular, domed or tubular (branched or unbranched) in shape. Interior non-septate or only partially subdivided but occasionally with complete partitions.

The following families are a selection of those placed in the Astrorhizacea by Loeblich and Tappan (1988). They are recognized on the basis of test morphology and wall characteristics. Most of these taxa are probably unnatural units; a thorough revision of astrorhizacean foraminifera, based on cytological organization as well as test features, is urgently required.

KEY TO THE FAMILIES

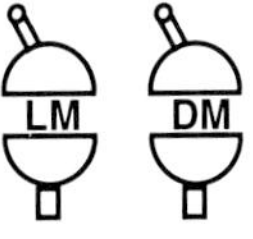

1. Test tubular... 2
1'. Test globular.. 5

2. Stolon-like tubes radiating from central area. .... **Astrorhizidae**
2'. Lacking central area...................................... 3

3. Test a straight, unbranched tube, usually non-septate.............................. **Bathysiphonidae**
3'. Test tubular, dichotomously branching or fanlike.. 4

4. Wall thin and flexible........... **Rhizamminidae**
4'. Wall thick, labyrinthic and canaliculate.........

.... **Schizamminidae**

5. Test usually free, typically globular............ 6
5'. Test free or attached, consisting of one or few subglobular chambers.. .... **Hemisphaeramminidae**

6. Test a single globular chamber with single or multiple apertures.............. **Saccamminidae**
6'. Test consisting of one to several globular chambers.. 7

7. Chambers single or loosely joined, aperture not recognizable............ **Psammosphaeridae**
7'. Chambers forming a linear series, aperture terminal rounded...... **Polysaccamminidae***

*-Family not illustrated in this volume

FAMILY ASTRORHIZIDAE Brady, 1881

Test free or attached, generally large, up to 15 mm in diameter, with one or more branches from a central area, apertures at the ends of arms.

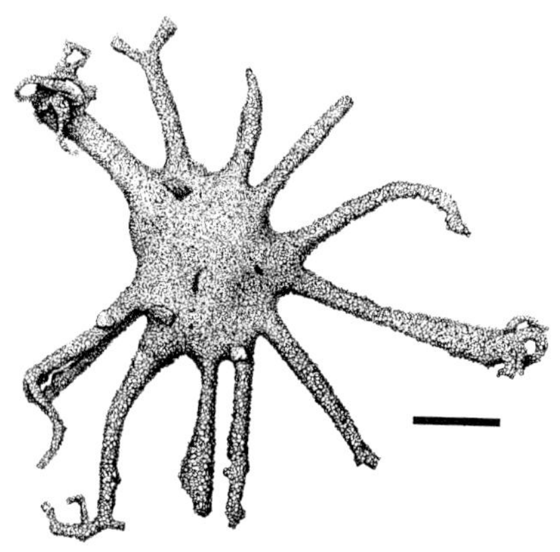

Fig. 18. *Astrorhiza limicola* Sandahl, 1858 (after Brady, 1884, Pl. 19, Fig. 1). Scale 2 mm.

KEY TO GENERA

1. Test agglutinated throughout.......................... 2
1'. Test flattened with agglutinated outer rim surrounding proteinaceous central area; rim gives rise to several short tubular necks .... ***Vanhoeffenella***

2. Test including tubular elements......................3
2'. Test rounded to elongate, fusiform; wall thick, fine-grained.................................... ***Pelosina***

3. Test flattened with stolonlike arms radiating from central area........................ ***Astrorhiza***
3'. Test consisting of thick tube that branches at irregular intervals, lacking the central area.................................... ***Astrorhizoidese***

Fig. 19. *Bathysiphon filiformis* G. O. & M. Sars, 1872, Lectotype (after Gooday, 1988b, new print). Scale 20 mm.

FAMILY BATHYSIPHONIDAE Avnimelech, 1952

Test forms unbranched, slender, tapered or untapered tube, open at both ends and either straight or curved. Apertural end may be domed and aperture constricted. Internal transverse partitions present in some species (ex. *Bathysiphon,* Fig. 19).

FAMILY RHIZAMMINIDAE WIESNER, 1931

Test large, forming long, narrow, open ended tube, >100 μm in diameter and >10 cm in length, with numerous branchings and no obvious starting point. Tube branches dichotomously and does not anasto-mose. Apertures probably simple although it is difficult to identify original apertures because of fragmentation. Wall has organic layer overlain by agglutinated particles. Lumen contains **stercomata** masses and protoplasm. Example: *Rhizammina* (Fig. 20) and *Rhabdammina* (Fig. 21) widely distributed in deep water in all major oceans.

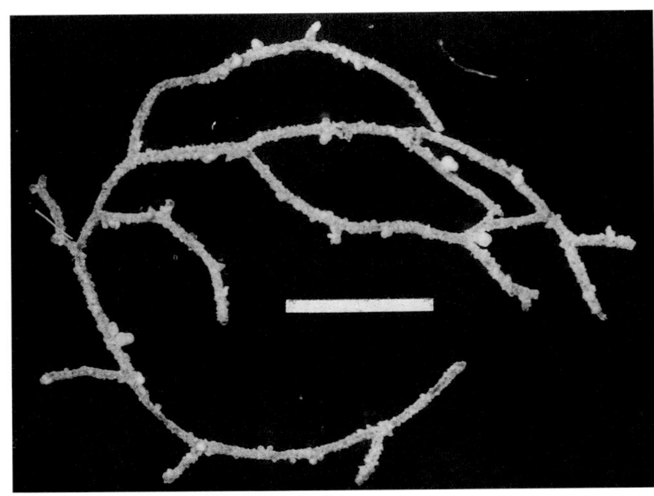

Fig. 20. *Rhizammina algaeformis* Brady, 1879 (after Gooday, 1986b). Scale 2 mm.

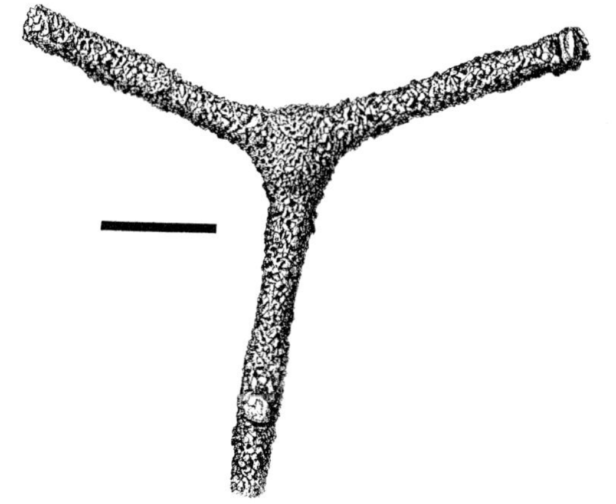

Fig. 21. *Rhabdammina abyssorum* M. Sars, 1869 (after Brady, 1884, Pl. 21, Fig. 1). Scale=2 mm.

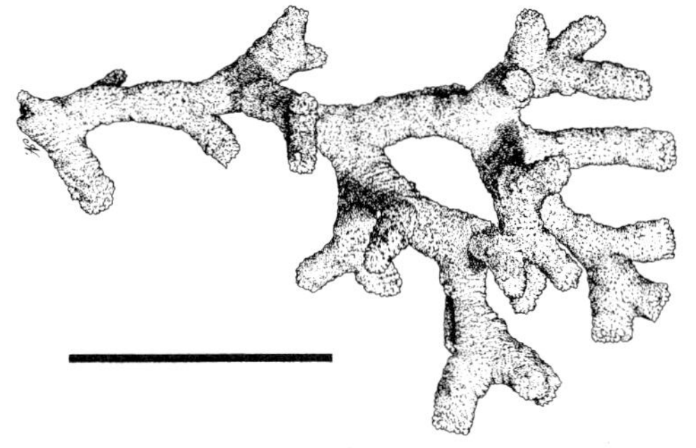

Fig. 22. *Schizammina labyrinthica* Heron-Allen & Earland, 1929. Scale=10 mm.

FAMILY SCHIZAMMINIDAE Norvang, 1961

Test free, large (2-10 cm maximum dimension)

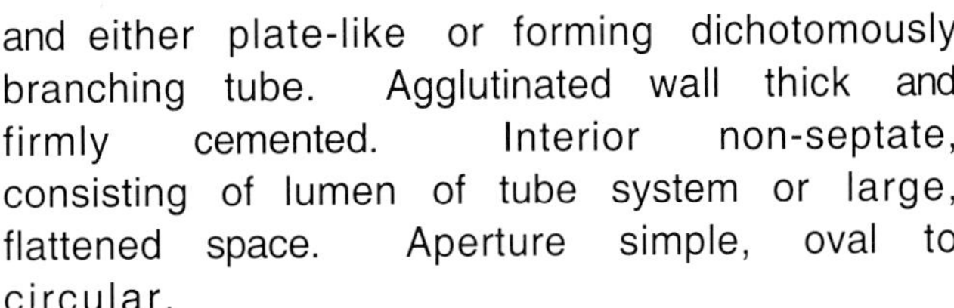

and either plate-like or forming dichotomously branching tube. Agglutinated wall thick and firmly cemented. Interior non-septate, consisting of lumen of tube system or large, flattened space. Aperture simple, oval to circular.

KEY TO REPRESENTATIVE GENERA OF THE SCHIZAMMINIDAE

1. Test tubular and dichotomously branching .... ***Schizammina***

1'. Test fan-shaped, lenticular to plate-like .... ***Jullienella***

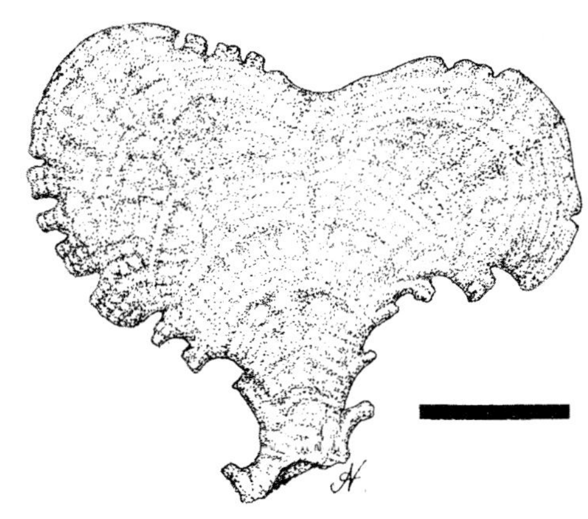

Fig. 23. *Jullienella foetida* Schlumberger, 1890, specimen from Atlantide Station 163. Scale=10 mm.

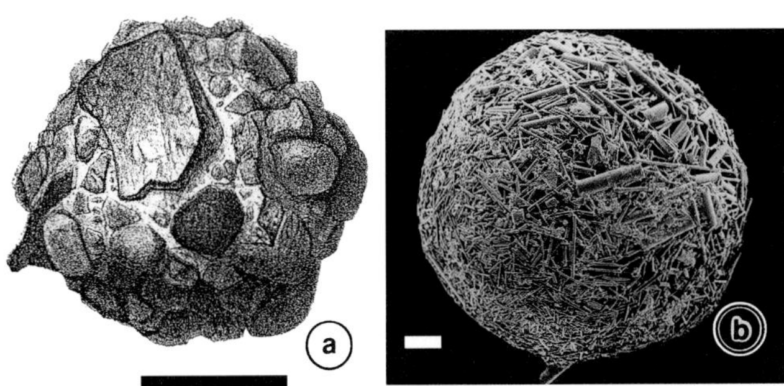

Fig. 24. *Psammosphaera fusca* Schultz, 1875 (after Brady, 1884, Pl. 18, Fig. 1). a, b. show different agglutination patterns. Scales=1 mm.

FAMILY PSAMMOSPHAERIDAE Haeckel, 1894

Test free, globular to irregular, large (up to 6 mm), consisting of one to several loosely joined spherical or subglobular chambers; aperture not recognizable but large interstitial pores occurring between grains probably allow communication with the exterior (ex. *Psammosphaera* [Fig. 24], *Sorosphaera*).

FAMILY SACCAMMINIDAE Brady, 1884

Test free or attached, consisting of single, globular to elongate chamber, aperture terminal, rounded or slitlike, in some genera multiple.

KEY TO REPRESENTATIVE GENERA

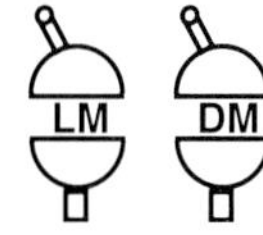

1. Test globular to elongate, aperture single 2
1'. Test large (>6mm), globular, with subconical or tubular arms projecting from the surface, apertures multiple, openings at the ends of the arms... ***Astrammina***

2. Aperture terminal, rounded, sometimes produced on a neck.. 3
2'. Aperture an elongate narrow slit, elevated on a low ridge; test globular, up to 3.5 mm in diameter... ***Pilulina***

3. Test ovoid to fusiform or elongate..................... 4
3'. Test globular, up to 3.5 mm in diameter........... .... ***Saccammina***

4. Test ovoid to fusiform, aperture terminal with inwardly projecting **entosolenian tube**; wall flexible, very finely agglutinated.... ***Ovammina***
4'. Test elongate, fusiform or cylindrical, up to 3 mm in length................................. ***Technitella***

FAMILY HEMISPHAERAMMINIDAE
Loeblich & Tappan, 1961

Test attached, or less commonly, free, consisting of one or few subglobular or hemispherical chambers; aperture single or multiple or may be unrecognizable; interior may be subdivided.

KEY TO REPRESENTATIVE GENERA

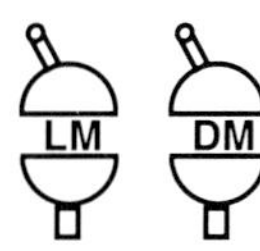

1. Test interior not subdivided............................... 2
1'. Test interior often partially subdivided by one or more variably developed septa; may be free or attached.................................... ***Crithionina***

2. Test consisting of a single chamber................... 3
2'. Test consisting of up to four hemispherical chambers; aperture single, rounded at the apex of each chamber........................ ***Ammopemphix***

3. Aperture unrecognizable.***Hemisphaerammina***
3'. Apertures at ends of tubular projections of the chamber; test basically organic but with some agglutinated particles forming outer layer......... .... ***Iridia***

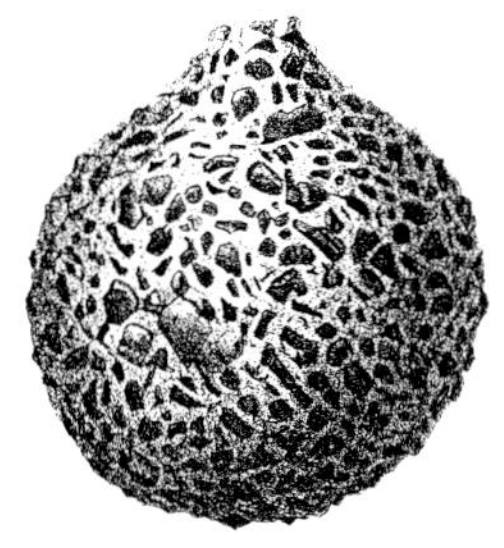
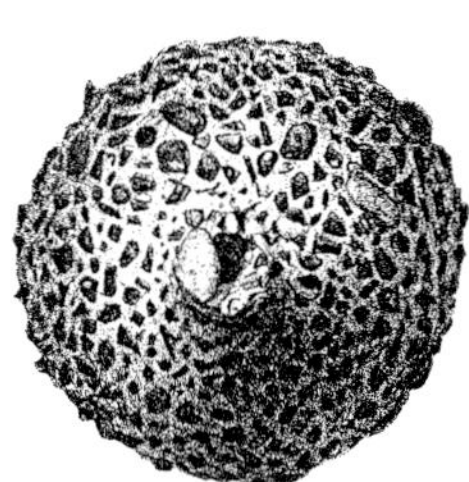

Fig. 25. *Saccammina sphaerica* Brady, 1871 (after Brady, 1884, Pl. 18, Figs. 11,13). Scale=1 mm.

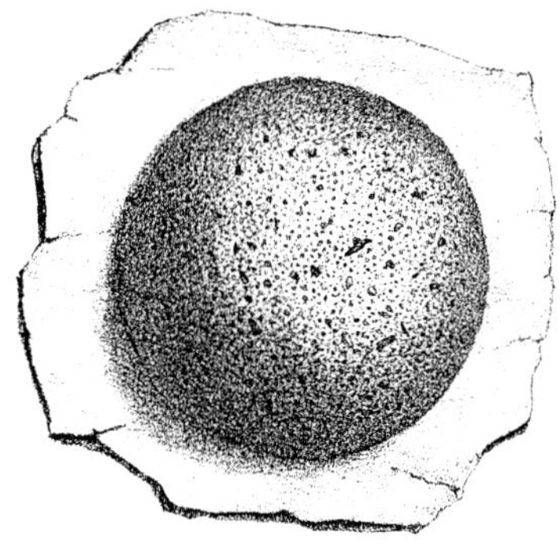

Fig. 26. *Hemisphaerammina bradyi* Loeblich & Tappan, 1957 (after Brady, 1884, Pl. 41, Fig. 11). Scale=1 mm.

SUPERFAMILY KOMOKIACEA
Tendal & Hessler, 1977

Test consists of more or less complex system of fine branched tubules of even diameter. Tubules stiff to flexible, basically cylindrical in form, infre-quently constricted, and with distance between branching points generally much longer than diameter. Wall simple, consisting of inner organic layer and thicker agglutinated outer layer of mainly argillaceous particles. **Stercomata** accumulate within tubules.

KEY TO THE FAMILIES

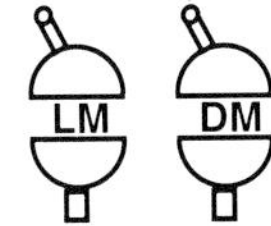

1. Test tree- or bush-like or forming tangled clumps, consisting of fine branched tubules, usually of even diameter........... **Komokiidae**
1'. Test a single shaft or clump of branching tubules with numerous short side-branches giving bead-like appearance... **Baculellidae**

FAMILY KOMOKIIDAE Tendal & Hessler, 1977

The test is tree- or bush-like in shape or forms a tangled clump. Tubules stiff to flexible, basically cylindrical in form, infrequently constricted, and with the distance between branching points gnerally much longer than the diameter.

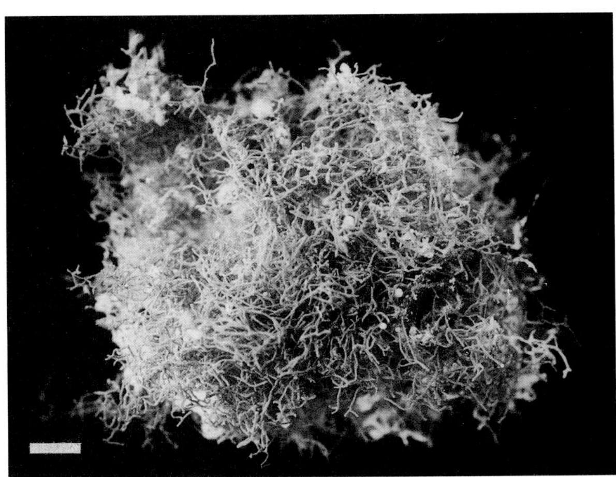

Fig. 27. *Lana neglecta* Tendal & Hessler, 1977. Scale=500 μm.

KEY TO REPRESENTATIVE GENERA

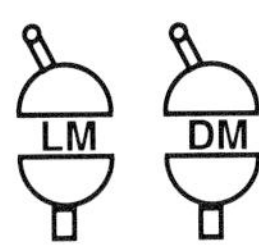

1. Test tree- or bush-like, tubules do not anastomose.. 2
1'. Test forming tangled clump of anastomosing tubules.. ***Lana***

2. Tubules decreasing markedly in diameter at each branching point.............................. ***Ipoa***
2'. Tubules of approximately uniform diameter throughout individual................................. 3

3. Test bush-like in overall form.................... 4
3'. Tubules radiate outwards from basal area and terminate in globular or club-like structures which may contain septa... ***Normanina***

4. Tubules non-septate......................... ***Komokia***
4'. Tubules divided internally by transverse septae with foramina...................... ***Septuma***

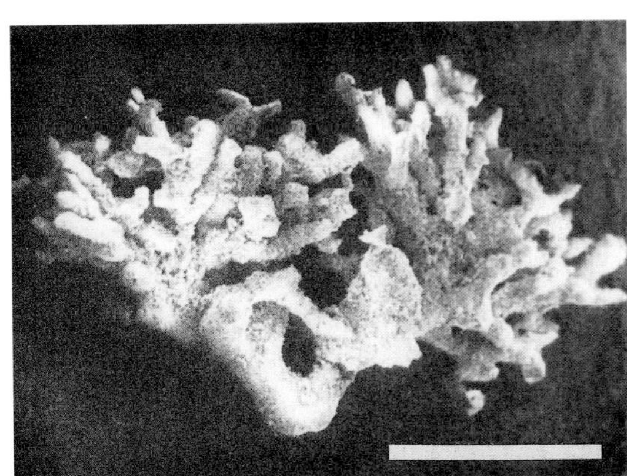

Fig. 28. *Ipoa fragilis* Tendal & Hessler, 1977 (after Tendal and Hessler, 1977). Scale=500 μm.

FAMILY BACULELLIDAE Tendal & Hessler, 1977

Test variously shaped; consisting either of a dense clump or straight, occasionally branched shaft with numerous short side branches. Fine morphology dominated by bead-like appearance arising from frequent, extremely short side branches, often constricted at base, or from frequent constriction of main tubules at regular short intervals.

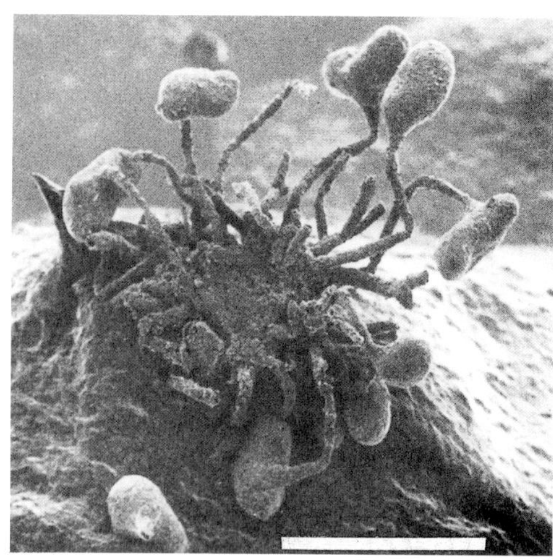

Fig. 29. *Normania tylota* Tendal & Hessler, 1977 (after Tendal and Hessler, 1977). Scale=500 μm.

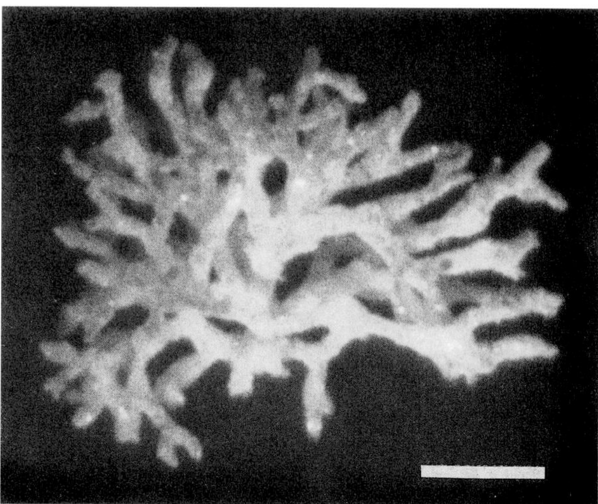

Fig. 30. *Komokia multiramosa* Tendal & Hessler, 1977 (after Tendal and Hessler, 1977). Scale=500 μm.

KEY TO REPRESENTATIVE GENERA

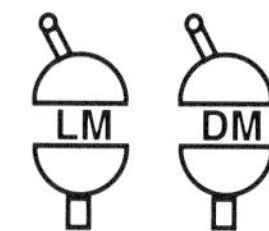

1. Test free.. 2

1'. Test attached, encrusting; constrictions near ends of tubules delimit terminal, bead-like sections which impart granular appearance... .... ***Chondrodapsis***

2. Test consists of single, sometimes dichotomously branched shaft, formed by non-septate tubule from which arise numerous short side branches, sometimes reduced to simple, bead-like lobes. .... ***Baculella***

2'. Test forms dense, rounded clump of branching tubules with numerous short side-branches and beads.................................. ***Edgertonia***

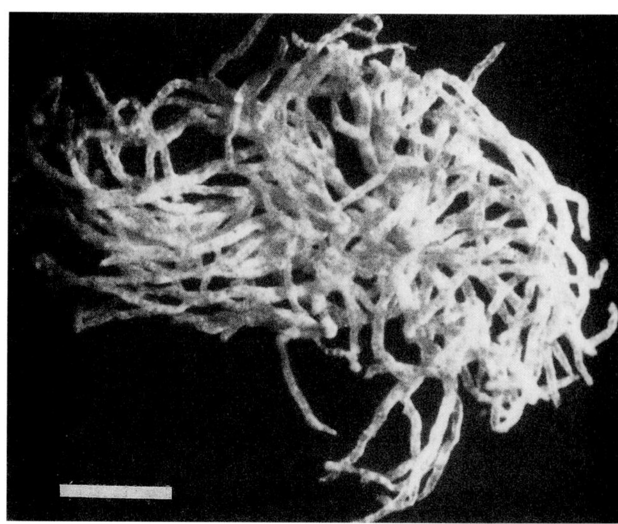

Fig. 31. *Septuma ocotillo* Tendal & Hessler, 1977 (after Tendal and Hesler, 1977). Scale=500 μm.

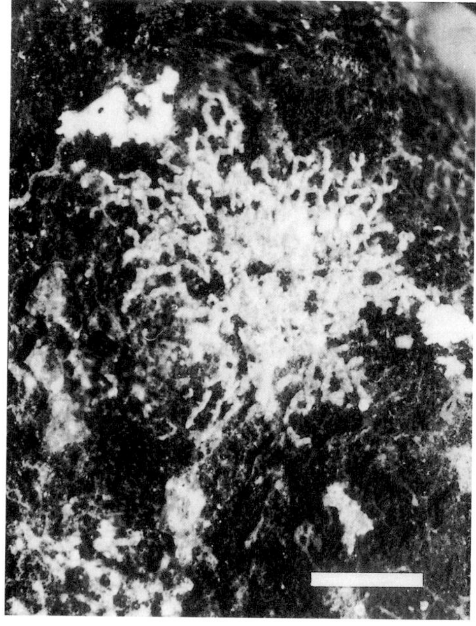

Fig. 32. *Chondrodapis hessleri* Mullineaux, 1988 (after Mullineaux, 1988). Scale=1 mm.

ORDER TEXTULARIIDA

Delage & Herouard, 1896

Test multilocular, wall agglutinated, foreign particles cemented using organic or calcareous cement; may be canaliculate. Various attempts (e.g. Loeblich and Tappan, 1989) have been made to divide this, the major group of agglutinating foraminifera, into several orders, based on presence or absence of "perforations" (**canaliculi**), cement-type (organic or calcareous), and type and layering of the wall (calcareous-agglut-inating or agglutinating with inner and/or outer organic layers. These schemes are, for the present, considered premature.

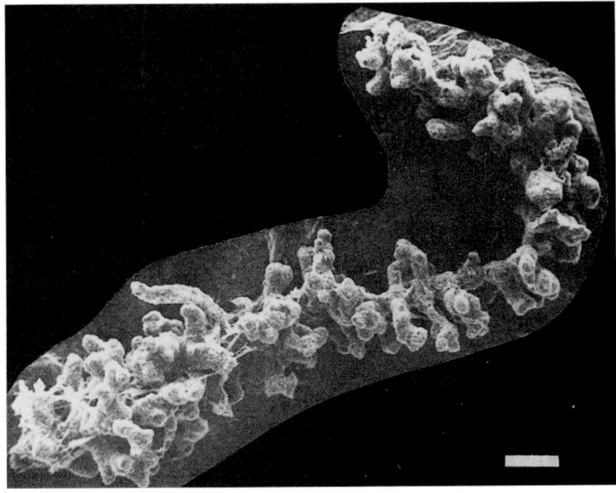

Fig. 33. *Baculella globofera* Tendal & Hessler, 1977 (after Tendal and Hessler, 1977). Scale=500 μm.

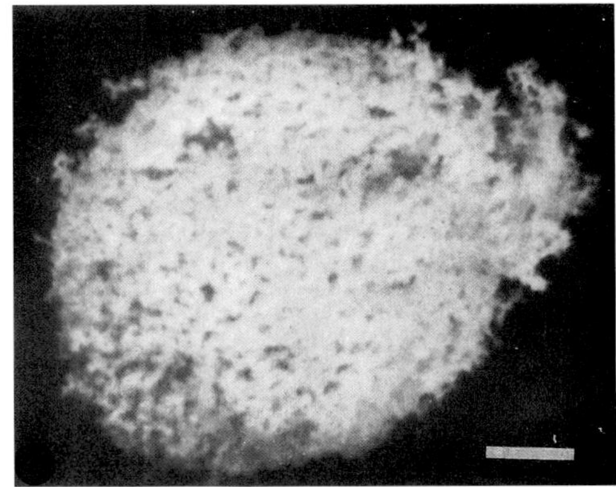

Fig. 34. *Edgertonia tolerans* Tendal & Hessler, 1977 (after Tendal and Hessler, 1977). Scale=500 μm.

KEY TO THE FAMILIES

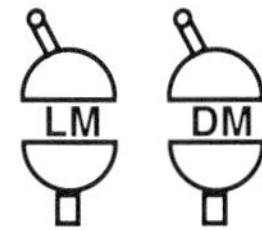

1. Wall agglutinated using organic cement, non-canaliculate.. 2

1'. Wall agglutinated using calcareous cement, typically canaliculate................................ 15

2. Test with chambers in uniserial arrangement. 3

2'. Test enrolled.. 4

3. Test free................................ **Hormosinidae**

3'. Test attached....................... **Telamminidae***

4. Test with chambers added in milioline coiling .... **Rzehakinidae**

4'. Test planispiral, **streptospiral,** or trochospiral.. 5

5. Test planispiral or streptospiral................ 6

5'. Test trochospiral....................................... 13

6. Test streptospiral....**Ammosphaeroidinidae**

6'. Test planispiral.. 7

7. Test planispiral throughout......................... 8

7'. Planispiral in early stage, later uncoiled.. 10

8. Wall simple, not alveolar............................ 9

8'. Wall with outer imperforate layer and inner alveolar layer.................... **Cyclamminidae**

9. Test involute to partially evolute......... .... **Haplophragmoididae**

9'. Test involute with embracing chambers .... **Sphaeramminidae***

10. Test free 11

10'.Test attached..................... **Placopsilinidae**

11. Test planispiral, later uncoiled, uniserial12

11'. Test planispiral, later biserial......... .... **Spiroplectamminidae**

12. Early stage with non-septate tubular chamber .... **Lituotubidae**

12'. Early stage with numerous chambers........... .... **Lituolidae**

13. Test trochospiral only in early stage, later triserial............................ **Verneuilinidae**

13'. Test trochospiral throughout 14

14. Low trochospiral coil..... **Trochamminidae**

14'. High trochospiral coil.**Globotextulariidae**

15. Test trochospiral or triserial, at least in early stage.. 16
15'. Test biserial .... **Textulariidae**

16. Early stage trochospiral, later triserial....... .... **Eggerellidae**
16'. Early stage **triserial** and triangular in section, later may be uniserial**Valvulinidae**

*-Families not illustrated in this volume.

FAMILY HORMOSINIDAE Haeckel, 1894

Test free, elongate, chambers arranged in rectilinear to arcuate series, each succeeding chamber attached near the base of the apertural neck of the preceding chamber; aperture terminal, rounded, produced on a slight neck; wall thin of a single layer of agglutinated grains of quartz, mica, sponge spicules, or foraminiferal tests held in a minimum of organic cement; 15 Recent genera, ex. *Reophax* (Fig. 35).

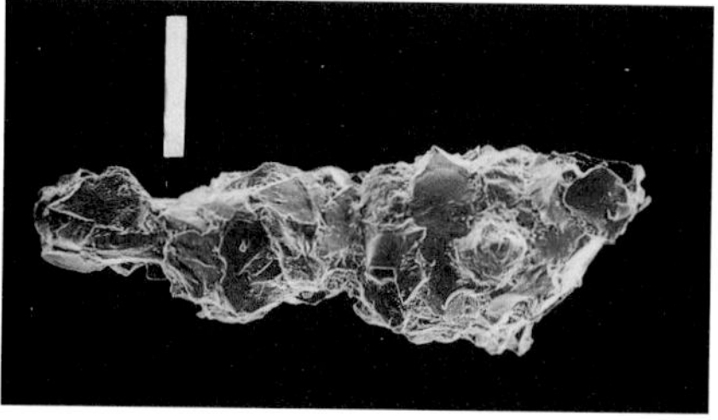

Fig. 35. *Reophax scorpiurus* de Montfort, 1808. Scale=150 μm.

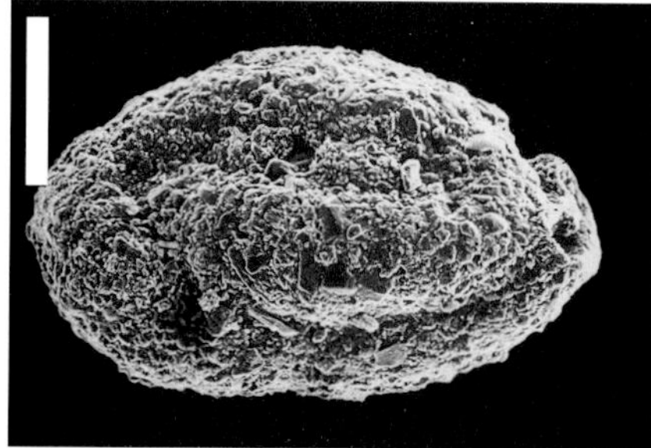

Fig. 36. *Miliammina fusca* (Brady, 1870). Scale 100μm.

FAMILY RZEHAKINIDAE Cushman, 1933

Test enrolled, with successive chambers added in varied planes as in the miliolines, wall not perforate, finely agglutinated, test ovate in outline; aperture at the end of the chamber; five Recent genera, ex. *Miliammina* (Fig. 36).

FAMILY AMMOSPHAEROIDINIDAE Cushman, 1927

Test subglobular, streptospirally enrolled, chambers few per whorl, only those of last whorl visible from exterior; aperture interiomarginal to areal; three Recent genera: *Adercotryma* (Fig. 37), *Cystammina* (Fig. 38), *Recurvoides* (Fig. 39).

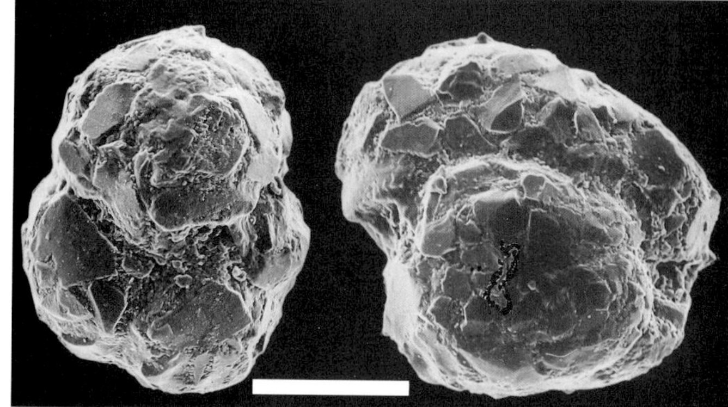

Fig. 37. *Adercotryma glomeratum* (Brady, 1878). Scale=100 μm.

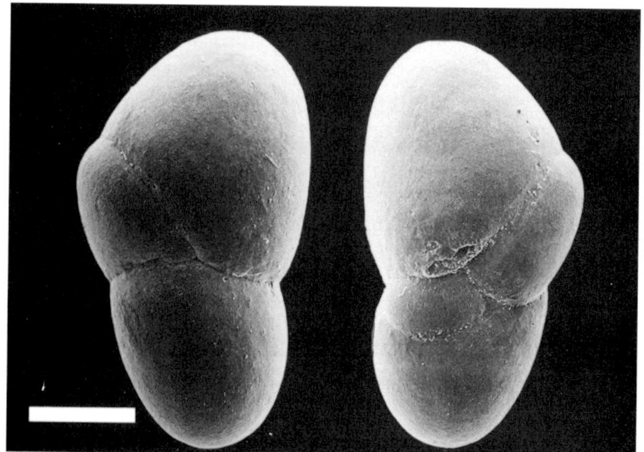

Fig. 38. *Cystammina pauciloculata* (Brady, 1879). Scale=100 μm.

FAMILY HAPLOPHRAGMOIDIDAE Maync, 1952

Test planispirally enrolled, involute to partially evolute, with septa formed by outer wall, few chambers per whorl; aperture equatorial in position, basal to areal; eight Recent genera, ex. *Cribrostomoides* (Fig. 40), *Haplophragmoides, Trochamminita.*

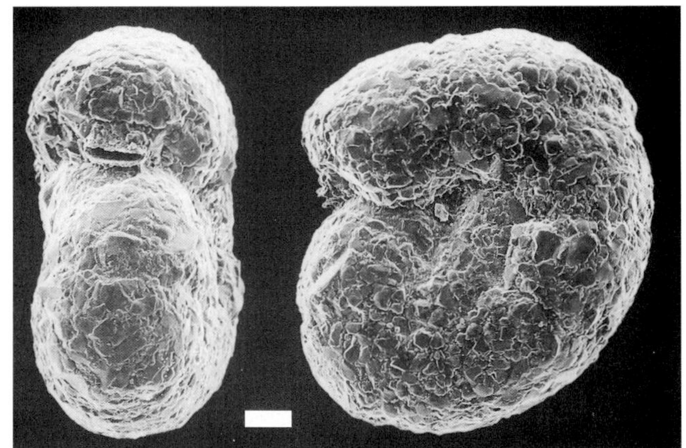

Fig. 39. *Recurvoides contortus* Earland, 1934. Scale=200 µm.

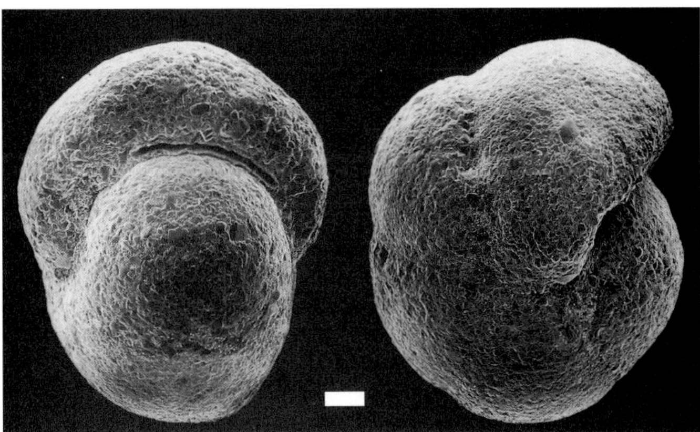

Fig. 40. *Cribrostomoides subglobosus* (Cushman, 1910). Scale=200 µm

FAMILY LITUOTUBIDAE Loeblich & Tappan, 1984

Test free, globular proloculus followed by enrolled, non-septate, tubular chamber, later elongate and irregularly septate chambers, few per whorl, finally becoming uncoiled and irregularly recti-linear; aperture rounded at the open end of the tubular chamber; two Recent genera: *Lituotuba* (Fig. 41), *Trochamminoides*.

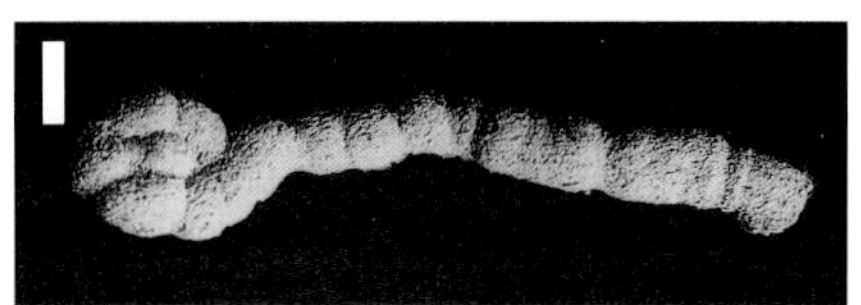

Fig. 41. *Lituotuba lituiformis* (Brady, 1879). Scale=500 µm.

FAMILY LITUOLIDAE de Blainville, 1827

Test free, early stage planispirally enrolled, later may be uncoiled and rectilinear, interior simple; aperture terminal, single or multiple; six Recent genera, ex. *Ammobaculites* (Fig. 42), *Ammotium, Ammoastuta* (Fig. 43), *Lituola*.

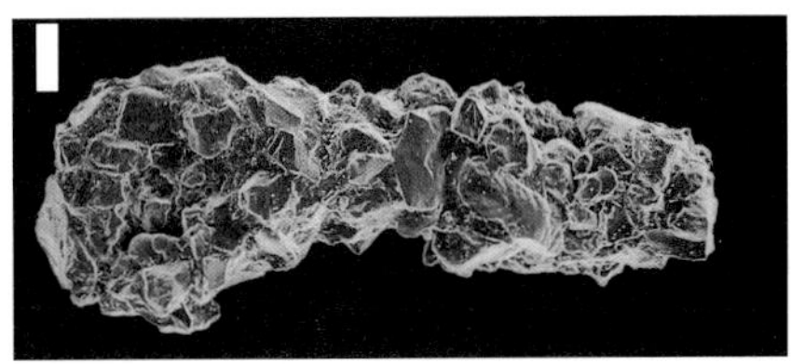

Fig. 42. *Ammobaculites balkwilli* Haynes, 1973. Scale=100 µm.

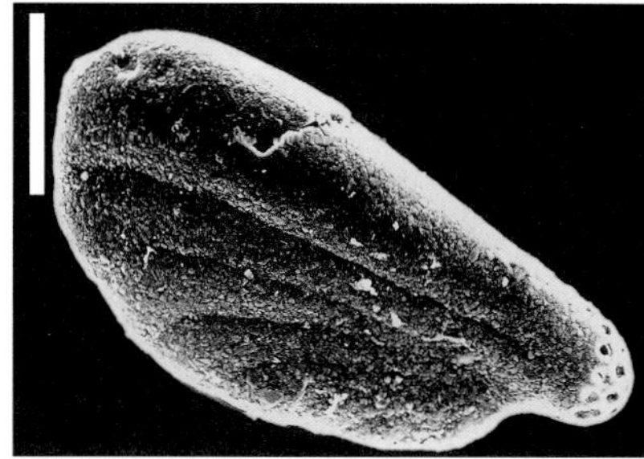

Fig. 43. *Ammoastuta salsa* Cushman & Bronnimann, 1948. Scale=100 µm.

Fig. 44. *Placopsilina cenomana* d'Orbigny, 1850. Scale=100 µm.

FAMILY PLACOPSILINIDAE Rhumbler, 1913

Test attached, early stage planispirally enrolled, later uncoiled and rectilinear, aperture terminal, rounded, may be bordered by a slight lip; three Recent genera, ex. *Placopsilina* (Fig. 44).

FAMILY CYCLAMMINIDAE Marie, 1941

Test planispirally coiled and involute, flattened, with numerous broad chambers per whorl; **aperture interiomarginal** slit and a series of round pores with elevated rims scattered over the face; wall with very thin outer imperforate layer and prominent inner alveolar layer; two Recent genera *Cyclammina* (Fig. 45), *Alveolophragmium*.

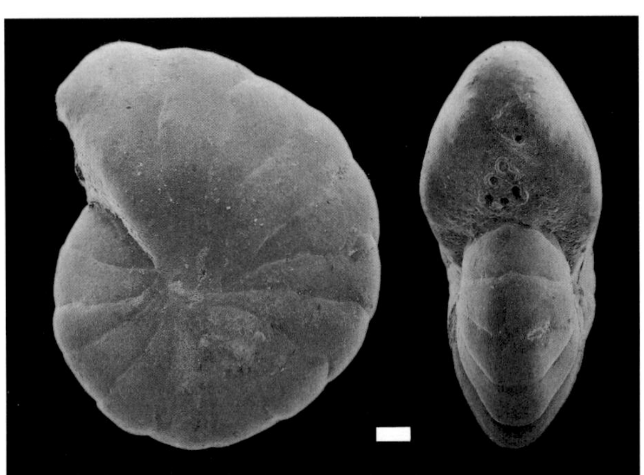

Fig. 45. *Cyclammina cancellata* Brady, 1879. Scale=200 μm.

FAMILY SPIROPLECTAMMINIDAE Cushman, 1927

Test free, planispirally coiled in early stage, later biserial; aperture interiomarginal.

KEY TO REPRESENTATIVE GENERA

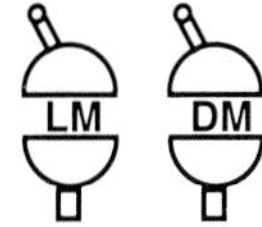

1. Chambers simple, undivided.......................... 2
1'. Chambers with internal partition.................... .... ***Spirotextularia***

2. Test elongate, narrow...... ***Spiroplectammina***
2'. Test broad and flattened................ ***Vulvulina***

FAMILY TROCHAMMINIDAE Schwager, 1877

Test multilocular, chambers in low trochospiral coil; aperture interiomarginal to areal, single or multiple, may have supplementary umbilical or areal openings; agglutinated wall with inner and outer organic layer and organic cement; non-canaliculate.

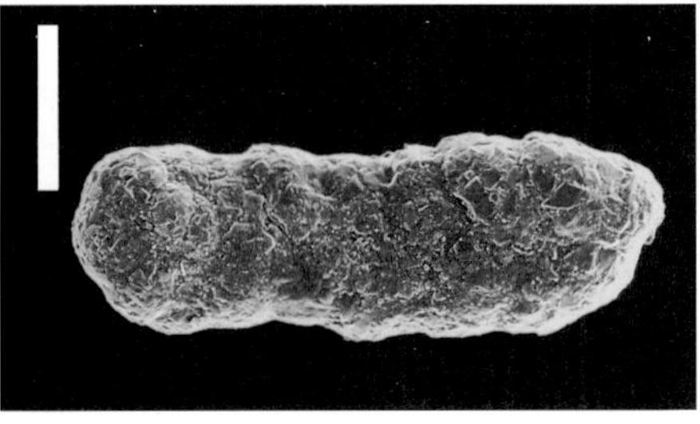

Fig. 46. *Spiroplectammina biformis* (Parker & Jones, 1865). Scale=100 μm.

Fig. 47. *Vulvulina capreolus* d'Orbigny, 1826. Scale=100 μm.

KEY TO REPRESENTATIVE GENERA

1. No supplementary openings............................ 2
1'. Supplementary openings................................ 5

2. Primary aperture interiomarginal................ 3
2'. Primary aperture areal, on umbilical side .... ***Trochaminella***

3. Umbilical flap............... ***Portatrochammina***
3'. No umbilical flap.. 4

4. Three chambers/whorl, aperture without lip .... ***Tritaxis***
4'. More than three chambers/whorl, aperture with narrow lip...................... ***Trochammina***

5. Supplementary openings areal....................... 6
5'. Supplementary openings umbilical, sutural..... .... ***Deuterammina***

6. Supplementary openings in the lower portion of the apertural face.................... ***Jadammina***
6'. Supplementary openings at the apex of the final chamber....................... ***Arenoparrella***

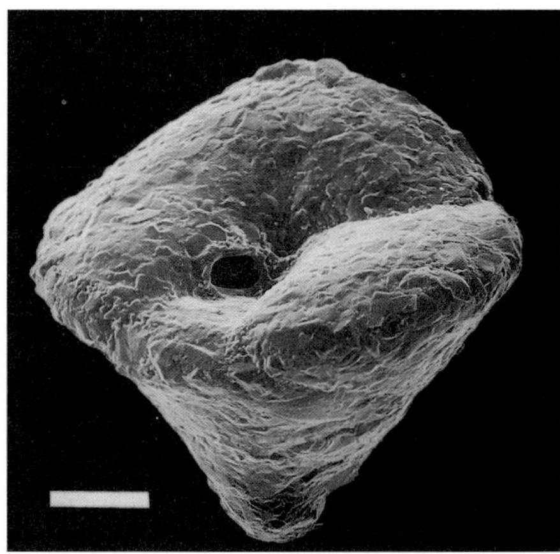

Fig. 48. *Trochamminella conica* (Parker & Jones, 1865). Scale=100 µm.

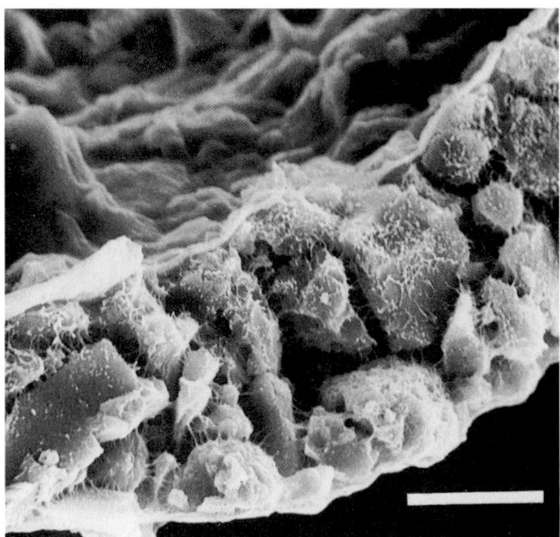

Fig. 49. Section through test of *Trochammina* showing "solid" agglutinating wall with inner and outer organic layers and organic cement. Scale=9 µm.

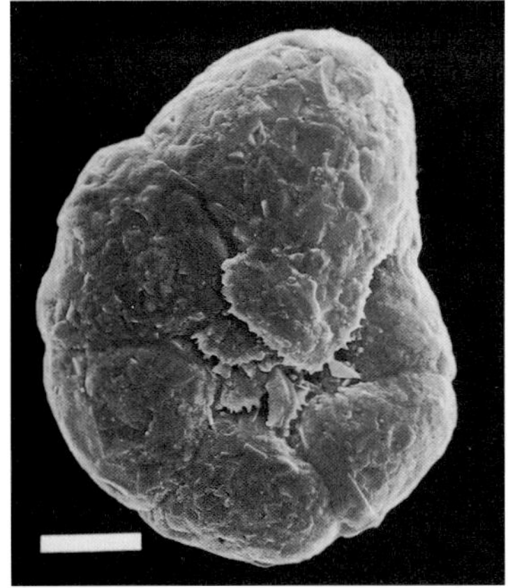

Fig. 50. *Portatrochammina bipolaris* Bronnimann & Whittaker, 1980. Scale=50 µm.

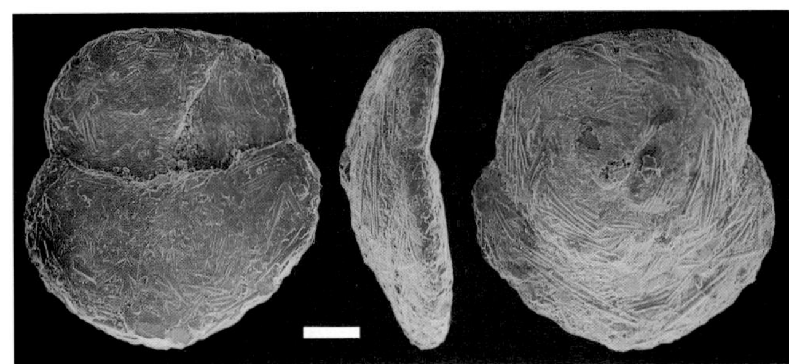

Fig. 51. *Tritaxis tusca* (Williamson, 1858). Scale=100 µm.

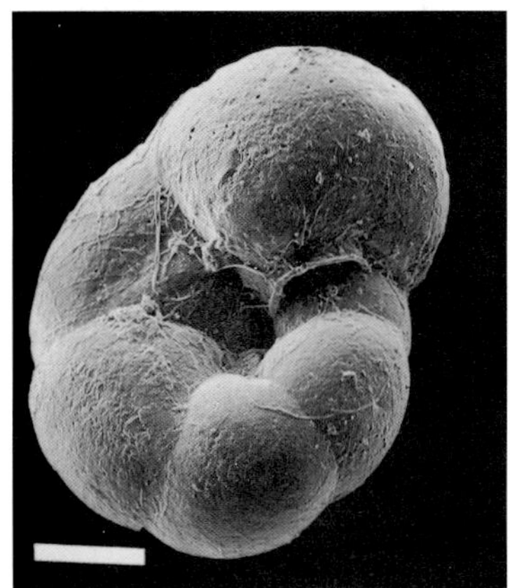

Fig. 52. *Trochammina inflata* (Montagu, 1808). Scale=100 µm.

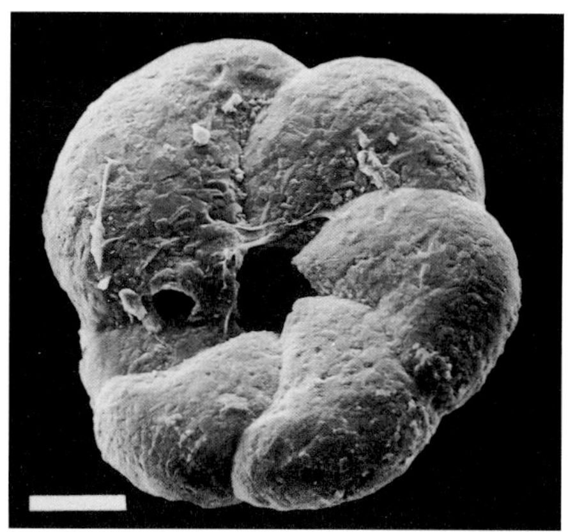

Fig. 53. *Deuterammina plymouthensis* Bronnimann & Whittaker, 1990. Scale=50 μm.

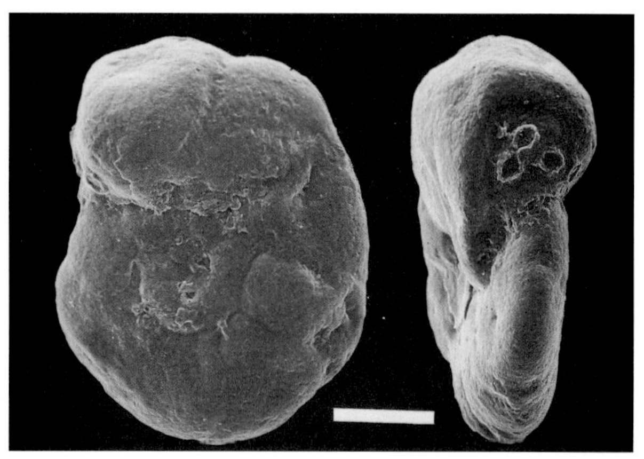

Fig. 54. *Jadammina macrescens* (Brady, 1870). Scale=100 μm.

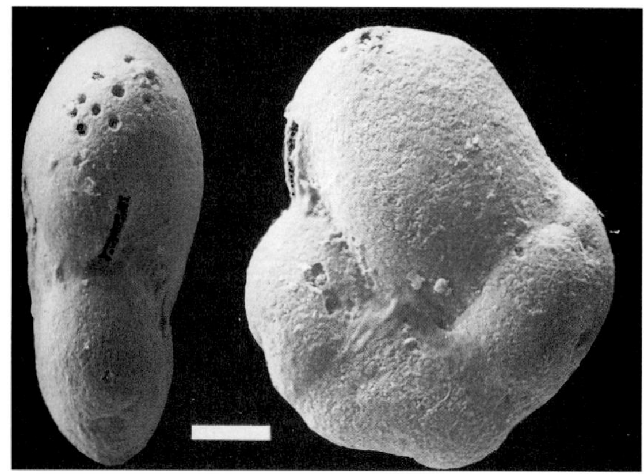

Fig. 55. *Arenoparrella mexicana* (Kornfield, 1931). Scale=50 μm.

FAMILY REMANEICIDAE Loeblich & Tappan, 1964

Test trochospiral, low, interior partially subdivided by infoldings of the umbilical wall and may have secondary septa; primary aperture interiomarginal, secondary openings umbilical; wall non-canaliculate.

KEY TO REPRESENTATIVE GENERA

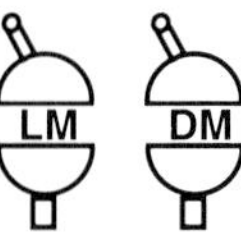

1. Secondary septula.......................... ***Remaneica***
1'. No secondary septa .... ***Asteroparatrochammina***

Fig. 56. *Remaneica plicata* (Terquem, 1876). Scale=100 μm.

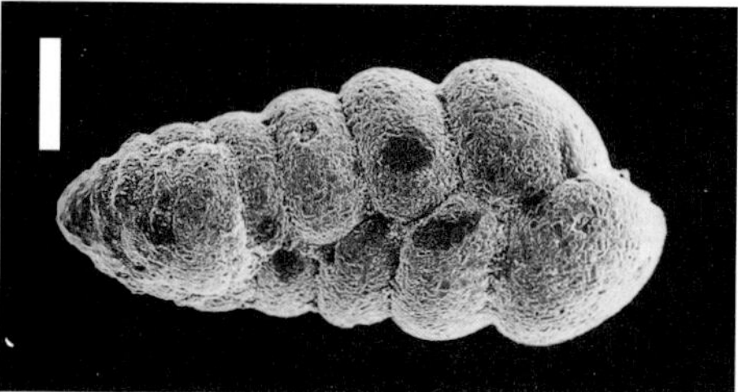

Fig. 57. *Gaudryina baccata* Schwager, 1866. Scale=100 μm.

FAMILY VERNEUILINIDAE Cushman, 1911

Test triserial and triangular in section, at least in the early stage, later may become biserial or uniserial, chambers angular; aperture an interio-marginal arch, may become terminal in later stage; wall non-canaliculate; three Recent genera, ex. *Gaudryina* (Fig. 57).

FAMILY GLOBOTEXTULARIIDAE Cushman, 1927

Test in high trochospiral coil, later stage with three to four chambers per whorl, chambers subglobular, inflated; aperture an interiomarginal umbilical arch; six Recent genera, ex. *Globotextularia* (Fig. 58).

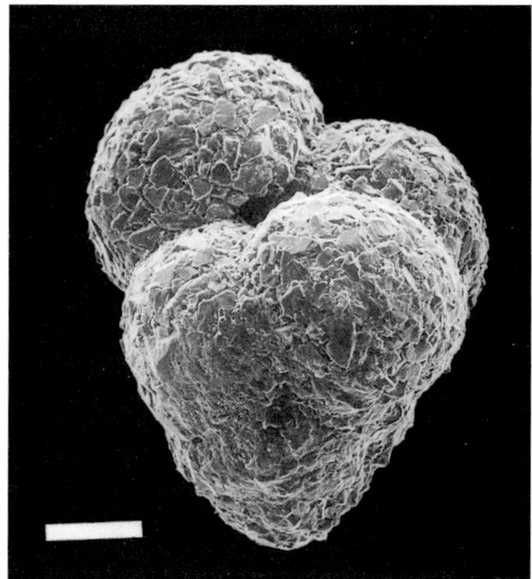

Fig. 58. *Globotextularia anceps* (Brady, 1884). Scale=100 μm.

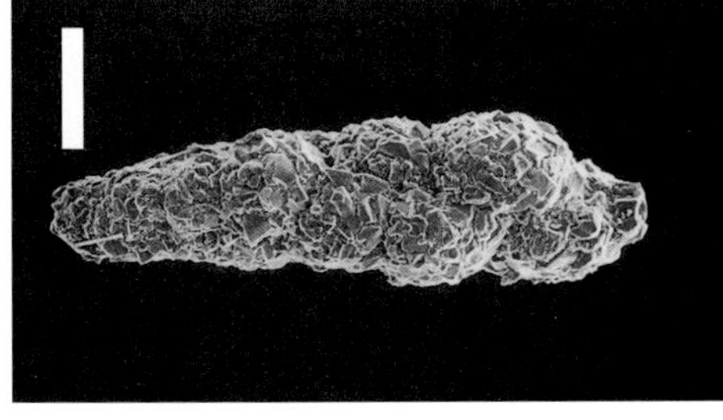

Fig. 59. *Karreriella apicularis* (Cushman, 1911). Scale=100 μm.

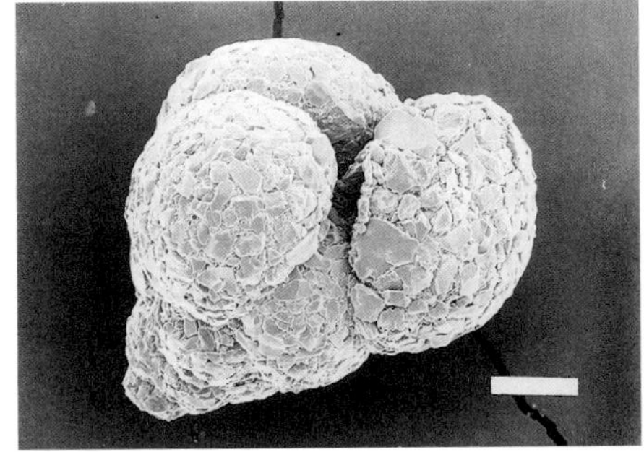

Fig. 60. *Eggerella* sp. Toulmin, 1941. Scale=100 μm.

FAMILY EGGERELLIDAE Cushman, 1937

Test trochospirally enrolled in the early stage, later may be reduced to triserial, biserial, or uniserial; aperture basal to **areal**, single or multiple, and bordered by a narrow lip; wall canaliculate.

KEY TO REPRESENTATIVE GENERA

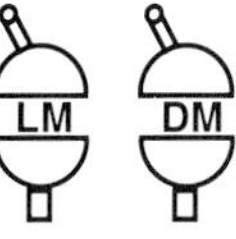

1. Test subconical or subfusiform, aperture basal.2

1'. Test elongate, aperture areal, rounded............... .... ***Karreriella***

2. Aperture a low slit near the base of the apertural face................................ ***Eggerella***

2'. Aperture a high arch in the center of the apertural face........................ ***Eggerelloides***

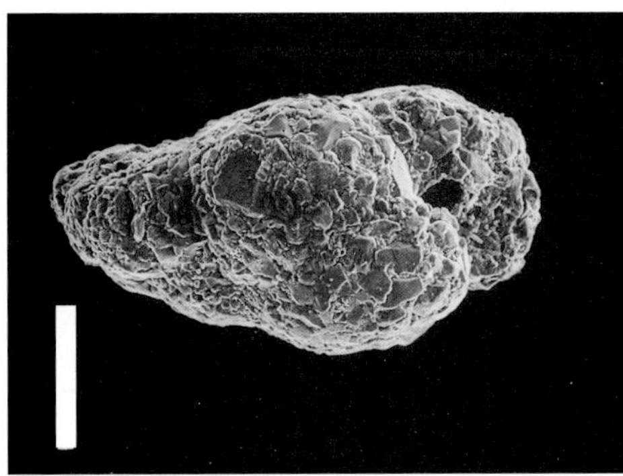

Fig. 61. *Eggerelloides scabra* (Williamson, 1858). Scale=100 μm.

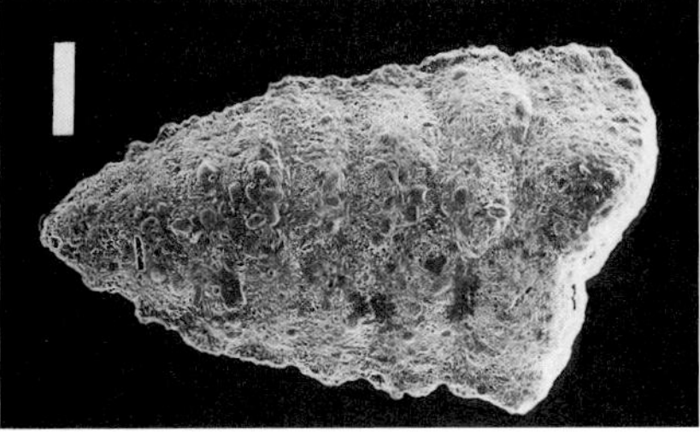

Fig. 62. *Textularia sagittula* Defrance, 1824. Scale=100 μm.

FAMILY TEXTULARIIDAE Ehrenberg, 1838

Test biserial, at least in the early stage, later may

be reduced to uniserial; aperture interiomarginal to **areal**, single or multiple; wall canaliculate.

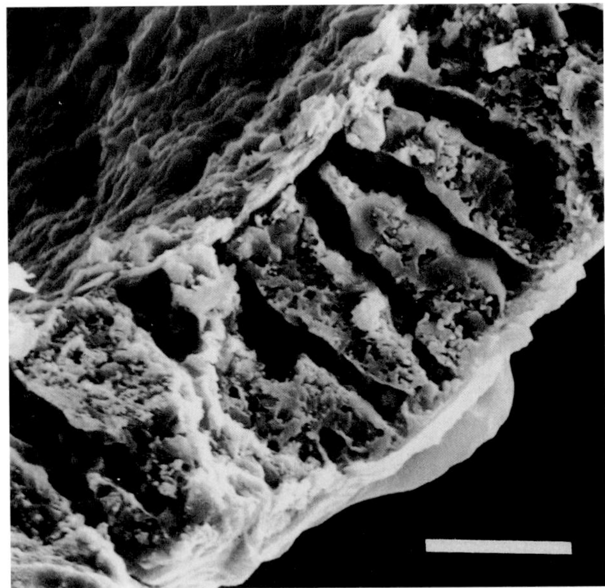

Fig. 63. Section through test of *Textularia* showing canaliculate wall. Unlike true pores, canaliculi end blindly against the inner and outer organic layers. Scale=9 μm.

KEY TO REPRESENTATIVE GENERA

1. Aperture interiomarginal, a slit at the base of the apertural face........................ ***Textularia***
1'. Aperture areal.. 2

2. Aperture areal, surrounded by a distinct lip or produced on a neck............ ***Siphotextularia***
2'. Aperture areal, cribrate......... ***Planctostoma***

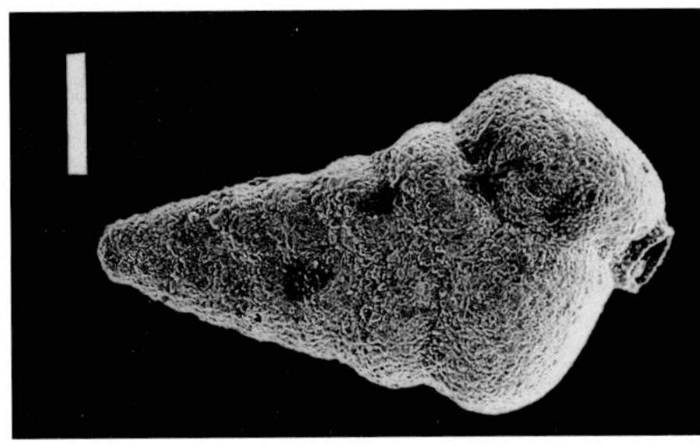

Fig. 64. *Siphotextularia heterostoma* (Fornasini, 1896). Scale=100 μm.

FAMILY VALVULINIDAE Berthelin, 1880

Test trochospiral to triserial in early stage, later uniserial; aperture interiomarginal in early stage, later terminal and rounded, with valvular tooth or flap; wall canaliculate.

KEY TO REPRESENTATIVE GENERA

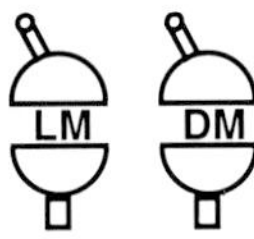

1. Test elongate, triserial in early stage, later uniserial .... ***Clavulina***
1'. Test triserial, triangular in section throughout...................................... ***Valvulina***

Fig. 65. *Clavulina*. a. *C. tricarinata* d'Orbigny, 1839. Scale=100 μm. b. *C. pacifica*. Scale=100 μm.

ORDER ROBERTINIDA
Loeblich & Tappan, 1984

Test trochospirally enrolled with chambers subdivided by internal partitions. Test wall of hyaline, finely perforate, ultrastructurally and optically radiate aragonite (orthorhombic crystal form of calcium carbonate). Mostly fossil, only few Recent species.

KEY TO THE FAMILIES

1. Single internal partition............................... 2
1'. Double transverse partition..... **Robertinidae**

2. Aperture an elongate slit extending up the apertural face............ **Ceratobuliminidae**
2'. Aperture a slit on peripheral margin........ .... **Epistominidae**

FAMILY ROBERTINIDAE Reuss, 1850

Test nearly planispiral to trochospiral, predominantly dextral, chambers divided by double trans-verse partition resulting from infolding of wall, and forming small supplementary chambers on one or both sides of the test; aperture consisting com-monly of two slit-like openings, one interiomarginal at the base of the chamber, the other extending up the center of the apertural face or areal; may have small supplementary openings on the spiral side.

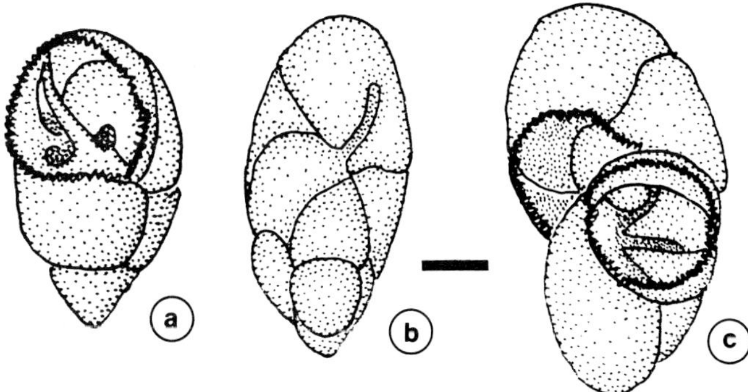

Fig. 66. *Robertinoides bradyi* (Brady). a. Dissected specimen with main chamber opened, showing broken toothplate and exposing the bulimine aperture from within. b. Apertural view. c. Dissected specimen in oblique view from below into last two chamberlets, connected with main chambers through foramina formed by previous primary apertures. (from McGowran 1968; Loeblich & Tappan 1988). Scale=10 μm.

KEY TO THE GENERA

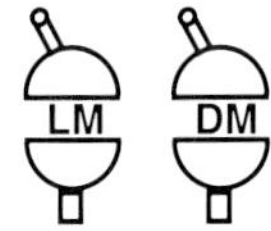

1. Test planispiral to low trochospiral 2
1'. Test a high trochospiral coil......................... 4

2. Test nearly planispiral................................ 3
2'. Test distinctly trochospiral.................... .... ***Pseudobulimina***

3. Supplementary sutural openings................ .... ***Cushmanella***
3'. No sutural openings........................ ***Alliatina***

4. Aperture single, extending up the face of the final chamber .... ***Robertina***
4'. Two slit-like apertures, one interiomarginal, other directed up the apertural face.... .... ***Robertinoides***

FAMILY CERATOBULIMINIDAE CUSHMAN, 1927

Test trochospirally coiled with rapidly enlarged chambers; internal partition attached to the posterior side of the apertural slit; aperture a long narrow slit within a groove that extends up the apertural face. Recent genera: *Ceratobulimina* (Fig. 67), *Ceratobuliminoides, Lamarckina*, cosmopolitan.

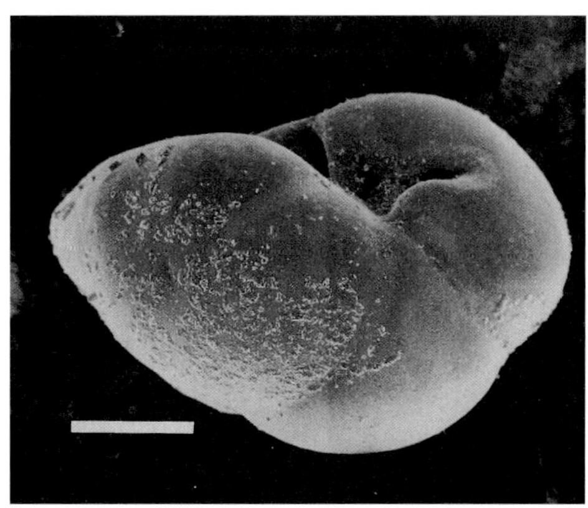

Fig. 67. *Ceratobulimina contraria* (Reuss). Scale=100 μm.

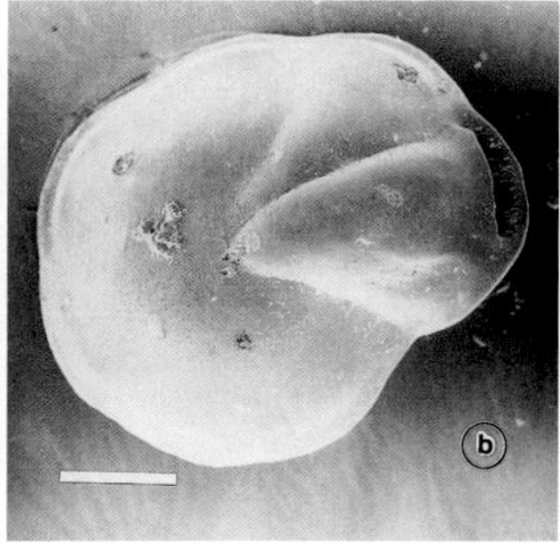

Fig. 68. *Hoeglundina elegans* (d'Orbigny). Scales=100 μm.

FAMILY EPISTOMINIDAE Wedekind, 1937

Test with predominantly sinistral low trochospire coiling, biconvex, internal partition flat, present only in the last chamber; aperture a slit on peripheral margin, closed by clear material in older chambers, may have a secondary interiomarginal aperture. Single Recent genus *Hoeglundina* (Fig. 68), cosmopolitan.

ORDER CARTERINIDA
Loeblich & Tappan, 1981

Test of secreted rodlike or fusiform calcitic spicules, each one crystallographically a single elongate crystal of calcite, commonly oriented parallel to periphery and embedded in organic matrix.

FAMILY CARTERINIDAE
Loeblich & Tappan, 1955

Test attached, trochospiral, large, flattened with numerous crescentic to irregularly coiled chambers, later chambers expanded, may be subdivided by secondary septula resulting from infolding of the wall; aperture umbilical single in the early chambers, multiple in the later chambers. Genera: *Carterina* (Fig. 69) and *Zaninettia*, common in tropical shallow waters.

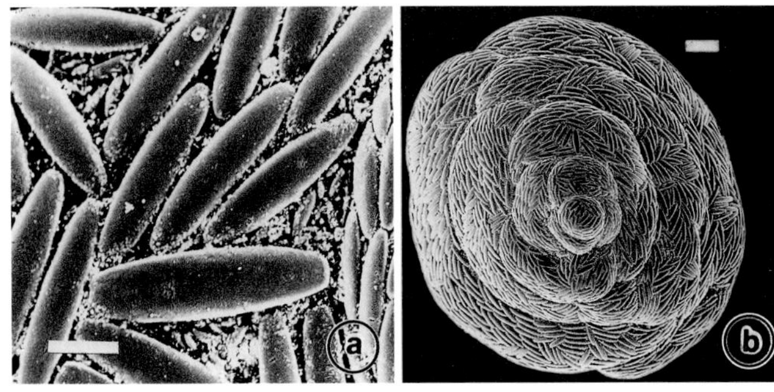

Fig. 69. *Carterina spiculotesta* (Carter). a. Close up of spicules. Scale=20 µm. b. Whole specimen. Scale=100 µm.

ORDER SPIRILLINIDA
Hohenegger & Piller, 1975

Test of calcite optically a single crystal or few to a mosaic of crystals. Wall formed by marginal accretion, not by calcification of an organic matrix.

KEY TO THE FAMILIES

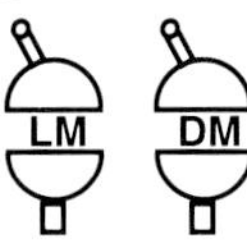

1. Test of undivided, enrolled tubular chamber .... **Spirillinidae**

1'. Test of enrolled tubular chamber in early stage, later stage with two chambers per whorl .... **Patellinidae**

FAMILY SPIRILLINIDAE Reuss & Fritsch, 1861

Test discoidal to conical, composed of early spherical portion "proloculus" followed by a non-septate tubular chamber, planispirally or trocho-spirally enrolled; aperture at open end of the tube. Reproduction gamontogamic. All nuclei of agamont stage identical (homokaryotic). Occur in shallow waters.

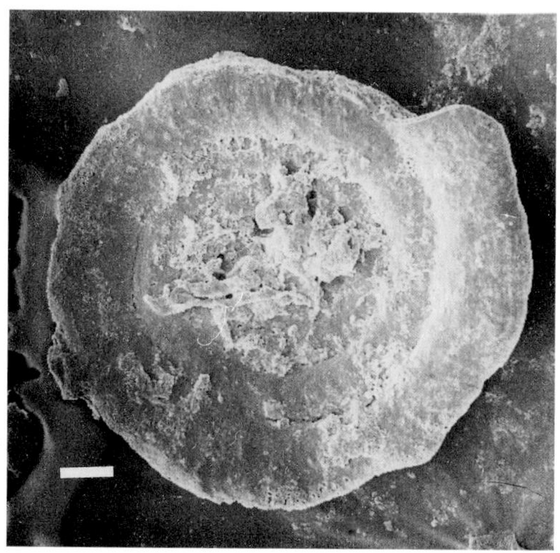

Fig. 70. *Spirillina vivipara* Ehrenberg. Scale=20 µm.

KEY TO THE GENERA

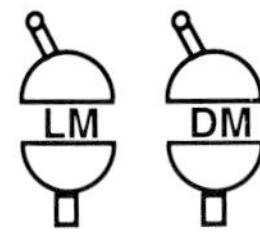

1. Test of closely appressed whorls 2
1'. Whorls separated by narrow plate, periphery carinate with long spines .... ***Sejunctella***

2. All whorls coil in the same plane 3
2'. Last whorls coil toward umbilicus, coarse perforations on both sides..... ***Mychostomina***

3. Test planispiral, 4-9 whorls, numerous pores or pseudopores.............................. ***Spirillina***
3'. Test conical, four whorls, fine perforations........................ ***Turrispirillina***

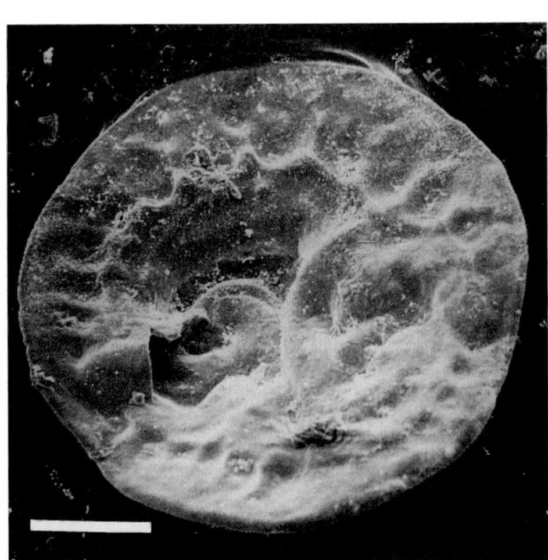

Fig. 71. *Patellina corrugata* Williamson. Scale=50 μm.

FAMILY PATELLINIDAE Rhumbler, 1906

Test low conical, planoconvex, spiral undivided tubular chamber in early stage, later with two crescentic chambers per whorl, only the final pair of the chambers visible on the flattened umbilical side; aperture a low opening at the end of the chamber directed toward the umbilicus, later may be covered by a broad apertural plate. Reproduction gamontogamic. All nuclei of agamont stage identical (homokaryotic). Few species living in shallow waters, cosmopolitan.

KEY TO REPRESENTATIVE GENERA

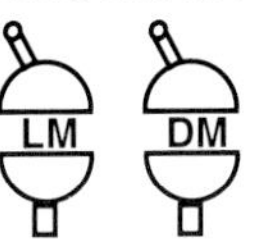

1. Chambers subdivided by radial septula, visible from the spiral side........................ ***Patellina***
1'. Chambers are not subdivided by radial septula .... ***Patellinoides***

ORDER MILIOLIDA
Delage & Herouard, 1896

Test of fine, randomly oriented, rodlike crystals of high magnesium calcite; the random crystal orientation refract light in all directions to result in the milky opacity or porcelaneous appearance; the test is **imperforate** in post embryonic stage; it may have added agglutinated material.

KEY TO THE FAMILIES

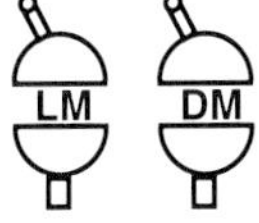

1. Test unilocular.................... **Squamulinidae***
1'. Test with two or more chambers.................... 2

2. Test tubular, later stage may be chambered and irregular.. 3
2'. Chambers in regular arrangement................. 6

3. Undivided tubular chamber............................ 4
3'. Tubular stage distinctly chambered 5

4. Coiling planispiral to streptospiral (later irregular growth)................. **Cornuspiridae**
4'. Coiling streptospiral becoming planispiral.... .... **Hemigordiopsidae**

5. Few chambers per whorl, planispiral, trochospiral, or streptospiral, may be uncoiled in later stage.............................. **Fischerinidae**
5'. Later stage uncoiled, may be spreading or branched............................. **Nubeculariidae**

6. Chambers partly subdivided by a few oblique secondary septa....................... **Riveroinidae**
6'. Chambers without internal sub-divisions..... 7

7. Two chambers per whorl in early stage, later may have more chambers per whorl or may be uncoiled.. 8
7'. More than two chambers per whorl in early stage 11

8. Proloculus followed by flexostyle................. 9
8'. **Flexostyle** generally absent..................... 10

9. Flexostyle well developed. **Ophthalmidiidae**
9'. Flexostyle reduced, no more than one whorl in length .... **Spiroloculinidae**

10. Wall imperforate...................... **Hauerinidae**
10'. Wall pseudopores numerous....... **Miliolidae***

11. Test globular or fusiform coiled about elongate axis 12
11'. Test planispiral, later serial, flabellulate or cyclical 13

12. Chambers subdivided by secondary partitions into chamberlets parallel to the direction of coiling...................................... **Alveolinidae**
12'. Chambers subdivided into irregular or meandriform chamberlets in spherical series .... **Keramosphaeridae***

13. Chambers not subdivided....... **Peneroplidae**
13'. Chambers subdivided into chamberlets . 14

14. Proloculus followed by cornuspirine coil.. .... **Discospirinidae**
14'. Cornuspirine coil reduced or absent..... .... **Soritidae**

*-Family not illustrated in this volume

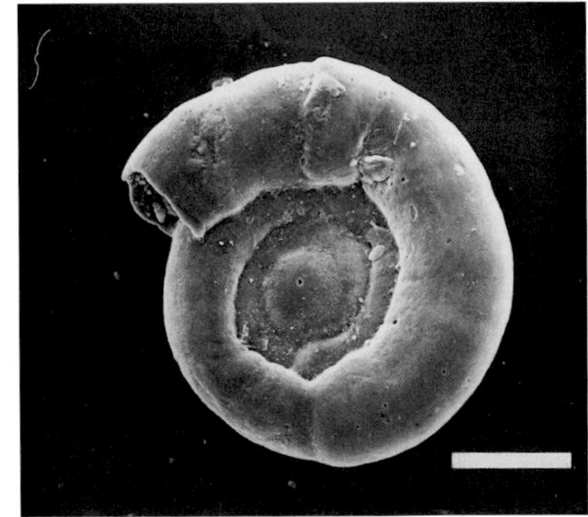

Fig. 72. *Cornuspira involens* (Royce). Scale=100 µm.

FAMILY CORNUSPIRIDAE Schultze, 1854

Test free or attached, composed of an undivided planispiral to streptospiral tubular chamber that may show later irregular growth. Occur in shallow waters.

KEY TO THE GENERA

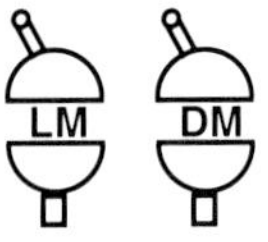

1. Tubular chamber zigzagging, involutely enrolled .... ***Meandrospira***
1'. Tubular chamber planispirally enrolled, at least in young stage...................................... 2

2. Tubular chamber rounded in section throughout .... ***Cornuspira***
2'. Tubular chamber flattened in last portion ... 3

3. Later portion flabelliform ***Cornuspiroides***
3'. Later portion irregularly spreading and branching .... ***Cornuspirella***

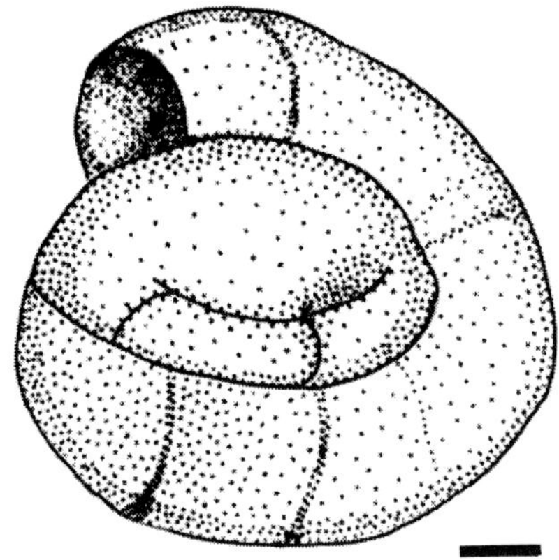

Fig. 73. *Gordiospira fragilis* Heron-Allen & Earland. Scale=35 µm.

FAMILY HEMIGORDIOPSIDAE A. Nikitina, 1969

Test free, composed of an undivided tubular chamber streptospirally coiled in early part, later becoming planispiral. The only Recent genus *Gordiospira* (Fig. 73) occurs in cold waters of South Atlantic, Arctic, and Antarctic.

FAMILY FISCHERINIDAE Millett, 1898

Test free, enrolled tubular portion divided into few chambers per whorl; may be uncoiled in later stage. Occur in tropical shallow waters.

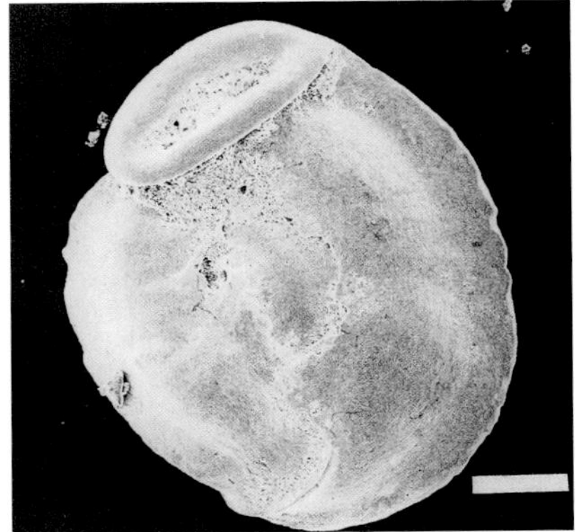

Fig. 74. *Wiesnerella auriculata* (Egger). Scale=100 μm.

KEY TO THE GENERA

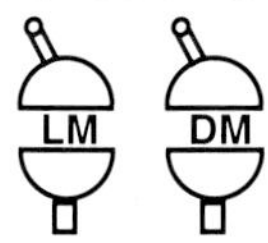

1. Planispiral, may be conical in early portion. 2
1'. Trochospiral or streptospiral, at least in the early stage.. 9

2. Aperture a terminal arch............................. 3
2'. Aperture terminal bordered by everted lip...7

3. Sutures radial... 4
3'. Sutures oblique.. 5

4. Up to eight chambers/whorl....... ***Fischerina***
4'. Three chambers/whorl, aperture with a thickened rim....................... ***Trisegmentina***

5. Two chambers/whorl in later coiling............ .... ***Subfischeri***
5'. Three chambers/whorl, increasing rapidly. 6

6. Test symmetrical bi-umbilicate........... .... ***Planispirina***
6'. Umbilicus obscured by addition of successive laminae. .... ***Planispirinella***

7. Test embracing on apertural face.............. .... ***Wiesnerella***
7'. Test symmetrical, last chamber uncoiled 8

8. Early stage planispiral.... ***Nodobaculariella***
8'. Early stage trochospiral, test covered by longi-tudinal ribs or stria...... ***Vertebralina***

9. Test trochospiral, up to four or five chambers/whorl .... ***Fischerinella***
9'. Early portion streptospiral....................... 10

10. Test discoidal, 12 chambers/whorl.............. .... ***Zoyaella***
10'. Test irregular, two chambers/whorl........... .... ***Glomulina***

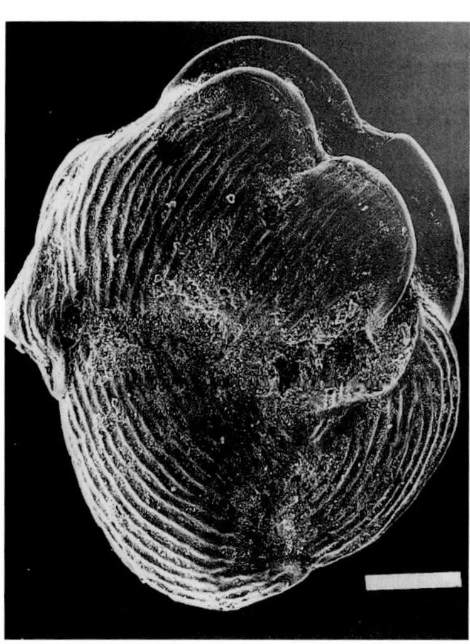

Fig. 75. *Vertebralina striata* d'Orbigny. Scale=100 μm.

FAMILY NUBECULARIIDAE Jones, 1875

Test free or attached, distinctly chambered, early stage planispiral or irregularly coiled, later may be uncoiled, spreading, or branching; wall may have outer agglutinated coating. Common in shallow tropical waters.

KEY TO THE GENERA

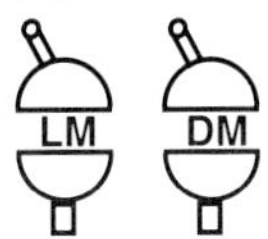

1. Test free.. 2
1'. Test attached.. 3

2. Test with agglutinated material... ***Nubeculina***
2'. Not agglutinated.............. ***Nodophthalmidium***

3. Chambers regularly arranged......................... 4
3'. Chambers irregularly arranged..................... 5

4. Two to three whorls of ovate chambers, the last ones uniserially arranged.. ***Nubeculinita***
4'. Chambers few in number.................. ***Webbina***

5. With agglutinated material............................. 6
5'. Without agglutinated material........................ 7

6. Incorporating fine sand grains.. ***Nubecularia***
6'. Occasionally coarse sand grains.... ***Gymnesina***

7. Chambers tending to uncoil...... ***Nubeculopsis***
7'. Chambers uniserially arranged or branching.8

8. Chambers tapering toward the aperture ....
.... ***Calcituba***
8'. Chambers cylindrical or pyriform...................
.... ***Cornuspiramia***

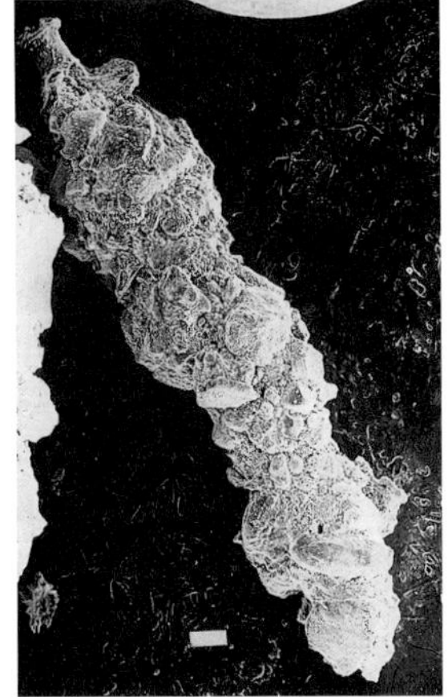

Fig. 76. *Nubeculina divariauta advena.* Scale=100 µm.

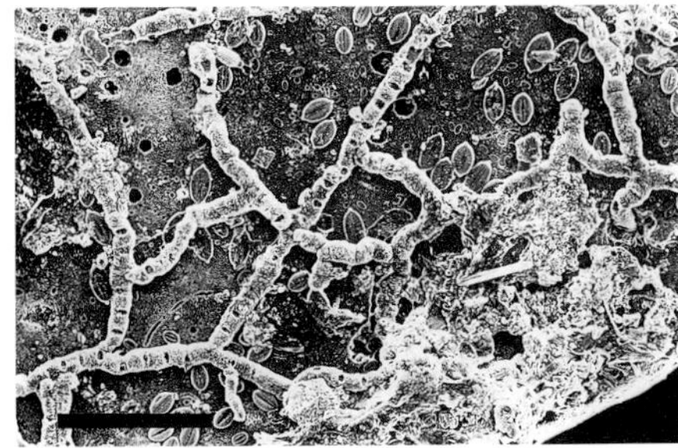

Fig. 77. *Cornuspiramia* sp. Cushman. Scale=100 µm.

FAMILY RIVEROINIDAE Saidova, 1981

Test planispiral, chambers one-half coil in length, partly subdivided by a few oblique secondary septa. Occur in tropical waters.

KEY TO THE GENERA

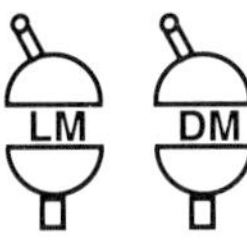

1. Planispiral, aperture curved slit...***Riveroina***
1'. Coiling quinqueloculine, becoming planispiral, aperture trematophore 2

2. Planispiral stage well developed ....
.... ***Pseudohauerina***
2'. Planispiral stage short . ***Pseudohauerinella***

Fig. 78. *Pseudohauerina occidentalis involi.* Scale=100 µm.

FAMILY OPHTHALMIDIIDAE Wiesner, 1920

Test free, proloculus followed by an undivided coiled second chamber (flexostyle) more or less developed, followed by chambers about one-half coil in length. Cosmopolitan.

KEY TO THE GENERA

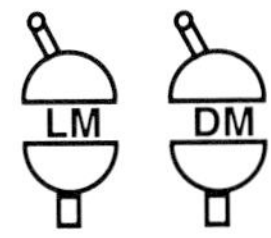

1. Chambers of adjacent whorls closely coiled, two chambers/whorl.................................. 2
1'. Chambers of adjacent whorls separated by a plate (former keel).................................. 3

2. Chambers elongated, flattened.................... .... ***Edentostomina***
2'. Chambers inflated.................... ***Ophthalmina***

3. Two chambers/whorl. ***Spirophthalmidium***
3'. Two and a half to three chambers per whorl .... ***Cornuloculina***

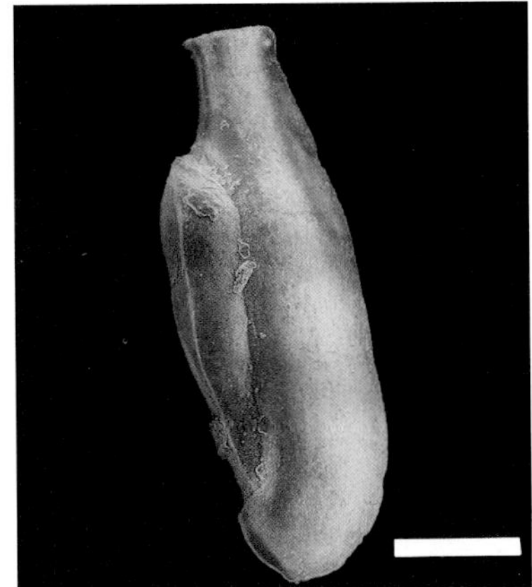

Fig. 79. *Edentostomina cultrata* (Brady). Scale=100 µm.

FAMILY SPIROLOCULINIDAE Wiesner, 1920

Test free, composed of a proloculus followed by reduced coiled second chamber, no more than one whorl in length, followed by two chambers per whorl; wall may be smooth, striate or costate. Cosmopolitan.

KEY TO THE GENERA

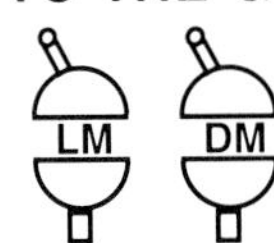

1. Last chambers uniserial.................................. .... ***Rectospiroloculina***
1'. All chambers enrolled. 2

2. Last chamber in **quinqueloculine** arrangement 3
2'. Last chambers arranged in 2-3 planes 4

3. Aperture with simple or **bifid tooth**....... .... ***Adelosina***
3'. Aperture cribrate in last stage .... ***Cribrolinoides***

4. Last chambers in triloculine arrangement .... ***Planispirinoides***
4'. Last chambers arranged in one or two planes. 5

5. Test dissymmetrical, chambers in two planes, 140 to 170° apart....................... ***Inaequalina***
5'. Last chambers in one plane......................... 6

6. Test dissymmetrical......................... ***Pippinia***
6'. Test symmetrical... 7

7. Test globular, only two last chambers visible .... ***Nummulopyrgo***
7'. Several chambers visible............................. 8

8. Peripheral keel in the middle........... ***Flintia***
8'. Flattened sides, may be keeled laterally...... 9

9. Aperturecribrate...... ***Cribrospiroloculina***
9'. Aperture simple... 10

10. Aperture without tooth ..***Neospiroloculina***
10'. With simple or bifid tooth.. ***Spiroloculina***

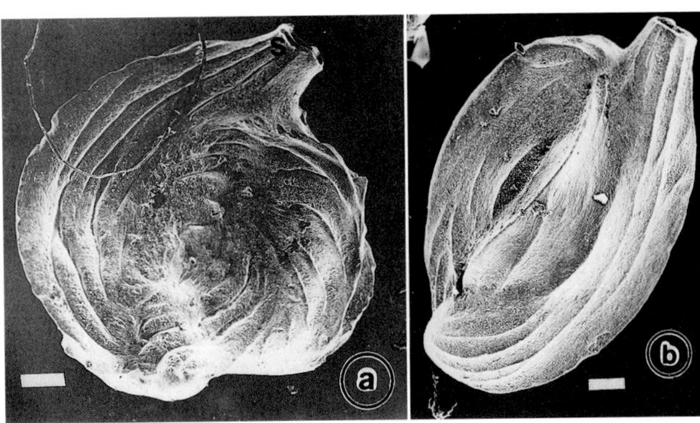

Fig. 80. *Adelosina* sp. d'Orbigny, 1826. a. "First stage." Scale=100 µm. b. Adult. Scale=100 µm.

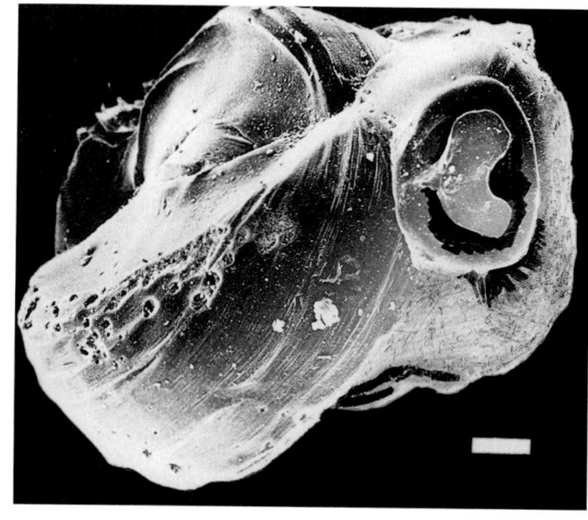

Fig. 81. *Cribrolinoides curtus* (Cushman). Scale=100 µm.

Fig. 82. *Spiroloculina excavata* d'Orbigny. Scale=100 µm.

FAMILY HAUERINIDAE Schwager, 1876

Test free; proloculus followed by two or less commonly more than two chambers per whorl; chambers added in one to five or more planes of coiling. The adult test may be uncoiled and rectilinear. Cosmopolitan.

KEY TO THE GENERA

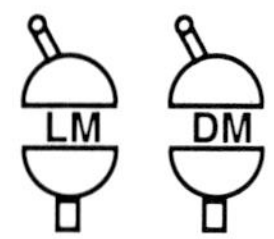

1. Test agglutinated... 2
1'. Test without agglutinated material 8

2. Quinqueloculine throughout 3
2'. Non-quinqueloculine, at least last stages.... 4

3. Aperture with simple tooth.. ***Siphonaperta***
3'. Aperture with bifid tooth and crenulate margin .... ***Dentostomina***

4. Chambers arranged in more than five planes throughout...................... ***Schlumbergerina***
4'. Chambers arranged in less than three planes 5

5. Last chambers arranged in two planes separated by increasing angle.. ***Sigmoilopsis***
5'. Early chambers quinqueloculine arranged.... 6

6. Later triloculine arrangement. ***Agglutinella***
6'. Later chambers in one plane 7

7. Two chambers/whorl, aperture cribrate...... .... ***Ammomassilina***
7'. Three chambers/whorl, aperture simple .... ***Pseudoflintina***
8. Last chambers uncoiled................................ 9
8'. All chambers coiled 13

9. First stage made of two ovoid chambers, later uniserial, cylindrical................... ***Tubinella***
9'. First stage of more than two chambers...... 10

10. First stage sigmoid. . ***Vertebrasigmoilina***
10'. First stage quinqueloculine...................... 11

11. Last stage irregular...................... ***Parrina***
11'. Last stage regular, uniserial................... 12

12. Quinqueloculine, later uniserial............... .... ***Articulina***
12'. Quinqueloculine, later planispiral, and finally uniserial .... ***Rectomassilina***

13. Only the last two chambers visible.......... 14
13'. More than two chambers visible............. 24

14. Aperture multiple.................................... 15
14'. Aperture simple....................................... 16

15. Numerous irregular pores.... ***Cribropyrgo***
15'. Six apertures separated by radiating encurved lamellae .... ***Nevillina***

16. Aperture without tooth or flap 17
16'. Aperture with tooth or flap 18

17. Aperture terminal................. ***Pseudopyrgo***
17'. Aperture on neck.............. ***Mesosigmoilina***

18. Aperture with flap.................................... 19
18'. Aperture with tooth.................................. 21

19. **Biloculine**........................... ***Biloculinella***
19'. Sigmoid arrangement in first stage........... 20

20. **Sigmoid** throughout................ ***Sigmoinella***
20'. Sigmoid, later planispiral .... ***Nummoloculina***

21. Tooth simple... 22
21'. Tooth short, bifid...................................... 23

22. Tooth small................................. ***Sigmoilina***
22'. Tooth large, triangular................ ***Pyrgoella***

23. Quinqueloculine, later biloculine....... ***Pyrgo***
23'. Sigmoid, later biloculine.......... ***Sigmopyrgo***

24. Last chambers in one plane....................... 25
24'. Last chambers in several planes................ 47

25. Chambers at least partially evolute........... 26
25'. Chambers involute, only three 45

26. Two chambers/whorl................................. 27
26'. More than two chambers/whorl................ 32

27. First stage sigmoid.................................... 28
27'. First stage milioline................................. 30

28. Surface of last chambers with transverse undulations .... ***Anchihauerina***
28'. Smooth or with longitudinal costae........... 29

29. Three or four chambers visible.................. .... ***Sigmoilinella***
29'. Chambers straight, numerous..................... .... ***Sigmoilinita***

30. Periphery truncated............. ***Massilinoides***
30'. Periphery rounded or keeled..................... 31

31. Periphery broadly rounded .... ***Proemassilina***
31'. Periphery keeled.......................... ***Massilina***

32. Aperture cribrate...................................... 33
32'. Aperture simple.. 37

33. First stage sigmoid.................................... 34
33'. First stage milioline.................................. 35

34. Three chambers/whorl..... ***Sigmoihauerina***
34'. More than three chambers/whorl, sutural pores .... ***Polysegmentina***

35. Aperture sievelike trematophore................ .... ***Hauerina***
35'. Aperture terminal, cribrate, secondary sutural apertures 36

36. Surface crenulate.................. ***Parahauerina***
36'. Surface smooth............. ***Parahauerinoides***

37. Aperture without tooth or flap.................... 38
37'. Aperture with tooth or flap........................ 40

38. Surface striate, broad aperture..... ***Sissonia***
38'. Surface non-striate.................................. 39

39. Four chambers/whorl, aperture a constricted slit .... ***Lorettaoides***
39'. Two and a half chambers/whorl, aperture a high slit with bordering lip .... ***Pseudomassilina***

40. Aperture with flap.................................... 41
40'. Aperture with bifid tooth........................... 42

41. Test symetrical......................... ***Miliolinella***
41'. Flattened at one side ...... ***Wellmanellinella***

42. Aperture a high arch.................... ***Steigerina***
42'. Aperture rounded...................................... 43

43. Test triangular in outline, last stage planispiral loosely coiled................ ***Ptychomiliola***
43'. Test rounded, chambers closely coiled....... 44

44. Test ribbed, small aperture........................ .... ***Mesopateoris***
44'. Test smooth, large aperture..... ***Neopateoris***

45. Aperture terminal, a sieve-like trematophore .... ***Involvohauerina***

45'. Aperture simple.. 46

46. Aperture strait arch, with lip, without tooth .... ***Flintinoides***
46'. Aperture large, ovate with bifid tooth becoming complex in adult................ ***Flintina***

47. Chambers in two planes about 180° apart.. 48
47'. Chambers in more than two planes........... 50

48. Lateral extensions of the chambers on one side of the test.......................... ***Subedentostomina***
48'. No lateral extensions................................. 49

49. Aperture rounded, tooth simple or bifid .... ***Pseudoschlumbergerina***
49'. Aperture ovate, tooth small .... ***Sigmella***

50. Test quinqueloculine, five chambers visible 51
50'. Test triloculine, three chambers visible.. 53

51. Aperture elongated, long tooth with bifid end .... ***Lachlanella***
51'. Aperture ovate or rounded........................ 52

52. Chambers without floor............ ***Cycloforina***
52'. Chambers with floor........ ***Quinqueloculina***

53. Last stage non triloculine.......................... 54
53'. Last stage triloculine................................ 55

54. Last stage planispiral, tooth bifid .... ***Pseudotriloculina***
54'. 3-5 chambers visible, tooth of varying shape... ***Varidentella***

55. Aperture complex 56
55'. Aperture simple, with tooth or flap.......... 57

56. Aperture irregularly triradiate .... ***Cribromiliolinella***
56'. Aperture cruciform slit....... ***Cruciloculina***

57. Flap or spatulate tooth.............................. 58
57'. Bifid or ringlike tooth.............................. 59

58. Test rounded, aperture flap.......................... .... ***Triloculinella***
58'. Test triangular, spatulate tooth .... ***Triloculinellus***
59. Tooth ringlike.................... ***Triloculinoides***
59'. Tooth bifid... 60

60. Aperture narrow, long tooth with an inflated end.. ***Affinetrina***
60'. Aperture rounded, short, bifid tooth .... ***Triloculina***

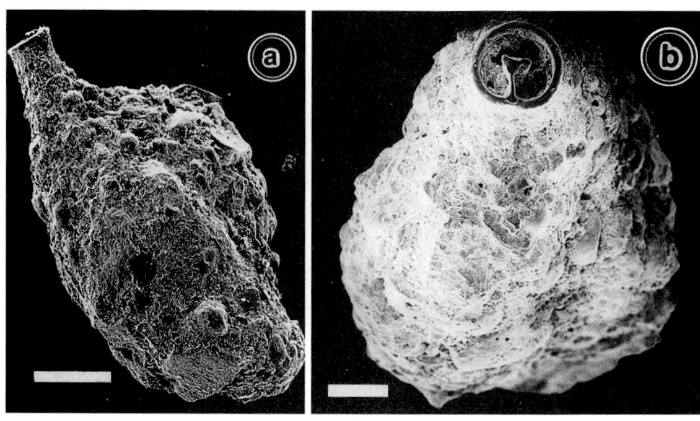

Fig. 83. *Siphonaperta anina excavata*. a. Spiral view. b. Apertural view. Scales=100 μm.

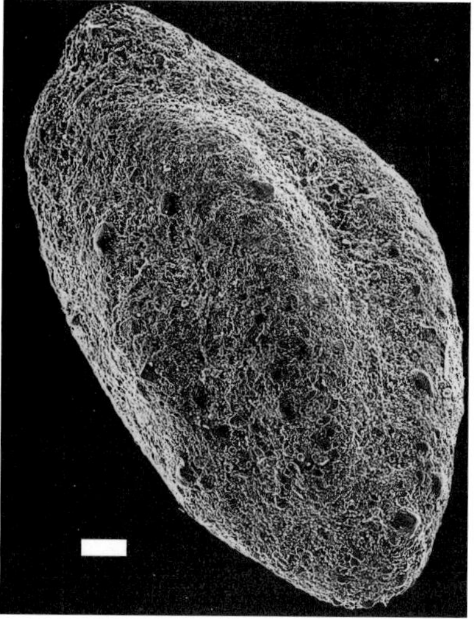

Fig. 84. *Schlumbergerina alveoliniformis* Brady. Scale=100 μm.

Fig. 85. *Ammomassilina alveoliniformis* (Millett). Scale=100 μm.

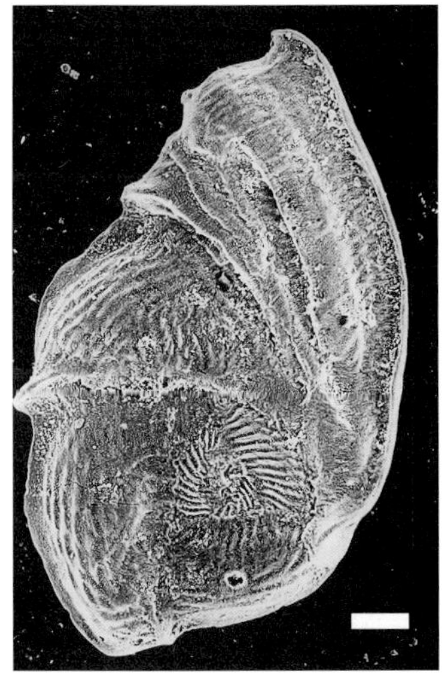

Fig. 86. *Vertebrasigmoilina* sp. Hofker, 1976. Scale=100 μm.

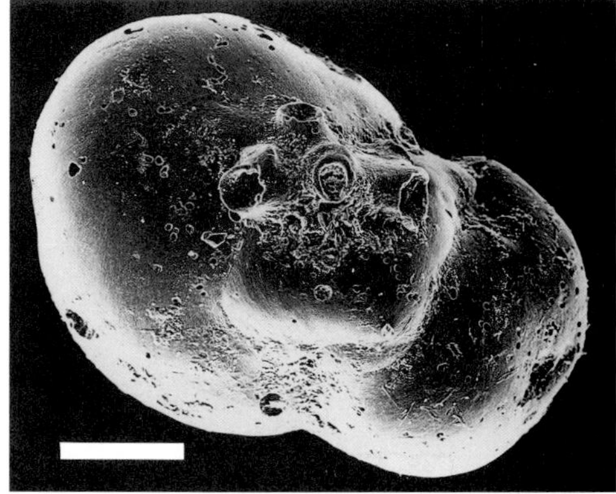

Fig. 87. *Parrina bradyi* (Millett). Scale=100 μm.

Fig. 88. *Pyrgo* sp. DeFrance, 1824. Scale=100 μm.

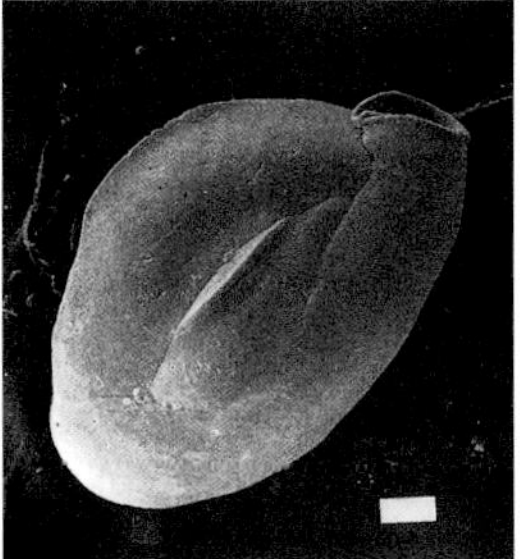

Fig. 89. *Massilina secans* (d'Orbigny). Scale=200 μm.

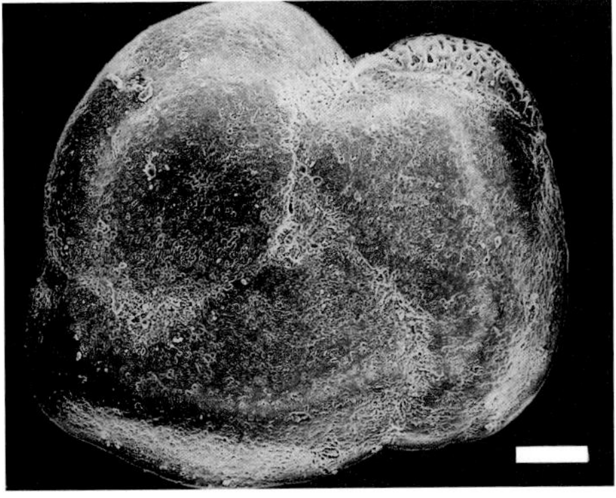

Fig. 90. *Hauerina diversa*. Scale=100 μm.

Fig. 91. *Pseudomassilina* sp. Lacroix, 1938. Scale=100 μm.

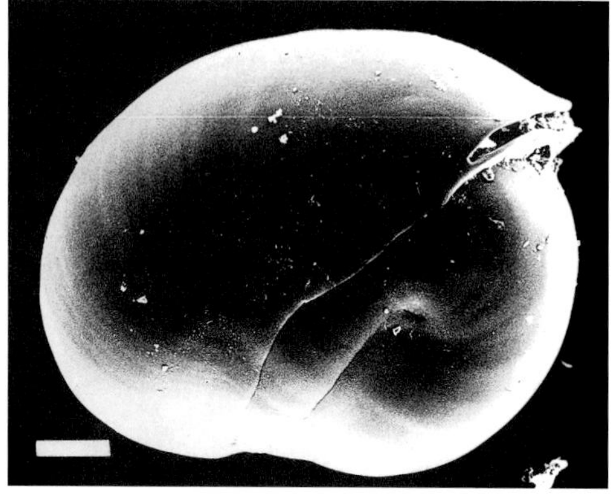

Fig. 92. *Miliolinella australis.* Scale=100 μm.

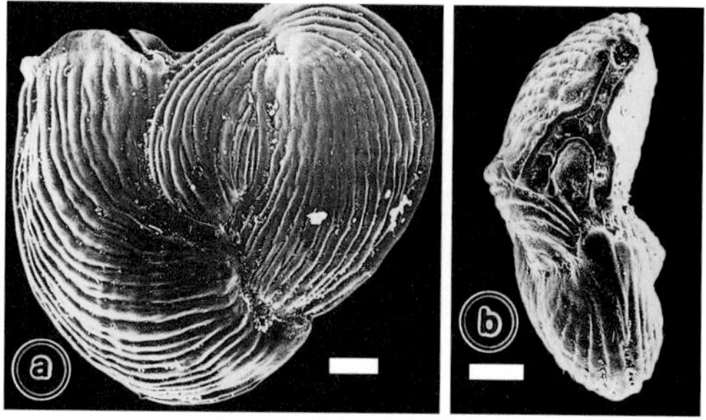

Fig. 93. *Wellmanellinella* sp. Cerif, 1970. a. Spiral view. b. Apertural view. Scales=100 μm.

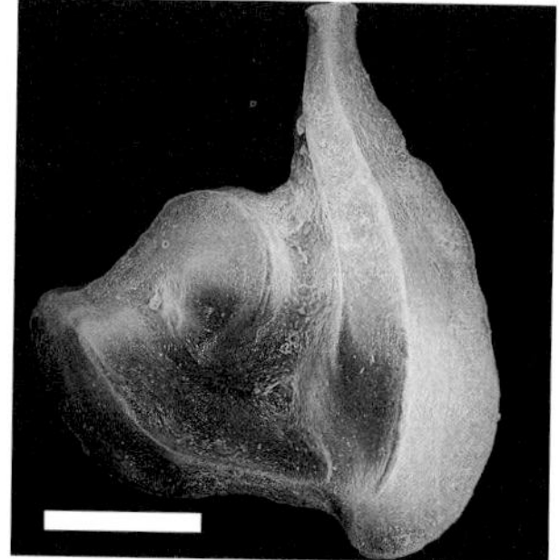

Fig. 94. *Ptychomiliola* sp. Eimer & Fickert, 1899. Scale=100 μm.

Fig. 95. *Flintina* sp. Cushman, 1921. Scale=100 μm.

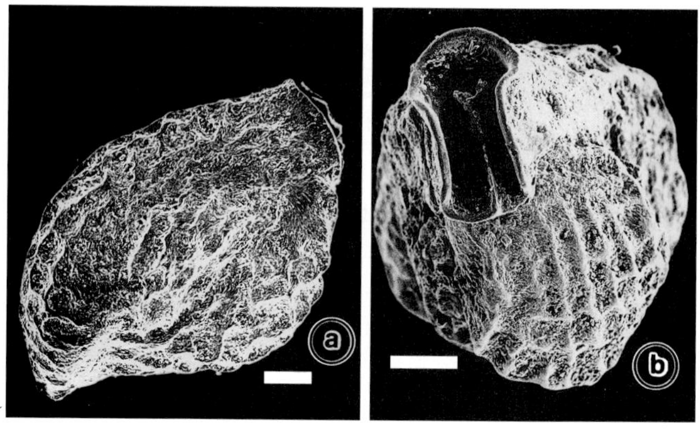

Fig. 96. *Lachlanella* sp. Vella, 1957. a. Spiral view. b. Apertural view. Scales=100 μm.

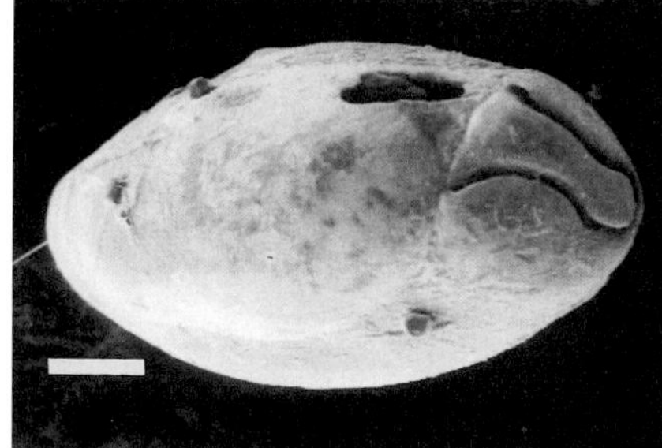

Fig. 97. *Affinetrina* sp. Luczkowska, 1972. Scale=50 μm.

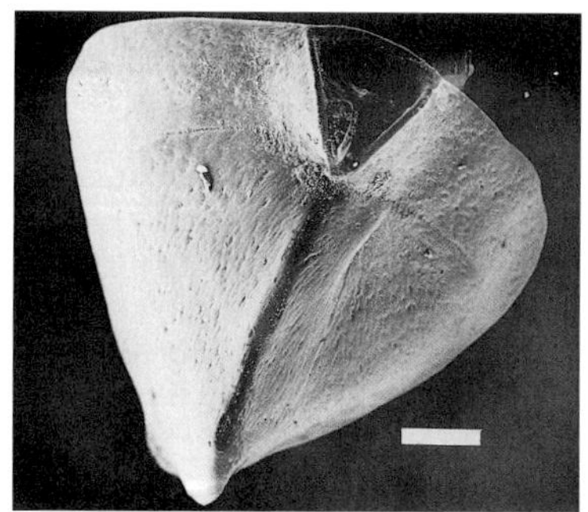

Fig. 98. *Triloculina* sp. d'Orbigny, 1826. Scale=100 μm.

FAMILY ALVEOLINIDAE Ehrenberg, 1839

Test free, large, subcylindrical to cigar-shaped, coiled about elongate axis; numerous adult planispiral chambers are divided by secondary partitions into one or more layers of chamberlets that parallel the direction of coiling. Stenohaline, common in coral reef.

Fig. 99. *Alveolinella quoyi* (d'Orbigny). Scale=200 μm.

KEY TO THE GENERA

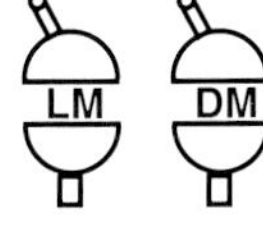

1. Apertures in one row........................ ***Borelis***
1'. Apertures in at least two rows....................... 2

2. Two rows of apertures............. ***Flosculinella***
2'. Several rows of apertures......... ***Alveolinella***

FAMILY PENEROPLIDAE Schultze, 1854

Test planispiral in the early stage, later may be uncoiled, flabellulate, or cyclical; chambers with simple interior, not subdivided. Cosmopolitan, common in coral reefs.

KEY TO THE GENERA

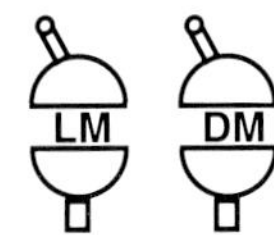

1. Test entirely planispiral, aperture dendritic .... ***Dendritina***
1'. Test uncoiled in later stage............................ 2

2. Uncoiled portion cylindrical........................... 3
2'. Uncoiled portion flabelliform......................... 5

3. Surface smooth, punctate aperture simple.... .... ***Monalysidium***
3'. Longitudinally ribbed or striate.................... 4

4. Aperture simple............................. ***Spirolina***
4'. Aperture cribrate.................. ***Coscinospira***

5. Surface smooth.................... ***Laevipeneroplis***
5'. Longitudinal striae or grooves..... ***Peneroplis***

FAMILY DISCOSPIRINIDAE
Wiesner, 1931

Test discoid, proloculus followed by a second chamber of several volutions, later chambers becoming annular, incompletely subdivided into chamberlets. Single genus *Discospirina* (Fig. 100) occurs in the Mediterranean Sea and Atlantic.

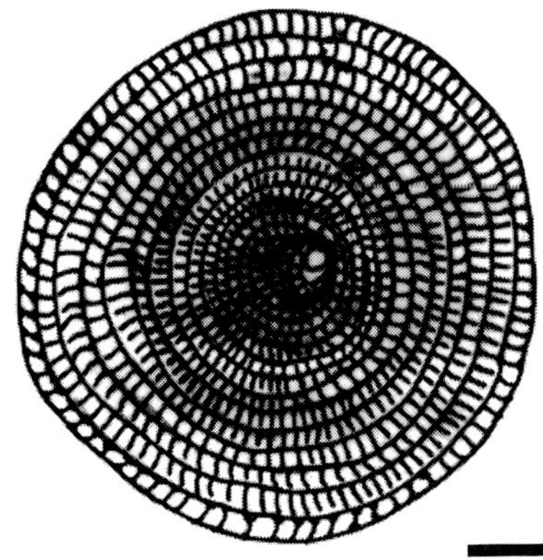

Fig. 100. *Discospirina italica* (Costa). Scale=120 μm.

FAMILY SORITIDAE Ehrenberg, 1839

Test free, early stage planispiral, later may be uncoiled, flabelliform, fusiform, or cylindrical; chambers subdivided by **interseptal pillars** or septula. Common in shallow tropical waters.

KEY TO THE GENERA

1. Test planispiral, last stage may be discoid..... 2

1'. Test discoidal, early stage may have a few chambers planispirally arranged.................... 6

2. Last chambers not annular............................. 3
2'. Last chambers completely annular................. 4

3. Chambers partly divided by flattened pillars and septula.................................... ***Androsina***
3'. Chambers partly divided by cylindrical pillars, no septula............................ ***Archaias***

4. Discontinuous vertical partition of the chambers projecting from the lateral faces ... 5
4'. Vertical partitions fusing in the central part of the chambers........................... ***Parasorites***

5. Short alternating partitions . ***Cycloputeolina***
5'. Opposite vertical partitions, pillars in the median part of the chamber.. ***Cyclorbiculina***

6. Aperture numerous pores scattered on the peripheral margin .... ***Marginopora***
6'. Aperture rows of openings............................. 7

7. One row of openings............................ ***Sorites***
7'. Two rows of openings...................***Amphisorus***

Fig. 101. *Marginopora*. Scale=1 mm.

ORDER LAGENIDA
Delage & Herouard, 1896

Test wall of optically and ultrastructurally radiate calcite, with crystal c-axes perpendicular to surface; primarily and secondarily lamellar.

KEY TO THE FAMILIES

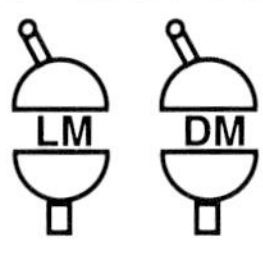

1. Test unilocular................................ **Lagenidae**
1'. Test multilocular.. 2

2. Entosolenian tube present....... **Glandulinidae**
2'. Entosolenian tube absent................................. 3

3. Test uniserial or biserial......... **Nodosariidae**
3'. Test encoiled.. 5

4. Planispiral at least in early stage.................. .... **Vaginulinidae**
4'. Spirally or sigmoidally coiled with overlapping chambers.............. **Polymorphinidae**

FAMILY LAGENIDAE Reuss,1862

Test unilocular, wall with one or two layers, aperture terminal, circular, ovate, slit-like, or radiate, may be produced on a neck, entosolenian tube present or absent. Taxonomic classification following Patterson and Richardson (1987).

KEY TO REPRESENTATIVE GENERA

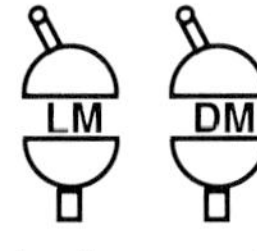

1. Wall with a single layer of calcite.................. 2
1'. Double wall connected by network of pillars..... .... ***Sipholagena***

2. Entosolenian tube absent................................ 3
2'. Entosolenian tube present.............................. 6

3. Test globular to ovate..................................... 4
3'. Test elongate, fusiform.................................. 5

4. Surface smooth, aperture multiple .... ***Cribrolagena***
4'. Surface with longitudinal striae or costae, aperture on a long or short neck........... ***Lagena***

5. Surface with costae................ ***Procerolagena***
5'. Surface smooth..................... ***Hyalinonetrion***

6. Test unilocular non-carinate........................ 7

6'. Test unilocular carinate................................ 11

7. Aperture without neck.................................... 8
7'. Aperture on neck.. 10

8. Aperture circular to radiate................ ***Oolina***
8'. Aperture slit-like.. 9

9. Aperture in the center of a fissurelike cavity at test apex .... ***Fissurina***

9'. Aperture within a hoodlike extension of the dorsal wall .... ***Parafissurina***

10. Surface hispid............. ***Pristinosceptrella***
10'. Row of punctae within longitudinal costae .... ***Cushmanina***

11. Carina with radiating tubules and struts....... .... ***Solenina***
11'. Carina without complex structures............. 7

12. Carina broad, thin, aperture subterminal, at the junction of test and carina..... .... ***Wiesnerina***
12'. Carina widening toward aperture enclosing upper portion of the neck .... ***Palliolatella***

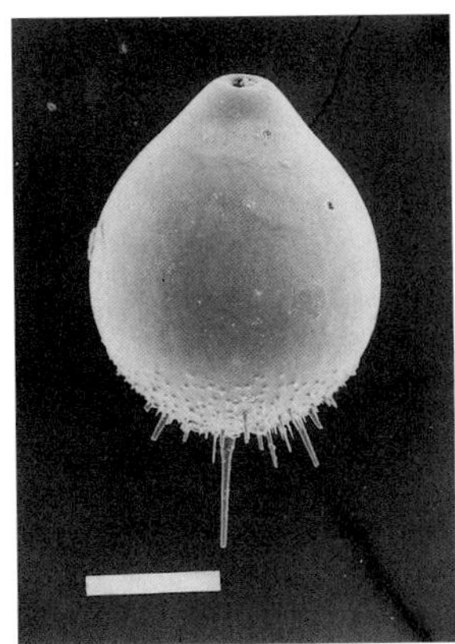

Fig. 102. *Oolina* sp. d'Orbigny, 1839. Scale 100 μm.

FAMILY GLANDULINIDAE Reuss, 1860

Test uniserial, biserial or **polymorphine**, with strongly overlapping chambers; aperture terminal, radiate or slit-like with entosolenian tube. Twenty genera.

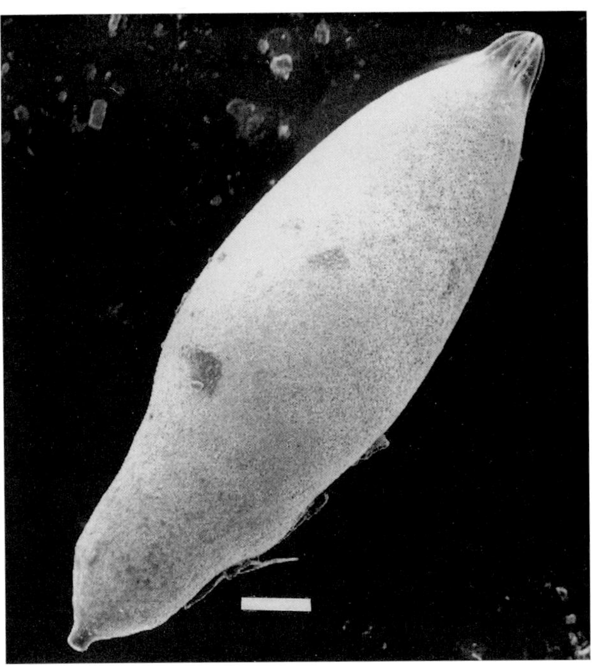

Fig. 103. *Glandulina* sp. d'Orbigny, 1839. Scale=50 μm.

KEY TO REPRESENTATIVE GENERA

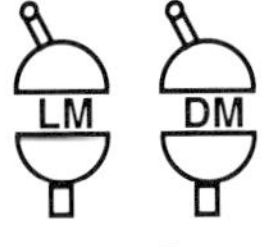

1. Aperture terminal, radiate............................. 2
1'. Aperture terminal, oval to slit-like.............. 4

2. Test uniserial, slightly arcuate..... ***Phlegeria***
2'. Test biserial at least in early stage................ 3

3. Test biserial to uniserial............. ***Glandulina***
3'. Test biserial and sigmoid........ ***Laryngosigma***

4. Test uniserial, rectilinear...... ***Entolingulina***
4'. Test compressed, each chamber completely enveloping earlier ones................. ***Seabrookia***

FAMILY NODOSARIIDAE Ehrenberg, 1838

Test elongate, multilocular, chambers arranged in curved or straight series, uniserial, or biserial in early stage; aperture terminal, located at top of the last chamber, commonly radiate, may be rounded, slit-like or multiple. Twenty-four modern genera, cosmopolitan.

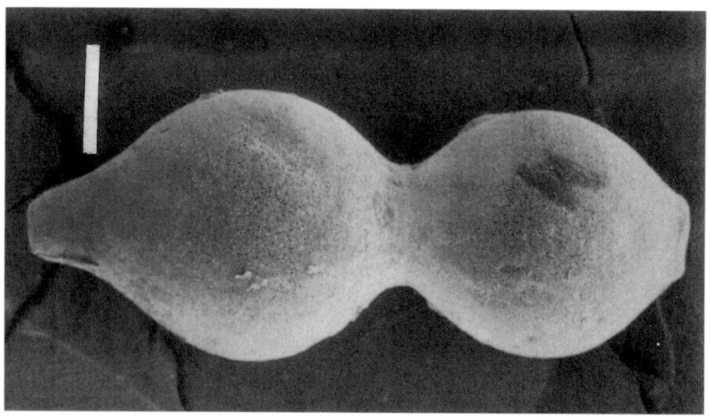

Fig. 104. *Nodosaria radicula* (Linne). Scale=200 μm

KEY TO REPRESENTATIVE GENERA

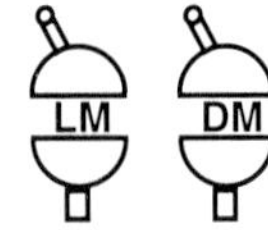

1. Uniserial.. 2
1'. Biserial in early stage, later uniserial.......... 6

2. Aperture rounded to radiate........................... 3
2'. Aperture an elongated, terminal slit, strongly overlapping chambers .... ***Lingulina***

3. Chambers globular to ovate........................... 4
3'. Chambers broad, test flattened to palmate, sutures highly arched............. ***Frondicularia***

4. Test rectilinear.. 5
4'. Test arcuate, longitudinal costae.... ***Dentalina***

5. Aperture radiate.......... (Fig. 104) ***Nodosaria***
5'. Aperture rounded, on the neck, test with distinct longitudinal costae...... ***Pyramidulina***

6. Test rounded in section, aperture cribrate.. .... ***Amphimorphina***
6'. Test strongly compressed, broadly palmate, aperture rounded .... ***Proxifrons***

FAMILY VAGINULINIDAE Reuss, 1860

Test lenticular to flattened, ovate to palmate; early stage planispiral, later uncoiled, or in slightly arcuate arrangement but lacking a distinct coil, or in nearly straight series with oblique sutures; aperture terminal, commonly radiate at the dorsal angle. Twenty Recent genera, cosmopolitan.

KEY TO REPRESENTATIVE GENERA

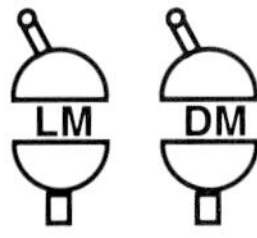

1. Test distinctly coiled 2
1'. Test lacking a distinct coil.............................. 3

2. Test lenticular, biumbonate, planispirally enrolled throughout..................... ***Lenticulina***
2'. Test planispiral only in the early stage, tending to become rectilinear .... ***Saracenaria***

3. Early chambers slightly curved 4
3'. Test uniserial, rectilinear.......... ***Vaginulina***

4. Surface smooth, aperture radiate, at the dorsal angle.. ***Astacolus***
4'. Surface longitudinally striate, aperture radiate at the end of a long neck........... .... ***Amphicoryna***

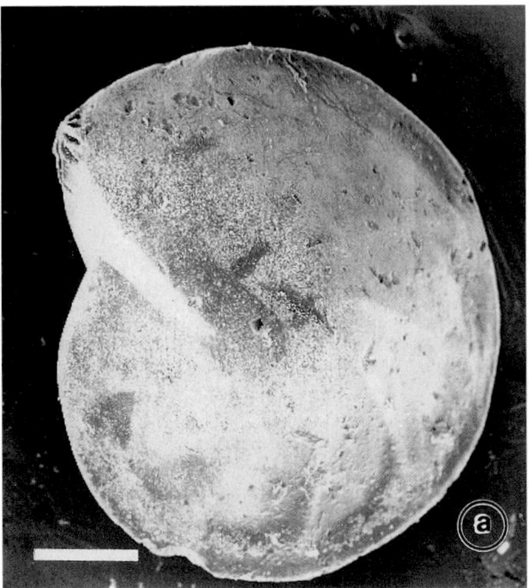

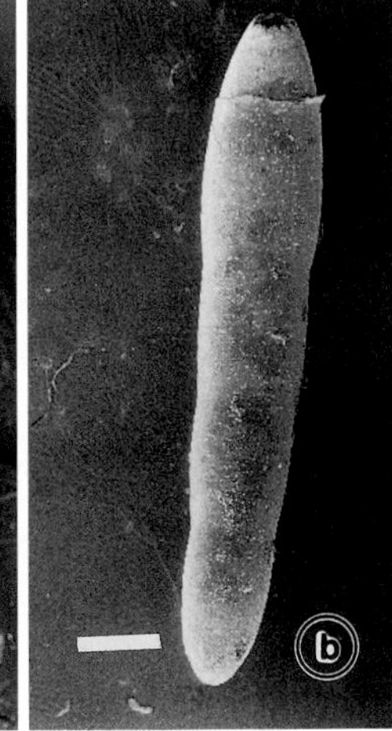

Fig. 105. a. *Lenticulina rotulata* (Lamarck). b. *Vaginulina legumen* (Linne). Scales=200 μm.

FAMILY POLYMORPHINIDAE d'Orbigny, 1839

Test free or attached, ovate to elongate, multilocular, chambers biserial or spirally arranged in three or more planes, strongly overlapping toward the early part of the test or irregular, aperture terminal, rounded, slit-like or radiate. eighteen Recent genera.

KEY TO REPRESENTATIVE GENERA

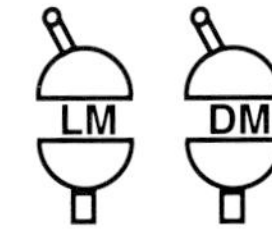

1. Chambers strongly overlapping........................ 2
1'. Chambers irregular, in branching series....... 7

2. Test free.. 3
2'. Test attached, early stage with chambers in polymorphine arrangemen .... ***Webbinella***

3. Aperture rounded to radiate............................. 4
3'. Aperture slit-like..... ***Fissuripolymorphina***

4. Chambers added in three to five planes 5
4'. Biserial, slightly twisted...... ***Polymorphina***

5. Chambers added in three planes, later biserial .... ***Pyrulina***
5'. Five planes at least in early stage.................. 6

6. Chambers strongly overlapping, added in five, later in three planes....................... ***Globulina***
6'. Chambers in five planes throughout................ .... ***Guttulina***

7. Test free, chambers connected by stolon-like tubes.. ***Ramulina***
7'. Test attached, with branches budded from a hemispheric proloculus...... ***Discoramulina***

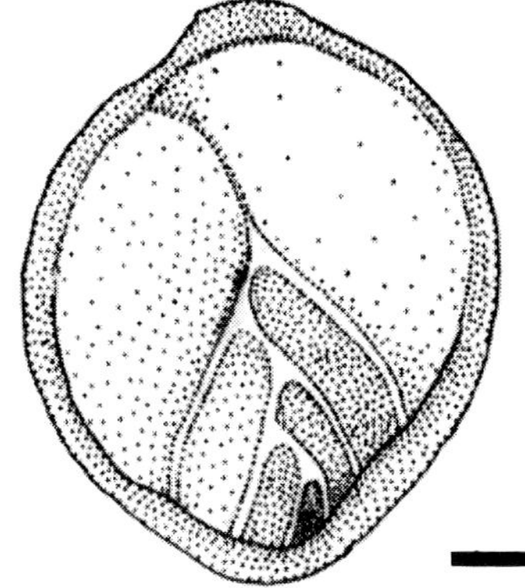

Fig. 106. *Webbinella crassa* (Saidova). Scale=75 μm.

ORDER ROTALIIDA
DELAGE & HEROUARD, 1896

Test wall lamellar of perforate, hyaline calcite; chambers multiple, simple or subdivided by secondary partitions, may have internal canal system.

KEY TO SUPERFAMILIES AND FAMILIES

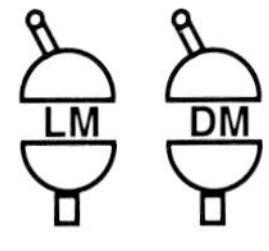

1. Test elongated, usually triserial to uniserial, internal toothplate common............................. 2
1'. Test discoidal, trochospiral to planispiral, some with internal structures.................................. 9

2. Internal toothplate absent ..... **Loxostomatacea**
2'. Internal toothplate present 3

3. Wall optically radial.. 4
3'. Wall optically granular 6

4. Uniserial throughout............. **Stilostomellacea**
4'. Biserial or triserial.. 5

5. Biserial throughout....................... **Bolivinacea**
5'. Triserial to biserial..................... **Buliminacea**

a. Aperture loop-shaped.................................... b
a'. Aperture terminal, commonly rounded........ d

b. Test elongate, fusiform...... **Stainforthiidae**
b'. Test elongate, ovate...................................... c

c. Test triserial........................... **Buliminidae**
c'. Numerous chambers per whorl..................... .... **Buliminellidae**

d. Test elongate, triserial to uniserial............. e
d'. Test pyramidal or flabelliform.................... g

e.Test biserial to uniserial............................... f
e'. Test triserial in early stage, aperture on neck, longitudinal costae..... **Uvigerinidae**

f. Chambers broad......... **Siphogenerinoididae**
f'. Chambers subcylindrical......... **Millettiidae**

g. Test pyramidal.. h
g'. Test flabelliform......................... **Pavonidae**

h. Aperture slit-like or cribrate...................
.... **Reussellidae**
h'. Aperture a broad opening...... **Trimosinidae**

6. Uniserial.......................... **Pleurostomellacea**
6'. Biserial, twisted or enrolled............................ 7

7. Biserial, in planispiral coil.... **Cassidulinacea**
7'. Biserial, fusiform.. 8

8. Wall with interlocular spaces and sutural openings............................. **Virgulinellacea**
8'. No complex wall structures.......................
.... **Fursenkoinacea**

9. Wall without internal structures................. 10
9'. Wall with internal structures...................... 18

10. Wall optically radial................................... 11
10'. Wall optically granular.............................. 16

11. Test free.. 12
11'. Test free in early stage, later attached........ 15

12. Aperture terminal, rounded, projected on a neck .... **Siphoninacea**
12'. Aperture an interiomarginal slit............... 13

13. Aperture umbilical.................................... 14
13'. Aperture equatorial or extraumbilical........15

14. Umbilical side with radial striae ....
.... **Glabratellacea**
i. Test a low trochospiral coil..................... j
i'. Test a high trochospiral coil.....................
.... **Buliminoididae**

j. Aperture may be covered by flap............. k
j'. Aperture closed by a later formed plate.................................. **Heronalleniidae**

k. Test minute, less than 100 µm ....
.... **Rotaliellidae**
k'. Test larger, commonly with rugose or reticulose surface............. **Glabratellidae**

14'. Umbilical side flattened with distinctive flap
.... **Rosalinacea**
l. Spiral side evolute, umbilical involute
.... m
l'. Test bi-involute or bi-evolute................ p

m. Supplementary sutural apertures ....
.... **Heleninidae**
m'. No sutural apertures.............................. n

n. Test distinctly perforate.......... **Rosalinidae**
n'. Test with imperforate or thickened areas.. o

o. Imperforate area near aperture..................
.... **Bagginidae**
o'. Thickened areas at the periphery on both sides .... **Mississippinidae**

p. Test bi-involute..................**Bueningiidae**
p'. Test bi-evolute.......... **Bronnimanniidae**

15. Chambers without internal partitions...........
.... **Discorbinellacea**

q. Test trochospiral, plano- to biconvex...... r
q'. Test nearly planispiral............................. s

r. Aperture a low slit bordered by a narrow lip....................................... **Parrelloididae**
r'. Aperture a vertical slit extending up the apertural face............... **Pseudoparrellidae**

s. Test biconcave, partly evolute on both sides................................ **Planulinoididae**
s'. Test planoconvex......... **Discorbinellidae**

15'. Chambers with internal partitions
.... **Asterigerinacea**
t. Internal toothplate forming supplementary chambers... u
t'. Supplementary chamberlets produced by umbilical coverplates... **Asterigerinatidae**

u. Secondary openings present..................... v
u'. No secondary openings.............................. w

v. Supplementary sutural and areal openings
.... **Epistomariidae**
v'.Multiple sutural openings and

bridges.................................. .**Alfredinidae**

w. Stellate pattern regular............................
.... **Asterigerinidae**
w'. Stellate pattern irregular.........................
.... **Amphisteginidae**

16. Chambers trochospirally coiled, becoming irregular...................... **Planorbulinacea**
A. Test free... B
A'. Test attached by spiral face...................... C

B. Test discoidal, partially evolute on both sides .... **Planulinidae**
B'. Test planispiral, bi-involute.................
.... **Bisacciidae**

C. Test trochospiral throughout....................
.... **Cibicididae**
C'. Test trochospiral in early stage, later chambers irregular.................................. D

D. Test irregularly discoid or conical........... E
D'. Chambers growing upward in spiral or irregular mass...................... **Victoriellidae**

E. Two apertures on the periphery at opposite ends of each chamber..................
.... **Planorbulinidae**
E'. Aperture multiple, small circular pores, gamont with float chamber ....
.... **Cymbaloporidae**

16'. Chambers irregular throughout, trocho-spiral only in early stage...................
.... **Acervulinacea**

F. Chambers encrusting, in one or more layers................................ **Acervulinidae**
F'. Chambers growing up in branching structure.......................... **Homotrematidae**

17. Test conical, uniserial................................
.... **Annulopatellinacea**
17'. Test planispiral or trochospiral............... 18

18.Test planispiral......................... **Nonionacea**

G. Secondary sutural or peripheral openings
.... H
G'. No secondary openings.......... **Nonionidae**

H. Sutural openings between the septal bridges.......................... **Spirotectinidae***
H'. Secondary slit-like aperture at peripheral margin of the chamber .... **Almaenidae**

18'. Test trochospiral with enveloping chambers
.... **Chilostomellacea**

I.Test free.. J
I'.Test attached....................... **Karreriidae***

J. Test trochospiral, chambers enveloping..... K
J'. Test lenticular, chambers not enveloping
.... L

K. Chambers strongly enveloping, without internal partition........... **Chilostomellidae**
K'. Chambers not strongly enveloping, with internal partition....... **Quadrimorphinidae**

L. Surface smooth... M
L'.Surface pustulose or granulose on the umbilical side..................... **Trichohyalidae**

M. Aperture interiomarginal.......................... N
M'. Aperture areal............. **Osangulariidae**

N. Aperture umbilical obscured by an infolded area of the apertural face..........
.... **Alabaminidae**
N'. Aperture equatorial................................ O

O. Aperture extending on the umbilical and spiral side........................ **Heterolepidae**
O'. Aperture extending on the umbilical side.
.... P

P. Aperture covered by a distinctive umbilical flap .... **Gavelinellidae**
P'. Small sutural openings on both sides ....
.... **Oridorsalidae**

19. Trochospiral to planispiral........... **Rotaliacea**

R. Test trochospiral throughout.................... S
R'. Test planispiral.................... **Elphidiidae**

S. Test with little or no differentiation of

spiral and umbilical surfaces, commonly with radial spines .... **Calcarinidae**
S'. Chambers well differentiated.................. T

T. Umbilical side with sinusoid lobes overhanging the internal coverplates and umbilical spiral canal.................. **Rotaliidae**
T'. Umbilical side with few or any sinusoid lobes.. U

V. Aperture with distinctive flap or lip.......... .... **Discorbidae**
V'. Aperture without flap............. **Eponididae**

19'. Test large, planispiral......... **Nummulitacea**

- Family not illustrated in this volume.

SUPERFAMILY LOXOSTOMATACEA
Loeblich & Tappan, 1962

FAMILY BOLIVINELLIDAE Hayward, 1980

Test biserial, chambers broad and palmate, aperture cribrate, apertural face with radial striae, no internal toothplate; reproduction gamontogamic. Single genus *Bolivinella* (Fig. 107), tropical to subtropical shallow waters.

Fig. 107. *Bolivinella folia* (Parker & Jones). a. Spiral view. b. Apertural view. Scale=50 µm.

FAMILY TORTOPLECTELLIDAE
Loeblich & Tappan, 1985

Test biserial, flattened, aperture an areal slit bordered by a lip, internal toothplate absent. Single genus *Tortoplectella* (108), tropical shallow waters.

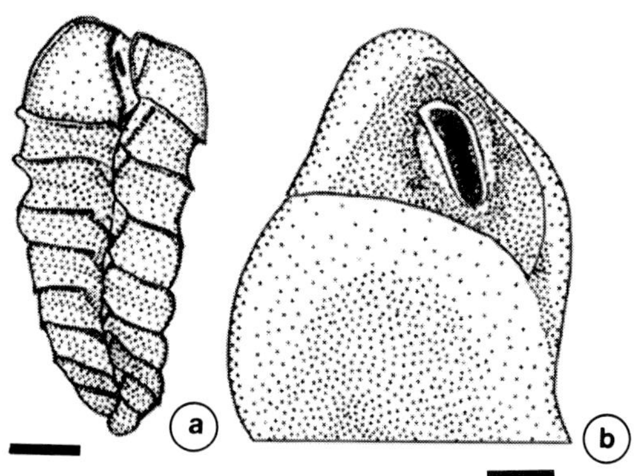

Fig. 108. *Tortoplectella crispata* (Brady). Scale=45 µm.

FAMILY TOSAIDAE Saidova, 1981

Test triserial, later become biserial; aperture a short slit on apertural face; internal toothplate absent; wall optically granular. Single genus *Tosaia* (Fig. 109), subtropical shallow waters.

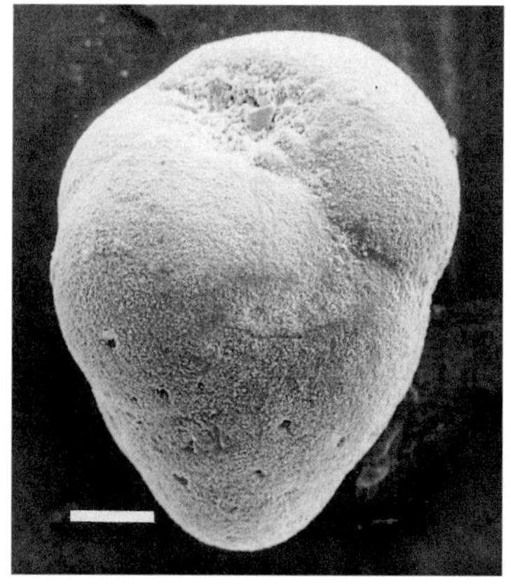

Fig. 109. *Tosaia hanzawai* Takayanagi. Scale=20 µm.

SUPERFAMILY STILOSTOMELLACEA Finlay, 1947

FAMILY STILOSTOMELLIDAE Finlay, 1947

Test uniserial, rectilinear or arcuate; aperture terminal, with **phialine lip** and with a distinct tooth projecting into the aperture.

KEY TO THE GENERA

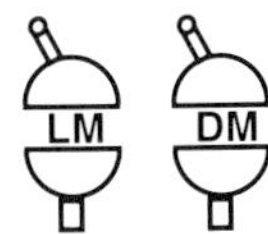

1. Test short and robust............. ***Stilostomella***
1'. Test narrow and elongate.............................. 2

2. Chambers subglobular........ ***Siphonodosaria***
2'. Chambers widest near the base, with a row of small spines at the edge .... ***Nodogenerina***

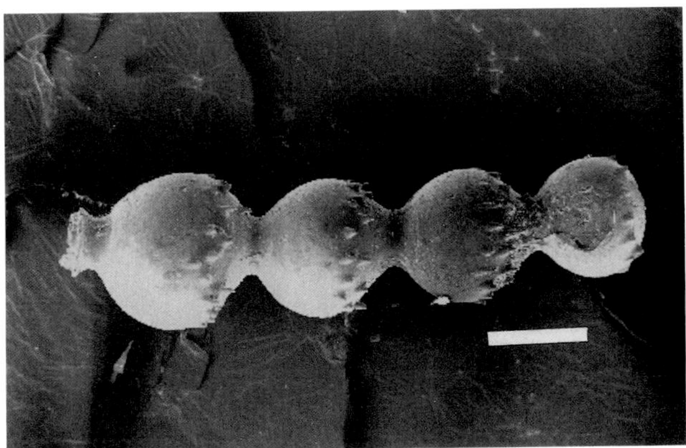

Fig. 110. *Nodogenerina adolphina* d'Orbigny. Scale=200 µm.

SUPERFAMILY BOLIVINACEA Glaessner, 1937

FAMILY BOLIVINIDAE Glaessner, 1937

Test biserial to uniserial, aperture a high loop-shaped opening with an internal toothplate. Five genera, cosmopolitan.

KEY TO REPRESENTATIVE GENERA

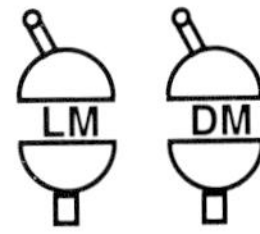

1. Free part of toothplate broad and short, test ovoid in outline............................... ***Bolivina***
1'. Free part of toothplate narrow and elongate 2

2. Test oval in section, wall coarsely perforate near sutures........................... ***Bolivinellina***

3. Periphery acute to carinate............ ***Brizalina***

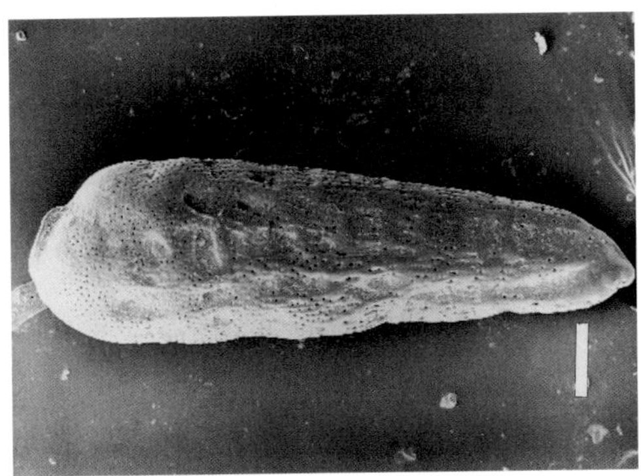

Fig. 111. *Bolivina plicata* d'Orbigny. Scale=100 µm.

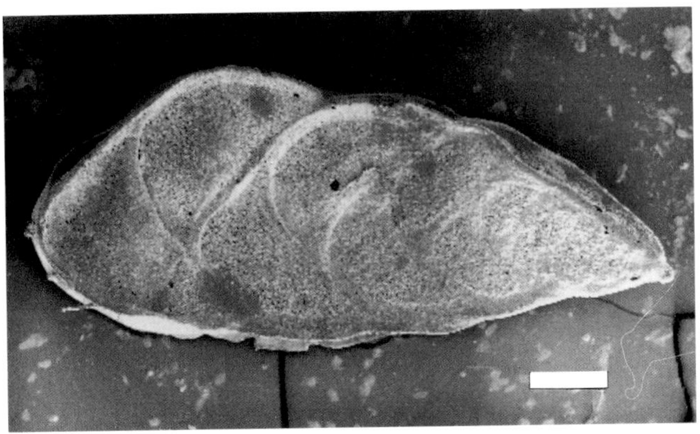

Fig. 112. *Bolivinita quadrilatera* (Schwager). Scale=200 µm.

FAMILY BOLIVINITIDAE Cushman, 1927

Test biserial, sides flattened and edges truncate, periphery carinate; aperture a high opening in the face of the final chamber.

KEY TO THE GENERA

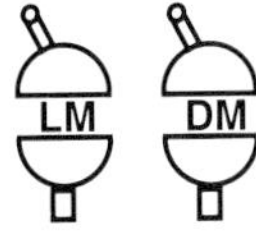

1. Surface highly ornamented..... ***Abditodentrix***
1'. Surface smooth, slightly nodose.... ***Bolivinita***

SUPERFAMILY FURSENKOINACEA Loeblich & Tappan, 1961

FAMILY FURSENKOINIDAE Loeblich & Tappan, 1961

Test elongate, biserial, may be twisted, aperture loop shaped, extending up the face of the final chamber, internal toothplate, wall optically granular. Cosmopolitan.

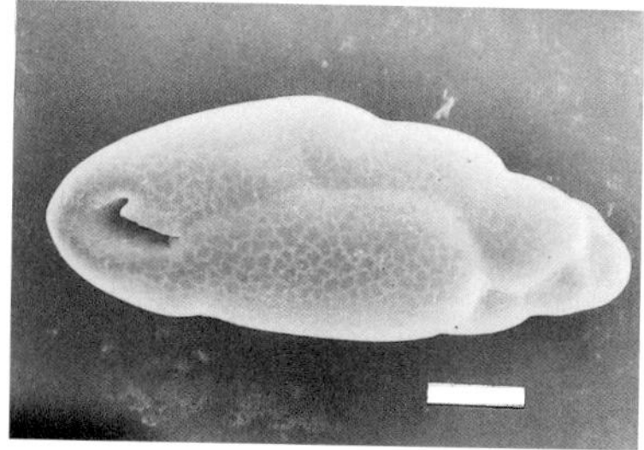

Fig. 113. *Fursenkoina persiformis.* Scale=20 µm.

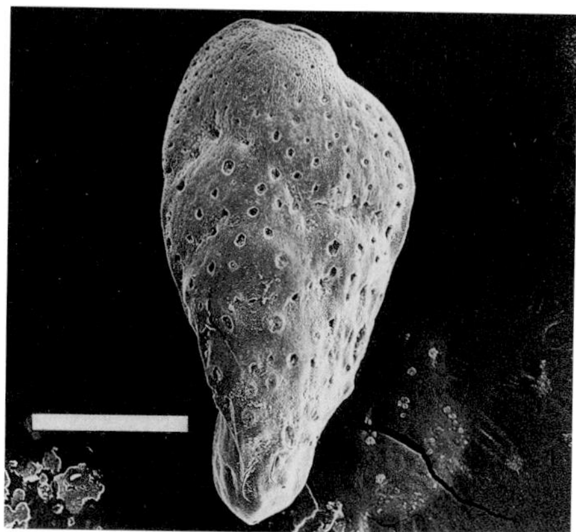

Fig. 114. *Sigmavirgulina tortuosa* (Brady). Scale=100 µm.

KEY TO THE GENERA

1. Test rounded in section.................................... 2
1'. Test flattened, palmate.................................... 4

2. Early chambers biserial, later becoming uniserial .... ***Coryphostoma***
2'. Chambers biserial throughout....................... 3

3. Chambers high and narrow........ ***Fursenkoina***
3'. Chambers strongly oblique and overlapping .... ***Rutherfordoides***

4. Early chambers in sigmoid alignment............ .... ***Sigmavirgulina***
4'. All chambers in a single plane........ ***Stuartia***

SUPERFAMILY VIRGULINELLIDAE
Loeblich & Tappan, 1984

FAMILY VIRGULINELLIDAE
Loeblich & Tappan, 1984

Test elongate, triserial in early stage, later biserial, aperture loop-shaped with internal toothplate, wall optically granular. Single genus *Virgulinella* (Fig. 115). (see Revets, 1991

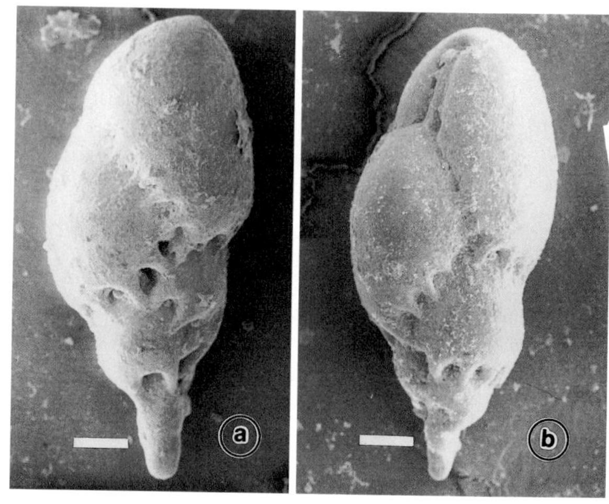

Fig. 115. *Virgulinella pertusa* (Reuss). a. Spiral view. b. Apertural view. Scales=100 µm.

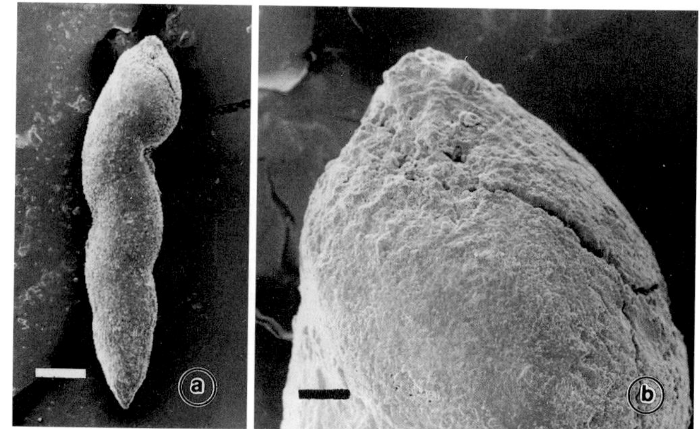

Fig. 116. a. *Pleurostomella* sp. Reuss, 1860. Scale=100 µm. b. Close-up of aperture. Scale=20 µm.

SUPERFAMILY PLEUROSTOMELLACEA
Reuss, 1860

FAMILY PLEUROSTOMELLIDAE Reuss, 1860

Test biserial in early stage, later reduced to uniserial, chambers cuneate, aperture a subterminal to terminal slit covered by a projecting hood, with internal siphon connecting apertures of adjacent chambers; wall optically granular.

KEY TO THE GENERA

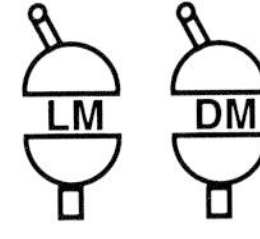

1. Chambers overlapping................................ 2
1'. Chambers rectilinear.................................. 3

2. Chambers completely overlapping, early stage biserial .... ***Ellipsobulimina***
2'. Chambers strongly overlapping, early stage uniserial .... ***Ellipsoglandulina***

3. Uniserial throughout................ ***Nodosarella***
3'. Early stage biserial.......... ***Pleurostomella***

SUPERFAMILY CASSIDULINACEA
d'Orbigny, 1839

FAMILY CASSIDULINIDAE d'Orbigny, 1839

Test with biserially arranged chambers enrolled in planispiral coil, at least in early stage, later may be uncoiled; aperture an interiomarginal slit, may have internal toothplate; wall optically radial or granular. Sixteen Recent genera.

.... KEY TO REPRESENTATIVE GENERA

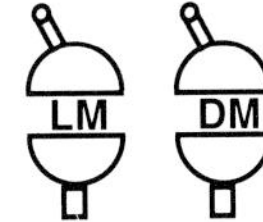

1. Chambers enrolled in planispiral coil............ 2
1'.Chambers tending to uncoil, compressed perpendicular to the plane of coiling............ 5

2.Wall optically radial.................... ***Islandiella***
2'. Wall optically granular............................... 3

3. Test globular..................... ***Globocassidulina***
3'. Test lenticular to flattened........................... 4
4. Surface smooth............................ ***Cassidulina***
4'.Surface reticulated .... ***Favocassidulina***

5. Test globular in section................ ***Burseolina***
5'. Test lenticular in section........ ***Ehrenbergina***

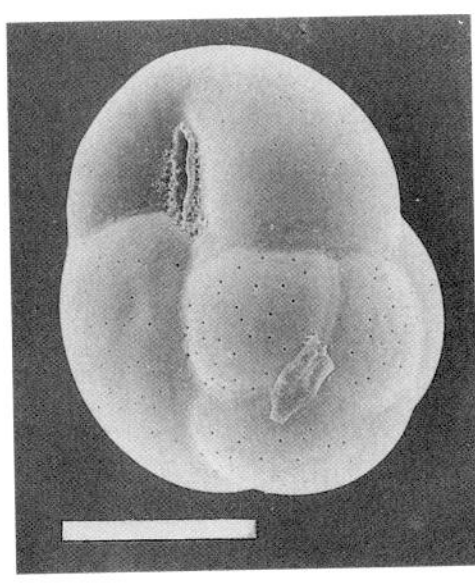

Fig. 117. *Globocassidulina subglobosa* (Brady). Scale=100 µm.

SUPERFAMILY BULIMINACEA Jones, 1875

Test highly trochospiral, triserial to uniserial in later stage; aperture a loop-shaped opening with internal toothplate; wall optically radial.

FAMILY STAINFORTHIIDAE Reiss, 1963

Test elongated, triserial to biserial; aperture a high loop in apertural face with internal toothplate; wall optically radial.

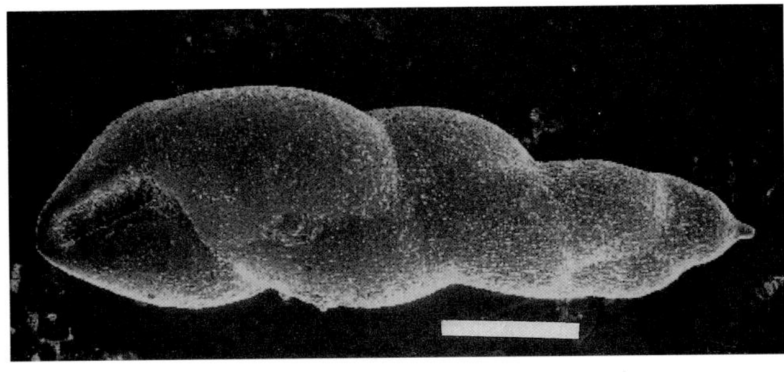

Fig. 118. *Stainforthia concava* (Hoeglund). Scale=100 µm.

KEY TO THE GENERA

1. Aperture broadly oval with a narrow lip, apical

spine 2
1'. Aperture elongate, no apical spine...................
.... ***Virgulinopsis, Virgulopsis***

2. Test twisted, biserial.................. ***Cassidelina***
2'. Early stage triserial, later twisted biserial
.... ***Stainforthia***

FAMILY BULIMINIDAE Jones, 1875

Test elongate, chambers triserially arranged, may be overlapping; aperture loop-shaped, with distinct platelike toothplate that extends backward from the aperture to the previous foramen. Cosmopolitan.

KEY TO THE GENERA

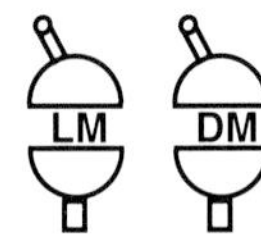

1. Chambers strongly overlapping.................. 2
1'. Chambers inflated, not overlapping..............
.... ***Bulimina***

2. Aperture loop-shaped...................................
.... (Fig. 119)***Globobulimina***
2'. Aperture a high narrow curved slit..............
.... ***Praeglobobulimina***

FAMILY BULIMINELLIDAE Hofker, 1951

Test highly trochospiral with numerous chambers per whorl. Aperture loop-shaped with internal toothplate. Single Recent genus *Buliminella* (Fig. 120).

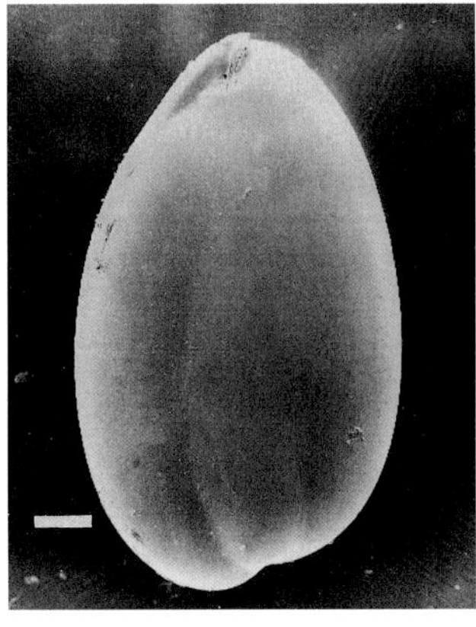

Fig. 119. *Globobulimina pacifica* Cushman. Scale=100 µm.

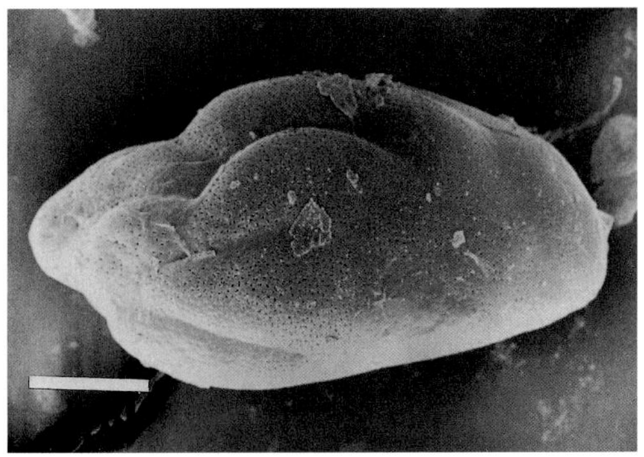

Fig. 120. *Buliminella elegantissima* (d'Orbigny). Scale=50 µm.

FAMILY UVIGERINIDAE Haeckel, 1894

Test elongate, triserial to biserial in early stage, later become uniserial; aperture terminal with a neck and phaline lip. Cosmopolitan, bathyal.

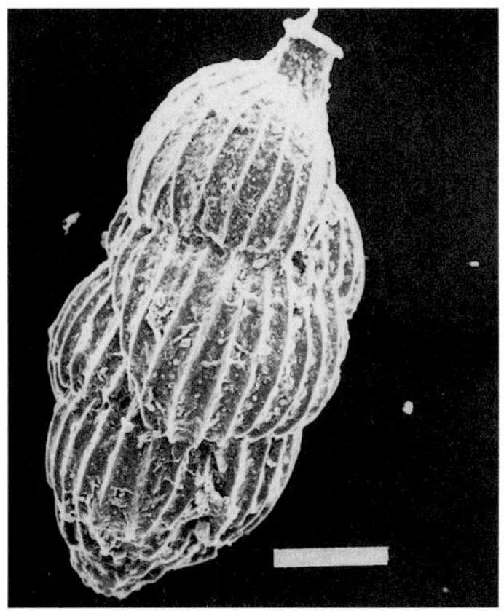

Fig. 121. *Uvigerina peregrina* Cushman. Scale=100 µm.

KEY TO THE GENERA

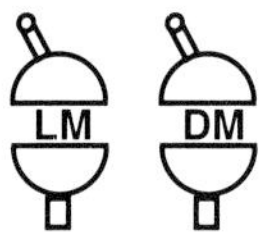

1. Test rounded in section................................ 2
1'. Test triangular in section, chambers with carinate angles.. 5

2. Triserial throughout.................................. 3
2'. Triserial to uniserial................................ 4

3. Surface with longitudinal costae ....
.... ***Uvigerina***
3'. Surface with numerous fine spines...............
.... ***Euuvigerina***

4. Triserial to biserial, chambers umbrella-like.................................. ***Siphouvigerina***
4'. Triserial to irregularly uniserial, surface finely hispid........................ ***Neouvigerina***

5. Triserial throughout............. ***Angulogerina***
5'. Early stage triserial, later distinctly uniserial and rectilinear............ ***Trifarina***

FAMILY SIPHOGENERINOIDIDAE Saidova, 1981

Test slightly compressed, biserial to uniserial; aperture terminal with bordering lip and internal siphon-like toothplate that generally changes in orientation by 180° (Siphogenerinoidinae) or by 120° (Tubulogenerininae) in successive chambers. Eleven Recent genera.

KEY TO REPRESENTATIVE GENERA

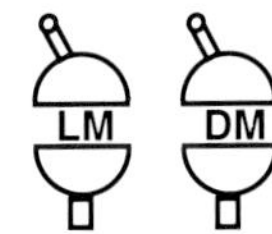

1. Test biserial in early stage, later may become uniserial, internal toothplate that changes orientation by 180°...................... 2
1'. Test triserial to uniserial, internal toothplate that changes orientation by 120° 6

2. Biserial throughout, aperture on a short neck.................................... ***Hopkinsinella***
2'. Biserial to uniserial.................................. 3

3. Chambers sharply angled......... ***Sagrinella***
3'. Chambers moderately inflated.................... 4

4. Aperture a broad circular opening with a lip .... ***Rectobolivina***
4'. Aperture oval or elliptical with a lip......... 5

5. Surface with longitudinal costae.............. .... ***Loxostomina***
5'. Surface smooth.................. ***Parabrizalina***

6. Early stage triserial, later biserial, chambers distinctly angular.......... ***Sagrina***
6'. Test triserial, later uniserial.................... 7

7. Aperture terminal, oval, with a lip, surface pustulose or spinose............... ***Sagrinopsis***
7'. Aperture terminal on a neck, surface with longitudinal costae........... ***Siphogenerina***

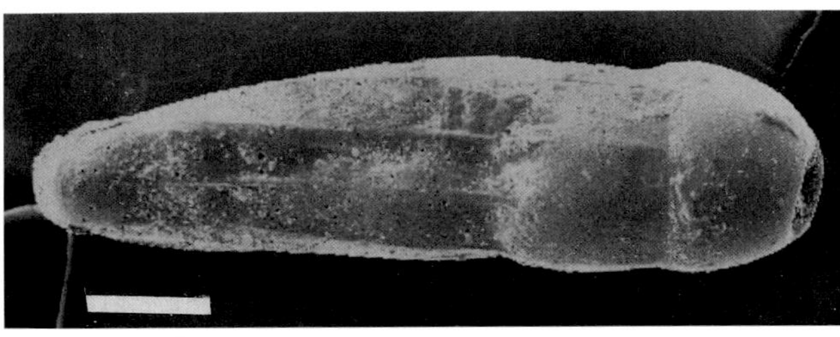

Fig. 122. *Rectobolivina bifrons* (Brady). Scale=100 µm.

FAMILY REUSSELLIDAE Cushman, 1933

Test elongate, pyramidal, triserial to uniserial; aperture terminal, slit-like or may be cribrate.

KEY TO THE GENERA

1. Test uniserial throughout................. ***Acostina***
1'. Test triserial at least in early stage............ 2

2. Triserial throughout..................................... 3
2'. Triserial, later biserial or uniserial.......... 4

3. Aperture a narrow slit and multiple openings at the apertural face.............................. ***Fijiella***
3'. Aperture a slit at the base of the final chamber .... ***Reussella***

4. Later chambers uniserial and rectilinear, aperture cribrate............. ***Chrysalidinella***
4'. Later chambers biserial................................ 5

5. Aperture terminal, ovate............................. .... ***Compressigerina***
5'. Aperture terminal, a narrow and elongate slit bordered by a rim .... ***Valvobifarina***

FAMILY TRIMOSINIDAE Saidova, 1981

Test elongate, pyramidal, triserial, later become biserial, chambers angled; aperture a broad opening, may be accompanied by a second areal opening. Two genera: *Trimosina* (Fig. 124), *Mimosina*, tropical shallow waters.

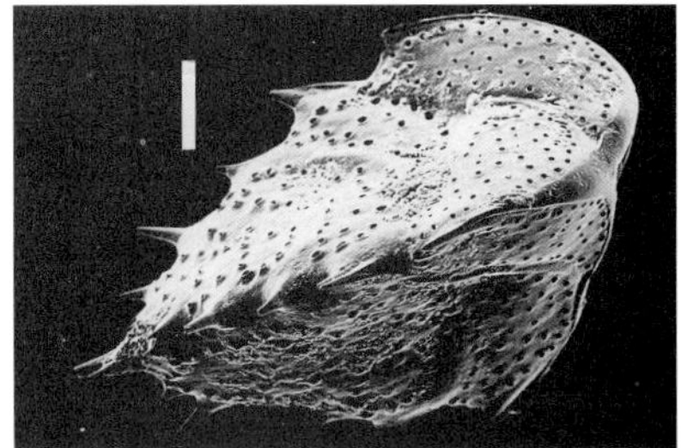

Fig. 123. *Reussella spinulosa* (Reuss). Scale=100 µm.

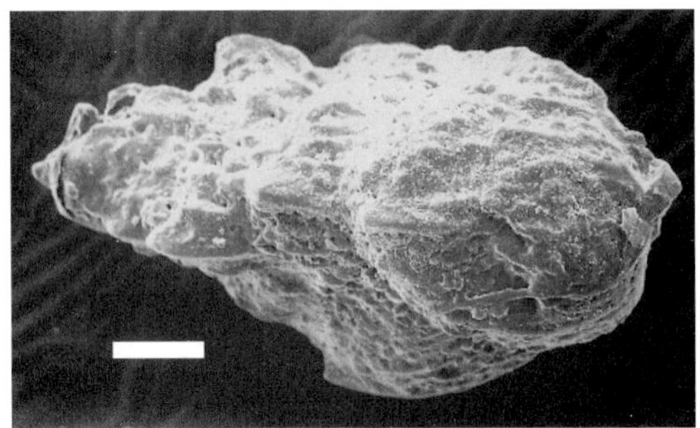

Fig. 124. *Trimosina spinulosa* Cushman. Scale=50 µm.

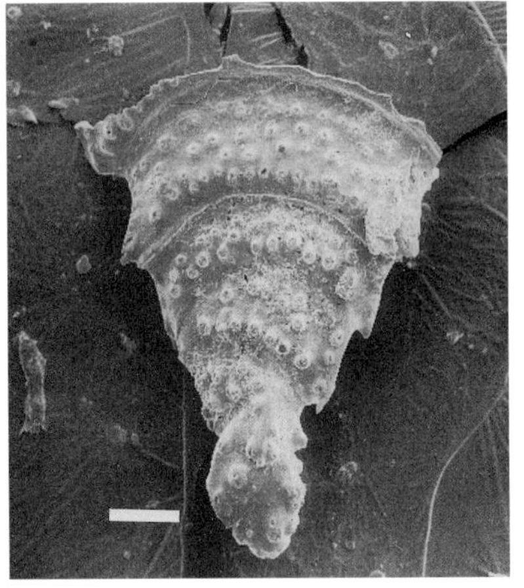

Fig. 125. *Bifarinella ryukyuensis* Cushman & Hanzawa. Scale=100 µm.

FAMILY PAVONINIDAE Eimer & Fickert, 1899

Test broadly palmate, triserial to biserial in early stage, later uniserial with semicircular, rapidly enlarging chambers; aperture a multiple coarse perforations on the apertural face. Two genera: *Bifarinella* (Fig. 125), *Pavonina* (Fig. 126), subtropical waters.

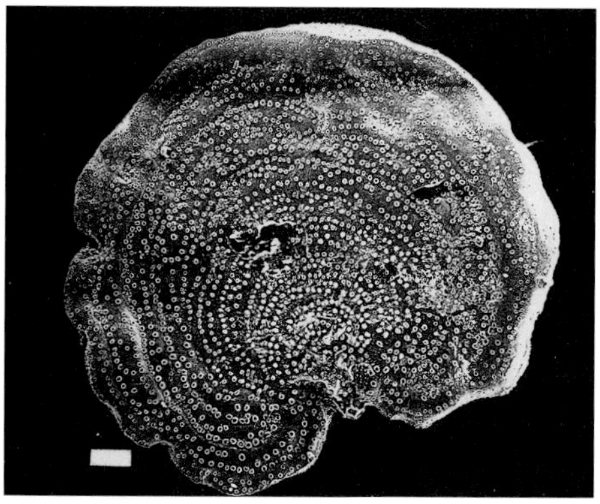

Fig. 126. *Pavonina flabelliformis* d'Orbigny. Scale=100 µm.

FAMILY MILLETTIIDAE Saidova, 1981

Test elongate, cylindrical, early chambers biserial, later uniserial; aperture terminal, surrounded by a collar and a narrow rim. Single genus *Millettia* (Fig. 127).

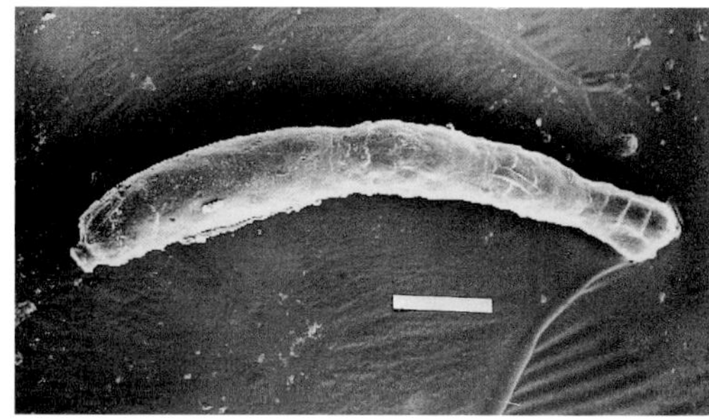

Fig. 127. *Millettia tessellata* (Brady). Scale=100 µm.

SUPERFAMILY SIPHONINACEA Cushman, 1927

FAMILY SIPHONINIDAE Cushman, 1927

Test lenticular, trochospiral in early stage, later may be uncoiled; aperture equatorial to terminal, produced on a distinct neck, with phialine lip.

KEY TO THE GENERA

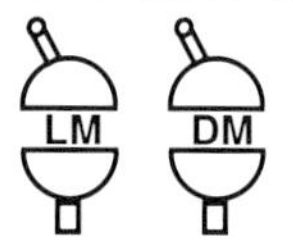

1. Aperture a small pore in a concave plate filling the apertural depression.. ***Siphoninoides***

1'. Aperture elliptical or slit-like.................. 2

2. Test lenticular, trochospiral...... ***Siphonina***
2'. Test lenticular in early stage, later stage uncoiling, uniserial.............. ***Siphoninella***

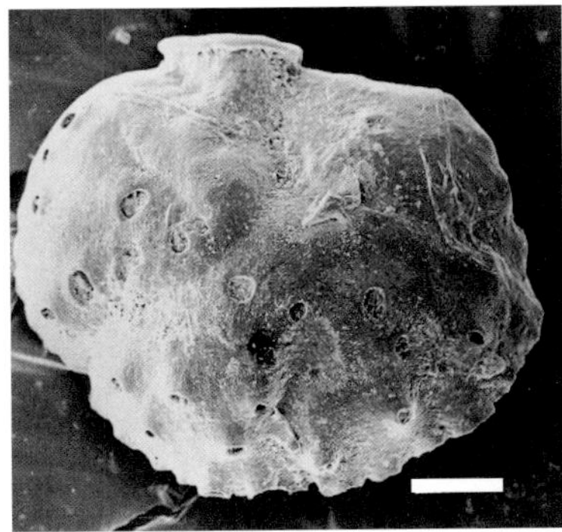

Fig. 128 *Siphonina tubulosa* Cushman. Scale=50 μm.

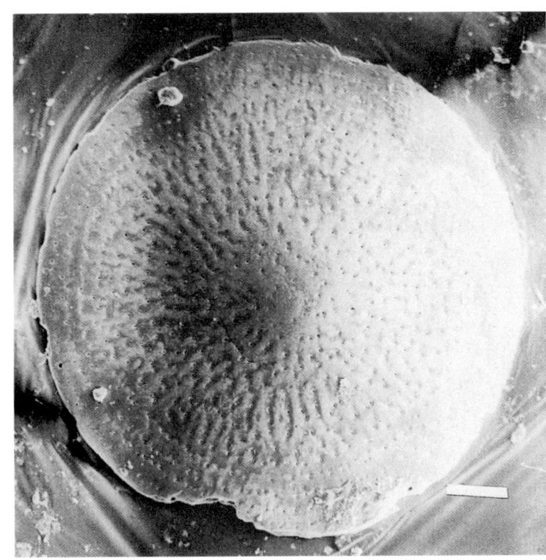

Fig. 129. *Annulopatellina annularis* (Parker & Jones). Scale=50μm.

SUPERFAMILY ANNULOPATELLINACEA
Loeblich & Tappan, 1964

FAMILY ANNULOPATELLINIDAE
Loeblich & Tappan, 1964

Test conical, chambers annular as viewed from the convex side, each chamber completely covering the concave side, chambers subdivided by radiating tubules; wall optically granular; reproduction gamontogamic. Single genus *Annulopatellina* (Fig. 129), tropical shallow waters.

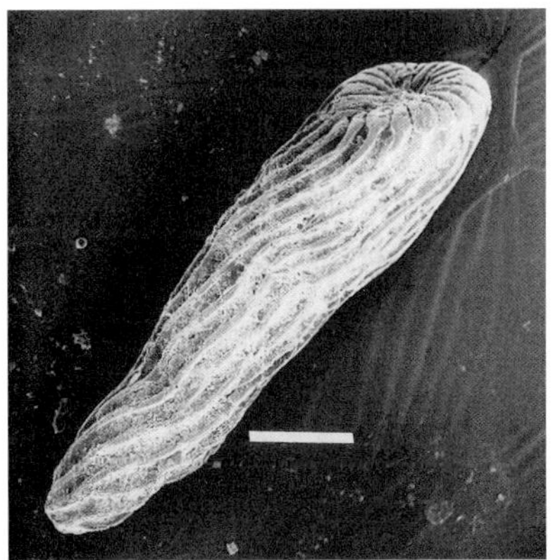

Fig. 130. *Buliminoides williamsoniana* (Brady). Scale 100μm.

SUPERFAMILY GLABRATELLACEA
Loeblich & Tappan, 1964

Test trochospiral, surface of umbilical side commonly with radial striae, costae, or nodes; aperture interiomarginal, umbilical; wall of hyaline, perforate, optically radial calcite.

FAMILY BULIMINOIDIDAE Seiglie, 1970

Test elongate, with high trochospiral coil, sutures may be obscured by prominent longitudinal costae; apertural face with radial striae; reproduction gamontogamic. Two genera: *Buliminoides* (Fig. 130), *Fredsmithia*, tropical shallow waters.

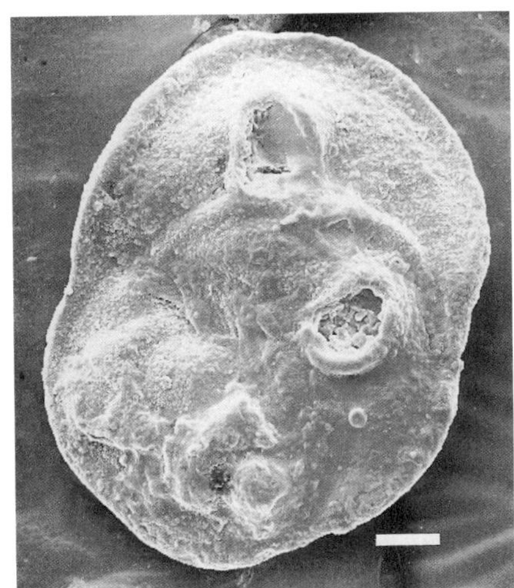

Fig. 131. *Heronallenia lingulata* . Scale 20μm.

FAMILY HERONALLENIIDAE
Loeblich & Tappan, 1986

Test trochospiral, planoconvex, with one or two whorls of rapidly enlarging chambers, umbilical side radially grooved; aperture secondarily closed by a later formed plate. Single genus *Heronallenia* (Fig. 131).

FAMILY ROTALIELLIDAE
Loeblich & Tappan, 1964

Test tiny, trochospiral, with few subglobular to arcuate chambers, umbilical side with few radial grooves and often with small denticles; aperture covered by umbilical flap; reproduction gamontogamic or autogamic; nuclear dimorphism observed (Grell, 1954; Pawlowski and Lee, 1992).

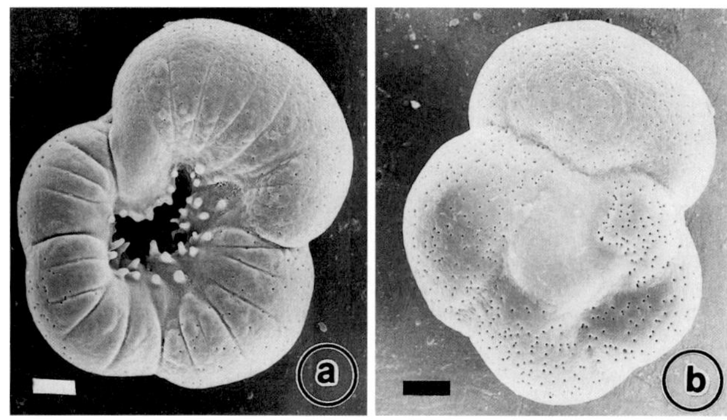

Fig. 132. *Rotaliella elatiana*. a. Apertural view. b. Spiral view. Scales=10 μm.

KEY TO THE GENERA

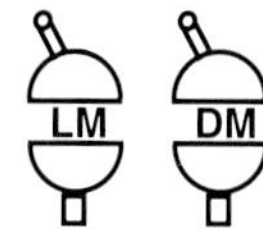

1. Embryonic pseudochamber, umbilical denticles .... ***Rotaliella***
1'. No embryonic pseudochamber, no denticles... 2

2. Umbilicus small with few umbilical grooves and ridges.............................. ***Metarotaliella***
2'. Umbilicus large with numerous grooves and ridges.. ***Rossyatella***

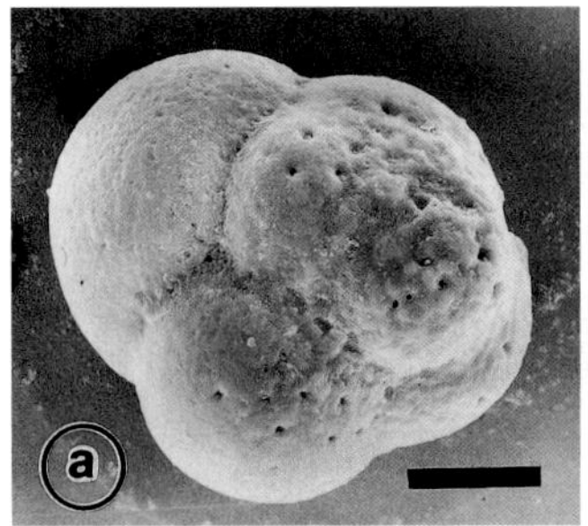
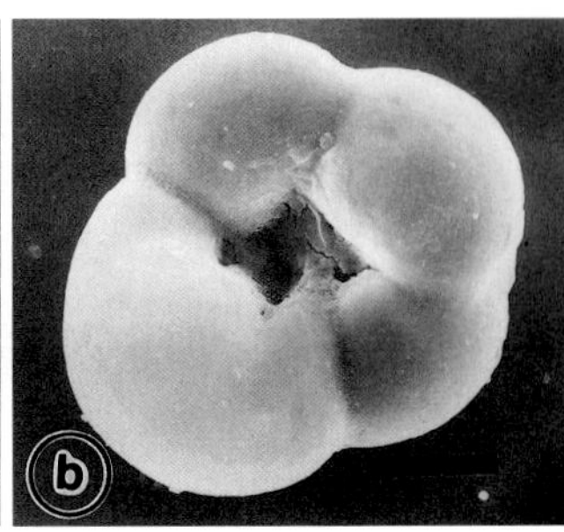

Fig. 133. *Metarotaliella simplex* Grell. a. Spiral view. b. Apertural view. Scales=10 μm.

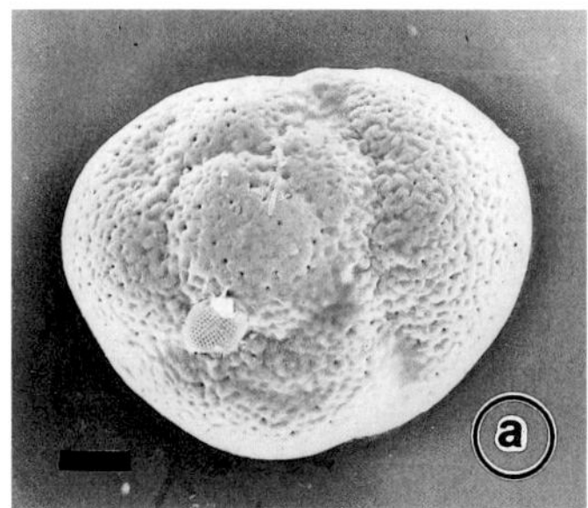
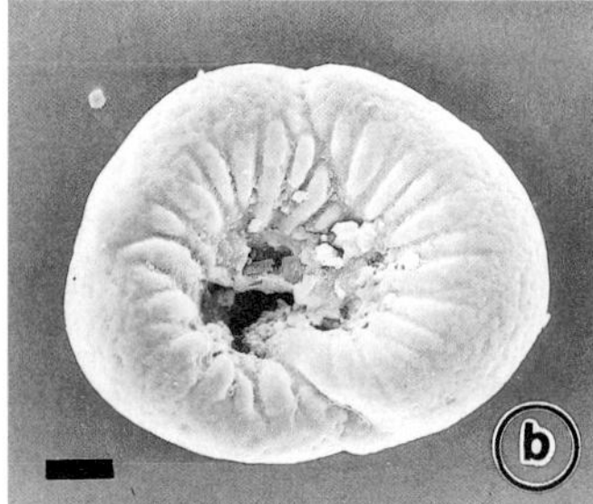

Fig. 134. *Rossyatella colonnensis*. a. Spiral view. b. Apertural view. Scales=50 μm.

FAMILY GLABRATELLIDAE
Loeblich & Tappan, 1964

Test a low trochospiral coil, umbilical side with radial ornamentation; aperture umbilical, may be covered by an umbilical flap; gamonts usually smaller and highly conical compared to the agamonts; reproduction gamontogamic; nuclear dimorphism observed in some species (Grell, 1958). Fifteen genera, shallow waters.

KEY TO REPRESENTATIVE GENERA

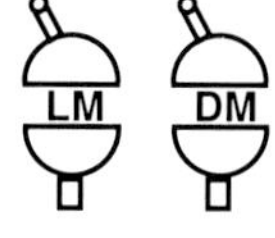

1. Test a low trochospiral coil.......................... 2
1'. Test pyramidal................. ***Angulodiscorbis***

2. Umbilicus open... 3
2'. Umbilicus closed............ ***Planoglabratella***

3. Umbilical flap........................ ***Glabratellina***
3'. No umbilical flap..................... ***Glabratella***

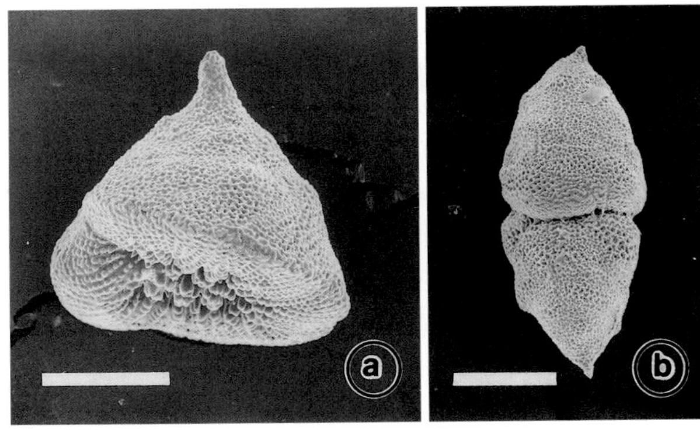

Fig. 135. *Glabratella erectia*. a. Gamont. b. Paired gamonts undergoing gamontogamy. Scales=100µm.

SUPERFAMILY ROSALINACEA Reiss, 1963
(Formally Discorbacea)

Test in low trochospiral coil; aperture interiomarginal on umbilical side of test, commonly covered by an umbilical flap; wall of perforate, hyaline, optically and ultrastructurally radiate calcite.

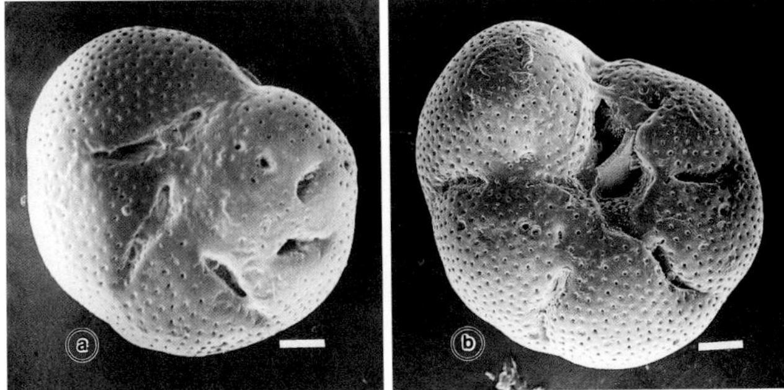

Fig. 136. *Pseudohelenia collinsi* (Parr). Scales=50 µm.

FAMILY HELENINIDAE
Loeblich & Tappan, 1988

Test a low trochospiral coil, both sides flattened, chambers enlarging gradually; aperture an umbilical-extraumbilical slit, covered by distinctive flap; supplementary sutural apertural slits present on both sides. Two genera: *Helenina, Pseudohelenina* (Fig. 136), subtropical shallow waters.

FAMILY ROSALINIDAE Reiss, 1963

Test trochospiral, commonly planoconvex, umbilicus may be closed by umbilical plug or flap; aperture a low arch at the base of final chamber with a narrow bordering lip. Eight genera, common in shallow waters.

KEY TO REPRESENTATIVE GENERA

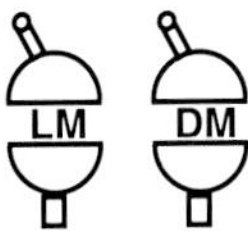

1. Umbilicus open, bordered by triangular flap...................... ***Rosalina, Neoconorbina***
1'. Umbilicus closed.. 2

2. Umbilicus closed by strongly overlapping chambers .... ***Planodiscorbis***
2. Umbilical plug........................... ***Gavelinopsis***

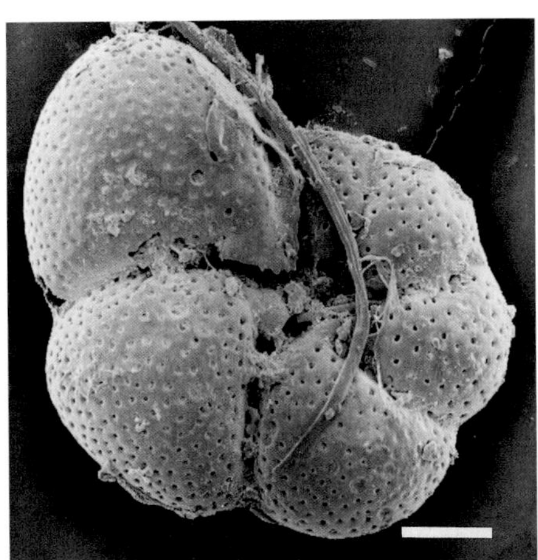

Fig. 137. *Rosalina globularis* d'Orbigny. Scale=50 µm.

FAMILY BAGGINIDAE Cushman, 1927

Test trochospiral, umbilical area closed; wall perforate except for a broad area of final chamber adjacent to aperture and umbilicus. Nine genera.

KEY TO REPRESENTATIVE GENERA

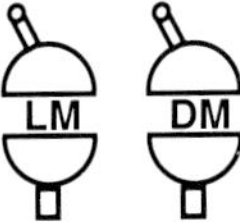

1. Umbilical flap.. 2
1'.Test sublobular, no umbilical flap...... ***Baggina***

2. Test lenticular, periphery carinate, chambers

rapidly enlarging................................ ***Cancris***

2'. Test rounded, chambers gradually enlarging .... ***Valvulineria***

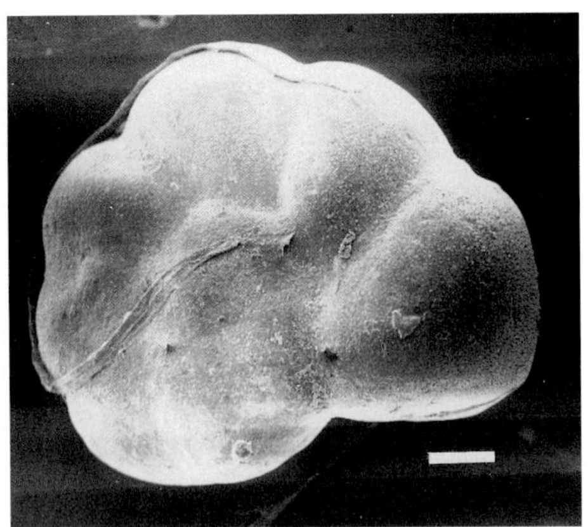

Fig. 138. *Valvulineria californica* Cushman. Scale=100 μm.

FAMILY BUENINGIIDAE Saidova, 1981

Test enrolled with both sides involute, spiral side inflated, umbilical side flattened with carinate margin; aperture umbilical with a small umbilical flap. Single genus *Bueningia* (Fig. 139).

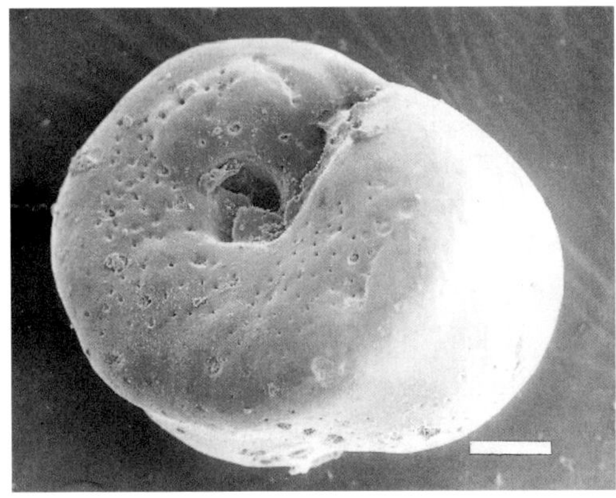

Fig. 139. *Bueningia creeki* Finlay. Scale=50 μm.

FAMILY BRONNIMANNIIDAE
Loeblich & Tappan, 1984

Test low trochospiral to planispiral, bi-evolute and biconcave with broad truncate periphery; aperture a low interiomarginal slit beneath umbilical flap. Single genus *Bronnimannia* (Fig. 140).

SUPERFAMILY DISCORBINELLACEA Sigal, 1952

Test trochospiral, at least in early stage; aperture interiomarginal, equatorial in position, a low arch to high slit up the apertural face; wall calcareous, optically radial.

FAMILY PARRELLOIDIDAE Hofker, 1956

Test trochospiral, biconvex, umbilicus closed; aperture a low equatorial slit bordered by a narrow lip; wall coarsely perforate on spiral side. Two genera: *Cibicidoides* (Fig. 141), *Parrelloides*, cosmopolitan.

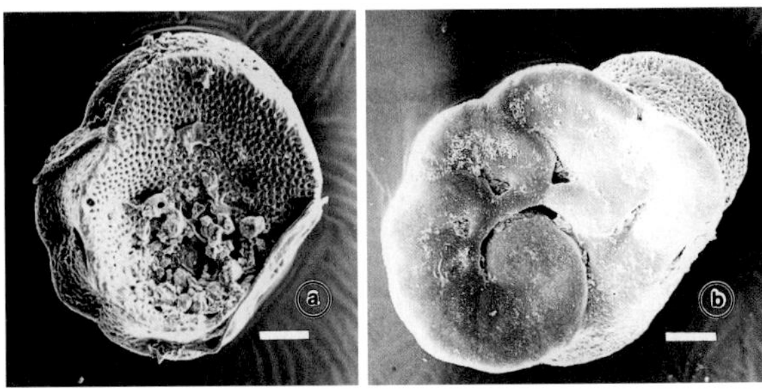

Fig. 140. *Bronnimannia haliotis* (Heron-Allen & Earland). a. Apertural view. b. Spiral view. Scales–50 μm.

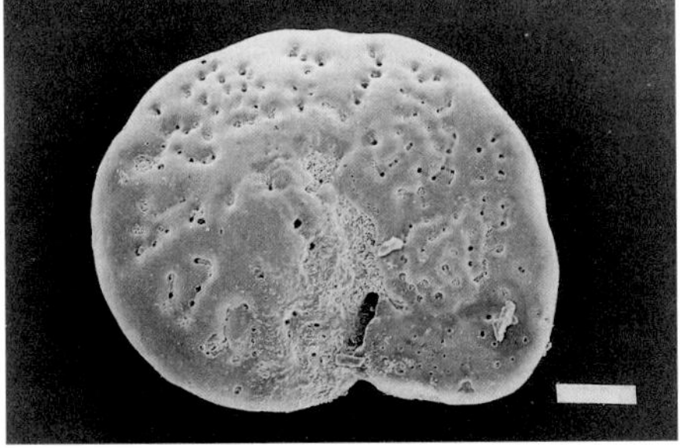

Fig. 141. *Cibicidoides muellersdorfi*. Scale=100 μm.

FAMILY PSEUDOPARRELLIDAE
Voloshinova, 1952

Test trochospiral, planoconvex to biconvex, may become involute on both sides; aperture a vertical slit extending up the face of the final chamber. Seven genera, bathyal.

KEY TO REPRESENTATIVE GENERA

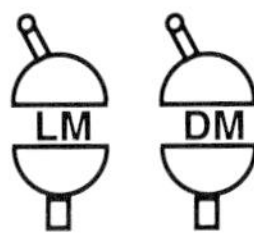

1. Spiral side evolute, umbilical side involute.. 2
1'. Final whorl involute on both sides...***Stetsonia***

2. Test planoconvex to biconvex.. ***Epistominella***
2'. Test biconvex....................... ***Pseudoparrella***

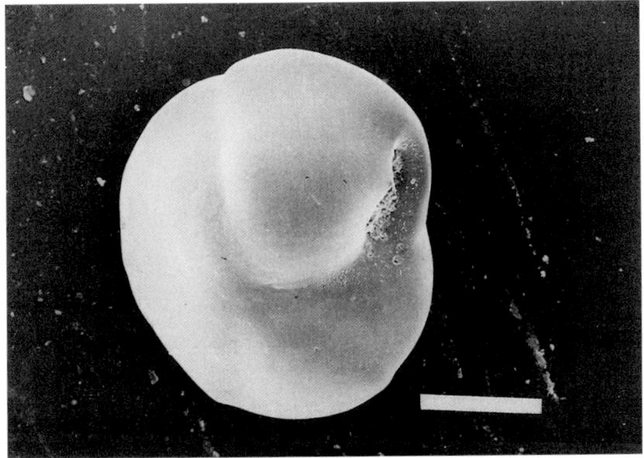

Fig. 142. *Epistominella exigua*. Scale=100 μm.

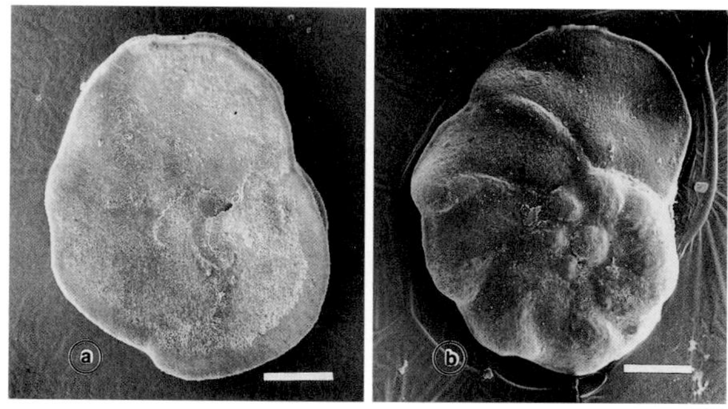

Fig. 143. *Planulinoides biconcavus* (Jones & Parker). a. Apertural view. b. Spiral view. Scales=100 μm.

FAMILY PLANULINOIDIDAE Saidova, 1981

Test near planispiral, biconcave, and partially evolute on both sides, periphery truncate, bicarinate; aperture areal and equatorial, an oval opening bordered by a distinct lip, **supplementary apertures** on the umbilical side; wall finely perforate. Single genus *Planulinoides* (Fig. 143).

FAMILY DISCORBINELLIDAE Sigal, 1952

Test nearly planispiral, with tendency to uncoil, may be carinate; aperture a small equatorial opening, supplementary openings at the posterior end of the umbilical flap. Seven genera.

KEY TO REPRESENTATIVE GENERA

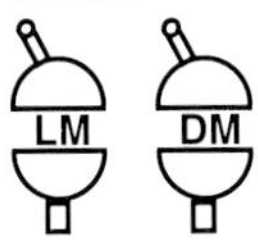

1. Chambers subdivided by secondary partitions .... ***Torresina***
1'. No secondary partitions............................... 2

2. Test bi-evolute, with very broad peripheral keel.. ***Laticarinina***
2'. Umbilical side involute................................ 3

3. Large, circular pores on the spiral side....................................... ***Discorbinita***
3'. Wall finely perforate............ ***Discorbinella***

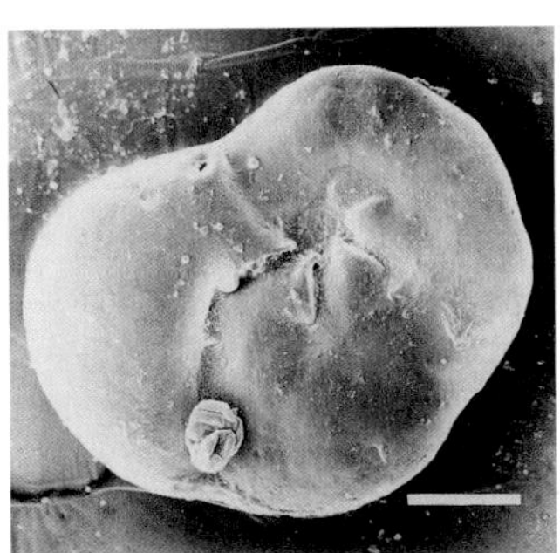

Fig. 144. *Discorbinella disparilis*. Scale=100 μm.

SUPERFAMILY PLANORBULINACEA Schwager, 1877

Test free or attached, trochospiral at least in early stage, later may be uncoiled and rectilinear or biserial, or may have many chambers per whorl, or chambers added irregularly; aperture interio-marginal, extraumbilical-umbilical to nearly equatorial; wall of perforate hyaline calcite, commonly optically radial, apertural face may be imperforate.

FAMILY PLANULINIDAE Bermudez, 1952

Test discoidal, very low trochospiral to nearly

planispiral, partially evolute on both sides; aperture a low equatorial arch extending beneath the umbilical folium. Genera: *Hyalinea* (Fig. 145; epifaunal, shelf, upper bathyal), *Planulina* (epifaunal, cosmopolitan).

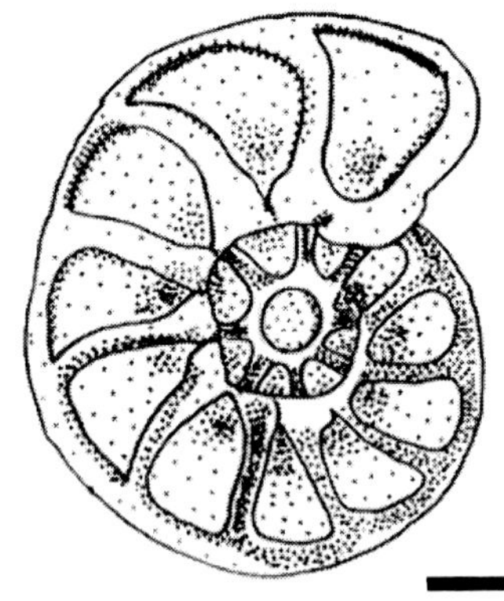

Fig. 145. *Hyalinea balthica* (Schröter). Scale=100 µm.

FAMILY BISACCIIDAE
Loeblich & Tappan, 1988

Test irregularly planispiral, periphery rounded, aperture equatorial covered by the plate of the apertural chamberlet; supplementary sutural openings present. Two genera (*Bisaccioides, Bisaccium*, Fig. 146).

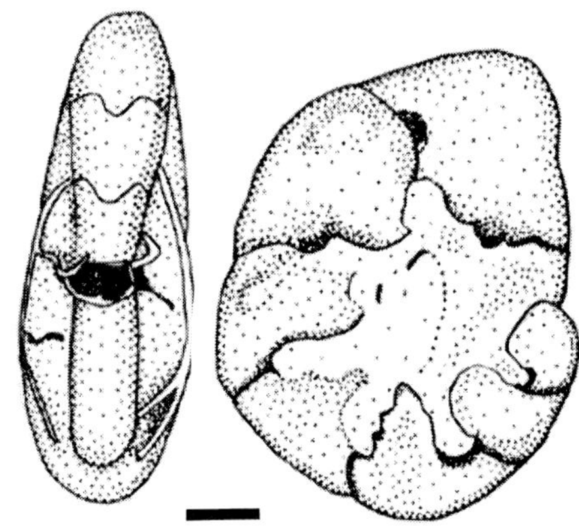

Fig. 146. *Bisaccium imbricatum* Anderson. Scale=80 µm.

FAMILY CIBICIDIDAE CUSHMAN, 1927

Test trochospirally enrolled at least in early stage, attached by flat spiral side, umbilical side convex; aperture a low equatorial opening extending along the spiral suture on the spiral side; wall coarsely perforate on the spiral side finely perforate on the umbilical side. Fifteen modern genera.

KEY TO REPRESENTATIVE GENERA

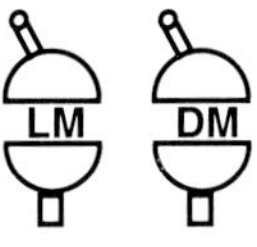

1. Test trochospirally enrolled throughout........ 2
1'. Test trochospiral in early stage only............ 4

2. Secondary apertures present........................ 3
2'. No secondary apertures.................. ***Cibicides***

3. Secondary apertures on the periphery............. .... ***Paracibicides***
3'. Secondary apertures beneath umbilical flap..... .... ***Montfortella***

4. Test biserial in the later stage. ***Dyocibicides***
4'. Later stage with cyclical chambers.................. .... ***Cyclocibicides***

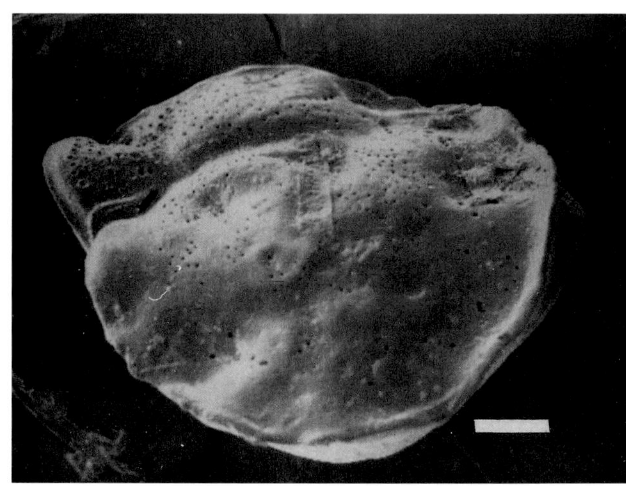

Fig. 147. *Cibicides refulgens* Montfort. Scale=100 µm.

FAMILY PLANORBULINIDAE Schwager, 1877

Test attached by spiral side, early stage trochospiral, later chambers proliferating more or less irregularly, forming discoid or conical test, each chamber with two apertures on periphery at opposite ends. Three genera (*Planorbulina* [Fig. 148], *Planorbulinella, Planorbulinopsis*).

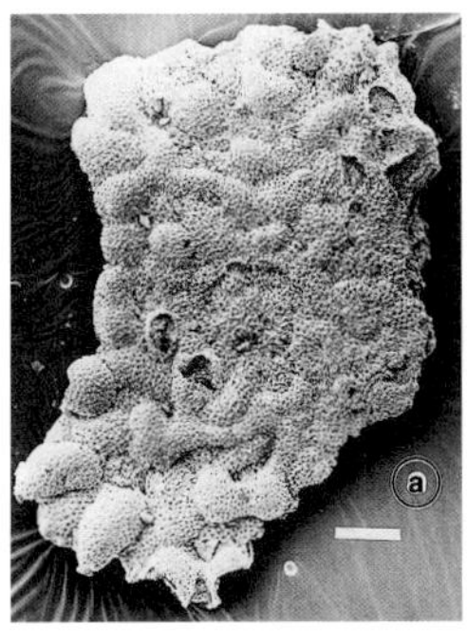
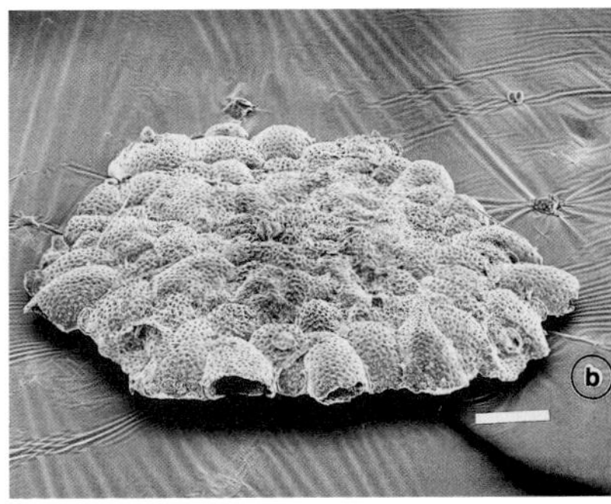

Fig. 148. *Planorbulina* sp. d'Orbigny, 1826. a. Spiral view. b. Apertural view. Scales 200µm.

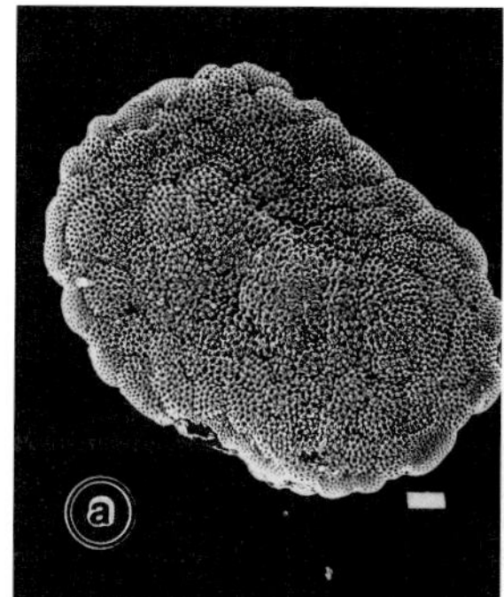
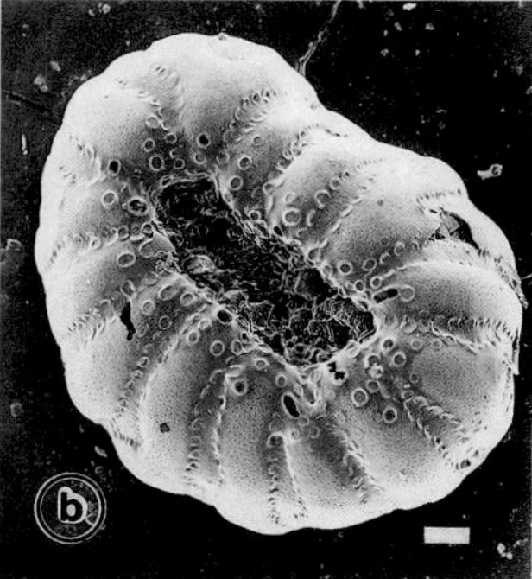

Fig. 149. *Cymbaloporella tabellaeformis* (Brady, 1884). a. Spiral view. b. Apertural view. Scales=100 µm.

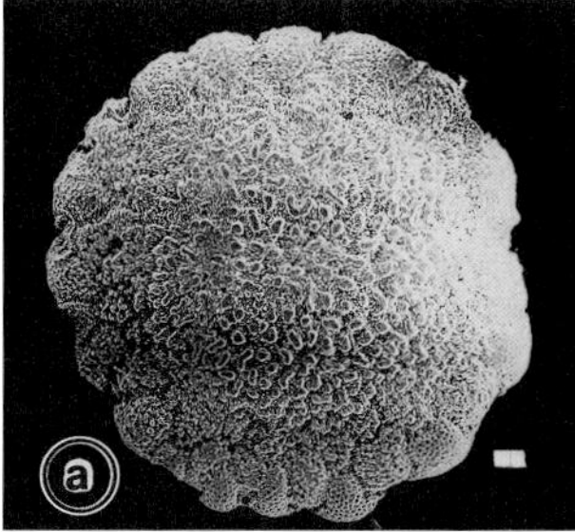
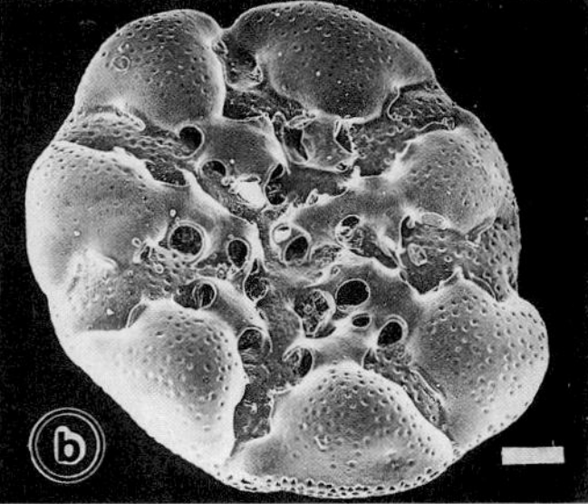

Fig. 150. *Cymbaloporetta squammosa* (d'Orbigny, 1826). a. Spiral view. b. Apertural view. Scales=100 µm.

FAMILY CYMBALOPORIDAE Cushman, 1927

Test trochospiral, later chambers in annular series in a single flat to conical layer; numerous apertures present as small circular pores; gamonts with floating reproductive chambers. Four modern genera (*Cymbaloporella* [Fig. 149], *Cymbalo-poretta* [Fig. 150], *Millettiana, Pyropilus*).

FAMILY VICTORIELLIDAE
Chapman & Crespin, 1930

Test attached, early chambers trochospiral, later may grow upward from the attachment in a loose spiral or irregular mass. Three genera.

KEY TO THE GENERA

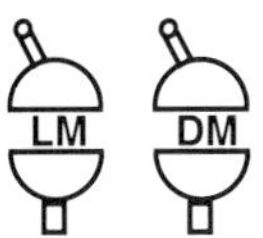

1. Test trochospiral throughout..... ***Carpenteria***
1'. Test attached by basal disc, growing up in a loose spiral........... ***Biarritzina, Rupertina***

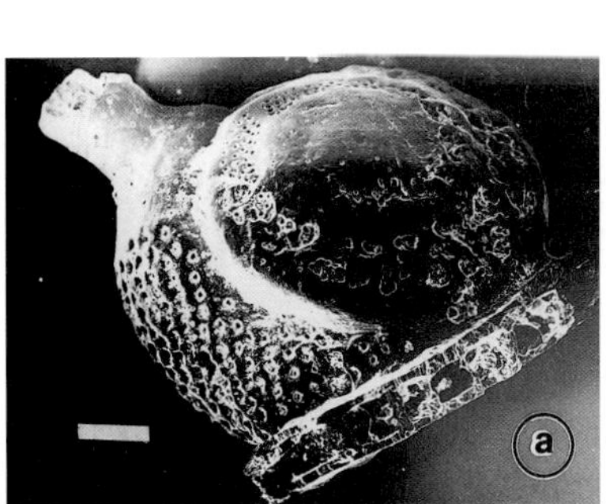
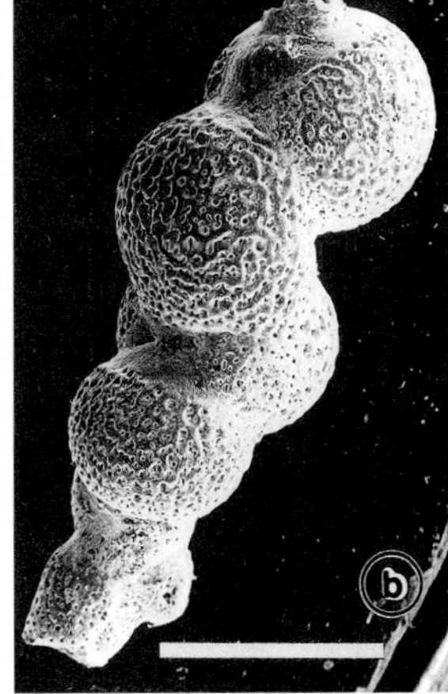

Fig. 151. *Carpenteria balaniformis*, proteiformis Goes, 1882. a. Juvenile. Scale=100 µm. b. Adult. Scale=1000 µm.

SUPERFAMILY ACERVULINACEA
Schultze, 1854

Test free or attached, early spiral stage followed by spreading or irregular chambers that form an irregular disc or branching structure; wall of hyaline, optically radial calcite, coarsely perforate.

FAMILY ACERVULINIDAE Schultze, 1854

Chambers spreading in one or more layers; no aperture other than mural pores.

KEY TO THE GENERA

1. Test globular....................... ***Sphaerogypsina***
1'. Test flattened.. 2

2. Test discoidal.. 3
2'. Chambers arranged irregularly.. ***Acervulina***

3. Chambers in very thin, single layer................ .... ***Planogypsina***
3'. Chambers inflated, in more than one layer.... 4

4. Successive layers of chambers alternating in position.. ***Gypsina***
4'. Successive layers not in alignment.................. .... ***Discogypsina***

FAMILY HOMOTREMATIDAE Cushman, 1927

Test attached, trochospiral in early stage, later numerous chambers forming a massive or branching test growing erect from the attachment; apertures may be large perforations or round openings at the ends of the branches.

KEY TO THE GENERA

1. Surface with areolae or pillar pores............ 2
1'. Surface lacking special structures, apertures terminal................................ ***Sporadotrema***

2. Surface with perforated plates surrounded by imperforate areolae.................. ***Homotrema***
2'. Wall with pillar-pores................ ***Miniacina***

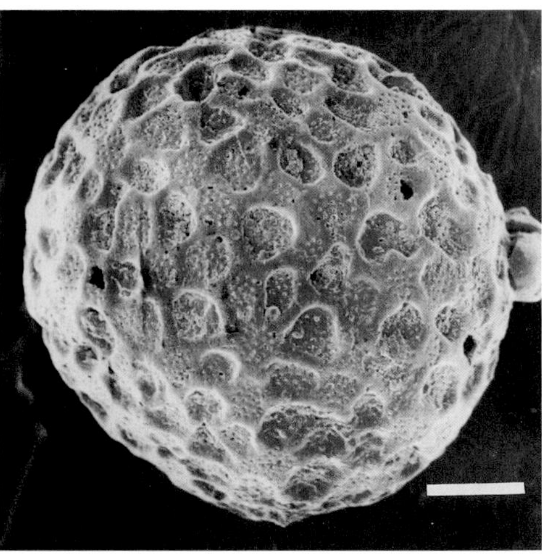

Fig. 152. *Gypsina globula* (Reuss, 1847). Scale 100µm.

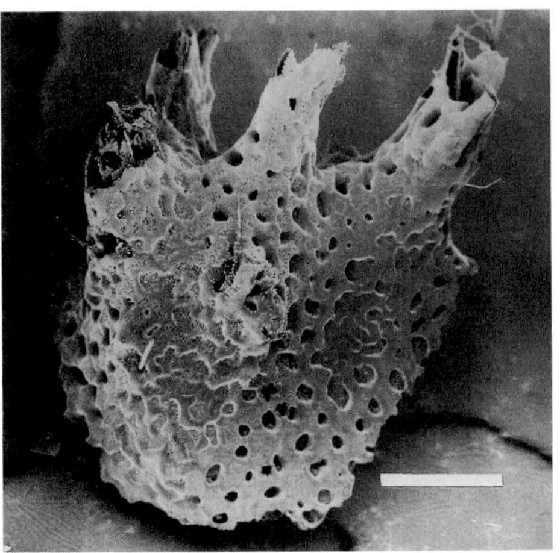

Fig. 153. *Homotrema* sp. Hickson, 1911. Scale=500 µm.

SUPERFAMILY ASTERIGERINACEA d'Orbigny, 1839

Test trochospiral to nearly planispiral, chambers with internal partitions that form supplementary chambers around the umbilicus, producing a stellate pattern on the umbilical side; primary aperture interiomarginal, equatorial to umbilical in position, secondary sutural or areal apertures; wall of hyaline perforate calcite, optically radial.

FAMILY EPISTOMARIIDAE Hofker, 1954

Test in a low trochospiral coil, biconvex, chambers divided by internal partitions that may form the supplementary chamberlets; numerous secondary openings on both sides of the wall.

KEY TO THE GENERA

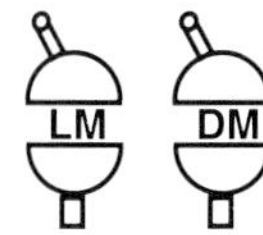

1. Prominent secondary chamberlets on the umbilical side.............................. ***Eponidella***
1'. Secondary chamberlets are not visible......... 2

2. Numerous supplementary sutural and areal openings.. 3
2'. No supplementary openings, aperture divided by a median plate..................... ***Palmerinella***

3. Internal partition not attached to opposite wall .... ***Nuttalides***
3'. Internal toothplate attached to the supplementary opening on the spiral side .... ***Pseudoeponides***

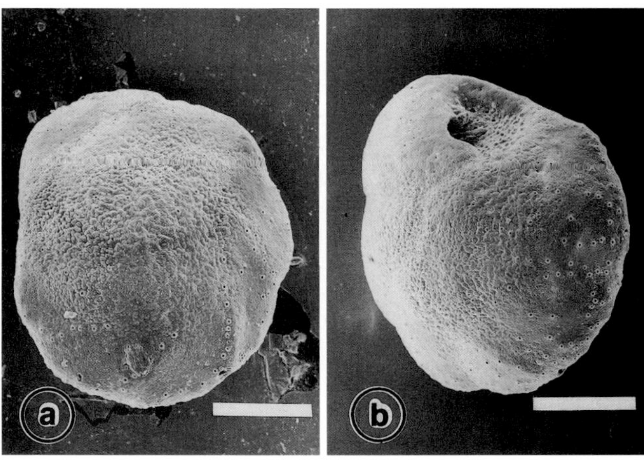

Fig. 154. *Nuttalides umboniferus* (Cushman, 1933). a. Spiral view. b. Apertural view. Scales=100 µm.

FAMILY ALFREDINIDAE Singh & Kalia, 1972

Test trochospiral, biconvex, supplementary chamberlets formed by a transverse internal partition, visible on the umbilical side, multiple sutural openings present on both sides of the test. Single genus *Epistomaroides* (Fig. 155).

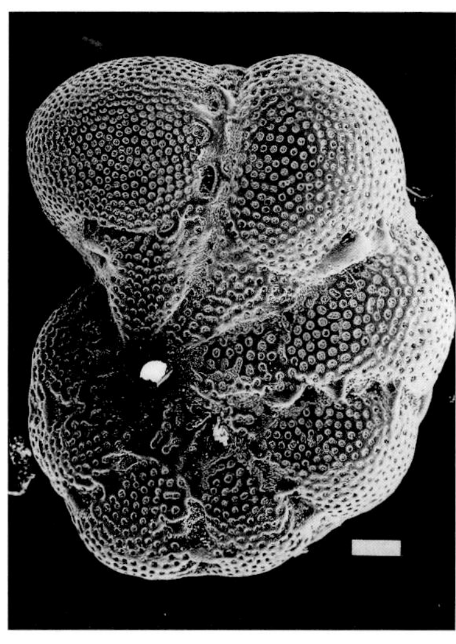

Fig. 155. *Epistomaroides polystomelloides* (Parker & Jones, 1865). Scale=100 µm.

FAMILY ASTERIGERINATIDAE Reiss, 1963

Test trochospiral, planoconvex, umbilical side with chambers partially covered by secondarily produced inflated coverplates forming a stellate pattern around the umbilicus, coverplates perforate, may cover all of the umbilical side and be partly visible on the spiral side (Pawlowski et al., 1992); aperture a broad umbilical arch.

KEY TO REPRESENTATIVE GENERA

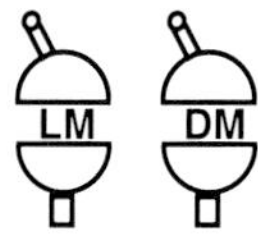

1. Umbilical plates large, extending beyond the periphery.................................. ***Rubratella***
1'. Umbilical plates small, forming a stellate pattern around the umbilicus........................ 2

2. Test small, chambers crescentic, aperture broad and large arch................. ***Eoeponidella***
2'. Test larger, chambers semilunate, aperture low arch, umbilical plug....... ***Asterigerinata***

FAMILY ASTERIGERINIDAE d'Orbigny, 1839

Test trochospiral, biconvex, with more elevated umbilical side, chambers divided by internal toothplates producing a stellate series of chamber-lets surrounding the closed umbilicus; aperture a low umbilical slit bordered by a distinct lip. Single genus *Asterigerina* (Fig. 158).

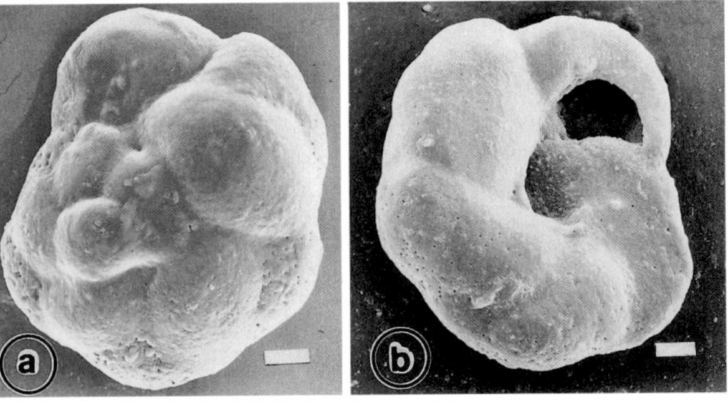

Fig. 156. *Rubratella* sp. Grell, 1956. a. Spiral view. b. Apertural view. Scales 10μm.

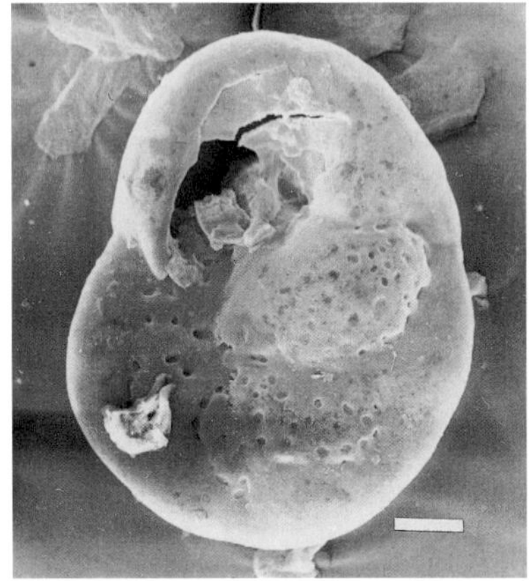

Fig. 157. *Eoeponidella pacifica.* Scale=20 μm.

Fig. 158. *Asterigerina carinata* d'Orbigny, 1839. Scale=100 μm.

FAMILY AMPHISTEGINIDAE Cushman, 1927

Test lenticular, biconvex, may be bi-involute or partially evolute on spiral side, with distinctive umbilical plug; internal toothplate dividing almost completely each chamber, producing an irregular stellate pattern on the umbilical side; aperture an umbilical slit bordered by a lip; surface smooth except for the apertural region covered with fine papillae. Single genus *Amphistegina* (Figs. 13a-e, 159).

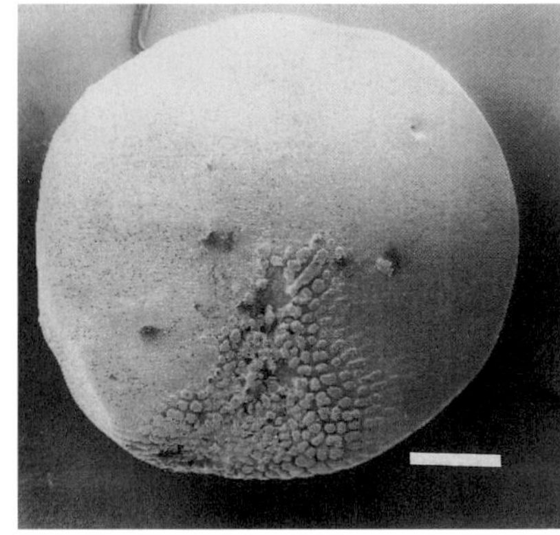

Fig. 159 *Amphistegina* sp. d'Orbigny, 1839. Scale=200 μm.

SUPERFAMILY NONIONACEA Schultze, 1854

Test enrolled, planispiral to slightly asymmetrical; wall of perforate hyaline calcite, appearing optically granular.

FAMILY NONIONIDAE Schultze, 1854

Test planispiral, involute to evolute, bi-umbilicate, umbilici filled with pustules, aperture a low equatorial slit at the base of the apertural face, extending laterally nearly to the umbilici.

KEY TO REPRESENTATIVE GENERA

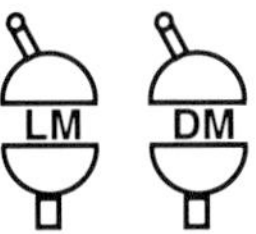

1. Test moderately compressed.......................... 2
1'. Test subglobular to globular......................... 6

2. Chambers with umbilical flap or flaplike projections.. 3
2'. No umbilical flap... 5

3. Umbilical flap extending from the umbilicus along the intercameral suture... ***Astrononion***
3'. Chambers with flaplike projection

overhanging the umbilicus.................................. 4

4. Surface with pustules bordering the umbilical rim... ***Nonionoides***
4'. Surface smooth without pustules or granules .... ***Nonionella***

5. Sutures with intercameral lacunae.......... .... ***Haynesina***
5'. No intercameral lacunae...................... ***Nonion***

6. Chambers numerous, umbilicus deeply open .... ***Melonis***
6'. Few chambers/whorl, umbilicus closed........... .... ***Pullenia***

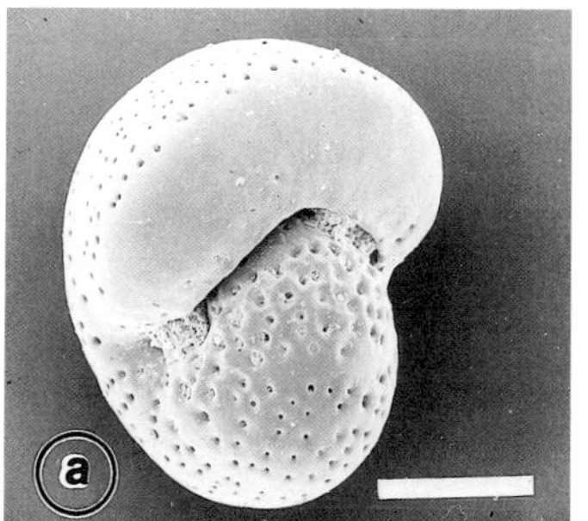
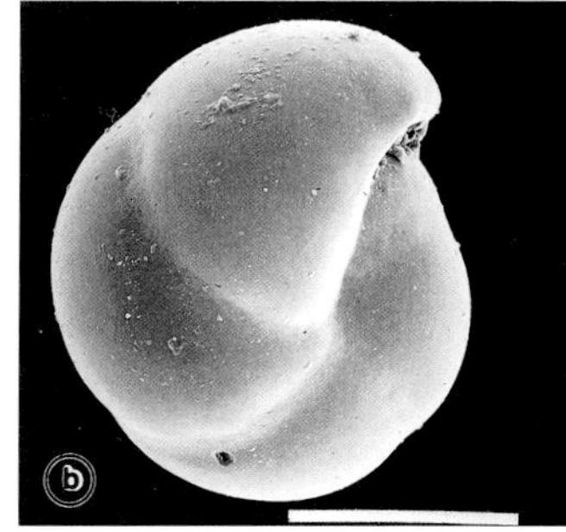

Fig. 160. a. *Melonis* sp. de Monfort, 1808. Scale=100 µm. b. *Pullenia bulloides* (d'Orbigny, 1846). Scale=100 µm.

FAMILY ALMAENIDAE Myatlyuk, 1959

Test lenticular, nearly planispiral, periphery carinate; aperture a low equatorial arch, secondary slit-like aperture at outer peripheral margin of chamber. Single Recent genus *Anomalinella* (Fig. 161)

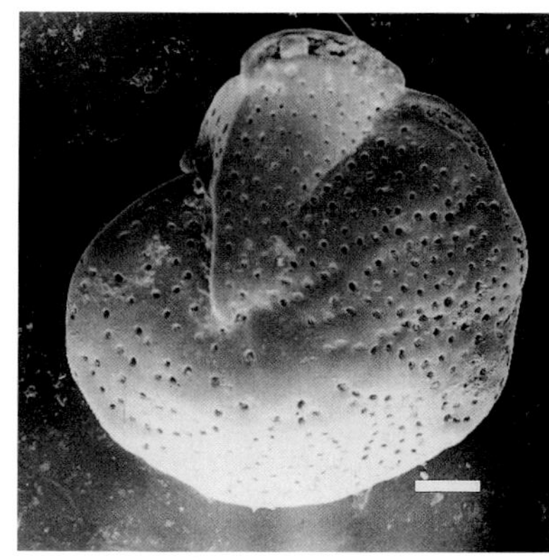

Fig. 161. *Anomalinella rostrata* (Brady, 1887). Scale=100 µm.

SUPERFAMILY CHILOSTOMELLACEA Brady, 1881

Test enrolled, low trochospiral, chambers enveloping; wall of perforate hyaline oblique calcite, appearing optically granular.

FAMILY CHILOSTOMELLIDAE Brady, 1881

Test ovoid, trochospiral to planispiral, two to three chambers per whorl, much enveloping, only those of final whorl visible externally; aperture a nar-row equatorial slit. Two Recent genera: *Allomorphina, Chilostomella* (Fig. 162).

Fig. 162. *Chilostomella ovidea* Reuss, 1850. Scale=100 µm.

FAMILY QUADRIMORPHINIDAE Saidova, 1981

Test low trochospiral, biconvex; aperture a low opening covered by a large lip with internal toothplate. Single genus *Quadrimorphina* (Fig. 163).

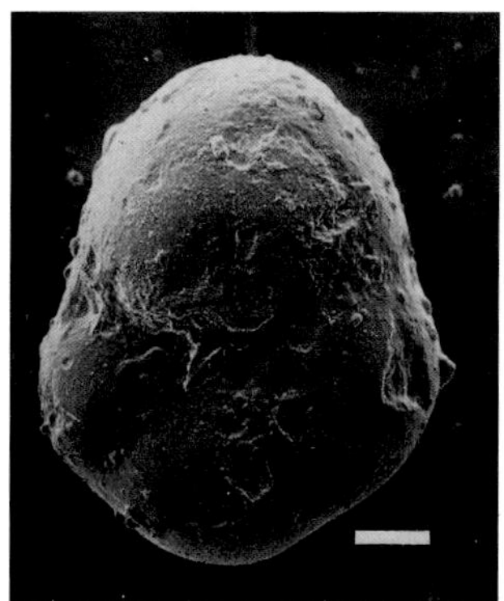

Fig. 163. *Quadrimorphina allomorphinoides* (Reuss, 1860). Scale=50 µm.

FAMILY ALABAMINIDAE Hofker, 1951

Test trochospiral, biconvex, apertural face deeply excavated; aperture an interiomarginal slit extending from periphery to open umbilicus. Two genera: *Alabamina* (Fig. 164), *Svratkina.*

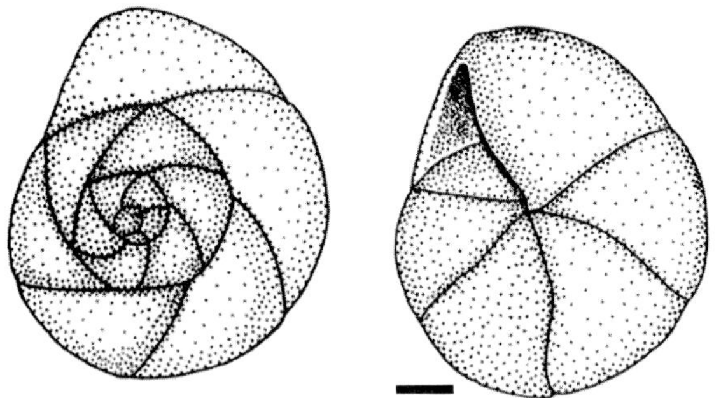

Fig. 164. *Alabamina* sp. Toulmin, 1941. Scale=42 µm.

FAMILY OSANGULARIIDAE Loeblich & Tappan, 1964

Test lenticular, periphery carinate, aperture areal extending up apertural face. Single Recent genus *Osangularia* (Fig. 165).

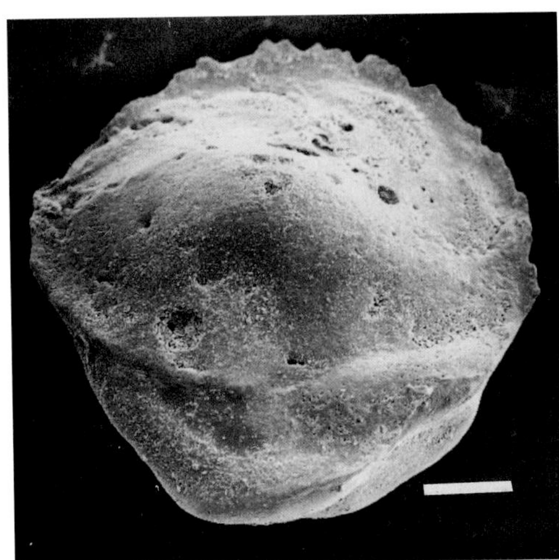

Fig. 165. *Osangularia culter* (Parker & Jones, 1865). Scale=50 µm.

FAMILY ORIDORSALIDAE Loeblich & Tappan, 1984

Test lenticular, chambers in low trochospiral coil; aperture at base of apertural face, extending from near the periphery to the umbilical region, small secondary sutural openings on both sides. Two genera: *Oridorsalis* (Fig. 166), *Schwantzia.*

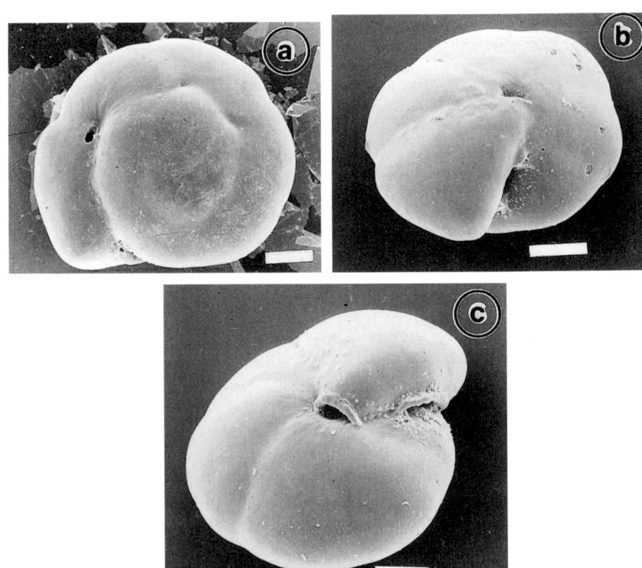

Fig. 166. *Oridorsalis umbonatus* (Reuss, 1851). a. Spiral view. b. Lateral view. c. Apertural view. Scales=50 µm.

FAMILY HETEROLEPIDAE Gonzalez-Donoso, 1969

Test lenticular, umbilicus closed; aperture a small slit on the umbilical side extending shortly onto the spiral side. Three genera: *Heterolepa* (Fig. 167), *Anomalinoides, Gemellides.*

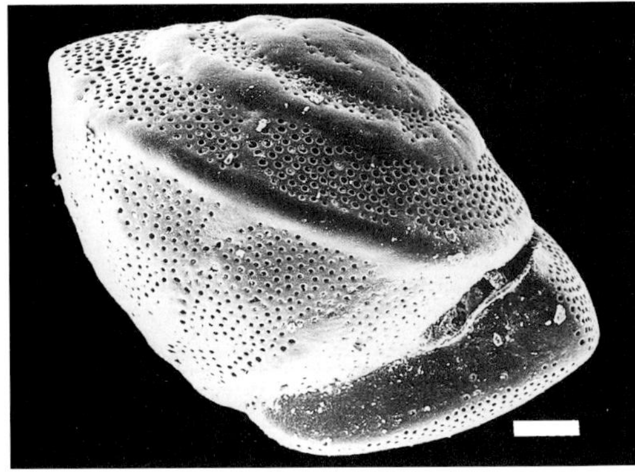

Fig. 167. *Heterolepa praecincta* (Karrer, 1868). Scale=100 µm.

FAMILY GAVELINELLIDAE Hofker, 1956

Test trochospiral, spiral side flat and evolute, umbilical side convex and involute, umbilicus open; aperture a low interiomarginal slit extending from the periphery to the umbilicus, may be partially covered by a distinctive flap. Eight genera: *Gyroid-inoides* (Fig. 168), *Anomalinulla, Discanomalina, Gyroidella, Gyroidina, Gyroidinopsis, Hansenisca, Hanzawaia*

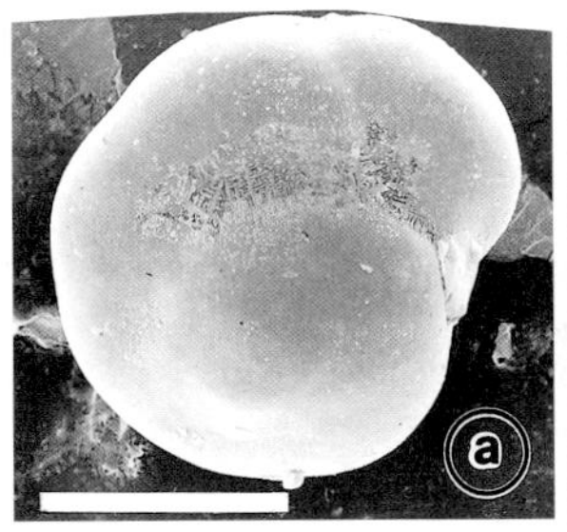
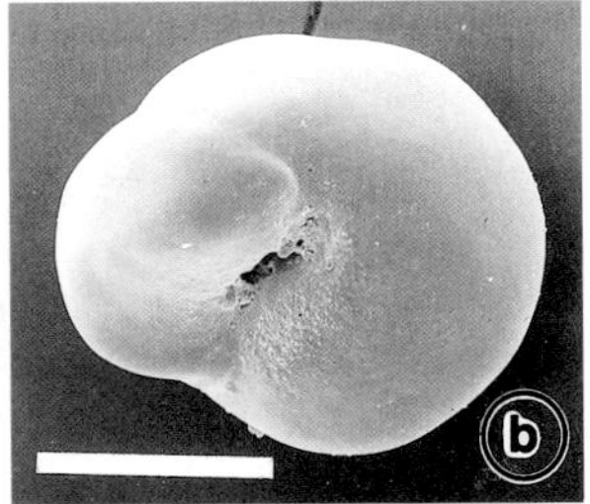

Fig. 168. *Gyroidinoides neosoldanii* (Brotzen, 1936). a. Spiral view. b. Apertural view. Scales=100 µm.

FAMILY TRICHOHYALIDAE Saidova, 1981

Test trochospiral, lenticular to planoconvex, umbilical side partly or wholly obscured by secondary growth of granules or pustules; aperture umbil-ical, obscured by the pustulose coating. Three genera: *Buccella* (Fig. 169), *Neobuccella, Trichohyalus*.

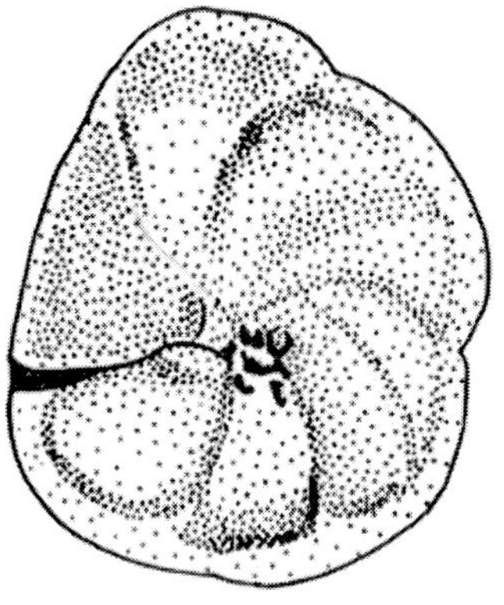
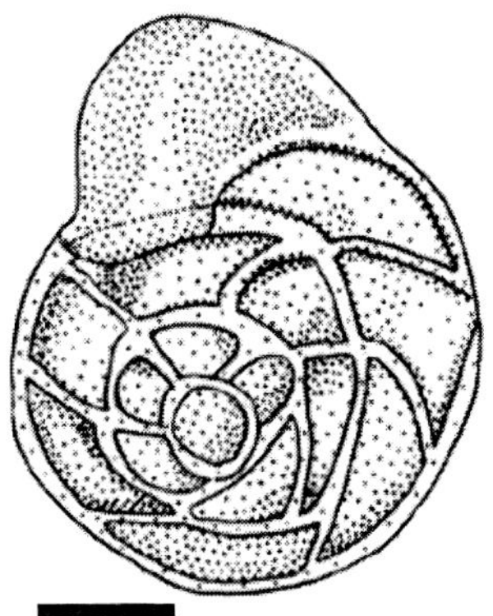

Fig. 169. *Buccella fastidiosa* (McCulloch, 1977). Scale=200 µm.

SUPERFAMILY ROTALIACEA Ehrenberg, 1839

Test enrolled, in low trochospiral coil, may be planispiral, involute to evolute, commonly with many small chambers in numerous whorls; usually more than one primary apertures; wall of perfor-ate, hyaline calcite, generally optically radiate, with internal structures including umbilical spiral canals, coverplates, hooks and gutter-like struc-tures. The families Discorbidae and Eponididae were transfered to Rotaliacea following the description of internal structures in *Discorbis* and *Eponides* (Hansen and Revets, 1992; Hottinger et al., 1992)

FAMILY ELPHIDIIDAE Galloway, 1933

Test planispiral, lenticular, bi-umbonate, sutural canal system opening into single or double row of sutural pores; aperture single or multiple.

KEY TO THE GENERA

1. Test planispiral, at least in early stage......... 2
1'. Test trochospiral to planispiral, umbilical side with chamber extensions 3

2. Test planispiral throughout........................... 4
2'. Test planispiral, later uncoiled......... ***Ozawaia***

3. Test periphery sharply angular. ***Notorotalia***
3'. Test periphery carinate.................................... .... ***Parrellina***

4. Sutural ponticuli and fossettes...................... 5
4'. No ponticuli and fossettes............ ***Elphidiella***
5. Periphery perforate, rounded.......................... .... ***Cribroelphidium***
5'. Periphery imperforate, carinate...................... .... ***Elphidium***

FAMILY CALCARINIDAE Schwager, 1876

Test enrolled, biconvex, showing little or no differentiation of chambers on spiral and umbilical side, commonly with prominent radial spines; canal system diffuse and confused with perforations.

KEY TO THE GENERA

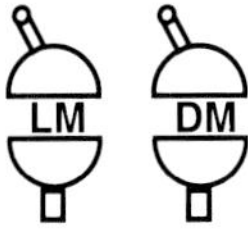

1. Test lenticular, biconvex, with prominent

radial spines.. 2

1'. Test globular with slightly projecting spines and tubercules................ ***Schlumbergerella***

2. Chambers trochospirally coiled throughout...... ***Calcarina***(Figs. 11b,13hi, ... ***Neorotalia***

2'. Chambers trochospirally coiled in early stage, later followed by a loose network of numerous chamberlets.......................... ***Baculogypsina***

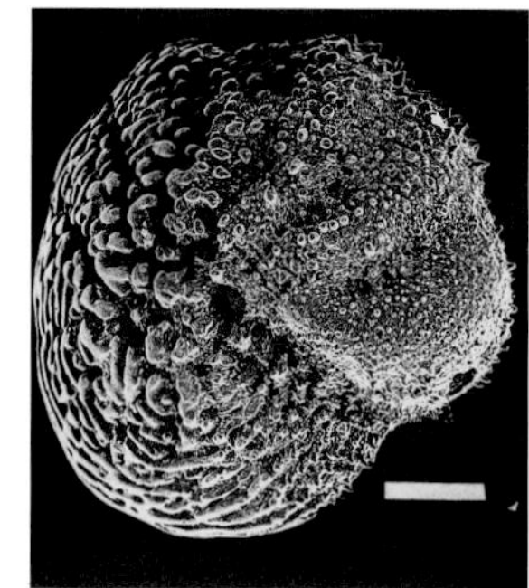

Fig. 170. *Parrellina hispidula* (Cushman, 1936). Scale=100 μm.

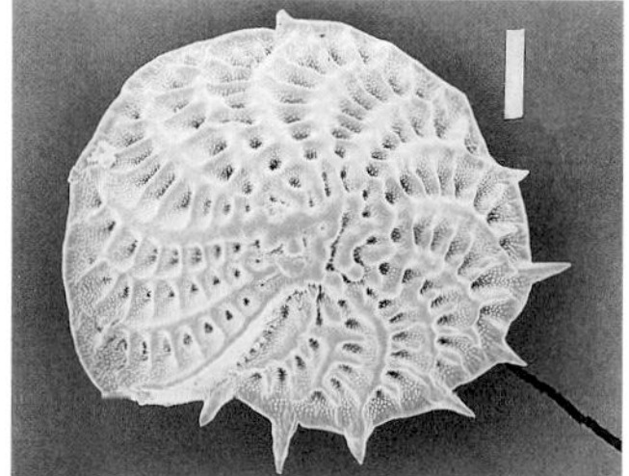

Fig. 171. *Elphidium aculeatum* (d'Orbigny, 1846). Scale=100 μm.

Fig. 172. *Neorotalia*. Scale=200 μm.

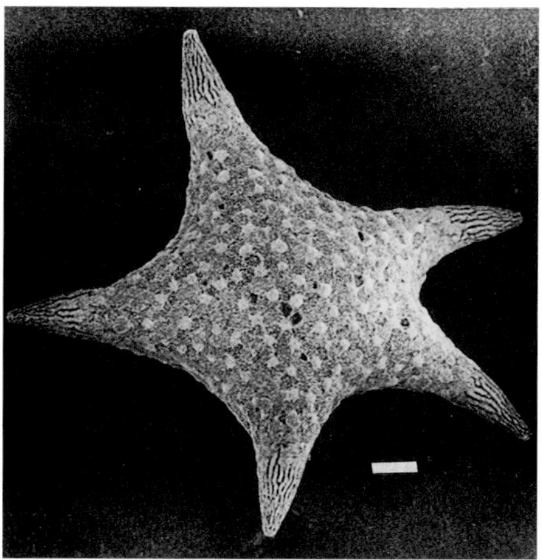

Fig. 173. *Baculogypsina sphaerulata* (Parker & Jones, 1860). Scale=200 μm.

FAMILY ROTALIIDAE Ehrenberg, 1839

Test trochospiral, biconvex to planoconvex, umbilical region secondarily closed by a foraminal coverplates, with **radial canals**, fissures, or umbilical cavities connecting the chambers; primary aperture a small umbilical slit.

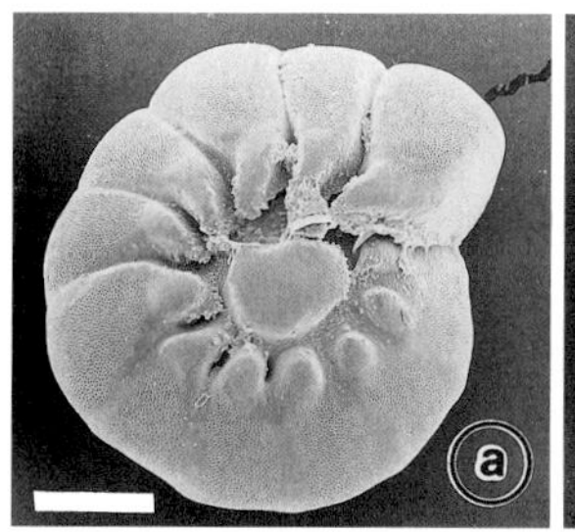

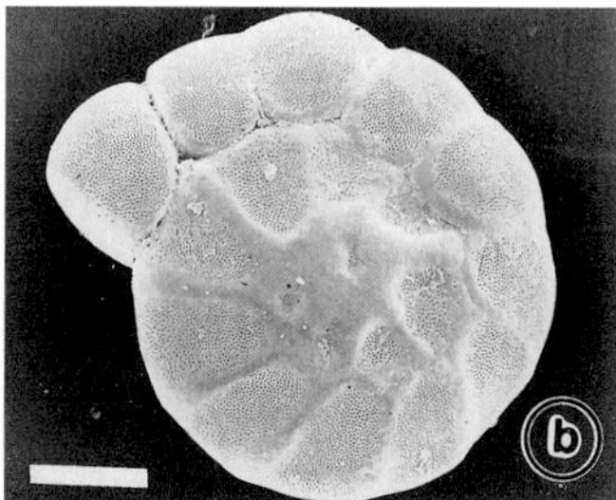

Fig. 174. *Ammonia parkinsoniana* (d'Orbiny, 1939). a. Spiral view. b. Apertural view. Scales=100 μm.

KEY TO THE GENERA

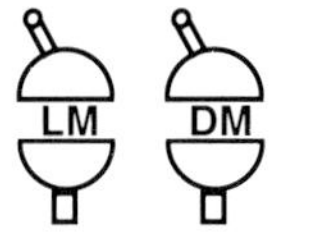

1. Test trochospiral, biconvex........................... 2

1'. Test flattened or conical................................. 4

2. **Spiral canal** present.................................. 3

2'. No spiral canal........ ***Ammonia, Pararotalia***

3. Spiral canal simple, around the umbilical plug .... ***Challengerella***

3'. Spiral canal broad, covered by umbilical extensions of the chambers......... ***Rotalidium***

4. Test flattened, nearly planispiral with three large solid spines.................... ***Asterorotalia***
4'. Test conical, planoconvex, with sutures elevated and limbate.............. ***Pseudorotalia***

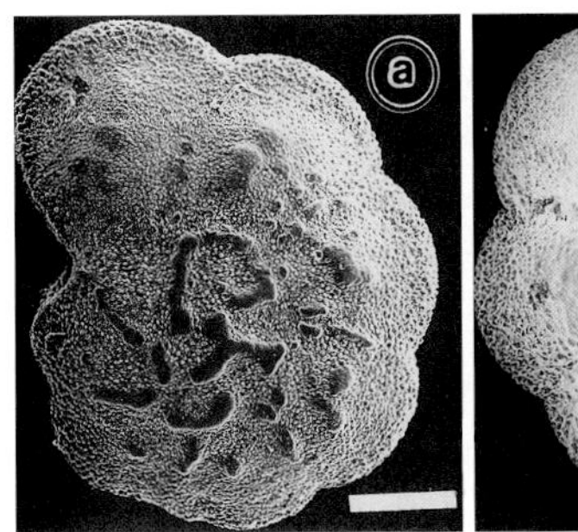
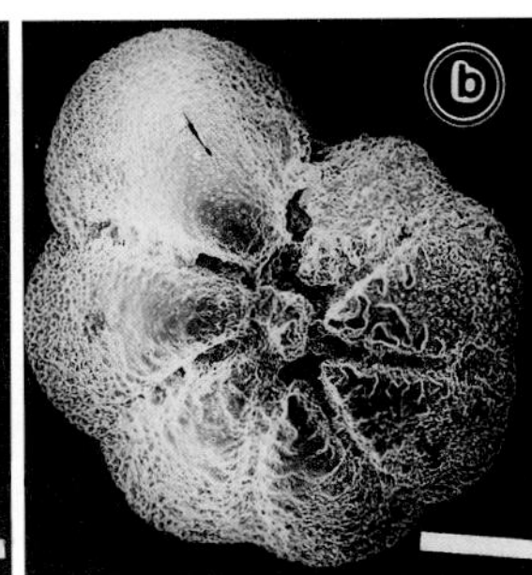

Fig. 175. *Pararotalia* sp. LeCalvez, 1949. a. Spiral view. b. Apertural view. Scales=100 µm.

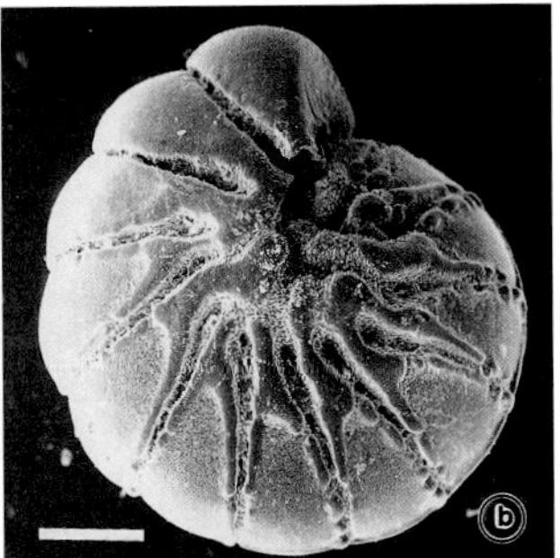

Fig. 176. *Challengerella* sp. Billman, Hottinger & Esterle, 1980. a. Spiral view. b. Apertural view. Scales=200 µm.

FAMILY DISCORBIDAE Ehrenberg, 1838

Test concavo- to planoconvex, chamber interior subdivided; aperture umbilical with distinct flap extending over the umbilical region.

KEY TO REPRESENTATIVE GENERA

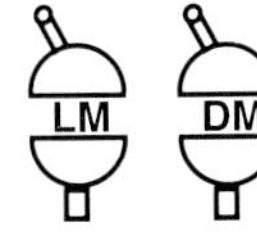

1. Aperture covered by a distinct flap................ 2
1'. Aperture with a trough-like lip.................... 3

2. Chambers narrow, curved............. ***Discorbis***
2'. Chambers ovoid, inflated............ ***Strebloides***

3. Spiral side distinctly convex, periphery angled and keeled... ***Rotorbis***
3'. Spiral side convex, surface ornamented by anastomosing network, periphery broadly rounded....................................... ***Rotorboides***

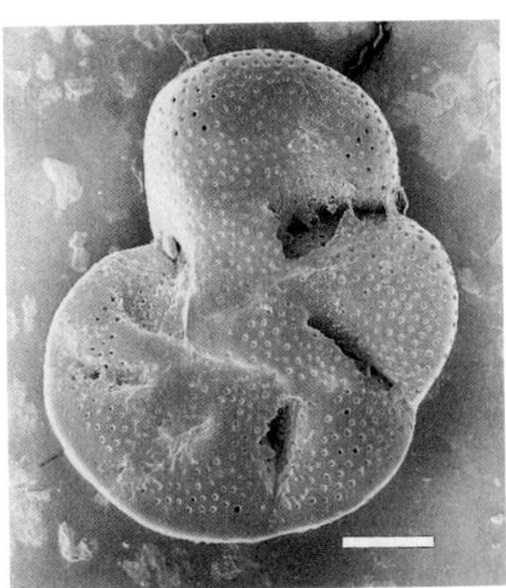

Fig. 177. *Discorbis vesicularis* (Lamarck, 1804). Scale=100 µm.

FAMILY EPONIDIDAE Hofker, 1951

Test trochospiral, biconvex, umbilicus closed; aperture a low arch extending from the umbilicus to the periphery, may have a few supplementary areal openings.

KEY TO THE GENERA

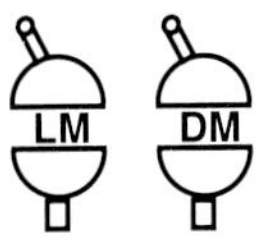

1. Aperture umbilical, extending onto the periphery, areal pores on umbilical side of the final chamber................... ***Poroeponides***
1'. Aperture umbilical, areal pores rare or absent..2

2. Trochospire high, umbilical side with stellate sutural ornamentation............... ***Neoeponides***
2'. Trochospire low, no sutural ornamentation...... .... ***Eponides***

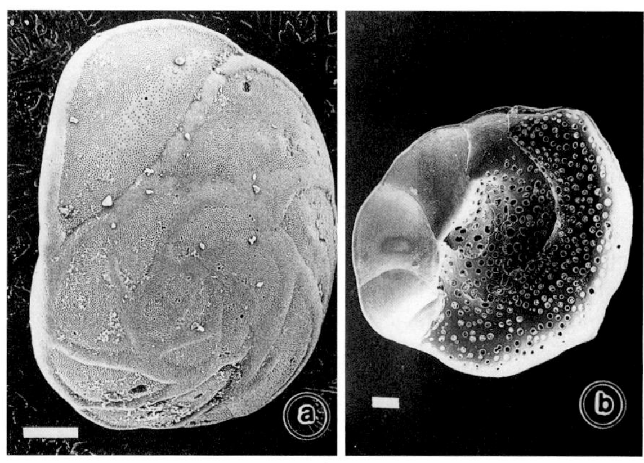

Fig. 178. *Poroeponides lateralis* (Terquem, 1878). a. Spiral view. b. Apertural view. Scales=100 μm.

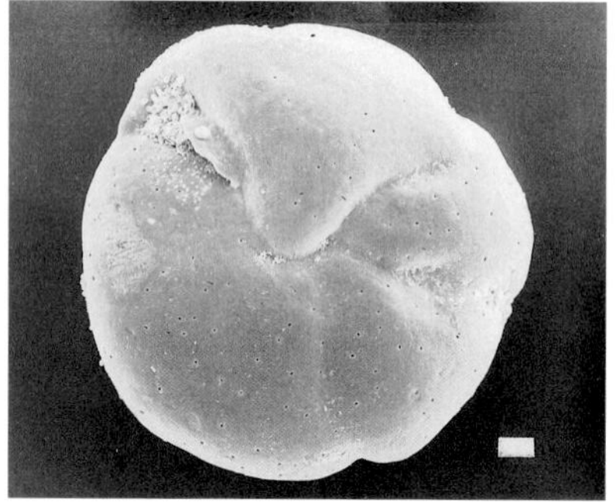

Fig. 179. *Eponides* sp. de Monfort, 1808. Scale=10 μm.

Fig. 180. *Heterostegina depress* d'Orbigny, 1826. a. Scale 400 μm. b. Scale=150 μm.

SUPERFAMILY NUMMULITACEA de Blainville, 1827

Test large, planispiral, evolute to involute; chambers numerous, may be subdivided into chamberlets, later chambers may be added in annular series; complex canal system of septal, marginal, and vertical canals; aperture an arched slit at the base of apertural face.

Fig. 181. *Operculina ammonoides* (Gronovius, 1781). Scale=100 μm.

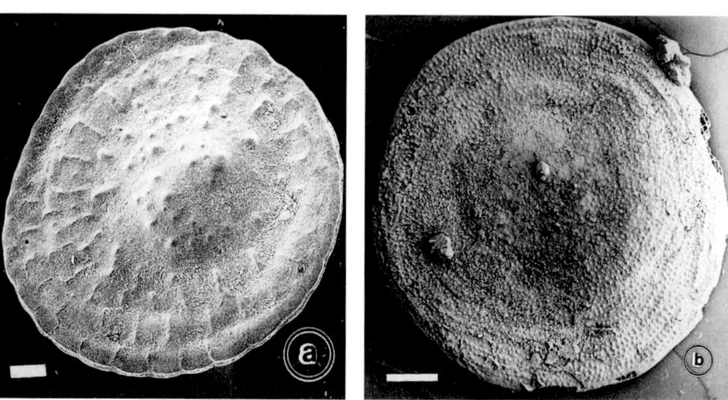

Fig. 182. *Cycloclypeus carpenteri* Brady, 1881. a. Juvenile. Scale=100 μm. b. Adult. Scale=1 mm.

FAMILY NUMMULITIDAE de Blainville, 1827

Characteristics as superfamily.

KEY TO THE GENERA

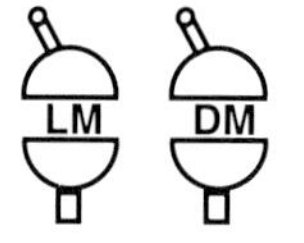

1. Chambers spirally coiled throughout............. 2
1'. Chambers annular in adult............................. 5

2. Chambers divided into chamberlets .... ***Heterostegina***
2'. Chambers undivided....................................... 3

3. Test lenticular or discoidal........ ***Nummulites***
3'. Test flattened, chambers narrow................... 4

4. Septal flap unfolded................................ ***Assilina***
4'. Septal flap moderately to strongly folded........ .... ***Operculina***

5. Test centrally umbonate, narrowed at the periphery .... ***Cycloclypeus***
5'. Test discoidal, thin....................... ***Heterocyclina***

ORDER GLOBIGERINIDA

Delage & Herouard, 1896

Planktonic in habit; test wall of perforate hyaline calcite, optically radiate, primary lamination bilamellar, with secondary lamination due to addition of material at formation of new chamber.

KEY TO THE SUPERFAMILIES:

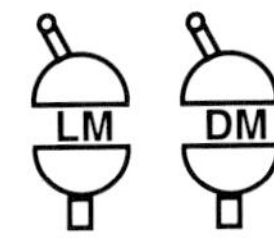

1. Test **microperforate**, non-spinose, without continous encrustation..................... **Candeinacea**
1'. Test macroperforate, adults spinose or non-spinose, may be encrusted or coated with smooth cortex...................................... **Globigerinacea**

SUPERFAMILY CANDEINACEA Cushman, 1927

Test trochospiral; wall microperforate in adult, with perforations irregularly scattered; surface unencrusted but smooth or with scattered pustules but no spines; aperture (or apertures) interio-marginal, with narrow lip (or lips) of uniform breadth.

FAMILY GLOBIGERINITIDAE Bermudez, 1961

Primary aperture single; dorsal sutural apertures absent or very few and sporadic. The species are characteristically highly eurythermal, thriving in surface and subsurface upper waters from the tropics to subpolar seas.

KEY TO GENERA OF THE GLOBIGERINITIDAE

1. Aperture intraumbilical... 2
1'. Aperture intra-extraumbilical 4

2. Lacking bulla................................ ***Tenuitellinata***
2'. With **umbilical bulla** 3

3. No accessory **bullate tunnels**........... .... ***Globigerinita***
3'. Infralaminal multiple apertures at end of bullate extension tunnels ...***Tinophodella***
4. Lacking bulla...................... ***Tenuitella***
4'. With bulla...................... ***Tenuitellita***

FAMILY CANDEINIDAE Cushman, 1927

Early growth stage as in the Globigerinitidae, but adults with umbilicus closed and the single primary aperture replaced by many small uniform apertures in each suture; surface smooth. Only one extant species (*Candeina nitida* [Figs. 183k-m]), rare in tropical waters.

SUPERFAMILY GLOBIGERINACEA

Carpenter, Parker & Jones, 1862

Test trochospiral, planospiral, or streptospiral; wall macroperforate, with perforations regularly and geometrically arranged; surface may be pitted, with or without spines in adult, but may be smooth in adult except for scattered pustules; aperture interiomarginal or areal.

KEY TO THE FAMILIES

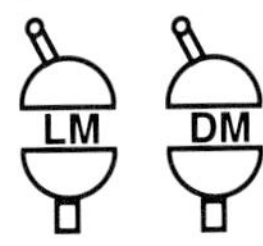

1. Test smooth except for distal, peripheral ends of chambers where thick spines, triradiate in section, are situated............ .... **Hastigerinidae**
1'. Test smooth or punctate; spines if present, circular in section, and scattered over the whole test 2

2. Adult test coated by cortical crust which narrows or blocks the exterior ends of the perforations .... **Sphaeroidinellidae**
2'. Test without a cortical crust 3

3. Test trochospiral, sometimes becoming planispiral or pseudoplanispiral during growth.. 4

3'. Test streptospiral.. 5

4. Surface of adult test spinose........................ 6
4'. Surface of adult test non-spinose.................... .... **Globorotaliidae**

5. Test smooth, adult wall without spines, aperture single, interiomarginal.......................... .... **Pulleniatinidae**
5'. Test punctate, adult thickly spinose, apertures multiple, areal............................ **Orbulinidae**

6. Test large, surface cancellate...................... .... **Globigerinidae**
6'. Test small, surface encrusted and smoothed....................... **Turborotalitidae**

FAMILY HASTIGERINIDAE
Bolli, Loeblich & Tappan, 1957

Test surrounded by a bubble-like capsule of vacuolated cytoplasm, a feature not known to occur in species of any other family. *Hastigerina* has no algal, dinoflagellate or diatom symbionts, but seems to be exclusively carnivorous. It also has a synodic lunar periodic reproduction cycle. These features are unknown in other families, and may prove to characterize all the Hastigerinidae. All are of low latitudinal distribution and commonly inhabits deep or intermediate water layers.

KEY TO THE GENERA

1. Adult chambers radially elongate, clavate, often partly distally divided into two or more lobes; test coiling, initially planispiral, becoming streptospiral....... ***Hastigerinopsis***
1'. Adult chambers not radially elongate........... 2

2. Coiling planispiral......................... ***Hastigerina***
2'. Coiling trochospiral.......................... ***Orcadia***

FAMILY SPHAEROIDINELLIDAE
Banner & Blow, 1959

Adult test coated in a thick, smooth calcitic cortex, which diminished or closed the perforations and thickened the apertural lips; surface of the younger whorls, below the cortex is punctate and spinose; primary aperture is single, interiomarginal, and intra-umbilical in position; supplementary dorsal, sutural apertures are present in only extant species (*Sphaeroidinella dehiscens* [Fig. 184c,d]). Rare in subsurface, tropical waters.

FAMILY GLOBIGERINIDAE
Carpenter, Parker & Jones, 1862

The trochospiral genera include both species which possess dinoflagellate endosymbionts and those which do not. *Globigerina* occurs in tropical to sub-polar latitudes, while *Globigerinoides* is confined to tropical and subtropical seas. *Beella* seems to be confined to the deeper waters of low latitudes. *Globigerinella* possesses two different endosymbi-onts, both of them being chrysophytes. It occurs in tropical and subtropical waters. *Bolliella* is much less common in these latitudes.

KEY TO THE GENERA

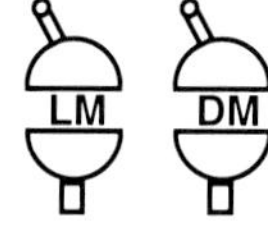

1. Adult test trochospiral................................. 2
1'. Adult test planispiral.................................... 3

2. With dorsal sutural supplementary apertures .... ***Globigerinoides***
2'. Without supplementary apertures................ 4
3. Adult chambers subglobular............................ .... ***Globigerinella***
3'. Adult chambers radially elongate..................... .... ***Bolliella***

4. Adult chambers subglobular.... ***Globigerina***
4'. Adult chambers radially elongate....... ***Beella***

FAMILY PULLENIATINIDAE Cushman, 1927

The initial and adult coils are streptospiral; the chambers increasingly embrace ventrally, so that the umbilicus becomes closed and the aperture becomes wholly extra-umbilical; in advanced forms, the chambers begin to embrace dorsally, so that the last whorl covers much of the preceding whorls; the surface of the neanic chambers is punctate, perforation-pitted and spinose, but the surface of the adult chambers is encrusted and smooth, except for tubercles around the aperture. There is only one extant genus, *Pulleniatina* (Fig. 185e,f), which occurs in tropical sea.

FAMILY ORBULINIDAE Schultze, 1854

Test coiling is intensely streptospiral, with the last, globular chamber completely embracing the ventral side (or all) of the juvenile test. The interiomarginal aperture of the nepionic and neanic stages is replaced by a multitude of rounded pores in the area of the last chamber. The surface of the adult test is covered with perforation-pits and spines. There is only one extant genus *Orbulina* (Fig. 185g), distributed from tropical to temper-ate seas, but is much commoner in lower latitudes. It is carnivorous, preying upon several species of copepods and other organisms. *Orbulina* is host for dinoflagellates of genus *Gymnodinium* (up to 23,000 symbiotic algae have been found inside one adult test).

FAMILY GLOBOROTALIIDAE Cushman, 1927

The peripherally carinate (keeled) species develop smooth tests which may possess scattered, small tubercles but which are not cancellate or punctate, -unlike the unkeeled genera, which never have smooth adult tests. *Paragloborotalia* species are wider spread latitudinally, being quite common in subpolar seas (e.g. *P. incompta*) to the tropics, but carinate *Globorotalia* is temperate to tropical only. None of genera is known to possess endosymbionts.

KEY TO THE GENERA

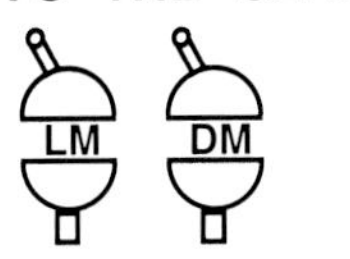

1. Equatorial periphery carinate.. ***Globorotalia***
1'. Equatorial periphery non carinate................. 2

2. Surface smooth, adult chambers often elongate .... ***Clavatorella***
2'. Surface rough, adult chambers not elongate.3

3. Principal aperture umbilical-extraumbilical throughout adult growth... ***Paragloborotalia***
3'. Principal aperture becomes intra-umbilical in last few chambers .... ***Neogloboquadrina***

FAMILY TURBOROTALITIDAE Hofker, 1976

The species of this family are all small, but their numbers of whorls and their calcitic crusts readily distinguish them from juvenile tests of taxa of other families.

KEY TO THE GENERA

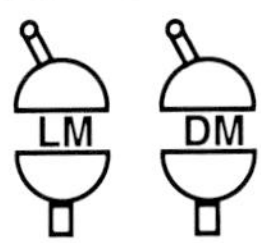

1. Apertural lip weak or absent........ ***Berggrenia***
1'. Apertural lip broad, making the final chamber ampullate, with small, accessory infra-laminal apertures at margin of ampulla, which may cover the umbilicus................ ***Turborotalita***

Species of *Turborotalita* carry endosymbionts (often chrysophytes) and live in the photic zone for part, at least, of their lives; it is possible that the adults live at greater depths and there develop thicker calcitic crusts to increase the smoothness and heaviness of their tests. It has a transglobal, tropical-subtropical distribution. *Berggrenia* is known only from tropical latitudes of the eastern Pacific.

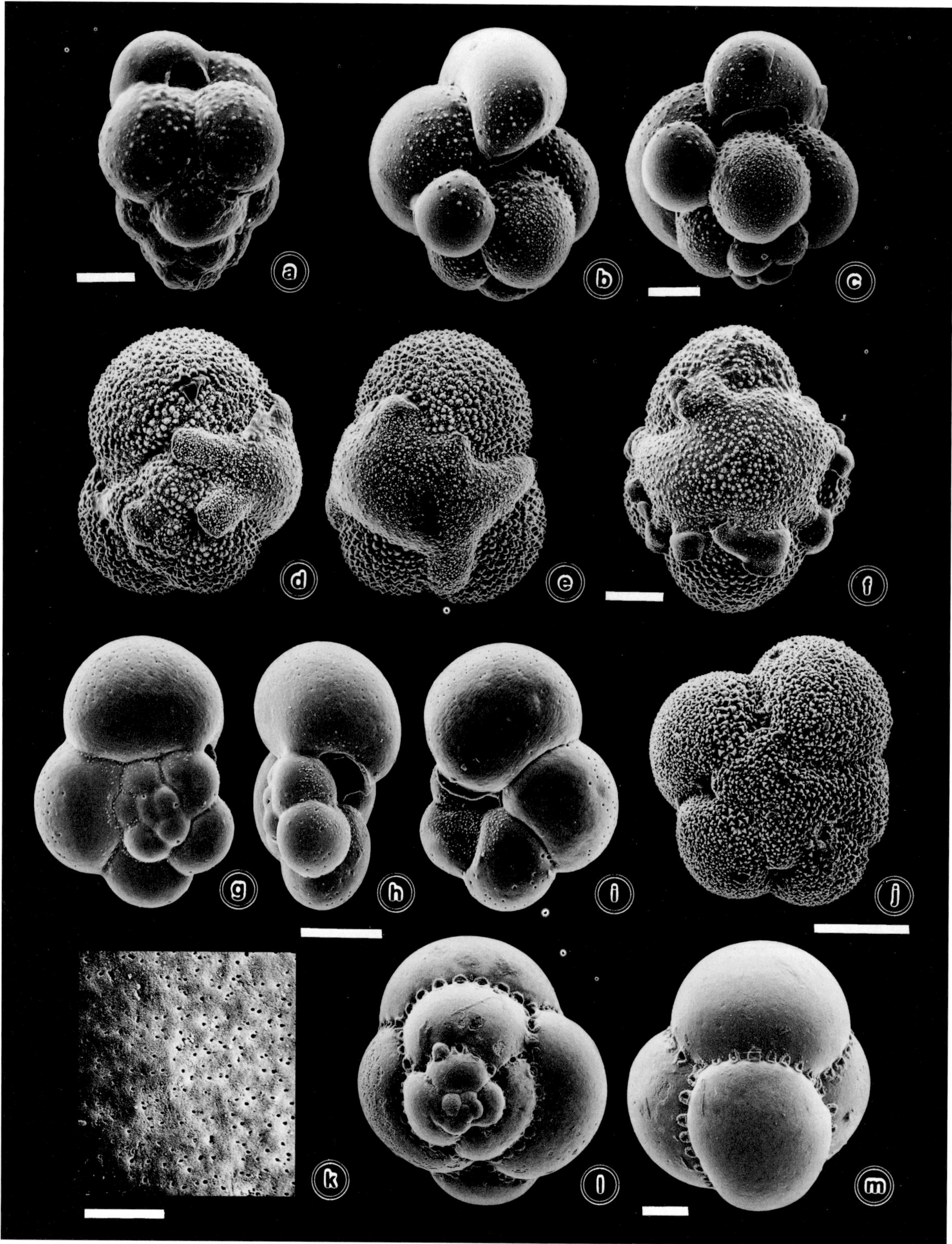

Fig. 183. a. *Tenuitellinata uvula* (Ehrenberg, 1861). Scale=25 μm. b,c. *Globigerinita minuta* (Natland, 1933). Scale=50 μm. d-f. *Tinophodella glutinata* (Egger, 1893). Scale=50 μm. g-i. *Tenuitella anfracta* (Parker, 1967). Scale=50 μm. j. *Tenuitellita iota* (Parker, 1962). Scale=50 μm. k-m. *Candeina nitida* d'Orbigny, 1839. Scale=100 μm (l,m) 100 μm (m). k. Microperforate wall. Scale=10 μm.

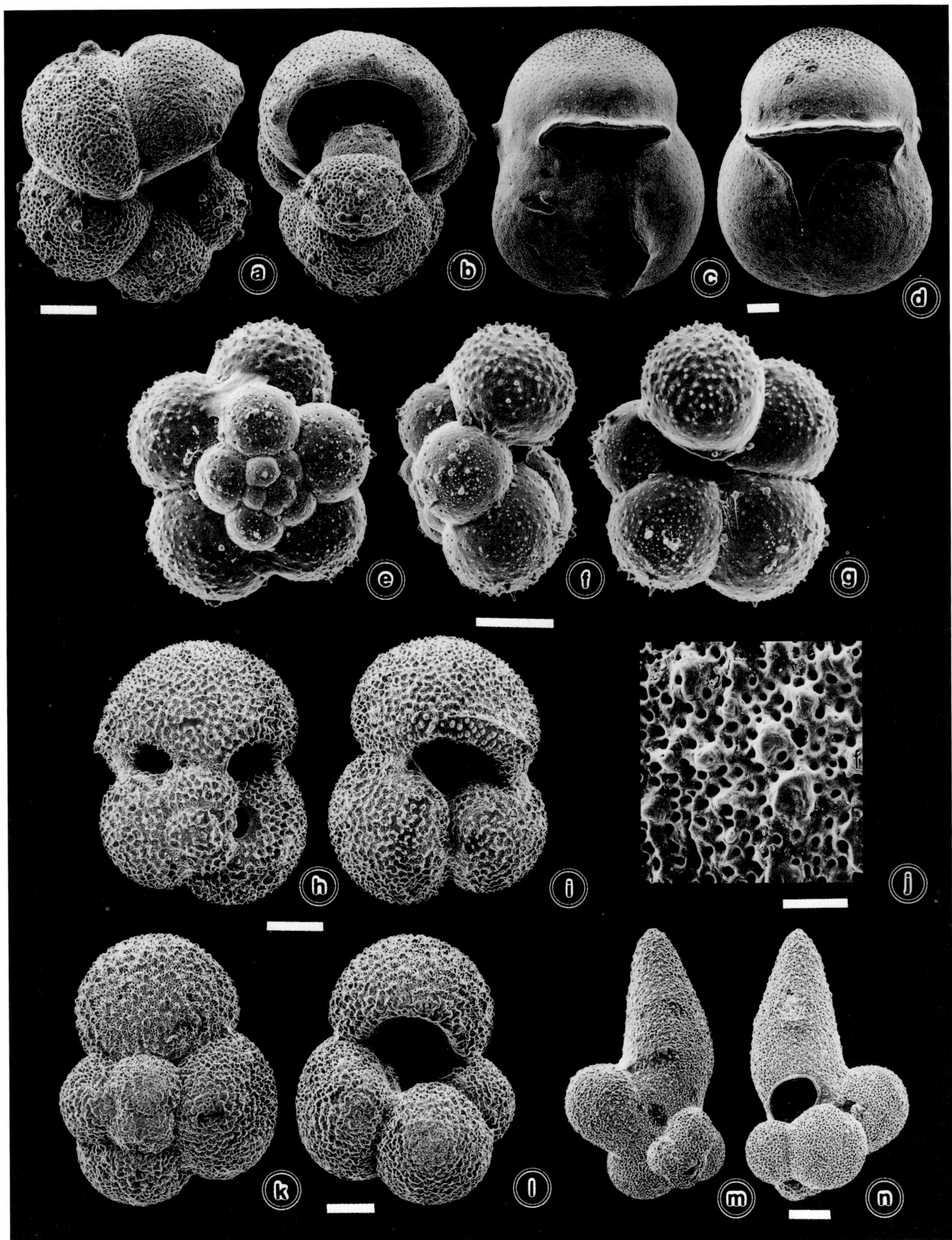

Fig. 184. a,b. *Hastigerina pelagica* (d'Orbigny, 1839). Scale=100 µm. c,d. *Sphaeroidinella dehiscens* (Parker & Jones, 1865). Scale=100 µm. e-g. *Orcadia riedeli* (Rögl & Bolli, 1973). Scale=50 µm. h-j. *Globigerinoides ruber* (d'Orbigny, 1839). (h,i) Scale=100 µm. j. Macroperforate wall. Scale=25 µm. k,l. *Globigerina bulloides* d'Orbigny, 1826. Scale=100 µm. m,n. *Beella digitata* (Brady, 1879). Scale=100 µm.

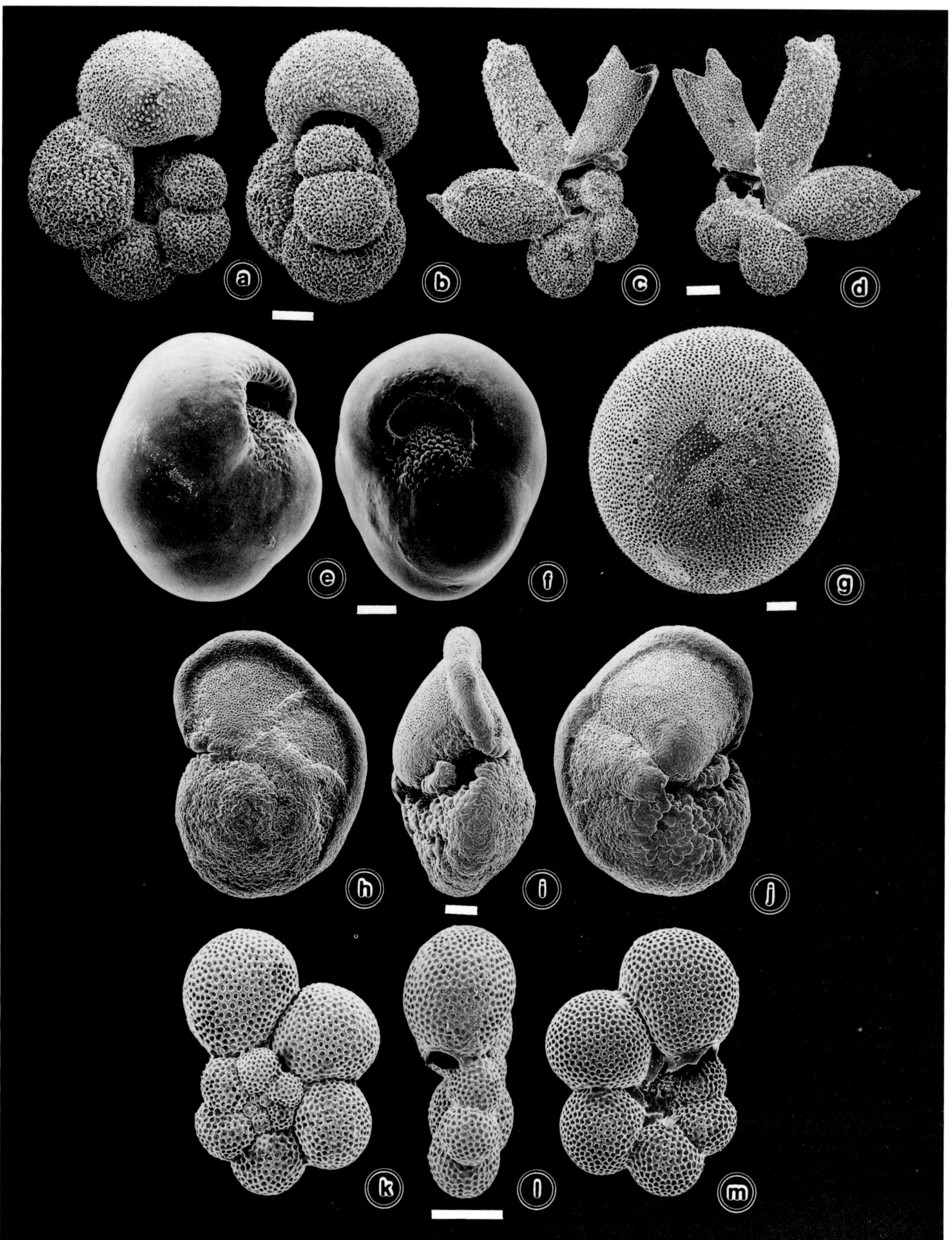

Fig. 185. a,b. *Globigerinella siphonifera* (d'Orbigny, 1839). c,d. *Bolliella adamsi* (Banner & Blow, 1959). e,f. *Pulleniatina obliquiloculata* (Parker & Jones, 1865). g. *Orbulina universa* d'Orbigny, 1839. h-j. *Globorotalia tumida* (Brady, 1877). k-m. *Clavatorella hexagona* (Natland, 1938). All scales=100 µm.

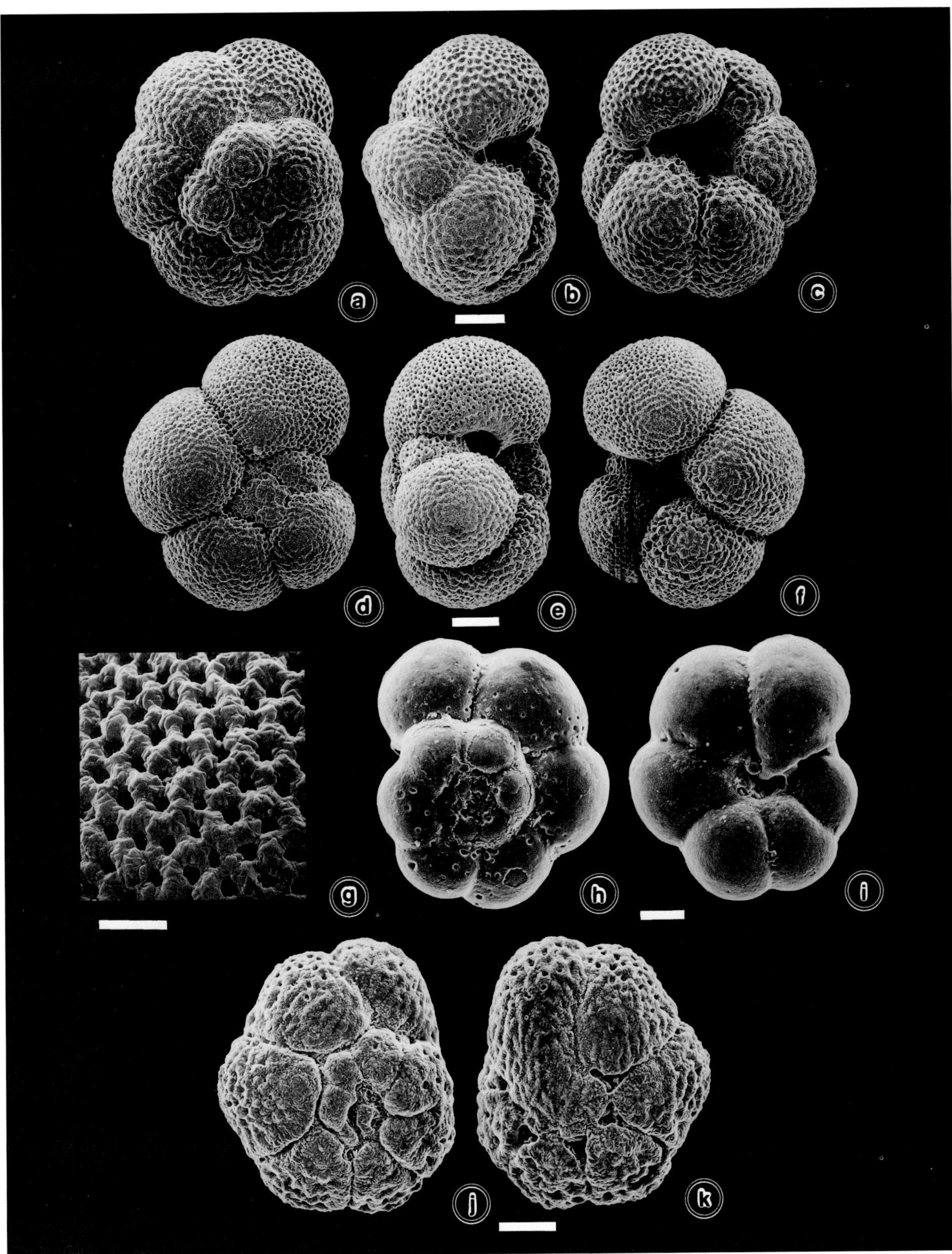

Fig. 186. a-c. *Neogloboquadrina dutertrei* (d'Orbigny, 1839). Scale 100 µm. d-g. *Paragloborotalia blowi* (Rögl & Bolli, 1973). (d-f) Scale=100 µm. g. Encrusted macroperforate wall. Scale=25 µm. h,i. *Berggrenia pumilio* (Parker, 1962). Scale=25 µm. j,k. *Turborotalita humilis* (Brady, 1884). Scale=50 µm.

ROTALIINA *INCERTAE SEDES*

FAMILY PLACENTULINIDAE Kasimova, Poroshina & Geodakchan, 1980

Test a low trochospiral coil with two to three overlapping chambers per whorl, chambers with peripheral septula as in *Patellina;* aperture an umbilical arch; wall finely perforate, optically radial. Three genera (*Ashbrookia, Patellinella* [Fig. 187), *Subpatellinella*).

Fig. 187. *Patellinella inconspicua* (Brady, 1884). Scale=20 μm.

FAMILY UNGULATELLIDAE Seiglie, 1964

Test trochospiral, composed of two or three chambers per whorl, early stage may be with an undivided tubular chamber; aperture umbilical, slit-like; three genera (*Ungulatella* [Fig. 188], *Ungulatelloides, Metapatellina*).

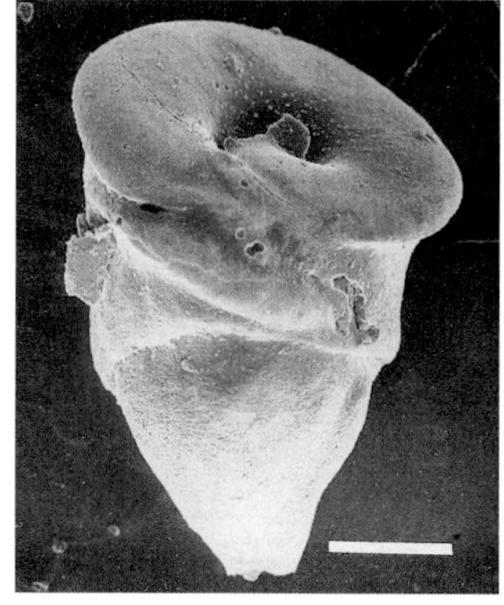

Fig. 188. *Ungulatella pacifica* Cushman, 1931. Scale=50 μm.

FAMILY PANNELLAINIDAE Loeblich & Tappan, 1984

Test tiny, planoconvex, 7-9 chambers per whorl, chambers aligned with those of the preceding whorl, sutures strongly elevated and costate on the spiral side; aperture umbilical. Single genus *Pannellaina* (Fig. 189).

Fig. 189. *Pannellaina byramensis* (Cushman, 1922). Scale=50 μm.

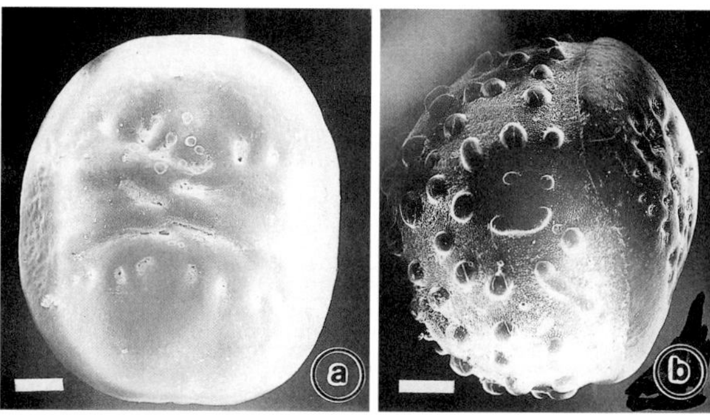

Fig. 190. *Pegidia dubia* (d'Orbigny, 1826). a. Spiral view. b. Apertural view. Scales=100 μm.

FAMILY PEGIDIIDAE Heron-Allen & Earland, 1928

Test globular with few partially enveloping chambers; aperture multiple. Three genera (*Sphaeridia, Pegidia* (Fig. 190), *Siphonidia*), subtropical.

FAMILY SPHAEROIDINIDAE Cushman, 1927

Test subglobular with strongly embracing,

chambers; aperture a crescentic opening near the base of the chamber; wall finely perforate. Single Recent genus *Sphaeroidina* (Fig. 191), deep sea.

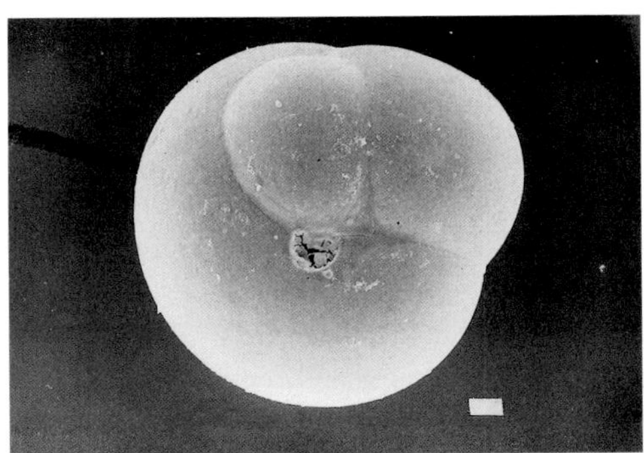

Fig. 191. *Sphaeroidina* sp. d'Orbigny, 1926. Scale=10 µm.

LITERATURE CITED

Grell, K. G. 1954. Der Generationswechsel der polythalamen Foraminifere *Rotaliella heterocaryotica*. *Arch. Protistenkd.*, 100:211-235.

Grell, K. G. 1958. Untersuchungen uber die Fortpflanz-ung und Sexualitat der Foraminiferen III. *Glabratella sulcata*. *Arch. Protistenkd.*, 102:449-472.

Hansen, H. J. & Revets, S. A. 1992. A revision and reclassification of the Discorbidae, Rosalinidae, and Rotaliidae. *J. Foraminiferal Res.*, 22:166-180.

Haynes, J. R. 1981. *Foraminifera*. John Wiley & Sons, New York. 433 pp

Hottinger, L. 1979. Araldit als Helfer in der Mikropal-äontologie. *Ciba-Geigy Aspekte*, 1979/3:1-10

Hottinger, L., Halicz, E. & Reiss, Z. 1992. Architecture of *Eponoises* and *Poroeponoides* reexamined. *Micropaleontology(NY)*, 37:60-75.

Loeblich, A. J. & Tappan, H. 1989. Implications of wall composition and structure in agglutinating foraminifers. *J. Paleont.*, 63:769-777.

Loeblich, A. J. & Tappan, H. 1988. *Foraminiferal Genera and their Classification*. Van Nostrand Reinhold, New York. Vol. 1. 970 pp. Vol 2, 212 pp. and 847 plates.

Patterson, R. T. & Richardson, R. H. 1987. A taxonomic revision of the unilocular foraminifera. *J. Foraminiferal Res.,* 17:212-226.

Pawlowski, J. & Lee, J. J. 1992. The life cycle of *Rotaliella elatiana* n. sp.: A tiny macro-algavorous foraminifer from the Gulf of Elat. *J. Protozool.*, 39:131-143.

Pawlowski, J., Zaninetti, L., & Lee, J. J. 1992. Systematic revision of tiny herokaryotic foraminfera described by Karl Grell in the years 1954-1979. *Rev. Paleobiol.*, 11:385-395.

Pawlowski, J., Bolivar, I., Guiard-Maffia, J. & Gouy, M. 1994a. Phylogenetic position of foraminifera inferred from LSU rRNA gene sequences. *Mol. Biol. Evol.*, 11:929-938.

Pawlowski, J., Bolivar, I., Fahrni, J., Cavelier-Smith, T. & Gouy, M. 1996. Early origin of foraminifera suggested by SSU rRNA gene sequences. *Mol. Biol. Evol.*, 13:445-450.

Revets, S. A. 1991. The nature of *Virgulinella* Cushman, 1932 and the implications for its classification. *J. Foraminiferal Res.*, 21:293-299.

Sogin, M. L. 1997. History assignment: when was the mitochondrion founded? *Current Opinion in Genetics & Development*, 7:792-799.

de Vargas, C., Zaninetti, L., Hilbrecht, H. & Pawlowski, J. 1997. Phylogeny and rates of molecular evolution of planktonic foraminifera: SSU rDNA sequences compared to the fossil record. *J. Mol. Evol.*, 45:285-294.

Wray, C. G., Langer, M. R., De Salle, R., Lee, J. J. & Lipps, J. H. 1995. Origin of the foraminifera. *Proc. Nat Acad. Sci. USA*, 92:141-145.

CLASS MYCETOZOA De Bary, 1859

By MICHAEL J. DYKSTRA and HAROLD W. KELLER

The Mycetozoa has three subclasses as described by Olive (1970): the Protosteliidae, the Guttulinia, and the Myxogastridae. The trophic phases of these organisms may be amoeboid, amoebo-flagellate, or plasmodial. Amoebae and amoebo-flagellate stages can encyst by secreting wall material when growth conditions such as food supply, water availability, and temperature become limiting. Under appropriate conditions of light, nutrition and humidity, members of this class form fruiting bodies that rise above the substratum and produce aerial spores. In some cases, the fruiting body stalks are cellular; others are acellular. **Sporangia** may exhibit ancillary sterile elements or may be merely an assemblage of spores contained within a sporangial wall. In other cases, spores may be formed in uniseriate chains covered with a slime sheath, with or without dichotomous branching (Guttulinia). Members of the Dictyosteliida have a mass of spores subtended by a cellular stalk and the entire **sorocarp** is covered with a sheath. No fossil record remains of this group, so their relationships are problematic. New research techniques using DNA comparisons should soon shed light on their origins and relationships to other members of this class and to other protists. These free-living, holozoic organisms are found almost anywhere organic material is located. They are readily observed on rotting logs, soil, living trees and herbaceous plants, and similar environments where dampness is coupled with assemblages of microorganisms on organic substrata. Mycetozoans have been classified historically with the fungi and given class, order, and family names with plant suffixes in accordance with the Code of Botanical Nomenclature, because fungi were considered to be plants due to their sporangial walls and stalked fruiting bodies. When Whittaker's landmark paper was published in 1969, creating five kingdoms and raising fungi to kingdom level, it reflected the fact that this group possessed characters unlike plants and fungi and should be placed with the protists because of animal-like amoebae, amoebo-flagellate, and plasmodial assimilative stages. In this chapter, the suffixes used for family names and higher taxonomic groupings agree by convention with the Zoological Code of Nomenclature. The vast majority of the members of this class cannot be grown in axenic culture, so many questions about their physiological and biochemical capabilities remain unanswered. Most mycetozoans have not been maintained successfully in culture under any conditions. Sexuality has been demonstrated in only one member of the Protosteliida (*Ceratiomyxa*), in several members of the Dictyosteliida, and in a number of the Myxogastria. The Myxogastria are most thoroughly discussed by Frederick (1990) and in *The Myxomycetes* (Martin and Alexopoulos, 1969). The cellular slime molds are treated by Cavender (1990), in *The Mycetozoans* (Olive, 1975), and in *The Dictystelids* (Raper, 1984). Olive (1975) gives the most thorough treatment of the Protosteliida. The economic and ecological significance of this class is uncertain, though the cellular slime mold *Dictyostelium discoideum* Raper and the plasmodial slime mold *Physarum polycephalum* Schw. have been used extensively in the study of genetics, developmental biology, and cellular physiology.

KEY TO THE SUBCLASSES

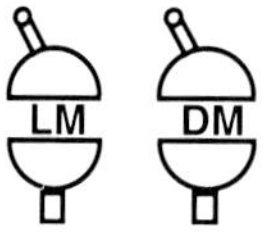

1. Spores not produced by endogenous cleavage; fruiting bodies typically minute; trophic stage uninucleate amoebae or amoebo-flagellates, or consisting of multinucleate plasmodia............ 2

1'. Spores formed endogenously by cytoplasmic cleavage within a typically multinucleate sporangium; trophic stage a **plasmodium**, amoeba or amoeboflagellate.. **Subclass Myxogastria**

2. Fruiting bodies arising from single amoeboid cells or aggregates of amoebae (pseudoplasmodia); stalk tube presen; trophic stage amoeboid or amoebo-flagellate with filose pseudopodia or multinucleate plasmodia .... **Subclass Protostelia**

2'. Fruiting bodies arising from pseudoplasmodia; no stalk tube formed; plasmodia not present; amoebae of limax type..**Subclass Guttulinia**

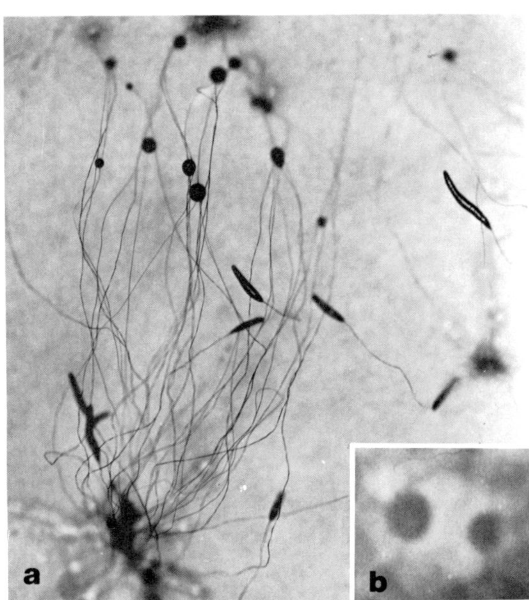

Fig. 1. *Acytostelium leptosomum*. a. Mature sporocarps (with globose sori and developing sorocarps with elongated cellular masses at the stalk tip. X 40. b. A nucleus showing two peripheral nucleoli, approximately X 2400. (Fig. 13, Hutner, 1985)

SUBCLASS PROTOSTELIA L. Olive, 1970

The trophic stage of these organisms consists of uninucleate amoebae or amoebo-flagellates and multinucleate plasmodia. During locomotion, the protoplasts form filose pseudopodia. Flagellate cells are found only in some members of the Protosteliida. One- to few-spored sporocarps are formed from the uninucleate or multinucleate protoplasts of the Protosteliida, while aggregates of amoebae form pseudoplasmodia in the Dictyosteliida and give rise to multi-spored sorocarps. When conditions are unfavorable, trophic stages of many species in the subclass can synthesize wall material and encyst on the substrate. Stalk tubes are formed during sporulation in all members of the subclass. Cellulose has been identified in the stalk tubes of both Dictyosteliida and Protosteliida species. Sexuality has been conclusively demonstrated only in *Ceratiomyxa* (Protosteliida) and the Dictyosteliida. Members of this subclass have been isolated from dung, soil, and plant parts where they are presumed to feed on bacteria, yeasts, and fungal spores identical or similar to those used to maintain them in culture. No practical or ecological importance has been ascribed to members of the subclass, though members of the Dictyosteliida have been used extensively for research on differentiation and chemotropism.

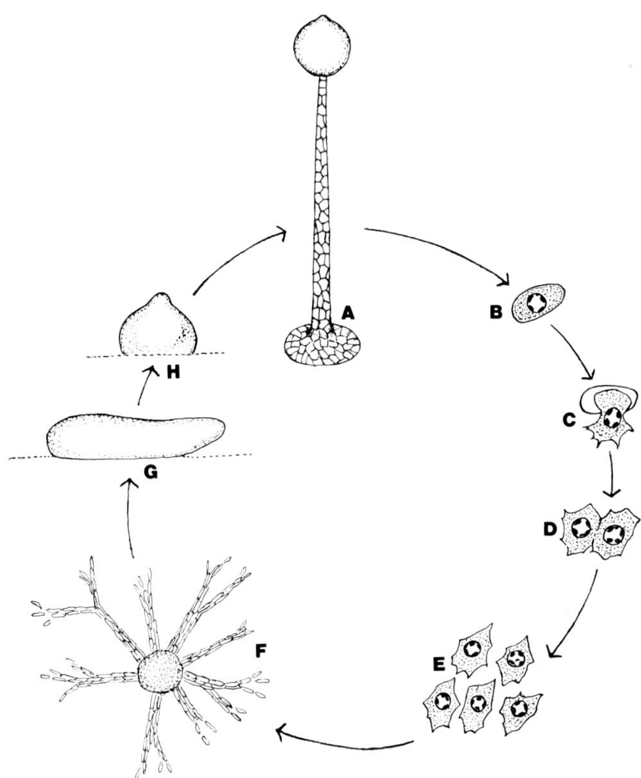

Fig. 2. Diagrammatic life cycle of *Dictyostelium discoideum* Raper. A. Sorocarp with sorus containing spores subtended by a cellular stalk. B. Spore. C. Amoebae emerging from spore. D. Cell division. E. Amoebae. F. Amoebae aggregating to form slug. G. Migrating slug (pseudoplasmodium). H. Beginning of sorogenesis (culmination). (Fig. 12, Hutner, 1985)

ORDER DICTYOSTELIIDA L. Olive, 1970

This order consists of one monotypic family (Acytosteliidae) and the Dictyosteliidae, containing three genera composed of about 70 species. The trophic stage is amoeboid and aflagellate, possessing filose pseudopodia during locomotion. In culture, once the food supply is exhausted, the amoebae can aggregate under the influence of chemoattractants to form a **pseudoplasmodium**. The pseudoplasmodium is an aggregation of individual amoebae that differentiate into pre-stalk and pre-spore cells. The pseudoplasmodium may or may not migrate extensively, but ultimately undergoes **soro-**

genesis, during which a stalk tube is formed by the stalk cells which, in the Dictyosteliidae become incorporated into the stalk tube. The remaining amoebae go to the top of the stalk tube, secrete a wall, and become spores. The mass of spores is referred to as a **sorus**. A slime sheath covers the entire **sorocarp**. Under appropriate conditions, amoebae of some species can fuse to form a diploid cell that becomes a macrocyst, wherein meiosis occurs. Useful taxonomic treatments include *Monograph of the Acrasieae* (E. W. Olive, 1902), *The Mycetozoa* (L. S. Olive, 1975), *The Dictyostelids* (K. B. Raper, 1984) and J. C. Cavender's chapter (1990) in the *Handbook of Protoctista* (Margulis et al., 1990).

KEY TO THE GENERA

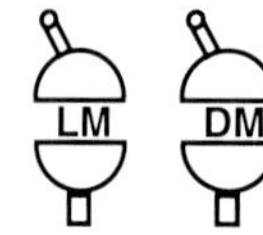

1. Pseudoplasmodia give rise to sorocarps with acellular stalk tubes supporting the spores....................... **Family Acytosteliidae**

1'. Pseudoplasmodia give rise to sorocarps with stalk tubes containing stalk cells...................... Family **Dictyosteliidae** 2

2. Sorocarp apex expanded into a denticulate cup bearing round spore.......................... ***Coenonia***

2'. Sorocarp apex not expanded into a denticulate cup 3

3. Unbranched sorocarp or sometimes with short, irregularly distributed lateral branches .... ***Dictyostelium***

3'. Sorocarp consisting of terminal sorus, beneath which arise whorls of secondary sori.......... .... ***Polysphondylium***

FAMILY ACYTOSTELIIDAE Raper & Quinlan, 1958

This family was described for *Acytostelium leptosomum* Raper & Quinlan, 1958 Cavender (pers. comm.) currently recognizes four or five species, not all of which have been formally described. A slender acellular stalk tube supports a mass of spores. The single nucleus of the amoeba has 1-3 peripheral nucleoli. Several sorocarps typically arise from one aggregate of amoebae. In culture, sorocarps with as few as two spores may be induced to form. Figure 1 shows sorocarps and the nucleus of an amoeba.

TYPE SPECIES: *Acytostelium leptosomum* Raper & Quinlan, 1958

FAMILY DICTYOSTELIIDAE Rostafinski, 1875

The family consists of three genera, one of which *Coenonia* has not been seen since its original description. After amoebal aggregation, the pseudoplasmodium may or may not migrate extensively during differentiation into pre-stalk and pre-spore cells. During sorogenesis, the stalk cells secrete a cellulosic stalk tube, crawl into the tube and synthesize a wall around themselves before dying. The remaining amoebae crawl up the outside of the cellular stalk tube and, upon reaching the apex, become spores following wall synthesis. The spores are presumed to be distributed by small organisms passing by as well as water and air currents. The life cycle of *Dictyostelium discoideum* Raper is shown in Fig. 2. Certain species (e.g. *Dictyostelium discoideum* and *Polysphondylium pallidum* E. W. Olive) have been grown in axenic culture.

Genus ***Coenonia*** van Tieghem, 1884

A single species reported only once from decaying beans.

TYPE SPECIES: *Coenonia denticulata* van Tieghem, 1884

Fig. 3. Migrating pseudoplasmodium of *Dictyostelium mucoroides* Brefeld forming stalk on agar substrate. X 52. (Fig. 14, Hutner, 1985)

Genus ***Dictyostelium*** Brefeld, 1869

Approximately 50 species represent the genus and are differentiated on the basis of sorocarp size, pigmentation, morphology, and spore shape. Figure 2 illustrates stages in the life cycle for *discoideum* Raper. The most commonly encountered species is *D. mucoroides* Brefeld (Fig. 3), which was also the first cellular slime mold isolated (Brefeld, 1869).

TYPE SPECIES: *Dictyostelium mucoroides* Brefeld, 1869

Stages in the life cycle of *P. pallidum* E. W. Olive are shown in Fig. 4. Approximately a dozen species are recognized by Cavender (pers. comm.), based on sorocarp size, pigmentation, morphology, and spore shape.
TYPE SPECIES: *Polysphondylium violaceum* Brefeld, 1884

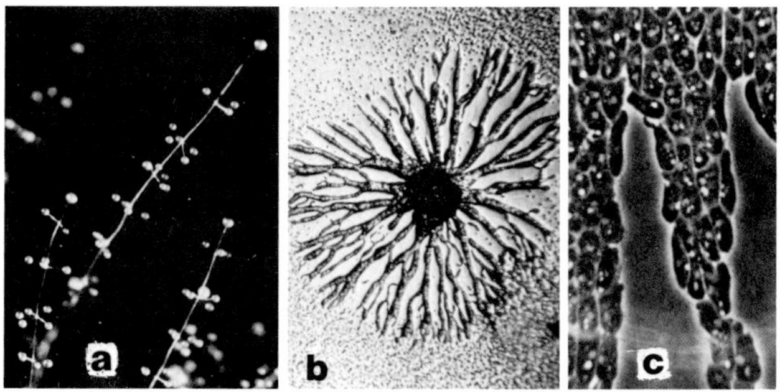

Fig. 4. *Polysphondylium pallidum* E. W. Olive. a. Sorocarps. X6. b. Aggregation. c. Amoebae streaming towards aggregation center. X280. (Fig. 17, Hutner, 1985)

SUBCLASS GUTTULINIA L. Olive, 1970
ORDER GUTTULINIDA L. Olive, 1970

This taxon represents a grouping of some potentially unrelated organisms that historically have been grouped together for purposes of convenience rather than phylogenetic relationship (Blanton, 1990). The trophic stage consists of limax-type amoebae in all cases, with only the species *Pocheina flagellata* described by Olive et al. (1983) possessing flagellate cells (see Blanton, 1990). All genera in the subclass form microcysts under conditions unfavorable to the trophic stage. Amoebae migrate singly or in small groups to form sorocarps, never exhibiting streams of aggregating cells typical for the Dictyosteliidae. The aggregated cells climb on top of each other, rising off of the substrate to produce a sorocarp that may consist of only one type of encysted cell (sorocysts) or may be differentiated into stalk cells and spores, as is the pattern in Dictyosteliidae. The species with differentiated stalk cells and spores have platelike mitochondrial cristae while those consisting entirely of sorocysts have tubular mitochondrial cristae (Dykstra, 1977). The only exception to this pattern is the species *Fonticula alba* described by Worley et al. (1979), which rapidly extrudes cells through the center of its stalk to form the sorus and has platelike mitochondrial cristae. This monotypic genus also forms filose pseudopodia on the advancing front of migrating amoebae. Unlike the dictyostelids, sorocarp stalks and sori formed by members of the Guttulinia contain viable cells at maturity and have no stalk tube. The family Fonticulidae has uncertain affinities but is included here as a monotypic genus, *Fonticula alba* Worley, Raper & Hohl. The subclass now consists of one order, four families, six genera, and 16 species, many of which are relatively rare. Our limited knowledge of their world-wide distribution probably reflects the fact that there have been very few individuals devoted to collecting these organisms. Less than 50 research papers have been published on this subclass, though the biology and taxonomy of this group are discussed by Raper (1984) and Blanton (1990). Some of these genera have been isolated exclusively from the bark of living trees and vines (*Pocheina*); some are isolated from dung (*Guttulinopsis*, *Copromyxella*, *Copromyxa*, *Fonticula alba*), while *Copromyxella* and *Guttulinopsis* have also been isolated from soil. *Acrasis rosea* Olive & Stoianovitch can be isolated from aerial inflorescenses of dead plants. *Guttulinopsis vulgaris* E. W. Olive is easily isolated by incubating horse or cow dung in a moist chamber for several days. *Acrasis rosea* is commonly isolated from moribund corn tassels collected in the fall, soaked in sterile water for a few hours, and put into a moist chamber where sorocarps can be seen within 4-5 days.

KEY TO THE FAMILIES AND GENERA OF THE ORDER GUTTULINIDA

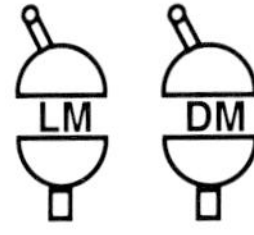

1. Sorocarp of encysted cells of only one type (sorocysts) without hila or markings in simple or branched columns; microcysts formed on the substrate.. Family **Copromyxidae** ... 2
1'. Sorocarp differentiated into stalk cells and spores.. 3

2. Amoebae with contractile vacuoles; yellow-brown macrocysts formed when different isolates are mixed........................ ***Copromyxa***
2'. Amoebae lacking contractile vacuoles and macrocysts............................... ***Copromyxella***

2'. Amoebae lacking contractile vacuoles and macrocysts.............................. ***Copromyxella***

3.. Stalk composed of viable stalk cells in intimate contact with each other; little or no cytoplasmic debris or matrix materials in stalk......Family **Acrasidae**........................... 4

3'. Stalk composed of large amounts of extra-cellular matrix material and cytoplasmic debris along with stalk cells not in intimate contact with each other.................................. 5

4. Spores in uniseriate chains, branched or not .... ***Acrasis***

4'. Spores contained in a globose sorus............. .... ***Pocheina***

5. Cells move individually up and outward from sorogen and then differentiate into spores...................... Family **Guttulinopsidae**

5'. Cells rapidly extruded from center of stalk to form sorus where they differentiate into spores........................ Family **Fonticulidae** ....(**=Fonticulaceae**)

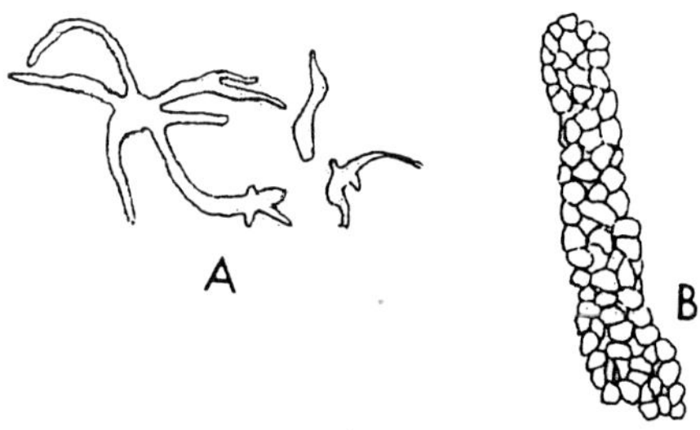

Fig. 5. *Copromyxa protea* (Fayod) Zopf. A. Branched sorocarps composed entirely of sorocysts. X8. B. Detail of tip of sorocarp branch tip, showing arrangement of sorocysts found throughout the sorocarp. X138. (Redrawn from Fig. 10-2, Raper, 1984)

FAMILY COPROMYXIDAE
OLIVE & STOIANOVITCH, 1975

Genus ***Copromyxa*** (Fayod, 1883) Zopf, 1885

Copromyxa protea (Fayod) Zopf (Fig. 5) has limax-type amoebae 25-35 µm in diameter that contain one or more contractile vacuoles that are normally uninucleate. Sorocarps are composed of sorocysts throughout. Sorocarps attain heights of 2.5-3.0 mm and may be sparsely to copiously branched. Raper (1984) considers this to be a monotypic genus.

TYPE SPECIES: *Copromyxa protea* (Fayod) Zopf, 1885

Fig. 6. *Copromyxella filamentosa* Raper, Worley & Kurzynski. A. Simple to occasionally branched sorocarps. X15. b. Detail of sorocarp tip, showing arrangement of sorocysts. X234. (Redrawn from Fig. 10-3, Raper, 1984)

Genus ***Copromyxella***
Raper, Worley & Kurzynski, 1979

Copromyxella filamentosa Raper, Worley, and Kurzynski, 1979 (Fig. 6) is representative of the genus. It has limax-type amoebae normally devoid of contractile vacuoles. The sorocarps may be simple or sparsely branched, measuring 25-40 µm in diameter and 1.0-1.5 mm in length. Near the tip of the sorocarp, the diameter may be as thin as 10-15 µm in diameter. The filamentous sorocarps appear as thread-like tangles in culture.

TYPE SPECIES: *Copromyxella filamentosa* Raper, Worley & Kurzynski, 1979

FAMILY ACRASIDAE POCHE, 1913

Genus ***Acrasis*** van Tieghem, 1880
emend. L. Olive, 1970

Acrasis granulata van Tieghem was described as the type species for the genus. It had a violaceous tint and was isolated from spent beer yeast. It has never been seen since. The other species in the genus, *Acrasis rosea* Olive & Stoianovitch, 1996 (Fig. 7), is easily isolated from dead plant inflorescences, particularly corn tassels, and feeds readily on the pink yeast, *Rhodotorula*. It forms sorocarps that are bright pink and dusty-appearing in culture. Microcysts are formed on the substratum and sorocarps are formed that can be up to 800 µm in height. The sorocarps have stalk cells that are viable at maturity and will germinate when placed on fresh agar with

Rhodotorula cells. The spores are round and have hila where they contact each other.
TYPE SPECIES: *Acrasis granulata* van Tieghem, 1880

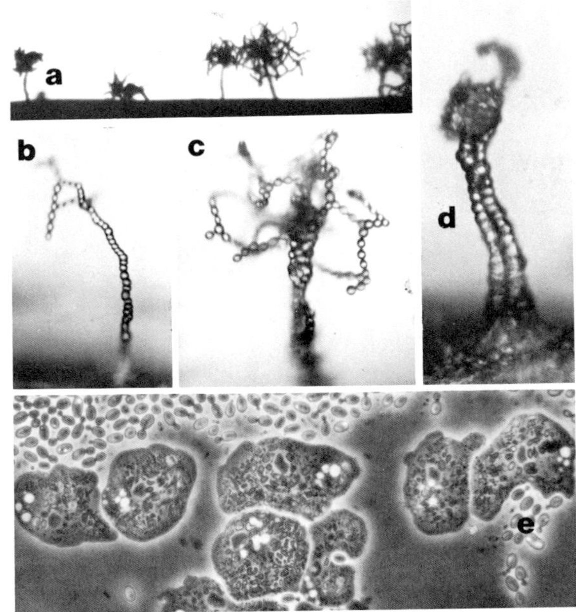

Fig. 7. *Acrasis rosea* Olive & Stoianovitch. a. Sorocarps. X60. b. Relatively simple sorocarp. X130. c. Typical dendroid sorocarp. X130. d. Sorocarp showing parallel rows of stalk cells. X300. e. Uninucleate amoebae ingesting yeast food source. X450. (Fig. 19, Hutner, 1985)

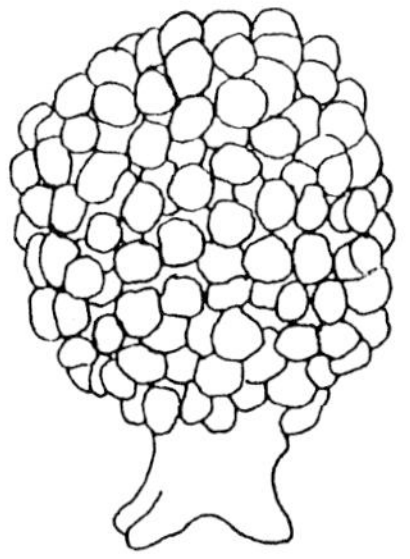

Fig. 8. *Pocheina* rosea (Cienk.) Loeblich & Tappan. Sorocarp consisting of cellular stalk region with globose mass of spores comprising sorus. All of the cells are pink in color. X442. (Redrawn from Fig. 13 Olive et al., 1983)

Genus ***Pocheina*** Loeblich & Tappan, 1961

This genus contains 2 or 3 species (see Raper, 1984). The organism is often isolated from bark wetted and placed in a moist chamber. The stalked sorocarps are pink, consisting of several rows of encysted cells capable of germinating and a globular sorus of spores (Fig. 8). The stalk is about 100 µm tall, as is the sorus. Uninucleate spores or stalk cells germinate to liberate limax-type amoebae. If a division occurs before germination, 2 biflagellate protoplasts or a binucleate protoplast may emerge, develop flagella, and undergo cytokinesis to produce 2 biflagellate amoebo-flagellate cells.
TYPE SPECIES: *Pocheina rosea* (Cienkowski) Loeblich & Tappan, 1961

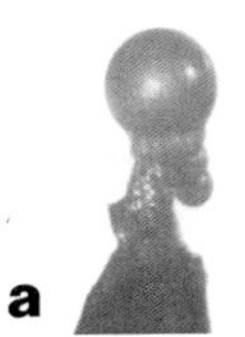

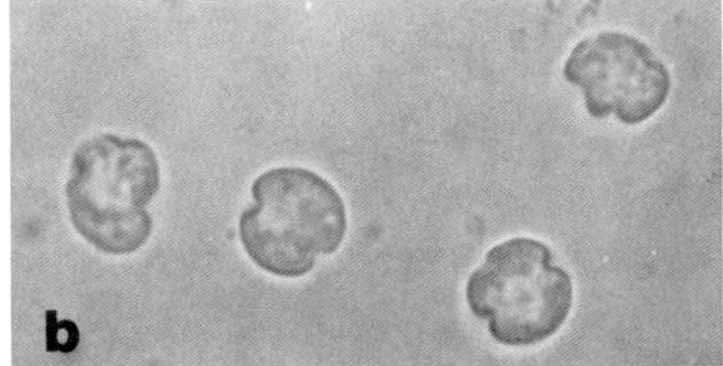

Fig. 9. *Guttulinopsis vulgaris* E. W. Olive. A. Sorocarp with globose sorus containing spores and subtending stalk region containing some stalk cells and a lot of acellular matrix materials. X45. b. Spores. X900. (Fig. 18, Hutner, 1985)

FAMILY GUTTULINOPSIDAE L. Olive, 1970

Genus **Guttulinopsis** E. W. Olive, 1901

The single genus *Guttulinopsis* E. W. Olive with two species is best represented by the ubiquitous *Guttulinopsis vulgaris* E. W. Olive (Fig. 9). A stalk may be prominent, reduced, or essentially absent, but if present, it consists of a mucilaginous matrix containing cellular debris and and stalk cells that are not in intimate contact with each other. Sorocarps may be up to 0.5 mm in diameter and can form small, lateral sori. The trophic stage is amoeboid with limax-type pseudopodia that undergo rapid locomotion by explosive extensions in the direction the amoebae are moving. Microcysts are not formed in this species but *G. nivea* Raper, Worley & Kessler does produce microcysts.
TYPE SPECIES: *Guttulinopsis vulgaris* E. W. Olive, 1901

FAMILY FONTICULIDAE (=FONTICULACEAE) Worley, Raper & Hohl, 1979

Genus ***Fonticula*** Worley, Raper & Hohl, 1979

A single genus with one species, *Fonticula alba* Worley, Raper & Hohl, was originally isolated from dog dung by L. S. Olive. The amoebae are 9 -

15 X 6-12 µm in diameter (Raper, 1984), uninucleate, and have filose pseudopodial extensions. Cells aggregate without forming streams of amoebae and then produce mounds of cells, which next form a tapered column of stalk cells. The tip of the column opens up with a quick extrusion of cells, which become the spores (Fig. 10).

TYPE SPECIES: *Fonticula alba* Worley, Raper & Hohl, 1979

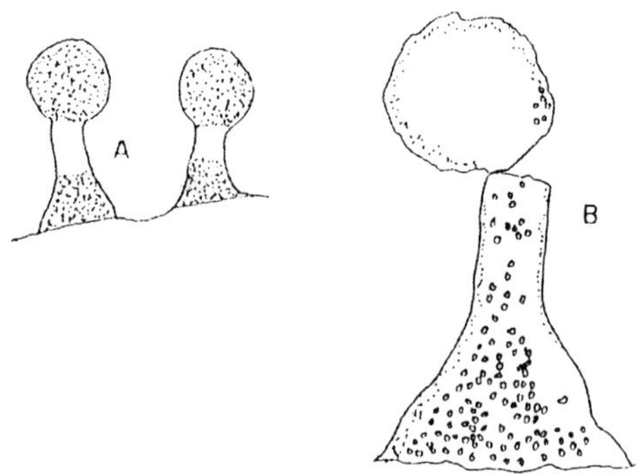

Fig. 10. *Fonticula alba* Worley, Raper & Hohl. A. Two sorocarps with inverted-funnel-shaped stalks subtending globose sori. X65. b. Detail of sorocarp, showing mass of spores extruded from stalk detaching from tip of stalk and cells still within the stalk. X156. (Redrawn from Fig. 14, Worley et al., 1979)

SUBCLASS MYXOGASTRIA, L. Olive, 1970

The Myxogastria make up a small class of organisms consisting of five orders, 57 genera, and approximately 600 species and have been known to biologists for over 300 years. They have been classified as animals (Mycetozoa), as plants (Myxomycophyta), as fungi (Myxomycetes), and as protists (Protista) by various taxonomists over the years, depending on what phase of the life cycle the taxonomist selected to emphasize. Ernst Haeckel (1879) suggested in his *Monographie der Moneren* that the Myxogastria should be considered neither plant nor animal, but should instead be placed in the Protista. One of the finest discussions of the general morphology and life cycle of the Myxomycetes is found in *The Myxomycetes* by G. W. Martin and C. J. Alexopoulos (1969). The serious student is strongly urged to read that authoritative text.

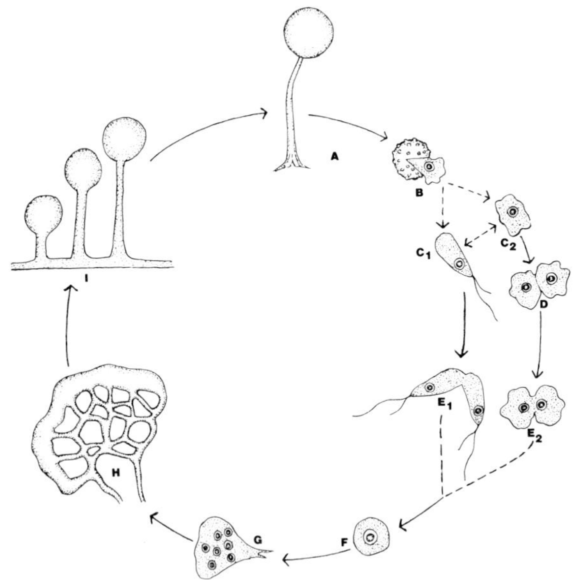

Fig. 11. Diagram of a generalized myxogastrian life cycle. A. Mature sporangium. B. Protoplast emerging from spore. C_1. Amoeba. C_2. Amoebo-flagellate cell. D. Cytokinesis of amoebae. E. Fusion of amoebo-flagellate (E_1) or amoebae (E_2). F. Diploid zygote resulting from cellular fusions. G. Young plasmodium after several divisions of nuclei after fusion of haploid cells. H. Large phaneroplasmodium. I. Sporogenesis. (Fig. 20, Hutner, 1985)

The reproductive phase is characterized by fruiting bodies containing few to many spores. The diploid (2n) plasmodium is the "slime" stage, and until recently the Myxogastria were referred to as "plasmodial" or "acellular" slime molds; however, recent work with members of the mycetozoan subclass Protostelia (Olive, 1975) has shown that they too possess a rudimentary plasmodial body known to be plurinucleate, and for that reason we have used the common name "true slime mold." The plasmodia may be found on moist logs in woods, on leaf litter, or even the moist bark of living trees and vines. The plasmodia may be colored red, orange, white, violet, or almost black, but most are yellowish, such as that of *Physarum polycephalum* Schw. If the plasmodium is found in the field, it may be transferred to a moist chamber for laboratory study, in the manner suggested by W. G. Camp (1936). Some species are more easily cultured than others in the laboratory, and the one most commonly used is *P. polycephalum*, which can be obtained from biological supply houses if it is not convenient to collect it from the field.

After the plasmodium has grown in the moist chamber, it is possible to dry it for storage, producing a sclerotium. The drying must be done in a special way, however, to be sure that it remains viable. Lonert (1965) developed a technique for producing sclerotia in the laboratory, which has proved quite reliable. These diploid sclerotia (2n) may be placed in an envelope and kept in the refrigerator for a year or more, and therefore can be stored for use at a later time. The sclerotia may be reactivated by wetting and placing them in a moist chamber and feeding them. When the plasmodium is revived under the proper conditions, it will feed and ultimately sporulate, producing haploid (n) spores.

Myxogastrid spores range in size from about 5 to 20 µm, but most are about 10 µm in diameter. Most are spherical, or nearly so, and have a spiny, warted, or reticulate surface, or are rarely perfectly smooth. The spores are released from the fruiting bodies when disturbed by animals, rain, or wind, and fall onto the substratum where, when water is present, they germinate, releasing protoplasts (Braun, 1971). The protoplast may develop into either an amoeba or an amoebo-flagellate "swarm" cell, both of which arc haploid and behave like gametes. The haploid gametes fuse in pairs to form diploid zygotes, which then divide mitotically without subsequent cell division, resulting in the formation of the multinucleated mass of cytoplasm called the plasmodium. A generalized life cycle for myxogastrids is shown in Fig. 11.

Myxogastria are universally distributed and live in moist dark places on decaying organic matter such as old decaying logs, dead twigs, and leaves on the forest floor. Moisture and temperature appear to be the most important factors governing distribution. Geographical distribution of myxogastrids is correlated, in part, with the collectors who hunt them. We know more about the temperate North American species because of the collecting activities of Thomas H. Macbride and George W. Martin in Iowa; Robert Hagelstein in New York and Pennsylvania; A. P. Morgan , Karl L. Braun in Ohio; Harold W. Keller in Ohio and Arkansas; Travis E. Brooks in Kansas; W. C. Sturgis and D. H. Mitchel in Colorado; C. J. Alexopoulos in Texas; and Donald T. Kowalski in California.

Myxogastria, as a group, begin to appear in early May and fruit throughout the summer season until October in the north termperate regions. Some species have seasonal fruiting patterns, occurring only in early spring, during the summer months, or the fall months until the first freeze. Most species appear to be independent of the substratum where they fruit, but many species show distinct affinity for decaying wood, decaying leaves, coniferous wood, living trees and vines, or dung of herbivorous animals. Some species appear to be confined to the tropics or subtropics, others to temperate regions, still others to alpine areas near melting snowbanks. Some are restricted to the Far East or to the New World.

Most species are collected from decaying organic matter on ground sites. Accumulated leaf litter under trees or shrubbery, piled mulching, or baled straw weathering in fields, all serve as prolific sources of Myxogastria, following lengthy periods of rain during the months of May to September throughout most of the United States. Exposed sites of decaying logs or stacked wood offer areas where myxogastrids are found in great abundance. Low, wet areas where water seepage keeps decaying leaves and logs continuously damp will yield Myxogastria throughout the warmer months. There are a number of man-made habitats such as living lawn grass, bark and leaf mulching around trees and shrubs, basal dead leaves of ornamental flowers in flower beds, and old tree stumps where interesting species may be found.

Corticolous Myxogastria occur on the bark of living trees and vines. Many of these species are restricted to the trunks or upper canopy of living trees. Trees that yield the greatest diversity of myxogastrid species are *Juniperus virginiana* (red cedar), *Ulmus americana* (American elm), *Malus* (apple), *Fraxinus* (ash), *Quercus alba* (white oak), and *Vitis* (grape), as well as trees heavily covered with mosses. There are many other tree species that also serve as productive habitats for Myxogastria. Collecting sites such as cemeteries and apple orchards are especially productive because of the number of junipers found in the former and apple trees in the latter (Keller, 1990).

It has become increasingly difficult to construct dichotomous keys to the orders and families of Myxogastria that will identify every genus. For example, the monotypic genera *Kelleromyxa*, *Protophysarum,* and *Trabrooksia* are atypical members of families and orders because they do not have the taxonomic suite of characters representative of their assigned rank. In the case of *Kelleromyxa*, it is a taxon of uncertain affinity and cannot be assigned to any order as presently circumscribed. This is further complicated by extreme variability in a given species that may result in the presence and absence of a character, such as calcium carbonate. These keys are based on the system of classification used by Martin and Alexopoulos (1969) and Martin, et al. (1983), and updated where appropriate to include new genera and nomenclatural changes since that time. We have included all genera on a worldwide basis. All the figures labeled Martin and Alexpoulos (M & A) have been photocopied from *The Myxomycetes* by Martin and Alexopoulos (1969) with the kind permission of the University of Iowa Press (Iowa City, IA). This excellent reference on the plasmodial slime molds is still in print and may be obtained directly from the University of Iowa Press.

KEY TO THE ORDERS

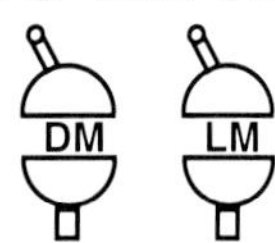

1. Fruiting bodies tiny, usually less than 0.5 mm in total height, rarely up to 1.5 mm, always forming stalked sporangia with a globose spore case, mostly on living trees and vines; spores white, pinkish, or light brown in mass .... **Echinosteliida**

1'. Fruiting bodies larger and of various types and shapes, either sessile or stalked; not with the above combination of characters.................... 2

2. Capillitium either typically absent or present as pseudocapillitium; true threads present in two monotypic genera, *Kelleromyxa* and *Listerella*; columella always absent.... **Liceida**

2'. Capillitium present as true threads; columella present or absent... 3

3. Spores bright colored—white, yellow, or red in mass; true capillitium present and columella always absent................................. **Trichiida**

3'. Spores dark colored—black, violet-brown, dark purple-brown or deep red in mass..........4

4. Calcium carbonate granules or crystals absent from fructification; stalk, when present, hollow or partially filled with strands......... .... **Stemonitida**

4.' Calcium carbonate present in fruiting bodies, (except in *Protophysarum* and *Trabrooksia*) either in the peridium, capillitium, or stalk; stalk when present filled with granular material or calcium carbonate....... **Physarida**

ORDER ECHINOSTELIIDA
Keller & Brooks 1976

The stalked sporangia are so small (less than 0.5 mm in diameter) that members of this order are frequently overlooked in the field and most specimens come from moist chamber cultures. Some species in the genus *Echinostelium* are common on the bark surface of living trees and vines. Students interested in keying out species of *Echinostelium* should consult the papers by Keller and Brooks (1976a) and Whitney (1980).

ORDER LICEIDA Jahn, 1928

The majority of species have spores brightly colored to dingy olivaceous in mass. Fruiting bodies are structurally simple, lacking characters found in other orders; no columella, no calcium carbonate, and no true capillitium associated with the mature fruiting body. Capillitial threads are present in two monotypic genera, *Kelleromyxa* and *Listerella*. The so-called pseudocapillitium found in some taxa in this group needs to be evaluated on the basis of ultrastructural and developmental studies. This order represents a heterogeneous assemblage of probably unrelated taxa. Modern methods of DNA analysis are needed to evaluate the phylogenetic relationships of members of this order.

ORDER TRICHIIDA Macbride, 1922

Spores in this group are light colored in mass. A capillitium is present and usually has distinctive surface markings. Fruiting bodies lack calcium carbonate and a columella.

ORDER STEMONITIDA Macbride, 1922

Spores in this group are dark colored, usually some shade of black or violet brown. A columella is usually present and this gives rise to the capillitium. Calcium carbonate is absent. This is the only order characterized by an **aphanoplasmodium**.

ORDER PHYSARIDA Macbride,1922

Spores in this group are dark, usually black in mass. Calcium carbonate is present in the peridium, capillitium, stalk, hypothallus, or in combinations of these. The capillitium is present as a system of noncalcareous threads, as calcareous tubules, or as non-calcareous threads interconnected by calcareous nodes. The assimilative stage is a **phaneroplasmodium**.

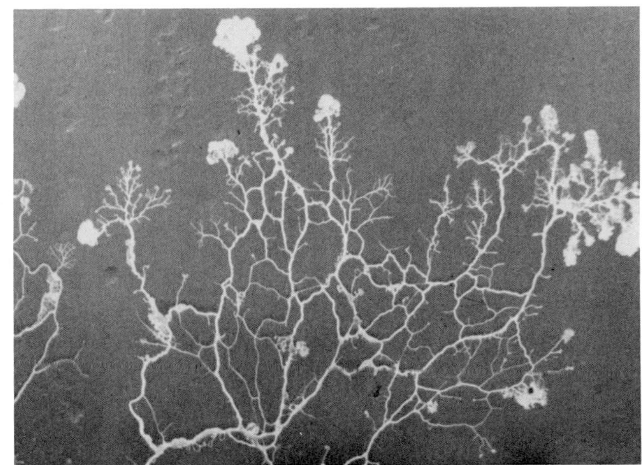

Figure 12. Photograph of reticulate phaneroplasmodium of a myxogastrian. 1X. (Fig. 21, Hutner, 1985)

KEY TO THE FAMILIES AND GENERA OF LICEIDA

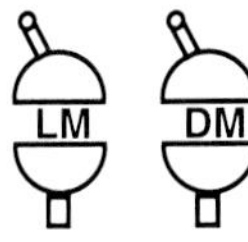

1. Capillitium present as threads, 2 μm or less in diameter, bearing annular or disc-shaped thickenings at regular intervals giving a beaded appearance; fructifications sporangiate, sessile, black, tiny, less than 0.3 mm in diameter; dehiscence by lobes along preformed lines.. **Listerellidae**

1'. Capillitium typically absent as threads; when present, lacking ornamentation, sometimes present as plates of pseudocapillitium; fructifications sporangiate, pseudoaethalioid, or aethalioid without the above combination of characters 2

2. Fructifications minute, usually sporangiate and less than 1 mm in diameter, or simple to branched plasmodiocarps larger in size, neither pseudocapillitium nor dictydine granules present; capillitium typically absent, present as capillitial threads in *Kelleromyxa* ....**Liceidae**.... 4

2'. Fructifications variable in size and type, sporangiate and minute or pseudoaethalioid to aethalioid and large, conspicuous; pseudocapillitium or dictydine granules may be present... 3

3. Dictydine granules present on the peridium, calyculus, and spores; peridium remaining as a surface net of threads with expanded nodes, the interstices fugacious except in *Lindbladia*.......5

3.' Dictydine granules absent; peridium persistent, intact, and complete.....**Enteridiidae** ...7

4. Capillitial threads absent....................... ***Licea***

4.' Capillitial threads present, simple and unbranched, or sometimes forming a sparsely branching network, attached to the peridial wall .... ***Kelleromyxa***

5. Fructifications aethalioid or pseudoaethalioid, often on an extensive, thickened hypothallus, rarely forming scattered sporangia; peridium mostly intact with a peridial network poorly developed or absent; dictydine granules sparse and pale.. ***Lindbladia***

5.' Fructifications sporangiate, usually free, sometimes aggregated or scattered; hypothallus delicate and inconspicuous; peridial network and dictydine granules distinct and conspicuous... 6

6. Peridium remaining as a network of threads with thickened or expanded nodes... ***Cribraria***

6.' Peridium remaining as thickened, dark, vertical threads connected by delicate, hyaline, transverse filaments..................... ***Dictydium***

7. Fructification a true aethalium without noticeable remains of individual sporangia.....8

7'. Fructification a pseudoaethalium with noticeable remains of individual sporangia..............9

8. Aethalium globose to conic or pulvinate, usually on a constricted base; pseudocapilli

tium consisting of colorless, branching, nearly smooth to sculptured tubules; spores in mass distinctly pinkish............................ ***Lycogala***

8'. Aethalium cushion-shaped or flattened on a broad base; pseudocapillitium consisting of perforated or frayed membranes; spores in mass brown, yellow, or olivaceous.................... ....***Enteridium*** (=***Reticularia***)

9. Pseudoaethalium of massed, tightly appressed sporangia, the outer surface with conspicuous polygonal areas, the tops of fused sporangia; sporangial walls disappearing at maturity below the apex except for thickened strands at the adjoining angles which persist as threads depending from the lids; hypothallus not massive or spongy............ ***Dictydiaethalium***

9.' Pseudoaethalium of fused sporangia with persistent sporangial walls; hypothallus conspicuous, massive, forming a stalk or cushion under a cluster of sporangia .... ***Tubifera***

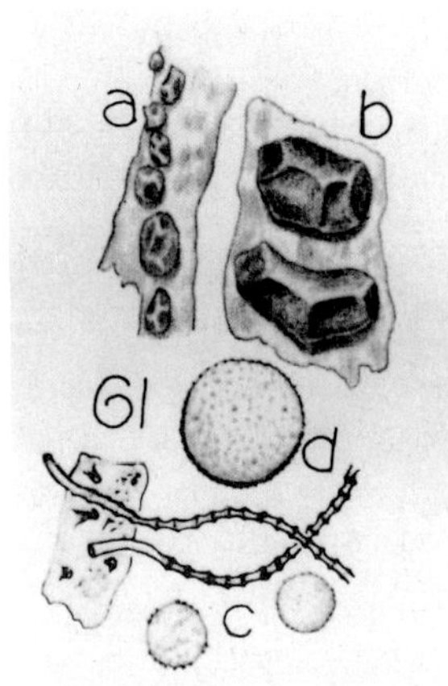

Fig. 13. *Listerella paradoxa* Jahn. a. Sporangial cluster. X12. b. Sporangia. X60. c. Capillitium and spores. X600. d. Enlarged spore. X1200. (Fig. 61, Martin and Alexopoulos, 1969)

FAMILY LICEIDAE Rostafinski, 1873

Genus *Licea* Schrader, 1797

A genus of at least 30 species (Keller and Brooks, 1977), most of which produce minute sporangia on the bark surface of living trees and vines in moist chamber culture (Fig. 12).
TYPE SPECIES: *Licea pusilla* Schrader, 1797

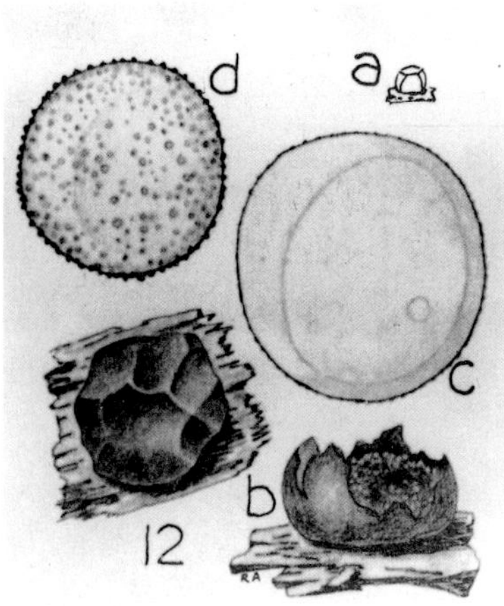

Fig. 14. *Licea pusilla* Schrad. a. Sporangium. X12. b. Sporangia in enlarged view. X30. c. Optical section of spore. X1200. d. Surface of spore. X1200. (Fig. 12, Martin and Alexopoulos, 1969)

Genus *Kelleromyxa* (Dearness & Bisby) Eliasson, 1991

Monotypic (Fig. 15). An atypical member of the family Liceidae (Eliasson et al., 1991).
TYPE SPECIES: *Kelleromyxa fimicola* (Dearness & Bisby) Eliasson, 1991

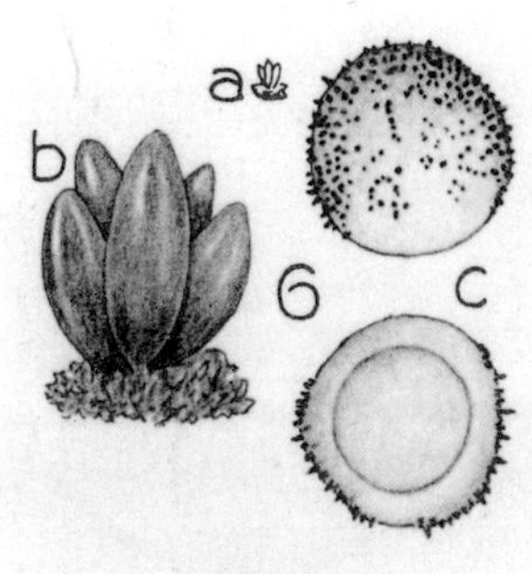

Fig. 15. *Kelleromyxa* (=*Licea*) *fimicola* (Dearness & Bisby) Eliasson. a. Sporangia. X6. b. Sporangia at higher magnification. X60. c. Spores in surface view (above) and optical section (below). X1200. (Fig. 6, Martin and Alexopoulos, 1969)

FAMILY LISTERELLIDAE Jahn, 1928

Genus *Listerella* Jahn, 1928

Monospecific(Fig. 13). Needs careful study to clarify its uncertain taxonomic position.
TYPE SPECIES: *Listerella paradoxa* Jahn, 1928

FAMILY ENTERIDIIDAE Farr, 1982

Genus ***Lindbladia*** Fries, 1849

Monotypic, cosmopolitan and common (Fig. 16). TYPE SPECIES: *Lindbladia tubulina* Fries, 1849

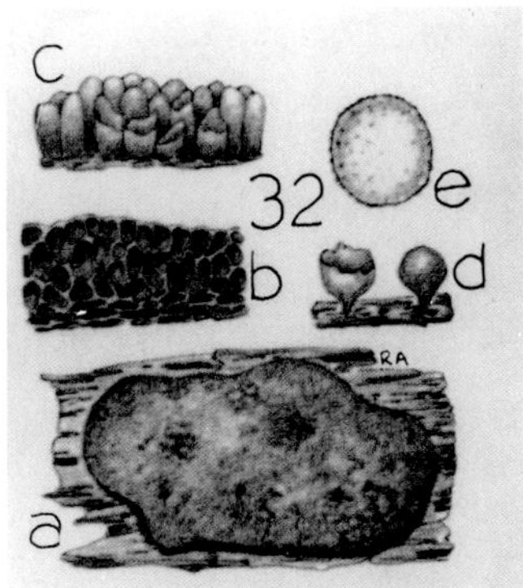

Fig. 16. *Lindbladia tubulina* Fries. a. Pseudoaethalium. X0.6. b. Cross-section of pseudoaethalium. X6. c. Massed sporangia. X6. d. Isolated, stipitate sporangia. X6. e. Spore with dictydine granules. X1200. (Fig. 32, Martin and Alexopoulos, 1969)

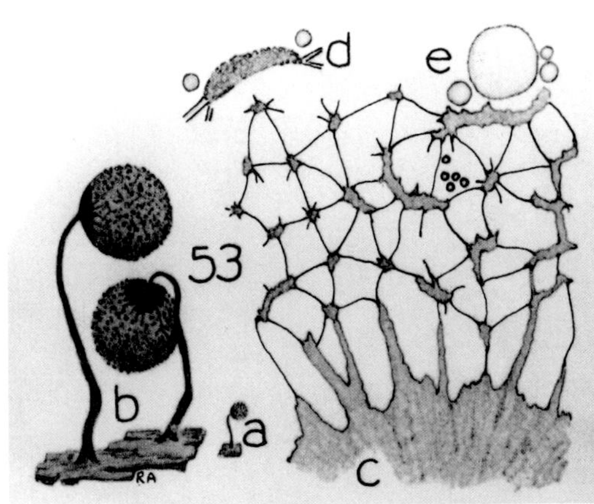

Fig. 17. *Cribraria aurantiaca* Schrad. a. Sporangium. X3.6. b. Sporangia. X18. c. Edge of cup (below) and net (above). X120. d. Node of net. X300. e. Spore with dictydine granules. X1200. (Fig. 53, Martin and Alexopoulos, 1969)

Genus ***Cribraria*** Persoon, 1794

Cribraria has over 30 species that are often difficult to identify and separate. *Cribraria aurantiaca* Schrader,1797 is shown in Fig. 17. TYPE SPECIES: *Cribraria rufescens* Persoon, 1794

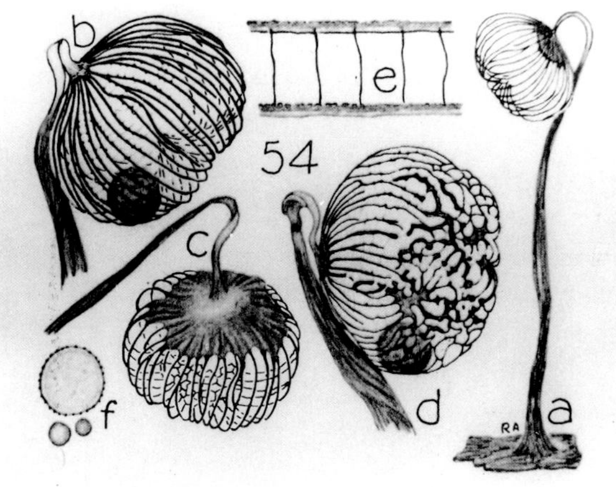

Fig. 18. *Dictydium cancellatum* (Batsch) Macbr. a. Sporangium. X24. b. Sporangium without cup. X48. c. Sporangium with cup. X48. d. Sporangium with cribrarioid net above. X48. e. Net detail. X300. f. Spore with dictydine granules. X1200. (Fig. 54, Martin and Alexopoulos, 1969)

Genus ***Dictydium*** Schrader, 1797

Dictydium cancellatum (Batsch) Macbride, 1899 (Fig. 18) is a cosmopolitan, common, and distinctive myxogastrid; 2 other species are rare. TYPE SPECIES: *Dictydium umbilicum* Schrader, 1797

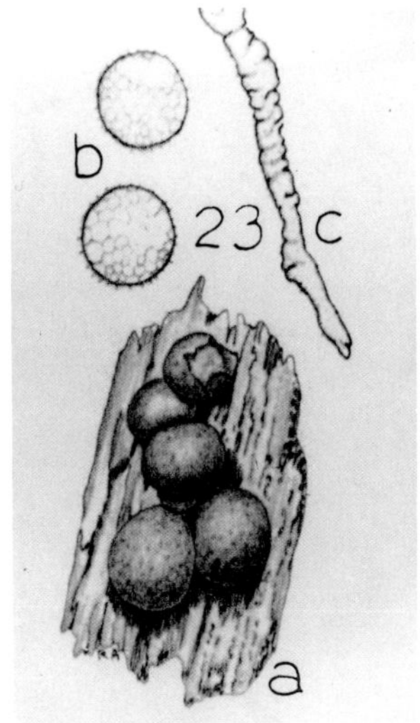

Fig. 19. *Lycogala epidendrum* (L.) Fries. a. Five aethalia. X1.2. b. Spores. X1200. c. Pseudocapillitial thread. X120. (Fig. 23, Martin and Alexopoulos, 1969)

Genus ***Lycogala*** Micheli, 1729

Lycogala epidendrum (L.) Fries, 1829 (Fig. 19) is the most common and cosmopolitan of 5 species.
TYPE SPECIES: *Lycogala epidendrum* (L.) Fries, 1829

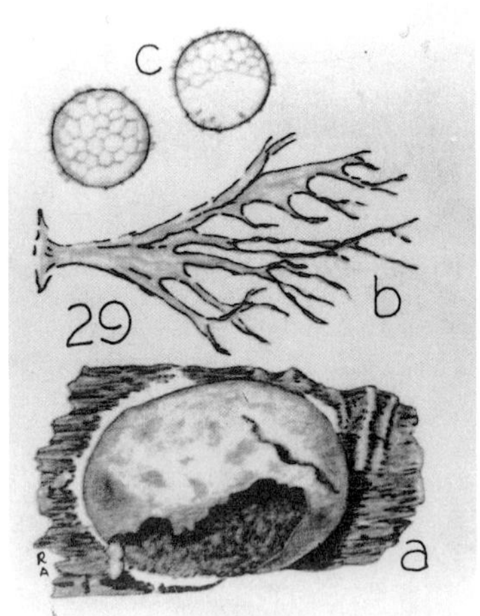

Fig. 20. *Enteridium* (=*Reticularia*) *lycoperdon* Bull. a. Aethalium. X0.6. b. Pseudocapillitium. X3.6. c. Spores. X1200. (Fig. 29, Martin and Alexopoulos, 1969)

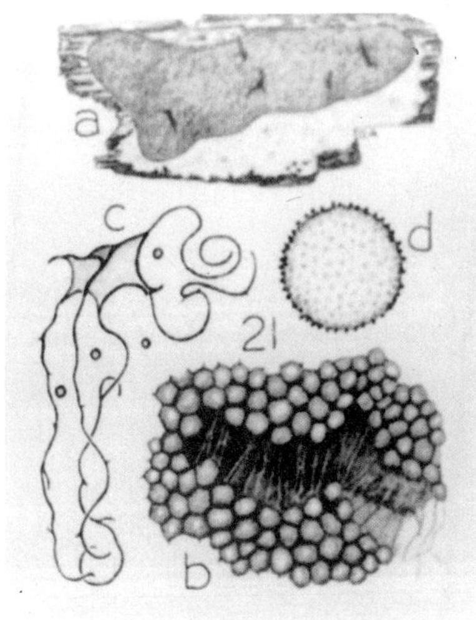

Fig. 21. *Dictydiaethalium plumbeum* (Schum.) Rost. a. Pseudoaethalium. X1.2. b. Detail of caps and threads. X24. c. Detail of threads attached to cap. X60. d. Spore. X1200. (Fig. 21, Martin and Alexopoulos, 1969)

Genus ***Enteridium*** Bulliard, 1791

Enteridium lycoperdon Bulliard (Fig. 20) and *E. splendens* Morgan are the most common and widely distributed of 7 species.
TYPE SPECIES: *Enteridium lycoperdon* (Bulliard, 1791)

Genus ***Dictydiaethalium*** Rostafinski, 1873

Dictydiaethalium plumbeum (Schumacher) Rostafinski, 1894 shown in Fig. 21 is the type species and the most common and widely distributed of 2 species.

Genus ***Tubifera*** Gmelin, 1791

Tubifera ferruginosa (Batsch) Gmelin, 1791. (Fig. 22) and *T. microsperma* (Berkeley & Curtis) Martin, 1947 are most common and widely distributed of 5 species.
TYPE SPECIES: *Tubifera ferruginosa* (Batsch) Gmelin, 1791

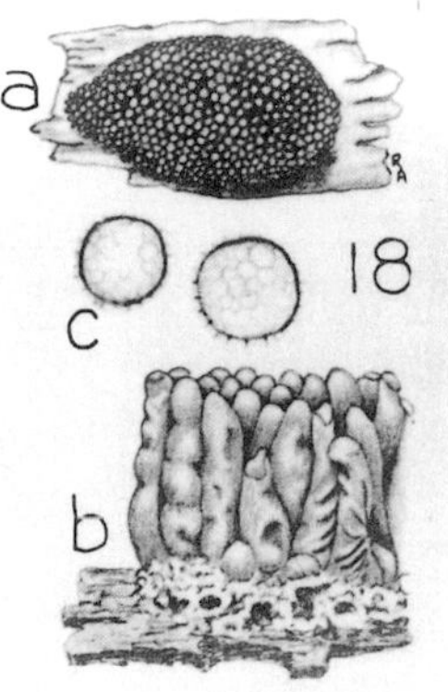

Fig. 22. *Tubifera ferruginosa* (Batsch) J. F. Gmel. a. Pseudoaethalium. X1.2. b. Sectional view of pseudoaethalium. X12. c. Spores. X1200. (Fig. 18, Martin and Alexopoulos, 1969)

KEY TO THE FAMILIES AND GENERA OF THE ORDER ECHINOSTELIIDAE

1. Spores in mass pallid, varying from white, cream-colored, yellowish, pinkish, to pinkish -brown................................ **Echinosteliedae**

1.' Spores in mass darkly colored shades of brown...........................**Clastodermidae**...... 2

2. Peridium tough, persistant, splitting into petal-like lobes remaining attached basally .... ***Barbeyella***
2.' Peridium delicate, evanescent, sometimes with small fragments that cling to the tips of the capillitial branches .... ***Clastoderma***

FAMILY ECHINOSTELIEDAE Rostafinski, 1873

Genus ***Echinostelium*** de Bary, 1873

A single genus of at least 12 species. *Echinostelium* has tiny stalked sporangia and is known mostly from the bark of living trees and vines. is shown in Fig. 23.
TYPE SPECIES: *Echinostelium minutum* de Bary, 1874

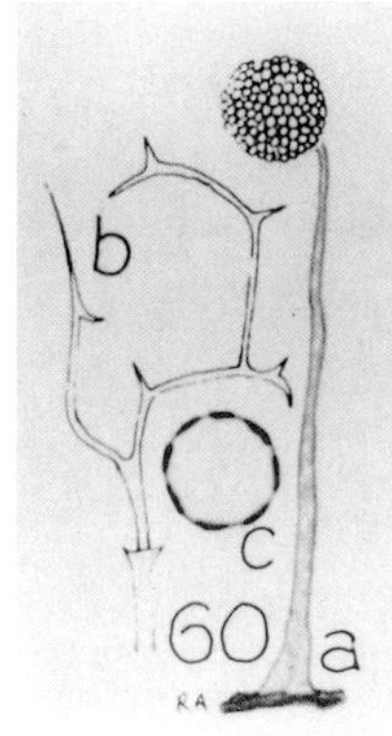

Fig. 23. *Echinostelium minutum* de Bary. a. Sporangium. X120. b. Capillitium. X300. c. Spore. X1200. (Fig. 60, Martin and Alexopoulos, 1969)

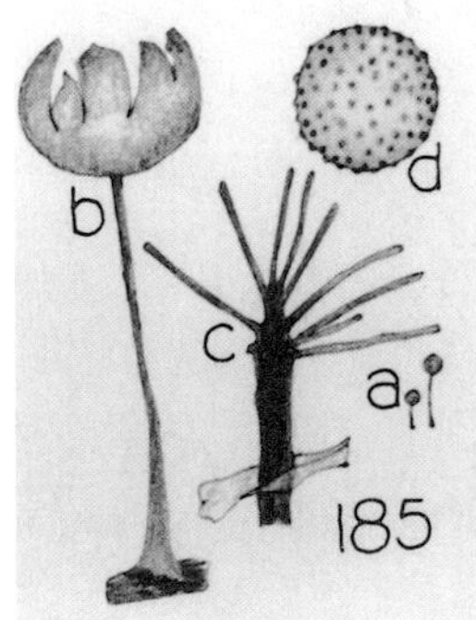

Fig. 24. *Barbeyella minutissima* Meylan. a. Sporangia. X6. b. Enlarged sporangium. X60. c. Basal collar of sporangium, columella, and capillitium arising from columella. X240. d. Spore. X1200. (Fig. 185, Martin and Alexopoulos, 1969)

FAMILY CLASTODERMIDAE
Alexopoulos & Brooks, 1971

Genus ***Barbeyella*** Meylan, 1914

Monotypic. Rare; found most frequently on bryophytes covering logs in the mountainous regions of the northwestern United States (Fig. 24).
TYPE SPECIES: *Barbeyella minutissima* Meylan, 1914

Genus ***Clastoderma*** Blytt, 1880

Clastoderma debaryanum var. *debaryanum* Blytt (Fig. 25) is a common and widely distributed taxon found mostly on decaying wood on ground sites. The other 3 species are restricted to the bark surface of living trees and vines.
TYPE SPECIES: *Clastoderma debaryanum* var. *debaryanum* Blytt, 1880

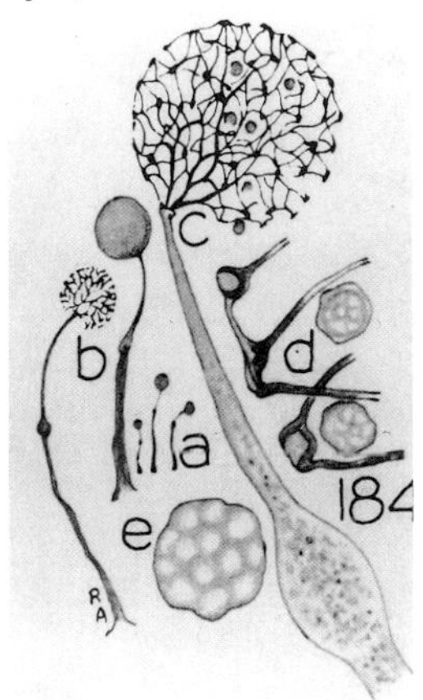

Fig. 25. *Clastoderma debaryanum* var. *debaryanum* Blytt. a. Sporangia. X6. b. Two sporangia, one without spores shed. X36. c. Sporangium, illustrating spores, capillitium, and upper part of stalk. X120. d. Capillitial tips with attached plates, and spores. X600. Spore. X1200. (Fig. 184, Martin and Alexopoulos, 1969)

KEY TO THE FAMILIES AND GENERA OF THE ORDER TRICHIIDA

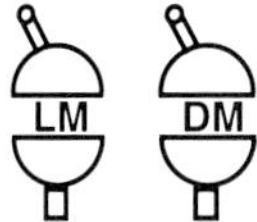

1. Capillitial threads appearing solid, thick-walled, lacking a hollow core, slender, usually less than 2 µm in diameter....**Dianemidae**.... 2

1'. Capillitial threads tubular with a hollow core, coarser, usually greater than 2 μm in diameter..**Trichiidae**.. 3

2. Capillitial threads slender, hairlike, coiled, with few attachments to the peridium... ***Calomyxa***
2'. Capillitial threads stout, mostly straight and vertically aligned, rarely forming a reticulum, with many of the tips attached to the peridium... ***Dianema***

3. Fructifications pseudoaethalioid to aethalioid..4
3'. Fructifications sporangiate, gregarious to sometimes heaped or crowded, varying to plasmodiocarpous... 5

4. Fructifications stalked, hanging in groups, aethalioid; peridium evanescent above, remaining below as a cup-like base.. ***Arcyriatella***
4'. Fructifications sessile, pseudoaethalioid to aethalioid; peridium fragmenting and falling away, without a cuplike base...... ***Minakatella***

5. Capillitium consisting of thick-walled, nearly solid, coiled threads, spirally twisted in bundles, becoming subdivided into penicillate tips attached to the sporangial walls.. ***Prototrichia***
5'. Capillitium consisting of thin-walled, hollow threads, not spirally coiled in bundles nor with penicillate tips.. 6

6. Capillitial threads bearing prominent, coarse, complete rings................................. ***Cornuvia***
6'. Capillitial threads almost smooth to variously ornamented with warts, spines, reticulations, half rings, cogs, or conspicuous raised spiral bands.. 7

7. Fructifications typically heaped or crowded, bright golden-yellow, shiny; peridium single, membranous, wholly persistent................... 8
7'. Fructifications sporangiate to plasmodiocarpous, usually scattered to gregarious without the above combination of characters.............. 9

8. Capillitium consisting of free elaters.. ***Oligonema***

8'. Capillitium consisting of threads united into an incomplete network.......................... ***Calonema***

9. Fructifications sessile sporangia, clustered or heaped, coppery to dull brownish-red or yellowish-brown; peridium persistent, especially below; capillitial threads bearing warts or spines... ***Arcyodes***
9'. Fructifications without this combination of characters of habit, color, peridium, and capillitial threads.. 10

10. Peridium evanescent above, leaving a well defined shallow or deep calyculus; capillitium often forming a network of elastic threads, expanding in length and marked with warts, spines, cogs, half rings, reticulations, or rarely spirals.................................... ***Arcyria***
10.' Peridium more persistent, sometimes thickened, cartilaginous, and double, especially below; capillitium consisting of elaters or forming a reticulum of threads, inelastic, conspicuosly marked with spiral bands or roughened with spines or reticulations........ 11

11. Peridium thickened, double, two-layered in most species, the outer layer dark, thick and charged with debris, the inner layer hyaline and membranous; capillitial threads lacking conspicuous, raised spiral bands..***Perichaena***
11'. Peridium membranous to thickened and cartilaginous but not obviously double-layered; capillitial threads marked with 2 - 6 conspicuous, raised, spiral bands................ 12

12. Peridium cartilaginous, shining, iridescent, opening by a preformed lid; sporangia deep maroon to black, often united into clusters resembling wasp nests; capillitial threads strongly spiny............................ ***Metatrichia***
12'. Peridium mostly membranous or thickened by accretion and then dull, opening irregularly or in lobate fashion. Capillitial threads spiny or smooth.. 13

13. Threads of capillitium united into an intricate net with few free ends............... ***Hemitrichia***
13'. Threads of capillitium broken into short, mostly unbranched, individual elaters with free ends.. ***Trichia***

FAMILY DIANEMIDAE Macbride, 1899

Genus ***Calomyxa*** Nieuwland, 1916

Calomyxa metallica (Berkeley) Nieuwland. (Fig. 26) is common and widespread occurring on living trees and decaying wood on ground sites. The only other species is known from alpine habitats.
TYPE SPECIES: *Calomyxa metallica* (Berkeley, 1836) Nieuwland, 1916

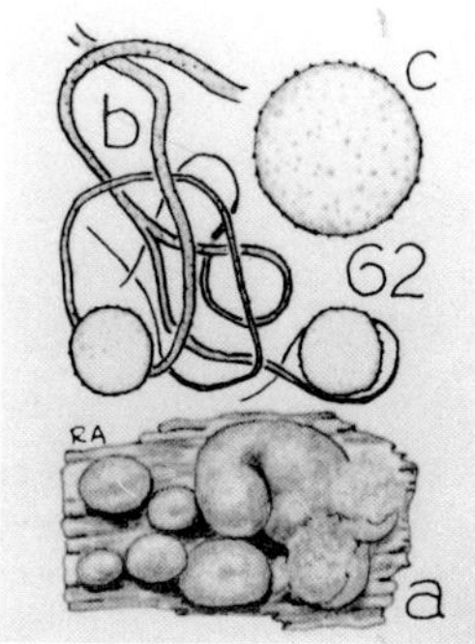

Fig. 26. *Calomyxa metallica* (Berk.) Nieuwl. a. Sporangia and plasmodiocarps. X12. b. Capillitium and spores. X600. c. Spore. X1200. (Fig. 62, Martin and Alexopoulos, 1969)

Genus ***Dianema*** Rex, 1891

A genus of 7 rarely collected species, *Dianema corticatum* A. Lister, 1894 (Fig. 27) is the most widespread.
TYPE SPECIES: *Dianema harveyi* Rex, 1891

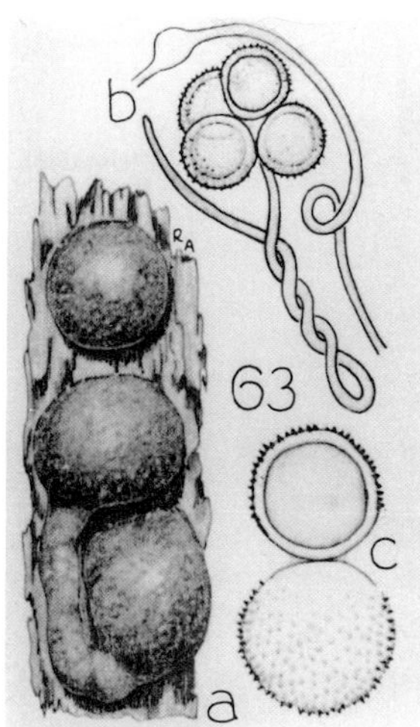

Fig. 27. *Dianema corticatum* A. Lister. a. Sporangium and plasmodiocarp. X12. b. Capillitium and spores. X600. c. Spore. X1200. (Fig. 63, Martin and Alexopoulos, 1969)

FAMILY TRICHIIDAE Rostafinski, 1873

Genus ***Arcyriatella***
Hochgesand & Gottsberger, 1989

Recently described in 1989 and is only known from Sao Paulo, Brazil. Monotypic (Fig. 28).
TYPE SPECIES: *Arcyriatella congregata* Hochgesand & Gottsberger, 1989

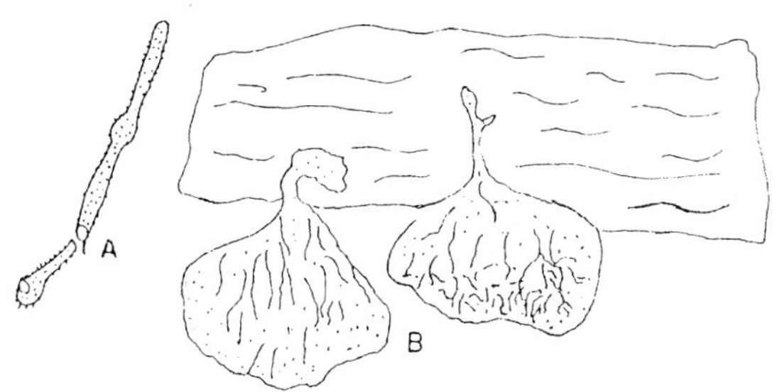

Fig. 28. *Arcyriatella congregata* Hochgesand & Gottsberger. a. Spinulose, hollow capillitial thread with expanded nodes. X3.4. b. Stipitate, pendent aethalia with profuse capillitial network. X240. (Redrawn from Figs. 1 and 2, Hochgesand and Gottsberger, 1989)

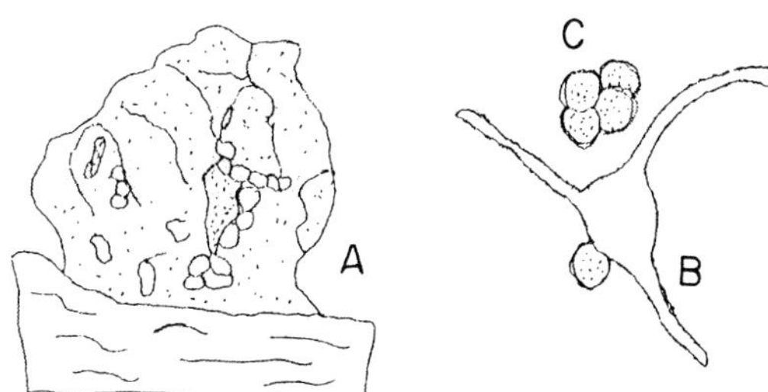

Fig. 29. *Minakatella longifila* G. Lister. a. Pseudoaethalium. X0.6. b. Capillitium with ornamented walls, nodal thickening, and one spore. X450. c. Subglobose spores, minutely spinulose, in cluster. X450. (Redrawn from Figs. 1 and 5, Keller et al., 1973)

Genus ***Prototrichia*** Rostafinski, 1876

Mostly montane in distribution. Monospecific (Fig. 30).
TYPE SPECIES: *Prototrichia metallica* (Berkeley, 1859) Massee, 1889

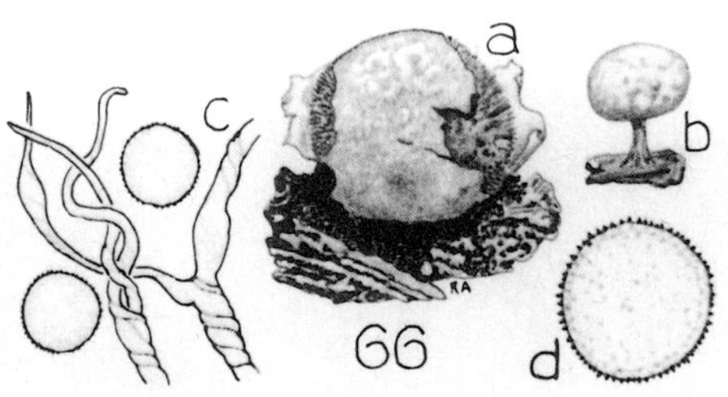

Fig. 30. *Prototrichia metallica* (Berk.) Massee. a. Sporangium. X12. b. Stalked sporangium. X12. c. Capillitium and spores. X600. d. Spore. X1200. (Fig. 66, Martin and Alexopoulos, 1969)

Genus ***Cornuvia*** Rostafinski, 1873

Monospecific; rare (Fig. 31).
TYPE SPECIES: *Cornuvia sepula* (Wigand,1863) Rostafinski, 1873

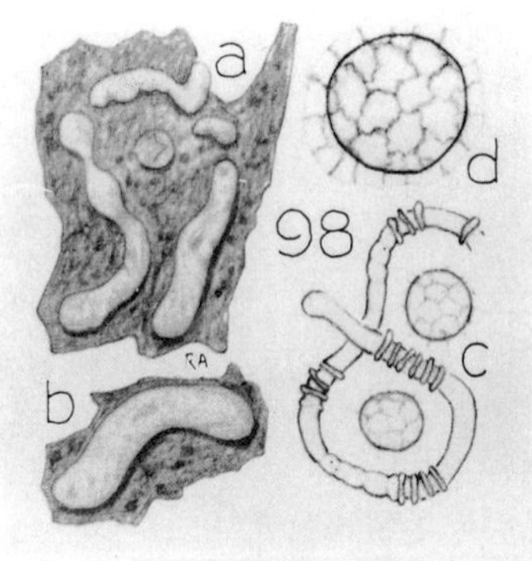

Fig. 31. *Cornuvia serpula* (Wigand) Rost. a. Plasmodiocarps. X12. b. Single plasmodiocarp. X24. c. Capillitium and spores. X600. d. Spore. X1200. (Fig. 98, Martin and Alexopoulos, 1969)

Genus ***Oligonema*** Rostafinski, 1875

Of the 3 species, *Oligonema schweinitzii* (Fig. 32) is the most common and widely distributed.
TYPE SPECIES: *Oligonema schweinitzii* (Berkeley, 1873) Martin, 1947

Genus ***Calonema*** Morgan,1893

Two rare species (Fig. 33).
TYPE SPECIES: *Calonema aureum* Morgan, 1893

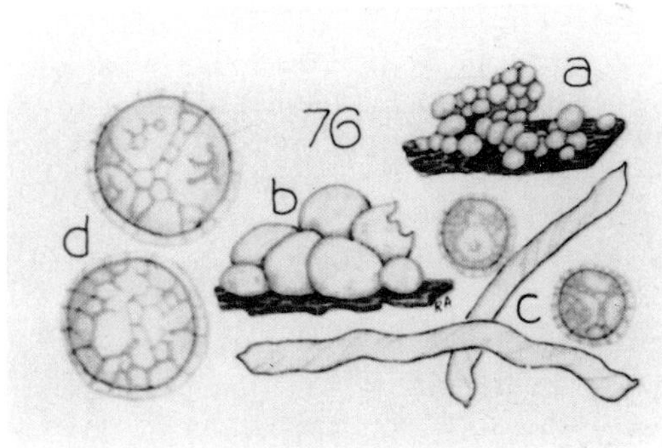

Fig. 32. *Oligonema schweinitzii* (Berk.) Martin. a. Heaped sporangia. X6. b. Sporangia. X24. c. Capillitial threads and spore. X600. Spores. X1200. (Fig. 76, Martin and Alexopoulos, 1969)

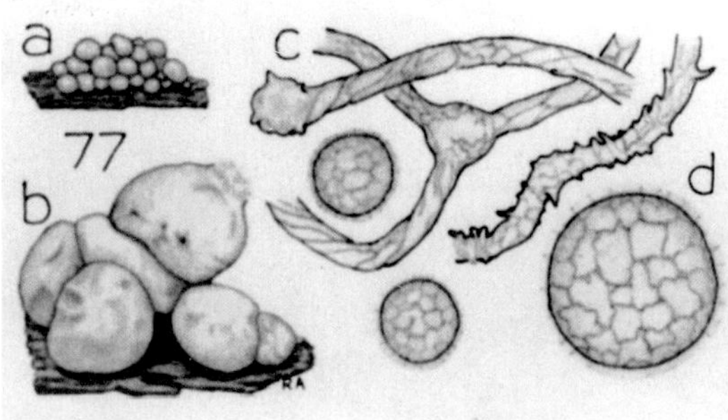

Fig. 33. *Calonema aureum* Morgan. a. Heaped sporangia. X6. b. Sporangia. X24. c. Capillitium and spores. X600. d. Spore. X1200. (Fig. 77, Martin and Alexopoulos, 1969)

Genus ***Arcyodes*** O. F. Cook, 1902

Monotypic species (Fig. 34).
TYPE SPECIES: *Arcyodes incarnata* (Albertini & Schweinitzii, 1805) O. F. Cook, 1902

Genus ***Arcyria*** Wiggers, 1780

Over 25 species, *Arcyria cinerea* (Bulliard) Persoon 1801 (Fig. 35), *A. denudata* (L.) Wettstein 1886, *A. incarnata* (Persoon) Persoon, 1746, and *A. nutans* (Bulliard) Greville, 1824 are cosmopolitan and among the most common myxogastrids.
TYPE SPECIES: *Arcyria denudata* (L., 1753) Wettstein, 1886

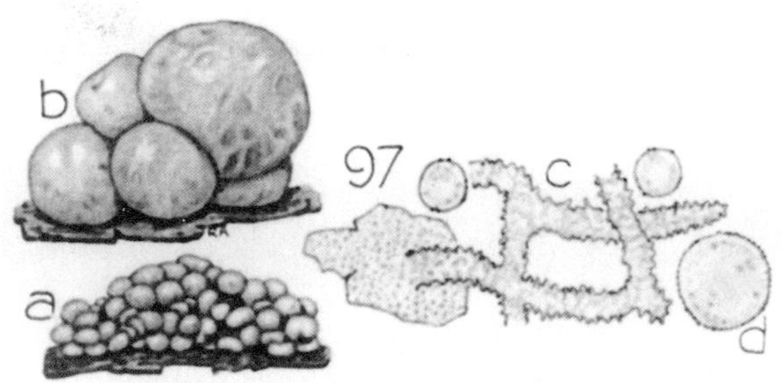

Fig. 34. *Arcyodes incarnata* (Alb. & Schw.) O. F. Cook. a. Sporangial cluster. X6. b. Sporangia. X24. c. Capillitium, partly attached to peridium, and spores. X600. d. Spore. X1200. (Fig. 97, Martin and Alexopoulos, 1969)

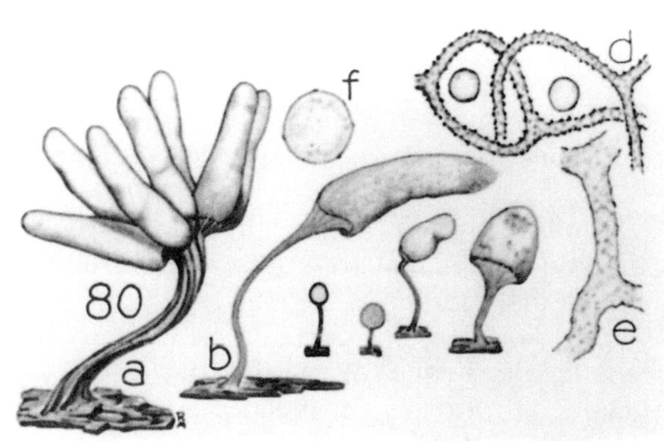

Fig. 35. *Arcyria cinerea* (Bull.) Pers. a. Digitate sporangia in cluster. X12. b. Isolated sporangia of various sizes, all shown at 12X. d. Capillitium and spores. X600. e. Capillitial thread. X600. f. Spore. X1200. (Fig. 80, Martin and Alexopoulos, 1969)

Genus ***Perichaena*** Freis, 1817

In the most recent key to the species, (Keller and Eliasson, 1992) there are 16 taxa recognized in the genus. *Perichaena depressa* Libert, 1837 is shown in Fig. 36.
TYPE SPECIES: *Perichaena populina* Fries, 1817

Genus ***Metatrichia*** Ing, 1964

Of the 2 species, *Metatrichia vesparium* (Fig. 37), the wasp nest slime mold, is cosmopolitan and one of the most common myxogastrids.
TYPE SPECIES: *Metatrichia horrida* Ing, 1964

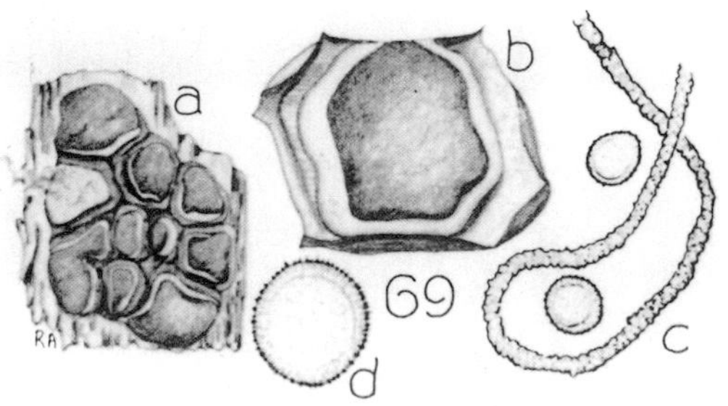

Fig. 36. *Perichaena depressa* Libert. a. Clustered sporangia. X6. b. Single sporangium with raised lid showing expanded spore mass. X24. c. Capillitium and spores. X600. d. Spore. X1200. (Fig. 69, Martin and Alexopoulos, 1969)

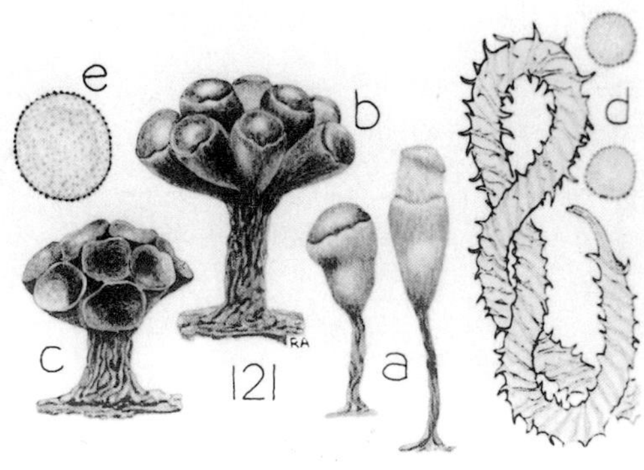

Fig. 37. *Metatrichia vesparium* (Batsch) Nann.-Brem. a. Two single sporangia. X12. b. Cluster of sporangia with fused stalks. X12. c. Group of empty sporangia. X12. d. Capillitial thread and spores. X600. e. Spore. X1200. (Fig. 121, Martin and Alexopoulos, 1969)

Genus ***Hemitrichia*** Rostafinski,1873

About 15 species, *Hemitrichia clavata* and *H. serpula* (Scopoli) Rostafinski, 1894 (Fig. 38) are common.
TYPE SPECIES: *Hemitrichia clavata* (Persoon 1794) Rostafinski, 1873

Genus ***Trichia*** Haller, 1768

Fourteen species are recognized in Martin and Alexopoulos (1969). *Trichia favoginea* (Batsch) Persoon, 1794 (Fig. 39) and *T. varia* (Persoon) Persoon, 1794 are probably the most common species.
TYPE SPECIES: *Trichia gregaria sessilis, piriformis flava* Haller, 1768

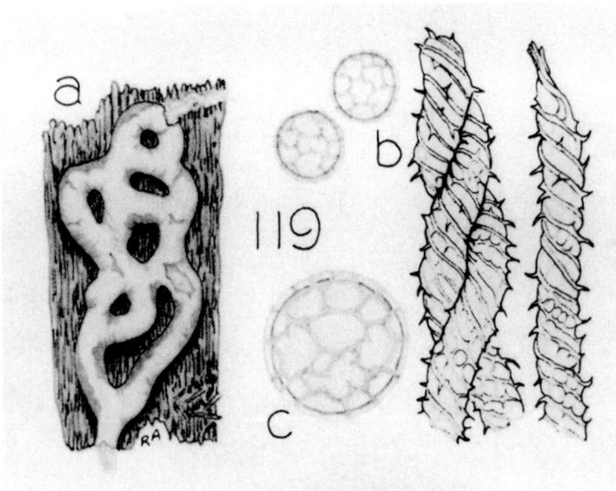

Fig. 38. *Hemitrichia serpula* (Scop.) Rost. a. Plasmodiocarp. X12. b. Capillitium and spores. X600. c. Spore. X1200. (Fig. 119, Martin and Alexopoulos, 1969)

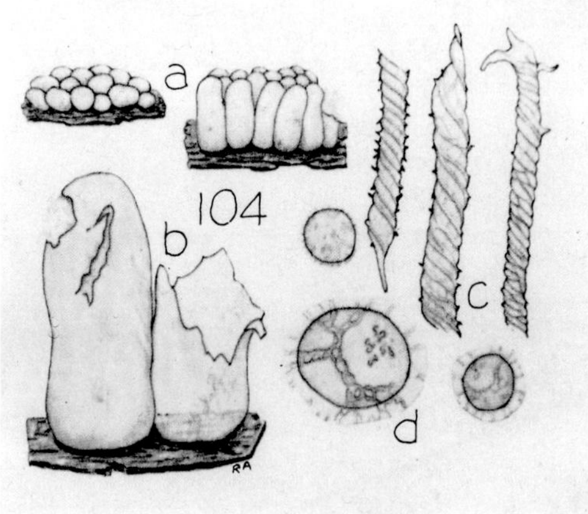

Fig. 39. *Trichia favoginea* (Batsch) Pers. a. Clusters of sporangia. X6. b. Two sporangia. X 24. c. Elater tips and spores. X600. Spore. X1200. (Fig. 104, Martin and Alexopoulos, 1969)

KEY TO THE GENERA OF THE ORDER STEMONITIDA

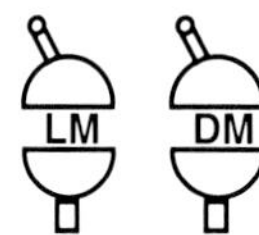

1. Fructification an aethalium............................ 2
1'. Fructification sporangiate or sometimes massed together into a pseudoaethalium.......... 3

2. Capillitial threads united into a network with many inflated, multicellular vesicles at the nodes .... ***Brefeldia***
2'. Capillitial threads dendroid, mostly branching from the base, without vesicle.***Amaurochaete***

3. Fructification a pseudoaethalium; peridium fugacious except for a basal cup and apical lid; capillitium composed of sparsely branched filaments coiled into columnar strands and attached to the peridial base and the peridial lid, then covered by a continuous membrane .... ***Schenella***
3'. Fructification sporangiate, sporangia free or clustered, sometimes united into a pseudoaethalium, but then the capillitium not in coiled spirals.. 4

4. Peridium double, consisting of a membranous inner layer and an outer layer becoming gelatinous when wet...................... ***Colloderma***
4'. Peridium single, not gelatinous when wet...... 5

5. Columella absent; peridium membranous, hyaline, iridescent throughout... ***Diacheopsis***
5'. Columella present; peridium variable but not as above... 6

6. Columella enlarged at the apex into a cupulate disk giving rise to the capillitium............. .... ***Enerthenema***
6'. Columella without a cupulate apical disk, the capillitium arising from the entire columella or from the sporangial base............................ 7

7. Fructification always sporangiate with stipe translucent, hollow, usually minute and often on the bark of living trees and vines................ .... ***Macbrideola***
7'. Fructification of various types, sessile or stalked, stalks not translucent, either hollow or with fibrous strands, usually larger than 0.5 mm in diameter 8

8. Peridium persistent, typically shiny and iridescent; capillitium arising mainly from tip of columella .... ***Lamproderma***
8'. Peridium early evanescent or, if present, thin, membranous, delicate............................ 9

9. Capillitial threads terminating in a peripheral surface net with few or no free ends .... ***Stemonitis***
9'. Capillitial threads not united into a surface net, sometimes with a subsurface network but with many free ends.................... ***Comatricha***

FAMILY STEMONITIDAE Rostafinski, 1873

Genus **Brefeldia** Rostafinski, 1873

Neither widespread nor common in the United States but is sometimes locally abundant in the northeastern states. Monospecific (Fig. 40).
TYPE SPECIES: *Brefeldia maxima* (Fries, 1825) Rostafinski, 1873

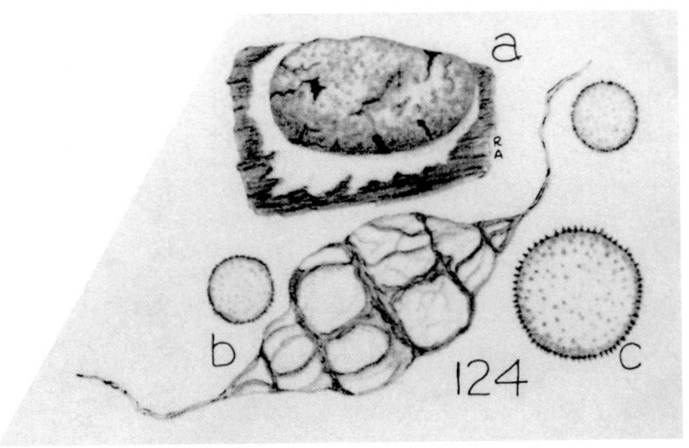

Fig. 40. *Brefeldia maxima* (Fries) Rost. a. Aethalium. X1.2. b. Vesicle and spores. X600. c. Spore. X1200. (Fig. 124, Martin and Alexopoulos, 1969)

Genus **Amaurochaete** Rostafinski,1873

Five species, none are common nor frequently collected. *Amaurochaete atra* is shown in Fig. 41.
TYPE SPECIES: *Amaurochaete atra* (Albertini & Schweinitz) Rostafinski, 1873

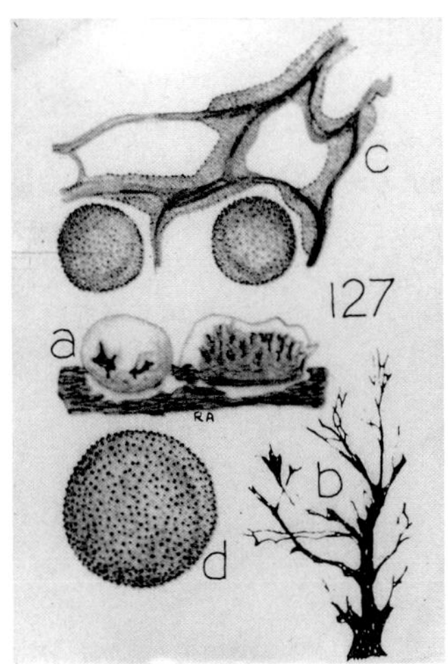

Fig. 41. *Amaurochaete atra* (Alb. & Schw.) Rost. a. Two aethalia. X1.2. b. Capillitium arising from stalk attached to base. X12. c. Capillitium and spores. X600. d. Spore. X1200. (Fig. 127, Martin and Alexopoulos, 1969)

Genus **Schenella** Macbride, 1911

An enigmatic genus of 2 rare species, *Schenella simplex* Macbride (Fig. 42) and *S. microspora* Martin, known only from their type localities in California. Their taxonomic fate will not be certain until additional specimens are found.
TYPE SPECIES: *Schenella simplex* Macbride, 1911

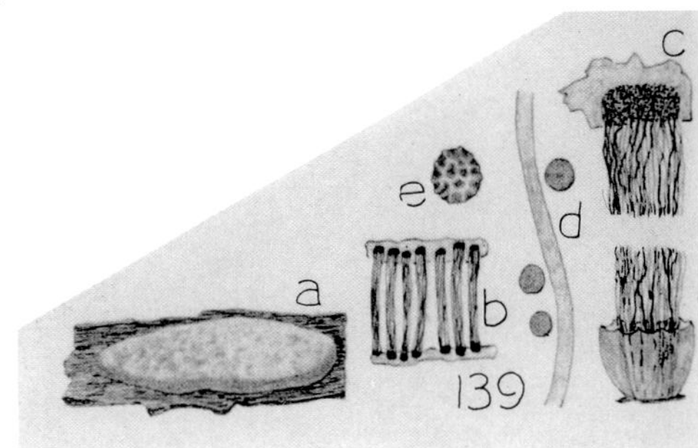

Fig. 42. *Schenella simplex* Macbride. a. Pseudoaethalium. X0.6. b. Longitudinal section, showing sporangia attached to base and cortex. X3.6. c. Base and cap of sporangium. X24. d. Capillitium and spores. X600. e. Spore. X1200. (Fig. 139, Martin and Alexopoulos, 1969)

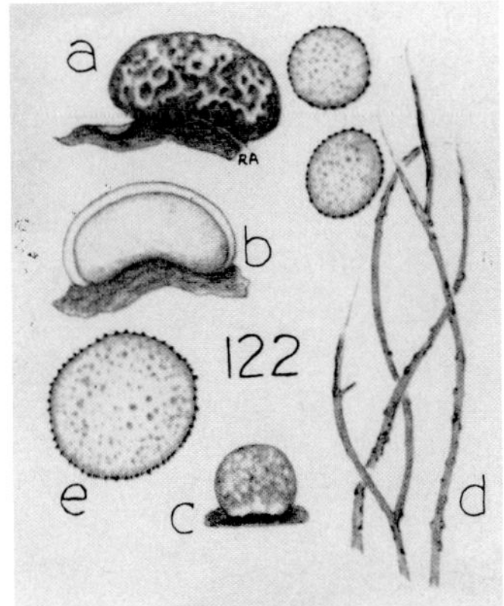

Fig. 43. *Colloderma oculatum* (Lippert) G. Lister. a. Dry sporangium. X24. b. Wetted sporangium with gelatinous outer wall. X24. c. Sporangium showing inner peridial layer. X24. d. Capillitium and spores. X600. e. Spore. X1200. (Fig. 122, Martin and Alexopoulos, 1969)

Genus **Colloderma** G. Lister, 1910

Only 2 species; *Colloderma oculatum* (Lippert) G. Lister, 1910 (Fig. 43) and *C. robustum* Meylan, 1933 are rarely collected in the United States. Usually associated with dead wood covered with mosses and lichens.

TYPE SPECIES: *Colloderma oculatum* (Lippert, 1894) G. Lister, 1910

Genus ***Diacheopsis*** Meylan, 1930

Kowalski (1975) recognized 6 species. *Diacheopsis effusa* Kowalski, *D. metallica* Meylan (Fig. 44), and *D. serpula* Kowalski are found near melting snow in alpine regions.
TYPE SPECIES: *Diacheopsis metallica* Meylan, 1930

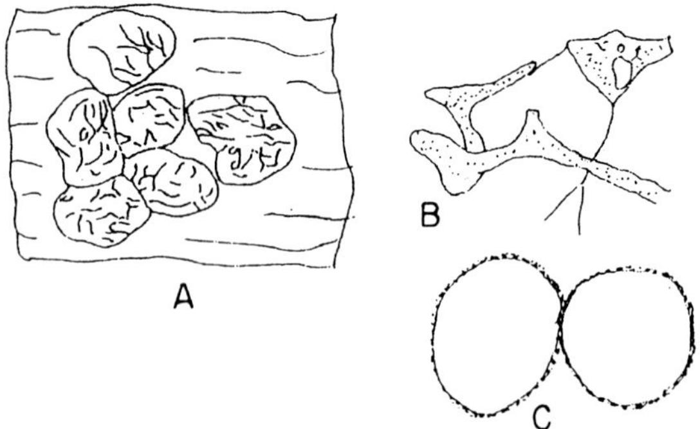

Fig. 44. *Diacheopsis metallica* Meylan. a. Sporangia. X7.8. b. Capillitium with expanded nodes. X403. c. Spores. X910. (Figs. 11-13, Kowalski, 1975)

Genus ***Enerthenema*** Bowman, 1830

A distinctive genus of 3 species with *Enerthenema papillatum* (Persoon) Rostafinski, 1876 (Fig. 45) the most common, on decayed wood and living trees and vines.
TYPE SPECIES: *Enerthenema elegans* Bowman, 1830

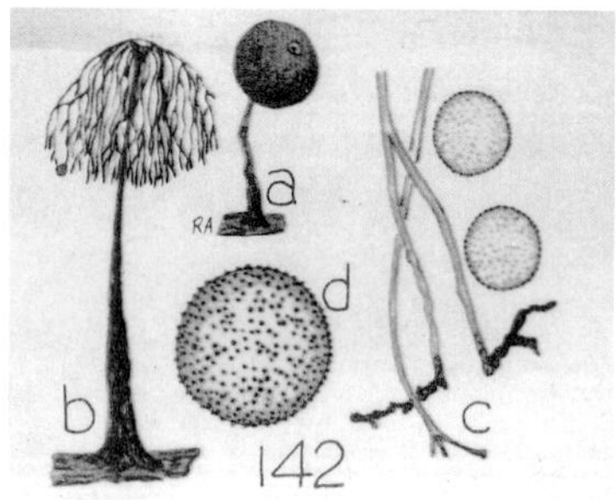

Fig. 45. *Enerthenema papillatum* (Pers.) Rost. a. Sporangium. X12. b. Sporangium without spores. X24. c. Capillitial tips and spores. X600. Spore. X1200. (Fig. 142, Martin and Alexopoulos, 1969)

Genus ***Macbrideola*** H. C. Gilbert, 1934 emend. Alexopoulos, 1967

At least 6 species; *M. decapillata* H. C. Gilbert, 1934 is common. Most members of the genus are found on the bark surface of living trees and vines (Fig. 46).
TYPE SPECIES: *Macbrideola scintillans* H. C. Gilbert, 1934

Genus ***Lamproderma*** Rostafinski, 1873

Kowalski (1970) recognized 21 species but over 25 species have now been described. Members of this genus are mostly montane, but the most cosmopolitan and common species, *L. arcyrionema* Rostafinski, 1874, *L. arcyrioides* (Sommerfield) Rostafinski, 1874, and *L. scintillans* (Berkeley & Broome) Morgan, 1894, are found in lowland areas. *Lamproderma columbinum* is shown in Fig. 47.
TYPE SPECIES: *Lamproderma columbinum* (Persoon, 1795) Rostafinski, 1873

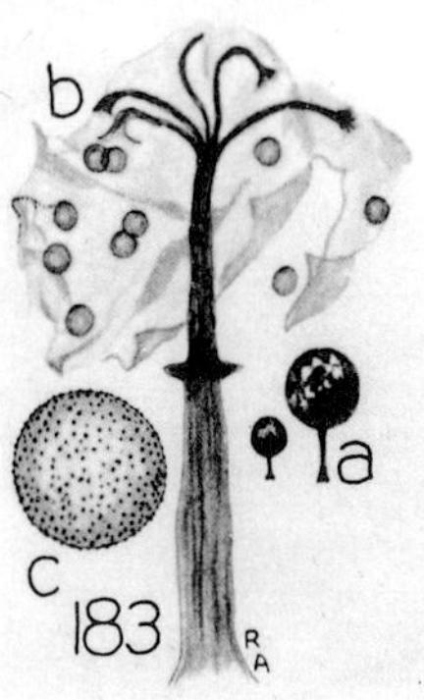

Fig. 46. *Macbrideola scintillans* H. C. Gilbert. a. Sporangia. X36. b. Capillitium, spores, and persistent peridium of sporangium. X180. c. Spore. X1200. (Fig. 183, Martin and Alexopoulos, 1969)

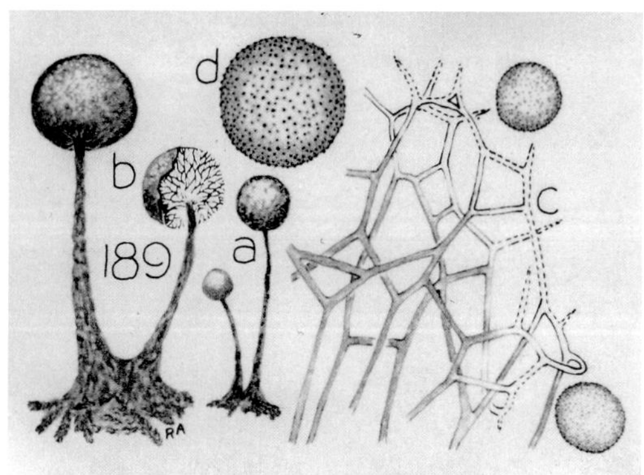

Fig. 47. *Lamproderma columbinum* (Pers.) Rost. a. Sporangia. X6. b. Sporangia. X12. c. Capillitial tips and spores. X600. d. Spore. X1200. (Fig. 189, Martin and Alexopoulos, 1969)

Genus ***Stemonitis*** Roth, 1787

A genus with over 20 species; the most common and cosmopolitan species are *S. fusca* Roth, 1787 (Fig. 48), *S. splendens* Rostafinski, and *S. axifera* (Bulliard) Macbride, 1889, which occur on decaying wood on ground sites.
TYPE SPECIES: *Stemonitis fusca* Roth, 1787

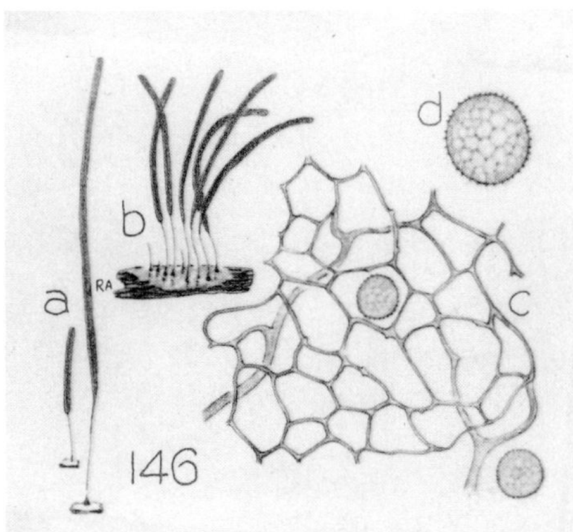

Fig. 48. *Stemonitis fusca* Roth. a. Sporangia of different sizes. X2.4. b. Sporangia. X6. c. Capillitium with surface net and spores. X600. Spore. X1200. (Fig. 146, Martin and Alexopoulos, 1969)

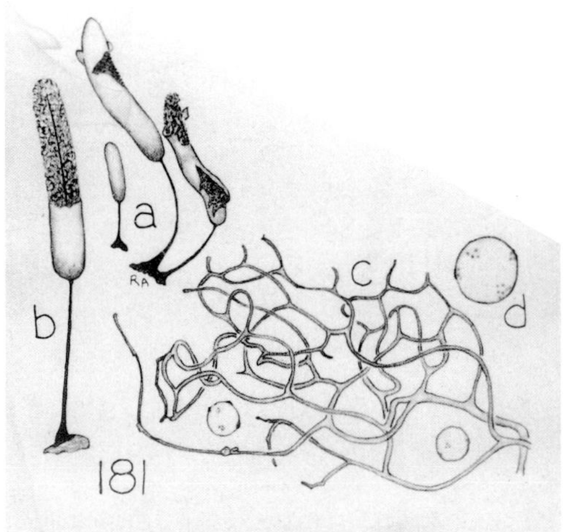

Fig. 49. *Comatricha typhoides* (Bull.) Rost. a. Sporangia. X6. b. Sporangium. X12. c. Capillitium and spores. X600. Spore. X1200. (Fig. 181, Martin and Alexopoulos, 1969)

Genus ***Comatricha*** Preuss, 1851

A genus of over 30 species sometimes merging with *Stemonitis*. *Comatricha typhoides* (Bulliard) Rostafinski, 1894 (Fig. 49) is cosmopolitan and probably the most common species in the genus.
TYPE SPECIES: *Comatricha obtusata* (Fries, 1829)

KEY TO THE FAMILIES AND GENERA OF THE PHYSARIIDA

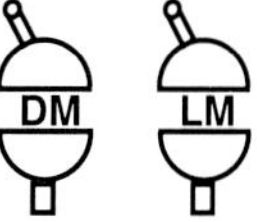

1. Fructifications with wax or oil in the stalk and often in other structural parts .... **Elaeomyxidae**
1'. Fructifications without wax or oil, usually calcium carbonate present as either granules or crystals.. 2

2. Capillitium calcareous, rarely absent, usually consisting of either calcareous tubules or non-calcareous threads interconnected by calcareous nodes; calcium carbonate granularin the fructifications.**Physaridae**.. 3
2'. Capillitium non-calcareous, a system of threads rarely with scattered calcareous inclusions; calcium carbonate granular or crystalline in the fructifications .**Didymiidae** 12

3. Calcium carbonate absent from sporangia; fruiting bodies sporangiate, tiny, not exceeding 0.75 mm in height and restricted to living trees, vines and desert plants; peridium membranous, shiny .... ***Protophysarum***
3'. Calcium carbonate present as granules in fructifications.. 4

4. Fructifications aethalioid, generally exceeding 1 cm across.. ***Fuligo***
4'. Fructifications sporangiate to plasmodiocarpous.. 5

5. Capillitium consisting of mostly simple, spike-like processes intruding from the upper peridium; a network of threads absent .... ***Badhamiopsis***
5'. Capillitium consisting of branching and anastomosing threads forming a network;

spike-like processes also present in *Physarella*... 6

6. Capillitium duplex, a calcareous portion interlaced with or connected to a mostly non-calareous network of hyaline threads............ 7
6'. Capillitium homogeneous, a single integrated system of threads either calcareous in part or throughout 9

7. Peridium of several layers, the outer one cartilaginous, brittle, smooth; sporangia ovoid.. ***Leocarpus***
7'. Peridium delicate to firm, but not brittle, polished and shell-like; fructifications either thimble-shaped or plasmodiocarpous............ 8

8. Fructifications thimble-shaped and when sporangiate deeply umbilicate; capillitium consisting of stout, spike-like trabeculae intruding from the peridium as well as a system of hyaline threads interconnected with fusiform calcareous nodes.......... ***Physarella***
8'. Fructifications plasmodiocarpous, internally divided into chambers by vertical calcareous plates; capillitium bearing free, hooked branches... ***Willkommlangia*** (=*Cienkowskia*)

9. Sporangia turbinate; dehiscence circumscissile, often by a preformed lid, the basal portion usually persisting as a deep cup .... ***Craterium***
9'. Sporangia of other shapes; dehiscence irregular or lobate but not by a preformed lid....... 10

10. Fructifications cylindrical, pendant............... .... ***Erionema***
10'. Fructifications sporangiate or plasmodiocarpous, rarely pendant.................................... 11

11. Capillitium a network of calareous tubes of nearly equal diameter, rarely with non-calcareous connective threads in some species... ***Badhamia***
11'. Capillitium a network of slender, non-calcareous threads interconnected by calcareous nodes .... ***Physarum***

12. Fructifications aethalioid............. ***Mucilago***
12'. Fructifications sporangiate to plasmodiocarpous... 13

13. Fructifications without visible calcareous deposits throughout the peridium, stalk, or hypothallus... 14
13'. Fructifications with visible calcareous deposits in or on the peridium, stalk, or hypothallus... 15

14. Fructifications plasmodiocarpous; capillitium consisting of simple, hyaline, subparallel threads vertically aligned.......... ***Trabrooksia***
14'. Fructifications sporangiate; capillitium consisting of dark threads branching and anastomosing but not vertically aligned.... .... ***Leptoderma***

15. Sporangial wall tuberculate, bearing numerous, blunt, calcareous, peg-like protuberances .... ***Physarina***
15'. Sporangial wall not roughened with peg-like protuberances, smooth or nearly so............. 16

16. Peridium non-calcareous, membranous, either iridescent or glossy...................***Diachea***
16'. Peridium typically with calcareous deposits, under certain conditions non-calcareous, and then the peridium neither iridescent nor glossy 17

17. Calcareous deposits granular, amorphous (a middle crystalline layer sometimes present in *Diderma trevelyani* [Grev.] Fries..... ***Diderma***
17'. Calcareous deposits in the form of stellate, angular, or platelike crystals either sprinkled on the peridium, united into scales, or forming a continuous crust... 18

18. Crystals either stellate or angular, scattered on the peridium, forming a crust, or embedded in an eggshell-like layer............... ***Didymium***
18'. Crystals united into distinct scales, scattered or massed on the peridium......... ***Lepidoderma***

FAMILY ELAEOMYXIDAE Hagelstein ex Farr & Keller, 1982

Genus *Elaeomyxa* Hagelstein, 1942

A single genus *Elaeomyxa*, with 2 rare species, *E. cerifera* (G. Lister) Hagelstein, 1942 and *E. miyazakiensis* (Fig. 50).
TYPE SPECIES: *Elaeomyxa miyazakiensis* (Emoto, 1935) Hagelstein, 1942

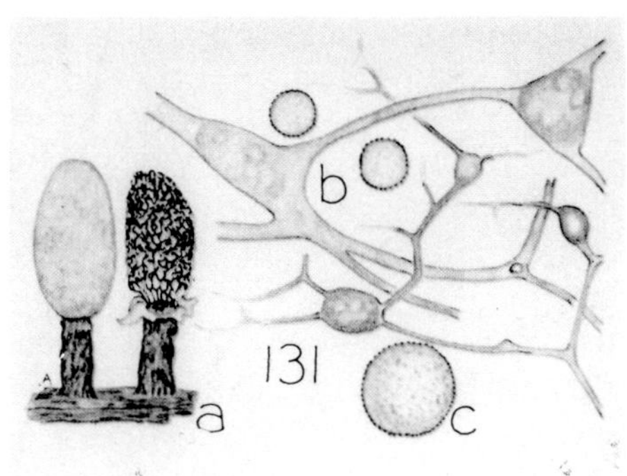

Fig. 50. *Elaeomyxa miyazakiensis* (Emoto) Hagelst. a. Sporangia. X18. b. Spores and capillitium. X600. c. Spore. X1200. (Fig. 131, Martin and Alexopoulos, 1969)

FAMILY PHYSARIDAE Rostafinski, 1873

Genus ***Protophysarum*** Blackwell & Alexopoulos, 1975

Monospecific, rare, (Fig. 51). Known only from living trees and desert plants in the United States.
TYPE SPECIES: *Protophysarum phloiogenum* Blackwell & Alexopoulos, 1975

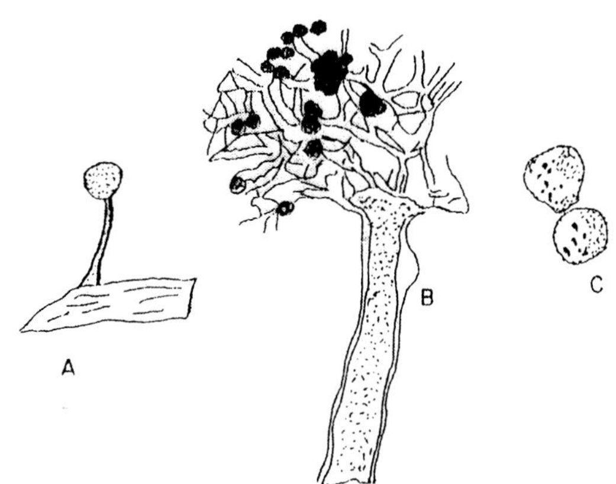

Fig. 51. *Protophysarum phloiogenum* Blackwell & Alexopolous. a. Sporangium. X39. b. Capillitial network and spores arising from tip of the stalk. X260. c. Spores. X650. (Redrawn from Figs. 1-3, Blackwell and Alexopoulos, 1975).

Genus ***Fuligo*** Haller, 1768

Martin and Alexopoulos (1969) recognized 5 species. *Fuligo septica* (L.) Wiggers, 1780 (Fig. 52) is cosmopolitan and the most common.

TYPE SPECIES: *Fuligo septica* (L., 1763) Wiggers, 1780

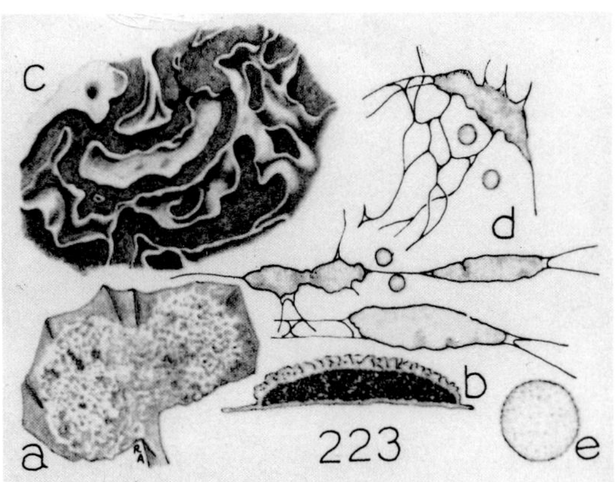

Fig. 52. *Fuligo septica* (L.) Wiggers. a. Aethalium. X1.2. b. Cross-section through aethalium. X1.2. c. Aethalium with surface removed. X24. d. Capillitium and spores. X300. e. Spore. X1200. (Fig. 223, Martin and Alexopoulos, 1969)

Genus ***Badhamiopsis*** Brooks & Keller, 1976

Monospecific; only known from the bark of living trees and vines, but is common there (Fig. 53).
TYPE SPECIES: *Badhamiopsis ainoae* (Yamashiro, 1936) Brooks & Keller, 1976

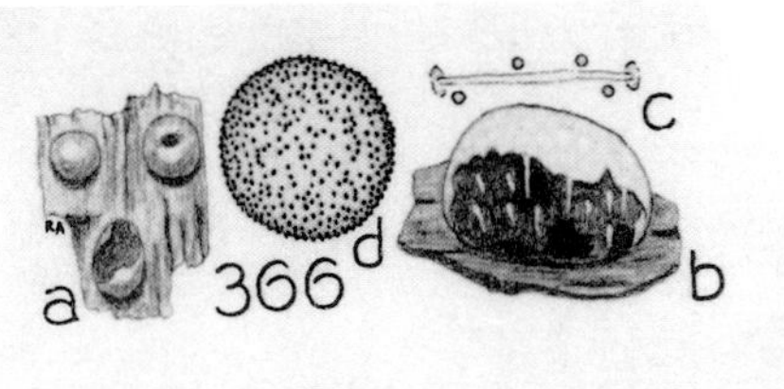

Fig. 53. *Badhamiopsis* (=*Badhamia*) *ainoae* (Yama) Brooks & Keller. a. Sporangia. X6. b. Partially opened sporangium showing limy columns. X24. c. Limy column with fragments of base and peridium attached, with spores. X60. d. Spore. X1200. (Fig. 366, Martin and Alexopoulos, 1969)

Genus ***Leocarpus*** Link, 1809

A single, distinctive species, cosmopolitan and common, is easily identified with the naked eye (Fig. 54).
TYPE SPECIES: *Leocarpus fragilis* (Dickson, 1785) Rostafinski, 1874

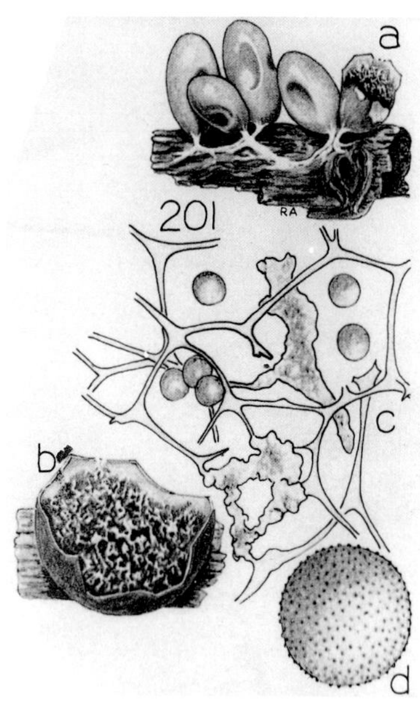

Fig. 54. *Leocarpus fragilis* (Dicks.) Rost. a. Sporangia. X6. b. Sporangium with top broken open. X24. c. Capillitium and spores. X300. d. Spore. X1200. (Fig. 201, Martin and Alexopoulos, 1969)

Genus *Physarella* Peck, 1882

A single, distinctive species, common and cosmopolitan, (Fig. 55), is easily identified with the naked eye; the stalked sporangia resemble tiny yellow daffodils.
TYPE SPECIES: *Physarella oblonga* (Berkeley & Curtis, 1873) Morgan, 1896

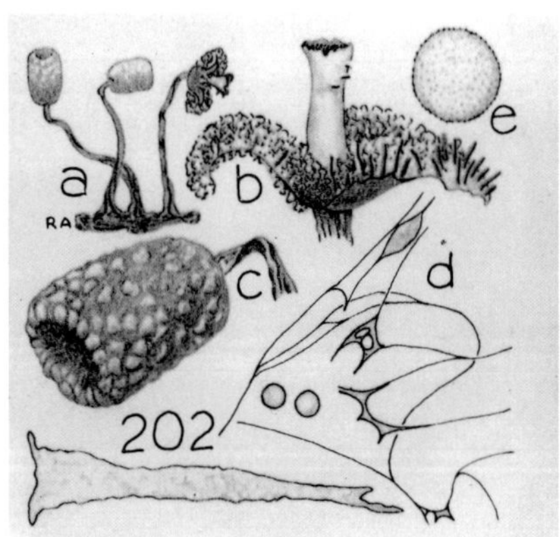

Fig. 55. *Physarella oblonga* (Berkeley & Curtis) Morgan. a. Sporangia. X6. b. Open sporangium, showing pseudo-columella and spikes extending from peridium. X24. c. Sporangium. X24. d. Capillitium and spores. X300. d. Spore. X1200. (Fig. 202, Martin and Alexopoulos, 1969)

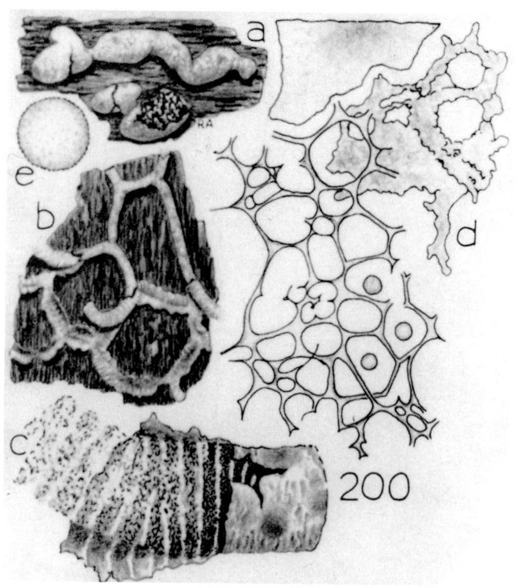

Fig. 56. *Willkommlangea* (=*Cienkowskia*) *reticulata* (Alb. & Schw.) Rost. a. Plasmodiocarp. X12. b. Portion of plasmodiocarp. X6. c. Plasmodiocarp showing plates. X60. d. Capillitium and spores. X300. e. Spore. X1200. (Fig. 200, Martin and Alexopoulos, 1969)

Genus *Willkommlangea* O. Kuntze, 1891

A single species (Fig. 56) is cosmopolitan but not frequently collected.
TYPE SPECIES: *Willkommlangia reticulata* (Albertini & Schweinitz, 1805) Rostafinski, 1874 (=*Cienkowskia* Rostafinski, 1873)

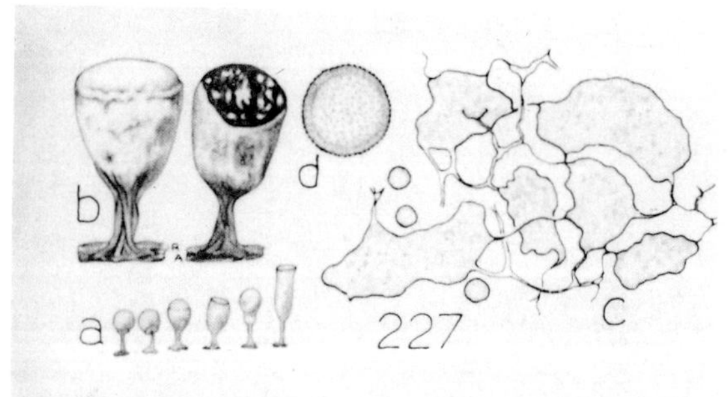

Fig. 57. *Craterium leucocephalum* (Pers.) Ditmar. a. Sporangia. X6. b. Sporangia. X24. c. Capillitium and spores. X300. d. Spore. X1200. (Fig. 227, Martin and Alexopoulos, 1969)

Genus *Craterium* Trentepohl, 1797

A distinctive genus based on the goblet shape of the sporangia. It merges with *Physarum* but most of

the 7 species recognized by Martin et al. (1983) have circumscissile dehiscence and separate by a distinct lid. *Craterium leucocephalum* (Persoon) Ditmar, 1813 (Fig. 57) is the most common species and is cosmopolitan.
TYPE SPECIES: Craterium pedunculatum Trentepohl, 1797

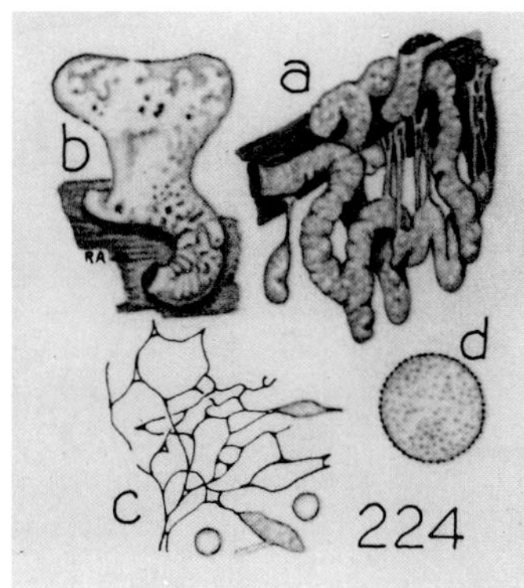

Fig. 58. *Erionema aureum* Penzig. a. Hanging plasmodiocarps. X12. b. Enlargement of plasmodiocarp. X30. c. Capillitium and spores. X300. d. Spore. X1200. (Fig. 224, Martin and Alexopoulos, 1969)

Genus ***Erionema*** Penzig, 1898

A single, rare species (Fig. 58), is confined to Southeast Asia.
TYPE SPECIES: *Erionema aureum* Penzig, 1898

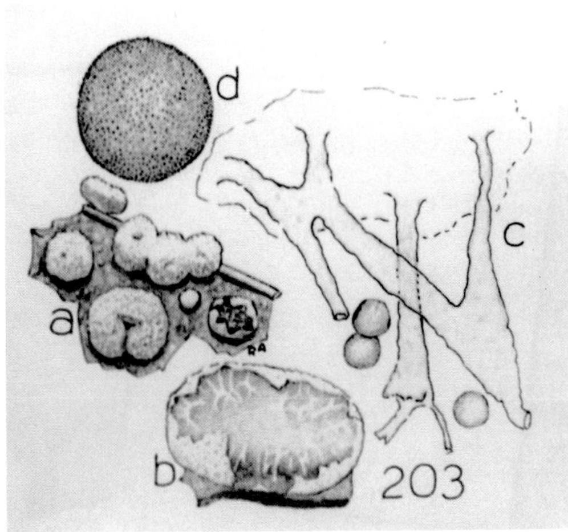

Fig. 59. *Badhamia affinis* Rost. a. Sporangia. X6. b. Sporangium with top broken open. X24. c. Capillitium attached to peridium with spores. X300. d. Spore. X1200. (Fig. 203, Martin and Alexopoulos, 1969)

Genus ***Badhamia*** Berkeley, 1853

A genus of 19 species as recognized by Martin and Alexopoulos (1969). Many new species have been described since that time. The genus is taxonomically difficult and merges with *Physarum. Badhamia affinis* Rostafinski, 1874 (Fig. 59) is cosmopolitan and one of the most common species on the bark of living trees and decaying logs on ground sites.
TYPE SPECIES: *Badhamia capsulifera* (Bulliard, 1791) Berkeley, 1853

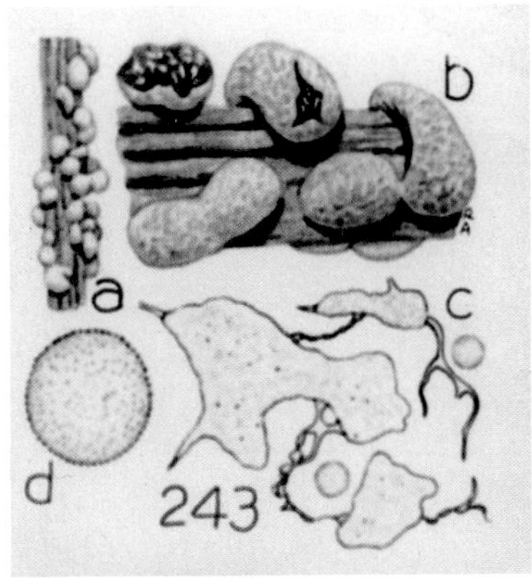

Fig. 60. *Physarum cinereum* (Batsch) Pers. a. Sporangia. X6. b. Sporangia exhibiting subplasmodiocarpous habit. X24. c. Capillitium and spores. X300. d. Spore. X1200. (Fig. 243, Martin and Alexopoulos, 1969)

Genus ***Physarum*** Persoon, 1794

This is the largest genus in the Myxogastria, with over 100 species. Many of the species are common and colorful. *Physarum cinereum* (Batsch) Persoon, 1794 (Fig. 60) is cosmopolitan and common, especially on living lawn grass blades.
TYPE SPECIES: *Physarum aureum* Persoon, 1794

FAMILY DIDYMIIDAE Rostafinski, 1873

Genus ***Mucilago*** Micheli ex Battara, 1755

A single species (Fig. 61) is common and cosmopolitan. It often fruits on living herbaceous plants on ground sites.
TYPE SPECIES: *Mucilago crustacea alba* Micheli, 1755

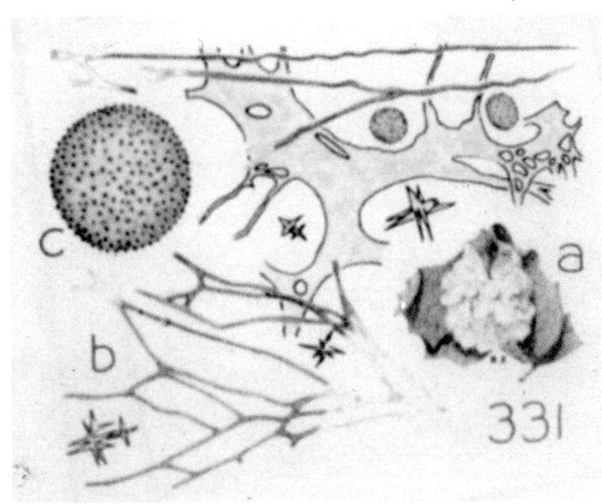

Fig. 61. *Mucilago crustacea* Wiggers. a. Aethalium. X0.6. b. Capillitium, pseudocapillitium, crystals, and spores. X300. c. Spore. X1200. (Fig. 331, Martin and Alexopoulos, 1969)

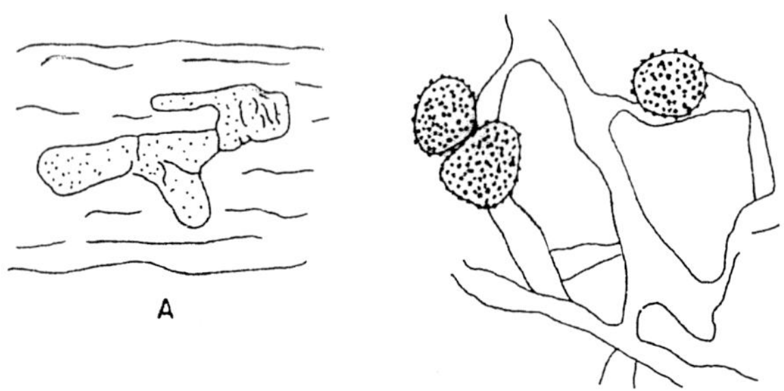

Fig. 62. *Trabrooksia applanata* Keller. a. Plasmodiocarp on wood. X4.5. b. Smooth capillitium and spinose spores. X780. (Redrawn from Figs. 1 and 11, Keller, 1980)

Genus ***Trabrooksia*** Keller, 1980

A single species (Fig. 62) is widespread and frequently collected on the bark of living trees and vines in the United States.
TYPE SPECIES: *Trabrooksia applanata* Keller, 1980

Genus ***Leptoderma*** G. Lister, 1913

A single, rare species (Fig. 63); sometimes is included in the Stemonitidae because of its iridescent peridium and network of blackish, non-calcareous threads.
TYPE SPECIES: *Leptoderma iridescens* G. Lister, 1913

Genus ***Physarina*** Höhn, 1909

Two rare, distinctive species, neither known from the United States. *Physarina echinospora* Thind & Manocha, 1964 is shown in Fig. 64.
TYPE SPECIES: *Physarina echinocephala* Höhn, 1909

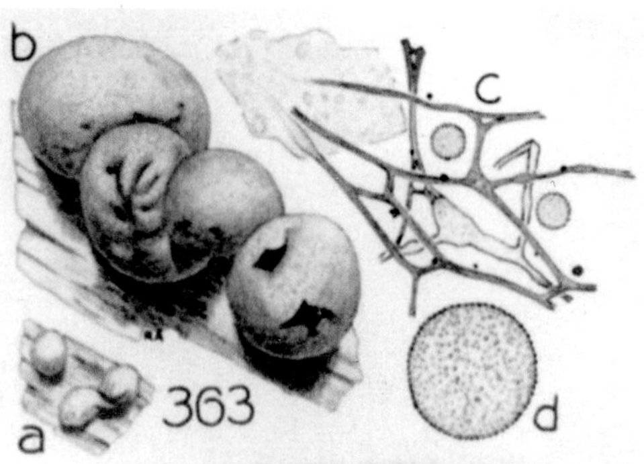

Fig. 63. *Leptoderma iridescens* G. Lister. a. Sporangia, two of them partially fused. X6. b. Sporangia. X24. c. Capillitium attached to peridium. with spores. X300. d. Spore. X1200. (Fig. 363, Martin and Alexopoulos, 1969)

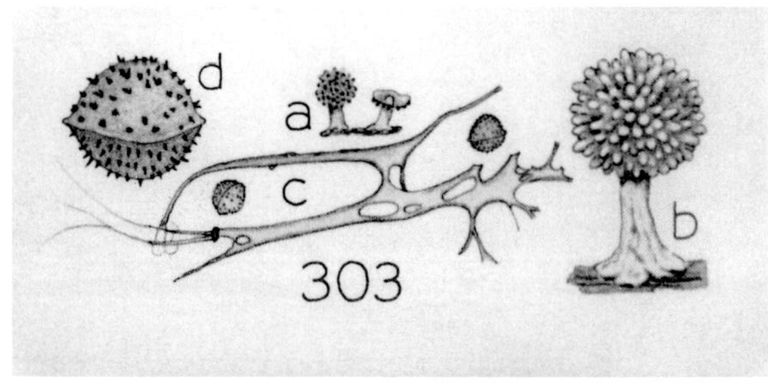

Fig. 64. *Physarina echinospora* Thind & Manocha. a. Sporangia. X6. b. Sporangium. X24. c. Capillitium and spores. X300. Spore. X1200. (Fig. 303, Martin and Alexopoulos, 1969)

Genus ***Diachea*** Fries, 1825

A distinctive genus of at least 10 species with *D. leucopodia* (Bulliard) Rostafinski, 1874 the most common and cosmopolitan (Fig. 65).
TYPE SPECIES: *Diachea leucopodia* (Bulliard, 1791) Rostafinski, 1874

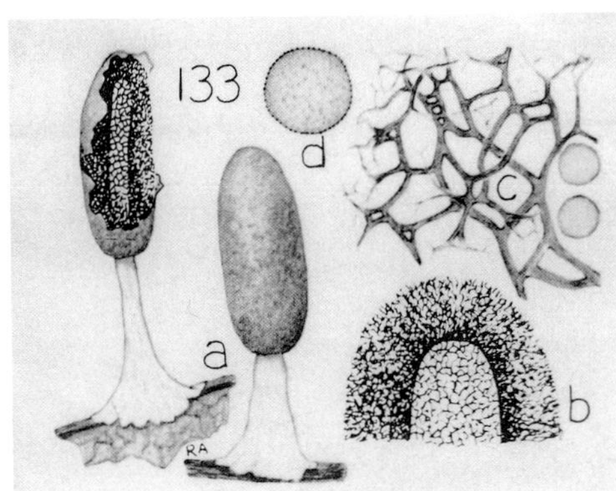

Fig. 65. *Diachea leucopodia* (Bull.) Rost. a. Sporangia. X24. b. Columella tip surrounded by capillitium. X60. c. Capillitium and spores. X600. d. Spore. X1200. (Fig. 133, Martin and Alexopoulos, 1969)

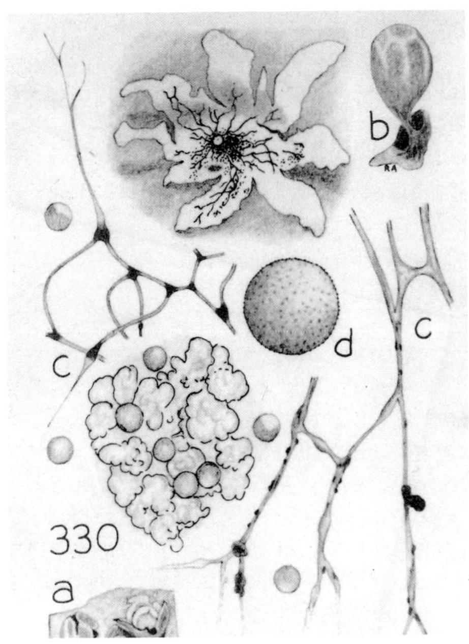

Fig. 66. *Diderma trevelyani* (Grev.) Fries. a. Sporangial cluster. X6. b, Sporangia, one closed and one open after spore discharge. X12. c. Capillitium, portion of peridium with lime crystals embedded, spores. X300. d. Spore. X1200. (Fig. 330, Martin and Alexopoulos, 1969)

Genus ***Diderma*** Persoon, 1794

Diderma has over 40 species; some are predominantly montane such as *D. niveum* (Rostafinski) Macbride, 1899. Others, namely *D. effusum* (Schweinitz) Morgan, 1894, *D. globosum* Persoon, 1794, and *D. testaceum* (Schrader) Persoon, 1801. are cosmopolitan and common on decaying leaves. *Diderma trevelyani* (Greville) Fries, 1829 is shown in Fig. 66.
TYPE SPECIES: *Diderma globosum*, Persoon, 1794

Genus ***Didymium*** Schrader, 1797

A genus of over 40 species, including *D. clavus* (Albertini & Schweinitz) Rabenhorst, 1844, *D. minus* (A. Lister) Morgan, 1894 and *D. squamulosum* (Albertini & Schweinitz) Fries, 1818, which are common and cosmopolitan. *Didymium iridis* (Ditmar) Fries, 1829 is shown in Fig. 67.
TYPE SPECIES: *Didymium farinaceum* Schrader, 1797

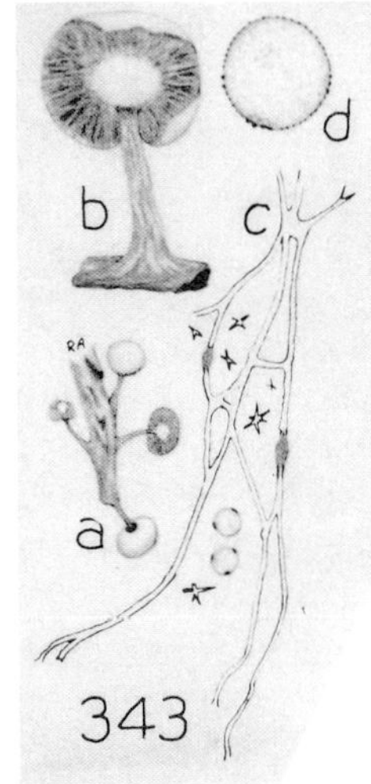

Fig. 67. *Didymium iridis* (Ditmar) Fries. a. Sporangia. X6. b. Sporangium. X24. c. Capillitium, with spores and crystals. X300. d. Spore. X1200. (Fig. 343, Martin and Alexopoulos, 1969)

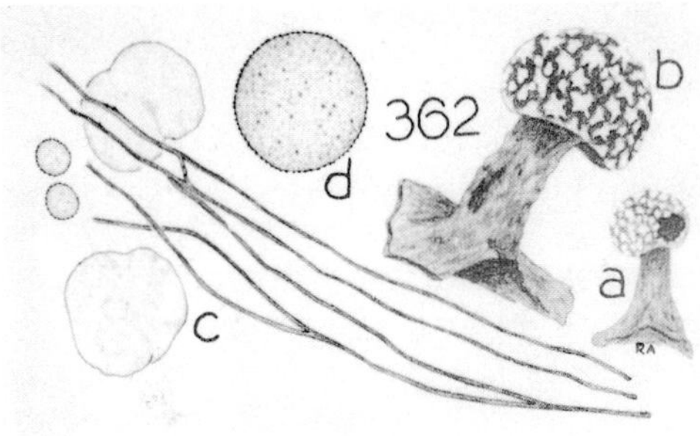

Fig. 68. *Lepidoderma tigrinum* (Schrad.) Rost. a. Sporangium. X6. b. Sporangium. X12. c. Capillitium.

Genus ***Lepidoderma*** de Bary, 1873

Kowalski (1971) recognized 6 species with most members predominantly montane (Fig. 68).
TYPE SPECIES: *Lepidoderma tigrinum* (Schrader, 1797) Rostafinski, 1873

FIGURES

The majority of figures shown have been copied, with the kind permission of the University of Iowa Press, from *The Myxomycetes* by Martin and Alexopoulos (1969). The black-and-white photographic copies do not do full justice to the hand-tinted originals. The figure legends still contain the original figure numbers used by Martin and Alexopoulos (1969). The other figures have been redrawn from several sources, all of which are cited in the figure legends.

REFERENCES

Alexopoulos, C. J. & Mims, C. J. 1979. *Introductory Mycology.* 3rd Edition. John Wiley and Sons, NY.

Blackwell, M. & Alexopoulos, C. J. 1975. Taxonomic studies in the Myxomycetes. IV. *Protophysarum phloiogenum*, a new genus and species of Physaraceae. *Mycologia,* 67:32-37.

Blanton, R. L. 1990. Phylum Acrasea. *In*: Margulis, L., Corliss, J. O., Melkonian, M. & Chapman, D. J. (eds.), *Handbook of Protoctista.* Jones and Bartlett Publishers, Boston. pp 75-87

Brefeld, O. 1869. *Dictyostelium mucoroides.* Ein neuer Organismus und der Verwandschaft der Myxomyceten. *Abh. Beckenberg. Naturforsch. Ges.*, 7:85-107.

Camp, W. G. 1936. A method of cultivating myxomycete plasmodia. *Bull. Torrey Bot. Club,* 63:205-210.

Cavender, J. C. 1990. Phylum Dictyostelida. *In*: Margulis, L., Corliss, J. O., Melkonian, M. & Chapman, D. J. (eds.), *Handbook of Protoctista.* Jones and Bartlett Publishers, Boston. pp 88-101

Demaree, R. S., Jr. & Kowalski, D. T. 1975. Fine structure of five species of Myxomycetes with clustered spores. *J. Protozool.*, 22:85-88.

Dykstra, M. J. 1977. The possible phylogenetic significance of mitochondrial configurations in the acrasid cellular slime molds with reference to members of the Mycetozoa and fungi. *Mycologia*, 69:579-591.

Eliasson, U. H., Keller, H. W. & Schoknecht, J. D. 1991. *Kelleromyxa*, a new generic name for *Licea fimicola* (Myxomycetes). *Mycol. Res.*, 95:1201-1207.

Frederick, L. 1990. Phylum plasmodial slime molds class Myxomycota. *In*: Margulis, L., Corliss, J. O., Melkonian, M. & Chapman, D. J. (eds.), *Handbook of Protoctista.* Jones and Bartlett Publishers, Boston. pp 467-483.

Haeckel, E. 1879. Monographie der moneren. Jena. *Z. Med. Naturwiss.*, 4:64-137.

Hochgesand, E. & Gottsberger, G. 1989. *Arcyriatella congregata*, a new genus and new species of the Trichiaceae (Myxomycetes). *Nova Hedwigia*, 48:485-489.

Hutner, S. H. 1985. Class Mycetozoea de Bary, 1859. *In*: Lee, J. J., Hutner, S. H. & Bovee, E. C. (eds.), *Illustrated Guide to the Protozoa.* Soc. Protozoologists, Lawrence, Kansas. pp 214-226

Keller, H. W. 1980. Corticolous Myxomycetes VIII: *Trabrooksia*, a new genus. *Mycologia*, 72:395-403.

Keller, H. W. 1990. Searching for Myxomycetes. *Mushroom, The Journal of Wild Mushrooms*, 9:32-33.

Keller, H. W., Aldrich, H. C. & Brooks, T. E. 1973. Corticolous Myxomycetes II: Notes on *Minakatella longifila* with ultrastructural evidence for its transfer to the Trichiaceae. *Mycologia* , 65:768-778.

Keller, H. W. & Brooks, T. E. 1976. Corticolous Myxomycetes IV: *Badhamiopsis*, a new genus for *Badhamia ainoae*. *Mycologia*, 68:834-841.

Keller, H. W. & Brooks, T. E. 1976a. Corticolous Myxomycetes V: Observations on the genus *Echinostelium*. *Mycologia*, 68:1204-1220.

Keller, H. W. & Brooks, T. E. 1977. Corticolous Myxomycetes VII: Contribution toward a monograph of *Licea,* five new species. *Mycologia*, 69:667-684.

Keller, H. W. & Eliasson, U. H. 1992. Taxonomic evaluation of *Perichaena depressa* and *P. quadrata* based on controlled cultivation, with additional observations on the genus. *Mycol. Res.*, 96:1085-1097.

Kowalski, D. T. 1970. The species of *Lamproderma*. *Mycologia*, 62:621-672.

Kowalski, D. T. 1971. The genus *Lepidoderma*. *Mycologia*, 63:490-516.

Kowalski, D. T. 1975. The genus *Diacheopsis*. *Mycologia*, 67:616-628.

Lonert, A. C. 1965. A high-yield method for inducing sclerotization in *Physarum polycephalum*. *Turtox News*, 43:98-102.

Margulis, L., Corliss, J. O., Melkonian, M. & Chapman, D. J. (eds.) 1990. *Handbook of Protoctista.* Jones and Bartlett Publishers, Boston.

Martin, G. W. & Alexopoulos C. J. 1969. *The Myxomycetes.* University of Iowa Press, Iowa City.

Martin, G. W., Alexopoulos, C. J. & Farr, M. L. 1983. *The Myxomycetes.* Revised version of the 1969 edition. University of Iowa Press, Iowa City.

Olive, E. W. 1902. Monograph of the Acrasieae. *Proc. Boston Soc. Natur. Hist.*, 30:451-513.

Olive, L. S. 1970. The Mycetozoa: a revised classification. *Bot. Rev.*, 36:59-87.

Olive, L. S. 1975. *The Mycetozoans.* Academic Press, NY.

Olive, L. S. & Stoianovitch, C. 1960. Two new members of the Acrasiales. *Bull. Torrey Bot. Club*, 87:1-20.

Olive, L. S., Stoianovitch, C. & Bennett, W. E. 1983. Descriptions of acrasid cellular slime molds: *Pocheina rosea* and a new species, *Pocheina flagellata. Mycologia*, 75:1019-1029.

Raper, K. B. & Quinlan, M. S. 1958. *Acytostelium leptosomum*: a unique cellular slime mould with an acellular stalk. *J. Gen. Microbiol.*, 18:16-32.

Raper, K. B. 1984. *The Dictyostelids.* Princeton University Press, Princeton, NJ.

Spiegel, F. W. 1990. Phylum plasmodial slime molds class Protostelida. *In*: Margulis, L., Corliss, J. O., Melkonian, M. & Chapman, D. J. (eds.), *Handbook of Protoctista.* Jones and Bartlett Publishers, Boston. pp 484-497

van Tiegham, M. P. 1880. Sur quelques Myxomycetes a plasmode agrege. *Bull. Soc. Bot. Fr.*, 27:317-322.

van Tieghem, M. P. 1884. *Coenonia*, genre noveau de Myxomcetes á plasmode agrege. *Bull. Soc. Fr.*, 31:303-306.

Whitney, K. D. 1980. The myxomycete genus *Echinostelium. Mycologia*, 72:950-987.

Whittaker, R. H. 1969. New concepts of kingdoms of organisms. *Science*, 163:150-160.

Worley, A. C., Raper, K. B. & Hohl, M. 1979. *Fonticula alba*: a new cellular slime mold (Acrasiomycetes). *Mycologia*, 71:746-760.

CLASS PHAEODAREA

by KOZO TAKAHASHI and O. R. ANDERSON

Phaeodaria are oceanic protists with porous skeletons composed of biogenic opal with organic substances and traces (to 1%) of Mg, Ca, and Cu. They are included in the artificial grouping of radiolaria in the older literature and are found in major oceanic locations, dwelling from the near surface to great depths (4000 to 8000 m) in the water column (e.g. Haeckel, 1887, Anderson, 1983; Takahashi, 1987, 1991). Skeletal sizes vary from approximately 50 µm for some genera to several hundred micrometers for many genera. Some large species are easily visible with the unaided eye. The skeletons of *Aulosphaera* (Fig. 20) and other large, geodesic, spherical genera may be several millimeters in diameter, and other genera such as *Coelographis* (Fig. 3) with long arm-like extensions can be tens of millimeters in over-all dimensions. Due to the porous structure and different chemical composition of the phaeodarian skeleton relative to the polycystines, they are less resistant to dissolution in marine sediments. Hence, the phaeodaria are less represented in the microfossil record (e.g. Takahashi et al., 1983).

Phaeodaria typically lack algal symbionts. Current knowledge of their role in marine food webs is meager although prey has been documented for some species. Phaeodaria appear to be generalists consuming a wide range of particulates including bacteria, *Chlorella*-like cells, other algae, diatoms, scale-bearing algae, tintinnids, crustacea, and "olive-green detrital matter" (Gowing, 1986, 1989; Nöthig and Gowing, 1991; Gowing and Garrison, 1992; Gowing and Wishner, 1986, 1992). *Coelographis* sp., a large phaeodarian with long skeletal styles, collected by SCUBA divers, contained ingested detrital matter, metazoans, flagellates, copepods, and gelatinous organisms (Swanberg et al., 1986). Additional information on phaeodarian ecology in relation to polycystines is presented in Anderson (1983, 1993).

Morphology. Much of the classification is based on the skeletal morphology. Major skeletal types include those with bilateral symmetry (e.g. bivalve shells resembling microscopic clams, Fig. 8), radial symmetry (forming geodesic spheres of remarkable complexity, Fig. 20), and multi-symmetry with polyhedral or more complex geometric forms (Figs. 13,14,35,47). The central capsular membrane (arrow, Fig. 25) is thick relative to other radiolaria and contains two major kinds of pores, each with a cytoplasmic strand that projects into the surrounding environment: 1) "**astropylum**" or **oral pore** with a large opening and massive cytoplasmic extension (A in Fig. 25) from the central capsular cytoplasm and 2) "**parapylae**" (P in Fig. 25), usually two pores, and smaller in diameter, at the opposite pole (Fig. 25). Digestion of prey can occur in the extracapsulum or within the **intracapsulum** by ingestion through the astropylum. A dense mass of darkly pigmented, undigested debris (**phaeodium**, Fig. 2a,b) is suspended in the **extracapsular cytoplasm**, usually in the vicinity of the oral region at one pole of the central capsule. The vegetative nucleus is large and spherical (only rarely cordiform) and occupies nearly all of the **central capsule** (Fig. 22). During reproduction, the cytoplasm becomes divided into numerous small swarmers, each containing a nucleus with eight chromosomes.

Some genera such as *Phaeodina* (Fig. 1) lack skeletal elements or are surrounded by shells of other protists. Other genera, such as *Aulacantha* (Fig. 2), contain only a loose arrangement of radially directed individual spicules embedded within the extracapsular cytoplasm, but many have more massive skeletal structures (Figs. 10,15,26,38). The skeleton can be bilaterally symmetrical with two halves forming a bivalve resembling a clam (Fig. 3), campanulate (Fig. 29), or nearly spherical (e.g. Figs. 15,16). Others have hollow tubular skeletal elements forming meshworks (Fig. 24), geodesic spheres (Fig. 20), and complex frameworks surrounding the central capsule. Skeletal surface structures include spines (Fig. 13) and verticillate extensions, i. e. ray-like extensions with many whorls of lateral spines (Fig. 22). A tubular elongation of the skeletal opening (**peristome**) near the mouth can be cylindrical (arrow, Fig. 18) or trumpet-shaped (arrow, Fig. 12) and may be connected to a circular arch (Fig. 11). Massive arm-like extensions of the skeletal framework occur in some genera. These may form paired **styles** (arrow, Fig. 3) supporting a meshwork or lattice mantle. Major projections known as horns (large arrow, Fig. 4), or in other genera, feet (arrow, Fig. 42) may extend from the periphery of the shell, and there may be a ridge-like keel as in *Conchidium* (arrow, Fig. 4). The surface of the mesh may be ornamented by numerous spine-like extensions with swollen or expanded tips (Figs. 3,20,23). The surface of the shell may be smooth (Fig. 26), dimpled (Fig. 37), or ridged (Fig. 4).

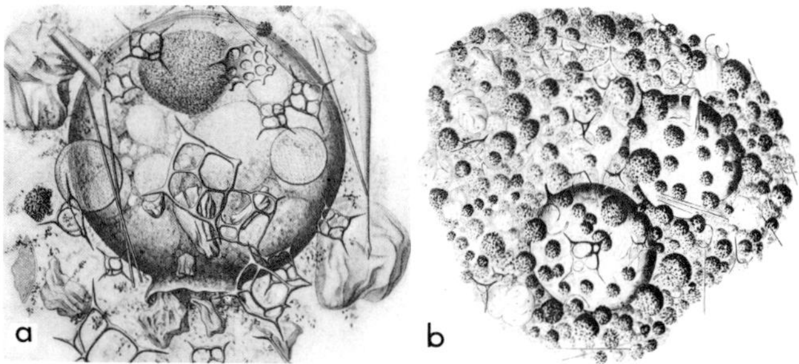

Fig. 1. Families Phaeosphaeridae and Phaeodinidae. 1a. *Phaeopyla spherica* Cachon-Enjumet, 1961. 1b. *Phaeodina valdivia* (Haecker, 1908).

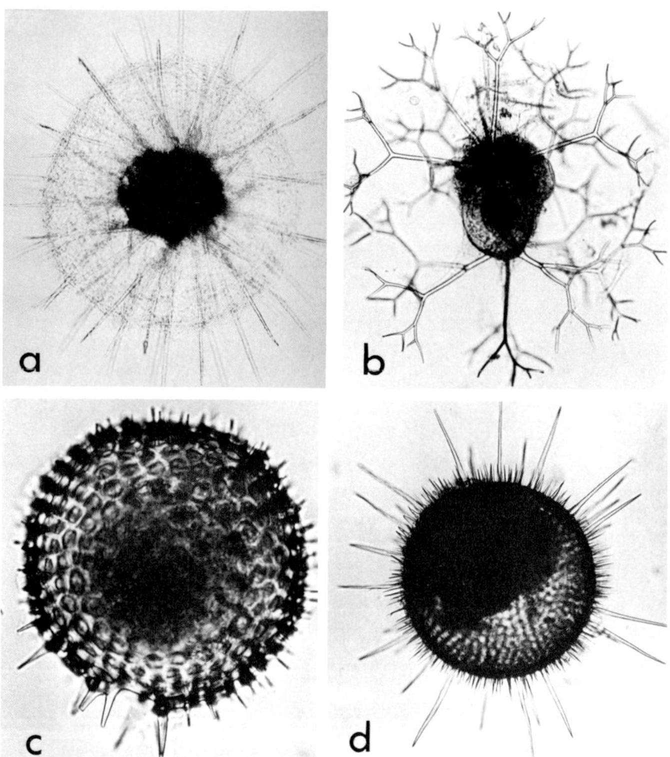

Fig. 2. Families Aulacanthidae, Coelodendridae, Castenellidae. 2a. *Aulacantha scolymantha* Haeckel, 1982. 2b. *Coelodendrum ramosissimum* Haeckel 1962. 2c. *Castanella wivillei* Haeckel, 1887. 2d. *Castanissa valdiviae* Haecher, 1907 (all alive).

Taxonomy. The taxonomy of the Phaeodaria needs extensive modern revision as is generally true for the "radiolaria." Much of our current taxonomy is based on the extensive systematics published by Haeckel (1887), but it is clearly artificial. A synopsis of current thought on taxonomy at the level of Family is presented here (Anderson, 1983, Cachon and Cachon, 1985, Takahashi, 1991) though it may change substantially with increasing evidence from fine structural and perhaps molecular taxonomic research.

The following key to some major living genera is not a natural grouping, but a practical guide to some commonly observed genera collected in plankton samples.

KEY TO MAJOR LIVING GENERA

LM

1. Skeleton lacking or composed of radiating loose spicules I
1'. Skeleton entire, robust or composed of rods in framework.. II

I. GENERA WITHOUT SKELETONS

1. No skeleton, loosely aggregated scales and spicules collected from other protists............ 2
1'. Radially arranged needle-like spicules, not connected .... ***Aulacantha***

2. Phaeodium extra- and intracapsular, with thick capsular membrane........................ ***Phaeopyla***
2'. Phaeodium extracapsular only........ ***Phaeodina***

II. GENERA WITH SKELETONS

1. With bivalves 2
1'. Without bivalves 6

2. With long paired styles, rhinocanna (nasal tube), and outer lattice-mantle. ***Coelographis***
2'. Without styles 3

3. With two horns, one on each valve at aboral hinge, lenticular, with keel .... ***Conchidium***
3'. Without horns... 4

4. Spherical bivalves 5
4'. Lenticular bivalves, and keel........ ***Conchopsis***
4". Lattice with circular pores and no hollow spines .... ***Conchellium***

5. With conical process (galea) and three or more branched spines........................ ***Coelodenrum***
5'. Lattice with rectangular pores.. ***Conchophacus***

6. With tubular arch and trumpet-shaped peristome .... ***Borgertella***
6'. Without tubular arch 7

7. Radial symmetry (spheres)............................ 8
7'. Bilateral symmetry (ovate, lenticular, campanulate) 17
7". Multilateral symmetry (polyhedral), smooth shell... 24

8. With aperture 9
8'. Without aperture 13

9. Porcellaneous shell, radial spines at center of stellate circles.............................. ***Haeckeliana***
9'. Non-porcellaneous, with lattice work 10
9". Trizonal meshwork (triangular pores), and tubular peristome........................ ***Porospathis***

10. With main radial spines............................... 11
10'. Without main radial spines, teeth on peristome... ***Castanella***

11. Main radial spines branched, teeth on peristome .... ***Castanea***
11'. Main radial spines unbranched 12

12. With teeth on peristome................ ***Castanissa***
12'. Without peristomal teeth........... ***Castanidium***

13. Large shells composed of tangential tubules, triangular or regular simple mesh, lacking pyramidal elevations......................................14
13'. Large partial skeletons, many radial tubules touching central capsule 15

14. With radial tubes........................***Aulosphaera***
14'. Without radial tubes...........................***Aularia***

15. Lateral, verticillate branches on radial tubes .... ***Aulospathis***
15'. Lacking lateral branches on radial tubes 16

16. Terminal branches simple......... ***Aulographis***
16'. Terminal branches forked.............. ***Auloceros***

17. Ovate or lenticular shells 18
17'. Campanulate shells 23

18. With pharynx 19
18'. Without pharynx... 20

19. With one or more oral teeth ...... ***Pharyngella***
19'. Lacking oral teeth or marginal spine................ .... ***Entocannula***

20. With marginal spines.................................. 21
20'. Without marginal spines 23

21. Shell surface basically smooth 22
21'. Shell surface alveolate...... ***Challengeranium***

22. With partial alveolate surface........................ .... ***Challengerosium***
22'. Without alveoli........................ ***Challengeron***

23. Smooth surface with oral teeth.... ***Protocystis***
23'. Ridged or furrowed surface, without oral teeth 25

24. Four equidistant articulate feet....... ***Medusetta***
24'. One large and three rudimentary articulate feet .... ***Euphysetta***

25. Octahedral shell........................... ***Circoporus***
25'. Icosahedral shell .... ***Circogonia***

CLASS PHAEODAREA Haeckel, 1879

Skeleton composed of biogenic opal with porous, sometimes hollow, structures; thick capsular wall with openings, a large astropyle and usually two smaller parapylae: one kind of axopodium associated with the parapylae and a different kind with the astropyle.

ORDER PHAEOGYMNOCELLIDA CACHON & CACHON, 1985

Skeleton absent or forming only a cup-shaped structure covering oral pole.

FAMILY PHAEOSPHAERIDAE Cachon-Enjumet, 1961

Skeleton lacking; peripheral cytoplasm contains shells of other protists (diatoms, silicoflagellates, dinoflagellates, etc.) and enclosed by much phaeodium. Intracapsular phaeodium present. (Fig. 1)

Genus ***Phaeopyla*** Cachon-Enjumet, 1961

Phaeodium extra- and intracapsular. Capsular membrane thick. Astropylum large, simple, widely open (Fig. 1a).

Genus ***Phaeodactylis*** Cachon-Enjumet, 1961

Phaeodium extracapsular only. Capsular membrane thick. Astropylum bordered with various long, finger-like appendages.

Genus ***Phaeosphaera*** Cachon-Enjumet, 1961

With characters of the family, intracapsular phaeodium present.

FAMILY PHAEODINIDAE Cachon-Enjumet, 1961

Skeleton lacking; central capsule surrounded by cytoplasm containing shells of other protists and numerous phaeodium-globules. No intracapsular phaeodium. (Fig. 1b)

Genus ***Phaeodina*** (Haecker) Cachon-Enjumet, 1961

With characters of the family. Generally with 2 central capsules (Fig.1b).

FAMILY ATLANTICELLIDAE Cachon-Enjumet, 1961

Skeleton and phaeodium usually absent, but when skeleton is present, it forms a cup-shaped structure covering the oral pore. Capsule always globular, usually with 3 openings. Nucleus usually adjacent to astropylum.

Genus ***Gymnocella*** Cachon-Enjumet, 1961

No cytoplasmic strand; parapylae in aboral hemisphere. Neither skeleton nor phaeodium.

Genus ***Halocella*** Borgert, 1907

Skeleton formed of spongy basket-like piece and 2 small wing-like rods.

Genus ***Lobocella*** Borgert, 1907

Saccular central capsule with finger-like processes.

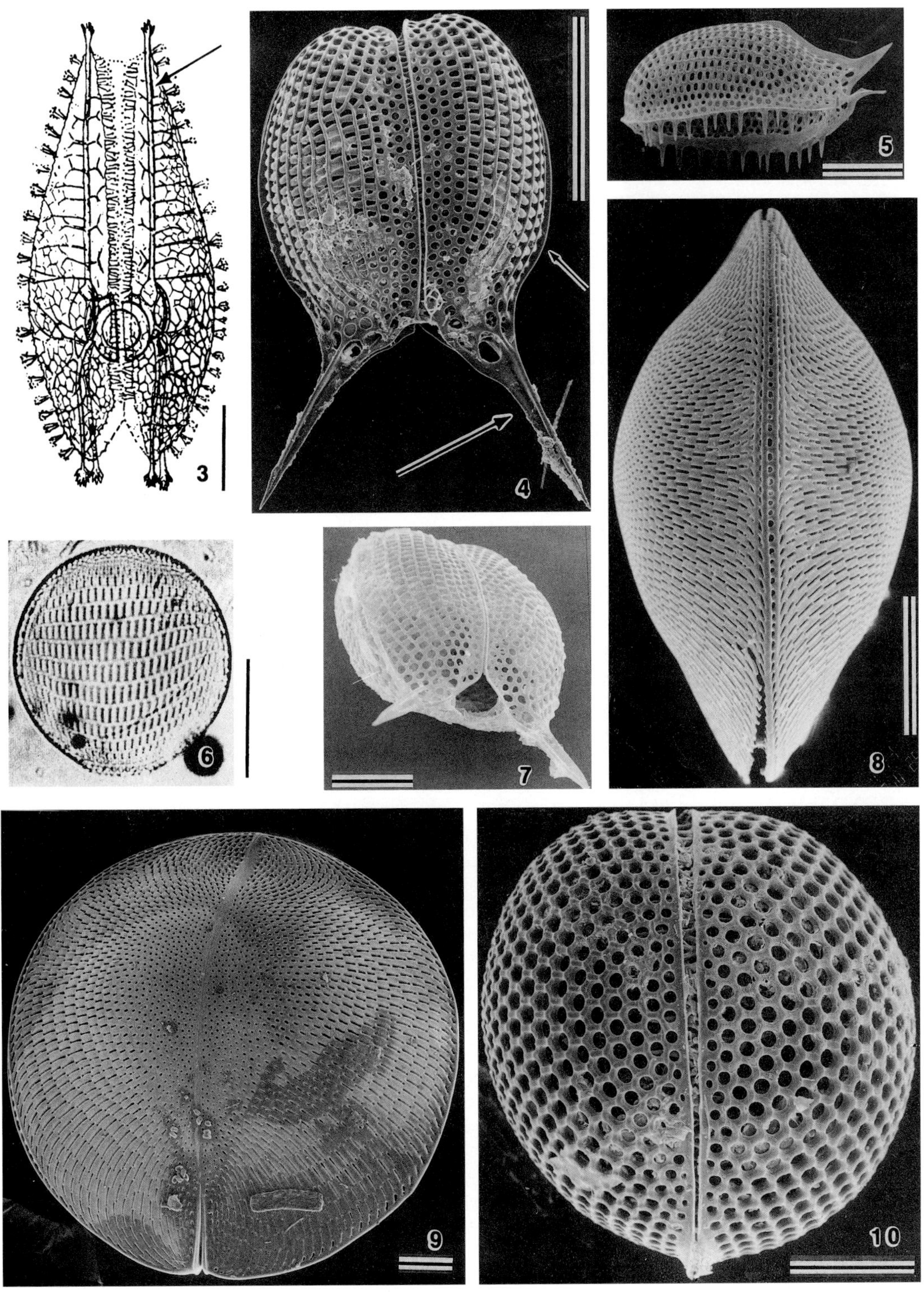

Families Coelodendridae and Concharidae. All scale bars are 100 µm except for Fig. 3, which is 1 mm, and Fig. 10, which is 50 µm. 3. *Coelographis regina* Haeckel, 1887, showing styles (arrow) (line drawing from Haeckel, 1887). 4,7. *Conchidium caudatum* (Haeckel, 1887), with keel (arrow) and horns (large arrow). 5. *Conchidium argiope* Haeckel, 1887. 6. *Conchophacus diatomeus* Haeckel, 1887. 8,9. *Conchopsis compressa* Haeckel, 1887. 10. *Conchellium capsula* Borgert, 1907.

Families Lirellidae, Castanellidae, and Porospathidae. Scale bars=10 μm for Figs. 11,12,19 and 100 μm for Figs. 13—18. 11,12. *Borgertella caudata* (Wallich, 1869), with trumpet-shaped peristome (arrow). 13. *Haeckeliana porcellana* Murray, 1885. 14. *Castanidium longispinum* Haecker, 1908. 15. *Castanella macropora* (Schmidt, 1908). 16. *Castanella aculeata* Schmidt, 1907. 17. *Castanissa circumvallata* Schmidt, 1907. 18,19. *Porospathis holostoma* (Cleve, 1899), with tubular peristome (arrow).

Genus ***Miracella*** Borgert, 1911

Parapylae at aboral pole. Skeleton absent or formed by foreign adhering matter.

Genus ***Planktonetta*** Borgert, 1902
emend. Cachon-Enjumet, 1961

Skeleton cup-shaped shell with articulate arms. Parapylae (2 or more) near astropylae (1 or more). Large central capsule with single, large vacuole.

ORDER PHAEOCYSTIDA Haeckel, 1887

Skeleton simple, composed of numerous hollow, thin and tangential needles at periphery or composed only of radial spines with proximal ends near the central capsule, or combinations of radial and tangential needles (Fig. 2).

FAMILY AULACANTHIDAE Haeckel, 1887

Skeleton simple, composed of numerous hollow, thin, and tangential needles at periphery, or composed only of radial spines with proximal ends near the central capsule; or combinations of radial and tangential needles (Fig. 2a).

Genus ***Aulacantha*** Haeckel, 1860

Tangential needles numerous, make an external interwoven veil; radial spines denticulate (Fig. 2a).

FAMILY: ASTRACANTHIDAE Haeckel, 1887

Skeleton of radial, hollow spines with proximal ends united at center, forming a hollow space, surrounding two enclosed central capsules.

Genus ***Astracantha*** Haecker, 1908

Spines with small, regularly disposed thorns.

Genus ***Castanella*** Haeckel, 1879

Dentate (toothed) mouth, without radial spines (Figs. 2c,15).

Genus ***Castanissa*** Haeckel, 1879

Large unbranched radial main spines scattered between short bristles; with dentate mouth.

ORDER PHAEOSPHAERIDA Haeckel, 1887

Shell, one or more, composed of hollow or solid rods enclosing the central capsule.

FAMILY AULOSPHAERIDAE Haeckel, 1887

Shell generally latticed, or sometimes spongy, composed of hollow rods that form a cortical network with triangular or polygonal meshes supporting radial by-spines (spines arising at the nodes of the meshwork) (Fig. 20).

Genus ***Aulosphaera*** Haeckel, 1887

Lattice shell with triangular meshes; smooth cylindrical, radial tubes, a verticil of 3 divergent terminal branches (Fig. 20).

Genus ***Aularia*** Haeckel, 1887

Like *Aulosphaera*, but shell surface smooth, lacks radial tubules (Fig. 25).

Genus ***Aulotractus*** Haeckel, 1887

Single shell elongate, ellipsoidal to spindle-shaped; radial tubules at nodal points.

FAMILY: CANNOSPHAERIDAE Haeckel, 1887

Two concentric shells united by numerous strands; external shell latticed with polygonal meshes and internal one massive with a pylum. Radial spines arise from nodal points of cortical network.

Genus ***Coelocantha*** Hertwig, 1879

Internal shell latticed, with 60 to 90 radial spines; external shell pentagonal meshed; from each nodal point emerges a smooth radial spine bearing a verticil of 3 by-spines.

FAMILY SAGOSPHAERIDAE Haeckel, 1887

Skeleton spherical with lattice work containing subregular triangular meshes and filiform solid rods. Internal shell, when present, has no pylum.

Genus ***Sagenoarium*** Borgert, 1891

Double lattice shell with numerous pyramidal elevations without axial rods and with radial spines.

Genus ***Sagenoscena*** Haeckel, 1887

Pyramidal tents or elevations formed by rodlets united at the apex, arising from the surface of the lattice work, with internal axial rods, in some species prolonged into a crowned radial spine.

Genus ***Sagoscena*** Haeckel, 1887

Pyramidal tents, arising from the surface of the lattice work, without internal axial rods.

ORDER PHAEOCALPIDA Haeckel, 1887

Spherical, polyhedral, or ovate shells, some with porcellanous quality.

FAMILY CASTANELLIDAE Haeckel, 1887

Shell spherical with round pores. Radial spines, arising from nodes of the lattice shell, cover the surface. Shell has a large mouth (Fig. 15).

Genus ***Castanea*** Haecker, 1906

Large solid shell; main radial spines branched; small smooth mouth, spines on peristome.

FAMILY CIRCOPORIDAE Haeckel, 1887

Shell spherical or polyhedral (generally large mouth) with either porcellanous structure (nearly polygonal network with crests), or tabulate (surface flattened and smooth at places, like a slate tablet). Hollow radial spines encircled at base by circle of radial pores (Fig. 46).

Genus ***Circoporus*** Haeckel, 1879

Spherical shell with 6 radial spines (Fig. 46).

Genus ***Circospathis*** Haeckel, 1879

Tetradecahedral shell with 9 radial spines.

Genus ***Haeckeliana*** Haeckel, 1887

Dimpled spherical shell without polygonal plates; unbranched radial spines often numerous but variable in number (Fig. 13).

FAMILY TUSCARORIDAE Haeckel, 1887

Shell spherical, ovate or spindle-shaped with a porcellanous surface, smooth or spiny, but not tabulate or paneled (with flattened tile-like segments). Few tubular spines regularly arranged around a large pore or around circle of small pores.

Genus ***Tuscarora*** Murray, 1879

Three equidistant radial legs.

Genus ***Tuscarilla*** Haeckel, 1887

Four crossed teeth.

Genus ***Tuscaretta*** Haeckel, 1887

Two oral teeth.

FAMILY POROSPATHIDAE Borgert, 1900

Shell spherical or ovate with a smooth or tabulate surface containing irregularly disposed tubular spines. Peristome prolonged as a tubule. (Fig. 18)

Genus ***Porospathis*** Haeckel, 1879

Single genus with characteristics of the family (Fig. 18).

FAMILY POLYPYRAMIDAE Reschetnjak, 1966

Shell spherical or polyhedric, loosely polygonal pores, covered by pyramids formed by 4-5 beams (rod-like segments) and from which arise radial spines.

Genus ***Polypyramis*** Reschetnjak, 1966

One genus, one species with characteristics of the family.

ORDER PHAEOGROMIDA Haeckel, 1887

Shell ovate, lenticular or cape-shaped, sometimes with spines.

FAMILY CHALLENGERIDAE Murray, 1885

Shell ovate or lenticular with fine pores composing hexagonal meshes. The peristome is prolonged as a tubular pharynx surrounded by spines. Shell may be covered by marginal or aboral spines (Fig. 34).

Genus ***Challengeria*** Murray, 1876

Shell with oral teeth, without marginal spines.

Genus ***Challengeron*** Murray, 1879

Like *Challengeria*, but has spines on sharp marginal edge of shell (Fig. 34).

FAMILY MEDUSETTIDAE Haeckel, 1887

Ovate, hemispherical or cape-shaped (ovate with a broad basal opening) shell of alveolated structure, with by-spines, and with or without apical spine. Peristome surrounded by cylindrical, hollow, articulated spines (Fig. 42).

Genus ***Euphysetta*** Haeckel, 1887

With apical spine; 1 long and 3 small teeth-spines.

Genus ***Gazelletta*** Haeckel, 1887

Shell hemispherical, no apical spine; 6 "feet" (very long) that are radiate without terminal branches.

Genus ***Medusetta*** Haeckel, 1887

Four branched feet, apex usually with horn (Figs. 42-43).

FAMILY LIRELLIDAE Loeblich & Tappan, 1961

Small, elliptical shell with longitudinal striae: apex with or without apical spine, or with an elliptical ring connecting apex and exterior of aperture. One of the most abundant families of deep-water dwellers (Fig. 40).

Genus ***Borgertella*** Dumitrica, 1973

Shell with 2 main parts: egg-shaped chamber closed at the aboral end and armed with a hollow spine, and a long, more or less curved, trumpet-like peristome. Inner cavity of 2 parts separated by a diaphragm and communicating only through a narrow tube entering the peristomal cavity (Fig. 11).

Genus ***Lirella*** Ehrenberg, 1872

Ovate or lenticular shell (not bivalved) without pharynx, ridged or furrowed surface, without oral teeth (Fig. 40).

ORDER PHAEOCONCHIDA Haeckel, 1887

Shell composed of 2 thick-walled valves resembling the 2 halves of a clam shell.

FAMILY CONCHARIDAE Haeckel, 1887

Shell composed of 2 thick-walled latticed valves, spherical or lenticular, perforated by many pores; valves with smooth or dentate edges and oral split between valves; horn on aboral hinge (Fig. 5).

Genus ***Conchidium*** Haeckel, 1887

Bivalve shell without styles, but with 2 horns, one on each valve at aboral hinge, lenticular, with keel (Figs. 4 ,5)

ORDER PHAEODENDRIDA Haeckel, 1887

Shell with 2 hemispherical, thin-walled valves.

FAMILY COELODENDRIDAE Haeckel, 1887

Shell composed of two thin-walled hemispherical valves (a dorsal and a ventral one) with many pores. Each valve with a conical process (galea)

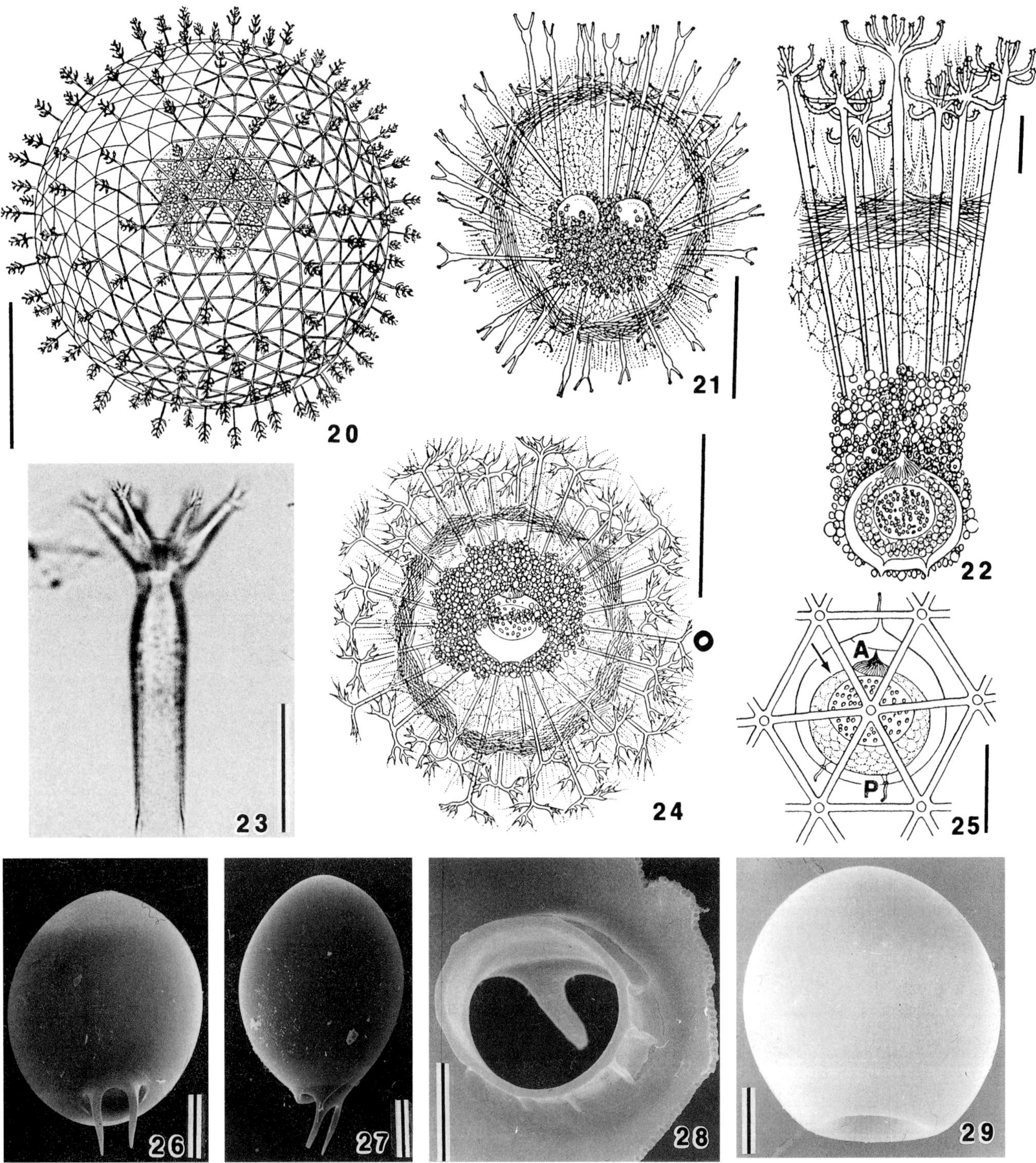

Families Aulacanthidae, Aulosphaeridae, and Challengeriidae. Scale bars=1 mm for Figs. 20-24, 100 µm for Figs. 25-27, 29, and 50 µm for Fig. 28. 20. *Aulosphaera dendrophora* Haeckel, 1887 (line drawing from Haeckel, 1887). 21. *Aulospathis bifurca* Haeckel, 1887 (line drawing from Haeckel, 1887). 22. *Aulographis candelabrum* Haeckel, 1887 (line drawing from Haeckel, 1887). 23. *Auloceros spathillaster* Haeckel, 1887. 24. *Auloceros elegans* Haeckel, 1887 (line drawing from Haeckel, 1887). 25. *Aularia ternaria* Haeckel, 1887 (line drawing from Haeckel, 1887). 26-27. *Pharyngella gastrula* Haeckel, 1887 with inside view (Fig. 28) showing detail of a pharynx. 29. *Entocannula infundibulum* Haeckel, 1887.

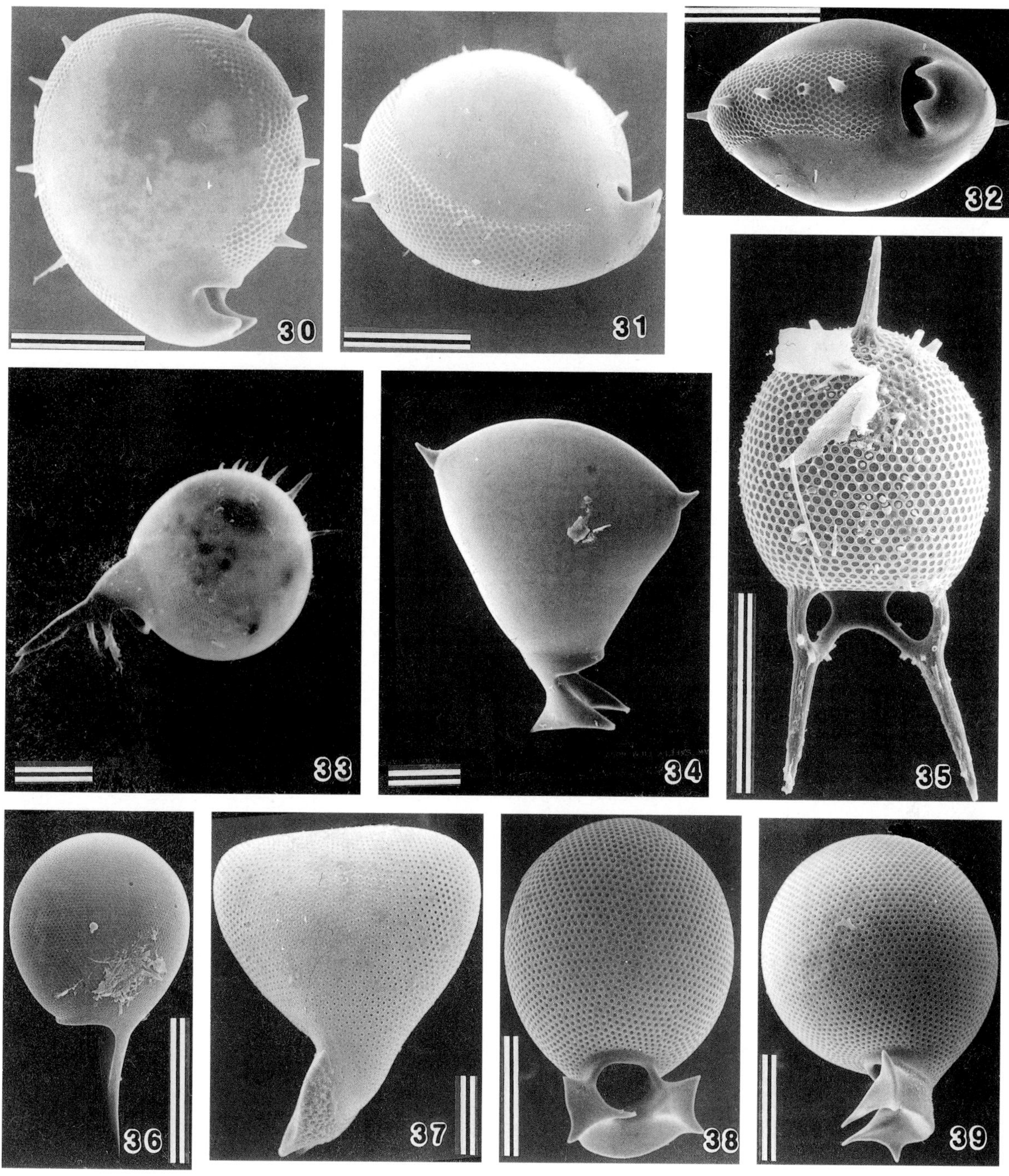

Family Challengeriidae. All scale bars are 100 µm except for Fig. 37, which is 50 µm. 30-32. *Challengerosium avicularia* Haecker, 1906. Lateral view (Fig. 30). Oblique lateral view (Fig. 31) and ventral view (Fig. 32). 33. *Challengeron willemoesii* Haeckel, 1887. 34. *Challengeron tizardi* (Murray, 1885). 35. *Challengeranium diodon* (Haeckel, 1887). 36. *Protocystis xiphodon* (Haeckel, 1887). 37. *Protocystis sloggetti* (Haeckel, 1887). 38-39. *Protocystis murrayi* (Haeckel, 1887). Ventral view (Fig. 38) and lateral view (Fig. 39).

Families Lirellidae, Medusettidae, and Circoporidae. Scale bars=10 µm for Figs. 40-41, 50 µm for Figs. 42-44; and 100 µm for Figs. 45-49. 40. *Lirella melo* (Cleve, 1899). 41. *Lirella baileyi* Ehrenberg, 1872. 42, 43. *Medusetta ansata* Borgert, 1902, showing feet (arrow, Fig. 42). 44. *Euphysetta elegans* Borgert, 1902. 45. *Euphysetta lucani* Borgert, 1892. 46. *Circoporus sexfuscinus* Haeckel, 1887. 47. *Circoporus oxyacanthus* Borgert, 1902. 48, 49. *Circogonia* sp.

from which three or more divergent, branched hollow spines arise whose branches may anastomose and form a spongy mantle (Fig. 3). Many genera.

Genus ***Coelodendrum*** Haeckel, 1860

Spherical bivalves with conical process (galea) and 3 or more branched spines (Fig. 2b).

Genus ***Coelographis*** Haeckel, 1887

Bivalved shell with long paired styles, rhinocanna (nasal tube) and outer lattice-mantle (Fig. 3).

FAMILY CHALLENGERIDAE Murray, 1885

Shell ovate or lenticular with fine pores composing hexagonal meshes. The peristome is prolonged as a tubular pharynx surrounded by spines. Shell may be covered by marginal or aboral spines. (Fig. 34)

FAMILY MEDUSETTIDAE Haeckel, 1887

Ovate, hemispherical, or cape-shaped (ovate with a broad basal opening) shell of alveolated structure, with by-spines, and with or without apical spine. Peristome surrounded by cylindrical, hollow, articulated spines. (Fig. 42)

ORDER PHAEOCONCHIDA Haeckel, 1887

Shell composed of two thick-walled valves resembling the two halves of a clam shell.

FAMILY CONCHARIDAE Haeckel, 1887

Shell composed of two thick-walled latticed valves, spherical or lenticular, perforated by many pores; valves with smooth or dentate edges and oral split between valves; horn on aboral hinge. (Fig. 5)

ORDER PHAEODENDRIDA Haeckel, 1887

Shell with two hemispherical, thin-walled valves.

FAMILY COELODENDRIDAE Haeckel, 1887

Shell composed of two thin-walled hemispherical valves, dorsal and ventral, each with many pores. Each valve with a conical process (galea) from which three or more divergent, branched hollow spines arise, the branches of which may anastomose and form a spongy mantle. (Figs. 2, 3)

Literature Cited

Anderson, O. R. 1983. *Radiolaria*.. Springer-Verlag, NY.

Anderson, O. R. 1993. The trophic role of planktonic foraminifera and radiolaria. *Mar. Microbial Food Webs*, 7:31-51.

Borgert, A. 1982. Vorbericht über einige Phaeodarien- (Trijpyleen-) Familien der Plankton-Expedition. *Ergebn. der Plankton-Expedition.*, 1A:176-184.

Borgert, A. 1900. Untersuchungen über die Fortzpflanzung der Tripyleen Radiolarien, speziell von *Aulacantha scolymantha. Teil I -- Zool. Jahrb.*, 14:203-276.

Borgert, A. 1902. Mittheilungen über die Tripyleen-Ausbeute der Plankton-Expedition. I. Neue Medusettidae, Circoporidae, und Tuscaroridae. *Zool. Jahrb.*, 2:566-577.

Borgert, A. 1907. Die Tripyleen Radiolarien der Plankton-Expedition. Concharidae. *Ergebnisse der Plankton Expedition*, 3, L.h.(5):195-232.

Cachon, J. & Cachon, M. 1985. III. Class Phaeodarea. *In*: Lee, J. J., Hutner, S. H. & Bovee, E. C. (eds.), *An Illustrated Guide to the Protozoa*. Society of Protozoologists, Lawrence, KS. pp 295-302

Cachon-Enjumet, M. 1961. Contribution à l'études Radiolaires Phaeodariés. *Arch. Zool. Exp. Gén.*, 100:152-273.

Cleve, P. T. 1899. Plankton collected by the Swedish Expedition to Spitzbergen in 1898. *K. Svenska Vetensk.-Akad., Handl.*, 32(3):1-51.

Ehrenberg, C. G. 1872. Nachtrag zur Übersicht der organischen Atmospharilen. *Kgh. Akad. Wiss. Berlin, Jahrg. 1871*. p 233

Gowing, M. M. 1986. Trophic biology of phaeodarian radiolarians and flux of living radiolarians in the upper 2000 m of the North Pacific central gyre. *Deep-Sea Res.*, 33:655-674.

Gowing, M. M. 1989. Abundance and feeding ecology of Antarctic phaeodarian radiolarians. *Mar. Biol.*, 103:107-118.

Gowing, M. M. & Coale, S. L. 1989. Fluxes of living radiolarians and their skeletons along a northeast Pacific transect from coastal upwelling to open ocean waters. *Deep-Sea Res.*, 36:561-576.

Gowing, M. M. & Garrison, D. L. 1992. Abundance and feeding ecology of larger protozooplankton in the ice edge zone of the Weddel and Scotia Seas during the austral winter. *Deep-Sea Res.*, 39:893-919.

Gowing, M. M. & Silver, M. W. 1985. Minipellets: a new and abundant size class of marine fecal pellets. *J. Mar. Res.*, 43:395-418.

Gowing, M. M. & Wishner, K. E. 1986. Trophic relationships of deep-sea calanoid copepods from the benthic boundary layer of the Santa Catalina Basin, California. *Deep-Sea Res.*, 33:939-961.

Gowing, M. M. & Wishner, K. E. 1992. Feeding ecology of benthopelagic zooplankton on an eastern tropical Pacific seamount. *Mar. Biol.*, 112:451-467.

Haeckel, E. 1862. *Die Radiolarien (Rhizopoda, Radiolaria). Eine Monographie.* Georg Reimer, Berlin. 586 pp

Haeckel, E. 1887. Report on Radiolaria collected by H. M. S. Challenger during the years 1873-1876. *In*: Thompson, C.W. & Murray, J. (eds.), *The Voyage of the H. M. S. Challenger.* Her Majesty's Stationery Office, London, V. 18:1-1803.

Haecker, V. 1906. Über Kenntis der Challengeriden; vierte Mitteilung über die Tripyleen-Ausbeute der deutschen Tiefsee-Expedition. *Arch. Protistenkd.*, 7(2):259-306.

Haecker, V. 1907. Zur Kenntis der Castanellide und Porospathiden. *Arch. Protistenkd.*, 8:52-65.

Haecker, V. 1908. Tiefsee Radiolarien Speziel Teil. L1, Aulacanthidae-Concharidae. *Deutsch. Tiefsee-Exped., Wiss. Ergeben.*, 14:1-336.

Murray, J. 1885. The Radiolaria. *In*: Tizard, T. H., Moseley, H. N., Buchanan, J. Y. & Murray, J. (eds.), *Narrative of the Cruise of the H.M.S. Challenger with a General Account of the Scientific Results of the Expedition. Rept. Voy. Challenger, Narrative*, 1(pt. 1):219-227.

Nöthig, E.-M. & Gowing, M. M. 1991. Late winter abundance and distribution of phaeodarian radiolarians, other large protozooplankton and copepod nauplii in the Weddell Sea, Antarctica. *Mar. Biol.*, 111:473-484.

Reschetnjak, B. B. 1966. Deep Sea Phaeodarian Radiolaria of the North-west Pacific. *Fauna SSSR. New Series*, 94:1-208.

Schmidt, W. J. 1907. Einige neue Castanellide-Arten. *Zool. Anz.*, 32:297-307.

Schmidt, W. J. 1908. Die Tripyleen Radiolarien der Plankton-Expedition. Castanellidae. *Ergebnisse der Plankton-Expedition der Humboldt-Stiftung*, Bd. III, L.h. 6:234-279.

Swanberg, N. R., Bennett, P., Lindsey, J. L. & Anderson, O. R. 1986. The biology of a coelodendrid: a mesopelagic phaeodarian radiolarian. *Deep-Sea Res.*, 33:15-25.

Takahashi, K. 1987. Radiolarian flux and seasonality: Climatic and *El Niño* Response in the Subarctic Pacific, 1982-1984. *Global Geochemical Cycles*, 1:213-231.

Takahashi, K. 1991. Radiolaria: flux, ecology, and taxonomy in the Pacific and Atlantic. *In*: Honjo, S. (ed.), *Ocean Biocoenosis*, Series No. 3. Woods Hole Oceanographic Institution Press. 303 pp

Takahashi, K., Hurd, D. C. & Honjo, S. 1983. Phaeodarian skeletons: their role in silica transport to the deep sea. *Science*, 222(4624):616-618.

Wallich, G. C. 1869. On some undescribed testaceous rhizopods from the North Atlantic deposits. *Monthly Microsc. Jour.*, 1:104-110.

CLASS POLYCYSTINEA

O. ROGER ANDERSON, CATHERINE NIGRINI, DEMETRIO BOLTOVSKOY, KOZO TAKAHASHI, and NEIL R. SWANBERG

The polycystines are solitary or colonial, actinopod-bearing, marine, planktonic protists falling into two main groups: 1) Spumellaria with a spherical, cell body plan (Figs. 1, 2, 4), though the skeletons may have very different symmetries; 2) Nassellaria with a non-spherical cell body plan and skeletons varying from simple spicules to complex helmet-shaped structures (Fig. 3). Sizes vary from about 30 μm to several hundred micrometers for many skeleton-bearing species or to several millimeters for larger gelatinous spumellarian species. Most known species occur in open-ocean environments, but some occur in coastal waters. Radiolarian siliceous skeletons contribute substantially to the microfossil record in marine sediments, providing one of the most continuous records of micro-organismic evolution (e.g. Nigrini, 1970, 1971; Kling, 1978; Sanfilippo and Riedel, 1985; Sanfilippo et al., 1985) extending back to the Cambrian, while many modern species can be clearly traced back at least to the early Mesozoic (e.g. Anderson, 1983a; Riedel and Sanfilippo, 1981).

General Morphology. The spherical body plan of the Spumellaria is particularly evident in the shape of the **central capsule** (nucleus-containing central mass of cytoplasm) which is surrounded by a spheroidal central capsular wall with numerous pores. **Actinopodia** radiate outward from the pores (Figs. 1, 4D). The shape of the surrounding skeletal matter, when present, can be very different, taking quadrangular, trigonal, bilocular, or other geometric shapes

(Figs. 2, 3). The central capsule of Nassellaria is typically a prolate spheroid with a pore plate at the base where several to many relatively robust actinopods extend outward into the surrounding environment. The polycystine central capsule contains reserve substances and major cytoplasmic organelles (nucleus, mitochondria, and other membranous organelles, exclusive of digestive vacuoles). The actinopods radiating from the pores in the central capsular wall produce a frothy or web-like extracapsular cytoplasm **(extracapsulum)** (Fig. 1A) where prey is sequestered within digestive vacuoles.

Skeletal matter is siliceous and is secreted within a thin cytoplasmic sheath or **cytokalymma** (Anderson, 1983a). Some species lack siliceous deposits and are among the largest of the solitary species (Fig. 1A). Algal symbionts, when present, are enclosed within peri-algal vacuoles usually in the extracapsulum (Figs. 1, 4). Colonial species (Fig. 4A-C) contain numerous central capsules, interconnected by a rhizopodial network, enclosed within a common gelatinous sheath. The colony can be spherical, spheroidal, or filiform forming elongate strands up to one meter or more in length.

Skeletal Morphology. Skeletal morphology is highly varied, often not based on a simple geometric plan, and can be as simple as a porous or latticed sphere or as elaborate as a multi-septate conical shell with pores and latticed wings, etc. (Figs. 2, 3). Some skeletal terminology relevant to use of the keys is presented here. Spicules in Spumellaria are spindle-shaped, curved, multiradiate or of more complex design, but always occurring as scattered, unconnected structures (Figs. 1D, 4E, G, H). The spumellarian skeleton can consist of spherical or spheroidal porous shells singly or in concentric array (Figs. 1B, 2A). Concentric spheres are joined by radial beams (Fig. 2A), which may or may not project outward as spines. Large spines, often in fixed number (2, 4, or 6) and regularly arranged, are known as primary spines. Smaller ones (sometimes bristle- or thorn-shaped) are known as secondary spines and are scattered irregularly about the shell's surface. When there is more than one shell, the outer one is the **cortical shell** and the innermost is the **medullary shell** (Fig. 2A). Skeletal structures may vary substantially across genera. These include spongiose ones, either spherical to spheroidal (Figs. 2D, 5B,C), flattened and square (Figs. 2M, 8J), discoidal (Figs. 2I,J, 8I), or trigonal (Y-shaped, Figs. 2N, 8K,L). In some cases the shell is a biconvex (lenticular) disk enclosing a spherical medullary shell (Fig. 2F,G). The skeleton may include latticed polar caps (Figs. 2H, 6E) or other ornate spongiose or latticed structures. In other genera, the skeleton can be augmented by arches of latticed skeletal matter or girdles which define openings or gates (Figs. 2Q, 7J,K), broad encircling latticed bands, etc., or consist entirely or partially of a spiraling latticed structure (Figs. 2O,P, 7H). A tube-like extension from the surface of the spumellarian skeleton (either entire, latticed, and/or ornamented with spines), or a funnel-like groove in the margin of the spongy disk, is known as a **pylome** (Figs. 2J,K, 11H,K). Sometimes forked or branched rods, **apophyses**, extend outward from the center of the skeleton (Figs. 2A, 5L).

In Nassellaria, the spicule is a complex structure, basically tripodal in design, and often forming the major scaffolding of the skeleton, either as the main skeletal element (Figs. 3A,B, 9C), as a central structure to which other structures are attached (Figs. 3B,C, 9G), or enclosed (hidden) within a porous (simple or segmented) shell (Fig. 3F,N). The shell may consist of only a basal tripod (Fig. 3K); the tripod and associated ring(s) (Fig. 3A,E); a single chamber known as the cephalis (Fig. 3D); or two or more chambers (Fig. 3H,S,T,U) connected serially, designated **cephalis, thorax, abdomen,** and **postabdominal segments.** The final chamber can be either open, without a closing plate (Fig. 3J), or closed by a porous plate (Fig. 3I). The organization of the cephalis varies from a simple helmet- or cap-shaped structure (Fig. 3J) to bilocular (two bilateral lobes, Fig. 3F) or multilocular (more than two lobate protrusions, Fig. 3O-Q) forms. The top of the cephalis can be augmented by a **galea** (cupola) (Fig. 3G) and may have a cylindrical or three-bladed horn (Fig. 3O,T) or lateral cephalic tube (protruding from the side, Fig. 3S). The thorax may have three latticed or solid radial apophyses or "wings" projecting from the proximal or middle part of the thoracic wall (Fig. 3D). The rim (**peristome**) of the terminal chamber, either the thorax or abdomen, may bear spine-like protrusions called "**teeth**," if they are numerous and short (Fig. 3 T), or "**feet,**" if there are three and they are relatively long (Fig. 3C,F,G). The term sagittal ring is used for a circular to D-shaped element that sometimes is the only structure present; or in other cases it reinforces the wall of a latticed shell in the medial sagittal plane (Fig. 3E). This region may be augmented by a sagittal beam.

Reproduction. The reproduction of many species of radiolaria is not known. In those that have been studied (e.g. Anderson, 1983a) reproduction can be by release of biflagellated swarmers, each with a vacuole bearing a large strontium sulfate crystal, or in some cases by fission of the central capsule. It is not known if the swarmers are asexual dissemules or motile gametes.

Distribution and Ecology. Polycystine radiolaria occur in most major oceanic sites with densities of tens to hundreds per cubic meter of seawater. Species bearing algal symbionts often occur in the upper strata of the water column in the well-illuminated portions of the photic zone. Others, preferring a cooler habitat, may extend to great depths at mid-latitudes, but these are often found nearer the surface in high latitudes (Nigrini and Moore, 1979; Boltovskoy, 1981, 1994; Anderson, 1983a; Steineck and Casey, 1990). In extrapolar areas, the layer of maximum radiolarian abundance is usually around 25-50 m for the polycystines, whereas the phaeodarians tend to inhabit deeper strata (Reshetnjak, 1966; Renz, 1976; Kling, 1979; Gowing, 1993; Kling and Boltovskoy [in press]). In polar waters, on the other hand, peak abundances can be located deeper, at 200-300 m (Boltovskoy, 1987; Boltovskoy and Alder, 1992).

Polycystines consume a wide variety of prey, spanning bacteria, algae, and other protists, to copepods, larvacea, and other small zooplankton. Among the species studied, many are omnivorous, taking phytoplankton and zooplankton prey, while others appear to be more algivorous (Anderson, 1983a, 1993; Caron and Swanberg, 1990). Algal symbionts, when present, secrete photosynthetic products that are assimilated by the host as a nutritional source (Anderson, 1983b, 1992).

Taxonomy. Although the first descriptions of radiolarian species date as far back as the beginning of the nineteenth century, the cornerstone of the taxonomy of this group was established in 1887, with Haeckel's impressive work based on collections of the HMS Challenger. In the three volumes of this monograph Haeckel described 3389 polycystine species (617 genera), 80% of them new, mostly on the basis of recent materials. It is now widely recognized, however, that Haeckel's excessively rigid classification system is plagued with synonyms (the estimated number of living polycystine species is probably around 700 to 1000). Its suprageneric categories, moreover, are artificial and unsatisfactory; however, advances in the development of a satisfactory alternative taxonomy have been modest at best, especially for the Spumellaria.

Revisions of Haeckel's spumellarian supra-generic systematics were undertaken by Riedel (1967a,b, 1971) and by French and Russian specialists (Hollande and Enjumet, 1960; Cachon and Cachon, 1972, 1985; Petrushevskaya, 1965, 1967, 1971, 1975, 1981, 1986; Petrushevskaya et al., 1976). In addition, selected groups were subject to detailed morphologic studies, and new classification schemes were suggested (e.g. Zhamoida and Kozlova, 1971; Dumitrica, 1988, 1989). Riedel's proposed system, based on skeletal features alone, is rather similar to Haeckel's and is also the one most widely used at present. By contrast, the revision by the French and Russian workers uses cytoplasmic features, such as the "nucleoaxopodial complex" (*sensu* Petrushevskaya, 1986) for suprageneric divisions, and the skeleton for generic and specific diagnoses. The revision by the French and Russian workers is more elaborate. Its arguments are probably stronger in biologic and phylogenetic terms, and it is based on criteria more closely comparable to those applied to other protista. Its appeal among radiolarian workers, however, has been very limited. The reasons for this are probably multiple, but at least three are more evident: 1) over 90% of radiolarian studies are carried out on the basis of sedimentary material devoid of any soft parts, which precludes applying classification systems centered on protoplasmic details, 2) information on the cytoplasmic morphology is mainly in the biological literature in French or in Russian, especially as applied to taxonomy, and 3) most of the taxonomy was done by English-speaking geologists for whom literature in other languages, particularly Russian, was not readily accessible. Riedel's (1971) spumellarian classification is restricted to 11 (8 extant) families, external skeletal morphology of which is usually quite distinctive; this simplicity is very attractive, especially considering that the French-Russian proposed divisions include orders, superfamilies, families, and subfamilies. The nassellarian classification inherited from earlier workers (Ehrenberg, 1847ab; Müller, 1858a; Haeckel, 1862, 1887; Hertwig, 1879; Popofsky, 1908, 1912, 1913) was profoundly changed both in whole (Riedel, 1967a,b, 1971; Petrushevskaya, 1981, 1986), and in selected groups (Petrushevskaya, 1965; Goll, 1968, 1969; Sanfilippo and Riedel, 1970; Foreman, 1973). Besides cytoplasmic traits (cf. Petrushevskaya, 1986), a feature of primary importance for the classification of this group, but often over-looked, is the internal skeleton.

This consists of a system of internal spines and connecting bars which allow close comparisons between homologous structures in forms differing widely in their external morphology. Analysis of these features is demanding, since it requires dedicated efforts at understanding the complex spatial relationships involved. In addition, observation of this internal skeleton is only feasible with well preserved individuals oriented in the right position, usually on permanent-mount slides. Nevertheless, family-level assignments are generally reasonably stable among publications, and these do take the structure of the internal skeleton into account. Again, the most widespread system is that proposed by Riedel (1971), which includes eight extant families.

In spite of the recognized importance of internal skeletal features, some recent publications on taxonomy of Spumellaria and Nassellaria at the generic and specific levels continue to base identifications solely on external traits of the skeletons. This is true even for new genus and species descriptions. It should be noted in passing, however, that among some larger spongiose Spumellaria (e.g. *Styptosphaera* and *Plegmosphaera*) phenotypic intergradation and variability of the internal structures make separation of the genera difficult, except for extreme extreme cases (Swanberg et al., 1990). The surging interest in ecologically and, especially, paleoecologically-oriented research places special emphasis on morphologically distinguishable indicators that can convey information on recent or past ecological settings. For this work, suprageneric and even generic traits are deemed of little use. Thus, current Linnean binomial denominations often have limited

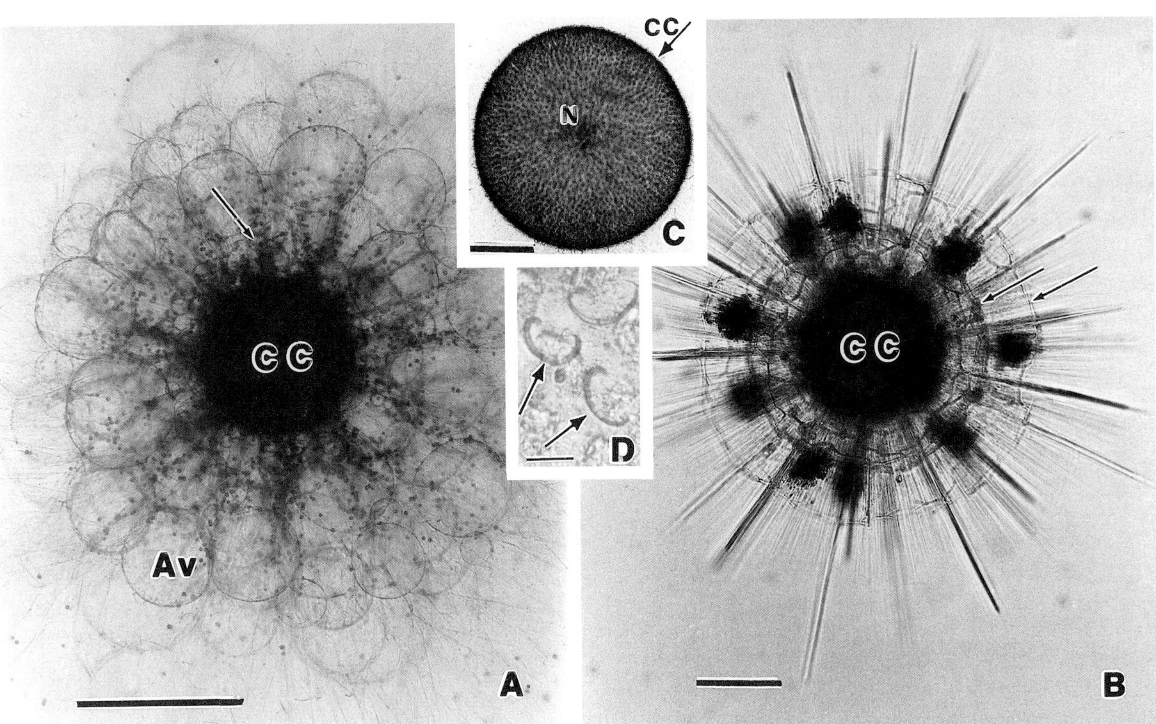

Fig. 1. Light microscopic views of living spumellarian radiolaria. A. *Thalassicolla nucleata* Huxley, 1851 (p. 433), a skeletonless species, with an opaque central capsule (CC) surrounded by an extracapsulum consisting of frothy alveoli (Av) and numerous algal symbionts (arrow). Scale=1 mm. B. A skeleton-bearing species showing two concentric latticed shells (arrows) with radial spines, surrounding the central capsule (CC). Scale=50 µm. C. *Physematium.* Scale =1 mm. D. C-shaped spicules (arrows) on the surface of the central capsular membrane of *P. mulleri*, Meyen, 1834 (p. 163), a large, gelatinous solitary spumellarian with scattered arcuate or variously curved spicules on the surface of a very large central capsule (arrow) enclosing radially arranged internal alveoli surrounding a central nucleus (N). Scale=100 µm. (Adapted from Anderson, 1983a, 1969; Sanfilippo and Riedel, 1970; Foreman, 1973).

value in so far as relationships between taxa are concerned. This is especially so at the generic level, because closely comparable morphotypes are widely recognized under different generic names. Likewise, some very different morphotypes share the same genus. Family-level divisions, on the other hand, while not necessarily correct, are more uniform and stable among publications. The following synopsis to family level for extant species (Riedel, 1971; Kling, 1973, 1978; Anderson, 1983a) is consistent with current perspectives, but is likely to change as more fine structural and evolutionary data are obtained.

Class Polycystina

Ehrenberg, 1838, emend. Riedel, 1967a

Central capsule with numerous actinopodia radiating from pores distributed over the entire surface or at a restricted region at one pole, skeletal matter when present composed of opal, some species with endosymbionts, but all lack a phaeodium.

ORDER SPUMELLARIA

Ehrenberg, 1875

Spherical body plan with actinopodia radiating uniformly from all over the surface of the central capsule.

KEY TO GENERA

The following key is not natural, since our knowledge of polycystine phylogenies is limited. It is a practical guide based on some fairly conspicuous characteristics observable with a light microscope

1. Skeleton lacking, or consisting of scattered spicules, etc...I
1.' Skeleton present II

I. SKELETONLESS OR SPICULAR SPUMELLARIA

1. Skeleton lacking.. 2
1'. Spicules or scattered siliceous product present.. 4

2. Colonial .... ***Collozoum***
2'. Solitary 3

3. Alveoli neither within or outside central capsule... ***Actissa***
3'. Alveoli within central capsule.. ***Thalassolampe***
3''. Alveoli surrounding central capsule .... ***Thalassicolla***

4. Colonial 5
4'. Solitary 6

5. Spicules double tri-radiate (three branches at each end)................................ ***Sphaerozoum***
5'. Spicules simple needles or radiate................ .... ***Rhaphidozoum***

6. Alveoli neither within or outside central capsule 7
6'. Alveoli present within or outside central capsule 8

7. Spicules unbranched............***Thalassosphaera***
7'. Spicules branched .... ***Thalassoxanthium***

8. Numerous alveoli within transparent central capsule .... ***Physematium***
8'. Bubble-like alveoli surrounding central capsule 9

9. Spicules unbranched............. ***Thalassoplancta***
9'. Spicules branched .... ***Lampoxanthium***

II. SKELETON PRESENT

1. Basic shell symmetry radial (includes spongiose and lattice shelled species).......................... **III**
1'. Basic shell symmetry not radial (includes tripodal skeletons or those with single or segmented helmet-shaped skeletons)............ **IV**

III. SKELETON-BEARING SPUMELLARIA

1. Radial symmetry, general shell shape spherical or ellipsoidal, single shell 2
1'. Radial symmetry, general shell shape spherical or ellipsoidal, multiple concentric shells... 13
1''. Radial symmetry, general shell shape cylindrical, quadrate, triaxial, or circular/lenticular 36

2. Shell spherical to ellipsoidal without annular constrictions. 3
2'. Shell subspherical with several annular constrictions forming dome-like protuberances.......................... ***Tholostaurus***

3. Outermost shell composed of a spongy framework 4
3'. Outermost shell composed of a latticed framework 5

4. Spongy sphere without a central cavity............ .... ***Styptosphaera***
4'. Spongy sphere with a central cavity.................. .... ***Plegmosphaera***

5. Pores irregular in size, arrangement, and spacing; shell wall very thin.......................... 6
5'. Pores more or less regular in size, arrangement and spacing, shell wall of variable thickness.. 11

6. Surface of shell smooth or lumpy, no spines or tubes.. ***Collosphaera***
6'. Shell surface with internal or external spines or tubes 7

7. Inwardly or outwardly directed tubular protuberances 8
7'. Numerous outwardly directed spines.............. .... ***Acrosphaera***

8. Inwardly directed tubes with perforated walls................................... ***Buccinosphaera***
8'. Outwardly directed tubes................................ 9

9. Tubes poreless...................... ***Siphonosphaera***
9'. Pored tubules... 10

10. Tube termination smooth .... ***Disolenia***
10'. Tube termination with one or more spines ....***Otosphaera***

11. Shell with prominent primary spines.........12
11'. Shell without prominent primary spines ....***Cenosphaera***

12. Primary spines branched.......... ***Cladococcus***
12'. Primary spines unbranched ....***Acanthosphaera***

13. Multiple shells, outermost shell spongy......14
13'. Multiple shells, outermost shell latticed.....18

14. Innermost (medullary) shell composed of rods which define the edges of a cube, spines radiating from corners of cube.....................15
14'. Innermost (medullary) shell(s) spherical to subspherical.. 16

15. Outer (cortical) shell in close proximity to cubic medullary shell .... ***Centrocubus***
15'. Outer (cortical) shell somewhat separated from cubic medullary shell....... ***Octodendron***

16. Outermost shell without equatorial constriction 17
16'. Outermost shell with equatorial constriction .... ***Spongoliva***

17. Primary spines present....... ***Spongosphaera***
17'. Primary spines absent.......... ***Spongoplegma***

18. Outer shell(s) ellipsoidal with equatorial constriction, with or without polar caps .... ***Didymocyrtis***
18'. Outer shell(s) spherical to ellipsoidal, no equatorial constriction............................ 19

19. Outermost shell composed of a series of girdles 33
19'. Outermost shell not composed of a series of girdles..20

20. Outer shell(s) form a series of concentric spheres/ellipsoids, no spiral structure visible 21
20'. Outer shell(s) partly or totally of a spiral structure 31

21. No primary spines 22
21'. Prominent primary spines......................... 23

22. Two concentric shells: one medullary and one cortical .... ***Carposphaera***
22'. Three concentric shells, two medullary and one cortical.............................. ***Thecosphaera***

23. Two primary spines, bipolar....................... 24
23'. More than two primary spines.................... 27

24. Two bipolar spines, not joined distally........25
24'. Two bipolar spines, joined distally so as to form a ring.................................. ***Saturnalis***

25. Two concentric shells, one medullary and one cortical.. 26
25'. Three concentric shells, two medullary and one cortical........................ ***Stylacontarium***

26. Spines similar.......................... ***Stylatractus***
26.' Spines dissimilar............... ***Druppatractus***

27. Four primary spines .... ***Staurolonche***
27'. Six primary spines arranged symmetrically 28
27". Eight or more primary spines.................. 29

28. Two concentric shells: one medullary and one cortical.. ***Hexalonche***
28'. Three concentric shells: two medullary and one cortical.............................. ***Hexacontium***

29. With pylome .... ***Cromyechinus***
29'. Without pylome 30

30. Two concentric shells, one medullary and one cortical.................................. ***Astrosphaera***
30'. Three concentric shells, two medullary and one cortical .... ***Actinomma***

31. Without pylome... 32
31'. With pylome................................ ***Larcopyle***

32. Spiral tightly wound....................... ***Lithelius***
32'. Spiral open................................ ***Larcospira***

33. Two systems of concentric girdles............... 34
33'. Three systems of concentric girdles........... 35

34. Four gates bisected by sagittal beam ***Octopyle***
34'. Four gates, simple, not bisected.....***Tetrapyle***

35. Six open gates................................ ***Hexapyle***
35'. Six closed gates............................... ***Pylolena***

36. General shell shape cylindrical................... 37
36'. Basic shell shape lenticular; overall shape circular to subcircular, quadrate, or trigonal... 38

37. Spongy cylinder with no visible structure but may vary in width .... ***Spongocore***
37'. Spongy cylinder, may have some structure, with or without pylome................ ***Spongurus***

38. Outer shell lenticular, latticed...***Heliodiscus***
38'. Outer shell spongy, with or without some visible structure 39

39. Shell spongy, but with some chambers or bands visible.. 40
39'. Shell spongy, no visible structure.............. 43

40. Outer meshwork forms two-chambered arm (one of which may bifurcate distally) radiating from central disc.***Amphirhopalum***
40'. No radiating arms, but with more or less concentric or spiral bands around central chamber... 41

41. With external radiating spines................. 42
41'. Without external radiating spines, but with a pylome...................................***Ommatodiscus***

42. With thin, porous equatorial girdle.......... .... ***Stylochlamydium***
42'. No equatorial girdle .... ***Stylodictya***

43. Lenticular shell without radiating arms......44
43'. Conspicuous arms or dense areas of spongy material radiating from central disc.........46

44 . With pylome .... ***Spongopyle***
44'. Without pylome 45

45. With external spines radiating from disc.......... .... ***Spongotrochus***
45'. Without external spines radiating from disc .... ***Spongodiscus***

46. Outer meshwork forms three radiating arms, sometimes bifurcated distally.........47
46'. Outer meshwork forms four or more radiating "arms" joined by meshwork, outline quadrangular.. ***Spongaster***

47. Arms disposed bilaterally, two paired arms, one unpaired (forming a Y)........ ***Euchitonia***
47'. Arms equally disposed (propeller-like)...... .... ***Hymeniastrum***

FAMILY OUTLINE WITH ILLUSTRATIVE GENERA

FAMILY THALASSICOLLIDAE Haeckel, 1862

Solitary Spumellaria without a skeleton, or in some genera containing spicules.

Genus ***Actissa*** Haeckel, 1887

Alveoli neither within nor without the central capsule. Nucleus spherical, sometimes ellipsoidal, not branched (Fig. 4D).

Genus ***Lampoxanthium*** Haeckel, 1887

Numerous large alveoli within the peripheral jelly calymma, not in the central capsule. Spicules branched.

Genus ***Physematium*** Meyen, 1834

Very thin capsular membrane surrounding a spherical nucleus suspended within cytoplasmic strands delineating alveolar spaces radiating toward the peripheral capsular membrane. Spicules smooth and spindle-shaped.

Genus ***Thalassicolla*** Huxley, 1851

Numerous large alveoli surrounding the central capsule within the gelatinous extracapsulum, nucleus spherical (Fig. 1A).

Genus ***Thalassolampe*** Haeckel, 1862

Numerous large alveoli within the central capsule, not in the extracapsulum, nucleus spherical.

Genus ***Thalassoplancta*** Haeckel, 1887

Numerous large alveoli within the extracapsulum, not in the central capsule, with simple, unbranched spicules scattered throughout the extracapsulum (Fig. 4G).

Genus ***Thalassosphaera*** Haeckel, 1862

Alveoli lacking (neither within or surrounding the central capsule), extracapsular spicules simple.

Genus ***Thalassoxanthium*** Haeckel, 1887

Alveoli absent, branched spicules scattered within the extracapsulum (Fig. 4E).

FAMILY COLLOZOIDAE Haeckel, 1862

Colonial Spumellaria without a skeleton or with scattered spicules in the gelatinous sheath.

Genus ***Collozoum*** Haeckel, 1862

Colonial species lacking spicules and containing numerous central capsules, spherical to irregular in shape depending on the species, enclosed by a gelatinous sheath (Fig. 4A).

Genus ***Rhaphidozoum*** Haeckel, 1862

Colonial species with simple or radiate spicules (most frequently with 4, or more rarely 3, 5, or 6 rays).

Genus ***Sphaerozoum*** Meyen, 1834

Colonial species with scattered, paired-triradiate spines bearing a central shaft and three rays at each pole (Fig. 4G).

FAMILY COLLOSPHAERIDAE Müller, 1858B

Colonial polycystines, each central capsule has a single, thin-walled, spherical or subspherical latticed shell. This is the only group of colonial polycystines with complete latticed shells. Colonies consist of a gelatinous mass in which hundreds to thousands of shells are immersed. The shape of the colony is not species-specific; it may be spherical, ellipsoidal, cylindrical, ribbon-shaped, etc., measuring up to several centimeters in length and a few millimeters in diameter. The siliceous shells are always represented by a single perforated sphere (internal spheres are never present), with or without centrifugal (external) or centripetal (internal) tubular projections and/or spines. Spines (when present) are conical (circular in cross-section) (Fig. 5D-H).

Genus **Acrosphaera** Haeckel, 1881

Central capsules widely spaced and each surrounded by a single latticed shell, the outer surface of which is covered with solid, conical, or radial spines perforated by pores at the base and distributed between the pores of the shell (Fig. 5E).

Genus ***Buccinosphaera*** Haeckel, 1887

Shell surrounding each central capsule appearing as crumpled sphere with shallow depressions from which extend short inwardly directed conical tubules (Fig. 5F).

Genus ***Collosphaera*** Müller, 1855

Each central capsule surrounded by a latticed shell, smooth-surfaced, lying at considerable distance from one another, and varying in shape among species from spherical, crumpled, polygonal, or in the form of a thin net of large and small polygonal meshes (Fig. 5D).

Genus ***Disolenia*** Ehrenberg, 1860

Shell smooth, thin-walled, subspherical, with numerous small irregular pores having no definite arrangement. Pored tubules, 3-8, approximately as long as broad, cylindrical, slightly expanded distally (Fig. 5I).

Genus ***Siphonosphaera*** Müller, 1858

Shell surrounding each central capsule spherical, rather thick-walled with somewhat rough or pitted surface and numerous, irregularly scattered, subcircular pores. Wall with 4 to 10 poreless, cylindrical tubules smoothly truncated tangentially or sometimes obliquely (Fig. 5H).

FAMILY ACTINOMMIDAE Haeckel, 1862
emend. Riedel, 1971, Sanfilippo and Riedel, 1980

Solitary species with latticed or spongy spherical, subspherical, or ovoid shells (not discoidal, nor equatorially constricted ellipsoids); with or without medullary shells. Surface of shell is often covered with spines, but not tubes. All actinommids posses either single or multiple, concentric spherical or ovoid shells. When several shells are present, they are connected to each other by radial beams which pierce the cell. (Figs. 2A-D, 7 A-F)

Genus ***Actinomma*** Haeckel, 1860

Without pylome, 3 concentric spheres and 10-20 unbranched spines of either uniform or irregular length (Fig. 7E).

Genus ***Astrosphaera*** Haeckel, 1881

As above, but 2 concentric shells joined by long radial spines (Fig. 7F).

Genus ***Hexacontium*** Haeckel, 1881

Shells composed of 3 spheres bearing 6 radial spines of equal size arranged in 3 mutually perpendicular axes (Fig. 7C).

Genus ***Saturnalis*** Haeckel, 1881

Spherical shell with 2 opposite spines joined by a ring encircling the shell (Fig. 6H).

Genus ***Staurolonche*** Haeckel, 1881

Spheroidal porous shell with 4 unbranched spines (Fig. 7A).

FAMILY PHACODISCIDAE Haeckel, 1881
emend. Campbell, 1954

A lenticular, cortical shell, not surrounded by spongy or chambered structures, encloses a much smaller, single or multiple medullary shell. The margin (but less commonly the surfaces) of the cortical shell may bear radial spines. (Figs. 2F,G)

Genus ***Heliodiscus*** Haeckel, 1862

Radial spines unbranched; medullary shell simple (Fig. 8B).

FAMILY: COCCODISCIDAE Haeckel, 1862,
emend. Sanfilippo and Riedel, 1980

Discoidal forms consisting of a lenticular cortical shell enclosing a small single or double medullary shell, and surrounded by an equatorial zone of spongy or concentrically-chambered structures. Or forms with ellipsoidal cortical shell, usually equatorially constricted and enclosing a single or double medullary shell. The opposite poles of the shell can bear spongy columns and/or single or multiple latticed caps. Fig. 2H, 6E.

Genus ***Didymocyrtis*** Haeckel, 1860a

Bilocular cortical shell with or without polar spongiose caps (Fig. 6E).

FAMILY: SPONGODISCIDAE Haeckel, 1862
emend. Riedel, 1967b, Petrushevskya and
Kozlova, 1972

Discoidal or cylindrical forms with or without a visible central chamber surrounded by concentric or spiral, continuous or interrupted, finely chambered bands, or partially or totally spongy with no visible structure. The surface of some forms may be partly or totally covered with a very thin, porous sieve-plate, which in lenticular forms may extend beyond the central spongy mass forming a delicate equatorial girdle around the periphery of the shell. Major part of the shell can be represented by 2-4 thickenings of the spongy mesh or separate arms radiating from the central disc in the equatorial plane. (Figs. 2I-N, 8J-L)

Genus ***Euchitonia*** Ehrenberg, 1860

Three undivided chambered arms (Y-shaped) with a patagium (spongiose veil) between the arms. Shell bilateral (Fig. 8K).

Genus ***Hymeniastrum*** Ehrenberg, 1847

Like *Euchitonia* but angles between arms equal (Fig. 8L).

Genus ***Spongaster*** Ehrenberg, 1860

Four spongy arms with patagium forming a quadrangular shape (Fig. 8J).

FAMILY PYLONIIDAE Haeckel, 1881
emend. Campbell, 1954

Shell consisting of an ellipsoidal inner chamber or microsphere of complicated construction and with one or two pairs of openings or gates, enclosed by a series of successively larger elliptical latticed girdles in three mutually perpendicular planes; with the major diameter of each girdle being the minor diameter of the next smaller one. Figs. 2Q, 7J)

Genus ***Octopyle*** Haeckel, 1881

Ellipsoidal shell with 2 perfect girdles and 4 gates bisected by sagittal septum (Fig. 7J).

FAMILY THOLONIIDAE Haeckel, 1887

Completely latticed shell, without larger openings, and with two or more annular constrictions or furrows separating dome-shaped protuberances (Fig. 2R).

Genus ***Tholostaurus*** Haeckel, 1887

Central chamber without medullary shell, 4 simple dome-shaped cupolas (Fig. 5A).

FAMILY LITHELIIDAE Haeckel, 1862

Ellipsoidal to spherical or lenticular shell in outer form, and of spiral organization, at least internally; initial chamber of basic pyloniid type (Figs. 2P, 7H).

Genus ***Larcospira*** Haeckel, 1887

Transverse girdle consisting of a simple or double spiral around a principal axis. At least in *L. quadrangula* shell consists of 2 open spirals arising from a common origin (Fig. 7I).

Genus ***Lithelius*** Haeckel, 1860b

Spirally arranged lattice shell with branched or unbranched radial spines (Fig. 7H).

Order Nassellaria Ehrenberg, 1875

Body plan typically not spherical; actinopodia arise from a pore plate at one pole of the central capsule, skeleton when present composed of spicules, rings, other geometric shapes or single or multi-segmented latticed chambers. A widely recognized, albeit seldom utilized, feature of primary importance for the classification of the Nassellaria is the internal skeleton. The internal skeleton consists of a complex set of spines and connecting bars enclosed in the cephalis, which allow comparison of homologous structures in forms differing widely in their external morphology.

KEY TO GENERA

The following key is not natural, since our knowledge of polycystine phylogenies is limited. It is a practical guide based on some fairly conspicuous characteristics observable with a light microscope

1. Shell composed of rings, bars and rods, lattice absent or very loose, open.............................. 2
1'. Latticed shell.. V

2. Basal tripod only.. 3
2'. Basal tripod plus sagittal ring........................ 5

3. Branches of basal tripod not connected distally .... ***Triplagia***
3'. Branches of basal tripod connected distally......4

4. Distal connections do not form a chamber....... .... ***Triplecta***
4'. Distal connections form an irregularly latticed chamber (cephalis)............. ***Phormacantha***

5. Simple vertical sagittal ring only. ***Zygocircus***
5'. Simple vertical ring with additional rings and/or appendages, open meshwork 6

6. Shell outline bilaterally symmetrical with two lobes clearly defined by a sagittal ring.......... 7
6'. No lobes evident 10

7. Lobes consisting of two crossed rings, verical a vertical meridional; additional horizontal basal ring.............................. ***Acanthodesmia***
7'. Lobes defined by open meshwork..................... 8

8. Lobes with large polygonal meshwork................ .... ***Lophospyris***
8'. Lobes kidney-shaped with irregular meshwork 9

9. Lobes formed by two plates joined only by sagittal ring .... ***Nephrospyris***
9'. Lobes complete, closed................. ***Lithocircus***

10. One complete vertical ring with additional half ring or arch between base and top .... ***Neosemantis***
10'. Two parallel polygonal rings of dissimilar size, connected by bars........ ***Pseudocubus***

size, connected by bars........ ***Pseudocubus***

V. NASSELLARIA WITH LATTICED SHELL

1. Lattice shell complete, cephalis simple, unilocular, spherical without constrictions or lobes... VI
1'. Lattice shell complete, cephalis lobed............ 2

2. Cephalis bilocular, lobes defined by a sagittal constriction... VII
2'. Cephalis multilocular..................................... 3

3. Cephalis trilocular VIII
3'. Cephalis apparently multilocular, composed of cephalis, ante- and post-cephalic lobes (cephalis simple).. IX

VI NASSELLARIA WITH LATTICED SHELL
(cephalis unilocular, spherical, without lobes)

1. Simple cephalis distinct from subsequent segments.. 2
1'. Cephalis rudimentary or submerged into thorax 16

2. Simple cephalis without lateral cephalic tube..3
2'. Simple cephalis with lateral cephalic tube; pores in precise horizontal rows................27

3. Simple cephalis with one additional segment (thorax); rudimentary, poorly defined abdomen may be present.............................. 4
3'. Simple cephalis with more than one additional segment 22

4. Thorax closed; no abdomen .... ***Lithopera***
4'. Thorax open 5

5. Cephalis and thorax approximately same width.6
5'. Thorax wider than cephalis.............................. 7

6. Without radial apophyses............. ***Lophophaen***
6'. With radial apophyses................ ***Peromelissa***

7. Thorax cylindrical, ovate or sharply conical...8
7'. Thorax widely conical (like a Chinese peasant hat.. 13

8. Thoracic ribs which may become external to form radial apophyses 9
8'. Apophyses form basal thoracic projections..20

9. Three thoracic ribs....................................... 10
9'. Three thoracic ribs, all joined with apical by a delicate meshwork..............***Arachnocorys***

10. Thoracic ribs do not become external........11
10'. Thoracic ribs and a large apical horn, all joined by a delicate meshwork...................12

11. Thoracic wall with regularly arranged sub–circular pores................ ***Cycladophora***
11'. Thoracic wall interrupted by large holes... .... ***Clathrocanium***

12. Meshwork continuous .... ***Callimitra***
12'. Meshwork interrupted by thoracic holes..... .... ***Clathrocorys***
13. Three radial apophyses 14
13'. Numerous sinuous ribs .... ***Sethophormis***

14. With skirt-like abdominal section......... 15
14'. Without abdominal section.. ***Lampromitra***

15. Cephalis with branched horns...................... .... ***Eucecryphalus***
15'. Cephalis with unbranched apical horn.......... ....***Theopilium***

16. Cephalis rudimentary 17
16'. Cephalis submerged into thorax 19

17. Thorax narrowly conical......................... 18
17'. Thorax widely conical (like a Chinese peasant hat or dau-li)...... ***Litharachnium***

18. Cephalis rudimentary, no thoracic ribs..... .... ***Cornutella***
18'. Cephalis rudimentary, numerous thoracic ribs...................................... ***Peripyramis***

19. Cephalis completely submerged into thorax .... ***Carpocanistrum***
19'. Cephalis partially submerged into thorax.... .... ***Antarctissa***

20. Three basal appendages are extensions of thoracic ribs.. 21
20'. More than three basal appendages................ .... ***Acanthocorys***

21. Appendages project from base of thorax as "feet"............................... ***Pterocanium***
21'. Appendages project distally from wall of thorax as "wings"............... ***Dictyophimus***

22. Simple cephalis with two additional segments (thorax,abdomen)...................................... 23
22'. Simple cephalis with more than two additional segments.. 24

23. Three external radial apophyses projecting from thorax................................ ***Lipmanella***
23'. No radial apophyses on thorax ....***Clathrocyclas***

24. Basal segment closed, wings on thorax ....***Cyrtopera***
24'. Basal segment open................................. 25

25. Radial apophyses (wings) on thorax ....***Stichopilium***
25'. No radial apophyses on thorax................ 26

26. Complete shell conical in outline ....***Lithostrobus***
26'. Complete shell spindle-shaped in outline ....***Eucyrtidium***

27. Cylindrical lateral cephalic tube 28
27'. Duck-billed lateral cephalic tube, pores quadrate, outline stepped; multisegmented ....***Spirocyrtis***

28. Tube short 29
28'. Tube elongate, lying along thorax................ .... ***Phormostichoartus***

29. Three-segmented, but third segment may have minor indentations alternating with single transverse pore rows.... ***Siphocampe***
29'. Multisegmented, constrictions well defined, several rows of pores per segment............. .... ***Botryostrobus***

VII. NASSELLARIA WITH LATTICED SHELL
(cephalis bilocular)

1. Lobed cephalis only 2
1'. Lobed cephalis plus galea and/or false thorax..3

2. Six feet .... ***Liriospyris***
2'. Three feet 4

3. Sagittal ring external............... ***Dendrospyris***
3'. Sagittal ring internal.............. ***Triceraspyris***

4. Galea present, three feet .... ***Tholospyris***
4'. False thorax present............... ***Phormospyris***

VIII. NASSELLARIA WITH LATTICED SHELL
(cephalis trilocular)

1. Trilocular cephalis with one additional segment (thorax)... 2
1'. Trilocular cephalis with two additiona segments (thorax, abdomen)... 3

2. Thorax ends in distinct poreless peristome with terminal "teeth"....... ***Anthocyrtidium***
2'. Thorax without peristome........ ***Lamprocyrtis***

3. Paired cephalic lobes lateral to larger unpaired lobe; abdomen wider than thorax.................... 4
3'. Paired cephalic lobes beneath larger unpaired lobe, forming a "neck"; abdomen and thorax of approximately equal width ... ***Theocorythium***

4. Abdomen ends in a distinct poreless peristome with terminal "teeth"............ ***Lamprocyclas***
4'. No differentiated peristome or terminal "teeth"... ***Pterocorys***

IX. NASSELLARIA WITH LATTICED SHELL
(cephalis apparently multilocular)

1. Multilocular cephalis, shell monothalamous; one cephalic lobe completely enclosed by the other.. ***Centrobotrys***

1'. Multilocular cephalis, shell consisting of more than one chamber.. 2

2. Multilocular cephalis, shell dithalamous (cephalis, thorax)....................................... 3
2'. Multilocular cephalis (two conspicuous lobes of similar size); shell trithalamous (cephalis, thorax, abdomen).. ***Botryocyrtis***

3. Two conspicuous larger lobes of similar size... .... ***Saccospyris***
3'. Lobes dissimilar in size................................. 4

4. With cephalic tubes .... ***Acrobotrys***
4.' Without cephalic tubes .... ***Botryopyle***

FAMILY OUTLINE WITH ILLUSTRATIVE GENERA

FAMILY PLAGONIIDAE Haeckel, 1881
emend. Riedel, 1967b

Simple "tripodal" nassellarian spicule or single latticed chamber (cephalis) lacking post-cephalic chambers. Wide variety of forms developed from elaboration of accessory spines and branches, including latticed chamber surrounding spicule. (Fig. 3A,B)

Genus ***Callimitra*** Haeckel, 1881

Small, dome-shaped cephalis provided with very long apical (directed upwards), dorsal, and main

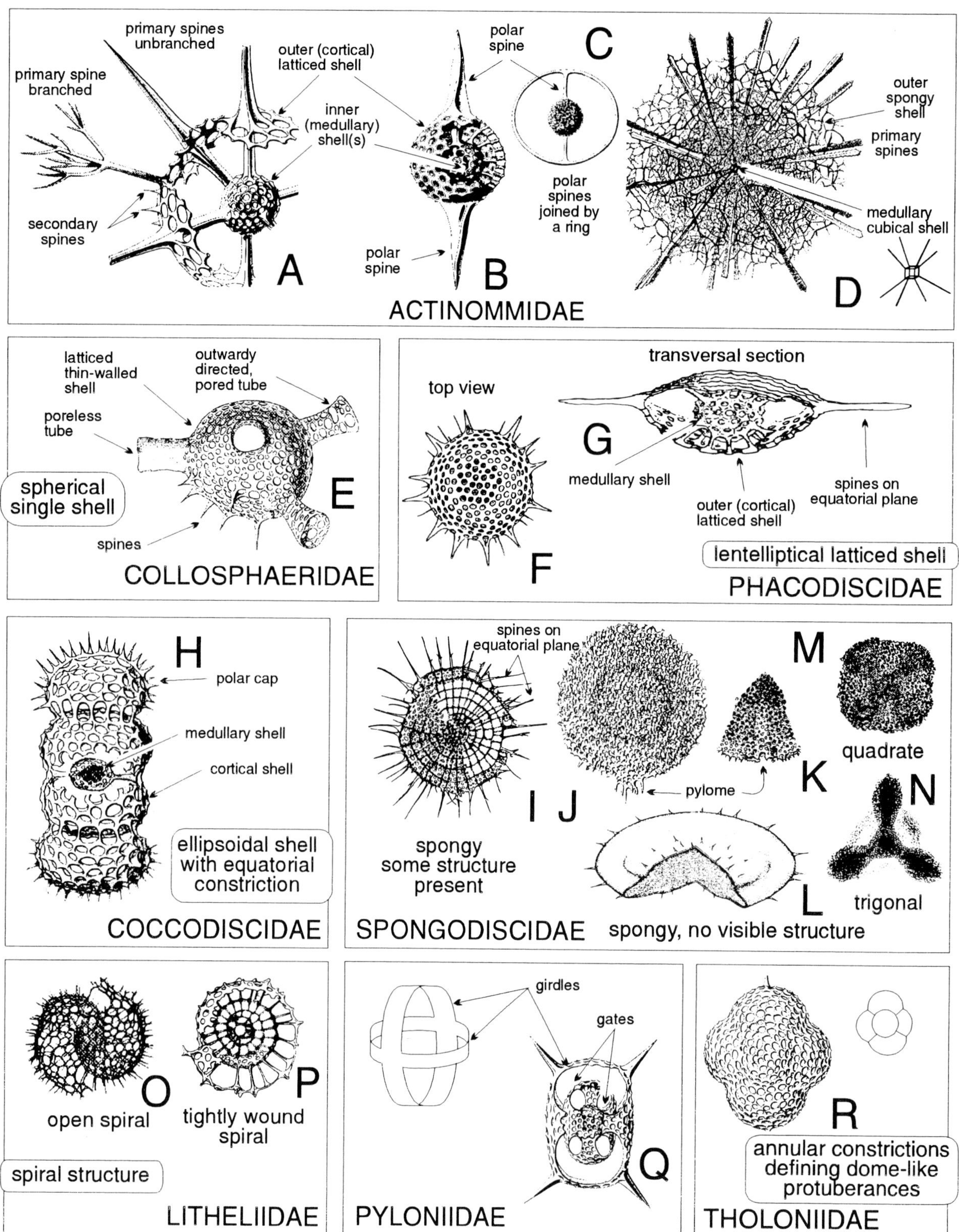

Fig. 2. Structural elements of the spumellarian skeleton and their nomenclature. A-D. Actinommidae Haeckel, 1862, *sensu* Riedel (1967b). E. Collosphaeridae Müller, 1858a,b. F,G. Phacodiscidae Haeckel, 1881. H. Coccodiscidae Haeckel, 1862. I-N. Spongodiscidae Haeckel, 1862, *sensu* Riedel, 1967b. O,P. Litheliidae Haeckel, 1862. Q. Pyloniidae Haeckel, 1881. R. Tholoniidae Haeckel, 1887. Sources. A-E, H, Q, R, modified from Haeckel (1887); F, G, M, modified from Boltovoskoy (1981); I, K, P, Petrushevskaya (1967); J, Dreyer (1889); L, modified from Riedel (1958); N, Nigrini and Moore (1979); and O, Takahashi (1991).

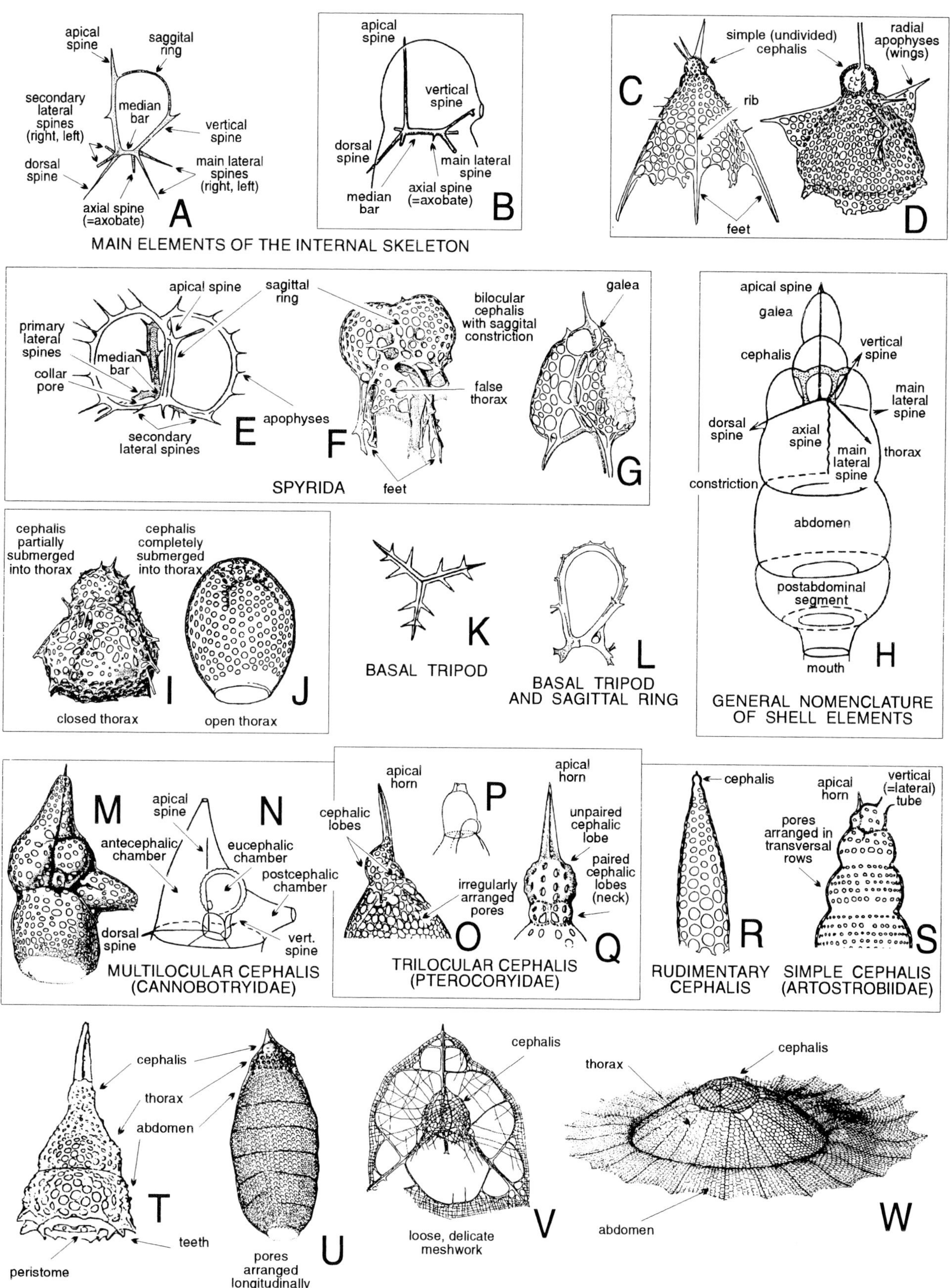

Fig. 3. Diversity and skeletal elements of Nassellaria. Sources. A, H, N, Petrushevskaya (1981); B, Riedel (1958); C-E, G, I, J, L, M, O, S, U, V, Petrushevskaya (1971); F, R, Petrushevskaya (1967); K, W, Haeckel (1887); P, Riedel (1957); Q, Nigrini (1967); T, Hays (1965).

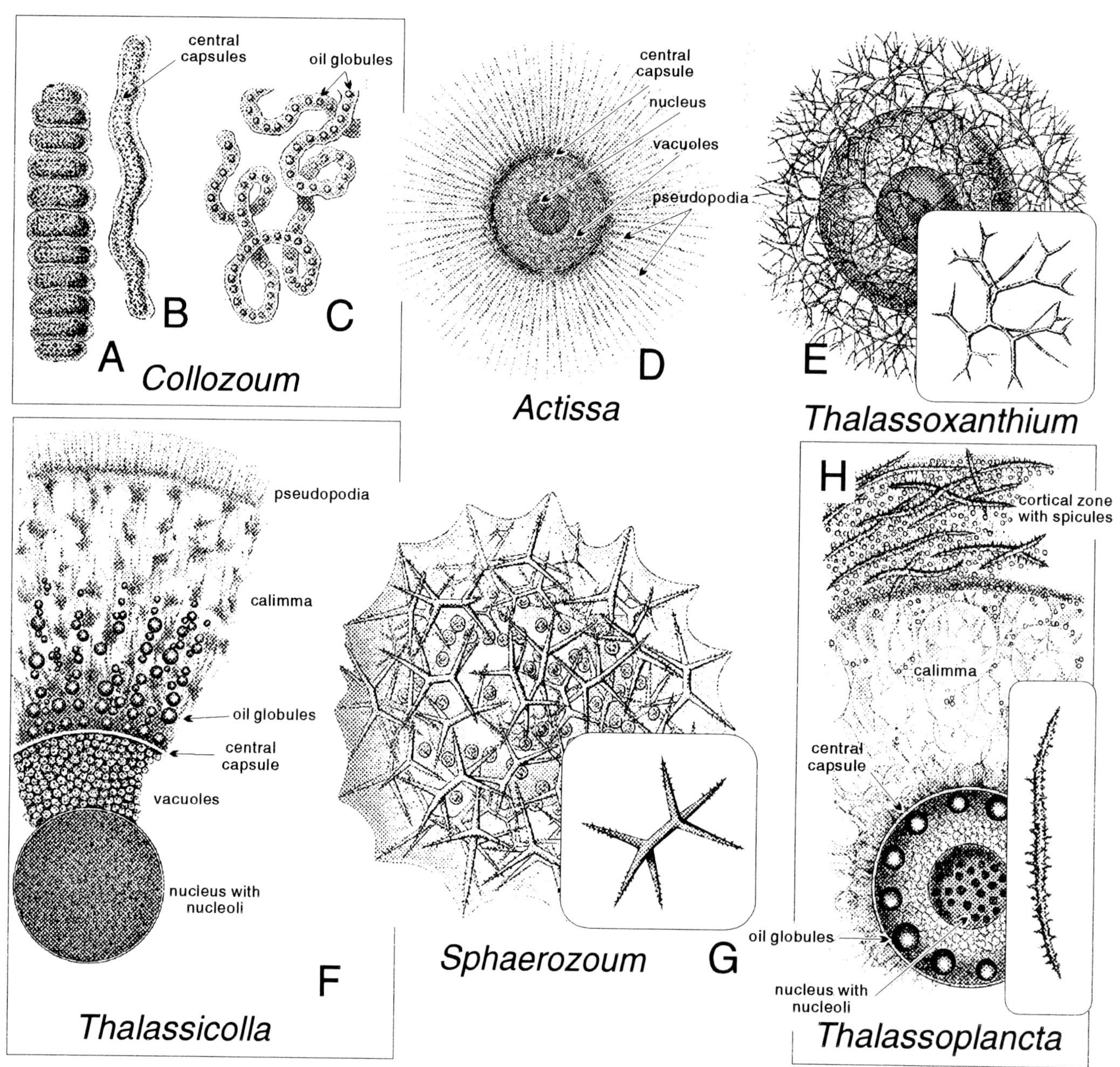

Fig. 4. Skeletonless and spiculated Spumellaria. A-C. *Collozoum* sp. Haeckel, 1862 (p 522), D. *Actissa* sp. Haeckel, 1887 (p 12), E. *Thalassoxanthium* sp. Haeckel, 1881 (p 470), F. *Thalassicolla* sp. Huxley, 1851 (p 433), G. *Sphaerozoum* sp. Meyen, 1834 (p 163), and G. *Thalassoplancta* sp. Haeckel, 1887 (p 29). (All adapted from Haeckel, 1887)

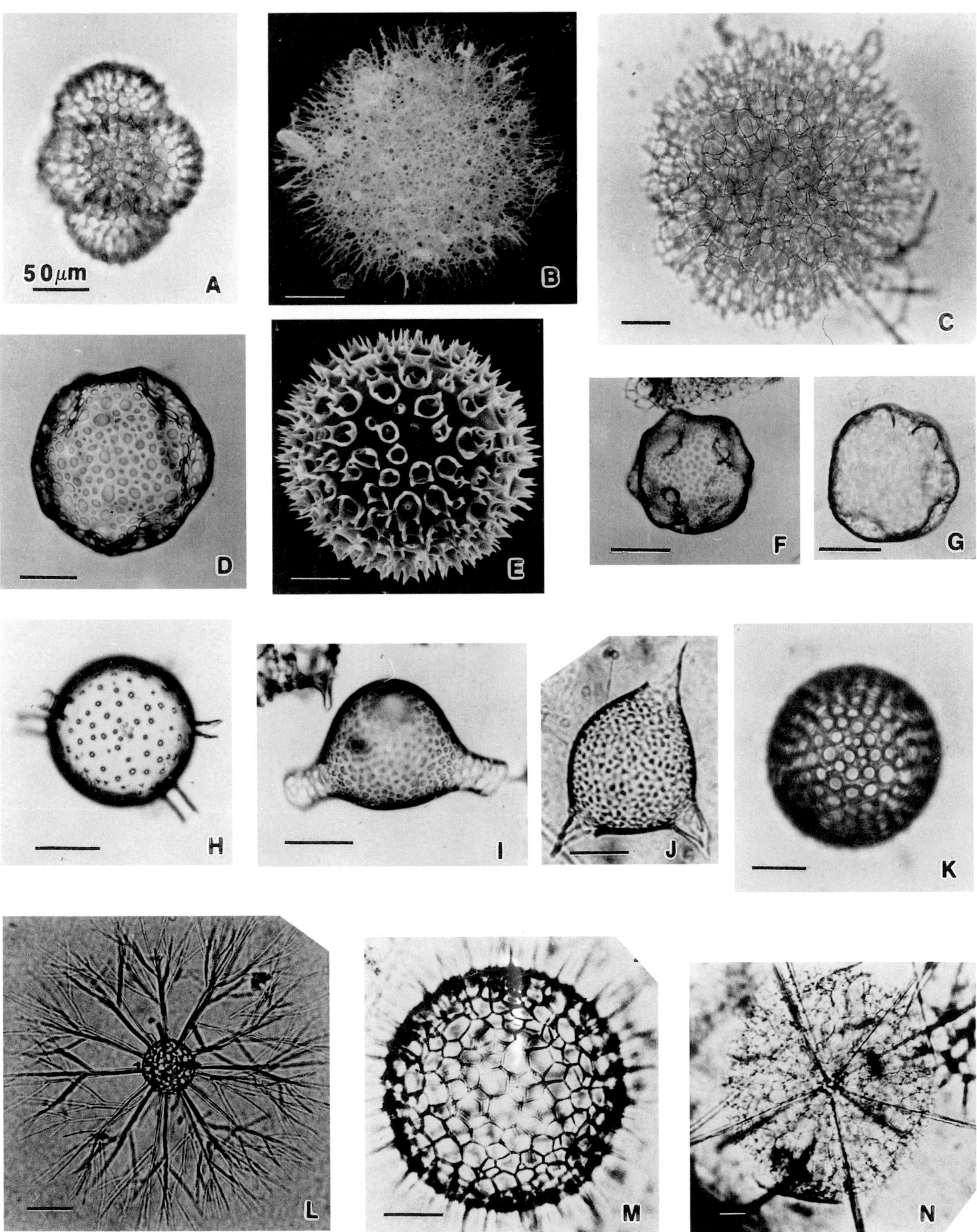

Fig. 5. Spumellaria. A. *Tholostaurus* sp. Haeckel, 1887 (p 670) (x225). B. *Styptosphaera spongiacea* Haeckel, 1881 (p 455 in Haeckel, 1887) (x210). C. *Plegmosphaera* sp. Haeckel, 1881 (p 455) (x180). D. *Collosphaera tuberosa* Müller, 1855 (p 238 in Haeckel, 1887) (x225). E. *Acrosphaera murrayana* Haeckel, 1881 (p 471in Haeckel, 1887) (x230). F. *Buccinosphaera invaginata* Haeckel, 1887 (p 99) (x215). G. *Buccinosphaera invaginata* (x215). H. *Siphonosphaera polysiphonia* Müller, 1858b (p 59 in Haeckel, 1887) (x233). I. *Disolenia quadrata* Ehrenberg, 1860 (p 831in Ehrenberg, 1872) (x250). J. *Otosphaera polymorpha* Haeckel, 1887 (p 116) (x210). K. *Cenosphaera* sp. Ehrenberg, 1854 (p 237) (x200). L. *Cladoccocus scoparius* Müller, 1857 (p 485 in Haeckel, 1887) (x160). M. *Acanthosphaera tunis* Ehrenberg, 1858 (p 12 in Haeckel, 1887) (x210). N. *Centrocubus octostylus* Haeckel, 1887 (p 277) (x104). (A,C) D. Boltovskoy; (D,F) J. Caulet; (K—J) Morley; (G-I) Nigrini (1967, 1971); (B,E,J,L-N) Takahashi (1991).

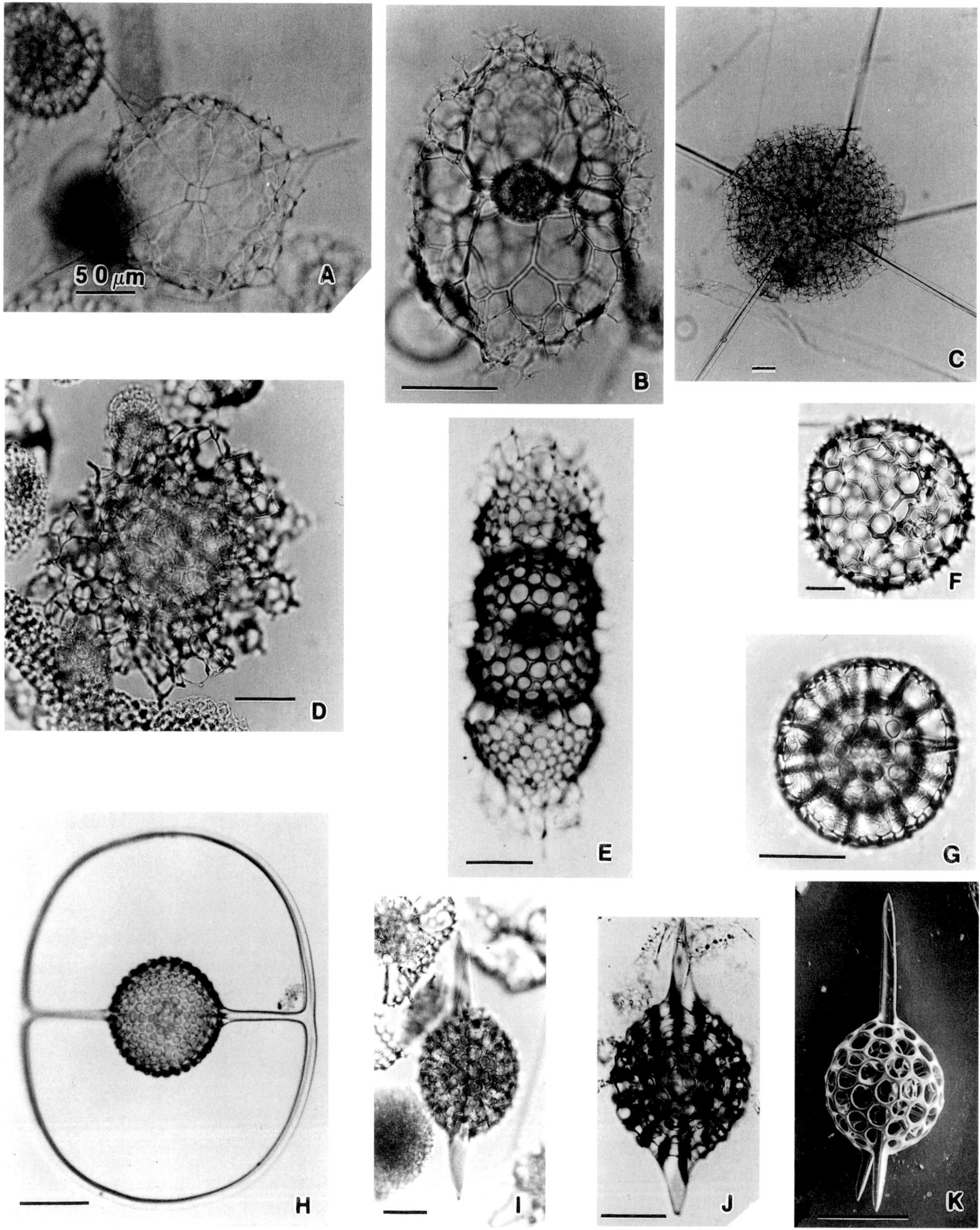

Fig. 6. Spumellaria. A. *Octodendron* sp. Haeckel, 1887 (p 279) (x200). B. *Spongoliva* sp. Haeckel, 1887, (p 351) (x360). C. *Spongosphaera* sp. Ehrenberg, 1847b (p 54) (x85). D. *Spongoplegma* sp. Haeckel, 1881 (p 455) (x225). E. *Didymocyrtis tetrathalamus* Haeckel, 1881 (p 816 in Haeckel, 1887) (x250). F. *Carposphaera* sp. Haeckel, 1881 (p 451) (x160). G. *Thecosphaera* sp. Haeckel, 1881 (p 452) (x320). H. *Saturnalis circularis* Haeckel, 1881 (p 450 in Haeckel, 1887) (x233). I. *Stylacontarium* sp. Popofsky, 1912 (p 90) (x160). J. *Stylatractus* sp. Haeckel, 1887 (p 328) (x233). K. *Druppatractus* sp. Haeckel, 1887 (p 324) (x325). (A) Boltovskoy (1987); (B,D,E) C. Nigrini; (F,G,I) D. Boltovskoy; (H) Nigrini (1967); (J, courtesy of J. Moore); (K, courtesy of J. Caulet).

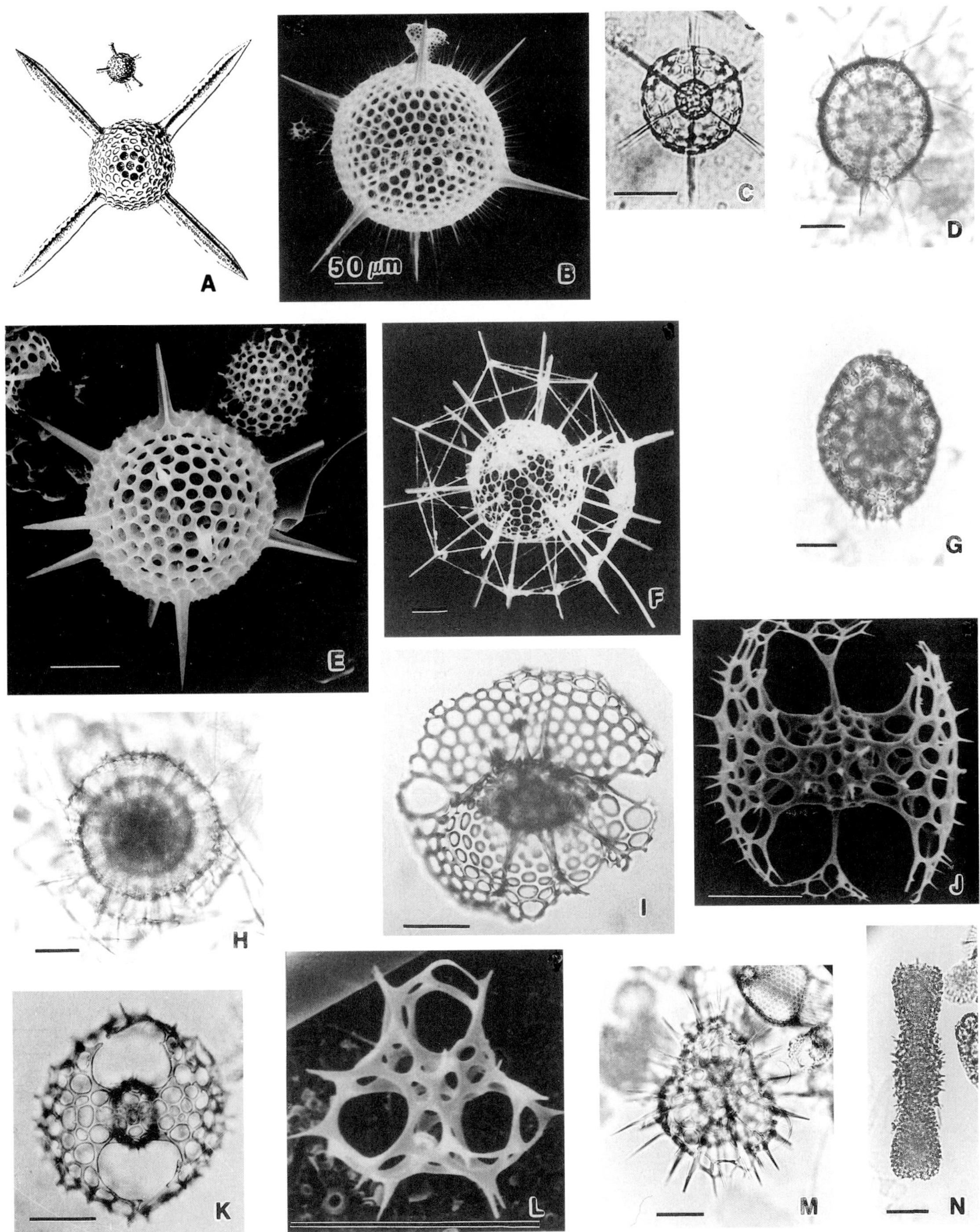

Fig. 7. Spumellaria. A. *Staurolonche pertusa* Haeckel, 1881 (p 451 in Haeckel, 1887) (mag. unknown). B. *Hexalonche amphisiphon* Haeckel, 1881 (p 451 in Haeckel, 1887) (x180). C. *Hexacontium axotrias* Haeckel, 1881 (p 452 in Haeckel, 1887) (x210). D. *Cromyechinus* sp. Haeckel, 1881 (p 453) (x160). E. *Actinomma* sp. Haeckel, 1860a (p 815) (X250). F. *Astrosphaera hexagonalis* Haeckel, 1887 (p 250) (x120). G. *Larcopyle* sp. Dreyer, 1889 (p 124) (x160). H. *Lithelius* sp. Haeckel, 1860b (p 843) (x160). I. *Larcospira quadrangula* Haeckel, 1887 (p 695) (x233). J. *Octopyle stenozona* Haeckel, 1881 (p 463, Haeckel, 1887) (x330). K. *Tetrapyle octachantha* Müller 1858a (p 154) (x230). L. *Hexapyle* sp. Haeckel, 1881 (p 463) (X960). M. *Pylolena* sp. Haeckel, 1887 (p 567) (x160). N. *Spongocore* sp. Haeckel, 1887 (p 345) (x160). (A) Haeckel (1887); (B,C,J) Takahashi (1991); (D,G,H,M,N) D. Boltovskoy; (E, K) J. Caulet; and (I) Nigrini (1970).

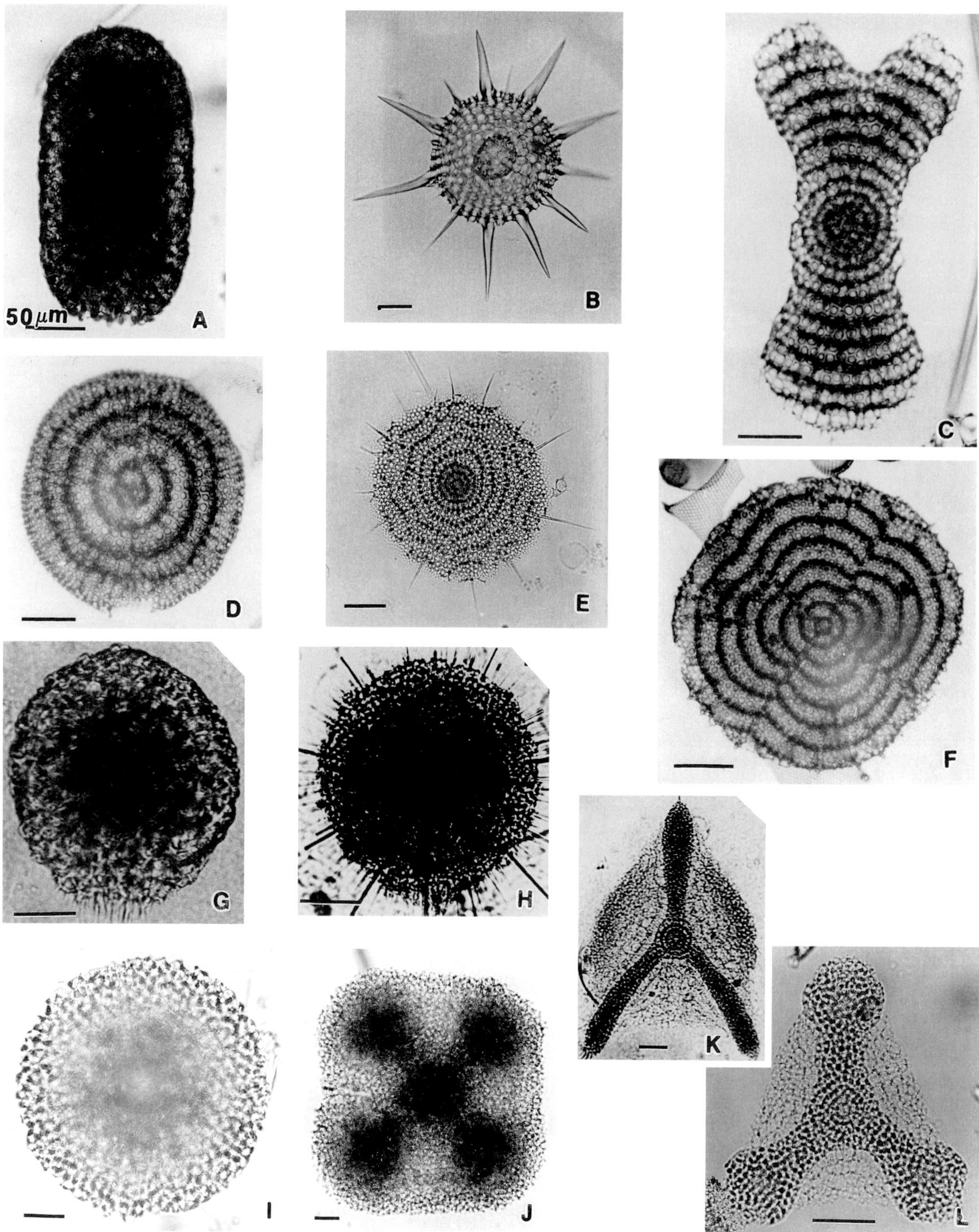

Fig. 8. Spumellaria. A. *Spongurus* sp. Haeckel, 1860b (p 844) (x200). B. *Heliodiscus asteriscus* Haeckel, 1862 (p 436 in Haeckel, 1887) (x136). C. *Amphirhopalum ypsilon* Haeckel, 1881 (p 460, Haeckel, 1887) (x233). D. *Ommatodiscus* sp. Stohr, 1880 (p 115) (x190). E. *Stylochlamydium* sp. Haeckel, 1881 (p 460) (x160). F. *Stylodictya* sp. Ehrenberg, 1847a (chart to p 385) (x200). G. *Spongopyle osculosa* Dreyer, 1889 (p 118) (x210). H. *Spongotrochus glacialis* Haeckel, 1860b (p 844, Popofsky, 1908) (x210). I. *Spongodiscus* sp. Ehrenberg, 1854 (p 237) (x160). J. *Spongaster tetras* Ehrenberg, 1860 (p 833) (x106). K. *Euchitonia elegans* Ehrenberg, 1860 (p 831in Ehrenberg, 1872) (x105). L. *Hymeniastrum euclidis* Ehrenberg, 1847a (chart to p 385 in Haeckel, 1887) (x225). (A) J. Morley; (B,C) C. Nigrini; (D) J. Caulet; (E,I,J,L) D. Boltovskoy; (G,H,K) Takahashi (1991).

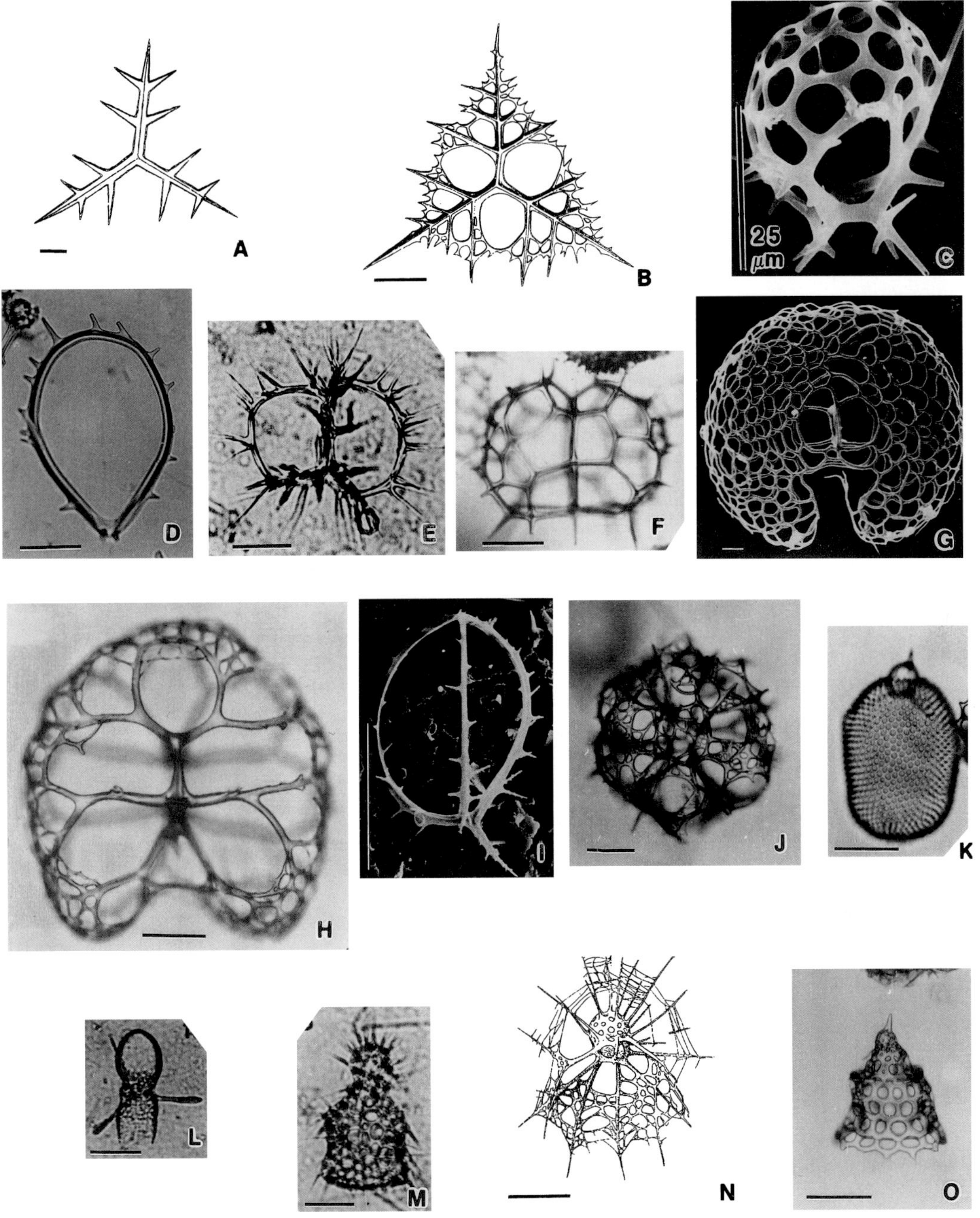

Fig. 9. Nassellaria. A. *Triplagia primordialis* Haeckel, 1881 (p 423 in Haeckel, 1887) (x100). B. *Triplecta triactis* Haeckel, 1881 (p 424 in Haeckel, 1887) (x180). C. *Phormacantha hystrix* Jörgensen, 1905 (p 132, Jörgensen 1900) (x1200). D. *Zygocircus productus* Butschli, 1882 (p 496) (x210). E. *Acanthodesmia vinculata* Müller, 1857 (x210). F. *Lophospyris pentagona* Haeckel, 1887 (p 1080 in Ehrenberg, 1872) (x233). G. *Nephrospyris renilla* Haeckel, 1887 (p 1100) (x80). H. *Lithocircus reticulata* Müller, 1857 (p 484, Ehrenberg, 1872) (x233). I. *Neosemantis distephanus* Popofsky, 1913 (p 298) (x520). J. *Pseudocubus warreni* Haeckel, 1887 (p 1010 in Goll, 1980) (x200). K. *Lithopera bacca* Ehrenberg, 1847a (chart to p 385 in Ehrenberg, 1872) (x233). L. *Lophophaena cylindrica* Ehrenberg, 1847b (p 54, Cleve, 1900) (x210). M. *Peromelissa phalacra* Haeckel, 1881 (p 433 in Haeckel, 1887) (x210). N. *Arachnocorys circumtexta* Haeckel, 1860b (p 837 in Haeckel, 1862) (x240). O. *Cycladophora davisiana* Ehrenberg, 1847a (chart to p 385 in Ehrenberg, 1861) (x230). Sources. (A,B) Haeckel (1887); (B,E,G,I,L,M) Takahashi (1991); (D,O) J. Caulet; (F,H,K) Nigrini (1967); (J) Nigrini and Caulet (1992); (N) Petrushevskaya (1971).

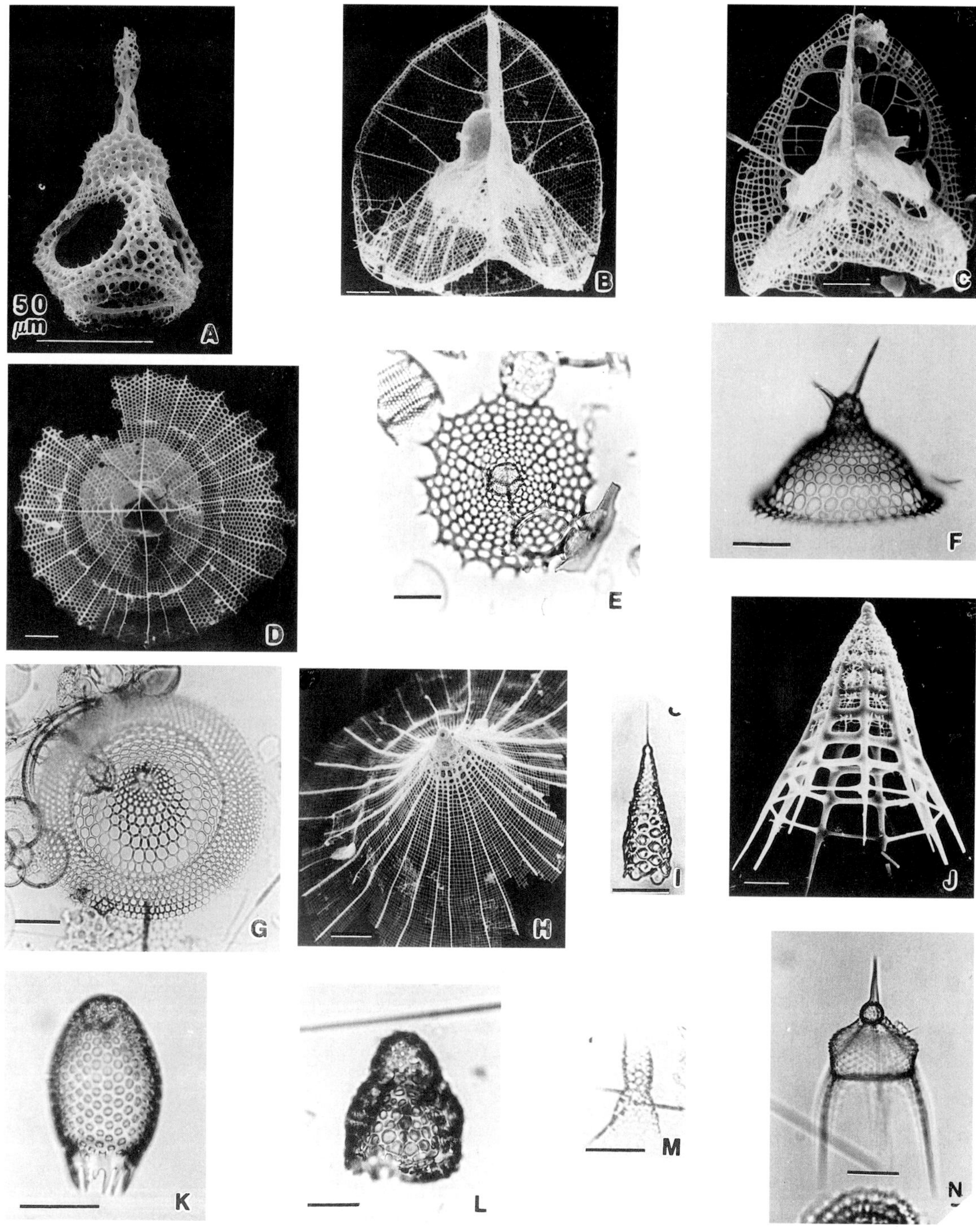

Fig. 10. Nassellaria. A. *Clathrocanium* sp. Ehrenberg, 1860 (p 829) (x440). B. *Callimitra annae* Haeckel, 1881 (p 431 in Haeckel, 1887) (x160). C. *Clathrocorys giltschii* Haeckel, 1881 (p 431 in Haeckel, 1887) (x190). D. *Sethophormis aurelia* Haeckel, 1887 (p 1243) (x120). E. *Lampromitra schultzei* Haeckel, 1881 (p 431 in Haeckel, 1862) (x180). F. *Eucecryphalus* sp. Haeckel, 1860b (p 836) (x200). G. *Theopilium* sp. Haeckel, 1881 (p 435) (x160). H. *Litharachnium tentorium* Haeckel, 1860b (p 835) (x150). I. *Cornutella profunda* Ehrenberg, 1838 (p 128 in Ehrenberg, 1854) (x210). J. *Peripyramis circumtexta* Haeckel, 1881 (p 428 in Haeckel, 1887) (x180), K. *Carpocanistrum* sp. Haeckel, 1887 (p. 1170) (x300), L. *Antarctissa* sp. Petrushevskaya, 1967 (p. 85) (x200). M. *Acanthocorys* cf. *variabilis* Haeckel, 1881 (p 432 in Popofsky, 1913) (x210). N. *Pterocanium praetextum* Ehrenberg, 1847a (chart to p 385 in Haeckel, 1872) (x175). Sources: (A,F,K) J. Caulet; (B-D,H-J,M) Takahashi (1991); (E) D. Boltovskoy; (L) J. Morley; (N) Nigrini (1970).

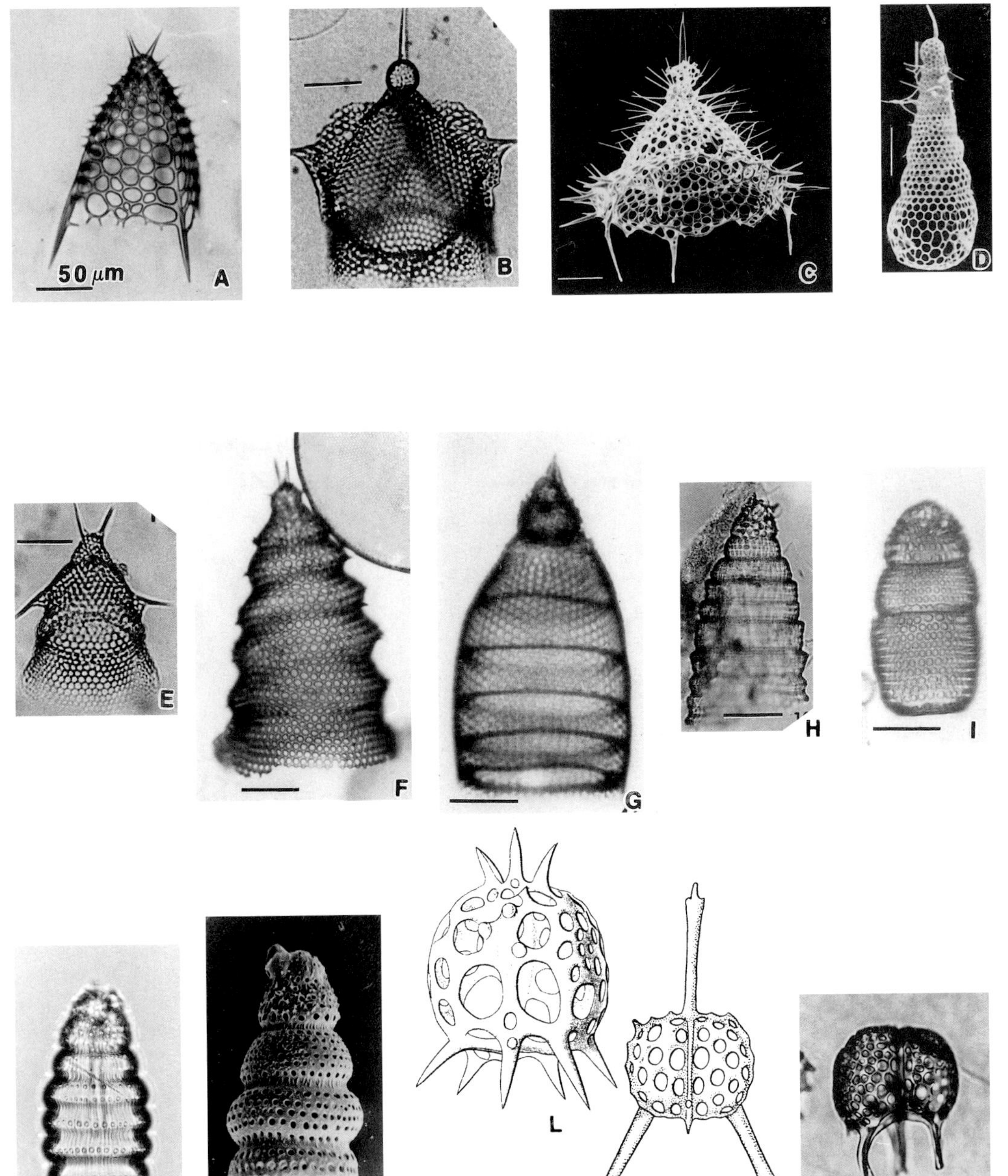

Fig. 11. Nassellaria. A. *Dictyophimus infabricatus* Ehrenberg, 1847a, (chart to p 383 in Nigrini, 1968) (x200). B. *Lipmanella virchowii* Loeblich & Tappan, 1961 (p 226 in Haeckel, 1862) (x210). C. *Clathrocyclas cassiopejae* Haeckel, 1881 (p 439 in Haeckel, 1887) (x150). D. *Cyrtopera laguncula* Haeckel, 1881 (p 434 in Haeckel, 1887) (x190). E. *Stichopilium bicorne* Haeckel, 1881 (p 439 in Haeckel, 1887) (x210). F. *Lithostrobus* cf *hexagonalis* Bütschli, 1882 (p 529 in Haeckel, 1887) (x200). G. *Eucyrtidium hexagonatum* Ehrenberg, 1847a (chart to p 385 in Haeckel, 1887) (x250). H. *Spirocyrtis scalaris* Haeckel, 1881 (p 438 in Haeckel, 1887) (x220). I. *Phormostichoartus corbula* Campbell, 1951 (p. 530, Harting, 1863) (x250). J. *Siphocampe nodosaria* Haeckel, 1881 (p 438 in Haeckel, 1887) (x320). K. *Botryostrobus auritus/australis* Haeckel, 1887 (p 1475 in Ehrenberg, 1844ab) (x400). L. *Liriospyris hexapoda* Haeckel, 1881 (p 441) (mag. unknown). M. *Dendrospyris* sp. Haeckel, 1881 (p 443 in Haeckel, 1887) (mag. unknown). N. *Triceraspyris* sp. Haeckel, 1881 (p 441) (x200). Sources. (A) Nigrini and Caulet (1992); (B-E) Takahashi (1991); (F, K) J. Caulet; (G) Nigrini (1967); (H) Nigrini (1977); (I) C. Nigrini; (J) D. Boltovskoy; (L) Haeckel (1887); (M) Petrushevskaya (1981); (N) J. Morley.

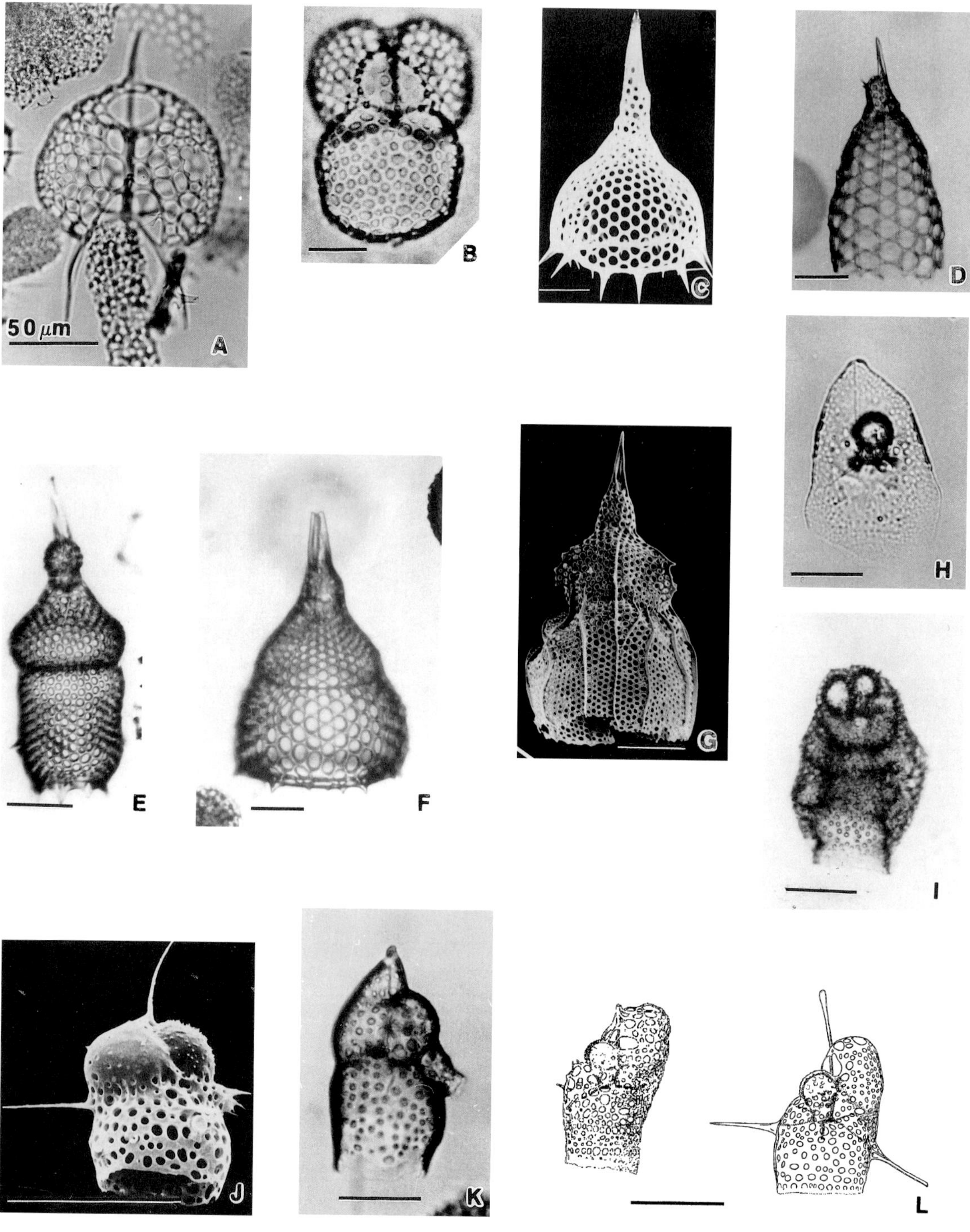

Fig. 12. Nassellaria. A. *Tholospyris* sp. Haeckel, 1881 (p 441) (x320). B. *Phormospyris stabilis* Haeckel, 1881 (p 442 in Goll, 1976) (x210). C. *Anthocyrtidium ophirense* Haeckel, 1881 (p 430 in Ehrenberg, 1872) (x200). D. *Lamprocyrtis nigriniae* Kling, 1973 (p 638 in Caulet, 1971) (x200). E. *Theocorythium trachelium* Haeckel, 1887 (p 1416 in Ehrenberg, 1872) (x233). F. *Lamprocyclas maritalis* Haeckel, 1881 (p 434 in Haeckel, 1887) (x200). G. *Pterocorys hertwigii* Haeckel, 1881 (p 435 in Haeckel, 1887) (x240). H. *Centrobotrys thermophila* Petrushevskya, 1965 (p 113) (x250). I. *Botryocyrtis scutum* Ehrenberg, 1860 (p 829 in Harting, 1863) (x250). J. *Saccospyris preantarctica* Haecker, 1907 (p 124 in Petrushevskya, 1975) (x630). K. *Acrobotrys* sp. (p 440 in Haeckel, 1881) (x300). L. *Botryopyle dictyocephalus* Haeckel, 1881 (p 440 in Haeckel, 1887) (x360). Sources: (A) D. Boltovskoy; (B,C,J) Takahashi (1991); (D) Nigrini and Caulet (1992); (E,I) Nigrini (1967); (F, G, K) J. Caulet; (H) C. Nigrini; (L) Petrushevskaya (1971).

lateral spines (directed down and sideways) interconnected by a delicate meshwork which forms 3 basal plates and 3 lateral plates (Fig. 10B).

Genus ***Clathrocanium*** Ehrenberg, 1860

Cephalis with a large, 3-bladed apical horn which may have lateral thread-like, anastomosing projections. The dorsal and 2 main lateral spines, directed down and sideways, are joined by narrow lattice plates which form a small thorax (Fig. 10A).

Genus ***Clathrocorys*** Haeckel, 1881

Terminal mouth of thorax as above and 3 radial ribs enclosed in the wall of the thorax that bears 3 large lateral holes between the 3 ribs. No frontal horn (Fig. 10C).

Genus ***Phormacantha*** Jörgensen, 1905

Primary spines with 3 arches and a strong ventral sagittal spine (Fig. 9C).

Genus ***Sethophormis*** Haeckel, 1887

Terminal mouth wide open. Radial ribs smooth, enclosed in the wall of the thorax. Shell flat, campanulate, or nearly discoidal. Cephalis without a horn (Fig. 10D).

Genus ***Triplagia*** Haeckel, 1881

Three radial spines simple or branched, radiating from a point (Fig. 9A).

Genus ***Triplecta*** Haeckel, 1881

Three regularly arranged spines in one plane, which anastomose to form a loose wickerwork (Fig. 9B).

Genus ***Zygocircus*** Bütschli, 1882

Pear-shaped or D-shaped, spiny, 3-bladed sagittal ring (Fig. 9D).

FAMILY ACANTHODESMIIDAE Haeckel, 1862

Skeleton represented by a well-developed D-shaped sagittal ring (median bar and anastomosed vertical and apical spines), either free or embedded into the latticed cephalic wall, in which case the cephalis is usually bilaterally lobed. Sometimes with thorax, abdomen always absent. The typical heteropolar nassellarian symmetry is often inconspicuous in the Spyridae (Fig. 3E, 9E).

Genus ***Acanthodesmia*** Müller, 1857

Four open lateral gates, partly latticed (Fig. 9E).

Genus ***Dendrospyris*** Haeckel, 1881

Shell composed of cephalis without apical cupola or dome or thorax. Feet branched like a tree, single apical horn (Fig. 11M).

Genus ***Liriospyris*** Haeckel, 1881

Shell as above, but with 6 basal feet and 3 apical horns (Fig. 11L).

Genus ***Lophospyris*** Haeckel, 1881

Shell with 7 to 12 or more basal feet; spines unbranched, meshes polygonal or within polygonal frames (Fig. 9F).

Genus ***Tholospyris*** Haeckel 1881

Cephalis with an apical cupola and horn, without thorax. Three basal feet (Fig. 12A).

FAMILY THEOPERIDAE Haeckel, 1881
emend. Riedel, 1967B

Small spherical or nearly spherical cephalis (smallest anterior chamber), often poreless or sparsely perforate and with apical spine, and one or more post-cephalic chambers. Cephalis contains a reduced internal spicule homologous with that of the plagoniids. Generally, cap- or helmet-shaped nassellarians or tending toward conical in overall plan (Fig. 9K)

Genus ***Eucyrtidium*** Ehrenberg, 1847
emend. Nigrini, 1967

Ovate or spindle-shaped shell with solid apical horn, multisegmented, with constricted mouth (Fig. 11G).

Genus ***Lithopera*** Ehrenberg, 1847

Shell 2-segmented with cephalis partly submerged into an oval thorax with a closed mouth. Three divergent ribs initially enclosed in the cavity of the thorax (Fig. 9K).

Genus ***Peripyramis*** Haeckel, 1881

Pyramidal shell with simple lattice and an outer net-like mantle, without apical horn; shell mouth open (Fig. 10J).

FAMILY CARPOCANIIDAE Haeckel, 1881 emend. Riedel, 1967b

Cephalis small and included in large second segment (thorax), nearly indistinguishable from latter, and reduced to a few bars that are homologous with nassellarian spicule in other groups. Abdomen absent or rudimentary (Fig. 10K).

Genus ***Carpocanistrum*** Haeckel, 1887

Shell subspherical to oval, cephalis completely submerged in thorax. Poreless peristome with or without terminal teeth (Fig. 10K).

FAMILY PTEROCORYTHIDAE Haeckel, 1881 emend. Riedel 1967b, emend. Moore, 1972

Cephalis large, elongate, divided into three lobes by two lateral furrows directed obliquely and downward from the apical spine to the base of the cephalis. The upper unpaired lobe is located above the two smaller paired ones; these basal paired lobes are not always conspicuous. Many pterocorythids are two or three-segmented, lacking postabdominal segments (Fig. 12C, G)

Genus ***Anthocyrtidium*** Haeckel, 1881

Two-segmented shell with apical horn, poreless peristome, mouth open, constricted, bearing terminal and subterminal teeth outside the mouth opening (Fig. 12C).

Genus ***Lamprocyclas*** Haeckel, 1881

Trilocular cephalis with 2 additional segments (thorax and abdomen). Shell mouth open, with terminal teeth. Peristome differentiated (Fig. 12F).

Genus ***Lamprocyrtis*** Kling, 1973

Lamprocyrtis hannai. Cephalis elongate, with a large 3-bladed apical horn. Thorax campanulate, thick-walled, with subregular, circular pores. Abdomen truncate-conical, with large, subregular, circular pores and usually with terminal and/or subterminal teeth (Fig. 12D).

Lamprocyrtis nigriniae. Cephalis elongated, usually open proximally, with a large 3-bladed horn. Thorax campanulate, thin-walled, with large, subregular, circular pores increasing in size distally; peristome absent or weakly developed. No abdomen.

Genus ***Pterocorys*** Haeckel, 1881

Shell with 3 segments, mouth open, with 3 solid thoracic ribs which may project as thorns or wings. No differentiated peristome or terminal feet (Fig. 12G).

Genus ***Theocorythium*** Haeckel, 1887

Shell with 2-part cephalis: (1) upper sphere and (2) 1 or 2 lobes directly beneath or slightly lateral to the upper sphere. Single apical horn. Abdomen subcylindrical with numerous terminal teeth and sometimes with subterminal teeth scattered irregularly over the distal half of the abdomen. Cephalic lobes form a "neck" structure between the upper cephalis and thorax, a feature peculiar to this genus (Fig. 12E).

FAMILY ARTOSTROBIIDAE Riedel 1967a emend. Foreman, 1973

Cephalis spherical with a lateral tubule, homologous with vertical spine, and one or more post-cephalic chambers. Pores usually in transverse rows encircling the chambers (Fig. 11H, K).

Genus ***Botryostrobus*** Haeckel, 1887 emend. Nigrini, 1977

Shell with more than 4 segments, vertical tube cylindrical, basal mouth open (Fig. 11K).

Genus ***Phormostichoartus*** Campbell, 1951, emend. Nigrini, 1977

Shell with 4 segments, cylindrical; mouth slightly constricted with well-developed peristome; vertical tube well developed, cylindrical lying along the thorax; no apical horn (Fig. 11I).

Genus ***Spirocyrtis*** Haeckel, 1882 emend. Nigrini, 1977

Shell with more than 4 segments, expanding distally; intersegmental constrictions sharply rounded to angular; vertical tube flared (duck-billed); apical horn or tube present (Fig. 11H).

FAMILY CANNOBOTRYIDAE Haeckel, 1881
emend. Riedel, 1967b

Cephalis, multilobed (sometimes appearing as irregular bulges), unpaired and asymmetrical; one lobe homologous with cephalis of theoperids (Fig. 12K).

Genus **Acrobotrys** Haeckel, 1881

Cephalis has tubules, thorax open (Fig. 12K).

Genus ***Botryopyle*** Haeckel, 1882

Lacks cephalic tubules, thorax is open (Fig. 12L).

Genus ***Saccospyris*** Haecker, 1907

Without apical horn, has corona of minute serrations around basal shell mouth (Fig. 12J).

ACKNOWLEDGEMENTS

We are very pleased to acknowledge the very generous contribution of drawings and photographs from Drs. J. Caulet, J. Moore, J. Morley, M. Petrushevskaya, and T. Takahashi. Some of the line drawings are adapted from Haeckel (1887).

LITERATURE CITED

Anderson, O. R. 1983a. *Radiolaria*. Springer-Verlag, New York. 355 pp

Anderson, O. R. 1983b. The radiolarian symbiosis. *In:* Goff, L. (ed.), *Algal Symbiosis: A Continuum of Interaction Strategies*. Cambridge Univ. Press, Cambridge. pp 69-89

Anderson, O. R. 1992. Radiolarian algal symbioses. *In:* Reisser, W. (ed.), *Algae and Symbioses: Plants, Animals, Fungi, Viruses, Interactions Explored*. Biopress Ltd., Bristol, England. pp. 93-109

Anderson, O. R. 1993. The trophic role of planktonic foraminifera and radiolaria. *Mar. Microb. Food Webs*, 7:31-51.

Boltovskoy, D. 1981. Radiolaria. *In:* Boltovskoy, D. (ed.), *Atlas del Zooplankton del Atlantico Sudoccidental. Public. esp. del INIDEP*. Mar del Plata, Argentina.

Boltovskoy, D. 1987. Sedimentary record of radiolarian biogeography in the equatorial to Antarctic western Pacific Ocean. *Micropaleontology,* 33:267-281.

Boltovskoy, D. 1994. The sedimentary record of pelagic biogeography. *Progress in Oceanography* (in press).

Boltovskoy, D. & Alder, V. A. 1992. Paleoecological implications of radiolarian distribution and standing stocks versus accumulation rates in the Weddell Sea. *The Antarctic Environment: a Perspective on Global Change, Antarctic Research Series*, American Geophysical Union, 56:377-384.

Bütschli, O. 1882. Beitrage zur Kenntnis der Radiolarienskelette, insbesondere der der Cyrtida. *Z. Wiss. Zool.*, 36:485-540.

Cachon, J. & Cachon, M. 1972. Le système axopodial des radiolaires sphaeroidés. II. Les Periaxoplastidiés, III - Cryptoaxoplastidiés (anaxoplastidiés), IV - Les fusules et le système rhèoplasmique. *Arch. Protistenkd.*, 114:291-307.

Cachon, J. & Cachon, M. 1985. II. Class Polycystinea. *In:* Lee, J. J., Hutner, S. H., & Bovee, E. C. (eds.), *An Illustrated Guide to the Protozoa*. Society of Protozoologists, Lawrence, Kansas. pp 283-295

Campbell, A. S. 1951. New genera and subgenera of Radiolaria. *J. Paleontol.*, 25(4):527-530.

Caron, D. A. & Swanberg, N. R. 1990. The ecology of planktonic sarcodines. *Aquatic Sciences*. CRC Press, Boca Raton, 3:147-180.

Caulet, J. P. 1971. Contribution à l'étude de quelques Radiolaires Nassellaires des boues de la Mediterrané et du Pacifique (Study of some nassellarian Radiolaria from Mediterranean and Pacific sediments). *Cahiers de micropaleontologie Serie 2. Archives originales Centre de Documentation C.N.R.S. No. 498*, 10:1-10.

Cleve, P. T. 1900. Notes on some Atlantic plankton-organisms. *Kongliga Svenska Vetenskapsakademiens handlingar*, 34(1):1-22.

Dreyer, F. 1889. Morphologischen Radiolarien studien. 1. Die Pylombildungen in vergleichend-anatomischer und entwicklungs geschichtlicher Beziehung bei Radiolarien und bei Protisten uberhaupt nebst System und Beschreibung neuer und der bis jetzt bekannten pylomatischen Spumellarien. *Jena. Z. Naturwiss.*, 23:1-138.

Dumitrica, P. 1988. New families and subfamilies of Pyloniacea (Radiolaria). *Rev. Micropaleontol.*, 31:178-195.

Dumitrica, P. 1989. Internal skeletal structures of the superfamily Pyloniacea (Radiolaria), a basis of a new systematics. *Rev. Esp. Micropaleontol.*, 21:207-264.

Ehrenberg, C. G. 1838. Über die Bildung der Kreidefelsen und des Kreidemergels durch unsichtbare Organismen. *K. Akad. Wiss. Berlin, Abh.*, 1838:59-147.

Ehrenberg, C. G. 1844a. Über 2 neue Lager von Gebirgsmassen aus Infusorien als Meeres-Absatz in Nord-Amerika und eine Vergleichung derselben mit den organischen Kreide-Gebilden in Europa und Afrika. *K. Akad. Wiss. Berlin, Abh.*, 1844:57-97.

Ehrenberg, C. G. 1844b. Einige vorlaufige Resultate seiner Untersuchungen der ihm von der Sudpolreise des Captain Ross, so wie von den Herren Schayer und Darwin zugekommenen Materialien uber das Verhalten des kleinsten Lebens in den Oceanen und den grossten bisher zuganglichen Tiefen des Weltmeeres. *K. Akad. Wiss. Berlin, Abh.*, 1844:182-207.

Ehrenberg, C. G. 1847a. Über eine halibiolithische, von Herrn R. Schomburgk entdeckte, vorherrschend aus mikroskopischen Polycystinen gebildete, Gebirgsmasse von Barbados. *K. Akad. Wiss. Berlin, Abh.*, 1846:382-385.

Ehrenberg, C. G. 1847b. Über die mikroskopischen kieselschaligen Polycystinen als machtige Gebirgsmasse von Barbados und über das Verhaltniss deraus mehr als 300 neuen Arten bestehenden ganz eigenthumlichen Formengruppe jener Felsmasse zu den jetzt lebenden Thieren und zur Kreidebildung. Eine neue Anregung zur Erforschung des Erdlebens. *K. Akad. Wiss. Berlin, Abh.*, 1847:40-60.

Ehrenberg, C. G. 1854. Die systematische Charakteristik der neuen mikroskopischen Organismen des tiefen atlantischen Oceans. *K. Akad. Wiss. Berlin, Abh.*, 1854:236-250.

Ehrenberg, C. G. 1858. Kurze Characteristik der 9 neuen Genera und der 105 neuen Species des agaischen Meeres und des Tiefgrundes des Mittel-Meeres. *K. Akad. Wiss. Berlin, Abh.*, 1858:10-40.

Ehrenberg, C. G. 1860. Über den Tiefgrund des stillen Oceans Inseln aus bis 15600' Tiefe nach Lieutenant Brooke. *K. Akad. Wiss. Berlin, Abh..*, 1860:819-833.

Ehrenberg, C. G. 1861. Über die Tiefgrund Verhaltnisse des Oceans am Eingange der Davisstrasse und bei Island. *K. Akad. Wiss. Berlin, Abh.*, 1861:275-315.

Ehrenberg, C. G. 1872. Mikrogeologische Studien als Zusammenfassung seiner Beobachtungen des kleinsten Lebens der Meeres-Tiefgrunde aller Zonen und dessen geologischen Einfluss. *K. Akad. Wiss. Berlin, Abh.*, 1872:265-322.

Ehrenberg, C. G. 1875. Fortsetzung der mikrogeologischen Studien als Gesammt-Uebersicht der mikroskopischen Palaeontologio. *K. Akad. Wiss. Berlin, Abh.*, 1875. 226 pp

Foreman, H. P. 1973. Radiolaria of Leg 10 with systematics and ranges for the families Amphipyndacidae, Artostrobiidae, and Theoperidae. *In:* Worzel, J. L., Bryant, W., et al. (eds.), *Initial Reports of the Deep Sea Drilling Project, vol. 10*. Washington, D.C.: U. S. Government Printing Office, 407-474.

Goll, R. M. 1968. Classification and phylogeny of Trissocyclidae (Radiolaria) in the Pacific and Caribbean Basins. Pt I. *J. Paleontol.*, 42:1409-1432.

Goll, R. M. 1969. Classification and phylogeny of Trissocyclidae (Radiolaria) in the Pacific and Caribbean Basins. Pt II. *J. Paleontol.*, 43:322-339.

Goll, R. M. 1976. Morphological intergradation between modern populations of *Lophospyris* and *Phormospyris* (Trissocyclidae, Radiolaria). *Micropaleontology*, 22(4):379-418.

Goll, R. M. 1980. Pliocene-Pleistocene Radiolaria from the East Pacific Rise and the Galapagos spreading center, Deep Sea Drilling Project Leg 54. *In:* Rosendahl, B. R., Hekinian, R., et al., (eds.) *Initial Reports of the Deep Sea Drilling Project, vol.54*. Washington, D.C.: U. S. Government Printing Office, 1980, 54:425-454.

Gowing, M., M. 1993. Seasonal radiolarian flux at the VERTEX North Pacific time-series site. *Deep-sea Res.*, 40:517-545.

Haeckel, E. 1860a Über neue, lebende Radiolarien des Mittelmeeres und die dazu gehorigen Abbildungen (Some new living Radiolaria from the Mediterranean Sea and their illustration). *K. Akad. Wiss. Berlin, Abh.*, 1860:794-817.

Haeckel, E. 1860b. Fernere Abbildungen und Diagnosen neuer Gattungen und Arten von lebenden Radiolarien des Mittelmeeres (Supplementary illustrations and diagnosis of new genera and species of living radiolarian of the Mediterranean Sea). *K. Akad. Wiss. Berlin, Abh.*, 1860:835-845.

Haeckel, E. 1862. *Die Radiolarien (Rhizopoda Radiaria). Eine Monographie.* Reimer, Berlin. 572 pp

Haeckel, E. 1881. Entwurf eines Radiolarien-Systems auf Grund von Studien der Challenger-Radiolarien (Basis for a radiolarian classification from the study of Radiolaria of the Challenger collection). *Jena. Z. Naturwiss.*, 15:418-472.

Haeckel, E. 1887. Report on Radiolaria collected by H. M. S. Challenger during the years 1873-1876. *In:* Thompson, C. W. & Murray, J. (eds.), *The Voyage of the H. M. S. Challenger*. Her Majesty's Stationery Office, London, V. 18, pp 1-1803.

Haecker, V. 1907. Altertümliche sphärellarien und cyrtellarien aus grossen Meerestiefen. *Arch. Protistenkd.*, 10:114-126.

Harting, P. 1863. Bijdrage tot de kennis der mikroskopische fauna en flora van de Banda-Zee (Contribution to the knowledge of microscopic fauna and flora from the Banda Sea). *Verh. K. Akad. Wet. Lett., Amsterdam*, 10(1):1-34.

Hays, J. D. 1965. Radiolarians and late Tertiary and quaternary history of Antarctic seas. *Biol. Ant. Seas II, Ant. Res. Ser. 5*, Amer. Geophys. Union. pp 125-184

Hertwig, R. 1879. *Der Organismus der Radiolarien.* G. Fischer, Jena. pp. 149 pp

Hollande, A. & Enjumet, M. 1960. Cytologie, évolution et systématique des Sphaeroidés (Radiolaires). *Arch. Mus. Nat. Hist. Natur., Paris, Ser. 7*, 7:1-134.

Huxley, T. H. 1851. Zoological notes and observations made on board H.M.S. Rattlesnake. III. Upon *Thalassicolla*, a new zoophyte. *Annals and Magazine of Natural History, ser. 2*, 8(48):433-442.

Jörgensen, E. 1900. Protophyten und Protozoen im Plankton aus der norwegischen Westküste. *Bergens Museums Aarbog (1899)*, 2 (6):112.

Jörgensen, E. 1905. The protist plankton and the diatoms in bottom samples. *Bergens Museums Skrifter, Bergen, 1905.* pp 49-151

Kling, S. A. 1973. Radiolaria from the eastern North Pacific, Deep Sea Drilling Project Leg 18. *In:* Kulm, L. D., Von Huene, R., et al., (eds.), *Initial Reports of the Deep Sea Drilling Project.* vol. 18. Washington, D.C.: U. S. Government Printing Office. pp 617-671

Kling, S. A. 1978. Radiolaria. *In:* Haq, B. U. & Boersma, A. (eds.), *Introduction to Marine Micropaleontology.* Elsevier, NY. pp 203-244

Kling, S. A. 1979. Vertical distribution of polycystine radiolarians in the central North Pacific. *Mar. Micropaleontol.*, 4:295-318.

Kling, S. A. & Boltovskoy, D. Radiolarian vertical distribution patterns across the southern California Current. *Deep-Sea Res.* (In press)

Loeblich, A. R. & Tappan, H. 1961. Remarks on the systematics of the Sarkodina (Protozoa), renamed homonyms and new validated genera. *Proc. Biol. Soc. Wash.*, 74:213-234.

Meyen, F. J .F. 1834. Über das Leuchten des Meeres und Beschreibung einiger Polypen und anderer niederer Tiere (On starfish and a description of some polyps and other inferior animals). Beitrage zur Zoologie, gesammelt auf einer Reise um die Erde, vol. 16. *Nova acta Academiae Caesareae Leopoldino Carolinae germanicae naturae curiosorum.* pp 125-216

Moore, T. C., Jr. 1972. Mid-Tertiary evolution of the radiolarian genus *Calocycletta. Micropaleontology*, 18(2):144-152.

Müller, J. 1855. Über *Sphaerozoum* und *Thalassicolla. K. Akad. Wiss. Berlin, Abh.*, 1855:229-253.

Müller, J. 1857. Über die Thalassicollen, Polycystinen und Acanthometren des Mittelmeeres (The Thalassicolla, Polycystina, and Acanthometra from the Mediterranean Sea). *K. Akad. Wiss. Berlin, Abh.*, 1856:474-503.

Müller, J. 1858a. Einige neue bei St. Tropez am Mittelmeer beobachtete Polycystinen und Acanthometren. *K. Akad. Wiss. Berlin, Abh..*, 1858:154-155.

Müller, J. 1858b. Über die Thalassicollen, Polycystinen und Acanthometren des Mittelmeeres. *K. Akad. Wiss. Berlin, Abh.*, 1858:1-62.

Nigrini, C. A. 1967. Radiolaria in pelagic sediments from the Indian and Atlantic oceans. *Bull. Scripps Inst. Oceanogr. Univ. Calif.*, 11:1-125.

Nigrini, C. A. 1968. Radiolaria from eastern tropical Pacific sediments. *Micropaleontol*ogy, 14(1):51-63.

Nigrini, C. A. 1970. Radiolarian assemblages in the North Pacific and their applications to a study of Quaternary sediments in core V20-130. *In:* Hays, J. D. (ed.) *Geological Investigations of the North Pacific.* Geol. Soc. Amer. Mem., 126:139-183.

Nigrini, C. A. 1971. Radiolarian zones in the Quaternary of the Equatorial Pacific Ocean. *In:* Funnell, B. M. & Riedel, W. R. (eds.), *Micropaleontology of Oceans.* Cambridge University Press, Cambridge. pp 443-461

Nigrini, C. A. 1977. Tropical Cenozoic Artostrobiidae (Radiolaria). *Micropaleontology*, 23:241-269.

Nigrini, C. A. & Caulet, J. P. 1992. Late Neocene radiolarian assemblages characteristic of Indo-Pacific areas of upwelling. *Micropaleontology*, 38:139-164.

Nigrini, C. A. & Moore, T. C., Jr. 1979. *A Guide to Modern Radiolaria.* Cushman Foundation for Foraminiferal Research, Sp. Publ. No. 16, Washington, D. C.

Petrushevskaya, M. G. 1965. Osobennosti i konstruktsii skeleta radiolyarii Botryoidae otr. (Nassellaria) (Peculiarities of the construction of the skeleton of radiolarians Botryoidae [Order Nassellaria]). *Tr. Zool. Inst. Akad. Nauk SSSR*, 35:79-118.

Petrushevskaya, M. G. 1967. Radiolyarii otryadov Spumellaria i Nassellaria Antarkticheskoi oblasti (po materialam Sovetskoi Antarkticheskoi Ekspeditzii). *Issledovaniya Fauny Morei IX(XII), Rezultaty Biologicheskikh Issledovanii Sovetskikh Antarkticheskikh Ekspeditsii 1955-1958*, 3:5-186.

Petrushevskaya, M. G. 1971. Radiolayarii Nassellaria v planktone Mirovogo Okeana. *Issledovaniya Fauny Morei IX(XVII)*, Nauka, Leningrad. pp 5-186

Petrushevskaya, M. G. 1975. Cenozoic radiolarians of the Antarctic, Leg 29, DSDP. *In:* Kennett, J., P., Houtz, R. E., et al., (eds.). *Initial Reports of the Deep Sea Drilling Project, vol. 29.* Washington, D.C.: U. S. Government Printing Office. pp 541-675

Petrushevskaya, M. G. 1981. Radiolyarii otryada Nassellaria Mirovogo Okeana. *Akad. Nauk SSSR, Zool. Inst.*, Nauka, Leningrad. pp 1-405

Petrushevskaya, M. G. 1986. Radiolyarevii analiz. *Akad. Nauk SSSR*, Nauka, Leningrad. pp 1-200

Petrushevskaya, M. G., Cachon, J. & Cachon, M. 1976. Comparative-morphological study of radiolarians: foundations of new taxonomy. *Zool. Zh.*, 55:485-496.

Popofsky, A. 1908. Die Radiolarien der Antarktis (mit Ausnahme der Tripyleen). *In:* Drygalski, E. (ed.), *Deutsche Sudpolar Expedition, 1901-1903*, vol. 10. Berlin, Germany: Georg Reimer. pp 183-306

Popofsky, A. 1912. Die Sphaerellarien des Warmwassergebietes. *In:* Drygalski, E. (ed.), *Deutsche Sudpolar-Expedition, 1901-1903*, vol. 13. Berlin, Germany: Georg Reimer. pp 73-159

Popofsky, A. 1913. Die Nassellarien des Warmwassergebietes. *In:* Drygalski, E. (ed.), *Deutsche Sudpolar-Expedition, 1901-1903*, vol. 14. Berlin, Germany: Georg Reimer. pp 217-416

Renz, G. W. 1976. The distribution and ecology of Radiolaria in the Central Pacific: plankton and surface sediments. *Bull. Scripps Inst. Oceanogrr*, 22:1-267.

Reshetnjak, V. V. 1966. Fauna SSSR. Radiolyarii. Akad. Nauk, SSSR, Nauka, Moskva. pp 1-208

Riedel, W. R. 1957. Radiolaria, a preliminary stratigraphy. *Rep. Swedish Deep Sea Exped. 1947-1948*, 6(3):61-96.

Riedel, W. R. 1958. Radiolaria in Antarctic sediments. *Rep. B.A.N.Z.A.R.E., B*, 6(10):219-255.

Riedel, W. R. 1967a. Some new families of Radiolaria. *Proc. Geol. Soc. Lond.*, 1640:148-149.

Riedel, W. R. 1967b. Subclass Radiolaria. *In:* Harland, W. B., Holland, C. H. & House, M. R. (eds.), *The Fossil Record. A Symposium with Documentation*. London, UK: Geological Society of London. pp 291-298

Riedel, W. R. 1971. Systematic classification of polycystine Radiolaria. *In:* Riedel, W. R. & Funnell, B. M. (eds.), *The Micropalaeontology of Oceans*, Cambridge Univ. Press., Cambridge. pp 649-660

Riedel, W. R. & Sanfilippo, A. 1981. Evolution and diversity of form in radiolaria. *In:* Simpson, T. L. & Volcani, B. E. (eds.), *Silicon and Siliceous Structures in Biological Systems.* Springer-Verlag, New York. pp 323-346

Sanfilippo, A. & Riedell, W. R. 1970. Post-Eocene "closed" theoperid radiolarians. *Micropaleontology,* 16:446-462.

Sanfilippo, A. & Riedel, W. R. 1980. A revised generic and suprageneric classification of the Artiscins (Radiolaria). *J. Paleontol.*, 54(5):1008-1011.

Sanfilippo, A. & Riedel, W. R. 1985. Cretaceous Radiolaria. *In:* Bolli, H. M., Saunders, J. B. & Perch-Nielsen, K. (eds.), *Plankton Stratigraphy.* Cambridge University Press, Cambridge, UK pp 573-630

Sanfilippo, A., Westberg-Smith, M. J. & Riedel, W. R. 1985. Cenozoic Radiolaria. *In:* Bolli, H. M., Saunders, J. B. & Perch-Nielsen, K. (eds.), *Plankton Stratigraphy.* Cambridge Univ. Press, Cambridge, UK. pp 631-712

Steineck, P. L. & Casey, R. E. 1990. Ecology and paleoecology of foraminifera and radiolaria. *In:* Capriulo, G. M. (ed.), *Ecology of Marine Protozoa.* Oxford Univ. Press, New York. pp 46-138

Stohr, E. 1880. Die Radiolarienfauna der Tripoli von Grotte, Provinz Girgenti in Sicilien (The radiolarian fauna of the Tripoli of Grotte, Girenti Province, Sicily). *Palaeontographica*, 26 (ser. 3, vol. 2):71-124.

Swanberg, N. S., Anderson, O. R. & Bennett, P. 1990. Skeletal and cytoplasmic variability of large spongiose spumellarian radiolaria (Actinopodea: Polycystina). *Micropaleontology,* 36:379-387.

Takahashi, K. 1991. Radiolaria: flux, ecology, and taxonomy in the Pacific and Atlantic. *Woods Hole Oceanogr. Inst., Ocean Biocoenosis Series No. 3.* 303 pp

Zhamoida, A. I. & Kozlova, G. E. 1971. Gootnochenie podotryadov i semeistv v otryade Spumellaria; radiolyarii. *Trudy Vsesoyuznogo Neftyanogo Nauchno-Issledovatelskogo Geologorazvedochnogo Instituta*, 291:76-82.

THE NAKED RAMICRISTATE AMOEBAE (GYMNAMOEBAE)

BY ANDREW ROGERSON and DAVID J. PATTERSON

The naked amoebae with branching tubular mitochondrial cristae are generally non-spore forming (dictyostelids excluded) heterotrophic protists with broad pseudopodia (lobopodia) of one form or another. In all cases the pseudopodia are used for locomotion and/or feeding; however, there is considerable variation within this grouping regarding pseuodopodial type. They can be thick or thin, long or short, single or numerous, and subpseudopodia can arise from the anterior hyaline zone of the cell (e.g. family Paramoebidae). Subpseudopodia can also be extremely fine, as in the case of echinopodia. These pseudopodia are without any discernable function and are distinct from the fine locomotory and predatory filopodia of naked amoebae previously assigned to the class Filosea.

Gymnamoebae are naked cells, that is they are not housed within tests, walls or tecta with distinct pores or openings for prey ingestion. Despite this seemingly straightforward categorization, the nature of the cell coat is varied as evidenced by transmission electron microscopy. The cell surface structure can be simple, with no discernable features, or can be covered to varying degrees in surface microfibrils or a thin mucoprotein layer termed a **glycocalyx.** It can also be elaborate with complex structures such as the **pentagonal glycostyles** of *Vannella*, the boat-shaped microscales of *Dactylamoeba*, or the thick cuticle of *Mayorella*. The Cochliopodiidae, sometimes classified with the Testacealobosia, are included here as their dorsal tectum of scales corresponds more closely to the system of glycostyles than it does to the rigid test with one or more discrete apertures—characteristic of the testate lobose and testate filose amoebae.

The best known of the naked amoebae is probably *Amoeba proteus*, which is a favorite demonstration cell in laboratory practicals. While *A. proteus* adequately displays the main features of the group, its large size is not representative of gymnamoebae likely to be encountered in field samples. Most isolates are much smaller, frequently less than 25 µm in length. There can be no doubt that many small forms (those less than 10 µm) have been overlooked and many new species in this size class remain to be described. For example, in recent studies *Flabellula demetica* (around 4.5 µm long) was one of the four most common amoebae on seaweed surfaces (Rogerson, 1991), and *Parvamoeba rugata*, with a mean length of only 3.9 µm, was frequently found in the plankton (Rogerson, 1993).

Amoebae have been found in all land masses and in all major habitats wherever moisture is present. Their ubiquity is related to the ability of many species to form resting stages, commonly cysts, which can be dispersed by wind. Cyst formation also accounts for the numerical dominance of amoebae in many soils where they often number several thousand per gram, making them the dominant bacterivores in this habitat (Clarholm, 1981). Recent research has shown that they are also numerically important in the water column, at interfaces such as the neustonic layer, and in the benthos (Butler and Rogerson, 1996; Anderson and Rogerson, 1995; Davis et al., 1978; Rogerson and Laybourn-Parry, 1992).

Many amoebae are facultative or obligate pathogens of man and animals. *Acanthamoeba*, for example, which is ubiquitous in soil and freshwater, contains known pathogenic species which can invade the cornea resulting in severe eye infections.

It is intended that this chapter be used, as far as possible, to identify amoebae by light microscopy (LM). To this end, when a new isolate is encountered, six key diagnostic features at the LM level should be considered. In most cases these will allow the identification of an isolate to genus. A few amoebae can only be identified to genus using ultrastructural features. For practical reasons, the use of these diagnostic features has been kept to a minimum in the present key.

1. *Locomotive form*. Characterizing the normal locomotive form of the cell is of paramount importance. To do this it is essential to observe amoebae under conditions that do not adversely affect their movement. Viewing cells under a coverglass is rarely satisfactory since O_2 levels soon decrease. Hanging drop preparations (Finlay et al., 1988) can be used or, better still, amoebae can be observed in dishes using an inverted microscope or in a drop of water using a water immersion lens.

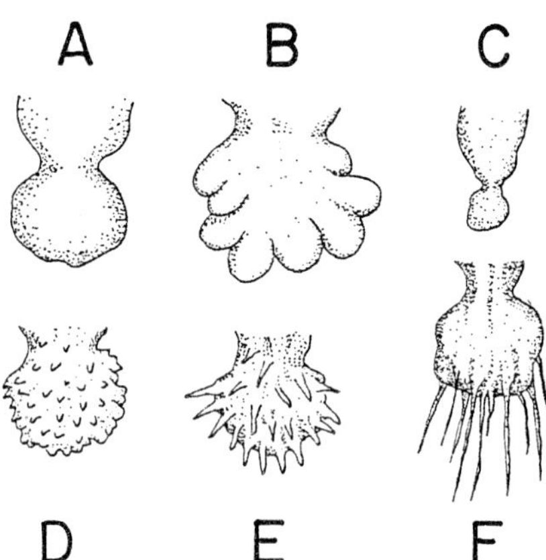

Fig. 1. Uroidal types. A. Bulbous. B. Morulate. C. Globular. D. Papillate. E. Villous. F. Knob-like with trailing filaments.

Amoebae can either be flattened or cylindrical during normal locomotion. Cylindrical amoebae can be **polypodial** or **monopodial**, the latter often being referred to as **limax**. An important dichotomy is whether locomotion is **steady** or **eruptive**; amoebae with markedly eruptive pseudopodia are frequently in the class Heterolobosea. The form of the pseuodopodia is important. Most amoebae advance by a single pseudopodium or by a succession of progressing pseudopodia. From the anterior edge, extending subpseudopodia are common. These can be short, long, slender or thick and may be distinctive in shape. Some are **digitiform** or **mamilliform** in shape, **furcate** (branched) or non-furcate. Subpseudopodia that branch usually do so close to their base. The anterior edge of leading pseudopodia often have a clear zone, called a **hyaline cap**. The extent of this zone is important for identification as it can be thin, broad, or even absent. At the posterior of some amoebae is the **uroid**. When morphologically differentiated, the shape of the uroid (e.g. bulbous, globular, morulate, villous, papillate, Fig. 1) is a useful diagnostic feature as is the presence or absence of trailing uroidal filaments.

2. *Floating form*. When identifying an unknown isolate the form of the suspended, or floating, amoebae should be observed. Some have long thin, tapering pseudopodia which radiate out from the cell body, while at the other extreme amoebae may have short, blunt radiating pseudopodia or none at all.

3. *Size*. Length and breadth dimensions of normal locomoting cells are essential for identification purposes. Where relevant, additional measurements should be recorded. These commonly include the lengths of subpseudopodia, radiating pseudopodia in floating amoebae, and any trailing uroidal filaments.

4. *Cytoplasmic inclusions*. The most obvious inclusions found in some amoebae are crystals. Truncate bipyramidal crystals, and sometimes plate-like crystals, are characteristic of many Amoebidae. The cytoplasm of other amoebae can be granular in appearance because of numerous irregular inclusions. Two species of amoebae, *Mayorella viridis* and *Parachaos zoochlorellae* are green in color because they harbor intracellular endosymbionts. The presence or absence of contractile vacuoles in the cytoplasm and their degree of activity can be used to aid some identifications (Patterson, 1981).

5. *Nucleus*. Most gymnamoebae are uninucleate but some are binucleate or multinucleate. The shape of the nucleus is frequently spherical with a central nucleolus (i.e. vesicular). But some amoebae have characteristic nucleolar patterns, for example, the nucleolus can be lobed or granular. Staining with the DNA-specific fluorochrome DAPI is useful for observing the configuration of nuclear DNA in amoebae (Rogerson, 1988). Mitotic pattern has been used by some as a diagnostic tool (Raikov, 1982), but this feature is not used in the present key. The diameter of the nucleus and nucleolus should also be noted, and DAPI staining is useful for observing small, or obscured nuclei (Rogerson, 1988). Some authors, notably Page (1983, 1988), have detailed the length:breadth ratio of

many species; however, this character is not used in the present key.

6. *Life-cycle.* Some amoeba have flagellate stages in their life-cyle, a feature characteristic of many members of the Vahlkampfiidae (see chapter on the Heterolobosea). Many amoebae form resistant cysts, and this ability, together with the size and morphology of the cyst, is an important diagnostic feature. *Acanthamoeba*, for example, can be identified solely on the basis of its distinctive cysts.

7. *Ultrastructural features.* Some identifications require the characterization of ultrastructural features and the use of transmission electron microscopy (TEM) to distinguish between genera (eg. *Vannella* from *Platyamoeba*); however, too few genera have been described fully at this level to make this a routine diagnostic feature for identifying amoebae. TEM is also time consuming and impractical unless cultured material is available. Full details on cultivation methods for marine and freshwater/soil amoebae are given in Page (1983, 1988). There are two major ultrastructural features used for identification: i) Mitochondrial cristae. The form of mitochondrial cristae in amoebae can be used to distinguish the class Heterolobosea from the gymnamoebae. The former have discoid or flattened cristae while the gymnamoebae have tubular mitochondrial cristae. ii) Cell coat. Many amoebae have a cell covering termed the **glycocalyx** lying external to the plasmamembrane. This can be thin and barely detectable in some genera or well developed as in the thick amorphous cuticle of *Mayorella*, or the complex cuticle of *Paradermamoeba* (Fig. 2). In some genera, the glycocalyx is differentiated into distinctive surface structures. The most common ones are the flexible glycostyles of *Vannella* with pentagonal symmetry, the hexagonal arrangements of *Platyamoba,* and the more rigid microscales on the surface of *Dactylamoeba.* Other ultrastructural features used in identification include the presence or absence of dictyosomes, cytoplasmic microtubules, centrosomes, filaments, unusual inclusions, and nuclear envelope structure.

The present identification scheme is a synthesis of previously published works. Freshwater and soil amoebae are dealt with in Page (1976, 1988) and Page and Siemensma (1991) and marine amoebae in Schaeffer (1926), Bovee and Sawyer (1979) and Page (1983). The reader should note that several genera of amoebae of uncertain affinity have been grouped in a chapter by Patterson, Simpson, and Rogerson and that many of the Heterolobosea have amoeboid stages.

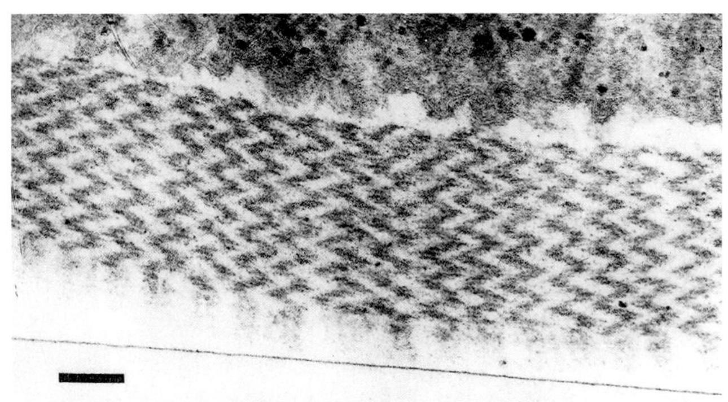

Fig. 2. Complex cuticle structure of *Paradermamoeba valamo.* Surace layer composed of regularly arranged helical glycostyles embedded in an electron-transparent matrix. (courtesy of A. V. Smirnov) Bar=0.1 µm.

Characters of the gymnamoebae

A heterogeneous group of amoebae unified by five features, two of which are discernible by light microscopy; there are no flagellate stages in the cell cycle and locomotion is not usually eruptive. At the ultrastructural level, mitochondria have branched tubular cristae, the cytoplasm usually has dictyosomes and the cell surface is often differentiated into a glycocalyx. Length of amoebae across the class is highly variable, from a few micrometers to 2 mm or more. Pseudopodial types are also variable and several genera produce subpseudopodia from their hyaline edge. Many form cysts but none form spores, which are usually formed in simple or branched chains. Ramicristates are naked amoebae without a test and with characters of the group. Any cell covering (glycocalyx) is flexible and without a discrete aperture.

Characters of the orders of the gymnamoebae

Locomotive form usually monopodial or polypodial with cylindrical pseudopodia. Flattened forms, regular in outline without branched (furcate) subpseudopodia. Long trailing uroidal filaments unusual. Active movement rarely eruptive. Identifiable glycocalyx in most.......... **Euamoebida**

Locomotive amoebae moderately flattened and irregular in outline. Many form cysts with pores closed by operculata. Lamellate centrosomes evident in the cytoplasm at the ultrastructural level... **Centramoebida**

Locomotive form usually flattened and irregular in outline. Amoebae can form fan-shaped unicells, long branches or reticulate sheet-like plasmodia. Active amoebae may develop temporary monopodial limax forms displaying eruptive motion.......**Leptomyxida**

ORDER EUAMOEBIDA Lepsi, 1960

KEY TO THE FAMILIES OF THE ORDER

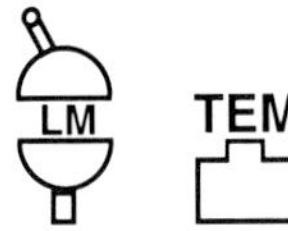

1. Locomotive form with subpseudopodia from anterior edge 2
1'. Locomotive form without subpseudopodia from anterior edge.. 3

2. Subspeudopodia short................ **Paramoebidae**
2'. Subpseudopodia long.................. **Vexilliferidae**

3. Locomotive form markedly compressed......... ....**Vannellidae**
3'. Locomotive form not markedly compressed.......4

4. Locomotive form with obvious surface folds, wrinkles or texture .... **Thecamoebidae**
4'. Locomotive form without obvious surface folds or texture... 5

5. Locomotive form usually monopodial and <80 µm in length .... **Hartmannellidae**
5'. Locomotive form usually polypodial and >80 µm in length... **Amoebidae**

FAMILY PARAMOEBIDAE Poche, 1913

Locomotive form with digitiform, blunt subpseudopodia frequently extending from the anterior hyaloplasm; temporarily lacking in some species. Floating form often with radiating pseudopodia. Uninucleate with central nucleolus. One genus of this family, *Paramoeba*, contains parasomes (DNA-rich intracellular inclusions situated adjacent to cell nucleus). These intracellular bodies can be stained with the DNA specific fluorochrome, DAPI (Rogerson, 1988). Family diagnosis emended by Page (1987b). At the ultrastructural level, two features are of diagnostic importance: a) Surface layer is complex and highly differentiated as a cuticle or as boat-shaped microscales discernable at the TEM level (e.g. Fig. 3) b) Pseudopodia are without a filamentous core (as in Vexilliferidae).

KEY TO GENERA

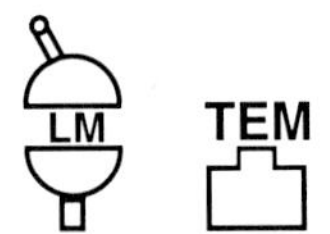

1. Surface layer as microscales discernable at EM level.. 2
1'. Surface layer as cuticle .... ***Mayorella***

2. Cell with parasome(s) .... ***Paramoeba***
2'. Cell without parasome(s) .... ***Dactylamoeba***

Genus *Mayorella* Schaeffer, 1926

Locomotive form with digitiform subpseudopodia extending from anterior hyaloplasm. Wide range of sizes, up to 350 µm in length. Floating form often with long pseudopodia, radiating from an irregular body. Nucleus usually spherical with a central nucleolus. Most mayorellae have prominent cytoplasmic crystals, truncated, bipyramidal, in pairs, or clustered with spheroids. Cell surface of those studied, as a cuticle some 200 nm thick. The genus is generally widespread in freshwater and marine habitats. *Mayorella* and *Dactylamoeba* are morphologically similar and require TEM methods to distinguish the surface cuticle of *Mayorella* from the surface microscales of *Dactylamoeba*. However, *Mayorella* appears to be the commoner of the 2 genera (Page, 1988) and is more likely to be

encountered in field samples. At least 28 species of *Mayorella* have been described, including *M. viridis* which contains numerous cytoplasmic zoochlorellae. Refs. Schaeffer (1926), Page (1983, 1988).
TYPE SPECIES: *Mayorella bigemma* (Schaeffer, 1918)

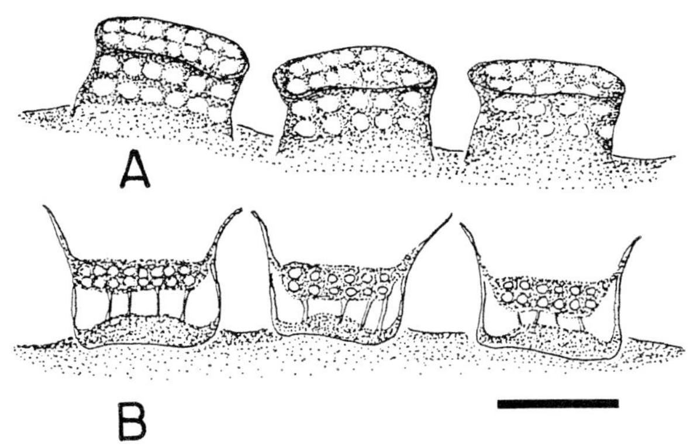

Fig. 3. Diagramatic representation of surface scales of *Dactylamoeba stella* (A) and *D. bulla* (B) (after Page). Bar=0.5 µm.

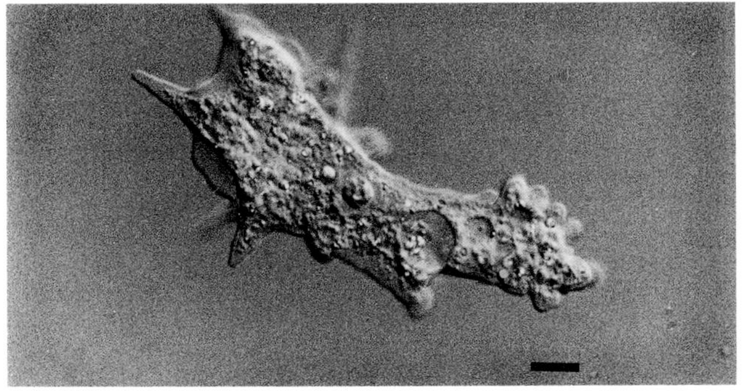

Fig. 4. *Mayorella cantabrigiensis.* Bar=10 µm.

Genus ***Paramoeba*** Schaudinn, 1896

Locomotive form with short blunt digitiform subpseudopdia extending from a narrow anterior hyaline margin. Length of cell commonly from 45 to 100 µm. Floating form with thin radiating pseudopdia. Surface coat with boat-shaped microscales. Always with one or more DNA-rich parasomes (=Nebenkörper) located close to the nucleus. Distinguish from parasome-containing amoebae of the genera *Neoparamoeba* and *Janickina*. Marine, monospecific. Refs. Cann and Page (1982), Grell (1961), Page (1970, 1973).

TYPE SPECIES: *Paramoeba eilhardi* Schaudinn, 1896

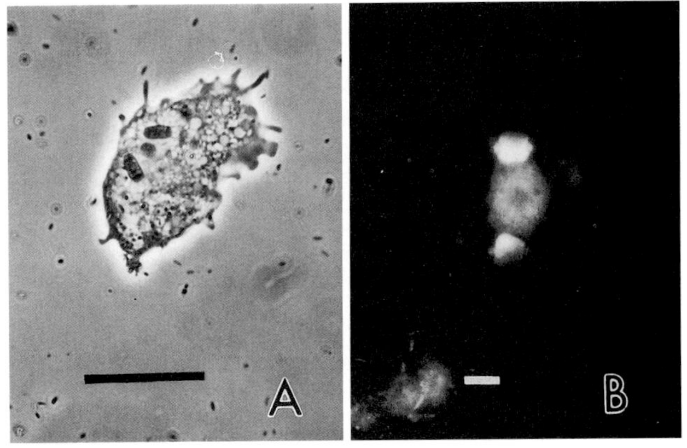

Fig. 5. *Paramoeba eilhardi*, locomotive cell (A). Bar=50 µm. DAPI-stained nucleus with two adjacent parasomes (B). Bar=10 µm.

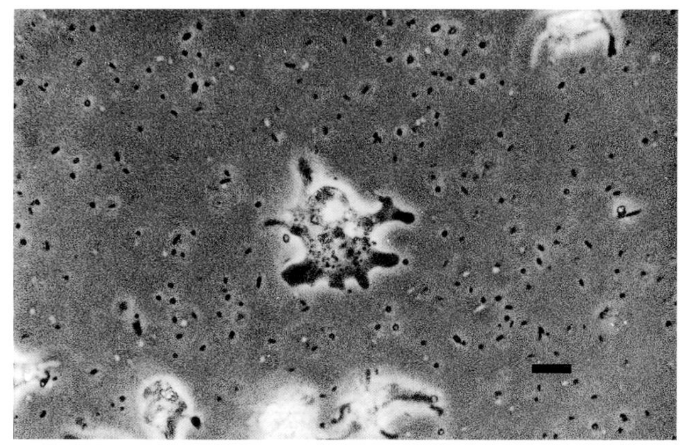

Fig. 6. *Dactylamoeba stella*. Bar=10 µm.

Genus ***Dactylamoeba*** Korotneff, 1880

Length of locomotive form up to 120 µm. Similar in appearance to *Mayorella* but less likely to lack blunt digitiform pseudodopodia extending from the anterior hyaline region. Floating form with slender radiating pseudopodia, frequently bent, extending from a more or less spherical central mass. Nucleus in some with lobed or irregular nucleolar material occasionally giving the impression of a binucleate cell. Cytoplasm without crystalline inclusions. Cell surface covered in complex boat-like microscales, up to 700 nm long, discernible with EM (Fig. 3). Some microscales with prominent terminal spikes. Freshwater with 3, possibly 4 described species if *Astramoeba*

tatianae De la Arene, 1955 is included in the genus (see Page, 1988). Refs. Korotneff (1880), Schaeffer (1926), Pennick and Goodfellow (1975).
TYPE SPECIES: *Dactylamoeba elongata* Korotneff, 1880.

FAMILY VEXILLIFERIDAE Page, 1987

Amoebae with long, slender subpseudopodia (usually more than one) frequently giving a spiny appearance to the cell. Subpseudopodia from hyaline region never branched. Locomotive form usually longer than broad. Floating form frequently with slender, radiating pseudopodia.

frequently with slender, radiating pseudopodia. By electron microscopy, the cell surface of those examined is seen as an amorphous glycocalyx, or more commonly, is differentiated into discrete hexagonal glycostyles. Pseudopodia of those species examined have a filamentous core. Included in this family are several genera for which no ultrastructural information is available. As new information is obtained, some genera may have to be relocated to other familial positions; for now all amoebae in this family share the feature of having one or more markedly slender subpseudopodia.

KEY TO THE GENERA

1. Frequently covered with slender pseudopodia, cell spiny in appearance.................................. 2
1'. Usually with slender pseudopodia at anterior edge only... 3

2. Cytoplasm colorless .... ***Vexillifera***
2'. Cytoplasm green, yellow or brown............... .... ***Dactylosphaerium***

3. Anterior pseudopodia usually less than half cell length... 4
3'. Anterior pseudopodia usually more than half cell length... 5

4. With intracellular parasomes....***Neoparamoeba***
4'. Without parasomes .... ***Pseudoparamoeba***

5. Amoebae with locomotive and floating forms.... 6
5'. Amoebae apparently only with a floating stage .... ***Boveella***

6. Several pseudopodia projecting from anterior edge.. 7
6'. Only one or two leading pseudopodia from anterior edge.. 8

7. Active locomotive form, usually club-shaped .... ***Striolatus***
7' Active locomotive form, usually trapezoidal with pseudopodia from a mound-like base .... ***Triaenamoeba***

8. Leading pseudopodium broadly tapering 9
8'. Leading pseudopodium flagellum-like............... .... ***Podostoma***

9. Leading pseudopodium with very fine tip........... .... ***Subulamoeba***
9'. Leading pseudopodium with blunt tip .... ***Oscillosignum***

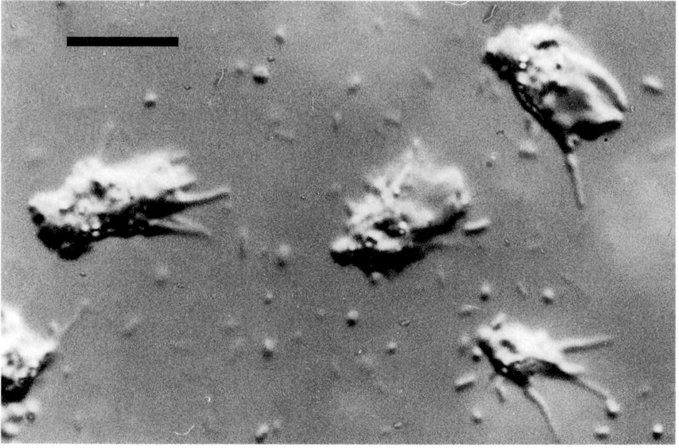

Fig. 7. *Vexillifera bacillipedes*. (courtesy F. C. Page and CCAP). Bar=10 µm.

Genus ***Vexillifera*** Schaeffer, 1926

Locomotive form frequently triangular in outline, often ranging from about 10 to 70 µm in length. Slender subpseudopodia, never branching, extend well beyond the anterior edge. In *Vexillifera*, the pseudopodia are sometimes carried posteriorly

giving the amoeba an overall spiny appearance. Cytoplasm sometimes containing a few crystals. Uninucleate, usually with a central nucleolus. No cysts reported. At the ultrastructural level, cell surface usually covered in hexagonal glycostyles. This surface structure distinguishes *Vexillifera* from amoebae in the morphologically similar family Paramoebidae. At least 16 named species from freshwater and marine habitats. Refs. Bovee and Sawyer (1979), Page (1976).
Type species: *Vexillifera lemani* Page, 1976.

Genus ***Dactylosphaerium*** Hertwig & Lesser, 1874

Shape of locomotive form variable, but cells usually ovoidal with numerous tapering subpseudopodia. Most of these are conical and extend from the hyaline edge, but they can also radiate out from surface of cell. Single spherical nucleus. Granular cytoplasm without cytoplasmic crystals. No posterior uroid. Cell very slow moving. Reported from both freshwater and marine muddy sediments. Presumably feeds on algae since the cytoplasm of the type species contained yellow/green/brown bodies thought to represent partly digested food particles.
TYPE SPECIES: *Dactylosphaerium polypodium* Schultze, 1854.

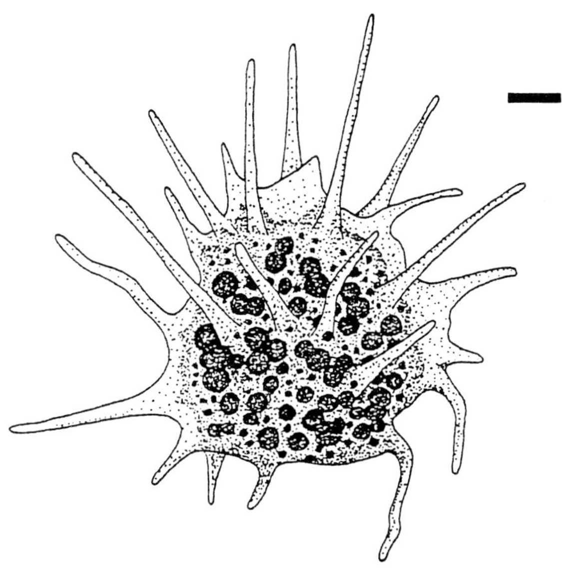

Fig. 8. *Dactylosphaerium acuum* during active locomotion. (after Schaeffer). Bar=10 μm.

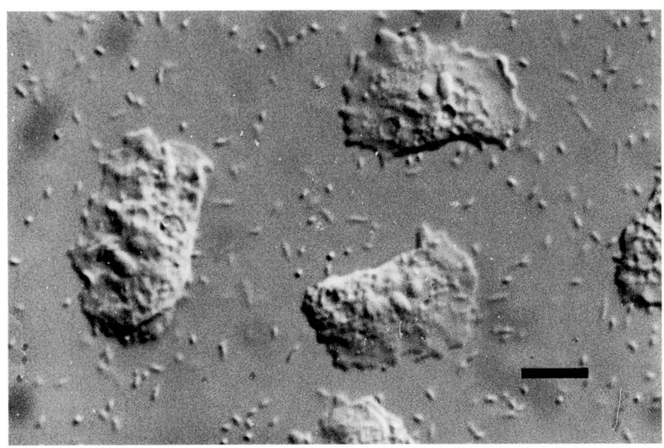

Fig. 9. *Neoparamoeba pemaquidensis*. Bar=10 μm.

Genus ***Neoparamoeba*** Page, 1987

Locomotive form usually longer than broad, often between 10 to 25 μm in length. Many have longitudinal ridges which may give rise to anterior, blunt, conical subpseudopodia. Floating form with radiating pseudopodia. Nucleus spherical with 1, 2, or 3 adjacent parasomes; supernumerary parasomes in some cells. Temporary morulate uroid. At ultrastructural level, surface coat as dense, amorphous glycocalyx, about 10 nm thick. Common in marine samples. One species found as a facultative parasite of sea urchins and another as a crustacean parasite. Refs. Page (1970, 1983, 1987b).
TYPE SPECIES: *Neoparamoeba pemaquidensis* (Page, 1970)

Genus ***Pseudoparamoeba*** Page, 1979

Locomotive form broad with an anterior hyaloplasmic region with a few, short, conical subpseudopodia. Some may elongate to long, narrow pseudopodia. Morphologically similar to *Neoparamoeba* but without intracellular parasomes. Floating form with a spherical central mass, frequently with 3-11 radiating pseudopodia tapering to fine points. At ultrastructural level, surface coat differentiated into blister-like structures with hexagonal bases. Monspecific, marine habitat. Ref. Page (1979).

TYPE SPECIES: *Pseudoparamoeba pagei* (Sawyer, 1975)

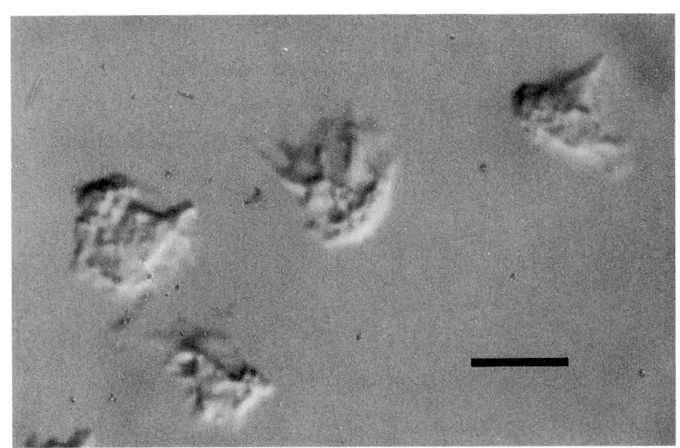

Fig. 10. *Pseudoparamoeba pagei* (courtesy F. C. Page and CCAP). Bar=10 μm.

Genus ***Boveella*** Sawyer, 1975

Floating form of the cell body spherical, about 20 μm across. Radiating pseudopodia, clear and conical, extending up to 50 μm. Pseudopodia retract quickly in the presence of light. Locomotive form of this genus unknown. Cytoplasm without crystals. Cell covered in organic detritus which obscures intracellular detail. Form of nucleus unknown. Marine, monospecific.
TYPE SPECIES: *Boveella obscura* Sawyer, 1975.

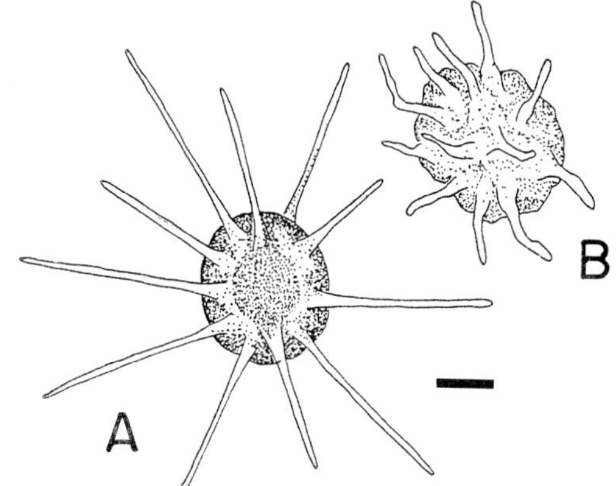

Fig. 11. *Boveella obscura*, radiate form (A) and contracted form (B). (after Sawyer). Bar=10 μm.

Genus ***Striolatus*** Schaeffer, 1926

Locomotive amoeba up to 60 μm in length; contracted form around 35 μm. Shape during locomotion mainly club-shaped with an occasional broad pseudopodium. The amoeba is very slow moving even in this 'active' phase taking several minutes to move one body-length. Shape at other times discoid with numerous long, slender pseuodopodia. Floating form with radiating pseudopodia. Cytoplasm filled with refractile granules and several vesicles. No posterior uroid or cysts formed. Mononucleate, or more commonly, binucleate. Monospecific, described from shallow coastal waters. Refs. Schaeffer (1926), Bovee and Sawyer (1979).
TYPE SPECIES: *Striolatus tardus* Schaeffer, 1926

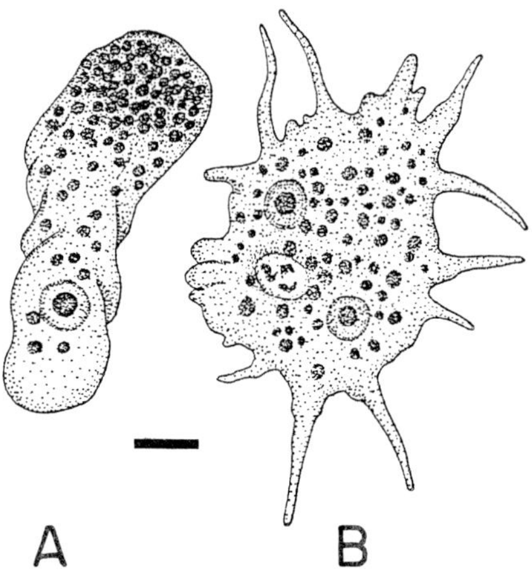

Fig. 12. *Striolatus tardus*, active form (A) and flattened form with conical subpseudopodia (B) (after Schaeffer). Bar=10 μm.

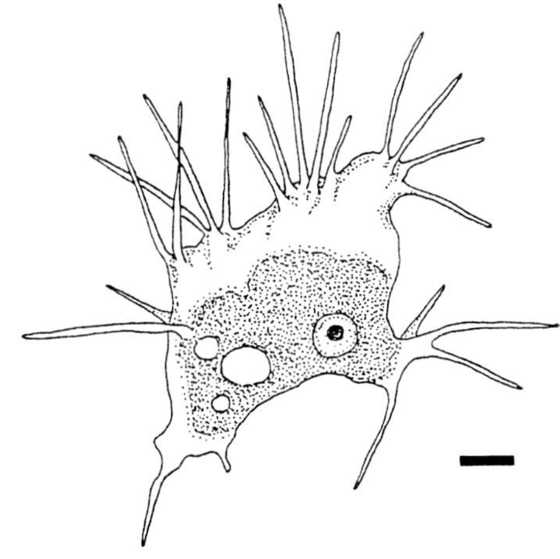

Fig. 13. *Triaenameba jackowskii*, locomotive form (after Sawyer). Bar=10 μm.

Genus ***Triaenamoeba*** Bovee, 1970.

Locomotive form generally trapezoidal with long, slender anterior subpseudopodia usually in groups of 2 or more produced from a common mound-like base. Length, including pseudodopodia, up to 60 µm although cell body usually only around 17 µm long. Floating form irregularly round with radiating pseudopodia. Spherical nucleus with central nucleolus. No uroid but cells may have a few temporary trailing filaments. Cytoplasm without crystals and often with a few vesicles. Two described species from coastal regions. Ref. Sawyer (1975c).

Genus ***Podostoma*** Claparède & Lachmann, 1858

Locomotive form usually between 30 and 60 µm not including the long, thin, active pseudopodium (40 to 50 µm) which can be supported by short conical pseudopodia. The long vibratile pseuodopodium extending from the body surface resembles (but is not) a flagellum. It can be extended and retracted rapidly and the cell can change form rapidly. With a temporary uroid. Granular cytoplasm with irregular crystals, a spherical nucleus and a contractile vacuole. Found in freshwater. Same as the genus *Flagellipodium* described by Bovee (1953). Ref. Claparède and Lachmann (1858/59). TYPE SPECIES: *P. filigerum* Claparède & Lachmann, 1858.

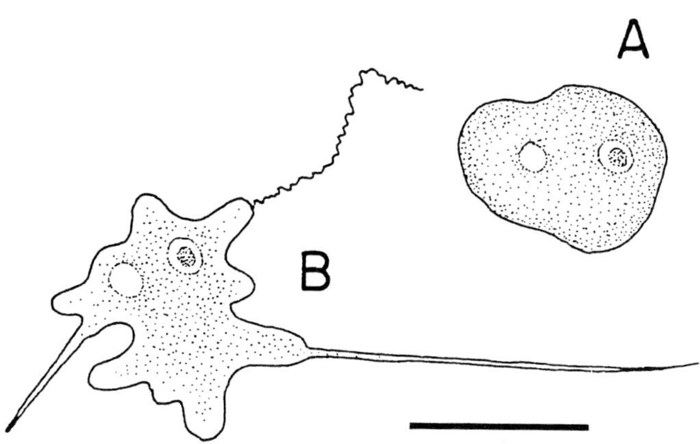

Fig. 14. *Podostoma filigerum*, stationary form (A) and active form with flagellum-like, prehensile pseudopodium (B) (after Claparède and Lachmann). Bar=50 µm.

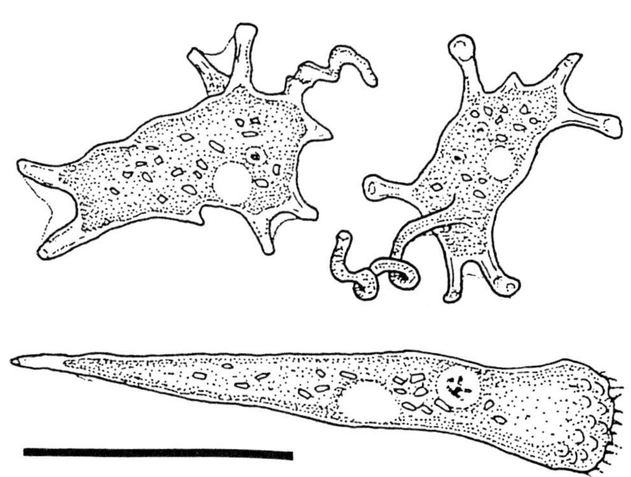

Fig. 15. *Subulamoeba saphirina* (after Bovee). Bar=50 µm.

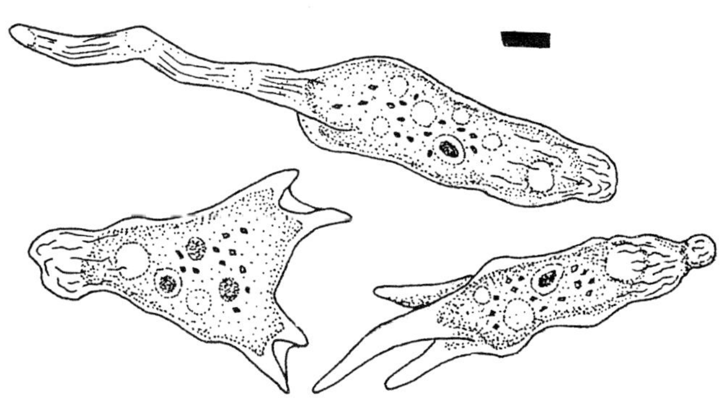

Fig. 16. *Oscillosignum proboscidium* (after Bovee). Bar=10 µm.

Genus ***Subulamoeba*** Bovee, 1953

Morphologically similar to *Oscillosignum* but with more pointed pseudopodia. Locomotive form normally between 30 and 70 µm. Pseuodopodia, usually paired, extending from cell periphery; sometimes these pseudopodia have flattened tips. During rapid locomotion the cell becomes awl-shaped, and longer, with one broad-based, active, conical pseudopodium. In this configuration, the posterior of the cell is wider than the anterior portion. No posterior uroid. Cytoplasm granular with a few crystals. Nucleus spherical with small endosomal granules. Rarely binucleate. Fresh-water. Ref. Page (1976).

TYPE SPECIES: *Subulamoeba saphirina* (Penard, 1902)

Genus *Oscillosignum* Bovee, 1953

Locomotive form with a long tapered pseudopodium which actively waves around. This leading pseudopodium frequently attaches to the substrate permitting the body of the cell to flow into it to form a flattened cell. These flattened forms have short pseudopodia which are paired and web-connected. Tapered pseudopodia granular except at tips. Floating form, spherical with blunt radiating pseudopodia. Cytoplasm with a single nucleus, contractile vacuole, and crystals. Four species described from fresh-water. Refs. Bovee (1970), Page (1976).
TYPE SPECIES: *Oscillosignum proboscidium* Bovee, 1953

FAMILY VANNELLIDAE Bovee, 1970

Amoebae have locomotive forms markedly flattened, fan-shaped, or linguiform with a prominant anterior hyaline zone without subpseudopodia. No posterior uroid. Floating form of most with radiating pseudopodia, either blunt or tapered. Nucleus of most with prominant central nucleolus, rarely with parietal lobes. Cytoplasmic crystals rare. Glycocalyx of the only two genera studied by TEM differentiated into either pentagonal glycostyles (*Vannella*) or hexagonal arrangements (*Platyamoeba*). Family diagnosis emended by Page (1987b) on the basis of this ultrastructural information. The present grouping of genera is based on morphological similarities at the LM level since ultrastructural details are unavailable for most.

KEY TO GENERA OF THE FAMILY

1. Hyaloplasm smooth.. 2
1'. Hyaloplasm with surface features 3

2. Hyaloplasm extends posteriorly as border .... ***Discamoeba***
2'. Hyaloplasm predominantly anterior....***Vannella***

3. Surface features faint 4
3'. Surface features obvious 5

4. Usually with faint longitudinal edge ridges .... ***Platyamoeba***
4'. Usually with faint lateral folds....... ***Clydonella***

5. Hyaloplasm with series of superimposed waves .... ***Unda***
5'. Hyaloplasm with conoidal bosses......***Pessonella***

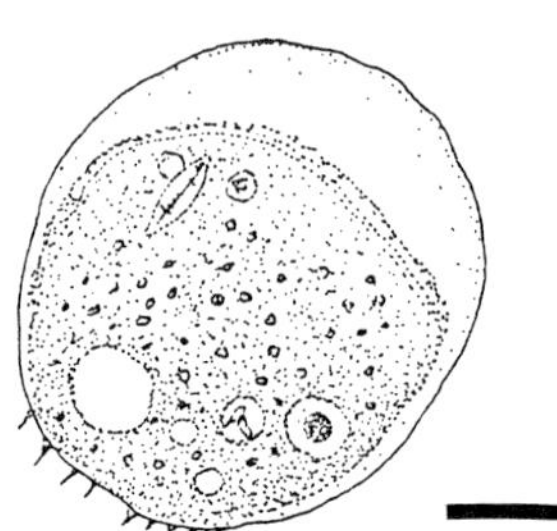

Fig. 17. *Discamoeba guttula* (after Bovee). Bar=10 µm

Genus *Discamoeba*
Jahn, Bovee & Griffith, 1979

Length and width of locomotive form similar. Clear zone extends around the endoplasmic mass. No posterior uroid. Cytoplasm contains a spherical nucleus and frequently a contractile vacuole and crystals. Floating form without radiating pseudopodia. Three species described from freshwater. Ref. Jahn et al. (1979).
TYPE SPECIES: *Discamoeba guttula* (Dujardin, 1835)

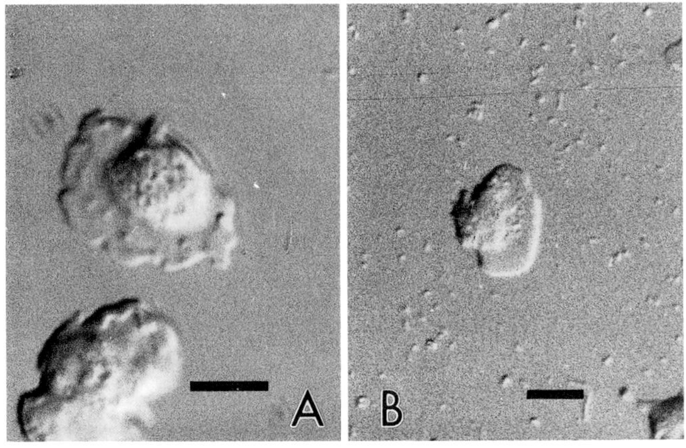

Fig. 18. *Vannella anglica* (A) and *V. denonica* (B). Bars=10 µm.

Genus ***Vannella*** Bovee, 1965

Locomotive form with a flattened anterior hyaloplasm occupying up to half the length of the cell. Breadth usually greater than length making the cell commonly fan-shaped. Some species less than 10 μm in length, others up to 75 μm long. All freshwater and most marine species have floating forms with tapering pseudopodia. Nucleus with central nucleolus. All certain species without cysts. At the ultrastructural level, the cell coat of those species examined has pentagonal glycostyles and simple filaments. Common in both freshwater and marine habitats with at least 15 well-documented species. Refs: Schaeffer (1926), Bovee (1965), Page (1987b).
TYPE SPECIES: *Vanella mira* (Schaeffer, 1926)

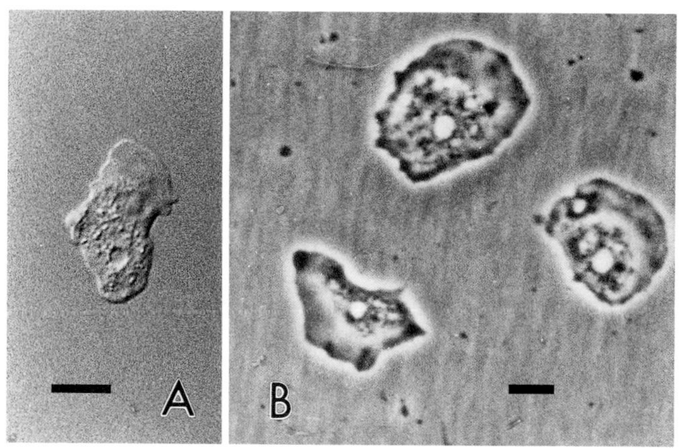

Fig. 19. *Platyamoeba placida* (A) and *P. calycinucleolus* (B). Bars=10 μm.

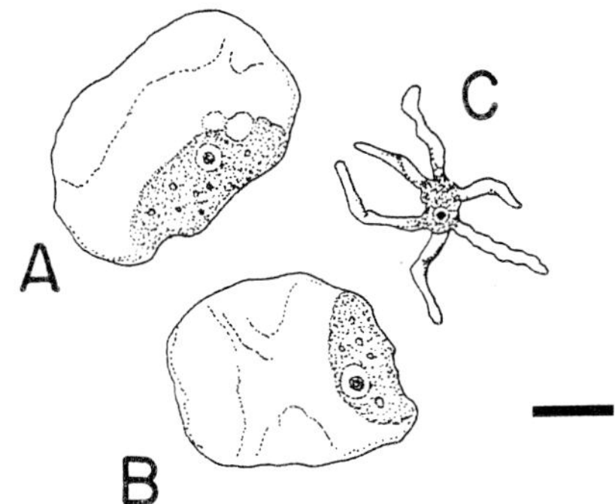

Fig. 20. *Clydonella sindermanni*, locomotive forms (A and B), floating form (C) (after Sawyer). Bar=10 μm.

Genus ***Platyamoeba*** Page, 1969

Locomotive form generally oval or linguiform but occasionally fan-shaped with extensive hyaline zone; some with faint surface folds. Length usually from 10 to 35 μm. Nucleus spherical usually with a central nucleolus but occasionally cup-shaped or parietal. Most platyamoebae have floating forms with blunt radiating pseudopodia. Cell coat of those studied by TEM as fuzzy glycocalyx with hexagonal arrangements of filaments (not glycostyles as in *Vannella*). Common in freshwater and marine habitats with at least 13 named species. Most described species are marine. Only freshwater and soil forms produce cysts. Page (1983) considers this genus to be synonymous with *Lingulamoeba* Sawyer, 1975. Refs. Page (1983, 1988), Sawyer (1975a).
TYPE SPECIES: *Platyamoeba placida* (Page, 1968)

Genus ***Clydonella*** Sawyer, 1975

Characters of the genus are intermediate between *Vannella* and *Platyamoeba*. Locomotive form with anterior hyaline region occupying 2/3 of the cell length. Hyaloplasm with faint lateral folds as ripples over anterior region. Locomotive form ovoid, usually 10 to 40 μm. No posterior uroid. Floating form with radiate, sometimes irregularly twisted pseudopodia; in some, radiating pseudopodia bent or extended in multiples of 2. Nucleus distinct with central nucleolus. Cytoplasm containing few vesicles. No cytoplasmic crystals or cysts. Fine structure of cell coat unknown. All 4 species marine. Refs. Schaeffer (1926), Sawyer (1975b), Bovee and Sawyer (1979).
TYPE SPECIES: *Clydonella vivax* (Schaeffer, 1926)

Genus ***Unda*** Schaeffer, 1926

Locomotive form as a flattened ovoid with length less than breadth. Length between 20 and 60 μm. Hyaloplasm characteristically advances in waves at anterior margin. No posterior uroid. Floating form with short cylindrical pseudopodia. Nucleus spherical. Cytoplasm often with small vesicles and conspicuous greenish or yellowish granules. No cysts known. Fine structure of surface coat

unknown. Marine, with 3 described species. Ref. Schaeffer (1926).
TYPE SPECIES: *Unda maris* Schaeffer, 1926.

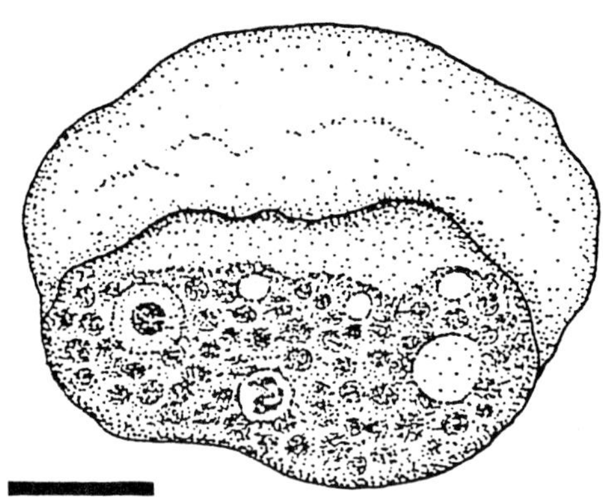

Fig. 21. *Unda maris* during rapid locomotion with new wave of cytoplasm beginning to roll out of granular ectoplasm (after Schaeffer). Bar=10 μm.

Genus ***Pessonella*** Pussard, 1973

Locomotive form flattened and broadly lingulate or fan-shaped. Anterior margin clear with many cone-like bosses on hyaloplasm. Floating form with blunt, cylindrical, and frequently bent pseudopodia. No posterior uroid. Cytoplasm finely granular without crystals. Nucleus spherical with 3 to 5 parietal nucleolar pieces. Reported from freshwater and compost. Monospecific. Ref. Pussard (1973).
TYPE SPECIES: *Pessonella marginata* Pussard, 1973

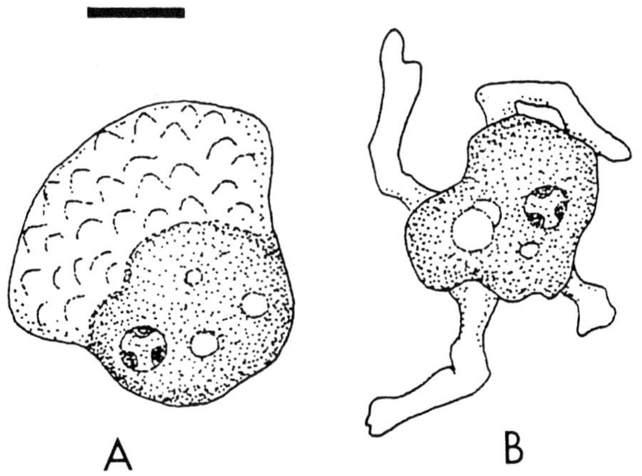

Fig. 22. *Pessonella marginata*, locomotive form (A) and floating form (B) (after Bovee). Bar=10 μm.

FAMILY THECAMOEBIDAE Schaeffer, 1926

Locomotive amoebae usually oblong with an anteriolateral crescent. Cells with a pellicle-like layer, or skin, which can be smooth, wrinkled, or ridged. Cell coat thick usually as an amorphous glycocalyx but can be as a cuticle or filamentous layer. Most uninucleate but one genus binucleate and one multinucleate. Most showing a diversity of nuclear configurations. Description based on the most recent statement of Page (1987a).

KEY TO GENERA

1. Less than 10 μm in length .... ***Parvamoeba***
1'. Greater than 10 μm in length.......................... 2

2. Uninucleate 3
2'. Multinucleate 4

3. Usually oval in outline during locomotion........ 5
3'. Often elongate in outline during locomotion...... 6

4. With numerous nuclei (100 to 200) .... ***Thecochaos***
4'. With one to four pairs of nuclei.......... ***Sappinia***

5. Usually with surface folds or wrinkles............ .... ***Thecamoeba***
5'. Usually with surface smooth ...... ***Dermamoeba***

6. Anterior hyaline cap without lateral extension .... ***Pseudothecamoeba***
6'. Anterior hyaline cap with lateral extension .... ***Paradermamoeba***

Genus ***Parvamoeba*** Rogerson, 1993

Oval in outline and never longer than 6.0 μm in length. Raised cell mass with random wrinkles and folds. Locomotion slow via extension of lobed hyaline pseudopodium, occasionally with fine radiating subpseudopodia. Floating form with finger-like pseudopodia. No cytoplasmic crystals. At the ultrastructural level, surface with an

amorphous glycocalyx some 30 nm across. Single nucleus with central nucleolus and *Thecamoeba*-like electron-dense layer adjacent to nuclear membrane. Marine. Monospecific. Ref. Rogerson (1993).
TYPE SPECIES: *Parvamoeba rugata* Rogerson, 1993

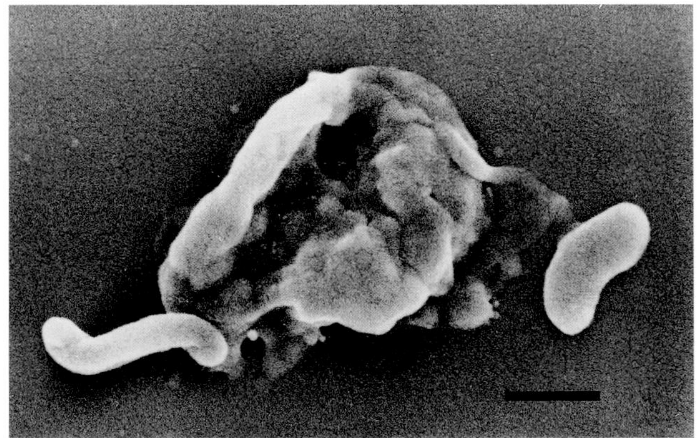

Fig. 23. *Parvamoeba rugata* adjacent to two bacterial cells. S.E.M. Bar=1 µm.

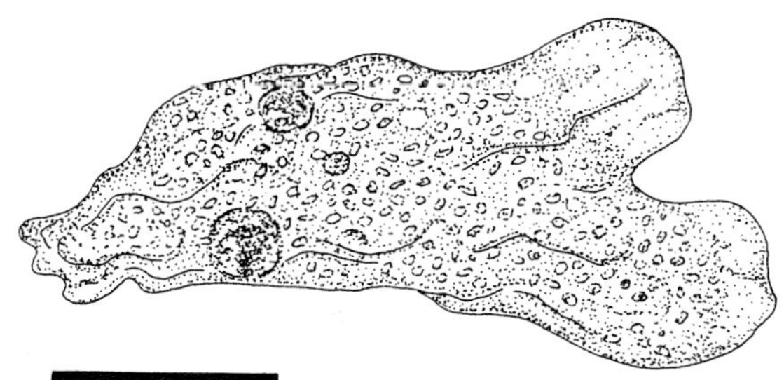

Fig. 24. *Thecochaos fibrillosum* (after Penard). Bar=100 µm.

Genus ***Thecochaos*** Page, 1981

Locomotive form irregular, but more or less oval between 160 and 360 µm in length. Branched only when changing direction. Surface with wrinkles. Anterior hyaloplasm as a crescentic cap. Numerous nuclei, around 100 to 200 per cell, each with a central nucleolus. Two species described from freshwater. Ref. Page (1981a).
TYPE SPECIES: *Thecochaos fibrillosum* (Greef, 1891)

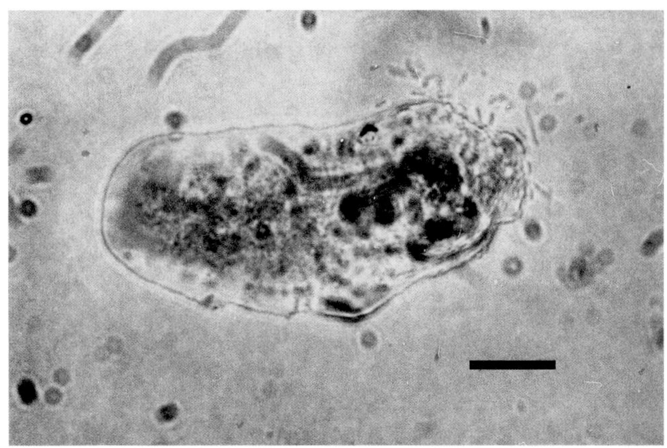

Fig. 25. *Sappinia diploidea* (courtesy F. C. Page and CCAP). Bar=10 µm

Genus ***Sappinia*** Dangeard, 1896

Locomotive form resembles a smooth *Thecamoeba*, some 63 µm in length. Tendency for cells to have 1 to 4 pairs of nuclei. Mature cysts uninucleate or binucleate; life cycle of this amoeba thought to involve sexual processes in the cyst phase. One species reportedly has stalked cysts or resting stages. At the ultrastructural level, surface coat as a thick glycocalyx. Widely distributed in freshwater, soil, and manure. Ref. Goodfellow et al. (1974).
TYPE SPECIES: *Sappinia pedata* Dangeard, 1896

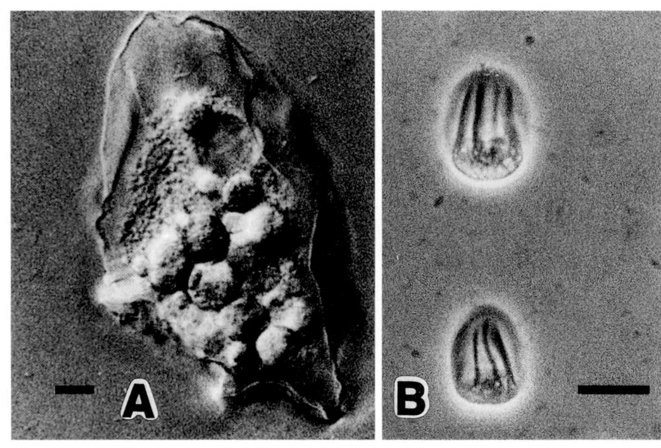

Fig. 26. *Thecamoeba sphaeronucleolus* (A) with surface wrinkles and *T. orbis* (B) with surface parallel folds. Bars=10 µm.

Genus ***Thecamoeba*** Fromentel, 1874

Locomotive form, usually between 30 and 350 µm long, flattened and oblong in outline, often with several parallel folds or wrinkles. Anterior edge semi-circular with an extensive hyaloplasm that can occupy more than half the cell. Posterior edge often slightly convex, sometimes with a knob uroid. Form of the nucleus is an important diagnostic feature for delineating species, and thecamoebae can have a single spherical or ovoidal nucleus with a central nucleolus, granular nucleolus, or nucleolus as 2 or 3 parietal pieces. Cytoplasm may have crystals and contractile vacuoles. No cysts known. At the ultrastructural level, nuclear envelope without an internal fibrous lamina and cell coat as a thick, dense, amorphous, glycocalyx. Common genus with 18 species described from freshwater, terrestrial, and marine habitats. Refs. Schaeffer (1926), Page (1983).
TYPE SPECIES: *Thecamoeba quadripartita* Fromentel, 1874

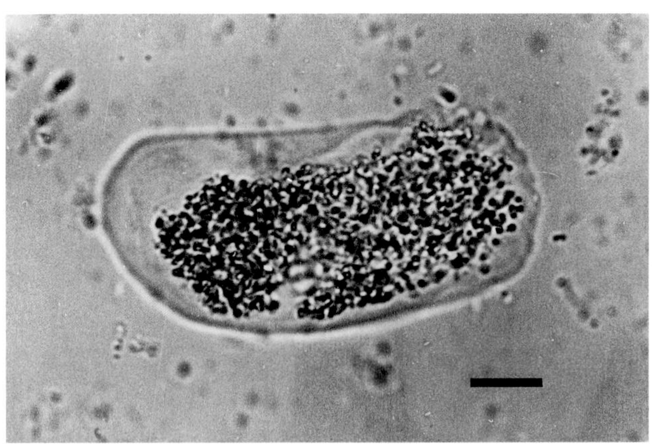

Fig. 27. *Dermamoeba granifera*. Bar=10 µm.

Genus ***Dermamoeba*** Page & Blakey, 1979

Locomotive form oval or elongate, up to 85 µm in length, with frequent narrowing of the anterior end. Extensive hyaline zone, often extending posteriorly along the margins. Surface of cell smooth, but, when present, morulate uroid gives rise to surface folds at the posterior. Nucleus frequently with a pair of central bodies. Cytoplasm can have yellow/brown ingesta; feeds on fungi and other microbes in the soil. Cysts in at least one of the 2 described species. Refs. Page (1976, 1983, 1988).
TYPE SPECIES: *Dermamoeba granifera* (Greef, 1966)

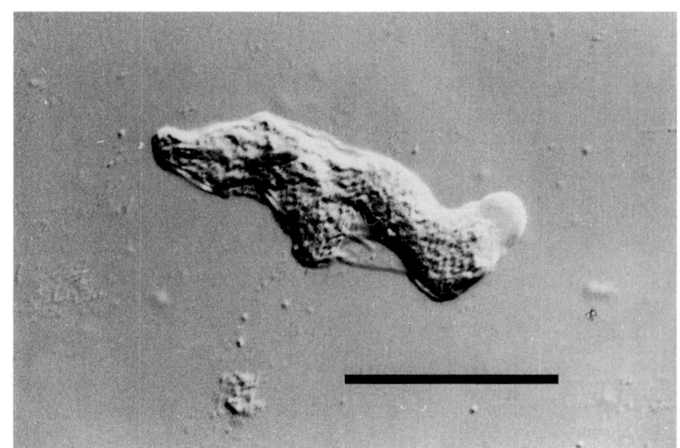

Fig. 28. *Pseudothecamoeba proteoides*. Bar=100 µm.

Genus ***Pseudothecamoeba*** Page, 1988

Locomotive form usually monopodial, as a flattened cylinder, about 170 µm in length but sometimes branched or polypodial. Surface with prominent wrinkles or folds. Cytoplasm without crystals. Nucleus elongate with a granular appearance due to numerous nucleolar fragments. At the ultrastructural level, the nucleus has an inner lamina, and the surface coat is composed of filaments (80 nm) extending from the plasma membrane. No cysts. Habitat: freshwater, Europe. Monospecific. Ref. Page (1988).
TYPE SPECIES: *Pseudothecamoeba proteoides* (Page, 1976)

Genus ***Paradermamoeba***
Smirnov & Goodkov, 1993

Locomotive form flattened and oblong to elongate, between about 50 to 100 µm in length. Distinct anterior hyaloplasm extending as flattened lateral extension towards posterior. Surface of cell without dorsal folds or wrinkles. At the ultrastructural level, unique glycocalyx (513 nm thick) composed of closely packed glycostyles (Fig. 2). Viewed longitudinally, glycocalyx appears as

spiral springs; in transverse sections the tips are as hollow pentagonal structures. Nucleus with prominant central nucleolus. Single species from freshwater and bottom sediments. Refs. Smirnov and Goodkov (1993, 1994).
TYPE SPECIES: *Paradermamoeba valamo* Smirnov & Goodkov, 1993

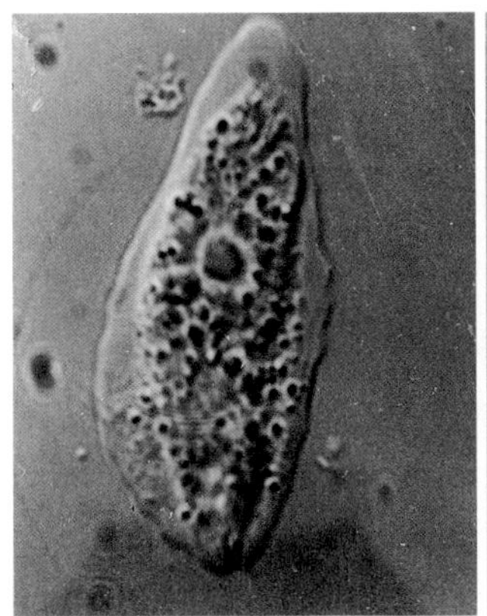
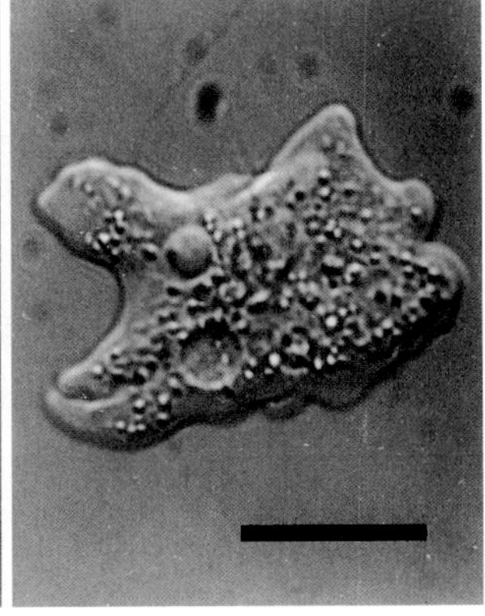

Fig. 29. *Paradermamoeba valamo* (courtesy of A. V. Smirnov). Bar=50 µm.

FAMILY HARTMANNELLIDAE Volkonsky, 1931

Uninucleate amoebae, monopodial with a cylindrical pseudopodium when active. Most floating forms without radiating pseudopodia. Nucleus with a prominant central nucleolus.

KEY TO GENERA OF THE FAMILY

1. Distinctive hyaline zone in locomotive form .... ***Hartmannella***
1'. Hyaline zone in locomotive form reduced or absent 2

2. Uroid, if present, reduced.................................3
2'. Uroid, often bulbous and villous...................... 4

3. Steady locomotion.................................... ***Cashia***
3'. Some eruptive activity.................... ***Nolandella***

4. Cytoplasm often with prominent crystals or inclusions .... ***Saccamoeba***
4'. Cytoplasm without crystals............... ***Glaeseria***

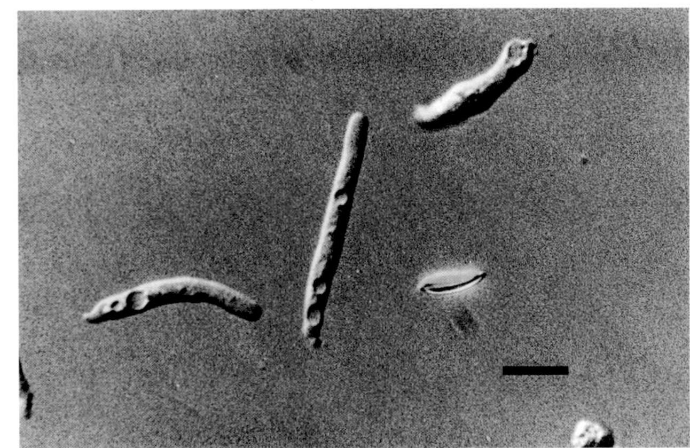

Fig. 30. *Hartmannella vermiformis*. Bar=10 µm.

Genus ***Hartmannella*** Alexeieff, 1912

Locomotive form monopodial with a prominent anterior hyaline zone. Locomotion steady, non-eruptive. Cytoplasm without prominent cytoplasmic crystals but frequently with many small crystals. Often with a contractile vacuole. Mitochondria, of those species examined, sometimes elongate. Forms round, or slightly oval, smooth bilaminar cysts. Glycocalyx thin, some with cup-like surface structures, around 12.5 nm in diameter. Habitat: widely distributed in freshwater, also marine species. Distinguish from other limax amoebae of the class Heterolobosea, and families Leptomyxidae and Amoebidae by comparing features of the groupings. Many inadequately described species in this genus including type species. Refs. Page (1986), Singh (1952).
TYPE SPECIES: *Hartmannella hyalina* (Dangeard, 1900)

Genus ***Cashia*** Page, 1974

Locomotive form monopodial with a wide, but reduced, anterior hyaline zone. Posterior of cell smooth or with a small uroid, never with a villous knob. Nucleus spherical with a central nucleolus. Mitochondria with helical cristae. No cytoplasmic crystals. No cysts. At the ultrastructural level, surface with cup-like structures some 30 nm in diameter. Habitat: freshwater, N. America and Europe. Refs. Page (1988), Page and Siemensma (1991).
TYPE SPECIES: *Cashia limacoides* (Page, 1967)

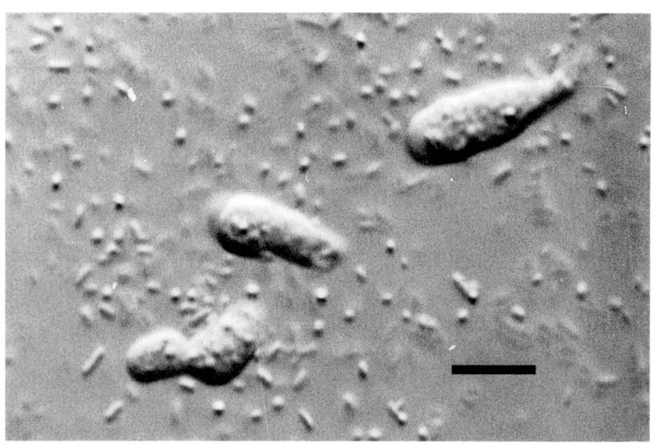

Fig. 31. *Cashia limacoides* (courtesy of F. C. Page and CCAP). Bar=10 μm.

Genus ***Nolandella*** Page, 1983

Locomotive form limax-like, usually around 12 μm long, displaying frequent changes in direction. Unusual among the hartmannellids in that some cells show eruptive activity. At the ultrastructural level, surface coat with hexagonal elements extending 30 nm above plasma membrane. Floating form with blunt radiating pseudopodia. Single marine species. Refs. Page (1980, 1983).
TYPE SPECIES: *Nolandella hibernica* (Page, 1980)

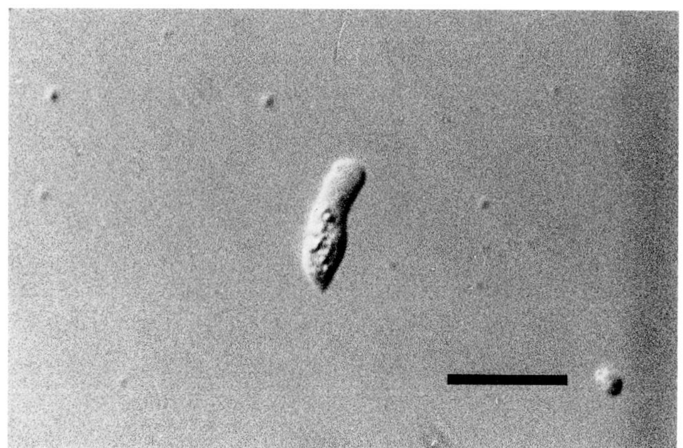

Fig. 32. *Nolandella hibernica*. Bar=10 μm.

Genus ***Saccamoeba*** Frenzel, 1892

Locomotive form active but not eruptive. Genus contains the largest of the hartmannellids, up to 175 μm in length, and shares features with members of the Amoebidae. For example, many species have cytoplasmic crystals and a villous or papillate knob uroid; however, floating form has irregularly rounded pseudopodia rather than the radiating pseudopodia found in most Amoebidae. Mitochondria often elongate. Frequently with bulging posterior contractile vacuole(s). Nucleus with prominent central nucleolus. Cysts reported in some species. Where studied, cell surface at ultrastructural level with cup-like structures. Seven species described from both freshwater and marine habitats. Refs. Bovee (1972), Page (1974).
TYPE SPECIES: *Saccamoeba lucens* Frenzel, 1892

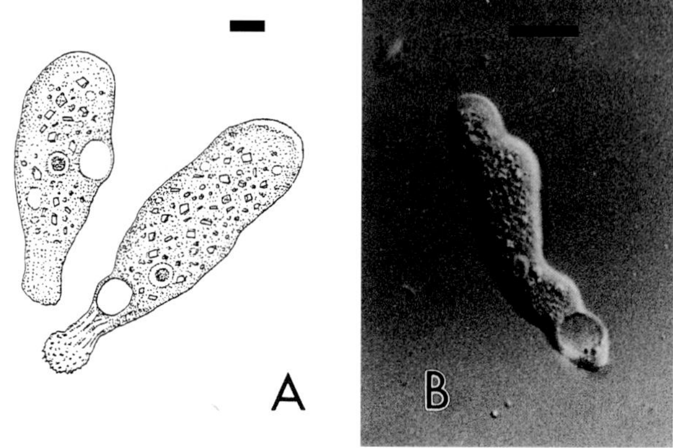

Fig. 33. *Saccamoeba lucens* (A, after Bovee) and *Saccamoeba stagnicola* (B). Bars=10 μm.

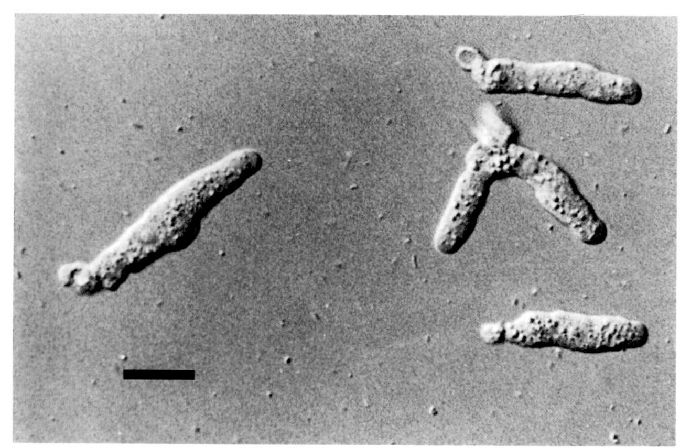

Fig. 34. *Glaeseria mira*. Bar=10 μm.

Genus ***Glaeseria*** Volkonsky, 1931

Locomotive form thin and up to 70 µm in length with a tendency to temporarily branch. Pseudopodia frequently with a thin hyaline cap. No cytoplasmic crystals. Posterior contractile vacuole(s) and knob-like uroid sometimes with a few trailing filaments. Cysts of free-living species with smooth inner wall and sticky outer coating, uninucleate to trinucleate. Distinguish from *Hartmannella* on the basis of ultrastructural features: larger (30 nm) surface cup-like structures and no elongate mitochondria. Freshwater and the intestine of *Testudo*. Ref. Page (1974).
TYPE SPECIES: *Glaeseria mira* (Glaser, 1912)

FAMILY AMOEBIDAE Ehrenberg, 1838

Locomotive amoebae usually greater than 80 µm in length, with cylindrical pseudopodia with hyaline caps. Some species monopodial but most polypodial. Many have cytoplasmic crystals. Nucleus often with nuclear material in small fragments.

KEY TO GENERA OF THE FAMILY AMOEBIDAE

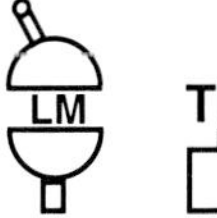

1. Amoebae always multinucleate.......................... 2
1'. Amoebae always uninucleate.............................. 3

2. Active locomotive form usually greater than 500 µm, cell without endosymbiotic zoochlorellae .... ***Chaos***
2'. Active cell usually less than 500 µm, frequently contains endosymbiotic zoochlorella. .... ***Parachaos***

3. Locomotive form usually monopodial................ 4
3'. Locomotive form usually polypodial................. 5

4. Amoeba free-living .... ***Trichamoeba***
4'. Amoeba parasitic on freshwater coelenterates .... ***Hydramoeba***

5. Locomotive amoeba with equal-sized pseudopodia radiating out from the cell....... ***Polychaos***
5'. Locomotive amoeba with one leading pseudopodium 6

6. Amoeba usually less than 200 µm in length .... ***Deuteramoeba***
6'. Amoeba usually more than 200 µm in length....7

7. With longitudinal pseudopodial ridges...***Amoeba***
7'. Without longitudinal pseudopodial ridges......... .... ***Metachaos***

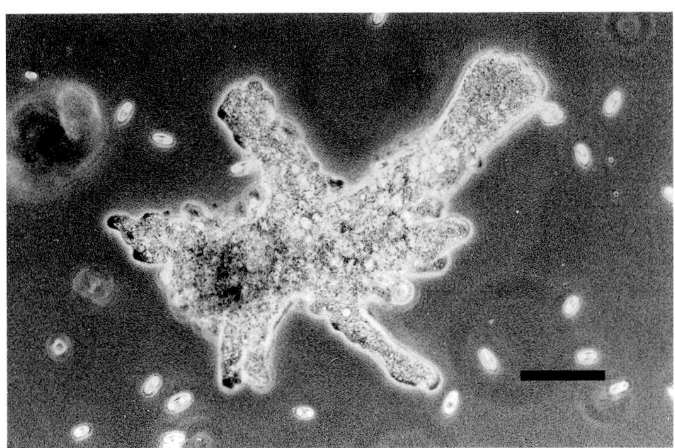

Fig. 35. *Chaos nobile*. Bar=100 µm.

Genus ***Chaos*** Linneaus, 1767

Locomotive amoebae polypodial or monopodial commonly between 1000 and 2000 µm in length, but up to 5000 µm. Pseudopodia cylindrical, frequently with ridges. Morulate uroid. Multinucleate with up to 1000 or more nuclei per cell. Nuclei biconvex or ellipsoidal discs, around 27 µm in length. Cytoplasm with crystals often bipyramidal. Inhabits swampy freshwater locales. Refs. Schaeffer (1926), Jeon (1973), Gromov (1986).
TYPE SPECIES: *Chaos chaos* (Linneaus, 1758)

Genus ***Parachaos***
Willumsen, Siemensma & Suhr-Jessen, 1987

Locomotive form commonly polypodial, 200 to 500 µm, or monopodial up to 900 µm in length. Multinucleate with 8 to 50 spherical, or slightly oval, nuclei per cell. Nucleolar material parietal as single layer or irregular fragments. Cytoplasm rich in bipyramidal or plate-like crystals and zoochlorellae in many isolates. Usually with

villous-bulb uroid. Similar to *Chaos* but at the ultrastructural level *Parachaos* can be distinguished by an amorphous cell coat without filaments and nuclei with no internal laminae. Freshwater. Refs. Willumsen (1982), Willumsen et al. (1987).
TYPE SPECIES: *Parachaos zoochlorellae* (Willumsen, 1982)

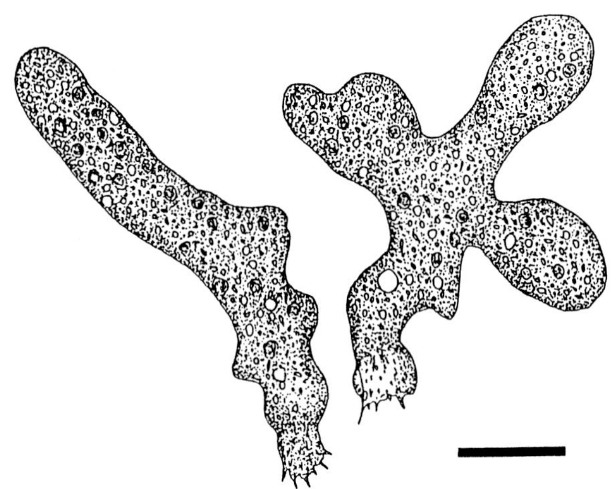

Fig. 36. *Parachaos zoochlorellae*, monopodial and polypodial forms (after Willumsen et al.). Bar=100 µm.

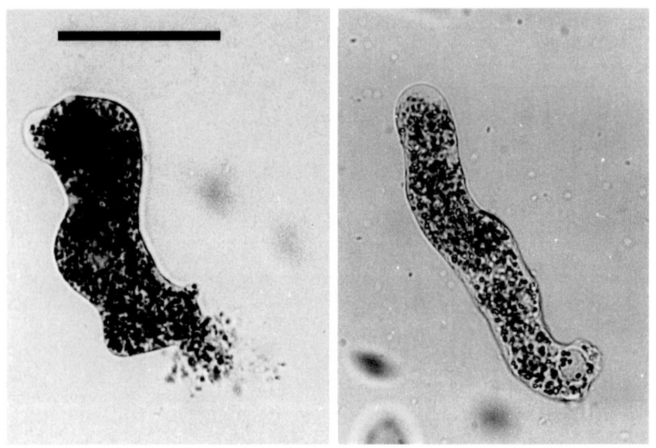

Fig. 37. *Trichamoeba sinuosa*. Bar=100 µm.

Genus ***Trichamoeba*** Fromentel, 1874

Active locomotive form regularly monopodial, slender to thick, most between 75 to 250 µm in length. Nucleus ovular. Cytoplasm often with bipyramidal crystals. Bulbous uroid smooth or with short villi; some with fine filaments. Floating form with radiating pseudopodia. May form cysts. Distinguish from *Saccamoeba*, which is similar but has a nucleus with distinct central nucleolus and a floating form without obvious radiating pseudopodia. Species found in both freshwater and marine habitats. Refs. Schaeffer (1926), Bovee (1972), Page (1988), Page and Siemensma (1991).
TYPE SPECIES: *Trichamoeba hirta* (Fromentel, 1974)

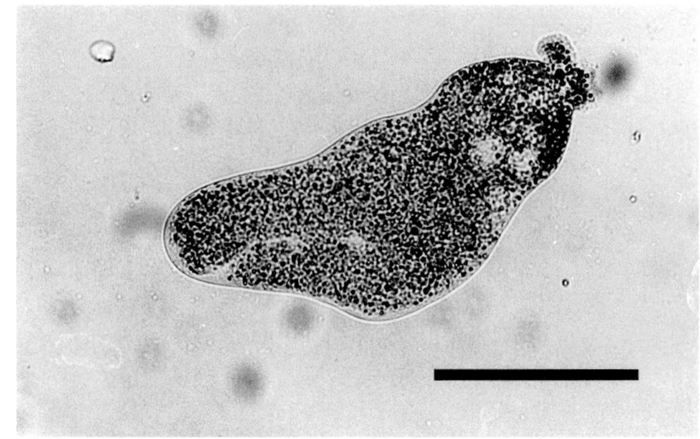

Fig. 38. *Hydramoeba hydroxena* (courtesy F. C. Page and CCAP). Bar=100 µm.

Genus ***Hydramoeba*** Reynolds & Looper, 1928

Monopodial amoeba, usually 150 to 250 µm in length, with a thin anterior hyaline zone. Usually has a single nucleus but can have up to 3 nuclei per cell. Nucleus spherical, around 15 µm diameter, with a central nucleolus and coarse nuclear fragments. Cytoplasm full of crystals. Uroid temporary, morulate. No cysts. Habitat: Parasitic, possibly obligate, on freshwater coelenterates. Monotypic genus. Ref. Page and Robson (1983).
TYPE SPECIES: *Hydramoeba hydroxena* (Entz, 1912)

Genus ***Polychaos*** Schaeffer, 1926

During active locomotion several pseudopodia may radiate out together. Posterior of cell often with a bundle of fused, remnant, pseudopodia. Some locomotive amoebae are monopodial with a bulbous uroid. These distinctive forms are useful in distinguishing *Polychaos* from *Amoeba*. Length usually 50 to 400 µm. Single nucleus is spherical, ovoid, or cup-shaped, often with granular nucleolar material. Cytoplasm with irregular

crystals, plate-like to bipyrimidal, some as paired bodies or clusters. Some species form cysts. Freshwater. Refs. Page (1976, 1987a), Schaeffer (1926).
TYPE SPECIES: *Polychaos dubium* (Schaeffer, 1916)

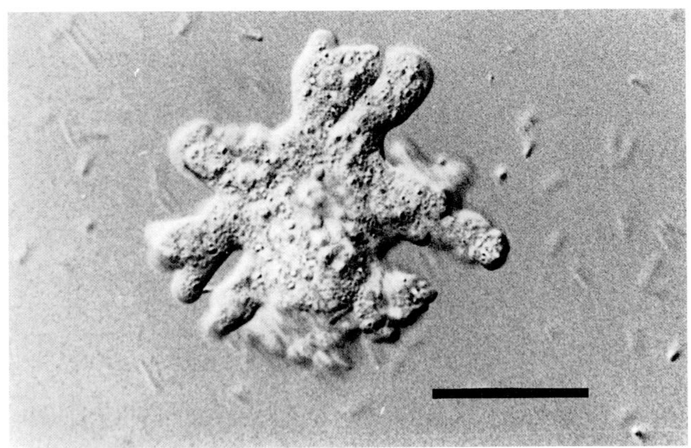

Fig. 39. *Polychaos fasciculatum.* Bar=50 µm.

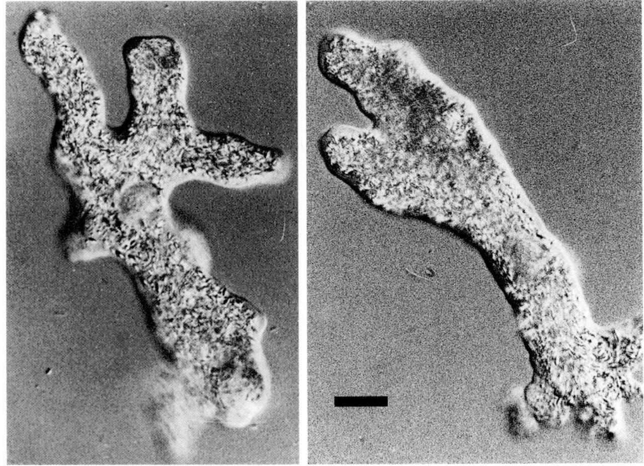

Fig. 40. *Deuteramoeba mycophaga*, locomotive forms. Bar=10 µm.

Genus **Deuteramoeba** Page, 1987

Locomotive form polypodial, although one species monopodial and only polypodial on occasion. Moving cells generally 70 to 180 µm in length with one pseudopodium dominant. Pseudopodia without ridges. Anterior hyaline caps frequently fill with advancing granuloplasm. Posterior morulate uroid. Floating form with short radiating pseudopodia. Cytoplasm with crystals, most as paired inclusions or as bipyramidal crystals. Nucleus spherical with nucleolar material in a few granular pieces or as a central endosome. At least one species forms cysts. At the ultrastructural level the cell coat has filaments. Freshwater and terrestrial. Two named species, one unusual (*D. mycophaga*) in that it feeds on fungi. Refs. Chakraborty and Old (1986), Baldock et al. (1983), Page (1987a, 1988).
TYPE SPECIES: *Deuteramoeba algonquinensis* (Baldock, Rogerson & Berger, 1983)

Genus **Amoeba** Bory de St. Vincent, 1822

Locomotive cells, often 220 to 760 µm in length, although *A. diminutiva* Bovee, 1972 is only 15 to 20 µm. Usually polypodial with one dominant cylindrical pseudopodium, often with ectoplasmic ridges. Resting, or well fed, cells irregularly rounded. Floating form with long, radiating pseudopodia. Cytoplasm granular in appearance often with truncated bipyramidal crystals. Posterior uroid morulate when present. Single nucleus, discoid or ovoidal, sometimes with peripheral granules. Nucleolar material often scattered. At the ultrastructural level, surface coat of *A. proteus* has filaments rising 230 nm above the plasma membrane whereas coat of *A. lenigradensis* amorphous and without filaments. Freshwater. References: Bovee (1972), Flickinger (1974), Friz (1987), Jeon (1973), Kalinina et al. (1987). Based on isozymic analysis, Friz (1992) has established a new genus *Metamoeba* with *M. leningradensis* as the type species.

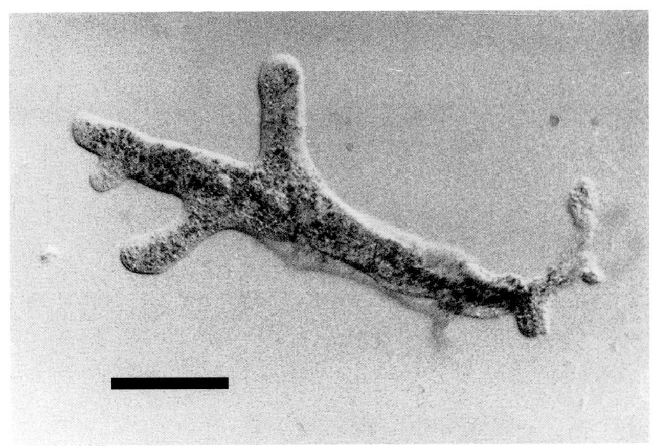

Fig. 41. *Amoeba proteus*. Bar=100 µm.

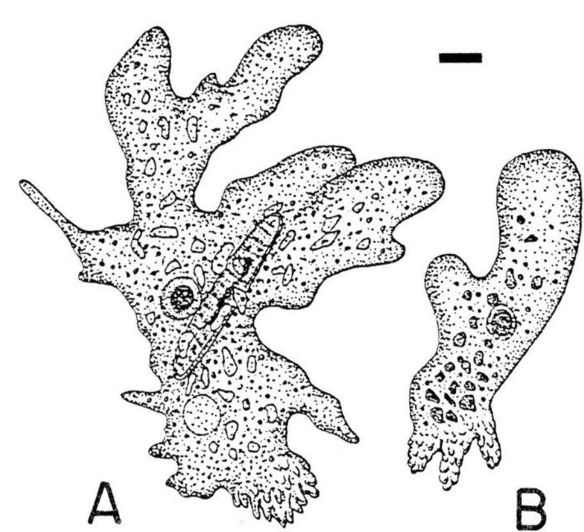

Fig. 42. *Metachaos gratum* (A) and *M. fulvuum* (B), both in active locomotion(after Schaeffer). Bar=10 µm.

Genus ***Metachaos*** Schaeffer, 1926

Locomotive form usually between 75 and 500 µm in length. Pseudopodia cylindrical and smooth with clear hyaline zone. Cytoplasm often very granular with bipyramidal or irregular crystals. Nucleus discoid or spherical. Uroid morulate and temporary. Distinguish from *Polychaos.* Freshwater and marine. Refs. Bovee (1972), Schaeffer (1926).
TYPE SPECIES: *Metachaos discoides* (Schaeffer, 1916)

ORDER CENTRAMOEBIDA

Ramicristate amoebae with extranuclear lamellate centrosomes (microtubule organizing centers).

KEY TO THE FAMILIES OF THE ORDER

1. With many tapering subpseudopodia .... **Acanthamoebidae**
1'. With long slender branched pseudopodia.................................. **Stereomyxidae**

FAMILY ACANTHAMOEBIDAE
Sawyer & Griffin, 1975

Amoebae more or less triangular to trapezoidal, somewhat flattened with flexible, slender, tapering subpseudopodia from the hyaline marigin. These subpseudopodia may branch near their base. Several to many tapering subpseudopodia (acanthopodia) produced from a broad hyaline zone. Cytoplasm frequently with small lipid globules. Cysts common, with or without pores.

KEY TO THE GENERA OF THE FAMILY

1. Cyst wall with a preformed pore through which trophozoite emerges, pore with opercula........... ....***Acanthamoeba***
1'. Cyst wall without pores or opercula, trophozoite breaks through cyst wall.***Protacanthamoeba***

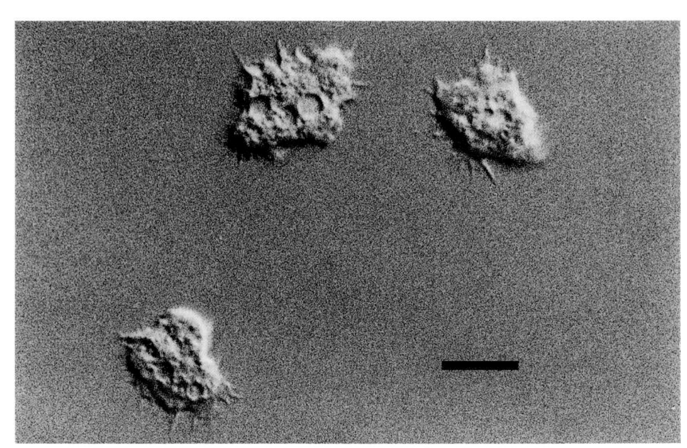

Fig. 44. *Acanthamoeba castellani.* Bar=10 µm

Genus ***Acanthamoeba*** Sawyer & Griffin, 1975

Locomotive form usually between 12 and 40 µm in length. Pseudopodia clear, slowly eruptive, and occasionally branching at base. With numerous, slender, and tapering subpseudopodia (acanthopodia) giving the cell a spiny appearance. Cytoplasm without crystals but frequently with small lipid globules and an obvious contractile vacuole. Nucleus spherical with a central nucleolus. Posterior uroid temporary. Cysts common, often with 2 layers, a thick, wrinkled

outer ectocyst and an inner polygonal endocyst and a wall pore with operculum. Possibly the most commonly isolated genus of all naked amoebae from freshwater and soil habitats. Acanthamoebae have also been isolated from salt water of low salinity, presumably from cysts. One author (AR) has never managed to maintain 'marine strains' on full salinity media for more than a few generations. Some *Acanthamoeba* have been implicated in infections of the human cornea. At least 18 named species which can be distinguished, with difficulty, on the basis of their cyst morphology. Non-morphological procedures (i.e. chemotaxonomic approaches) have been applied to the genus and readers should consult Byers et al. (1983), Costas and Griffiths (1985), McLaughlin et al. (1988), Johnson et al. (1990).
TYPE SPECIES: *Acanthamoeba castellanii* (Douglas, 1930)

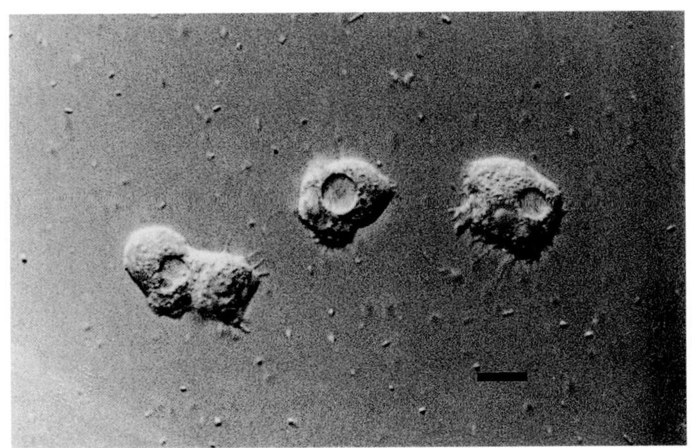

Fig. 45. *Protacanthamoeba caledonica*. Bar=10 µm.

Genus ***Protacanthamoeba*** Page, 1981

Trophic amoebae indistinguishable from *Acanthamoeba*. Locomotive form usually between 20 and 40 µm. Nucleus with central nucleolus. Cysts smooth, circular to oval, about 16.5 µm across. Can be distinguished from *Acanthamoeba* on basis of cyst wall which is without pores and opercula. Freshwater and soil. Refs. Page (1981b, 1988).
TYPE SPECIES: *Protacanthamoeba caledonica* Page, 1981

FAMILY STEREOMYXIDAE Grell, 1966

Amoebae with long slender branched pseudopodia. In some genera, these fuse to form reticulate plasmodia. Some species been shown to have microtubular organizing centers (Page, 1983).

KEY TO GENERA OF THE FAMILY

1. Attached amoebae, uninucleate... ***Stereomyxa***
1'. Attached amoebae, multinucleate.. ***Corallomyxa***

Genus ***Stereomyxa*** Grell, 1966

Amoebae branched, often with long slender branches. Never anastomosing. When actively moving, amoebae extend up to 300 µm or more as in *S. ramosa*. Uninucleate, with an obvious central nucleolus. Floating form rounded with a few to several radiating, blunt-ended, pseudopodia. Marine. Ref. Grell (1966).

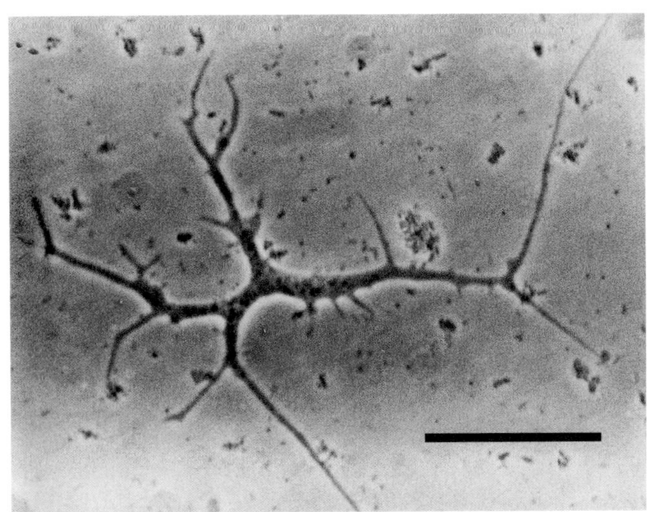

Fig. 46. *Stereomyxa ramosa*. Bar=50 µm.

Genus ***Corallomyxa*** Grell, 1966

Multinucleate amoeba forms reticulate plasmodium when attached. Releases multinucleate or uninucleate 'buds' under adverse conditions, such as prey depletion. In some, centriole-like bodies have

been observed by TEM suggesting possible allegiance with the Acanthamoebidae. Two marine species. Refs. Grell (1966), Grell and Benwitz (1978).
TYPE SPECIES: *Corallomyxa mirabilis* Grell, 1966

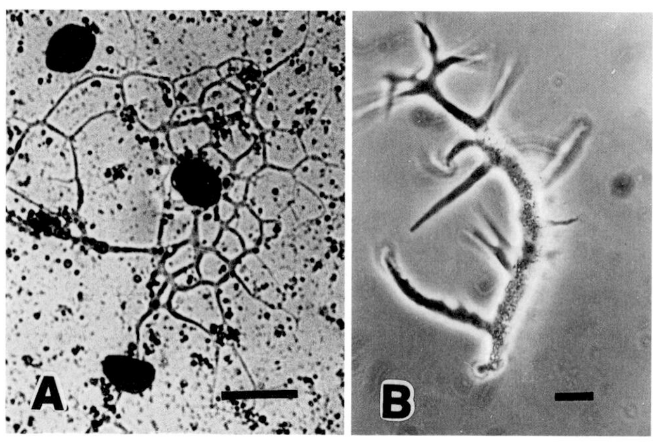

Fig. 47. *Corallomyxa chattoni*, reticulate plasmodium (A). Bar=100 µm. Uninucleate 'bud' (B). Bar=10 µm. (courtesy of F. C. Page and CCAP)

ORDER LEPTOMYXIDA
Pussard & Pons, 1976

Amoebae commonly flattened. Wide range of forms within the order, from uninucleate elongate forms to multinucleate plasmodial amoebae which can be reticulate or branched. Some amoebae produce monopodial stages which are eruptive, otherwise locomotion generally slow and steady. Diagnosis emended in line with Page (1987b).

KEY TO THE FAMILIES OF THE ORDER

1. Amoebae always uninucleate.**Gephyramoebidae**
1'. Amoebae can be multinucleate.......................... 2

2. Often fan-shaped or spatulate......**Flabellulidae**
2'. Often monopodial or plasmodial. **Leptomyxidae**

FAMILY GEPHYRAMOEBIDAE

Amoebae often much branched but not plasmodial. Although amoebae can be spatulate in outline, they lack the mass of trailing filaments sometimes found in the Leptomyxidae. Only one genus is known in this family.

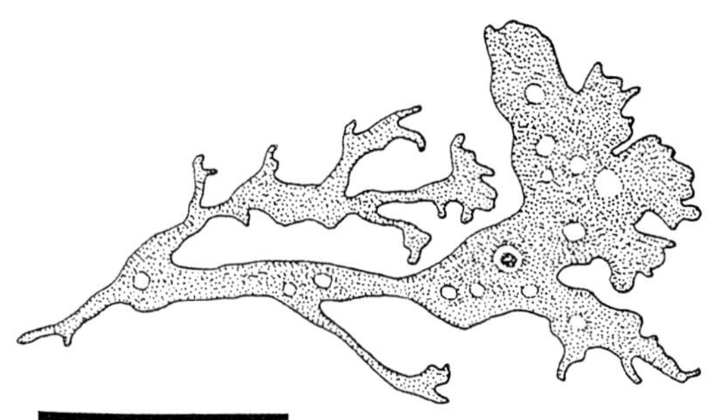

Fig. 48. *Gephyramoeba delicatula* (after Page). Bar=100 µm.

Genus ***Gephyramoeba*** Goodey, 1914

Locomotive form 30 to 300 µm. Often fan-shaped with a long posterior tail, sometimes highly branched but not anastomosing. Does not form plasmodia. Uninucleate. Cysts spherical or ovoid. Monospecific, from soil. Ref. Pussard and Pons (1976).
TYPE SPECIES: *Gephyramoeba delicatula* Goodey, 1914

FAMILY FLABELLULIDAE Bovee, 1970

Flattened amoebae with extensive anterior hyaline zones, with or without subpseudopodia. Often with posterior trailing uroidal filaments.

KEY TO GENERA OF THE FAMILY

1. Anterior hyaline with clefts rather than subpseudopodia .... ***Flabellula***
1'. Anterior hyaline with short, round-tipped subpseudopodia....................... ***Paraflabellula***

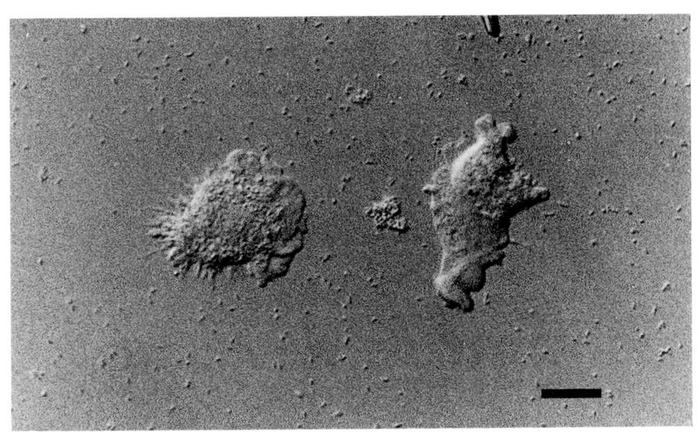

Fig. 49. *Flabellula citata*, fan-shaped and elongate forms. Bar=10 µm.

Genus ***Flabellula*** Schaeffer, 1926

Size of locomotive cell can range between 3 and 7 5 µm. Amoebae flattened with an obvious hyaloplasm that can separate into anterior clefts. Shape variable from fan-shaped to elongate with a tendency towards eruptive motion when changing direction. Some with uroidal filaments. Floating form usually irregular without radiating pseudopodia. Five species from marine habitats. Ref. Bovee (1965).
TYPE SPECIES: *Flabellula citata* Schaeffer, 1926

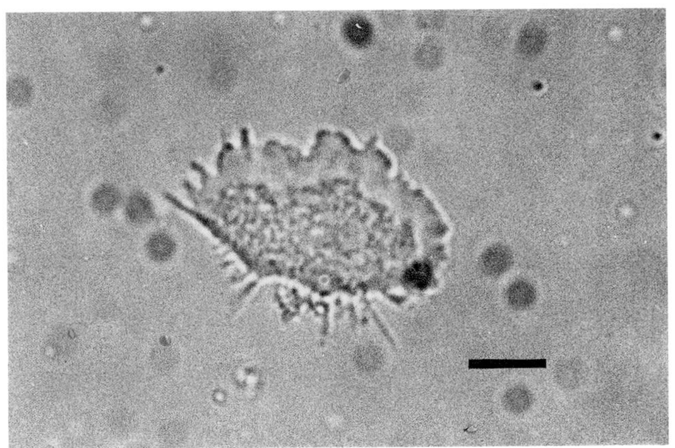

Fig. 50. *Paraflabellula reniformis*. Bar=10 µm.

Genus ***Paraflabellula*** Page & Willumsen, 1983

Locomotive form usually between 8 and 45 µm, flattened, fan-shaped to elongate with an extensive hyaloplasm with short subpseudopodia. Trailing filaments common, some branched. Floating form typically with radiating pseudopodia which terminate in a fine tip. Two marine and one freshwater species. Ref. Page and Willumsen (1983).
TYPE SPECIES: *Paraflabellula reniformis* (Schmoller, 1964)

FAMILY LEPTOMYXIDAE Pussard & Pons, 1976

Uninucleate monopodial amoebae with a tendency to form multinucleate plasmodia which can be flabellate, multilobed, or highly branched and reticulate. Amoebae never have anterior subpseudopodia.

KEY TO GENERA OF THE FAMILY

1. Active amoebae commonly limax-like... ***Rhizamoeba***
1'. Active amoebae never limax-like... ***Leptomyxa***

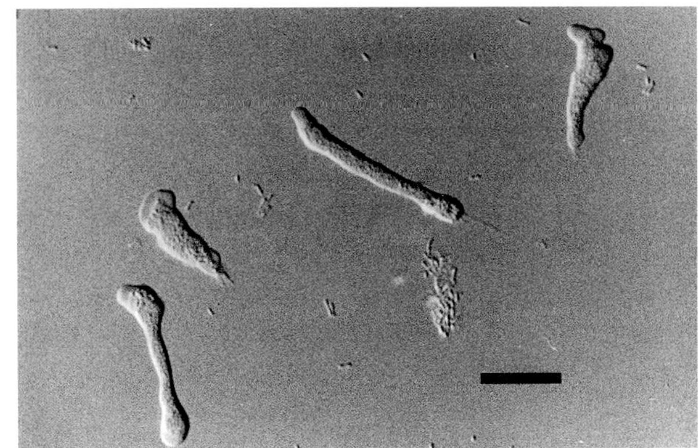

Fig. 51. *Rhizamoeba saxonica*. Bar=10 µm.

Genus ***Rhizamoeba*** Page, 1972

Limax-like in rapid motion, often eruptive. Anterior with a hyaline zone, posterior usually with conspicuous trailing filaments. Less active forms are flattened, sometimes with peripheral filaments. Most uninucleate, but some species multinucleate. Nucleus difficult to observe in living cells. Marine, freshwater, and soil species; some form cysts. Refs. Page (1972, 1988).
TYPE SPECIES: *Rhizamoeba polyura* Page, 1972

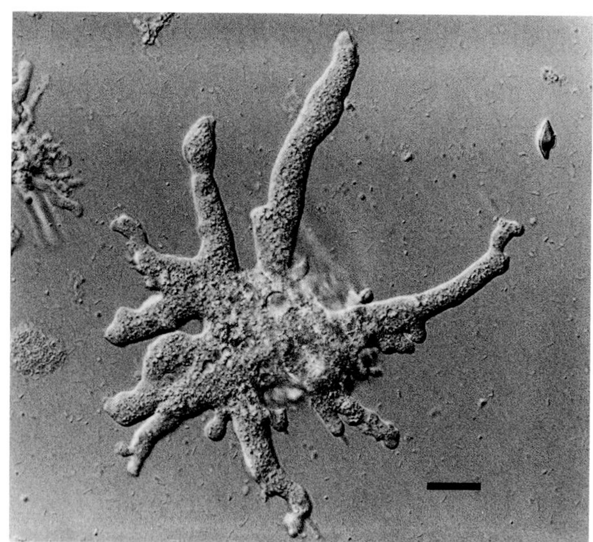

Fig. 52. *Leptomyxa reticulata.* Bar=10 µm.

Genus ***Leptomyxa*** Goodey, 1914

Flattened plasmodium from around 100 µm to over 1 mm in length, often reticulate, multi-lobed, or branched. Locomotion can be eruptive and trailing filaments are common. Multinucleate. Forms cysts. Widespread in soil and reported in freshwater. Ref. Page (1988).
TYPE SPECIES: *Leptomyxa reticulata* Goodey, 1914

FAMILIES OF UNCERTAIN AFFINITIES

FAMILY ECHINAMOEBIDAE Page, 1975

Flattened amoebae with many short, extremely fine subseudopodia which may be furcate. Subpseudopodia with a central filamentous core at the ultrastructural level.

KEY TO THE GENERA OF THE FAMILY

1. Amoebae usually less than 15 µm, pseudodopodia short as spine-like projections, cysts with thin cell walls.................................... ***Echinamoeba***
1'. Amoebae usually larger than 15 µm, cysts with thick walls... 2

2. Many filose-like pseudopodia.......... ***Filamoeba***
2'. With a few short spine-like echinopodia........... .... ***Comandonia***

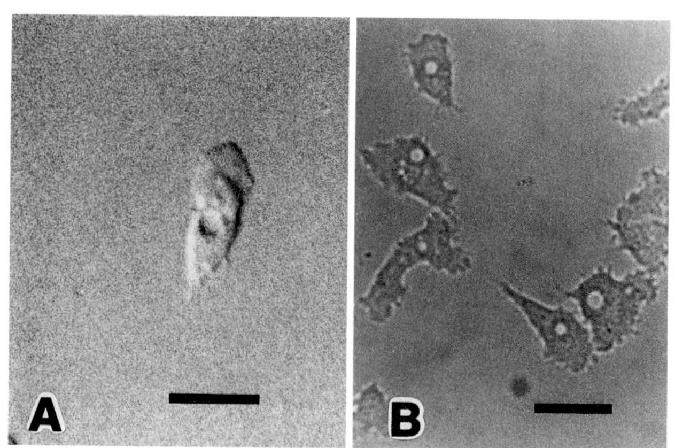

Fig. 53. *Echinamoeba sylvestris* (A) and *E. exudans* (B). Bars=10 µm.

Genus ***Echinamoeba*** Page, 1975

Locomotive cell small, around 12 µm, usually triangular, fan-shaped, or elongate. With a few short spiny pseudopodia (echinopodia up to 1.5 µm long) extending from the hyaline region. Single nucleus. Cysts smooth and spherical. Thin cyst wall with a narrow space between the outer and inner layers, although one species has cysts without layers. Two species described from leaf litter. Ref. Page (1975).
TYPE SPECIES: *Echinamoeba exudans* (Page, 1967)

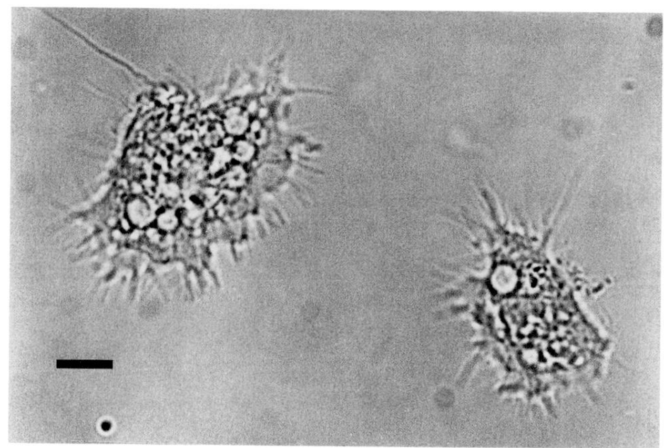

Fig. 54. *Filamoeba nolandi* (courtesy F. C. Page and CCAP). Bar=10 µm.

Genus ***Filamoeba*** Page, 1967

Cells flattened, fan-shaped, and sometimes with an elongate posterior giving the cell a spatulate outline, between 15 and 50 µm in length. Occasionally with 2 or more hyaline lobes. Clear, thin, filopodia, up to 10 µm long, extend from the hyaline edge. These can be branched at their base but are never anastomosing. Cytoplasm with 15 or more small contractile vacuoles. Cysts smooth and round, oval, or irregular in shape with a relatively thick wall. Monospecific, described from freshwater. Ref. Page (1967).
TYPE SPECIES: *Filamoeba nolandi* Page, 1967

Genus ***Comandonia*** Perin & Pussard, 1979

Irregularly shaped cell, usually around 40 µm in length, with a narrow anterior hyaline margin. Frequently with a few, fine, short subpseudopodia (echinopodia) arising from the margin giving the cell a spine-like appearance. Posterior often with adhesive uroidal filaments. Cytoplasm with numerous contractile vacuoles and a spherical nucleus. No centriole-like bodies at the ultrastructural level. Cysts usually irregularly polyhedral with a thick inner wall and very thin outer layer. Only genus in the family to have cysts with pores closed by operculum. Monospecific, freshwater. Ref. Perin and Pussard (1979).
TYPE SPECIES: *Comandonia operculata* Perin & Pussard, 1979

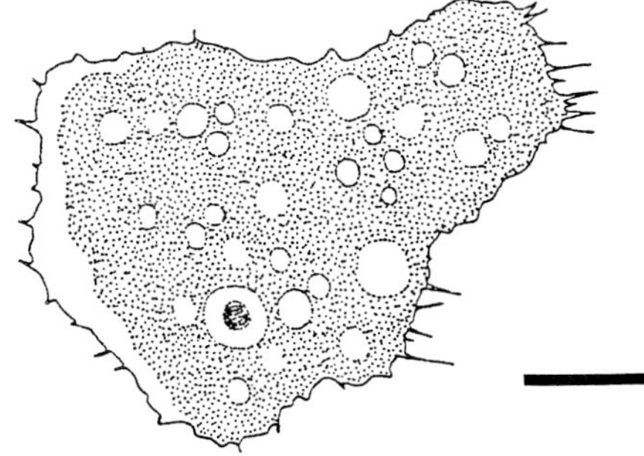

Fig. 55. *Comandonia operculata* (after Perin and Pussard). Bar=10 µm.

FAMILY HYALODISCIDAE Poche, 1913

Locomotive forms regularly discoid, ovoid, or fan-shaped. ,With a prominent granular cytoplasmic mass surrounded either entirely or anteriorly and laterally by a flattened hyaline margin. With or without slender conical subseudopodia from the clear border. Amoebae frequently move with a distinctive rolling motion. Family description emended by Page (1976).

KEY TO THE GENERA OF THE FAMILY

1. Central granular mass; hyaline border often surrounds cell 2
1'. Posterior granular mass; hyaline border usually anterior and lateral only.................... 3

2. With subspeudopodia from hyaloplasm .... ***Gibbodiscus***
2'. Without subpseudopodia from hyaloplasm .... ***Ovalopodium***

3. With extremely fine, anterior subspeudopodia ....***Hyalodiscus***
3'. With conical, anterior, subpseudopodia........ .... ***Flamella***

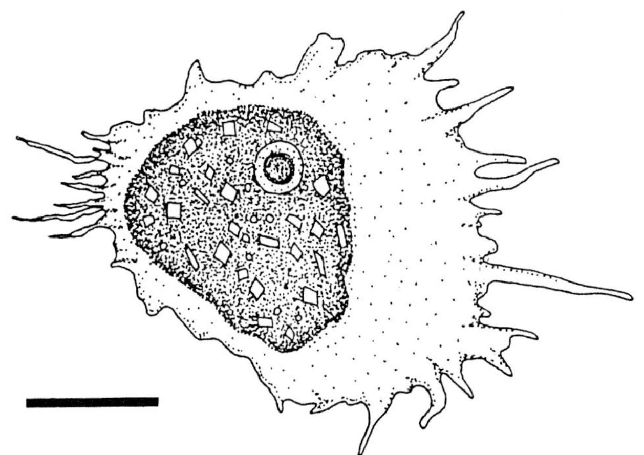

Fig. 56. *Gibbodiscus gemma* (after Schaeffer). Bar=10 µm.

Genus ***Gibbodiscus*** Schaeffer, 1926

Length in locomotion usually 20 to 35 µm. Cell as a flattened ovoid with a prominent sub-hemispherical post-central hump. A peripheral zone, broadest at the anterior, is always present. During locomotion, numerous slender sub-

pseudopodia are extended and retracted from the anterior edge. Dense granular cytoplasm without conspicuous streaming when cell in motion. Instead the entire hump is dragged along by the clear, anterior peripheral zone. Cytoplasm may contain numerous cubical or ovoid crystals. Spherical nucleus with a central nucleolus. Uroid of slender, trailing elements usually present. Two marine species. Refs. Schaeffer (1926), Sawyer (1975a), Bovee and Sawyer (1979).
TYPE SPECIES: *Gibbodiscus gemma* Schaeffer, 1926

Genus ***Ovalopodium*** Sawyer, 1980

Cell broadly ovoid with a central or post-central hump and spine-like posterior uroidal filaments. Anterior and lateral margins smooth or rippled. Hyaline margin, without subpseudopodia, encircles the cell. Cytoplasm granular with numerous clear vesicles and a distinct nucleus with a central nucleolus. Sawyer (1980) considers *Ovalopodium* to be in the family Cochliopodiidae; however, it differs from other members in that it lacks scales or a cuticle and does not form cysts. Marine, monospecific, from oyster holding trays. Ref. Sawyer (1980).
TYPE SPECIES: *Ovalopodium carrikeri* Sawyer, 1980

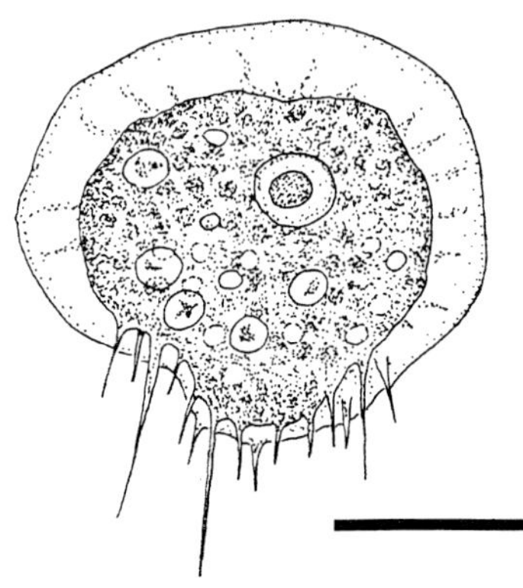

Fig. 57. *Ovalopodium carrikeri* (after Sawyer). Bar=10 μm.

Genus ***Hyalodiscus*** Hertwig & Lesser, 1874

Cells discoid with breadth greater than length. Extensive, flattened hyaline margin surrounding the cytoplasmic hump anteriorly and laterally and sometimes posteriorly. Locomotion frequently rapid with a distinctive rolling motion. Fine, short subpseudopodia extend from the hyaloplasm. Floating form with radiate pseudopodia. Cytoplasmic mass sometimes reddish brown or greenish in color with 1 or 2 nuclei. At least 4 described species from freshwater and marine habitats. Refs. Schaeffer (1926), Page and Willumsen (1980).
TYPE SPECIES: *Hyalodiscus rubicundus* Hertwig & Lesser, 1874

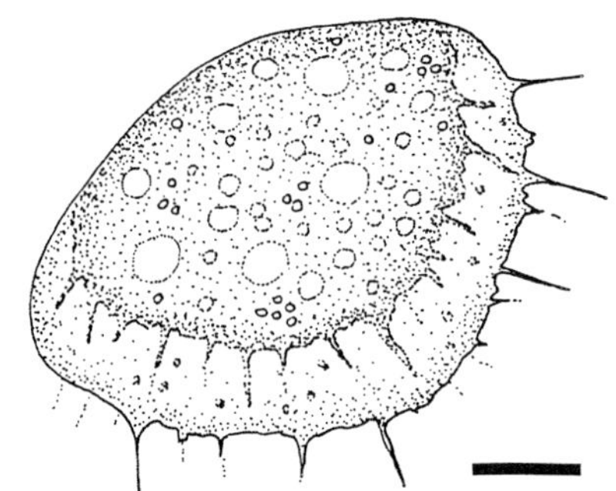

Fig. 58. *Hyalodiscus rubicundus* (after Cash and Willumsen). Bar=10 μm.

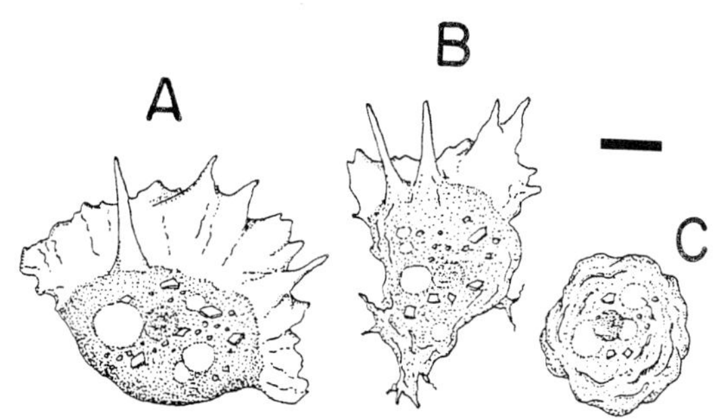

Fig. 59. *Flamella citrensis,* active forms (A, B), floating form prior to attachment (C) (after Bovee). Bar=10 μm.

Genus ***Flamella*** Schaeffer, 1926.

Shape during locomotion variable due to rapid form changes, but usually oval, around 30 to 60 μm in length. Anterior 2/3 of the cell as a flattened hyaline margin; posterior third as a granular cytoplasmic mound. Anterior edge with numerous, long, slender, conical subpseudopodia, often in clusters of 2 or 3 emerging from pyramidal bases. Some pseudopodia may become fused with the hyaloplasm. No posterior uroid but cells can have a

few, fine, short filaments. Cytoplasm with crystals. Nucleus obscure. Reported from marine, coastal waters and from freshwater citrus pulp wastes. At least one species (*F. citrensis*) is anaerobic. Refs. Schaeffer (1926), Bovee (1956).
TYPE SPECIES: *Flamella magnifica* Schaeffer, 1926

FAMILY COCHLIOPODIIDAE de Saedeleer, 1934

Amoebae partially enclosed within a flexible cuticle or wall (tectum) usually covered with microscales. The covering is open along the region of attachment to the substratum; with no well defined aperture. Amoebae in this family have been classified as testate amoebae by Page (1987b) but are included in this chapter because they resemble euamoebae at the light microscope level. Typically they have a distinct hyaline zone often with short subpseudopodia, and a prominant central granular hump.

KEY TO GENERA OF THE FAMILY

1. Tectum with surface microscales .... ***Cochliopodium***
1'. Tectum without surface microscales................. 2

2. Edge of hyaline margin usually irregular in outline... ***Gocevia***
2'. Edge of hyaline margin usually regular in outline.. ***Paragocevia***

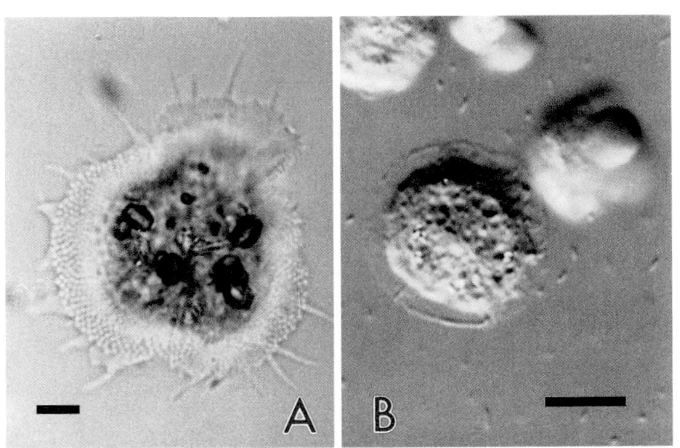

Fig. 60. *Cochliopodium bilimbosum* (A) and *C. minus* (B). Bars=10 μm.

Genus ***Cochliopodium*** Hertwig & Lesser, 1874

Discoid or spherical in shape with a central granular hump partly or completely surrounded by a distinctive hyaline margin. At the LM level, circular microscales can be seen as regular punctation on the surface of the margin. Occasional furcate subpseudopodia produced from the anterior hyaline zone and trailing filaments from the posterior. Granular mass contains cytoplasmic crystals. Some species form cysts. Freshwater habitats. Refs. Bark (1973), Yamaoka et al. (1984).

Genus ***Gocevia*** Valkanov, 1932

Similar in shape to *Cochliopodium* but without microscales. Most species have a hyaline border with adhering refractile particles such as mineral grains. Hyaline zone often irregular in outline with few subpseudopodia. Cells uninucleate or binucleate. Freshwater habitats. Ref. Pussard (1965).

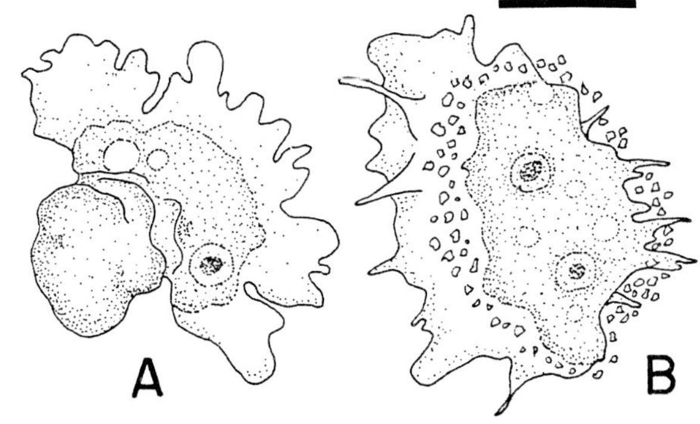

Fig. 61. *Gocevia fonbrunei* (A, after Pussard) and *G. binucleata* (B, after de Saedeleer). Bars=10 μm.

Genus ***Paragocevia*** Page, 1987

Discoid cell around 100 μm in length with a flattened hyaline margin and slightly raised granular mass. Frequently with numerous small (5 to 10 μm) conical subpseudopodia from the margin. Cuticle without microscales or foreign particles. Freshwater habitats. Ref. Page and Willumsen (1980).

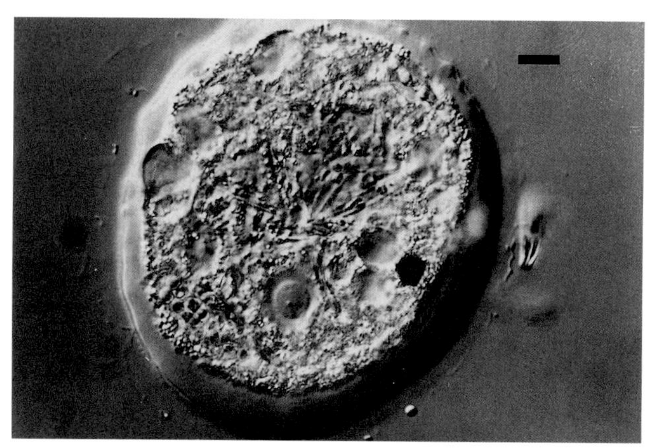

Fig. 62. *Paragocevia placopus* (courtesy F. C. Page and CCAP).

LITERATURE CITED

Anderson, O. R. & Rogerson, A. 1995. Annual abundances and growth potential of gymnamoebae in the Hudson estuary with comparative data from the Firth of Clyde. *Europ. J. Protistol.*, 31:223-233.

Baldock, B. M., Rogerson, A. & Berger, J. 1983. A new species of freshwater amoeba: *Amoeba algonquinensis* n. sp. (Gymnamoebia: Amoebidae). *Trans. Am.. Microsc. Soc.*, 102:113-121.

Bark, A. W. 1973. A study of the genus *Cochliopodium* Hertwig & Lesser 1874. *Protistologica*, 9:119-138.

Bovee, E. C. 1956. Some observations on the morphology and activities of a new ameba from citrus wastes, *Flamella citrensis* n. sp. *J. Protozool.*, 3:151-153.

Bovee, E. C. 1965. An emendation of the ameba genus *Flabellula* and a description of *Vannella* gen. nov. *Trans. Am. Microsc. Soc.*, 84:217-227.

Bovee, E. C. 1970. The lobose amebas. I. A key to the suborder Conopodina Bovee & Jahn, 1966, and descriptions of thirteen new or little known *Mayorella* species. *Arch. Protistenkd.*, 112:178-227.

Bovee, E. C. 1972. The lobose amebas. IV. A key to the order Granulopodia Bovee & Jahn, 1966 and descriptions of some new and little known species in the order. *Arch. Protistenkd.*, 114:371-403.

Bovee, E. C. & Sawyer, T. K. 1979. Marine Flora and Fauna of the Northeastern United States. Protozoa: Sarcodina: Amoebae. NOAA Technical Report NMFS circular 419, 1-55.

Butler, H. & Rogerson, A. 1996. Temporal and spatial abundance of naked amoebae (Gymnamoebae) in marine benthic sediments of the Clyde Sea area, Scotland. *J. Euk. Microbiol.*, in press.

Byers, T. J., Bogler, S. A. & Burianek, L. L. 1983. Analysis of mitochondrial DNA variation as an approach to systematic relationships in the genus *Acanthamoeba*. *J. Protozool.*, 30:198-203.

Cann, J. P. & Page, F. C. 1982. Fine structure of small free-living *Paramoeba* (Amoebida) and taxonomy of the genus. *J. Mar. Biol. Ass. UK.*, 62:25-43.

Chakraborty, S. & Old, K. M. 1986. Ultrastructure and description of a fungus-feeding amoeba, *Trichamoeba mycophaga* n. sp. (Amoebidae, Amoebae), from Australia. *J. Protozool.*, 33:64-569.

Clarholm, M. 1981. Protozoan grazing of bacteria in soil - impact and importance. *Microb. Ecol.*, 7:343-350.

Claparède, É. & Lachmann, J. 1858/59. Études sur les Infusoires et les Rhizopodes. Extracit des Tomes V et VI de L'Institut Genevois, vol. 4, 441-442.

Costas, M. & Griffiths, A. J. 1985. Enzyme composition and the taxomomy of *Acanthamoeba*. *J. Protozool.*, 32:604-607.

Davis, P. G., Caron, D. A. & Sieburth, J. McN. 1978. Oceanic amoebae from the North Atlantic: culture, distribution, and taxonomy. *Trans. Am. Microsc. Soc.*, 97:73-86.

Finlay, B. J., Rogerson, A. & Cowling, A. J. 1988. *A Beginner's Guide to the Collection, Isolation, Cultivation and Identification of Freshwater Protozoa.* CCAP Publication, Titus Wilson & Son Ltd, Kendal, England.

Flinkinger, C. J. 1974. The fine structure of four 'species' of *Amoeba*. *J. Protozool.*, 21:59-68.

Friz, C. T. 1987. Taxonomic analyses of the free-living amoeba using protein polymorphism. *Arch. Protistenkd.*, 133:165-171.

Friz, C. T. 1992. Taxonomic analyses of seven species of family Amoebidae by isozymic characterization of electrophoretic patterns and the descriptions of a new genus and a new species: *Metamoeba* n. gen. *Amoeba amazonas* n. sp. *Arch. Protistenkd.*, 142:29-40.

Goodfellow, L. P., Belcher, J. H. & Page, F. C. 1974. A light- and electron-microscopical study of *Sappinia diploidea*, a sexual amoeba. *Protistologica*, 10:207-216.

Grell, K. G. 1961. Über den Nebenkörper von *Paramoeba eilhardi* Schaudinn. *Arch. Protistenkd.*, 105:303-312.

Grell, K. G. 1966. Amöeben der Familie Stereomyxidae. *Arch. Protistenkd.*, 109:147-154.

Grell, K. G. & Benwitz, G. 1978. Ultrastruktur mariner Amöben. IV. *Corallomyxa chattoni* n. sp. *Arch. Protistenkd.*, 120:287-300.

Gromov, D. B. 1986. Ul'trastruktura yader mnogoyadernoy ameby *Chaos carolinense*. (The ultrastructure of the nuclei of the multinucleate amoeba *Chaos carolinense*). *Tsitologiya*, 28:446-447.

Jahn, T. L., Bovee, E. C. & Griffith, D. 1979. Taxonomy and evolution of Sarcodina: a reclassification. *Taxon,* 23:483-496.

Jeon, K. W. 1973. *The Biology of Amoeba*. Academic Press, New York.

Johnson, A. M., Fielke, R., Christy, P. E., Robinson, B & Baverstock, P. R. 1990. Small subunit ribosomal RNA evolution in the genus *Acanthamoeba*. *J. Gen. Microbiol.*, 136:1689-1698.

Kalinina, L. V., Afon'kin, S. Yu., Gromov, D. B., Khrebtukova, I. A. & Page, F. C. 1987. *Amoeba borokensis* n. sp., a rapidly dividing organism especially suitable for experimental purposes. *Arch. Protistenkd.*, 132:343-361.

Korotneff, A. 1880. Etudes sur les Rhizopodes. *Arch. Zool. Exp. Gén.*, 8:467-482.

McLaughlin, G. L., Brandt, F. H. & Visvesvara, G. S. 1988. Restriction fragment length polymorphisms of the DNA of selected *Naegleria* and *Acanthamoeba*. *J. Clin. Microbiol.*, 26:1655-1658.

Page, F. C. 1967. *Filamoeba nolandi* n. g., n. sp., a filose amoeba. *Trans. Am. Microsc. Soc.*, 86:405-411.

Page, F. C. 1970. Two new species of *Paramoeba* from Maine. *J. Protozool.*, 17:421-427.

Page, F. C. 1972. *Rhizamoeba polyura* n. g. n. sp., and uroidal structures as a taxonomic criterion for amoebae. *Trans. Am. Microsc. Soc.*, 91:502-513.

Page, F. C. 1973. *Paramoeba*: a common marine genus. *Hydrobiologia*, 41:183-188.

Page, F. C. 1974. A further study of taxonomic criteria for limax amoebae, with descriptions of new species and a key to genera. *Arch. Protistenkd.*, 116:149-184.

Page, F. C. 1975. A new family of amoebae with fine pseudopodia. *Zool. J. Linn. Soc.*, 56:73-89.

Page, F. C. 1976. *An Illustrated Key to Freshwater and Soil Amoebae*. Freshwater Biological Association, Ambleside.

Page, F. C. 1979. Two genera of marine amoebae (Gymnamoebia) with distinctive surface structures: *Vannella* Bovee, 1965 and *Pseudoparamoeba* n.gen., with two new species of *Vannella*. *Protistologica*, 15:245-257.

Page, F. C. 1980. A light- and electron- microscopical comparison of marine limax and flabellate amoebae belonging to four genera. *Protistologica*, 16:57-78.

Page, F. C. 1981a. Eugène Penard's slides of Gymnamoebia; re-examination and taxonomic evaluation. *Bull. Br. Mus. Nat. Hist. (Zool.)*, 40:1-32.

Page, F. C. 1981b. A light- and electron-microscopical study of *Protacanthamoeba caledonica* n. sp., type species of *Protacanthamoeba* n. g. (Amoebida, Acanthamoebidae). *J. Protozool.*, 28:70-78.

Page, F. C. 1983. *Marine Gymnamoebae*. Institute of Terrestrial Ecology, Cambridge.

Page, F. C. 1986. The limax amoebae; comparative fine structure of the Hartmannellidae (Lobosea) and further comparisons with the Vahlkampfiidae (Heterolobosea). *Protistologica*, 21:361-383.

Page, F. C. 1987a. The general and possible relationships of the family Amoebidae, with special attention to comparative ultrastructure. *Protistologica*, 22:301-316.

Page, F. C. 1987b. The classification of 'naked' amoebae (Phylum Rhizopoda). *Arch. Protistenkd.*, 133:199-217.

Page, F. C. 1988. *A New Key to Freshwater and Soil Gymnamoebae.* Freshwater Biological Association, Ambleside.

Page, F. C. & Robson, E. A. 1983. Fine structure and taxonomic position of *Hydramoeba hydroxena* (Entz, 1912). *Protistologica*, 19:41-50.

Page, F. C. & Siemensma, F. J. 1991. *Nackte Rhizopoda und Heliozoea.* Gustav Fisher Verlag, Stuttgart, New York.

Page, F. C. & Willumsen, N. B. S. 1980. Some observations on *Gocevia placopus* (Hülsmann, 1974), an amoeba with a flexible test, and on *Gocevia*-like organisms from Denmark, with comments on the genera *Gocevia* and *Hyalodiscus*. *J. Nat. Hist.*, 14:413-431.

Page, F. C. & Willumsen, N. B. S. 1983. A light- and electron-microscopical study of *Paraflabellula reniformis* (Schmoller, 1964), type species of a genus of amoebae (Amoebida, Flabellulidae) with subpseudopodia. *Protistologica,* 19:567-575.

Patterson, D. J. 1981. Contractile vacuole complex behaviour as a diagnostic character for free-living amoebae. *Protistologica*, 17:243-248.

Pennick, N. C. & Goodfellow, L. P. 1975. Some observations on the cell surface structures of species of *Mayorella* and *Paramoeba*. *Arch. Protistenkd.*, 117:41-46.

Perin, P. & Pussard, M. 1979. Etude en microscopie photonique et électronique d'une Amibe voisine du genre *Acanthamoeba*: *Comandonia operculata* n. gen., n. sp. (Amoebida, Acanthamoebidae). *Protistologica*, 15:87-102.

Pussard, M. 1965. Cuticle et caryocinvése de *Gocevia fonbrunei* n. sp. (Cochliopodiidae - Testacealobosa). *Arch. Zool. Exp. Gén.*, 105:101-117.

Pussard, M. 1973. Description d'une amibe de type flabellulien: *Pessonella marginata* n. g., n. sp. (Mayorellidae, Amoebaea). *Protistologica*, 9:175-185.

Pussard, M. & Pons, R. 1976. Etudes des genres *Leptomyxa* et *Gephyramoeba* (Protozoa, Sarcodina). III - *Gephyramoeba delicatula* Goodey, 1915. *Protistologica*, 12:351-383.

Raikov, I. B. 1982. *The Protozoan Nucleus. Morphology and Evolution.* Wein. Springer.

Rogerson, A. 1988. DAPI-staining for the rapid examination of nuclei and parasomes in marine gymnamoebae. *Arch. Protistenkd.*, 135:289-298.

Rogerson, A. 1991. On the abundance of marine naked amoebae on the surfaces of five species of macroalgae. *FEMS Microb. Ecol.*, 85:301-312.

Rogerson, A. 1993. *Parvamoeba rugata* n. g., n. sp., (Gymnamoebia, Thecamoebidae): an exceptionally small marine naked amoeba. *Eur. J. Protistol.*, 29:446-452.

Rogerson, A. & Laybourn-Parry, J. 1992. The abundance of marine naked amoebae in the water column of the Clyde estuary. *Estuar. Coast Shelf Sci.*, 34:187-196.

Sawyer, T. K. 1975a. Marine amoebae from surface waters of Chincoteague Bay, Virginia: one new genus and eleven new species within the families Thecamoebidae and Hyalodiscidae. *Trans. Am. Microsc. Soc.*, 94:305-323.

Sawyer, T. K. 1975b. *Clydonella* n. g. (Amoebida: Thecamoebidae) proposed to provide an appropriate generic home for Schaeffer's marine species of *Rugipes*, *C. vivax* (Schaeffer, 1926) n. comb. *Trans. Am. Microsc. Soc.*, 94:395-400.

Sawyer, T. K. 1975c. Marine amoebae from surface waters of Chincoteague Bay, Virginia: two new genera and nine new species within the families Mayorellidae, Flabellulidae, and Stereomyxidae. *Trans. Am. Microsc. Soc.,* 94:71-92.

Sawyer, T. K. 1980. Marine amoebae from clean and stressed bottom sediments of the Atlantic Ocean and Gulf of Mexico. *J. Protozool.*, 27:13-32.

Schaeffer, A. A. 1926. Taxonomy of the amebas. *Pap. Dep. Mar. Biol. Carnegie Instn. Wash.*, 24:1-116.

Singh, B. N. 1952. Nuclear division in nine species of small free-living amoebae and its bearing on the classification of the order Amoebida. *Phil. Trans. R. Soc.*, B, 36:405-461.

Smirnov, A. V. & Goodkov, A. V. 1993. *Paradermamoeba valamo* gen. n., sp. n. (Gymnamoebia, Thecamoebidae) - a freshwater amoeba from bottom sediments. *Zool. Zh.,* 72:5-11. (in Russian)

Smirnov, A. V. & Goodkov, A. V. 1994. Freshwater gymnamobae with a new type of surface structure *Paradermamoeba valamo* and *P. levis* sp. n. (Thecamoebidae), and notes on the diagnosis of the family. *Acta Protozool.*, 33:109-115.

Vodkin, M. H., Howe, D. K., Visvesvara, G. S. & McLaughlin, G. L. 1992. Identification of *Acanthamoeba* at the generic and specific levels using the polymerase chain reaction. *J. Protozool.*, 39:378-385.

Willumsen, N. B. S. 1982. *Chaos zoochorellae* nov. sp. (Gymnamoebia, Amoebidae) from a Danish freshwater pond. *J. Nat. Hist.*, 16:803-813.

Willumsen, N. B. S., Siemensma, F. & Suhr-Jessen, P. 1987. A multinucleate amoeba, *Parachaos zoochlorellae* (Willumsen 1982) comb. nov., and a proposed division of the genus *Chaos* into the genera *Chaos* and *Parachaos* (Gymnamoebia, Amoebida). *Arch. Protistenkd.*, 134:303-313.

Yamaoka, I., Kawamura, N., Mizuno, M. & Nagatani, Y. 1984. Scale formation in an amoeba, *Cochliopodium* sp. *J. Protozool.*, 31:267-272.

ACKNOWLEDGEMENTS

We are grateful to the Culture Collection of Algae and Protozoa (CCAP), Windermere, for allowing access to examine and photograph some of their living 'type' strains. We also thank Dr. F. C. Page and Mr. A. V. Smirnov for some of the photomicrographs and Dr. J. Gunderson for helpful discussion.

INCERTAE SEDIS

Malamoeba Taylor & King, 1937.

Amoeba parasitizing insect malpighian tubules or gut epithelia, 2 species. Trophozoites with up to 4 nuclei, cysts are uninucleated. There is some uncertainty as to whether the 2 species assigned to this genus are congeneric. The type species was originally placed in *Malpighamoeba*, but

transferred out by Taylor and King (1937). Refs. Larsson et al. (1992), Purrini (1980), Taylor and King (1937)
TYPE SPECIES: *Malamoeba locustae* Taylor & King, 1937

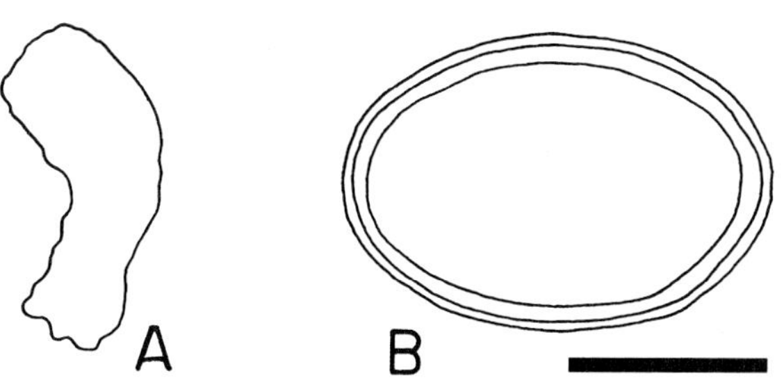

Fig. 63. *Malamoeba locustae* (after King and Taylor, 1936). Scale bar=5 µm.

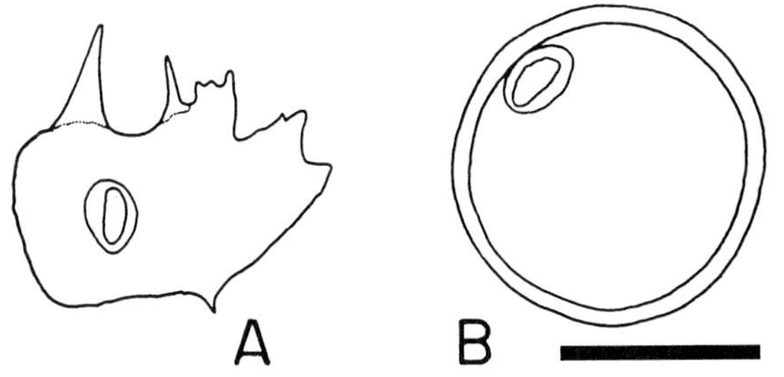

Fig. 64. *Malpighamoeba mellificae* (after Prell, 1926). Scale bar=5 µm.

Malpighamoeba Prell, 1926.

One species of amoeba parasitizing insect Malpighian tubules or digestive system of insects, transmission by ovoid cysts. Ultrastructure studied by Schwantes and Eichelberg (1984). Ref. Harry and Finlayson (1975, 1976).
TYPE SPECIES: *Malpighamoeba mellificae* Prell, 1926

References

Harry, O. G. & Finlayson, L. H. 1975. Histopathology of secondary infections of *Malpighamoeba locustae* (Protozoa, Amoebidae) in the Desert Locust *Schistocerca gregaria* (Orthoptera, Acrididae). *J. Invert. Pathol.*, 25:25-33.

Harry, O. G. & Finlayson, L. H. 1976. The life-cycle, ultrastructure and mode of feeding of the locust amoeba *Malpighamoeba locustae*. *Parasitology*, 72:127-135

King, R. L. & Taylor, A. B. 1936. *Malpighamoeba locustae*, n. sp. (Amoebidae), a protozoon parasitic in the malpighian tubes of grasshoppers. *Trans. Am. Microsc. Soc.*, 55:6-10

Larsson, J. I. R., de Roca, C. B. & Gaju-Ricart, M. 1992. Fine structure of an amoeba of the genus *Vahlkampfia* (Rhizopoda, Vahlkampfiidae), a parasite of the gut epithelium of the bristletail *Promesomachilis hispanica* (Microcoryphia, Machilidae). *J. Invert. Pathol.*, 59:81-89.

Prell, H. 1926. Beiträge zur Kenntnis der Amöbenseuche der erwachenen Honigbiene. *Arch. Bienenkunde*, 7:113-121.

Purrini, K. 1980. *Malamoeba scoltyi* sp. n. (Amoebidae, Rhizopoda, Protozoa) parasitizing the bark beetles, *Dryocoetes autographus* Ratz., and *Hyalurgops palliatus* Gyll. (Scolytidae, Coleoptera). *Arch. Protistenkd.*, 123:358-366.

Taylor, A. B. & King, R. L. 1937. Further studies on the parasitic amoebae found in grasshoppers. *Trans. Am. Microsc. Soc.*, 56:172-176.

SCHIZOCLADIDAE

by

TOMAS CEDHAGEN and STEFAN MATTSON

Reticulate plasmodium, with few, large nuclei, embedded in a matrix consisting of endogeneous and exogeneous particles in a clear, mucilaginous substance. No solid wall structures. Longitudinal supportive rods run through the plasmodium and the matrix. Body, up to 10 mm, branched dichotomously 3-4 times, bush-like, funnel-shaped (Fig. 1). Branching dichotomous, achieved by splitting of the branches of the latest order simultaneously along their entire length. Eleven distal branches in holotype; branches nearly circular in cross-section.

Body surface even and smooth. Brown-yellow grey with an olive tinge. A mainly longitudinal arrangement of the plasmodium and matrix makes the body look longitudinally striated (Fig 1). Body very soft, non-elastic, droops even when submerged in water. A pressure mark made on the body remains without any tendency of reversion.

The plasmodium is formed as a network of irregularly anastomosing sheets and strings, mainly running longitudinally. Few nuclei, 33-44 µm in diameter; each contains a large number (>1,000) of small bodies, possibly nucleoli, ca. 0.5 µm in diameter. The plasmodium also contains some scattered particles similar to those in the matrix (see below).

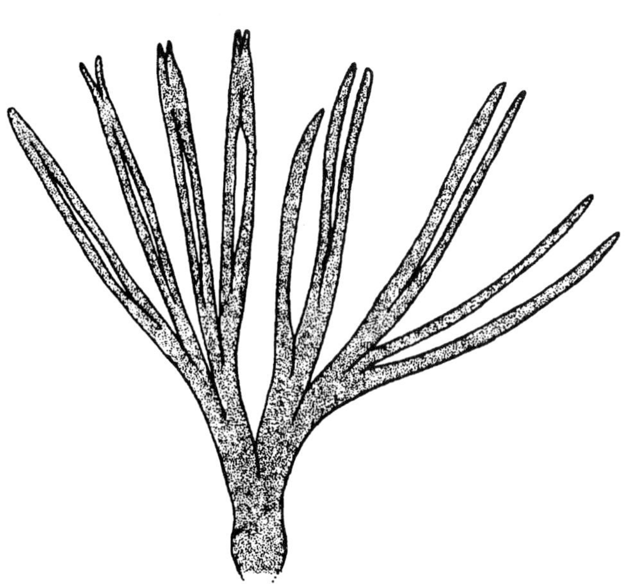

Fig. 1. *Schizocladus sublittoralis.* Diagrammatic sketch showing the branching pattern. Length=10 mm.

Plasmodium interwoven with a reticulate matrix. Matrix semitransparent, consisting of endogeneous and exogeneous particles in a clear, mucilaginous substance and of the same consistency as the plasmodium. Relative volume of matrix largest near base of body and decreasing towards branch ends. There are two kinds of endogeneous particles, the most numerous being flattened, rounded, clear crystals of 4-24 µm in diameter, which are probably calcareous. The other kind are rounded, yellowish brown xanthosomes (11-35 µm in diameter). The exogeneous particles are small detritus particles, such as mineral grains, fragments of sponge spicules and crustacean exoskeletons, diatom tests, etc. Each branch contains at least one longitudinal, unbranched, elastic, supportive rod running through the plasmodium and the matrix. The rods are >1mm, circular in cross-section, 3.5-8 µm thick, of constant diameter, and have a smooth surface. They are composed of longitudinal fibers, but the fibers are only visible as such in splintered rods.

Schizoclada differ from the Xenophyophorea and the Foraminifera primarily in its lack of solid wall structures. Its basic organization can be likened to that of a xenophyophore without test and organic lining. Distribution: On clay bottoms at 55-75 m depth in Gullmarsfjorden, Swedish west coast. Monospecific.

Type species: *Schizocladus sublittoralis* Cedhagen & Mattson, 1992

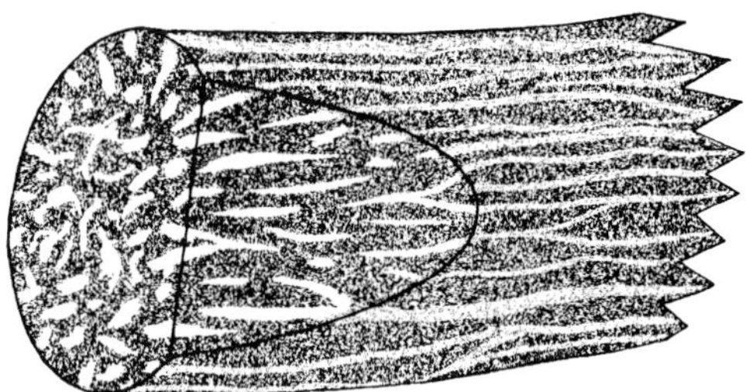

Fig. 2. *Schizocladus sublittoralis.* Diagram of branch fragment, cut transversely and obliquely, showing organization of plasmodium (light) and matrix (dark).

LITERATURE CITED

Cedhagen, T., Mattson, S. 1992. *Schizocladus sublittoralis* gen. et sp. n. (Protozoa: Sarcodina: Schizocladea classis n.) from the Scandinavian sublittoral. *Sarsia*, 76:279-285.

TESTATE AMOEBAE WITH FILOPODIA

by RALF MEISTERFELD

Amoebae with filopodia, and cell body in a test.

Species with these characters are traditionally grouped in the subclass Testaceafilosia de Saedeleer, 1934 or the order Gromiida Claparède & Lachmann, 1858 (e.g. Lee et al., 1985). It is now generally accepted that they do not form a natural assemblage. *Gromia* with its complex life cycle, gametes and test structure probably has closer affinities to the foraminifera and can no longer be considered a member of the Filosea. The ordinal name Gromiida should not be used for filose testate amoebae (Page, 1987).

Morphology of the shell. Filose testate amoebae have **proteinaceous** or **agglutinated** shells. The former can either be flexible (Chlamydophryidae) and sometimes very thin (*Rhogostoma*), or rigid

(*Archerella*). Organic tests built from single units, as in the lobose testate amoebae such as the Arcellidae, have not been described from the filose testate amoebae. The amount of organic cement varies and may dominate an agglutinated shell, or it may be limited to small cement strands to glue the mineral particles together. In all genera studied it is sheet-like.

The most homogenous group are the Euglyphida with their siliceous test composed of secreted plates. These plates differ in shape, size, and arrangement from one genus to another and also from one species to another. They are often used as a character for identification.

Morphology of the cell. Only a few genera have been studied by electron microscopy: *Chlamydophrys* (Peshkov and Akinshina, 1963), *Rhogostoma* (Simitzis and Le Goff, 1981), *Archerella* (Bonnet et al., 1981b), *Corythion* (Ogden, 1991c), *Euglypha* (Hedley and Ogden, 1973, 1974b; Ogden, 1979c), *Tracheleuglypha* (Ogden and Coûteaux; 1987), *Trinema* (Hedley and Ogden, 1974a) and *Paulinella* (Kies, 1974). The cells are usually divided into a highly vacuolar anterior region rich in food vacuoles, a dark band (pigment zone) with cement vesicles and excretion crystals, and a dense compact posterior region, which contains the nucleus surrounded by a dense rough endoplasmatic reticulum. The spherical nucleus is usually single, although in cultures of *Euglypha,* cells may have two or three. The nuclei can be classified either ovular or vesicular (Raikov, 1982). In the former there are many nucleoli while in the latter the nucleolus is single and situated in a finely granular nuclear matrix with dispersed small concentrations of chromatin. In many genera the type of nucleus is a constant character.

Golgi complexes (dictyosomes) participate in the production of siliceous building elements (Euglyphida) and in the organic cement. In well nourished cells there are large electron-dense vesicles situated anterior to this zone. **Reserve plates** are stacked at the margins of the posterior part of the parental cell prior to formation of daughter cells.

Reproduction. All species reproduce by binary fission. In the Chlamydophryidae a transition exists from genera with longitudinal division to those with division by budding. *Clypeolina* has rigid shells composed of two valves, one of which is replaced by a new one during longitudinal fission. In the few species studied by transmission electron microscopy (Ogden, 1979c; Ogden and Coûteaux, 1987), nuclei divide by closed **orthomitosis** and have intranuclear MTOC's. When the new shell is formed, some cytoplasm is extruded, and reserve plates are sequentially arranged to form the test, beginning with the apertural plates. The plates are moved and held in position with microfilaments which are connected with adhesion plaques (Hedley and Ogden, 1974b). When the often genus-specific pattern is finished, cement vesicles discharge their fibrillar content into the inner casing and the plates are fixed onto a sheet of cement. In genera like *Euglypha* this cement cannot be seen from the outer surface of the test (although exceptions do occur, e.g. *E. penardi* Meisterfeld, 1979a).

Ecology. Filose testate amoebae have a world-wide distribution. They are common in the same habitats as the lobose testate amoebae and have similar habitat demands as them. Usually they are studied together with the same methods. For general information about their ecology see the corresponding section of the Arcellida chapter.

Most of the Chlamydophryidae have not been reported recently and little ecological information is available. Species of (as examples) *Chlamydophrys* or *Capsellina* seem to prefer microhabitats rich in organic material substrates, which are not rare in our polluted environment. On the other hand *Penardeugenia*, *Diaphoropodon*, and *Clypeolina* have only been found in oligotrophic, neutral water bodies, a habitat type which is increasingly rare due to eutrophication and atmospheric acidification (at least in Central Europe).

Filose testate amoebae do not fossilize very well. Different genera of Euglyphida have been found in Pleistocene peats and sedimentary deposits. Usually only shell fragments (plates, spines) are preserved and used as indicators of bog development in the catchment area of lakes (Douglas and Smol, 1987).

Phylogeny and Classification. Only a few testate amoebae with filopodia have been studied with modern methods. A recent ribosomal RNA study (Bhattacharya et al.,1995) places two members of the Euglyphida (*Euglypha rotunda* and *Paulinella chromatophora*) and the chlorarachniophyte algae together. Our own sequence data of several genera (Wylezich, Meisterfeld and Schlegel in prep.) show that the Euglyphida are monophyletic and support the view of Cavalier-Smith (1997) and Cavalier-Smith & Chao (1997) that the euglyphid testate amoebae form a sister group to flagellates like *Cercomonas*. So far this lineage is isolated and the affinities to other filose amoebae are not clear. Whether one or more of the types of filose amoebae have evolved from a testate state (Cavalier-Smith, 1993) and whether the siliceous building elements of the Euglyphida are an ancestral character is still speculative given our present knowledge. Alternatively the Chlamydophryidae could be placed as the most primitive members, at the base of this group.

The siliceous body-plates are considered to be homologous in all Euglyphida. Our molecular data lead to a first picture of the relation of the families and genera in this order. According to this data, *Paulinella* branches earliest followed by *Tracheleuglypha*, the next group is formed by *Trinema* and *Trachelocorythion* and the most highly evolved genus is *Euglypha*. This branching sequence is also reflected in a more and more complex shell morphology. This is in contrast to the scheme proposed by Schönborn (1989). Based on the topophenetic analysis in which he correlates morphological trends of the shell with habitat structure, Schönborn derives other genera from *Euglypha*.

During the last twenty-five years a new and rich association of genera with unclear affinities was discovered in marine interstitial habitats (Golemansky, 1970a, b; 1974; 1991; Chardez, 1977). Psammonobiotic (living in interstitial habitats) taxa have funnel shaped apertures or large apertural collars. These features are considered to be adaptations to the life on sand grains or to the interstices between them and therefore do not necessarily indicate that all species are closely related. Genera with similar test shape but different types of shell construction are placed in different families. Further information on ultrastructure and molecular biology may help to group the taxa correctly.

Due to the lack of a generally accepted phylogeny, the testate amoebae with filose pseudopodia are grouped here in the order Euglyphida and the families Chlamydophridae, Psammonobiotidae, Volutellidae, Pseudodifflugiidae, Amphitremidae, and a few genera *incertae sedis*. The Euglyphida are traditionally subdivided into four families, but with the availability of more sequence data their composition will change. Sudzuki (1979a) has proposed a family Pseudocorythionidae for psammobiont genera such as *Pseudocorythion* or *Messemvriella* which have funnel-shaped apertures and large apertural collars. In the present classification these genera are placed with the Cyphoderiidae because no ultrastructure data are available and intermediate genera like *Campascus* partly close the gap between *Cyphoderia* and (for example) *Pseudocorythion*. Genera with tests built with external material belong to the Psammonobiotidae.

Genera, such as *Ampullataria*, *Euglyphidion*, *Heteroglypha*, which are based only on descriptions of empty shells and which have no clear affinities to any of the families of Euglyphidae (although they are often placed with them) are placed here *incertae sedis*.

Observation and Identification. Many of the filose testate amoebae with firm shells are small. To study their test morphology on which the classification is based, scanning electron microscopy is the most appropriate technique. Such microscopes are not always readily available and for ecological studies are not practical. Until now only the more common species have been studied by scanning electron microscopy, and it has not always been easy to coordinate these new findings with earlier descriptions. These problems are partly caused by the limitations of light microscopy but more often by inadequate diagnoses. Although electron microscopy should be used for modern descriptions, most species can be identified using light microscopy, but often phase-contrast or interference-contrast is necessary. For bright field microscopy, it can help to mount empty tests of Euglyphida in media with high refractive indices such as Hyrax®, Naphrax® or Styrax®. Identification is often based on characters such as the shape of the shell, the kind and pattern of body plates, and sometimes the type of nucleus (e.g. number of nucleoli). Chlamydophryidae are best detected and observed in fresh field or raw cultured material where pseudopods help to distinguish them from other microbiota. It is usually not possible to see filopods in fixed cells and the organisms are easily overlooked in samples made for ecological studies.

A promising approach for the identification of clones of lobose as well filose testate amoebae is RAPD-PCR (Wanner et al., 1997).

Literature for identification. There is no modern monograph of the filose testate amoebae available. Penard (1902), Cash et al. (1915), Grospietsch (1972a) and Bonnet and Thomas (1960) are useful. Decloitre has published a series of compilations (*Euglypha* 1962, 1976b, 1979b, 1982, 1986; *Trinema*: 1981). Although uncritical and without any discussion, these papers can help to find an access to original descriptions.

KEY TO THE ORDERS, FAMILIES AND GENERA *INCERTAE SEDIS*

1. Test rigid, with secreted siliceous scales often arranged regularly........... Order **Euglyphida**
1'. Test either organic or agglutinate (foreign material in a cement matrix) or if covered with triangular scales and short spines, then flexible apertural region.............................. 2

2. Test with one apertural opening..................... 3
2'. Test with two apertural openings..................... .... Family **Amphitremidae**

3. Whole test or aperture flexible or shell bilayered with two outer valves and an inner flexible sac, enclosing the cell........................ .... Family **Chlamydophryidae**
3'. Test rigid.. 4

4. Test, a hemispherical cup with open ventral face, cell in a membranous sac................. .... *Frenzelina*
4'. Test not a hemispherical cup.......................... 5

5. Aperture surrounded by thickened lobes......... ....*Feuerbornia*
5'. Aperture without lobes.................................. 6

6. Shell half-coiled.. 9
6'. Shell not half-coiled...................................... 7

7. Aperture terminal, or simple opening.............. .... Family **Pseudodifflugiidae**
7'. Aperture with collar or funnel-like............. 8

8. Aperture like a funnel or with disk-shaped collar .... Family **Psammonobiotidae**
8'. Aperture with scalloped collar........................ .... Genus *Rhumbleriella*

9. Shell composed of small sausage-like plates... .... *Lesquerella*
9'. Shell organic or with few attached mineral particles....................... Family **Volutellidae**

FAMILY CHLAMYDOPHRYIDAE de Saedeleer, 1934

With a more or less flexible test or **tectum** which may be distorted by the contraction of the cell. Some species with attached foreign particles, scales, spines, or spicules; division often longitudinal.

KEY TO THE GENERA OF THE FAMILY

1. Aperture terminal.. 2
1'. Aperture internalized................................... 7

2. Test with smooth surface.............................. 3
2'. Test with scales or attached particles............ 5

3. Plasma fills the test completely, refractile granules not restricted to a median band, tectum flexible .... *Lecythium*
3'. Plasma does not fill the test completely, refractile granules concentrated in a median band...4

4. Cell divide by budding, test more or less firm. .... *Chlamydophrys*
4'. Longitudinal division, test at aperture flexible .... *Leptochlamydophrys*

5. Test with triangular embedded plates, and short spines which may be difficult to observe.......... .... *Penardeugenia*
5'. Test with attached particles, some with spines 6

6. Test compressed, composed of two valves.......... .... *Clypeolina*
6'. Test ovoid, not composed of two valves, but with thin spines................... *Diaphoropodon*

7. Tectum with external layer of debris................ .... *Capsellina*
7'. Tectum smooth............................ *Rhogostoma*

Genus ***Chlamydophrys*** Cienkowsky, 1876

<u>Test</u>: clear in some species difficult to observe, ovoid, longish, or broad pyriform, round in cross-section, more or less firm. ,Plasma nearly fills shell with transversal layer of brilliant granules (Fig. 1). ,Nucleus: vesicular. ,In culture most species form rosette-like colonies. Division: budding, closed **orthomitosis**. Pseudopods may anastomose. All species small (<55 µm). Some species form round brownish cysts, in collapsed test. Feeding: bacteria. Habitat: freshwater, one endozooic (gut of lizards), and 2 marine species. Eight species. Refs. Belar (1921), de Saedeleer (1934), Peshkov and Akinshina (1963) (TEM). A very similar genus is *Leptochlamydophrys* Belar,

1921. In contrast to *Chlamydophrys*, longitudinal division (Belar, 1921). Ref. Breuer (1917).
TYPE SPECIES: *Chlamydophrys stercorea* Cienkowsky, 1876.

Genus ***Penardeugenia*** Deflandre, 1958

Test: round; with triangular, embedded siliceous plates, 2 µm long, in rows; with siliceous spines 10 µm long projecting between plates (Fig. 2). Vesicular nucleus. Size: 25 to 35 µm diameter. **Pseudostome**: flexible. Feeding: herbivorous. Habitat: deep oligotrophic lakes, on bottom. Monospecific. Other names: *Pamphagus bathybioticus, Eugenia bathybioticus* (homonym, was replaced by Deflandre [1958]). Ref. Penard (1904).
TYPE SPECIES: *Penardeugenia bathybiotica* (Penard, 1904).

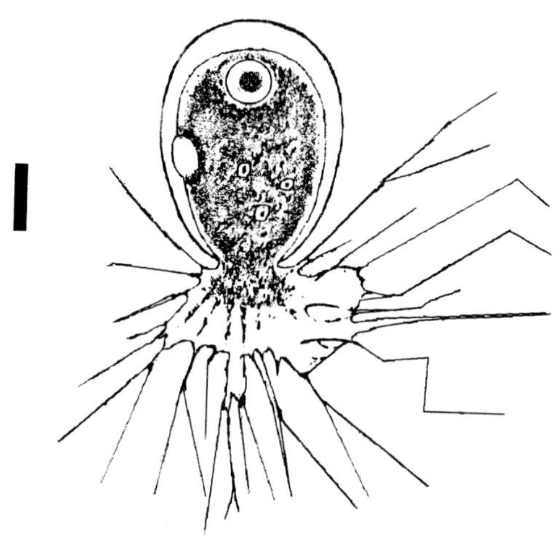

Fig. 1. *Chlamydophrys stercorea*. Scale bar=10 µm. (After Hoogenraad and De Groot, 1940)

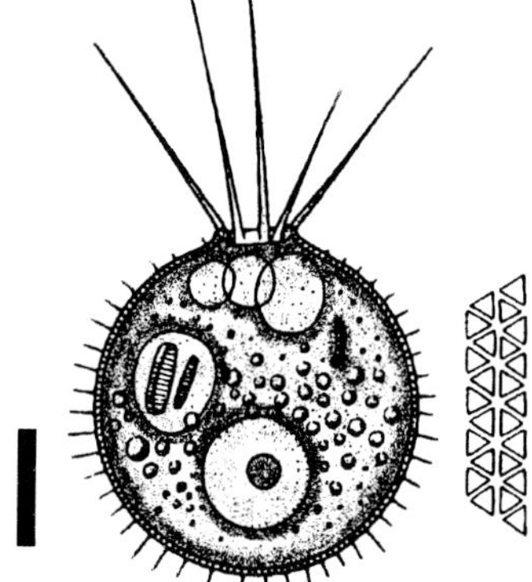

Fig. 2. *Penardeugenia bathybiotica*. Scale bar=10 µm. (After Deflandre, 1953)

Genus ***Lecythium*** Hertwig & Lesser, 1874

Test: round, lenticular or pyriform, flexible membrane, follows size changes of plasma during feeding or starvation (Fig. 3). Aperture often between folds. Plasma fills the test completely, nucleus vesicular or ovular. Division longitudinal, closed orthomitosis, nucleolus disintegrates (*L. hyalinum*). Habitat: freshwater; in culture the individuals are found in groups. Feeding: herbivorous and bacterivorous. Other names: *Pamphagus,* Bailey (a homonym!); *Baileya* Awerintzew (unjustified replacement name). About 6 species. Refs. Belar (1921), de Saedeleer (1934).

TYPE SPECIES: *Lecythium hyalinum* (Ehrenberg, 1838).

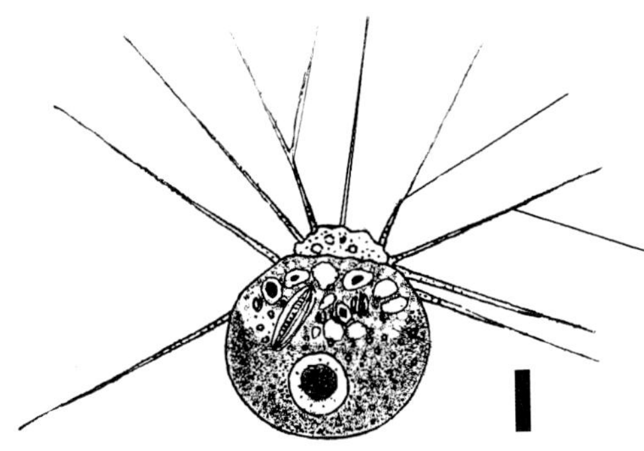

Fig. 3. *Lecythium hyalinum*. Scale bar=10 µm. (After Penard, 1902)

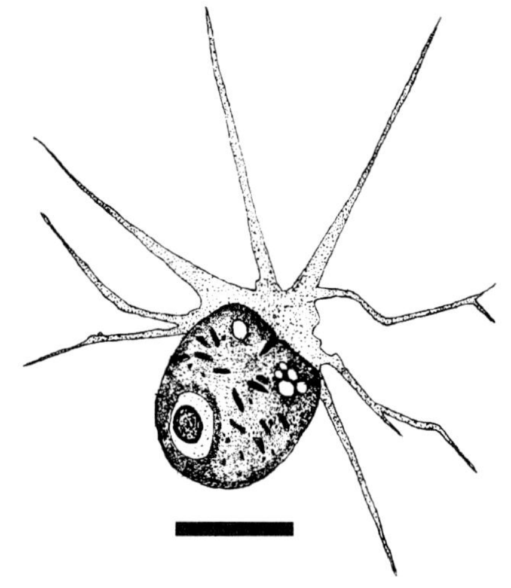

Fig. 4. *Rhogostoma schuessleri*. Scale bar=10 µm. (After Simitzis and Le Goff, 1981)

Genus ***Rhogostoma*** Belar, 1921

Test: thin, flexible, transparent membrane, subspherical, compressed, no attached material. Pseudostome: a slit at end of oval invagination of test, perpendicular to the compression (Fig. 4). Plasma fills the test completely, vesicular nucleus. Division longitudinal, open orthomitosis, nucleolus disintegrates. Feeding: bacteria. Habitat: freshwater debris, mosses, manure. Two species: *Rhogostoma schuessleri* Belar, 1921. According to Simitzis and Le Goff (1981) a species complex. Refs. Belar (1921), for ultrastructure: Simitzis and Le Goff (1981).

Genus ***Capsellina*** Penard, 1909

Test: ovoid, greyish, shaped like *Rhogostoma*; but with attached brown debris. Pseudostome a slit at base of a deep invagination (Fig. 5). Plasma fills the test completely; nucleus spherical with several nucleoli. Division longitudinal. Contractile vacuole 10 to 12 µm. Feeding: herbivorous. Monospecific. Note: *C. timida* Brown, 1911 belongs to *Rhogostoma* (Simitzis & Le Goff, 1981). Habitat: sphagnum mosses. Ref. Penard (1909).
TYPE SPECIES: *Capsellina bryorum* Penard, 1909

Genus ***Diaphoropodon*** Archer, 1869

Test: ovoid, clear, membranous, flexible; attached particles. The key character is the hair-like, 8 to 10 µm long spines (Fig. 6). Pseudostome terminal, of variable shape. Endoplasm grey, granular, 1 or 2 contractile vacuoles. Nucleus according to De Saedeleer (1934) ovular. Longitudinal division? Feeding: herbivorous. Habitat: freshwater plants, rare. Two species. Ref. Penard (1902).
TYPE SPECIES: *Diaphoropodon mobile* Archer, 1869

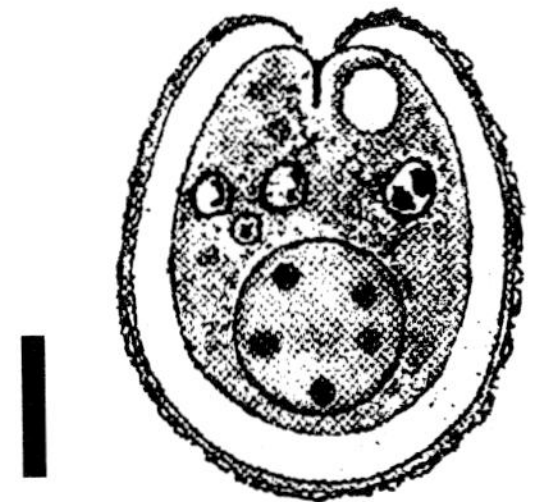
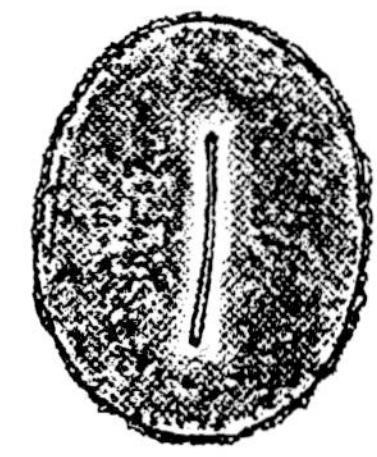

Fig. 5. *Capsellina bryorum*. Lateral and top view. Scale bar=10 µm. (After Penard, 1909)

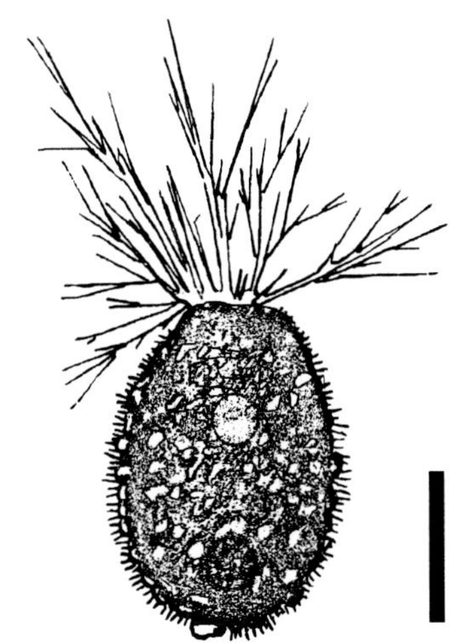

Fig. 6. *Diaphoropodon mobile*. Scale bar=50 µm. (After Deflandre, 1953)

Genus ***Clypeolina*** Penard, 1902

Test ovoid, 2-layered, outer of opposed chitinoid valves covered with small scales, forms keel; inner a thin flexible sac closely investing the cell (Fig. 7). Pseudostome terminal, wide, oval, or linear. Endoplasm sometimes with zoochlorellae. Pseudopods extremely thin. Nucleus spherical, irregularly shaped endosome (nucleolus). Reproduction by longitudinal fission, one valve is replaced by a new one. Feeding: algivorous. Habitat: oligotrophic freshwater, sediments or on plants. Monospecific, rare. Refs. Penard (1907), Pateff (1926).
TYPE SPECIES: *Clypeolina marginata* Penard, 1902

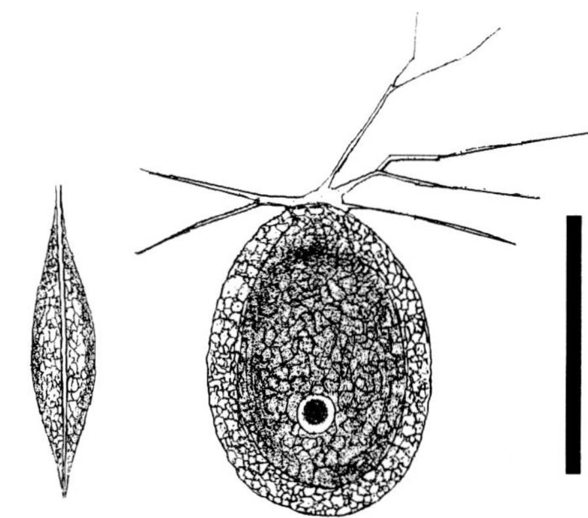

Fig. 7. *Clypeolina marginata*. Frontal and lateral view. Scale bar=100 µm. (After Penard, 1907)

FAMILY PSEUDODIFFLUGIIDAE de Saedeleer, 1934

Test rigid, agglutinated. Cell does not fill test. Division by budding. De Saedeleer (1934) included *Clypeolina* and *Frenzelina* in this family. *Clypeolina* is grouped here with the

Chlamydophryidae because both have longitudinal division. *Frenzelina* is so different that it has no clear affinities to any of the families and is placed *incertae sedis*. Recently Anderson et al. (1996) have described the marine genus *Ovulina*, which is provisionally placed in this family.

KEY TO THE GENERA OF THE FAMILY

1. Organic shell, <20 µm.................... ***Ovulinata***
1'. Test with xenosomes............ ***Pseudodifflugia***

Genus ***Ovulinata***
Anderson, Rogerson & Hannah, 1996

Test ovoid, small (~15 µm), organic, lacking scales or mineral particles. Aperture terminal, round to oval, sometimes with a slightly thickened border (Fig. 8a). Surface somewhat granular (SEM). Nucleus large, vesicular. Feeding: bacterivorous. Habitat: sandy marine sediments. The amoebae grow in a broad range of salinities (fresh to seawater). Distribution: known only from collection in Scotland. Possibly overlooked elsewhere because of its small size. Monospecific. Ref. Anderson, Rogerson & Hannah (1996).
TYPE SPECIES: *Ovulina parva* Anderson, Rogerson & Hannah, 1996

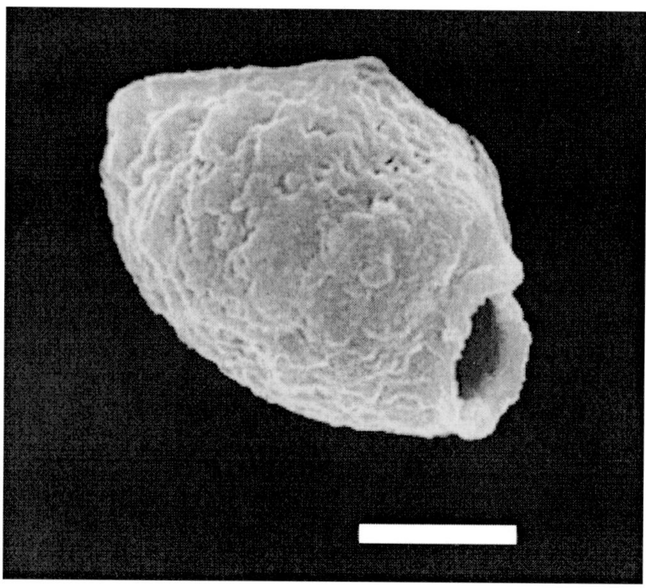

Fig. 8a. *Ovulinata parva.* Scale=5 µm. (modified from Anderson et al., 1996)

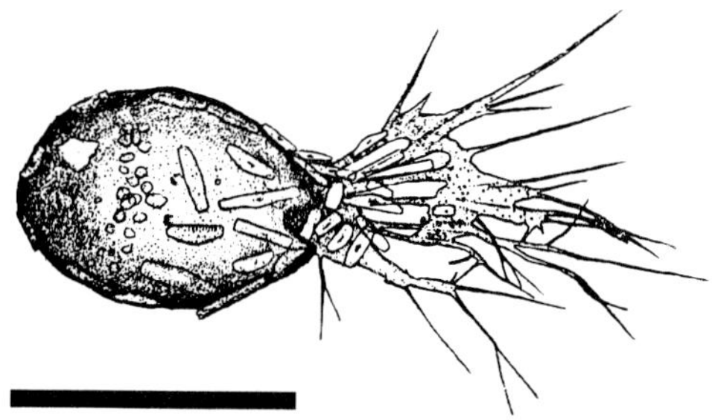

Fig. 8b. *Pseudodifflugia gracilis.* Scale bar=50 µm. (After de Saedeleer, 1932)

Genus ***Pseudodifflugia*** Schlumberger, 1845

Small species with agglutinate test, in general appearance similar to *Difflugia* but with filopodia (Fig. 8b). Test firm, chitinoid or completely covered with attached quartz grains. Shape often variable ovoid to subspherical. Feeding: herbivorous. Species of this genus can be found in freshwater, soil, and marine littoral sands. About 20 species, often with incomplete descriptions. Refs. Cash et al. (1915), Ogden and Hedley (1980).
TYPE SPECIES: *Pseudodifflugia gracilis* Schlumberger, 1845.

FAMILY PSAMMONOBIOTIDAE Golemansky, 1974

Test more or less shaped like a Greek vase, with funnel-shaped or disk-like apertural collar; aperture may be terminal or more or less bent to one side; clear or with attached or embedded particles. The concept of the Family applied here differs from that of Golemansky (1974) in that genera with round or oval siliceous **idiosomes** are excluded and placed with the Euglyphida; those with a terminal collar like *Micramphora*, *Nadinella,* and *Ogdeniella* are included. Whether these genera form a natural group is open to dispute. The striking similarity of the apertural structures is perhaps only a convergence as a result of the selective pressure in the interstitial environment. The genera of this family are separated mainly by characters of the shell. None of the species has been established in clonal culture; a critical evaluation of the diagnostic characters is lacking. The distinctions between the genera have been partly lost by recent species descriptions. This also may make identification to genus difficult.

KEY TO THE GENERA OF THE FAMILY

1. Aperture terminal 2
1'. Aperture bent to one side................................... 4

2. Shell circular in cross-section or often compressed, apertural collar disk-shaped.......... 3
2'. Shell pot-shaped, aperture funnel-shaped....... .... ***Micramphora***

3. Apertural collar completely hyaline, freshwater species................................ ***Nadinella***
3'. Apertural collar encrusted with mineral particles, marine species .... ***Ogdeniella***

4 . Test organic, clear, no structure visible by light microscopy 8
4'. Test with mineral particles.......................... 5

5. Shell with lateral and terminal spines........... .... ***Alepiella***
5'. Fundus without lateral appendices................ 6

6. Apertural collar large disk, thin; marine or freshwater species........... ***Psammonobiotus***
6'. Apertural collar short, angle with long axis usually less than 90°.................................... 7

7. Marine species.................... ***Micropsammella***
7'. Soil species.......................... ***Edaphonobiotus***

8. Shell with caudal spine.................. ***Chardezia***
8'. Fundus rounded, collar short .... ***Propsammonobiotus***

Genus *Micramphora* Valkanov, 1970

Test rounded, pot-shaped, with somewhat constricted neck and a funnel-shaped terminal pseudostome (Fig. 9). Shell transparent, chitinoid. Most species small (<30 µm). Habitat: marine interstitial. Six species, all marine. A similar genus with an irregular apertural funnel but lacking a neck is *Micramphoraeopsis* Sudzuki (1979b). Whether it is a valid genus is yet to be confirmed. Refs. Valkanov (1970), Chardez (1977).

TYPE SPECIES: *Micramphora pontica* Valkanov, 1970

Genus *Ogdeniella* Golemansky, 1982

Test: ovoid elongate, in contrast to *Micramphora* laterally compressed, in lateral view, on the narrow side fundus lanceolate, broad side with round base; constricted neck and disk-shaped apertural collar. **Pseudostome** round to oval (Fig. 10). Shell composed of flat amorphous mineral particles. It is assumed that these elements are endogenous, but this has yet to be proven. Habitat: marine interstices, salinity 0.035% to 3.08%. Eight species. First described as *Amphorellopsis* Golemansky, 1970. Ref. Golemansky (1982).

TYPE SPECIES: *Ogdeniella elegans* (Golemansky, 1970)

Fig. 9. *Micramphora pontica*. Scale bar=10 µm.

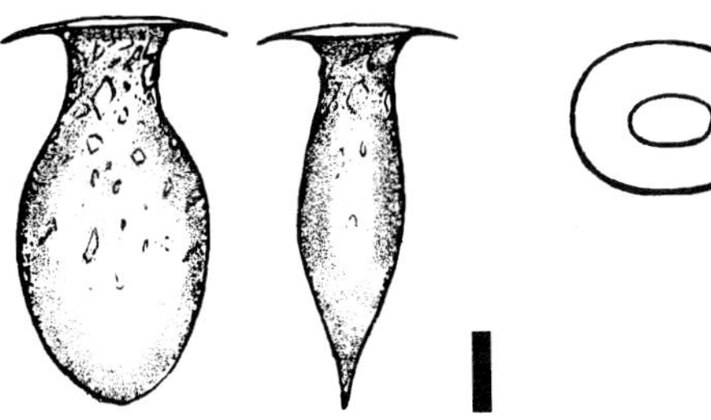

Fig. 10. *Ogdeniella elegans*. Scale bar=10 µm. (After Golemansky, 1970a)

Genus *Nadinella* Penard, 1902

Test pyriform, greyish yellow, chitinoid, fundus globular, neck compressed, becomes narrower towards the pseudostome; attached siliceous **xenosomes** (Fig. 11). Pseudostome elliptic, with large completely hyaline rim, flared at right angle to aperture, slightly recurved. Endoplasm granular; does not completely fill test, one contractile vacuole. Few filopods, rapidly moving. Nucleus spherical. Feeding: herbivorous. Habitat: freshwater, benthos and Aufwuchs. Monospecific. Ref. Penard (1902).

TYPE SPECIES: species: *Nadinella tenella* Penard, 1902

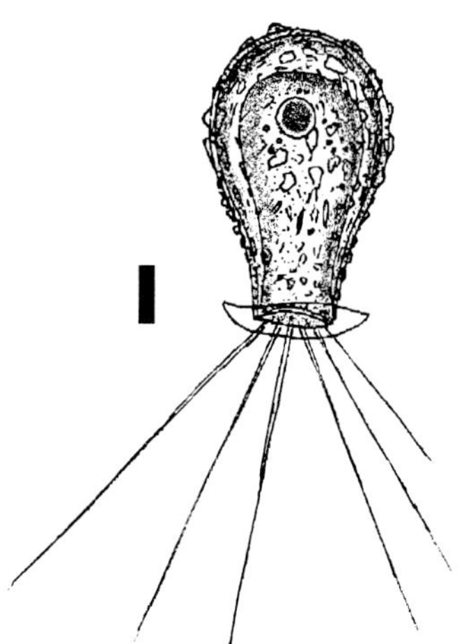

Fig. 11. *Nadinella tenella.* Scale bar=10 μm. (After Penard, 1902)

Genus ***Micropsammella*** Golemansky, 1970

Test ovate elongate, dorso-ventrally compressed, ventral face flat, constricted neck, pseudostome at 45-90° angle to long axis, with chitinoid collar (Fig. 12). Shell transparent, composed of small angular mineral particles. Habitat: marine interstitial. Two species. Refs. Golemansky (1970a), Golemansky and Coûteaux (1982), SEM Ogden and Coûteaux (1989).
TYPE SPECIES: *Micropsammella retorta* Golemansky, 1970

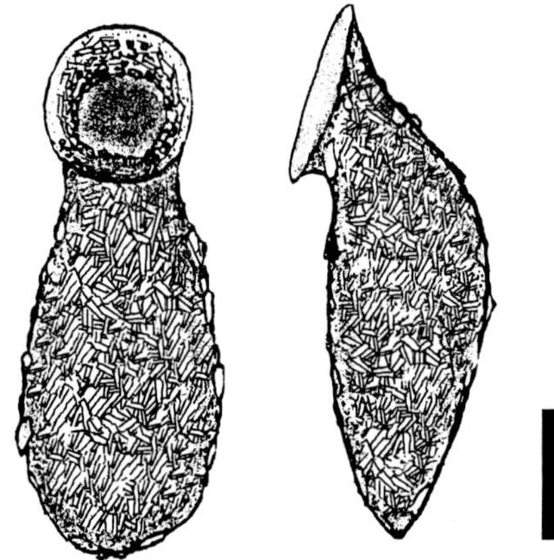

Fig. 12. *Micropsammella retorta.* Scale bar=10 μm.

Genus ***Chardezia*** Golemansky, 1970

Test clear, structureless, flask-shaped, with apertural collar, **pseudostome** bent at ≈45° angle to long axis; base of test drawn out as spine. Size: 38 to 50 μm long (Fig. 13). Filopodia. Ecology: marine, sandy beaches, algivore. Monospecific. Ref. Golemansky (1970c).
TYPE SPECIES: *Chardezia caudata* Golemansky, 1970

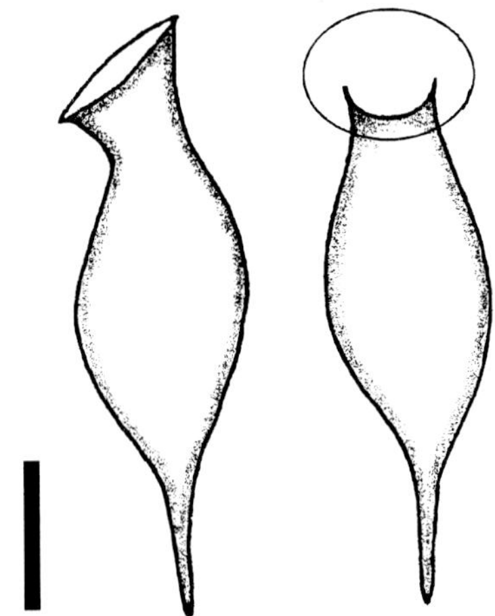

Fig. 13. *Chardezia caudata.* Scale bar=10 μm. (After Golemansky, 1970c)

Genus ***Alepiella*** Golemansky, 1970

Shell dorso-ventrally compressed, with 1 terminal and 2 forward curved, blunt, lateral spines. Pseudostome with funnel-shaped collar at ≈60° angle to long axis of test. Shell composed of calcareous and siliceous mineral particles bound by a structured organic cement. Pseudopodia unknown. Habitat: marine, sandy beaches. Monospecific (Fig. 14). Ref. Ogden and Coûteaux (1989).
TYPE SPECIES: *Alepiella tricornuta* Golemansky, 1970.

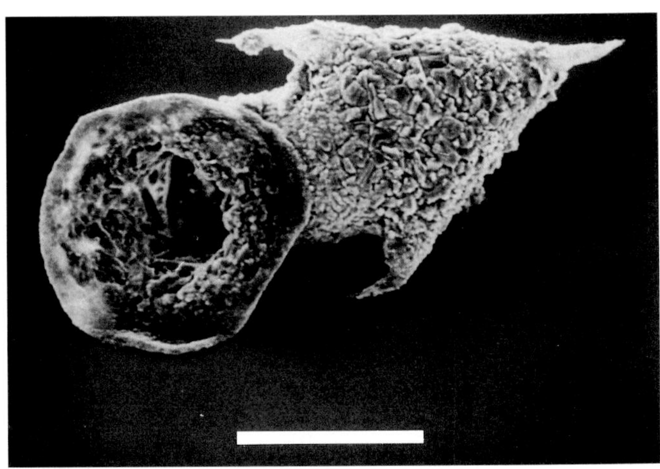

Fig. 14. *Alepiella tricornuta.* Scale bar=20 μm. (From Ogden and Coûteaux, 1986)

Genus ***Propsammonobiotus*** Golemansky, 1991

Shell ovoid, cross-section round, pseudostome bent with an angle of app. 45°, short collar. Test:

organic, transparent (Fig. 15). Pseudopodia unknown. Plasma does not fill the shell. Habitat: marine interstitial. Monospecific. Ref. Golemansky, 1991.
TYPE SPECIES: *Propsammonobiotus atlanticus* Golemansky, 1991.

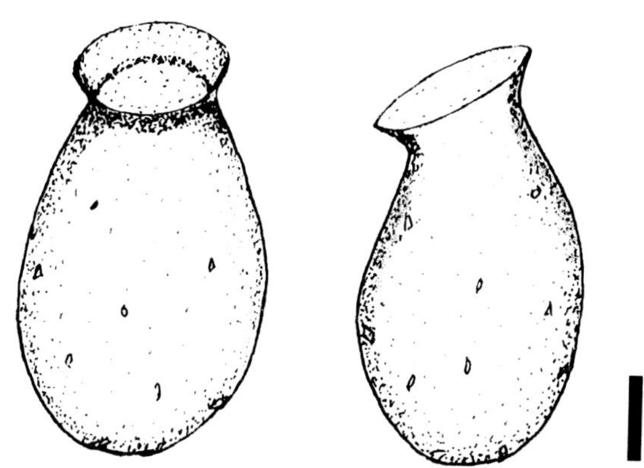

Fig. 15. *Propsammonobiotus atlanticus*. Scale bar=10 µm. (After Golemansky, 1991)

Genus ***Psammonobiotus*** Golemansky, 1967

Test flask-shaped, oval in cross-section with hemispherical base, pseudostome at ≈ 90° angle to long axis, large hyaline collar. Filopodia, alveolar nucleus (*P. minutus*). Shell transparent, composed of flat and a few larger attached mineral particles. Habitat: fresh water and marine interstitial. Seven species (Fig. 16). Refs. Golemansky (1967), Chardez (1977); for SEM, Ogden and Coûteaux (1989).
TYPE SPECIES: *Psammonobiotus communis* Golemansky, 1967

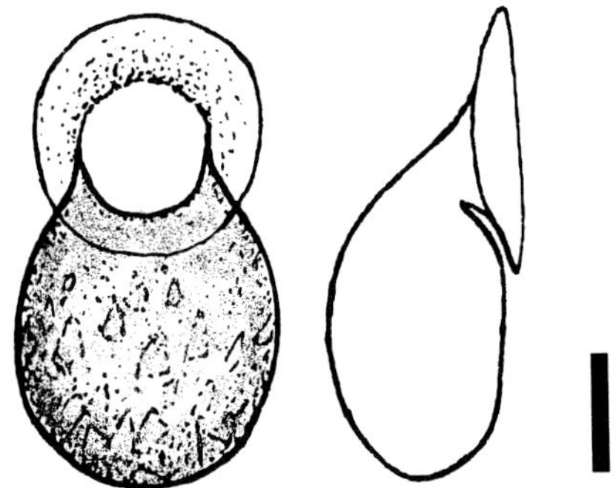

Fig. 16. *Psammonobiotus communis*. Scale bar 10 µm. After Golemansky (1967).

Genus ***Edaphonobiotus***
Schönborn, Foissner & Meisterfeld, 1983

Shell flask-shaped, curved neck which terminates in an expanded elliptical collar, cross-section circular, fundus rounded, test composed of xenosomes, fragile (Fig. 17). Pseudopodia not yet reported. Similar to *Campascus* but collar always present, differs from the marine genera *Psammonobiotus* Golemansky and *Micropsammella* Golemansky in the structure of the test, and most importantly, the biotope. Monospecific. The agglutinate test is formed by very delicate, mostly irregular platelets; the margins stain intensively with protargol. Plasma transparent, no zonation, one vesicular nucleus. Ecology: humus. Refs. Schönborn et al. (1983), Wanner and Funke (1986).
TYPE SPECIES: *Edaphonobiotus campascoides* Schönborn, Foissner & Meisterfeld, 1983

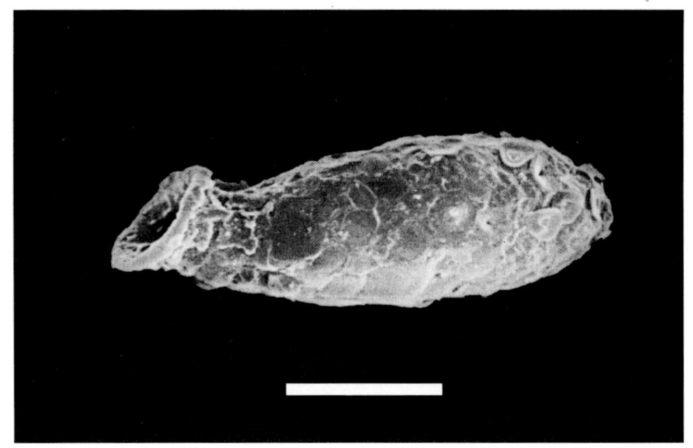

Fig. 17. *Edaphonobiotus campascoides*. Scale bar=10 µm.

FAMILY AMPHITREMIDAE Poche, 1913

Test has 2 pseudostomes; shell either organic or with attached extraneous material. Due to an error by de Saedeleer (1934) this family has been sometimes included with the Granuloreticulosea, but there is no doubt that *Amphitrema* and *Archerella* have filopodia. The pseudopodia of *Paramphitrema* differ significantly, and it is unclear whether the presence of two apertures is sufficient to place this genus in the Amphitremidae. *Diplophrys* and *Microcometes* have previously been included in this area, but *Diplophrys* has since been shown to have affinities with the Labyrinthulids (Dykstra & Porter, 1984) and *Microcometes* is a flagellate (see chapter on unassigned flagellates).

KEY TO THE GENERA OF THE FAMILY

1. Only one type of pseudopodia: numerous thin filopodia from each aperture......................... 2
1'. Two types of pseudopodia: either thin filopodia or one flagellum-like pseudopodium from each opening .... ***Paramphitrema***

2. Test with attached extraneous particles........... .... ***Amphitrema***
2'. Test with smooth surface.............. ***Archerella***

Genus ***Amphitrema*** Archer, 1869

Test elliptical, compressed, attaching a mixture of siliceous particles, diatom frustules, or flagellate cysts; a patterned cement network can be seen sometimes by scanning electron microscopy (Fig. 18). Pseudostomes: 2, at opposite ends, elliptical, with or without short collar. Only *A. wrightianum* and *A. stenostoma* have been studied in detail. In these species the protoplast nearly fills the test; many filopods from protoplast at pseudostomes, often branched, to 150 µm long. Nucleus: spherical, with few small nucleolar globules. Symbionts: zoochlorellae. Habitat: wet sphagnum, peat bogs; indicator species in palaeolimnology (Tolonen, 1985). Four species. Refs. Penard (1902), for SEM, Ogden (1984) (*A. stenostoma*?), Meisterfeld (1977), Tolonen et al. (1992).
TYPE SPECIES: *Amphitrema wrightianum* Archer, 1869

Fig. 18. *Amphitrema wrightianum*. The arrows mark the two apertures. Scale bar=20 µm.

Genus ***Archerella*** Loeblich & Tappan, 1961

Test thick, rigid, compressed, with parallel sides, rounded ends; wall organic, 3 layers, no xenosomes, brown due to iron accumulation (Fig. 19). Two pseudostomes at opposite ends, each elliptical, slightly thickened rim, very thin, short collar (at the limits of the resolution of the light microscope). Protoplast nearly fills test, green from zoochlorellae. Vesicular nucleus. Filopodia very thin, emerging from an internal ectoplasmic stalk at both apertures, rarely branching. Ecology: *A. flavum* is an important character species of upper green parts of wet sphagnum. (Schönborn, 1965b; Meisterfeld, 1977). Two species. *Ditrema mikrous* and *D. marina* de Saedeleer, 1934, which also have been assigned to this genus, have granuloreticulose pseudopodia and belong to the genus *Pseudoditrema* Deflandre, 1953. Ref. for TEM, Bonnet et al. (1981b).
TYPE SPECIES: *Archerella flavum* (Archer, 1877) (synonyms=*Amphitrema flavum, Ditrema flavum*)

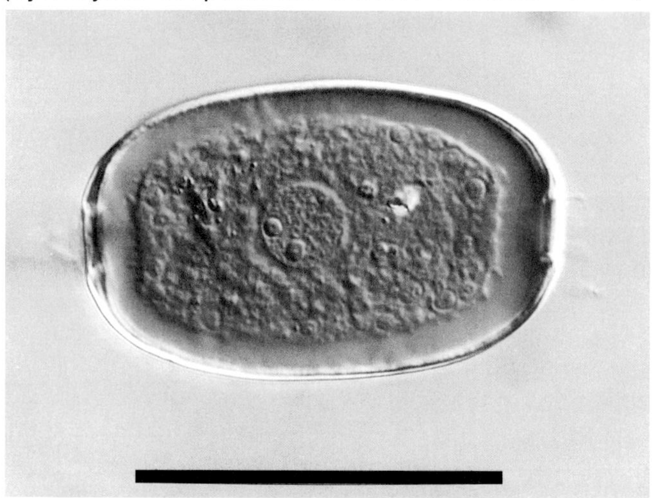

Fig. 19. *Archerella flavum*. The arrows mark the two apertures. Scale bar=50 µm.

Genus ***Paramphitrema*** Valkanov, 1970

Shell elliptical, compressed with 2 apertures at opposite poles more or less ending in 2 tubes (Fig. 20). Two types of filopodia, either thin and branching or thick with only 1 per pseudostome during rapid locomotion. Tubes usually long sometimes with terminal collar, deformable during feeding. Test insoluble in concentrated hydrochloric acid, covered with small mineral particles. Plasma fills the shell completely; bright granules and rusty brown pigment from the ingested diatoms. Ecology: marine and freshwater, algivorous. Three species. Two of the species were originally assigned to *Amphitrema* (Penard,

1903). It is not clear whether this genus is related to *Amphitrema* or *Archerella*. Ref. Valkanov (1970).
TYPE SPECIES: *Paramphitrema pontica* Valkanov, 1970

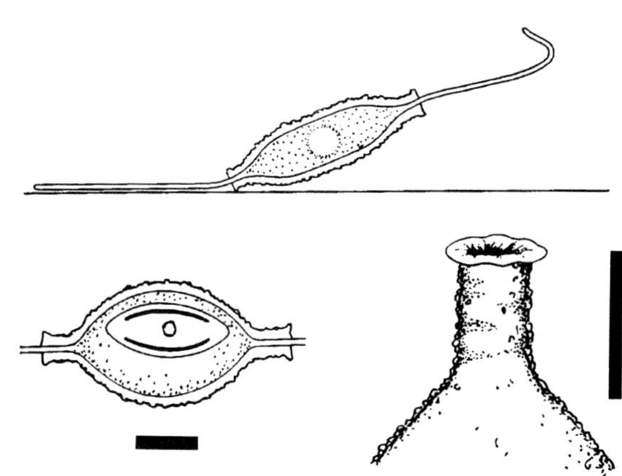

Fig. 20. *Paramphitrema pontica*. Scale bar=10 µm. (After Valkanov, 1970)

GENERA *INCERTAE SEDIS*

Genus ***Frenzelina*** Penard, 1902

Test clear, membranous, hemispherical cup with completely open ventral face, attached sand granules, in dorsal view often elliptic (Fig. 21). In contrast to Penard (1902) the cup is not gelatine-filled and does not hold a cytoplasmic sac (Hoogenraad and De Groot, 1940; Schönborn (pers. com.). Pseudopods numerous, very thin. Feeding: herbivorous. Habitat: freshwater, sewage. Two species. Refs: Penard (1902), Hoogenraad and De Groot (1940).
TYPE SPECIES: *Frenzelina reniformis* Penard, 1902

Genus ***Feuerbornia*** Jung, 1942

Test oval, aperture surrounded by thickened yellow to brownish lobes, in many specimens bent like an ampoule (Fig. 22). The shell is transparent with a very fine **areolation**. Often ring of mineral particles at the base of the apertural lobes. Plasma filled with green symbionts. Pseudopodia not yet observed. Habitat: moss, rain forest, southern Chile and New Zealand (Meisterfeld unpbl.). Monospecific. Ref. Jung (1942a).
Type species: *Feuerbornia lobophora* Jung, 1942

Genus ***Rhumbleriella*** Golemansky, 1970

Test oval to pouch-shaped, ventral surface flat, dorsal surface rounded. Pseudostome oval, antero-ventral, with thin, clear anterior, scalloped **collarette** (Fig. 23). Shell composed of small siliceous grains, in organic matrix, colorless, transparent. Pseudopods unknown. Habitat: marine littoral sands. Two species. Refs. Golemansky (1970b), Chardez (1977).
TYPE SPECIES: *Rhumbleriella filosa* Golemansky, 1970

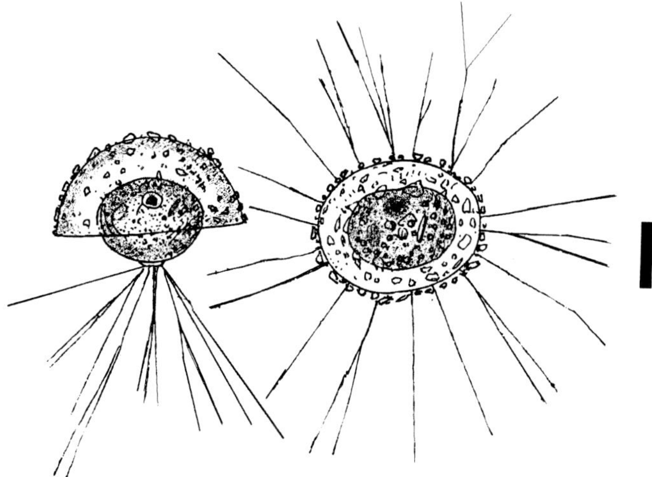

Fig. 21. *Frenzelina reniformis*. Scale bar=10 µm. (After Penard, 1902)

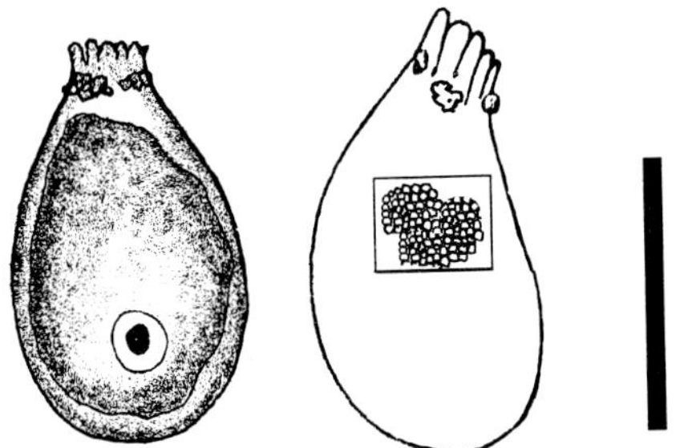

Fig. 22. *Feuerbornia lobophora*. Scale bar=50 µm. (After Jung, 1942a)

Genus ***Lesquerella*** Chardez & Thomas, 1980

Test half-coiled with asymmetrical neck, attached to the body. Aperture circular (Fig. 24). Test built of small rod- or sausage-like elements. Filopodia! Habitat: marine littoral sands. Monospecific. Ref. Chardez and Thomas (1980).
TYPE SPECIES: *Lesquerella mesopsammophila* Chardez & Thomas, 1980

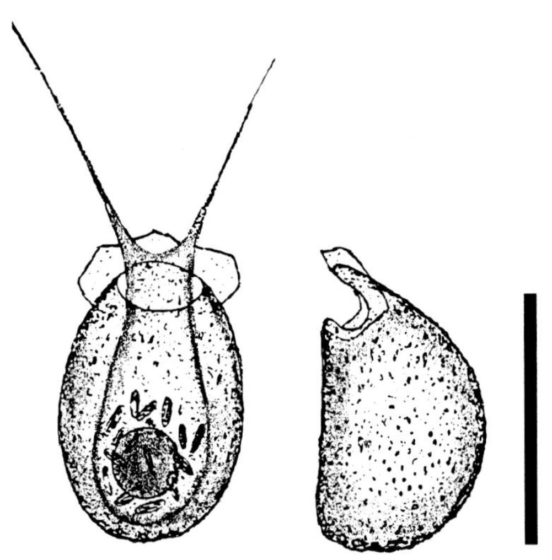

Fig. 23. *Rhumbleriella filosa*. Scale bar=50 µm. (After Golemansky, 1970b)

Fig. 24. *Lesquerella mesopsammophila*. Scale bar=50 µm. (After Chardez and Thomas, 1980)

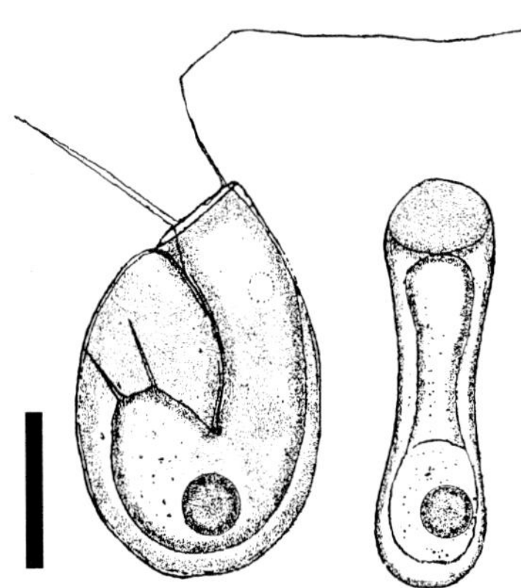

Fig. 25. *Volutella hemispiralis*. Scale bar=10 µm. (After Chardez, 1972)

FAMILY VOLUTELLIDAE Sudzuki, 1979

Shell half-coiled, organic, sometimes with few attached particles. Marine species.

NOTE: *Volutella* was described as having a transparent completely organic test, sometimes with a few encrusted mineral particles (Chardez, 1972). Sudzuki (1979a) described circular plates which indicate affinities to the Euglyphida. Golemansky (1992) places the family under the Monothalamida. Sudzuki (1979a) described *Pseudovolutella* in which the anterior part of the neck is separated by a septum.

Genus ***Volutella*** Chardez, 1977

Test tubular, half-coiled. Aperture circular. Shell organic, hyaline. One or 2 long filopodia (Fig. 25)! Two species. Habitat: marine; beach sands, common. Ref. Chardez (1972).
TYPE SPECIES: *Volutella hemispiralis* (Chardez, 1972)
NOTE: This genus was originally described by Chardez (1972) as *Voluta*, a homonym of a marine mollusc, and wascorrected by Chardez (1977).

ORDER EUGLYPHIDA Copeland, 1956

Test covered with internally formed, variously shaped, siliceous plates, mostly transported to and arranged in definitive patterns, bound by or sometimes embedded in an unstructured organic sheet-type cement.

KEY TO THE FAMILIES OF THE ORDER AND TO GENERA *INCERTAE SEDIS.*

1. Shell with long straight cylindrical neck.......... .... ***Ampullataria***
1'. Without neck or with bent neck.................... 2

2. Body plates not adjacent................................ 3
2'. Body plates overlapping or adjacent............. 4

3. Cross-section elliptical.............***Matsakision***
3'. Cross-section round................ ***Euglyphidion***

4. Body plates long, appear hexagonal, long axes at right angles to pseudostome.......................... .... Family **Paulinellidae**
4'. Body plates round or slightly elliptical....... 5

5 Aperture terminal, scales round to elliptical, overlapping or adjacent .... Family **Euglyphidae**
5'. Aperture eccentric... 6

6. Aperture bordered by denticulate apertural plates .... Family **Trinematidae**
6'. Aperture at the end of bent neck, sometimes with a funnel-like collar.. Family **Cyphoderiidae**

FAMILY EUGLYPHIDAE Wallich, 1864

Scales round to elliptical, thin, overlapping, adjacent or irregularly scattered.

KEY TO THE GENERA OF THE FAMILY

1. Test with denticulate apertural plates.. ***Euglypha***
1'. Aperture with collar of small plates in organic matrix.. 2
1''. Aperture with organic lip or rim, sometimes dentate.. 3

2. Shell circular in cross-section, collar distinct .... ***Sphenoderia***
2'. Shell compressed, collar small.. ***Trachelocorythion***

3. Thick organic lip..................... ***Heteroglypha***
3'. Thin organic rim.. 4

4. Shell circular in cross-section..................... 5
4'. Shell compressed... 6

5. Body plates large (6-9 µm), circular or elliptical.......................... ***Tracheleuglypha***
5'. Body plates small (<3.5 µm), caudal horn... ***Pareuglypha***

6. Aperture elongate slit, often with spines.. ***Placocista***
6'. Aperture circular or oval, never spines, often brownish...... .... ***Assulina***

Genus ***Pareuglypha*** Penard, 1902

Test yellowish, ovoid, tapered toward pseudostome; fundus with scaly spine of variable length (Fig. 26). Pseudostome circular. Pseudopodia, endoplasm, inclusions like *Euglypha.* Nucleus spherical, ovular. Feeding: herbivorous. Habitat: freshwater plants, sapropel. Two species. Refs. Penard (1902), for SEM, Grospietsch (1982).
TYPE SPECIES: *Pareuglypha reticulata* Penard, 1902

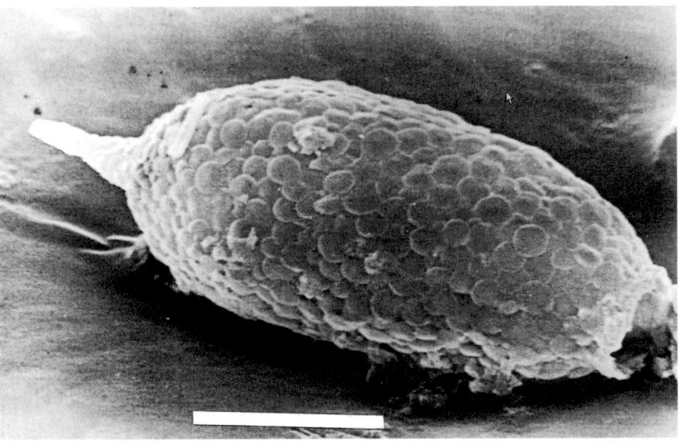

Fig. 26. *Pareuglypha reticulata.* Scale bar=20 µm. Photo Grospietsch unpubl.

Genus ***Euglypha*** Dujardin, 1841

Test elongate ovoid or pyriform, built of overlapping plates usually arranged in longitudinal rows. Aperture always with denticulate mouth plates! This large and common genus contains more than 40 species and twice as much sub- and infrasubspecific taxa. Only a few have been studied with modern methods. Refs. Decloitre (1962, 1976b, 1979b), Hedley and Ogden (1973), Hedley et al. (1974), Ogden (1981), Ogden and Hedley (1980).
TYPE SPECIES: *Euglypha tuberculata* Dujardin, 1841

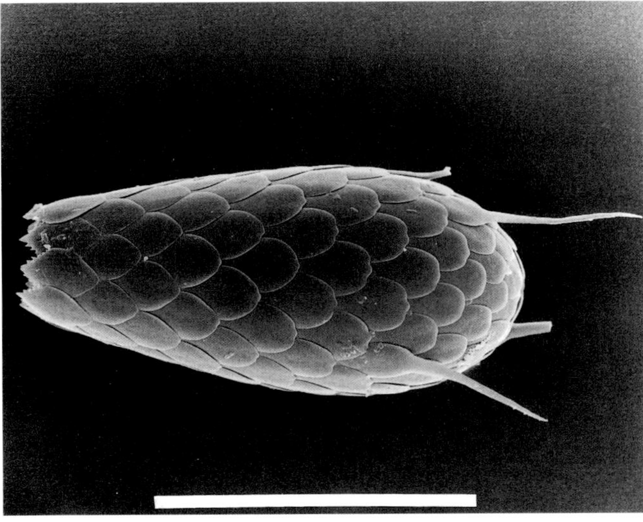

Fig. 27. *Euglypha acanthophora.* Scale bar=50 µm.

Traditionally this genus is subdivided into 3 artificial groups:
<u>Group I</u>
Shell in cross-section and aperture circular, spines, if present, appendix of body plates, e.g.

Euglypha acanthophora (Ehrenberg, 1841) (Fig. 27). Similar but without spines: *Euglypha tuberculata* Dujardin, 1841 (Fig. 28).

Group II

Shell in cross-section elliptical, aperture ± circular, spines not appendix of body plates, e.g. *Euglypha strigosa* (Ehrenberg, 1871) (Fig. 29). Habitat: common in mosses, sphagnum, and organic soils. Feeding: herbivorous.

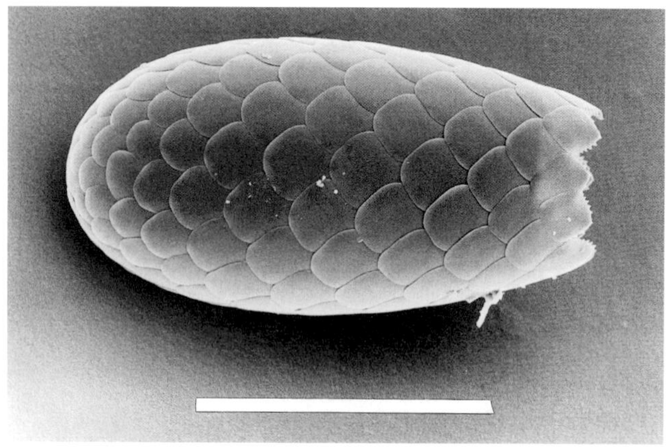

Fig. 28. *Euglypha tuberculata.* Scale bar=50 µm.

Fig. 29. *Euglypha strigosa.* Scale bar=20 µm.

Group III

Shell compressed, aperture elliptical, spines not modified body plates (e.g. *Euglypha compressa* Carter, 1869) (Fig. 30). Feeding: herbivorous. Habitat: aquatic vegetation, sphagnum, acid soils.

Fig. 30. *Euglypha compressa.* Scale bar=20 µm.

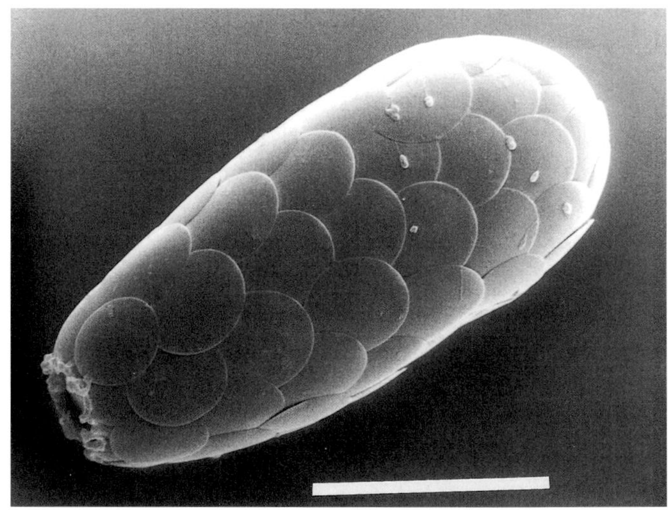

Fig. 31. *Tracheleuglypha dentata.* Scale bar=20 µm.

Genus ***Tracheleuglypha*** Deflandre, 1928

Test clear, elliptical, round cross-section; with round, overlapping scales; no spines, no dentate apertural plates (Fig. 31). Pseudostome terminal, circular with short, sometimes dentate, organic collar, which is produced during fission to hold the opposed shells together. Ogden and Coûteaux (1988) have demonstrated that this feature is highly variable! Ovular nucleus. Feeding: herbivorous. Habitat: mosses, soil, and aquatic

vegetation. Two species and a few subspecies. Refs. Thomas and Gauthier-Lièvre (1959b), Ogden and Coûteaux (1987, 1988), Coûteaux and Ogden, 1988), Bonnet (1991b).
TYPE SPECIES: *Tracheleuglypha dentata* (Penard, 1890)

Genus ***Placocista*** Leidy, 1879

Test ovoid, flattened, with acute lateral border; edge has in some species many short, often paired, lanceolate spines, cemented between scales (Fig. 32). Pseudostome large, narrowly biconvex, with pointed edges, in lateral view incised and surrounded by a thin rim of organic cement (only visible by SEM). Pseudopodia, endoplasm and inclusions like *Euglypha*, sometimes with zoochlorellae. Nucleus spherical, one nucleolus. Feeding: bacteria, algae, and fungi; others depend on their symbionts. Habitat: mainly *Sphagnum*. Eight species and several subspecies. Refs. Cash et al. (1915), Thomas and Gauthier-Lièvre (1959b), for SEM, Ogden and Hedley (1980).
TYPE SPECIES: *Placocista spinosa* (Carter, 1865)

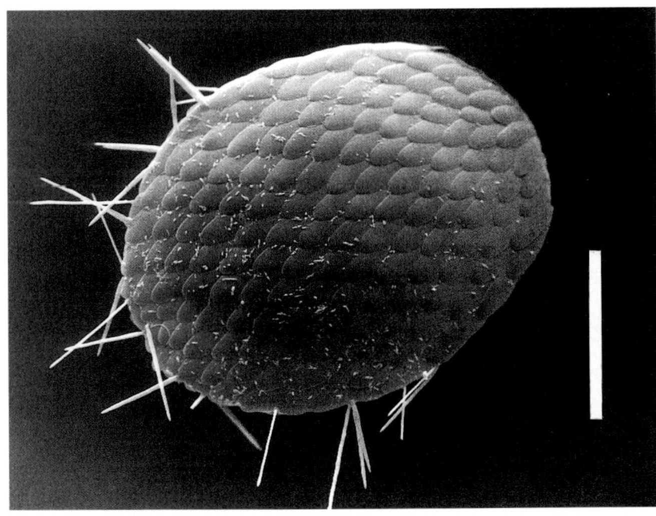

Fig. 32. *Placocista spinosa*. Scale bar=50 μm.
eumi-iqp2-p1069-f032

Genus ***Assulina*** Leidy, 1879

Test clear to brown (in culture the color is lost often), ovoid, flattened, biconvex in cross-section, with overlapping elliptical scales (Fig. 33). Pseudostome narrow oval, more or less crenulate, organic margin. Pseudopodia like *Euglypha*. Vesicular nucleus. Feeding: herbivorous. Habitat: mosses, soil. For ecological data for the soil species, see Bonnet (1991b). Four species. Refs. Cash et al. (1915), Ogden and Hedley (1980), Lüftenegger et al. (1988).
TYPE SPECIES: *Assulina seminulum* (Ehrenberg, 1848)

NOTE: *Valkanovia* Tappan, 1966 is closely related, being colorless with a more circular aperture, but there are transitions between the genera so the status of the genus is uncertain. *Valkanovia* has 4 species. Cyst spherical. Feeding: mainly bacteria, yeasts. Habitat: mosses, humus. Refs. Tappan (1966), Chardez (1967c), Schönborn and Peschke (1990).
TYPE SPECIES: *Valkanovia delicatula* (Valkanov, 1962) (Synonym=*Euglyphella delicatula*)

Fig. 33. *Assulina seminulum*. Scale bar=50 μm.

Fig. 34. *Sphenoderia lenta*. Scale bar=10 μm.

Genus ***Sphenoderia*** Schlumberger, 1845

Test clear, circular or ovoid; shell plates overlapping, elliptical or nearly circular (Fig. 34). Pseudostome more or less terminal, a narrow, oval slit at the end of a broad, clear collar covered with numerous small (1-2 μm) plates.

Nucleus spherical, few small endosomes (nucleoli). Feeding: herbivorous, mainly bacteria. Habitat: mosses, *Sphagnum*. To date 16 species and several subspecies have been described, which differ mainly in size, shape, and arrangement of body scales. Only a few have been studied with modern methods. Refs. Thomas and Gauthier-Lièvre (1959b), Ogden (1984).
TYPE SPECIES: *Sphenoderia lenta* Schlumberger, 1845

Genus ***Trachelocorythion*** Bonnet, 1979

Test clear, compressed, in lateral view lenticular, aperture at the end of a short collar, dorsal lip slightly longer than the ventral one (Fig. 35). The shell is composed of oval siliceous platelets while the apertural collar is covered with much smaller but not denticulated (cf. *Sphenoderia*) scales. Vesicular nucleus. Ecology: sphagnum mosses, acid skeletal soils (Bonnet, 1991b). This genus has been erected to accommodate *Corythion pulchellum*, which is quite different to other species in the genus *Corythion.* Monospecific. *Trachelocorythion pulchellum* is size polymorphic. Dimensions of the normal small morph: length 20-35 µm, the larger morph is 47-61 µm long. Refs. Penard (1902), Bonnet (1979b).
TYPE SPECIES: *Trachelocorythion pulchellum* (Penard, 1890)

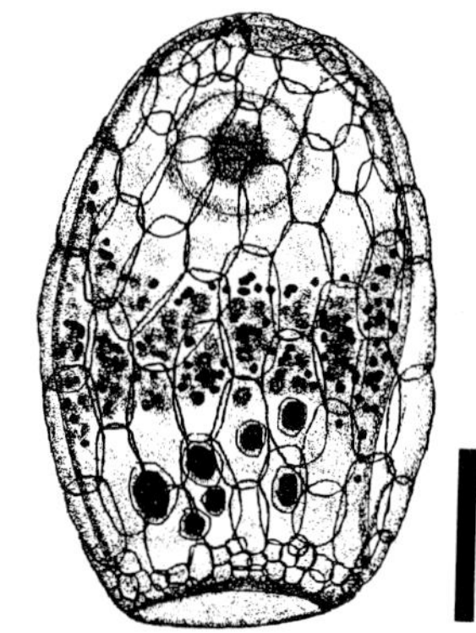

Fig. 35. *Trachelocorythion pulchellum.* Scale bar=10 µm.

FAMILY TRINEMATIDAE
Hoogenraad & De Groot, 1940

Test with bilateral symmetry and an eccentric, often invaginated pseudostome.

KEY TO THE GENERA

1. Aperture invaginated (lateral view)............. 2
1'. Aperture not invaginated, or with visor........ 3

2. Body plates circular, usually two sizes, the smaller plates filling the gaps between the larger ones...................................... ***Trinema***
2'. Body plates elongate oval, aperture only posteriorly invaginated, anterior part like a spoon .... ***Corythion***

3. Aperture with visor.................. ***Deharvengia***
3'. No visor.. 4

4. Body plates large oval, forming bosses with small more circular plates in between..........
.... ***Pileolus***
4'. Body plates circular, two sizes, shell surface smooth....................................... ***Playfairina***

Genus ***Trinema*** Dujardin, 1841

Two types of tests are common in this genus:
1. pouch-shaped, ends round, oval in cross-section, dorsum arched, ventrum flat or concave (e.g. *T. enchelys*, Fig. 36).
2. in ventral view broad oval, sides parallel; in lateral view tapering sharply towards the aperture (e.g. *T. complanatum*, Fig. 37).

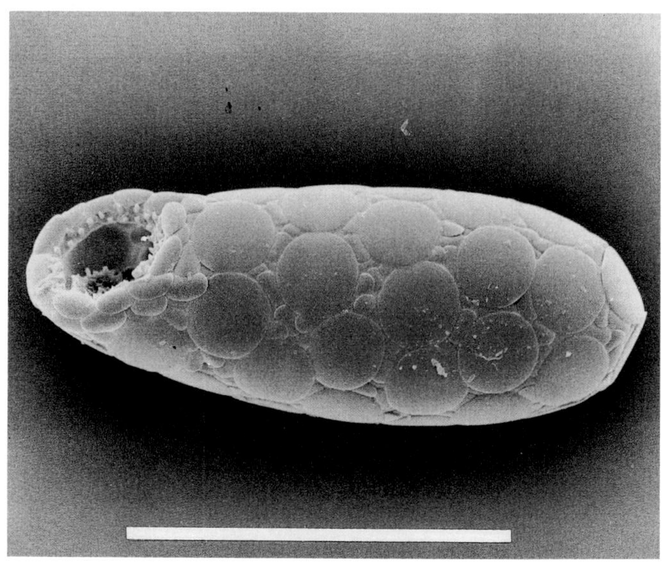

Fig. 36. *Trinema enchelys.* Scale bar=50 µm.

Always 2 sizes of body scales, large circular plates with smaller circular or broad elliptical scales in between, overlapping. Pseudostome sub- terminal, ventral, round or oval, invaginated, with rows of toothed apertural plates; apertural rim with oval small scales. Vesicular nucleus. Feeding: herbivorous. Habitat: common in freshwater, moss, and soil. This genus houses more than 2 dozen species, a similar number of subspecies, varieties, and forms. Most of the descriptions are insufficient and should be taken as *nomina nuda*. Considering the large variability observed in clonal cultures (Hedley and Ogden, 1974a), others are probably synonyms. Most species are size polymorphic. (Fig. 36). Refs. Bonnet (1970, 1991a), Ogden and Hedley (1980), Lüftenegger et al. (1988), Schönborn (1992).
TYPE SPECIES: *Trinema enchelys* (Ehrenberg, 1838)
NOTE: Closely related (synonym?) to *Trinema* is the genus *Puytoracia* Bonnet, 1970; here the aperture is not invaginated.

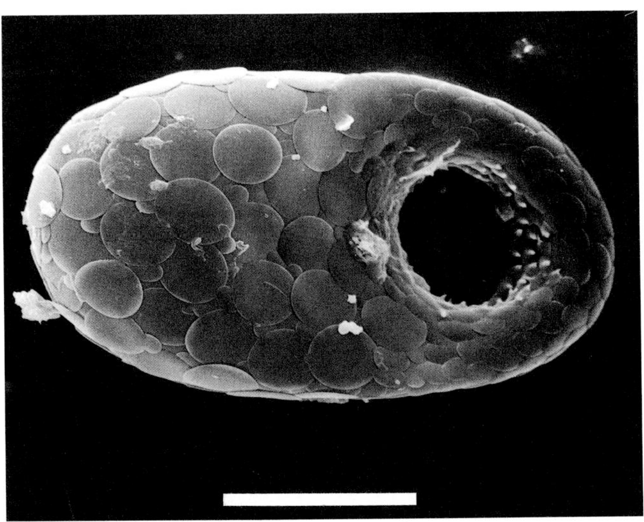

Fig. 37. *Trinema complanatum*. Scale bar=20 µm.

Genus **Corythion** Taranek, 1881

Test ovoid, asymmetrically compressed, cross-section lenticular, lateral margins often slightly acute; aperture invaginated, oblique, oval, semicircular, or round, anterior part like a spoon, only one type of body plates: rounded rectangular, imbricated, irregularly distributed but at lateral margins in rows (Fig. 38). Apertural plates with one central tooth. Some species have a terminal horn or are decorated with short organic or siliceous spines. Vesicular nucleus. Feeding: mainly bacterivore. Habitat: mosses, sphagnum, soils. Seven species, several forms and varieties. Refs. Cash et al. (1915), Cowling (1986), Lüftenegger et al. (1988), Bonnet (1991a), Ogden (1991c).
TYPE SPECIES: *Corythion dubium* Taranek, 1881

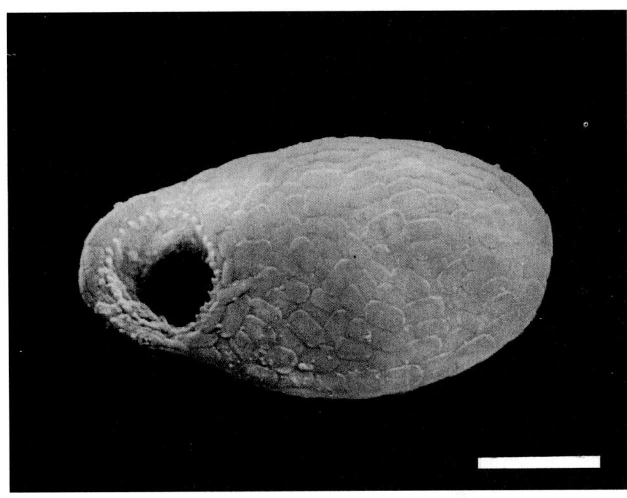

Fig. 38. *Corythion dubium*. Scale bar=10 µm.

Fig. 39. *Playfairina caudata* specimen from Mt. Buffalo, Australia. Scale bar=20 µm. Inset: Apertural view.

Genus **Playfairina** Thomas, 1961

Test sub-cylindrical, tapering a little towards the aperture, fundus rounded or terminating in a horn (Fig. 39). Pseudostome and shell in transversal section circular. Shell composed of endogenous circular siliceous platelets, arrangement in most cases as in *Trinema*. Aperture evaginated, one row of denticulated marginal plates. Pseudopodia unknown, but due to the homologies in shell morphology *Playfairina* is placed here. Habitat:

sphagnum or humus. Two species. Refs. Thomas (1961), Golemansky (1966).
TYPE SPECIES: *Playfairina caudata* Thomas, 1961

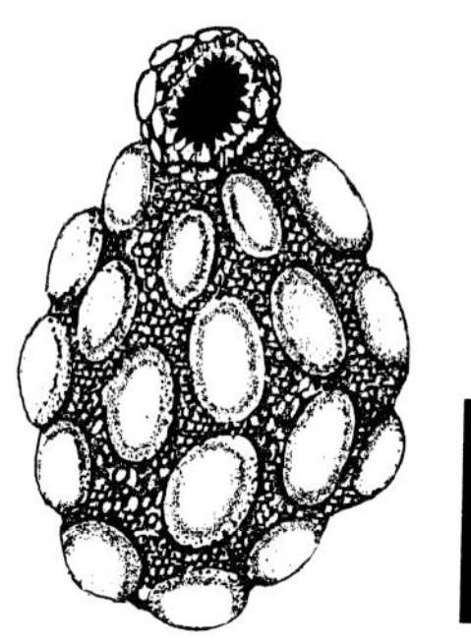

Fig. 40. *Pileolus tuberosus.* Scale bar=10 µm. (After Coûteaux and Chardez, 1981)

Genus ***Pileolus*** Coûteaux & Chardez, 1981

Shell with larger oval plates forming bosses and smaller more circular scales filling the space between the larger ones (Fig. 40). Aperture with a collar turned up towards the exterior and surrounded by small elongated plates, and inside, a circle of plates with a single tooth. Monospecific. Habitat: rhizosphere of Bromeliaceae, French Guyana. Ref. Coûteaux and Chardez (1981).
TYPE SPECIES: *Pileolus tuberosus* Coûteaux & Chardez, 1981

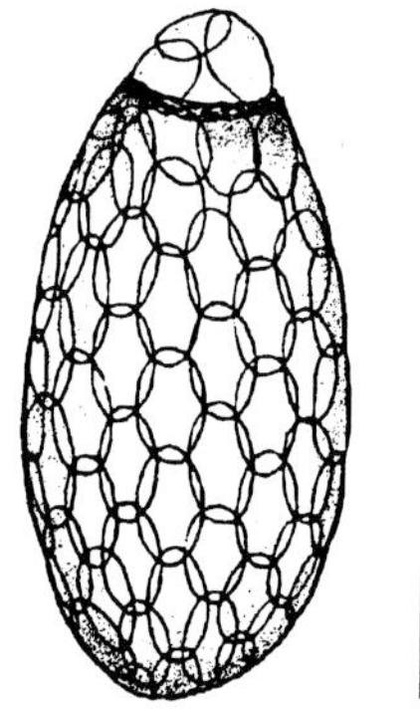

Fig. 41. *Deharvengia papuensis*. Scale bar=10 µm. (After Bonnet, 1979b)

Genus ***Deharvengia*** Bonnet, 1979

Shell large (89-115 µm), oval, very regular, transparent. Aperture elliptic, bordered by a yellowish lip-like thickening and surmounted by a small visor formed by a few siliceous platelets (Fig. 41). Shell composed of large oval plates arranged regularly similar to *Tracheleuglypha*; close to the aperture the scales are smaller and disposed irregularly. Ecology: tropical rain forest soils, Indonesia. Monospecific. Ref. Bonnet (1979b)
TYPE SPECIES: *Deharvengia papuensis* Bonnet, 1979

FAMILY CYPHODERIIDAE de Saedeleer, 1934

Scales circular or oval, adjacent or overlapped, test usually with aperture bent to one side. The marine forms have usually an apertural funnel or collar.

KEY TO GENERA OF THE FAMILY

1. Aperture with funnel or distinct collar......... 3
1'. Aperture without funnel distinct collar......... 2

2. Shell in cross-section circular or laterally compressed.................................. ***Cyphoderia***
2'. Shell in cross-section triangulate.................. .... ***Schaudinnula***

3. Apertural collar hyaline, fragile, disappears in empty shells rapidly................ ***Campascus***
3'. Apertural collar with siliceous scales........... 4

4. Shell with caudal spine...... ***Pseudocorythion***
4'. Fundus rounded... 5

5. Shell plates oval, shell dorso-ventrally compressed............................... ***Corythionella***
5'. Shell plates circular, cross-section circular .... ***Messemvriella***

Genus ***Cyphoderia*** Schlumberger, 1845

Test clear, yellowish, ovoid, round in cross-section, base round or tapered to mammillate tip; plates round, small, depending on species either adjoining or overlapping on organic matrix; short cylindrical neck, bent; pseudopods long, thin (Fig. 42). Endoplasm with refractive grains. Nucleus spherical, ovular. Feeding: herbivorous. Habitat: freshwater, aquatic mosses, sphagnum, and sediment. This genus comprises a dozen species and an equal number of varieties. *Cyphoderia ampulla* has adjoining body scales but is often confused with *C. trochus*, with regularly imbricated scales. Refs. Ogden and Hedley (1980), Chardez (1991).
TYPE SPECIES: *Cyphoderia ampulla* (Ehrenberg, 1840)

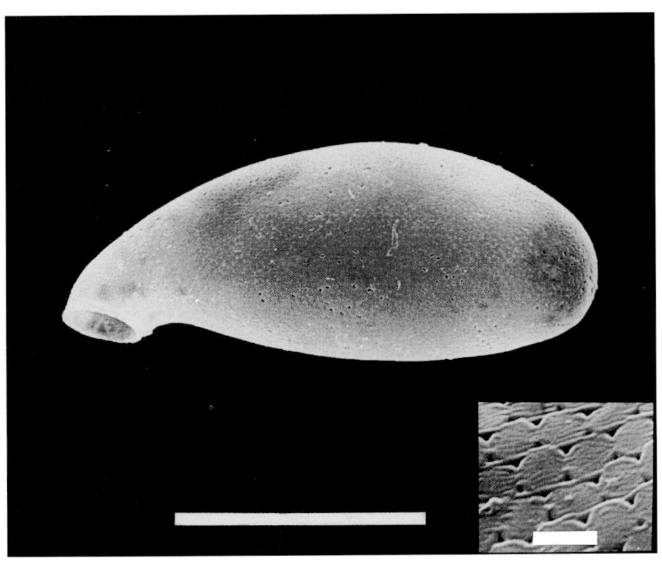

Fig. 42. *Cyphoderia ampulla*. Inset shows shell surface. Scale bars=50 μm and 2 μm.

Genus ***Schaudinnula*** Awerintzew, 1907

Test like *Cyphoderia*; pointed tip; but triangular in cross-section (Fig. 43). Differs from *Campascus* in lacking the hyaline collar. Round scales irregularly overlapping, appear hexagonal. Pseudostome round. Pseudopods and cytoplasm like *Cyphoderia*. Nucleus vesicular. Feeding: herbivorous. Habitat: freshwater plants; very rare. Monospecific. Ref. Schönborn (1965c).
TYPE SPECIES: *Schaudinnula arcelloides* Awerintzew, 1907

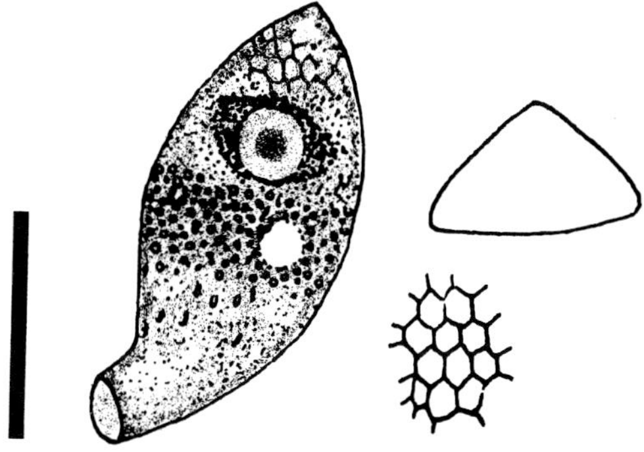

Fig. 43. *Schaudinnula arcelloides*. Lateral view, cross-section of fundus and shell plates. Scale bar=50 μm. (After Schönborn, 1965c)

Genus ***Campascus*** Leidy, 1879

Test broadly ovoid; cross-section ovoid or triangular; neck cylindroid, curved (Fig. 44). The base is rounded or has up to 3 projections. Aperture round. The distinctive character is the hyaline collar. The composition of the shell is variable. Some species have circular, others amorphous, siliceous scales sometimes mixed with mineral particles, usually not regularly arranged. Nucleus ovular. Feeding: herbivorous. Habitat: freshwater plants and sediments, one marine species. There are similarities with some genera of the family Psammonobiotidae. Due to the lack of information it is unclear if these similarities warrant a new classification. Seven species. In this genus marine and freshwater species are combined. Refs. Leidy (1879), Penard (1902).
TYPE SPECIES: *Campascus cornutus* Leidy, 1879

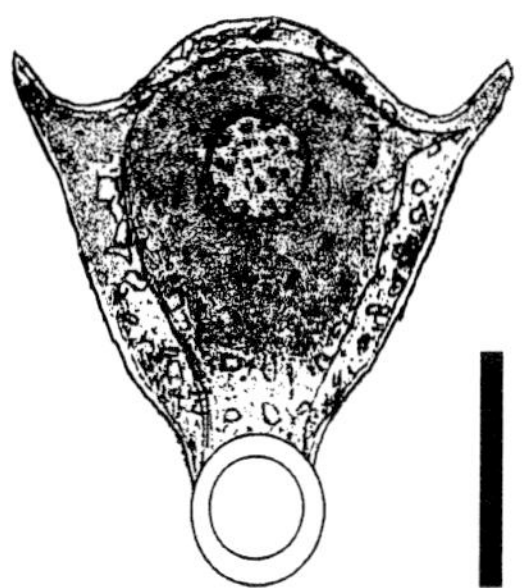

Fig. 44. *Campascus cornutus*. Scale bar=50 μm. (After Leidy, 1879)

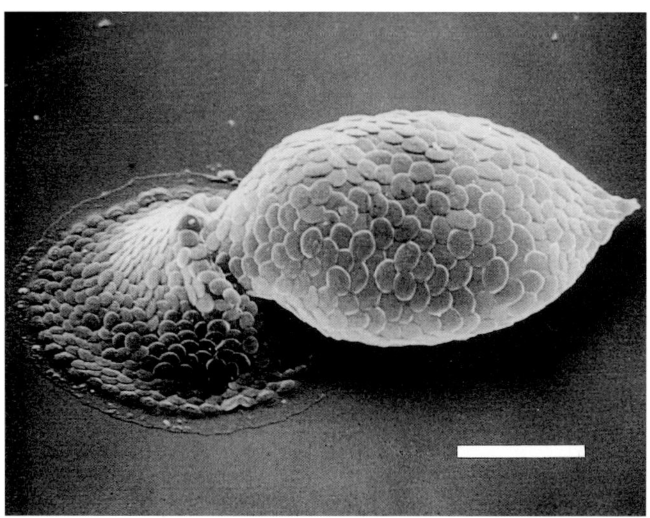

Fig. 45. *Pseudocorythion wailesi*. Scale bar=10 μm. (From Ogden and Coûteaux, 1989)

Genus ***Pseudocorythion*** Valkanov, 1970

The shell is ovoid in ventral view, tapering at both ends, slightly flattened dorso-ventrally (Fig. 45). Aperture with a large circular collar of the same size as the maximum breadth; posteriorly the test carries a small pointed spine of variable length. In

lateral view, the fan-like collar is constricted to form a ridge at the rear junction with the body (*P. wailesi*). Outer limit of the collar marked by a very thin organic rim. This collar can have a wavy outline. The shell is composed of small circular or oval overlapping plates which are randomly arranged. The cell almost fills the test. The spherical nucleus is surrounded by a dense granular area. Habitat: marine interstitial. Four species. Similar genera are *Micropsammelloides* and *Corythionelloides*, Sudzuki, 1979a. Refs. Golemansky (1971), for SEM, Ogden and Coûteaux (1989).
TYPE SPECIES: *Pseudocorythion acutum* (Wailes, 1927).

Genus ***Corythionella*** Golemansky, 1970

Test ovoid, dorso-ventrally compressed, large circular collar, fundus rounded, with elongate, siliceous, internally formed plates, irregularly arranged as in *Corythion*; the plates of the collar are smaller (Fig. 46). Habitat: marine interstitial. Five species. A similar genus is *Pseudowailesella* Sudzuki (1979a). Refs. Golemansky (1970b), for SEM, see Ogden and Coûteaux (1989).
TYPE SPECIES: *Corythionella pontica* Golemansky, 1970

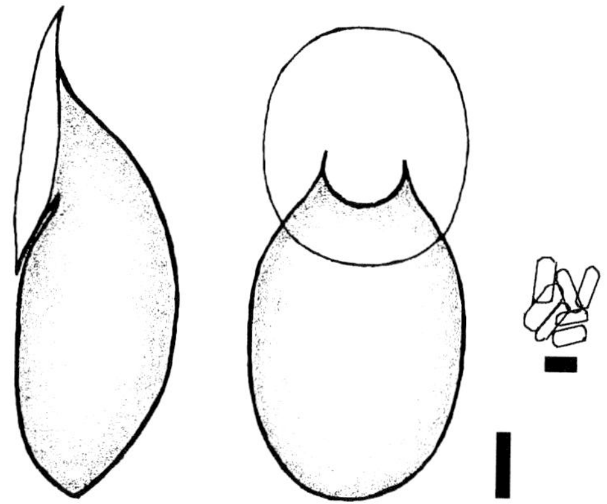

Fig. 46. *Corythionella pontica*. Frontal and lateral view, shell scales. Scale bar=10 µm and 1 µm. (After Golemansky, 1970b)

Genus ***Messemvriella*** Golemansky, 1973

Shape resembles *Corythionella*, but circular in transverse section, with circular, regularly overlapping siliceous scales (Fig. 47). Apertural funnel less developed than in *Corythionella*. This genus differs from *Pseudocorythion* by the lack of a caudal horn. Typical ovular nucleus. Few thin filopodia. Ecology: marine, littoral sands, bacterivorous and algivorous. Two species. Ref. Golemansky (1973).
TYPE SPECIES: *Messemvriella filosa* Golemansky, 1973

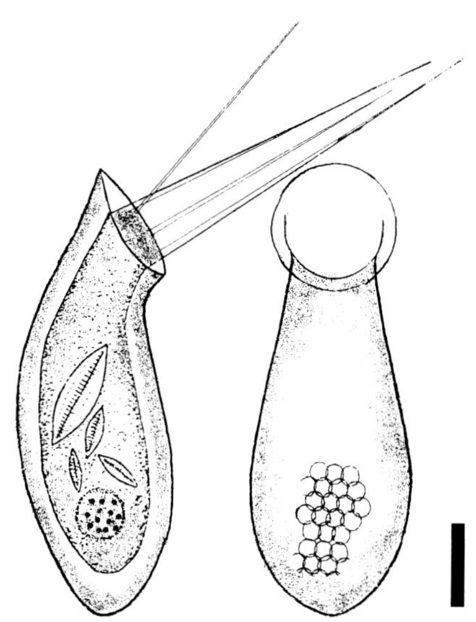

Fig. 47. *Messemvriella filosa*. Scale bar=10 µm. (After Golemansky, 1973)

FAMILY PAULINELLIDAE de Saedeleer, 1934

Scales long, with rounded ends, long axes at right angles to aperture. One genus.

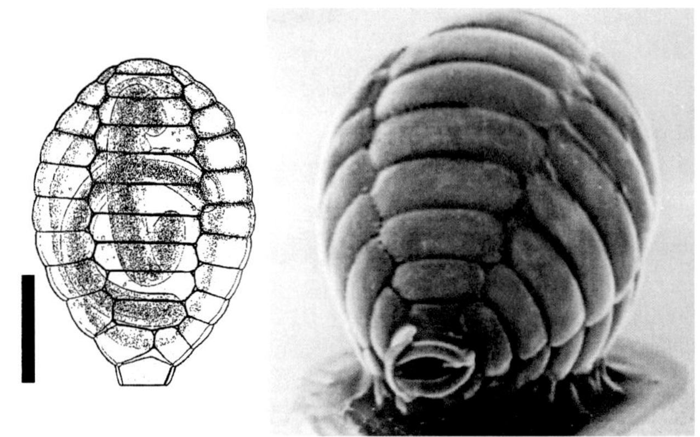

Fig. 48. *Paulinella chromatophora*. Scale bar=10 µm. After de Saedeleer (1934). Photo Grospietsch unpubl.

Genus ***Paulinella*** Lauterborn, 1895

Test ovoid, made of siliceous rectanguloid plates with slightly rounded, scarcely overlapping ends, arranged in transversely circumferential longitudinal rows; in apertural view the plates form a counter-clockwise spiral, one pentagonal scale at the aboral pole (Fig. 48). All known species small

(<45 μm). Ultrastructure of plates that have a smooth surface but a porous subsurface layer or internal channels with rows of pores differs from that of other *Euglyphida* (Kies, 1974). Pseudostome on short neck. *Paulinella chromatophora* has 2 sausage-like curved, short blue-green cyanobacteria as symbionts, no food vacuoles, 1 to 3 rapidly moving filopods. Habitat: freshwater plants, plankton, but also brackish water and marine sediments (Pankow, 1982; Hannah et al., 1996). Four species. Refs. Kies (1974), Kies and Kremer (1979), Johnson et al. (1988), Vørs (1993).
TYPE SPECIES: *Paulinella chromatophora* Lauterborn, 1895

Genera *incertae sedis*

Genus *Ampullataria* Van Oye, 1956

Test ellipsoid, round in cross-section, cylindrical neck; scales oval, overlap, appear polygonal; neck-scales irregular at aperture (Fig. 49). Size: 110 μm long, 75 μm diameter; neck 20 μm long, 20 μm basal diameter. Pseudostome round, 17 μm. Habitat: sphagnum, Venezuela (alt. 3800 m), rare. Monospecific. Ref. Van Oye (1956).
TYPE SPECIES: *Ampullataria rotunda* Van Oye, 1956

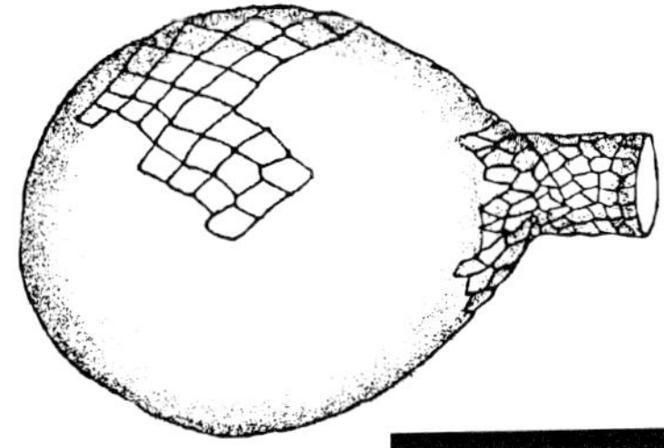

Fig. 49. *Ampullataria rotunda.* Scale bar=50 μm. (After van Oye, 1956)

Genus *Euglyphidion* Bonnet, 1960

Test clear, ovoid, round in cross-section, sides taper; scales not adjacent, not overlapping (Fig. 50). Size: 30-31 μm long, scales of 2 sizes, 6 by 3.5 μm and 1.0 by 1.5 μm. Pseudostome terminal, round, 10 μm. Feeding: herbivorous. Habitat: humid mor or moder type soils, rarely reported, acidophilous (Bonnet, 1991b). Monospecific. Ref. Bonnet (1960b).
TEST SPECIES: *Euglyphidion enigmaticum* Bonnet, 1960.

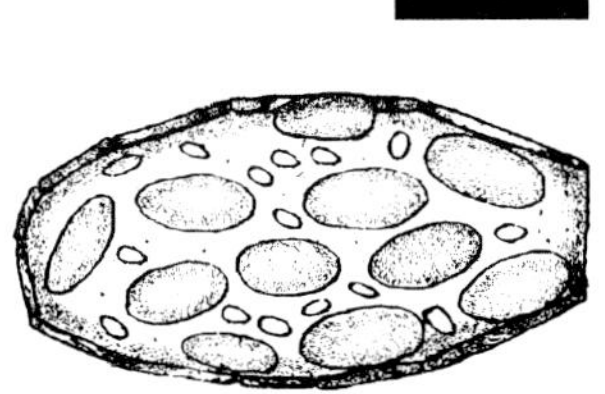

Fig. 50. *Euglyphidion enigmaticum.* Scale bar=10 μm. (After Bonnet, 1960b)

Genus *Heteroglypha* Thomas & Gauthier-Lièvre, 1959

Test transparent, wedge-shaped, in frontal view fundus rounded, sides sub-parallel, anterior end only little smaller than fundus (Fig. 51). Aperture slit-like or slightly arched; at right angle to the broad diameter, 22 to 23 μm wide, with clear organic rim. In side view test tapered almost to point at apertural end. Scales elliptical, arranged as in *Euglypha*. Habitat: acid freshwater, between plants; tropical Africa. Monspecific. Ref. Thomas and Gauthier-Lièvre (1959b).
TYPE SPECIES: *Heteroglypha delicatula* Thomas & Gauthier-Lièvre, 1959

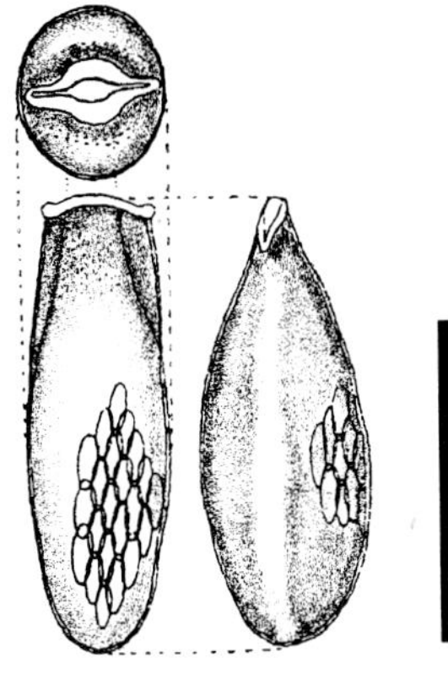

Fig. 51. *Heteroglypha delicatula.* Scale bar=50 μm. (After Thomas and Gauthier-Lièvre, 1959b)

Genus *Matsakision* Bonnet, 1967

Shell transparent, ovoid elongate, very slightly compressed, perpendicularly truncated near the oval aperture; small organic collar, hardly visible; test covered with elliptical intrinsic scales, not overlapping, bound by a clear yellow cement (Fig. 52). Pseudopodia not reported. Ecology: terrestrial mosses. Two species. Ref. Bonnet (1967).

TYPE SPECIES: *Matsakision cassagnaui* Bonnet, 1967

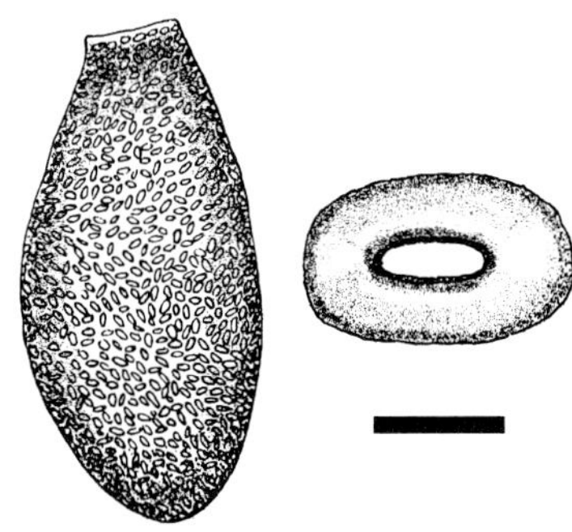

Fig. 52. *Matsakision cassagnaui.* Scale bar=10 µm. (After Bonnet, 1967)

Literature cited:

Aescht, E. & Foissner, W. 1989. *Catalogus Faunae Austriae.* Ein systematisches Verzeichnis aller auf österreichischem Gebiet festgestellten Tierarten. Ia: Stamm: Rhizopoda. *Verl. Österr. Akad. Wiss. Wien*, 1-79.

Aescht, E. & Foissner, W. 1992. Effects of mineral and organic fertilizers on the microfauna in a high-altitude reafforestation trial. *Biol. Fertil. Soils*, 13:17-24.

Anderson, O. R. 1987. Fine structure of a silica-biomineralizing testate amoeba, *Netzelia tuberculata. J. Protozool.*, 34:302-309.

Anderson, O. R. 1989. Some observations of feeding behavior, growth, and test particle morphology of silica-secreting testate amoeba *Netzelia tuberculata* (Wallich) (Rhizopoda, Testacea) grown in laboratory culture. *Arch. Protistenkd.*, 137:211-221.

Anderson, O. R. 1990. Effects of silicate deficiency on test morphology, cytoplasmic fine structure, and growth of the testate amoeba *Netzelia tuberculata* (Wallich) Netzel (Rhizopoda, Testacea) grown in laboratory culture. *Arch. Protistenkd.*, 138:17-27.

Anderson, O. R., Rogerson, A. & Hannah, F. 1996. A description of the testate amoeba *Ovulina parca gen. nov., sp. nov.* from coastal marine sediments. *J. Mar Biol. Assoc. U.K*, 76:851-865.

Belar, K. 1921. Untersuchungen über Thecamöben der *Chlamydophrys*-Gruppe. *Arch. Protistenkd.*, 43:287-354.

Bhattacharya, D., Helmchen, T. & Melkonian, M. 1995. Molecular evolutionary analyses of nuclear-encoded small subunit ribosomal RNA identify an independent rhizopod lineage containing the Euglyphina and the Chlorarachniophyta. *J. Euk. Microbiol.*, 42:65-69.

Bonnet, L. 1959. Nouveaux thécamoebiens du sol. *Bull. Soc. Hist. Nat. Toulouse*, 94:177-188.

Bonnet, L. 1960a. Thécamoebiens des sols d'Angola (I). *Pub. Cultural. Compan. Diamant. Angola*, 1:79-86.

Bonnet, L. 1960b. Nouveaux thécamoebiens du sol. III. *Bull. Soc. Hist. Nat. Toulouse*, 95:209-211.

Bonnet, L. 1961. L'émission pseudopodique, chez les thécamoebiens endogés (I). *Bull. Soc. Zool. France*, 86:17-28.

Bonnet, L. 1962. Thécamoebiens du sol. *Biol. Amer. Australe*, 1:43-47.

Bonnet, L. 1963. L'émission pseudopodique chez les thécamoebiens endogés. II. *Bull. Soc. Zool. France,* 88:57-63.

Bonnet, L. 1964. Le peuplement thécamoebien des sols. *Rev. Ecol. Biol. Sol,* 1:123-408.

Bonnet, L. 1965. Nouveaux thécamoebiens du sol V. *Bull. Soc. Hist. Nat. Toulouse,* 100:330-332.

Bonnet, L. 1967. Le peuplement thécamoebien des sols de Grèce. *Biol. Gallo-Hell.,* 1:7-26.

Bonnet, L. 1970. Nouveaux thécamoebiens du sol (VI). *Bull. Soc. Hist. Nat. Toulouse,* 106:328-332, plate XVIII.

Bonnet, L. 1974a. A propos de *Geopyxella sylvicola* et de *Pseudawerintzewia calcicola* (Rhizopodes Thécamoebiens édaphiques). *Rev. Ecol. Biol. Sol*, 10:509-522.

Bonnet, L. 1974b. *Lamtoquadrula* gen. nov. et la structure plagiostome chez les thécamoebiens nébéliformes. *Bull. Soc. Hist. Nat. Toulouse*, 110:297-299.

Bonnet, L. 1974c. Protistologie. - Les Lamtopyxidae fam. nov. et la structure propylostome chez les thécamoebiens (Rhizopoda, Testacea). *C. R. Acad. Sci. Paris*, 278:2935-2937.

Bonnet, L. 1975a. Nouveaux Thécamoebiens du sol (VIII). *Bull. Soc. Hist. Nat. Toulouse*, 111:300-302.

Bonnet, L. 1975b. Types morphologiques, écologie et évolution de la thèque chez les thécamoebiens. *Protistologica*, 11:363-378.

Bonnet, L. 1977. Nouveaux thécamoebiens du sol. IX. *Bull. Soc. Hist. Nat. Toulouse*, 113:152-156.

Bonnet, L. 1979a. Origine et biogéographie des Distomatopyxidae (Rhizopodes Thécamoebiens des sols, Lobosia, Arcellinida). *C. R. Acad. Sci. Paris*, 288:775-778.

Bonnet, L. 1979b. Nouveaux thécamoebiens du sol. X. *Bull. Soc. Hist. Nat. Toulouse*, 115:106-118.

Bonnet, L. 1984a. Les *Plagiopyxis* à structure Callida (Thécamoebiens des sols). *Protistologica*, 20:475-489.

Bonnet, L. 1984b. Nouvelles données sur le genre *Planhoogenraadia* (Thécamoebiens). *Bull. Soc. Hist. Nat. Toulouse*, 120:117-122.

Bonnet, L. 1985. Le signalement écologique des thécamoebiens des sols. *Bull. Soc. Hist. Nat. Toulouse*, 121:7-12.

Bonnet, L. 1988. Ecologie du genre *Plagiopyxis* (Thécamoebiens des sols). *Bull. Soc. Hist. Nat. Toulouse*, 124:13-21.

Bonnet, L. 1989a. Données écologiques sur quelques Centropyxidae (Thécamoebiens) des sols. *Bull. Soc Hist. Nat, Toulouse*, 125:7-16.

Bonnet, L. 1989b. Données écologiques sur quelques Hyalospheniidae et Paraquadrulidae (Thécamoebiens) des sols (Première partie). *Bull. Soc. Hist. Nat. Toulouse*, 125:17-22.

Bonnet, L. 1989c. Attracteurs et morphologie de la thèque des Thécamoebiens (Rhizopoda Testacea). *Bull. Soc. Hist. Nat. Toulouse*, 125:23-26.

Bonnet, L. 1990. Données écologiques sur quelques Hyalospheniidae et Paraquadrulidae (Thécamoebiens) des sols (Deuxième partie: genre *Nebela*). *Bull. Soc. Hist. Nat. Toulouse*, 126:9-17.

Bonnet, L. 1991a. Ecologie de quelques Euglyphidae (Thécamoebiens, Filosea) des milieux édaphiques et paraédaphiques (Première partie: genres *Corythion* et *Trinema*). *Bull. Soc. Hist. Nat. Toulouse*, 127:7-13.

Bonnet, L. 1991b. Ecologie de quelques Euglyphidae (Thécamoebiens, Filosea) des milieux édaphiques et paraédaphiques (Deuxième partie). *Bull. Soc. Hist. Nat., Toulouse*, 127:15-20.

Bonnet, L. & Gomez-Sanchez, M. S. 1984. Note préliminaire sur le peuplement thécamoebien des sols des Asturies (Espagne). *Bull. Soc. Hist. Nat. Toulouse*, 120:111-116.

Bonnet, L. & Thomas, R. 1958. Une technique d'isolement des Thécamoebiens (Rhizopoda Testacea) du sol et ses résultats. *C. R. Acad. Sci. Paris*, 247:1901-1903.

Bonnet, L. & Thomas, R. 1960. Faune terrestre et d'eau douce des Pyrénées-Orientales. Thécamoebiens du sol. *Vie Milieu*, 11:suppl.,1-103.

Bonnet, L. & Vandel, .M. A. 1970. Les Distomatopyxidae fam. nov. et la structure diplostome chez les Thécamoebiens (Rhizopoda, Testacea). *C. R. Acad. Sci. Paris*, 271:1189-1191.

Bonnet, L., Brabet, J., Comoy, N. & Guitard, J. 1981a. Observations sur l'ultrastructure de *Nebela marginata* (Rhizopoda, Testacea, Lobosia, Hyalospheniidae). *Protistologica*, 17:235-241.

Bonnet, L., Brabet, J., Comoy, N. & Guitard, J. 1981b. Nouvelles données sur le thécamoebien filosia *Amphitrema flavum* (Archer 1877) Penard 1902. *Protistologica* 17:225-233.

Brabet, J., Comoy, N., Guitard, J. & Bonnet, L. 1982. Quelques données sur l'ultrastructure d'*Heleopera petricola amethystea* (Rhizopoda, Testacea). *Bull. Soc. Hist. Nat. Toulouse*, 118:161-167.

Breuer, R. 1917. Fortpflanzung und biologische Erscheinungen einer *Chlamydophrys*-Form auf Agarkulturen. *Arch. Protistenkd.*, 37:65-92.

Cash, J. & Hopkinson, J. 1905. *The British Freshwater Rhizopoda and Heliozoa I.* Ray Soc. London.

Cash, J. & Hopkinson, J. 1909. *The British Freshwater Rhizopoda and Heliozoa II.* Ray Soc. London.

Cash, J., Wailes, G. H. & Hopkinson, J. 1915. *The British Freshwater Rhizopoda and Heliozoa III.* Ray Soc. London.

Cavalier-Smith, T. 1993. Kingdom Protozoa and its 18 Phyla. *Microbiol. Rev.*, 57:953-995.

Cavalier-Smith, T. 1997. Amoeboflagellates and mitochondrial cristae in eukaryote evolution: megasystematics of the new protozoan subkingdoms Eozoa and Neozoa. *Arch. Protistenkd.*, 147:237-258.

Cavalier-Smith, T. & Chao, E. E. 1997. Sarcomonad ribosomal RNA sequences, rhizopod phylogeny, and the origin of euglyphid amoebae. *Arch. Protistenkd.*, 147:227-236.

Chardez, D. 1961. Note sur les thécamoebiens d'Otrange (Hesbaye) et découverte d'un genre nouveaux. *Ann. Soc. Zool. Belg.*, 91:39-43.

Chardez, D. 1966. Etudes monographiques sur quelques genres de Thécamoebiens. *Bull. Rech. Agron. Gembloux* N.S., 1:177-189.

Chardez, D. 1967a. Monographie du genre *Quadrulella* Cockerell (Protozoa, Rhizopoda, Testacea). *Bull. Rech. Sta. Agron. Gembloux* N.S., 2:230-241.

Chardez, D. 1967b. Le genre *Geopyxella* Bonnet et Thomas. *Bull. Rech. Sta. Agron. Gembloux* N.S., 2:596-599.

Chardez, D. 1967c. Histoire naturelle des protozoaires thécamoebiens. *Les Naturalistes Belges*, 48:484-576.

Chardez, D. 1969. Le genre *Phryganella* Penard. *Bull. Rech. Agr. Gembloux*, 4:315-322.

Chardez, D. 1972. *Voluta hemispiralis*. Thécamoebien psammophile nouveau. *Rev. Verv. Hist. Nat.*, 29:10-12.

Chardez, D. 1977. Thécamoebiens du mesopsammon des plages de la Mer du Nord. *Rev. Verv. Hist. Nat.*, 34:18-34.

Chardez, D. 1984. Notes sur les *Microcorycia* (Protozoa Sarcomastigophora). *Rev. Verv. Hist. Nat.*, 3:38-42.

Chardez, D. 1991. The Genus *Cyphoderia* Schlumberger, 1845 (Protozoa, Rhizopoda, Testacea). *Acta Protozool.*, 30:49-53.

Chardez, D. & Krizelj, S. 1970. Protozoaires thécamoebiens et ciliés du sol. *Bull. Inst. R. Sci. Nat. Belge*, 46:1-19.

Chardez, D. & Thomas, R. 1980. Thécamoebiens du mesopsammon des plages Lacanau et Leporge Ocean (Gironde, France), (Protozoa, Rhizopoda testacea). *Acta Protozool.*, 19:277-285.

Chardez, D., Delecour, F. & Weissen, F. 1972. Evolution des populations thécamoebiennes de sols forestiers sous l'influence de fumures artificielles. *Rev. Ecol. Biol. Sol*, 9:185-196.

Charret, R. 1964. Contribution à l'étude cytologique et biologique de *Hyalosphenia papilio* (Leidy), Rhizopode Testacé. *Bull. Biol. France Belg.*, 98:369-390.

Collin, B. 1914. Notes protistologiques. *Arch. Zool.*, 54:85-97.

Coûteaux, M.-M. 1976a. Dynamisme de l'équilibre des Thécamoebiens dans quelques sols climaciques. *Mem. Mus. Nat. Hist. Nat., Ser. A, Zool.*, 96:1-183.

Coûteaux, M.-M. 1976b. Le peuplement Thécamoebien du sol et la nature de l'eau disponible. *Bull. Ecol.*, 7:197-206.

Coûteaux, M.-M. 1978. Quelques Thécamoebiens du sol du Japon. *Rev. Ecol. Biol. Sol*, 15:119-128.

Coûteaux, M.-M. 1979. L'effet de la déforestation sur le peuplement thécamoebien en Guyane Française: étude préliminaire. *Rev. Ecol. Biol. Sol*, 16:403-413.

Coûteaux, M.-M. & Chardez, D. 1981. Thécamoebiens édaphiques et muscicoles de Guyane Française. *Rev. Ecol. Biol. Sol*, 18:193-208.

Coûteaux, M.-M. & Ogden, C. G. 1988. The growth of *Tracheleuglypha dentata* (Rhizopoda: Testacea) in clonal cultures under different trophic conditions. *Microb. Ecol.*, 15:81-93.

Cowling, A. J. 1986. Culture methods and observations of *Corythion dubium* and *Euglypha rotunda* (Protozoa, Rhizopoda) isolated from maritime antarctic moss peats. *Protistologica*, 22:181-191.

Decloitre, L. 1960. Perfectionnements apportés à un appareil pour une technique d'isolement des microorganismes du sol, des mousses, des eaux. *Int. Rev. Ges. Hydrobiol.*, 45:169-171.

Decloitre, L. 1961. Le genre *Paraquadrula* (Thecamoebina). *Int. Rev. Ges. Hydrobiol.*, 46:321-330.

Decloitre, L. 1962. Le genre *Euglypha* Dujardin. *Arch. Protistenkd.*, 106:51-100.

Decloitre, L. 1964. Matériaux pour une faune thécamoebienne du Maroc. (4. note). *Bull. Soc. Hist. Nat. Phys. Maroc*, 44:3-12.

Decloitre, L. 1976a. Le genre *Arcella* Ehrenberg. Compléments à jour au 31. Décembre 1974 de la monographie du genre parue en 1928. *Arch. Protistenkd.*, 118:291-309.

Decloitre, L. 1976b. Le genre *Euglypha*. Compléments à jour au 31. Décembre 1974 de la monographie du genre parue en 1962. *Arch. Protistenkd.*, 118:18-33.

Decloitre, L. 1977a. Le genre *Cyclopyxis*. Compléments à jour au 31. Décembre 1974 de la monographie du genre parue en 1929. *Arch. Protistenkd.*, 119:31-53.

Decloitre, L. 1977b. Le genre *Nebela*. Compléments à jour au 31. Décembre 1974 du genre parue en 1936. *Arch. Protistenkd.*, 119:325-352.

Decloitre, L. 1978. Le genre *Centropyxis* I. Compléments à jour au 31.12.1974 de la monographie du genre parue en 1929. *Arch. Protistenkd.*, 120:63-85.

Decloitre, L. 1979a. Le genre *Centropyxis* II. Compléments à jour au 31. Décembre 1974 de la monographie du genre parue en 1929. *Arch. Protistenkd.*, 121:162-192.

Decloitre, L. 1979b. Mises à jour au 31.12.1978 des mises à jour au 31.12.1974 concernant les genres *Arcella, Centropyxis; Cyclopyxis; Euglypha* et *Nebela*. *Arch. Protistenkd.*, 122:387-397.

Decloitre, L. 1981. Le genre *Trinema* Dujardin, 1941. Révision à jour au 31. XII.1979. *Arch. Protistenkd.*, 124:193-218.

Decloitre, L. 1982. Compléments aux publications précédentes. Mise à jour au 31. XII. 1981 des genres *Arcella*, *Centropyxis*, *Cyclopyxis*, *Euglypha*, *Nebela* et *Trinema*. *Arch. Protistenkd.*, 126:393-407.

Decloitre, L. 1986. Compléments aux publications précédentes. Mise à jour au 31. XII. 1984 des genres *Arcella*, *Centropyxis*, *Cyclopyxis*, *Euglypha* et *Nebela*. *Arch. Protistenkd.*, 132:131-136.

Deflandre, G. 1928a. Le genre *Arcella* Ehrenberg. Morphologie-Biologie. Essai phylogénetique et systématique. *Arch. Protistenkd.*, 64:152-189.

Deflandre, G. 1928b. Deux genres nouveaux de rhizopodes testacés. 1. *Wailesella* gen. nov. 2. *Tracheleuglypha* gen. nov. *Ann. Protistol.*, 1:37-43.

Deflandre, G. 1929. Le genre *Centropyxis* Stein. *Arch. Protistenkd.*, 67:322-375.

Deflandre, G. 1931. Thécamoebiens nouveaux ou peu connus, I. *Ann. Protistol.*, 3:81-95.

Deflandre, G. 1932. *Paraquadrula* nov. gen. *irregularis* (Archer). Conjugation et enkystment. *C. R. Soc. Biol.*, 109:1346-1347.

Deflandre, G. 1936. Etude monographique sur le genre *Nebela* Leidy (Rhizopoda-Testacea). *Ann. Protistol.*, 5:201-286.

Deflandre, G. 1953. Ordres des Testacealobosa (De Saedeleer, 1934), Testaceafilosa (De Saedeleer, 1934), Thalamia (Haeckel, 1862) ou thécamoebiens (Auct.) (Rhizopoda Testacea). *In*: Grassé, P.-P. (ed.) *Traité de Zoologie*, 1:97-149.

Deflandre, G. 1958. Eugene Penard, 1855-1954. Correspondance et souvenirs. Bibliographie et bilan systématique de son œuvre. *Hydrobiologia*, 10:2-37.

Dekhtyar, M. N. 1993. New species of the family Difflugiidae (Lobosea, Rhizopoda) with remarks on validity of the genus *Protocucurbitella*. *Zool. Zhurnal*, 72, 6:5-15.

De Saedeleer, H. 1934. Beitrag zur Kenntnis der Rhizopoden: morphologische und systematische Untersuchungen und ein Klassifikationsversuch. *Mem. Mus. R. Hist. Nat. Belge*, 60:1-112.

Doflein, F. 1916. Studien zur Naturgeschichte der Protozoen. VIII. *Pyxidicula operculata* (Agardh). *Zool. Jb.*, 39:585-647.

Douglas, S. V. & Smol, J. P. 1987. Siliceous protozoan plates in lake sediments. *Hydrobiologia*, 154:13-23.

Dykstra, M. J. & Porter, D. 1984. *Diplophrys marina*, a new scale-forming marine protist with labyrinthulid affinities. *Mycologia*, 76:626-632.

Foissner, W. 1987. Soil protozoa: fundamental problems, ecological significance, adaptations in ciliates and testacea, bioindicators, and guide to the literature. *Progr. Protistol.*, 2:69-212.

Foissner, W. & Adam, H. 1981. Die Gemeinschaftsstruktur und Produktion der terricolen Testaceen (Protozoa, Rhizopoda) in einigen Böden der Österreichischen Zentralalpen (Hohe Tauern, Glocknergebiet). *Veröff. Österr. MaB-Programms*, 4:53-78.

Foissner, W., Franz, H. & Adam, H. 1987. Untersuchungen über das Bodenleben in ökologisch und konventionell bewirtschafteten Acker- und Grünlandböden im Raum Salzburg. *Verh. Ges. Ökol. (Graz 1985)*, 15:333-339.

Freeman, C. J. 1974. Shell formation, cell division and longevity in *Lesquereusia spiralis* (Ehrenberg) Penard and *Pontigulasia vas* (Leidy) Schouteden. *PhD Thesis, Univ. S.C.*, 66 pp.

Gauthier-Lièvre, L. 1953. Les genres *Nebela, Paraquadrula* et *Pseudonebela* (Rhizopodes testacés). *Bull. Soc. Hist. Nat. Afrique Nord*, 44:324-366.

Gauthier-Lièvre, L. & Thomas, R. 1958. Les genres *Difflugia, Pentagonia, Maghrebia* et *Hoogenraadia* (Rhizopodes testacés) en Afrique. *Arch. Protistenkd.*, 103:241-370.

Gauthier-Lièvre, L. & Thomas, R. 1960. The genus *Cucurbitella* Penard. *Arch. Protistenkd.*, 104:569-602.

Golemansky, V. 1966. *Playfairina valkanovi* nov. sp.—une nouvelle Rhizopode d'eau douce (Rhizopoda, Testacea). *C. R. Acad. Bulg. Sci.*, 19:57-59.

Golemansky, V. 1967. Matériaux sur la systématique et l'écologie des Thécamoebiens (Protozoa, Rhizopoda) du lac d'Ohrid. *Sect. Sci. Nat. Univ. Skopje*, 14:3-26.

Golemansky, V. 1968. Matériaux sur la faune thécamoebienne (Rhizopoda, Testacea) de Cuba. *Acta Protozool.*, 6:335-340.

Golemansky, V. 1970a. Contribution à la connaissance des Thécamoebiens (Rhizopoda, Testacea) des eaux souterraines littorales du Golf de Gdansk (Pologne). *Bull. Inst. Zool. Mus. (Sofia)*, 31:77-87.

Golemansky, V. 1970b. Rhizopodes nouveaux du psammon littoral de la Mer Noire. *Protistologica*, 6:365-371.

Golemansky, V. 1970c. *Chardezia caudata* gen. n. sp. n. et *Rhumbleriella filosa* gen. n. sp. n.—deux Thécamoebiens nouveaux du psammon littoral de la Mer Noire. *Bull. Inst. Zool. Mus. (Sofia)*, 32:121-125.

Golemansky, V. 1971. Taxonomische und zoogeographische Notizen über die Thekamöbe Fauna (Rhizopoda, Testacea) der Küstengrundgewässer der sowjetischen Fernostküste (Japanisches Meer) und der Westküste Kanadas (Stiller Ozean). *Arch. Protistenkd.*, 113:235-249.

Golemansky, V. 1973. *Messemvriella filosa* n. gen. n. sp.—une nouvelle thécamoebienne psammobionte (Rhizopoda: Testacea) des eaux souterraines littorales de la Mer Noire. *Zool. Anz.*, 190:302-304.

Golemansky, V. 1974. Psammonobiotidae fam. nov.—une nouvelle famille de thécamoebiens (Rhizopoda, Testacea) du psammal supralittoral des mers. *Acta Protozool.*, 8:137-141.

Golemansky, V. 1982. Révision du genre *Ogdeniella* nom. n. (=*Amphorellopsis* Golemansky, 1970) (Rhizopoda, Gromida) avec considérations sur son origine et évolution dans le milieu interstitiel. *Acta Zool. Bulg.*, 19:3-12.

Golemansky, V. 1991. Thécamoebiens mésopsammiques (Rhizopoda: Arcellinida, Gromida & Monothalamida) du sublittoral marin de l'Atlantique dans la région de Roscoff (France). *Arch. Protistenkd.*, 140:35-43.

Golemansky, V. & Coûteaux, M.-M. 1982. Etude en microscopie électronique à balayage de huit espèces de thécamoebiens interstitiels du supralittoral marin. *Protistologica*, 18:473-480.

Golemansky, V. & Ogden, C. G. 1980. Shell structure of three littoral species of testate amoebae from the Black Sea (Rhizopodea, Protozoa). *Bull. Br. Mus. Nat. Hist. (Zool.)*, 38:1-6.

Golemansky, V., Skarlato, S. O. & Todorov, M. T. 1987. A light- and electron-microscopical (SEM and TEM) study of *Microchlamys sylvatica* n. sp. (Rhizopoda: Arcellinida). *Arch. Protistenkd.*, 134:161-167.

Green, J. 1975. Freshwater ecology in the Mato Grosso, Central Brazil, IV: Associations of testate Rhizopoda. *J. Nat. Hist.*, 9:545-560.

Griffin, J. L. 1972. Movement, fine structure and fusion of pseudopods of an enclosed amoeba, *Difflugiella* sp. *J. Cell Sci.*, 10:563-583.

Grospietsch, T. 1954. Studien über die Rhizopodenfauna von Schwedisch-Lappland. *Arch. Hydrobiol.*, 49:546-580.

Grospietsch, T. 1964. Die Gattungen *Cryptodifflugia* und *Difflugiella* (Rhizopoda, Testacea). *Zool. Anz.*, 172:243-257.

Grospietsch, T. 1965. Monographische Studie der Gattung *Hyalosphenia* Stein, (Rhizopoda, Testacea). *Hydrobiologia*, 26:211-241.

Grospietsch, T. 1972a. *Wechseltierchen* (*Rhizopoden*). Franckh. Verl., Stuttgart.

Grospietsch, T. 1972b. Testacea und Heliozoa. *In*: Elster, H.-J. & Ohle, W. (eds.), *Das Zooplankton der Binnengewässer, I.* Schweizerbarth, Stuttgart. pp 1-30

Grospietsch, T. 1982. Untersuchungen über die Thekamöbenfauna (Rhizopoda, Testacea) im Murnauer Moos. *Entomofauna*, 1:57-88.

Hallas, T. E. 1975. Notes on the encystation in *Microcorycia radiata* (Testacealobosa, Protozoa). *Pedobiologia*, 15: 149-150.

Hannah, F., Rogerson, A. & Anderson, O. R. 1996. A description of *Paulinella indentata* N. Sp. (Filosea: Euglyphina) from subtidal coastal benthic sediments. *J. Euk. Microbiol.,* 43:1-4.

Harnisch, O. 1958. Rhizopoda. *In*: Brohmer, P. Ehrmann, P. & Ulmer, G. (eds.), *Tierwelt Mitteleurop.*, Vol. 1, 1b:1-75.

Harrison, F. W., Dunkelberger, D., Watanabe, N. & Stump, A. B. 1975. The cytology of the testaceous rhizopod *Lesquereusia spiralis* (Ehrenberg) Penard. I. Cytochemistry. *Acta Histochem.*, 54:71-77.

Harrison, F. W., Dunkelberger, D., Watanabe, N. & Stump, A. B. 1976. The cytology of the testaceous rhizopod *Lesquereusia spiralis* (Ehrenberg) Penard. II. Ultrastucture and shell formation. *J. Morph.*, 150:343-358.

Harrison, F. W., Dunkelberger, D., Watanabe, N. & Stump, A. B. 1981. Ultrastructure and deposition of silica in rhizopod amoebae. *In*: Simpson, T. L. & Volcani, B. E. (eds.), *Silicon and Siliceous Structures in Biological Systems*, Springer-Verlag, New York. pp 281-294

Hedley, R. H. & Ogden, C. G. 1973. Biology and fine structure of *Euglypha rotunda* (Testacea; Protozoa). *Bull. Br. Mus. Nat. Hist. Zool.*, 25:121-137.

Hedley, R. H. & Ogden, C. G. 1974a. Observations on *Trinema lineare* Penard (Testacea: Protozoa). *Bull. Brit. Mus. Nat. Hist. Zool.*, 26:185-199.

Hedley, R. H. & Ogden, C. G. 1974b. Adhesion plaques associated with the production of a daughter cell in *Euglypha* (Testacea: Protozoa). *Cell Tiss. Res.*, 153:261-268.

Hedley, R. H., Ogden, C. G. & Krafft, J. I. 1974. Observations on clonal cultures of *Euglypha acanthophora* and *Euglypha strigosa* (Testacea: Protozoa). *Bull. Br. Mus. Nat. Hist. Zool.*, 27:103-111.

Hedley, R. H., Ogden, C. G. & Mordan, N. J. 1977. Biology and fine structure of *Cryptodifflugia oviformis* (Rhizopodea: Protozoa). *Bull. Br. Mus. Nat. Hist. Zool.*, 30:313-328.

Hoogenraad, H. R. & De Groot, A. A. 1940. Zoetwaterrhizopoden en Heliozoen. *In*: Boschma, H., Beaufort, L. F., Redeke, H. C. & Roepke, W. (eds.), *Fauna van Nederland*, Leiden, 9: 1 - 302.

Jennings, H. S. 1916. Heredity, variation and the results of selection in the uniparental reproduction of *Difflugia corona*. *Genetics*, 1:407-534.

Johnson, P. W., Hargraves, P. E. & Sieburth, J. M. 1988. Ultrastructure and ecology of *Calycomonas ovalis* Wulff, 1919, (Chrysophyceae) and its redescription as a testate rhizopod, *Paulinella ovalis* N. Comb. (Filosea: Euglyphina). *J. Protozool.*, 35:618-626.

Joyon, M. L. & Charret, R. 1962. Sur l'ultrastructure du Thécamoebien *Hyalosphenia papilio* (Leidy). *C. R. Acad. Sci. Paris*, 255:2661-2663.

Jung, W. 1942a. Südchilenische Thekamöben. (aus dem südchilenischen Küstengebiet, Beitrag 10). *Arch. Protistenkd.*, 95:253-356.

Jung, W. 1942b. Illustrierte Thekamöben-Bestimmungstabellen. I. Die Systematik der Nebelinen. *Arch. Protistenkd.*, 95:357-390.

Kies, L. 1974. Elektronenmikroskopische Untersuchungen an *Paulinella chromatophora* Lauterborn, einer Thekamöbe mit blau-grünen Endosymbionten (Cyanellen). *Protoplasma*, 80:69-89.

Kies, L. & Kremer, B. P. 1979. Function of cyanelles in the Thecamoeba *Paulinella chromatophora. Naturwissenschaften*, 66:578-579.

Laminger, H. 1971. Sedimentbewohnende Schalenamöben (Rhizopoda, Testacea) der Finstertaler Seen (Tirol). *Arch. Hydrobiol.*, 69:106-140.

Leidy, J. 1879. Freshwater rhizopods of North America. *U.S. Geol. Surv. Terr.*, 12:1-324.

Lee. J. J., Hutner, S. H. & Bovee, E. C. (eds.) 1985. *An Illustrated Guide to the Protozoa.* The Society of Protozoologists, Lawrence, Kansas.

Lee, J. J. & Soldo, A. T. (eds.) 1992. *Protocols in Protozoology*. The Society of Protozoologists, Lawrence, Kansas.

Loeblich, A. R., Jr. & Tappan, H. 1961. Remarks on the systematics of the Sarkodina (Protozoa), renamed homonyms and new and validated genera. *Proc. Biol. Soc. Wash.*, 74:213-234.

Loeblich, A. R., Jr. & Tappan, H. 1964. Sarcodina chiefly "Thecamoebians" and Foraminiferida. *In*: Moore, R. C. (ed.), *Treatise on Invertebrate Paleontology*, Geol. Soc. Am. and Univ. Kansas Press, Lawrence, Kansas. Part C, Protista 2:1-54.

Lousier, J. D. 1974. Effects of experimental soil moisture fluctuations on turnover-rates of Testacea. *Soil Biol. Biochem.*, 6:19-26.

Lousier, J. D. 1984a. Population dynamics and production studies of species of Euglyphidae (Testacea, Rhizopoda, Protozoa) in an aspen woodland soil. *Pedobiologia*, 26:309-330.

Lousier, J. D. 1984b. Population dynamics and production studies of *Phryganella acropodia* and *Difflugiella oviformis* (Testacea, Rhizopoda, Protozoa) in an aspen woodland soil. *Pedobiologia*, 26:331-347.

Lousier, J. D. 1984c. Population dynamics and production studies of species of Nebelidae (Testacea, Rhizopoda) in an aspen woodland soil. *Acta Protozool.*, 23:145-159.

Lousier, J. D. 1985. Population dynamics and production studies of species of Centropyxidae (Testacea, Rhizopoda) in an aspen woodland soil. *Arch. Protistenkd.*, 130:165-178.

Lousier, J. D. & Parkinson, D. 1984. Annual population dynamics and production ecology of Testacea (Rhizopoda Protozoa) in an aspen woodland soil. *Soil Biol. Biochem.*, 16:103-114.

Lüftenegger, G. & Foissner, W. 1991. Morphology and biometry of twelve soil testate amoebae (Protozoa, Rhizopoda) from Australia, Africa, and Austria. *Bull. Br. Mus. Nat. Hist. (Zool.)*, 57:1-16.

Lüftenegger, G., Petz, W., Berger, H., Foissner, W. & Adam, H. 1988. Morphologic and biometric characterization of twenty-four soil testate amoebae (Protozoa, Rhizopoda). *Arch. Protistenkd.*, 136:153-189.

Meisterfeld, R. 1977. Die horizontale und vertikale Verteilung der Testaceen (Rhizopoda, Testacea) in *Sphagnum*. *Arch. Hydrobiol.*, 79:319-356.

Meisterfeld, R. 1978. Die Struktur von Testaceenzönosen (Rhizopoda, Testacea) in *Sphagnum* unter besonderer Berücksichtigung ihrer Diversität. *Verh. Ges. Ökol. (Kiel 1977)*, 7:441-450.

Meisterfeld, R. 1979a. Zur Systematik der Testaceen (Rhizopoda, Testacea) in *Sphagnum*. Eine REM-Untersuchung. *Arch. Protistenkd.*, 121:246-269.

Meisterfeld, R. 1979b. Clusteranalytische Differenzierung der Testaceenzönosen (Rhizopoda, Testacea) in *Sphagnum*. *Arch. Protistenkd.*, 121:270-307.

Meisterfeld, R. 1986. The importance of protozoa in a beech forest ecosystem. *Symp. Biol. Hung.*, 33:291-299.

Meisterfeld, R. 1989. Die Bedeutung der Protozoen im Kohlenstoffhaushalt eines Kalkbuchenwaldes. (Zur Funktion der Fauna in einem Mullbuchenwald 3). *Verh. Ges. Ökol. (Göttingen 1987)*, 17:221-227.

Meisterfeld, R., Dohmen, C., Meyer, A., Panfil, C. & Wienand, J. 1992. Influence of food quality and quantity on growth and feeding rates of testate amoebae (Testacealobosia). *Europ. J. Protistol.*, 28:351.

Mignot, J. P. & Raikov, I. B. 1990. New ultrastructural data on the morphogenesis of the test in the Testacea *Arcella vulgaris*. *Europ. J. Protistol.*, 26:132-141.

Mignot, J. P. & Raikov, I. B. 1992. Evidence for meiosis in the testate amoeba *Arcella*. *J. Protozool.*, 39:287-289.

Netzel, H. 1972. Die Bildung der Gehäusewand bei der Thekamöbe *Centropyxis discoides* (Rhizopoda, Testacea). *Z. Zellforsch. Microsk. Anat.*, 135:45-54.

Netzel, H. 1975a. Die Entstehung der hexagonalen Schalenstruktur bei der Thekamöbe *Arcella vulgaris var. multinucleata* (Rhizopoda, Testacea). *Arch. Protistenkd.*, 117:321-357.

Netzel, H. 1975b. Morphologie und Ultrastruktur von *Centropyxis discoides* (Rhizopoda, Testacea). *Arch. Protistenkd.*, 117:369-392 + Tab. 58-69.

Netzel, H. 1976. Die Ultrastruktur der Schale von *Difflugia oviformis* (Rhizopoda, Testacea). *Arch. Protistenkd.*, 118:321-329+ T. 63-70.

Netzel, H. 1983. Gehäusewandbildung durch mehrphasige Sekretion bei der Thekamöbe *Netzelia oviformis* (Rhizopoda Testacea). *Arch. Protistenkd.*,127:351-381.

Netzel, H. & Grunewald, B. 1977. Morphogenesis in the shelled rhizopod *Arcella dentata*. *Protistologica*, 13:299-319.

Ogden, C. G. 1979a. Siliceous structures secreted by members of the subclass Lobosia. *Bull. Br. Mus. Nat. Hist. (Zool.)*, 36:203-207.

Ogden, C. G. 1979b. Comparative morphology of some pyriform species of *Difflugia*. *Arch. Protistenkd.*, 122:143-153.

Ogden, C. G. 1979c. An ultrastructural study of division in *Euglypha* (Protozoa: Rhizopoda). *Protistologica*, 15:541-556.

Ogden, C. G. 1980. Shell structure in some pyriform species of *Difflugia* (Rhizopodea). *Arch. Protistenkd.*, 123:455-470.

Ogden, C. G. 1981. Observations of clonal cultures of Euglyphidae (Rhizopoda, Protozoa). *Bull. Br. Mus. Nat. Hist. (Zool.)*, 41:137-151.

Ogden, C. G. 1983a. Observations on the systematics of the genus *Difflugia* in Britain (Rhizopoda, Protozoa). *Bull. Br. Mus. Nat. Hist. (Zool.)*, 44:1-73.

Ogden, C. G. 1983b. The significance of the inner dividing wall in *Pontigulasia* Rhumbler and *Zivkovicia* gen. nov. (Protozoa: Rhizopoda). *Protistologica*, 19:215-229.

Ogden, C. G. 1984. Shell structure of some testate amoebae from Britain (Protozoa Rhizopoda). *J. Nat. Hist.*, 18:341-361.

Ogden, C. G. 1985. The flexible shell of the freshwater amoeba *Microchlamys patella* (Claparède & Lachmann, 1859) (Rhizopoda: Arcellinida). *Protistologica*, 21:141-152.

Ogden, C. G. 1987a. The fine structure of the shell of *Pyxidicula operculata*, an aquatic testate amoeba (Rhizopoda). *Arch. Protistenkd.*, 133:157-164.

Ogden, C. G. 1987b. The taxonomic status of the genera *Pontigulasia*, *Lagenodifflugia* and *Zivkovicia* (Rhizopoda: Difflugiidae). *Bull. Br. Mus. Nat. Hist. (Zool.)*, 52:13-17.

Ogden, C. G. 1988. Fine structure of the shell wall in the soil testate amoeba *Cyclopyxis kahli* (Rhizopoda). *J. Protozool.*, 35:537-540.

Ogden, C. G. 1989. The agglutinate shell of *Heleopera petricola* (Protozoa, Rhizopoda), factors affecting its structure and composition. *Arch. Protistenkd.*, 137:9-24.

Ogden, C. G. 1990. The structure of the shell wall in testate amoebae and the importance of the organic cement matrix. *In*: Claugher, D. (ed.), *Scanning Electron Microscopy in Taxonomy and Functional Morphology*. Syst. Ass. Special Vol., Clarendon Press, Oxford, 41:235-257.

Ogden, C. G. 1991a. The biology and ultrastructure of an agglutinate testate amoeba *Difflugia geosphaira* sp. nov. (Protozoa, Rhizopoda). *Arch. Protistenkd.*, 140:141-150.

Ogden, C. G. 1991b. The ultrastructure of *Heleopera petricola* an agglutinate soil amoeba with comments on feeding and silica deposition. *Europ. J. Protistol.*, 27:238-248.

Ogden, C. G. 1991c. Ultrastructure of the vegetative organisation and initial stages of silica plate deposition in the soil testate amoeba *Corythion dubium*. *Protoplasma*, 163:136-144.

Ogden, C. G. & Coûteaux, M.-M. 1987. The biology and ultrastructure of the soil testate amoeba *Tracheleuglypha dentata* (Rhizopoda: Euglyphidae). *Europ. J. Protistol.*, 23:28-42.

Ogden, C. G. & Coûteaux, M.-M. 1988. The effect of predation on the morphology of *Tracheleuglypha dentata* (Protozoa: Rhizopoda). *Arch. Protistenkd.*, 136:107-115.

Ogden, C. G. & Coûteaux, M.-M. 1989. Interstitial marine rhizopods (Protozoa) from littoral sands on the east coast of England. *Europ. J. Protistol.*, 24:281-290.

Ogden, C. G. & Ellison, R. L. 1988. The value of the organic cement matrix in the identification of the shells of fossil testate amoebae. *J. Micropalaeontol.*, 7:233-240.

Ogden, C. G. & Hedley, R. H. 1980. *An Atlas of Freshwater Testate Amoebae.* Oxford University Press, Oxford.

Ogden, C. G. & Meisterfeld, R. 1989. The taxonomy and systematics of some species of *Cucurbitella*, *Difflugia* and *Netzelia* (Protozoa: Rhizopoda); with an evaluation of diagnostic characters. *Europ. J. Protistol.*, 25:109-128.

Ogden, C. G. & Meisterfeld, R. 1991. The biology and ultrastructure of the testate amoeba, *Difflugia lucida* Penard (Protozoa, Rhizopoda). *Europ. J. Protistol.*, 26:256-269.

Ogden, C. G. & Pitta, P. 1989. Morphology and construction of the shell wall in an agglutinate soil testate amoeba *Phryganella acropodia* (Rhizopoda). *J. Protozool.*, 36:437-445.

Ogden, C. G. & Pitta, P. 1990. Biology and ultrastructure of the mycophagous, soil testate amoeba, *Phryganella acropodia* (Rhizopoda, Protozoa). *Biol. Fertil. Soils*, 9:101-109.

Ogden, C. G. & Zivkovic, A. 1983. Morphological studies on some Difflugiidae from Yugoslavia (Rhizopoda, Protozoa). *Bull. Br. Mus. Nat. Hist. (Zool.)*, 44:341-375.

Page, F. C. 1966. *Cryptodifflugia operculata* n. sp. (Rhizopodea: Arcellinida, Cryptodifflugiidae) and the status of the genus *Cryptodifflugia*. *Trans. Am. Microsc. Soc.*, 85:506-515.

Page, F. C. 1987. The classification of "naked" amoebae (Phylum, Rhizopoda). *Arch. Protistenkd.*, 133:199-217.

Pankow, H. 1982. *Paulinella chromatophora* Lauterborn, eine bisher nur im Süßwasser nachgewiesene Thekamöbe, in den Boddengewässern des Darß und des Zingst (südliche Ostsee). *Arch. Protistenkd.*, 126:261-263.

Pateff, P. 1926. Fortpflanzungserscheinungen bei *Difflugia mammillaris* Penard und *Clypeolina marginata* Penard. *Arch. Protistenkd.*, 55:516-544.

Penard, E. 1890. Études sur les rhizopodes d'eau douce. *Mem. Soc. Phys. Hist. Genève*, 31:1-230.

Penard, E. 1902. *Faune Rhizopodique du Bassin de Léman.* Kündig, Genève.

Penard, E. 1903. Sur quelques Protistes voisin des Héliozoaires ou des Flagellates. *Arch. Protistenkd.*, 2:283-304.

Penard, E. 1904. Quelques nouveaux rhizopodes d'eau douce. *Arch. Protistenkd.*, 3:391-422.

Penard, E. 1906. Notes sur quelques sarcodinés. *Rev. Suisse Zool.*, 14:109-141.

Penard, E. 1907. Etude sur la *Clypeolina marginata*. *Arch. Protistenkd.*, 8:66-85.

Penard, E. 1909. Sur quelques rhizopodes des mousses. *Arch. Protistenkd.*, 17:258-296.

Penard, E. 1910. Rhizopodes nouveaux. *Rev. Suisse Zool.*, 18:929-940.

Penard, E. 1912. Notes sur quelques Sarcodinés. *Rev. Suisse Zool.*, 20:1-29.

Peshkov, M. A. & Akinshina, G. T. 1963. Cytological studies of the rhizopod *Chlamydophrys major*. 1. Fine structure of the *Chlamydophrys* trophont based on electron microscopy of ultrathin sections and in vivo observations. *Tsitologiya*, 5:554-564.

Poinar, G. O., Waggoner, B. M. & Bauer, U. 1993. Terrestrial soft-bodied protists and other microorganisms in Triassic amber. *Science*, 259:222-224.

Rauenbusch, K. 1987. Biologie und Feinstruktur (REM-Untersuchungen) terrestrischer Testaceen in Waldböden (Rhizopoda, Protozoa). *Arch. Protistenkd.*, 134:191-294.

Raikov, I. B. 1982. *The Protozoan Nucleus. Morphology and Evolution.* Springer-Verlag, Berlin.

Raikov, I. B. & Mignot, J. P. 1991. Fine-structural study of mitosis in the Testacea *Arcella vulgaris* Ehrbg. *Europ. J. Protistol.*, 26:340-349.

Raikov, I. B., Karadzhan, B. P., Kaur, R. & Mignot, J. P. 1989. Nuclear fine structure at interphase and during encystment in 2 forms of the testacean *Arcella vulgaris*. *Europ. J. Protistol.*, 24:369-380.

Schönborn, W. 1962a. Die Ökologie der Testaceen im oligotrophen See, dargestellt am Beispiel des Großen Stechlinsees. *Limnologica*, 1:111-182.

Schönborn, W. 1962b. Die Stratigraphie lebender Testaceen im Sphagnetum der Hochmoore. *Limnologica*, 1:315-321.

Schönborn, W. 1964. Bodenbewohnende Testaceen aus Deutschland. I. Untersuchungen im Naturschutzgebiet Serrahn (Mecklenburg). *Limnologica*, 2:105-122.

Schönborn, W. 1965a. Studien über die Gattung *Difflugiella* Cash (Rhizopoda, Testacea). *Limnologica*, 3:315-328.

Schönborn, W. 1965b. Untersuchungen über die Zoochlorellen-Symbiose der Hochmoor-Testaceen. *Limnologica*, 3:173-176.

Schönborn, W. 1965c. Die sedimentbewohnenden Testaceen einiger Masurischer Seen. *Acta Protozool.*, 3:297-309.

Schönborn, W. 1966. *Beschalte Amöben.* A. Ziemsen-Verlag, Wittenberg-Lutherstadt.

Schönborn, W. 1967. Taxozönotik der beschalten Süßwasser-Rhizopoden. Eine raumstruktur-analytische Untersuchung über Lebensraum-erweiterung und Evolution bei der Mikrofauna. *Limnologica*, 5:159-207.

Schönborn, W. 1977. Production studies on Protozoa. *Oecologia*, 27:171-184.

Schönborn, W. 1982. Estimation of annual production of Testacea (Protozoa) . in mull and moder (II). *Pedobiologia*, 23:383-393.

Schönborn, W. 1989. The topophenetic analysis as a method to elucidate the phylogeny of testate amoebae (Protozoa, Testacealobosia and Testaceafilosia). *Arch. Protistenkd.*, 127:223-245.

Schönborn, W. 1992. Adaptive polymorphism in soil-inhabiting testate amoebae (Rhizopoda)—its importance for delimitation and evolution of asexual species. *Arch. Protistenkd.*, 142:139-155.

Schönborn, W. & Peschke, T. 1990. Evolutionary studies on the *Assulina-Valkanovia* complex (Rhizopoda, Testaceafilosia) in *Sphagnum* and soil. *Biol. Fertil. Soils*, 9: 95-100.

Schönborn, W., Flössner, D. & Proft, G. 1966. Die limnologische Charakterisierung des Profundals einiger norddeutscher und masurischer Seen mit Hilfe von Testaceen-Gemeinschaften. *Verh. Int. Verein. Limnol.*, 16:251-257.

Schönborn, W., Foissner, W. & Meisterfeld, R. 1983. Licht- und Rasterelektronenmikroskopische Untersuchungen zur Schalenmorphologie und Rassenbildung bodenbewohnender Testaceen (Protozoa: Rhizopoda) sowie Vorschläge zur biometrischen Charakterisierung von Testaceen-Schalen. *Protistologica*, 19:553-566.

Schönborn, W., Petz, W., Wanner, M. & Foissner, W. 1987. Observations on the morphology and ecology of the soil-inhabiting testate amoeba *Schoenbornia humicola* (Schönborn, 1964) Decloitre, 1964 (Protozoa, Rhizopoda). *Arch. Protistenkd.*, 134:315-330.

Simitzis, A. M. & Le Goff, F. 1981. Observations on testate Amoebae of the group *Capsellina* Penard 1909 *Rhogostoma* Belar 1921 (Gromiidae, Gromiida, Filosca). *Protistologica*, 17:99-111.

Stepanek, M. 1963. Die Rhizopoden aus Katanga (Kongo-Afrika). *Annls. Mus. R. Afr. Cent.(Zool.)*, Serie IN 8,117:8-91 with 20 plates.

Stepanek, M. 1967. Testacea des Benthos der Talsperre Vranov am Thayafluss. *Hydrobiologia*, 29:1-67.

Stump, A. B. 1959. Mitosis in the Rhizopod *Lesquereusia spiralis*. *J. Protozool.*, 6:185-189.

Sudzuki, M. 1979a. Marine Testacea of Japan. *Sesoko Mar. Sci. Lab. Tech. Rep.*, 6:51-61.

Sudzuki, M. 1979b. Marine interstitial Testacea from Plau Pinang, Malaysia. *Annot. Zool. Japon.*, 52:50-53.

Tappan, H. 1966. *Valkanovia*, a new name for the thecamoebian *Euglyphella* Valkanov, 1962 non Warthin, 1934 (Protozoa: Rhizopodea). *Proc. Biol. Soc. Wash.*, 79:89.

Thomas, R. 1957. Remarques sur deux Rhizopodes: *Paralieberkuehnia gracilis* (Möbius) comb. nov. et *Amphizonella violacea* Greeff. *P. V. Soc. Linn. Bordeaux*, 96:1-7.

Thomas, R. 1958. Le genre *Plagiopyxis* Penard. *Hydrobiologia*, 10:198-214.

Thomas, R. 1961. Note sur quelques rhizopodes de France. *Cah. Nat.*, *Bull. Naturalistes Parisiens. n.s.*, 17:74-80.

Thomas, R. & Gauthier-Lièvre, L. 1959a. Le genre *Lesquereusia* Schlumberger 1845 (Rhizopodes testacés). *Bull. Soc. Hist. Nat. Afr. Nord*, 50:34-83 + Pl. I, II.

Thomas, R. & Gauthier-Lièvre, L. 1959b. Note sur quelques Euglyphidae d'Afrique. *Bull. Soc. Hist. Nat. Afr. Nord*, 50:204-221.

Thomas, R. & Gauthier-Lièvre, L. 1961. Addenda et corrigenda le genre *Lesquereusia* Schlumberger. *Bull. Soc. Hist. Nat. Afr. Nord*, 52:162-163.

Tolonen, K. 1986. Rhizopod analysis. *In*: Berglund, B. E. (ed.), *Handbook of Holocene Palaeoecology and Palaeohydrology*. John Wiley & Sons N.Y. pp 645-666

Tolonen, K., Warner, B. G. & Vasander, H. 1992. Ecology of Testacea (Protozoa, Rhizopoda) in Mires in Southern Finland. 1. Autecology. *Arch. Protistenkd.*, 142:119-138.

Valkanov, A. 1970. Beitrag zur Kenntnis der Protozoen des Schwarzen Meeres. *Zool. Anz.*, 184:241-290.

Van Oye, P. 1956. Rhizopoda Venezuelas mit besonderer Berücksichtigung ihrer Biogeographie. *Hydrobiologia*, 8:329-360.

Velho, L. F. M. & Lansac-Tôha, F. A. 1996. Testate amoebae (Rhizopodea-Sarcodina) from zooplankton of the High Paraná River floodplain, State of Mato Grosso do Sul, Brazil: II. Family Difflugidae. *Stud. Neotrop. Fauna & Environm.*, 31:179-192.

Vucetich, M. C. 1974. Comentarios criticos sobre *Argynnia* Jung, 1942 (Rhizopoda, Testacea). *Neotropica*, 20:126-128.

Vørs, N. 1993. Marine heterotrophic amoebae, flagellates and heliozoa from Belize (Central America) and Tenerife (Canary Island), with descriptions of new species, *Luffisphaera bulbochaete* n. sp., *L. longihastis* n. sp., *L. turriformis* n. sp. and *Paulinella intermedia* n. sp. *J. Euk. Microbiol.*, 40:272-287.

Wailes, G. H. & Hopkinson, J. 1919. *The British Freshwater Rhizopoda and Heliozoa IV.* Ray Soc. London.

Wailes G. H. & Penard, E. 1911. Rhizopoda, Clare Island Survey, pt.65. *Proc. R. Ir. Acad. Dublin*, 33:1-64 + pl. 1-6.

Wanner, M. 1991. Zur Ökologie von Thekamöben (Protozoa: Rhizopoda) in süddeutschen Wäldern. *Arch. Protistenkd.*, 140:237-288.

Wanner, M. & Funke, W. 1986. Rasterelektronen-mikroskopische Untersuchungen an *Edaphonobiotus campascoides* (Protozoa: Testacea). *Arch. Protistenkd.*, 132:187-190.

Wanner, M., Nähring, J. M. & Fischer, R. 1997. Molecular identification of clones of testate amoebae using single nuclei for PCR amplification. *Europ. J. Protistol.*, 33:192-199.

Zivkovic, A. R. 1975. Nouvelles et rares espèces de Testacea (Rhizopoda) dans la fauna de Danube. *Bull. Mus. Hist. Nat. Belgrad, Ser. B.*, 30:119-123.

FAMILY TRICHOSIDAE Möbius, 1889

by F. L. SCHUSTER

Multinucleate marine amebae with alternation of schizont (asexual) and gamont (sexual) generations, the former with calcite spicules on the surface and the latter without. Page (1983) suggests use of terms "fuzzy" and "smooth" for **schizont** and **gamont** individuals, respectively, until the life-cycle can be better defined. Gamonts rare in cultured material. Spicule morphology and size of amebae useful characters for speciation, though spicule morphology is varied. Organism covered by a complex test or glycocalyx consisting of spicules embedded in mucin layer on top of a fibrillar layer (Sheehan and Banner, 1973). Test contains multiple apertures through which unbranched, probably sensory, dactylopods extend. **Dacytlopods** probably tactile in function. Movement by means of **lobopodia**. Wide distribution in littoral zone, and frequently isolated from marine aquaria. Euryhaline and eurythermal according to Sheehan and Banner (1973). Gametogenesis reported by Schaudinn (1899) and Angell (1976). *Pontifex maximus* is a large marine amoeba described by Schaeffer (1926). Because it resembles in its general characteristics the gamont or smooth form, Griffin and Spoon (1977) and Page (1983) consider it to be *Trichosphaerium*.

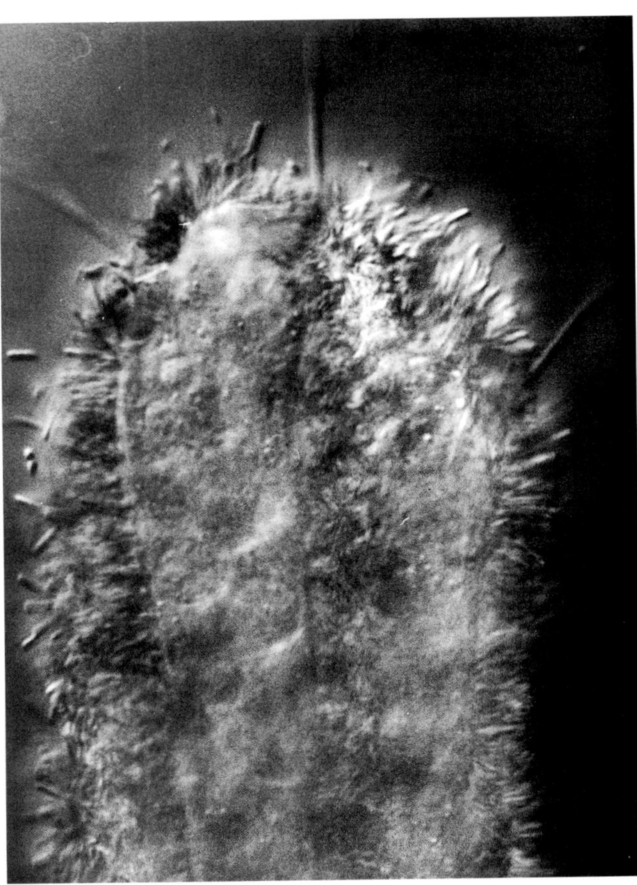

Fig. 1. Schizont stage of *Trichosphaerium sieboldi* with spicules covering the cell surface and dactylopods projecting beyond the spicule layer. X2700.

Genus ***Trichosphaerium*** Schneider, 1878

Trichosphaerium sieboldi Schneider, 1878
Multinucleate marine amoebae occurring as schizonts or gamonts, with characteristics of the the Order. Test of schizont individuals covered with spicules having triangular aspect in cross-section. Nuclear divisions synchronous and occur within the nuclear envelope. Ref. Page (1983), Schaudinn (1899), Schuster (1976).

KEY TO SPECIES

1. Calcareous spicules triangular in cross-section. Large amoebae up to 2 mm in diameter ......***Trichosphaerium sieboldi***
2. Amebae average 36 µm in diameter.......... ....***Trichosphaerium micrum***
3. Calcareous spicules flat and blade-shaped...... ....***Trichosphaerium platyxyrum***

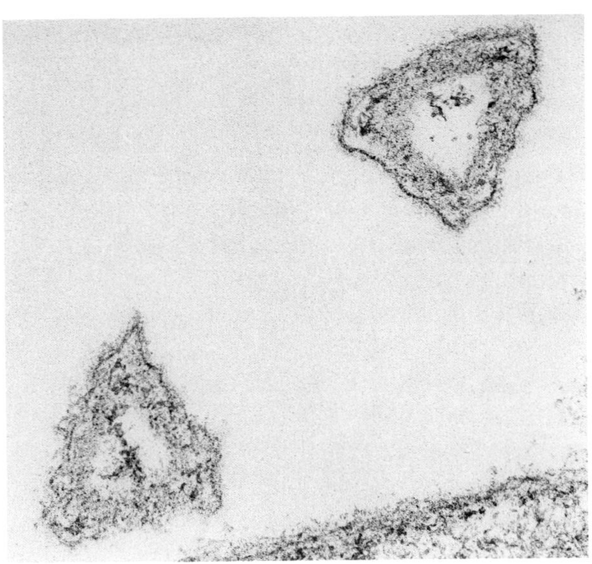

Fig. 2. Transverse section (TEM) through two spicules showing triangular shape and hollow center. X94,000.

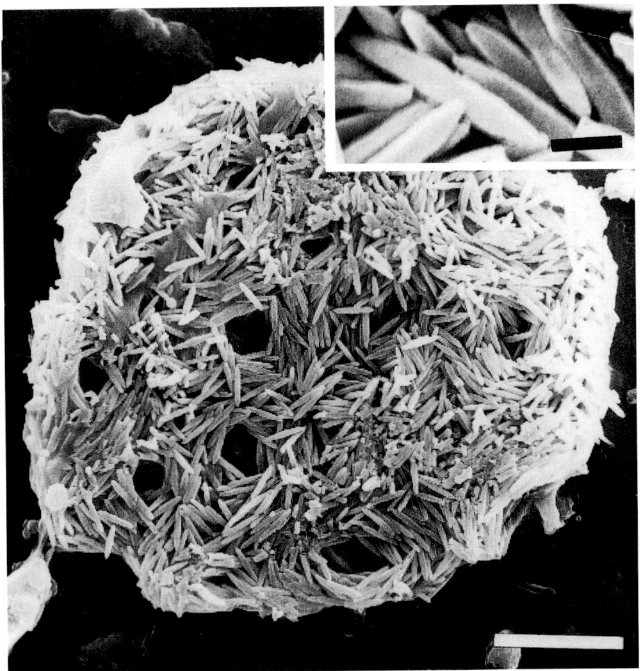

Fig. 3. SEM micrograph of entire schizont of *Trichospherium* with dense layer of spicules. Apertures can be seen in the test. Scale=10 µm. Inset: spicules at higher magnification. Scale=1 µm.

Trichosphaerium micrum Angell, 1975. Size range of amebae 22-50 μm with average at 36 μm. Nuclei range from 2-24 in number per amoeba. Spicules on test are elongate (1-3 μm) and hollow, with blunt ends. Gamonts not observed in cultured material. Ref. Angell (1975).

Trichosphaerium platyxyrum Angell, 1976. Schizont amoebae with size range of 47-93 μm, average 67 μm, with 23-53 nuclei per amoeba, with average of 38. Spicules of schizont are thin and blade-shaped, with squared ends (3-10 μm in length). Multinucleate gamonts lack spicules on test surface. Nuclei number 9-56 with average of 23. Size range of gamonts 37-110 μm (average is 63 μm) but some individuals measure over 1 mm in size. Nuclear division asynchronous in gamont. Early stages of gametogenesis observed. Ref. Angell (1976).

LITERATURE CITED

Angell, R. W. 1975. Structure of *Trichosphaerium micrum* sp. n. *J. Protozool.*, 22:18-22.

Angell, R. W. 1976. Observations on *Trichosphaerium platyxyrum* sp. n. *J. Protozool.*, 23:357-364.

Griffin, J. L., Spoon, D. M. 1977. Ecological interactions of *Trichosphaerium* sp., a multinucleate marine rhizopod. *Int. Congr. Protozool., 5th*, abstract 202.

Page, F. C. 1983. *Marine Gymnamoebae*. Institute of Terrestrial Ecology. Culture Centre of Algae and Protozoa, Cambridge.

Schaeffer, A. A. 1926. Taxonomy of the Amebas. *Carnegie Inst. Wash. Publ.* 345:1-116.

Schaudinn, F. 1899. Untersuchungen über den Generationswechsel von *Trichosphaerium sieboldi* Schn. *Abh. Konigl. Preuss. Akad. Wiss.*, Berlin, 1-93.

Schuster, F. L. 1975. Fine structure of the schizont stage of the testate marine ameba, *Trichosphaerium* sp. *J. Protozool.*, 23:86-93.

Sheehan, R. & Banner, F. T. 1973. *Trichosphaerium*—An extraordinary testate rhizopod from coastal waters. *Estuarine Coastal Mar. Sci.*, 1:245-260.

CLASS XENOPHYOPHOREA Schulze, 1904

ANDREW J. GOODAY and OLE S. TENDAL

INTRODUCTION

Xenophyophores are marine rhizopod plasmodial protozoans which construct an agglutinated test. They range from a few millimeters to 25 cm in size, making them among the largest known protists. Initially described more than a century ago as either sponges or foraminifera, xenophyophores were first recognized as a distinct taxon within the Rhizopoda by Schulze (1907a). The monograph of Tendal (1972) provides the starting point for our modern understanding of the taxonomy, distribution, and biology of the group. Recent reviews by Tendal (1989), Levin (1991, 1994), and Levin and Gooday (1992) have dealt with various aspects of xenophyophore biology, ecology, and taxonomy.

Xenophyophores are one of the few higher taxa to be restricted to the deep sea. They occur at depths between about 50O m and >8000 m (Tendal, 1972; Tendal et al., 1982) and are particularly abundant in regions which experience a high nutrient flux, for example, under upwelling areas and on sloped topography associated with deep sea trenches, submarine canyons, the flanks of seamounts, and continental slopes (Levin and Gooday, 1992). A report by Schepotieff (1912) on xenophyophores from 1-20 m depth in tropical localities is considered unreliable. Tendal (1972) chose to accept it but has since found evidence that the records may represent mistakes, mislabellings, or even fraud. Being confined to the deep sea, xenophyophores are difficult and expensive to collect. The early material was captured in trawls and other towed gears and, as a result, is often fragmented (Tendal, 1972). During the last decade, numerous intact specimens have been recovered in box cores or using submersibles (Levin and Thomas, 1988; Gooday, 1991). *In situ* photographs obtained from submersibles or remote camera systems have provided valuable ecological information (Tendal and Gooday, 1981; Levin and Thomas, 1988; Gooday et al., 1993).

Most xenophyophores live epifaunally on soft sediments but one genus, *Occultammina, is* infaunal. Species belonging to several genera sometimes occur on hard substrates and *Semipsammina* is restricted to hard substrates. Xenophyophores have never been cultured and little is known about their biology. They accumulate large volumes of fine sediment which provide a source of food and of particles for test construction. Digestion of food (sediment) particles probably occurs extracellularly, the remnants being packaged into faecal pellets (stercomata) which are retained within the test. There are indications that sexual reproduction occurs by gametogamy and involves biflagellate gametes which give rise to an amoeboid stage. Recent evidence indicates that some xenophyophores grow episodically with phases of rapid growth, lasting 2-3 days, being separated by much longer periods of quiescence (Gooday et al., 1993).

The Xenophyophorea is a small class which presently includes 14 genera, 50 described species and at least half that number of undescribed species known from specimens or *in situ* photographs. Several possible xenophyophore body and trace fossils have been identified, but the group has no confirmed fossil record (Levin, 1994; Maybury and Evans, in press).

KEY CHARACTERS

The main features of xenophyophore organization are illustrated in Fig. 1.

1) Most xenophyophores possess an agglutinated test composed of foreign particles ("**xenophyae**") held together with an organic cement. Morphologies are very diverse and include spheres, discs, irregular plates, plates which are folded into complex flower-like forms, tubes, networks of tubes, and reticulate, branched, and irregularly lumpy forms. The tests are generally fragile and can be either hollow or solid structures ranging from soft and flaccid to hard and brittle in consistency. Many are large enough to be studied with the naked eye although a binocular microscope is needed to view smaller specimens and details of test structure.

2) The protoplasm is organized as a branched, multinucleate plasmodium surrounded by a tubular, transparent organic sheath. These two elements together constitute the **granellare system** which occupies an estimated 0.1-5% of the test volume (Levin and Gooday, 1992). Rose Bengal staining can be used to make the granellare system more easily visible.

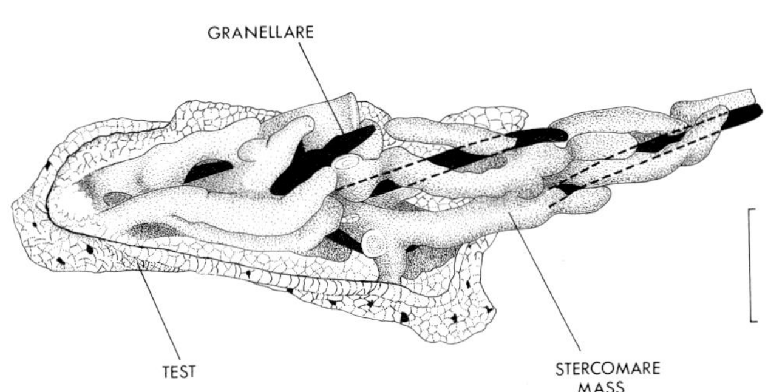

Fig. 1. Contents of the test of *Aschemonella ramuliformis* showing the relation between the granellare strands (black) and the stercomare masses (stippled) (slightly modified after Gooday and Nott, 1982). Scale bar=l mm.

3) An exceptional feature is the accumulation in the protoplasm of numerous 2-5 μm barium sulphate crystals called **granellae**. These and other cytological structures (e.g. nuclei) are often visible in squash preparations.

4) The granellare is closely associated, and often intertwined with, dark colored strings and masses of **stercomata** (the stercomare) which form an integral and occasionally dominant part of the test structure. The granellare and stercomare together ramify through the test interior and are easily visible when live tests are broken open (Fig. 1).

The class Xenophyophorea is included in the Phylum Granuloreticulosea despite the lack of definite information regarding the nature of the pseudopodia. This placement is adopted because xenophyophores seem to be closely related to foraminifera, as indicated particularly by the presence of an agglutinated test and the accummulation of stercomata within the test. The main features distinguishing xenophyophores from foraminifera are the branching plasma/granellare system, the presence of intracellular granellae and the retention of stercomata in large masses enveloped within an organic membrane (Tendal, 1972).

The class is divided into two orders according to whether long fibers (**linellae**) composed of a silk-like protein (Hedley and Rudall, 1974) are present (Order Stannomida) or absent (Order Psamminida). Families are distinguished according to the arrangement of xenophyae within the test, particularly the degree to which they form layered structures, and the relative amount of cement. Genera are distinguished largely on the test morphology.

CLASS XENOPHYOPHOREA

KEY TO FAMILIES AND GENERA

1. Test made of xenophyae and cement only; body rigid or at most slightly elastic ....Order **Psamminida**......3
1'. Test made up of xenophyae, cement, and linellae; body clearly flexible ..... Order **Stannomida**, ...Family **Stannomidae**.......2

2. Body composed of one or more leaflike parts....... .... ***Stannophyllum***
2'. Body bush- or tree-like with rounded branches .... ***Stannoma***

3. Xenophyae cemented together at random points of contact and rather uniformly distributed throughout test 11
3'. Xenophyae either cemented together to form tubes or plates, or some of them form a distinct, coherent surface layer covering a looser mass of internal xenophyae 4

4. Test composed of plate-like elements, or more massive with a distinct surface layer ....Family **Psamminidae**.....7
4'. Test composed of tubular elements; xenophyae found only in tube walls ....Family **Syringamminidae**.......5

5. Tubular elements unbranched or branched in one plane only 6
5'. Test forms a 3-dimensional network of tubes .... ***Syringammina***

6. Test composed of tubular chambers or inflated chambers joined by short tubes; branching may occur... ***Aschemonella***
6'. Tubular elements of uniform diameter; often with dichotomous branching and anastomosing .... ***Occultammina***

7. Test with distinct, cemented surface layer of xenophyae, massive or reticulate in form 10
7'. Test composed of platelike parts that may be convoluted 8

8. Test composed of more than one thin plate between which run the granellare and stercomare 9
8'. Test attached to solid substrate and composed of one thin plate that covers the granellare and stercomare .... ***Semipsammina***

9. Test consists of two parallel plates; sometimes the whole test is convoluted but the two plates always lie parallel to each other..... ***Psammina***
9'. Test consists of numerous layers arranged in a stack; overall form rounded and lumpy .... ***Cerelpemma***

10. Test massive, forming rounded lumps or lumps with a few thick undivided branches....... .... ***Galatheammina***
10'. Test reticulate with anastomosing branches...... .... ***Reticulammina***

11. Xenophyae have cement only at points of contact ....**Family Psammettidae**................ 12
11'. Xenophyae surrounded by conspicuous masses of cement; test consistency doughy..................... ....**Family Cerelasmidae** ...... ***Cerelasma***

12. Test strongly constructed and of stony or felt-like consistency; granellae of rather uniform diameter.. 13
12'. Test fragile, friable, very loosely cemented;

strands look like strings of pearls of rather uniform diameter.................... ***Homogammina***

13. Test very strong and stony in appearance, irregularly lumpy...................... ***Maudammina***

13'. Test felt-like and slightly elastic; at least some species with a special kind of xenophyae in the oldest (central) part of the test .... **Psammetta**

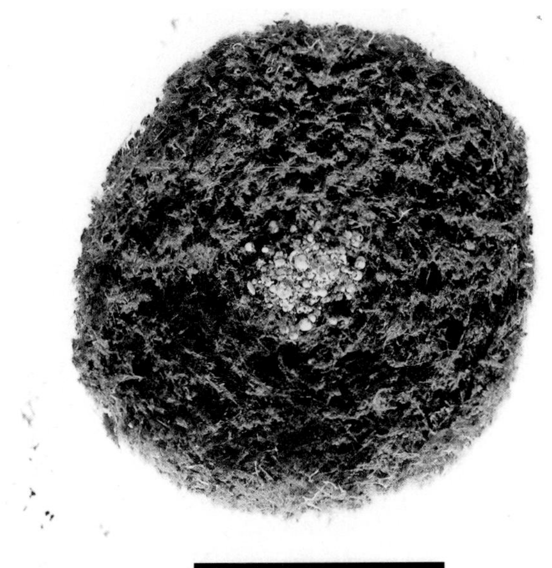

Fig. 2. *Psammetta globosa* Schulze, 1906. Cross-sectional view of test (after Tendal, 1972, new print). Scale bar=lO mm.

ORDER PSAMMINIDA TENDAL, 1972

Xenophyophores with rigid test and xenophyae usually arranged in some kind of order. Linellae absent.

FAMILY PSAMMETTIDAE Tendal, 1972

Test normally forms massive lump. Xenophyae arranged haphazardly, without any obvious order, and cemented only at random points of contact; no specialized surface layer. Apertures and other test openings not developed.

Genus *Psammetta* Schulze, 1906

Test free, regular in shape with no excrescences and forming either spherical or ellipsoidal lump or circular disc. Test free, regular in shape with no excrescences and forming either spherical or ellipsoidal lump or circular disc. Consistency felt-like and surface slightly shaggy. Most species have 2 types of xenophyae; central part of test contains one kind (primary xenophyae) remainder of the test contains another kind (secondary xenophyae). In all known species secondary xenophyae are siliceous sponge spicules. Occurrence: Indian Ocean, depth range 1158-4730 m. *Psammetta*-like organisms have been photographed at 7665-7710 m depth in the New Britain Trench (SW Pacific). Refs. Schulze (1906. 1907a), Tendal (1972), Lemche et al. (1976).

TYPE SPECIES: *Psammetta globosa* Schulze, 1906 (Fig. 2): Test spherical, 5-25 mm diameter, dark brown in color and of soft to firm consistency with either a slightly shaggy or felt-like surface. Primary xenophyae foraminiferal tests, secondary xenophyae sponge spicules. Yellow granellare branches sometimes conspicuous on test surface. Other species: *P. arenocentrum* Tendal, 1972; *P. ovale* Tendal, 1972; *P. erythrocytomorpha* Schulze, 1907.

Genus ***Homogammina*** Gooday & Tendal, 1988

Test free, fragile and of brittle to friable consistency. Irregular in shape but tending to be approximately rounded or plate-like in general form. Short, root-like structures and lobate protuberances sometimes developed. Xenophyae are mixture of different-sized planktonic foraminiferal tests, with overall nature and size distribution homogeneous throughout test. Stercomare strands consist of thicker sections connected by narrow necks. Occurrence: Northeast Atlantic, eastern equatorial Pacific; bathymetric range 2963-4550 m.

TYPE SPECIES: *Homogammina lamina* Gooday & Tendal, 1988 (Fig. 3). Test whitish and fragile, up to at least 7 mm maximum dimension. Fragments usually plate-like and irregular in outline, less commonly lumpy. Xenophyae consist of small juvenile and scattered larger planktonic globigerinacean shells distributed homogeneously throughout test. Stercomare strands arranged in more or less parallel pattern. Other species: *Homogammina maculosa* Gooday & Tendal, 1988, in which test forms rounded lump (Fig. 4); *H. crassa* Gooday, 1991. Several undescribed species occur in the eastern equatorial Pacific. Refs. Gooday and Tendal (1988), Gooday (1991).

Fig. 3. *Homogammina lamina* Gooday & Tendal, 1988. Holotype (after Gooday and Tendal, 1988, new print). Scale bar=2 mm.

Fig. 4. *Homogammina maculosa* Gooday & Tendal, 1988. Holotype (after Gooday and Tendal, 1988, new print). Scale=lO mm.

Genus ***Maudammina*** Tendal, 1972

Test free, flattened and irregular in form, with or without ridge-like crests and of very firm consistency. Xenophyae all of same type and cemented only at random points of contact; test space consists of numerous small interstices between grains. Occurrence: one locality off southeast Africa, 5110 m depth. Only known species *Maudammina* arenaria Tendal, 1972 (Fig. 5): Test yellowish-brown in color, very firmly constructed, 45 mm long and 4-8 mm thick. Flattened and more or less irregularly bent with a few plate-like ridges on one side. Xenophyae sand grains. Reference: Tendal (1972).

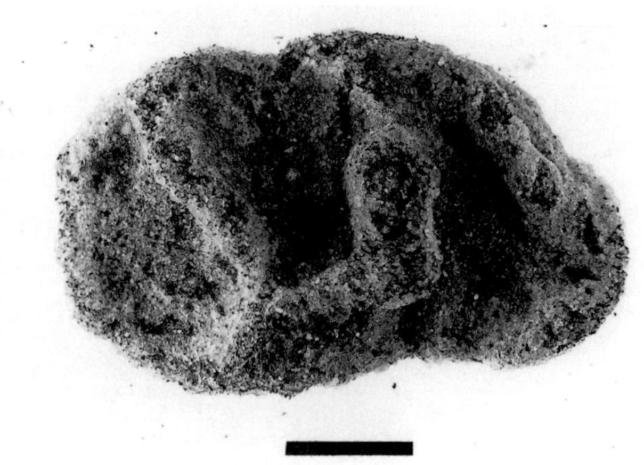

Fig. 5. *Maudammina arenaria* Tendal, 1972. Holotype (after Tendal, 1972, new print). Scale bar=lO mm.

FAMILY PSAMMINIDAE Haeckel, 1889

Test usually solid, often fragile and varying in form from plate-like to lumpy, branched or reticulate. External xenophyae arranged in surface layer; occasionally test consists of a stack of layers. There is very little cement.

Fig. 6. *Psammina nummulina* Haeckel, 1889. Fragmentary specimens, largest specimen 6.5 mm long (after Tendal, 1972, new print). Scale=lO mm.

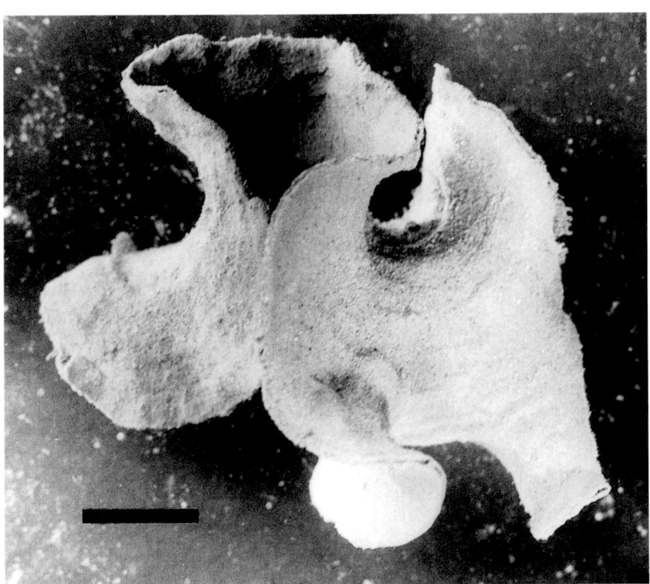

Fig. 7. Large, folded, undescribed species of *Psammina* from eastern Pacific. Scale=IO mm.

Genus ***Psammina*** Haeckel, 1889

Test fragile, brittle, basically plate-like and typically discoidal but sometimes folded into more complex shapes. Circular apertures developed along margin in some species. External xenophyae firmly cemented to form upper and lower plates; internal xenophyae sparse, commonly forming pillar or bar-like structures between plates. Granellare branches and well-developed stercomare strings run between plates and branch around pillars. Occurrence: Northeast Atlantic, equatorial and west Pacific, 1158-6059 m.
TYPE SPECIES: *Psammina nummulina* Haeckel, 1889 (Fig.6): Test light grey, very brittle and irregularly circular in shape, measuring 12-15 mm in diameter and 0.6-1.4 mm thick. Edge either swollen or tapered and perforated by small circular apertures. Upper and lower plates also have small, scattered pores. Plates easily detached, revealing yellowish granellare strands and dark, radially organized, sometimes anastomosing stercomare masses. Occasional connecting pillars occur between plates. Xenophyae mainly radiolarian tests. Other species: *P. globigerina* Haeckel, 1889; *P. plakina* Haeckel, 1889; *P. delicata* Gooday & Tendal, 1988, *P. fusca* Gooday & Tendal, 1988, *P. sabulosa* Gooday & Tendal, 1988. Several large undescribed species with folded morphologies occur in the east Pacific (Fig. 7). Refs. Schulze (1906); Tendal (1972; in press); Gooday and Tendal (1988).

Genus ***Semipsammina*** Tendal, 1975

Test attached, consisting of a single fragile layer of xenophyae covering granellare and stercomare which are attached directly to substrate. Occurrence: Puerto Rico Trench, 6000-5890 m; eastern equatorial Pacific, 4200 m.
TYPE SPECIES: *Semipsammina fixa* Tendal, 1975 Test layer irregularly rounded in outline and conforms to shape of substrate. Yellow-brown in color, up to 5.5 mm diameter and about 0.5 mm thick. Xenophyae are sponge spicules, mineral particles and fine-grained interstitial material. Other species: an undescribed species occurs in the eastern Pacific. Refs. Tendal (1975), Mullineaux (1987).

Fig. 8. *Cerelpemma radiolarium* (Haeckel, 1889). Lectotype (after Tendal, 1972, new print). Scale bar=IO mm.

Genus ***Cerelpemma*** Laubenfels, 1936

Generic diagnosis: Test of friable consistency, lumpy and composed of numerous thin layers of xenophyae arranged in wafer-like stack. Granellare and stercomare occupy flat spaces between these layers. Occurrence: Central Pacific, 4438-5353 m. Monospecific.
TYPE SPECIES: *Cerelpemma radiolarium* (Haeckel, 1889) (Fig. 8) Test light grey in color, forming more or less irregular, rounded mass 5-25 mm diameter. Xenophyae exclusively radiolarian tests cemented into numerous thin layers each consisting of a single layer of particles. Intervening spaces are of approximately same thickness as layers themselves. Ref. Tendal (1972).

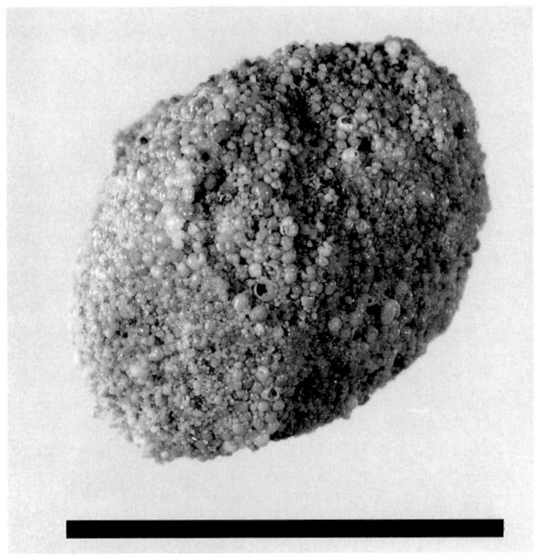

Fig. 9. *Galatheammina calcarea* (Haeckel, 1889) (after Tendal, 1972, new print). Scale bar=lO mm.

Genus ***Galatheammina*** Tendal, 1972

Test free and of hard to friable consistency. All described species have totally different test morphologies. Type species has 3-dimensional star-shaped test with 4 branches. In other species test forms compact, rounded lump, irregular plate (sometimes with smaller side plates), or subtriangular plate with short, root-like bars at base. Interior consists of loose accumulation of xenophyae between which granellare and stercomare interweave. Surface xenophyae form firmly cemented layer. Occurrence: Atlantic, Indian, and Pacific Oceans, bathymetric range 1320m-5353 m. Illustrated species (Fig.9) *G. calcarea* (Haeckel, 1889): Test massive and lumpy, forming rounded, occasionally somwhat irregular masses 6-25 mm diameter. Color white and consistency hard. Xenophyae exclusively calcareous foraminiferal tests. Other species: *G discovery* Gooday & Tendal, 1988; *G. microconcha* Gooday & Tendal, 1988; *G. irregularis* Gooday, 1991; *G. erecta* Gooday, 1991. Several undescribed species occur in the eastern Pacific. Refs. Tendal (1972j; Levin and Thomas (1988); Gooday and Tendal (1988); Gooday (1991).
TYPE SPECIES: *Galatheammina tetraedra* Tendal, 1972

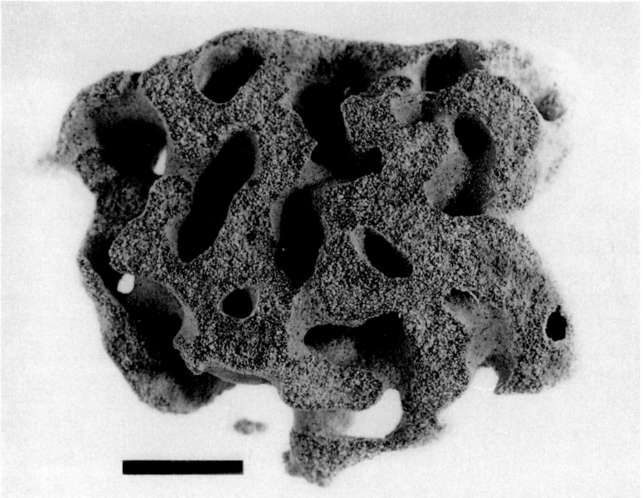

Fig. 10. *Reticulammina novazealandica* Tendal, 1972. Scale bar=lO mm.

Genus ***Reticulammina*** Tendal, 1972

Test free and of generally friable consistency. Hemispherical or subrectangular in overall shape and consisting of anastomosing branches and plate-shaped elements which delimit open spaces. Surface xenophyae form more or less firmly cemented layer. Interior contains loose accumulations of xenophyae, between which stercomare and xenophyae interweave. Occurrence: Northeast Atlantic, Indian Ocean, west and east Pacific, Antarctica; bathymetric range 743-6059 m.
TYPE SPECIES: *Reticulammina novezealandica* Tendal, 1972 (Fig. 10): Test light grey in color, rounded, somewhat flattened and up to 60 mm maximum diameter, consisting of anastomosing plate-like branches mostly about 5 mm wide. Open

spaces circular to oval and 2-lO mm wide. Internal xenophyae are planktonic foraminiferal tests; surface layer consists of well-cemented, fine-grained calcareous material with scattered planktonic foraminiferal tests. Other species: *R. labyrinthica* Tendal, 1972 (Fig. 11); *R. lamellata* Tendal, 1972; *R. cretacea* (Haeckel, 1889); *R. maini* Tendal & Lewis, 1978; *R. antarctic* Riemann & Gingele, 1993. Several undescribed species occur in the eastern Pacific. References: Tendal (1972); Rice et al. (1979); Tendal and Gooday (1982); Levin and Thomas (1988); Gooday and Tendal (1988); Gooday (1991); Riemann et al. (1993).

Fig. 11. *Reticulammina labyrinthica* Tendal, 1972. Specimens photographed on the seafloor off NW Africa at 4000 m bathymetric depth. Largest specimen about 5 cm diameter.

FAMILY SYRINGAMMINIDAE Tendal, 1972

Test fragile, consisting of system of interconnected tubes or, less commonly, a single tube. There are no internal xenophyae.

Genus ***Syringammina*** Brady, 1883

Test usually free, occasionally attached, hemispherical in overall shape and consisting either of numerous radiating tubes connected by side branches or of more irregular networks of tubes. Tube walls consist of firmly cemented xenophyae. Interior occupied by granellare and stercomare. Occurrence: North Atlantic, Southern Ocean, southwest Pacific, 740-4795 m.

TYPE SPECIES: *Syringammina fragilissima* Brady, 1883 (Fig. 12): Test fragile, greyish in color, rounded and up to 38 mm in size. Tubes organized irregularly in central region, towards periphery they are more radial and anastomosing side branches form consecutive layers 1.3-2.5 mm apart. Tube diameter varies from 0.5 mm to 1.O mm and wall thickness is about 0.13 mm. Xenophyae are fine sand grains and small planktonic foraminiferal tests. Each tube contains 3 - 6 longitudinally running granellare strands and one thick, centrally located stercomare string. Other species: *S. minute* Pearcey, 1914; *S. tasmanensis* Lewis, 1966, *S. reticulata* Gooday, 1996, (Fig. 14). A large, undescribed species has been photographed on the seafloor in the NE Atlantic (Tendal and Gooday, 1981, Fig 1. Gooday and Bett. unpubl.). Refs. Brady (1883); Lewis (1966); Tendal (1972); Tendal and Lewis (1978); Tendal and Gooday (1981); Levin (1991).

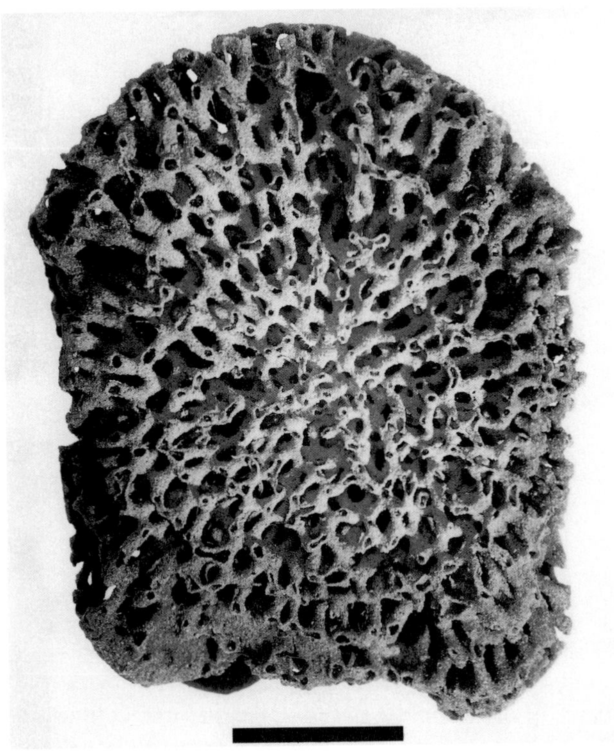

Fig. 12. *Syringammina fragilissima* Brady, 1883. Lectotype (after Tendal, 1972, new print). Scale bar=lO mm.

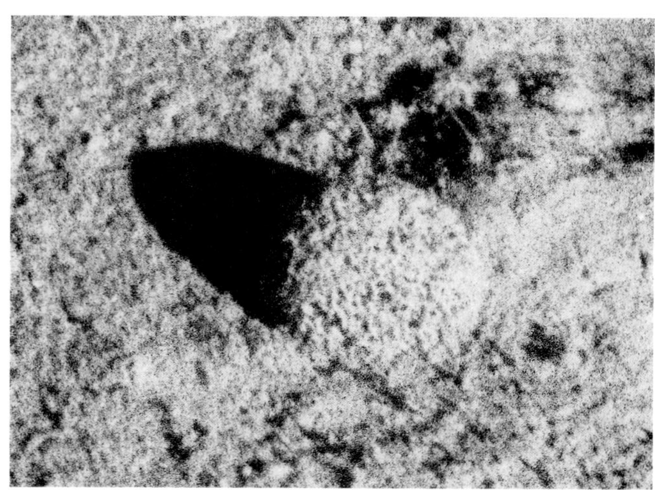

Fig. 13. Undescribed species of *Syringammina* photographed on the seafloor off SW Ireland at 1745 m bathymetric depth. Specimen is about 7.5 cm wide.

Fig. 14. *Syringammina reticulata*, Gooday, 1996 from NE Atlantic. Scale= l0 mm.

Genus *Occultammina*
Tendal, Swinbanks &Shirayama, 1982

Tubular test is unbranched, or bifurcates, or anastomoses to form polygonal network. Tube wall consists of 1 or 2 distinct layers of xenophyae. Tube contains single, sometimes bifurcating granellare strand and dichotomously branching, sometimes anastomosing stercomare strings. Occurrence: Northeast Atlantic, 4550-4845 m, western Pacific, 6440-8260 m.

TYPE SPECIES: *Occultammina profunda* Tendal, Swinbanks & Shirayama, 1982: Test forms straight tube, 0.5-1.O mm diameter and light grey in color, which sometimes bifurcates and fragments very readily. Wall composed of poorly cemented xenophyae with thicker inner layer composed of silt-sized mineral grains and outer, finer-grained layer; 1-4 ridges run longitudinally on inside of wall. Other species: There is at least one undescribed species in which the tube branches and anastomoses to form a polygonal network (Swinbanks, 1982). Ref. Tendal et al., (1982).

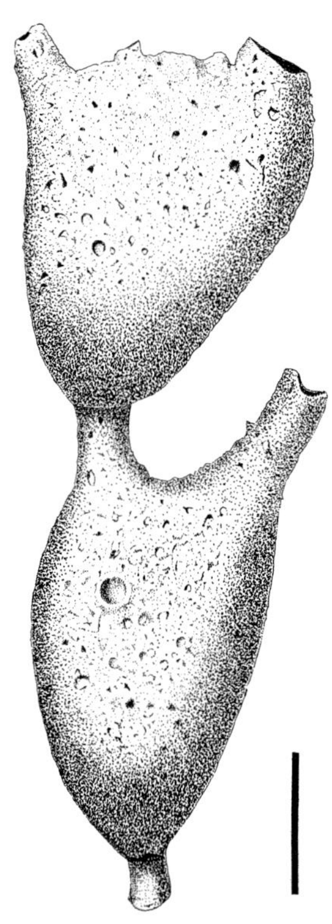

Fig. 15. *Aschemonella scabra* Brady, 1879 (after Brady, 1884). Scale bar=l mm.

Genus *Aschemonella* Brady, 1879

Test free, relatively small (a few millimeters long), of fragile, brittle consistency, and consisting of tubular or inflated chambers, sometimes with short side branches. Chambers sometimes arranged in series although isolated chambers often occur through fragmentation. Apertures located at ends of tubular or approximately conical processes. Interior occupied by anastomosing stercomare strings and branching granellare strands. Wall

thin, composed mainly of mineral grains. Occurrence: widely distributed in deepwater.
TYPE SPECIES: *Aschemonella scabra* Brady, 1879 (Fig. 15): Test a few millimeters in size, consisting of several inflated, often approximately spherical chambers, joined by short stolons. Apertures located at ends of short tubular processes. Other species: *A. ramuliformis* Brady, 1884. Refs. Brady (1879); Gooday and Nott (1982), Schroder (1986); Schroder et al. (1988).

FAMILY CERELASMIDAE Tendal, 1972

Test fairly soft, and agglutinating cement present in large quantities. Xenophyae occur in varying amounts and arranged in no obvious order. Each particle surrounded by cement so that there is no contact with adjacent particles.

Fig. 16. *Cerelasma gyrosphaera* Haeckel, 1889. Lectotype (after Tendal, 1972, new print). Scale=lO mm.

Genus ***Cerelasma*** Haeckel, 1889

Test free, of firm to doughy consistency and easily broken. Lumpy, irregularly rounded in form and composed of numerous anastomosing branches. Occurrence: Atlantic, Pacific, and Indian Oceans, 3660-4829 m.
TYPE SPECIES: *Cerelasma gyrosphaera* Haeckel, 1889 (Fig. 16): Test gray-brown to very dark brown in color, of firm but slightly elastic consistency, and up to 70 mm diameter. Test has convoluted surface and is composed of anastomosing branches 1.5-5 mm diameter which delimit open spaces 1-6 mm diameter. Xenophyae are exclusively radiolarian tests. Other species: *C. lamellosa* Haeckel, 1889; *C. masse* Tendal, 1972. Refs. Schulze (1907a); Tendal (1972, 1980a).

ORDER STANNOMIDA TENDAL, 1972

Xenophyophores with flexible test and poorly organized xenophyae. Linellae present in large numbers.

FAMILY STANNOMIDAE HAECKEL, 1889

Test is branched and treelike, or plate-like and of varying degrees of softness.

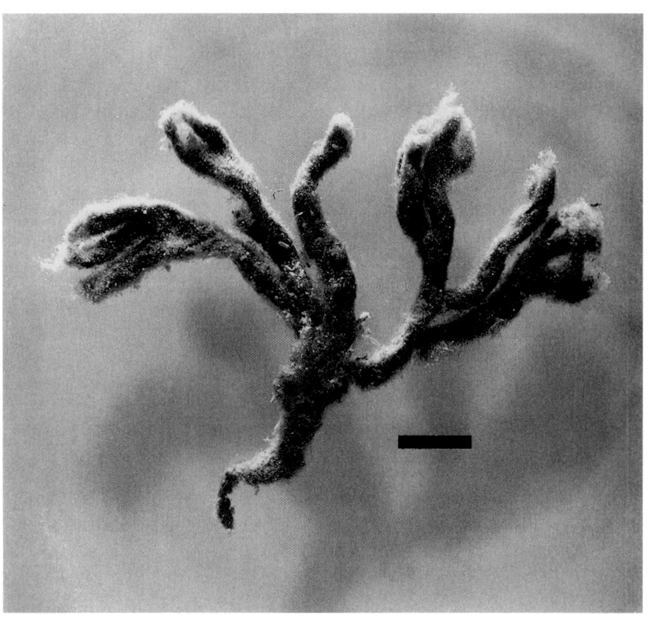

Fig. 17. *Stannoma dendroides* Haeckel, 1889 (after Tendal, 1972, new print). Scale=lO mm (natural size).

Genus ***Stannoma*** Haeckel, 1889

Test free, soft, flexible, and tree- or bush-like in form with few or many branches. Occurrence: Indian and Pacific Oceans, 3814-4930 m.
TYPE SPECIES: *Stannoma dendroides* Haeckel, 1889 (Fig. 17): Tree-like test is up to 80 mm in length, yellowish to greenish-brown in color, soft and very flexible. It branches dichotomously 1-3 times, in one plane. Branches are 2-5 mm in

diameter, circular or slightly flattened in cross-section, and do not anastomose. Linellae simple and without ramifications. Xenophyae are mainly radiolarian tests. Other species: *S. coralloides* Haeckel, 1889. Refs. Schulze (1907a,b); Tendal (1972).

Fig. 18. *Stannophyllum zonarium* Haeckel, *1889*. Scale=lO mm (natural size).

Genus ***Stannophyllum*** Haeckel, 1889

Test free, more or less flexible and varying in consistency from soft and flaccid to stiff but somewhat elastic. Composed of one or more plate-like parts. Occurrence: One species has been recorded from 3 closely spaced localities in the abyssal northeast Atlantic but the vast majority of records are from the Indian and Pacific Oceans, 981-6380 m. Refs. Schulze (1907a,b); Tendal (1972, 1980a,b).

TYPE SPECIES: *Stannophyllum zonarium* Haeckel, 1889 (Fig. 18): Test up to 190 mm long and 1 - 2 mm thick, brown in color, plate-like with an approximately kidney-shaped outline. Consistency stiff but somewhat elastic. Surface with distinct concentric zones, each 35 mm broad. Linellae strongly developed, unbranched, 3-4 m thick, forming a distinct surface layer and often occuring in bundles. Xenophyae are mainly radiolarian tests. Other species: There are 13 other described species, some of which are poorly known.

LITERATURE CITED

Brady, H. B. 1879. Notes on some of the Reticularian Rhizopoda of the "Challenger" Expedition. Part 1. On new or little known arenaceous types. *Q. J. Microsc. Sci.,* 19:20-63, pls III-V.

Brady, H. B. 1883. Syringammina, a new type of arenaceous Rhizopoda. *Proc. R. Soc.* Lond., 35:155-161.

Brady, H. B. 1884. Report on the foraminifera dredged by H.M.S. *Challenger* during the years 1873-1876. *Report on the Scientific Results of the Voyage of H.M.S. Challenger during the years 1873-1876, Zoology* 9: 1814, pls 1-115.

Gooday, A. J. 1991. Xenophyophores (Protista, Rhizopoda) in box-core samples from the abyssal northeast Atlantic (BIOTRANS area): their taxonomy, morphology, and ecology. *J. Foram. Res.*, 21:197-212.

Gooday, A. J. 1996. Xenophyophores (Protista) including two new species, from abyssal sites in the northeast Atlantic Ocean. *J. Foram. Res.*, 26:197-212

Gooday, A. J. & Nott, J. A. 1982. Intracellular barite crystals in two xenophyophores, *Aschemonella ramuliformis* and *Galatheammina sp.* (Protozoa: Rhizopoda) with comments on the taxonomy of *A. ramuli*formis. *J. Mar. Biol Ass. UK*, 62:595-605.

Gooday, A. J. & Tendal, O. S. 1988. New xenophyophores (Protista) from the bathyal and abyssal north-east Atlantic Ocean. *J. Nat. Hist.*, 22:413-434.

Gooday, A. J., Bett, B. J. & Pratt, D. N. 1993. Direct observation of episodic growth in an abyssal xenophyophore. *Deep-Sea* Res., 40:2123-2143.

Haeckel, E. 1889. Report on the deep-sea Keratozoa. *Report on the Scientific Results of the Voyage of H.M.S. Challenger during the years 1873-1876, Zoology,* 32:1-92, pls 1-8.

Hedley, R. H. & Rudall, K. M. 1974. Extracellular silk fibres in *Stannophyllum* (Rhizopdea: Protozoa). *Cell. Tiss. Res.,* 150:107-111.

Lemche, H., Hansen, B., Madsen, F. J., Tendal, O. S. & Wolff, T. 1976. Hadal life as analysed from photographs. *Vidensk. Meddr. Dansk Naturh. Foren.*, 139:263-336.

Levin, L. A. 1991. Interactions between metazoans and large agglutinating protozoans: implications for the community structure of the deep-sea benthos. *Am. Zool.*, 31:886-900.

Levin, L. A. 1994. Ecology and paleoecology of xenophyophores. *In*: Miller, W., (ed.), *Recent Advances in Deep-Sea Paleoecology. Palaios* Theme Issue.

Levin, L. A. & Gooday, A. J. 1992. Possible roles for xenophyophores in deepsea carbon cycling. *In*: Rowe, G. T. & Pariente, V. (eds.), *Deep-Sea Food Chains and the Global Carbon Cycle.* Kluwer Academic Publishers, Dordrecht, The Netherlands. pp 93-104

Levin, L. L. & Thomas, C. L. 1988. The ecology of xenophyophores (Protista) on east Pacific seamounts. *Deep-Sea Res.,* 35:2003-2027.

Lewis, K. B. 1966. A giant foraminifer: a new species of *Syringammina* from the New Zealand region. *N. Z. J. Sci.,* 9:114-123.

Maybury, C. A. & Evans, K. R. A new interpretation of certain Pennsylvanian phylloid algae as shallow water xenophyophores. *Lethaia* (in press)

Mullineaux, L. 1987. Organisms living on manganese nodules and crusts: distribution and abundance at three North Pacific sites. *Deep-Sea Res.,* 34:165-184.

Rice, A. L., Aldred, R. G., Billett, D. S. M. & Thurston, M. H. 1979. The combined use of an epibenthic sledge and a deep-sea camera to give quantitative relevance to macro-benthos samples. *Ambio Special Report,* **6**:59-72.

Schroder, C. J. 1986. Deep-water arenaceous Foraminifera in the Northwest India Ocean. *Canadian Technical Report of Hydrography and Ocean Sciences,* 71:1-191.

Schroder, C. J., Medioli, F. S. & Scott, D. B. 1989. Fragile abyssal foraminifera (including new Komokiacea) from the Nares Abyssal Plain. *Micropaleontology N.Y.,* 35:10-48.

Schulze, F. E. 1906. Die Xenophyophoren der 'Siboga'-Expedition. *Siboga-Exped.* Monographie IV:1-18, pls I-III.

Schulze, F. E. 1907a. Die Xenophyophoren, eine besondere Gruppe der Rhizopoden. Wiss. *Ergebn. dt. Tiefsee-Exped.,* 11:1-55, pls I-VIII.

Schulze, F. E. 1907b. Die Xenophyophora. *Bull Mus. Comp. Zool. Harv.,* 51:205-229.

Swinbanks, D. D. 1982. *PaleodicLyon:* The traces of infaunal xenophyophores? *Science*, 218:47-49.

Tendal, O. S. 1972. A monograph of the Xenophyophora. *Galathea Report,* 12:7-103, pls 1-17.

Tendal, O.S. 1975. A new xenophyophore (Rhizopoda, Protozoa), living on solid substratum, and its significance. *Deep-Sea Res.*, 22:45-48.

Tendal, O. S. 1980a. Xenophyophores from the French expeditions "INCAL" and "BIOVEMA" in the Atlantic Ocean. *Cah. Biol. Mar.,* 21:303-306.

Tendal, O. S. 1980b. *Stannophyllum setosum* sp. n., a remarkable xenophyophore (Rhizopoda, Protozoa) from the eastern Pacific. *Cah. Biol. Mar.*, 21:383-385.

Tendal, O. S. 1989. Phylum Xenophyophora. *In*: Margulis, L., Corliss, J. O., Melkonian, M. & Chapman, D. J. (eds.), *Handbook of the Protoctista.* Jones and Bartlett Publishers, Boston. pp 135-138

Tendal, O. S. & Lewis, K. B. 1978. New Zealand xenophyophores: upper bathyal distribution, photographs of growth position, and a new species. N.Z. *J. Mar. Freshwat. Res.,* 12:197-203.

Tendal, O. S., Swinbanks, D. D. & Shirayama, Y. 1982. A new infaunal xenophyophore (Xenophyophorea, Protozoa) with notes on its ecology and possible trace fossil analogues. *Oceanol. Acta,* 5:325-329.

Tendal, O. S. & Gooday, A. J. 1981. Xenophyophoria (Rhizopoda, Protozoa) in bottom photographs from the bathyal and abyssal NE Atlantic. *Oceanol. Acta,* 4:415-422.

*Supported partly by the European Community under contracts 0037C(EDB) and MAS2-CT92-0

ORDER PELOBIONTIDA Page, 1976

by
GUY BRUGEROLLE and DAVID PATTERSON

The order Pelobiontida includes four genera of mostly free-living heterotrophic protists. They are usually found in anaerobic or microaerophilic habitats such as stagnant water which is rich in organic matter. They occur in freshwater and marine habitats. One species is endocommensal. The most usual trophic form in all but *Pelomyxa* is a flagellated cell, but amoeboid and encysted forms are known from some species and may be characteristic of all species. Flagellated cells have one or many flagella and move by flagellar motion and/or by (a sometimes distinctive type of) amoeboid streaming.

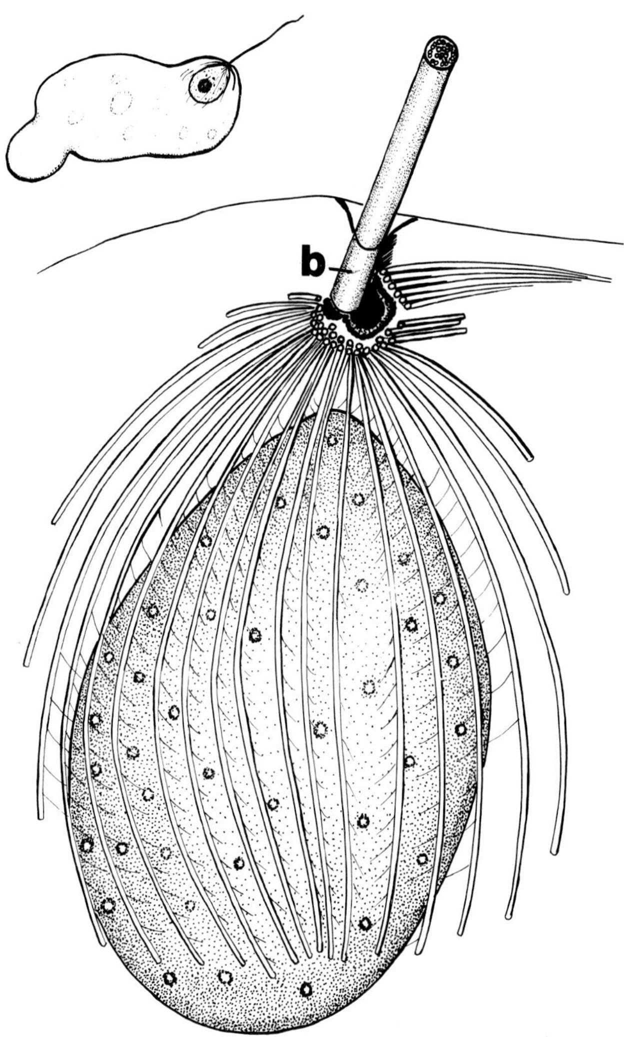

Fig. 1. Basic organization of a mastigamoebid (from Brugerolle, 1991a). Upper left: typical uniflagellated cell. The single flagellum/basal body (b) is connected to the nucleus by a cone of microtubules. The microtubules arise from a microtubule-organizing-center structure that is attached to the base of the basal body. A lateral band of microtubules extends towards the cell surface.

The order has been more clearly circumscribed after the discovery of flagella in *Pelomyxa* by Griffin (1988) and as a result of ultrastructural studies of various species (Brugerolle 1991a, 1991b; Mylnikov, 1991; Simpson et al., 1997). Light and electron microscopic studies have shown that flagella/basal bodies occur singly. In some cases the flagella have an 'n+n' substructure rather than the usual '9+2' arrangements of microtubules, and there may also be abnormalities in the basal body. Flagella are often attached to the nucleus by a cone or cape of microtubules radiating from the basal body (Figs. 1, 2a,b,c). The microtubules arise from a region at the base of or around the basal body. This has been demonstrated in *Mastigina* (Brugerolle, 1982), *Pelomyxa* (Griffin, 1988), *Mastigamoeba* (Brugerolle, 1991a, 1991b; Simpson et al., 1997), and *Mastigella* (A. G. B. Simpson and C. Bernard, unpubl. observ.). The cytoplasm is rich in microfilaments, and cytoplasmic movements may lead to locomotion (especially in *Mastigina* and *Pelomyxa*) or to the ingestion of food. Most species move by the slow, undulating actions of the flagella.

The endoplasmic reticulum is sometimes concentrated in large stacks of cisternae. There are no dictyosomes nor mitochondria; glycogen is used as a reserve material. Cells contain one to many nuclei. These contain large nucleoli. Mitosis involves an intranuclear spindle; the nucleoli are persistent and divide, but without ultrastructural studies of division it is unclear if the nuclear envelope is persistent or if it breaks down.

Because of the absence of mitochondria and Golgi apparatus and because of the simplicity of the mastigont system, pelobionts have been regarded as an early offshoot in eukaryotic evolution and have also been classified as Archeozoa (Cavalier-Smith, 1991). The term Archezoa was previously used by Perty and Haeckel with different meanings and therefore is not adopted here. The primitive status of the Pelobionts is not supported by current molecular studies (Hinkle et al., 1994). General reviews of the group are found in Lemmermann, 1914 (species diversity), Mylnikov, 1991 (generic diversity), and Brugerolle, 1991a, 1991b (ultrastructure).

KEY CHARACTERS

All pelobionts have an amoeboid body which may give rise to pseudopodia and have one (or more, e.g. in *Pelomyxa*) flagella. The body gives rise to blunt pseudopodia, the form of which may have some diagnostic value—although several species are known to have variable pseudopodial form. Amoeboid motion occurs in *Mastigina* and *Pelomyxa*. This motion is distinctive with the cytoplasm flowing along the center of the body and spreading out, fountain-like, at the anterior end. The posterior end of the cell has a bulbous or filose uroid, and food particles may be ingested here.

In most cells of genera other than *Pelomyxa* a flagellum is located anteriorly in moving cells. The relationship between the nucleus and flagellum is used to segregate the genera *Mastigamoeba* and *Mastigella*—the latter genus being said to have a greater relative difference between the two organelles. The flagellum is often very long and beats languidly, typically with a single wave passing from base to tip—the beat pattern is distinctive.

Members of this genus may be confused with some other taxa (e.g. *Cercomonas*), and electron microscopy is desirable to confirm that the organism is a pelobiont. Taxa are assigned to genus depending on the number of flagella, the connections between flagella and nuclei, and the extent of amoeboid motion.

KEY TO FAMILIES AND GENERA

1. Amoeboid, one large pseudopod, posterior **uroid**, one to many nuclei, flagella numerous, but hard to see................................ ***Pelomyxa***
1'. With prominent apical flagellum when gliding/ swimming (some cells may have more than one flagellum) 2

2. With marked cytoplasmic streaming; one species endocommensal.................... ***Mastigina***
2'. Cytoplasmic streaming not marked 3

3. Flagellum closely linked to the conical anterior nucleus .... ***Mastigamoeba***
3'. Flagellum and nucleus distant from each other, cytoskeletal material linking the two visible or not visible................................ ***Mastigella***

FAMILY PELOMYXIDAE Schulze, 1877

Flagellated monopodial amoebae, thickly cylindrical during amoeboid motion, sometimes with fountain streaming and with food being taken up posteriorly. Cell with one to many nuclei with associated flagella (**karyomastigonts**). Two genera.

Genus ***Pelomyxa*** Greeff, 1874

Free-living from anaerobic or micro-aerophilic habitats, cells amoeboid reaching 3 mm long with a large, anterior pseudopod and a posterior uroid. With one to several nuclei. Short flagella have been demonstrated at some stages of the life cycle. Cytoplasm contains 2 to several "glycogen bodies" as large as the nuclei (10 µm). A complex life cycle has been described. In spring, cysts release small binucleate amoebae, which grow and become multinucleate and acquire **endosymbiotic** bacteria. They become elongate with a posterior uroid, and flagella are evident at this stage. The cells later become spherical, and endosymbiotic bacteria congregate around the nuclei. These cells fragment into rosettes giving rise either to cysts in winter or to small amoebae which may undergo another cycle of development (Whatley and Chapman-Andresen, 1990). Electron-microscopical studies have shown that the basal bodies and flagella are connected to a cone of microtubules. The microtubules are not in contact with the nuclei when the nuclei are surrounded by endosymbiotic bacteria. Three types of bacteria have been observed: one is a large, cylindrical bacterium with an internal cleft, the other is a long rod-shaped bacterium, the third is a **methanogen** (Stumm and Zwart,

1986; Griffin, 1988; Whatley and Chapman-Andresen, 1990). Many species have been assigned to this genus, but all may be assignable to the type species *P. palustris* Greeff, 1874.

endosome, and mitosis includes division of the endosome. *Mastigina hylae,* which is endocom-

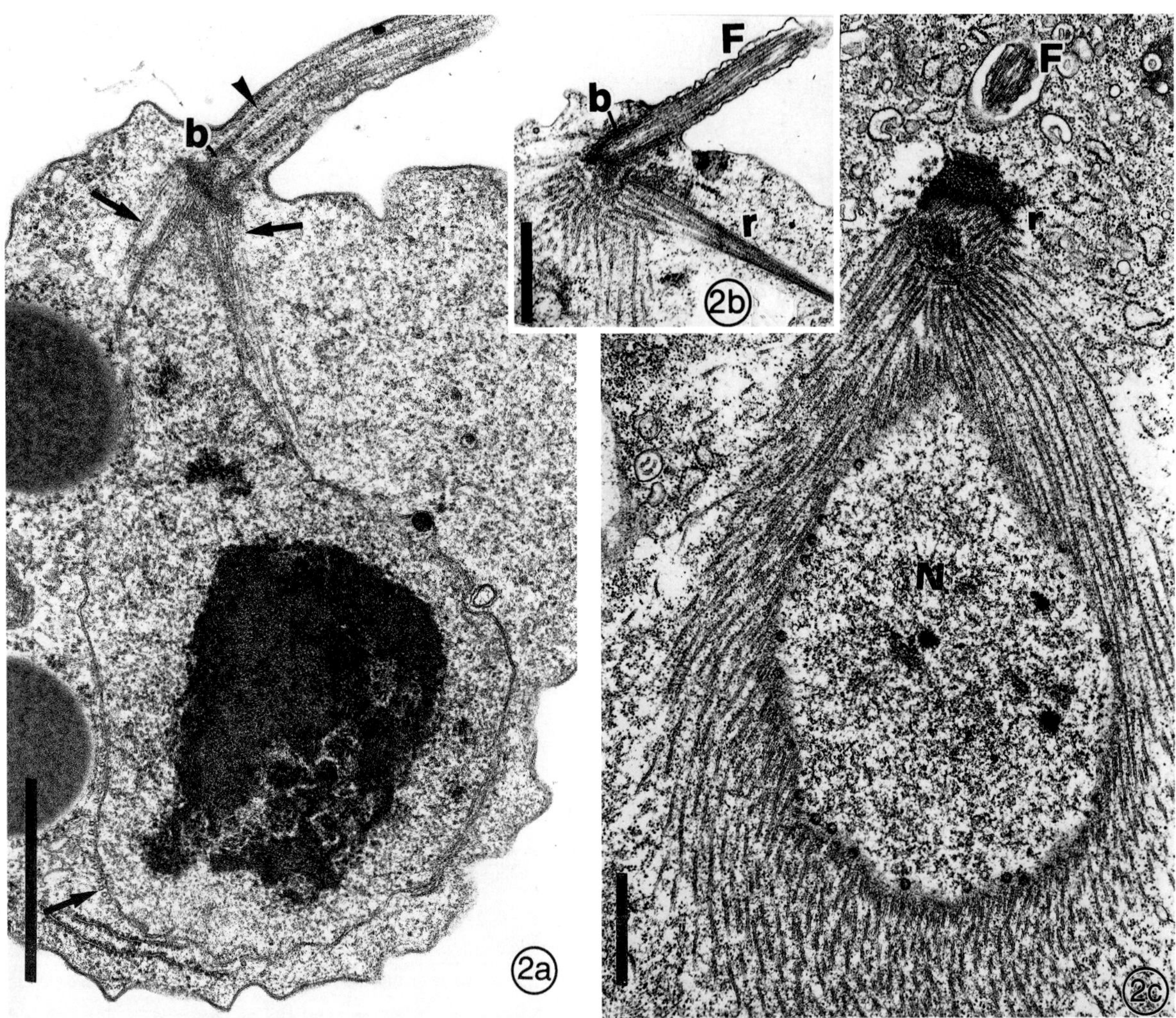

Fig. 2a. Section of *Mastigamoeba* showing the cone of microtubules (arrows) arising from the basal body (b). These microtubules link the flagellum to the nucleus. Note the helix-like structure around the axoneme of the flagellum (arrowhead) and the large nucleolus. Bar=1 μm.

Fig. 2b,c. In *Mastigella hylae*, the microtubules of the cape covering the nucleus (N) are numerous; they arise from a MTOC at the base of the basal body (b) and some form a lateral ribbon (r). The flagellum (F) is composed of microtubules not arranged in a 9+2 configuration. Bar=1 μm. (from Brugerolle, 1991a)

Genus ***Mastigina*** Frenzel, 1897

Amoeboid monopodial cells with one flagellum, long or short, active or inactive, and always attached to an anterior nucleus. There is no posterior uroid, and there are several nuclei in one species. The nucleus has a large, central

mensal in intestine of toads, has solitary basal bodies connected to the nucleus by a cape or cone of microtubules (Figs. 2b,c and Fig. 5). The single flagellum may not have the usual 9+2 arrangement of microtubules in the axoneme. A lateral ribbon of microtubules extends from the

basal body to the cell surface. Secondary nuclei, which occur in large cells, bear one basal body without a flagellum and have a few associated microtubules. Some species may have a **polymorphic life cycle** (Goldschmidt, 1907). Electron microscopy has been carried out by Brugerolle (1982, 1991a). About 6 species have been reported. Refs. Brugerolle, 1982, 1991a; Collin, 1913; Goldschmidt, 1907; Penard, 1909.
TYPE SPECIES: *Mastigina chlamys* Frenzel, 1897

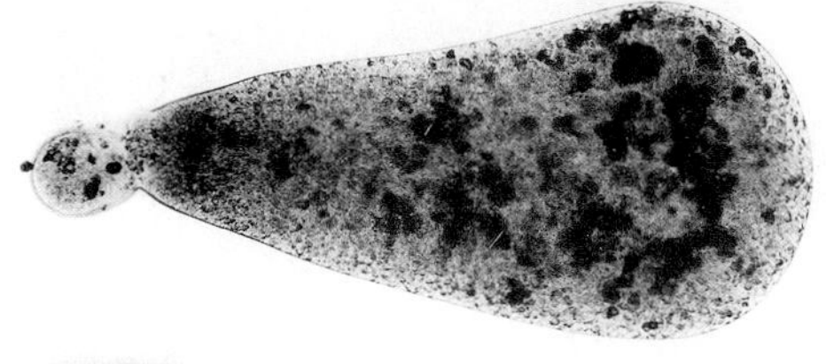

Fig. 3. *Pelomyxa palustris,* showing the anterior, large pseudopod and the small, posterior uroid. Bar=250 μm. (courtesy of C. K. Stumm)

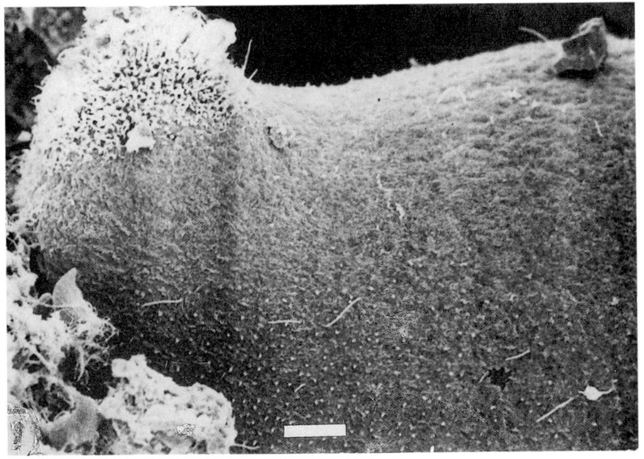

Fig. 4. *Pelomyxa palustris* showing flagella on the surface and the posterior uroid. Bar=10 μm. (from Griffin, 1988)

FAMILY MASTIGAMOEBIDAE Goldschmidt, 1907

Flagellated cells with or without simple or branched, hyaline pseudopodia during amoeboid motion. Swimming is erratic with the single, languidly beating flagellum directed forward. Species with one or several nuclei, with a life cycle which may include amoebae without flagella, and cysts.

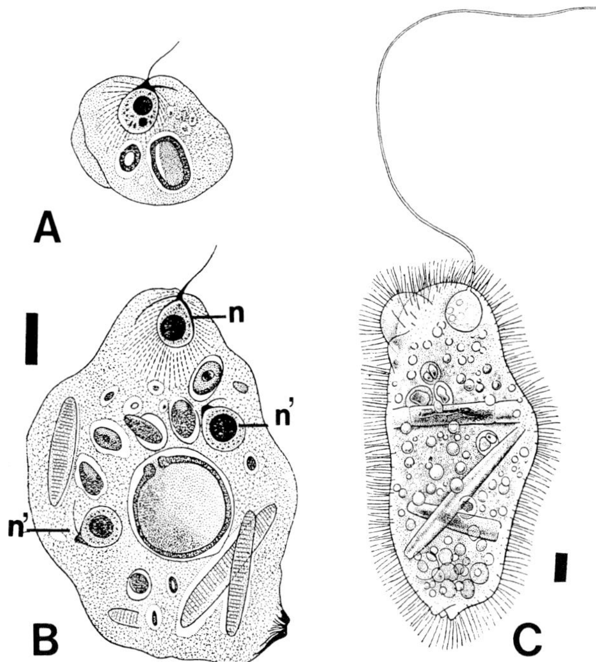

Fig. 5. A. *Mastigina hylae* with one nucleus and associated flagellum. B. With two secondary nuclei (n') without flagella. Bar=10 μm. (after Collin, 1913) C. *Mastigina setosa* (after Goldsdhmidt, 1907). Bar=10 μm.

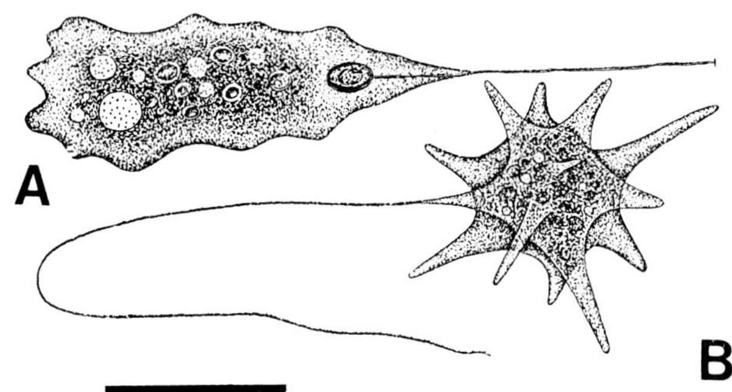

Fig. 6. *Mastigamoeba scholaia* with flagella connected to the nucleus; cell swimming (A), gliding (B). Bar = 10 μm. (from Klug 1936).

Genus ***Mastigamoeba*** Schulze, 1875

Amoeboid cells with a flagellated basal body located immediately adjacent to the anterior nucleus. The nucleus may be pulled out anteriorly to have a conical appearance (Fig. 6). The connection may be seen with the light microscope. The flagellum is long and directed forward, the pseudopodia are simple, branched,

or absent. The cells may swim or glide. Life cycles may include amoeboid organisms, multinucleated cells, or cysts (Simpson et al., 1997). Electron-microscopical studies (e.g. Brugerolle et al., 1991a) have shown that a cone of microtubules connects the short basal body to the nucleus (Fig. 2a) and that there is a ribbon of microtubules extending from the basal body into the cytoplasm. The flagellum has the conventional 9+2 arrangement of microtubules within the axoneme, but variations on the normal organization of the basal body have been reported (Simpson et al., 1997). The monotypic genus *Phreatamoeba* was erected by Chavez et al. (1986) for organisms with a complex life cycle. *Phreatamoeba* cannot be distinguished from *Mastigamoeba* and thus is a synonym of *Mastigamoeba*, (Simpson et al., 1997). *Dinamoeba mirabilis* Leidy, 1874, the type species of that genus, also appears to be an incomplete description of *Mastigamoeba aspera* (Simpson et al., 1997). With about 40 nominal species, many of which will prove to be incomplete descriptions of other species. References: Brugerolle, 1991a; Klug, 1936; Lemmermann, 1914; Penard, 1909; Schulze, 1875; Simpson et al., 1997.
TYPE SPECIES: *Mastigamoeba aspera* Schulze, 1875

Genus ***Mastigella*** Frenzel, 1897

Amoeboid cells with an apical flagellum (Fig. 7) and the nucleus removed some distance from the flagellum. Reportedly without direct connection between the two, although unpublished ultrastructural studies indicate that connections are present. The cells are naked but usually with a distinct cortical region or periplast, and some species may have adhering bacteria. Pseudopodia are simple or absent and there may be one 1 or many contractile vacuoles. Swimming is disorderly, cells may glide but rarely. One species at least described with aflagellated amoebae and cyst in life cycle. The distinction between this genus and *Mastigamoeba* is probably not sustainable.

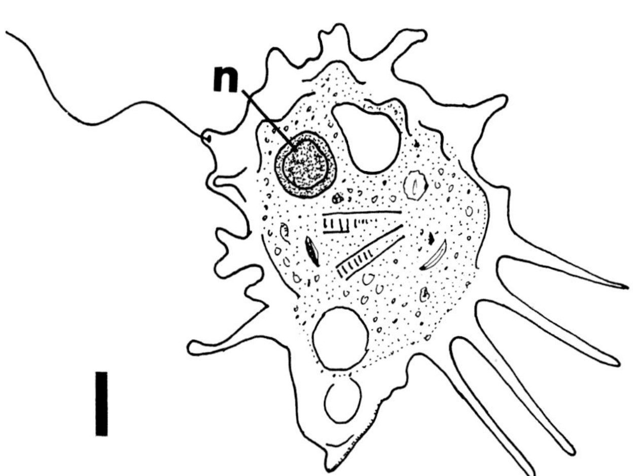

Fig. 7. *Mastigella nitens* with a flagellum not directly connected to the nucleus (n). Bar=10 μm. (after Penard, 1909)

With about 150 nominal species, many of which have been reported by Skvortzkov and Noda (1975) and are probably not distinct species. Published ultrastructure is restricted to an examination of symbiotic methanogens. Refs. Goldschmidt, 1907; Klug, 1936; Lemmermann, 1914; Penard, 1909; Skvortzkov and Noda 1975; van Bruggen et al., 1986.
TYPE SPECIES: *Mastigella polymastix* Frenzel, 1897

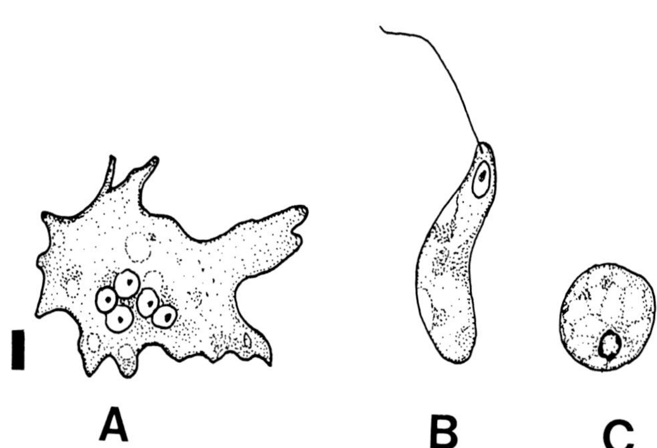

Fig. 8. *Mastigamoeba balamuthi*, multinucleated amoeboid stage (A), flagellate (B), and cyst (C). Bar=10 μm. (after Chavez et al., 1986)

Genus ***Phreatamoeba***
Chavez, Balamuth & Gong, 1986 (Fig. 8)

See *Mastigamoeba*.

References

Brugerolle, G. 1982. Caractères ultrastructuraux d'une mastigamibe: *Mastigina hylae* (Frenzel). *Protistologica*, 18:227-235.

Brugerolle, G. 1991a. Flagellar and cytoskeletal systems in amitochondrial flagellates: Archamoebae, Metamonada and Parabasala. *Protoplasma*, 164:70-90.

Brugerolle, G. 1991b. Cell organization in free-living amitochondriate heterotrophic flagellates. *In*: Patterson, D. J. & Larsen, J. (eds.), *The Biology of Free-living Heterotrophic Flagellates.* The Systematics Association, Clarendon Press, Oxford, 45:133-158.

Chavez, L. A., Balamuth, W. & Gong, T. 1986. A light and electron microscopical study of a new, polymorphic free-living amoeba, *Phreatamoeba balamuthi* n. g., n. sp. *J. Protozool.*, 33:397-404.

Cavalier-Smith, T. 1991. Cell diversification in heterotrophic flagellates. *In*: Patterson, D. J. & Larsen, J. (eds.), *The Biology of Free-living Heterotrophic Flagellates.* The Systematic Association, Clarendon Press, Oxford, 45:113-131.

Collin, B. 1913. Sur un ensemble de protistes parasites des Batraciens. *Arch. Zool. Exp. Gén.* (Notes Rev.), 52:67-76.

Goldschmidt, R. 1907. Lebensgeschichte der mastigamöben *Mastigella vitrea* n. sp. u. *Mastigina setosa* n. sp. *Arch. Protistenkd.*, 1: (suppl.1) 83-168.

Griffin, J. L. 1988. Fine structure and taxonomic position of the giant amoeboid flagellate *Pelomyxa palustris. J. Protozool.*, 35:300-315.

Hinkle, G., Leipe, D. D., Nerad, T. A. & Sogin, M. L. 1994. The unusually long small subunit ribosomal RNA of *Phreatamoeba balamuthi. Nucl. Acids Res.*, 22:465-469.

Klug, G. 1936. Neue oder wenig bekannte Arten der Gattung *Mastigamoeba*, *Mastigella*, *Cercobodo*, *Tetramitus* und *Trigonomonas. Arch. Protistenkd.,* 87:97-116.

Lemmermann, E. 1914. Flagellata I. *In*: Pascher, A. (ed.), *Die Süsswaterflora Deutschlands, Osterreichs und der Schweiz*, vol. 1., G. Fisher, Jena. pp 30-51

Mylnikov, A. P. 1991. Diversity of flagellates without mitochondria. *In*: Patterson, D. J. & Larsen, J. (eds.), *The Biology of Free-living Heterotrophic Flagellates.* The Systematic Association, Clarendon Press, Oxford. pp 149-158

Penard, E. 1909. Sur quelques mastigamibes des environs de Genève. *Rev. Suisse Zool.*, 17:405-440.

Schulze, F. E. 1875. Rhizopodienstudien. *Arch. Mikrosk. Anat.,* 11:583-596.

Simpson, A. G. B., Bernard, C., Fenchel, T. & Patterson, D. J. 1997. The organisation of *Mastigamoeba schizophrenia* n. sp.: more evidence of ultrastructural idiosyncrasy and simplicity in pelobiont protists. *Eur. J. Protistol.*, 33:87-98.

Skvortzkov, B. V. & Noda, M. 1975. Flagellates of clean and polluted waters. V. A short description of colourless flagellates with one swimming flagellum— genus *Mastigella* (Frenzel) nob (Rhizomastigaceae) reported during 1962 - 1975 in many parts of the world and especially from subtropics in Brasil, South America. *In*: *Prof. Mitsuzo Noda commemorative publication on his retirement from the Faculty of Science, Niigata Universit.* Niigata University. pp 25-66

Stumm, C. K. & Zwart, K. B. 1986. Symbiosis of protozoa with hydrogen utilizing methanogens. *Microb. Sci.*, 3:100-105.

van Bruggen, J. J. A., Stumm, C. K., Zwart, K. B. & Vogels, G. D. 1985. Endosymbiotic methanogenic bacteria of the sapropelic amoeba *Mastigella. FEMS Microb. Ecol.,* 31:187-192.

Whatley, J. M. & Chapman-Andresen, C. 1990. Phylum Karyoblastea. *In*: Margulis, L., Corliss, J. O., Melkonian, M. & Chapman, D. J. (eds.), *Handbook of Protoctista*. Jones & Bartlett, Boston. pp 167-185

CLASS HETEROLOBOSEA

DAVID J. PATTERSON, A. ROGERSON
and NAJA VØRS

This grouping of the schizopyrenid amoebae and amoeboflagellates and the acrasid (cellular) slime moulds was created by Page and Blanton (1985) on the basis of ultrastructural characteristics (see also Page, 1985). Initially, the schizopyrenid organisms were thought of as being predominantly amoeboid with some species having distributive flagellated stages. Several taxa have been added that are known only as flagellates (Fenchel and Patterson, 1986; Larsen and Patterson, 1990). As the flagellated state probably evolved only once in eukaryotes, the ancestors of the schizopyrenids must have been flagellated organisms. The group contains some parasitic organisms. Divisions among some genera are subtle and often require examination of life cycle characteristics. These characteristics may only be explored with any confidence using pure cultures. Note that some species are facultative pathogens. The group has been reviewed by Page (1988) and Page and Siemensma (1991) and the genera with flagellates in Patterson and Larsen (1991). The organism referred to as a vahlkampfiid by Larsson et al. (1992) is not accepted as such here. This account concentrates on those characteristics which can be determined by light microscopy of living cells. Details on techniques for isolation and culturing these flagellates are given in Page (1988) and in Patterson and Larsen (1991). Primary characters are (i) the number of phenotypes in the life cycle (amoebae, flagellates, cysts), (ii) the nature of the cysts (with or without excystment pores and gelatinous coat), and (iii) the flagellated forms (number of flagella, with or without cytostome, and curving ridge near flagella). Family level distinctions depend on the aspect or the behavior of the nucleus during mitosis.

CLASS HETEROLOBOSEA
Page & Blanton, 1985

Protists with flattened cristae in mitochondria, mitochondria associated with endoplasmic reticulum, dictyosomes not well developed, nuclear division **promitotic** (persistent nucleoli and endonuclear division spindle). Amoebae with eruptive pseudopodia, cell length usually less than 65 µm. Species may exhibit amoeboid and/or flagellated and/or encysted forms. Some taxa (acrasids) produce stalked fruiting bodies in which encysted cells are referred to as spores. With two orders, Schizopyrenida and Acrasida.

The Schizopyrenids contain facultative pathogens. The group is abundant, but any cells should be regarded initially as potentially pathogenic until full identification has been carried out.

ORDER SCHIZOPYRENIDA Singh, 1952

Heterolobosea without **fruiting bodies**. With two families (Vahlkampfiidae and Gruberellidae).

FAMILY VAHLKAMPFIIDAE Jollos, 1917

Schizopyrenids in which the nucleolus forms two **polar masses** during mitosis (**promitosis**). Amoeboid form usually cylindrical and usually uninucleate. Most species with a flagellated stage, one genus (*Percolomonas*, *incertae sedis*) so far not known to have an amoeboid stage. *Vahlkampfia* lacks a flagellated stage.

KEY TO GENERA OF THE FAMILY VAHLKAMPFIIDAE

TEM

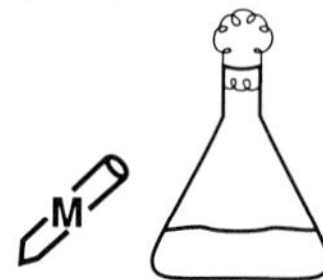

To determine generic identities it is necessary to establish pure cultures of each isolate. The Heterolobosea can have amoeboid and/or flagellated and/or encysted stages in their life cyle, and the presence or absence and appearance of these are essential diagnostic criteria. In practice this may be difficult since, depending upon culture conditions used, some isolates may fail to display all stages in the laboratory. Moreover, older amoeboid cultures frequently lose their ability to form flagellates or cysts. Viewing several protozoa in a culture is required to detail the features of a 'typical' vahlkampfiid. For example, some flagellates have four flagella, but occasionally three. For the purposes of this key, the typical form would be four. A phenotypically diverse group of protists which are being studied by molecular phylogeneneticists (e.g. Hinkle and Sogin, 1993). One genus, *Lyromonas*, is amitochondriate and requires TEM for identification.

1. With mitochondria... 2
1'. Without mitochondria..................... ***Lyromonas***

2. Without flagellate stage.................................... 3
2'. With a flagellate stage (or known only as a flagellate).. 4

3. Amoebae always uninucleate .... ***Vahlkampfia***
3'. Amoebae form multinucleate plasmodia on agar .... ***Pseudovahlkampfia***

4. With two flagella 5
4'. With more than two flagella 9

5. With anterior beak-like rostrum......................... .... ***Paratetramitus***
5'. Without beak-like rostrum 6

6. With cytostome 7
6'. Without cytostome 8

7. Cytostome obvious, as deep groove....................... .... ***Heteramoeba***
7'. Cytostome indistinct, as circular apical depression.. ***Adelphamoeba***

8. Cysts with pores .... ***Naegleria***
8'. Cysts without pores .... ***Didascalus***

9. With three flagella .... ***Trimastigamoeba***
9'. With four or more flagella 10

10. With four flagella 11
10'. With four sets of four flagella .... ***Psalteriomonas***

11. Cysts with pores .... ***Willaertia***
11'. Cysts without pores 12

12. With cytostomal groove 13
12'. Without cytostomal groove .... ***Tetramastigamoeba***

13. With lip-like anterior rostrum... ***Tetramitus***
13'. Without anterior rostrum...... ***Percolomonas***

Genus ***Lyromonas*** Cavalier-Smith, 1993

Amitochondriate genus with one species described as *Psalteriomonas vulgaris* (Broers et al. 1993), the flagellated stage only is known. Closely related to *Tetramitus*.

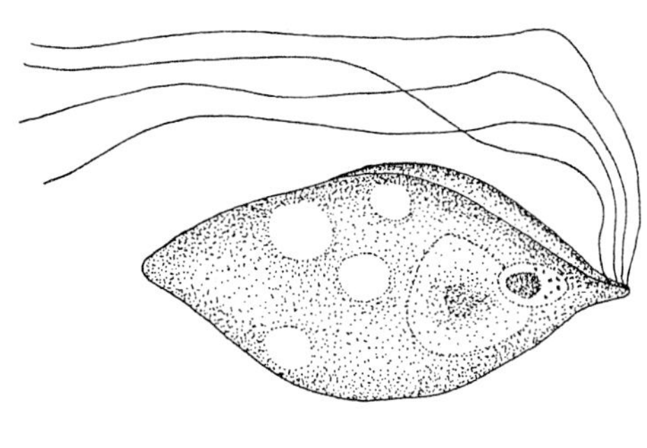

Fig. 1. *Lyromonas vulgaris* (after Broers et al.). Bar=5 µm.

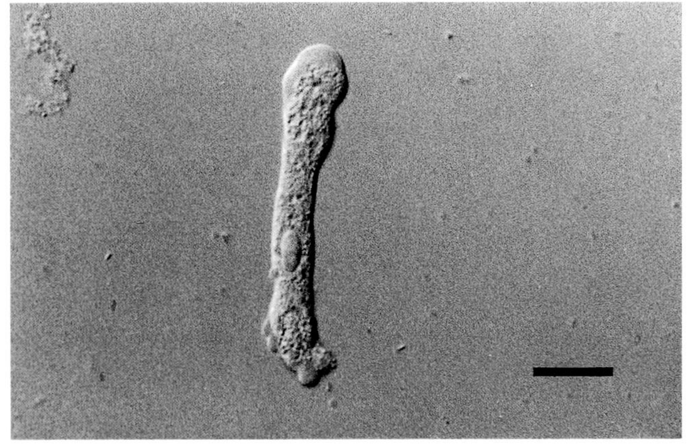

Fig. 2. *Vahlkampfia ustiana*. Active cell. Bar=10 µm

Genus ***Vahlkampfia*** Chatton & LaLung-Bonnaire, 1912 (Fig. 2)

Heterolobosea without flagellated stage described in life cycle. Cysts with or without gelatinous coat. One species only reported to have excystment pores. Many species (Page 1988; Page and Siemensma).

Genus ***Pseudovahlkampfia*** Sawyer, 1980 (Fig. 3)

Uninucleate amoeboid locomotive cells 18-30 µm long with the tendency to form multinucleate plasmodia up to 245 µm when cultured on agar. A vahlkampfiid without described flagellate stage. Monotypic genus. Marine, commensal in digestive tract and on gills of crabs (Bovee and Sawyer, 1979).

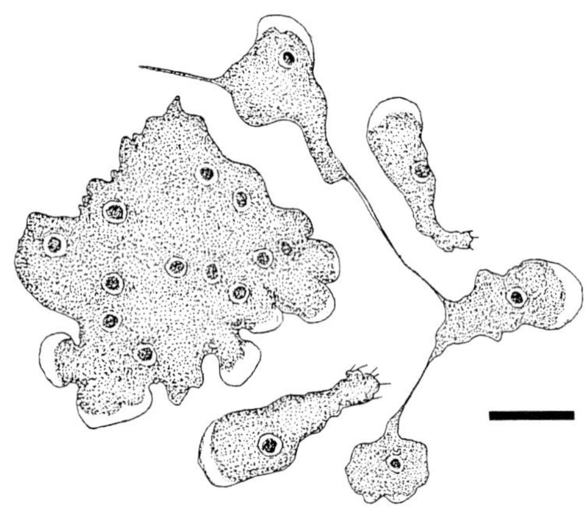

Fig. 3. Different forms of *Pseudovahlkampfia emersoni* (after Sawyer). Bar=10 µm.

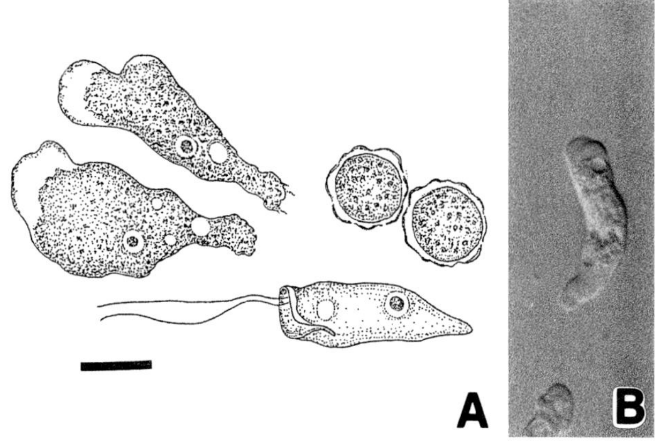

Fig. 4. *Paratetramitus jugosus*. (A) Line drawings of flagellate, cysts, and amoeboid stages. (after Page). (B) Bocomotive amoeboid cells. Bar=10 µm.

Genus ***Paratetramitus*** Darbyshire, Page & Goodfellow, 1976 (Fig. 4)

Monotypic vahlkampfiid genus, with cyst, amoeba, and flagellate phenotypes. Flagellates measure 1 5 - 20 µm, usually with 2 flagella but up to 6 reported; an anterior curved lip is present but ingestion not observed in original study. Flagellates may divide. *Paratetramitus* is most similar to *Heteramoeba*, being distinguished principally by chromatin distribution in the nucleus. Cyst walls without excystment pores (Darbyshire et al., 1976).

Genus ***Heteramoeba*** Droop, 1962 (Fig. 5)

The flagellated form is large (30 µm), 2 flagella, an elongate cytostome curving around the anterior of the cell and forming a groove. Nucleus with peripheral chromatin. Probably feeds and divides as a flagellate. One species. This genus is most like *Paratetramitus* from which it can be distinguished by peripheral location of chromatin material. Cysts without pores, excystment through a weak region of wall (Page, 1983). Marine.

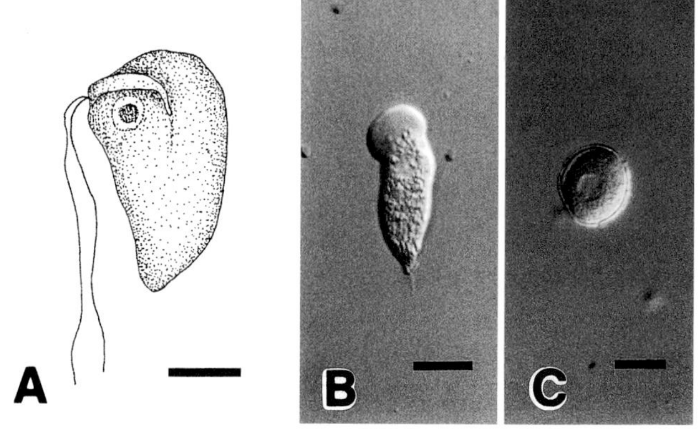

Fig. 5. *Heteramoeba clara*. (A) Flagellate stage. (B) Locomotive amoeba. (C) cyst. (after Page) Bars=10 µm.

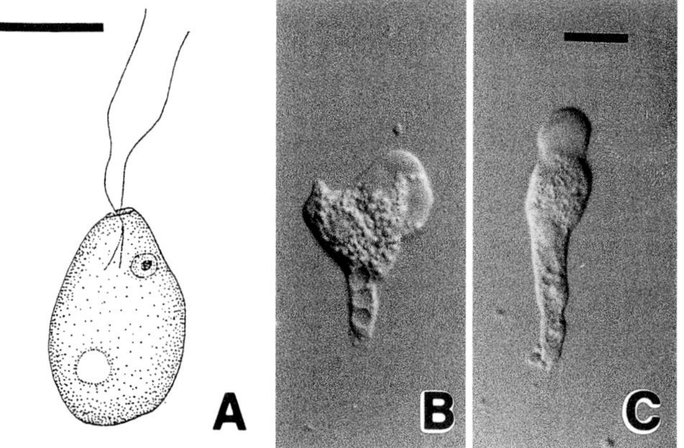

Fig. 6. *Adelphamoeba galeacystis*. (A) Flagellate stage (after Page) and (B, C) amoeboid stages. Bars=10 µm.

Genus ***Adelphamoeba***
Napolitano, Wall & Ganz, 1970 (Fig. 6)

Monospecific vahlkampfiid genus, with cyst, amoeboid, and flagellated stages. Flagellate about 10-20 µm long, 2 anterior flagella insering apically, with circular cytostome adjacent to flagellar insertion, feeds on bacteria. Cysts without pores, and no gelatinous sheath. Found in soil.

Genus ***Naegleria*** Aléxéieff, 1912 (Fig. 7)

Vahlkampfiid with cyst, amoeboid, and flagellated forms. Flagellate with 2 anteriorly inserting flagella, without cytostome, and without division. Cysts with plugged pores. Includes pathogenic species.

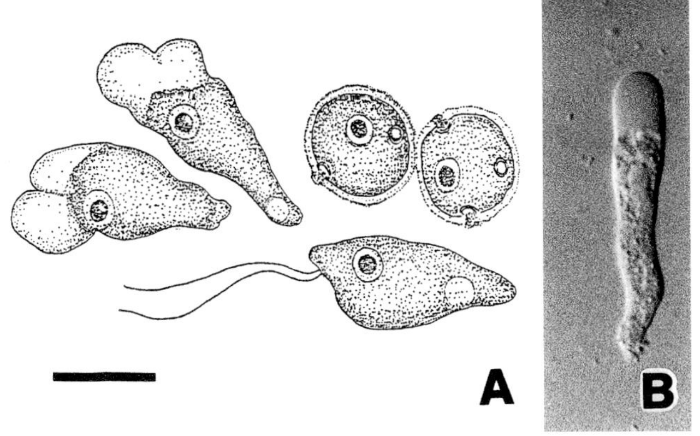

Fig. 7. *Naegleria gruberi* (A) Flagellate, cyst, and amoeboid stages (after Page). Note the plugged pores in the cysts. (B) N. jadini, active amoeboid stage. Bar=10 µm.

Genus **Didascalus** Singh, 1952 (Fig. 8)

Vahlkampfiid with amoeba, cyst, and biflagellated stage resembling the flagellate stage of *Naegleria*, the genus being distinguished by lack of excystment pores in cyst wall and by the presence of a mucus sheath around the cyst. Monotypic genus found in soil.

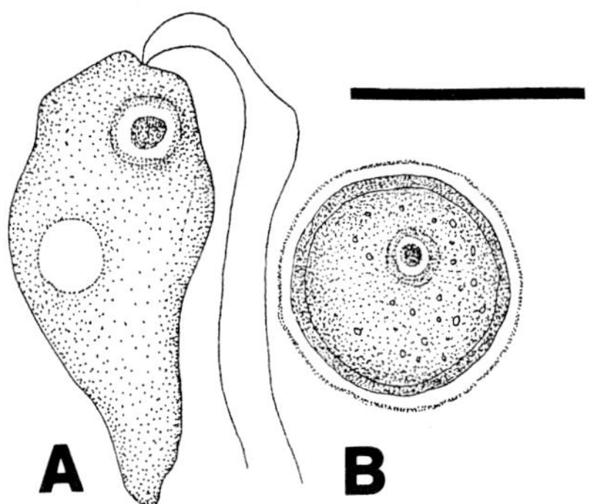

Fig. 8. *Didascalus thorntoni*. (A) Flagellate stage and (B) cyst (after Singh). Bar=10 µm.

Genus ***Trimastigamoeba*** Whitmore, 1911 (Fig. 9)

Vahlkampfiid with cysts, amoeba, and flagellate stage. Flagellate measures 16-22 µm. According to the original description normally with 3 flagella inserting in a deep depression, but no mouth—although the flagellated form has been reported as dividing. Cyst with pores. The nature of the taxon is unclear but probably related to the vahlkampfiids (Page, 1988).

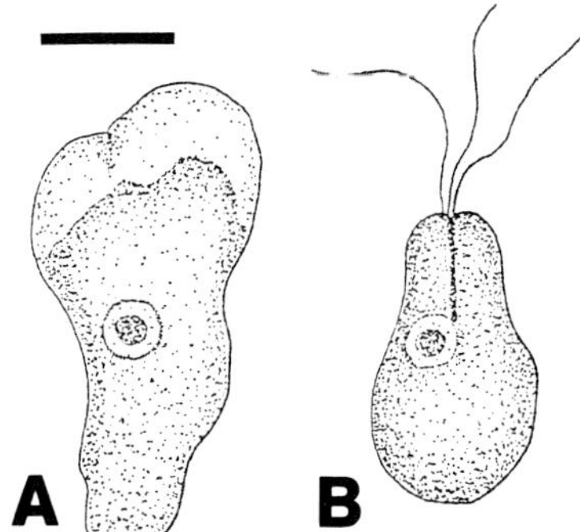

Fig. 9. *Trimatigamoeba philippenensis*. (A) Amoeboid stage. (B) Flagellated stage. Bar=10 µm.

Genus ***Psalteriomonas***
Broers, Stumm, Vogels & Brugerolle, 1990 (Fig. 10)

Vahlkampfiid, the flagellated stage of which has 4 nuclei and 4 sets of 4 flagella, giving rise to mononucleated amoeboid stage. From anoxic habitats. With methanogenic bacteria. Type species, *P. lanterna* (Broers et al., 1990). This genus is probably most closely related to *Tetramitus* because of similarly complex microtubular and fibrous roots derived from the flagellar basal bodies.

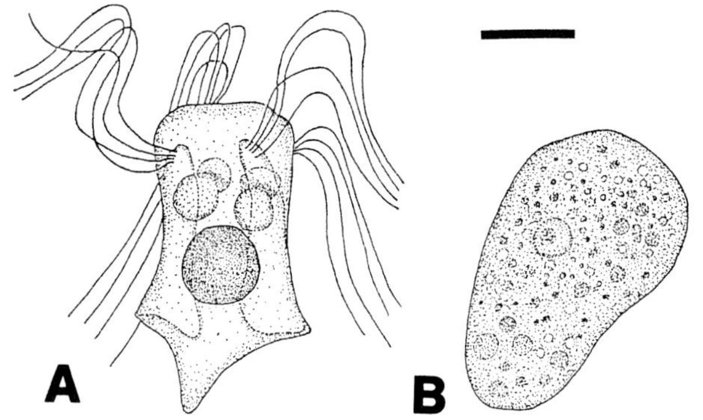

Fig. 10. *Psalteriomonas lanterna*. (A) Flagellate with prominent central globule and (B) amoeboid stage (after Broers et al.). Bar=10 µm.

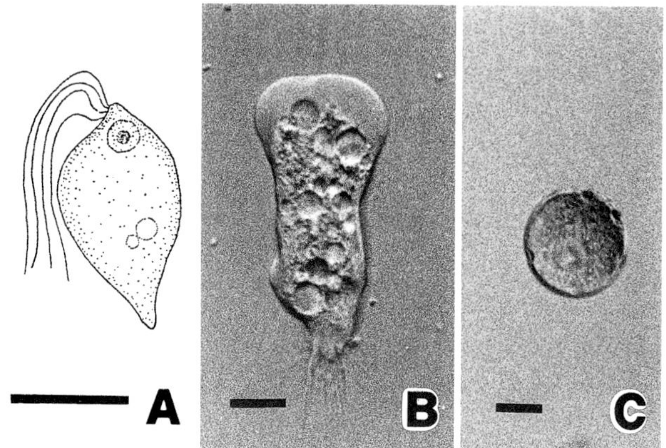

Fig. 11. *Willaertia magna*. (A) Flagellate, (B) locomotive amoeboid cell, (C) cysts with plugged pores. Bars=10 µm.

Genus ***Willaertia***
de Jonckheere, Dive, Pussard & Vickerman, 1984 (Fig. 11)

Vahlkampfiid with amoeboid stage, multi-pored cyst, and flagellated stage which can divide. Flagellate 19-20 µm in length with 4 flagella and no cytostome. Synonymy with *Protonaegleria* (Michel and Raether, 1985) suggested by Page (1988). *Protonaegleria* was distinguished by having a flagellated stage in the life cycle, but this was subsequently confirmed for *Willaertia* (Robinson et al., 1989). There is a rhizoplast and curving microtubular rootlets as in *Tetramitus* (Michel et al., 1987). Division of flagellate reported in Page (1988). Distinguishable from *Tetramastigamoeba* by having plugged pores in the cyst.

Genus ***Tetramastigamoeba***
Singh & Hanumaiah, 1977 (Fig. 12)

Monospecific genus of vahlkampfiids; with amoeboid and flagellated stages, plus cyst without pores. Flagellate with 4 flagella and no cytostome or ridge; the flagellates divide. Page (1988) doubts that the cyst is without pores. If pores are present, then this is the same as *Willaertia* and *Protonaegleria*, and *Tetramastigamoeba* is the senior synomym.

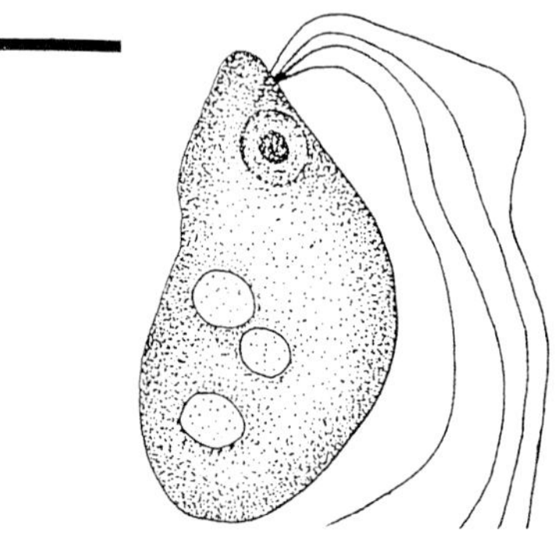

Fig. 12. *Tetramastigamoeba hoarei* (after Page). Bar=10 µm.

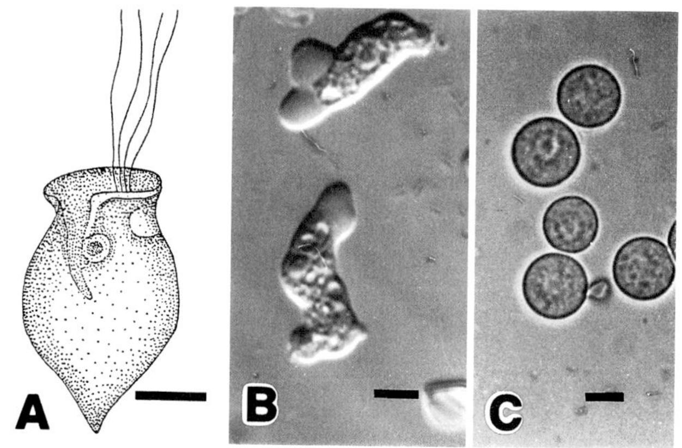

Fig. 13. *Tetramitus rostratus*. (A) Flagellate (after Page). (B) Amoeboid cells. (C) Cysts without pores. Bars=5 µm.

Genus ***Tetramitus*** Perty, 1852 (Fig. 13)

With 4 equal flagella arising in a flattened slit at the front end of the cell, surrounded by a distinct anterior lip-like rostrum. Substantial fibrous root associated with flagellar bases. Cysts without pores. Many of the flagellates assigned to this genus are likely to be more closely related to *Percolomonas cosmopolitus* than to *T. rostratus*, the type of this genus. Reported from soil, freshwater, and feces.

Genus ***Percolomonas*** Fenchel & Patterson, 1986. *incertae sedis* (Fig. 14)

Protists known only as flagellates; with 4 flagella, usually 1 used for adhesion, others for feeding on small suspended particles which are ingested within a ventral grove. Ultrastructural features suggest affinities with the Heterolobosea (Fenchel and Patterson, 1986).

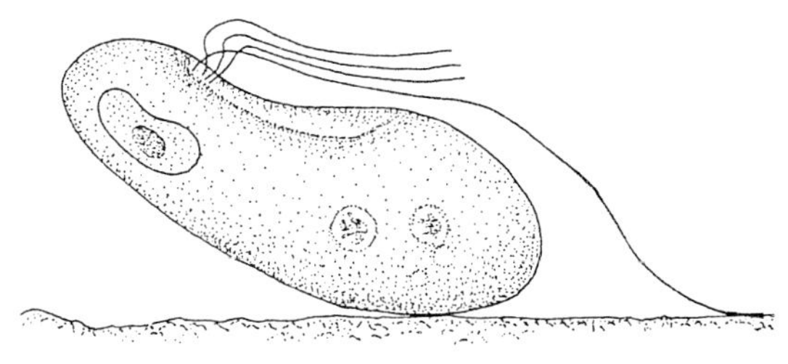

Fig. 14. *Percolomonas cosmopolitus*, attached (after Fenchel and Patterson). Bar=5 μm.

FAMILY GRUBERELLIDAE Page & Blanton, 1985

Nucleolus disintegrates during mitosis, amoeboid form flattened or limax and often multinucleate. No flagellated stage known. Amoebae may have fine **subpseudopodia.**

KEY TO THE GENERA OF THE FAMILY GRUBERELLIDAE

1. Normally uninucleate; from soil or freshwater .... ***Stachyamoeba***
1'. Normally multinucleate, marine.. ***Gruberella***

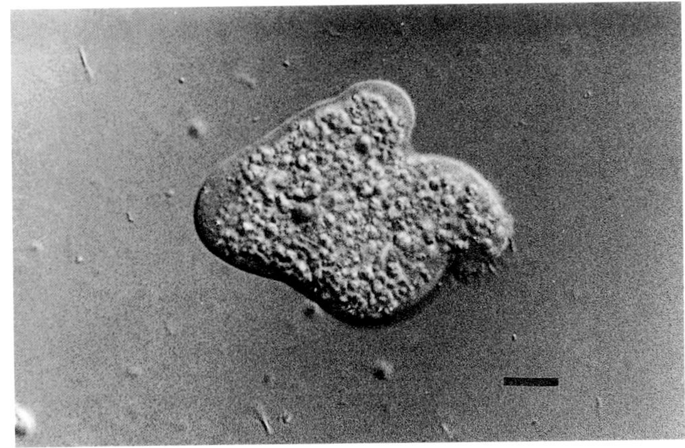

Fig. 15. *Gruberella flavescens*. Bar=10 μm.

Genus ***Gruberella*** Gruber, 1889 (Fig. 15)

Marine amoeba with rapidly moving locomotive form as a thick cylinder around 60 μm in length. Produces eruptive hyaline bulges. Multinucleate with around 6 nuclei but up to 37 per cell. Posterior adhesive filaments a common feature. At the ultrastructural level surface coat as a thin glycocalyx. Monotypic genus. *Euhyperamoeba*, a **limax** marine amoeba, is similar to *Gruberella* (see Chapter on Lobosea).

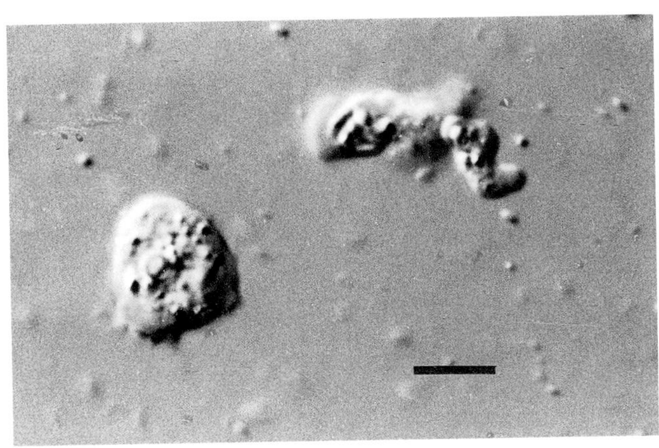

Fig. 16. *Stachyamoeba lipophora*, limax and flattened forms. Bar=10 μm.

Genus ***Stachyamoeba*** Page, 1975 (Fig. 16)

Amoeboid form commonly flattened during locomotion, sometimes fan-shaped, occasionally limax. Pseudopodia eruptive; some may have a few, short, fine subpseudopodia. Cytoplasm with single nucleus with parietally arranged nucleolar material. Often with at least one prominent lipid inclusion. Cysts spherical with crenulated wall, but without excystment pores. Monotypic genus found in soil.

GENERA WITH UNCERTAIN FAMILIAL POSITION

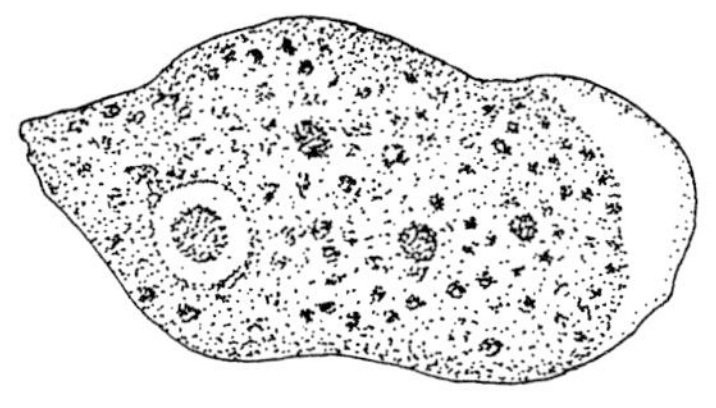

Fig. 17. *Pernina chaumonti* (after Kadiri et al.). Bar=10 µm.

Genus ***Pernina*** Kadiri, Joyon & Pussard, 1992 (Fig. 17)

Locomotive amoeba **limax** with rapid, eruptive monopodial locomotion. Lobed pseudopodia produced from anterior hyaloplasm. Posterior uroidal filaments common. Usually uninucleate, nucleus with a conspicuous central nucleolus. At the ultrastructural level, mitochondria with flattened cristae and cell surface without coating. Forms spherical cysts with double walls, with pores and plugs. This strain possesses characters intermediate between the families Gruberellidae and Vahlkampfiidae (Kadiri et al., 1992). Monotypic genus, marine. Type species: P. chaumonti Kadiri, Joyon & Pussard, 1992.

Genus ***Rosculus*** Hawes, 1963 (Fig. 18)

Protists known only as an amoeboid stage, reportedly with discoid mitochondrial cristae (Page, 1988). Morphologically very similar to *Guttulinopsis* (subclass Guttulinia) but fruiting bodies have not been seen in culture. Produces smooth, oval or spherical cysts. Monotypic genus. From freshwater, forest litter, and the rectum of *Natrix natrix*.

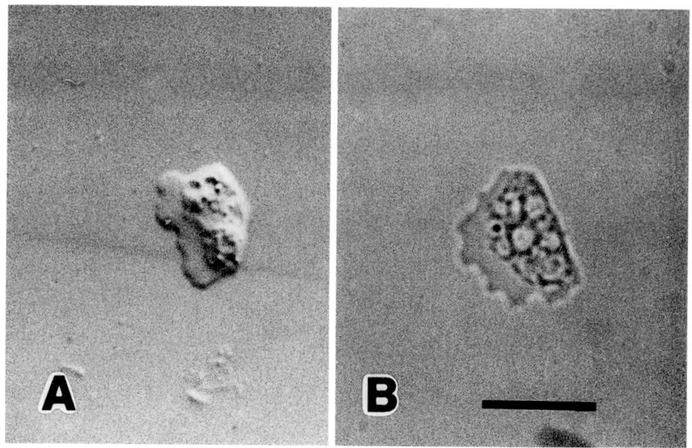

Fig. 18. *Rosculus ithacus.* Locomotive cell by interference contrast (A) and phase contrast (B) microscopy. Bar=10 µm. Courtesy of Dr. F. C. Page and CCAP.

LITERATURE CITED

Bovee, E. C. & Sawyer, T. K. 1979. Marine Flora and Fauna of the Northeastern United States. Protozoa: Sarcodina: Amoebae. *NOAA Technical Report NMFS* 419:1-55.

Broers, C. A. M., Stumm, C. K., Vogels, G. D. & Brugerolle, G. 1990. *Psalteriomonas lanterna* gen. nov., sp. nov., a free-living amoeboflagellate isolated from freshwater anaerobic sediments. *Europ. J. Protistol.*, 25:369-380.

Broers, C. A. M., Meyers, H. H. M., Symens, J. C., Brugerolle, G., Stumm, C. K. & Vogels, G. D. 1993. Symbiotic association of *Psalteriomonas vulgaris* n. spec. with *Methanobacterium formicicum*. *Europ. J. Protistol.*, 29:98-105.

Darbyshire, J. F., Page, F. C. & Goodfellow, L. P. 1976. *Paratetramitus jugosus*, an amoeboflagellate of soils and freshwater, type-species of *Paratetramitus* nov. gen. *Protistologica*, 12:375-387.

Fenchel, T. & Patterson, D. J. 1986. *Percolomonas cosmopolitus* (Ruinen) n.gen., a new type of filter feeding flagellate from marine plankton. *J. Mar. Biol. Assoc. U.K.*, 66:465-482.

Hinkle, G, & Sogin, M. L. 1993. The evolution of the vahlkampfiidae as deduced from 16S-like ribosomal RNA analysis. *J. Euk. Microbiol.*, 40:599-603.

Kadiri, G. E., Joyon, L. & Pussard, M. 1992. *Pernina chaumonti*, n.g., n.sp., a new marine amoeba (Rhizopoda, Heterolobosea). *Europ. J. Protistol.*, 28:43-50.

Larsson, J. I. R., De Roca, C. B. & Gaju-Ricart, M. 1992. Fine structure of an amoeba of the genus

Vahlkampfia (Rhizopoda, Vahlkampfiidae), a parasite of the gut epithelium of the bristletail *Promesomachilis hispanica* (Microcoryphia, Machilidae). *J. Invert. Pathol.*, 59:81-89.

Larsen, J. & Patterson, D. J. 1990. Some flagellates (protista) from tropical marine sediments. *J. Nat. Hist.*, 24:801-937.

Michel, R. & Raether, W. 1985. *Protonaegleria westphali* gen.nov., sp.nov. (Vahlkampfiidae), a thermophilic amoebo-flagellate isolated from freshwater habitat in India. *Z. Parasitenkd.*, 71:705-713.

Michel, R., Raether, W. & Schupp, E. 1987. Ultrastructure of the amoebo-flagellate *Protonaegleria westphali*. *Parasitol. Res.*, 74:23-29.

Page, F. C. 1983. *Marine Gymnamoebae.* Institute of Terrestrial Ecology, Cambridge, England.

Page, F. C. 1985. The limax amoebae: comparative fine structure of the Hartmannellidae (Lobosea) and further comparisons with the Vahlkampfiidae (Heterolobosea). *Protistologica*, 21:361-383.

Page, F. C. 1988. *A New Key to Freshwater and Soil Gymnamoebae.* Freshwater Biological Association, Ambleside, England.

Page, F. C. & Blanton, R. L. 1985. The Heterolobosea (Sarconina:Rhizopoda), a new class uniting the Schizopyrenida and the Acrasidae (Acrasida). *Protistologica*, 12:37-53.

Page, F. C. & Siemensma, F. J. 1991. *Nackte Rhizopoda und Heliozoea.* Gustav Fischer Verlag, Stuttgart, New York.

Patterson, D. J. & Larsen, J. (eds.) 1991. *The Biology of Free-living Heterotrophic Flagellates.* Clarendon Press, Oxford.

Robinson, B. S., Christy, P. E. & DeJonckheere, J. F. 1989. A temporary flagellate (mastigote) stage in the vahlkampfiid amoeba *Willaertia magna* and its possible evolutionary significance. *BioSystems*, 23:75-86.

ORDER CRYPTOMONADIDA Senn, 1900

by PAUL KUGRENS, ROBERT E. LEE and DAVID R. A. HILL

The cryptomonads are a relatively small but distinct taxonomic group of unicellular flagellates that are cosmopolitan in distribution and are important primary producers in both freshwater and marine habitats (Gillott, 1990; Klaveness 1988). Cryptomonads are common in many phytoplankton samples, but because their cells are extremely delicate, they rupture when fixatives are added or when temperatures are elevated; their numbers generally are underestimated in preserved samples. In sub-ice habitats and in early spring and late fall populations of temperate lakes and reservoirs, they often become dominant. In fact, the structural diversity discovered by using specialized electron-microscopic techniques (Hill, 1990, 1991b; Hill and Wetherbee, 1986, 1988, 1989, 1990; Kugrens and Lee, 1987, 1991, Kugrens et al., 1986a, 1987; Lee and Kugrens, 1986) strongly suggest that the numbers of species in this group of organisms probably has been significantly underestimated (Andersen, 1992).

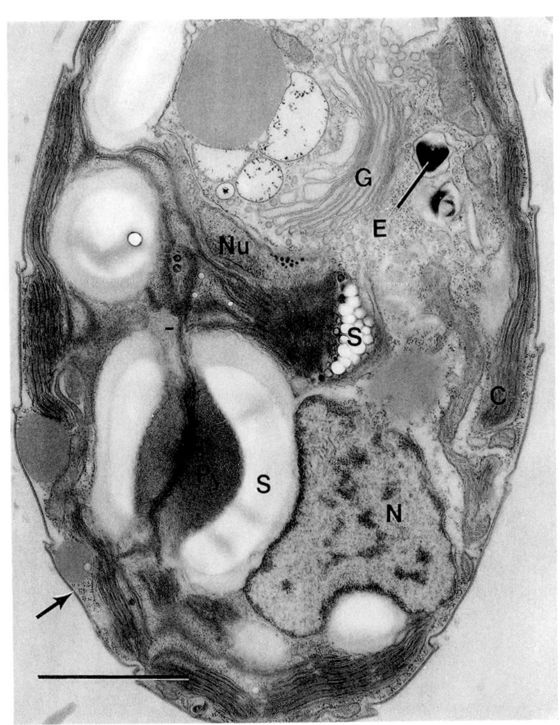

Fig 1. TEM of *Chroomonas coerulea* showing the general features of a cryptomonad cell and the relative locations of cellular structures. The periplast plates impart a serated appearance to the periplast (arrow). Important cell structures include a chloroplast (C) with a pyrenoid (Py) traversed by a pair of thylakoids and a stigma (S); a nucleus (N); a nucleomorph (Nm); a golgi apparatus (G); ejectosomes (E); and starch in the periplastidal space. Scale=1 µm.

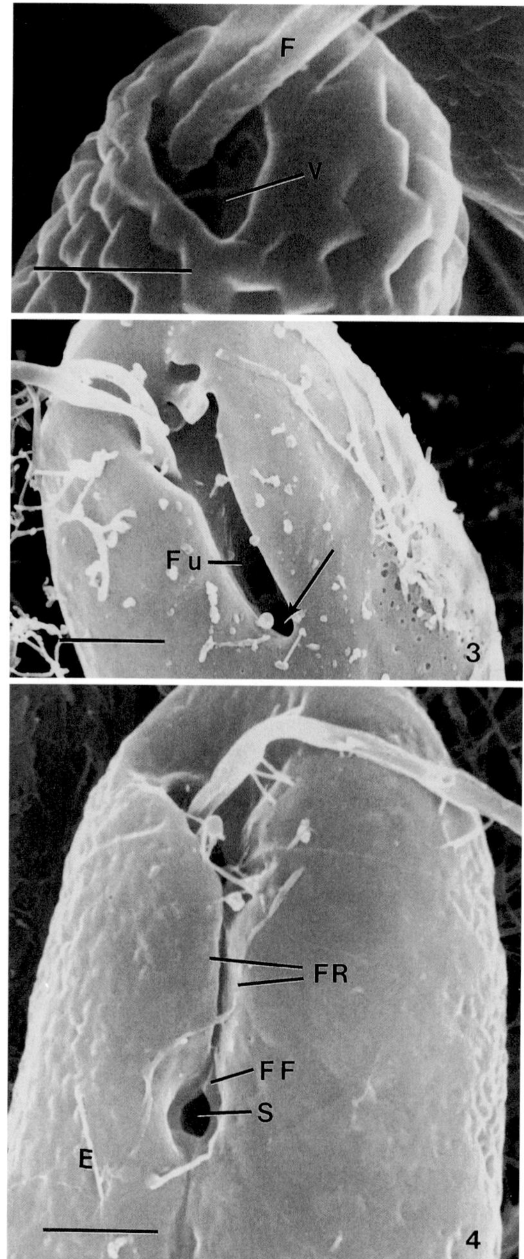

Fig. 2. Anterior of cell of *Chroomonas oblongata* showing vestibulum (V) and absence of a furrow. Flagella (F) are inserted on the right side of the vestibulum. Scale=1 µm. (From Kugrens et. al., 1986) Fig. 3. SEM of the anterior end of *Campylomonas reflexa* showing the simple furrow (Fu) and a gullet posterior to the furrow (arrow). At the anterior end and on the dorsal side of the vestibulum is a vestibular ligule (VL). Scale=1 µm. (Kugrens et al., 1986). Fig. 4. SEM of *Cryptomonas tetrapyrenoidosa* showing a complex furrow which consists of furrow ridges (FR), furrow folds (FF), and a posterior stoma (S). Scale=1 µm. (Kugrens et al., 1986)

Cryptomonads are quite complex (Fig. 1), and their fine structure reveals, that except for *Goniomonas*, their cells are derived from a eukaryotic endosymbiosis as well as the common prokaryotic (mitochondria and chloroplasts) endosymbioses (Douglas et al., 1991; Gillott, 1990; Ludwig and Gibbs, 1985, 1989; McKerracher and Gibbs, 1982; Anderson, 1992). The host nuclear encoded rRNA appears to be related to the *Acanthamoeba* and green algal lineage, whereas the **nucleomorph** encoded rRNA appears to be related to red algae (Douglas et al., 1984, 1987, 1991; Sespenwol, 1973). Owing to their unicellularity and small size, current and future classification schemes for cryptomonads must continue to rely upon ultrastructural features.

By using information from the structures discussed in the following section, specifically the furrow/gullet complex and periplast types, it has been possible to delineate 15 genera of cryptomonads. It is significant that the number of genera has increased since the review articles of Santore (1984,1987) and the systematic treatment by Gillott (1990), when only six genera were recognized. Features recognized by freeze-fracture in the TEM and in the SEM are important characters for species and some generic distinctions (Brett and Wetherbee, 1986; Dwarte and Vesk, 1983; Grim and Staehelin, 1984; Hill, 1990, 1991b; Hill and Wetherbee, 1986, 1988, 1989; Klaveness, 1985; Kugrens and Lee, 1987, 1991; Kugrens et al., 1986, 1987; Munawar and Bistricki, 1979; Wetherbee et al., 1986, 1987). For information on earlier classification schemes, the publications by Bourrelly (1970), Huber-Pestalozzi (1950), and Skuja (1939, 1948) are recommended for freshwater cryptomonads, whereas the treatise by Butcher (1967) deals with the marine forms.

The external architecture of the cell is asymmetrical due to a **furrow/gullet complex**. All cells, whether they are oval, compressed, lunate, caudate, acute, elongate, sigmoid, or contorted, possess an anterior, outwardly facing depression called a **vestibulum**, from which the flagella arise on the right side (Fig. 2).

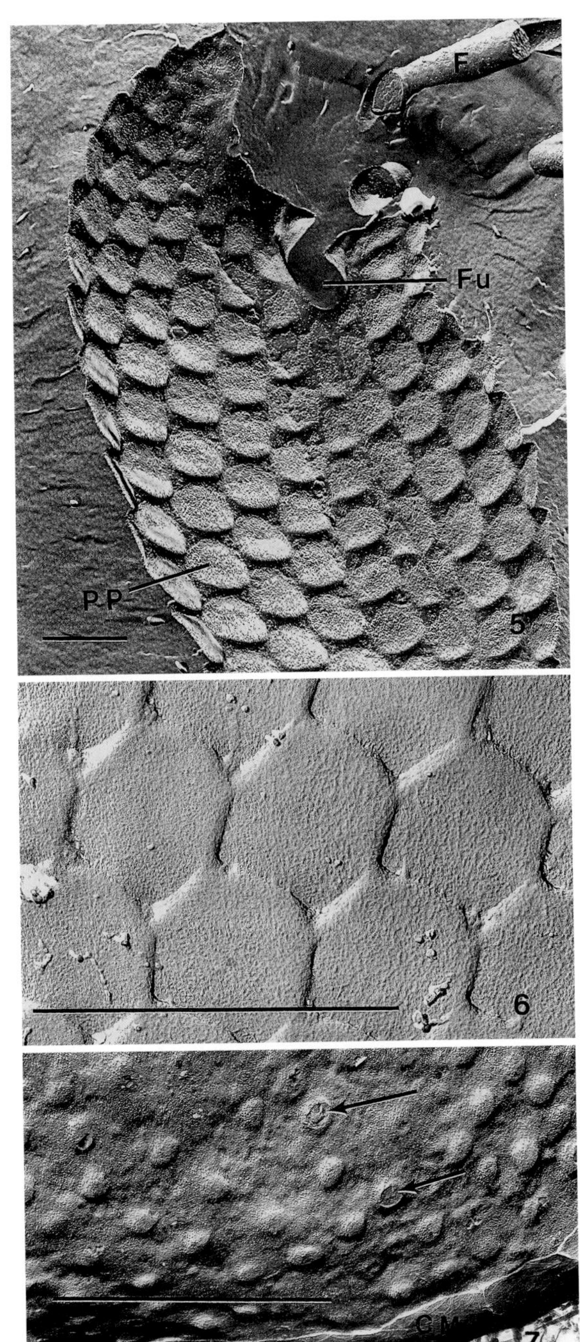

Fig. 5. Freeze-fracture replica of the periplast of *Rhodomonas ovalis*. Periplast plates (PP) appear round with tapered anteriors. Also shown is the furrow (Fu) and one of the flagella (F). Scale=1 µm. Fig. 6. Hexagonal surface periplast plates of *Komma* sp. Scale=1 m. (From Kugrens and Lee, 1991). Fig. 7. Inner periplast sheet of *Storeatula rhinosa* with pores (arrows) through which ejectisomes project. A portion of the cell membrane (CM) is visible toward the outside of the cell. Scale=1 µm.

In addition, there may be a **furrow**, a **gullet,** or combination of both. Furrows are ventral grooves originating in the vestibulum extending posteriorly, but terminating in the anterior half of the cell. A gullet is a tubular invagination which extends posteriorly from the vestibulum or the end of the furrow. Variations in these features are illustrated in Figs. 2-4. **Ejectisomes** flank the furrow and/or gullet. The **contractile vacuole** discharges into a predetermined site in the dorsal portion of the vestibulum. In some genera, a **vestibular ligule** covers the discharge site of the contractile vacuole (Kugrens and Lee, 1991).

Periplasts (Figs. 5-7) are unique conformation to cryptomonads. Periplasts consist of the plasma membrane and the inner and surface components (Hill, 1990, 1991b; Hill and Wetherbee, 1986, 1988, 1989; Kugrens and Lee, 1987, 1991; Wetherbee et al., 1987, 1986) and both can be variable in their composition. The inner periplast component (IPC) is protein (Gantt, 1971; Faust, 1974) and may be composed of many variously shaped plates or a sheet. The surface periplast component (SPC) may consist of plates, heptagonal scales, mucilage, or combinations of these. The plates of the inner periplast component are connected to the cell membrane by intramembrane particles or proteins (Brett and Wetherbee, 1986; Kugens and Lee, 1987; Wetherbee et al., 1986,1987). The arrangement of these IMP domains conform to the plate shapes which have been used to characterize genera. The periplast has pores (Fig. 7) through which the ejectisomes dock with the cell membrane (Grim and Staehelin, 1984).

With the exception of *Goniomonas,* cells have two subapically inserted subequal flagella (Fig. 8) with tubular hairs on at least one of them. There are at least five variations in the arrangement of the tubular and non-tubular hairs on the flagella. In the most common, the longer dorsal flagellum bears two laterally opposed rows of tubular hairs and the shorter ventral flagellum bears a single row of hairs (Fig. 8). Tubular hairs on the dorsal flagellum have one terminal filament and tubular hairs of the ventral flagellum have two unequal terminal filaments. In addition flagella may bear heptagonal scales (Lee and Kugrens, 1986; Pennick, 1981). *Goniomonas* has a unilateral row of curved spikes on one of its flagella instead of tubular hairs and fine non-tubular hairs on both flagella (Kugrens and Lee, 1987, 1991). In addition, the flagella of *Goniomonas* are inserted on the dorsal side of the vestibulum (Fig 13). The flagellar transition region is unique in cryptomonads and consists of a doublet system of septa (Fig. 9) (Grain et al., 1988; Kugrens and Lee, 1991). A **rhizostyle** is an intergral component of the flagellar apparatus in most

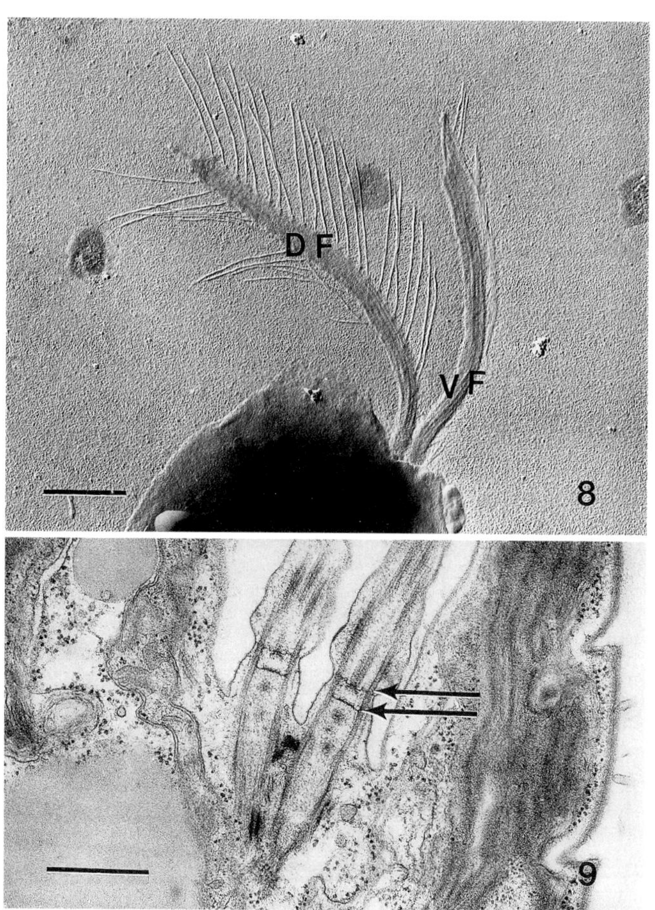

Fig. 8. Shadowed replica of the flagella of *Komma* sp. The dorsal flagellum (DF) bears two rows of tubular hairs, whereas the ventral flagellum (VF) has only one row of tubular hairs. Scale=1 μm. Fig. 9. TEM of transition regions of both flagella of *Chroomonas coerulea* displaying the double septate nature of this region (arrows). Scale=1 μm.

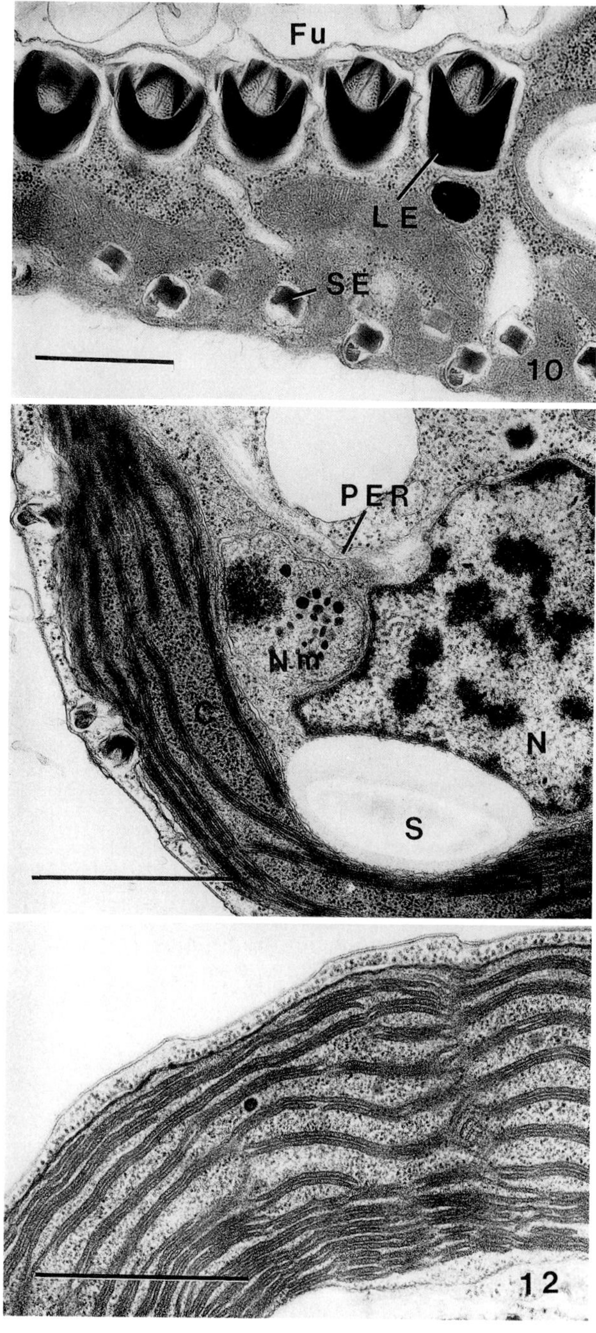

Fig. 10. Large ejectisomes of *Chilomonas paramecium* (LE) are located near the furrow (Fu), and the small ejectosomes (SE) are situated beneath the cell membrane. Scale=1 μm. Fig. 11. Portion of *Komma caudata* showing the location of the nucleomorph (Nm) near the nucleus (N) and chloroplast (C) in the periplastidal compartment. (PER) periplastiidal endoplasmic reticulum; (S) starch. Scale=1 μm. Fig. 12. Portion of a chloroplast of *Komma caudata* showing the paired arrangement of thylakoides. Scale=1 μm.

cryptomonads, and it consists of microtubules that originate near the basal bodies and then extend posteriorly into the cell. One type of rhizostyle, found in *Chilomonas* (Roberts et al., 1981; Kugrens and Lee, 1991) *Cryptomonas phi* (Gillott, 1990; Gillot and Gibbs, 1983) *Teleaulax* (Hill, 1990), *Storeatula* (Hill, 1990), *Geminigera* (Hill, 1990), and *Proteomonas* (Hill and Wetherbee, 1986) passes close to the nucleus and terminates near the posterior end of the cell. Each microtubule has a wing-like extension (**lamella**) and the length of these lamellae may vary with the respective microtubule (Gillot and Gibbs, 1983). A second type of rhizostyle, reported for *Cryptomonas ovata* (Roberts, 1984; Hill, 1990) and *Cryptomonas theta* (=*Guillardia theta*)(Gillot and Gibbs, 1983), lacks wings on the microtubules. This rhizostyle terminates anterior to the nucleus (Roberts, 1984).

Ejectisomes are the extrusive organelles of all cryptomonads, and they appear to be identical in all genera investigated. Two sizes of ejectisomes (Fig. 10) are found in cryptomonad cells (Dodge, 1969; Schuster, 1968, 1970; Morall and Greenwood, 1980; Santorc, 1982a, 1984). Large ejectisomes are associated with the furrow/gullet and small ejectisomes are found beneath the periplast in other regions of the cell. Ejectisomes consist of two different-sized, tightly coiled and tapered ribbons, which are joined, and both are surrounded by a membrane (Fig. 10). Upon discharge the ribbons roll up to form a tube. A crystalline substructure, similar to some prasinophyte ejectosomes has been reported by Morrall and Greenwood (1980) and Grim and Staehelin (1981).

Cryptomonads have a single reticulate mitochondrion with flattened cristae (Santore, 1977; Roberts et al., 1981; Kugrens and Lee, 1991). Chloroplasts may be olive green, brown, blue-green, or red, depending upon the pigments present. Pigments found in cryptomonads are chlorophylls *a* and c_2, and β carotene, alloxanthin, diadinoxanthin, and several forms of blue and red phycobilliproteins called Cr-phycocyanin and Cr-phycoerythrin to differentiate them from cyanobacterial and red algal phyco-biliproteins. With the exception of a marine endosymbiont (Hibberd, 1977), pigmented cells have only one or two chloroplasts. Either Cr-phycocyanin or Cr-phycoerythrin is located in the intrathylakoidal lumina of photosynthetic lamellae (Faust and Gantt, 1973; Gantt, 1979, 1980; Gantt et al., 1971; Ludwig and Gibbs, 1989). Chloroplasts are surrounded by a double-membraned endoplasmic reticulum (Fig.11) called the **periplastidial envelope**, **periplastidal complex**, or **chloroplast endoplasmic reticulum**, which originates from an evagination of the outer membrane of the nuclear envelope and surrounds the chloroplast, starch grains, and a reduced nucleus known as the **nucleomorph** (Gillott and Gibbs, 1980; Ludwig and Gibbs, 1985; Santore, 1982c). Starch grains are formed within the periplastidal compartment, not within the chloroplast, and they generally are associated with a pyrenoid, if present. The number of thylakoids penetrating the pyrenoid has been used as a taxonomic character (Santore, 1984).

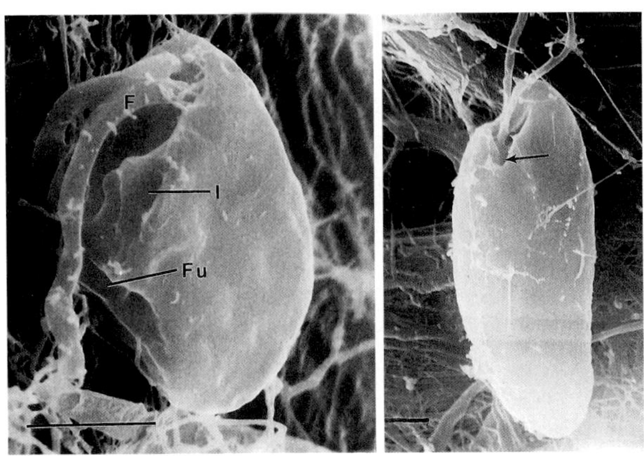

Fig.13. SEM of *Goniomonas truncata* showing the relationships between the dorsal flagella (F), furrow (Fu), and infundibulum. Scale=1 µm. Fig. 14. Comparatively short furrow (arrow) of *Chilomonas paramecium.* Scale=1 µm.

Thylakoids in the chloroplasts usually are arranged in pairs (Fig. 12) (Dwarte and Vesk, 1982, 1983; Gantt et al., 1971), sometimes in groups of three (Klaveness, 1981; Hill, 1991b), or in stacks of variable number (Hill, 1991b).

Chilomonas has a reduced chloroplast which lacks pigments and is called a **leucoplast**. *Goniomonas* lacks plastids and a nucleomorph; consequently it also lacks the periplastidial compartment.

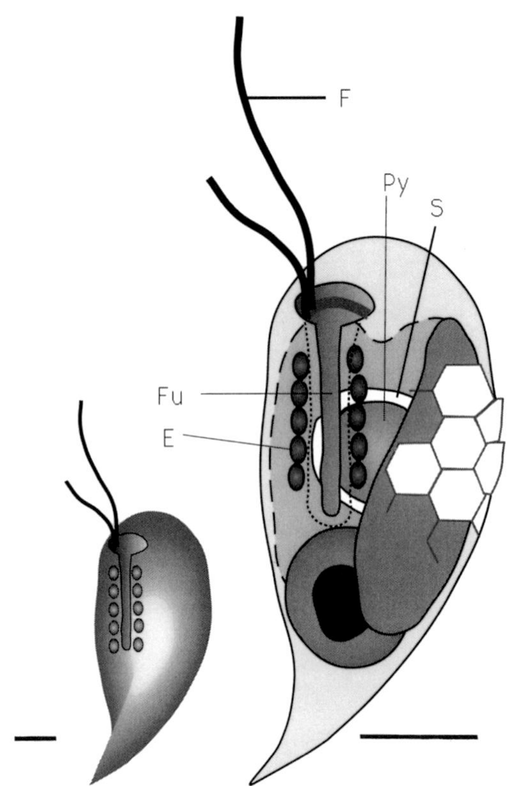

Fig. 15. Diagram of a typical cryptomonad (*Falcomonas daucoides*). (E) ejectisome; (F) flagellum; (Fu) furrow; (Py) pyrenoid; (S) starch. Scale=1 μm.

Nucleomorphs (Figs. 1,11) are located in the periplastidial compartment (Gillott and Gibbs, 1980; Kugrens and Lee, 1989, 1991; McKerracher and Gibbs, 1982; Ludwig and Gibbs, 1985; Morrall and Greenwood, 1982; Santore, 1982c, 1984, 1987). They represent a vestigial nucleus which remains from an ancestral red algal endosymbiont. (Douglas et al., 1991). The nucleomorph is small, is limited by a double membrane, and contains DNA (Ludwig and Gibbs, 1985; Douglas et al., 1991). In addition, the nucleomorph contains a fibrillar-granular region and dense bodies (Fig. 11). Its location within the compartment may have systematic applications (Santore, 1984; Hill and Wetherbee, 1989). In *Rhodomonas* the nucleomorph is located in the pyrenoidal bridge (Hill and Wetherbee, 1989).

With the exception of *Goniomonas* and *Chilomonas*, which are heterotrophs, all cryptomonads are obligate phototrophs. *Goniomonas* is phagocytic, whereas *Chilomonas* is strictly osmotrophic, and it does not ingest particulate materials as was previously assumed. One species of *Chroomonas* is mixotrophic (Kugrens and Lee, 1991), but there is no evidence that other cryptomonads are also mixotrophic.

Reproduction usually occurs by mitotic divisions, although sexual cycles have been documented for *Proteomonas* (Hill and Wetherbee, l986) and *Chroomonas* (Kugrens and Lee, 1988). Cysts may be produced to withstand adverse conditions.

KEY CHARACTERS

1. Two apically inserted subequal flagella with tubular hairs on at least one of them. Most commonly, the dorsal flagellum bears two laterally opposed rows of tubular hairs, while the ventral flagellum bears only a single row of hairs.
2. Cells asymmetrical with an anterior vestibular depression and gullet/furrow complex.
3. Unique cell covering, the periplast, comprised of plates, protein sheets, heptagonal scales, mucilage, or some combination of these.
4. Most have a rhizostyle consisting of microtubules that originate near the basal bodies and pass to the posterior close to, or anterior to, the nucleus.
5. A single reticulate mitochondrion with flattened cristae.
6. Two sizes of tightly coiled tapered ribbon-like ejectisomes associated with the gullet/furrow complex and periplast.
7. May be pigmented or colorless. One or two chloroplasts per cell, may be olive green, brown, blue-green, or red depending upon the pigments present.
8. In addition to a usual eukaryotic nucleus, most possess in the periplastidal compartment a vestigial nucleus known as a nucleomorph.

KEY TO GENERA

LM SEM TEM

1. Cells colorless.. 2
1'. Cells pigmented.. 3

2. Cells lack plastids.......................... ***Goniomonas***
2'. Cells with leucoplasts.................... ***Chilomonas***

3. Chloroplasts blue-green.................................. 4
3'. Chloroplasts red, olive-green, or brown..........6

4. Cells with a gullet only................................... 5
4'. Cells with a furrow...................... ***Falcomonas***

5. Retangular periplast plates.......... ***Chroomonas***
5'. Hexagonal periplast plates........................ ***Koma***

6. Life history with two dissimilar phases............ .... ***Proteomonas***
6'. Life history with one phase 7

7. Cells with a furrow or gullet only 8
7'. Cells with a furrow/gullet complex 12

8. Cells with a gullet only 9
8'. Cells with a furrow only .... ***Teleaulax***

9. Periplast components of inner & outer hexagonal plates 10
9'. Inner periplast component a sheet................. 11

10. Nucleomorph located in a cytoplasmic evagination into pyrenoid...................... ***Rhinomonas***
10'. Nucleomorph located anterior to pyrenoid and not located in pyrenoid.............. ***Hemiselmis***

11. Outer periplast component of large, irregular, longitudinally oriented plates ..... ***Guillardia***
11'. Outer periplast component of coarse fibrils..... .... ***Storeatula***

12. Cells with a simple furrow........................ 13
12'. Cells with a complex furrow... ***Cryptomonas***

13. Inner periplast component consists of a sheet. 14
13'. Inner periplast component consists of multiple plates.. ***Rhodomonas***

14. Surface periplast component consists of heptagonal plates.......................... ***Geminigera***
14'. Surface periplast component lacking or comprised of diffuse fibrils....... ***Campylomonas***

Genus ***Goniomonas*** Stein (Syn. *Cyathomonas*)

Cells laterally compressed, colorless, phagocytic; flagella inserted on the dorsal side of the vestibulum, one flagellum with recurved spines, the other flagellum with fine fibrillar hairs; the vestibulum connects to a ventral furrow, a gullet is absent, but the furrow has a posterior stoma; an infundibulum is located on the left side of the cell, presumably for ingestion of particulates; large ejectisomes form a band around the vestibulum; the periplast is comprised of inner and outer rectangular plates, which are not offset (Fig. 13); the periplastidial compartment lacking; fresh water and marine. Refs. Hill (1991a), Kugrens and Lee (1991), Mignot. (1965), Schuster (1968).

Genus ***Chilomonas*** Ehrenberg

Colorless cells with a rostrate anterior, not phagocytic; furrow/gullet complex consisting of a vestibulum, a short furrow, and long tubular gullet (Fig. 14). A vestibular ligule covers the area of contractile vacuole discharge. Both flagella with a unilateral row of tubular hairs. The periplast has an inner single sheet with numerous ejectisome pores and a surface periplast component consisting primarily of fibrils. The periplastidal compartment contains leucoplasts lacking thylakoids, numerous large starch grains, and several nucleomorphs located in the periplastidal space near the nucleus. Freshwater (Anderson, 1962; Grim and Staehelin, 1984; Heywood, 1988; Kugrens and Lee, 1987, 1991; Kugrens et al., 1986; Roberts et al., 1981; Schuster, 1970; Sespenwol, 1973).

Genus *Falcomonas* Hill

The anterior end of cells rounded, gradually tapering to an acute, pointed, and slightly curved posterior end which gives it a comma to teardrop shape (Fig 15). A furrow extends posteriorly from a vestibulum for approximately half the cell length, and a short sac-like gullet extends beyond the furrow. The periplast appears serrated due to ejectosome chambers located beneath the anterior corners of the raised plates. The inner and outer periplast components consist of hexagonal plates with the outer plates being crystalline and composed of minute subunits. Sometimes heptagonal scales lie on the superficial component plates. The periplastidal compartment contains a single chloroplast with a centrally situated pyrenoid bisected by a periplastidial cytoplasmic tongue, and a nucleomorph located anterior to the pyrenoid. Chloroplasts are blue-green and contain Cr-phycocyanin 569. Common in marine habitats (Hill, 1990).

Genus *Chroomonas* Hansgirg

Cells subovate and lack a furrow. Subequal flagella with typical hairs inserted dorso-ventrally on the right side of the cell. Cells lack a furrow but a tubular gullet extends posteriorly from the vestibulum. The inner and outer components of the periplast consist of offset retangular plates (Hill, 1990) with the anterior edges raised, due to rows of intramembrane particles in the cell membrane adhering tightly to the plates in the posterior end of each plate. Scales or fibrils may be part of the superficial component in some species. The periplastidial compartment contains a single chloroplast with a pyrenoid and a nucleomorph, which is usually located near the pyrenoid. The golgi apparatus and contractile vacuole (in freshwater species) are anterior, while the nucleus is posterior. Cells are blue-green due to the Cr-phycocyanin 645 or 630 in the chloroplasts. Some species have a stigma. Freshwater and marine (Antia et al., 1973; Dodge, 1969; Gantt, 1971; Hill, 1990; Kugrebs and Lee, 1987; Kugrens et al., 1986; Meyer and Pienaar, 1984).

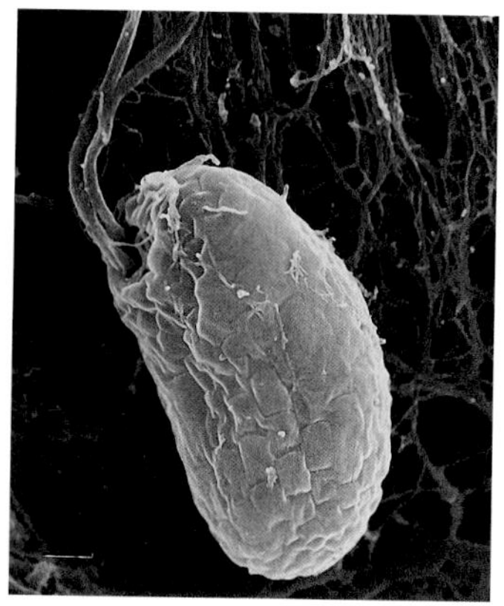

Fig. 16. SEM of *Chroomonas coeulea.* Scale=1 µm. (From Kugrens et al., 1986)

Genus *Komma* Hill, 1990

Cells are comma-shaped, or acuminate, with a rounded anterior end, tapering to a pointed or acutely rounded posterior. A furrow is lacking but a tubular gullet extends posteriorly from the base of the vestibulum. The periplast consists of relatively small internal and surface hexagonal plates, with the surface plates being crystalline in composition and occasionally, rosulate, heptagonal scales lie on the surface of these plates. The periplastidial compartment contains a single blue-green chloroplast with a central pyrenoid lacking traversing thylakoids and projecting from the chloroplast, the chloroplast occupying a dorso-central position of the cell, and a nucleomorph situated at the level of the pyrenoid. The chloroplast contains C-phycocyanin 645. Freshwater (Hill 1990).

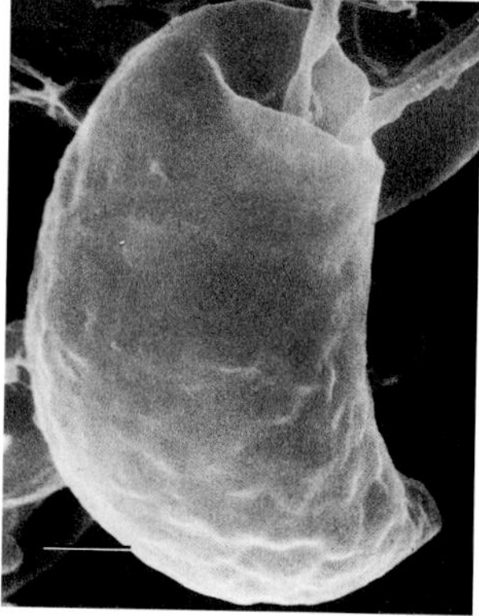

Fig. 17. SEM of *Komma caudata.* Scale=1 µm.

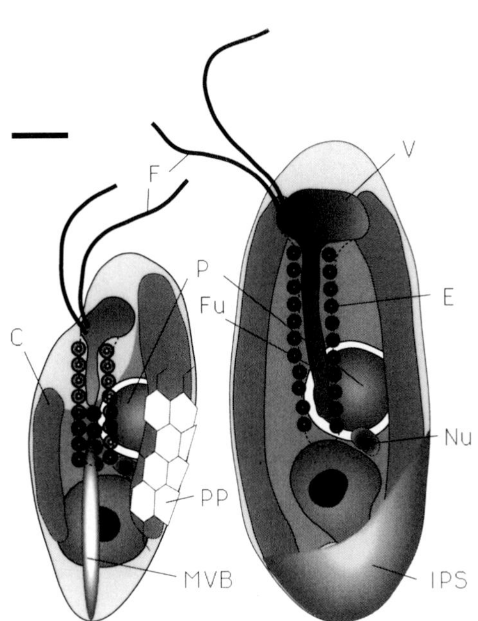

Fig. 18. *Proteomonas sulcata*. Diagram of the haplomorph is on the left side and diplomorph on the right. The haplomorph has hexagonal periplast plates (PP) and the diplomorph has an inner periplast sheet (IPS). (C) Chloroplast; (E) ejectisome; (Fu) furrow; (MVB) midventral band; (Nu) nucleomorph; (P) pyrenoid. Scale=1 μm.

Genus ***Proteomonas*** Hill & Wetherbee, 1986

Cells constantly swimming, occasionally alternating between dissimilar haploid and diploid forms. Haplomorph is slightly compressed with dorsal side convex. The anterior and posterior ends rounded in lateral view. A furrow arises from the posterior of the vestibulum and a sac-like gullet extends posteriorly from the end of the furrow. The periplast consists of inner and outer hexagonal to ovovate plates. The periplastidial compartment contains a single parietal chloroplast (pigment - Cr-phycoerythrin 545) with a pyrenoid, lacking traversing thylakoids, and a nucleomorph in the posterior of the cell positioned between the pyrenoid and nucleus. The diplomorph is larger and circular in transverse section, narrowly elliptic to obovate in lateral view. The vestibulum is situated anteriorly on the ventral surface from which the furrow runs at a slight angle posteriorly for about one third the cell length. A short, sac-like gullet is located posterior to the furrow. The periplast consists of an inner sheet and a surface component of an imbricate layer of heptagonal rosette scales. The periplastidial compartment has a single parietal chloroplast with a pyrenoid not traversed by thylakoids, and a nucleomorph situated between the nucleus and pyrenoid. Chloroplast contains Cr-phycoerythrin 545. Marine (Hill and Wetherbee, 1986).

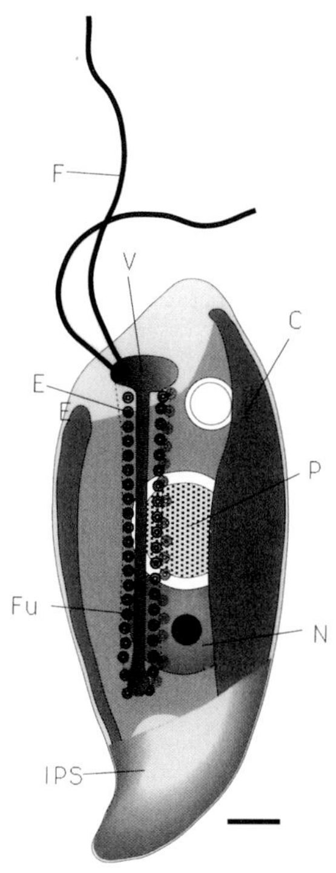

Fig. 19. Diagram of *Teleaulax acuta*. (C) chloroplast; (E) ejectisome; (F) flagellum; (Fu) furrow; (IPS) inner periplast sheet; (N) nucleus; (P) pyrenoid. Scale=1 μm.

Genus ***Teleaulax*** Hill, 1991

Cells with arcuate ends, possessing a long furrow but lacking a gullet. The inner periplast component is a sheet, whereas the surface component consists of an imbricate layer of heptagonal rosulate scales complemented with overlaying and interdispersed fibrils. The periplastidial compartment has a single boat-shaped chloroplast with thylakoids arranged in loose aggregates of three, with a dorsal central pyrenoid, which is not traversed by thylakoids and a nucleomorph. Chloroplasts contain phycoerythrin Cr-545. Marine (Hill, 1991b).

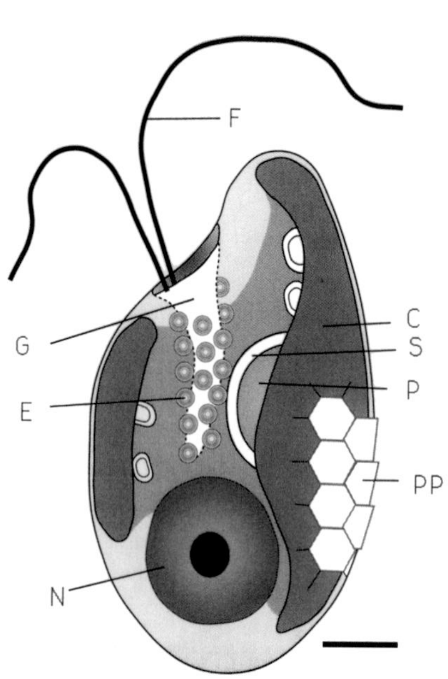

Fig. 20. Diagram of a left-ventral view of *Rhinomonas pauca.* Note the absence of a furrow and the presence of hexagonal periplast plates (PP). (C) chloroplast; (E) ejectisome; (F) flagellum; (G) gullet; (IPS) inner periplastic sheet; (N) nucleus; (P) pyrenoid. Scale=1 µm.

Genus ***Rhinomonas*** Hill & Wetherbee, 1988

Cells with rhinote anterior. Vestibulum located about one third of the cell length from the apex; tubular gullet posterior. Periplast with internal and external hexagonal plates elevated anteriorly to accomodate underlying ejectisomes. Internal plates attached peripherally and posteriorly. The surface plates composed of short rod-like fibrils in a crystalline arrangement with coalescent borders. The periplastidial compartment has a single chloroplast with a single pyrenoid lacking traversing thylakoids and a nucleomorph in a periplastidial cytoplasmic evagination into the pyrenoid. Cells red; chloroplast with Cr-phycoerythrin 545. Marine (Hill and Wetherbee, 1988; Novarino, 1991).

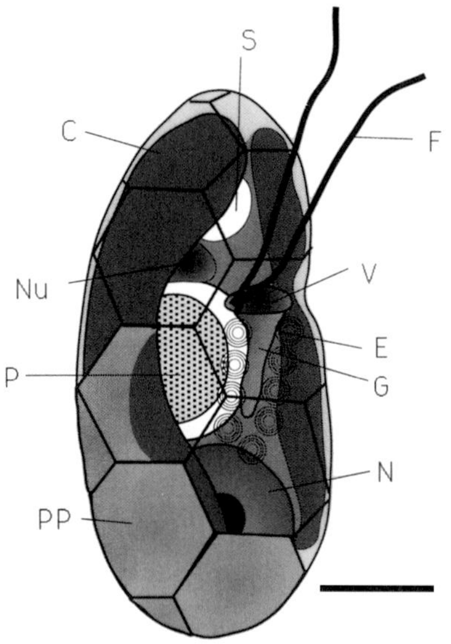

Fig. 21. Diagram of *Hemiselmis brunnescens.* (PP) periplast plates; (C) chloroplast; (E) ejectisome; (F) flagellum; (Fu) furrow; (IPS) inner periplast sheet; (N) nucleus; (Nu) nucleomorph; (P) pyrenoid; (V) vestibulum. Scale=1 µm.

Genus ***Hemiselmis*** Parke

Cells rounded anteriorly and posteriorly, dorso-ventrally flattened, appearing somewhat bean-shaped in lateral view; vestibulum and gullet located approximately one third of the cell length from the anterior, a furrow is absent; inner and surface periplast components comprised of hexagonal plates. The periplastidial compartment contains a single dorsal, boat-shaped chloroplast with a centrally situated, stalked pyrenoid traversed by one thylakoid and a nucleomorph located anterior to the pyrenoid. Salmon pink to orange in color (Cr-phycoerythrin 555). Marine (Santore, 1977, 1982b; Kugrens and Lee 1987; Pennick, 1981, 1982; Wetherbee et al., 1986).

Genus ***Guillardia*** Hill & Wetherbee, 1990

Cells actively motile; slightly dorso-ventrally compressed, ovate in lateral view with both anterior and posterior rounded ends, elliptic in ventral view; a furrow is lacking, a tubular gullet extends posteriorly from the vestibulum. Periplast consists of an inner sheet and a surface component of large, longitudinally oriented and irregular, crystalline plates. The periplastidial compartment contains a single chloroplast with a pyrenoid not traversed by thylakoids and a nucleomorph situated anterior to the pyrenoid. Chloroplast contains Cr-phycoerythrin 545. Marine (Hill and Wetherbee, 1990).

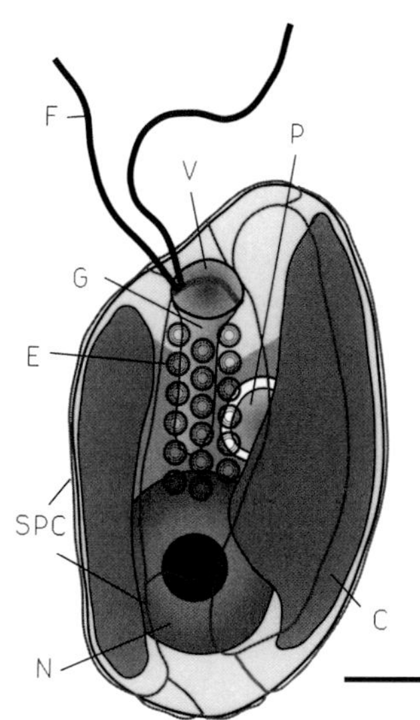

Fig. 22. Diagram of *Guillardia theta*. Note the irregular surface components of the periplast (SPC) and the absence of a furrow. (C) chloroplast; (E) ejectisome; (F) flagellum; (Fu) furrow; (G) golgi; (N) nucleus; (P) pyrenoid; (V) vestibulum. Scale=1 μm.

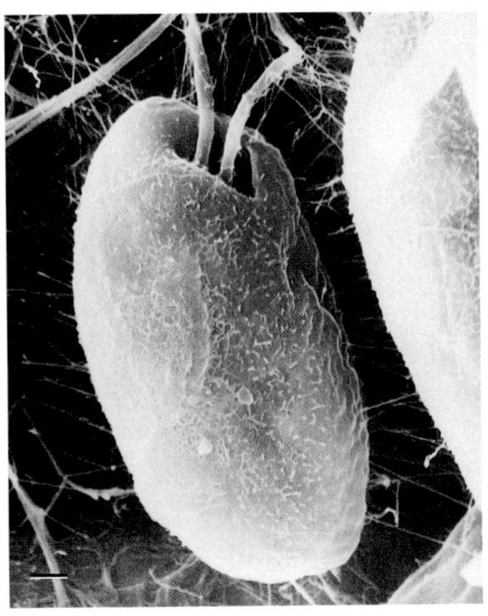

Fig. 23. SEM of *Storeaulata hanrattyi*. Note absence of a furrow. Scale=1 μm.

Genus ***Storeatula*** Hill, 1991

Cells are ellipsoid with a slightly pointed anterior; a furrow is lacking, the tubular gullet subtending the posterior margin of the vestibulum, extending to approximately the middle of the cell and lined with four rows of ejectisomes; the periplast has an inner sheet and an outer component of a thick mat of coarse fibrils. The periplastidial compartment contains a single, lobed chloroplast, a pyrenoid and a nucleomorph located in an anterior groove or depression in the pyrenoid. Chloroplast contains the biliprotein Cr-phycoerthrin. Marine (Hill, 1991b).

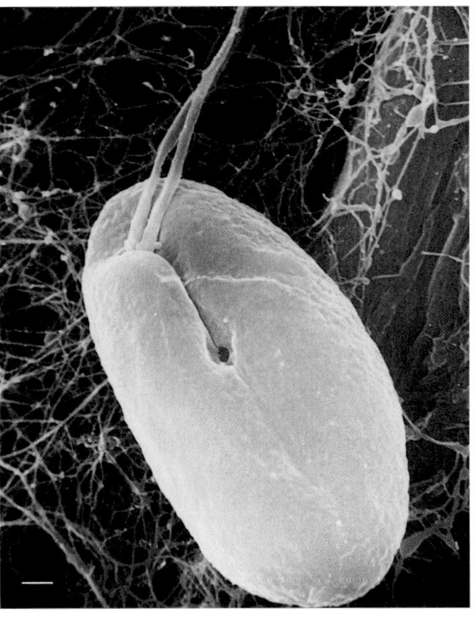

Fig. 24. SEM of *Cryptomonas tetrapyrenoidsa* showing the complex furrow of this genus. Scale=1 μm. (From Kugrens et al., 1986)

Genus ***Cryptomonas*** Ehrenberg

Cells often forming mucilaginous pallmelloid colonies. Motile cells possess 2 flagella, with typical tubular flagellar hairs, that originate from the right side of the vestibulum. Complex type of vestibular/furrow/gullet (Fig. 4) with furrow consisting of furrow ridges, folds, and a persistent opening called the stoma located near the posterior one third of the furrow. Periplast consisting of an inner component of round to oval-shaped plates and a surface component of a thin layer of fibrils. The periplastidial compartment has 2 chloroplasts with 2 pyrenoids, not transversed by thylakoids, and 2 nucleomorphs. Chloroplasts possess CR-phycoerythrin 566. Freshwater only (Brett et al., 1986; Hill, 1991b; Kugrens and Lee, 1987;

Kugrens et al., 1986; Munawar and Bistricki, 1979; Roberts, 1984; Santore, 1977, 1984).

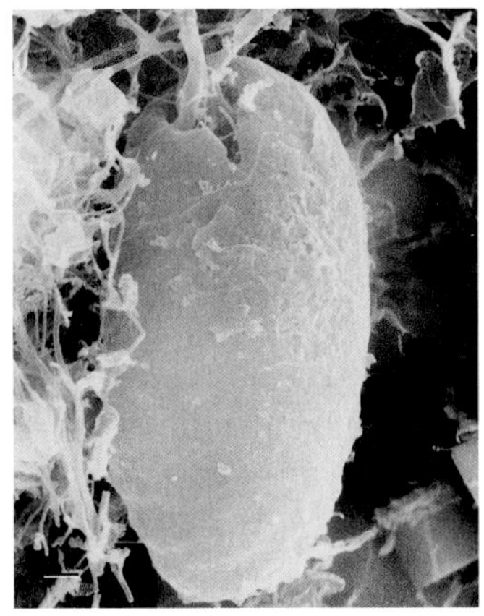

Fig. 25. SEM of *Rhodomonas ovalis.* Scale=1 µm.

Genus ***Rhodomonas*** Karsten (=***Pyrenomonas*** Santore)

Cells with a variable furrow and a short tubular gullet. Periplast consisting of inner, more or less square plates which taper slightly toward the anterior and a surface component of intertwining fibrils. The periplastidial compartment usually has a single, bilobed chloroplast with a pyrenoid situated between the 2 lobes of the chloroplast. Thylakoids do not traverse the pyrenoid. Nucleomorph located in an evagination of the periplastidal cytoplasm into the pyrenoid. Cells not necessarily colored red; chloroplasts contain Cr-phyco-erythrin 545. Marine and freshwater (Erata and Chihara, 1989; Novarino, 1993; and Santore, 1984).

Genus ***Geminigera*** (Taylor & Lee) Hill, 1991

Cells are laterally compressed; a medium length furrow extends posteriorly from the vestibulum and a short sac-like gullet is posterior to the furrow; the periplast consists of an inner sheet and a surface component of imbricate heptagonal scales and a fibrillar layer. The periplastidial compartment contains a single, somewhat lobed, peripheral chloroplast with 2 stalked kidney-shaped pyrenoids not traversed by thylakoids, with the thylakoids in the chloroplast arranged singly or stacked in variable numbers, and a nucleomorph located in a tongue of periplastidial cytoplasm in an ivagination of the nucleus. Chloroplast contains biliprotein Cr-phycoerythrin 545; reported only from marine habitats (Hill, 1991).

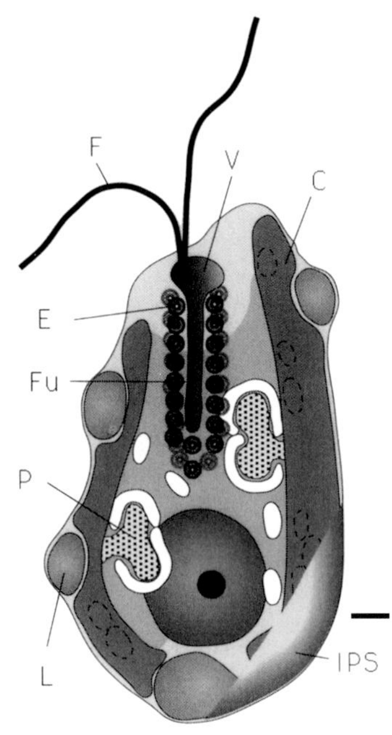

Fig. 26. Diagram of *Geminigera cryophila* showing characteristic lipid accumulation (L) and a portion of the inner periplast sheet (IPS). (C) chloroplast; (E) ejectosome; (F) flagellum; (Fu) furrow; (N) nucleus; (P) pyrenoid; (V) vestibulum. Scale=1 µm.

Genus ***Campylomonas*** Hill, 1991

Cells with a characteristic recurved posterior, imparting a sigmoid shape in lateral view. A keeled or winged rhizostyle extends the length of the cell acting as a cytoskeleton to maintain the cell shape. A simple furrow extends posteriorly from the vestibulum to a sac-like gullet. The periplast consists only of an inner sheet which is not associated with the cell membrane; a surface peri-plast component is lacking. The periplastidal compartment contains 2 chloroplasts, each with its own pyrenoid not transversed by thylakoids, and 2 nucleomorphs located near each pyrenoid. Chloroplasts possess Cr-phycoerythrin 566. Freshwater (Hill, 1991b; Klaveness, 1985; Kugrens and Lee, 1987; Kugrens et al., 1986; Munawar and Bistricki, 1979).

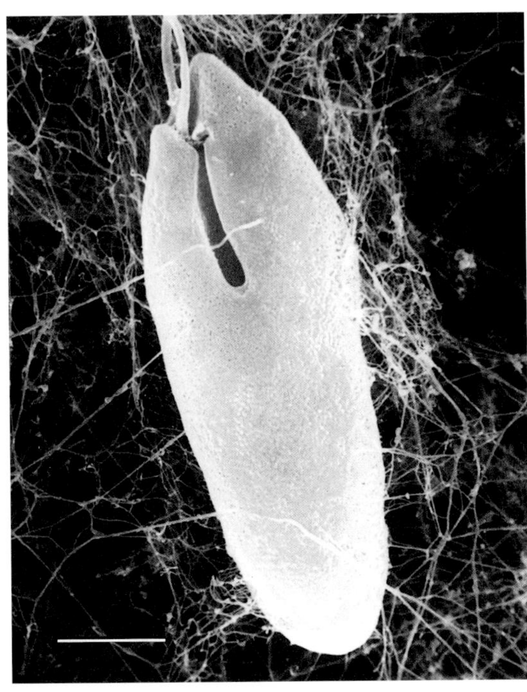

Fig. 27. SEM of *Campylomonas rostratiformis*. Scale=1 µm. (From Kugrens et al., 1986)

LITERATURE CITED

Andersen, R. A. 1992. Diversity of eukaryotic algae. *Biodiversity Conserv.*, 1:267-292.

Anderson, E. 1962. A cytological study of *Chilomonas paramaecium* with particular reference to the so-called trichocysts. *J. Protozool.*, 9:380-395.

Antia, N. J., Kalley, J. P., McDonald, T. & Bisalputra, T. 1973. Ultrastructure of the marine cryptomonad *Chroomonas salina* cultured under conditions of photoautotrophy and glyceroheterotrophy. *J. Protozool.*, 20:377-385.

Bourrelly, P. 1970. *Les Algues d'Eau Douce. Tome III: Les Algues Bleues et Rouges. Les Eugleniens, Peridiniens et Cryptomonadines.* Editions N. Boubee & Cie, Paris. 512 pp

Brett, S. J. & Wetherbee, R. 1986. A comparative study of periplast structure in *Cryptomonas cryophila* and *C. ovata* (Cryptophyceae). *Protoplasma*, 131:23-31.

Butcher, R. W. 1967. *An Introductory Account of the Smaller Algae of British Coastal Waters IV. Cryptophyceae.* Fishery Invest., Ser. 4, London, pp 1-54

Dodge, J. D. 1969. The ultrastructure of *Chroomonas mesostigmatica* Butcher (Cryptophyceae). *Arch. Mikrobiol.*, 69:266-280.

Douglas, S. E., Murphy, C. A., Spencer, D. F. & Gray, M. W. 1991. Cryptomonad algae are evolutionary chimeras of two phylogenetically distinct unicellular eukaryotes. *Nature*, 350:148-151.

Dwarte, D. & Vesk, M. 1982. Freeze-fracture thylakoid ultrastructure of representative members of chlorophyll *c* algae. *Micron*, 13:325-326.

Dwarte, D. & Vesk, M. 1983. A freeze-fracture study of cryptomonad thylakoids. *Protoplasma*, 117:130-141.

Erata, M. & Chihara, M. 1989. Re-examination of *Pyrenomonas* and *Rhodomonas* (Class Cryptophyceae) through ultrastructural survey of red pigmented cryptomonads. *Bot. Mag. Tokyo*, 102:429-442.

Faust, M. A. 1974. Structure of the periplast of *Cryptomonas ovata* var. *palustris*. *J. Phycol.*, 10:121-124.

Gantt, E. R. 1971. Micromorphology of the periplast of *Chroomonas* sp. (Cryptophyceae). *J. Phycol.*, 7:177-184.

Gantt, E. 1979. Phycobiliproteins of Cryptophyceae. *In*: Levandowsky, M. & Hutner, S. H. (eds.), *Biochemistry and Physiology of the Protozoa*, Vol. 1. Academic Press, New York and London. pp 121-138

Gantt, E. R. 1980. Photosynthetic cryptophytes. *In*: *Phytoflagellates, Developments in Marine Biology,* Vol. 2. Elsevier, North Holland Amsterdam. pp 381-405

Gantt, E., Edwards, M. R. & Provasoli, L. 1971. Chloroplast structure of the Cryptophyceae, evidence for phycobiliproteins within intrathylakoidal spaces. *J. Cell Biol.*, 48:280-290.

Gillot, M. 1990. Phylum Cryptophyta (Cryptomonads). *In*: Margulis, L., Corliss, J. O., Melkonian, M. & Chapman, D. J. (eds.), *Handbook of Protoctista*. Jones & Bartlett Publishers, Boston. pp 139-151

Gillot, M. A. & Gibbs, S. P. 1980. The cryptomonad nucleomorph: its structure and evolutionary significance. *J. Phycol.,* 16:558-568.

Gillott, M. A & Gibbs, S. P. 1982. Comparison of the flagellar rootlets and periplast in two marine cryptomonads. *Can. J. Bot.*, 61:1964-1980.

Grain, J., Mignot, J. P. & Puytorac, P. 1988. Ultrastructures and evolutionary modalities of flagellar and ciliary systems in protists. *Biol. Cell.,* 63:219-237.

Grim, J. N. & Staehelin, L. A. 1984. The ejectisomes of the flagellate *Chilomonas paramecium*:

visualization by freeze-fracture and isolation techniques. *J. Protozool.,* 3:259-267.

Heywood, P. 1988. Ultrastructure of *Chilomonas paramecium* and the phylogeny of cryptoprotists. *Biosystems,* 21:293-298.

Hibberd, D. J. 1977. Observations on the ultra-structure of the cryptomonad endosymbiont of the red-water ciliate *Mesodinium rubrum. J. Mar Biol. Assoc. U. K.,* 57:45-61.

Hibberd, D. J., Greenwood, A. D. & Griffiths, H. B. 1971. Observations on the ultrastructure and flagella and periplast in the Cryptophyceae. *Br. Phycol. J.,* 6:61-72.

Hill, D. R. A. 1990. *Chroomonas* and other blue-green cryptomonads. *J. Phycol.,* 26:133-145.

Hill, D. R. A. 1991a. Diversity of heterotrophic cryptomonads. *In*: Patterson, D.J. & Larsen, J , (eds.), *The Biology of Free-Living Heterotrophic Flagellates.* Systematics Association Special Volume 45. pp 235-240

Hill, D. R. A. 1991b. A revised circumscription of *Cryptomonas* (Cryptophyceae) based on examination of Australian strains. *Phycologia,* 30:170-188.

Hill, D. R. A. & Rowan, K. S. 1989. Biliproteins of the Cryptophyceae. *Phycologia*, 28:455-463.

Hill, D. R. A. & Wetherbee, R. 1986. *Proteomonas sulcata* gen. et sp. nov. (Cryptophyceae) a crypto-monad with two morphologically distinct and alter-nating forms. *Phycologia,* 27:521-543.

Hill, D. R. A. & Wetherbee, R. 1988. The structure and taxonmy of *Rhinomonas pauca* gen. et sp. nov. (Cryptophyceae). *Phycologia,* 27:355-365.

Hill, D. R. A. & Wetherbee, R. 1989. A reappraisal of the genus *Rhodomonas* (Cryptophyceae). *Phycologia*, 28:143-158.

Hill, D. R. A. & Wetherbee, R. 1990. *Guillardia theta* gen. et sp. nov. (Cryptophyceae). *Can. J. Bot.*, 68:1873-1876

Huber-Pestalozzi, G. 1950. Das Phytoplankton das Süsswaters, Tiel 3., Cryptophyceen, Chloromonadinen, Peridineen. *In*: Thienenemann, A. (ed.), Stuttgart, pp2-78.

Klaveness, D. 1981. *Rhodomonas lacustris* (Pascher & Ruttner) Javornicky (Cryptomonadida): ultra structure of the vegetative cell. *J. Protozool.,* 28:83-90.

Klaveness, D. 1985. Classical and modern criteria for determining species of Cryptophyceae. *Bull. Plankton Soc. Jap.*, 32:111-128.

Klaveness, D. 1988. Ecology of the Cryptomonadida: A first review. *In*: Sandgren, C, D. (ed.), *Growth and Reproductive Strategies of Freshwater Phyto-plankton.* Cambridge Univ. Press, New York. pp 105-133

Kugrens, P. & Lee, R. E. 1987. An ultrastructural survey of cryptomonad periplasts using quick-freezing freeze-fracture techniques. *J. Phycol.*, 23:365-376.

Kugrens, P. & Lee, R. E. 1988. Ultrastructure of fertilization in a cryptomonad. *J. Phycol.*, 24:510-518.

Kugrens, P. & Lee, R. E. 1990. Ultrastructural evidence for bacterial incorporation and mixotrophy in the photosynthetic cryptomonad *Chroomonas pochmanni* Huber-Pestalozzi (Cryptomonadida). *J. Protozool.*, 37:263-267.

Kugrens, P. & Lee, R. E. 1991. Organization of cryptomonads. *In*: Patterson, D. J. and Larsen, J. (eds.), *The Biology of Free-Living Heterotrophic Flagellates.* The Systematics Association, Special Volume 45. pp 219-233

Kugrens, P., Lee, R. E. & Andersen, R. E. 1986. Cell form and surface patterns in *Chroomonas* and *Cryptomonas* cells (Cryptophyta) as revealed by scanning electron microscopy. *J. Phycol.*, 22:512-522.

Kugrens, P., Lee, R. E. & Andersen, R. E. 1987. Ultrastructural variations in cryptomonad flagella. *J. Phycol.*, 23:511-518.

Lee, R. E. & Kugrens, P. 1986. The occurrence and structure of flagellar scales in some freshwater cryptophytes. *J. Phycol.*, 22:549-552.

Lucas, I. A. N. 1970a. Observations on the ultrastructure of representatives of the genera Hemiselmis and *Chroomonas* (Cryptophyceae). *Br. Phycol. J.*, 5:29-37.

Lucas, I. A. N. 1970b. Observation on the fine structure of the Cryptophyceae. I. The genus *Cryptomonas. J. Phycol.*, 6:30-38.

Lucas, I. A. N. 1982. Observation on the fine structure of the Cryptophyceae. II. The eyespot. *Br. Phycol. J.*, 17:113-119.

Ludwig, M. & Gibbs, S. P. 1985. DNA is present in the nucleomorph of cryptomonads: further evidence that the chloroplast evolved from a eukaryotic endosymbiont. *Protoplasma*, 127:9-20.

Ludwig, M. & Gibbs, S. P. 1989. Localization of phycoerythrin at the luminal surface of the thylakoid membrane in *Rhodomonas lens. J. Cell Biol.*, 108:875-884.

McKerracher, L. & Gibbs, S. P. 1982. Cell and nucleomorph division in the alga *Cryptomonas. Can J. Bot.*, 60:2440-2452.

Meyer, S. R. & Pienaar, R. N. 1984. The microanatomy of *Chroomonas africana* sp. nov. (Cryptophyceae). *S. Afr. J. Bot.*, 3:306-319.

Minot, J. P. 1965. Etude ultrastructurale de *Cyathomonas truncata* From (flagelle Cryptomonadine). *J. Microsc.*, 4:239-252.

Morrall, S. & Greenwood, A. D. 1980. A comparison of the periodic substructure of the trichocysts of the Cryptophyceae and Prasinophyceae. *BioSystems*, 12:71-83.

Munawar, M. & Bistricki, T. 1979. Scanning electron microscopy of some nanoplankton cryptomonads. *Scanning Electron Microsc.*, 3:247-252.

Novarino, G. & Lucas, I. A. N. 1993. Some proposals for a new classification system of the Cryptophyceae. *Bot. J. Lin. Soc., London*, 111:3-21.

Pennick, D. L. 1981. Flagellar scales in *Hemiselmis brunnescens* Butcher and *H. virescens* Droop (Cryptophyceae). *Arch. Protistenkd.*, 124:267-270.

Pennick, D. L. 1982. Observations on the fine structure of *Hemiselmis brunnescens* Butcher. *Arch. Protistenkd.*, 126:241-245.

Roberts, K. R. 1984. Structure and significance of the cryptomonad flagellar apparatus. I. *Cryptomonas ovata* (Cryptophyta). *J. Phycol.*, 20:159-167.

Roberts, K. R., Stewart, K. D. & Mattox, K. R. 1981. The flagellar apparatus of *Chilomonas paramecium* (Cryptophyceae) and its comparison with certain zooflagellates. *J. Phycol.*, 17:159-167.

Santore, U. J. 1977. Scanning electron microscopy and comparative micromorphology of the periplast of *Hemiselmis rufescens*, *Chroomonas* sp., *Chroomonas salina* and members of the genus *Cryptomonas* (Cryptophyceae). *Br. Phycol. J.*, 12:255-270.

Santore, U. J. 1982a. Comparative ultrastructure of two members of the Cryptophyceae assigned to the genus *Chroomonas* - with comments on their taxonomy. *Archiv. Protistenkd.*, 125:5-29.

Santore, U. J. 1982b. The ultrastructure of *Hemiselmis brunnescens* and *Hemiselmas virescens* with additional observations on *Hemiselmas rufescens* and comments about the Hemiselmidaceae as a natural group of the Cryptophyceae. *Br. Phycol. J.*, 17:81-89.

Santore, U. J. 1982c. The distribution of the nucleomorph in the Cryptophyceae. *Cell Biol. Int. Reports*, 6:1055-1063.

Santore, U. J. 1984. Some aspects of taxonomy in the Cryptophyceae. *New. Phytol.*, 98:627-646.

Santore, U. J. 1987. A cytological survey of the genus *Chroomonas* - with comments on the taxonomy of this natural group of the Cryptophyceae. *Arch. Protistenkd.*, 134:83-114.

Santore, U. J. & Greenwood, A. D. 1977. The mitochondrial complex in Cryptophyceae. *Arch. Microbiol.*, 112:207-218.

Schuster, F. L. 1968. The gullet and trichocysts of *Cyathomonas truncata*. *Exper. Cell Res.*, 49:277-284.

Schuster, F. L. 1970. The trichocysts of *Chilomonas paramecium*. *J. Protozool.*, 17:521-526.

Sespenwol, S. 1973. Leucoplast of the cryptomonad *Chilomonas paramecium*; evidence for the presence of a true plastid in a colorless flagellate. *Exp. Cell Res.*, 76:395-409.

Skuja, H. 1939. Beitrag zur Algenflora Lettlands. II. *Acta. Horti. Bot. Univ. Latviensis.*, 11/12:41-168.

Skuja, J. 1948. Taxonomie des Phytoplanktons einiger Seen in Uppland Schweden. *Symb. Bot. Upsaliensis*, 9:1-399.

ORDER DIPLOMONADIDA

by

GUY BRUGEROLLE and JOHN J. LEE

Diplomonads are small (less than 30 μm) unicellular heterotrophic flagellates which are free-living or parasitic in animals (Fig. 1). They are anaerobic or microaerophilic, and the cell has no mitochondria and although there may be Golgi function, there is no Golgi apparatus (dictyosome).

The cell basically has a **flagellar (mastigont) system** with four basal bodies and flagella closely associated with one nucleus forming one **karyomastigon**t in "**monozoic forms**" (Enteromonadinae) or two karyomastigonts in "**diplozoic forms**" (Diplomonadinae) (Fig. 1). In each karyomastigont one flagellum is always directed backward and is generally associated with a ventral **cytostome**. The four basal bodies are arranged in two pairs and one basal body is generally located near a cup-like depression on the nucleus (Fig. 2a). Three microtubular fibers originating from the basal body periphery have been distinguished: the **supra-nuclear fiber** above the nucleus, the **infra-nuclear fiber** under the nucleus and the **cytostomal fiber** along the recurrent flagellum. In genera forming the Diplomonadinae such as *Trepomonas, Hexamita, Spironucleus, Octomitus, Giardia*, the cell has two karyomastigonts arranged in an axial binary symmetry (Fig. 2b). In these "diplozoic forms" the cell is like a double organism with two nuclei, two sets of four flagella, two cytostomal apparatuses. They reproduce by longitudinal binary fission and the mitosis is of the semi-opened type, the nuclear envelope persists and the two hemi-spindles are extranuclear with kinetochore microtubules penetrating into the nucleus by pores (Brugerolle, 1974). They form resistant cysts which ensure the transmission in parasitic species. General description could be found in Grassé (1952) for light microscopy, Kulda and Nohynkovà (1978) for parasitic species of vertebrates and in Vickerman (1990) and Brugerolle (1991a) for electron microscopy.

The free-living species are found in water sites rich in organic matter and deficient in oxygen such as sediments, stagnant reservoirs, marshes, water treatment plants, and also in brackish or salt water. They swim actively and feed on bacteria and also on dead cells of other protists, plants, and animals which are engulfed by the cytostomal apparatus. Information can be found in Klebs (1892), Lemmerman (1910), Klug (1936), Grassé (1952), Skuja (1956), Hänel (1979), Brugerolle (1991b), Mylnikov (1985), 1991.

The parasitic species are mostly found in vertebrate intestine and cloaca, but they can invade other organs and sometimes the blood. Some species live in invertebrates such as molluscs or insects. Most of them are endocommensal; they generally feed on bacteria and on food digested by the host. Some are pathogenic such as *Hexamita nelsoni* in oysters (Schlicht and Mackin, 1968), *Hexamita salmonis* in salmonid fishes (Becker, 1977), *Spironucleus muris* for mice, *S. meleagridis* in turkeys, or *Giardia intestinalis* in man. Species parasitic in vertebrates are dealt with by Kulda and Nohynkovà 1978.

Two suborders: the Enteromonadina with one karyomastigont and the Diplomonadina with two karyomastigonts arranged in axial binary symmetry (Fig. 1).

Key Characters

1. Cells with one or two karyomastigonts, each with 1-4 flagella including a recurrent flagellum.
2. Cells with two karyomastigonts ("diplozoic forms") organized in axial binary symmetry.
3. Cup-like depression on the nucleus surface in front of one basal body and three basal body-associated microtubular fibers: supra-nuclear, infra-nuclear, and cytostomal or recurrent fiber.
4. No Golgi, mitochondria, hydrogenosomes, or axostyle.

KEY TO SUBORDERS, FAMILIES, SUBFAMILIES AND GENERA

1. Monozoic organism (with a single karyomastigont).................... Suborder Enteromonadina ....Family **Enteromonadidae**.................2

1'. Diplozoic organisms (with two karyomastigonts)................. Suborder Diplomonadina.... ....Family **Hexamitidae**...............4

2. With four flagella:................... ***Enteromonas***

2'. With three or less flagella............................. 3

3. With three flagella (one recurrent and two anterior flagella).................................. ***Trimitus***

3'. With one flagellum...................... ***Caviomonas***

4. With two posterior oral grooves (cytostomes) 5

4'. With two posterior cytostomal canals 6

4." No posterior oral groove or cytostomal pocket 7

5. Oral grooves containing three recurrent flagella .... ***Trepomonas***

5'. Oral grooves with only two flagella............. .... ***Trigonomonas***

5". Oral grooves with only one flagellum......... .... ***Gyromonas***

6. Flagella inserted on the external side of the nuclei, endoplasmic reticulum associated with the cytostomal canals.................... ***Hexamita***

6'. Flagella inserted at the apex, S-shaped nuclei, no endoplasmic reticulum associated with the canals.. ***Spironucleus***

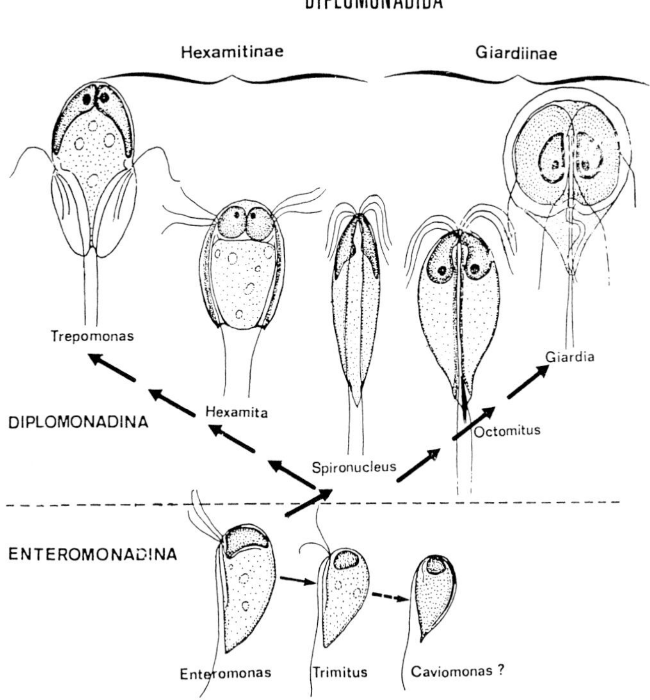

Fig. 1. Major types, systematics, and possible evolutionary relationships of diplomonads (from Brugerolle, 1999)

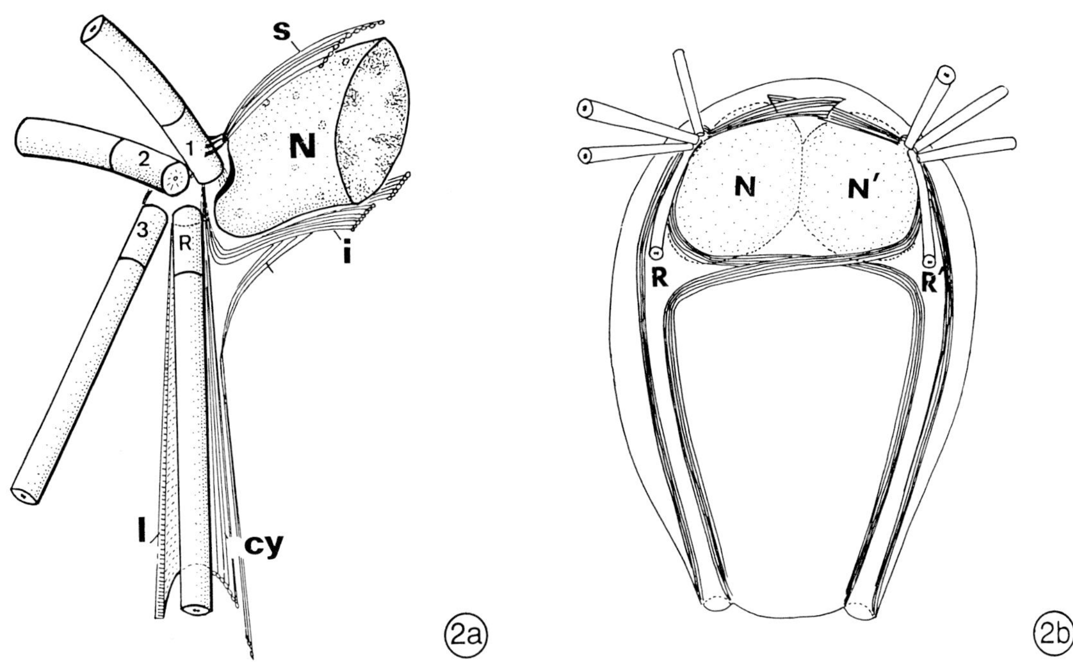

Fig. 2. General organization of diplomonads. Fig. 2a. Arrangement of the four basal bodies in two pairs close to one pole of the nucleus which presents a cup-like depression. The basal bodies are associated with three microtubular fibers: supranuclear (s), infranuclear (i), and cytostomal fiber (cy); a dense lamina (l) reinforces the cytostomal membrane. Fig. 2b. Organization of the two sets of basal bodies and associated fibers in relation to the nuclei in "diplozoic" forms of diplomonads such as *Hexamita,* where the two recurrent flagella (R,R') traverse the cell in two canals opened posteriorly. (from G. Brugerolle, 1991a)

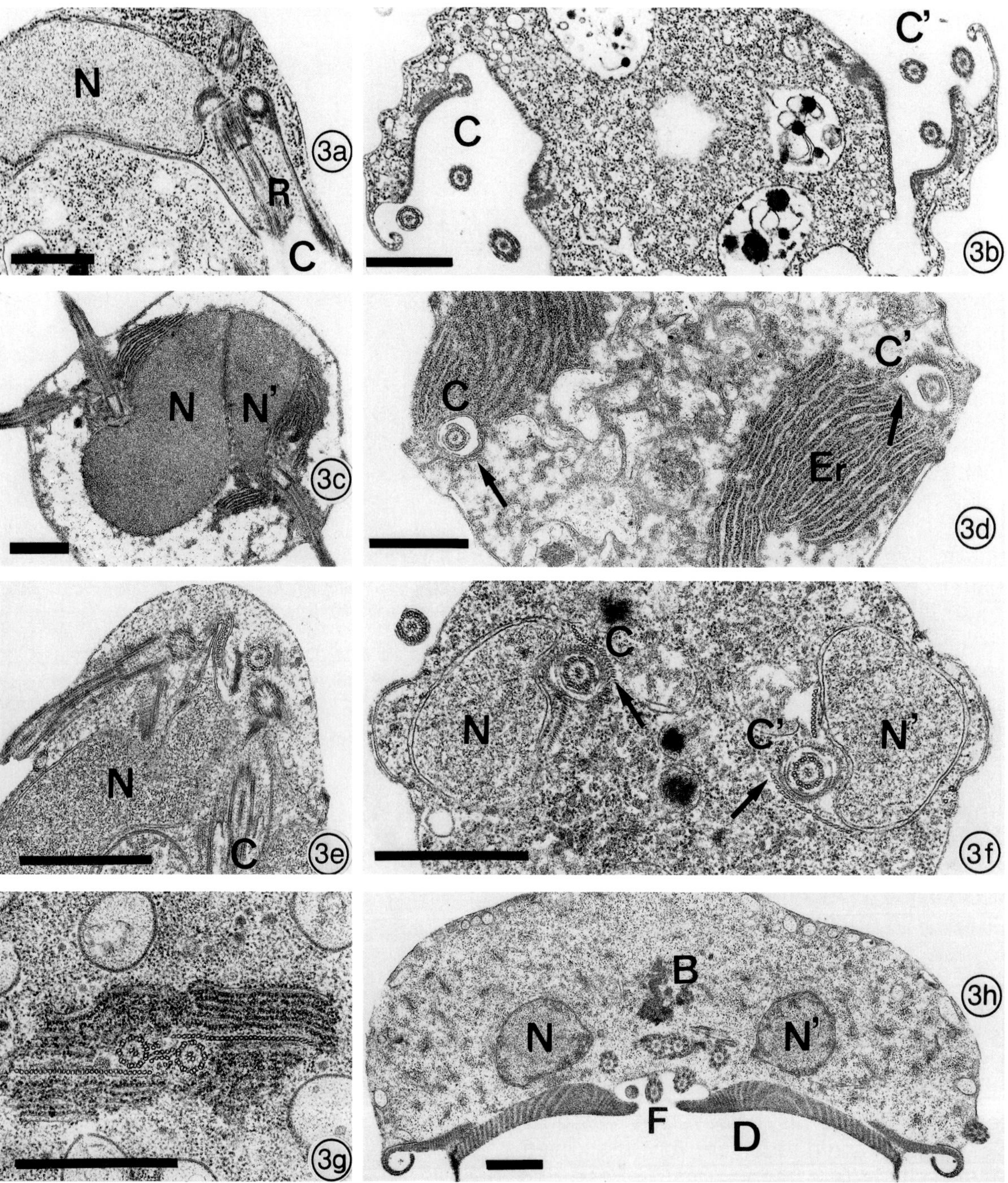

Fig. 3. Ultrastructutral features of diplomonad genera. Fig. 3a. *Enteromonas*: basal bodies at one pole of the nucleus (N), base of the cytostomal ridge (C) with the recurrent flagellum (R). Fig. 3b. *Trepomonas*: cross section of the two opposed cytostomal pockets (C-C') containing three flagella. Fig. 3c,d. *Hexamita*: insertion of flagella on the external face of the nuclei (N,N') and recurrent flagella into canals (C,C') lined by microtubular fibers (arrow) and endoplasmic reticulum (Er). Fig. 3e,f. *Spironucleus*: flagella inserted near the anterior knob of the S-shaped nucleus (N). Recurrent flagella into cytostomal canals (C,C') between the two nuclei (N,N'). Fig. 3g. *Octomitus*: central axis with two recurrent axonemes, microtubular fibers, and endoplasmic reticulum. Fig. 3h. *Giardia*: cross section of the ventral disk (D) axoneme, and flagella (F) between the nuclei (N,N') and median body (B). Bars=1 μm. (from G. Brugerolle)

7. Central axis formed by two recurrent axonemes, microtubular fibers and endoplasmic reticulum: .... ***Octomitus***
7'. With a ventral disk:........................... ***Giardia***

FAMILY ENTEROMONADIDAE

Genera of this family have one karyomastigont; they are monozoic forms.

Genus ***Enteromonas*** da Fonseca

Pyriform cell of small size (3-8 µm) with an anterior nucleus transversely positioned and 4 flagella inserted on one side of the nucleus (Fig. 4a). Convex dorsal face and flattened ventral face marked by a ridge. Three flagella are anteriorly directed; the fourth is recurrent and associated with the ventral cytostomal ridge. This ridge does not extend to the posterior end, and the recurrent flagellum continues freely backward (Nie, 1950). Electron microscopic study has shown the diplomonad characters of *Enteromonas*. The vental narrow cytostomal ridge is supported by 2 microtubular fibers and one lip is more prominent than the other (Fig. 3a). Phagocytosis of bacteria occurs on the bottom of this ridge. Incompletely divided cells with 3 nuclei, each associated with a set of flagella have been observed. Resistant cysts are the transmission forms. Among the 8 species described, *Enteromonas hominis* is found in human intestine, and also in monkeys (Dobell, 1935) and rodents such as rabbits and probably other animals. *Enteromonas caviae* from guinea-pigs has been correctly described after hematoxylin staining by Nie (1950). The list of described species is given in Kulda and Nohynkovà (1978); electron microscopy by Brugerolle (1975a).

Genus ***Trimitus*** Alexeieff

Very small flagellate (3-8 µm) with similar characters to *Enteromonas*; however, it has only 2 anterior flagella and a recurrent one associated with the ventral ridge extending to the posterior extremity (Fig. 4b). The 4 species described live in the intestine of fishes (Alexeieff, 1910), frogs, tortoises, insects (Pérez-Reyès, 1964), and snake. Electron microscopic study of flagellates assigned to the genus *Trimitus* (Brugerolle, 1986) has shown they have the characters of Enteromonadidae and differ from the tiny *Tricercomitus* which are trichomonads. The recurrent flagellum is adherent to the ventral ridge, which shows ultrastructural differences with that of *Enteromonas*. However, 4 basal bodies and flagella are present suggesting either a confusion with *Enteromonas* species or an erroneous light microscopic description of the genus *Trimitus*.

Genus ***Caviomonas*** Nie

Very small cell (2-6µm) which has a single free flagellum inserted on one side of the anterior nucleus (Fig. 4c). A short ventral ridge possibly homologous to the cytostomal ridge of *Enteromonas* is observed. The cell has also a thick dorsal fiber or peristyle which extends to the posterior end. No ultrastructural study to ascertain the diplomonad characters of this genus (Kulda & Nohynkovà 1978). *C. mobilis* of the coecum of the guinea-pig (Nie 1950) is the only described species.

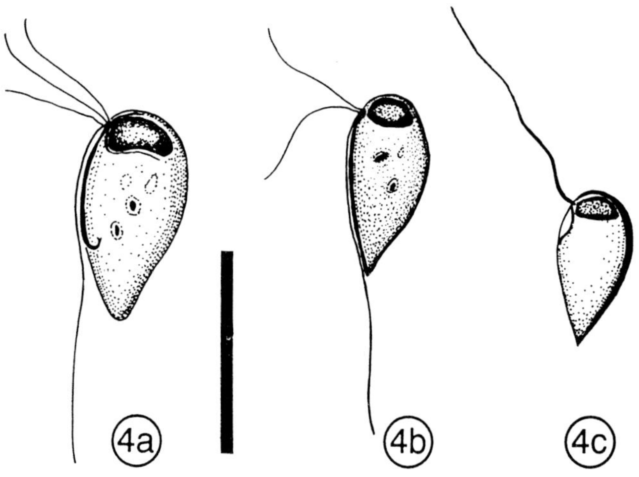

Fig. 4a. *Enteromonas intestinalis* from rabbits (after Brugerolle, 1975). Fig. 4b. *Trimitus ranae* from amphibians (after Péres-Reyès, 1964). Fig. 4c. *Caviomonas mobilis* from guinea-pigs (after Nie, 1950). Bar=10 µm.

FAMILY HEXAMITIDAE

Genera of this family have two karyomastigonts arranged in a binary axial symmetry (diplozoic forms).

SUBFAMILY HEXAMITINAE

Genera of this subfamily have two opened and functional cytostomes.

Genus ***Trepomonas*** Dujardin

Trepomonas are ovoid, bilaterally compressed cells (5-30 µm) with 2 lateral locomotory flagella, 2 anterior nuclei, and 2 posterior grooves largely opened on each opposed side of the cell (Fig. 5). The anterior part of the cell contains the 2 crescent-shaped nuclei, which abut on top and possess anterior nucleoli. Each set of 4 flagella are inserted on each side of the cell body near the equator at the base of each nucleus. The posterior half of the cell is grooved by 2 widely open cytostomal depressions or pockets, each containing 3 recurrent flagella, 1 of them being longer than others. The 2 locomotory flagella beat on each side and propel the cell, which generally moves by step-wise rotation. The bottom of the cytostomal pockets are the sites of the ingestion of food, mostly bacteria. One contractile vacuole forms in the middle part and discharges at the posterior end. Electron microscopic studies have shown the diplomonad characters: arrangement of the basal bodies and of the associated microtubular fibers. A transverse section (Fig. 3b) shows the cytostomal pockets opened on the flattened opposite faces of the cell, demonstrating the axial binary symmetry. Two cytostomal microtubular ribbons, 2 striated fibers, and a striated lamina border the faces of the cytostomal pockets. Cysts have been described. The 6 species described are free-living, but *T. agilis* which is a mesosaprobic species also occurs in intestine of amphibians, a marine fish, and tortoises. Light microscopic studies have been published by Klebs (1892), Lemmerman (1910), Grassé (1952), Skuja (1956), Calaway and Lackey (1962), Mylnikov (1991); electron microscopy in Brugerolle et al. (1973); Eyden and Vickerman (1975).

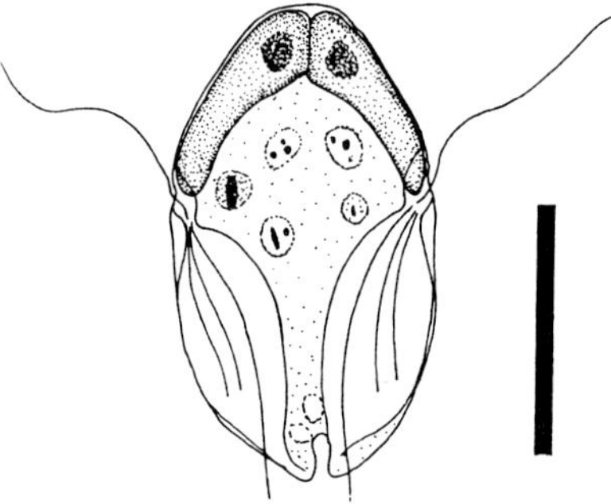

Fig. 5. *Trepomonas agilis* var. *communis* (after Brugerolle et al., 1973), with two anterior nuclei, two lateral locomotory flagella, two cytostomal pockets opened on opposite sides and containing three recurrent flagella. Bar=10 µm.

Genus ***Trigonomonas*** Klebs

Triangular or spindle-shaped cell (8-30 µm), often twisted spirally, with the posterior end of the body flattened (Fig. 6). The 2 nuclei are located anteriorly, and only 3 flagella are inserted at the base of each nucleus. One flagellum is longer than others and used for locomotion. The 2 largely opened cytostomal grooves extend from the flagella emergence to the posterior of the cell. They have a moving contractile vacuole. While swimming, the cell rotates about the longitudinal axis. Cysts have a smooth wall. They may feed osmotrophically and several species live in freshwater rich in organic matter. At least 7 species described. Light microscopic descriptions are given in Klebs (1892), Lemmermann (1910), Klug (1936), Liebmann (1938), Skuja (1956), van Meel (1982), Mylnikov (1985, 1991); no ultrastructural study.

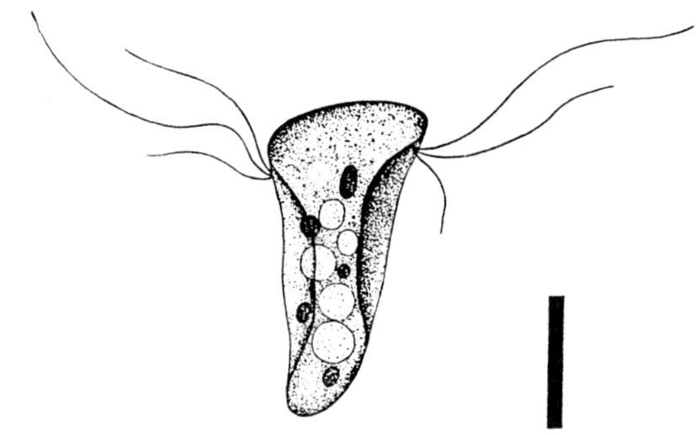

Fig. 6. *Trigonomonas compressa* (from Klug, 1936, and Klebs, 1892). The cell has only three flagella inserted at the top of two lateral cytostomal grooves. Bar=10 µm.

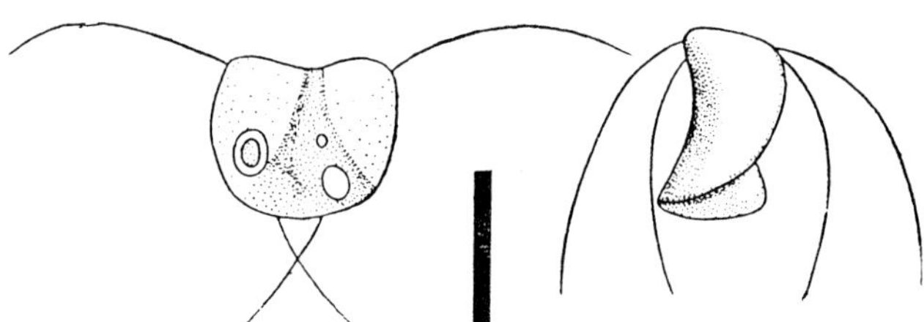

Fig. 7. *Gyromonas ambulans* (from Lemmermann, 1910) with only two flagella on each side; front and lateral views. Bar=10 µm.

Genus ***Gyromonas*** Seligo

Small cell (6-10 µm) with 2 nuclei and 2 pairs of flagella inserted on each side mostly occupied by 2 posterior and widely opened grooves (Fig. 7). Two species; light microscopic description of *G. ambulans* in Klebs (1892), Lemmerman (1910), Hänel (1979) and of *G. salinus* in Ruinen (1938); no electron microscopic study.

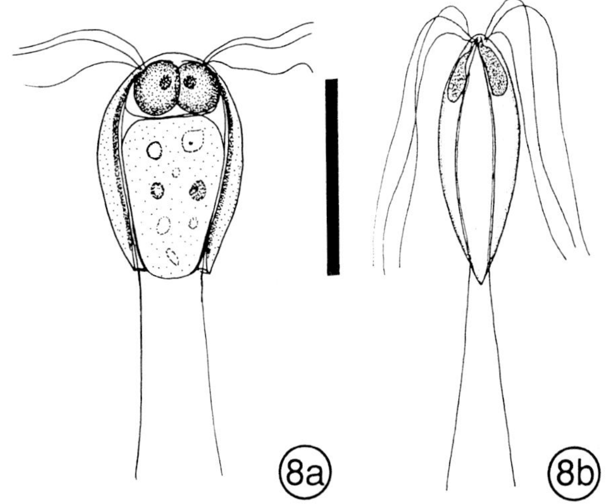

Fig. 8a. *Hexamita inflata* (after Lemmermann, 1910, and Brugerolle, 1974). Two rounded nuclei, two lateral canals opened posteriorly in two cytostomes containing the recurrent flagella. Bar=10 µm. Fig. 8b. *Spironucleus muris* from mice (after Brugerolle et al., 1980). Two sets of flagella inserted anteriorly, two S-shaped elongated nuclei, two cytostomal canals opening posteriorly. Bar=10 µm.

Genus ***Hexamita*** Dujardin

Hexamita (Syn=*Hexamitus, Urophagus*) have an oval or pyriform cell body (6-35 µm), truncated or tapered posteriorly, bearing 6 anterior locomotory flagella and 2 posterior trailing ones (Fig. 8a). They have 2 rounded nuclei closely apposed each other at the anterior end. Each set of flagella is inserted anteriorly on the external side of each nucleus (Fig. 3c). The recurrent flagella traverse the cell to emerge posteriorly as trailing flagella. They are lying in longitudinal canals which open posteriorly forming 2 apertures which are the cytostome of the cell, and the canals are equivalent to cytopharynges (Brugerolle, 1975c). The fine structure study of *Hexamita inflata* has shown the diplomonad characters of this genus. The cytostomal canals are lined by stacks of rough endoplasmic reticulum and by 2 microtubular fibers, and the membrane is also reinforced by a dense lamina (Fig. 3d). Around the posterior cytostomal openings the fibers lining the canals spread or recurve on each side forming a complex architecture. Additional dense striated fibers are present in this area in some species (Brugerolle, 1991a). At least 16 species described. Many species live in freshwater and saltwater rich in organic matter and bacteria. They prefer low oxygen sites. They form cysts. Light microscopic descriptions of various free-living species are given in Klebs (1892), Lemmerman (1910), Skuja (1956), Calaway and Lackey (1962), van Meel (1982), Mylnikov (1991). Parasitic species occur in insects: *H. cryptocerci* (Cleveland et al., 1934), in oysters *H. nelsoni* (Schlicht and Mackin, 1968), in salmonid fishes *H. salmonis* (Moore, 1922, Poynton and Morrison, 1990), in the cloaca of reptiles, in the coecum of rodents *H. teres* (Kirby and Honigberg, 1949) in monkeys *H. pitheci* (Wenrich, 1933). The distinction between parasitic *Hexamita* species and *Spironucleus* species is difficult. Description and list of species can be found in Grassé (1952) and for the species parasitic in vertebrates in Kulda and Nohynkovà (1978). Electron microscopic study of *Hexamita inflata* (Brugerolle, 1974) and of *Hexamita nelsoni* (Papayanni and Vivarès, 1987).

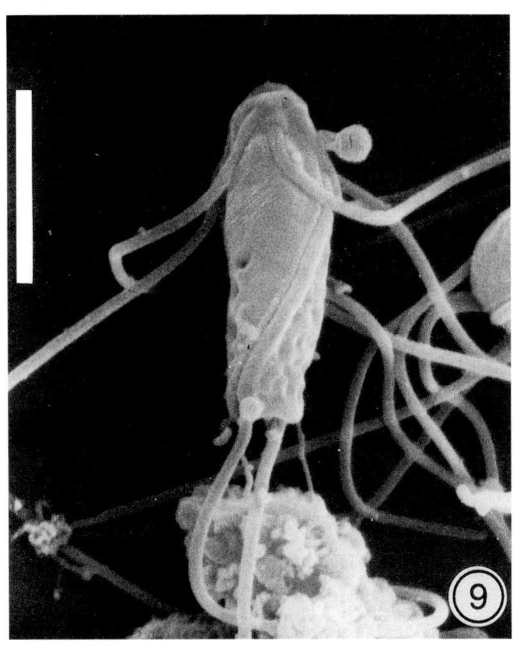

Fig. 9. *Spironucleus elegans* from amphibians, showing the anterior flagella, the two trailing flagella, and an oblique lateral ridge along the cell body. Bar=10 µm. (photograph from G. Brugerolle)

Genus ***Spironucleus*** Lavier

Ovoid or cigar-shaped cell body (5-10 µm) tapered posteriorly bearing 6 anterior flagella and 2 posterior trailing ones (Figs. 8b, 9). *Spironucleus* was distinguished from *Hexamita* and *Octomitus* on the basis that its nuclei were S shape or slightly spiralled and more elongated than *Hexamita* (Lavier, 1936). Electron microscopy has confirmed the shape of the nuclei which present an anterior knob. The basal bodies of the flagella are inserted at the base of these knobs which are surrounded by the supra-nuclear fibers (Fig. 3e). The 2 recurrent flagella in their canal run between the 2 nuclei (Fig. 3f) in contrast to *Hexamita* where they are located on the external face of the nuclei. There are 2 to 3 microtubular fibers around the canals and a dense lamina but not much rough endoplasmic reticulum. Complex structures around the posterior cytostomal openings have been observed. These organisms are intestinal parasites; they feed on digested food. They form ovoid cysts with a thick cyst wall. At least 3 species identified. They live in the intestine of various vertebrates such as *S. elegans* from amphibians (Grassé, 1952) and fishes (Becker, 1977). Some are pathogenic such as *S. muris* for mice and *S. meleagridis* for turkeys. A list of species is given in Kulda and Nohynkovà (1978). Several species have been studied by electron microscopy: *S. elegans* from amphibians (Brugerolle et al., 1974a), *S. intestinalis* from mice (Brugerolle et al., 1980), *S. torosa* from fishes (Poynton and Morrison, 1990; Poynton et al., 1995; Sterud, 1998; Sterud et al., 1997).

SUBFAMILY GIARDIINAE

Genera of this subfamily have no cytostome.

Genus *Octomitus* Prowazek

Octomitus (syn=*Syndyomita*) has a cell body broadly pyriform, tapered posteriorly bearing 6 anterior flagella and 2 posterior trailing ones (Fig. 10). The 2 anterior nuclei are bean-shaped; they face up and adjoin each other in their anterior part. A large endosome is present in the posterior lobe of the nuclei. Two sets of 3 flagella emerge on each side of the anterior part of the body. The 2 recurrent flagella traverse the cell forming a central axis before emerging as trailing flagella. The study of their ultrastructure has shown that the basal bodies are located between the 2 nuclei. In each karyomastigont, 2 anterior flagella emerge on one side and the third anterior flagellum on the opposite side of the cell. The supra-nuclear fibers are reduced. The 2 axonemes of the recurrent flagella are accompanied by microtubular fibers and by stacks of rough endoplasmic reticulum (Fig. 3g). Posteriorly the fibers spread around the flagella emergence, and there are 2 additional dense fibers which form 2 terminal spikes. There are no cytostomal openings or cytopharyngeal canals; these structures are cryptic. Endocytosis takes place in any part of the cell surface. The 2 species identified are parasitic in the intestine and transmitted by cysts. A general description and a list of species and synonyms is provided by Kulda and Nohynkovà (1978). *Octomitus neglecta* from amphibians and *O. intestinalis* from rodents have been studied by electron microscopy (Brugerolle et al., 1974b).

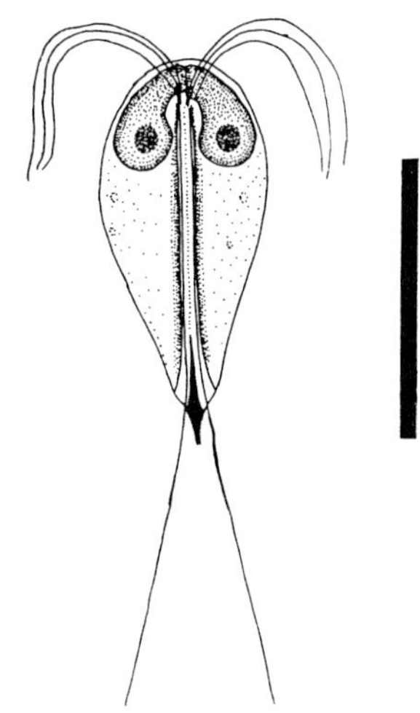

Fig. 10. *Octomitus intestinalis* from the caecum of mice (from Brugerolle et al., 1974b). The axonemes of the two trailing flagella are axial. Bar=10 µm.

Genus *Brugerolleia* Desser, Hong & Sidell, 1993

Created for a species invading the blood of *Rana* in Canada with the generic characters of *Octomitus* and some more specific ones involving the architecture of the posterior end (Desser et al., 1993).

Genus *Giardia* Kunstler

The cell body looks like a split pear with its flattened ventral face occupied by the adhesive disk, which attaches the parasite to the intestinal mucosa of its host (Figs. 11, 12). The cell tapers posteriorly forming a tail prolongated by 2 caudal flagella. The basal bodies of the 4 pairs of flagella are situated between the 2 nuclei in the middle of the cell. All flagella are directed posteriad. The axoneme of the 2 lateral flagella are first directed forward then they cross each other and follow the rim of the disk before emerging laterally. The 2 ventral flagella emerge ventrally in a groove posterior to the disk. The 2 ventro-lateral flagella emerge on each side of the tail. The axonemes of the 2 caudal flagella follow the midline of the body and the flagella prolong the tail. A pair of median bodies which looks like commas are situated in the median part of the cell. The transmittal forms are cysts with 4 nuclei and internal cytoskeletal structures (Fig. 11).

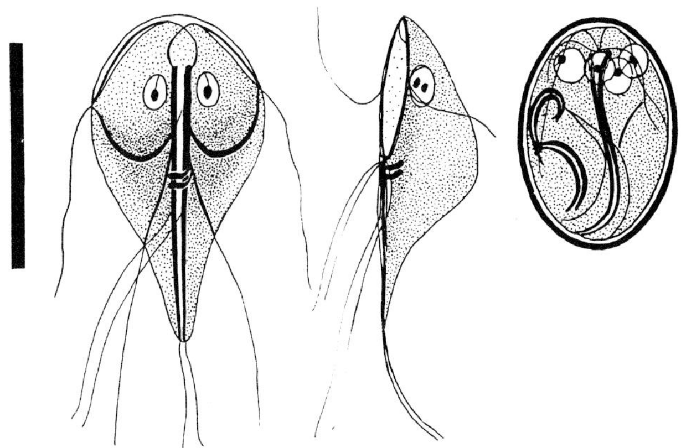

Fig. 11. *Giardia intestinalis* from man (after Kofoid and Swezy, 1922). Dorsal and lateral views of the flagellate and of the cyst with four nuclei. Bar=10 µm.

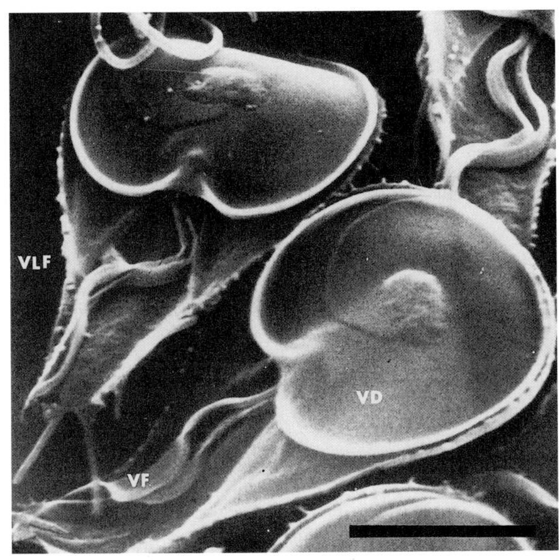

Fig. 12. *Giardia intestinalis* from human jejunum, showing the spiralled sucking disk (D), the vento-lateral flange (VLF), and flattened flagella in ventral channel (VF). Bar=10 µm. (Courtesy of K. Vickerman)

Electron microscopic studies have revealed the details of the structure of the ventral disk (Fig. 3h). Its central part is composed of a spiral of interlinked microtubules forming microribbons (Figs. 3h, 12). The external rim of the disk or ventro-lateral flange contains a striated or paracrystalline lamina. The median body is composed of disorganized microribbons. Examination of basal body arrangement has shown the 2 sets of basal bodies are in axial binary symmetry as in other "diplozoic forms" of diplomonads (Brugerolle, 1975b, 1991a). The apparent bilateral symmetry is secondary and consecutive to the development of the adhesive disk. The cell has no cytostome and the parasite feeds by pinocytosis occurring on the dorsal face. The cell adheres on the brush border of the intestinal epithelial cells, and the sucking force is generated by the beating of the ventral enlarged flagella, which produce a ventral depression (Holberton, 1973). The 41 species described live in vertebrates and are distributed in 3 morphological groups corresponding to the species *G. intestinalis* from man and other animals, *G. muris* from rodents and *G. agilis* from amphibians. The light microscopic description is given by Hegner (1922), Kofoid and Swezy (1922), Filice (1952) and Kulda and Nohynkovà (1978), who provided also a list of species. Electron microscopy is found in Holberton (1973), Brugerolle (1975b), Kulda and Nohynkovà (1978) and in a book devoted to *Giardia* and giardiasis (Erlandsen and Meyer, 1984).

LITERATURE CITED

Alexeieff, A. 1910. Sur les flagellés intestinaux des poissons marins. *Arch. Zool. Exp. Gén. Notes Rev.,* 5° Série, VI:1-20.

Becker, C. D. 1977. Flagellate parasites of fish. *In:* Kreier, J. P. (ed.), *Parasitic Protozoa*, Vol. I. Academic Press, New York. pp 357-416

Brugerolle, G. 1974. Contribution à l'étude cytologique et phylétique des Diplozoaires (Zoomastigophorea, Diplozoa, Dangeard 1910). III Etude ultrastructurale du genre *Hexamita* (Dujardin 1838). *Protistologica,* 10:83-90.

Brugerolle, G. 1975a. Etude ultrastructurale du genre *Enteromonas* da Fonseca (Zoomastigophorea) et révision de l'ordre des Diplomonadida Wenyon. *J. Protozool.*, 22:468-475.

Brugerolle, G. 1975b. Contribution à l'étude cytologique et phylétique des Diplozoaires (Zoomastigophorea, Diplozoa, Dangeard 1910). V.

Nouvelle interprétation de l'organisation cellulaire de *Giardia*. *Protistologica*, 11:99-109.

Brugerolle, G. 1975c. Contribution à l'étude cytologique et phylétique des Diplozoaires (Zoomastigophorea, Diplozoa, Dangeard 1910). VI. Caractères généraux des Diplozoaires. *Protistologica*, 11:111-118.

Brugerolle, G. 1986. Séparation des genres *Trimitus* (Diplomonadida) et *Tricercomitus* (Trichomonadida) d'après leur ultrastructure. *Protistologica*, 22:31-37.

Brugerolle, G. 1991a. Flagellar and cytoskeletal systems in amitochondrial flagellates: Archamoebae, Metamonada and Parabasala. *Protoplasma,* 164:70-90.

Brugerolle, G. 1991b. Cell organization in free-living amitochondriate heterotrophic flagellates. *In*: Patterson, D. J. & Larsen J. (eds.), *The Biology of Free-living Heterotrophic Flagellates.* The Systematics Association, Clarendon Press, Oxford, 45:133-158.

Brugerolle, G, Joyon, L. & Oktem, N. 1973. Contribution à l'étude cytologique et phylétique des Diplozoaires (Zoomastigophorea, Diplozoa Dangeard 1910). I. Etude ultrastructurale du genre *Trepomonas*. *Protistologica*, 9:339-348.

Brugerolle, G., Joyon, L. & Oktem, N. 1974a. Contribution à l'étude cytologique et phylétique des Diplozoaires (Zoomastigophora, Diplozoa, Dangeard 1910). II. Etude ultrastructurale du genre *Spironucleus* (Lavier 1936). *Protistologica,* 9:495-502.

Brugerolle, G., Joyon, L. & Oktem, N. 1974b. Contribution à l'étude cytologique et phylétique des Diplozoaires (Zoomastigophorea, Diplozoa, Dangeard 1910). IV. Etude ultrastructurale du genre *Octomitus* (Prowazek 1904). *Protistologica,* 4:457-463.

Brugerolle, G., Kunstyr, I., Senaud, J. & Freidhoff, K. T. 1980. A fine structure study of the pathogenic intestinal diplomonad flagellate *Spironucleus muris* (ex. *Hexamita muris)* trophozoite and cyst. *Z. Parasitenkd.*, 62:47-61.

Calaway, W. T. & Lackey, J. B. 1962. Waste treatment Protozoa. Flagellata. *Florida Engineering Series, Gainsville,* 3: 1-140.

Cleveland, L. R., Hall, S. R., Sanders, E. P. & Collier, L. 1934. The wood-feeding roach *Cryptocercus,* its Protozoa and the symbiosis between Protozoa and roach. *Mem. Amer. Acad. Arts Sci.,* 17:185-342.

Desser, S. S., Hong, H. & Sidall, M. E. 1993. An ultrastructural study of *Brugerolleia algonquinensis* gen. nov., sp. nov. (Diplomonadina; Diplomonadina), a flagellate parasite in the blood of frogs from Ontario, Canada. *Europ. J. Protistol.*, 29:72-80.

Dobell, C. 1935. Researches on the intestinal Protozoa of monkeys and man. VII. On the *Enteromonas* of macaques and *Embadomonas intestinalis. Parasitology*, 27:564-592.

Erlandsen, S. L. & Meyer, E. A. (eds.) 1984. *Giardia and giardiasis.* Plenum Press, New York.

Eyden, B. P. & Vickerman K. 1975. Ultrastructure and vacuolar movements in the free-living diplomonad *Trepomonas agilis* Klebs. *J. Protozool.,* 22:52-66.

Filice, F. P. 1952. Studies on the cytology and life history of a *Giardia* from the laboratory rat. *Univ. Calif. Publ. Zool.*, 57:53-146.

Grassé, P.-P. 1952. Ordre des Distomatinés ou Diplozoaires. *In*: Grassé, P.-P. (ed.), *Traité de Zoologie*. Vol. I. Masson et Cie, Paris. pp 963-982

Hänel, K. 1979. Systematik und Ökologie dere farblosen Flagellaten des Abwassers. *Arch. Protistenkd.*, 121: 73-137.

Hegner, R. W. 1922. A comparative study of the giardias living in man, rabbit, and dog. *Am. J. Hyg.,* 2:442-454.

Holberton, D. V. 1973. Fine structure of the ventral disk apparatus and the mechanism of attachment in the flagellate *Giardia muris*. *J. Cell Sci.*, 13:11-41.

Klebs, G. 1892. Flagelatenstudien I. *Z. Wiss. Zool.*, 55:265-351.

Klug, G. 1936. Neue oder wenig bekannte Arten der Gattungen *Mastigamoeba, Mastigella, Cercobodo, Tetramitus* und *Trigonomonas*. *Arch. Protistenkd.,* 87:97-116.

Kirby, H. & Honigberg, B. M. 1949. Flagellates of the caecum of ground squirrels, *Univ. Calif. Publ. Zool.,* 53:30-36.

Kofoid, C. A. & Swezy, O. 1922. Mitosis and fission in the active and encysted phases of *Giardia enterica* (Grassi) of man, with a discussion of the method of origin of bilateral symmetry in polymastigote flagellates. *Univ. Calif. Publ. Zool.,* 20:199-234.

Kulda, J. & Nohynkovà, E. 1978. Flagellates of the human intestine and of the intestines of other species. *In:* Kreier, J. P. (ed.), *Parasitic Protozoa*. Vol 2. Academic Press, New York. pp 2-138

Lavier, G. 1936. Sur la structure des Flagellés du genre *Hexamita* Duj. *C. R. Soc. Biol., Paris*, I:1177-1180.

Lemmermann, E. 1910. Algen I. *In:* Lemmermann, E. (ed.), *Kryptogamenflora der mark Brandenburg*, Leipzig. 1910. pp 406-415

Liebmann, H. 1938. Weitere Beiträge zur Kenntnis des Protozoenfauna des Faulschlammes des

Bleilochtalsperre. *Arch. Protistenkd.*, 90:272-291.

Moore, E. 1922. *Octomitus salmonis*, a new species of intestinal parasite of trout. *Trans. Am. Fish. Sci.*, 52:74-97.

Mylnikov, A. P. 1985. A key to free-living flagellates of order Diplomonadida (Wenyon, Brugerolle). *In*: Jakovlev V. N. (ed.), *Aquatic Communities and Hydrobiont Biology*. Nauka, Leningrad. pp 174-198 (in Russian)

Mylnikov, A. P. 1991. Diversity of flagellates without mitochondria. *In:* Patterson, D. J. & Larsen, J. (eds.), *The Biology of Free-Living Heterotrophic Flagellates.* Clarendon Press, Oxford. pp 149-158

Nie, D. 1950. Morphology and taxonomy of the intestinal protozoa of the guinea-pig, *Cavia porcella. J. Morphol.,* 86:381-493.

Papayanni, P. & Vivarès C. P. 1987. Etude cytologique de *Hexamita nelsoni* Schlicht and Mackin 1968 (Flagellata, Diplomonadida) parasite des huîtres méditerranéennes. *Aquaculture,* 67:171-177.

Pérez-Reyès, R. 1964. Estudios sobre protozoarios intestinales. I. Los flagelados del género *Trimitus* Alexeieff, 1910. *An. Esc. Nac. Cien. Biol., Mexico City*, 13:59-66.

Poynton, S. L. & Morrison, C. L. 1990. Morphology of diplomonad flagellates: *Spironucleus torosa* n. sp. from atlantic cod *Gadus morhua* L. and haddock *Melanogrammus aeglefinus* L. and *Hexamita salmonis* Moore from brook trout *Salvenius fontinalis* (Mitchill). *J. Protozool.*, 37:369-383.

Poynton, S. L., Fraser, W., Francis-Floyd, R. Rutledge, P. Reed, P. & Nerad, T. A. 1995. *Spironucleus vortens* n. sp. from the freshwater angelfish *Pterophyllum scalare:* morphology and culture. *J. Euk. Microbiol.*, 42:731-742.

Ruinen, J. 1938. Notizen über Salzflagellaten. II. Über die Verbreiten der Salzflagellaten. *Arch. Protistenkd.*, 90:210-258.

Schlicht, F. G. & Mackin, J. C. 1968. *Hexamita nelsoni* sp. n. (Polymastigina Hexamitidae) parasite of oysters. *J. Invertebr. Pathol.*, 11:35-39.

Skuja, H. 1956. Nova Taxonomische und Biologische Studien über das Phytoplankton schwedischer Binnengewässer. *Nova Acta Regiae Societatis Scientiarum Upsaliensis.* Ser. IV, 16:1-400.

Sterud, E. 1998. Ultrastructure of *Spironucleus torosa* Poynton & Morrison, 1990. (Diplomonadida: Hexamitidae), in cod *Gadus morhua* (L.) and saithe (*Pollachius virens* (L.) from south-eastern Norway. *Europ. J. Protistol.,* 34: 69-77.

Sterud, E., Mo, T. A. & Poppe, T. T. 1997. Ultrastructure of *Spironucleus barkhanus* n. sp. (Diplomonadida: Hexamitidae) from grayling *Thymallus thymallus* (L.) (Salmonidae) and atlantic salmon *Salmo salar* (L.) (Salmonidae). *J. Euk. Microbiol.,* 44:399-407.

Van Meel, L. I. J. 1982. Les eaux saumâtres de Belgique. *Institut Royal des sciences naturelles de Belgique, Mémoires,* 179:1-396.

Vickerman, K. 1990. Phylum Zoomastigina, Class Diplomonadida. *In*: Margulis, L., Corliss, J. O., Melkonian, M. & Chapman, D. J. (ed.), *Handbook of Protoctista.* Jones & Bartlett, Boston. pp 167-185

Wenrich, D. H. 1933. A species of *Hexamita* (Protozoa, Flagellata) from the intestine of a monkey (*Macacus rhesus). J. Parasitol.,* 19:225-228.

PHYLUM EUGLENOZOA
CAVALIER-SMITH, 1981

GORDON F LEEDALE and KEITH VICKERMAN

The Euglenozoa was created by Cavalier-Smith (1981) to formalize the grouping of the euglenids and kinetoplastids. Two minor groups are currently included within the Euglenozoa (*Postgaardi* and diplonemids). Others may join this group in due course (e.g. *Anehmia*), and others (Hemimastigphora, *Stephanopogon*, *Bordnamonas*) have recently been excluded (Simpson, 1997). Flagellates in the Euglenozoa have one or two flagella (rarely four or more) arising from a pocket-like invagination. Flagellar bases are anchored by a system which includes **three-microtubular roots**; flagella usually have **paraxonemal rods** accompanying the **axoneme**; mitochondria have discoid (rarely flattened or tubular) cristae; peroxisomes or glycosomes (not both) may be present; the cytoskeleton includes a regular array of cortical microtubules; the contractile vacuole, if present, empties into the **flagellar pocket**; distinctive cruciate **extrusomes** are present in some members of all four constituent subtaxa and may be an ancestral character; chloroplasts, if present, have chlorophylls *a* and *b* and an envelope of three membranes, but they lack starch and are located in the cytosol.

Composition of and ranks within the Euglenozoa are in dispute (Simpson, 1997). Euglenozoa is here ranked as a phylum. The four subtaxa (Simpson, 1997) are euglenids, kinetoplastids, diplonemids and *Postgaaardi* (Simpson, 1997). Evolutionary relationships among euglenids are still uncertain (Dawson and Walne, 1994; Simpson, 1997).

Euglenids are distinguished by the presence of interlocking and sometimes fused proteinaceous **pellicular strips** which run helically from cell anterior to posterior; these can be seen in some species by light microscopy. Kinetoplastids are distinguished by the presence of the **kinetoplast**, a mass of DNA within the mitochondrion. Diplonemids and *Postgaardi* lack both of these characters.

CLASS EUGLENOIDEA Bütschli, 1884

ORDER EUGLENIDA Bütschli, 1884

GORDON F. LEEDALE

The euglenids are a cosmopolitan group of green or colorless, free-living or parasitic, freshwater or marine flagellates. Cells are elongate, spherical, ovoid, fusiform, cork-screwed, or leaf-like, varying in length from 15 μm to 500 μm. Most genera are flagellated and motile but a few are sessile, including the colonial *Colacium*. The total number of valid euglenoid species is probably about 1000. Leedale (1967a, 1967b) divided the order into six suborders on a cytological basis (dispensing with older classifications which separated green and colorless forms). Reviews of Euglenida include Hollande (1952), Huber-Pestalozzi (1955), Leedale (1967a, 1967b, 1999), Starmach (1983), and Bourrelly (1985). The heterotrophs are reviewed by Larsen and Patterson (1991). The biology of *Euglena*, by far the most studied genus, is surveyed in Buetow (1967a, 1967b, 1982, 1989).

Autotrophic euglenids are mostly freshwater organisms, often producing blooms in waters contaminated by animals or decaying organic matter. Farmyards, middens, greenhouse tanks, agricultural ditches, and the muddy edges of streams and ponds are all worth sampling for common species; larger bodies of purer water such as lakes, reservoirs, and rivers often have sparse populations of rarer planktonic species. Some forms favor very acidic locations such as peaty pools in sphagnum bogs or sulfur lakes at pH 3-4. Among the more bizarre habitats from which euglenids have been recorded are mucus secretions of aquatic animals, barks and leaves of tropical trees, bladders and pitchers of insectiverous plants, snowfields, swimming pools, bird baths, and civic water supplies. Marine autotrophic forms are less common; *Eutreptia*, *Eutreptiella*, and *Klebsina* are exclusively marine or brackish; some other genera have one or two marine species in open sea, tidal zones among seaweeds, or on sandy beaches. Brackish species of *Euglena* can color estuarine mudflats green in dull light, the color vanishing in full sunlight as cells creep away from the mud surface. Heterotrophic forms are diverse and abundant components of the microbiota of insolated and other sediments. Parasitic genera are known from flatworms, copepods, annelids, and tadpoles. Fossil records are very sparse.

The classification is currently somewhat arbitrary. For example, Sphenomonadina and Heteronematina are distinguished by the absence of a cytostome in the former group. Triemer and Farmer (1991b) and Larsen and Patterson (1991) argue that revelation of a feeding apparatus by electron microscopy in some of the Sphenomonadina, or the presence of food particles, or possession of a cytostome visible by light microscopy (e.g. in *Notosolenus ostium*, Larsen & Patterson, 1990), removes the justification for a distinction between the two suborders. An alternative is that the suborder Sphenomonadina be used to house phagotrophic euglenids with no feeding apparatus or a simple one lacking ingestion rods or siphons. That approach is used here.

In our main text some genera are dealt with in detail and figured (in the case of the common and much studied genus *Euglena*, three species are described). Other genera are mentioned briefly in the appropriate taxonomic position but not figured, though they may be included in the key. Michajlow (1969a, b, 1972, 1978 inter alia) assigned a number of endozooic or parasitic species to euglenoid genera *Anisonema*, *Astasia*, *Conradnema*, *Dinema*, *Dinemula*, *Eutreptia*, *Naupliicola*, *Ovicola*, *Embryocola*, *Mesastasia*, *Mononema*, *Paradinemula*, *Paradistigma*, *Paradistigmoides*, *Parastasia*, *Parastasiella*, *Petalomonas*. In many cases the generic assignment is dubious. These taxa are not included here.

Key Characters

1. Flagella arise within an anterior invagination of the cell (a tubular **canal** and pyriform chamber, the so-called "**reservoir**"). The basic number of flagella is two per cell; both may emerge from the canal as locomotory organelles, but in some forms only one is emergent and the other is so short that it ends within the reservoir; one genus (*Scytomonas*) has no second flagellum; a few parasitic species have more than two flagella. In most genera the two flagella are of different lengths, even when both are emergent; equal-length flagella occur in a few genera but they are always **heterodynamic** and usually of different thickness. The diameter of the emergent flagella is increased in most species by **paraxonemal (paraflagellar) rods**. Emergent flagella of all species bear complex arrays of helically organized fibrous material, including a unilateral row of fine hairs (2-3 μm long, visible only in the electron microscope).
2. Chloroplast-containing and colorless forms are both common. Chloroplasts are diverse in shape (discs, plates, or ribbons), number per cell (1 or 2 to several hundreds), size (2 μm diameter to 10 x 20 (+) μm), and pyrenoids (naked, sheathed, projecting, immersed, or absent), even within one genus (e.g. *Euglena*). The chloroplast lamellae have three or more thylakoids except for the two-thylakoid lamellae that enter the pyrenoid matrix.
3. Euglenoid chloroplasts are grass-green in color and contain chlorophylls *a* and *b*, ß-carotene, diadinoxanthin, neoxanthin, and small amounts of other pigments.
4. All green species and a few colorless ones have an orange-red **eyespot** that is completely independent from the chloroplasts and ensheathes the base of the canal. The eyespot (or "stigma") contains ß-carotene derivatives and other carotenoids.
5. All forms with an eyespot also have a crystalline swelling near the base of the long flagellum, within the reservoir and opposite the eyespot; the **flagellar swelling** is presumed to be a photoreceptor.
6. Euglenoid cells have helical symmetry, usually with an imposed bilateral symmetry, with or without cell flattening. Many species are elongated, others are spherical, ovoid, or leaf-shaped. All genera are monads except for *Ascoglena*, which is sessile, and *Colacium*, which forms bunches, sheets, or dendroid colonies.
7. The euglenoid **pellicle** consists of interlocking proteinaceous strips that spiral along the cell and intussuscept at both ends or fuse along their entire lengths; the strips lie within the cell membrane and are flexible and elastic in some species, rigid in others; the cell is naked except for a thin layer of mucilage secreted from subpellicular muciferous bodies; some species eject large lumps of mucilage on irritation; in others, copious secretion of mucilage results in cysts or **palmellae**. In some genera (e.g. *Trachelomonas*), ferric hydroxide precipitates on the mucilage to form a dense envelope.
8. All euglenids, except for a few marine and parasitic species, have a contractile vacuole near the reservoir.
9. Mitosis in euglenids is intranuclear, with persistent dividing nucleoli, staggered and untidy chromatid segregation, kinetochores subtending tiny numbers of microtubules, and continuous microtubules (nucleoplasmic only, none from the cytoplasm) that associate with the chromosomal microtubules to form sparse spindle fibers.
10. Reproduction is by longitudinal fission; records of sexual reproduction are rare and unsubstantiated except for *Scytomonas* .
11. Euglenids have a range of diagnostic ultrastructural features, including details of flagellar roots, pellicular structure, Golgi bodies, mitochondria, endoplasmic reticulum, and contractile vacuole (Leedale, 1967a; Triemer and Farmer, 1991b; Simpson, 1997).
12. Locomotion occurs by rapid swimming with highly mobile flagella and helical rotation of the cell, by gliding with outstretched anteriorly

directed flagellum (and a trailing flagellum in some genera), or by creeping over a surface. Some species exhibit **euglenoid movement** (rapid changes of body shape) when swimming ceases; others are completely rigid. Swimming is the more common form of movement in autotrophs (e.g. in *Euglena*); gliding is more common among the heterotrophic genera.

13. All green euglenids are **phototrophic**; some are facultatively heterotrophic; none has been observed to ingest particulate material although some authors have argued for the presence of a residual cytostome in green euglenids. Colorless euglenids are heterotrophs, being **osmotrophic** and/or **phagotrophic**. Some phagotrophic genera have complex **ingestion organelles.**

14. The carbohydrate storage material is **paramylon**, a ß-1:3-linked glucan deposited as solid grains with helical organization; it gives no color reaction with iodine in KI. Ageing cells become charged with brown or orange droplets containing lipids, carotenoid pigments, and cyclic metaphosphates.

ARTIFICIAL KEY TO SELECTED GENERA

1. Swimming plastic/metabolic cells with two emergent flagella.. 2
1'. Swimming cell with one or many emergent flagella, or not metabolic, or cells normally gliding, skidding, or stationary....................... 5

2. Cell green (with chloroplasts)......................... 3
2'. Cell colorless 4

3. Flagella equal in length....................... ***Eutreptia***
3'. Flagella not equal.......................... ***Eutreptiella***

4. Free-living... ***Distigma***
4'. Isolated from horse-dung........... ***Distigmopsis***

5. Cell with chloroplasts.. 6
5'. Cell without chloroplasts................................ 15

6. Cell normally attached....................................... 7
6'. Cell always free-swimming.............................. 9

7. Cell normally attached to substratum by anterior end; usually grouped with other cells in bunches, sheets, or dendroid colonies .... ***Colacium***
7'. Cell solitary.. 8

8. Cell attached by posterior point to base of open muciferous lorica.......................................11
8'. Cell embedded in mucus, sometimes free-swimming.................................... ***Ascoglena***

9. Cell with envelope.. 10
9'. Cell without envelope...................................... 12

10. Cell enclosed by external envelope except for apical pore, usually with a neck, through which emerges the flagellum..............***Trachelomonas***
10'. Cell with envelope tapered anteriorly or widely open anteriorly.................................. 11

11. Envelope widely open anteriorly...... ***Klebsina***
11'. Envelope tapering anteriorly.......................... .... ***Strombomonas***

12. Cross-sectional profile rounded or slightly flattened...13
12'. Cross-sectional profile flattened, leaflike, cell may be twisted.. ***Phacus***

13. Cell plastic or semi-rigid, usually with pronounced euglenoid movement, or at least flexing of the cell; with one or no emergent flagellum; cell usually elongated............................ ***Euglena***
13'. Cell rigid.. 14

14. Cell with evenly spaced longitudinal folds or ridges.. ***Lepocinclis***
14'. Cell with occasional folds or ridges................ .... ***Cryptoglena***

15. Cell typically swimming................................ 16
15'. Cell typically gliding.................................. 27

16. Cell with multiple emergent flagella, parasitic 17
16'. Cell with one or two flagella......................... 18

17. Cell with four emergent flagella...................... .... ***Euglenamorpha***

17'. Cell with more than four emergent flagella.. ***Hegneria***

18. Cell with one emergent flagellum; eyespot present... 19
18'. Cell with one or two emergent flagella; eyespot absent.. 20

19. Cell plastic and rounded in cross-section, with pronounced euglenoid movement.. ***Khawkinea***
19'. Cell not plastic, flattened, may be twisted, leaflike .... ***Hyalophacus***

20. Cell metabolic or at least capable of bending 21
20'. Cell rigid.. 22

21. Cell very metabolic............................ ***Astasia***
21'. Cell long, thin, cylindrical; change of shape restricted to bending; canal opening apical.. ***Cyclidiopsis***

22. Cell rounded or slightly flattened in cross-section...23
22'. Cell very flattened in cross-section............. 26

23. Cell with two emergent flagella.***Sphenomonas***
23'. Cell with one emergent flagellum................ 24

24. Cell not flattened, ridged............................ 25
24'. Cell smooth, if flattened, only slightly so.. ***Rhabdospira***

25. Cell slightly ridged.................. ***Rhabdomonas***
25'. Cell markedly keeled.................. ***Gyropaigne***

26. Canal opening apical.................... ***Menoidium***
26'. Canal opening subapical................ ***Parmidium***

27. Cell very flexible, metabolic........................ 28
27'. Cell rigid or nearly so................................ 34

28. Cell with anterior flare..................... ***Urceolus***
28'. Cell with no anterior flare.......................... 29

29. Cell elongated.. 30
29'. Cell not elongated....................................... 32

30. Cell with large ingestion apparatus............. 31
30'. Cell without large ingestion apparatus......... 33

31. Recurrent flagellum adhering to cell, usually lying in a groove.............................. ***Peranema***
31'. Recurrent flagellum free from cell.. ***Heteronema***

32. Ingestion apparatus can be extruded.. ***Entosiphon***
32'. Ingestion apparatus cannot be extruded.. ***Ploeotia***

33. Cell golden in color, no ingestion organelle .... ***Calkinsia***
33'. Cell not golden.. 34

34. Cell with two emergent flagella..................... 35
34'. Cell with one emergent flagellum.................. 38

35. Recurrent flagellum with thickened basal hook .... ***Anisonema***
35'. Recurrent flagellum without hook................ 36

36. Cell gliding with anterior flagellum projected forwards.. 37
36'. Cell gliding/skidding with anterior flagellum beating.. ***Metanema***

37. Cell with strong helical keels.. ***Tropidoscyphus***
37'. Cell without strong helical keels.. ***Notosolenus***

38. Cell covered in bacteria.............. ***Dylakosoma***
38'. Cell without attached bacteria...................... 39

39. Cell flattened.............................. ***Petalomonas***
39'. Cell not flattened.. 40

40. Cell keeled................................ ***Calycimonas***
40'. Cell not keeled............................ ***Scytomonas***

SUBORDER EUTREPTIINA Leedale, 1967

Swimming euglenids with two emergent flagella.

Genus ***Distigma*** Ehrenberg, 1838

Colorless, osmotrophic, elongated, unflattened cells, with pronounced euglenoid movement; 2 emergent flagella, the shorter directed laterally during swimming with a cilium-like beat, the longer directed anteriorly; no eyespot or flagellar swelling; cysts known; 7 species, all freshwater. The genus ***Distigmopsis*** Hollande, 1944 (not figured), resembling a small *Distigma* (2 unequal flagella, heterotrophic) but with '**leucostigma**', was isolated from horse dung and is possibly parasitic. See Pringsheim (1942), Larsen and Patterson (1991).

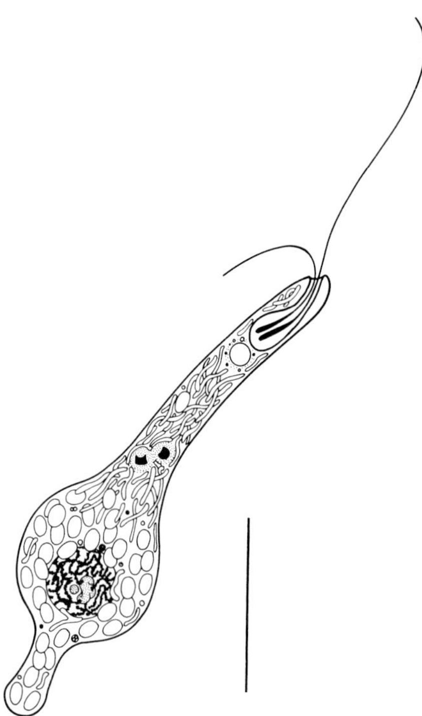

Fig. 1. *Distigma proteus.* Scale bar=20 µm.

Distigma proteus Ehrenberg emend. Pringsheim (Fig. 1). Cell spindle-shaped to cylindrical, 40-120 x 5-20 µm; during swimming the long flagellum lashes about wildly, with waves travelling from base to tip, while the shorter has a straight stroke backwards and a relaxed recovery stroke; euglenoid movement violent, even when the cell is swimming (a rare feature); 2 or 3 very large Golgi bodies and numerous mitochondrial threads are clearly visible in the anterior half of the living cell; paramylon ovals pack the cell posteriorly around the nucleus; common species worldwide in acidic freshwater habitats.

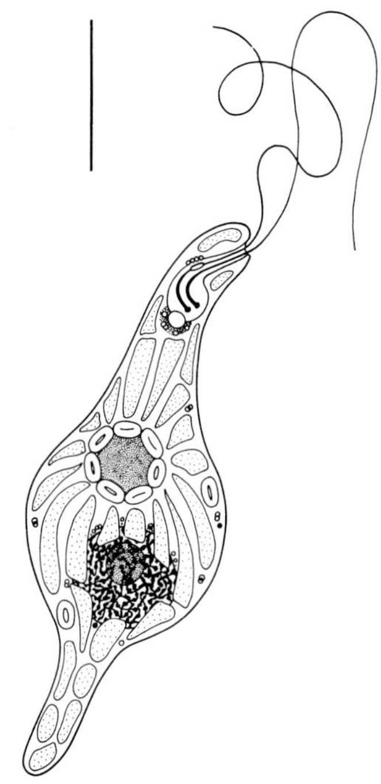

Fig. 2. *Eutreptia pertyi.* Scale bar=20 µm.

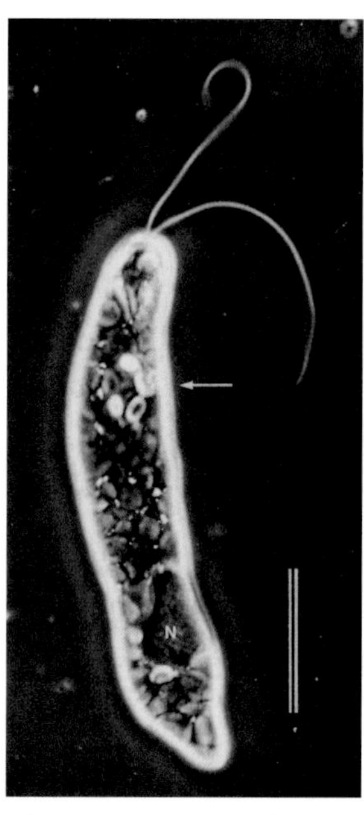

Fig. 3. *Eutreptia pertyi.* Living cell showing two flagella, paramylon center (arrow), numerous chloroplasts in the discoid state, posteriorly placed nucleus (N). Anoptral contrast light microscopy. Scale bar=20 µm.

Genus ***Eutreptia*** Perty, 1852

Green, spindle-shaped, unflattened cells, with pronounced euglenoid movement; 2 emergent flagella, equal in length and thickness, heterodynamic, both highly mobile during swimming; numerous ribbon chloroplasts, usually radiating from a paramylon center (pyrenoid surrounded by a rigid sheath of paramylon grains); eyespot present, flagellar swelling on one flagellum; 6 species, typically brackish or marine. See Huber-Pestalozzi (1955).

Eutreptia pertyi Pringsheim (Figs. 2,3). Cell 30-60(80) x 5-15 µm; chloroplast ribbons (Fig. 2) can become discoid (Fig. 3); nucleus large; euglenoid movement violent in cells that have just stopped swimming; copious mucilage production results in extensive palmellae in which cells divide repeatedly; quite common in brackish and marine eutrophic waters.

Genus ***Eutreptiella*** da Cunha, 1913

Small, green, elongated cells with 2 flagella of unequal length; reported from marine habitats only; 4 or 5 species (Fig. 4). See Huber-Pestalozzi (1955).

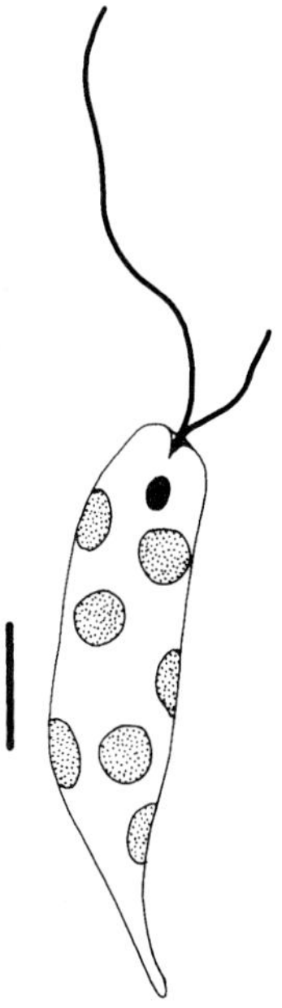

Fig. 4. *Eutreptiella.*

SUBORDER EUGLENINA Bütschli, 1884

Swimming euglenids, one emergent flagellum; with or without chloroplasts, no ingestion apparatus.

Genus ***Euglena*** Ehrenberg, 1830

Green, elongate or ovoid cells, some cylindrical, others flattened and cork-screwed; only 1 flagellum emergent as a locomotory organelle; (a few species lack emergent flagella); swimming locomotion involves helical rotation of the cell; most species exhibit euglenoid movement when swimming stops; gliding occurs in some taxa; a few are almost rigid; eyespot and flagellar swelling present; canal opening subapical; about 125 species, mostly freshwater, a few marine. See especially Pringsheim (1956); also Dangeard (1901), Chadefaud (1937), Hollande (1942), Chu (1946), Conrad and van Meel (1952), Gojdics (1953), Buetow (1968a, 1968b, 1982, 1989).

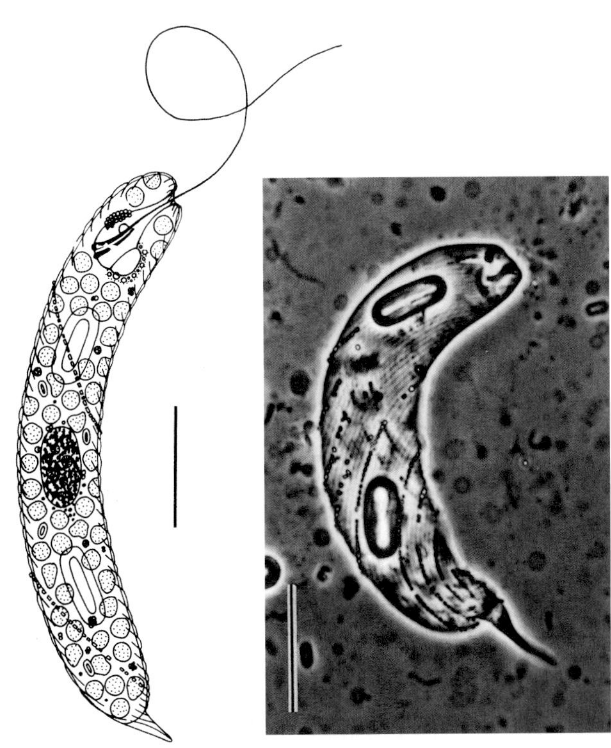

Figs. 5, 6. *Euglena spirogyra.* Scale-bars=20 µm. Fig. 6. Phase contrast micrograph with pellicle showing helical striations and rows of warts; two large paramylon granules lie within the cell.

Euglena spirogyra Ehrenberg (Figs. 5-7). Cell cigar-shaped, 45-250 x 5-35 µm; slightly flattened and twisted; almost rigid but shows slight bending movements in response to heat or chemical stimuli; many small discoid chloroplasts (2-3 µm diameter); pyrenoids absent; 2 large paramylon grains typically present (Fig. 6), one anterior to and one posterior to the nucleus; pellicle ornamented with rows of warts (Fig. 6) consisting of mucilage impregnated with ferric hydroxide (ornamentatiion being heavier in habitats rich in iron); thin mucilaginous sheath produced but cysts and palmellae unknown; common fresh-water species, cosmopolitan in ditches, pools, and lakes. See Leedale et al. (1965).

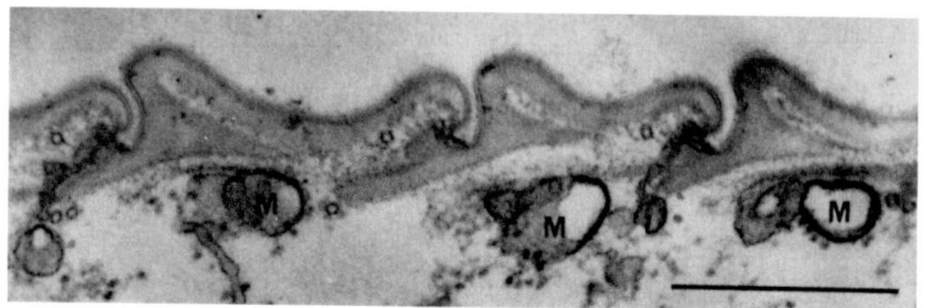

Fig. 7. *Euglena spirogyra*. Electron micrograph of a section of pellicle showing elaborate cross-sectional shape of several pellicular strips and their articulation one with another; muciferous bodies (M) associate with each strip. Scale bar=0.5 µm.

Euglena gracilis Klebs (Fig. 8). Tapering cigar-shaped cell, 35-65 x 5-15 µm; swimming rapid, with anterior end of the gyrating cell tracing a wide circle; euglenoid movement pronounced when swimming stops; chloroplasts 6-12 per cell, large, flat, shield-shaped, each with a central pyrenoid covered on both sides by a watchglass-shaped paramylon cap; freshwater species, in ponds and ditches containing rotting leaves; palmellae form on mud; can be grown axenically in undefined media such as 0.2% beef extract; widely used research organism (see below).

Numerous physiological strains of *Euglena gracilis* have been isolated by E. G. Pringsheim and are maintained in culture collections (especially at the FBA, England; Austin, Texas, USA & Göttingen, Germany); strains grow phototrophically or heterotrophically on acetates, organic acids, alcohols, or sugars, with an absolute requirement for cobalamin (vitamin B12), therefore being used in hospitals to assay blood for B12, lack of which causes pernicious anemia in man.

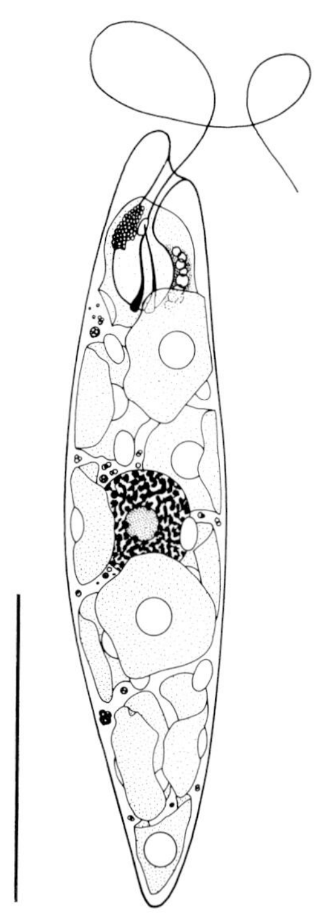

Fig. 8. *Euglena gracilis*. Scale bar=20 µm.

Euglena gracilis, though relatively rare in nature, is the dominant research organism of euglenology and many areas of physiology and biochemistry. Cells grown in the dark reduce their chloroplasts to proplastids that divide and survive for years and then, given light, will expand, synthesize chlorophyll, and resume photosynthesis within 24 hours, thus providing ideal material for research on chloroplast physiology and development. Permanently colorless races without chloroplasts can be produced by treatment with heat (Pringsheim and Pringsheim, 1952), ultraviolet irradiation (Lyman et al., 1959), streptomycin (Provasoli et al., 1948), other antibiotics, antihistamines, and a range of other chemicals (see Leedale, 1967a). Some of the bleached races are identical to naturally occurring species of *Astasia*, especially *A. longa* Pringsheim. Strains of *E. gracilis*, in particular *E. gracilis* Z and *E. gracilis* var. *bacillaris*, are used in research on DNA species (plastidial and nuclear), RNA species (plastidial and ribosomal), photoinduction of enzyme synthesis, lipid metabolism, heterotrophic nutrition, respiratory pathways, flagellar movement, and photoreception; these studies are summarized in Buetow (1968a,b, 1982, 1989).

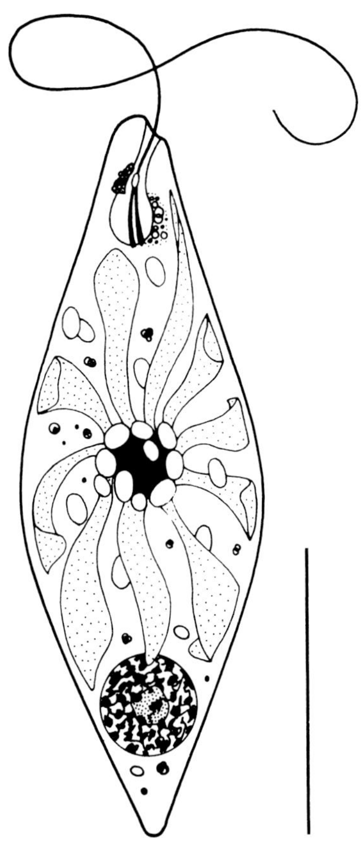

Fig. 9. *Euglena viridis.* Scale bar=20 µm.

Euglena viridis Ehrenberg (Fig. 9). Cell fusiform, 40-80 x 12-20 µm; swimming rapid and jerky; euglenoid movement also jerky, not flowing; chloroplasts arranged as ribbons radiating from a central pyrenoid surrounded by a shell of paramylon granules; under poor conditions the ribbons round off into apparently discrete discoid chloroplasts; mucilage production copious from small spherical muciferous bodies; cysts and palmellae often form on mud surfaces at pond edges (and in culture); common freshwater species of polluted habitats, frequently producing green blooms in farmyards. The genus ***Euglenocapsa*** Steinecke, 1931 (not figured) has small green cells within a mucus sheath and is probably just a palmelloid stage of a species of *Euglena.*

Genus ***Khawkinea*** Jahn & McKibben, 1937

Colorless, osmotrophic, fusiform or elongated, unflattened cells; flagella as *Euglena*; canal opening subapical; eyespot and flagellar swelling present; euglenoid movement pronounced; c. 10 species, fresh-water, widespread; some recorded as parasitic in fresh-water animals (flatworms, copepods, annelids); almost certainly derived from species of *Euglena* by loss of chloroplasts (Pringsheim 1956, 1963). See Jahn and McKibben (1937), Pringsheim (1956).

Khawkinea quartana (Moroff) Jahn & McKibben (Fig. 10). Cell fusiform, 40-60 x 15 µm; locomotory flagellum often shed, followed by violent euglenoid movement; contractile vacuole large; mitochondrial threads easily seen in the living cell; palmellae known; recorded from freshwater habitats with high sulfur content.

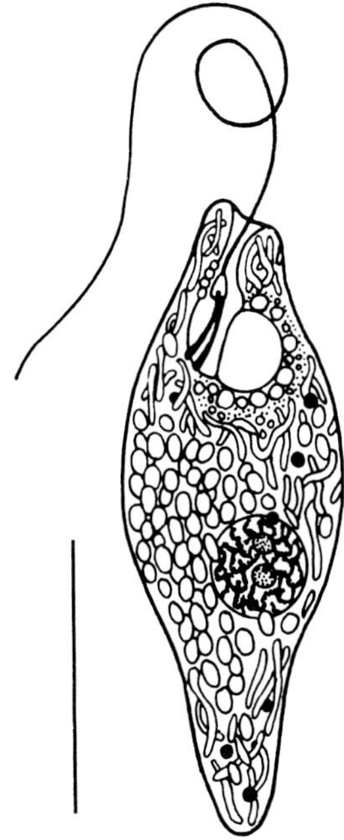

Fig. 10. *Khawkinea quartana.* Scale bar=20 µm.

Genus *Astasia* Dujardin,1830

Colorless, osmotrophic, fusiform or elongated cells, a few species slightly flattened; flagella as *Euglena*; canal opening apical or subapical (subgeneric difference in Christen's system, 1963); no eyespot, no flagellar swelling; euglenoid movement often pronounced; about 40 species, occurring worldwide, mostly in freshwaters rich in organic matter, some marine. See Christen (1958, 1963).

Astasia klebsii Lemmermann (Fig. 11). Cell elongated but always with a bulge that moves from posterior to anterior as euglenoid movement occurs (even when cell is swimming), 50-75 x 10-20 µm; canal opening apical; large Golgi body can be seen in the living cell, immediately behind the contractile vacuole; large mitochondria usually occur in the anterior half of the cell, oval paramylon grains in posterior, nucleus central, all churned about during euglenoid movement; cosmopolitan species of polluted freshwater habitats.

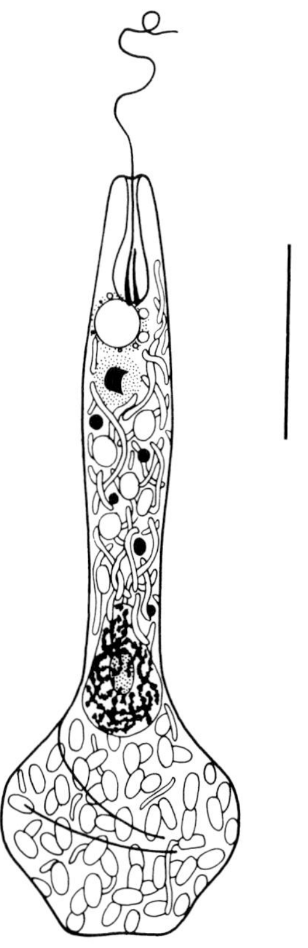

Fig. 11. *Astasia klebsii.* Scale bar=20 µm.

Genus *Cyclidiopsis* Korschikov, 1917

Colorless, osmotrophic, elongated, tapering, cylindrical cells; canal opening apical; eyespot and flagellar swelling present (suggesting derivation from *Euglena* by loss of chloroplasts); cell semi-rigid but can bend and loop; paramylon grains are long needles and small ovals; 4 species, all freshwater

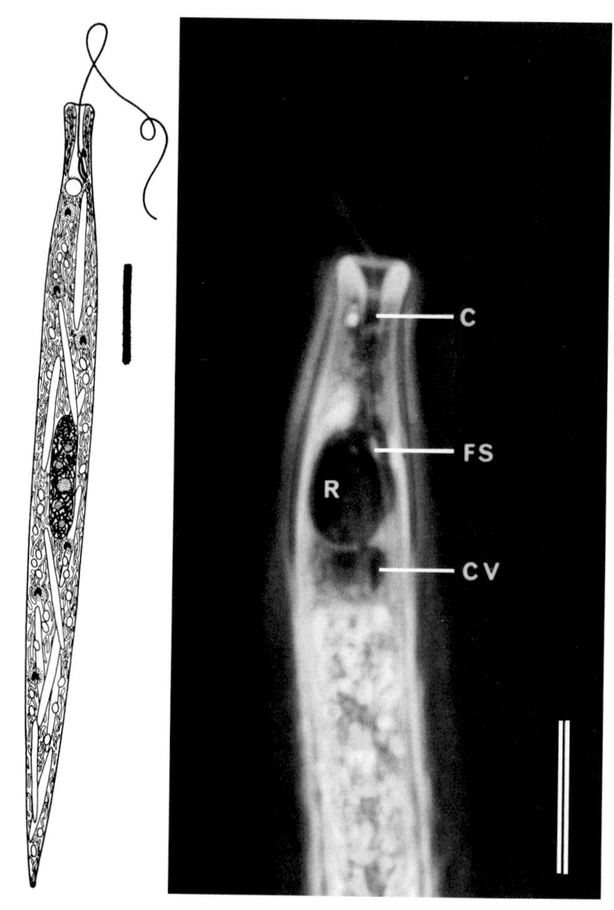

Figs. 12,13. *Cyclidiopsis acus.* Scale bar=20 µm. Fig. 13. Anoptral contrast light microscopy of anterior end of living cell showing canal (C) with apical opening, flagellar swelling (FS), reservoir (R) with flagellar bases, contractile vacuole (CV).

Cyclidiopsis acus Korschikov (Figs. 12,13). Cell long, thin and tapering, 120-210 x 5-15 µm; swimming rapid with anterior end tracing a wide circle, cell glides when not swimming; no euglenoid movement but cell can coil into loops; canal opening funnel-shaped and truly apical (Fig. 12); eyespot pale orange, flagellar swelling small (Fig. 12); no cysts or palmellae known but heavy coatings of mucilage cause cells to stick to substratum or to one another by their posterior ends; long paramylon needles always present; numerous mitochondrial threads and large Golgi bodies visible in living cell; widespread species of acidic habitats; the superficially similar green species *Euglena acus* Ehrenberg has a subapical canal opening, and close relationship is doubtful. See Pringsheim (1963).

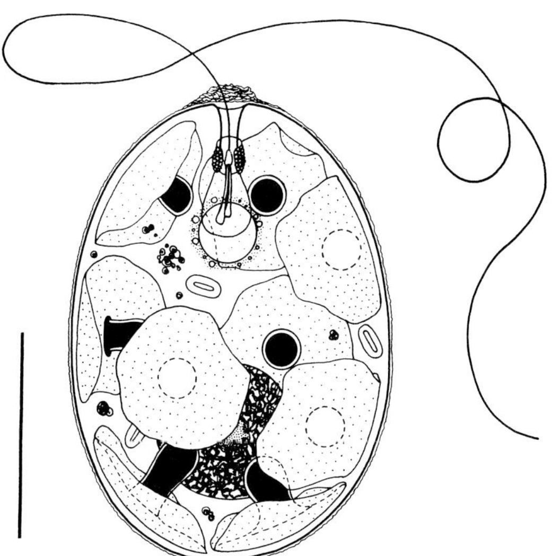

Fig. 14. *Trachelomonas grandis.* Scale bar=20 μm.

Genus ***Trachelomonas*** Ehrenberg,1833

Green *Euglena*-like cells, solitary and free-swimming but enclosed in an envelope with sharply defined neck or collar surrounding an apical pore (through which emerges the long locomotory flagellum); naked cells escape from the envelope during reproduction (and at other times) but immediately secrete a new envelope of species-specific shape and size (spherical, ovoid, ellipsoid, or elongated, 20-50 x 10-35 μm), which is colorless and smooth at first but soon becomes brown, ornamented (pores, punctae, spines, warts, or ridges), and brittle with ferric hydroxide and manganese salts; one emergent flagellum; chloroplasts of several distinct types (numerous, small, discoid, without pyrenoids; flat plates with double-sheathed pyrenoids; flat plates with inwardly-projecting pyrenoids (Fig. 15); or flat plates with naked pyrenoids); a few species are colorless and osmotrophic; eyespot and flagellar swelling present; euglenoid movement and cell rotation occur within the envelope; canal opening sub-apical even though the envelope pore is apical; over 300 species; exclusively freshwater, common in acidic to neutral waters (pH 4.5-7), often in peaty pools and other habitats rich in reduced iron and manganese, cosmopolitan; many species known only from their distinctive envelopes. See Deflandre (1926), Pringsheim (1953b); also Leedale (1975) for ultrastructural details, e.g *T. oblonga* (Fig. 15).

Other solitary *Euglena*-like genera with envelopes are ***Strombomonas*** Deflandre, 1930 with more than 50 species (see Conrad and van Meel, 1952) with a a pale or colorless envelope with no collar (Fig. 17); ***Ascoglena*** Stein, 1878 with 3 species in which the flask-shaped envelope is attached posteriorly to filamentous algae (Fig. 18); and the monotypic ***Klebsina*** Silva, 1961 (renamed from *Klebsiella* Pascher, 1931, since that name had been used twice previously), a marine form which is half enclosed in a brown cup-shaped envelope (Fig. 19).

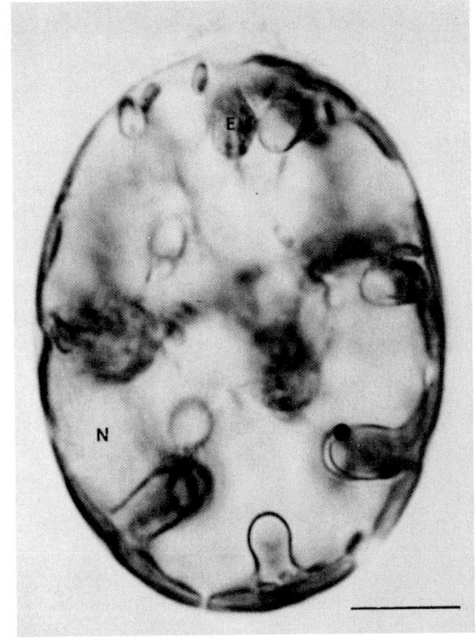

Fig. 15. *Trachelomonas grandis.* Light micrograph of living cell showing peripheral plate chloroplasts, each subtending an inwardly projecting pyrenoid with paramylon sheath, eyespot (E), nucleus (N). Scale bar=20 μm.

Trachelomonas grandis Singh (Figs. 14,15). Cell larger than most *Trachelomonas* spp., 40-60 x 35-40 µm; swimming slow, locomotory flagellum very long (150-200 µm); chloroplasts 8 - 12(15) per cell (10 µm diameter), flat, with inwardly-projecting cylindrical pyrenoids, each covered with a paramylon sheath (Fig. 15); eyespot large, bright red; envelope thin, pale brown, never becoming as thick, dark, and elaborate as in other species; recorded only once, from a quarry pond in India, but available in culture for study and research.

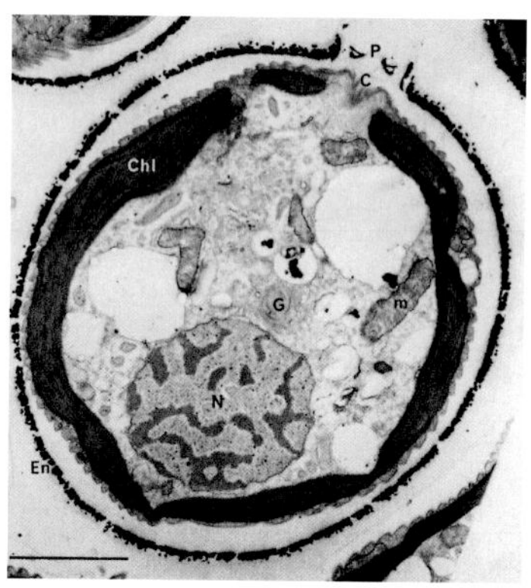

Fig. 16. *Trachelomonas oblonga* var. *punctata*. L.S. cell within its envelope (En), showing canal opening (C), chloroplasts (Chl), Golgi bodies (G), mitochondria (m), nucleus (N), envelope pore (P). Electron micrograph. Scale bar=4 µm.

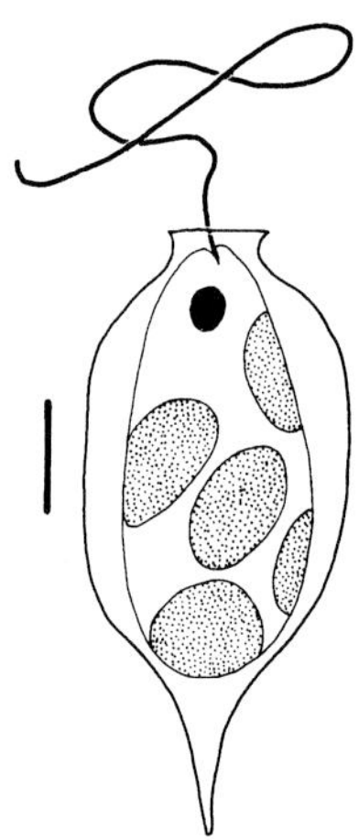

Fig. 17. *Strombomonas*.

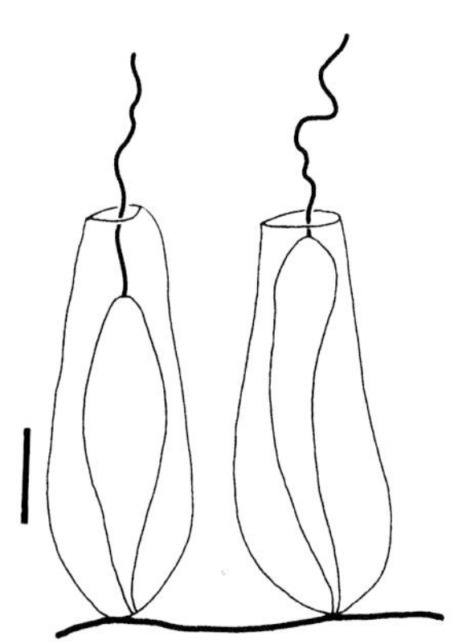

Fig. 18. *Ascoglena*.

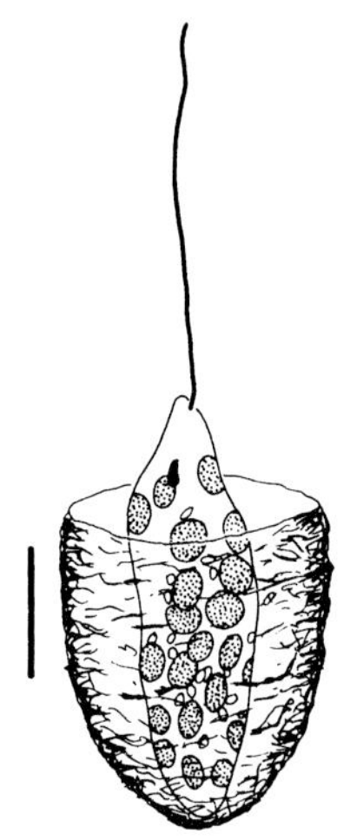

Fig. 19. *Klebsina*.

Genus ***Colacium*** Ehrenberg, 1833

Green *Euglena*-like cells, normally attached to a substratum by mucilaginous secretions from the canal; cell division occurs to produce bunches, sheets, or dendroid colonies of a few to hundreds of cells with thin sheaths, joined together by mucilaginous stalks; settled cells shed the locomotory flagellum but retain the basal portions of their 2 flagella within the reservoir, including the (reduced) flagellar swelling; any cell can regrow the long flagellum, escape from the colony as a free-swimming organism (indistinguishable from *Euglena*), settle elsewhere on its anterior end and secrete a new stalk and sheath of mucilage;

eyespot and flagellar swelling present; cell plastic, euglenoid movement only slight in colonial state; cell not flattened; canal opening subapical; about 10 species, freshwater, growing attached to filamentous algae, aquatic angiosperms, *Cyclops*, *Daphnia* and other aquatic animals, including fish; fairly common, world-wide. See Johnson (1934), Pringsheim(1953c), Huber-Pestalozzi (1955).

Colacium mucronatum Bourrelly & Chadefaud (Fig. 20). Large dendroid colonies formed when undisturbed, 1-2 mm across; cell ovoid, 25-35 x 12-20 µm; mucilage stalks apical but covering canal opening, 2-5 µm thick, transversely ridged and longitudinally striated, ~30 µm long from one branch to the next; flagellar swelling and eyespot small, both becoming larger in motile cells; released cells grow the locomotory flagellum (20-30 µm long) in about one hour; chloroplasts 12-20 per cell, 5 µm diameter, flat, with inwardly projecting ovoid pyrenoids, each one covered with a paramylon sheath; muciferous bodies small but copious mucilage produced for stalks and sheath; other species produce smaller colonies, often only 2 or 3 cells with short (2 µm) stalks.

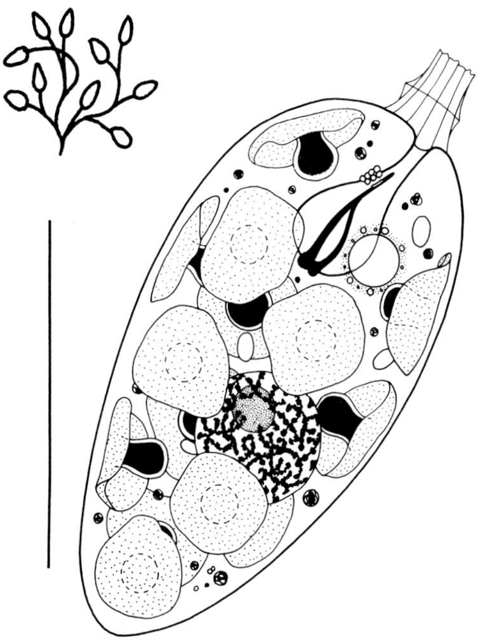

Fig. 20. *Colacium mucronatum*. Scale bar=20 µm. Inset shows small portion of a colony.

Genus ***Phacus*** Dujardin, 1841

Green, rigid, compressed cells, most species very flat and leaf-shaped, often with ridges, folds, or grooves running helically or longitudinally, giving an irregular or triradiate cross-section; many species with a long posterior spine, shaped like flattened spinning tops; some species twisted into flat corkscrews; flagella, eyespot, and flagellar swelling as in *Euglena*; chloroplasts usually small, discoid, numerous, without pyrenoids; a few species (e.g. *P. splendens*) have large flat chloroplasts with pyrenoids; paramylon is typically deposited as a few large granules (often rings) together with many small ones; canal opening subapical; no cysts; palmelloid stages rare; about 150 species, freshwater, worldwide. See Huber-Pestalozzi (1955).

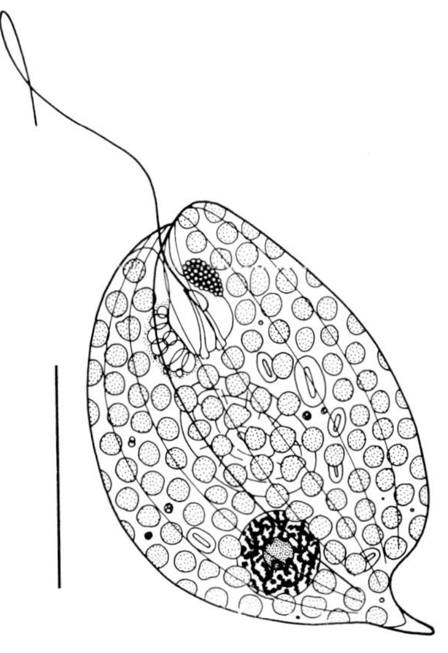

Fig. 21. *Phacus triqueter*. Scale bar=20 µm.

Colorless *Phacus*-like cells can be placed in the genus, ***Hyalophacus*** Pringsheim, 1936 (not figured). The genus ***Lepocinclis*** Perty, 1849 with approximately 50 species, also has rigid cells, discoid chloroplasts and large paramylon rings, but differs from *Phacus* in being unflattened (Fig. 23).

The 6 species of the genus ***Cryptoglena*** Ehrenberg, 1831 have small cells with a single chloroplast but are difficult to separate from *Phacus* (or *Euglena*).

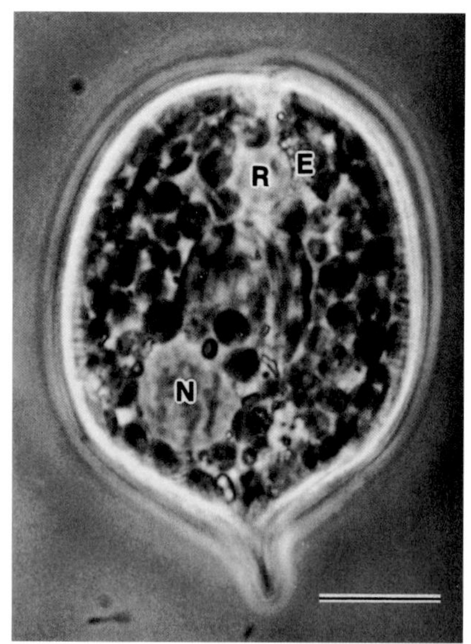

Fig. 22. *Phacus triqueter.* Living cell showing many discoid chloroplasts, eyespot (E), nucleus (N), reservoir (R). Phase contrast light microscopy. Scale bar=10 µm.

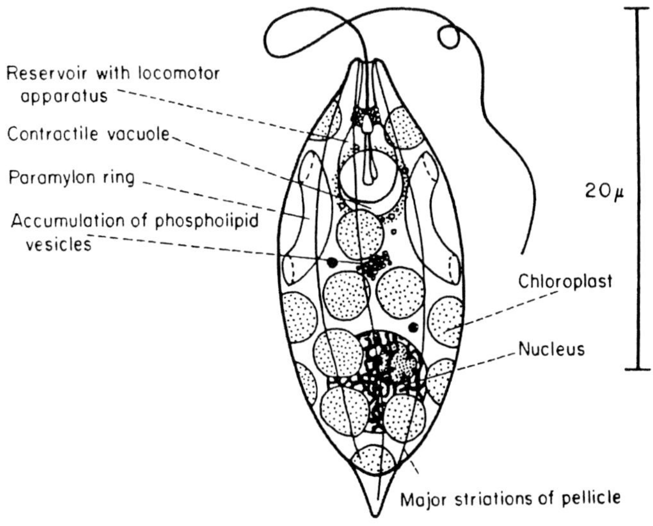

Fig. 23. *Lepocinclis.*

Phacus triqueter (Ehrenberg) Dujardin (Figs. 21, 22). Cell compressed, with lateral ridge, 35-70 x 25-45 µm; completely rigid, no euglenoid movement; swimming rapid, with flagellum held in front of the gyrating cell; chloroplasts numerous, small, discoid (Figs. 17,18); no pyrenoids; usually a few large paramylon grains and many small ones (Figs. 17,18); pellicular striations (lines of overlap between pellicular strips) very clear, with transverse struts running between them; eyespot large, orange; cosmopolitan freshwater species, in lakes and ponds.

SUBORDER RHABDOMONADINA Leedale, 1967

Swimming rigid cells with a single emergent flagellum, all heterotrophic; no clear evidence of phagotrophy.

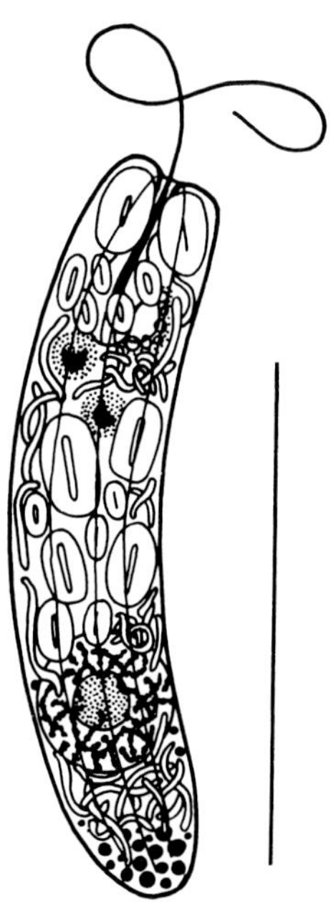

Fig. 24. *Rhabdomonas costata.* Scale bar=20 µm.

Genus ***Rhabdomonas*** Fresenius, 1858

Colorless (osmotrophic, never phagotrophic), rigid, cylindrical, curved cells with rounded ends; periplast shallowly fluted, with 6-8 steeply helical ridges; one emergent flagellum, mobile throughout its length during swimming, held straight when cell is stationary; second flagellum non-emergent; canal opening subapical; paramylon and lipid abundant; no cysts or palmellae; 6 species, freshwater, cosmopolitan. See Pringsheim

(1942,1963), Christen (1963) Leedale and Hibberd (1974). The monotypic genus ***Rhabdospira*** Pringsheim, 1963 (not figured) is similar except that the small cell is irregularly spiralled, with 2 major twists at right-angles to the long axis, slightly flattened and tapering posteriorly, not fluted.

Rhabdomonas costata (Korschikov) Pringsheim (Fig. 24). Cell 23-30 x 7-10 μm; completely rigid, no euglenoid movement; emergent flagellum 10-20 μm long; large Golgi bodies visible in the living cell, just posterior to the reservoir; nucleus in posterior half of the cell; lipid droplets characteristically grouped in tail region; common in peaty waters.

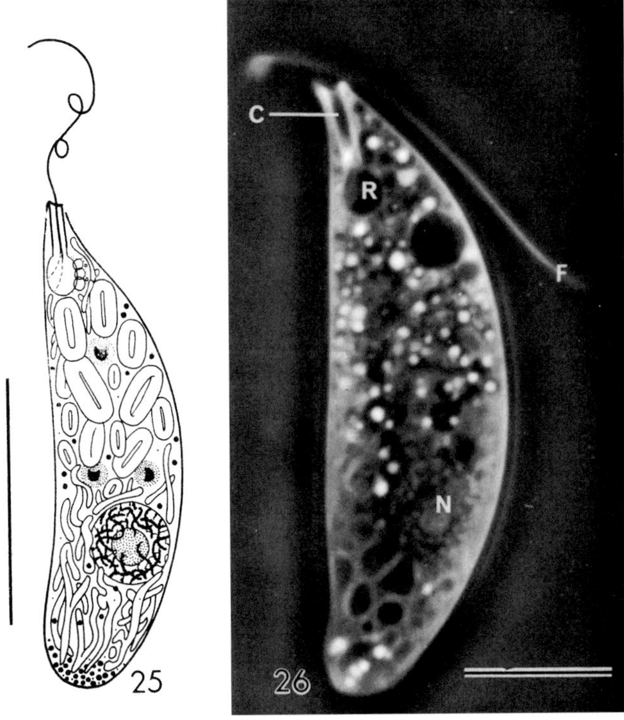

Figs. 25,26. *Menoidium bibacillatum*. Fig. 25. Scale bar=20 μm. Fig. 26. Anoptral contrast light microscopy of living cell showing canal (C), reservoir (R), nucleus (N); the emergent portion of the locomotory flagellum (F) has been shed. Scale bar=10 μm.

Genus ***Menoidium*** Perty, 1852

Colorless (osmotrophic, never phagotrophic), rigid, elongated, curved, flattened cells, narrowly triangular or elliptical in cross-section; pellicle with delicate striations (shown by electron microscopy to consist of pellicular strips fused into a single unit, Fig. 27) but not keeled; flagella as for *Rhabdomonas*; canal opening apical, anterior end of cell protruded into a narrow neck; cell contents include large paramylon links and typically show cytoplasmic streaming; cysts and palmellae unknown; about 10 species, freshwater, worldwide. See Pringsheim (1942, 1963), Christen (1963), Leedale and Hibberd (1974).

The genus ***Gyropaigne*** Skuja, 1939 contains 4 species with ovoid unflattened cells with pronounced helical keels (Fig. 28). ***Parmidium*** Christen, 1963 (not figured) has 2 species with ovoid, greatly compressed cells.

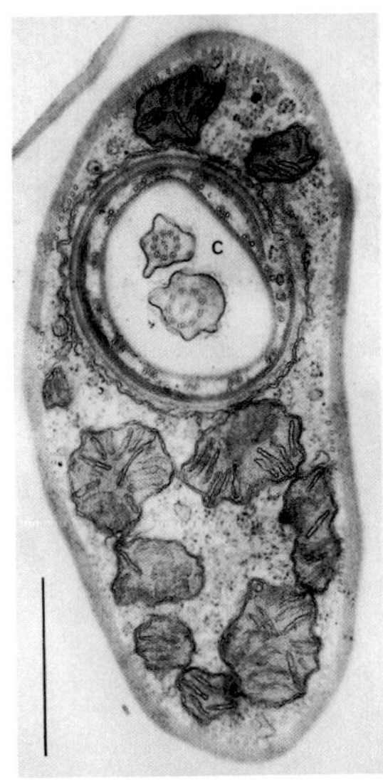

Fig. 27. *Menoidium bibacillatum*. Section through cell anterior in region of the canal (C) showing two flagella in T.S., skeletal material and associated microtubules around the canal, pellicle of fused strips and underlying microtubules, mitochondria. Electron micrograph. Scale bar=1 μm.

Menoidium bibacillatum Pringsheim (Figs. 25-27). Cell shaped like a flattened banana, narrowly triangular in cross-section, 30-40 x 10 x 5 μm; locomotory flagellum short, flickering in front of the cell during swimming; cell gyrates during rapid swimming, oscillates from side to side during slow swimming; several large Golgi bodies are visible, dispersed throughout the living cell; recorded only once, from a bog.

SUBORDER SPHENOMONADINA Leedale, 1967

Cells with one or two emergent flagella, no visible ingestion apparatus, fairly rigid.

Genus *Sphenomonas* Stein, 1859

Colorless, osmotrophic, cylindrical or ovoid, usually unflattened cells, characterized by 1 or 2 longitudinal ridges giving an angular cross-section, usually rigid but capable of some change of shape under extreme stimuli, easily squashed; 2 emergent, unequal, heterodynamic flagella, the longer directed anteriorly, the shorter curving laterally; canal opening almost apical; hyaline inclusion (possibly leucosin) often in the posterior half of the cell; 10 species, freshwater and marine, cosmopolitan but several species known only from Australia; little known genus with no modern studies.

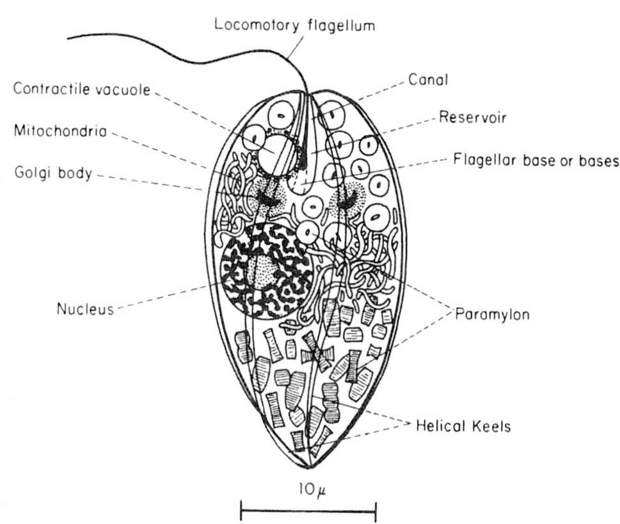

Fig. 28. *Gyropaigne*.

The monotypic genus ***Atraktomonas*** Christen, 1962 (not figured), with rigid fusiform unflattened cells, is similar but has only 1 emergent flagellum; Christen (1962) regards it as derived from *Sphenomonas* by reduction of the latter's short flagellum. Monotypic *Calkinsia* Lackey, 1960 has insufficient characters to separate it from *Sphenomonas* (apart from its golden color!).

Sphenomonas laevis Skuja (Fig. 29). Shape resembles that of *Menoidium* but cell not flattened, 25-35 x 7-8 μm; rigid but delicate; long flagellum held straight in front of cell during swimming, with shallow waves passing from base to tip; short flagellum relatively inactive; rare freshwater species.

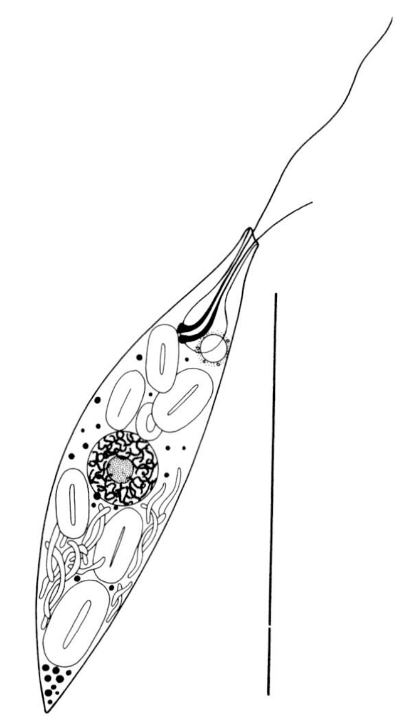

Fig. 29. *Sphenomomas laevis*. Scale bar = 20 μm.

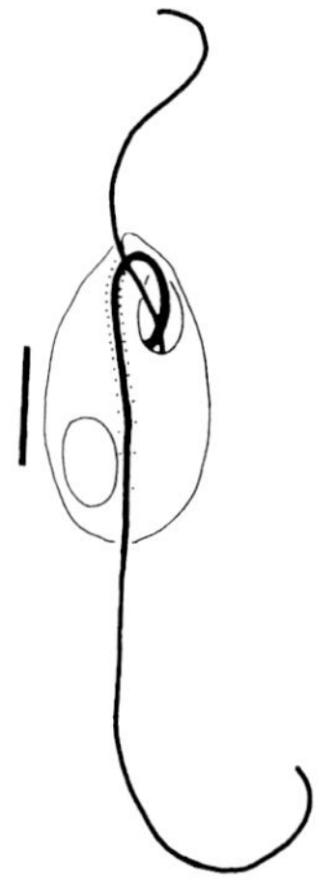

Fig. 30. *Anisonema*.

Genus ***Anisonema*** Dujardin, 1841

Almost rigid, gliding cells with 2 emergent flagella; the recurrent flagellum is well developed basally, with a 'hook'; gliding is interrupted by sudden jerks; cytostome not visible by light-microscopy (Fig. 30). Marine and freshwater, benthic, about 25 species. See Larsen and Patterson (1990, 1991).

Electron microscopy by Triemer (1985) shows his "*Anisonema* sp." to have a feeding apparatus of vanes and rods as found in *Peranema* and *Urceolus*. Triemer and Farmer (1991a, 1991b) argue that this and other observations on *Calycimonas* and *Petalomonas* (see below) remove the justification for separation of the suborders Sphenomonadina and Heteronematina; however, existence of ingestion rods in *Anisonema acus* and the numerous other described species has not yet been demonstrated and it can reasonably be suggested that Triemer's "*Anisonema* sp." is a *Heteronema* !

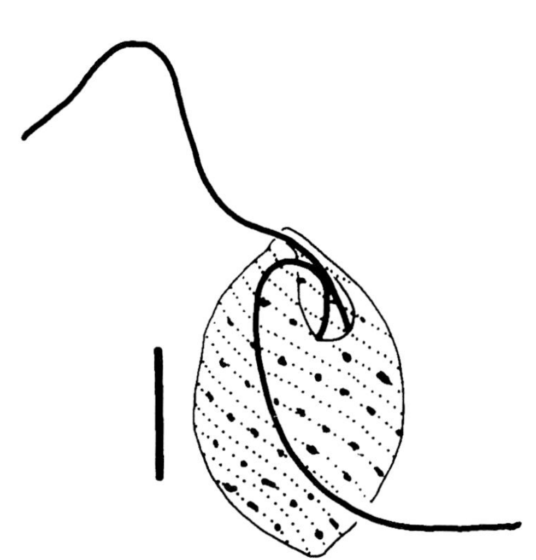

Fig. 31. *Metanema.*

Genus ***Metanema*** Senn, 1900

Flattened cells with a limited capacity to squirm and equal flagella which often bend in opposite directions when skidding-gliding (Fig. 31). Species of *Metanema* have previously been referred to *Anisonema*, but the characters above serve to differentiate the genera. No ingestion apparatus is visible by light microscopy although cells with ingested diatoms have been observed. Freshwater and marine, often common in marine sediments, less than 10 species. See Larsen (1987), Larsen and Patterson (1990, 1991).

Genus ***Calycimonas*** Christen, 1959

Colorless, phagotrophic, rigid, ovoid, unflattened cells, with 5-8 pronounced helical keels; food vacuoles observed in all species by light microscopy but no ingestion organelle; electron microscopy of 2 species has revealed a simple feeding apparatus consisting of a cytoplasmic pocket reinforced by microtubules; canal opening subapical with 1 emergent flagellum directed anteriorly, straight, during swimming; second flagellum non-emergent; 4 species, freshwater, European. See Christen (1959), Larsen and Patterson (1991).

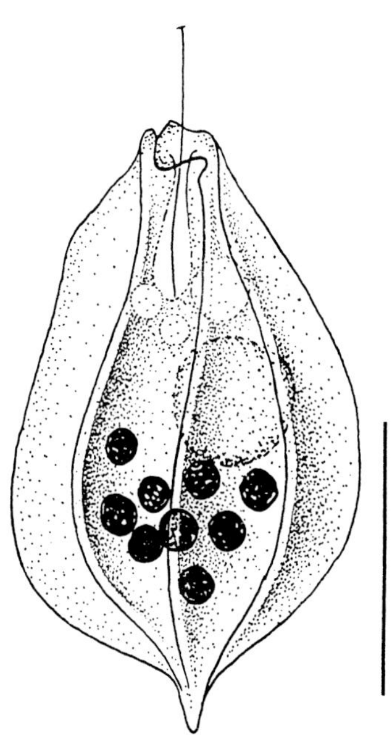

Fig. 32. *Calycimonas physaloides.* From Christen (1959). Scale bar=20 μm.

The taxonomic position of this genus, near to *Petalomonas* except for the latter's flattened cell (see below) and to *Tropidoscyphus* except for the latter's 2 flagella, is discussed by Christen (1959). However, Triemer and Farmer (1991) argue that the presence of a feeding apparatus removes the main distinction between Sphenomonadina and Heteronematina and these suborders should be merged. Alternatively, it can reasonably be suggested that the suborder Sphenomonadina be retained for phagotrophic euglenids with a simple feeding apparatus lacking ingestion rods or siphons and that line has been taken here.

Calycimonas physaloides Christen (Fig. 32). Cell 30 x 20 µm with delicate striations between the keels; swimming slow; paramylon said to be absent; mud-dwelling species in eutrophic ponds.

Genus ***Petalomonas*** Stein, 1859

Colorless, phagotrophic and/or osmotrophic, rigid, fusiform or triangular, flattened cells, usually very flat and leaf-shaped, mostly with strong ribs or keels; no special ingestion organelle visible in the light microscope, but at least some species have a simple feeding apparatus as in *Calycimonas*; canal opening subapical with 1 emergent flagellum, directed anteriorly, straight, during swimming; second flagellum non-emergent; paramylon abundant; c. 60 species, freshwater, mainly benthic, in mud sediments, cosmopolitan. See Shawan and Jahn (1947), Christen (1962).

The 2 species of the genus ***Dylakosoma*** Skuja, 1964 (not figured) are distinguished from *Petalomonas* only by being covered by a single-layer mosaic of symbiotic bacteria and by a modest capacity for euglenoid movement. The 3 species of ***Scytomonas*** Stein, 1878 are unique among euglenids in having only a single flagellum (no reduced second flagellum in the reservoir); electron microscopy shows the colorless phagotrophic cells to have a simple feeding pocket with microtubules, as in *Petalomonas*; cells are 7 - 15 µm long, pyriform, not flattened, rigid, with 5 pellicular strips; the genus is also unique in being the only euglenid where sexuality is known; 2 cells fuse as isogametes, 1 flagellum is lost and the zygote swims by means of the other; the gamete nuclei fuse (Mignot, 1961) but meiosis and further stages have not been seen. ***Dolium*** Larsen & Patterson, 1990 (not figured) is a sessile, monotypic genus, with one emergent flagellum and no visible cytostome although it is phagotrophic; recorded only from tropical and subtropical marine sediments. See Huber-Pestalozzi (1955), Mignot (1961), Christen (1962), Larsen and Patterson (1990, 1991), Lee and Patterson (1999).

Petalomonas tricarinata Skuja (Fig. 33). Cell 40-50 x 30 x 10-15 µm, much compressed, with 3 keels on 1 face; swims smoothly, with slow gyrations; large reservoir and contractile vacuole; mud-dwelling form of polluted waters.

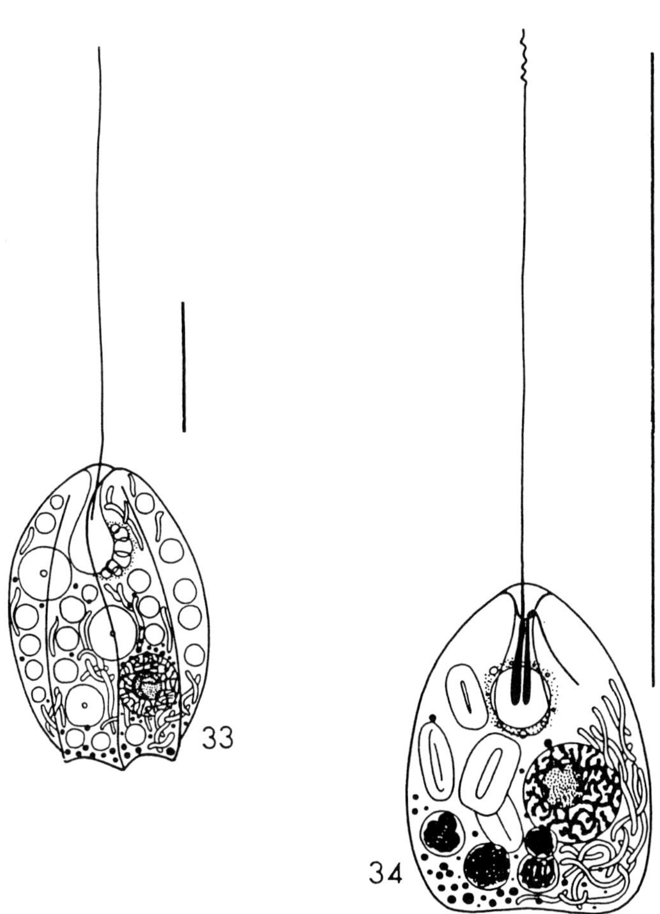

Fig. 33. *Petalomonas tricarinata*. Scale bar=20 µm.
Fig. 34. *Notosolenus apocamptus*. Scale bar=20 µm.

Genus ***Notosolenus*** Stokes, 1884

Colorless, phagotrophic, rigid, ovoid, flattened (usually very flat) cells, concavo-convex in cross-section or with keels; no ingestion organelle visible in the light microscope; canal opening subapical, with 2 emergent flagella, unequal, heterodynamic, the longer directed anteriorly during swimming, the shorter trailing; mostly small; c. 20 species; freshwater, characteristically in bog-pools and sediments; Europe, Asia, and America. The genus ***Tropidoscyphus*** Stein,1878 with 7 species, differs only in being less flattened and more obviously keeled. See Skuja (1939).

Notosolenus apocamptus Stokes (Fig. 34). Cell 6 - 10 x 5-6 x 3-4 µm, much compressed, unusual in being flattened dorsoventrally (most compressed euglenoids are flattened laterally), ventral face concave; long flagellum straight during swimming, with minute waves at the tip only; uncommon species of bog-pools and mud.

SUBORDER HETERONEMATINA Leedale, 1967

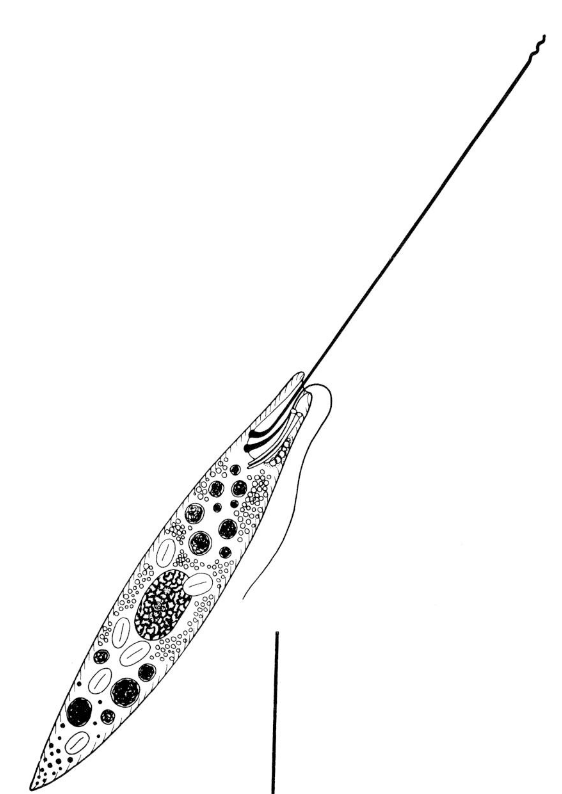

Fig. 35. *Heteronema acus.* Scale bar=20 µm.

Genus *Heteronema* Dujardin, 1841

Colorless, phagotrophic, non-rigid, elongated, tapering, unflattened cells, with moderate euglenoid movement; pellicular striations coarse; ingestion organelle composed of separate rods associated with a subapical cytostome independent from the canal and reservoir; both flagella emergent, unequal in length and thickness, the longer (thicker) one directed anteriorly during swimming, with a coiling or flickering motion of the tip only, giving a characteristic gliding locomotion, the shorter one curving posteriorly, lying free from the cell; canal opening subapical; muciferous bodies large; cysts and palmellae unknown; c. 25 species; common and widespread genus in fresh and marine waters; cosmopolitan.

Heteronema acus (Ehrenberg) Stein (Fig. 35). Cell 40-50 x 10-20 µm, elongated, cylindrical; feeds on bacteria, coccoid algae, etc.; muciferous bodies spherical; common species in freshwater and brackish ponds.

Genus *Peranema* Dujardin, 1841

Colorless, phagotrophic, elongated, blunt-ended, non-rigid cells, slightly flattened when swimming, rounding up and with violent euglenoid movement when swimming stops; pellicular striations coarse; ingestion apparatus as in *Heteronema*; flagella as in *Heteronema* except that the trailing flagellum is usually pressed closely to the cell, lying in a groove in some species; one form of swimming is a gliding locomotion with the long (very thick) flagellum held straight while the beating tip traces a funnel shape; a second is swimming with a cilium-like beat and recovery stroke of most of the flagellar length; a third is gliding against a substratum with a snapping action of the anterior third of the thick flagellum; canal opening subapical; muciferous bodies large but cysts and palmellae unknown; different species vary from 8 to 200 µm in length but the type species, *P. trichophorum,* is also very variable (see below); c. 20 species; freshwater and marine, common in ponds, ditches, marshes, sediments, and marine sands; cosmopolitan. The genus ***Urceolus*** Mereschkovsky, 1877 (not figured), has approximately 10 species with ovoid cells with a funnel-shaped collar and one very long emergent flagellum. Some species have a small ingestion apparatus opening into the canal, others have a larger apparatus opening into the collar;

reported from freshwater and marine habitats, especially intertidal sediments. See Huber-Pestalozzi (1955), Nisbet (1974), Larsen and Patterson (1990, 1991).

Peranema trichophorum (Ehrenberg) Stein (Fig. 36). Cell with many size varieties, 20-80 x 10-25 µm, slight lateral flattening; leading flagellum more than 1 µm thick; trailing flagellum very thin, lying in a groove and apparently stuck to the cell; feeds by engulfing particles and whole living organisms (bacteria, yeasts, algae, other euglenids) into food vacuoles; muciferous bodies fusiform; cosmopolitan flagellate of ponds and ditches; species studied in detail by Chen (1950) and, earlier, by Dangeard (1901), Hall and Powell (1928), Lackey (1929, 1933), and Chadefaud (1938). Electron microscopic studies include those by Mignot (1966) and Nisbet (1974).

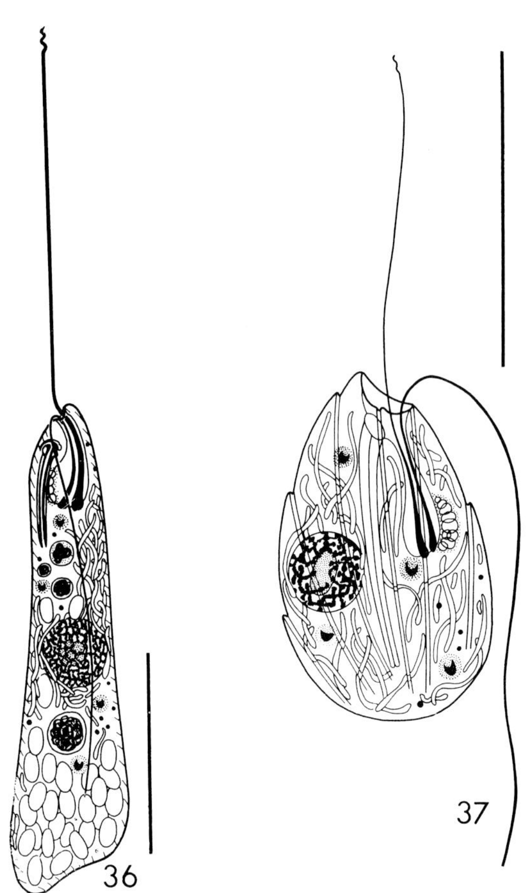

Fig. 36. *Peranema trichophorum*. Scale bar=20 µm.
Fig. 37. *Entosiphon sulcatum*. Scale bar=20 µm.

Genus ***Entosiphon*** Stein, 1878

Colorless, phagotrophic, rigid, slightly flattened cells, with helical fluting and ridges; ingestion organelle composed of rods fused together to make a protrusible tube ("siphon"); cytostome separate from canal opening, but both lie in a common apical depression; both flagella emergent, unequal in length and thickness, the longer (thinner) one directed anteriorly during swimming, with a cilium-like beat and recovery stroke, the longer one trailing posteriorly as a non-propulsive skid during gliding locomotion; 5 species; cosmopolitan and quite common genus, planktonic in freshwater ponds and lakes. Very similar to ***Ploeotia*** Dujardin, 1841 (not figured) from which it is distinguished by the protrusibility of the siphon. See Mignot (1963, 1964, 1966).

Entosiphon sulcatum (Dujardin) Stein (Fig. 37). Cell 20-25 x 10-15 µm, ovoid, slightly compressed; swims with slow gyration, anterior flagellum not rigid but with oscillating tip during gliding; siphon dorsal to reservoir, almost as long as the cell; cell with 12 helical grooves (pellicle composed of 12 articulating strips); paramylon not recorded; Golgi bodies of moderate size visible in the living cell; common species. See Mignot (1963, 1964).

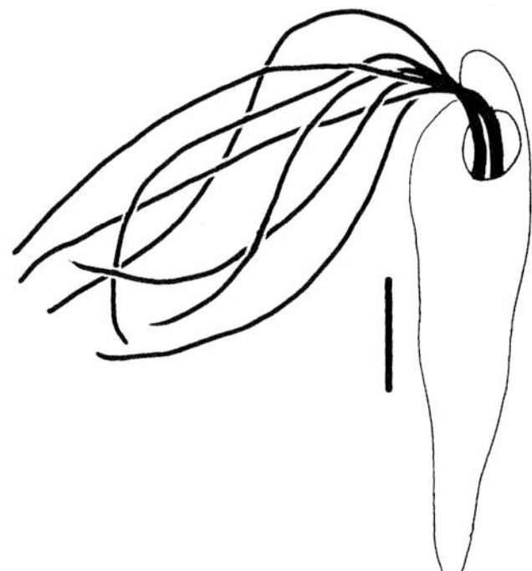

Fig. 38. *Euglenamorpha*.

SUBORDER EUGLENAMORPHINA Leedale, 1967

This order was established to contain the monotypic parasitic genera *Euglenamorpha* and *Hegneria*, included in the euglenids by some authorities

(Hollande, 1952; Leedale. 1967; Bourrelly, 1970) but not others (Huber-Pestalozzi, 1955). ***Euglenamorpha*** Wenrich, 1924 (Fig. 38) has green, cigar-shaped, non-rigid cells, 50 x 10 µm, with numerous discoid chloroplasts and 3 emergent flagella; it was recorded once only within the digestive tract of *Rana* tadpoles in Pennsylvania. ***Hegneria*** Brumpt & Lavier, 1924 (Fig. 39) has colorless cells with 6 or 7 emergent flagella; it was recorded once only from the rectums of tadpoles in Brazil. See Leedale (1967a).

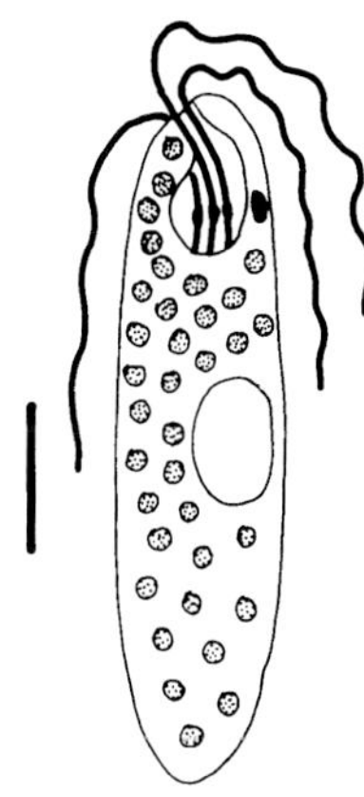

Fig. 39. *Hegneria*.

LITERATURE CITED

Bourrelly, P. 1970. Les Algues d'Eau Douce. Initiation à la Systématique. Tome 3: Les Algues Bleus et Rouges, les Eugléniens, Peridiniens et Cryptomonadines. N. Boubee et Cie, Paris. pp 115-184

Bourrelly, P. 1985. Ibid. 2nd ed. N. Boubee et Cie, Paris. pp 115-184, 514-532

Buetow, D. E. (ed.) 1967a. *The Biology of* Euglena. Vol. 1. Academic Press, New York.

Buetow, D. E. (ed.) 1967b. *The Biology of* Euglena. Vol. 2. Academic Press, New York.

Buetow, D. E. (ed.) 1982. *The Biology of* Euglena. Vol. 3. Academic Press, New York.

Buetow, D. E. (ed.) 1989. *The Biology of* Euglena. Vol. 4. Academic Press, New York.

Chadefaud, M. 1930. Anatomie comparée des Eugléniens. *Botaniste*, 28:85-185.

Chadefaud, M. 1938. Nouvelles recherches sur l'anatomie comparée des Eugléniens: les Péranémines. *Rev. Algol.*, 11:189-220.

Chen, Y. T. 1950. Investigations of the biology of *Peranema trichophorum* (Euglenineae). *Q. J. Microsc. Sci.*, 91:279-308.

Christen, H. R. 1958. Farblose Euglenalen aus dem Hypo-limnion des Hausersees. *Schweiz. Z. Hydrol.*, 20:141-176.

Christen, H. R. 1959. New colorless Euglenineae. *J. Protozool.*, 6:292-303.

Christen, H. R. 1962. Neue und wenig bekannte Eugleninen und Volvocalen. *Rev. Algol. Sér.* 2, 6:162-202.

Christen, H. R. 1963. Zur Taxonomie der farblosen Eugleninen. *Nova Hedwigia*, 4:437-464.

Chu, S. P. 1946. Contributions to our knowledge of the genus *Euglena*. *Sinensia*, 17:75-134.

Conrad, W. & van Meel, L. 1952. Matériaux pour une monographie de *Trachelomonas* Ehrenberg, C., 1834, *Strombomonas* Deflandre, G., 1930 et *Euglena* Ehrenberg, C., 1832, genres d'Euglénacées. *Inst. Roy. Sci. Nat. Belg.*, 124:1-176.

Dangeard, P.-A. 1901. Recherches sur les Eugléniens. *Botaniste*, 8:97-357.

Deflandre, G. 1926. *Monographie du Genre* Trachelomonas *Ehr.* Imprimerie André Lesot, Nemours.

Deflandre, G. 1926. Monographie du Genre *Trachelomonas* Ehr. Imprimerie André Lesot, Nemours.

Ekebom, J., Patterson,. D. J. & Vørs, N. 1996. Heterotrophic flagellates from coral reef sediments (Great Barrier Reef, Australia). Archiv Protistenkd., 146:251-272

Farmer, M. A. & Triemer, R. E. 1988. A redescription of the genus *Ploeotia* Dujardin (Euglenophyceae). *Taxon*, 37:319-25.

Farmer, M. A. & Triemer, R. E. 1994. An ultrastructural study of *Lentomonas applanatum* (Preisig) N. G. (Euglenida). *J. Euk. Microbiol.*, 41:112-119.

Gojdics, M. 1953. *The Genus* Euglena. Univ. Wisconsin Press, Madison.

Hall, R. P. & Powell, W. N. 1928. Morphology and binary fission of *Peranema trichophorum* Ehrbg. Stein. *Biol. Bull., Woods Hole*, 54:36-65.

Hollande, A. 1942. Étude cytologique et biologique de quelques flagellés libres. Volvocales, Cryptomonadines, Eugléniens, Protomastigines. *Arch. Zool. Exp. Gen.*, 83:1-268.

Hollande, A. 1952. Classe des Eugléniens (Euglenoidina Bütschli 1884). *In*: Grassé, P.-P. (ed.), *Traité Zoologie*. Vol. I(1). Masson et Cie, Paris. pp 238-284

Huber-Pestalozzi, G. 1955. *Das Phytoplankton des Süsswassers. 4. Euglenophyceen E.* Schweizerbart'sche Verlagsbuchhandlung, Stuttgart.

Jahn, T. L. & McKibben, W. R. 1937. A colorless euglenoid flagellate, *Khawkinea halli* N. Gen., N. Sp. *Trans. Am. Microsc. Soc.*, 56:48-54.

Johnson, D. F. 1934. Morphology and life history of *Colacium vesiculosum* Ehrbg. *Arch. Protistenkd.*, 83:241-263.

Lackey, J. B. 1929. Studies on the life histories of Euglenida. II. The life cycles of *Entosiphon sulcatum* and *Peranema trichophorum. Arch. Protistenkd.*, 67:128-156.

Lackey, J. B. 1933. Studies on the life histories of Euglenida. III. The morphology of *Peranema trichophorum* Ehrbg., with special reference to its kinetic elements and the classification of the Heteronemidae. *Biol. Bull., Woods Hole*, 65:238-248.

Lackey, J. B. 1960. *Calkinsia aureus* gen. et sp. nov., a new marine euglenid. Trans. Am. Microsc. Soc., 79:105-107.

Larsen, J. (1987). Algal studies of the Danish Wadden Sea. IV. A taxonomic study of the interstitial euglenoid flagellates. Nord. J. Bot., 7:589-607.

Larsen, J. & Patterson, D. J. 1990. Some flagellates (protista) from tropical marine sediments. J. Nat. Hist., 24:801-937.

Larsen, J. & Patterson, D. J. 1991. The diversity of heterotrophic euglenids. *In*: Patterson, D. J. & Larsen, J. (eds.), *The Biology of Free-living Heterotrophic Flagellates.* Clarendon Press, Oxford. pp 205-217

Lee, W. J. & Patterson, D. J. 1999. Heterotrophic flagellates of Botany Bay, Australia. J. Nat. Hist. (in press)

Leedale, G. F. 1967a. *Euglenoid Flagellates* Prentice-Hall Inc., Englewood Cliffs, New Jersey.

Leedale, G. F. 1967b. Euglenida/ Euglenophyta. *Ann. Rev. Microbiol.*, 21:31-48.

Leedale, G. F. 1975. Envelope formation and structure in the euglenoid genus *Trachelomonas. Br. Phycol. J.*, 10:17-41.

Leedale, G. F. 1995. Euglenophyceae. *In*: Parker, B. (ed.) *Encyclopedia of Algal Genera.*

Leedale, G. F. & Hibberd, D. J. 1974. Observations on the cytology and fine structure of the euglenoid genera *Menoidium* Perty and *Rhabdomonas* Fresenius. *Arch. Protistenkd.*, 116:319-345.

Leedale. G. F., Meeuse, B. J. D. & Pringsheim, E. G. 1965. Structure and physiology of *Euglena spirogyra*. I, II. *Arch. Mikrobiol.*, 50:68-102.

Lyman, H., Epstein, H. T. & Schiff, J. A. 1959. Ultraviolet inactivation and photoreactivation of chloroplast development in *Euglena* without cell death. *J. Protozool.*, 6:264-265.

Massart, J. 1920. Recherches sur les organismes inférieurs. VIII. - Sur la motilité des flagellates. Bull. Acad. Roy. Belg. Classe des Sci., Ser. 5, 6:116-141.

Michajlow, W. 1969a. Taxonomic problems in the classification of Euglenoidina parasitizing Copepoda. Bull. Acad. Polon. Sci. (Ser. Sci. Biol.), 17603-605.

MIchajlow, W. 1969b. The position of flagellata-parasites of Copepoda in the systematics of Euglenoidina. Bull. Acad. Sci. Polon. (Ser. Sci. Biol.), 17:499-501.

Michajlow, W. 1972. *Euglenoidina Parasitic in Copepoda. An Outline Monograph.* PWN Poliszh Scientific Publishers, Warsaw.

Michajlow, W. 1978. *Eutreptra parasitica* sp. n. and other Euglenoidina-parasites of Australian copepoda. *Bull. Acad. Polon. Sci. (Ser. Sci. Biol.,* 26:269-272.

Mignot, J.-P. 1961. Contribution à l'étude cytologique de *Scytomonas pusilla* (Stein) (Flagellé euglénien). *Bull. Biol. Fr. Belg.*, 95:665-678.

Mignot, J.-P. 1963. Quelques particularités de l'ultrastructure d'*Entosiphon sulcatum* (Duj.) Stein, flagellé euglénien. *C. R. Hebd. Seanc. Acad. Sci. Paris* , 257:2530-2533.

Mignot, J.-P. 1964. Observations complémentaires sur la structure des flagelles d'*Entosiphon sulcatum* (Duj.) Stein, flagellé euglénien. *C. R. Hebd. Seanc. Acad. Sci. Paris*, 258:3360-3363.

Mignot, J.-P. 1966. Structure et ultrastructure de quelques Euglenomonadines. *Protistologica*, 2:51-117.

Nisbet, B. 1974. An ultrastructural study of the feeding apparatus of *Peranema trichophorum. J. Protozool.*, 21:39-48.

Pringsheim, E. G. 1942. Contributions to our knowledge of saprotrophic algae and flagellata. III. *Astasia, Distigma, Menoidium and Rhabdomonas. New Phytol.*, 41:171-205.

Pringsheim, E. G. 1953a. Salzwasser-Eugleninen. *Arch. Mikrobiol.*, 18:149-164.

Pringsheim, E. G. 1953b. Observations on some species of *Trachelomonas* grown in culture. *New Phytol.*, 52:93-113, 238-266.

Pringsheim, E. G. 1953c. Notiz über *Colacium* (Euglenaceae). *Öst. Bot. Z.*, 100:270-275.

Pringsheim, E. G. 1956. Contributions towards a monograph of the genus *Euglena. Nova Acta Leopold.*, 18:1-168.

Pringsheim, E. G. 1963. *Farblose Algen: Ein Beitrag zur Evolutionsforschung.* Gustav Fischer, Stuttgart.

Pringsheim E. G. & Pringsheim, O. 1952. Experimental elimination of chromatophores and eyespot in *Euglena gracilis. New Phytol.*, 51:65-76.

Provasoli, L., Hutner, S. H. & Schatz, A. 1948. Streptomycin-induced chlorophyll-less races of *Euglena. Proc. Soc. Exp. Biol. Med.*, 69:279-282.

Shawan, F. M. & Jahn, T. L. 1947. A survey of the genus *Petalomonas* Stein (Protozoa: Euglenida). *Trans Am. Microsc. Soc.*, 66:182-189.

Simpson, A. G. B. 1997. The identity and composition of the Euglenozoa. Arch. Protistenkd., 148:318-328

Skuja, H. 1939. Beitrag zur Algenflora Lettlands II. *Acta Hort. Bot. Univ. Latv.*, 11/12:41-169.

Skuja, H. 1948. Taxonomie des Phyto-planktons einiger Seen in Uppland, Schweden. *Symb. Bot. Upsal.*, 9:1-399.

Starmach, K. 1983. *Flora Slodkowodna Polski.* Tom. 3. Polska Academia Nauk, Warsaw.

Triemer, R. E. 1985. Ultrastructural features of mitosis in *Anisonema* sp. (Euglenida). *J. Protozool.*, 32:683-690.

Triemer, R. E. & Farmer, M. A. 1991a. An ultrastructural comparison of the mitotis apparatus, feeding apparatus, flagellar apparatus and cytoskeleton in euglenoids and kinetoplastids. *Protoplasma* , 164:91-104.

Triemer, R. E. & Farmer, M. A. 1991b. The ultrastructural organization of the heterotrophic euglenoids and its evolutionary implications. *In*: Patterson, D. J. & Larsen, J. (eds.), *The Biology of Free-Living Heterotrophic Flagellates.* Syst. Ass. Spec. Vol. 45: 185-204. Clarendon Press, Oxford.

DIPLONEMIDS
(CLASS: DIPLONEMEA Cavalier Smith, 1993)

By KEITH VICKERMAN

The diplonemids contain two genera (Simpson, 1997) and have been classified up to the level of Class (Class Diplonemea Cavalier-Smith, 1993 (Cavalier-Smith, 1993), although a family is a sufficient rank to contain the total hierarchy of two genera. They are relatively rarely reported, recorded from plankton and benthos, freshwater and marine, sometimes associated with other protists (diatoms), aquatic plants, or animals (Crustacea) as scavengers, epibionts, or parasites. They have a characteristic gliding/creeping trophic phase with flagella barely emerging from the flagellar pocket and not beating; this stage also exhibits pronounced squirming ("euglenoid") movement. A fully flagellated dispersive phase, bearing some resemblance to the kinetoplastids *Bodo* and *Hemistasia* is present in some species.

Key Characters

Diplonemids are phagotrophic or osmotrophic flagellates. All species which have been studied by electron-microscopy have a mouth. There are two usually equal flagella lacking paraxial rods; the flagellar pocket has a posterior diverticulum. The mitochondrion is a cortical branched network with extensive flattened lamellar cristae; it lacks a kinetoplast, but its matrix contains knotted fibers of an unknown nature. Chloroplasts are absent, as are paramylum granules; energy reserves have not been identified; peroxisomes are not glycosomes (see Kinetoplastea, below). Pellicular tubules are evenly spaced but a plicate cortex (with folds/strips as in euglenoids) is absent, yet the feeding apparatus includes a plicate cytostome (a cytostome supported by two rods composed of microtubules and a series of folds or vanes, as in some euglenoids). In *Diplonema* the chromosomes, unlike those of euglenoids and kinetoplasteans form a metaphase plate during nuclear division. Locomotion is as described above. All diplonemeans show a central nucleus, a prominent cylindrical ingestion apparatus alongside the flagellar pocket, and a cytoplasm packed with refractile granules, many of them lipid globules. The genus *Diplonema* was described by Griessman (1913). Ultrastructural features of *Diplonema* (Triemer and Ott, 1990) are virtually identical with those of *Isonema* as described by Schuster et al. (1968) and later by Porter (1973) and Kent et al. (1987); Triemer and Ott (1990) have synonymised the two genera. *Rhynchopus* was described by Skuja (1948), and its ultrastructure (Schnepf, 1994) would appear also to be similar to that of *Diplonema*.

Diplonemids are easy to grow in axenic culture (Porter, 1973; Triemer and Ott, 1990). A study of three epibiont isolates from crustacea (Vickerman and Appleton, unpubl.) has recently demonstrated the presence of a dispersive flagellate phase, unrecorded from the numerous isolates of *Diplonema* now in culture (ATCC lists 4). Preliminary SSU rRNA gene sequencing data (Triemer, unpubl.) suggest that the Diplonemea form a sister group to the Euglenoidea and that the Diplonemea/Euglenoidea clade is a sister group to the Kinetoplastea. Only a single family, the Diplonemidae, need be recognized, which forms part of an unresolved polychotomy with the euglenids, kinetoplastids, and *Postgaardi* (Simpson, 1997)

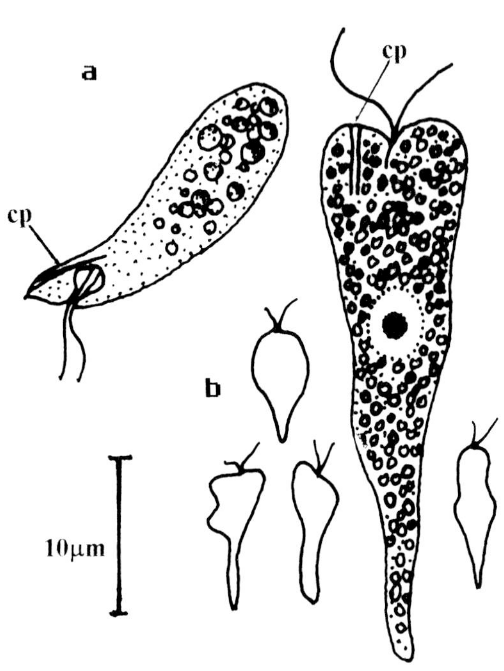

Fig. I. (a) *Diplonema ambulator* showing protruding flagella used in crawling. (b) *D. metabolicum* with outlines showing variation in shape. (cp) cytopharynx. (After Larsen and Patterson, 1990) Scale bar (except for outlines) =10 μm.

Genus ***Diplonema*** Griessmann, 1913

The type species, *D. breviciliata* Griessmann, has two short (3-4 μm), projecting non-motile flagella while in the similar-sized (14-23 μm) *D. ambulator* (Larsen and Patterson, 1990; Fig 1a) these short flagella flex slowly at their bases to give the appearance of walking legs during the organism's creeping movements. *Diplonema metabolicum* (Fig.1b), another sediment-dwelling species (30-48 μm; Larsen and Patterson, 1990), lacks the broader posterior end characteristic of most diplonemeans and moves only by writhing, not crawling or swimming. Triemer and Ott (1990) found *D. ambulator* associated with decaying leaves of an aquarium plant (*Cryptocoryne* sp.) and Porter found *D. papillatum* associated with eelgrass (*Zostera marina*). The protists enter plant cells and scavenge degenerating cytoplasm. Refs. Larsen and Patterson (1990); Triemer and Ott (1990).
TYPE SPECIES: *Diplonema breviciliata* Griessmann, 1913.

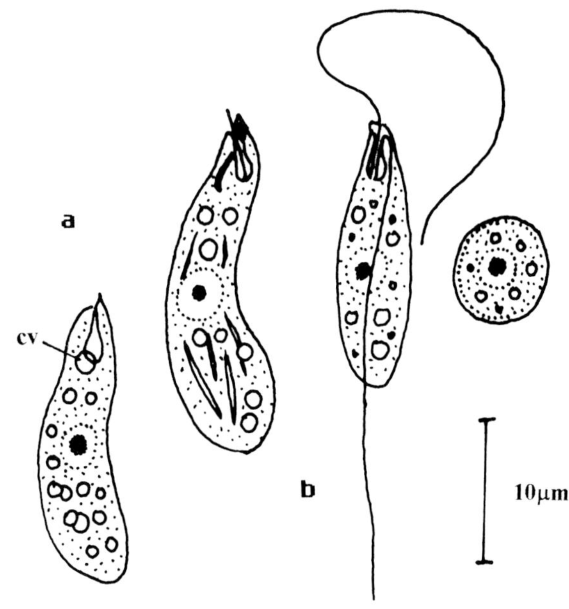

Fig. 2. (a) *Rhynchopus amitus*. (cv) Contractile vacuole. (Based on Skuja, 1948) (b) *Rhynchopus* sp. from gills of *Nephrops norvegicus*; creeping, enflagellated, and encysted stages in culture. (K. Vickerman, unpubl.) Scale bar=10 μm.

Genus ***Rhynchopus*** Skuja, 1948

Rhynchopus amitus, the type species, described from Baltic plankton by Skuja (1948) had an elongate pear-shaped body, often more concave along one side than the other. An anterior papilla separated the ingestion apparatus from the flagellar pocket (Fig. 2a), and the 2 flagella barely emerged from the pocket of the creeping cell. Schnepf (1994) found a similar form (*R. coscinodiscivorus*) feeding on the cytoplasm of the planktonic diatom *Coscinodiscus.* Bodammer and

Sawyer (1981) described an organism with diplonemean ultrastructure on the gills of the crab *Cancer irroratus*, and, more recently, isolates from the blood and gills of the Norway lobster, *Nephrops norvegicus* (Vickerman and Appleton, unpubl.) have proved to be *Rhynchopus* species. When starved, these isolates readily produce *Bodo*-like motile flagellates (Fig. 2b), and it is suggested that the presence of a fully flagellated dispersive phase in the life-cycle serves to distinguish *Rhynchopus* from *Diplonema*. Encysted stages are also produced. Refs. Bodammer and Sawyer (1981); Schnepf (1994); Skuja (1948).
TYPE SPECIES: *Rhynchopus amitus* Skuja, 1948

References

Bodammer, J. E. & Sawyer, T. K. 1981. Aufwuchs protozoa and bacteria on the gills of the rock crab, *Cancer irroratus* Say: a survay by light and electron microscopy. *J. Protozool.*, 28:35-46.

Griessmann, K. 1914. Uber marine Flagellaten. *Arch. Protistenkd.*, 32:1-78.

Kent, M. L., Elston, R. A., Nerad, T. A. & Sawyer, T. K. 1987. An *Isonema*-like flagellate (Protozoa: Mastigophora) infection in larval geoduck clams, *Panope abrupta*. *J. Invertebr. Pathol.*, 50:221-229.

Larsen, J. & Patterson, D. J. 1990. Some flagellates (Protista) from tropical marine sediments. *J. Nat. Hist.*, 24:801-937.

Porter, D. 1973. *Isonema papillatum* sp. n., a new colorless marine flagellate: a light and electron microscopic study. *J. Protozool.*, 20:351-356.

Schnepf, E. 1994. Light and electron microscopical observations in *Rhynchopus coscinodiscivorus* spec. nov., a colorless phagotrophic euglenozoan with concealed flagella. *Arch. Protistenkd.*, 144:63-74.

Schuster, F. L., Goldstein, S. & Hershenov, B. 1968. Ultrastructure of a flagellate, *Isonema nigricans* nov. gen., nov. sp. from a polluted marine habitat. *Protistologica*, 4:141-149.

Simpson, A., G. B. 1997. The identity and composition of the Euglenozoa. Arch. Protistenkd., 148:319-329.

Skuja, H. 1948. Taxonomie des Phytoplanktons einiger Seen in Uppland, Schweden. *Symb. Bot. Ups.*, 10:1-399.

Triemer, R. E. & Ott, D. W. 1990. Ultrastructure of *Diplonema ambulator* Larsen & Patterson (Euglenozoa) and its relationship to *Isonema*. *Europ. J. Protistol.*, 25:316-320.

ORDER KINETOPLASTEA Honigberg, 1963

By KEITH VICKERMAN

Kinetoplastid flagellates are abundant in nature either as free-living consumers of bacteria in freshwater, marine, and terrestrial ecosystems or as parasites of animals, flowering plants, and other protists. They are best known as the causative agents of important diseases of humans and their domestic animals, the trypanosomiases and leishmaniases. The free-living kinetoplastids all belong to the family Bodonidae whose members are characterized by the presence of two **heterodynamic flagella** (Fig. 1).

The parasitic kinetoplastids belong largely to the uniflagellate Trypanosomatidae (Fig. 2), which has no free-living members though three bodonid genera are composed of parasitic species, some of which cause disease in fishes. The trypanosomatids have the broadest host range of any major group of parasitic eukaryotes after the nematodes. Two-host life cycles have evolved in both families, and several morphologically and metabolically different developmental stages may occur in each host; multiplicative and non-multiplicative stages often alternate (Vickerman, 1994). Altruistic cooperation between different phenotypes in the same host environment may be a feature of such parasite life cycles (Vickerman, 1989, 1993), for example, the laying down of life by some trypanosome antigenic variants in the face of the host's immune response, so that the stock itself can survive.

Although they have no fossil history, molecular-sequencing techniques, especially of ribosomal RNA genes (Schlegel, 1994), have confirmed that the kinetoplastids represent an extremely ancient lineage that separated from other eukaryotes not long after the origin of the eukaryotic cell, and that they were among the first mitochondria-containing cells. They have modified their single mitochondrion by amplifying the mitochondrial DNA to form a stainable structure, the

kinetoplast, readily revealed by traditional Romanowsky- or Feulgen-staining procedures, or, where kinetoplasts are multiple and submicroscopic, by fluorescence microscopy using the fluorochrome DAPI (Vickerman, 1978). Gene sequencing techniques are rapidly transforming our view of the evolutionary pathway of the kinetoplastids. Comparison of sequences suggests that the two-host (**digenetic**) life cycle has evolved repeatedly from the one-host (**monogenetic**) life cycle on different branches of their phylogenetic tree (Fernandes et al., 1993; Maslov et al., 1994).

The difference in size between kinetoplastids and euglenoids is striking. Most kinetoplastids are less than 30 µm in length while most euglenids are larger than this; however, the haematozoic (blood-dwelling stages) in the life cycle of some fish trypanosomes may reach 130 µm. The **amastigote** stages of *Leishmania* spp. at 2-3 µm are among the smallest eukaryotes known. The free-living, phagotrophic kinetoplastids can usually be maintained in monoxenic culture (with e.g the bacterium *Klebsiella aerogenes*) using a weakly nutritive medium (e.g. cerophyll infusion, soil extract) to support growth of the bacterial food organism; axenic growth has not yet been achieved. Most trypanosomatids can be isolated and maintained in axenic culture, usually on blood-containing agar media or insect tissue culture media, but different stages in the life cycle are not equally amenable. Thus the sleeping sickness parasite, *Trypanosoma brucei*, has for several decades been cultivated as the tsetse fly vector midgut (**procyclic**) stage, but only relatively recently have the mammalian stages been maintained in vitro, and the salivary gland stages of the vector are still proving difficult to grow in culture (Taylor and Baker, 1987). The intracellular stages of *Leishmania* spp. are now routinely grown without the macrophage host cell (Bates, 1994), and the pathogenic *Phytomonas* spp. are slowly accommodating to life outside plant phloem sieve tubes (Menara et al., 1988). Many bodonids and monogenetic kinetoplastids have bacterial endosymbionts and, in the case of the latter, a role in providing essential nutrients, especially haemin, has been demonstrated (Roitman and Camargo, 1985). Doubling time in vitro is usually of the order of 6-12 hours. Reproduction is almost invariably by binary fission, though delayed cytokinesis results in multiple fission in rare cases.

There is now good genetic evidence for hybrid formation during the development of *Trypanosoma brucei* in the tsetse fly (Jenni et al., 1986), though meiosis and **syngamy** have not been demonstrated cytologically and mating is not obligatory for completion of the life cycle (reviewed Gibson, 1995). Sexual processes in other kinetoplastids are even more of a mystery, but synaptonemal complexes (suggesting meiosis) have been clearly demonstrated in one isolate of the bodonid *Dimastigella trypaniformis* (Vickerman, 1991) yet not in another (Breunig et al., 1993). The kinetoplastids may largely have abandoned sex.

Accounts of the pathogenic Trypanosomatidae loom large in texts on tropical medicine and veterinary medicine. Wenyon's (1926) textbook is still a mine of detailed and useful information on the kinetoplastid flagellates, but the Kinetoplastida were not monographed as an Order until Lumsden and Evans (1976, 1979) published their two important volumes. Hoare's (1972) monograph on the mammalian trypanosomes is a classic. The symposium edited by Patterson and Larsen (1991) has the most recent general reviews on free-living kinetoplastids. Wallace's reviews (1966, 1979) are an invaluable source of information on the monogenetic trypanosomatids. Updating reviews on parasitic kinetoplastids are to be found in Parasitology Today and Advances in Parasitology; the latter includes one of the few reviews on the cryptobias and trypanoplasms (Woo, 1987). Vickerman (1990) has given a general review of kinetoplastid biology.

Key Characters

The kinetoplastids are set apart from all other eukaryote organisms by distinctive organizational features which include: (1) presence of a single, often branched mitochondrion in which the DNA is amplified to form a stainable mass (or masses), the kinetoplast, usually being located close to the base of the flagellar apparatus (Fig.3); the circular DNA molecules composing the kinetoplast encode not only mitochondrial enzymes and mitochondrial ribosomal RNAs, but also "guide RNAs" involved in editing transcripts of the mitochondrial enzyme genes (reviewed Stuart and Feagin, 1992); in trypanosomatids the circular molecules are catenated together to form a network linked via the mitochondrial membrane to the flagellum base (Robinson and Gull, 1991), but in bodonids the kDNA molecules are not always catenated (Hajduk et al., 1986) - hence their ability to disperse within the mitochondrion (Fig.4); (2) segregation of enzymes of the glycolytic chain inside a membrane-bounded organelle, the glycosome (reviewed by Opperdoes, 1991); (3) one or two flagella in which a lattice-like paraxial rod runs alongside the axoneme in the flagellar shaft; the flagellum may be modified for attachment to surfaces as well as locomotion (reviewed Vickerman and Tetley, 1990); (4) a unique cytoskeleton reinforcing the body surface and composed of microtubules which have their own idiosyncratic form of nucleation on microtubule-associated proteins of adjacent (parallel) microtubules (Sherwin and Gull, 1989); (5) restriction of endocytosis and exocytosis of macromolecules, and insertion of new membrane into the body surface, to the flagellar pocket (Fig. 5) (reviewed Webster and Russell, 1993) or to that site and, for uptake, to an associated microtubule-lined cytopharynx (Figs. 3, 4); bodonids with extrusomes may be an exception to the pocket-only exocytosis rule; (6) a nuclear genome consisting of a large number of chromosomes that do not condense at mitosis (Van der Ploeg et al., 1984) when they are partitioned on an intranuclear spindle not associated with the flagellar basal bodies ; the chromosomes contain polycistronic transcription units; control of gene expression is largely at the post-transcriptional level (reviewed Pays, 1993).

The kinetoplastids share with the euglenids (1) a basically biflagellate condition with an anterior locomotory and a posterior recurrent flagellum that may be attached to the body; (2) presence of paraxial rods in the flagella; (3) both flagella arise from a pocket into which the contractile (pulsatile) vacuole empties and from which pinocytosis may occur; a separate microtubule-reinforced cytopharynx may be present in both; (4) extrusomes, when present, have a similar

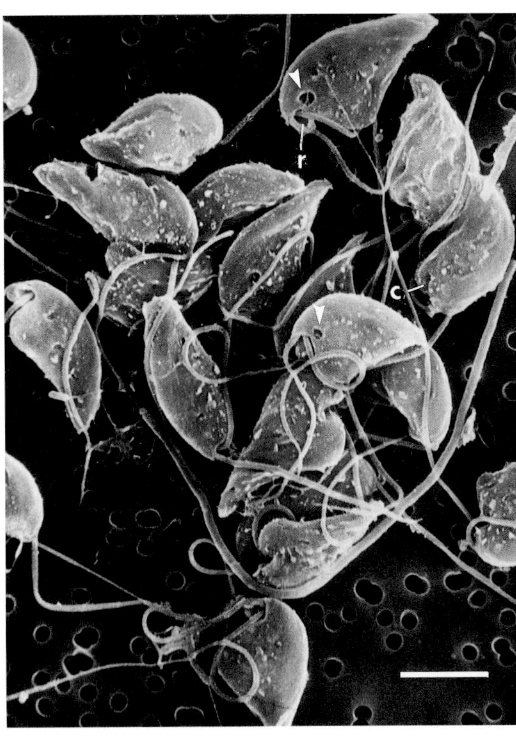

Fig. 1. Scanning electron micrograph showing several individuals of *Bodo caudatus*, each with two flagella emerging from a flagellar pocket at the anterior end of the flattened pear-shaped body. The rostrum (r) is evident in those seen from the left hand side, the cytostome opening (c) in those seen from the right. Arrowheads show artifactual pit formed by rupture of contractile vacuole through surface membrane. Scale bar=5 µm. (Micrograph: K. Vickerman)

structure; (5) similar transcriptional arrangements for gene expression. There are notable differences, however: the euglenids possess a more labile mitochondrial network which lacks kinetoplasts, and glycolytic enzymes are not organelle-associated; kinetoplastids lack all trace of chloroplasts and, usually, of storage carbohydrates, being dependent upon exogenous sugars or amino acids as a source of energy; the strip-like composition of the euglenid cortex with its less complete microtubule cytoskeleton enables exocytosis to take place away from the flagellar pocket. Although both groups exhibit a vesicular nucleus with a prominent central nucleolus that persists within the endonuclear spindle apparatus during division, the permanently condensed chromosomes of the euglenids contrast with the persistently uncondensed chromosomes of the Kinetoplastida (for more detailed comparisons see Kivic and Walne, 1984; Triemer and Farmer, 1991).

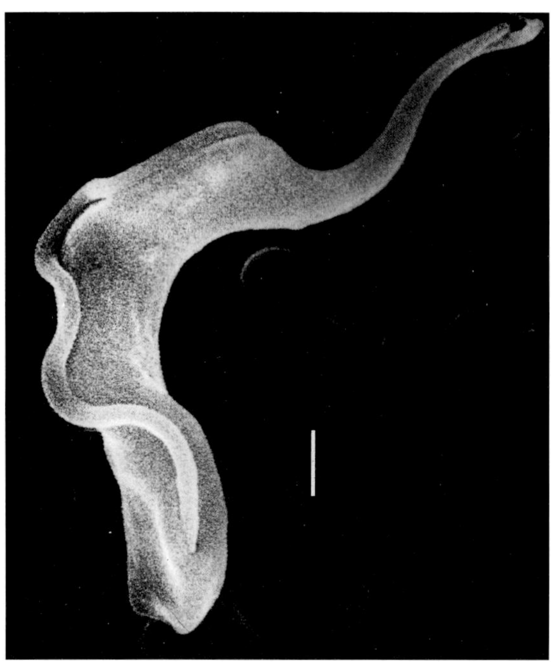

Fig. 2. Scanning electron micrograph of stumpy bloodstream form of *Trypanosoma brucei* showing the single flagellum emerging from the flagellar pocket at the posterior end of the flagellate and its attachment along the length of the body to form the undulating membrane. Scale bar=1 μm. (Micrograph: K. Vickerman)

Some 800 reasonably well characterized species of kinetoplast-bearing flagellates have been described and named, about 700 of which are trypanosomatids and of these about 500 are species of *Trypanosoma*. Trypanosomatids have often been named after the host in which they were found; a comprehensive list of species and hosts has been published by Podlipaev (1990). Neither generic nor specific distinctions are easy to draw. In bodonids, the distribution of kinetoplast DNA (kDNA)—whether it is massed close to the flagellar bases (eukinetoplasty, Figs. 3,5) or dispersed through the mitochondrion as several identical bodies (polykinetoplasty) or unevenly spread as diffuse masses (pankinetoplasty, Fig. 4)—may be used to define specific genera (Fig. 7). Other useful characters are the degree of development of the rostrum/oral apparatus and the extent of attachment of either or both flagella to the body.

Trypanosomatid genera have long been distinguished on the spectrum of morphological stages evident in the life cycle. These stages are identified by the topological relations of the single eukineto-plast/flagellum base to the nucleus and to the anterior end of the body, also by whether the emergent flagellum is free or attached to the body to form an undulating membrane (Hoare and Wallace, 1966); thus **amastigote**, **promastigote**, **choanomastigote**, **opisthomastigote**, **endomastigote**, **epimastigote**, and **trypomastigote stages** can be recognized (Fig. 6). Another consideration in defining genera has been whether the life cycle involves one or two hosts; the nature of the hosts may also be taken into account, e.g. all two-host trypanosomatids parasitizing plants are referred to the genus *Phytomonas*. Wallace et al (1983) have suggested guidelines for distinguishing species in the lower trypanosomatids. Species distinctions in free-living bodonids may depend upon motility patterns and kinetoplast shape (e.g. see Fig. 8); in parasitic kinetoplastids, kinetoplast size and shape may be important, but host species and preferred location in it are also important.

Molecular sequencing techniques suggest that genetic distances between so-called species within a morphologically defined genus may vary enormously. Thus *Trypanosoma brucei* and *T. cruzi*, both parasitizing humans may be as far apart genetically as sea urchin and man (Hecker, 1993), while the *Leishmania* species are genetically very close, not only to one another but also to the genus *Endotrypanum* (Fernandes et al., 1993). The taxonomy of the so-called monogenetic trypanosomatids in particular is currently in a state of flux: reliance on the traditional characters listed above is crumbling as closer affinities at the molecular level are revealed between some members of different genera than between members of the same putative genus. Even the sacrosanct distinction between monogenetic and digenetic genera is breaking down as some trypanosomatids assigned on morphological grounds to the former are found to be able to infect plants or vertebrates (see *Herpetomonas*, *Phytomonas*, and *Leishmania* below).

KEY TO FAMILIES AND GENERA

1. Flagellates with a single anteriorly directed flagellum, sometimes attached to body to form undulating membrane; kinetoplast compact, usually small and invariably associated with base of flagellar apparatus; cytostome, if present, lacks associated oral structures; non-phagotrophic; all parasitic..**Family Trypanosomatidae**...... 10

1'. Flagellates with two heterodynamic flagella, one projecting anteriorly and never attached to body, the other trailing posteriorly, sometimes attached to body; kinetoplast large or diffuse, either associated with flagellar bases or multiple and spread throughout body as several masses; cytostome with prominent preoral ridge developed as rostrum in phagotrophic forms; free-living or parasitic..**Family Bodonidae**............... 2

2. Kinetoplasts multiple, but discrete uniformly sized bodies, often minute................................ 3

2'. Kinetoplast single or present as diffuse masses of DNA.. 5

3. Kinetoplasts prominent; flagellates ectoparasitic on fish; rostrum developed for insertion into epidermal cell, reduced in free-swimming dispersive phase of life cycle....... ***Ichthyobodo***

3'. Kinetoplasts minute and only demonstrable by fluorescence or electron microscopy............... 4

4. Trailing flagellum not attached to body; predatory flagellates with cytostome at tip of well-developed snout.................. ***Rhynchobodo***

4'. Trailing flagellum attached to body; anterior and posterior flagella originate at an obtuse angle to one another; bacterivorous.. ***Dimastigella***

5. Recurrent flagellum free from body over greater part of its length.................................. 6

5'. Recurrent flagellum attached to body for more than one third its length................................ 7

6. Rostrum small and independent of anterior flagellum... ***Bodo***

6'. Rostrum forms prominent mobile snout attached to anterior flagellum which is same length as snout........................ ***Rhynchomonas***

7. Flagellate sedentary, cluster of individuals anchored to substratum by communal secreted stalk attached to their posterior extremitities .... ***Cephalothamnium***

7'. Flagellates free swimming............................... 8

8. Recurrent flagellum not attached along entire length of body; lacks ventral groove alongside recurrent flagellum; phagotrophic... ***Procryptobia***

8'. Recurrent flagellum attached along entire length of body; ventral groove present alongside recurrent flagellum; osmotrophic....................9

9. Monogenetic parasites; flagellum does not form prominent undulating membrane; kinetoplast oval, elongate, or dispersed through body as series of irregular masses of DNA... ***Cryptobia***

9' Digenetic parasites in poikilotherms and leeches; recurrent flagellum forms prominent undulating membrane; kinetoplast rod-like .... ***Trypanoplasma***

10. Trypanosomatids with flagellum emerging from flagellar pocket at anterior end of body and never attached to body................................... 11

10'. Trypanosomatids with flagellum emerging from flagellar pocket laterally and attached to body along part or most of its length......................17

11. Trypanosomatids with flagellar pocket elongate and extending behind the nucleus in some stages of life cycle...................................... 12

11'. Trypanosomatids with flagellar pocket entirely prenuclear (except in rare individuals).....13

12. Trypanosomatids with flagellar pocket extending to posterior end of cell (opisthomastigote form) in some stages of life cycle.................... .... ***Herpetomonas***

12'. Trypanosomatids with stage in life cycle in which flagellar pocket extends to rear of cell and then curves up forward towards nucleus, with flagellar base and kinetoplast lying beside nucleus (endomastigote form)............... .... ***Wallaceina***

13. Monogenetic parasites, largely of insects......14

13'. Digenetic parasites of insects and their food source organisms (plants, vertebrates, other insects); promastigote and amastigote stages only in life cycle... 15

14. Monogenetic parasites with promastigote stage dominant in life cycle; amastigotes or cysts may also be present................... ***Leptomonas***

14'. Monogenetic parasites with choanomastigote stage dominant in life cycle............ ***Crithidia***

15. Digenetic parasites alternating between plants and sap-sucking bugs (Heteroptera); promastigote stage dominant at all stages in life cycle.................................... ***Phytomonas***

15'. Digenetic parasites alternating between vertebrates and sand flies (Diptera: Phlebotominae); promastigote stages in the latter, amastigote stage in the former........ 16

16. Vertebrate host a reptile (lizard); amastigotes in blood cells (monocytes, thrombocytes) and not in tissues.................... ***Sauroleishmania***

16'. Vertebrate host a mammal; amastigotes in mononuclear phagocyte system of tissues...... .... ***Leishmania***

17. Monogenetic parasites of insects (rarely ticks) 18

17'. Parasites of bloodstream of vertebrates as trypomastigote (kinetoplast and base of undulating membrane post-nuclear) stage; usually digenetic with a bloodsucking invertebrate vector (arthropod, leech).........19

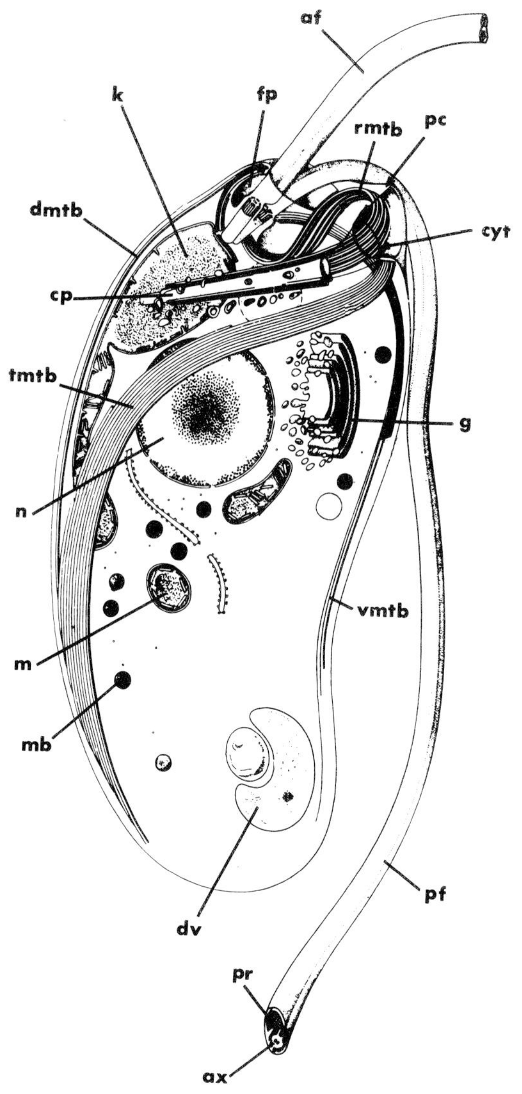

Fig. 3. Structural features of *Bodo caudatus* seen in right lateral view showing basic bodonid cytoskeletal system. The two flagella (af, anterior flagellum; pf, posterior flagellum) arise from the flagellar pocket (fp). The basal bodies are apposed to the kinetoplast (k) region of the unitary mitochondrion which is represented as several mitochondrial profiles (m). The cytoskeleton consists of four microtubular bands, the dorsal (dmtb), ventral (vmtb), transverse (tmtb) and reinforcing (rmtb) microtubular bands. This last reinforces the preoral crest (pc) and is deflected inwards at the cytostome to support the cytopharynx (cp). (ax) flagellar axoneme; (dv) digestive vacuole; (g) Golgi apparatus; (mb) microbodies (probably glycosomes); (n) nucleus; (pr) paraxial rod. The contractile vacuole is seen behind cp and tmbt. (Adapted from Brugerolle et al., 1979)

18. Epimastigote stage dominant in life cycle; cysts also formed .... ***Blastocrithidia***
18'. Trypomastigote stage dominant in life cycle...... .... ***Rhynchoidomonas***
19. Intraerythrocytic trypomastigotes and epimastigotes in mammalian (edentate) host, promastigotes and amastigotes in phlebotomine insect vector .... ***Endotrypanum***
19'. Exoerythrocytic bloodstream trypomastigotes in vertebrate host (though multiplicative stage may be epimastigote or amastigote in tissues), epimastigotes among vector developmental stages.. ***Trypanosoma***

SUBORDER BODONINA Hollande, 1952

Kinetoplastid flagellates with two subapically inserted heterodynamic flagella, one projecting anteriorly and never attached to body, the other trailing posteriorly, sometimes attached to body; **kinetoplast** large and associated with flagellar bases or multiple and spread through body as several masses; kDNA molecules not necessarily catenated into a network; cytostome with prominent preoral ridge developed as a **rostrum** in phagotrophic forms; free living or parasitic

FAMILY BODONIDAE Bütschli, 1887

With characters of the Suborder. The recognition of two families within the suborder—1) Bodonidae, phagotrophic and with free recurrent flagellum, and 2) Cryptobiidae, osmotrophic and with attached recurrent flagellum (cf Vickerman, 1976)—has become difficult to sustain and therefore abandoned. Differences between genera are often poorly drawn. The genera of free-living bodonids have been reviewed by Zhukov (1991) and their cellular organization by Vickerman (1991).

Genus *Bodo* Ehrenberg, 1830

Solitary phagotrophic flagellates with a single discrete, usually ellipsoidal eukinetoplast and the recurrent flagellum free (or mainly free) from the body, which is ovoid, bean- or comma-shaped (Figs. 1,3). The rostrum (preoral ridge) is not usually prominent, and bacteria (more rarely larger particles) are ingested into a **cytostome** alongside the **flagellar pocket**. Cysts are commonly found (but not in *B. saltans*) and may be **polykinetoplastic**.

Most *Bodo* spp. are 5 µm long and free-swimming. The anterior flagellum beats actively, and the recurrent serves as a skid during movement; differences in flagellar beat and motility pattern are helpful in distinguishing different species. In the type species, *Bodo saltans*, however, rapid swimming takes place intermittently between bouts of saltatory activity effected by rapid contraction of the posterior flagellum, and during these the flagellate ingests bacteria. Although some 200 nominal species have been ascribed to this genus, most descriptions are inadequate and only a handful of species are moderately well characterized. These include the common coprophile *B. caudatus* Dujardin (Figs. 1,3,7a), *B. designis* Skuja ubiquitous in fresh and saltwater habitats, with characteristic rotatory movement, *B. saliens* Larsen & Patterson and *B. curvifilus* Griessman—both common in marine habitats, the latter differing from the former in its slower creeping behavior and sweeping pattern of flagellar beating; all these species are available from the ATCC. In view of its very different feeding habits, behavior, and flagellar and oral structure (Brooker, 1971a), *B. saltans* should probably be placed in a different genus from the above named species.
Russian workers in particular (c.f. Zhukov, 1991) have adopted the name *Pleuromonas jaculans* Perty for *B. saltans*, while, confusingly, referring to *B. designis* as *B. saltans*. The genus *Parabodo* Skuja was distinguished from *Bodo* by a flattened body, polar emergence of the anterior flagellum and lateral emergence of the posterior flagellum, but these differences are less marked than those between the above species.

A thorough comparison of the motility patterns and ingestion apparatus of different species of *Bodo* is urgently needed in order to establish reliable species indentification. Species of *Bodo* appear to tolerate wide ranges of temperature and salinity and are widespread in both planktonic and benthic habitats, as well as being extremly common in infusions of decaying matter. Although parasitic species have been described, none has been authenticated

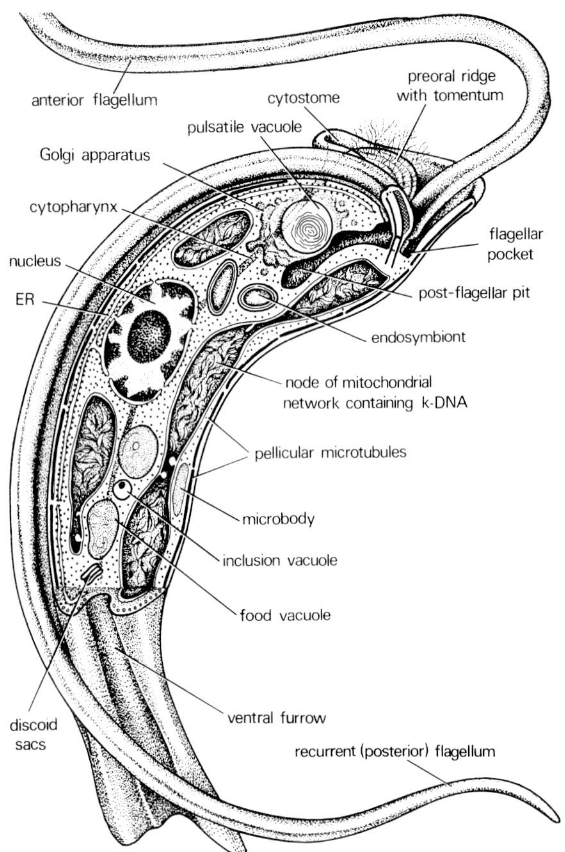

Fig. 4. Structural features of *Cryptobia vaginalis* from the leech as seen in left ventral view (with anterior two-thirds of body removed) to show relationships of flagellar pocket, postflagellar pit, preoral crest (ridge), and cytostome. The kinetoplast DNA is dispersed throughout the mitochondrial network in this species. (From Vickerman, 1977)

Genus ***Rhynchomonas*** Klebs, 1893

Solitary, usually creeping phagotrophic flagellates with a single large oval eukinetoplast and the rostrum developed as the left lobe, the flagellar pocket wall as the right lobe of a relatively large mobile snout that lies alongside and is attached along the length of the short anterior flagellum (Fig. 7b). During movement along the substratum, the snout moves from side to side and when occasionally its tip presses against an attached bacterium or detrital particle, this is ingested via the cytostome at the tip of the snout. The long posterior flagellum trails behind during locomotion, but may wrap around the body during rest. *Rhynchomonas nasuta* (3-11 µm) is the only well-characterized species (see Swale, 1973), though several others have been inadequately described. It is common in mesosaprobic waters, both freshwater and marine, and as a coprophile. Available from the ATCC.

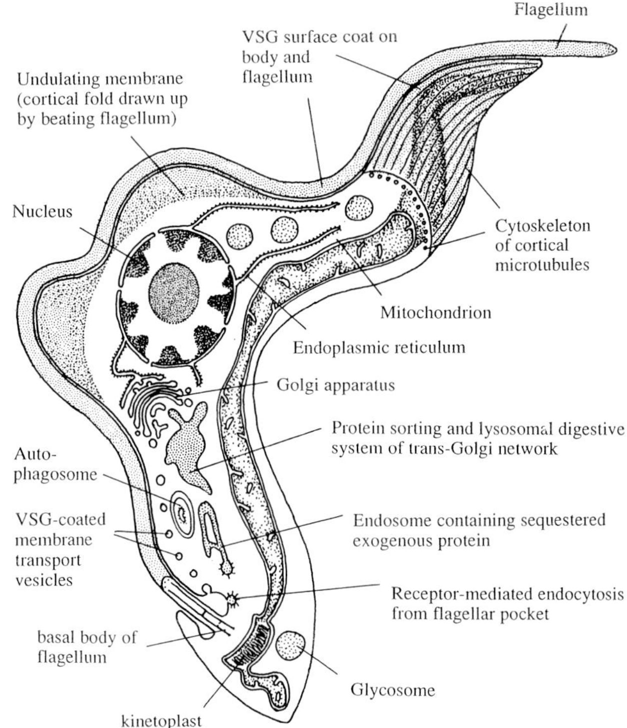

Fig. 5. Structural features of *Trypanosoma brucei* (bloodstream intermediate form) in partial longitudinal section. The cytoskeletal microtubules form an enveloping corset underlying the entire surface membrane of the body, except for the flagellar pocket. Insertion of new surface membrane bearing the variant-specific glycoprotein (VSG) coat, receptor-mediated endocytosis (uptake of proteins) and exocytosis of autophagosomes can only occur via the flagellar pocket membrane. A cytostome/cytopharynx is present in *T. cruzi* and trypanosomes of lower vertebrates, but not in the African trypanosomes. (From Vickerman et al., 1993)

Genus ***Rhynchobodo*** Lackey, 1940
emend. Vørs, 1992

Fast, free-swimming, phagotrophic, polykinetoplastid flagellates with a short anterior flagellum and a long trailing posterior flagellum (Fig. 7c). The apex of the body is drawn out to form a massive snout, free from the flagella which appear to be inserted further back than in most bodonids, and used in seizing prey for ingestion through a terminal cytostome. The numerous kinetoplasts are submicroscopic. Unique among bodonids in possessing extrusomes of the lattice-tube type as in euglenids. *Rhynchobodo armata* (Brugerolle, 1985) appears to be similar to Lackey's *R. agilis* and possibly identical with Mylnikov's (1986) *Phyllomitus apiculatus*. Vørs (1992) has shown that *Cryptaulax taeniata* Skuja, 1948 is a similar organism, but has a prominent spiral grove running along the body, and Elbrachter et al. (1996) have described as *Hemistasia phaeocysticola* another *Rhynchobodo*-like flagellate that is scavenging rather than predatory. For a discussion of the problematic name of this genus see Larsen and Patterson (1990) and Vørs (1992). Marine and freshwater. One species in ATCC.

Genus ***Ichthyobodo*** Pinto, 1928
(=***Costia*** Leclercq, 1890)

Ectoparasitic flagellates on fishes; polykinetoplastic with several prominent kinetoplasts dispersed within the reticular mitochondrion (Fig. 7e). Attached to epithelial cells by a hypertrophied **proboscis** derived from the rostrum and adjacent flagellar pocket and a wall attachment plaque between parasite and host cell at the surface of the latter, the flagellate ingests host cell cytoplasm through the terminal cytostome (for ultrastructure and pathogenesis see Joyon and Lom, 1969;

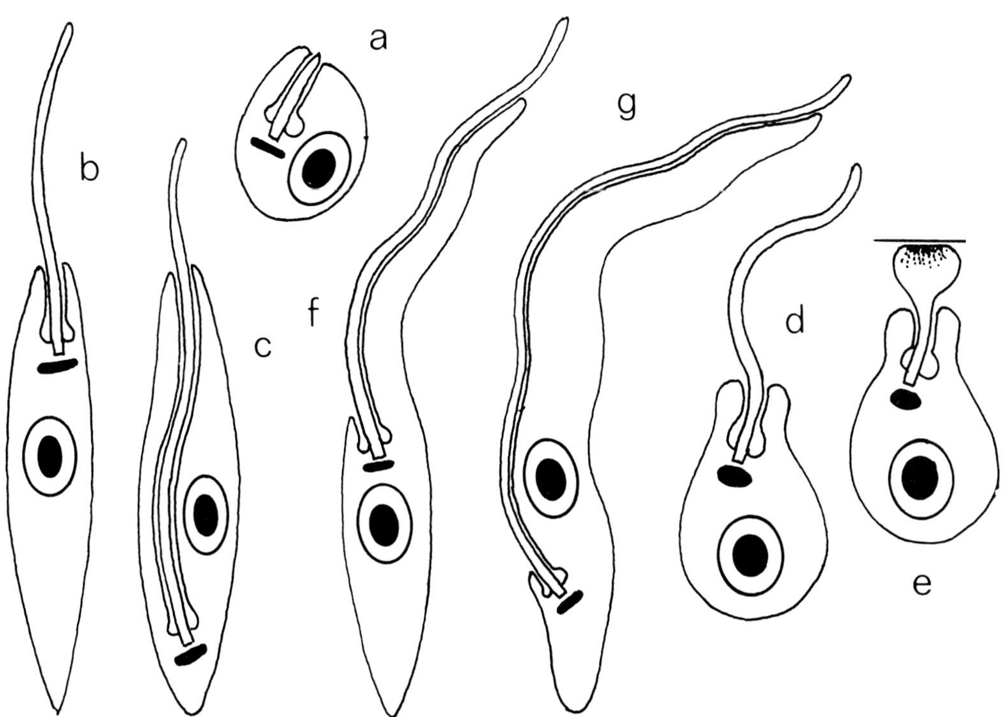

Fig. 6. Diagrams of developmental stages of trypanosomatids named according to the schemes of Hoare and Wallace (1966). Fbk=flagellum base/kinetoplast complex. Flagellar pocket (FP) opening anterior. (a) Amastigote (no emergent flagellum, body oval/rounded). (b) Promastigote (Fbk anterior, prenuclear). (c) Opisthomastigote (Fbk postnuclear, FP long and narrow). (d, e) Choanomastigotes (Fbk prenuclear, FP broad). (d) Nectomonad (free-swimming form). (e) Haptomonad (flagellum-attached form). FP opening laterally, flagellum attached to body for most of its length. (f) Epimastigote (Fbk prenuclear). (g) Trypomastigote (Fbk postnuclear). For endomastigote stage see Fig. 11.

Diamant, 1987). A free-swimming dispersive phase which lacks the hypertrophied proboscis is also present in the life cycle (Fig. 7f). *Ichthyobodo* (spelled*Ichtyobodo* by French workers) *necator* (fomerly called *Costia necatrix*, but this generic name is invalid) is an important cosmopolitan parasite of both freshwater and marine fishes, especially in hatcheries where heavy mortalities of fry may arise as a result of destruction of the gill epithelium. The disease is still called costiasis. *Ichthyobodo necator* on migratory chum salmon (*Onchorhynchus keta*) can adapt their salinity tolerance to the migratory behavior of the host, but cross-infection experiments suggest that on obligate marine fishes, e.g. the Japanese flounder (*Paralychthys olivaceus*) the parasite may represent a separate marine species (Urawa and Kusakari, 1990).

Genus ***Procryptobia*** Vickerman, 1978

Free-swimming phagotrophic eukinetoplastic flagellates. Bacterivorous and resembling *Bodo* in ingestion apparatus, but with recurrent flagellum attached to body along most of its length. Lacks the prominent ventral furrow of *Cryptobia* spp. but displays independent cortical contractility and changes shape while swimming; flattened when creeping, elongate and sinuous when moving among bacterial aggregates. Three species described, all free-living and differing in shape of kinetoplast (spheroidal to dagger-shaped, Fig. 8) (Vickerman, 1978). Cysts may be polykinetoplastic (Fig. 7d).

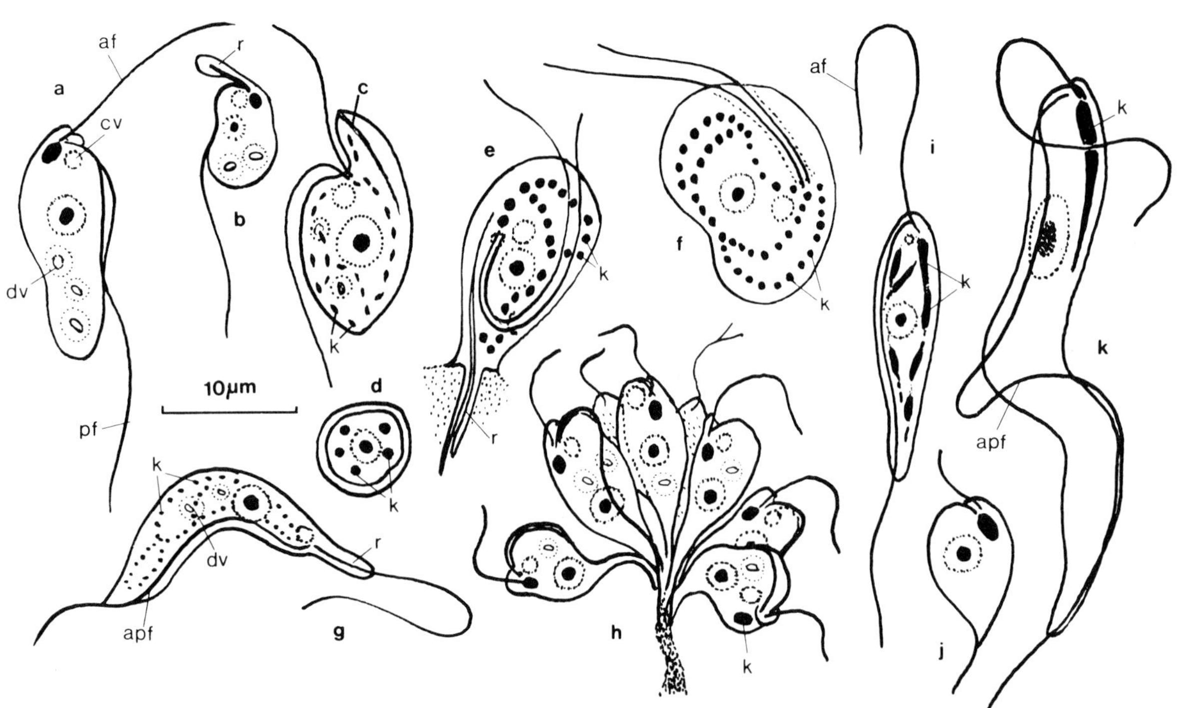

Fig. 7. Light microscope morphology of bodonid flagellates: *parasitic/endocommensal species; k, kinetoplast; n, nucleus, cv, contractile vacuole; dv, digestive vacuole; af, anterior flagellum; pf, posterior flagellum; apf, attached posterior flagellum; r, rostrum; EUK, eukinetoplastic; PLK, polykinetoplastic; PNK, pankinetoplastic. (a) *Bodo caudatus* (EUK). (b) *Rhynchomonas nasuta* (EUK); (c) *Rhynchobodo armata* (PLK); (d) *Procryptobia glutinosa* cyst (PLK); (e, f) *Ichthyobodo necator** (PLK); (e) attached phase on fish skin; (f) free-living migratory phase; (g) *Dimastigella trypaniformis* (PLK); (h) *Cephalothamnium cyclopum* (EUK) colony with secreted stalk; (i,j) *Cryptobia vaginalis** from leech (i, PNK phase; j EUK phase); (k) *Trypanoplasma keysselitzi* (EUK) from tench. (Adapted from Vickerman, 1990)

Genus ***Dimastigella*** Sandon, 1928

Elongate, fusiform, phagotrophic flagellates with the flagellar bases disposed at an obtuse angle to one another and the flagella emerging from opposite ends of an elongate flagellar pocket (Fig. 7g). The long anterior flagellum is initially attached alongside a proboscis that bears a broad cytostome for ingestion of bacteria; the posterior flagellum is attached along the entire length of the body becoming free at its posterior extremity; sinuous movement; polykinetoplastic, the minute kinetoplasts visible only with fluorescence or electron microscopy (Vickerman, 1978; Breunig et al., 1993). Polykinetoplastic cysts are formed.

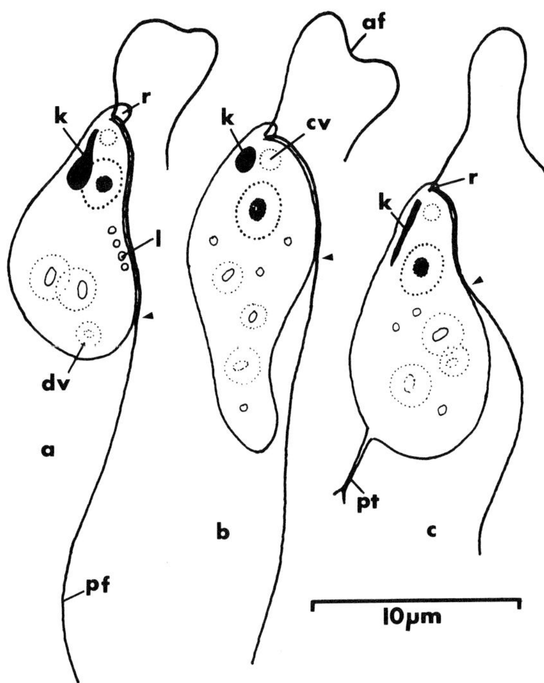

Fig. 8. Morphological differences between three species of *Procryptobia*. (a) *P. vorax* with tear-shaped kinetoplast (k), prominent rostrum (r), and posterior flagellum (pf) attached along length of body (arrowhead shows limit); (b) *P. tremulans* with subspherical kinetoplast, short rostrum and pf attached over half body length; (c) *P. glutinosa* with dagger-shaped kinetoplast, insignificant rostrum, and posterior flagellum attached one third to over half body length. Other differences include relative lengths of flagella (af, anterior flagellum), presence of paraflagellar lipid globules in *P. vorax* and sticky adherent periplast (pt) in *P. glutinosa.* (Based on Vickerman, 1991)

The type species is *D. trypaniformis* Sandon, common in soil and freshwater habitats and as a contaminant of urine samples or urine-impregnated bedding from animal cages. If from the latter situation it contaminates smeared tail blood from rodents under examination for experimental trypanosome infections, then it may be mistaken for a trypanosome (Vickerman, 1978). The identification of an isolate from termite gut contents as a parasite (Berchtold et al., 1994) is almost certainly erroneous, as the termite hindgut is anoxic. Several isolates listed by ATCC.

Genus ***Cephalothamnium*** Stein, 1878
(=***Cephalothamnion*** Lemmerman, 1913)

Sedentary phagotrophic flagellates forming colonies of 20-30 individuals (5-15 µm) attached to a common secreted stalk. Single eukinetoplast. One species described, *C. cyclopum* (Fig. 7h), epizoic on freshwater copepods (*Cyclops* spp). The anterior flagellum seems to thrust bacterial food towards the cytostome, the posterior flagellum is attached to the body, its tip and that of the body embedded in the stalk. Hitchen's (1974) ultrastructural study (Fig. 9) does not give details of the oral region.

Genus ***Cryptobia*** Leidy, 1846

Solitary, monogenetic (single host), bodonid flagellates in which the recurrent flagellum is attached along the entire length of the body but does not form a prominent undulating membrane; eukinetoplastic (Fig. 7j) or, more rarely, **pankinetoplastic** (Fig. 7i) with kinetoplast DNA irregularly disposed in masses in the mitochondrion; microtubule cytoskeleton reinforces most of plasma membrane except for ventral groove alongside the **recurrent flagellum** (Fig. 4). All parasitic, in some cases live as commensals; free-living species that have been described (e.g. by Ruinen, 1938) are most probably not kinetoplastids. Feeding is by pinocytosis through a cytostome located at the end of a long preoral ridge (Fig. 4). Cysts are unknown. This monogenetic genus and the digenetic (two host) *Trypanoplasma* are commonly united into a single genus, *Cryptobia* (as in Vickerman, 1976, 1990).

The case for a single genus has been admirably reviewed by Woo (1987). Candidate species of both genera are in need of considerably more comparative study, but Brugerolle et al. (1979) and Woo (1987) have made a start. The type species is *Cryptobia helicis* Leidy, 1846 (16-27 µm) from various land snails.

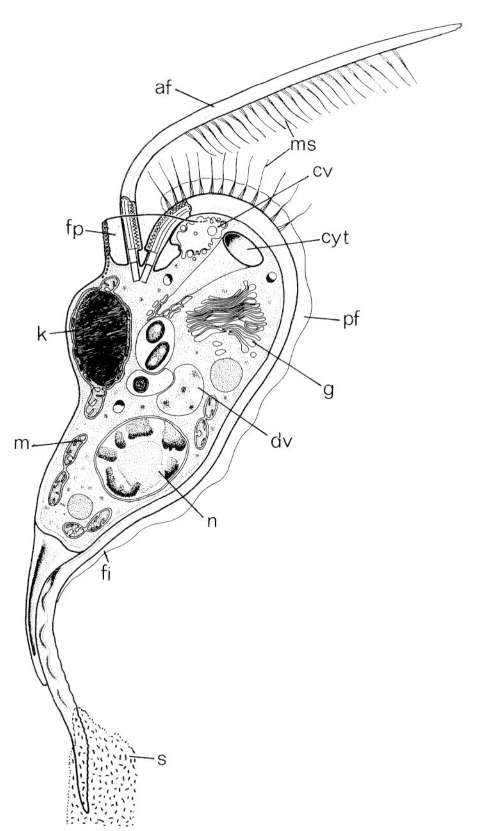

Fig. 9. Structural features of *Cephalothamnium cyclopum*, epibiont on the copepod *Cyclops*. (Af) anterior flagellum; (cv) contractile vacuole; (cyt) cytostome; (dv) digestive vacuole; (fi) flagellar fin; (fp) flagellar pocket; (g) golgi; (k) kinetoplast; (m) mitochondrion; (n) nucleus; (ms) mastigonemes; (pf) posterior flagellum; (s) stalk. (Based on Hitchen, 1974)

Monogenetic cryptobias include parasites of: (1) the reproductive system of molluscs (e.g. the **eukinetoplastic** *C. helicis* of gastropods [Current, 1980], and leeches e.g. the polykinetoplastic *C. vaginalis* of *Hirudo* and *Haemopis* [Vickerman, 1977; Figs. 4,7i,j]); (2) the gut of freshwater planarians (e.g. *C. dendrocoeli*); (3) the gut of fishes (e.g. *C. intestinalis* from the marine *Box boops*); *C. jubilans* of the gut of freshwater cichlids can spread to other organs (e.g. spleen, liver) where it appears to multiply inside macrophages (Nohynkova, 1984); (4) the gut of lizards (Bovee and Telford, 1962). Transmission is probably during copulation in (1), through surrounding water in (2) and (3), while in (4) the mode of transmission is unknown. The *Cryptobia*-like kinetoplastids found on the gills of fishes e.g. *C. branchialis* and *C. eilatica* (Diamant, 1990) reputedly feed phagotrophically on bacteria and dead cells in the gill mucus and may merit segregation in a separate genus, *Bodomonas* Davis, 1947 (but see *Trypanoplasma* below). These species attach to host epithelium by the recurrent flagellum, while *C. helicis* attaches to the microvilli lining the snail seminal vesicle by extravagant outgrowths of its anterior flagellum (Current 1980). As yet no monogenetic cryptobias have been cultivated in vitro.

Genus ***Trypanoplasma*** Laveran & Mesnil, 1901

Solitary, digenetic (two host) bodonidid flagellates in which the recurrent flagellum is attached along the entire length of the body to form a prominent undulating membrane in the **haematozoic** (blood-dwelling) phase; eukinetoplastic with elongate kinetoplast (Fig. 7k); microtubule cytoskeleton reinforces most of plasma membrane except for ventral groove alongside recurrent flagellum. All parasitic, alternating between blood of fish or, more rarely, terrestrial salamander as host and crop and proboscis of leech vector; feeding is by pinocytosis; the cytostome at the end of a long preoral ridge is located about one third way down the body. The size of the hematozoic forms ranges from 10-53 µm long.

The type species, *Trypanoplasma borreli* Laveran & Mesnil, 1901, which infects carp and several european cyprinid fishes has been most studied from the point of view of development in the leech vector (*Piscicola geometra* or *Hemiclepsis marginata*; see Lom, 1979; Kruse et al., 1989) and has been cultivated in vitro on biphasic blood agar medium supplemented with B vitamins or in a monophasic medium with red blood cell lysate

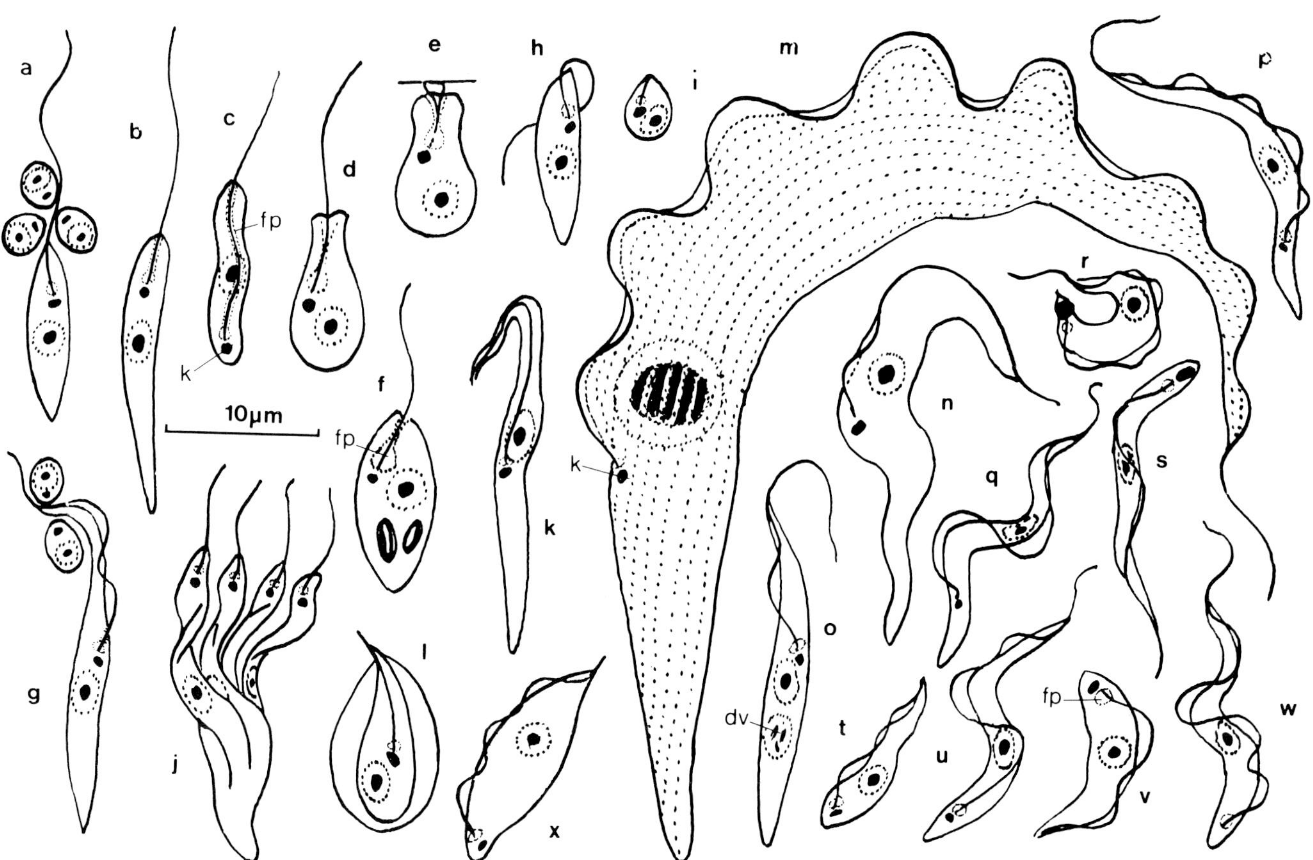

Fig. 10. Light microscope morphology of trypanosomatid flagellates: fp, flagellar pocket; PM, promastigote; AM, amastigote; OPM, opisthomastigote; CHM, choanomastigote; EPM, epimastigote; TPM, trypomastigote; other abbrev-iations as in Fig. 7. (a) *Leptomonas oncopelti* PM, with straphanger cysts; (b) *Herpetomonas muscarum* PM; (c) *H. muscarum* OPM; (d) *Crithidia fasciculata* CHM, nectomonad; (e) *C. fasciculata* CHM haptomonad; (f) *C. oncopelti* CHM, with endosymbionts (es); (g) *Blastocrithidia familiaris* EPM, with cysts; (h) *Leishmania major* PM; (i) *L. major* AM; (j) *Phytomonas elmassiani* PM, multiple fission stage in plant latex; (k) *Rhynchoidomonas drosophilae* TPM; (l) *Endotrypanum schaudinni* EPM in sloth red blood cell; (m) *Trypanosoma grayi* TPM, from crocodile blood; (n) *T.* (*Megatrypanum*) *cyclops* TPM from blood of macacque; (o) *T.* (*M.*) *cyclops* EPM with pigment in digestive vacuole (dv), from culture; (p) *T.* (*Herpetosoma*) *musculi* TPM from mouse blood; (q) *T.*(*Tejeiraia*) *rangeli* TPM from human blood; (r) *T.* (*Schizotrypanum*) *dionisii* TPM from pipistrelle bat blood; (s) *T.* (*Duttonella*) *vivax* TPM and (t) *T.* (*Nannomonas*) *congolense* TPM, both from cattle blood; (u) *T.* (*Trypanozoon*) *brucei* TPM slender bloodstream form; (v) *T.* (*T.*) *brucei* TPM short stumpy bloodstream form; (w) *T.* (*T.*) *evansi* TPM, dyskinetoplastic, from camel blood; (x) *T.* (*Pycnomonas*) *suis* TPM from pig blood. (From Vickerman, 1990)

(reviewed Peckova and Lom, 1990). Blood stages are pleomorphic, showing changes in body shape and respective positions of nucleus and kinetoplast. Parasitaemia may be determined by temperature and after an initial rise may be followed by a slow fall. Development in the leech initially involves a multiplicative short broad form in the crop giving rose later to long, slim, metacyclic forms, which migrate to the proboscis ready to infect a new fish host when the leech bites. Infection of the vector is terminated with completion of digestion of the blood meal. The salmonid parasite *T. salmositica* has also been extensively studied (Becker, 1977; Woo, 1987). In this species, in addition to cyclical transmission by the leech, direct transmission may occur by migration of the parasite from the blood to the gill and body surfaces, then contact between uninfected and infected fishes (references in Woo, 1987). Such transfer is important in leech-free housing of farmed fish and obviously confounds identification of gill-commensal *Cryptobia* spp and *Trypanoplasma* spp. (see under *Cryptobia* above).

Woo (1987) has reviewed the pathogenicity of trypanoplasms to their fish hosts. Anemia, exophthalmia, anorexia, and loss of normal reactions may precede death of infected fishes in some stocks from breakdown of osmoregulation. Trypanoplasmosis is preferable to cryptobiasis as a name for the disease as the latter may be confused with cryptobiosis—a state of suspended animation or dormancy observed in a variety of organisms.

Over 50 species of *Trypanoplasma* have now been described (Lom, 1979; Woo, 1987), mostly from freshwater fishes and anadromous salmonids, but a few have been reported from marine fishes (e.g. see Burreson [1982] on the life cycle of *T. bullocki* from various flatfish and the leech *Calliobdella vivida* and references in Woo [1987]). Limited cross infection experiments have as yet beencarried out, however, and many synonymies may eventually be demonstrated.

SUBORDER TRYPANOSOMATINA Kent, 1880

Kinetoplastid flagellates with a single locomotory flagellum (corresponding to anterior flagellum of bodonines) emerging from the flagellar pocket either at the anterior end as a free structure or laterally and attached along the body to form an **undulating membrane** during beating; kinetoplast relatively small and usually compact with kDNA in form of a network of catenated circles invariably associated with the flagellum base and usually in a distinct capsular region of the mitochondrion (Fig. 5); kDNA composed of 25-50 homogeneous **maxicircles** (each 20-38 kb) and 5,000 to 27,000 usually heterogeneous minicircles (each 0.46-2.5 kb)—the latter encoding guide RNAs responsible for editing the transcripts of the **maxicircle genes** into a translatable message. The cytostome, if present lacks associated oral structures. All parasitic, non-phagotrophic species.

Admittedly, relatively few species have been examined as yet, but these phylogenetic conclusions are supported by studies on RNA editing and on the nature of the parasite surface. In this account, however, the traditional sequence will be followed with the monogenetic genera that have simpler life cycles considered first.

FAMILY TRYPANOSOMATIDAE Doflein, 1901

With characters of the Suborder. The family Trypanosomatidae contains genera of outstanding medical (*Tryapanosoma, Leishmania*), veterinary (*Trypanosoma*), or agricultural (*Phytomonas*) importance. Traditionally the trypanosomatids have been divided into monogenetic (MG; one host) and **digenetic** (DG; two host) genera, and it has been tacitly assumed that in evolution the former preceded the latter. Molecular sequencing techniques are now radically reshaping previous ideas on the phylogeny of these organisms (Fernandes et al., 1993; Maslov et al., 1994; Vickerman, 1994; Maslov and Simpson, 1995). The tsetse fly-transmitted African trypanosomes (DG; causative agents of human sleeping sickness and of nagana in cattle and horses), previously viewed as the most advanced trypanosomatids on account of their complex life cycles involving multiple differentiation events (reviewed Vickerman, 1985), are now emerging as the most ancient lineage, followed by *Trypanosoma cruzi* (causative agent of Chagas' disease in humans) and other trypanosomes species (DG), then *Blastocrithidia* (MG), *Herpetomonas* (MG) and *Phytomonas* (DG), with *Leptomonas* (MG), *Crithidia* (MG), *Leishmania* (DG), and *Endotrypanum* (DG) forming the crown of the evolutionary tree.

Genus ***Leptomonas*** Kent, 1880

Monogenetic trypanosomatids with promastigote forms dominating in the life cycle; amastigotes and cysts may also occur (Fig. 10a). Parasites mainly of insects (Hemiptera, Diptera, Hymenoptera, Blattoidea, Lepidoptera, Siphonaptera, Anoplura) where they are found usually in the gut and associated organs, rarely (and doubtfully) of other invertebrates, but present in some ciliate protozoa.

The type species, *L. buetschlii* Kent from the nematode *Tobrilus gracilis* has not been rediscovered and the best known life cycles are those of the flea parasites *L. ctenocephali* (Molyneux et al., 1981) and *Leptomonas pulexsimulantis* (Beard et al., 1989), and the plant sap-sucking hemipteran parasite, *L. oncopelti* (McGhee and Hanson, 1962; Lauge and Nishioka, 1977). The life cycle of the ciliate parasites is also well documented.

In the American dog flea, *Pulex simulans*, *L. pulexsimulantis* may be found as a free swimming (**nectomonad**) promastigote in the midgut and malpighian tubules, but the parasites are mainly found attached by their flagella (**haptomonad** stage) to the cuticle of hindgut and rectum. In culture, long, slender, twisted, and short stubby promastigotes are produced and amastigotes with a corrugated surface but no cyst wall. The amastigotes nevertheless survive desiccation in the feces of the flea and serve to transmit the infection to larval fleas, which consume adult feces as an important dietary supplement (Beard et al., 1989). *Leptomonas oncopelti* (Fig.10a) produces amastigote "cysts" as "**straphangers**" attached to the flagellum as a result of unequal division of the parent flagellate: transmission is the result of contamination of the asclepiad food plant with cysts to be consumed during probing by uninfected bugs (McGhee and Hanson, 1962).

One of the few pathogenic species of *Leptomonas* is that occurring in silkworm larvae (Abe, 1980). Promastigotes multiply in the midgut lumen and in the haemocoel where they are said to encyst; encystation appears to occur after fusion of 2 promastigotes whose kinetoplasts and nuclei reportedly fuse, suggesting sexuality in this species.

Leptomonas ciliatorum (Gortz and Dieckmann, 1987) infects the macronucleus of the brackish water hypotrichous ciliate *Paraholosticha sterkei* and carries out its entire life cycle there. The ciliate host can divide, encyst, and excyst while infected. After being ingested by a new host, the infecting promastigote invades the macronucleus and divides as an amastigote before giving rise to large and then small promastigotes. These are farmed out to daughter hosts during macronuclear division; under conditions of host starvation parasite multiplication may result in death of the host and release of infective promastigotes. Unless a new host can be found within a few hours the parasite dies.

Without doubt, the genus *Leptomonas* is a dumping ground for trypanosomatids not yet shown to have additional morphological stages (e.g. opisthomastigote, endomastigote) in the life cycle or more than one host, and it is not surprising that marked heterogeneity has been demonstrated among its species with respect to restriction enzyme cleavage sites in ribosomal RNA genes and in kinetoplast DNA, and with respect to enzymes for arginine-citrulline metabolism (Camargo et al., 1992). The ATCC offers 8 species, mostly from Hemiptera, but including the exquisitely long and twisted *Leptomonas mirabilis* from the fly *Cynomyopsis cadaverina*; some authorities recognize a separate subgenus *Cercoplasma* to hold this and other elongate species of *Leptomonas* (Wallace, 1976).

Genus *Herpetomonas* Kent, 1880

Monogenetic trypanosomatids with a dominant promastigote stage (Fig.10b) in the life cycle, but also producing opisthomastigotes (Fig.10c) in which the flagellum courses through a long pocket running the length of the flagellate's body (Fig.6), and intermediate forms described as paramastigotes. Parasites of Diptera and possibly other groups of insects (see remarks under *Phytomonas* below). Of the ornithine cycle enzymes, arginine deiminase is always present and arginase absent. The type species is *Herpetomonas muscarum* (Leidy) Kent, from house flies and other muscids. Two different organisms have been confused under this name and designated as subspecies *H. muscarum muscarum* and *H. m. ingenoplastis* by Rogers and Wallace (1971). The former lacks bacterial symbionts and has a **Type A kinetoplast** (Vickerman and Preston, 1976), i.e. the kDNA mass is separated from the mitochondrial capsule envelope and is compact. *Herpetomonas m. ingenoplastis* has bacterial symbionts and a **Type**

B kinetoplast, i.e. the kDNA is loosely packed and entirely fills the very large teardrop or pear-shaped mitochondrial capsule, the wider end of the kDNA mass being anterior. A type B kinetoplast is common among symbiont-bearing trypanosomatids. Both subspecies inhabit the midgut endotrophic lumen and hindgut, but whereas *H. m. muscarum* exhibits conventional aerobic respiration and has a cristate mitochondrion, *H. m. ingenoplastis* has an acristate mitochondrion and is facultatively anaerobic in it respiration (Coombs, 1988). The two would appear to represent quite different species and their subspecific status is inappropriate.

The opisthomastigote stage is produced in the stationary phase of cultures, under cultivation at higher than optimal temperature, at increased pH or osmolality, or in the presence of urea or hydroxyurea; it is usually non-dividing, but division does occur among the abundant opisthomastigotes of *H. roitmani* in young cultures (Faria-e-Silva et al., 1996). The ATCC has 5 named species.

Genus ***Crithidia*** Leger, 1902

Monogenetic trypanosomatids possessing the characteristic choanomastigote form, i.e. barley-corn-shaped and with the kinetoplast large and lateral, close to the nucleus in both nectomonad (Fig. 10d) and haptomonad (Fig. 10e) phases; flagellar pocket wide and funnel-shaped, its mouth occupying most of the truncate anterior end. Parasites in the gut of Diptera, Hemiptera, Trichoptera, and Hymenoptera. Contain arginase and arginosuccinate lyase but not arginine deiminase. Require either arginine or citrulline. The type species is *C. fasciculata* Leger, 1902 of *Anopheles maculipennis* and other mosquitoes. Its nectomonads are 6-8 μm long, haptomonads which may cover the rectal wall are only 3-4 μm long (Brooker, 1971b).

The occurrence of monogenetic trypanosomatids with an undulating membrane (genus *Blastocrithidia*) alongside barleycorn-shaped *Crithidia* lacking this structure led to much confusion in early descriptions of species of this genus. Over 30 *Crithidia* species have been described (Podlipaev, 1990) of which the ATCC holds 9 in culture. Crithidias are readily obtained in culture, often overgrowing other trypanosomatids if taken from a mixed infection. Several subspecies of *C. fasciculata* and *C. luciliae* (a parasite of calliphorid and muscid flies) have been described on the basis of nutritional requirements, and these 2 species have been widely used in experimental studies. *Crithidia* (=*Strigomonas*) *oncopelti* (Fig. 10f), a symbiont-bearing laboratory strain of uncertain origin, but probably derived from a hemipteran, has also been favored in this respect. Like many symbiont-bearing trypanosomatids, it lacks a hemin requirement and grows in relatively simple media. Although most crithidias are non-pathogenic, *C. bombi* is known to regulate bumble bee population size (Durrer and Schmid-Hempel, 1994); transmission of the parasite occurs when bees share nectar and pollen sources.

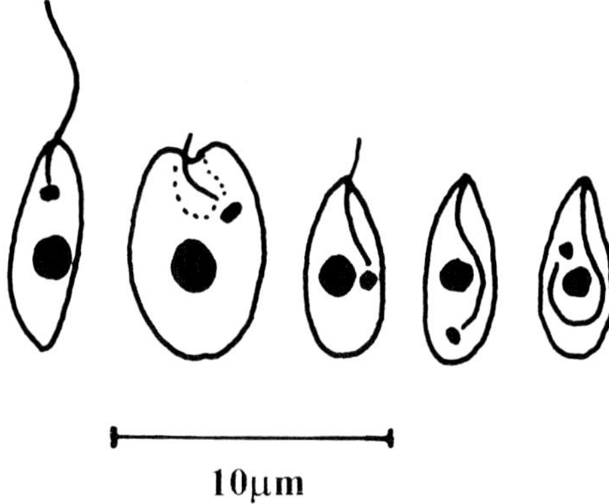

Fig. 11. *Wallaceina inconstans*. Stages in development (promastigote, choanomastigote, paramastigote, opisthomastigote, and endomastigote) from the gut of the bug *Calocoris sexguttatus*. (After Podlipaev et al., 1990)

Genus ***Wallaceina*** Podlipaev, Frolov & Kolesnikov, 1990 nom. nov for ***Proteomonas*** Podlipaev, Frolov & Kolesnikov, 1990

Monogenetic trypanosomatids with promastigote (or choanomastigote) and endomastigote stages in the life cycle. Parasites of Hemiptera and Diptera. This genus is distinguished by retreat of the flagellum base and kinetoplast behind the nucleus and around it, so that the flagellum emerges from the body facing in the posterior end of the flagellate, but remains within the convoluted flagellar pocket and does not emerge from the cell itself. On account of this character, Podlipaev et al. (1990) transferred two former *Crithidia* spp from Hemiptera to the genus *Proteomonas*, but this name proved to be preoccupied by a cryptomonad flagellate *Proteomonas* Hill & Weetherbee, so the authors have agreed to change the generic name to *Waiiaceina* in memory of the late F. G. Wallace. The type species is *P. inconstans* (Fig. 11) from the gut of *Calocoris sexguttatus*. *Herpetomonas mariadeanei* (Yoshida et al., 1978) from the fly *Muscina stabulans* also displays endomastigotes and should therefore be transferred to the genus. Endomastigotes may represent a special type of transmission phase, resistant to desiccation (Frolov and Malysheva, 1992).

Genus ***Blastocrithidia*** Laird, 1959

Monogenetic trypanosomatids with the characteristic epimastigote form in nectomonad and haptomonad phases. Amastigotes may be produced as flagellar cysts (as in *Leptomonas*) in some species (Fig. 10g). Parasites of Diptera, Hemiptera, Siphonaptera, possibly Hymenoptera, and ixodid ticks, usually in the gut, occasionally in the salivary glands; transovarian transmission of some *Blastocrithidia* spp. flagellates has been reported, as well as contaminative trasnmission by coprophagy. Predatory bugs may acquire their infections from dipterous prey. About 40 named species have been described (Podlipaev, 1990).

The type species is *Blastocrithidia gerridis* Patton (18-23 μm long) of the water strider bug *Gerris fossarum* and other hosts in the hemipteran families Gerridae and Veliidae, where it may coat the lining of the intestine with haptomonad forms. So far this symbiont-lacking species, along with many others of the genus, has not proved amenable to continuous culture. *Blastocrithidia culicis*, a symbiont-bearing species that reputedly does not form haptomonads in its culicid host, is readily cultured and available from the ATCC (both natural and aposymbiotic forms), as is the "straphanger" cyst producing *B. leptocoridis*. One of the most important and well-studied species is *B. triatomae*, which also produces flagellar cysts and is pathogenic in *Triatoma infestans*, one of the vectors of Chagas' disease. Indeed, this *Blastocrithidia* is a good candidate for biological control of the bug; pathogenicity appears to depend upon interaction of the parasites with the mycetome symbionts of the host insect (Schaub, 1994).

Genus ***Rhynchoidomonas*** Patton, 1910

Monogenetic trypanosomatids with a dominant trypomastigote stage (Fig. 10k) in the life cycle but lacking the conspicuous undulating membrane of trypomastigotes of *Trypanosoma* spp. Epimastigotes and amastigotes also produced. Parasites of Diptera and Lepidoptera, often in the malpighian tubules as well as the gut. This is a poorly known genus. Of the 6 species named, none has been cultured and only one, *R. operophterae* (Page et al., 1986) from a single lepidopteran host (the winter moth *Operophtera brumata*), has been studied since 1935. Its trypomastigotes had a free flagellum, previously regarded as absent from the genus. The type species, *R. luciliae*, from the malpighian tubules of various calliphorid and muscid flies may be as long as 50 μm. *Drosophila* spp.host 2 of the described species.

Genus ***Phytomonas*** Donovan, 1909

Digenetic trypanosomatids alternating between a plant host and a phytophagous hemipteran. Cyclical development of the parasite occurs in the bug; after transformation in the midgut of the vector, the flagellates migrate to the salivary glands via the hemocoel and transmission occurs via the anterior station; the parasite retains the promastigote form throughout this cycle, and twisting of the body is a common feature in this genus (Fig. 10j). Multiple fission occurs in some latex flagellates (Fig. 10j, see below). Amastigotes have also been reported but their role in the cycle is not clear. About a dozen species have been named but several isolates now grown in vitro in laboratories are named informally after the plant host from which they were taken (e.g. *Phytomonas* sp from *Euphorbia characias*). The habit of assigning all plant trypanosomatids to a single genus may soon have to be abandoned as enormous heterogeneity is being discovered amongst them. The type species is

Phytomonas davidi Lafont (15-20 μm) from the latex of *Euphorbia pilulifera* and transmitted by the bug *Stenocephalus agilis*. The parasite maintained under that name in the ATTC but isolated from American *E. pilulifera* (*E. hirta*) and transmitted by the bug *Pachybrachius*, however, produces opisthomastigote forms occasionally and has more biochemical features in common with *Herpetomonas* than with other *Phytomonas* spp. (Wallace et al., 1992).

From the plant pathologist's viewpoint, 3 groups of trypanosomatids have been referred to the genus: (1) parasites living in the latex tubes of laticiferous plants of the families Euphorbiaceae and Asclepiadaceae principally, but also in Apocynaceae, Cecropiaceae, Asteraceae (Compositae), Moraceae, Urticaceae, and Sapotaceae; transmission is by bugs of the families Coreidae and Lygeidae; examples include *P. davidi* of euphorbias and *P. elmassiani* of asclepiads; with the exception of *P. francai* of manioc, these species are not associated with a wilt of the host plant; (2) parasites living in the phloem sieve tubes of non-laticiferous plants of the Families Arecaceae (Palmae), Rubiaceae, and Zingiberaceae; transmission is by pentatomid bugs of the genus *Lincus*; these species are pathogenic causing a fatal wilt, possibly by blocking sieve tubes and preventing the transport of photoassimilates; examples include *P. staheli* pathogenic in oil and coconut palms and *P. leptovasorum* pathogenic in coffee plants; (3) parasites restricted to fruits and unable to live in other parts of the plant; damage to the host is localized to the fruit, and families affected include the Anacardiaceae, Oxalidaceae, Passifloraceae, Punicaceae, Rosaceae, Rutaceae, Solanaceae, and Poaceae (Gramineae); an example is *P. serpens* from tomatoes (Jankevicius et al., 1989). The fruit pericarp or seed aril may serve as a culture medium, not only for *Phytomonas* spp., but also for *Leptomonas* and other insect trypanosomatids (Conchon et al., 1989). Frolov and Malysheva (1993) have described as *Phytomonas nordicus* a promastigote parasite of the gut and salivary glands of the predatory pentatomid bug *Troilus luridus*. Transmission from bug to bug occurs contaminatively via the feces but also via the saliva when infected and uninfected bugs feed simultaneously on the hemolymph of the same dipteran larva; a plant host is not involved in the cycle.

Ease of cultivation of *Phytomonas* spp. varies. The latex parasites can usually be grown in insect tissue culture media, but the pathogenic species require insect feeder cells for some time after isolation (see Wallace et al., 1992). Strangely, no *Phytomonas* species harbors bacterial symbionts, yet neither *P. serpens* nor the phloem parasites require hemin as a growth factor.

Genus *Leishmania* Ross, 1903 emend. Saf'janova, 1982

Digenetic trypanosomatids multiplying in the mononuclear phagocyte cells of mammals and in the gut lumen of sandflies (Diptera: Psychodidae; Phlebotominae); characterized by intracellular (intravacuolar) amastigotes in the mammal (Fig. 10i) and free or attached promastigotes (Fig. 10h) (and paramastigotes) in the vector. Transmission is by bite, the infective metacyclic promastigotes being deposited in mammalian skin through parasite-induced defective pumping of blood by the cibarium. Primates, rodents, carnivores, edentates, hyracoids, and marsupials are the mammalian orders infected. The type species is *Leishmania donovani* (Laveran & Mesnil, 1903) Ross, 1903, causative agent of visceral leishmaniasis (kala azar) in humans; its amastigotes (2-3 μm diameter) live in the lysosomal system of macrophages of liver, spleen, and bone marrow. As is the case for other Old World species of *Leishmania*, the vectors are species of the sandfly genus *Phlebotomus*, whereas New World species are transmitted by species of *Lutzomyia*.

Leishmanias are important causative agents of disease in man in the tropics and subtropics, the human disease, leishmaniasis, having reservoirs in other mammals. The different species are distinguished primarily on their clinical manifestations and geographical distribution, as all are morphologically alike except for *L. major, L. enriettii* of guinea pigs (both with larger amastigotes), and *L. hertigi* (tiny dot-like kinetoplast) of porcupines. Behavior of the parasite in the sandfly and in the mammalian host, the latter's response in terms of a variety of serological and immunological reactions, and biochemical characters—especially DNA analysis and isoenzyme profiles have also been used to refine the classification of the genus into species and species complexes. For details the reader is referred to the reviews of Lainson and Shaw (1987) and Rioux et al. (1990). Lainson and Shaw (1987) have distinguished 2 subgenera on the basis of behavior of the parasite in the vector.

Subgenus ***Leishmania*** Saf'janova, 1982

Development in the insect gut limited to the midgut and foregut of the alimentary tract. Old and New World. Type species *L.* (*Leishmania*) *donovani* (Laveran & Mesnil, 1903). Other species include: *L.* (*L.*) *infantum* infecting mainly children in the Mediterranean area and *L.*(*L.*) *chagasi* infecting infants in S. America, both with a reservoir host in dogs; *L.*(*L.*) *tropica* causing human dry cutaneous leishmaniasis (oriental sore) in Asia (no known reservoir host); *L.*(*L.*) *major* causing moist oriental sore in Asia with a reservoir in rodents; *L. aethiopica* causing moist sores (occasionally becoming diffuse) in eastern Africa, with a reservoir host in hyracoids; *L.*(*L.*) *mexicana* (Chiclero's ulcer, Central America) and *L.*(*L.*) *amazonensis* (Brazil) causing mild cutaneous leishmaniasis, in the latter sometimes spreading to result in diffuse sores; reservoir hosts are rodents and opossums. *Leishmania* (*L.*) *enriettii* and *L. hertigi* (see above) are not known to infect humans.

Subgenus ***Viannia*** Lainson & Shaw, 1982

Development in the insect gut attached to the wall of the hindgut with later migration of flagellates to the midgut and foregut. New World only. Type species. *L.* (*Viannia*) *brasiliensis* Vianna, 1911 emend. Mata, 1916, causative agent of facially disfiguring mucocutaneous leishmaniasis (espundia, metastasizing to the nasopharynx) in Brazil and the forest areas east of the Andes; reservoir hosts are rodents, opossums, and sloths. Other species include: *L.*(*V.*) *guyanensis* causing forest yaws in northern S. America, with lesions metastasizing along the lymphatics and reservoir hosts in edentates, opossums, rodents; *L.*(*V.*) *panamenis* causing similar disease with limited lymphatic metastasis, in Panama, reservoir hosts in sloths, procyonid carnivores, and rodents; *L.*(*V.*) *peruviana* causing cutaneous leishmaniasis (uta) along the western slopes of the Peruvian Andes. This last species is the only non-forest form in the subgenus and it has a reservoir host in the dog. For a comprehensive survey of the leishmanias see Peters and Killick-Kendrick (1987).

Genus ***Sauroleishmania*** Ranque, 1973

Digenetic trypanosomatids alternating between the blood of lizards or, more rarely, snakes and the gut of a sandfly vector (*Sergentomyia* spp.). Blood stages are either amastigotes in macrophages and their precursors, thrombocytes, or erythrocytes, or promastigotes free in the blood. In the vector, they develop either entirely in the hindgut, so that transmission is contaminative, or initially there, progressing to the midgut and foregut for transmission by bite. The blood promastigotes are morphologically identical with those of *Leishmania*; the amastigotes have only rarely been seen. Evidence for multiplication in the reptile host is wanting. The type species is *Sauroleishmania tarentolae* (Wenyon, 1920) according to Killick-Kendrick et al. (1986), who list all known species. It has been much studied as a molecular biologist's pet and safe substitute (in the promastigote form) for the pathogenic leishmanias.

Sauroleishmania spp. are known mostly as culture promastigote forms, easily isolated from heart blood and maintained on blood agar media, but rarely seen in the reptile host. Their serological characters and isoenzyme and DNA restriction fragment profiles indicate that they are genetically distant from the mammalian leishmanias, and separation at generic level was advocated by Lainson and Shaw (1987) and by Rioux et al. (1990). Promastigotes found in the cloacal glands of reptiles and referred to as leishmanias by Wenyon (1926) have been dismissed as monogenetic trypanosomatids derived from insect prey (Vickerman, 1976) using the cloacal secretions as a fortuitous culture medium. Second thoughts on this question have been aroused, however, by the finding that the digenetic *Trypanosoma cruzi* in the opossum can quit the circulatory system and set up a mimic vector cycle of development in the anal glands of its mammalian host (Deane et al., 1986; see below)!

Genus ***Endotrypanum*** Mesnil & Brimont, 1908

Digenetic trypanosomatids with trypomastigote and epimastigote stages inside erythrocytes (Fig. 10l) of edentate mammals (sloths) and promastigotes and amastigotes in the sandfly (phlebotomine) vector. The type species is *Endotrypanum schaudinni* Mesnil & Brimont, 1908 from *Choleopus didactylus*; it also occurs in *Bradypus* the Brazilian sloth. A second species, *E. monterogei* Shaw, 1969, described from the same host, appears to be identical in its serological, isoenzyme, and DNA buoyant density characteristics (Croft et al., 1980).

This genus provides the only known intraerythrocytic trypanosomatid. Despite its trypanosome-like morphology, small subunit ribosomal RNA gene sequencing suggests that it is closely related to leishmanias of the subgenus *Viannia* with which it shares sandfly (*Lutzomya* spp) and sloth hosts; the similarity of its promastigote vector stage to that of *Leishmania* may lead to misidentification in leishmaniasis vector surveys (Christensen and Herrer, 1979). *Endotrypanum* has a homologue of the surface protease (gp63) found on all leishmanias and in some monogenetic trypanosomatids, but not in trypanosomes. Yet it possesses surface-anchored sialidase and trans-sialidase enzymes that, respectively, remove sialic acid residues from host glycoconjugates and transfer them to the parasite's own surface—activities important in entry into non-phagocytic cells in *Trypanosoma cruzi*. Leishmanias lack these enzymes—and incidentally the ability to enter non-phagocytic cells (Medina-Acosta et al., 1994).

Genus ***Trypanosoma*** Gruby, 1843

Digenetic trypanosomatids with a cycle of multplicative and developmental stages in the blood of vertebrates and in the gut of leeches or arthropods; trypomastigote and epimastigote stages common to nearly all life-cycles, amastigotes more unusual and promastigotes rare. A blood-dwelling trypomastigote form is common to all species in the vertebrate host. Intracellular stages (usually amastigote) occur in some life-cycles. Exceptions to cyclical development in 2 hosts occur in the life cycles of some mammalian trypanosomes (see below). The type species is *Trypanosoma rotatorium* Mayer, 1843, a leech-transmitted trypanosome of frogs.

Trypanosomes are found as blood parasites of all vertebrate classes. The species parasitizing elasmobranchs, bony fishes, certain amphibians, and chelonians are leech-transmitted, the cycle of development terminating with the production of **metacyclics** which (like the trypanoplasms) migrate to the leech's proboscis sheath; trypanosomes of terrestrial poikilotherms are transmitted by haematophagous arthropods as are trypanosomes of birds and mammals. Anuran hosts present an interesting transition. Thus, *Trypanosoma inopinatum* of the frog *Rana esculenta* is transmitted by leeches of the genus *Helobdella*, while *T. bocagei* of the toad *Bufo bufo* is vectored by a sandfly *Phlebotomus squamirastris* (reviewed Bardsley and Harmsen, 1973), the frog having a long aquatic larval phase during which infection can take place; the toad conversely having a very short larval phase.

Of the trypanosomes of poikilotherms, *Trypanosoma mega* from African toads has been much used as an experimental trypanosome in culture. *Trypanosoma grayi* (Fig. 10m) of African crocodiles is of interest in that its insect hosts are tsetse flies, the trypanosome undergoing its development cycle in the gut; epimastigotes and metacyclics occur in the hindgut of an infected fly, however, not in the mouthparts (as in *T. congolense*) or salivary glands (as in *T. brucei*; see below). The ATCC has trypanosomes from anurans (*T. chattoni*, *T. mega*, *T. ranarum*), and from birds (*T. avium*, *T bennetti*) as well as those listed below from mammals.

The literature on the classification and different species of trypanosomes of mammals has been monographed by Hoare (1972) and what follows is simply an updating summary. The introduction of subgenera for the mammalian trypanosomes (Hoare, 1964) met with wide approval and depended upon adequate knowledge of the pattern of development of the parasites in the insect host. The subgenera are arranged in 2 sections, the *Stercoraria* and *Salivaria*, which are based on the site of production of metacyclic trypanosomes in the insect host and the consequent method of infection of the mammalian host. The trypanosomes of lower vertebrates have not yet been fitted into the subgenera, except that *Trypanosoma* Gruby is the nominate sub-genus of the generic type species, *T. rotatorium* Mayer, 1843, from frogs. In the account below, species marked * are in the ATCC.

Section **STERCORARIA**.

Developmental cycle in the insect host is completed in the hind-gut, metacyclics are present in the vector's feces and transmission is contaminative. Remarkably, in *T.* (*Schizotrypanum*) *cruzi* and *T.* (*Megatrypanum*) *freitasi* in opossums, the vector cycle can occur in the mammal's anal glands so that metacyclics are deposited on feces and transmission occurs via the sniffing nose to another opossum (Deane and Jansen, 1986).

Subgenus ***Megatrypanum*** Hoare, 1964

Large trypanosomes with kinetoplast located near to nucleus and far from posterior end of body; reproduction as epimastigotes in the mammalian host. Type species: *Trypanosoma theileri* Laveran, 1902* of cattle and tabanid flies. Other species include *T. melophagium* (sheep), *T. ingens* (antelopes), *T. conorhini** (rodents), *T. cyclops** (Fig. 10n,o) (monkeys), also parasites of monotremes, marsupials, insectivores, bats, edentates, and carnivores (see Wells, 1976).

Subgenus: ***Herpetosoma*** Doflein, 1901

Medium-sized trypanosomes with kinetoplast sub-terminal but well away from posterior end of body; reproduction in mammal in amastigote/ epimastigote stages. Type species: *T.* (*Herpetosoma*) *lewisi* Kent, 1880* of the rat and rat flea. ,Other species include *T.* (*H.*) *musculi** (Fig. 10p) (mice), *T. rabinowitschae* (hamsters), *T. nabiasi* (rabbits), also parasites of a wide range of rodents, insectivores, bats, edentates, and primates (see Molyneux, 1976).

Subgenus ***Schizotrypanum*** Chagas, 1909

Relatively small trypanosomes, typically C-shaped in blood smears (Fig. 10r); large kinetoplast lying close to posterior end of body; reproduction in mammalian host in intracellular amastigote form. Type species: *Trypanosoma* (*Schizotrypanum*) *cruzi* Chagas, 1909, causative agent of South American trypanosomiasis or Chagas' disease in man and transmitted by triatomine bugs (e.g. *Rhodnius*, *Triatoma*). Morphologically identical forms from marsupials, edentates, rodents, carnivores, and monkeys. Other species include *T.* (*S.*) *vespertilionis*, *T. dionisii* (Fig. 10r) and several other species from bats (see Marinkelle, 1976; Miles, 1979).

Subgenus ***Tejeraia*** Anez, 1982

Trypanosomes in mammalian blood resemble those of s.g. *Herpetosoma* but division forms from mammalian blood are rare; moreover, produces metacyclics in the salivary glands of its triatomine bug vector after migration from the gut via the haemocoel and so is not strictly-speaking stercorarian. The only species is *Trypanosoma* (*Tejeraia*) *rangeli** (Fig. 10q) widespread in Central and South America as a non-pathogenic parasite of humans, dogs, opossums, raccoons, and edentates. Despite its salivarian-type life cycle, molecular evidence indicates a close relationship with *T. cruzi* with which it shares invertebrate and vertebrate hosts. *Trypanosoma rangeli* is the only trypanosome overtly pathogenic in its vector (see Anez, 1982).

Section **SALIVARIA**

Developmental cycle in the insect host is completed in the mouthparts or salivary glands so that metacyclics are present in the saliva and transmission is inoculative; reproduction is in trypomastigote form in mammal, in epimastigote and trypomastigote forms in the tsetse fly vector (*Glossina* spp.). In one species, *Trypanosoma* (*Trypanozoon*) *evansi*, which was clearly derived from *T. brucei* when this trypanosome left the tsetse belt of Africa (Hoare, 1972), there is no cycle of development in the invertebrate (tabanid fly) host which acts purely as a mechanical (as opposed to cyclical) transmitter betwen ungulates. Vampire bats (*Desmodus* spp.) may also act in transmission of this trypanosome in S. America (Hoare, 1965). A closely-related species, *T.* (*Trypanozoon*) *equiperdum* relies on coitus for transmission between equine hosts and so has eliminated the second host entirely. With lack of selection for cyclical development in the fly, both of these trypanosomes have undergone maxicircle deletions (Artama et al., 1982) or dyskinetoplasty mutations (Fig. 10w) in which the minicircle DNA becomes dispersed throughout the mitochondrion. Transmission from ungulates to the large cats (e.g. lion, leopard) by carnivory occurs readily among the salivarian trypanosomes.

Subgenus ***Duttonella*** Chalmers, 1908

Trypanosomes with large and usually terminal kinetoplast; free flagellum always present; vector development takes place only in proboscis. Type species: *T.* (*Duttonella*) *vivax* Ziemann, 1905 (Fig. 10s) of ungulates, causative agent of souma in cattle. Mechanically transmitted outside Africa (S. America, islands of Indian Ocean). Other species: *T.* (*D.*) *uniforme* of Bovidae.

Subgenus ***Nannomonas*** Hoare, 1964

Small trypanosomes with medium-sized marginal kinetoplast; free flagellum never present; vector development takes place in midgut and proboscis. Type species: *T.* (*Nannomonas*) *congolense* Broden, 1904 (Fig. 10t) of ungulates, causative organism of nagana in cattle. Other species: *T.* (*N.*) *simiae* of ungulates and monkeys, causative agent of acute porcine trypanosomiasis.

Subgenus ***Trypanozoon*** Luhe, 1906

Trypanosomes with small subterminal kinetoplast, with or without free flagellum; live in connective tissues and central nervous system as well as in blood of mammalian host; vector development is in midgut and salivary glands of tsetse fly (*T.* [*T.*] *brucei*). No cyclical development occurs in *T.* (*T.*) *evansi* or *T.*(*T.*) *equiperdum* (see above).

Type species: *T.* (*Trypanozoon*) *brucei* Plimmer & Bradford, 1899 (Fig. 10u,v) of ungulates, causative agent of nagana in cattle. The causative agents of human sleeping sickness are genetic variants of this species which is characterized by pleomorphism (i.e. presence of long, slender, dividing forms and short, stumpy, non-dividing forms in the blood). Medical parasitologists have long given the sleeping sickness trypanosomes separate specific names, viz. *T. gambiense* and *T. rhodesiense*, though most protozoologists now prefer to refer to these two organisms as subspecies of *T. brucei*, the subspecies *T. brucei brucei* denoting variants not infective to man. Hoare (1972) designates both the sleeping sickness trypanosomes *T. b. gambiense* regarding the Gambian and Rhodesian parasites as nosodemes. Molecular biological techniques, however, suggest that in macromolecular composition, as in the ecology of the insect vectors, *T. b. gambiense sensu stricto* stands apart from *T. b. rhodesiense* and *T. b. brucei* and may merit separation at the specific level (Hide et al., 1990; Kanmogne et al., 1996) Other species: *T.* (*T.*) *evansi* (Figs. 6,17,18) causative agent of surra in camels and horses outside the tsetse fly belt of Africa and in Asia. An indistinguishable parasite, *T.* (*T.*) *hippicum* causes murrina in S. America, and a dyskinetoplastic variant (*T.* [*T.*] *equinum*) causes mal de caderas in the same continent. Trypanosomes of this "*T. evansi* complex" lack a true insect host, the vector performing mechanical transmission. *Trypanosoma* (*T.*) *equiperdum* causes dourine in horses.

Subgenus ***Pycnomonas*** Hoare, 1964

Stout monomorphic trypanosomes with small subterminal kinetoplast and short free flagellum; vector development in midgut and salivary glands of *Glossina*. Type species: *T.* (*P.*) *suis* Ochmann, 1905 (Fig. 10x) causative agent of chronic porcine trypanosomiasis; known only to infect Suidae.

References

Abe, Y. 1980. On the encystation of *Leptomonas* sp. (Kinetoplastida, Trypanosomatidae), a parasite of the silkworm, *Bombyx mori* Linnaeus. *J. Protozool.*, 22:372-374.

Anez, N. 1982. Studies on *Trypanosoma rangeli* Tejera 1920. IV. A reconsideration of its systematic position. *Mem. Inst. Oswaldo Cruz, Rio De J.*, 77:405-415.

Artama, W.,T., Agey, M. W. & Donelson, J. E. 1992. DNA comparisons of *Trypanosoma evansi* (Indonesia) and *Trypanosoma brucei* spp. *Parasitology*, 104:67-74.

Bardsley, J. E & Harmsen, R. 1973. The trypanosomes of Anura. *Adv. Parasitol.*, 11:1-73.

Bates, P. 1994. Complete developmental cycle of *Leishmania mexicana* in axenic culture. *Parasitology*, 108:1-9.

Beard, C. B., Butler, J. F. & Greiner, E. C. 1989. In vitro growth characterisation and host-parasite relationship of *Leptomonas pulexsimulantis* n.sp., a trypanosomatid flagellate of the flea, *Pulex simulans*. *J. Parasitol.*, 75:658-668.

Becker, C. D. 1977. Flagellate parasites of fish. *In*: Kreier, J. P. (ed.), *Parasitic Protozoa.* Vol.1. Academic Press, London. pp 358-416

Berchtold, M., Phillipe, H., Breunig, A., Brugerolle, G. & Konig, H. 1994. The phylogenetic position of *Dimastigella trypaniformis* within the parasitic kinetoplastids. *Parasitol. Res.*, 80:672-679.

Bovee, E. C. & Telford, S. R. 1962. Protozoan inquilines from Florida reptiles. 3. *Rigidomastic scincorum* sp.n., *Cercobodo stilosomorum* n.sp. and *Cryptobia geccorum* n.sp. *Q. J. Fl. Acad. Sci.*, 25:180-191.

Breunig, A., Konig, H., Brugerolle, G., Vickerman, K. & Hertel, H. 1993. Isolation and ultra-structural features of a new strain of *Dimastigella trypaniformis* Sandon 1928 (Bodonina, Kinetoplastida) and comparison with a previously-isolated strain. *Eur. J. Protistol.*, 29:416-424.

Brooker, B. E. 1971a. Fine structure of *Bodo saltans* and *Bodo caudatus* (Zoomastigophora, Protozoa) and their affinities with the Trypanosomatidae. *Bull. Brit. Mus.* (*Nat.Hist.*), 22:82-102.

Brooker, B. E. 1971b. Flagellar attachment and detachment of *Crithidia fasciculata* in the gut wall of *Anopheles gambiae*. *Protoplasma*, 73:191-202.

Brugerolle, G. 1985. Des trichocystes chez les bodonides, un charactere phylogenetique supplementaire entre Kinetoplastida et Euglenida. *Protistologia*, 21:339-348.

Brugerolle, G., Lom, J., Nohynkova, E. & Joyon, L. 1979. Comparaison et evolution des structures cellulaires chez plusiers espèces de Bodonides et Cryptobiides, appartenant aux genres *Bodo*, *Cryptobia* et *Trypanoplasma* (Kinetoplastida. Mastigophora). *Protistologia*, 15:197-221.

Burreson, E. M. 1982. The life cycle of *Trypanoplasma bullocki* (Zoomastigophorea: Kinetoplastida). *J. Protozool.*, 29:72-77.

Camargo, E. P., Sbravate, C., Teixeira, M. M. G., Uliana. S. R. B., Soares, M. B. M., Affonso, H. T. & Floeter-Winter, I. 1992. Ribosomal DNA restriction analysis and synthetic oligonucleotide probing in the identification of genera of lower Trypanosomatidae. *J. Parasitol.*, 78:40-48.

Cavalier-Smith, T. 1993. Kingdom Protozoa and its 18 phyla. *Microbiol. Rev.*, 57:953-994.

Cavalier-Smith, T. 1987. Eukaryotic kingdoms: seven or nine? *BioSystems*, 14:461-481.

Christensen, H. A. & Herrer, A. 1979. Susceptibility of sand flies (Diptera, Psychodidae) to Trypanosomatidae from two-toed sloths (Edentata: Bradypodidae). *J. Med. Entomol.*, 16:424-427.

Conchon, I., Campaner, M., Sbravate, C. & Camargo, E. P. 1989. Trypanosomatids other than *Phytomonas* spp. isolated and cultured from fruit. *J. Protozool.*, 36:412-414.

Coombs, G. H. 1988. *Herpetomonas muscarum ingenoplastis*: an anaerobic kinetoplastid flagellate. *In*: Lloyd, D., Coombs, G. H. & Paget, T. A. (eds.), *Biochemistry and Molecular Biology of Anaerobic Protozoa*. Harwood Academic Publishers, London. pp 254-266

Croft, S., Chance, M. L. & Gardiner, P. J. 1980. Ultrastructual and biochemical characterisation of stocks of *Endotrypanum*. *Ann. Trop. Med. Parasitol.*, 74:584-589.

Current, W. L. 1980. *Cryptobia* sp. in the snail *Triadopsis multilineata* (Say). Fine structure of the attached flagellates and their mode of attachment ot the spermatheca. *J. Protozool.*, 27:278-287.

Deane, M. P., Lenzie, H. L. & Jansen, A. M. 1986. Double developmental cycle of *Trypanosoma cruzi* in the opossum. *Parasitol. Today*, 2:146-147.

Diamant, A. 1987. Ultrastructure and pathogenesis of *Ichthyobodo* sp. from wild common dab *Limanda limanda* L in the North Sea. *J. Fish Dis.*, 10:41-247.

Diamant, A. 1990. Morphology and ultrastucture of *Cryptobia eilatica* n. sp. (Bodonidae, Kinetoplastida), an ectoparasite from the gills of marine fish. *J. Protozool.*, 37:482-489.

Durrer, S. & Schmid-Hempel, P. 1994. Shared use of flowers leads to horizontal pathogen transmission. *Proc. Roy. Soc. Ser. B*, 258:299-302.

Elbrachter, M., Schnepf, E. & Balzer, I. 1996. *Hemistasia phaeocysticola* (Scherffel) comb. nov., redescription of a free-living, marine, phagotrophic kinetoplastid flagellate. *Arch. Protistenkd.*, 147:125-136.

Faria E Silva, P. M., Soares, M. J. & De Souza, W. 1996. Proliferative opisthomastigote forms in *Herpetomonas roitmani* (Kinetoplastida: Trypanosomatidae). *Parasitol. Res.*, 82:125-129.

Fernandes, A.,P., Nelson, K.,& Beverley, S. M. 1993. Evolution of nuclear ribosomal RNAs in kinetoplastid protozoa: perspectives on the age and origin of parasitism. Proc. Nat. Acad. Sci. USA, 90:11608-11612.

Frolov, A. O. & Malysheva, M. N. 1992. The endomastigote, a special type of transmission stage of trypanosomatids of the genus *Proteomonas. Parasitologia*, 26:97-107. (in Russian)

Frolov, A. O. & Malysheva, M. N. 1993. Description of *Phytomonas nordicus* sp.n. (Kinetoplastida, Trypanosomatidae) from the predatory bug *Troilus luridus* (Hemiptera, Pentatomidae). *Parasitologia,* 27:227-232. (in Russian)

Gibson, W. C. 1995. The significance of genetic exchange in trypanosomes. *Parasitol. Today*, 11:465-468.

Görtz, H.-D. & Dieckmann, J. 1987. *Leptomonas ciliatorum* n. sp. (Kinetoplastida, Trypanosomatidae) in the macronucleus of a hypotrichous ciliate. *J. Protozool.*, 34:259-263.

Hajduk, S. L., Siqueira, A. M. & Vickerman, K. 1986. Kinetoplast DNA of *Bodo caudatus. Mol. Cell. Biol.*, 6:4372-4378.

Hecker, H. 1993. Man and sea urchin: more closely related than African and South American trypanosomes? *Parasitol. Today*, 9:57.

Hide, G., Cattand, P., Le Ray, D., Barry, J. D. & Tait, A. 1990. The identification of *Trypanosoma brucei* subspecies using repetitive DNA sequences. *Mol. Biochem. Parasitol.*, 39:213-226.

Hitchen, E. T. 1974. The fine structure of the colonial kinetoplastid flagellate *Cephalothamnium cyclopum* Stein. *J. Protozool.*, 21:221-231.

Hoare, C. A. 1965. Vampire bats as vectors and hosts of equine and bovine trypanosomiasis. *Acta Trop.*, 22:204-216.

Hoare, C.,A. 1972. *The Trypanosomes of Mammals. A Zoological Monograph.* Blackwell Scientific Publications, Oxford.

Hoare, C. A. & Wallace, F. G. 1966. Developmental stages of trypanosomatid flagellates: a new terminology. *Nature,* 212:1385-1386.

Jankevicius, J. V., Jankevicius, S. I., Campaner, M., Conchon, I., Maeda, L. A., Teixeira, M. M. G., Freymuller, E. & Camargo, E. P. 1989. Life cycle and culturing of *Phytomonas serpens* (Gibbs), a trypanosomatid parasite of tomatoes. *J. Protozool.*, 36:265-271.

Jenni, L., Marti, S., Sweizer, J., Betschart, B., Le Page, R.,W.,F., Wells, J.,M., Tait, A., Pandavoine, P., Pays, E. & Steinert, M. 1986. Hybrid formation between African trypanosomes during cyclical transmission. *Nature*, 322:173-175.

Joyon, L. & Lom, J. 1969. Sur la structure de *Costia necatrix* Leclercq (Zooflagelle); place systematique de ce protiste. *C. R. Hebd. Seances Acad. Sci.*, 262:660-663.

Kanmogne, G. D., Stevens, J. R., Asonganyi, & Gibson, W. C. 1996. Genetic heterogeneity in the *Trypanosoma brucei gambiense* genome analysed by random amplification of poly-morphic DNA. *Parasitol. Res.*, 82:535-541.

Killick-Kendrick, R., Lainson, R., Rioux, J.-A. & Saf'janova, V. M. 1986. The taxonomy of *Leishmania*-like parasites of reptiles. *In*: Leishmania: *Taxonomie et Phylogenese. Applications Eco-epidemiologiques. Colloquia Internationales* CNRS/INSERM 1984. IMEEE Montpellier. pp 143-148

Kivic, P. A. & Walne, P. L. 1984. An evaluation of a possible phylogenetic relationship between the Euglenophyta and Kinetoplastida. *Origins of Life* 13:269-288.

Kruse, P., Steinhagen, D. & Korting, W. 1989. Development of *Trypanoplasma borreli* (Mastigophora, Kinetoplastida) in the leech vector *Piscicola geometra* and its infectivity for the common carp, *Cyprinus carpio. J.* Parasitol., 75:527-531.

Lainson, R. & Shaw, J. J. 1987. Evolution, classification and geographical distribution. *In*: Peters, W. & Killick-Kendrick, R. (eds.), *The Leishmaniases in Biology and Medicine.* Academic Press, London. pp 1-120

Larsen, J. & Patterson, D. J. 1990. Some flagellates (Protista) from tropical marine sediments. *J. Nat. Hist.*, 24:801-937.

Lauge, G. & Nishioka, R. S. 1977. Ultrastructural study of the relations between *Leptomonas oncopelti* (Noguchi & Tilden) Protozoa, Trypanosomatidae, and the rectal wall of adults of *Oncopeltus fasciatus* Dallas, Hemiptera, Lygaeidae. *J. Morph.*, 154:291-305.

Lom, J. 1979. Biology of trypanosomes and trypanosplasms of fish. *In*: Lumsden, W. H. R & Evans, D. A. (eds.), *Biology of the Kinetoplastida* Vol. 2. Academic Press, New York. pp 270-337

Lumsden, W. H. R. & Evans D. A. (eds.) 1976. *Biology of the Kinetoplastida*. Vol. 1. Academic Press, London.

Lumsden, W. H. R & Evans, D. A. (eds.) 1979. *Biology of the Kinetoplastida*. Vol. 2. Academic Press, London.

Maslov, D. A. & Simpson, L. 1995. Evolution of parasitism in kinetoplastid protozoa. *Parasitol. Today*, 11:30-32.

Maslov, D. A., Avila, H. A., Lake, J. A. & Simpson, L. 1994. Evolution of RNA editing in kinetoplastid protozoa. *Nature*, 368:345-348.

Mcghee, R. B. & Hanson, W. L. 1962. Growth and reproduction of *Leptomonas oncopelti* in the milkweed bug, *Oncopeltus fasciatus. J. Protozool.,* 9:488-493.

Medina Acosta, E., Paul, S., Tomlinson, S. & Pontes de Carvalho, L. C. 1994. Combined occurrrence of trypanosome sialidase/transsialidase activities and leishmanial metalloproteinase gene homologs in *Endotrypanum* sp. *Molec. Biochem. Parasitol.*, 64:273-282.

Menara, A., Dollet, M., Gargani, D. & Louise, C. 1988. Culture in vitro sur cellules d'invertebres de *Phytomonas* sp. (Trypanosomatidae) associes au hartrot, maladie du cocotier. *C. R. Seances Acad. Sci., Paris Ser.* 3, 307:597-602.

Molyneux, D. H. 1976. Biology of trypanosomes of the subgenus *Herpetosoma*. *In*: Lumsden, W. H. R & Evans, D. A. (eds.), *Biology of the Kinetoplastida*. Vol. 1. London, Academic Press. pp 285-326

Molyneux, D. H., Croft, S. L. & Lavin, D. R. 1981. Studies on the host-parasite relationships of *Leptomonas* species (Protozoa: Kinetoplastida) of Siphonaptera. *J. Nat. Hist.*, 15:395-406.

Mylnikov, A. P. 1986. Ultrastructure of a colourless flagellate *Phyllomitus apiculatus* Skuja 1948 (Kinetoplastida). *Arch. Protistenkd.*, 132:1-10.

Nohynkova, E. 1984. A new pathogenic *Cryptobia* from freshwater fishes: a light and electron microscopy study. *Protistologica*, 20:181-195.

Opperdoes, F. R. 1991. Glycosomes. *In*: Coombs, G. H. & North, M. J. (eds.), *Biochemical Protozoology*. Taylor & Francis: London. pp 134-144

Page, A. M., Canning, E. U., Barker, R. J. & Nicholas, J. P. 1986. A new species of *Rhynchoidomonas* Patton 1910 (Kinetoplastida, Trypanosomatina) from *Operophtera brumata* (Lepidoptera, Geometridae). *Syst. Parasitol.*, 8:101-105.

Patterson, D. J. & Larsen, J. (eds.) 1991. *The Biology of Free-Living Heterotrophic Flagellates*. Systematics Association Special Volume No. 45. Clarendon Press, Oxford.

Pays, E. 1993. Genome organisation and control of gene expression in trypanosomatids. *In*: Brod, P., Oliver, S. G. & Sims, P. (eds.), *The Eukaryotic Genome*. Cambridge University Press. pp 127-160

Peckova, H & Lom, J. 1990. Growth, morphology and division of flagellates of the genus *Trypanoplasma* (Protozoa, Kinetoplastida) in vitro. *Parasitol. Res.*, 76:553-558.

Peters, W. & Killick-Kendrick, R. (eds.) 1987. *The Leishmaniases in Biology and Medicine*. Vols. 1 & 2. Academic Press, London.

Podlipaev, S. A. 1990. Catalogue of World Fauna of Trypanosomatidae (Protozoa). *Proc. Zool. Inst.*, 217:1-177. USSR Academy of Sciences, Leningrad. (In Russian)

Podlipaev, S. A, Frolov, A. O. & Kolesnikov, A. A. 1990. *Proteomonas inconstans* n. g., n. sp. (Kinetoplastida, Trypanosomatidae), a parasite of the bug *Calocoris sexguttatus* (Hemiptera, Miridae). *Parasitologiya (Leningr.*), 24:339-346. (In Russian).

Rioux, J. A., Lanotte, G., Serres, E., Pratlong, F., Bastien, P. & Perieres, J. 1990. Taxonomy of *Leishmania*. Use of isoenzymes. Suggestions for a new classification. *Ann. Parasitol. Hum. Comp.*, 65:111-125.

Robinson, D. R. & Gull, K. 1991. Basal body movements as a mechanism for mitochondrial genome segregation in the trypanosome cell cycle. *Nature, London*, 352:731-733.

Rogers, W. E. & Wallace, F. G. 1971. Two new subspecies of *Herpetomonas muscarum* (Leidy 1856) Kent 1880. *J. Protozool.*, 18:645-649.

Roitman, I. & Camargo, E. P. 1985. Endosymbionts of Trypanosomatidae. *Parasitol. Today*, 1:143-144.

Ruinen, L. 1938. Notizen uber Saltzflagellaten. II Uber die Verbreitung der Saltzflagellaten. *Arch. Protistenkd.*, 90:210-258.

Schaub, G. A. 1994. Pathogenicity of trypanosomatids on insects. *Parasitol. Today*, 10:463-468.

Schlegel, M. 1994. Molecular phylogeny of eukaryotes. *Trends Ecol. Evol.,* 9:330-335.

Sherwin, T. & Gull, K. 1989. The cell division cycle of *Trypanosoma brucei*; timing of event markers and cytoskeletal modifications. *Phil. Trans. Roy. Soc. London B*, 323:573-588.

Simpson, A. G. B. 1997. The identity and composition of the Euglenozoa. *Arch. Protistenkd.*, 148:318-328.

Stuart, K. & Feagin, J. E. 1992. Mitochondrial DNA of kinetoplastids. *Int. Rev. Cytol.*, 141;65-88.

Swale, E. M. F. 1973. A study of the colourless flagellate *Rhynchomonas nasuta* (Stokes) Klebs. *Biol. J. Linn. Soc.*, 5:255-264.

Taylor, A. E. R. & Baker, J. R. 1987. *In Vitro Methods for Parasite Cultivation.* Academic Press, London.

Triemer, R. E. & Farmer, M. A. 1991. The ultrastructural organisation of the heterotrophic euglenids and its evolutionary implications. *In*: Patterson, D. J. & Larsen, J. (eds.), *The Biology of Free-Living Heterotrophic Flagellates.* Systematics Assocation Special Volume No. 45. Clarendon Press, Oxford. pp 185-204

Urawa, S. & Kusakari, M. 1990. The survivability of the freshwater ectoparasitic flagellate *Ichthyobodo necator* on chum salmon fry (*Oncorhynchus keta*) in sea water and comparison to *Ichthyobodo* sp. on Japanese flounder (*Paralichthys olivaceus*). *J. Parasitol.*, 76:33-40.

Van der Ploeg, L.,H.,T., Cornelissen, A. W. C. A., Barry, J. D. & Borst, P. ,1984. ,Chromosomes of the Kinetoplastida. ,*EMBO J.*, **3**:3109-3115.

Vickerman, K. 1976. The diversity of the kinetoplastid flagellates. *In*: Lumsden, W. H. R. & Evans, D. A. (eds.), *Biology of the Kinetoplastida.* Vol. 1. Academic Press, London. pp 1-34

Vickerman, K. 1977. DNA throughout the single mitochondrion of a kinetoplastid flagellate: observations on the ultrastructure of *Cryptobia vaginalis* (Hesse 1910). *J. Protozool.*, 24:221-233

Vickerman, K. 1978. The free-living trypanoplasms. Descriptions of three species of the genus *Procryptobia* N.G. and redescription of *Dimastigella trypaniformis* Sandon, with notes on the microscopical diagnosis of disease in man and animals. *Trans. Am. Microsc. Soc.*, 97:485-502.

Vickerman, K. 1985. Developmental cycles and biology of pathogenic trypanosomes. *Brit. Med. Bull.*, 41:105-114.

Vickerman, K. 1989. Trypanosome sociology and antigenic variation. *Parasitology*, 99:537-547.

Vickerman, K. 1990. Phylum Zoomastigina: Class: Kinetoplastida. *In*: Margulis, L., Corliss, J. O., Melkonian, M. & Chapman, D. (eds.), *Handbook of Protoctista.* Jones & Bartlett: Boston. pp 200-210

Vickerman, K. 1991. Organisation of the bodonid flagellates. *In*: Patterson, D. J. & Larsen, J. (eds.), *The Biology of Free-Living Heterotrophic Flagellates.* Systematics Association Special Volume No 43. Clarendon Press, Oxford. pp 159-176

Vickerman, K. 1993. Natural Selection and the life cycles of protozoan parasites. *Proc. Zool. Soc., Calcutta.* Haldane Commemorative Volume. pp 41-52

Vickerman, K. 1994. The evolutionary expansion of the trypanosomatid flagellates. *Int. J. Parasitol.*, **24**: 1317-1331.

Vickerman, K. & Preston, T. M. 1976. Comparative cell biology of kinetoplastid flagellates. *In*: Lumsden, W. H. R. & Evans, D. A. (eds.), *Biology of the Kinetoplastida.* Vol. 1. Academic Press, London. pp 36-130

Vickerman, K. & Tetley, L. 1990. Flagellar surfaces of parasitic protozoa and their role in attachment. *In*: Bloodgood, R. A. (ed.), *Flagellar and Ciliary Membranes.* Wiley, New York. pp 267-304

Vickerman, K., Myler, P. J. & Stuart, K. D. 1993. African Trypanosomiasis. *In*: Warren, K. S. (ed.), *Imunology and Molecular Biology of Parasitic Infections.* Blackwell Scientific Publications, Oxford. pp 170-212

Vørs, N. 1992. Heterotrophic amoebae, flagellates and Heliozoa from the Tvarminne area, Gulf of Finland, in 1988-90. *Ophelia,* 36:1-109.

Wallace, F. G. 1966. The trypanosomatid parasites of insects and arachnids. *Exper. Parasitol.*, 18:124-193.

Wallace, F. G. 1979. Biology of the Kinetoplastida of arthropods. *In*: Lumsden, W. H. R. & Evans, D. A. (eds.), *Biology of the Kinetoplastida.* Vol 2. Academic Press, London.

Wallace, F. G., Camargo, E. P., McGhee, R. B. & Roitman, I. 1983. Guidelines for the description of new species of lower trypanosomatids. *J. Protozool.,* 30:308-313.

Wallace, F. G., Roitman, I. & Camargo, E. P. 1992. Trypanosomatids of Plants. *In*: Kreier, J. P. &

Baker, J. R. (eds.), *Parasitic Protozoa.* 2nd Ed., Vol II. Academic Press, New York. pp 55-84

Webster, P. & Russell, D. G. 1993. The flagellar pocket of trypanosomatids. *Parasitol. Today*, 9:201-206.

Wells, E. A. 1976. Subgenus *Megatrypanum*. *In*: Lumsden, W. H. R. & Evans, D. A. (eds.), *Biology of the Kinetoplastida.* Vol. 1. Academic Press, London. pp 257-284

Wenyon, C. M. 1926. Protozoology: a Manual for Medical Men, Veterinarians and Zoologists. Vols. I & II. Balliere, Tindall & Cox, London.

Woo, P. T. K. 1987. *Cryptobia* and cryptobiasis in fishes. *Adv. Parasitol.*, 26:199-237.

Yoshida, N., Freymuller, E. & Wallace, F. G. 1978. *Herpetomonas mariadeanei* sp. n. (Protozoa, Trypanosomatidae) from *Muscina stabulans* (Fallen 1816) (Diptera, Muscidae). *J. Protozool.*, 25:421-425.

Zhukov, B. F. 1991. The diversity of bodonids. *In*: Patterson, D. J. & Larsen, J. (eds.). *The Biology of Free-Living Heterotrophic Flagellates.* Systematics Association Special Volume No. 45. Clarendon Press, Oxford. pp 176-184

HEMIMASTIGOPHORA

By WILHELM FOISSNER and ILSE FOISSNER

The Hemimastigophora are a small, but still growing group of heterotrophic flagellates having a unique constellation of characters. Thus, they have been classified as distinct phylum, the Hemimastigophora, by Foissner et al. (1988), containing the single family, Spironemidae Doflein. The diagonal (rotational) symmetry of the cortical plates (Fig. 1a) and the conspicuous ability of some species to perform **metaboly** (Fig. 1b) indicate a relationship with euglenids (Foissner and Foissner, 1993).

General characterization: small to medium sized (10-60 µm long), vermiform to ovoid, with slight anterior constriction producing head-like "**capitulum**". Two slightly spiraled rows of flagella, shorter or as long as body, in more or less distinct furrows located at sites where cortical plates abut. Basal bodies single, each

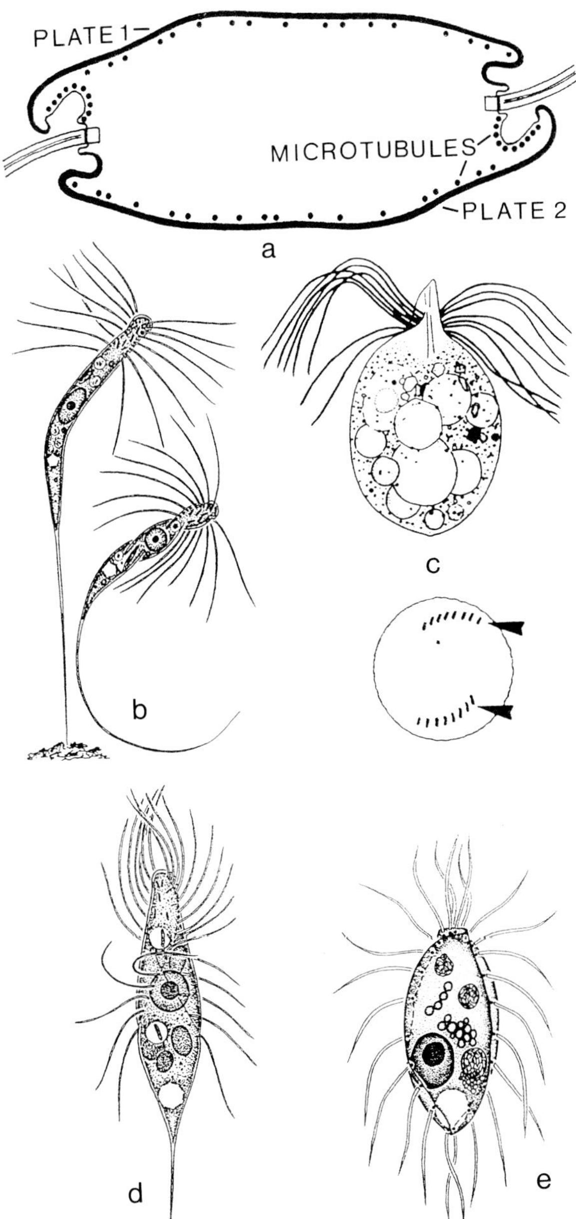

Fig. 1a-e. Hemimastigophoran flagellates. a. Schematized transverse section showing that the cortex is composed of two plicate plates having diagonal (rotational) symmetry (modified from Foissner and Foissner, 1993). b. *Spironema terricola* extended and contracted, length 40 µm (from Foissner and Foissner, 1993). c. *Paramastix conifera*, lateral and frontal view showing two flagellar rows (arrowheads), length 15 µm (from Zölffel and Skribbe, 1996). d. *Stereonema geiseri*, length 25 µm (from Foissner and Foissner, 1993). e. *Hemimastix amphikineta*, length 17 µm (from Foissner et al., 1988).

associated with a membranous sac, a short microtubule ribbon, a long microtubule ribbon, and nine filamentous arms (transitional fibers) forming a distinct basket. Cortex composed of two folded plates with diagonal (rotational) symmetry, supported by granular layer (epiplasm) in flagellated region and by microtubules either evenly spaced or in discrete groups (Fig. 1a). Single nucleus with prominent central nucleolus persisting throughout division. Contractile vacuole near posterior end of body. Mitochondrial cristae tubular to saccular. Complex, bottle-shaped **extrusomes** consisting of cylindroid posterior and rod-like anterior compartment. Food uptake at anterior end, defined oral structures, however, not recognizable. Fission in free-swimming condition, symmetrogenic. In freshwater and soil.

Four genera, as characterized in the key, with a total of eight reliable species are known; all genera, except *Paramastix*, have been confirmed by electron microscopy.

KEY TO GENERA

1. With euglenoid metaboly; flagella rows terminating in or distinctly above mid-body; body fusiform; cortex soft; freshwater and soil .... ***Spironema***

1'. Without euglenoid metaboly; flagella rows as long as body or distinctly shorter; body fusiform or globular; cortex rigid or soft; soil or freshwater... 2

2. Flagella rows extending whole body length; broadly fusiform and distinctly flattened; cortex rigid; soil .... ***Hemimastix***

2'. Flagella rows terminating in or distinctly above mid-body; body fusiform or globular; cortex soft; freshwater 3

3. Flagella rows terminating in mid-body; fusiform..................................... ***Stereonema***

3'. Flagella rows terminating distinctly above mid-body, i.e. restricted to rounded anterior end; globular.............................. ***Paramastix***

LITERATURE CITED

Foissner, I. & Foissner, W. 1993. Revision of the family Spironemidae Doflein (Protista, Hemimastigophora), with description of two new species, *Spironema terricola* n. sp. and *Stereonema geiseri* n. g., n. sp. *J. Euk. Microbiol.*, 40:422-438.

Foissner, W., Blatterer, H. & Foissner, I. 1988. The Hemimastigophora (*Hemimastix amphikineta* nov. gen., nov. spec.), a new protistan phylum from Gondwanian soils. *Europ. J. Protistol.*, 23:361-383.

Zölffel, M. & Skribbe, O. 1997. Rediscovery of the multiflagellated protist *Paramastix conifera* Skuja 1948 (Protista incertae sedis*)*. *Nova Hedwigia*, 65:443-452.

ORDER OXYMONADIDA

by GUY BRUGEROLLE and JOHN J. LEE

Oxymonads are intestinal parasitic flagellates with a cell basically comprising one **karyomastigont** composed of one nucleus, four flagella arranged in two pairs, a preaxostylar lamina, and a **paracrystalline axostyle** (several genera have two to several karyomastigonts)(Fig. 1). One or more of the anterior flagella is recurrent or adhering to the body surface. The pairs of basal bodies are separated by a **preaxostylar lamina** of a composite paracrystalline structure (Figs. 2,3,4a,b). The crystalline axostyle, originating from or near the anterior preaxostylar lamina, is composed of parallel rows of interlinked microtubules (Fig. 4c). An anterior row of microtubules or **pelta** is generally present (Fig. 2). Several species have developed an anterior expansion named holdfast (microfibrillar) and **rostellum** (microtubular) to attach to the chitinous intima of the insect host intestine (Figs. 1,3,4d,e). No cytostome, nutrition by phagocytosis and pinocytosis; several **xylophagous** species; glycogen is the reserve. They reproduce by binary fission and the mitosis is of the closed type with an intranuclear spindle (Cleveland, 1938, Hollande anb Carruette-Valentin, 1970b). Parental axostyle depolymerizes during division. Sexual

processes are described in some species. All species are endocommensal or symbiotic and live in the gut of insects except one *Monocercomonoides* which also lives in the gut of vertebrates. Transmission by cysts or trophic form.
Five families reported: Polymastigidae, Saccinobaculidae, Oxymonadidae, Pyrsonymphidae, Streblomastigidae, (Grassé, 1952, Vickerman, 1982). The Polymastigidae have the basic characters of the group and seem more primitive. The species of the other four families live in termites and in the wood-eating roach *Cryptocercus*, and they have developed adaptations for host attachment (holdfast, rostellum), contractile axostyle, complex life cycle including sexual processes. The list of oxymonad species living in termites and in the roach *Cryptocercus punctulatus* is given by Yamin (1979a).

The light microscopic coverage of the group is found in Grassé (1952); ultrastructural features in Brugerolle (1991).

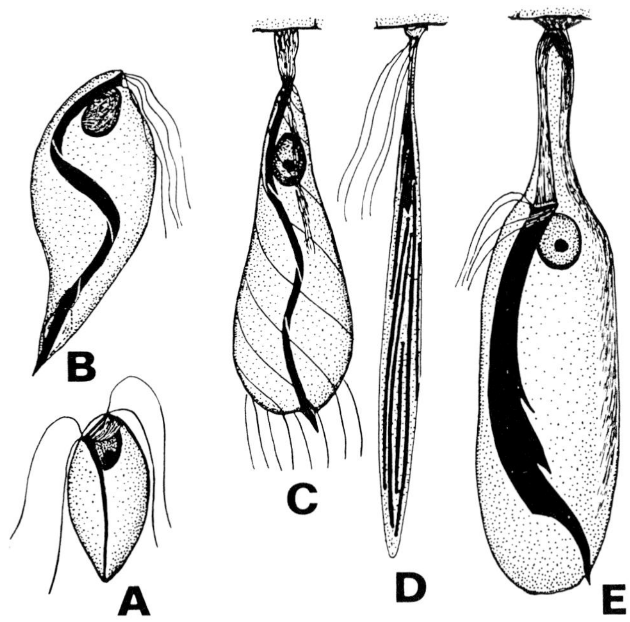

Fig. 1. General view of the genera representing each subfamily: A. *Polymastix*. B. *Saccinobaculus* C. *Pyrsonympha*. D. *Streblomastix*. E. *Oxymonas*. (G. Brugerolle)

Key Characteristics

1. Motile cells contain one or more karyomastigonts each with four flagella typically arranged in two separated pairs.
2. The two pairs of basal bodies separated by a preaxostylar lamina.
3. Axostyle composed of parallel rows of microtubules; contractile in some genera.
4. No cytostome; nutrition by phagocytosis or pinocytosis.
5. Closed mitosis with intranuclear spindle.

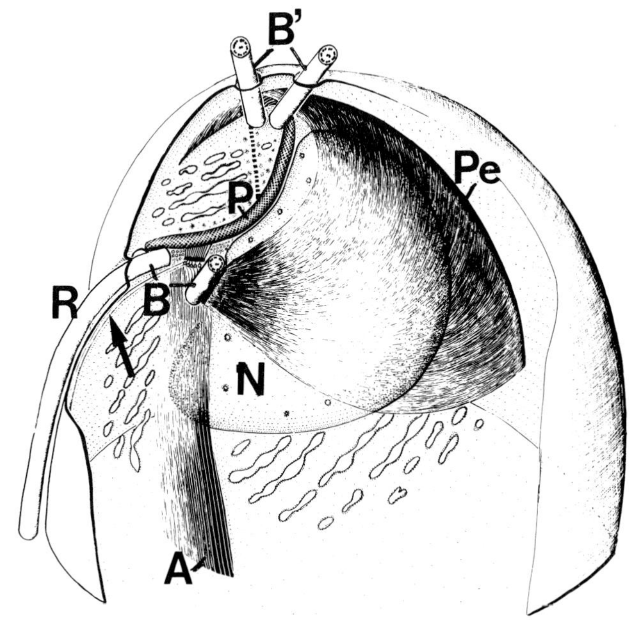

Fig. 2. Organization of the mastigont system of *Monocercomonoides*: the two pairs of basal bodies/flagella (B,B') are separated by the preaxostylar lamina (P) from which arise the axostyle (A). The pelta (Pe) caps the nucleus, and a row of microtubules (arrow) is adjacent to the recurrent flagellum (R). (From Brugerolle and Joyon, 1973)

ARTIFICIAL KEY TO FAMILIES AND SELECTED GENERA

1. No attachment organelle (holdfast or rostellum), no attached stage... 2
1'. With an attachment organelle (holdfast or rostellum) and attached phase to the host intestine... 3

2. With a slender axostyle, non-contractile....... ...Family **Polymastigidae**.......... ***Polymastix***, ...***Monocercomonoides***, ***Paranotila***

2'. With a large and contractile axostyle ....Family **Saccinobaculidae**... ***Saccinobaculus*, *Notila***

3. With an anterior rostellum.................................. ....Family **Oxymonadidae**.............5
4. With an anterior holdfast and recurrent adhering flagella...................... Family **Pyrsonymphidae** .... ***Pyrsonympha***
4'. With an holdfast and non-adhering flagella......... Family **Streblomastigidae**.... ***Streblomastix***

5. With a rostellum and uninucleate..... ***Oxymonas***
5'. With a rostellum and multinucleate.. ***Baroella***, ... ***Microrhopalodina*, *Sauromonas***

FAMILY POLYMASTIGIDAE

Three genera, *Monocercomonoides*, *Polymastix,* and *Paranotila*, of small flagellates with four free anterior flagella, one being recurrent. The cell has an anterior nucleus, a slender axostyle, and a row of microtubules or pelta covering the anterior end.

Genus ***Monocercomonoides*** Travis

Small oval to pyriform flagellates (5-15 µm) with 4 flagella extended in 2 pairs, one flagellum recurrent or trailing (Fig. 5A). The preaxostylar lamina is applied against the spherical anterior nucleus, which contains a central endosome. The large pelta caps the anterior part, and the slender axostyle originating from the preaxostyle traverses the cell and protrudes at the posterior end. A thin microtubular fiber follows the adhering zone of the recurrent flagellum (Figs. 2,4a). Electron microscopy has confirmed the oxymonad characters of this genus. The flagellate phagocytizes bacteria or wood and also feeds by pinocytosis. Among the 30 species reported, many occur in the posterior part of the gut of xylophagous insect larvae such as *Tipula* or coleoptera *Lygirodes, Cetonia* (Travis, 1932) and in orthoptera such as roaches, *M. globus* of *Cryptocercus* (Cleveland et al., 1934); a list of species of insects is given in Grassé (1952). Many species live in the digestive tract of vertebrates: cattle, rodents (Kirby and Honigberg, 1949, Nie, 1950), reptiles, amphibians; the list of vertebrate species is given by Kulda and Nohynkovà (1978), ultrastructure in Brugerolle and Joyon (1973) and in Radek (1994).

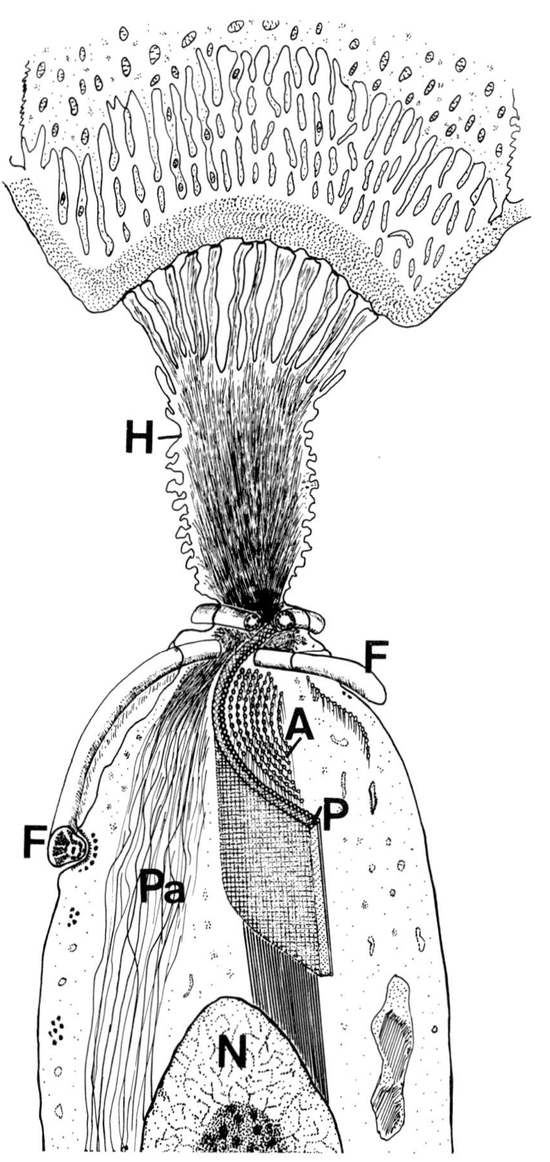

Fig. 3. Organization of *Pyrsonympha*: the two pairs of basal bodies/flagella are inserted close to the preaxostylar lamina (P) from which arises the crystalline axostyle (A). A bundle of non-organized microtubules or paraxostyle (Pa) develops parallel to the large axostyle. The flagella (F) are adherent to the cell body. The cell is attached to the intima of the intestine of the termite host by mean of the multifid anterior holdfast (H) containing a bundle of microfibrils. (From G. Brugerolle)

Genus ***Polymastix*** Bütschli

Almond or spindle-shaped cell with 4 anterior free flagella arranged in 2 pairs (Fig. 5B). Flagella deflected backward, but no adherent recurrent flagellum. Nucleus pointed anteriorly with a posterior endosome. Reduced anterior pelta and slender axostyle not protruding posteriorly. Electron microscopy has shown the basic oxymonad characters: basal bodies arranged in 2 pairs and attached on each side of the preaxostylar lamina (Brugerolle, 1980). The anterior tip of the nucleus is linked by a striated fiber to one pair of basal bodies. The cell is often covered by adhering rod-shaped bacteria such as *Fusiformis*. The 5 species reported occur in the lower intestine of coleoptera. insect larvae such as *Melolontha*, *Phyllophaga,* and in Myriapods such as *Glomeris*; light microscopic description and list of species in Grassé, 1952; for EM, Brugerolle, 1980.

Genus ***Paranotila*** Cleveland

Paranotila is a cell (15-25 µm) longer than *Monocercomonoides* and smaller than *Notila* (Figs. 6A,B). It bears 4 anterior flagella deflected backward, a slender axostyle not protruding posteriorly and not contractile. The large nucleus containing 2 nucleoli lies anteriorly near the origin of the flagella and the axostyle. There is no parabasal and this genus is an oxymonad. It presents a sexual cycle close to that of *Notila*: the cell undergoes gametogenesis forming 8 male and 8 female pronuclei which fuse producing zygotic nuclei and after division 8 diploid cells (Fig. 6B). Considering the sexual cycle, this genus is closer to *Notila* but the cytological features indicate a similarity with Polymastigidae. One species described: *P. lata* in *Cryptocercus punctulatus* (Cleveland, 1966). No EM study.

FAMILY SACCINOBACULIDAE

These oxymonad flagellates have a contractile axostyle but no evident attachment apparatus as in *Pyrsonympha* and *Oxymonas*. The flagella are not adherent to the cell body except in their proximal part. The two described genera *Saccinobaculus* and

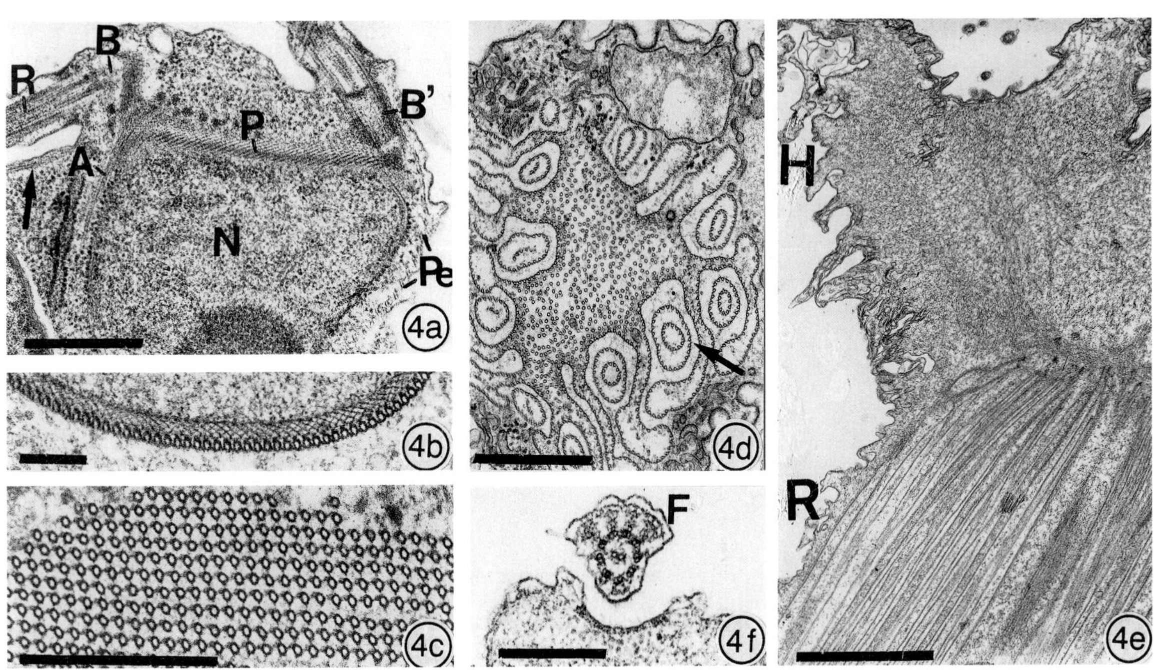

Figs. 4. Ultrastructural characters of oxymonads. Fig. 4a. Section of *Monocercomonoides* showing the two pairs of basal bodies/flagella (B, B') separated by the preaxostylar lamina (P) from which arise the microtubules of the axostyle (A) near the nucleus (N); microtubules of the pelta (Pe), microtubular fiber (arrow) underlying the recurrent flagellum (R) (bar=1 µm). Fig. 4b. Transverse section showing the composite structure of the preaxostylar lamina (bar=0.2 µm). Fig. 4c. Cross section of the crystalline axostyle in *Oxymonas* showing the interlinked rows of microtubules (bar=0.5 µm). Fig. 4d. Cross section of the rostellum of *Oxymonas* showing central microtubules and circonvoluted microtubular ribbons at the periphery (arrow) (bar=1 µm). Fig. 4e. Longitudinal section of the rostellum (R) containing microtubules and the holdfast (H) containing microfilaments in *Oxymonas.* (bar=1 µm.) Fig. 4f. Cross section of a modified flagellum adhering on gutters on the cell surface of *Pyrsonympha* (bar=0.5 µm). Photographs from G. Brugerolle.

Notila live in the gut of the roach *Cryptocercus* and have sexuality.

Genus ***Saccinobaculus*** Cleveland

Pyriform cell tapered posteriorly of 15-170 µm long according to the species (Fig. 7A). Four flagella inserted apically, directed backward and only adherent to the body in their proximal part. Nucleus anterior, close to the 2 pairs of basal bodies connected to the preaxostylar lamina. Large ribbon-like axostyle protruding posteriorly and surrounded by a sheath at the posterior end. The parallel rows of microtubules (up to 60) forming the axostyle begin to assemble at the anterior end close to the preaxostylar lamina. The axostyle is contractile and undulates in the cytoplam with bending waves originating at the anterior end and passing backwards (McIntosh et al., 1973). A row of microtubules corresponding to the pelta covers the anterior end. At least 2 species live in the roach *Cryptocercus punctulatus*. They swim free in the lumen and they have no developed attachment apparatus. They feed by pinocytosis and divide during the intermolt of the insect. Encysted cells also ensure host to host transmission (Cleveland et al., 1934). A sexual process comprising gametogenesis, fertilization, and meiosis in relation with the molting of the insect host has been described (Cleveland, 1950a), partial electron microscopic study in McIntosh et al. (1973).

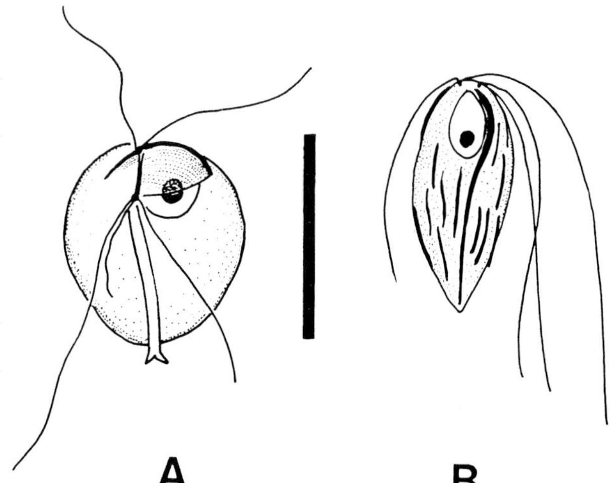

Fig. 5A. *Monocercomonoides caviae* from guinea-pigs (after Nie, 1950). Bar=10 µm. 5B. *Polymastix melolonthae* from Coleoptera larvae (after Grassé, 1952). Bar=10 µm.

Genus ***Notila*** Cleveland

Notila is a genus very close to *Saccinobaculus*; it has 4 flagella adhering to the cell body in their proximal portion and a ribbon-like contractile axostyle (Fig. 7B). It differs from the latter by its axostyle, which is not protruding and has no terminal sheath and contains many granules. The only species lives freely in the gut of *Cryptocercus punctulatus* and reproduces by asexual mitosis between the molts of the host. Sexual processes including the sequence—gametogenesis, first fertilization, meiosis, second fertilization—is correlated with the host molting (Cleveland, 1950b). No EM study.

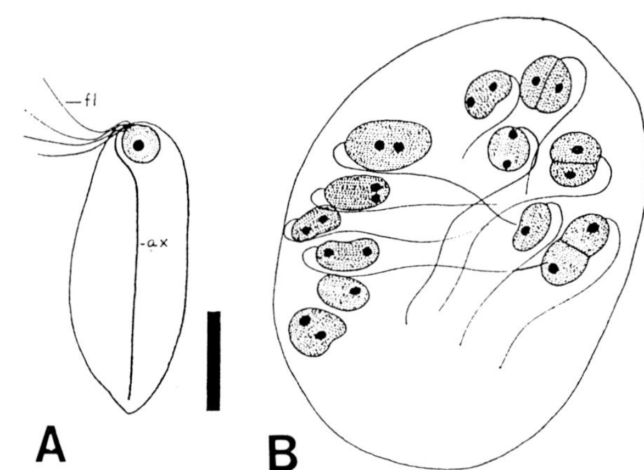

Fig. 6. *Paranotila lata* from *Cryptocercus punctulatus* (after Cleveland, 1966). A. Interphasic cell. B. Gametocyst with several nuclei undergoing fusion. Bar=10 µm.

FAMILY PYRSONYMPHIDAE

Oxymonads symbiotic in the termite *Reticulitermes* with flagella adhering along the cell body, contractile axostyle, cell attached to the intestinal wall by a microfibrillar holdfast at some stages of the life cycle. One genus *Pyrsonympha* (=*Dinenympha*).

Genus ***Pyrsonympha*** Leidy

Flagellates with a spirally twisted and contractile cell body of 20-150 µm depending of the species (Figs. 8,9). The 4 or 8 flagella are wrapped around the cell body and have a posterior trailing

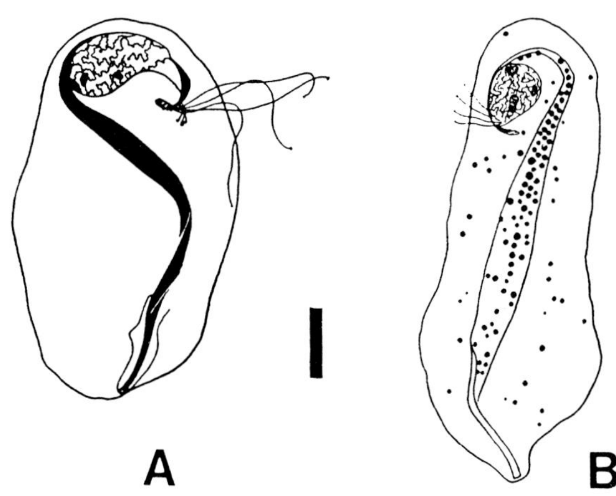

Fig. 7A. *Saccinobaculus ambloaxostylus* from *Cryptocercus punctulatus* (after Cleveland, 1950a). Bar=10 µm. Fig. 7B. *Notila proteus* from *Cryptocercus punctulatus* (after Cleveland, 1950b). Bar=10 µm.

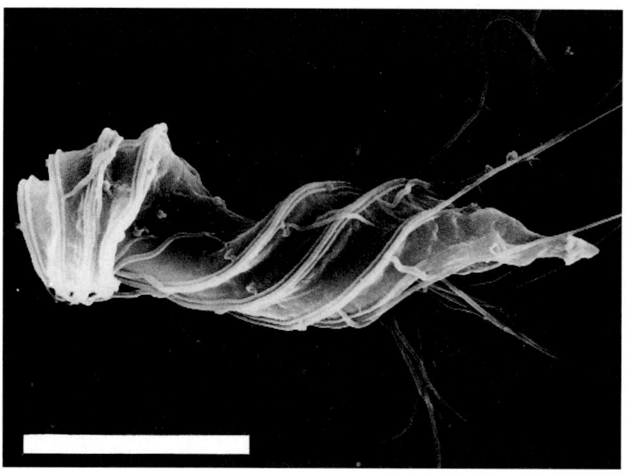

Fig. 9. *Pyrsonympha vertens* showing the four adhering flagella, several rod-shaped bacteria, and the posterior axostyle protrusion. Bar=10 µm. (G. Brugerolle)

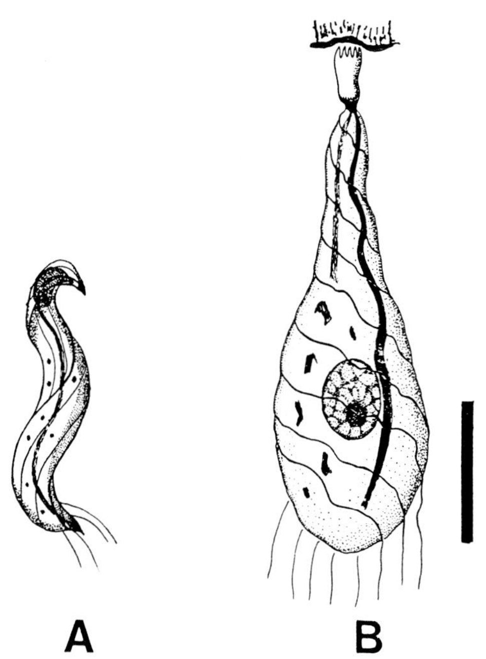

Fig. 8. *Pyrsonympha vertens* from *Reticulitermes lucifugus*. A. Motile *"Dinenympha"* form. B. Attached "*Pyrsonympha*" form. Bar=10 µm. (G. Brugerolle)

portion. The small and ribbon-like cells (12-30 µm), also called "*Dinenympha*" forms, have 4 flagella and swim in the intestinal fluid. The large and pyriform cells (150-200 µm), the "*Pyrsonympha*" forms, have 8 flagella and are generally attached by a holdfast to the intestinal wall of the termite. Electron microscopy has permitted the recognition of the oxymonad characters (Fig. 3). The basal bodies are separated into 2 pairs and linked to the preaxostylar lamina. The axostyle, originating close to the preaxostylar lamina, is composed of parallel rows (up to 50) of microtubules, which enroll at the posterior end, where the axostyle is surrounded by a sheath. The central axostyle is contractile and undulates in the cytoplasm. The flagella rest on gutters and adhere to the cell membrane forming undulating membranes (Fig. 9). The axoneme of the flagella present additional paraxonemal striated fibers alongside the 4 external triplets (Fig. 4f). All cells have an anterior holdfast, which is filled with longitudinal microfibrils originating from a dense centrosomal zone in contact with one pair of basal bodies. The multifid holdfast serves to attach the cell to the intestinal intima of the termite but does not reach the epithelial cells. Paraxostylar microtubules (Fig. 3) arising from the centrosomal zone form a non-organized bundle, which parallels the central crytalline axostyle. The cell surface is covered by a microfibrillar coat and by salpingoid and dendroid scales (Smith and Arnot, 1973). Flagellates phagocytize and digest wood and bacteria and also feed by pinocytosis. They live in the rectal pouch of the Rhinotermitidae of the genus *Reticulitermes.* During their life cycle young forms (*Dinenympha*) with 4 flagella transform to large fixed forms (*Pyrsonympha*) with 8 flagella by incomplete divisions while they increase their nuclear ploïdy (32 times) by successive endomitosis (Hollande eand Carruette-Valentin, 1970b). When the termite molts *Pyrsonympha* cells detach and divide several times restoring their normal ploïdy. Synaptonemal complexes occur in the first mitosis of the series and seem not related to a sexual process. The nuclear division is of the closed type with an intra-nuclear spindle (Cleveland, 1938, Hollande and Carruette-Valentin, 1970b). No cyst has been described; after the molt, the termite is reinfected by proctodeal feeding. Light microscopy of the 2 species is found in Grassé, 1952; Koidzumi, 1921; Powells, 1928; Cleveland, 1938; and electron microscopy by Brugerolle, 1970, Hollande and Carruette-Valentin, 1970a,b; Smith and Arnot, 1973; Bloodgood et al., 1974; Cochrane et al., 1979.

FAMILY STREBLOMASTIGIDAE

Oxymonad with 4 non-adhering flagella, a holdfast, a slender axostyle; only one genus *Streblomastix.*

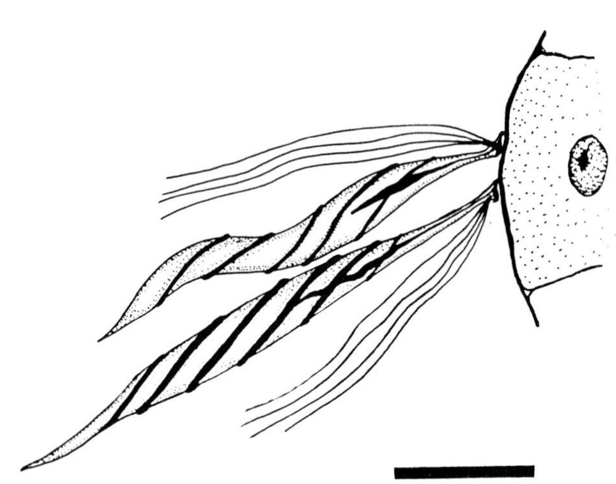

Fig. 10. *Streblomastix strix* from *Zootermopsis angusticollis* (after Kidder, 1929) attached and bearing rod bacteria. Bar=10 µm.

Genus ***Streblomastix*** Kofoid & Swezy

Long rod shape flagellate (15-530 µm), *Streblomastix strix*, the only species, lives in the hindgut of termopsid termites such as *Zootermopsis angusticollis* (Fig. 10). The body surface is carved in deep spiral ridges (4 to 8) and covered with long rod-shaped epibiont bacteria. The 4 flagella are inserted subapically at the base of the holdfast and are not adherent to the cell body. The 2 pairs of basal bodies are connected to the preaxostylar lamina. The axostyle is composed of few microtubular rows in the anterior part and reduced to one row around the elongated nucleus and to a thin bundle in the posterior part. Axostyle contraction is probably at the origin of the flexing movements of the cell body. Paraxostylar and pelta microtubules are present in the anterior part. The holdfast develops in cells attached to the intestinal wall. They probably feed by pinocytosis. They reproduce by longitudinal binary fission. During the molt of the termite, they are eliminated and then reintroduced with the proctodeal fluid. Light microscopy in Grassé, 1952; Kofoid and Swezy, 1919; Kidder, 1929; STM views in Yamin, 1979b; electron microscopy in Hollande and Carruette-Valentin, 1970a.

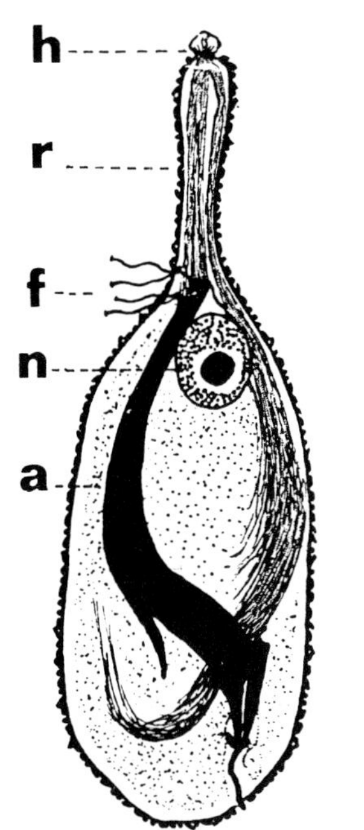

Fig. 11. *Oxymonas megakaryosoma* from *Glyptotermes latus* (after Cross, 1946) showing the anterior rostellum (r) terminated by a small holdfast (h) and containing the fibrous structure which penetrates into the cell body, the 4 flagella (f), the nucleus (n), and the axostyle (a). Bar=10 µm.

FAMILY OXYMONADIDAE

Uni- or multinucleated oxymonad flagellates composed of 1 to several karyomastigonts. They occur in alternate motile unattached forms and sedentary forms attached to the chitinous intima of the host intestine by an anterior expansion, the rostellum. The fibrous structure of the rostellum is mostly microtubular, differing from the microfibrillar holdfast of Pyrsonymphidae. Four genera: *Oxymonas, Barroella, Microrhopalodina, Sauromonas.*

Genus ***Oxymonas*** Janicki

Club-shaped cell (5-165 µm) with an anterior extensile rostellum containing fibers which arise at the base of the attachment point (Fig. 11). In attached forms a terminal nodule like a holdfast insures the fixation to the intestinal intima. The 4 flagella arise in 2 pairs in the shoulder region. The crystalline axostyle originates at the base of the rostellum, traverses the cell, and protrudes at the posterior end where it is surrounded by a sheath. The nucleus is generally anterior, at the base of the rostellum. Electron microscopic study has shown the oxymonad characters: presence of a preaxostylar lamina connected to the 2 pairs of basal bodies and a crystalline axostyle composed of parallel rows of microtubules (Fig. 4c). The rostellum is composed of a central bundle of non-organized microtubules surrounded by circonvoluted microtubulear ribbons arising at tne base of the holdfast and penetrating into the cell body (Fig. 4d). The rostellum is terminated by a holdfast containing microfibrils (Fig. 4e). Attached flagellates are non-motile, and unattached ones swim in the intestinal fluid and partly loose their rostellum. Twenty-two species living in Kalotermitidae (Kofoid and Swezy, 1926; Connell, 1930; Zeliff, 1930; Cross, 1946) and 2 in the roach *Cryptocercus* (Cleveland, 1950c). Hormone-induced sexual cycle comprising gametogenesis, fertilization, and meiosis has been described in species living in the roach *Cryptocercus* (Cleveland, 1950c). Ultrastructure study in Brugerolle and König, 1997; Rother et al., 1999).

Genus ***Barroella*** Zeliff

Multinucleate organism (25-220 µm) (Fig. 12A). Attached or mature forms are club-shaped cells with 1 rostellum, several nuclei (2-114), and axostyles scattered throughout the cell body and without flagella. Immature forms occurring from plasmotomy of large forms contain several karyomastigonts, with poorly developed flagella, arranged in one or several coronas (Figs. 12B,C). They phagocytize wood. Two described species, one occurring in *Calcaritermes brevicollis* of Panama and one in *Neotermes howa* from Mauritius island (Cross 1946). No ultrastructural study.

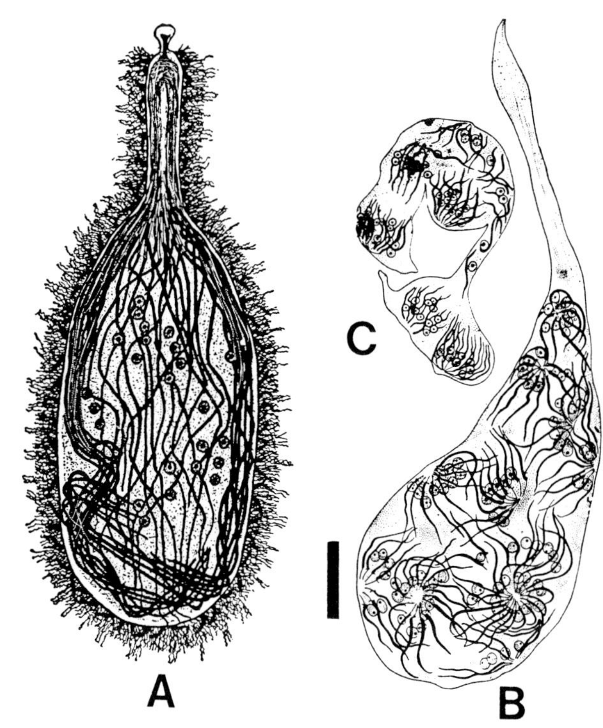

Fig. 12. *Barroella coronaria* from *Neotermes howa* var. *mauritiana* (after Cross, 1946). A. Attached form. B,C. Dividing forms undergoing plasmotomy. Bar=30 µm.

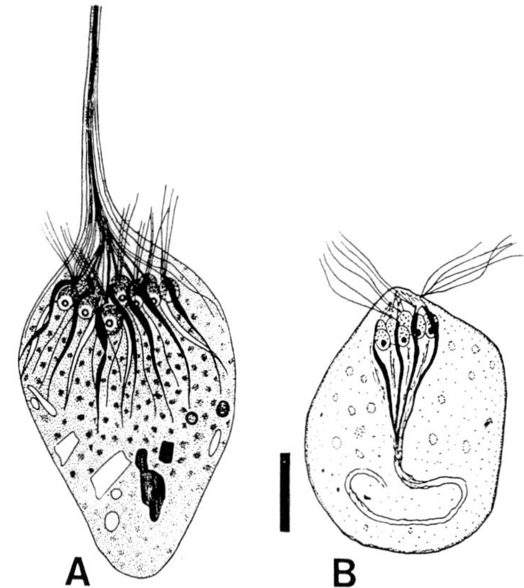

Fig. 13. *Microrhopalodina multinucleata* from *Cryptotermes dudleyi*. A. Attached form (after Kofoid and Swezy, 1926). B. Detached form without rostellum (after Kirby, 1929). Bar=20 µm.

Genus ***Microrhopalodina*** Grassi & Foa

Multinucleated organism (20-165 µm) (Fig. 13A). Attached forms are funnel-shaped cells composed of several karyomastigonts (1-26) forming a corona of nuclei, axostyles, and flagella at the base of the rostellum. Unattached forms lose their rostellum (Fig. 13B). Partial ultrastructural study has shown the oxymonad characters of the genus. Four species described in Kalotermitidae (Kirby, 1929; Cross, 1946; Grassé, 1952). The genera *Proboscidiella* (Kofoid and Swezy, 1926) and *Opisthomitus* (Grassé, 1952) are probably synonyms of *Microrhopolodina.* Ultrastructure study by Lavette (1973); Rother et al. (1999).

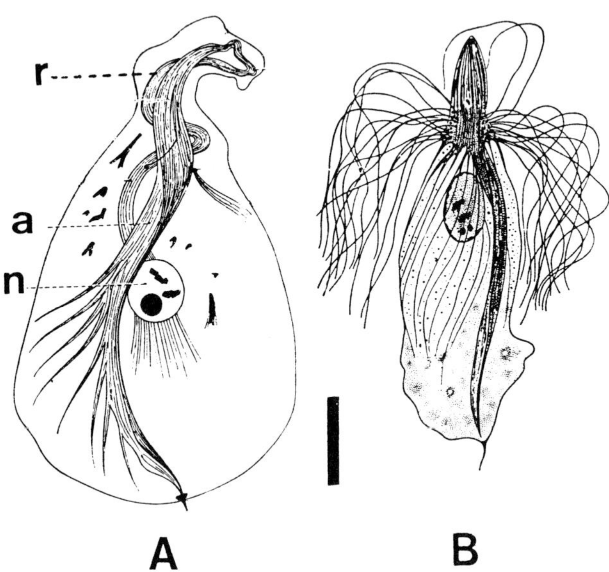

Fig. 14. *Sauromonas m'baikiensis* from *Glyptotermes boukoko* (after Grassé & Hollande, 1952). A. Attached form showing the anterior rostellum (r), the axostyle (a), the nucleus (n). B. Multiflagellated cell undergoing transformations during the molt of the termite. Bar=20 µm.

Genus ***Sauromonas*** Grassé & Hollande

The trophic form has one karyomastigont and is organized like an *Oxymonas* with 4 flagella, 1 anterior nucleus, a large axostyle protruding posteriorly, a rostellum containing a recurvent bundle of fibers (Fig. 14A). At the molting period of the termite, the flagellate detaches from the intestine wall and undergoes a series of transformations comprising the multiplication of flagella (Fig. 14B) and an encystment phase where the flagella regress. The only species, *S. m'baikiensis*, a symbiont of *Glyptotermes boukoko* with a complex life cycle is incompletely known (Grassé, 1952).

LITERATURE CITED

Bloodgood, R. A., Miller K. R., Fitzharris, T. P., McIntosh, J. R. 1974. The ultrastructure of *Pyrsonympha* and its associated microorganisms. *J. Morphol.*, 143:77-106.

Brugerolle, G. 1970. Sur l'ultrastructure et la position systématique de *Pyrsonympha vertens* (Zooflagellata Pyrsonymphina). *C. R. Acad. Sc.*, Paris, 270:3474-3478.

Brugerolle, G. 1980. Etude ultrastructurale du flagellé parasite *Polymastix melolonthae* (Oxymonadida). *Protistologica*, 17:139-145.

Brugerolle, G. 1991. Flagellar and cytoskeletal systems in amitochondrial flagellates: Archamoebae, Metamonada and Parabasala. *Protoplasma*, 164:70-90.

Brugerolle, G. & König, H. 1997. Ultrastructure and organization of the cytoskeleton in *Oxymonas*, an intestinal flagellate of termites. *J. Euk. Microbiol.*, 44:305-313.

Brugerolle, G. & Joyon, L. 1973. Ultrastructure du genre *Monocercomonoides* (Travis). Zooflagellata, Oxymonadida. *Protistologica*, 9: 1-80.

Cleveland, L. R. 1938. Mitosis in *Pyrsonympha. Arch. Protistenkd.,* 91:452-455.

Cleveland, L. R. 1950a. Hormone-induced sexual cycles of flagellates. II. Gametogenesis, fertilization, and one-division meiosis in *Saccinobaculus. J. Morphol.,* 86:215-228.

Cleveland, L. R. 1950b. Hormone-induced sexual cycles of flagellates. IV. Meiosis after syngamy and before fusion in *Notila. J. Morphol.,* 87:317-348.

Cleveland, L. R. 1950c. Hormone-induced sexual cycles of flagellates. II. Gametogenesis, fertilization, and one-division meiosis in *Oxymonas. J. Morph.,* 86:185-214.

Cleveland, L. R. 1966. Nuclear division without cytokinesis followed by fusion of pronuclei in *Paranotila lata* gen. et sp. nov. *J. Protozool.,* 13:132-136.

Cleveland, L. R., Hall, S. R., Sanders, E. P. & Collier, J. 1934. The wood-feeding roach *Cryptocercus*, its protozoa and the symbiosis betweeen protozoa and roach. *Mem. Am. Acad. Arts Sci.,* 17:155-342.

Cochrane, M., Smith, H. E., Buhse, Jr., H. E. & Scammell, J. G. 1979. Structure of the attached stage of *Pyrsonympha* in the termite *Reticulitermes flavipes* Kollar. *Protistologica,* 15:259-270.

Connell, F. H. 1930. The morphology and life-cycle of *Oxymonas dimorpha* sp. nov., from *Neotermes simplicicornis* (Banks). *Univ. Calif. Publ. Zool.*, 36:51-66.

Cross, J. B. 1946. The flagellate subfamily Oxymonadidae. *Univ. Calif. Publ. Zool.,* 53:67-162.

Grassé, P. P. 1952. Famille des Polymastigidae, Ordre des Pyrsonymphines, Ordre des Oxymonadines. *In:* Grassé P.- P. (ed.), *Traité de Zoologie*, Vol. I. Masson et Cie, Paris. pp 780-823

Hollande, A. & Carruette-Valentin, J. 1970a. La lignée des Pyrsonymphines et les caractères infrastructuraux communs aux autres genres *Opisthomitus, Oxymonas, Saccinobaculus, Pyrsonympha* et *Streblomastix*. *C. R. Acad. Sci.*, Paris, D, 270:1587-1590.

Hollande, A. & Carruette-Valentin, J. 1970b. Appariement chromosomique et complexes synaptonématiques dans les noyaux de dépolyploïdisation chez *Pyrsonympha flagellata*: le cycle évolutif des Pyrsonymphines symbiotiques de *Reticulitermes lucifugus*. *C. R. Acad. Sci.,* Paris, D, 270:2250-2255.

Kidder, G. W. 1929. *Streblomastix strix*, morphology and mitosis. *Univ. Calif. Publ. Zool.*, 33:109-124.

Kirby, H. 1929. A species of *Proboscidiella* from *Kalotermes (Cryptotermes) dudlei* Banks, a termite of Central America, with remarks on the oxymonad flagellates. *Q. J. Microsc. Sci.*, 72:355-386.

Kirby, H. & Honigberg, B. M. 1949. Flagellates of the caecum of ground squirrels. *Univ. Calif. Publ. Zool.,* 53:315-366.

Kofoid, C. A. & Swezy, O. 1919. Studies on the parasites of the termites. I. On *Streblomastix strix*, a polymastigote flagellate with a linear plasmodial phase. *Univ. Calif. Publ. Zool.,* 20:1-20.

Kofoid, C. A. & Swezy, O. 1926. On *Oxymonas*, a flagellate with an extensile and retractile proboscis from *Kalotermes* from British Guiana *Univ. Calif. Publ. Zool.,* 28:285-300.

Kofoid, C. A. & Swezy, O. 1926. On *Proboscidiella multinucleata* gen. nov., from *Planocryptotermes nocens* from the Phillipine islands, a multinucleate flagellate with a remarkable organ of attachment. *Univ. Calif. Publ. Zool.,* 28:301-316.

Koidzumi, M. 1921. Studies on the intestinal Protozoa found in the termites of Japan. *Parasitology,* 13:235-309.

Kulda, J. & Nohynkovà, E. 1978. Flagellates of the human intestine and of the intestine of other species. *In:* Kreier J. P. (ed.), *Parasitic Protozoa.* Vol. II. Academic Press New York. pp 2-138

Lavette, A. 1973. Sur l'ultrastructure et les affinités systématiques de *Microrhopalodina inflata*, flagellé symbiotique du termite à cou jaune *Calotermes flavicollis*. *C. R. Acad. Sci.* Paris, D, 276:1307-1311.

McIntosh, R., Ogata, E. S. & Landis, S. C. 1973. The axostyle of *Saccinobaculus*. I. Structure of the organism and its microtubule bundle. *J. Cell Biol.*, 56:304-323.

Nie, D. 1950. Morphology and taxonomy of the intestinal Protozoa of the guinea-pig, *Cavia porcella*. *J. Morphol.,* 86:381-493.

Powells, W. N. 1928. On the morphology of *Pyrsonympha* with a description of 3 new species from *Reticulitermes hesperus* Banks. *Univ. Calif. Publ. Zool.,* 31:179-200.

Radek, R. 1994. *Monocercomonoides termitis* n. sp., an oxymonad from lower termite *Kalotermes sinaicus*. *Arch. Protistenkd.*, 144:373-382.

Rother, A., Radek R. & Hausmann, K. 1999. Characterization of surface structures covering termite flagellates of the Family Oxymonadidae and ultrastructure of two oxymonad species, *Microrhopalodina multinucleata* and *Oxymonas* sp. *Europ. J. Protistol.*, 35:1-16.

Smith, H. S. & Arnot, H. J. 1973. Scales associated with the external surface of *Pyrsonympha vertens*. *Trans. Am. Microsc. Soc. USA,* 92:670-677.

Travis, B. W. 1931-32. A discussion of synonymy in the nomenclature of certain insect flagellates, with the description of a new flagellate from the larvae of *Lygirodes relictus* Say (Coleoptera-Scarabeidae). *Iowa State College J. Sc.*, VI: 317-323.

Vickerman, K. 1982. Mastigophora. *In*: Parker S. P. (ed.), *Synopsis and classification of Living Organisms*. Vol. I. McGraw-Hill Book Company, New York. pp 496-508

Yamin, M. A. 1979a. Flagellates of the orders Trichomonadida Kirby, Oxymonadida Grassé, and Hypermastigida Grassi and Foà reported from lower termites (Isoptera families Mastotermitidae, Kalotermitidae, Hodotermitidae, Termopsidae, Rhinotermitidae, and Serritermitidae) and from the wood-feeding *Cryptocercus* (Dictyoptera: Cryptocercidae). *Sociobiology,* 4:3-117.

Yamin, M. A. 1979b. Scanning electron microscopy of some symbiotic flagellates from the termite *Zootermopsis*. *Trans. Am. Microsc. Soc.,* 98:276-279.

Zeliff, C. C. 1930. A cytological study of *Oxymonas*, a flagellate, including description of new species. *Am. J. Hyg.,* 11:714-739.

PHYLUM PARABASALIA

By GUY BRUGEROLLE and JOHN J. LEE

Parabasalids are anaerobic flagellates without mitochondria; most of them live as parasites in the alimentary or urogenital tract of vertebrates and invertebrates. Half of the genera are symbiotic in xylophagous insects such as termites and roaches (*Cryptocercus* spp.) (Yamin, 1979a); however, two free-living species are found in stagnant water, ponds, and water rich in organic matter. Some species are pathogenic in the urogenital or intestinal tract and are able to invade other organs, e.g. *Trichomonas vaginalis* for man, *Tritrichomonas foetus* for cattle, and *Trichomonas gallinae* and *Histomonas meleagridis* for birds.

Light and electron microscopic studies have revealed that parabasalids constitute a monophyletic evolutionary lineage and share numerous homologous characters: 1.) flagellar apparatus/cytoskeletal system; 2.) type of mitosis; and 3.) the presence of **hydrogenosomes**; (Brugerolle, 1975-76, 1991a). The simpler parabasalids are the trichomonads (order Trichomonadida) from which have probably evolved the larger and more complex hypermastigids of the order Hypermastigida (Fig. 1). Several homologous characters involve the flagellar apparatus. The basic unit is four flagella, and their basal bodies are typically arranged with the basal body of the recurrent flagellum (R) perpendicular to the basal bodies of the anteriorly-directed flagella (Figs. 2,3,4a). These basal bodies bear special appendages which make them recognizable: basal bodies 1 and 3 bear a hooked lamina, F1 and F3, respectively, and basal body 2 bears **sigmoid fibers**, F2, (Figs. 2,3,4a,b). This typical arrangement is found from the small monocercomonads to the hypermastigids, such as *Joenia* and *Lophomonas*, and is partially conserved in other hypermastigids. These basal bodies are designated as privileged basal bodies (Fig. 2).

Another common feature is the existence of a **parabasal body** composed of **parabasal fibers** supporting the Golgi cisternae or dictyosome (Figs. 3,4c). Two parabasal fibers are normally attached to the basal bodies; however, in a species deprived of basal bodies (*Dientamoeba*) these fibers and the parabasal apparatus are always present. In hypermastigids the parabasal fibers subdivide into several parabasals.

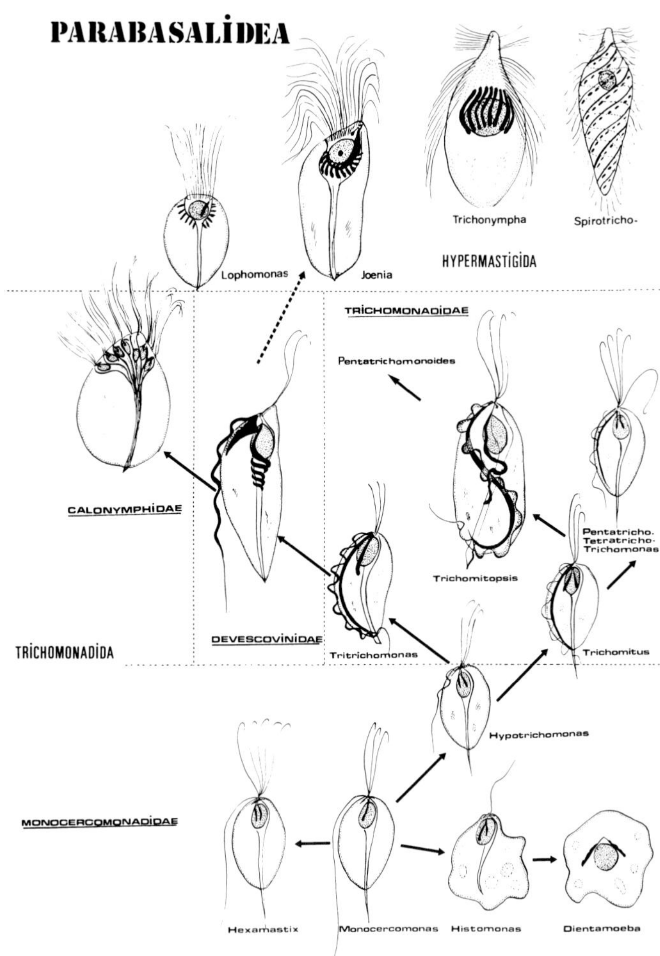

Fig. 1. Systematic and classic evolutionary relationships of parabasalids: two classes, Trichomonada and Hypermastigia. Class Trichomonada with one order Trichomonadida containing five families: Monocercomonadidae, Trichomonadidae, Devescovinidae, Calonymphidae, and Cochlosomatidae (not shown). (From G. Brugerolle)

Another homologous structure is the nearly general existence of the **pelta-axostyle complex** composed of one row of microtubules which inrolls to form a hollow tube through the axis of the cell (Figs. 3,4a,e,f). The microtubules arise from an organizing center which is close to the sigmoid fibers F2, also named **preaxostylar fibers**. This **axostyle** differs by its organization and origin from the crystalline axostyle of the oxymonads. In some hypermastigids the pelta-axostyle becomes more complex (*Joenia*) or reduced (*Trichonympha*), but it remains recognizable.

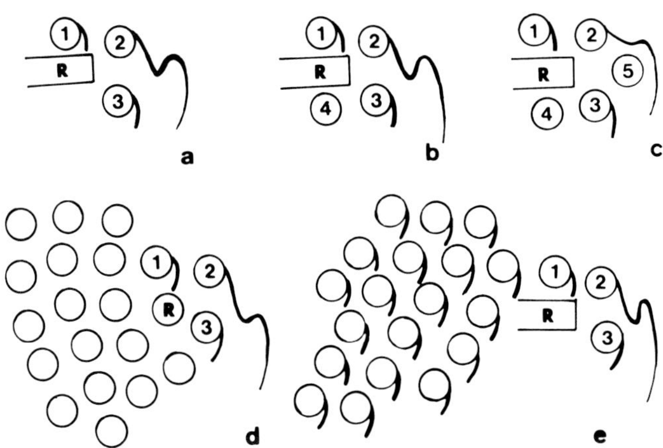

Fig. 2. Basal body arrangement in parabasalids. The basal bodies 1, 2, and 3 of the anteriorly directed flagella are orthogonal to the basal body of the recurrent flagellum (R). These four basic or privileged basal bodies found in *Monocercomonas* (a) are augmented by additional basal bodies in *Tetratrichomonas* (b), in *Hexamastix* and *Pentatrichomonas* (c), in *Lophomonas* (d), and in *Joenia* (e). (From Brugerolle, 1991a)

All these organisms divide in a closed mitosis with an external spindle (**pleuromitosis**) (Fig. 4g). The poles of the spindle are occupied by atractophores (dumb-bell-shaped bodies equivalent to centrioles) which give rise both to pole-to-pole microtubules (forming a bundle or paradesmose) and to chromosomal microtubules attached to the centromeres inserted in the nuclear envelope (Cleveland, 1961; Hollande and Carruette-Valentin, 1971, 1972; Brugerolle, 1975).

These cells also contain dense granules of about 1 µm in diameter named hydrogenosomes, which are limited by two closely applied membranes (Fig. 4d). They have no cristae and are different from mitochondria. They transform pyruvate to acetate, producing energy by phosphorylation and generating molecular hydrogen (Müller, 1988).

Parabasalids are divided into two classes: Trichomonada and Hypermastigia. Class Trichomonada contains a single order, Trichomonadida. Species in the order generally have only four to six flagella, except in species which have a reduced mastigont system, such as *Histomonas* or *Dientamoeba*, and those which have several **karyomastigonts**: Calonymphidae. Class Hypermastigia comprises species with hundreds or thousands of flagella and bearing many associated fibers forming a more complex cytoskeleton. General characters of the two classes can be found in Vickerman (1982) and Brugerolle (1991a). There is a single order in the Trichomonada.

CLASS TRICHOMONADA

ORDER TRICHOMONADIDA

These flagellates generally have three to five anteriorly directed flagella and a recurrent one, which is free or associated with the cell body, forming an undulating membrane (Fig. 5a). Basal bodies are typically arranged, and they bear the typical basal body appendages (Figs. 2,3,4). There are some exceptions, however, in the genera *Ditrichomonas*, *Histomonas*, and *Parahistomonas*, the number of flagella is reduced but basal bodies are still present; in the genus *Dientamoeba*, basal bodies and flagella are absent. Additional appendages occur in some genera: there is an infrakinetosomal body in *Monocercomonas* and *Tritrichomonas* (Figs. 3,5e) and a supra-kinetosomal body in the latter (Fig. 3). Basal body and fiber organization in trichomonads is presented in Brugerolle (1975-76, 1991a), Kulda et al. (1986), and Honigberg and Brugerolle (1990).

All genera have a parabasal apparatus composed of two parabasal fibers supporting one or two dictyosomes (Fig. 4c). These fibers are sometimes very long and in several Devescovinidae are twisted around the axostyle. The axostyle-pelta complex occupies the central axis of the cell and generally protrudes at the posterior end (Fig. 4f); however, some genera have a reduced (*Histomonas*) or no axostyle (*Dientamoeba*), and in *Pentatrichomonoides* the axostylar row of microtubules is not axial but peripheral. The axostyle forms a tube-like trunk which tapers and protrudes posteriorly and has an anterior capitulum, which is spoon-shaped and cups the surface of the nucleus, and is prolonged by the pelta. Hydrogenosomes are the spherical or bacilliform granules which concentrate around the axostyle and under the costa (Fig. 4d). They were also called paraxostylar, paracostal, or chromatic granules by light microscopists.

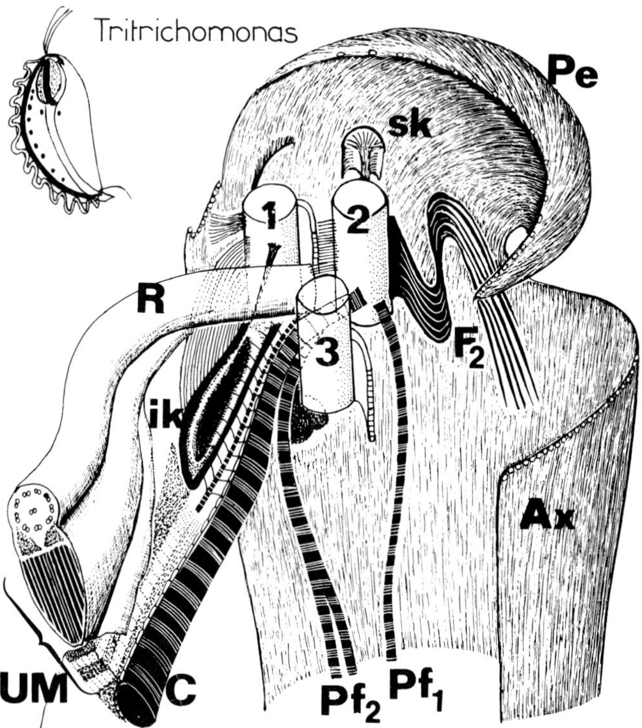

Fig. 3. Organization of basal bodies and associated fibers in *Tritrichomonas muris*. Basal bodies 1, 2, and 3 of the anterior flagella are arranged in a fixed position around the basal body R of the recurrent flagellum. Basal bodies 1 and 3 bear a hooked lamina and basal body 2 a sigmoid fiber F2 and a supra-kinetosomal body (sk). The two parabasal fibers (Pf1, Pf2) are attached to the basal bodies as well as the costa (C), which supports the undulating membrane (UM) and the infrakinetosomal body (ik). The two microtubular rows, pelta (Pe) and axostyle (Ax), are also positioned in relation to the basal bodies. (From Brugerolle, 1991a).

Trichomonads divide by a pleuromitosis where the paradesmose is the major distinguishable structure (Brugerolle, 1975). During division basal bodies separate into two sets and move apart as the paradesmose elongates and the pelta-axostyle microtubules depolymerize and reform in each sister cell (Grassé, 1952).

These flagellates have various striated roots attached to the basal bodies; the parabasal fibers are present in all species, and the costa is generally present in genera that have an **undulating membrane**, i.e. the Trichomonadidae (Fig. 5e). These fibers are composed of longitudinal parallel microfibrils which generate a periodic striation (period=42 nm). Two types of patterns of striation have been recognized. In the A-type pattern, the dense bands alternate with clear bands, and the fiber is massive in cross section (Figs. 5e,f). This pattern is found in all parabasal fibers and in the costae of *Tritrichomonas* and *Trichomitus*, and since it is more general, it is probably ancestral. In the B-pattern, thin dense bands alternate with clear bands showing a lattice structure (Fig. 6b), and the fiber is striated in cross section (Fig. 6c). This pattern occurs in the **costa** of *Trichomonas*, *Tetratrichomonas*, *Pentatrichomonas*, *Pentatrichomonoides*, and *Trichomitopsis*; all these genera also have a lamellar undulating membrane. A comparative ultrastructural study of the striated fiber pattern can be found in Brugerolle (1975-76) and in Honigberg and Brugerolle (1990).

The presence of an undulating membrane (UM) is more specific to the family Trichomonadidae. There are two general types of UMs. In the rail-type occuring in *Tritrichomonas,* the recurrent flagellum is modified by additional fibers along the axoneme (**para-axonemal fibers**) and adheres to the top of an elevation of the cell surface (Figs. 3,5a,b,c,d). This elevation contains microfibrils linked to the subjacent costa (Fig. 3). In the lamellar-type, the recurrent flagellum is normal or has additional structures (*Trichomitopsis*, *Pentatricho-monoides*) and adheres laterally to a lamellar elevation of the cell surface (Fig. 6a,d,e). This

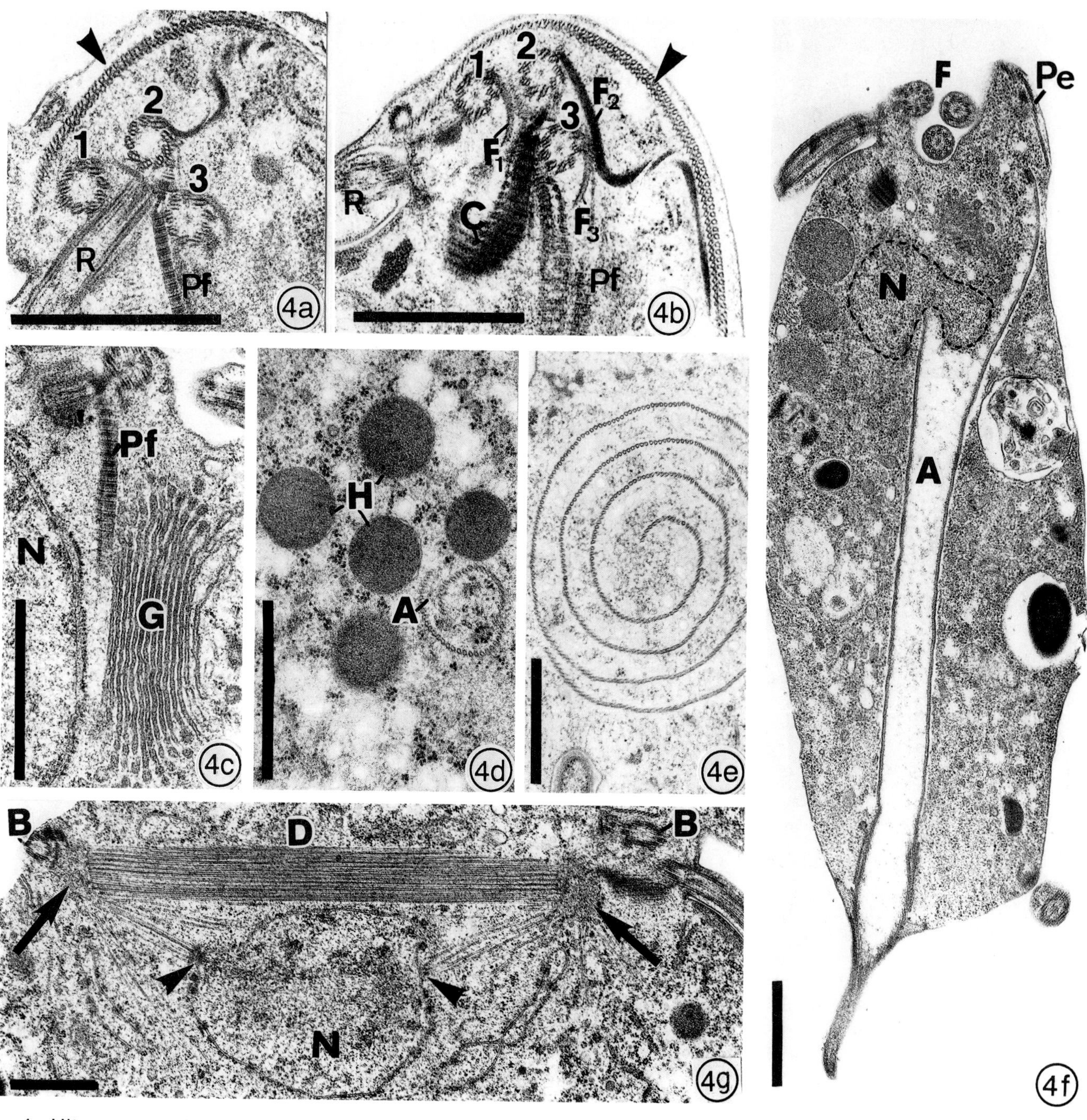

Fig. 4. Ultrastructural characters in class Trichomonada. Bars=1 μm (from G. Brugerolle). 4a,b. Basal body arrangement 1, 2, 3, and R and associated fibers F1, F2, F3, parabasal fibers (Pf), and costa (C), and axostyle and pelta microtubular rows (arrowhead). 4c. Parabasal apparatus composed of a parabasal fiber (Pf) and Golgi cisternae (G) close to the nucleus (N). 4d,e. Hydrogenosomal granules (H) around the hollow axostyle (A) in *Trichomonas vaginalis* (4d) and inrolled microtubular row of the axostyle in *Devescovina* (4e). 4f. Longitudinal section of the axostyle (A), anterior pelta (Pe), flagella (F), nucleus (N) in *Tritrichomonas minuta*. 4g. Pleuromitosis in *Trichomonas vaginalis*: a paradesmose (D) is stretched between the opposed poles or atractophores (arrows), attached to the basal bodies (B) of the two forming mastigont systems. Chromosomal microtubules are connected to the kinetochores of chromosomes (arrowheads) inserted in the persistent nuclear envelope (N).

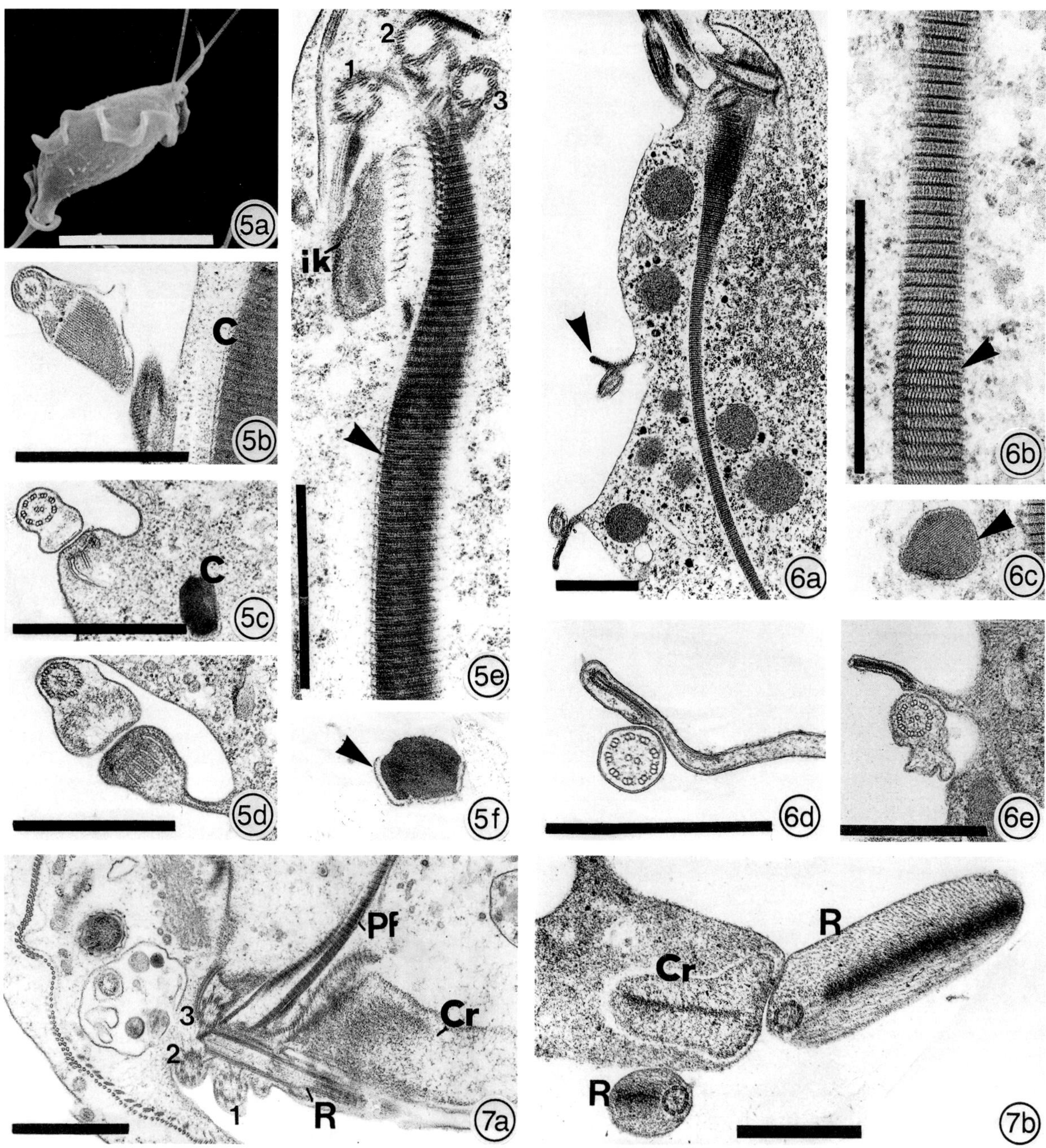

Figs. 5,6,7. Characteristics of the order Trichomonadida. Bars=1 µm (from G. Brugerolle). 5. Undulating membrane and costa in *Tritrichomonas*. 5a,b. Scanning view of the UM and costa in *T. muris* (5a) and cross section of the modified recurrent flagellum forming a rail-like UM attached to the underlain costa (C) (5b). 5c,d. UM in *T. foetus* and in *T. augusta.* 5e,f. Costa of *T. muris* attached to basal bodies 1, 2, and 3 and to the infrakinetosomal body (ik). The costa of the A-type shows its striation pattern and the external arched side (arrowhead) in longitudinal and cross section (Fig. 5f). Fig. 6. Lamellar-type UM and costa in Trichomonadidae. 6a. The costa underlies the lamellar UM in *Trichomonas vaginalis.* 6b,c. Costa of the B-type in *Pentatrichomonas hominis* showing the striation and a lattice structure (arrowhead), and cross section showing the stack of lamellae (arrowhead). 6d,e. Recurrent flagellum adhering to the lamellar UM containing a dense structure in *T. vaginalis.* Modified recurrent flagellum associated with a lamellar UM in *Pentatrichomonoides* (Fig. 6e). Fig. 7. Characters of Devescovinidae. Bar=1 µm. (from G. Brugerolle and J. P. Mignot) 7a. A dense structure named the cresta (Cr) develops under the adhering portion of the recurrent flagellum (R) and is attached to a comb-like fiber close to the parabasal fiber (Pf2) in *Foaina*. 7b. Cross section of the cresta (Cr) and of the recurrent flagellum (R) which has an axoneme associated with a bundle of microfibrils in *Devescovina*.

contains a dense/striated structure, and there is no evident connection with the supporting costa. This type of UM occurs in *Trichomitus, Trichomitopsis, Trichomonas, Tetratrichomonas, Pentatrichomonas*, and *Pentatrichomonoides* and also in two genera which have no costa, *Ditrichomonas* and *Pseudotrichomonas*. lamella
A third type of modification of the recurrent flagellum occurs in Devescovinidae, but it does not form a true undulating membrane. The cord-like or ribbon-like recurrent flagellum contains a bundle of microfibrils associated with the axoneme (Fig. 7b). It adheres to the cell body surface where it is underlain by a dense fibrous structure named the **cresta** (Fig. 7a). The length of the cresta and of the adhering portion is variable depending upon the species. Scanning views of the UM are found in Fig. 13 and in Warton and Honigberg (1979), and the ultrastructure and drawings in Brugerolle (1975-76, 1986b).

Trichomonads do not have a cytostome; they feed by phagocytosis and pinocytosis. Some symbiotic species in insects digest wood particles. The two free-living species and some parasitic species form resistant cysts (Brugerolle, 1973), but most are transmitted in the flagellated form.

There are four families in the order: the Monocercomonadidae have the simplest flagella /cytoskeleton, the Trichomonadidae are equipped with an undulating membrane supported by a costa, the Devescovinidae occur and have diversified in termites as have the Calonymphidae, which are polymastigotes related to the Devescovinidae. Genera of the family Calonymphidae are multinucleated with several karyomastigonts. Systematics based on light microscopic characters has been reviewed by Honigberg (1963) and, using additional ultrastructural characters, by Brugerolle (1975-76). A partial molecular phylogeny has been established by comparing the sequences of the ribosomal RNA (Viscogliosi et al, 1993; Keeling et al., 1998). General ultrastructural characters of trichomonads can be found in Brugerolle (1986b), Kulda et al. (1986), Honigberg and Brugerolle (1990), and Brugerolle (1991a).

KEY CHARACTERS OF THE PHYLUM

Motile cells typically with 4-6 flagella per mastigont system, one of which is recurrent (R) and free or associated with an undulating membrane (with some exceptions); with a hollow axostyle and a parabasal apparatus (Golgi body with a supporting fiber) associated with each mastigont system; basal bodies typically arranged (R orthogonal to 1, 2, and 3) and bearing typical fiber appendages; pleuromitosis with an external spindle and large paradesmosis.

KEY TO FAMILIES, SUBFAMILIES, AND SELECTED GENERA

(partly from Honigberg, 1963)

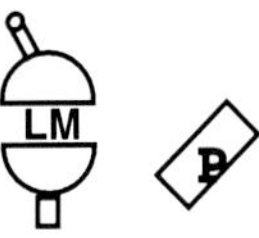

1. Costa absent.. 2
1.' Costa present.. Family **Trichomonadidae**
2. Cresta absent; undulating membrane absent in most genera; if present, poorly developed .. Family **Moncercomonadidae**
2'. Cresta present; no undulating membrane 3

3. Normally one mastigont per organism .... Family **Devescovinidae**
3'. Permanent polymonad organization............ .... Family **Calonymphidae**

FAMILY MONOCERCOMONADIDAE

1. Undulating membrane absent 2
1'. Undulating membrane present...................... .... Subfamily **Hypotrichomonadinae**.. 3

2. Flagella originating in funnel-shaped antero-ventral depression......Subfamily **Chilomitinae**........................ ***Chilomitus***
2'. Flagella not originating in funnel-shaped antero-ventral depression ....Subfamily **Monocercomonadinae**......... 5

3. Free posterior flagellum absent . ... ***Pseudotrichomonas***
3'. Free posterior flagellum present 4

4. With three anterior flagella.......................... .... ***Hypotrichomonas***
4'. With two anterior flagella. ***Ditrichomonas***

5. Three anterior flagella in mature organisms 6
5'. More than three anterior flagella in mature organisms... 10

6. Recurrent flagella present 9
6'. Axostyle not emergent 7

7. Recurrent flagellum presumably absent, body irregular spheroid..***Protrichomonas***
7'. Amoeboid body characteristically one anterior flagellum...................................... 8

8. Axostyle reduced.................... ***Histomonas***
8'. Not flagellated; lacks axostyle .... ***Dientamoeba***

9. Recurrent flagellum free or with proximal part adherent to body for some distance, typically not much longer than anterior flagella (not over 3x); trunk of axostyle of varying diameter; no modified forms in life cycle; no "normal" young forms with two anterior flagella .... ***Monocercomonas***
9'. Recurrent flagellum stout and long (2-10x longer than anterior flagella), with proximal part adherent to nearly entire length of body; trunk of axostyle very slender; modified forms present in life cycle; young individuals with two anterior flagella in many populations.............. .... ***Tricercomitus***

10. Four anterior flagella in mature organisms .... ***Tetratrichomastix***
10'. Five anterior flagella in mature organisms .... ***Hexamastix***

FAMILY TRICHOMONADIDAE

1. Trunk of axostyle hyaline, usually not stout, not tube-like in appearance, without periaxostylar rings or pointed terminal bulbous expansion; undulating membrane varying in degree of development and length 2
1'. Trunk of axostyle stout, tube-like in appearance in all but one genus; typically with periaxostylar rings or pointed terminal bulbous expansion; no typical pelta; undulating membrane well developed, as long as, or longer than body .....Subfamily **Tritrichomonadinae**..... 6

2. Trunk of axostyle usually slender or of moderate diameter projecting from posterior body surface; typical pelta; lamellar undulating membrane varying in degree of development and length; recurrent flagellum normal....................... ......Subfamily **Trichomonadinae**.......... 3
2'. Trunk of axostyle very slender, without projecting part; undulating membrane well developed, spiralling around the body from anterior to posterior end; five anterior flagella of single type ....Subfamily **Pentatrichomonoidinae** .... ***Pentatrichomonoides***

3. Free posterior flagellum present 4
3'. Free posterior flagellum absent............... .... ***Trichomonas***

4. Three anterior flagella; parabasal body usually V-shaped, occasionally rod-shaped; undulating membrane varying in degree of development and length................ .... ***Trichomitus***
4'. More than three anterior flagella 5

5. Four anterior flagella in mature individuals; parabasal body usually discoid, associated with at least one long filament; undulating membrane typically about as long as body. ***Tetratrichomonas***

5'. Five anterior flagella in at least 75% of every population, four grouped together at base, one independent; parabasal body small granule(s), which may be surrounded by clear, faintly outlined area associated with filament; undulating membrane about as long as body ***Pentatrichomonas***

6. Trunk of axostyle stout, tube-like in appearance; free posterior flagellum present 7
6'. Trunk of axostyle of moderate diameter, not tube-like in appearance; free posterior flagellum absent; costa very stout; parabasal body very long, typically branched; four anterior flagella................ .... ***Pseudotrypanosoma***

7. Typically three anterior flagella; axostyle typically with periaxostylar rings, tapers to point immediately after leaving body; costa of moderate width or stout and of "A" striation pattern .... ***Tritrichomonas***
7'. Four anterior flagella; axostyle typically without rings, often with terminal, pointed, bulbous expansion; costa stout Subfamily Trichomitopsiinae.................... .... ***Trichomitopsis***

FAMILY DEVESCOVINIDAE

1. Flagellate phase with fully developed organelles the only phase in life cycle... ....Subfamily **Devescovininae**.... 2
1'. Flagellate form with large and long cresta associated with the recurrent flagellum and enlarged, frequently binucleate, amoeboid phases, with reduced mastigont organelles in the life cycle........ Subfamily **Gigantomonadinae**........ ***Gigantomonas***

2. Parabasal single 3
2'. Parabasal branched or otherwise 11

3. Parabasal not coiled, short 4
3'. Parabasal coiled around axostyle 5

4. Parabasal comma-shaped, inserted in concavity of a ring-shaped nucleus............ .... ***Achemon***
4'. Parabasal not longer than nucleus, no emergent parabasal filament ...... ***Foaina***

5. Emergent axostyle 6
5'. Non-emergent axostyle 7

6. Short axostyle with membranous capitulum .... ***Metadevescovina***
6'. Long, thin, emergent axostyle. Capitulum extended in a cape .... ***Hyperdevescovina***

7. Elongate cell body with three anterior flagella and a recurrent one; aflagellate stage in life cycle......... ***Polymastigoides***
7'. Cell body width 1/2 to 1/3 length 8

8. Axostyle very thin, completely buried in cytoplasm .... ***Caduceia***
8'. Axostyle clearly visible 9

9. Parabasal U-shaped in early development. .... ***Bullanympha***
9'. Parabasal never U-shaped 10

10. Short, falx-shaped cresta... ***Devescovina***
10'. Conspicuous, large, long cresta along recurrent flagellum. ***Macrotrichomonas***

11. Parabasals very numerous throughout cytoplasm, difficult to distinguish by light microscopy .... ***Mixotricha***
11'. Parabasals branched.............................. 12

12. Parabasals comb-like (numerous parabasals) 13
12'. Parabasals with one or more rami......... 14

13. Parabasals around the nucleus.. ***Evemonia***
13'. Parabasals below the nucleus.................... .... ***Pseudodevescovina***

14. Parabasals with long, central, diagonal part and one or more rami..... ***Parajoenia***

14'. Parabasal ramified, branches usually associated with nuclear surface, attached stage with internal flagella; sac-like polymastigote stage in life cycle....... ***Kirbynia***

FAMILY CALONYMPHIDAE

1. Karyomastigonts with four flagella, one thicker; cresta present 2
1'. Karyomastigonts with two to fourflagella, no cresta 3

2. Eight or 16 karyomastigonts arranged in a circle at anterior of cell, axostyles independent.......................... ***Coronympha***
2'. 150 individual karyomastigonts arranged in a spiral, axostyles independent............ .... ***Metacoronympha***

3. Karyomastigonts with a single flagellum arranged in a spiral, axostyles grouped in a central bundle .... ***Stephanonympha***
3'. Many karyomastigonts, each with two flagella and 1-3 nuclei ..... ***Diplonympha***

4. With apical akaryomastigonts and subapical karyomastigonts ....***Calonympha***
4'. Many akaryomastigonts with four flagella, arranged nearly all around the body; nuclei independent .... ***Snyderella***

FAMILY COCHLOSOMATIDAE

1. Flagellates with an anterior adhesive disc...... 2
2. A single karyomastigont with six flagella, an axostyle, and a parabasal body........................... 3
3. One recurrent flagellum associated with a lamellar undulating membrane in a lateral groove... 4
4. Costa fiber with a B striation pattern 5
5. Flat adhesive disc.......................... ***Cochlosoma***
6. Cylindrical adhesive disc.......... ***Psychostoma***

FAMILY MONOCERCOMONADIDAE Kirby, 1944

Small flagellates, generally with three to five anteriorly directed flagella and a recurrent one free for its entire length or partly adhering to the body surface. Undulating membrane (UM) present in two genera, but no costa under the UM. Some genera have a reduced flagellar apparatus or no flagella. Four subfamilies.

SUBFAMILY MONOCERCOMONADINAE Kirby, 1944

Genus ***Monocercomonas*** Grassi, 1879

Monocercomonas (syn=*Eutrichomastix*) are small flagellates (5-15 µm) with 3 anterior free flagella and a recurrent one adhering to the cell body only on its proximal part. Well developed axostyle protruding posteriorly. Parabasal body rod-, disc- or V-shaped. Electron microscopy has shown the trichomonad characters and particularly the presence of an infrakinetosomal body similar to that of *Tritrichomonas*. About 20 species living in the intestinal tract of vertebrates, such as *M. colubrorum* in fishes, amphibians, reptiles (Moskowitz, 1951); birds and mammals, e.g. *M. caviae* from guinea pigs (Nie, 1950). Other species live in the gut of invertebrates, especially arthropods such as termites and roaches, coleoptera, tipulid larvae, and myriapods (Grassé, 1952b). At least one species forms cysts (Brugerolle, 1973). Diagnoses can be found in Honigberg (1963), a list of species in Grassé (1952a), electron microscopy by Mattern et al. (1972) and Brugerolle (1975-76).
TYPE SPECIES: *Monocercomonas colubrorum* (Hammerschmidt, 1844)
HOSTS: Fish, squamate reptiles, birds, mammals, arthropods

Genus ***Tricercomitus*** Kirby, 1930

Tricercomitus are very small flagellates (3-8 µm) with 3 anterior flagella and a recurrent one very long and adhering for most of its length along the body, thus differing from *Monocercomonas*. Slender axostyle and parabasal body present. In *T. divergens* of *Kalotermes flavicollis*, double forms with two sets of flagella per nucleus and polymorphism occur in relation to the molt of the termite. They could be interpreted either as part of a sexual process or stages of division (Grassé, 1952b). *Tricercomitus* has been wrongly put in synonymy with *Trimitus* by Grassé (1952).

Electron microscopy of *T. divergens* has shown its trichomonad characters and differences with *Trimitus* and *Enteromonas* which are "monozoic" forms of diplomonads. The double forms have 2 flagellar apparatus and no mitotic spindle. Diagnosis in Honigberg (1963), ultrastructural study by Brugerolle (1986a).
TYPE SPECIES: *Tricercomitus termopsidis* Kirby, 1930
TYPE HOST: *Zootermopsis angusticollis*

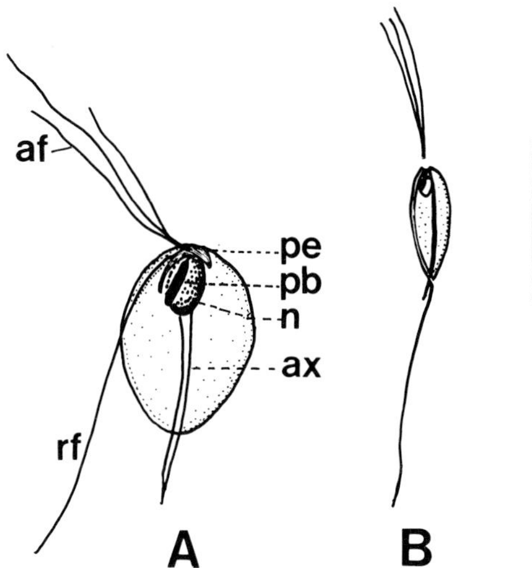

Fig. 8A. *Monocercomonas verrens* (after Honigberg, 1947, 1963). Bar=10 μm. 8B. *Tricercomitus termosidis* (after Kirby, 1930).

Genus ***Tetratrichomastix*** Mackinnon, 1913

Tetratrichomastix (syn.=*Trichomastix*) are small flagellates (8-15 μm) with 4 anterior flagella and a recurrent, non-adherent one. The only species described is *T. parisii* from *Tipula* larvae (Mackinnon, 1913; Ludwig, 1946) (Fig. 9A). No ultrastructural study.
TYPE SPECIES: *Tetratrichomastix parisii* Mackinnon, 1913
TYPE HOST: *Tipula abdominalis* Say

Genus ***Hexamastix*** Alexeieff, 1912
(Figs. 2, 9B)

Hexamastix are small (10-30 μm) trichomonad flagellates with 5 grouped, anterior free flagella and a non-adherent recurrent one. Axostyle well developed and parabasal V-shaped. Electron microscopy has shown the basal body arrangement and other trichomonad features. Among the 11 species described, several live in invertebrates; e.g., *H. claviger* in termites (Kirby, 1930) and other species in hosts such as the mole cricket, tipulid larvae, and a leech. In vertebrates, *H. batrachorum* is found in salamanders (Honigberg and Christian, 1954), *H. caviae* in rodents (Nie, 1950). Light microscopic studies in Grassé (1952a), diagnosis in Honigberg (1963), ultrastructure in Brugerolle (1975-76).
TYPE SPECIES: *Hexamastix batrachorum* (Alexieff, 1911)
HOSTS: Insects, leeches, caudate amphibians, rodents

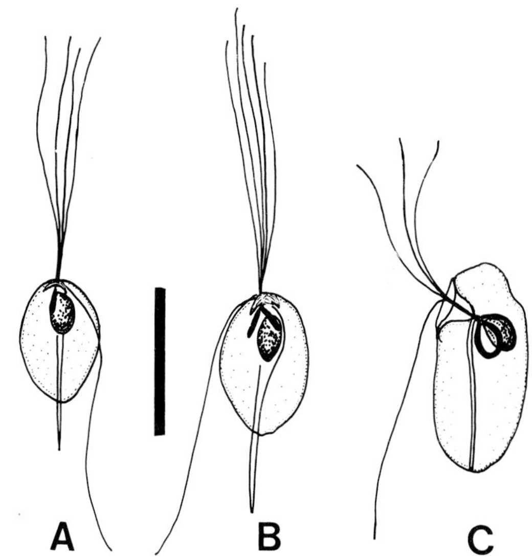

Fig. 9A. *Tetratrichomastix parisii* (after Mackinnon, 1913, in Honigberg, 1963). Bar=10 μm. 9B. *Hexamastix kirbyi* (after Honigberg, 1963). 9C. *Chilomitus* (after Nie, 1950, in Honigberg, 1963).

SUBFAMILY CHILOMITINAE Honigberg, 1963

Genus ***Chilomitus*** da Fonseca, 1915

Chilomitus are small flagellates (6-14 μm) with 3 anterior flagella and a recurrent, non-adhering one. The flagella arise from a funnel-shaped depression of the antero-ventral face. Axostyle slender and not protruding at the rear; parabasal ring-shaped. Not yet studied by electron microscopy. The 3 described species live in rodents: *C. caviae* and *C. conexus* from the guinea pig (Nie, 1950), diagnosis in Honigberg (1963). No EM study.

TYPE SPECIES: *Chilomitus caviae* da Fonseca, 1915
TYPE HOST: Guinea pigs

SUBFAMILY PROTRICHOMONADINAE

Genus ***Protrichomonas*** Alexeieff, 1911

Small flagellates (10-15 µm) which have 3 anterior flagella and no external recurrent flagellum. The capitulum of the axostyle is broad and the axostyle of the trunk slender and not protruding. The posterior part of the cell is enlarged and amoeboid. Electron microscopy of the species *P. legeri* living in the esophagus of the fish *Box boops* has shown the trichomonad features and also that the axoneme of the recurrent flagellum is intracytoplasmic. Another species has been reported in ducks. Light microscopy in Alexeieff (1910), diagnosis in Honigberg and Kuldova (1969), EM by Brugerolle (1980).
TYPE SPECIES: *Protrichomonas legeri* (Alexeieff, 1910)
TYPE HOST: *Box boops*

Genus ***Parahistomonas*** Lund, 1963

Spheroid or amoeboid flagellates (10-22 µm) with 4 free flagella and 1 flagellum separated from the other 3. Axostyle with a broad capitulum and a slender trunk, not protruding. Parabasal apparatus with a rod-shaped body and a long parabasal fiber. No tissue-dwelling phase in the life cycle. The only species described, *P. wenrichi*, lives in the cecum of fowl (Lund, 1963; Honigberg and Kuldova, 1969). Ultrastructure not yet studied.
TYPE SPECIES: *Parahistomonas wenrichi* Lund, 1963.
TYPE HOST: fowl

Genus ***Histomonas*** Tyzzer, 1920

Spheroidal and amoeboid flagellates (8-28 µm) with a single flagellum. Axostyle with a broad capitulum and a slender trunk not protruding; V-shaped parabasal. A flagellated tissue-dwelling phase in the cycle. Ultrastructural study has confirmed the trichomonad features; 3 basal bodies are barren. One species, *H. meleagridis*, causes enterohepatitis ("black-head") in fowl. The infection of cecal mucosa reaches the liver. Transmitted by the eggs of the nematode *Heterakis gallinae* (Lund and Chute, 1973). Description by Tyzzer (1920), Wenrich (1943), Honigberg and Bennett (1971), McDougald and Reid (1978); electron microscopy by Schuster (1968), Rybicka et al. (1972).
TYPE SPECIES: *Histomonas meleagridis*
TYPE HOSTS: Fowl

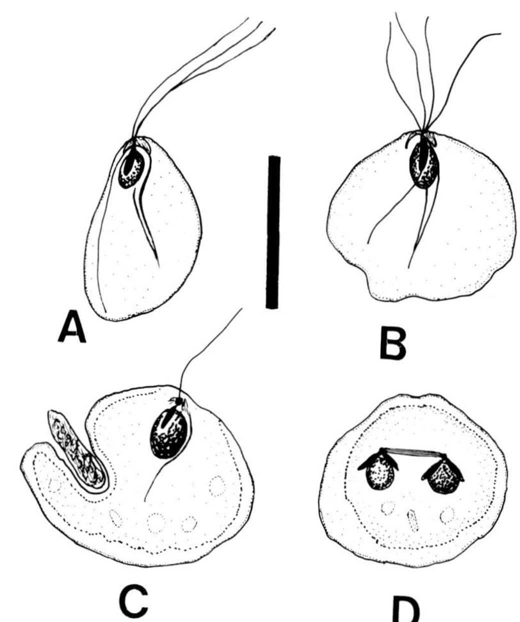

Fig. 10A. *Protrichomonas legeri* (after Alexeieff, 1910, Brugerolle, 1980). Bar=10 µm. Fig. 10B. *Parahistomonas wenrichi* (after Honigberg and Kuldova, 1969). Fig. 10C. *Histomonas meleagridis* (after Wenrich, 1943). Fig. 10D. *Dientamoeba fragilis* (after Dobell, 1940; Wenrich, 1944a).

SUBFAMILY DIENTAMOEBINAE Grassé

Genus ***Dientamoeba*** Dobell, 1940

Small binucleate amoebae (3.5-22 µm) in arrested telophase predominant in populations. Mononucleate individuals with neither flagella nor axostyle throughout the life cycle. V-shaped parabasal body close to the nucleus. Paradesmose stretched between the 2 polar centers or origin of the parabasal fibers. Ultrastructural studies demonstrated the trichomonad affinities: presence

of a parabasal apparatus, a paradesmose as part of an extranuclear spindle originating from atractophores, hydrogenosomes. Basal bodies and axostyle are lacking. The only species known, *D. fragilis*, is a harmless amoeba, common in the human colon but less numerous than other intestinal amoebae. It forms small spherical cysts. Light microscopy by Dobell (1940) and Wenrich (1944a), diagnosis and ultrastructure in Camp et al. (1974), Honigberg and Brugerolle (1990).
TYPE SPECIES: *Dientamoeba fragilis*
TYPE HOST: man

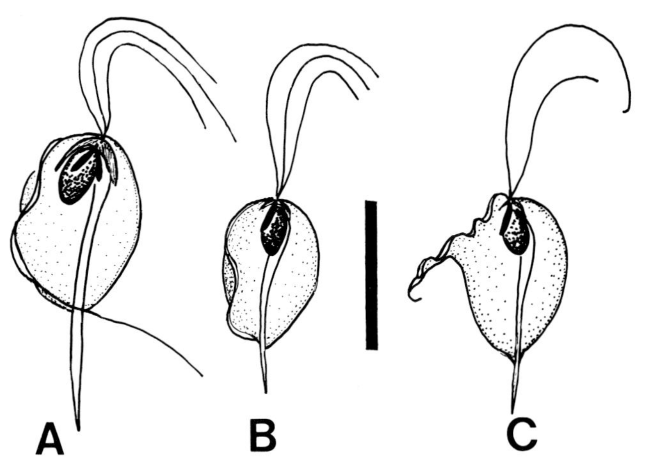

Fig. 11A. *Hypotrichomonas acosta* (after Lee 1960). Bar=10 µm. Fig. 11B. *Pseudotrichomonas keilini* (after Bishop, 1939; Brugerolle, 1991b). Fig. 11C. *Ditrichomonas honigbergi* (after Farmer, 1993).

SUBFAMILY HYPOTRICHOMONADINAE
Honigberg, 1963

Trichomonads of this subfamily have two to three anteriorly directed flagella and a recurrent one associated with an undulating membrane of variable length. Costa absent. Axostyle generally protruding posteriorly.

Genus ***Hypotrichomonas*** Lee, 1960

Small flagellates (10-35 µm) with 3 anterior free flagella and a recurrent one associated with a shallow, undulating membrane (UM), shorter than the body. Recurrent flagellum with a posterior free end. Axostyle protruding posteriorly; rod or V-shaped parabasal apparatus. Electron microscopy has shown the trichomonad characters and the shape of the UM looks like the lamellar-shape UM. Three or more species such as *H. acosta* living in intestine of squamate reptiles and chelonians (Moskowitz, 1951; Lee, 1960). Diagnosis in Honigberg (1963), electron microscopy by Mattern et al. (1969) and Brugerolle (1975-76).
TYPE SPECIES: *Hypotrichomonas acosta* (Moskowitz, 1951)
HOSTS: squamate reptiles and chelonians

Genus ***Pseudotrichomonas*** Bishop, 1939

Small flagellates (10-14 µm) with 3 anterior flagella and a recurrent one associated with a shallow, undulating membrane nearly as long as the body; no free posterior recurrent flagellum. Costa absent, slender axostyle, generally protruding posteriorly. Ultrastructural study has shown the trichomonad features, the lamellar-type structure of the UM, and the absence of the costa. The only species, *P. keilini* (Bishop, 1939) lives in ponds. Forms cysts. Electron microscopic study by Brugerolle (1991b).
TYPE SPECIES: *Pseudotrichomonas keilini* Bishop, 1939

Genus ***Ditrichomonas*** Farmer, 1993

Small flagellates (10-19 µm) with 2 anterior flagella and a recurrent one associated with an undulating membrane shorter than the body; recurrent flagellum free posteriorly. Axostyle protruding at the rear. It forms spherical cysts. Electron microscopy has shown the trichomonad characters; the basal body number 3 is barren; the UM is of the lamellar type and the costa is absent. Scanning and transmission electron microscopy by Farmer (1993).
TYPE SPECIES: *Ditrichomonas honigbergii* Farmer, 1993

FAMILY TRICHOMONADIDAE
Chalmers & Pekkola, 1918 emend. Kirby, 1946

Trichomonad genera with three to five anteriorly directed free flagella and a recurrent one associated with an undulating membrane supported by a striated fiber, the costa (Fig. 5e); four subfamilies.

SUBFAMILY TRITRICHOMONADINAE
Honigberg, 1963

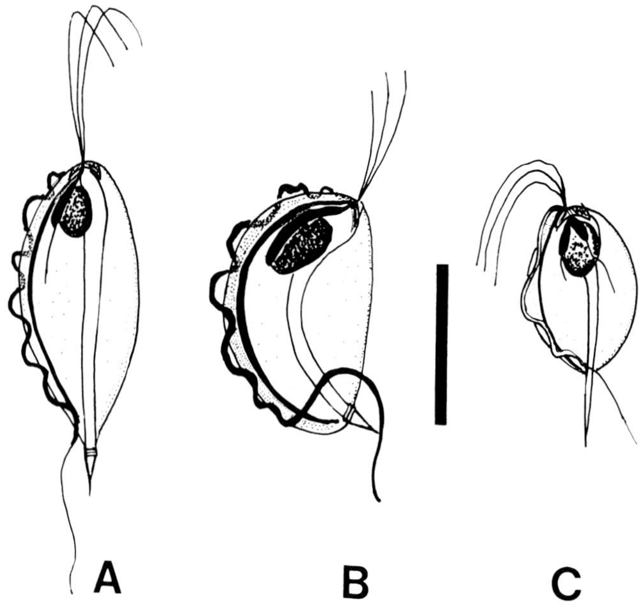

Fig. 12A. *Tritrichomonas augusta* (after Honigberg, 1963). Bar=10 µm. Fig. 12B. *Tritrichomonas muris* (after Honigberg, 1963). 12C. *Trichomitus batrachorum* (after Samuels, 1957; Honigberg, 1963)

Genus ***Tritrichomonas*** Kofoid, 1920

Small flagellates (8-22 µm) with 3 free anterior flagella and a recurrent one forming a well-developed undulating membrane (UM); recurrent flagellum free posteriorly. Costa slender or stout; axostyle well developed, protruding at the rear, but tapering abruptly. Periaxostylar rings near the extremity of the axostyle in some species. Rod- or sausage-shaped parabasal apparatus. Electron microscopy has shown the trichomonad features. There is an infra-kinetosomal body. The UM is of the rail-type, and the recurrent flagellum is modified by paraaxonemal fibers. In different species there are variations in the UM. The cell body part of the UM is always linked by microfibrils to the supporting costa, which has the A-type pattern. The 20 described species live in the intestinal tract of rodents, e.g., *T. muris* (Wenrich, 1921; Nie, 1950; Gabel, 1954); the intestine of swine, e.g., *T. suis* (Hibber et al., 1960); in the intestine of birds e.g., *T. eberthi* (McDowell, 1953); in the intestine of reptiles and amphibians; e.g., *T. augusta* (Buttrey, 1954); in the genital tract of bovines where *T. foetus* causes serious outbreaks and abortion in cows (Wenrich and Emerson, 1933; Kirby, 1951); or in squirrel monkeys e.g., *T. mobilensis* (Culberson et al., 1986). Ultrastructural studies by Daniel et al. (1971), Honigberg et al. (1971), Brugerolle (1975-76), Brugerolle (1981), and Honigberg and Brugerolle (1990).

TYPE SPECIES: *Tritrichomonas augusta* (Alexeieff, 1911)

HOSTS: Rodents, swine, cattle, birds, caudate and acaudate amphibians, saurian reptiles

SUBFAMILY TRICHOMONADINAE
Chalmers & Pekkola, emend., 1918

Genus ***Trichomitus*** Swezy, 1915

Small flagellates (10-15 µm) with 3 free anterior flagella and a recurrent one adhering to the undulating membrane developed along the entire length of the cell body; with a recurrent flagellum free posteriorly. Costa slender or well developed according to the species. Axostyle protruding posteriorly and V-shaped parabasal body. Electron microscopic study has shown the trichomonad features with a UM of lamellar type and a costa of A-type pattern. Since this genus has 3 anterior flagella and a recurrent one associated with an undulating membrane, which also are common characters of the genus *Tritrichomonas*, the 2 genera sometimes have been confused. Among the 5 species reported, *T. batrachorum* lives in amphibians and squamate reptiles. It has been described sometimes under the name of *Tritrichomonas batrachorum* (Honigberg, 1953; Samuels, 1957) but it is distinguished from *Tritrichomonas augusta* from the same hosts by distinct differences in the shape of the UM. It forms pseudocysts (Mattern et al., 1973) and true cysts (Brugerolle, 1973). One species living in a termite, *T. trypanoides*, has been identified by electron microscopic study (Boykin et al., 1986). The other species placed in this

genus live in rodents e.g., *T. marmotae* (Gabel, 1954); *T. wenyoni* (Wenrich and Nie, 1949); in swine, *T. rotunda* (Hibber et al., 1960); or in the feces of man, *T. fecalis* (Cleveland, 1928), but this identification has not been confirmed by an ultrastructural study. Diagnosis by Honigberg (1963); ultrastructure study by Brugerolle (1971, 1975-76), Honigberg et al. (1972), and Boykin et al. (1986).
TYPE SPECIES: *Trichomitus batrachorum* (Perty, 1852)
HOSTS: amphibians, squamate reptiles, rodents, swine, man, termites

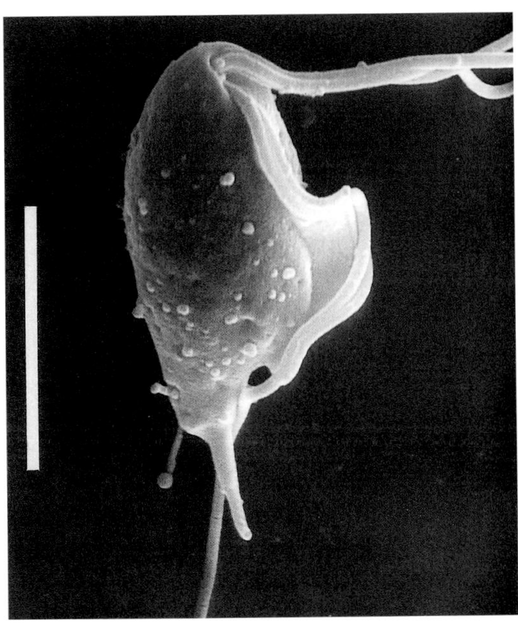

Fig. 13. *Trichomitus batrachorum* showing the recurrent flagellum associated with the lamellar UM (from Brugerolle). Bar=10 µm

Genus ***Tetratrichomonas*** Parisi, 1910

Flagellate of small size (10-15 µm) with 4 free anterior flagella and a recurrent one associated with an undulating membrane extending along the entire length of the body; free recurrent posterior flagellum. Costa well developed and axostyle protruding posteriorly; disc- or V-shaped parabasal. Ultrastructure studies have shown the *Trichomonas* features and the lamellar-type of the UM and the B-type pattern of the costa. Among the 10 species described, one, *T. limacis*, occurs in limacine mollusks (Kozlov, 1945); *T. prowazeki* occurs in fishes, amphibians, and squamate reptiles (Alexeieff, 1910, Honigberg, 1951); *T. gallinarum* in fowl (McDowell, 1953); *T. guttula* in rodents (Kirby and Honigberg, 1950); and *T. buttreyi* in swine (Hibber et al., 1960). Diagnosis in Honigberg (1963); electron microscopy (Saleudin, 1971, Brugerolle, 1975-76).
TYPE SPECIES: *Tetratrichomonas prowazeki* (Alexeieff, 1909)
HOSTS: Mollusks, fish, amphibians and squamate reptiles, birds, rodents, swine

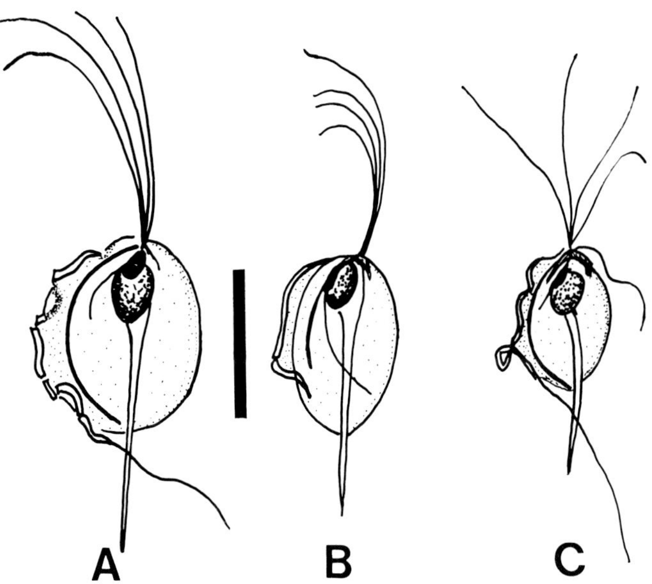

Fig. 14A. *Tetratrichomonas prowazecki* (after Honigberg, 1951). Bar=10 µm. Fig. 14B. *Trichomonas tenax* (after Honigberg, 1963). Fig. 14C. *Pentatrichomonas hominis* (after Kirby, 1945a; Honigberg, 1963).

Genus ***Trichomonas*** Donné, 1836

Polymorphic flagellates (4.5-30µm) with 4 free anterior flagella and 1 recurrent flagellum associated with an undulating membrane shorter than the body; no free posterior recurrent flagellum. Costa relatively slender; capitulum of the axostyle spatulate, trunk of the axostyle slender; parabasal body rod- or V-shaped, with one or several long parabasal filaments. Amoeboid and polymastigote forms present in natural or culture conditions. Electron microscopic studies have shown the trichomonad characters, the lamellar-type of the UM and the B-type pattern of the costa, microfibrillar structure of phagocytic

areas, and division. Three species described: *T. vaginalis* inhabits the urogenital tract of humans, causing venereal diseases (Wenrich, 1944a; Honigberg and Brugerolle, 1990); *T. tenax* lives in the mouth of humans (Wenrich, 1944b; Honigberg and Lee, 1959); *T. gallinae* is pathogenic in pigeons and other birds, invading the cecum and liver (Stabler, 1947; Honigberg, 1978). Diagnosis by Honigberg (1963). Scanning electron microscopy in Warton and Honigberg (1979). Ultrastructural studies of *T. vaginalis* in Nielsen et al. (1966), Nielsen (1972), Brugerolle et al. (1974), and Brugerolle (1975, 1975-76); *T. tenax* in Honigberg and Brugerolle (1990); *T. gallinae* in Mattern et al. (1967) and Honigberg (1978).
TYPE SPECIES: *Trichomonas vaginalis* Donné, 1836, emend. Ehrenberg
TYPE HOST: man

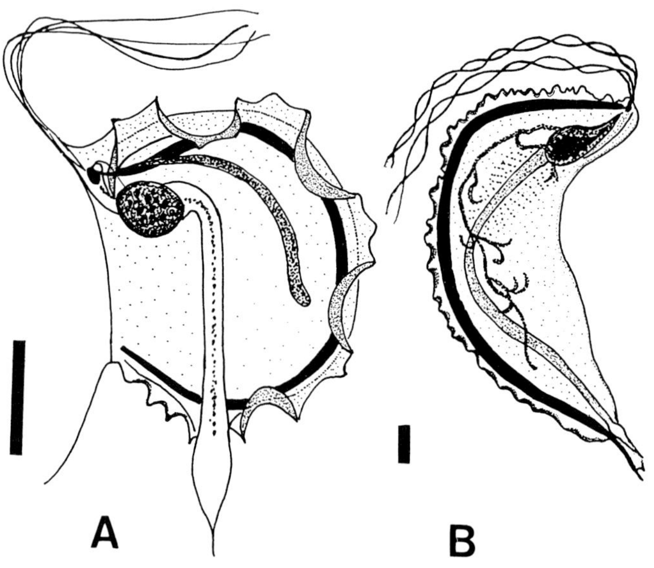

Fig. 15A. *Trichomitopsis termopsidis* from *Zootermopsis angusticollis* (after Kirby, 1931). Bar=10 µm. Fig. 15B. *Pseudotrypanosoma giganteum* from *Porotermes adamsoni* (after Cleveland, 1961).

Genus ***Pentatrichomonas*** Mesnil, 1914

Small flagellates (6-14 µm) with 5 anterior free flagella in normal organisms and 1 recurrent flagellum associated with an undulating membrane extending to the posterior end; free posterior recurrent flagellum. One of the anterior flagella originates independently from others. Costa and axostyle well developed, parabasal apparatus rather small. Ultrastructure study has confirmed the trichomonad features and that the basal body of the independent anterior flagellum is oriented opposite to the basal body of the recurrent flagellum. The UM is of the lamellar type and the costa has the B-pattern of striation. The only species described is *P. hominis* which lives in the intestine of humans and in other mammals (monkeys, cats, dogs, rats) and possibly in birds; it is not a pathogen (Wenrich, 1944c; Kirby, 1945a). Diagnosis by Honigberg (1963), ultrastructural study by Honigberg et al. (1968), Brugerolle (1975-76), and Honigberg and Brugerolle (1990).
TYPE SPECIES: *Pentatrichomonas hominis* (Davaine, 1860)
TYPE HOST: man

SUBFAMILY TRICHOMITOPSINAE Brugerolle 1977

Genus ***Trichomitopsis*** Kofoid & Swezy, 1919

Flagellates ranging in size from 11-150 µm, living in termites, with 4 free anterior flagella and a recurrent one associated with a well developed undulating membrane; short free posterior recurrent flagellum. Costa stout; axostyle stout with a terminal segment often expanded into a pointed bulbous enlargement. Parabasal long and sausage-shaped. Electron microscopy of *T. termopsidis* has shown the trichomonad features, the lamellar-type UM, and the associated recurrent flagellum enlarged by a paraxonemal fiber. The costa is of the B-type pattern with a pronounced lattice structure. By the costa and UM structure, this genus seems more closely related to genera of Trichomonadinae. A special study has shown that the costa of *T. termitidis* is contractile and undulates in the cytoplasm (Amos et al., 1979). The best known species is *T. termopsidis* from *Zootermopsis angusticollis* (Kirby, 1931); other species with 4 anterior flagella and living in termites could be included in this genus: *T. termitis, T. barbouri, T. cartagoensis* (Honigberg, 1963). Light microscopy in Kirby (1931) and Cleveland (1961), scanning views in Yamin

(1979b), electron microscopy in Hollande and Valentin (1968) and Amos et al. (1979).
TYPE SPECIES: *Trichomitopsis termopsidis* (Cleveland, 1925)
TYPE HOST: *Zootermopsis angusticollis* (Hagen)

Genus ***Pseudotrypanosoma*** Grassi, 1917
(Fig. 15B)

The only species known, *P. giganteum* is very large (55-205 µm) and lives in the termite *Porotermes adamsoni* from Australia. The cell has 4 free anterior flagella and a recurrent one associated with a well-developed undulating membrane; no free posterior recurrent flagellum. Costa very stout and contractile axostyle with a moderate diameter and a terminal bulbous expansion. Parabasal body very long and branched. Light microscopy by Grassi (1917), Kirby (1931), Sutherland (1933), and Cleveland (1961); EM partly in Amos et al. (1979).
TYPE SPECIES: *Pseudotrypanosoma giganteum* Grassi, 1917
TYPE HOST: *Porotermes adamsoni* (Froggatt)

SUBFAMILY PENTATRICHOMONOIDINAE
Honigberg, 1963

Genus ***Pentatrichomonoides*** Kirby, 1931

Flagellates (18-45 µm) with 5 free anterior flagella and a recurrent one incorporated into a well-developed undulating membrane, no free posterior recurrent flagellum. Cell shape changeable; cigar-shaped and truncated posteriorly or broadly triangular. Costa rather slender; costa and undulating membrane larger than the body, often recurved posteriorly. No central axostyle; parabasal of variable shape associated with a parabasal filament. Scanning and transmission electron microscopy has shown the trichomonad characters. Flagella arise from a gullet; the recurrent flagellum is associated with a prominent lamellar UM and a para-axononemal fiber. The microtubules of the axostyle develop under the plasma membrane all around the cell except at the UM. The UM is bordered by 2 deep grooves and is underlain by a costa of B-type pattern. *Pentatrichomonoides scroa*, has been described from several species of termites in the genus *Cryptotermes* (Kirby, 1931; Grassé, 1952a) and *P. darwiniensis* has been described from *Mastotermes darwiniensis*. Diagnosis by Honigberg (1963) and Kirby (1931); electron microscopy by Brugerolle et al. (1994).
TYPE SPECIES: *Pentatrichomonoides scroa* Kirby, 1931
TYPE HOST: *Cryptotermes* (formerly *Kalotermes*) *dudleyi* Banks

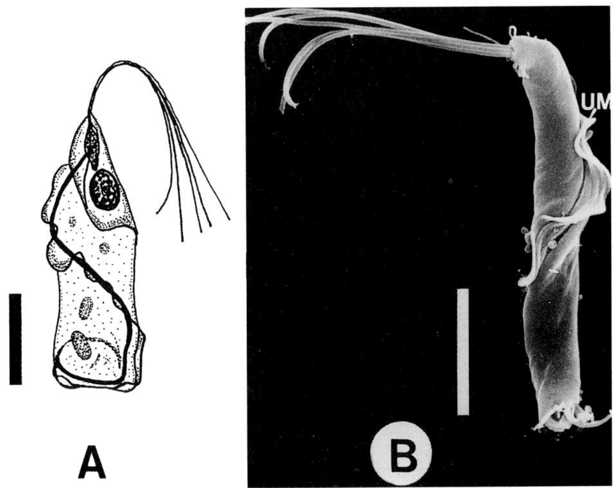

Fig. 16A. *Pentatrichomonoides darwiniensis* from *Mastotermes darwiniensis* (after Kirby, 1931). Bar=10 µm. Fig. 16B. *Pentatrichomonoides darwiniensis* showing its undulating membrane (from Brugerolle et al., 1994). Bar=10 µm.

FAMILY DEVESCOVINIDAE Doflein, 1911
emend. Kirby, 1931

Trichomonad flagellates with three free anterior flagella and a thick cord-like or ribbon-like recurrent flagellum. The recurrent flagellum adheres to the cell body along a variable length which is underlain by a cresta. There is no real undulating membrane. The pelta-axostyle complex is well developed. The parabasal apparatus is rodlike or branched or inrolled around the axostyle and these differences are used to distinguish the genera and species. All species occur in the termite gut and are xylophagous or bacterivorous. Two subfamilies: Devescovininae and Gigantomonadinae.

SUBFAMILY DEVESCOVININAE Doflein, 1911 emend. Kirby, 1946

Genus **Foaina** Janicki, 1915

Small devescovinids (6-54 µm) with 3 long anterior flagella and a trailing flagellum thicker than normal and longer than the body. Anterior papilla linked to the proximal part of the three anterior flagella. Cresta of variable length, reaching the end of the body in one species. Parabasal body ellipsoid or rod-shape applied on the nucleus, never coiled around the axostyle. Axostyle well developed, usually projecting at the rear, capitulum containing the nucleus. Ultrastructural study of the 2 species of *Foaina* in *Kalotermes flavicollis* have shown the specific structures of the recurrent flagellum, cresta, and parabasal found in the Devescovinidae (Brugerolle, 1975-76). The 25 species described occur in Kalotermitidae (Kirby, 1942b; Grassé, 1952a). The genera *Crucinympha* Kirby, *Janickiella* Duboscq & Grassé, and *Paradevescovina* Kirby have been put in synonymy with *Foaina* by Kirby (1942b). EM study by Lavette, 1971a.
TYPE SPECIES: *Foaina gracilis* Janicki, 1915
TYPE HOST: *Neotermes connexus* Snyder

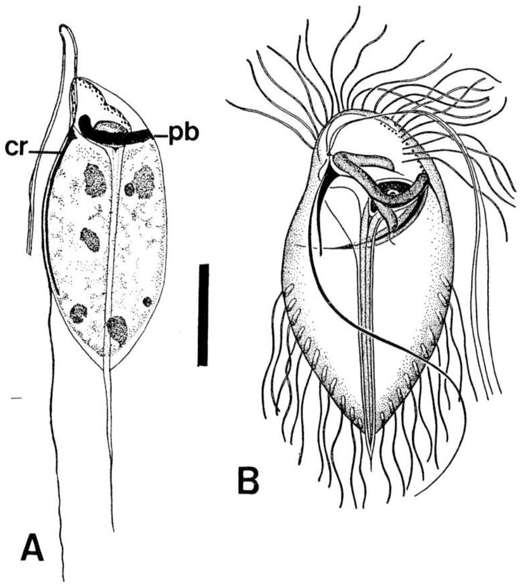

Fig. 17A. *Foaina dogieli* from *Kalotermes flavicollis* (after Kirby, 1942b). Bar=10 µm. Fig. 17B. *Parajoenia grassii* from *Neotermes connexus* bearing anterior and posterior spirochetes (after Kirby, 1942b).

Genus **Parajoenia** Janicki, 1911

Moderately sized devescovinids (29-50 µm) with a cell body rounded anteriorly with 3 long anterior flagella and a recurrent one, cord-like, slightly longer than the body. Moderate sized cresta; parabasal highly developed consisting of a long, slanting main part and one or more rami. Stout axostyle with a short pointed posterior projection and a leaf-like capitulum; anterior and posterior spirochetes. Only one species, *P. grassii* from *Neotermes connexus* from Hawaii (Kirby, 1942b). No EM study.
TYPE SPECIES: *Parajoenia grassii* Janicki, 1911
TYPE HOST: *Neotermes connexus* Snyder

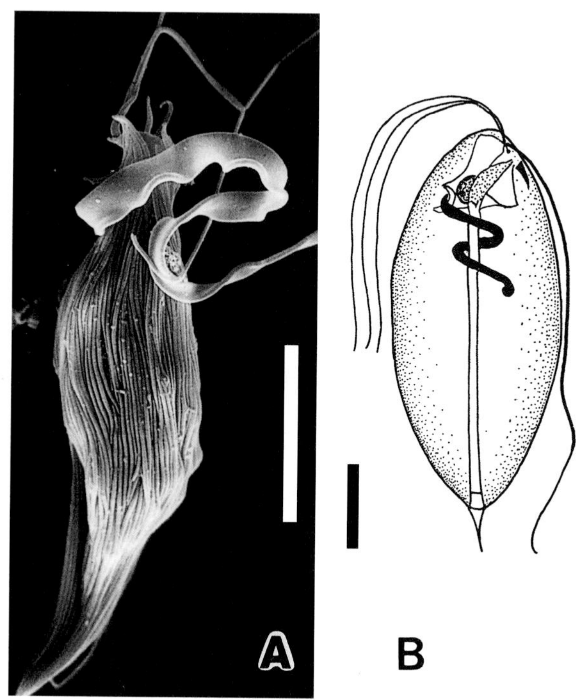

Fig. 18A. *Devescovina striata* from *Cryptotermes longicollis* (after Kirby, 1941). Bar=10 µm. Fig. 18B. *Metadevescovina debilis* from *Marginitermes* (formerly *Kalotermes*) *hubbardi* (after Kirby, 1945b).

Genus **Devescovina** Foà, 1905

Flagellates with an elongated body pointed posteriorly (20-80 µm) with 3 anterior free flagella and a recurrent trailing one forming a slender cord or ribbon. Short falx-shaped cresta; parabasal body spiraled around the trunk of the axostyle in 1/2 to 5 turns, not branched.

Axostyle tapering posteriorly with end enclosed in the cytoplasm. Electron microscopy has shown the trichomonad features and the structure of the recurrent flagellum containing an axoneme associated with many microfibrils. The cresta is a microfibrillar structure which develops under the membrane along the adhesive portion of the recurrent flagellum as in *Foaina.* The axostyle is composed of an inrolled sheet of microtubules. Among the 20 species reported in Kalotermitidae, *D. striata* lives in *Cryptotermes brevis* (Kirby, 1941; Grassé, 1952a). Electron microscopy in Mignot et al. (1969), Joyon et al. (1969), and Brugerolle (1975-76).
TYPE SPECIES: *Devescovina striata* Foa, 1905
TYPE HOST: *Cryptotermes brevis* (Walker)

Genus ***Metadevescovina*** Light, 1926

Devescovinids with a stout body (9.5-90 µm) in comparison to *Devescovina*, from which it is distinguished with difficulty (this genus has been put in synonymy with *Devescovina* by Grassé [1952a]). It bears 3 anterior free flagella and a trailing one, cord-like or occasionally ribbon-like, as long as the body. Cresta variable in length (4-24 µm). Parabasal body spiraled around the axostyle (1-7 turns), occasionally branched. Axostyle stout with a marked posterior projection and a capitulum with membranous extensions around the nucleus. They occur only in Kalotermitidae except for one species in *Mastotermes darwiniensis*. Among the 16 species described, *M. debilis* lives in *Marginitermes* (formerly *Kalotermes*) *hubbardi* (Kirby, 1945b). Light microscopy and diagnosis in Kirby (1945b). No ultrastructural study.
TYPE SPECIES: *Metadevescovina debilis* Light, 1926
TYPE HOST: *Marginitermes hubbardi* (Banks)

Genus ***Caduceia*** França, 1918

Large devescovinids (35-138 µm) with 3 long anterior flagella and a trailing flagellum slightly thicker than normal, usually shorter than the body. Cresta short; parabasal body coiled around the axostyle (2-20 turns); axostylar trunk tapered with posterior enclosed in the cytoplasm as in *Devescovina*. A rotatory movement of the head of the cell has been described (Tamm, 1978). Body covered with short spirochetes and short *Fusiformis*-like bacteria adhering in some areas. Among the 6 species described in Kalotermitidae, 4 occur in *Neotermes* such as *C. theobromae* which lives in *N. gestri* (Kirby, 1942a; Grassé, 1952a). No EM study.
TYPE SPECIES: *Caducia theobromae* França, 1918.
TYPE HOST: *Neotermes gestri* Silvestri

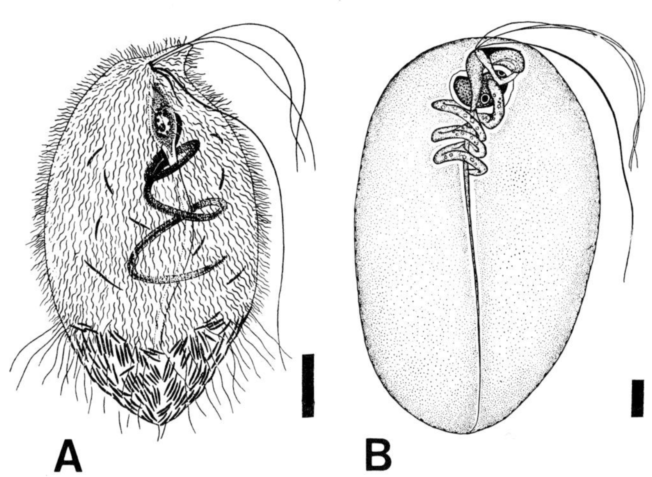

Fig. 19A. *Caduceia theobromae* from *Neotermes aburiensis* (after Grassé, 1952a). Bar=10 µm. Fig. 19B. *Bullanympha silvestrii* from *Neotermes erythraeus* (after Kirby, 1949b).

Genus ***Bullanympha*** Kirby, 1938

Large devescovinids (50-138 µm) with 3 anterior flagella and a recurrent flagellum cord-like in its middle part and shorter than the body. Cresta relatively long. Parabasal apparatus with a U-shaped proximal part on the nucleus and a distal part turned around the axostylar trunk (0.5-3.5 times). Capitulum and trunk of the axostyle slender, tapering gradually with the end enclosed in the cytoplasm. One species described: *B. silvestrii* occurring in *Neotermes erythraeus* (Kirby, 1949a). No EM study.
TYPE SPECIES: *Bullanympha sylvestrii* Kirby, 1938
TYPE HOST: *Neotermes erythraeus* Silvestri

Genus ***Hyperdevescovina*** Kirby, 1947

Large devescovinids (33-139 µm) with 3 short anterior flagella and a trailing one slightly thicker than normal, generally shorter than the body. Cresta very small, equilateral in shape, typically excavated posteriorly. Parabasal body without longitudinal part, turned in helix around the axostyle (2-19 turns). Capitulum of the axostyle extended in a cape; trunk of the axostyle stout in anterior part with a slender posterior projection of variable length. The 8 species reported occur in Kalotermitidae of the Australian region, e.g. *H. calotermitis* from *Kalotermes* (formerly *Proglyptotermes*) *brouni* (Kirby, 1949a), reported in Grassé, 1952a). No EM study.

TYPE SPECIES: *Hyperdevescovina calotermitis* (Nurse, 1945)

TYPE HOST: *Kalotermes* (formerly *Proglyptotermes*) *brouni* Froggatt

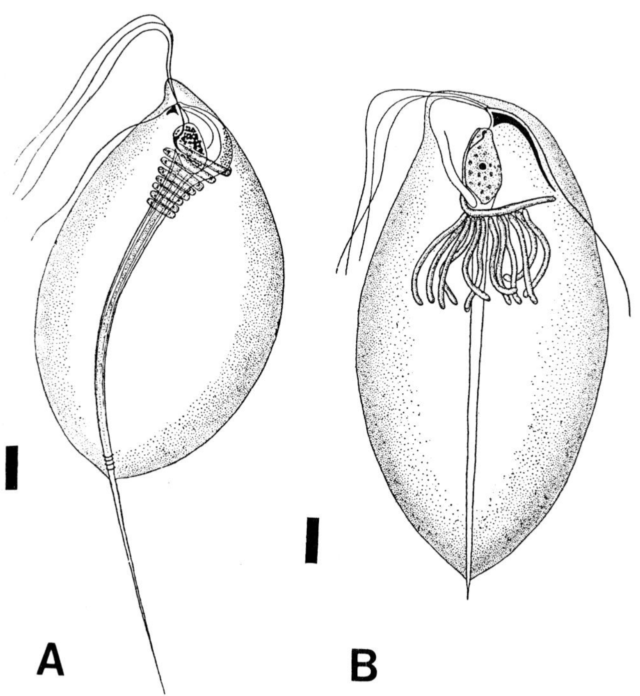

Fig. 20A. *Hyperdevescovina calotermitis* from *Proglyptotermes brouni* (after Kirby, 1949b). Bar=10 µm. Fig. 20B. *Pseudodevescovina uniflagellata* from *Neotermes insularis* (after Kirby, 1945b).

Genus ***Pseudodevescovina*** Sutherland, 1933

Relatively large devescovinids (40-140 µm) with 3 short anterior flagella and a trailing flagellum slightly thicker than normal, shorter than the body. Cresta well developed. Parabasal apparatus L-shaped supporting 7-19 appended parabasals not coiled around the axostyle. Axostyle is stout with a posterior projection and a large capitulum. One species, *P. uniflagellata,* from *Neotermes insularis*. Description in Sutherland (1933), Kirby (1945b), and Grassé (1952a). No EM study.

TYPE SPECIES: *Pseudodevescovina uniflagellata* Sutherland, 1933

TYPE HOST: *Neotermes insularis* (White)

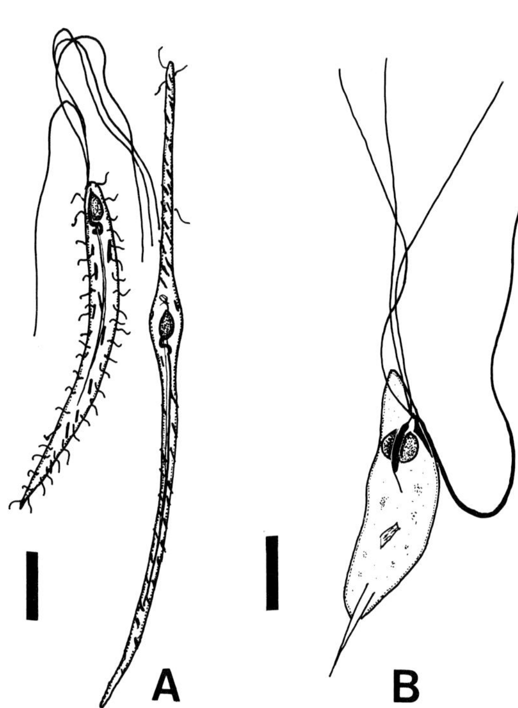

Fig. 21A. *Polymastigotoides elongata* from *Anacanthotermes ochraceus* (after Grassé and Hollande, 1951). Bar=10 µm. Fig. 21B. *Achemon platycaryon* from *Cryptotermes lamanianus* (after Grassé and Hollande, 1950). Bar=10 µm.

Genus ***Polymastigoides*** Grassé & Hollande,1951

Devescovinid with a long and slender body (30-37 µm) bearing 3 long anterior flagella and a recurrent one longer than the body. Anterior

papilla associated with the proximal part of the flagella. Short, narrow cresta; slender axostyle not protruding posteriorly. Parabasal inrolled around the axostylar trunk (1 turn). *Fusiformis* and spirochetes on the surface. Aflagellate forms with the nucleus and the mastigont system situated in the mid length of the body, mostly occurring in winged termites. One species, *P. elongata*, lives in *Anacanthotermes murgabicus* and *A. ochraceus* (Grassé and Hollande, 1951; Grassé, 1952a) and was described under the name *Devescovina elongata* in Bernstein, (1928). No EM study.
TYPE SPECIES: *Polymastigoides elongata* (Bernstein, 1928)
TYPE HOST: *Anacanthotermes murgabicus* Vasiljev

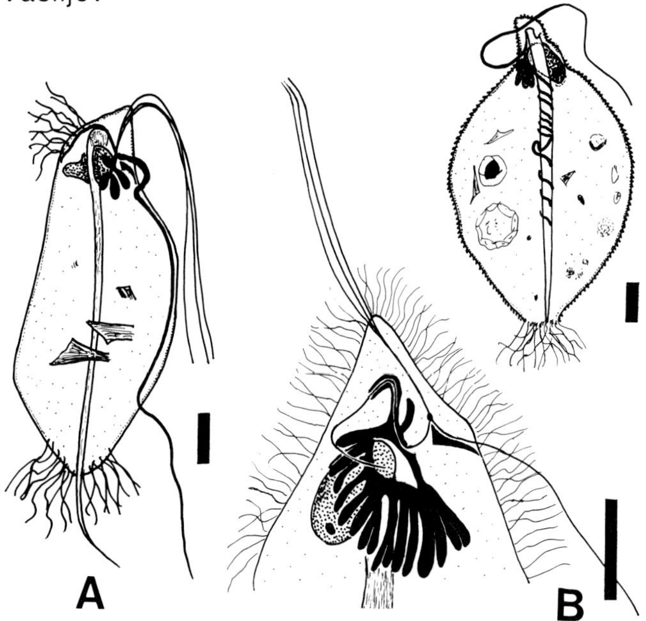

Fig. 22A. *Astronympha punctata* from *Cryptotermes lamanianus* (after Grassé and Hollande, 1950). Bar=10 μm. Fig. 22B. *Evemonia punctata* from *Neotermes aburiensis* (after Grassé and Hollande, 1950).

Genus ***Achemon*** Grassé & Hollande, 1950

Small devescovinids (30 μm) with 3 anterior flagella and a long cord-like recurrent flagellum (60 μm). Small cresta; axostyle projecting out of the body. Comma-shaped parabasal apparatus inserted in the concavity of the nucleus, which has an open ring shape. The only species, *A. platycaryon*, occurs in the gut of *Cryptotermes lamanianus* (Grassé and Hollande, 1950; Grassé, 1952a).
TYPE SPECIES: *Achemon playcaryon* Grassé & Hollande, 1950
TYPE HOST: *Cryptotermes havilandi* (Sjostedt)

Genus ***Astronympha*** Grassé, 1952

Devescovinids (28-45 μm) with 3 anterior flagella and a thick recurrent flagellum, cord-like, longer than the body. Cresta slender; nucleus like a transverse open ring; parabasal with a proximal V-shaped part supporting 4 to 7 small branches or parabasalies. Axostyle traversing the nucleus, with a terminal projection. Spirochetes on the anterior and posterior ends. One species, *A. nucleoflexa*, occurs in several species of *Cryptotermes* (Grassé, 1952a) (described under the name *Stellaria* in Grassé and Hollande, 1950 and as *Foaina nucleoflexa* in Kirby, 1942b).
TYPE SPECIES: *Astronympha nucleoflexa* (Kirby, 1942)
TYPE HOST: *Cryptotermes havilandi* (Sjostedt)

Genus ***Evemonia*** Grassé & Hollande, 1950

Devescovinid (50-70 μm) with 3 short anterior flagella and a thick recurrent one shorter than the body and non-adherent. Small cresta. Comb-like parabasal apparatus typically forming several branches or parabasalies around the nucleus. Axostyle stout, passing in the concavity of the flattened nucleus, no posterior projection. Three species, *E. punctata*, *E. ramosa*, and *E. brevirostris* described from *Neotermes aburiensis* (Grassé and Hollande, 1950; Grassé, 1952a), were also described under the name *Pseudodevescovina punctata* and *P. ramosa* by Kirby (1945b).
TYPE SPECIES: *Evemonia punctata* (Kirby, 1945)
TYPE HOST: *Neotermes aburiensis* Sjostedt

Genus ***Mixotricha*** Sutherland,1933

Large devescovinids (300-500 μm) with 4 anterior flagella inserted below an anterior papilla. Three flagella are anteriorly directed, and one is recurrent and not adherent to the body.

The small and elongated nucleus is anterior and appears connected to the basal bodies of the flagella and enclosed in the capitulum of the axostyle, which develops in a slender axostylar trunk to the posterior of the cell. All the cell surface except the anterior cone and the posterior end is covered by regularly arranged bacteria and spirochetes, which were identified as cilia by the first observer. The cell glides, propelled by the undulations of the spirochetes. Electron microscopy has shown a hexagonal fibrillar network under the surface where bacteria and spirochetes are attached. In the cytoplasm about 500 dictyosomes are present, each supported by a parabasal fiber. The cytoplasm is filled with vacuoles containing wood which is phagocytosed at the posterior end. In spite of the incomplete description, this flagellate seems to belong to the Devescovinidae. One species, *M. paradoxa*, described in *Mastotermes darwiniensis* by Sutherland (1933), electron microscopy in Cleveland and Grimstone (1964).
TYPE SPECIES: *Mixotricha paradoxa*, Sutherland, 1933
HOST: *Mastotermes darwiniensis* Froggatt

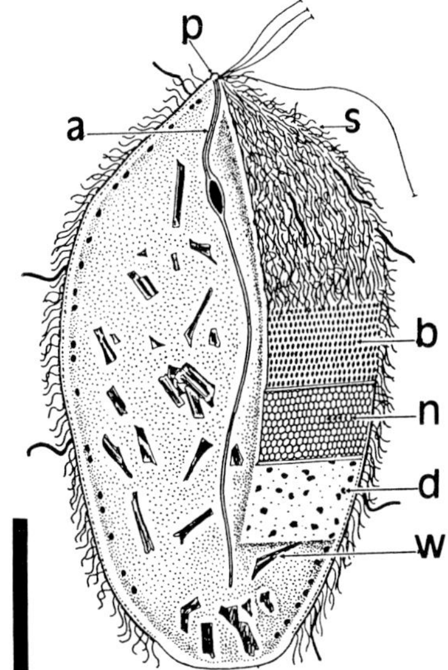

Fig. 23. *Mixotricha paradoxa* from *Mastotermes darwiniensis* (after Cleveland and Grimstone 1964). Bar=100 μm.

Genus *Kirbynia* Grassé & Hollande, 1950

Devescovinids with a body enclosing 1 or several karyomastigonts, free or fixed to the gut of the termite. The flagellated forms (50-70 μm) have 3 anterior flagella and a thick recurrent one shorter than the body. The cresta is short and the slender axostyle protrudes posteriorly. The parabasal is divided into several branches or parabasalies around the nucleus. Polymastigote cells of *K. pectinata*, comprising up to 70 karyomastigonts and 30 akaryomastigonts, with flagella at one pole and axostyles grouped at the opposite pole, are found swimming free in the intestinal fluid (Grassé and Hollande, 1951a).

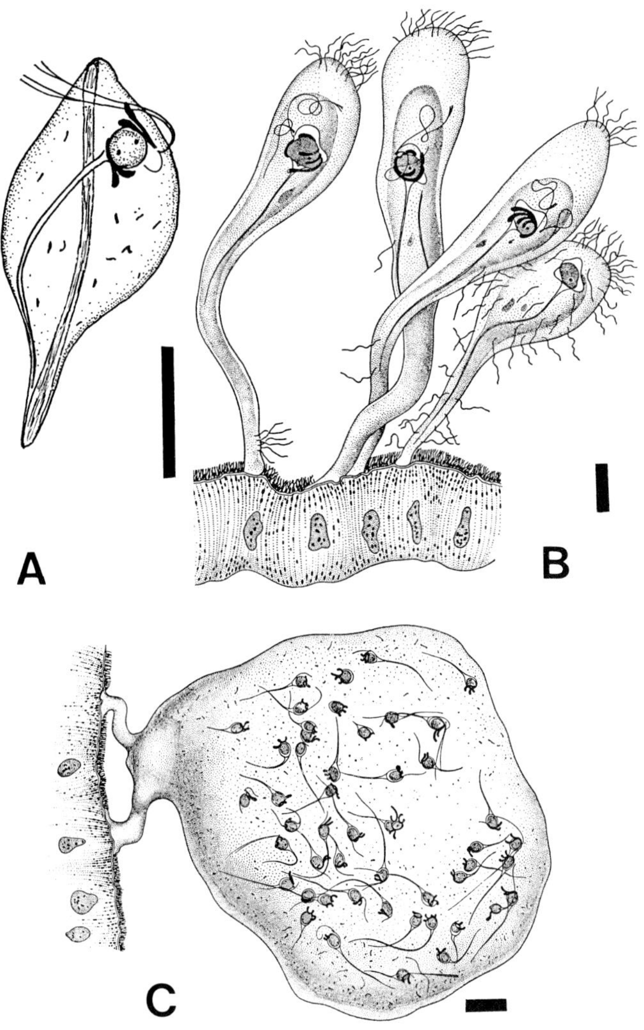

Fig. 24A,B,C. *Kirbynia pulchra* from *Anacanthotermes ochraceus*. A. Flagellate form containing a piece of wood. B. Fixed form with one karyomastigont and internal flagella. C. Fixed form with several karyomastigonts (from Grassé and Hollande, 1951a). Bar=10 μm.

The fixed forms (250 μm) with 1 karyomastigont and internal flagella are attached by their

posterior end to the gut of the termite. They can detach and acquire external flagella. In *K. pulchra*, large plasmodia containing many karyomastigonts (300) with regressed flagella are also fixed by a kind of stalk to the gut. Three species occur in Hodotermitidae: *K. pectinata* in *Microhodotermes wasmanni* and *Anacanthotermes ochraceus, K. pulchra* in *Anacanthotermes ochraceus* of North Africa, and *K. dogieli* in *A. murgabicus* (Grassé and Hollande, 1951a; Grassé, 1952a). No EM study.
TYPE SPECIES: *Kirbynia dogieli* (Bernstein, 1928)
HOSTS: *Anacanthotermes murgabicus* (Vasiljev) and *Anacanthotermes ochraceus* (Burmeister)

SUBFAMILY GIGANTOMONADINAE Kirby, 1946

Devescovinids occurring in both flagellated and amoeboid multinucleated forms.

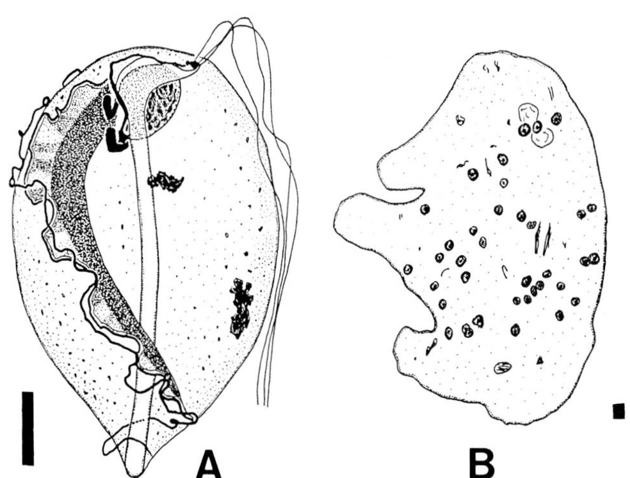

Fig. 25. *Gigantomonas herculea* from *Hodotermes mossambicus.* A. Flagellate form. B. Amoeboid form containing 36 nuclei (after Kirby, 1946). Bar=10 µm.

Genus ***Gigantomonas*** Dogiel,1916

Devescovinid flagellates (35-50 µm) with 3 long anterior flagella and a long and slender recurrent one, weakly adhering to the cell body along a large cresta; short, free posterior portion. No real undulating membrane; cresta developed almost all along the body with an outer edge thinner and undulating. Stout rod-like axostyle with a capitulum covering the nucleus and with a blunt or pointed end, not projecting from cell. Parabasal body near the nucleus, not coiled around the axostyle.

Amoeboid forms (80-300 µm) uni- or bi-nucleated, with reduced mastigont system and paradesmoses between 2 dividing nuclei. Large amoeboid forms with many nuclei have been described. Reproduction both by binary and multiple fission has been reported by Cleveland (1966c). One species, *G. herculea*, occurring in *Hodotermes mossambicus* and in additional hosts such as *Microhodotermes viator* (Kirby, 1946; Grassé, 1952a; Cleveland, 1966c). No EM study.
TYPE SPECIES: *Gigantomonas herculea* Dogiel, 1916
TYPE HOST: *Hodotermes mossambicus* (Hagen)

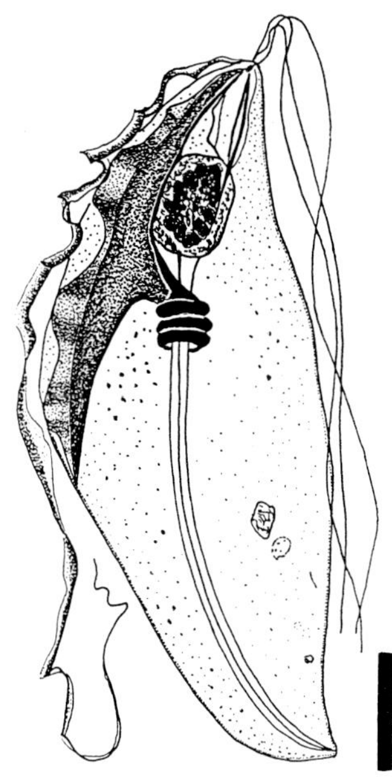

Fig. 26A. *Macrotrichomonas pulchra* from *Glyptotermes parvulus* (formerly *dubius*) (after Kirby, 1942a). Bar=10 µm.

Genus ***Macrotrichomonas*** Grassi, 1917

Large devescovinids (26-91µm) with 3 anterior flagella and a well developed trailing one, band-formed in some species (1-1.5 times the length of the body). Cresta developed into a broad internal membrane reaching from the nucleus to the periphery of the body. Parabasal body coiled around the trunk of the axostyle (1-13 turns). Axostyle tapering posteriorly with a pointed end

generally with a posterior projection. Among the 7 species occurring in 23 termite hosts, *M. pulchra* has been found in termites of the genus *Glyptotermes* and other Kalotermitidae (Kirby, 1942a; Grassé and Hollande, 1950; Grassé, 1952a); ultrastructural study by Hollande and Valentin (1969a).
TYPE SPECIES: *Macrotrichomonas pulchra* Grassi, 1917
TYPE HOST: *Glyptotermes parvulus* (Sjostedt)

FAMILY CALONYMPHIDAE Grassi & Foa, 1911

Multinucleated trichomonad flagellates having a permanent polymonad organization. The cell is composed of several karyomastigonts, each comprising 2-4 flagella, an axostyle-pelta complex, a parabasal, 1 nucleus, sometimes 2 or 3. The presence of a thick recurrent flagellum and a cresta in some of them indicates that they have probably evolved from devescovinids by multiplication of the karyomastigonts. All are xylophagous and bacterivorous symbionts in the gut of termites of the family Kalotermitidae. Six genera described showing a progressive integration of the karyomastigonts: *Coronympha, Metacoronympha, Stephanonympha, Diplonympha, Calonympha*, and *Snyderella*.

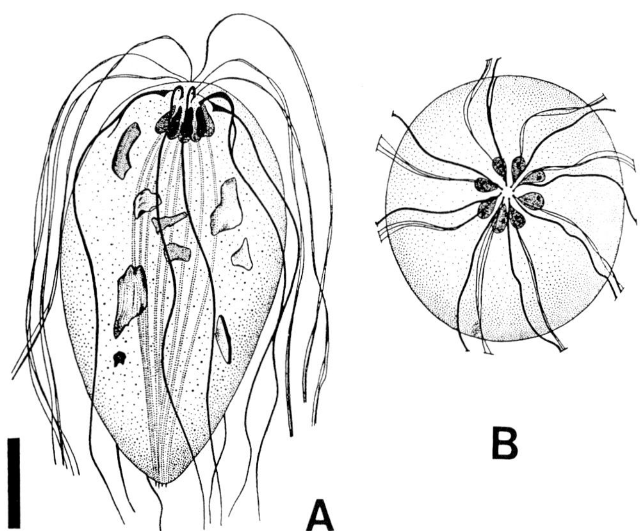

Fig. 27. *Coronympha octonaria* from *Incisitermes* (formerly *Kalotermes*) *emersoni.* A. Side view. B. Apical view (after Kirby, 1929). Bar=10 µm.

Genus ***Coronympha*** Kirby, 1929

Multinucleated ovoid cell (30-53 µm) with 8 or 16 karyomastigonts arranged in a single anterior circle. Each karyomastigont comprises 3 anterior flagella, a thick recurrent flagellum with an adherent proximal portion underlain by a cresta, an axostyle-pelta complex, and a drop-shaped parabasal close to the nucleus. The axostylar trunks are independent and only meet posteriorly, forming a caudal projection. Among the species described, *C. octonaria* has 8 karyomastigonts and lives in *Incisitermes* (formerly *Kalotermes*) *emersoni* (Mexico), and *C. clevelandi* with 16 karyomastigonts occurs in *Incisitermes immigrans* (formerly *Kalotermes clevelandi*) (Panama). Light microscopy by Kirby (1929) and Grassé (1952a). No EM study.
TYPE SPECIES: *Coronympha octonaria* Kirby, 1929
TYPE HOST: *Incisitermes* (formerly *Kalotermes*) *emersoni* (Light)

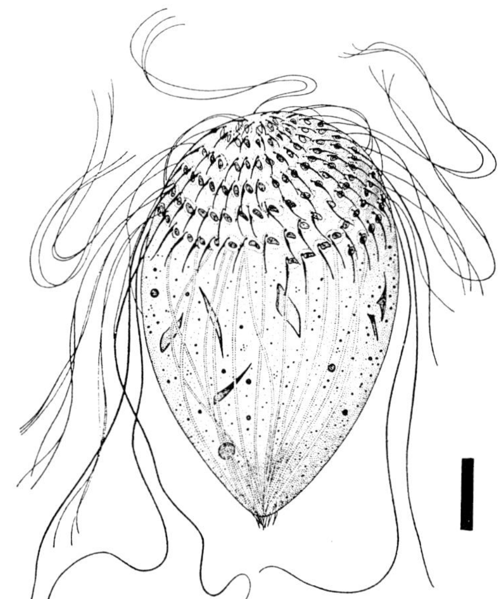

Fig. 28. *Metacoronympha senta* from *Kalotermes emersoni* (after Kirby, 1939). Bar=10 µm.

Genus ***Metacoronympha*** Kirby, 1939

Multinucleate flagellates (22-92 µm) composed of 66-345 karyomastigonts arranged along spiral lines from the apex. Each karyomastigont comprises 3 anterior flagella and a recurrent thick flagellum adhering along a cresta in its

proximal part. Axostyles independent, grouped at the posterior end. Parabasal bodies dot-shaped applied against the nuclei. The only species, *M. senta*, occurs in several termites of the genus *Incisitermes* (formerly *Kalotermes*) in America. Light microscopy by Kirby (1939) and Grassé (1952a).
TYPE SPECIES: *Metacoronympha senta* Kirby, 1939
TYPE HOST: *Incisitermes* (formerly *Kalotermes*) *emersoni* (Light)

Genus ***Stephanonympha*** Janicki, 1911

Multinucleate cells (70-130 µm) composed of numerous karyomastigonts with 4 flagella (sometimes reduced to as few as 2 flagella) concentrated at the apex in a complex spiral. Axostyles forming a central bundle not protruding at the rear. This genus is very close to *Metacoronympha*. Among the 12 species reported by Janicki (1915), *S. silvestrii* lives in *Neotermes connexus* (Honolulu), *Cryptotermes havilandi* (Nigeria), and *Neotermes erythraeus* (East Africa). Another description for *S. nelumbium* in *Cryptotermes domesticus* (formerly *hermsi*) was given by Kirby (1926). Light microscopic studies by Janicki (1915) and Grassé (1952a). EM study in Lavette, 1971b.
TYPE SPECIES: *Stephanonympha silvestrii* Janicki, 1911
TYPE HOST: *Neotermes connexus* Snyder

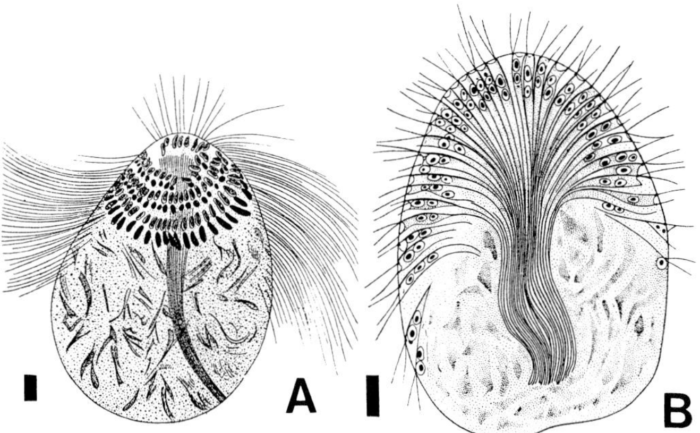

Fig. 29A. *Stephanonympha silvestrii* from *Neotermes connexus* (after Janicki, 1915). Bar=10 µm. Fig. 29B. *Diplonympha foae* from *Glyptotermes parvulus* (after Grassi, 1917). Bar=10 µm.

Genus ***Diplonympha*** Grassi, 1917

Multinucleated cells composed of several karyomastigonts, each comprising 2 flagella in Grassi's representation, an axostyle and generally 1-3 accompanying nuclei. Axostyles form a central bundle not protruding at the rear. The one species, *D. foae*, lives in *Glyptotermes parvulus* and was found again by Grassé in *G. bookoko* in West and Central Africa. Studied by light microscopy (Grassi, 1917; and Grassé, 1952a). No EM study.
TYPE SPECIES: *Diplonympha foae* Grassi, 1917
TYPE HOST: *Glyptotermes parvulus* (Sjostedt)

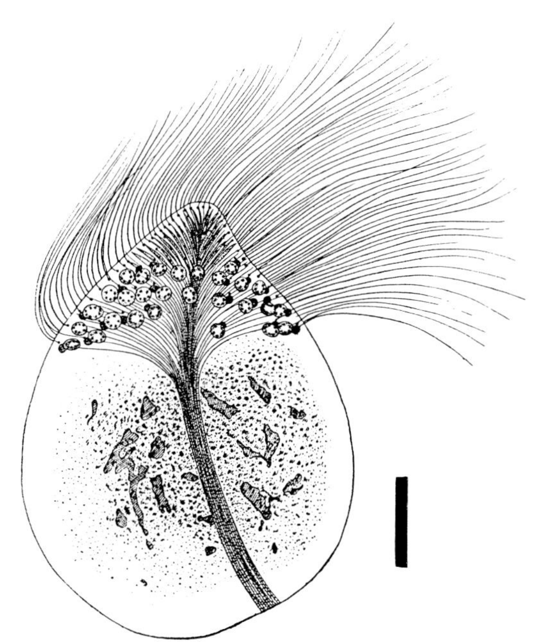
Fig. 30. *Calonympha grassii* from *Cryptotermes brevis* (after Janicki, 1915). Bar=10 µm.

Genus ***Calonympha*** Foà, 1905

Multinucleate flagellates (70-80 µm) with several coronas of akaryomastigonts anterior to 2-3 coronas of karyomastigonts. Each karyomastigont comprises 4 anteriorly directed flagella, no recurrent flagellum or cresta. The axostyles meet together forming a central bundle, which reaches the posterior end. A dot-shaped parabasal is present in each mastigont close to the nucleus. Electron microscopy has shown the trichomonad and devescovinid organization of the

mastigont system. One of the 3 species, *C. grassii*, lives in *Cryptotermes brevis* (formerly *grassii*) of Chile. Light microscopic studies by Janicki (1915) and Grassé (1952a), electron microscopic study by Joyon et al. (1969).
TYPE SPECIES: *Calonympha grassii* Foa, 1905
TYPE HOST: *Cryptotermes brevis* (Walker)

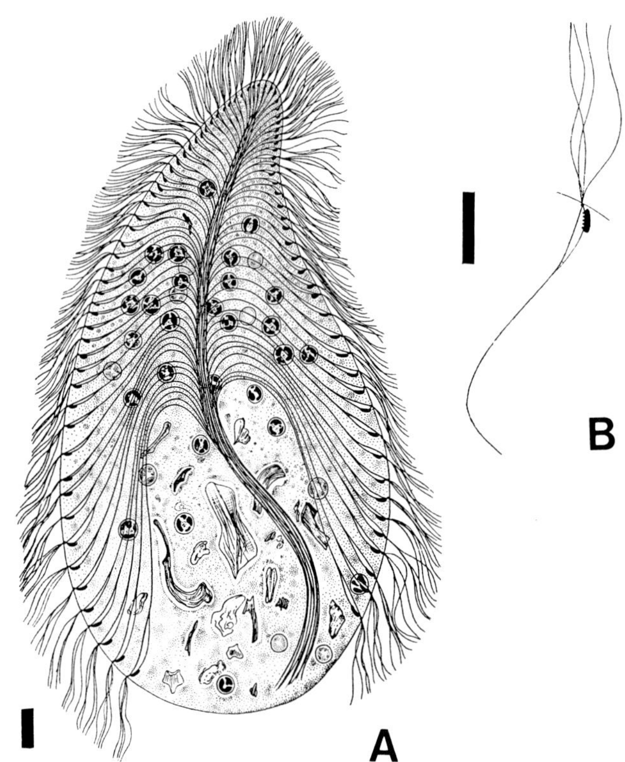

Fig. 31. *Snyderella tabogae* from *Kalotermes longicollis*. A. Side view of the whole organism. B. One karyomastigont with four flagella, one axostyle, and one parabasal (after Kirby, 1929). Bar=10 µm.

Genus ***Snyderella*** Kirby, 1929

Multinucleate flagellates (77-150 µm) composed of many akaryomastigonts (100-3000). Each karyomastigont has 4 flagella, a parabasal body, and an axostyle. The mastigonts are close together and extend through the major part of the peripheral region. The axostyles are collected into a bundle posterior to the middle of the body. Nuclei are mostly spread in the central part. The type species, *S. tabogae*, occurs in *Cryptotermes* (formerly *Kalotermes*) *longicollis* (Kirby, 1929; Grassé, 1952a). No EM study.
TYPE SPECIES: *Snyderella tabogae* Kirby, 1929
TYPE HOST: *Cryptotermes* (formerly *Kalotermes*) *longicollis* Banks

FAMILY COCHLOSOMATIDAE Tyzzer, 1930

Uninucleate flagellate with an anterior adhesive disk, a lateral groove, and a mastigont system of six flagella, an axostyle, and a parabasal body. The species of the three genera reported are intestinal parasites of birds.

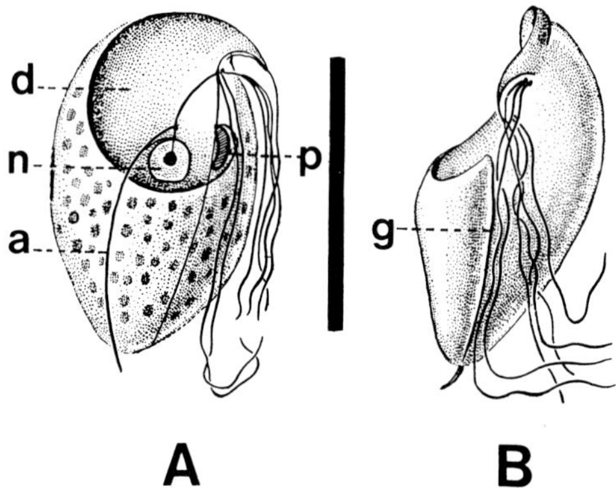

Fig. 32. *Cochlosoma anatis*. A. Anterior view showing the sucking disc, flagella, nucleus, parabasal, axostyle. B. Side view showing the lateral groove and associated flagella (after Kimura, 1934). Bar=10 µm

Genus ***Cochlosoma*** Kotlàn, 1923

Small flagellates (6-18 µm) anteriorly truncated by a spiraled adhesive disk. From a cleft on the disk, a lateral groove develops along the length of the body. The 6 flagella arise from the cleft and are deflected backward. A recurrent flagellum is associated with the dorsal left margin of the lateral groove, as in undulating membranes, and terminates by a free posterior portion. Four flagella are free and the fifth emerges independently or dorsally. An axostyle originating near the basal bodies traverses the cell and protrudes posteriorly. A parabasal body situated near the anterior nucleus has been described. Among the 7 species reported, 5 occur in the intestine of birds, such as ducks, turkeys, American magpies, eastern robins, and *Turdus*, and possibly one in bats (Kotlan, 1923; Kimura, 1934; Travis, 1938; Kulda and Nohynkovà, 1978). The scanning electron microscopic study (Watkins et al., 1989) has confirmed the external morphology of the disk and the groove, the number of flagella, the recurrent flagellum attached to the rim of the lateral groove, and the posterior axostyle. An EM study clearly shows the trichomonad characters of this genus and of the family (Pecka et al., 1996).

Genus ***Cyathosoma*** Tyzzer, 1930

This genus created by Tyzzer, (1930) for a species found in the cecum of the ruffed grouse, *Bonasa umbellus*, has the same features as *Cochlosoma*, except the lateral groove does not reach the posterior end, which is rounded and without a protruding axostyle. This genus has been put in synonymy with *Cochlosoma* under the name *C. striatum* Tyzzer by Kulda and Nohynkovà (1978) but not by Grassé (1952).
TYPE SPECIES: *Cyathosoma striatum* Tyzzer, 1930
TYPE HOST: *Bonasa umbellus*

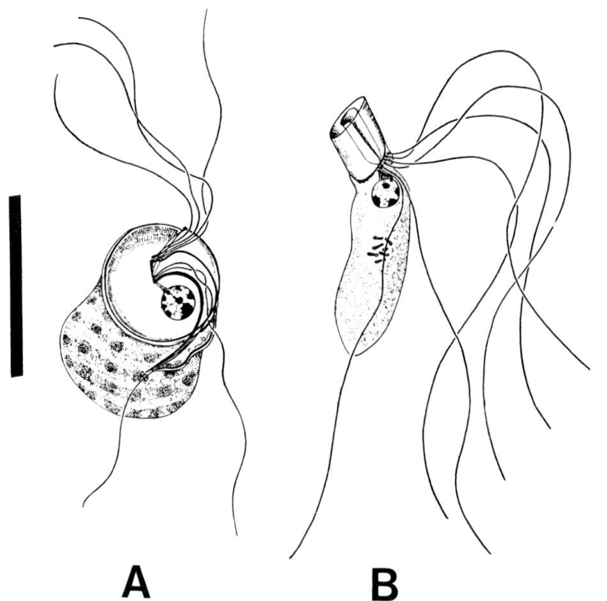

Fig. 33A. *Cyathosoma striata* (after Tyzzer, 1930). Bar=10 µm. Fig 33B. *Ptychostoma bonasae* (after Tyzzer, 1930).

Genus ***Ptychostoma*** Tyzzer, 1930

Cochlosomid flagellate with an elongated cell body (9-15 µm) bearing 6 flagella and a tube-like anterior sucker inrolled like a scroll. There is no apparent lateral groove, but the recurrent flagellum seems attached to the cell body. Multiplication by unequal binary fission. The species, *P. bonasae*, occurs in the cecum of the ruffed grouse (*Bonasa umbellus*), and flagellates are attached to the epithelium surface by their sucker (Tyzzer, 1930). This genus has been put in synonymy with *Cochlosoma* by Kulda and Nohynkovà (1978), but not by Grassé (1952). No EM study.
TYPE SPECIES: *Ptychostoma bonasae* Tyzzer, 1930
TYPE HOST: *Bonasa umbellus*

CLASS HYPERMASTIGIA

Hypermastigotes live in the intestine of termites, cockroaches, and woodroaches (Fig. 34). They have numerous flagella in their mastigont systems, branched and multiple parabasal bodies, and a single nucleus. Basal bodies of the flagella are arranged in a complete or partial circle, in one or more plates, or in longitudinal or spiral rows meeting in a centralized structure (**rostrum**). Axostylar supporting structures (microtubular complexes) are present but variable. The cytoplasm contains hydrogenosomes (no mitochondria) and stored glycogen. Reproduction is by binary fission, with the formation of an extranuclear spindle between elaborate centrosomal structures (**atractophores**) in a typical **pleuromitosis** (Cleveland, 1958a; Hollande and Carruette-Valentin, 1971). In some genera the large chromosomes remain condensed and visible throughout the cell cycle (Cleveland, 1949b). A sexual cycle in relation to the molting and hormone (ecdysone) production of the insect occurs in several genera (Cleveland, 1956a, 1957). Hypermastids phagocytose and digest wood and are associated with symbiotic bacteria. A general description by light microscopy is found in Cleveland et al. (1934), Grassé (1952b), and Vickerman (1982); the biology of hypermastigotes is reviewed in Grassé (1952b) and Honigberg (1970); a list of species and hosts in Yamin (1979a).

Electron microscopic studies have demonstrated that the hypermastigids share the basic characters of the trichomonads and belong to the parabasalid lineage. The four **privileged basal bodies** have been found in *Joenia* (Fig. 35b) (Hollande and Valentin, 1969b), *Lophomonas* (Hollande and Carruette-Valentin, 1972), and *Deltotrichonympha* (Hollande, 1986a) and form two sets of one, two, or three privileged basal bodies in Trichonymphida and Spirotrichonymphida (Hollande and Carruette-Valentin, 1971). They polarize the parabasals, the pelta-axostyle(s), and the two atractophores (**batachio**) equivalent to centrosomes. At division they divide into two sets, which are at the origin of the new mastigont systems in *Joenia, Lophomonas,* and *Deltotrichonympha* or at the origin of the symmetrical mastigont system in Hypermastigida and Spirotrichonymphida (Hollande and Carruette-Valentin, 1971).

In the order Lophomonadida, families Joeniidae, Lophomonadidae, and Deltotrichonymphidae, there is a single set of basal bodies and an associated fiber, which regresses at division and reconstitutes in each daughter cell. In the orders Trichonymphida and Spirotrichonymphida, there are two symmetrical sets of basal bodies and fibers which separate equally at division; then each daughter cell reconstitutes the symmetry. In Joeniidae (Fig. 35a,c) (Hollande and Valentin, 1969b) and Lophomonadidae (Hollande and Carruette-Valentin, 1972), there are **pelta-axostyle complexes** very similar to those of trichomonads. In trichonymphids, pelta-axostyle complexes are multiple and do not form a single axial bundle. The situation is mixed in spirotrichonymphids.

Parabasals are basically organized like those of trichomonads, striated parabasal fibers originating from the privileged basal bodies support the parabasal bodies or dictyosomes (Figs. 35d, 36e). In many hypermastigotes the number of parabasal bodies is multiplied, and in some the dictyosomes seem independent. Hypermastigotes have hydrogenosomes like trichomonads (Figs. 35a, 36e).

All divide in a pleuromitosis with an extranuclear spindle comprising pole to pole microtubules forming a paradesmose and with kinetochore or chromosomal microtubules attached to kinetochores of the chromosomes and inserted into the nuclear envelope. Electron microscopic studies of division may be found in Hollande and Valentin (1967, 1969b), Hollande and Carruette-Valentin (1971), and Hollande (1979, 1986a).

KEY CHARACTERS

1. Mastigont system with many flagella.
2. Basal bodies and flagella distributed as either an anterior tuft, as plate(s), or as longitudinal or spiral rows meeting anteriorly in a specialized structure.
3. Multiple parabasal bodies.
4. Motile cells with a single nucleus with one exception (*Rhizonympha*).

Three orders have been distinguished: Lophomonadida, Trichonymphida, and Spirotrichonymphida (Fig. 34) (Kirby, 1949a; Grassé, 1952b; Hollande and Carruette-Valentin, 1971; Vickerman, 1982; Hollande, 1986a). Six families compose the order Lophomonadida: Joeniidae, Lophomonadidae, Rhizonymphidae, Kofoidiidae, Microjoeniidae, and Deltotrichonymphidae. All have a single set of flagella which regresses at division. The Joeniidae have an anterior area of basal bodies and flagella in longitudinal rows. The basal bodies in Lophomonadidae are in an anterior ring; the flagella project as a distinctive anterior tuft. In Kofoidiidae there are several anterior tufts. In Rhizonymphidae the plasmodium contains 300-400 lophomonad karyomastigonts dispersed in the cytoplasm. The Microjoeniidae are the least known and seem to have a radial organization of flagella and parabasals. The Deltotrichonymphidae share more characters with this order than with the other two orders: they have a single set of flagella which disassembles and reconstitutes at division, but the basal bodies are separated into

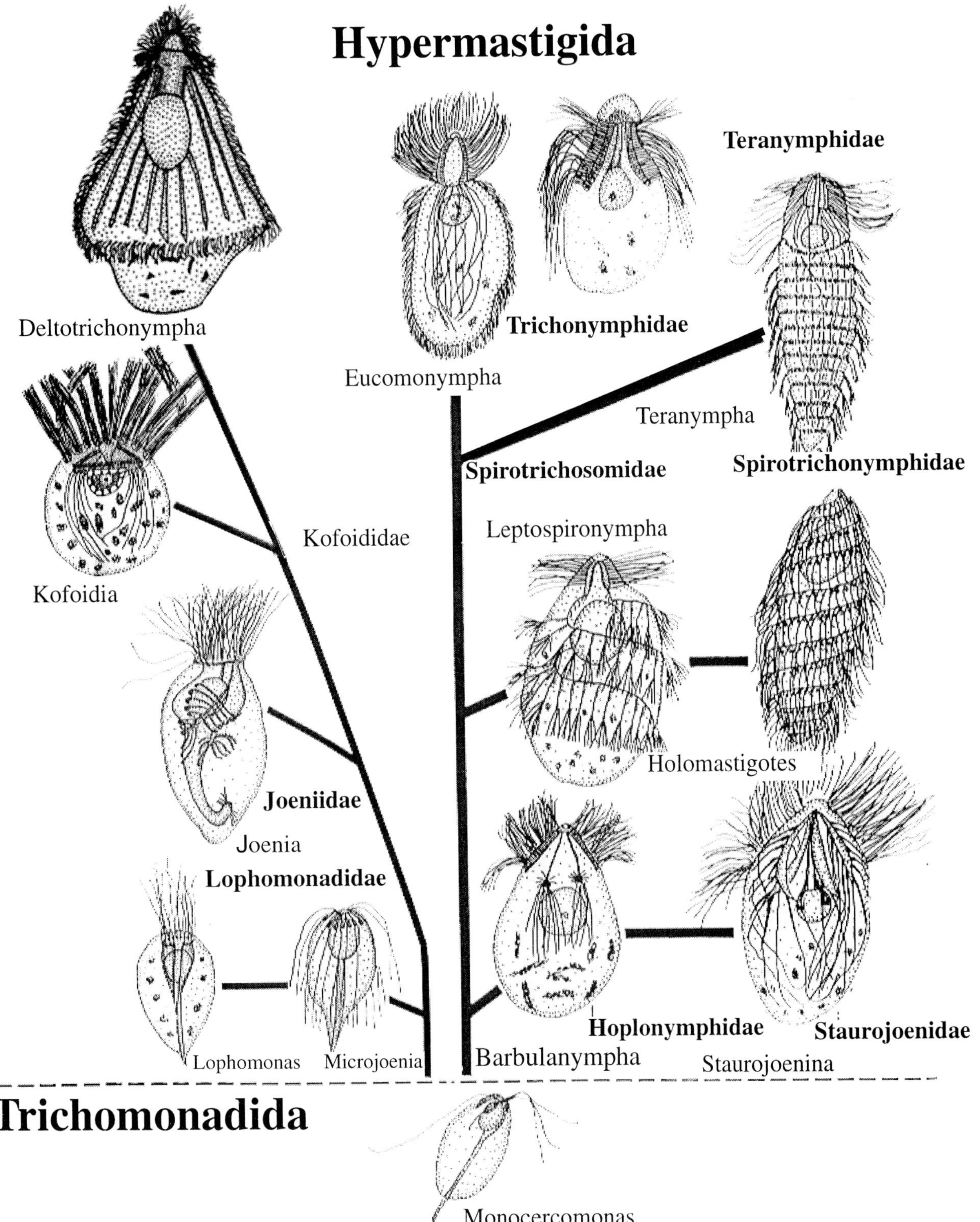

Fig. 34. Chart shows structural types, possible relationships in families of Hypermastigia, with suggested origin from Trichomonada. On the left is the order Lophomonadida, on the right members of the orders Trichonymphida and Spirotrichonymphida. (Based on figure by Kirby [1949] with modifications suggested by the papers of Hollande and Valentin [1969b] and Hollande and Carruette-Valentin [1971].)

two regions: rostral and post-rostral as in trichonymphids.

The order Trichonymphida contains six families: the Hoplonymphidae, Staurojoeninidae, Trichonymphidae, Eucomonymphidae, Teranymphidae, and Spirotrichosomidae: all have a bilateral symmetry and two sets of flagella, which separate at division. In Hoplonymphidae flagella are borne at the rostrum and separated into two groups by ectoplasmic lobes. The Staurojoenidae have thesame organization, but there are four sets of flagella separated by four lobes. Flagella in Trichonymphidae never trail over the postrostral region, and postrostral flagellar rows are independent of those in the rostrum.

Eucomonymphidae are flagellated over the entire surface, and the rostral flagella are longer than the postrostral ones; rostral flagellar rows are nearly continuous with the postrostral ones but are underlain by parabasal bodies. In Teranymphidae the flagella of the postrostral region form parallel circles. In Spirotrichosomidae there are only two bands of flagella spiraling along the body, but these bands can divide into longitudinal flagellar rows.

The order Spirotrichonymphida contains three families: the Spirotrichonymphidae, Holomastigotoididae, and the Holomastigotidae. All have two or more spiraled flagellar rows, which separate equally at division. The Spirotrichonymphidae have a **pseudorostrum** from which arise the rows of basal bodies and also a central axostyle. The Holomastigotoididae have no anterior pseudorostrum, but a rounded pole, and all flagellar bands do not arise from the apex. The Holomastigotidae are not well known; spiraled flagellar rows arise from the anterior pole, and they seem to lack an axostyle.

ARTIFICIAL KEY TO ORDERS, FAMILIES AND SELECTED GENERA

1. Flagella and fibers forming only one area and regressing at division; pseudorostrum without a tube and cap.................. Order **Lophomonadida**
2. Flagella and fibers forming two bilaterally symmetrical areas in the rostrum; each separates into two at division; rostrum separated from the body with a rostral tube and a cap; several axostyles.. Order **Trichonymphida**
3. Flagella and fibers forming two to several spiraled rows, dividing in two longitudinally at division; pseudorostrum without a tube and cap; dot-shaped parabasals between flagellar rows; axostyle central when present.. Order **Spirotrichonymphida**

ORDER LOPHOMONADIDA Light

1. Flagella forming an anterior oblique area, parabasal bi- or multiramous around the nucleus.................................. Family **Joeniidae:** .... ***Joenia, Placojoenia, Joenina,*** .... ***Joenoides, Joenopsis, Cyclojoenia***
2. Flagella forming an anterior tuft; parabasal bodies under the basal bodies; one central axostyle.....Family **Lophomonadidae**................ .... ***Lophomonas***
3. Flagella forming several anterior tufts, several axostyles... Family **Kofoididae**.......... ***Kofoidia***
4. Plasmodium with several karyomastigonts, each with a tuft of flagella, an axostyle, and a parabasal................. Family **Rhizonymphidae** .... ***Rhizonympha***
5. Flagella and parabasal radially arranged around the apex. .... Family **Microjoenidae** .... ***Microjoenia***
6. Flagella forming an anterior rostrum without a rostral tube and a cap.. Family **Deltotrichonymphidae**...................... ....***Koruga, Deltotrichonympha***

Order TRICHONYMPHIDA Grassé

1. Only two rostral flagellar areas separated by ectoplamic zones..... Family **Hoplonymphidae**: .... ***Barbulanympha, Hoplonympha,*** .... ***Urinympha, Rhynchonympha***
2. Only four rostral flagellar areas separated by ectoplasmic zones...

Family **Staurojoeninidae**............................
.............***Staurojoenina***, ... ***Idionympha.***

3. Rostral flagella short; body flagella in longitudinal rows only on half of the body; parabasals around the nucleus; no axostyle... Family **Trichonymphidae**... ***Trichonympha***
4. Rostral flagella long; body flagella cover the body in longitudinal or slightly spiral rows; parabasals under the flagellar rows; several axostyles.............. Family **Eucomonymphidae** ..... ***Eucomonympha, Pseudotrichonympha***
5. Rostral flagella long; body flagella arranged in circular rows on all the body.............................Family **Teranymphidae**, ***Teranympha***
6. Rostral flagella forming two spiraled bands continuing along the body ....Family **Spirotrichosomidae**................ ***Spirotrichosoma, Macrospironympha, Leptospironympha, Apospironympha, Colospironympha***.

ORDER SPIROTRICHONYMPHIDA Light

1. Spiraled flagellar rows arising on a pseudo-rostrum and continuing around the body; central axostyle; dot-shaped parabasals between the flagellar rows...Family **Spirotrichonymphidae**......... ***Spirotrichonympha, Micromastigote, Spirotrichonymphella***, ***Spironympha***
2. Spiraled rows arising on a flat apex continuing on the body; central axostyle. ....Family **Holomastigotoididae** ....***Holomastigotoides, Rostronympha***

3. Body completely covered by spiraled flagellar rows; no axostyle described..Family **Holomastigotidae** ....***Holomastigotes, Spiromastigotes***

ORDER LOPHOMONADIDA Light

Hypermastigid flagellates with one flagellar area comprising four privileged basal bodies, numerous.

FAMILY JOENIIDAE Janicki

Cells with an apical flagellar area forming a kind of rostrum; axostyle of trichomonad type, comprising a capitulum or calyx enveloping the anterior area and nucleus, and an axostylar trunk. Parabasal biramous at origin, attached to the privileged basal bodies. Flagellar area and axostyle resorbed during division. All symbiotic in termites.

Genus *Joenia* Grassi, 1917

Flagellate (50-300 µm) with a conical rostrum partly covered by the flagellar area (Fig. 37A); axostyle forming a calyx enveloping the nucleus and the flagellar area; trunk of axostyle stout, protruding posteriorly in young cells or ending within the body. The fine structure of all components of the mastigont system are known: The flagellar area comprises basal bodies all bearing a hooked lamina (Fig. 35b). The 4 privileged basal bodies with a barren orthogonal R-basal body, are present at the anterior edge of the flagellar area (Fig. 35b); the 2 major parabasal fibers inroll around the axostyle near the nucleus and divide into branches supporting a parabasal body (Fig. 35a,d); a striated lamina underlies the basal bodies of the flagellar area. The axostylar trunk is composed of several encased rows of microtubules (Fig. 35 a,c). Atractophores at the origin of the division spindle have been observed as well as the regression of the parental flagellar area during division. Among the 4 species living in Kalotermitidae, *J. annectens* from *Kalotermes flavicollis* and *J. duboscqui* in *K. praecox* are the best known. General descriptions can be found in Duboscq and Grassé (1933) and Grassé (1952b) and ultrastructure in Hollande and Valentin (1969) and Hollande (1979).

Genus *Placojoenia* Radek & Hausmann, 1994

Flagellate (200-104 µm) with a clavate cell body and a horse-shoe shaped tuft of flagella slightly shifted to ventral side; 5-7 tube-like parabasal bodies and fibers wound around the axostyle just under the nucleus; axostylar capitulum surrounding the flagellar area and the nucleus; axostylar trunk stout, protruding posteriorly. Nuclear envelope with branched extensions. One species described by light and EM: *Placojoenia sinaica* from *Kalotermes sinaicus* (Radek and Hausmann, 1994) (Fig. 37B).

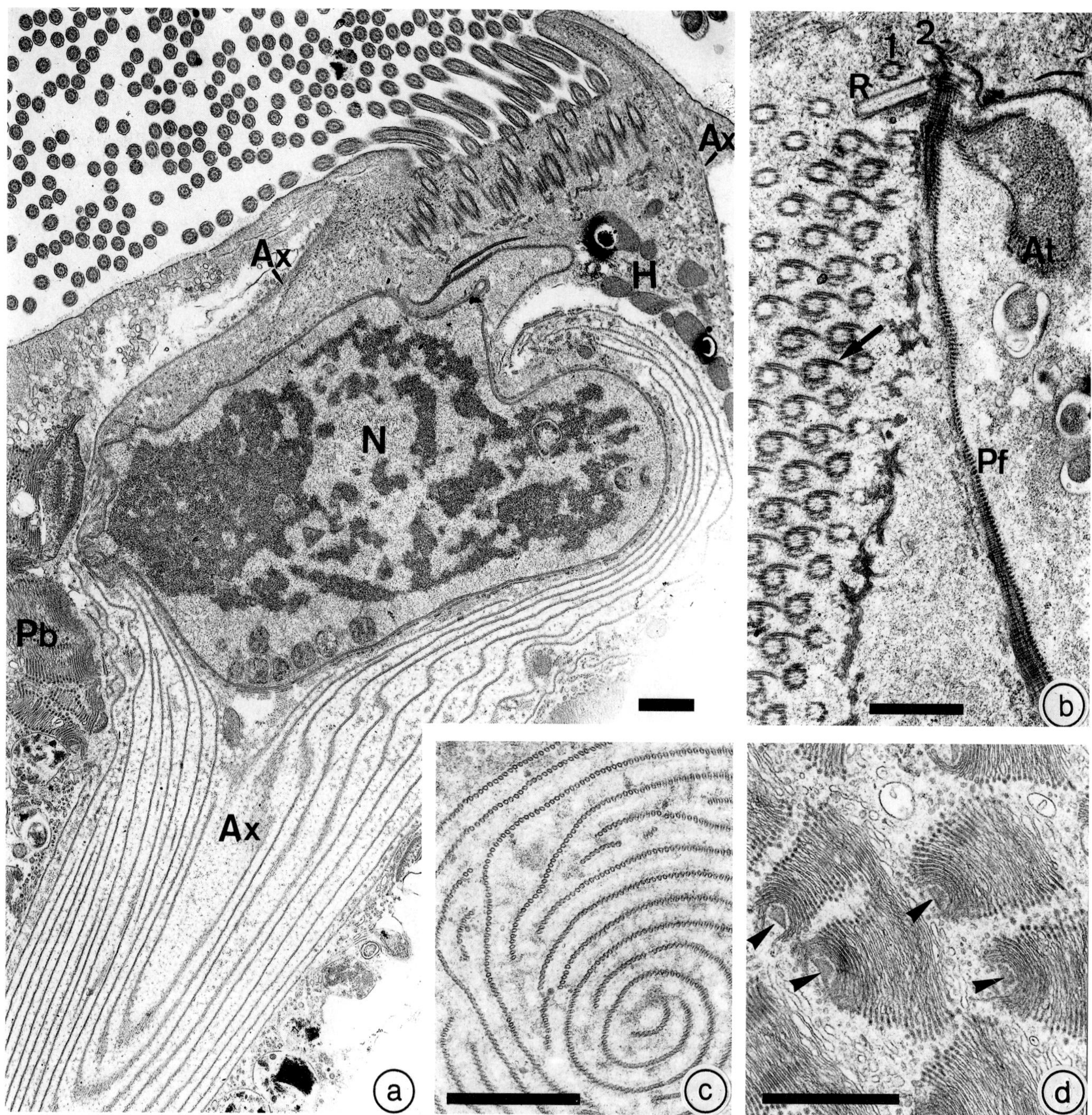

Fig. 35. Ultrastructure of *Joenia*. Bars=1 μm. (from G. Brugerolle) Fig. 35a. General view showing a part of the flagellar area on top, the nucleus (N), the microtubular rows of the axostyle (Ax), some parabasal bodies (Pb), hydrogenosomes (H). Fig. 35b. Privileged basal bodies (1, 2, R) on one side of the flagellar area. Notice the hooked lamina (arrow) borne by all basal bodies of the flagellar area, the parabasal fiber (Pf), and one atractophore (At). Fig. 35c. Transverse section of the axostyle. Fig. 35d. Transverse section of the parabasal bodies, where each dictyosome is supported by a parabasal fiber (arrowheads).

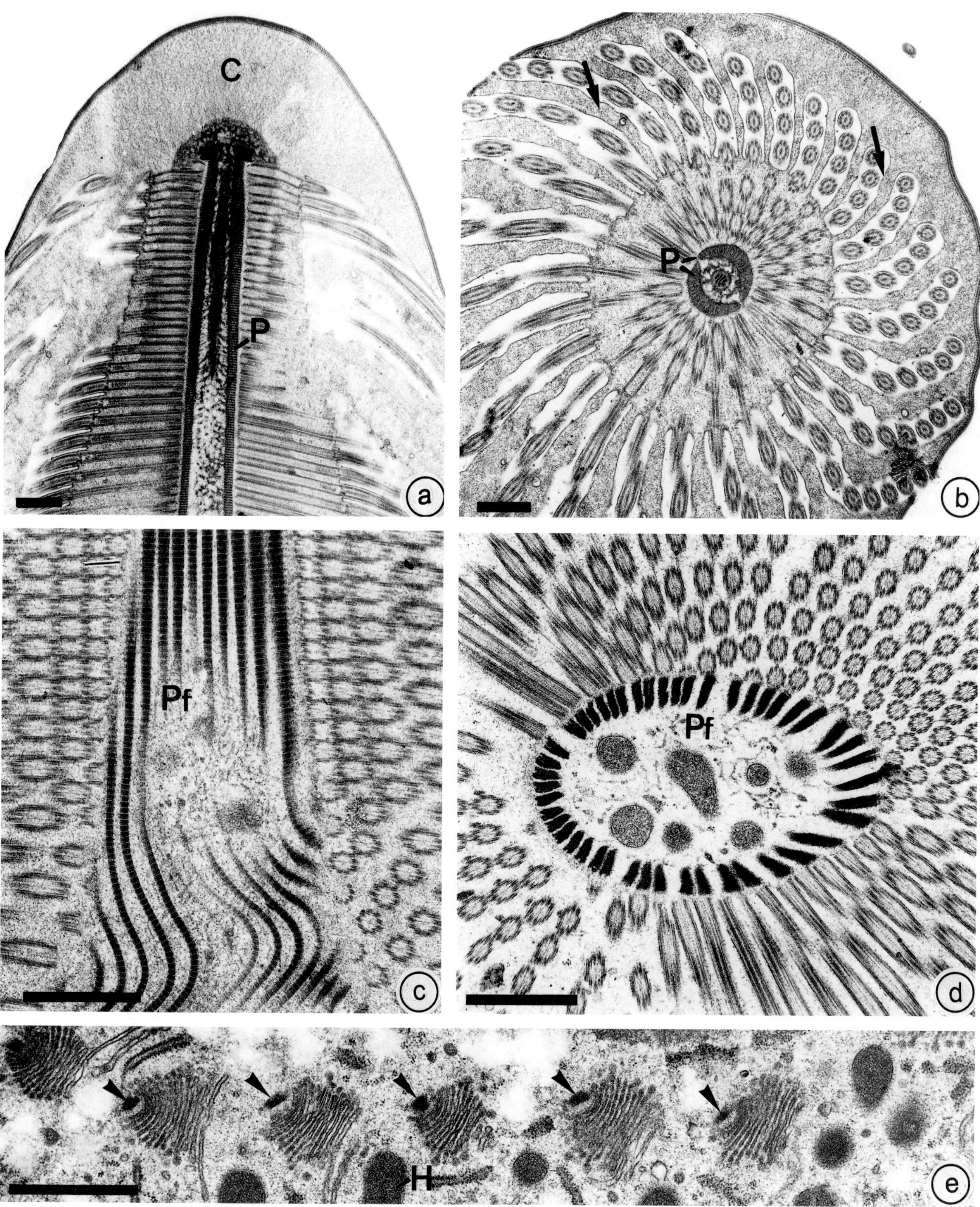

Fig. 36. Ultrastructure of *Trichonympha.* Bar=1 μm; (from G. Brugerolle). Longitudinal sections (Fig. 36a,c) and transverse sections (Fig. 36b,e). Two plates (P) composed of parabasal fibers (Pf) form the rostral tube, surmounted by a cap (C). Basal bodies are inserted perpendicular to the rostral tube or plates and give rise to flagellar rows separated by lanes of ectoplasm (arrows) (Fig. 36a,b). The two plates subdivide in several parabasal fibers (Pf) (Fig. 36c,d). Fig. 36e. Transverse section at the level of the nucleus, showing the parabasal bodies comprising a dictyosome supported by a parabasal fiber (arrow); hydrogenosome (H).

TYPE SPECIES: *Placojoenia sinaica* Radek & Hausmann, 1994
TYPE HOST: *Kalotermes sinaicus*

Genus ***Joenina*** Grassi, 1885

Flagellate (80-120 µm) with a semi-circular flagellar area (Fig. 38A). Axostylar capitulum enveloping the flagellar area. Trunk of the axostyle formed anterior to the nucleus; biramous parabasal bearing parabasalies around the nucleus. Monospecific. Reported in Grassé (1952b). No EM study.
TYPE SPECIES: *Joenina pulchella* Grassi, 1917
TYPE HOST: *Porotermes adamsoni*

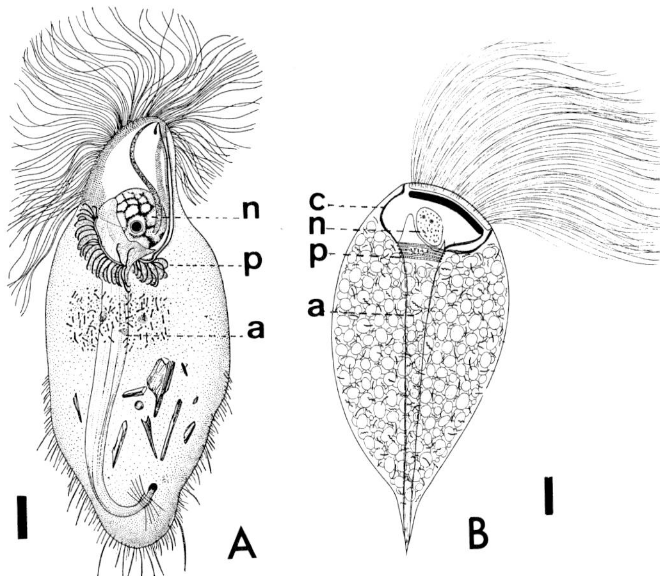

Fig. 37A. *Joenia annectens* from *Kalotermes flavicollis* showing the rostrum covered by the flagellar area, the nucleus (n), and the axostylar trunk (a) surrounded by parabasal bodies (p). (after Duboscq and Grassé, 1933) Fig. 37B *Placojoenia sinsica* Radek & Hausmann. Bars=10 µm.

Genus ***Joenoides*** Grassé, 1952

Flagellate (60-220 µm) with a semi-crescent-shaped flagellar area (Fig. 38B). Axostylar capitulum enveloping the nucleus and the flagellar area. Splinter-like parabasal bodies not encircling the nucleus. Monspecific. Reported in Grassé (1952b). No EM study.

TYPE SPECIES: *Joenoides intermedia* (Dogiel)
TYPE HOST: *Hodotermes mossambicus* (Hagen)

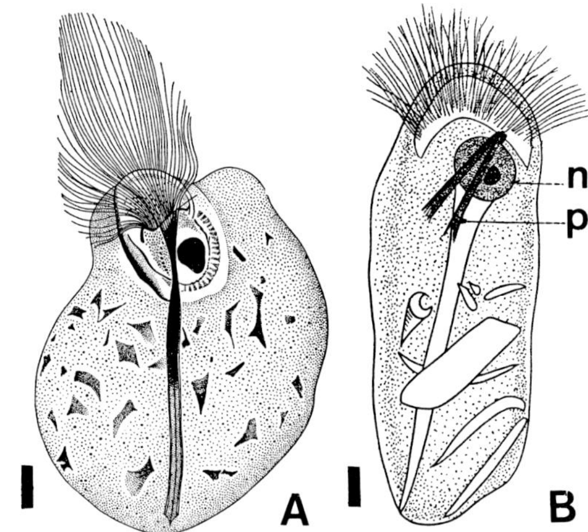

Fig. 38A. *Joenina pulchella* from *Porotermes adamsoni* (after Grassi, 1917). Bar=10 µm. Fig. 38B. *Joenoides intermedia* from *Hodotermes mossambicus* (after Dogiel in Grassé [1952b]). Bar=10 µm.

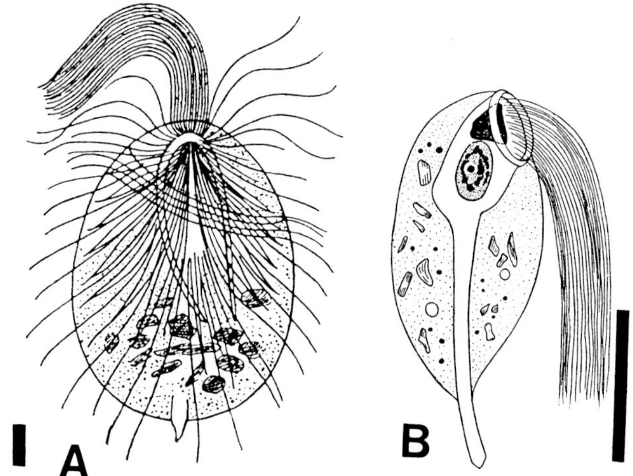

Fig. 39A. *Joenopsis polytricha* from *Archotermopsis wroughtoni* (after Cutler, 1920). Bar=10 µm. Fig. 39B. *Cyclojoenia australis* from *Stolotermes ruficeps* (after Nurse, 1945). Bar=10 µm.

Genus ***Joenopsis*** Cutler, 1920

Flagellate (65-160 µm) with a flagellar area shaped like an inverted U, prolonged distally by lines of flagella extending into the anterior half of the body (Fig. 6A). Capitular axostyle not apparent; trunk of the axostyle slender; biramous

pinnate parabasal bodies not encircling the nucleus. Two species described, *J. polytricha* and *J. cephalotricha* in *Archotermopsis wroughtoni* of India, by Cutler (1920); reported in Grassé (1952b). No EM study.
TYPE SPECIES: *Joenopsis polytricha* Cutler, 1920.
TYPE HOST: *Archotermopsis wroughtoni* (Desneux).

Genus ***Cyclojoenia*** Nurse, 1945
(*incertae sedis*)

Flagellate (18-23 µm) with a tuft of about 20 flagella as long as the body, arising from a circular depression. Axostylar capitulum enveloping the nucleus, the parabasal body, and the flagellar area; trunk of the axostyle protruding posteriorly. Triangular parabasal body situated between the nucleus and the flagellar area. Monospecific. The species, *C. australis*, has been found by Nurse (1945) in *Stolotermes ruficeps* of New Zealand and could belong to *Joenia*; reported in Grassé (1952b).
TYPE SPECIES: *Cyclojoenia australis* Nurse, 1945
TYPE HOST: *Stolotermes ruficeps* Brauer

Genus ***Projoenia*** Lavette, 1970

Projoenia was created for the species *P. sawayai*, a symbiont of a unidentified termite in Brazil. The flagellate of 45 to 200 µm, only known by EM photographs, has a flagellar area typical of Joeniidae and a recurrent flagellum and cresta typical of Devescovinidae and Joeniidae (Lavette 1970).

FAMILY LOPHOMONADIDAE Stein

Genus ***Lophomonas*** Stein

Flagellate (20-60 µm) with an apical tuft of flagella and an axostyle forming an anterior calyx enveloping the flagellar area, the nucleus, and a part of the parabasal bodies (Fig. 40). The calyx is opened along a longitudinal slit, and the trunk of the axostyle is slender and points posteriorly. When the cell divides by a typical pleuromitosis, flagella and axostyle regress and 2 new flagellar areas reform (Bêlâr, 1926). Ultrastructural studies have shown the 4 privileged basal bodies on one side of the flagellar area. They polarize the 2 parabasal fibers and atractophores (Hollande and Carruette-Valentin, 1972). Several longitudinal rows of microtubules form the wall of the calyx and the axostylar trunk. In *L. blattarum* a ring of sausage-shaped dense bodies surround the calyx and correspond to the extra-calycial bodies observed by light microscopy (Beams and Sekhon, 1969). The 2 species known, *L. blattarum* and *L. striata* (bearing epibiotic *Fusiformis*), are endocommensal parasites living in the colon of omnivorous roaches such as *Blatta, Periplaneta,* and *Blatella.* They form resistant cysts insuring host-to-host transmission. Light microscopic studies are found in Janicki (1910), Kudo (1926), Hirschler (1927), and Grassé (1952b). Electron microscopy by Beams and Sekhon (1969) and Hollande and Carruette-Valentin (1972). The genus *Prolophomonas*, created for the species *P. topocola* living in *Cryptocercus punctulatus* (Cleveland et al., 1934), could belong to *Lophomonas* (Grassé, 1952b).

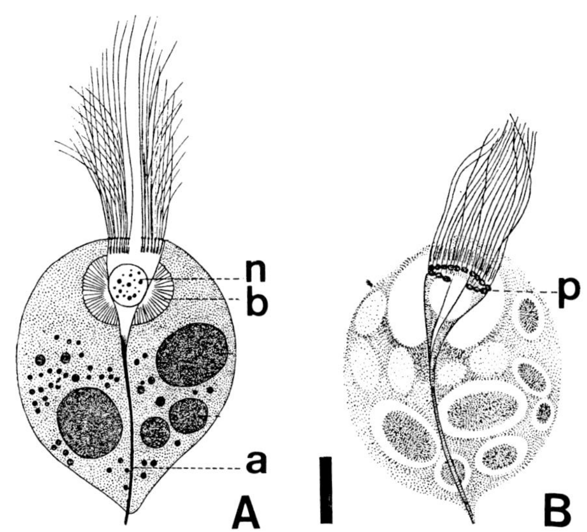

Fig. 40A. *Lophomonas blattarum* with the tuft of flagella, the extra-calycial bodies (b) around the nucleus (n), the axostyle (a) (from Janicki, 1910). Fig. 40B. *L. blattarum* with the crown of parabasal bodies (p) (from Hirschler, 1927). Bar=10 µm.

FAMILY RHIZONYMPHIDAE

Genus ***Rhizonympha*** Grassé & Hollande, 1951

Plasmodial flagellate, variable in size, attached to the intestine of the termite *Anacanthotermes ochraceus.* It contains 300-400 karyomastigonts embedded in the plasmodial cytoplasm. Each karyomastigont comprises an apical tuft of flagella, an axostyle with a calyx enveloping the nucleus, the parabasal bodies, and the flagellar area. The parabasal forms 2 branches applied to the nucleus. When a karyomastigont divides by pleuromitosis, the flagellar area and axostyle regress and reform in each sister mastigont system. Monospecific. Reported in Grassé (1952b). No EM study.
TYPE SPECIES: *Rhizonympha jahieri* Grassé & Hollande, 1951.
TYPE HOST: *Anacanthotermes ochraceus* (Burmeister)

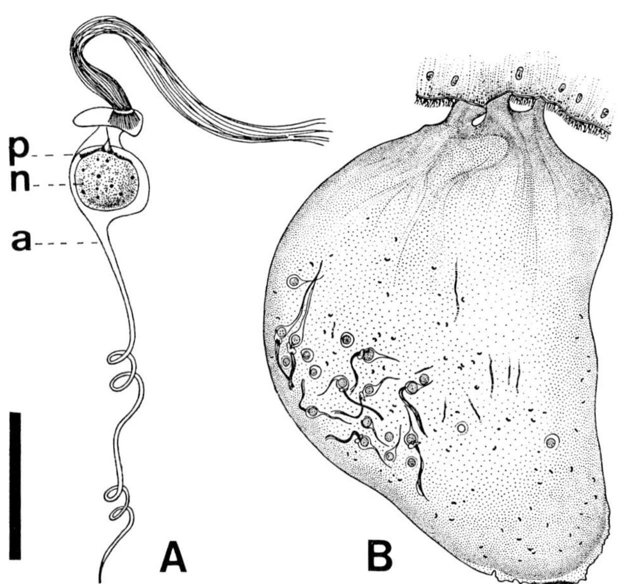

Fig. 41. *Rhizonympha jahieri* from *Anacanthotermes ochraceus.* A. Karyomastigont, B. Plasmodial organism attached to the gut of the termite (from Grassé and Hollande, 1951a). Bar=100 µm.

FAMILY MICROJOENIIDAE

Genus ***Microjoenia*** Grassi, 1892

Flagellates which have different sizes related to their stages of development (3.5-20 µm). Flagella arise around an apical bare pole and are deflected posteriorly; parabasals arranged in a crown around the pole and the anterior nucleus; axostylar trunk tube-like, protruding posteriorly (Fig. 42). Division and morphogenesis not decribed. Xylophagous, occurring in termites of the genus *Reticulitermes.* Brown (1930) distinguished *Microjoenia* from a new genus *Torquenympha* in *Reticulitermes hesperus,* and Grassé (1952b), arguing that *Microjoenia* presents stages and forms like *Torquenympha,* put the latter genus in synonymy with *Microjoenia.* This genus could have been confused with young *Spirotrichonympha,* and more precise studies are desirable to separate the genera. Four species, such as *M. fallax* and *M. hexamitoides* of *Reticulitermes lucifugus,* are described by Grassi (1917), Duboscq and Grassé (1928), and Grassé (1952b); a somewhat different species is *M. axostylis* from *Archotermopsis wroughtoni* (Cutler, 1920); reported in Grassé (1952b). No EM study.
TYPE SPECIES: *Microjoenia hexamitoides* Grassi, 1892
TYPE HOST: *Reticulitermes lucifugus* (Rossi)

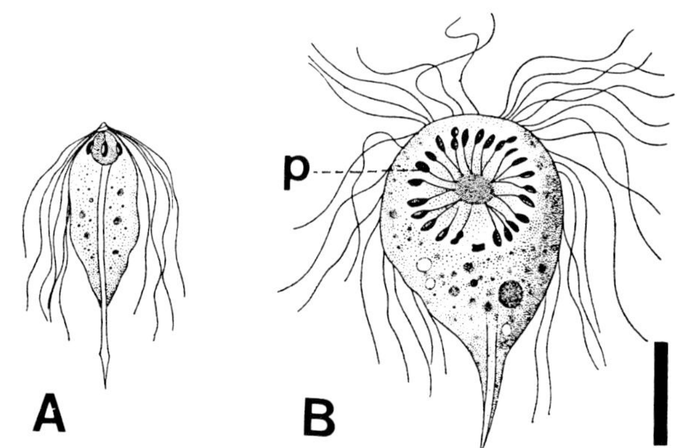

Fig. 42. *Microjoenia fallax* from *Reticulitermes lucifugus.* A. Young stage. B. Apical view of an adult stage, parabasal (p). (from Duboscq and Grassé, 1928, in Grassé [1952b]). Bar=10 µm.

FAMILY KOFOIDIIDAE

Monogeneric family which has the characters of the suborder.

Genus ***Kofoidia*** Light, 1927

Hypermastigid with a spherical shape (60–140 µm) bearing 8–16 anterior cylindrical flagellar bundles or "loricula" arranged in a curved spiral series (Fig. 43). Flagellar area surrounded by an incomplete chromatic collar; several threads or a "suspensorium" connect the base of the flagellar bundles to the flattened anterior face of the nucleus; from the outer edge of the collar, filaments (axostyles ?) run in the cytoplasm toward the rear of the cell; parabasal not identified; nucleus large with a peripheral alveolar zone. At division the nucleus migrates to the opposite side, and the parental mastigont system disintegrates; 2 new mastigont systems arise *de novo,* and nuclear mitosis is a typical pleuromitosis with a large paradesmose. Single species described by Light (1927) from *Kalotermes* (*Paraneotermes*) *simplicicornis* in Arizona; reported in Grassé (1952b).
TYPE SPECIES: *Kofoidia loriculata* Light, 1927.
TYPE HOST: *Paraneotermes* (formerly) *Kalotermes simplicicornis* (Banks).

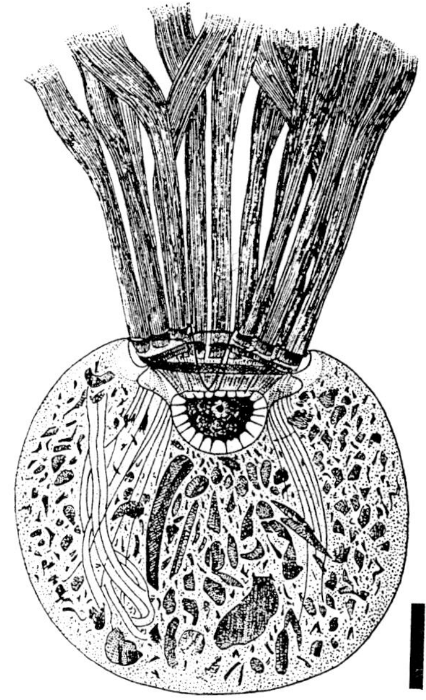

Fig. 43. *Kofoidia loriculata* from *Kalotermes simplicicornis* (after Light, 1927). Bar=10 µm.

FAMILY DELTOTRICHONYMPHIDAE

Genus ***Deltotrichonympha*** Sutherland, 1933

Large hypermastigid (450–500 µm), living in *Mastotermes darwiniensis.* Body broadly triangular in shape with an anterior rostrum completely flagellated and the cell body covered by longitudinal rows of flagella except the amoeboïd posterior area, which is separated by a girdle and used for ingestion (Fig. 44). The nucleus is situated in the anterior region and axostylar fibers surround the nucleus and some of them group to form a central axostyle posterior to the nucleus.

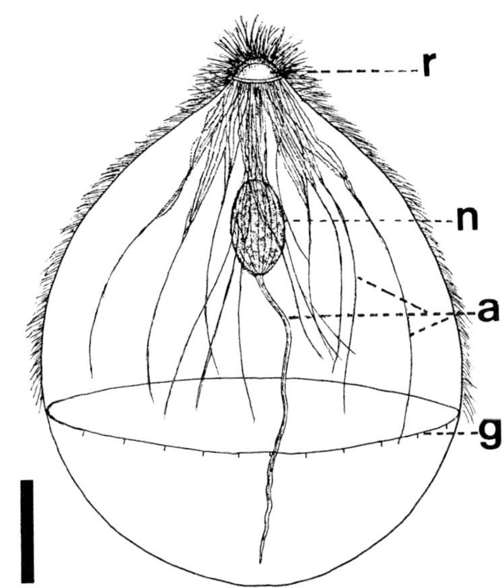

Fig. 44. *Deltotrichonympha operculata* from *Mastotermes darwiniensis;* rostrum (r), nucleus (n), axostyles (a), girdle (g), (after Cleveland, 1966a). Bar=100 µm.

Electron microscopic studies have shown that the flagellar area of the rostrum is composed of long radiating basal bodies. This zone is underlain by several layers of striated fibers connected to basal bodies. There is no rostral tube and cap. The flagellar area posterior to the rostrum is composed of basal bodies prolonged by 3 to 4 short radiating striated roots very characteristic among hypermastigid striated fibers. Inside the rostrum 3 privileged basal bodies have been recognized; they polarize the parabasal fibers, the atractophores, and the axostylar fibers.

Parabasal bodies supported by parabasal fibers are scattered in a central zone anterior to or around the nucleus. At division the whole flagellar area of the rostrum regresses and 2 new areas are probably formed from the numerous basal bodies which have accumulated in a central column from the nucleus to the rostrum. This type of morphogenesis is similar to that of *Joenia* or *Lophomonas*, and these 3 genera also share other characters such as the presence of a tubular axostyle and a single set of privileged basal bodies. Two species: *D. operculata,* the type species, and *D. nana,* a smaller one, occur in *Mastotermes darwiniensis*. Several other species reported under the name *Deltotrichonympha* such as *D. turkestanica,* which have a bilaterally symmetrical rostrum, belong to the genus *Trichonympha* and must be removed from this genus and family. Light microscopy by Sutherland (1933) and Cleveland (1966a,b); electron microscopy by Tamm (1972), Tamm and Tamm (1973a,b), and Hollande (1986a).
TYPE SPECIES: *Deltotrichonympha operculata* Sutherland, 1933.
TYPE HOST: *Mastotermes darwiniensis* Froggatt

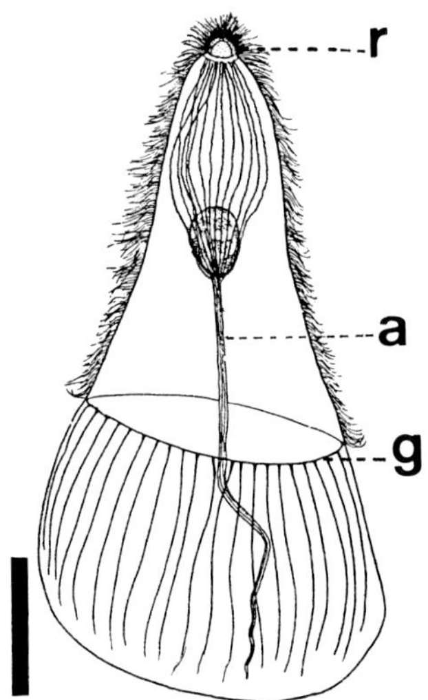

Fig. 45. *Koruga bonita* from *Mastotermes darwiniensis*; rostrum (r), axostyle (a), girdle (g), (after Cleveland, 1966c). Bar=100 µm.

Genus ***Koruga*** Cleveland, 1966

The only species, *Koruga bonita,* from *Mastotermes darwiniensis* has been reported by Cleveland (1966d). It has the same characters as *Deltotrichonympha* and is difficult to distinguish from it. The 2 genera have been intermixed by Tamm (1972) and Tamm and Tamm (1973a); Sutherland (1933) reported only the genus *Deltotrichonympha,* and Hollande (1986a) was unable to distinguish the 2 genera reported by Cleveland (1966a,d). No EM study.
TYPE SPECIES: *Koruga bonita* Cleveland, 1966
TYPE HOST: *Mastotermes darwiniensis*

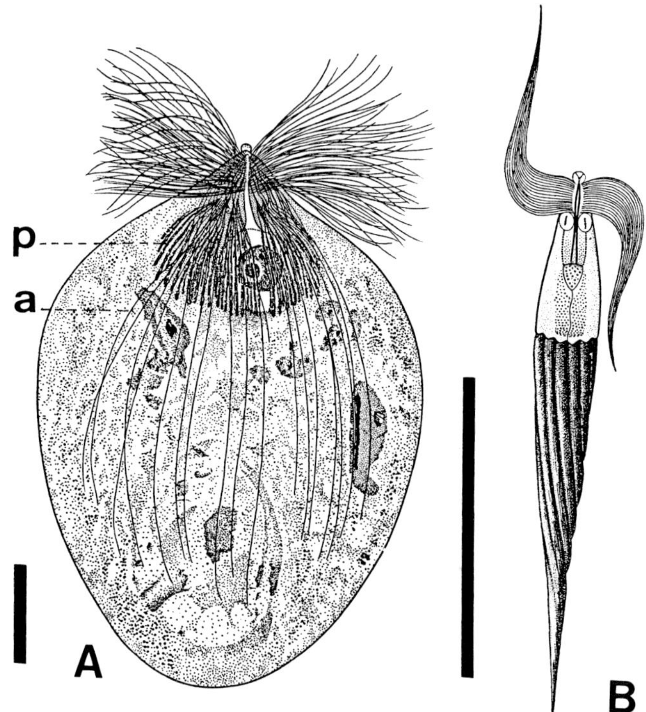

Fig. 46A. *Barbulanympha sp.* from *Cryptocercus punctulatus* with the two flagellar areas, the parabasals (p), and the axostyles (a), (after Cleveland, 1953). Bar=50 µm. Fig. 46B. *Hoplonympha natator* from *Paraneotermes* (formerly *Kalotermes*) *simplicicornis* (after Light, 1926). Bar=50 µm.

ORDER TRICHONYMPHIDA

Hypermastigids bilaterally symmetrical sometimes with a superimposed tetraradiate symmetry. Body composed of rostral and post-rostral zones. Rostrum divided into two hemi-rostra and terminated by an anterior cap. The internal rostral structures are cleaved or well separated by a band of cytoplasm free of flagella. Each hemi-rostrum comprises a set of privileged basal bodies, a flagellar area, a parabasal lamina and fibers, a set of axostylar fibers, and an atractophore. The post-rostral region is bare or flagellated, and flagella are arranged in meridional lines, concentric circles, or spiral lines. Axostylar and parabasal filaments originating in the rostrum penetrate into the post-rostral part of the cytoplasm. Dictyosomes are generally supported by the parabasal fibers or disseminated in the cytoplasm.

FAMILIES HOPLONYMPHIDAE and STAUROJOENINIDAE

These two families are distinguished from others by the absence of a rostral tube.

FAMILY HOPLONYMPHIDAE

Hypermastigid with flagella restricted to the anterior rostrum without a rostral tube. Two symmetrical flagellar areas are separated by cytoplasmic bands. In each flagellar area, longitudinal rows of flagella are not separated by cytoplasmic lanes as in Trichonymphidae. Parabasal fibers do not form a rostral tube. One set of privileged basal bodies is present on the top of each flagellar area and is connected to a set of pelta-axostylar fibers, parabasal fibers, and to one atractophore. Parabasal bodies are short and localized in the nuclear region. Four genera: *Barbulanympha, Hoplonympha, Urinympha, Rhynchonympha.*

Genus ***Barbulanympha*** Cleveland, Hall, Sanders & Collier, 1934

Large hypermastigid (113–200 µm) with a cell body rounded and a flagellated obtuse anterior rostrum terminated by a bare cap (Fig. 46A). The 2 triangular flagellar areas are separated by ectoplasmic bands. Thin parabasal fibers form a lamella lying under the basal bodies. These fibers, beyond the rostrum, support parabasal bodies and are arranged around the nucleus. Axostylar fibers formed by rows of 6–7 microtubules originate from the apex and follow the course of the parabasal fibers, but extend beyond the nucleus. One set of privileged basal bodies is attached on the top of each flagellar area. Two long atractophores are appended to each flagellar area and polarize the external spindle at division. When the 2 hemi-rostra separate, each daughter cell regenerates a new hemi-rostrum and thus symmetry is restored. *Barbulanympha* is the largest hypermastigid in *Cryptocercus punctulatus* and *B. laurabuda* is the type species of the 4 known species. Light microscopy by Cleveland et al. (1934) and Cleveland (1953); EM by Hollande and Valentin (1967) and Hollande and Carruette-Valentin (1971).
TYPE SPECIES: *Barbulanympha laurabuda* Cleveland, Hall, Sanders & Collier, 1934
TYPE HOST: *Cryptocercus punctulatus* Scudder

Genus ***Hoplonympha*** Light, 1926

Cigar-shaped and slender cell body (60–120 µm) which tapers posteriorly (Fig. 46B). The cell seems to have a pellicle thrown up into a series of about 12–15 conspicuous linear ridges (probably elongate bacteria) separated by narrow grooves. The flagella arise from 2 subapical triangular bundles, each bearing about 30 flagella and moving independently. On top there is a bare cap. Two rhizoplasts underlie each flagellar area and join the nucleus. Chromatic granules lie between the flagellar area and the nucleus, which has the shape of an inverted cone. At division the 2 flagellar areas separate, and each reconstitutes its opposite. Monospecific; reported in Grassé (1952b). No EM study.
TYPE SPECIES: *Hoplonympha natator* Light, 1926.
TYPE HOST: *Paraneotermes* (formerly *Kalotermes*) *simplicicornis*

Genus ***Urinympha*** Cleveland, Hall, Sanders & Collier, 1934

Hypermastigid with a spindle-shaped elongated cell body (75–300 µm) and a posterior end tapering to a fine point. The 600 flagella are distributed in 2 areas confined to the anterior

rostrum (Fig. 47A). Unlike *Barbulanympha* and *Rynchonympha,* each flagellar area moves as a unit and independently. A transverse section of the rostrum shows it has a dumb-bell shape with 2 parallel flagellar areas opposed and joined by lobes of ectoplasm. The 2 flagellar areas are triangular, and in longitudinal and perpendicular section they form an angle like a roof. Basal bodies of each flagellar area are underlain by longitudinal parabasal fibers (1.5 µm in diameter). These fibers originate from an anterior lamina and extend slightly beyond the nucleus, bearing parabasal bodies arranged around the nucleus. Two sets of 3 privileged basal bodies are present and polarize 2 atractophores. Two groups of pre-axostylar fibers form an internal dome into the cap of the rostrum. Axostylar fibers (about 24) extend to the middle of the cell. Rod-shaped bacteria covering the cell body. Four species occurring in *Cryptocercus punctulatus,* e.g. *U. talea*. Light microscopy by Cleveland et al. (1934) and Cleveland (1951a); EM by Hollande and Carruette-Valentin (1971).
TYPE SPECIES: *Urinympha Talea* Cleveland, Hall, Sanders & Collier, 1934.
TYPE HOST: *Cryptocercus punctulatus* Scudder

Genus ***Rhynchonympha***
Cleveland, Hall, Sanders & Collier, 1934

Hypermastigid with an elongated cell body (average length 173 µm) and a relatively long and slender rostrum terminated by a cap (Fig. 47B). The 2400 flagella are distributed in 2 opposite flagellar areas. There are as many rows of flagella at the anterior margin of the flagellar area as at the posterior. Organization is very similar to *Urinympha.* Parabasal bodies form a basket-like structure around the nucleus and axostylar fibers extend more than half the body. A nuclear sleeve forms a column from the rostrum toward the nucleus, which it encircles. Non-flagellated surface covered by epibiotic bacteria. Monospecific; from the Pacific coast of the USA; described by Cleveland et al. (1934) and Cleveland (1952); reported in Grassé (1952b). No EM study.
TYPE SPECIES: *Rhynchonympha tarda* Cleveland, Hall, Sanders & Collier, 1934
TYPE HOST: *Cryptocercus punctulatus* Scudder

FAMILY STAUROJOENINIDAE

Hypermastigids with flagellar areas localized in the rostrum. There are four flagellar areas separated by large lobes of ectoplasm. The longitudinal rows of flagella are not separated by ridges of ectoplasm as in Trichonymphidae. Parabasal lamina do not form a rostral tube. A tetra-radial symmetry, which corresponds to the dividing plane of the cell, is superimposed on the bilateral symmetry. Each basic hemi-rostrum has two flagellar areas, two parabasal plates and fibers, two sets of axostylar fibers, but one atractophore. At division each daughter cell receives two flagellar areas corresponding to one hemi-rostrum and reconstitutes the rostral symmetry.

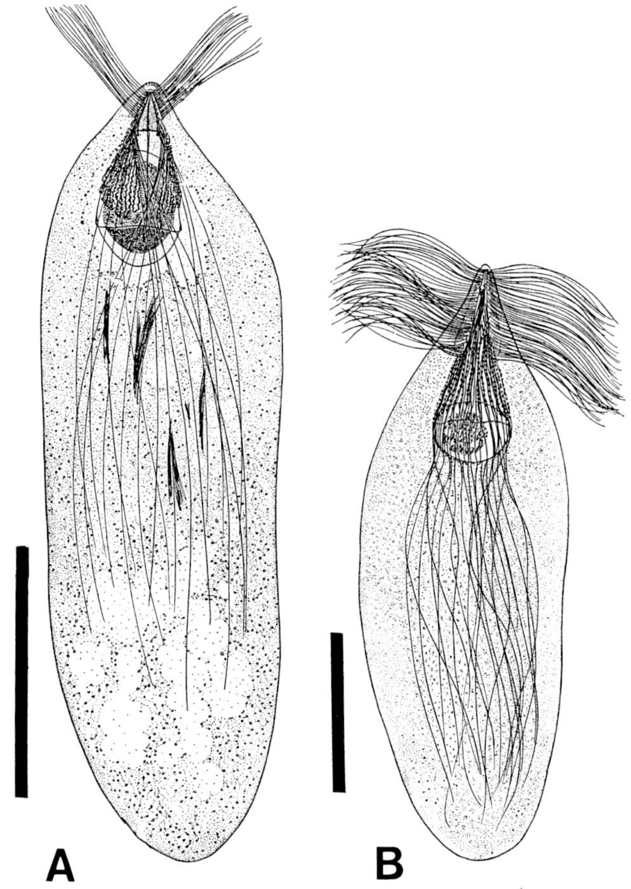

Fig. 47A *Urinympha talea* from *Cryptocercus punctulatus* (after Cleveland, 1951a). Bar=50 µm. Fig. 47B. *Rhynchonympha tarda* from *Cryptocercus punctulatus* (after Cleveland, 1952). Bar=50 µm.

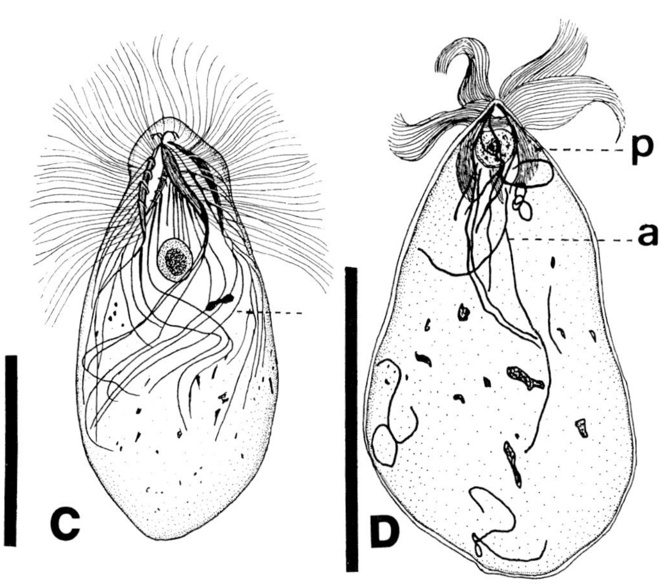

Fig. 47C. *Staurojoenina caulleryi* with four flagellar areas, parabasals (p) from *Postelectrotermes* (formerly *Neotermes*) *praecox* (after Grassé and Hollande, 1947, in Grassé, 1952b). Bar=50 µm. Fig. 47D. *Idionympha perissa* in division, with four flagellar areas, parabasals (p), axostyles (a), from *Cryptocercus punctulatus* (after Cleveland et al., 1934). Bar=50 µm.

Genus ***Staurojoenina*** Kirby, 1926

Hypermastigid with an amphora-shaped cell body (150–170 µm) and with an anterior broad rostrum bearing flagella and terminated by a cap (Fig 47A). The 4 flagellar areas alternate with prominent ectoplasmic lobes. Each flagellar area is composed of 200 flagella arranged in about 15 longitudinal rows, not separated by ectoplasmic ridges. Each flagellar area is underlain by a striated parabasal plate, which divides in a fan-like manner to give 12–14 parabasal fibers. Part of the parabasal fibers support a parabasal body and extend into the posterior cytoplasm. On the top of each flagellar area there is 1 privileged basal body bearing a pre-axostylar fiber from which arise 4 axostylar tubes composed of an inrolled sheet of microtubules. Each privileged basal body is also linked to 1 atractophore. Only the 2 major atractophores (1 in each hemi-rostrum) reach the nucleus and polarize the spindle. Five species or more: *S. caulleryi* lives in *Postelectrotermes* (formerly *Neotermes*) *praecox* of the island of Madeira and has been described by Grassé and Hollande (1947) and Hollande (1986b); *S. assimilis* is described in *Incisitermes* (formerly *Kalotermes*) *minor* by Kirby (1926a); EM study by Hollande and Carruette-Valentin (1971).
TYPE SPECIES: *Staurojoenina mirabilis* (Grassi, 1917)
TYPE HOST: *Epicalotermes* (formerly *Kalotermes*) *aethiopicus* Silvestri

Genus ***Idionympha***
Cleveland, Hall, Sanders & Collier, 1934

Hypermastigid with a pyriform shape (160–275 µm) and an anterior flagellated rostrum. Flagella arising from 4 flagellar areas which are separated by flat ectoplasmic bands (Fig. 47B). In each area a parabasal plate underlies the basal bodies and gives rise to the parabasal fibers. About 90 parabasal bodies surround and extend beyond the nucleus. Several fibers, described as axostyles originate from the rostrum and develop irregularly and variably in the cytoplasm toward the nucleus. Two atractophores originating at the top of the flagellar area meet the nucleus situated at the base of the rostrum and polarize the spindle. At division the hemi-rostra and the 2 flagellar areas separate, and each reconstitutes its symmetry in the daughter cells. Monospecific. No EM study.
TYPE SPECIES: *Idionympha perissa* Cleveland, Hall, Sanders & Collier, 1934
TYPE HOST: *Cryptocercus punctulatus* Scudder

FAMILIES TRICHONYMPHIDAE,
EUCOMONYMPHIDAE, TERANYMPHIDAE

These three families have common characters: flagella are found on the rostrum and over almost all the surface of the cell. In the rostrum, the two flagellar areas are juxtaposed, not separated by broad bands of ectoplasm. Parabasal fibers are associated in two plates to form the rostral tube. Longitudinal rows of flagella are separated by lanes of ectoplasm. One set of privileged basal bodies, independent from those of the flagellar areas; atractophores present or absent in interphasic cells; they form at division.

Dictyosomes are supported by parabasal fibers or free in the cytoplasm. Diverse patterns of pelta-axostylar fibers.

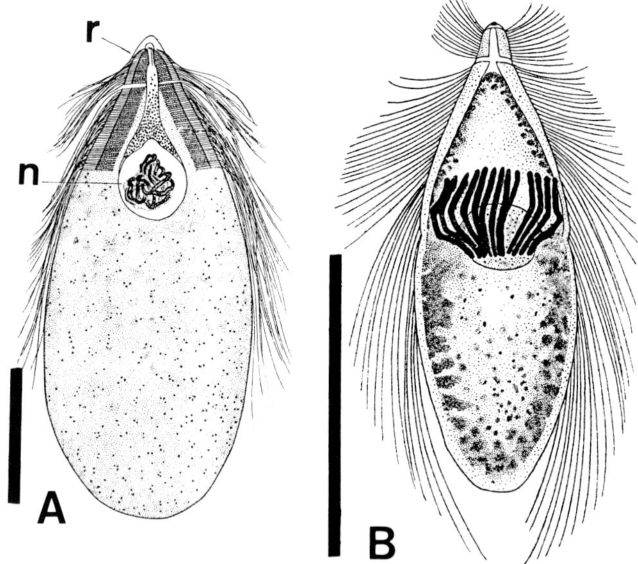

Fig. 48A. *Trichonympha okolona* from *Cryptocercus punctulatus* with the anterior rostrum (r), nucleus (n) (from Cleveland, 1949a). Bar=50 µm. Fig. 48B. *Trichonympha agilis* from *Reticulitermes lucifugus* showing the parabasal bodies arranged around the nucleus (silver staining) (from Duboscq and Grassé, 1933, in Grassé, 1952b). Bar= 50 µm.

FAMILY TRICHONYMPHIDAE

Genus ***Trichonympha*** Leidy, 1877

Hypermastigid with a spindle-shaped body (75–150 µm) in length. Rostral kineties are independent from kineties on the rest of the cell (Figs. 48 A,B). Rostral tube divided into 2 hemi-rostral plates composed of the association of parabasal fibers (Fig. 36a,b,c,d). Parabasal bodies regularly arranged around the nucleus (Figs. 48B,36e). Ultrastructural study has shown 2 privileged basal bodies on the top of the flagellar area of the rostrum and pelta-axostylar rows underlying the membrane of the cap. At division the 2 hemi-rostra separate and reconstitute their symmetry in each daughter cell. Seven species live in *Cryptocercus punctulatus* e.g. *T. okolona* (Cleveland et al., 1934; Cleveland, 1949a) and about 13 species occur in termites such as *T. agilis* in *Reticulitermes lucifugus* (Grassi, 1917; Koidzumi, 1921; Kirby, 1932; Grassé, 1952b; Cleveland, 1960). EM by Grimstone and Gibbons (1966) and Hollande and Carruette-Valentin (1971).
TYPE SPECIES: *Trichonympha agilis* Leidy, 1877
TYPE HOST: *Reticulitermes flavipes* Kollar

FAMILY EUCOMONYMPHIDAE

Radially symmetrical in appearence but bilateral in the rostral tube which is composed of two plates which separate at division as do the two atractophores. All or most of the body covered by flagella. Each flagellar row in the post-rostral part is underlain by parabasal fibers composed of two to several thin striated ribbons.

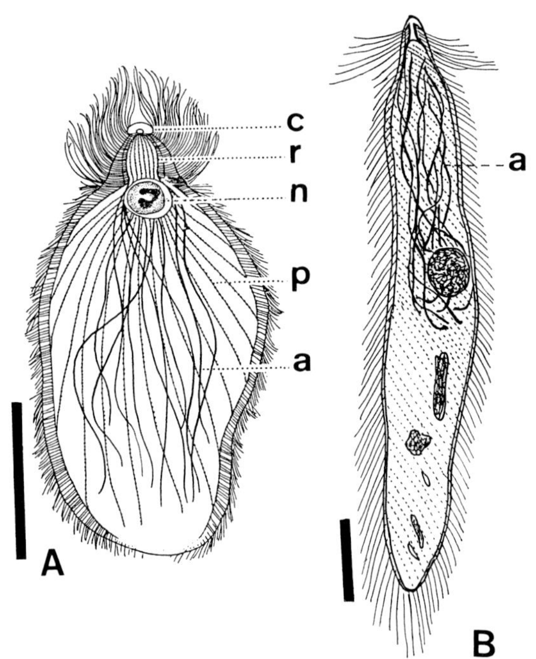

Fig. 49A. *Eucomonympha imla* from *Cryptocercus punctulatus* with the anterior cap (c), rostrum (r), nucleus (n), parabasals (p), and axostyles (a) (from Cleveland, 1950). Bar=50 µm.Fig. 49B. *Pseudo-trichonympha sp.*, from *Coptotermes travians*, axostyle (a), (after Cleveland et al., 1934). Bar=50 µm.

Genus ***Eucomonympha***
Cleveland, Hall, Sanders & Collier, 1934

Hypermastigid with an acorn-shaped body (100–165 µm) living in *Cryptocercus punctulatus.* Anterior rostrum only separated

from the post-rostral portion by a slight constriction (Fig. 49A). Rostrum terminated by a cap not limited by a pelta-axostylar microtubular row and without inner cap. Flagella covering all the body, those of the rostrum longer than others (different than *Trichonympha*). There are as many rows in the rostrum as in the post-rostral portion, but the rows are not continuous. The rostral tube is ampulla-shaped, and the 2 hemi-rostral plates are thin. These plates are striated and divide, giving parabasal fibers, which follow the rows of basal bodies extending into the post-rostral region. The ectoplasm of the cell body contains: long basal bodies, parabasal fibers associated with the proximal end of the basal bodies, dictyosomes between the basal bodies, and numerous hydrogenosomes. Axostylar fibers follow the rows of basal bodies in the rostrum and form long bundles around the nucleus and extend posteriorly in the endoplasm. At division 2 atractophores form and polarize the spindle. The rostrum splits into 2 hemi-rostra, each of which grows to a complete rostrum. Monospecific. Light microscopy by Cleveland et al. (1934) and Cleveland (1950). EM by Hollande and Carruette-Valentin (1971).
TYPE SPECIES: *Eucomonympha imla* Cleveland, Hall, Sanders & Collier, 1934
TYPE HOST: *Cryptocercus punctulatus* Scudder

Genus ***Pseudotrichonympha*** Grassi & Foà,1911

Hypermastigids living in several species of Rhinotermitidae such as *Heterotermes*, *Coptotermes*, *Prorhinotermes*, *Rhinotermes*, and *Schedorhinotermes.* They have a rostrum with long flagella separated from the post-rostral body, which is covered by flagella arranged in longitudinal and slightly spiraled rows. Axostylar filaments are scattered in the endoplasm of the anterior half of the body (Fig 49B). *Pseudotrichonympha* differs from *Eucomonympha* by several characters: the 2 hemi-rostral plates forming the rostral tube are thicker and composed of a thin, internal dense layer and a thick external one in contact with the basal bodies of the rostral area. There is only a thin undulated parabasal ribbon underlying the post-rostral basal bodies; the dictyosomes are scattered in the endoplasm and are not situated between the basal bodies. Light microscopic studies by Grassi (1917), Cleveland et al., (1934), and Grassé and Hollande (1951b). Light and EM studies of *P. grassei* from *Psammotermes* by Hollande and Carruette-Valentin (1971).
TYPE SPECIES: *Pseudotrichonympha hertwigi* (Hartmann, 1910)
TYPE HOST: *Coptotermes hartmanni* Holmgren

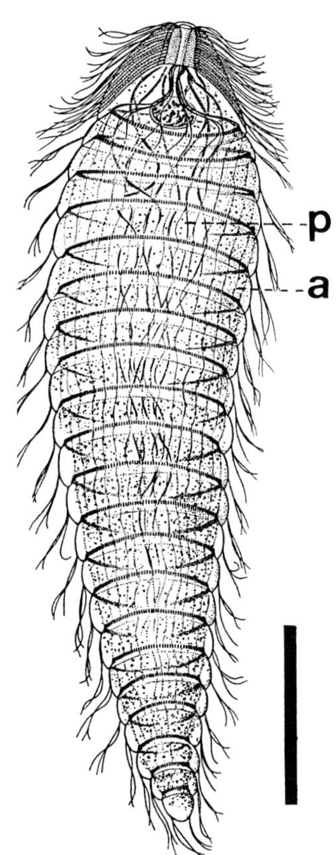

Fig. 50. *Teranympha mirabilis* from *Reticulitermes speratus*, axostyles (a), parabasals (p), (from Cleveland, 1938a). Bar=50 µm.

FAMILY TERANYMPHIDAE

Genus ***Teranympha*** Koidzumi, 1917

Teranympha (=*Teratonympha, Cyclonympha*) *mirabilis* is a large hypermastigid (90–270 µm) with a body divided into 2 regions: the rostrum and the post-rostral part. The rostrum is crowned by a low cap and covered by about 40 longitudinal rows of flagella. The post-rostral area is flagellated, but flagella are inserted on parallel circles separated by ectoplasmic bands. The rostral tube is broad and composed of an outer lamella adjacent to the rows of basal bodies. At division it splits in 2, forming 2 semi-circular rostra; as they grow, they become circular and complete as in *Pseudotrichonympha*. Nuclear sleeve highly developed and extended around the nucleus. Nuclear supporting strands (14–16) arise from the base of the rostral tube, join the nucleus, and bifurcate toward the periphery of the ectoplasm; they could be related to axostyles. Another set of 6–8 fibers originating from the base of the rostrum develop in the central cytoplasm and could be parabasals. Post-rostral flagellar rings (16–24) seem to have no connections, thus differing from the Spirotrichosomidae; they give the cell body its segmented appearance. No permanent atractophores; they develop only at mitosis. Light microscopy of *T. mirabilis* of *Reticulitermes speratus* in Japan by Koidzumi (1921) and Cleveland (1938a). No EM study.

TYPE SPECIES: *Teranympha mirabilis* Koidzumi, 1917

TYPE HOST: *Reticulitermes speratus* (Kobe)

FAMILY SPIROTRICHOSOMIDAE

The rostrum looks like that of Trichonymphidae, but flagella are in two bands lying side by side. Post-rostral flagellar rows are spiraled as in Spirotrichonymphidae, but in a clockwise direction. Parabasal threads and parabasal bodies are in relation to the flagellar bands. Axostyles are long, thin, and free, occasionally grouped in bundles.

Genus ***Macrospironympha***
Cleveland, Hall, Sanders & Collier, 1934

Large hypermastigid (110–160 μm) with a rostrum separated from the cell body. Flagella inserted in 2 broad flagellar bands arising on the top of the rostrum and continuing in 10–20 spirals on the body. There are 8 rows of flagella in a band on the body and 16 rows in the rostrum. The rostrum is composed of 2 rostral plates, not exactly opposed (somewhat S-shaped in transverse section), which form a rostral tube. Each rostral plate gives rise to parabasal fibers, which underlie the flagellar bands in the cell body. Dictyosomes are adjacent to this structure. Axostylar fibers (30–50) originate from the rostrum and develop in the central endoplasm beyond the middle of the body. At division the nucleus migrates posteriorly, carrying the 2 atractophores (centrioles) and divides by pleuromitosis. The 2 hemi-rostra separate and each reconstitutes a small daughter flagellar band and a complete rostrum; finally, each nucleus moves again toward its rostrum. Encystation and sexual processes occur in relation to the molting of the roach. Monospecific from the Pacific coast of the USA. Light microscopy in Cleveland et al. (1934) and Cleveland (1956b); EM in Hollande and Carruette-Valentin (1971).

TYPE SPECIES: *Macrospironympha xylopletha* Cleveland, Hall, Sanders & Collier, 1934

TYPE HOST: *Cryptocercus punctulatus* Scudder

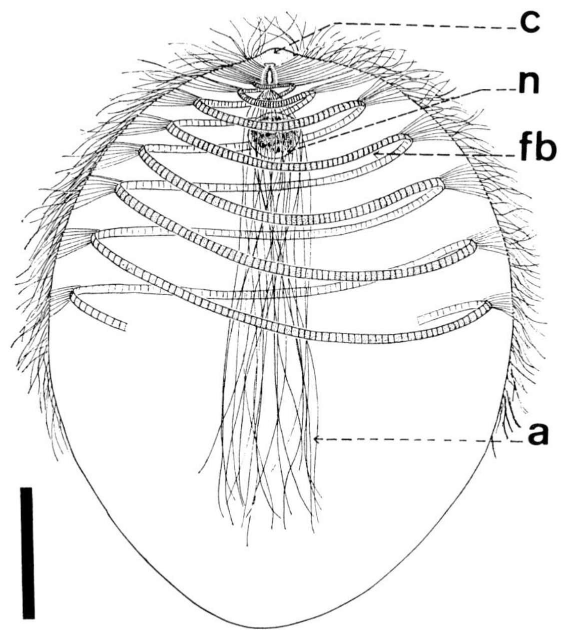

Fig. 51. *Macrospironympha xylopletha* with anterior cap (c), nucleus (n), two flagellar bands (fb), axostyles (a), from *Cryptocercus punctulatus* (after Cleveland, 1956b). Bar=50 μm.

Genus ***Leptospironympha*** Cleveland, Sanders, Hall & Collier, 1934

Species vary in size (53–250 µm) with a rostrum relatively longer than *Macrospironympha.* The 2 flagellar bands are underlain by parabasal fibers and parabasal bodies (Fig. 52). During division each centriole follows each hemirostrum and continuing flagellar band for a short distance, but the nucleus does not migrate posteriorly. Sexuality present. Three species occur in *Cryptocercus punctulatus*, e.g. *L. wachula* (Cleveland, 1951b); 6 other species have been described in the termite *Stolotermes* from Australia (Cleveland and Day, 1958). Partial EM study in Hollande and Carruette-Valentin (1971).
TYPE SPECIES: *Leptospironympha eupora* Cleveland, Sanders, Hall & Collier, 1934
TYPE HOST: *Cryptocercus punctulatus* Scudder

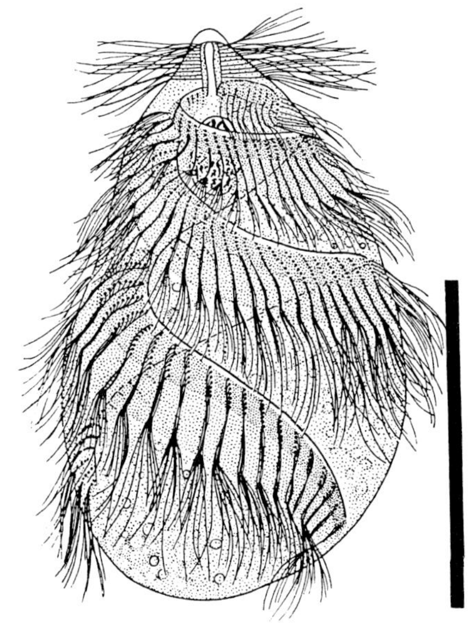

Fig. 52. *Leptospironympha wachula* from *Cryptocercus punctulatus* (from Cleveland, 1951b). Bar=50 µm.

Genus ***Spirotrichosoma*** Sutherland, 1933

Species vary in size (55–395 µm); with 2 types of flagellar bands, primary and secondary. The 2 parabasal bands, originating from the rostrum, form 2 large spirals supporting the flagellar rows in the anterior part of the body. These bands become thinner and continue in the posterior part of the body. Parabasal bodies are in relation to these bands; axostyles form long fibrils in the endoplasm. Nucleus situated close to the rostrum. At division, secondary bands and flagella disappear and primary bands unwind. Asexual multiplication. They occur in all *Stolotermes;* 9 species described in *Stolotermes* of Australia. Light microscopic studies by Sutherland (1933), Helson (1935), and Cleveland and Day (1958). No EM study.
TYPE SPECIES: *Spirotrichosoma capitatum* Sutherland, 1933.
TYPE HOST: *Stolotermes victoriensis* Hill

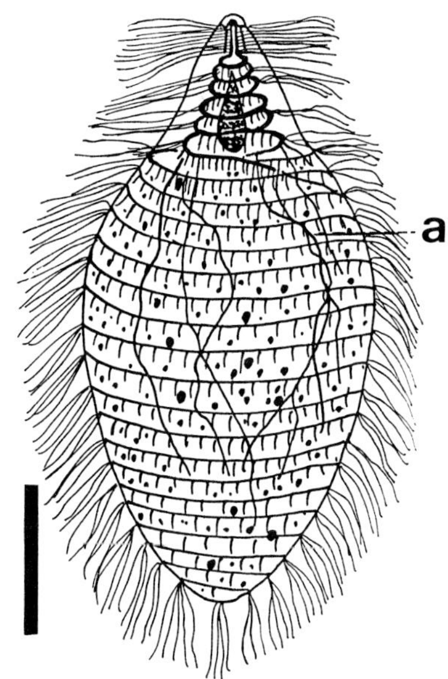

Fig. 53. *Spirotrichosoma capitatum* from *Stolotermes ruficeps,* axostyles (a) (after Cleveland and Day, 1958). Bar=50 µm.

Genus ***Apospironympha*** Cleveland & Day, 1958

Hypermastigid (58–180 µm) with 2 types of flagellar bands, primary and secondary. The 2 flagellar bands, originating from the rostrum, form 2 spirals in the anterior region of the body. Thin longitudinal secondary bands bearing the flagella arise from the anterior part of these bands and extend to the posterior. Parabasal bodies not exactly situated and long axostylar fibrils present in the central cytoplasm. Four species in *Stolotermes africanus*. Light microscopy by Cleveland and Day (1958). No EM study.
TYPE SPECIES: *Apospironympha lata* Cleveland & Day, 1958

TYPE HOST: *Stolotermes africanus* Emerson

Genus ***Colospironympha***
Cleveland & Day, 1958

Hypermastigid (60–120 µm) with 2 types of flagellar bands, primary and secondary. The 2 flagellar bands originate from the rostrum, form 2 spirals in the anterior region of the body. Secondary thin longitudinal bands arise from them which subdivide. Secondary bands bear flagella in regularly spaced segments and reach the posterior end. Parabasal bodies not recognized; axostylar fibrils are in the central cytoplasm. Asexual reproduction only. Three species occurring in *Stolotermes africanus*. Light microscopic study by Cleveland and Day (1958). No EM study.
TYPE SPECIES: *Colospironympha cateia* Cleveland & Day, 1958
TYPE HOST: *Stolotermes africanus* Emerson

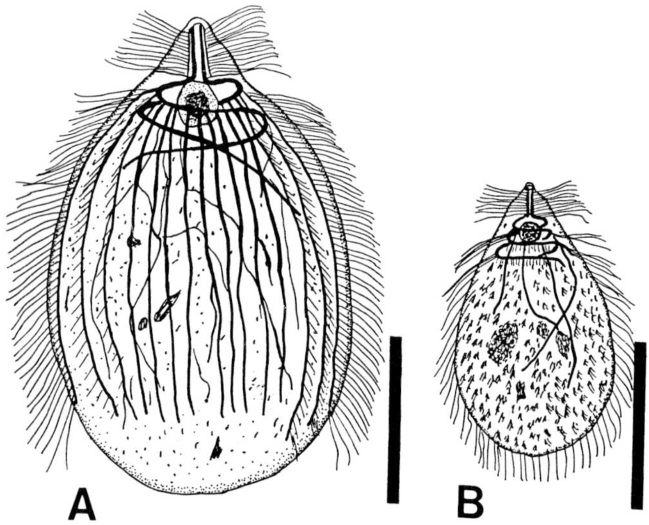

Fig. 54A. *Apospironympha lata* from *Stolotermes africanus* (after Cleveland and Day, 1958). Bar=50 µm. Fig. 54B. *Colospironympha cateia* from *Stolotermes africanus* (after Cleveland and Day, 1958). Bar=50 µm.

ORDER SPIROTRICHONYMPHIDA

Spirotrichonymphids have two or more flagellar rows composed of basal bodies and associated parabasal fibers. The autonomous flagellar rows pursue a counterclockwise course around the body. There is no typical rostrum, but the flagellate has a basic bilateral symmetry and has 2 groups of flagellar rows, which separate in 2 at division. Parabasal bodies (dictyosomes) form patches lining the flagellar rows. The polar basal body in each flagellar row is privileged and bears a preaxostyle at the origin of the pelta-axostylar complex. Axostylar rows pass around the nucleus and continue independently, or group in a bundle. At division, 2 flagellar rows on each side of the cleavage plane play the role of an atractophore in polarizing the spindle. Three families, Spirotrichonymphidae, Holomastigotoididae, and Holomastigotidae (Hollande and Carruette-Valentin, 1971).

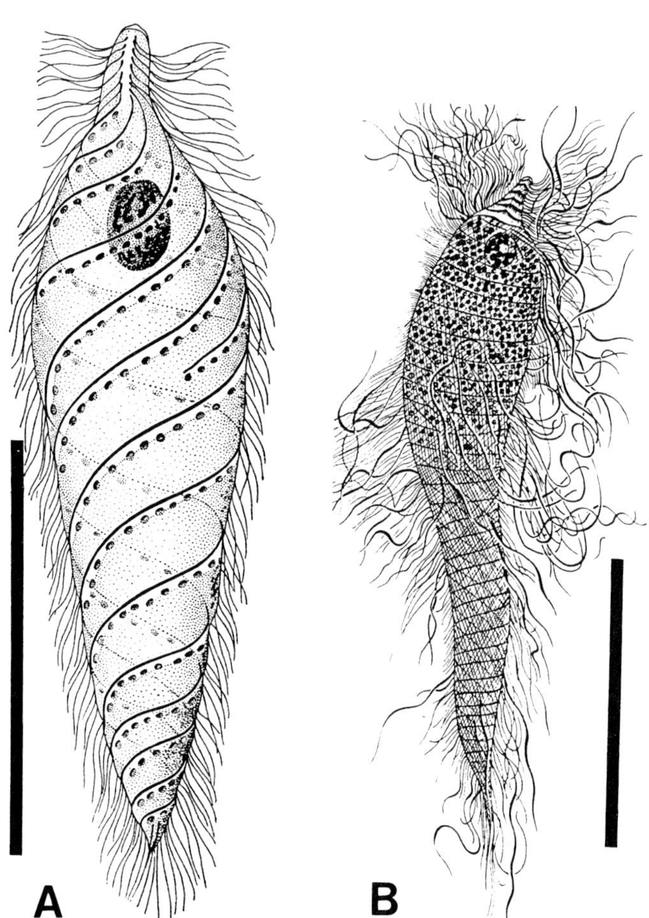

Fig. 55A. *Spirotrichonympha flagellata* from *Reticulitermes lucifugus* (after Duboscq and Grassé, 1933, in Grassé, 1952b). Bar=50 µm. Fig. 55B. *Spirotrichonymphella pudibunda* from *Porotermes adamsoni* (from Grassi, 1917, in Grassé, 1952b). Bar=50 µm.

FAMILY SPIROTRICHONYMPHIDAE

Comprses genera with a pseudo-rostrum containing a central columella. Parabasal striated lamina sandwiches the proximal part of the basal bodies, which are slightly interconnected by microfibrils and underlain by dictyosomes. At division, flagellar rows separate equally, or not (Cleveland, 1938b). Xylophagous flagellates, living in the termite genera: *Reticulitermes, Anacanthotermes, Hodotermes, Heterotermes, Coptotermes, Schedorhinotermes, Paraneotermes, Prorhinotermes, Psammotermes, Postelectrotermes,* and *Porotermes.* Comprises 3–4 genera *Spirotrichonympha, Spirotrichonymphella, Spironympha,* and *Micromastigotes.*

Genus ***Spirotrichonympha***
Grassi & Foà,1911

Spirotrichonymphids with a spindle-shaped body (11–200 µm) and a rostrum with a columella (Fig. 55A). Spiraled flagellar rows, in some reaching to the posterior end, which are underlain by dot-shaped dictyosomes. Axostylar fibrils form 1 bundle posterior to the nucleus to a terminal projection. Among 11 species described, *S. flagellata* occurs in *Reticulitermes lucifugus, S. decipiens* in *Anacanthotermes ochraceus*, *S. fallax* in *Coptotermes sjöstedti,* and *S. bispira* and *S. polyspira* in *Paraneotermes* (formerly *Kalotermes*) *simplicicornis* (Cleveland, 1938b). Light microscopic studies in Grassi (1917), Duboscq and Grassé (1928, 1933, 1943) Cleveland (1958b), Grassé (1952b); EM study in Hollande and Carruette-Valentin (1971), Radek (1997).
TYPE SPECIES: *Spirotrichonympha flagellata* Grassi, 1892)
TYPE HOST: *Reticulitermes lucifugus* (Rossi)

Genus ***Spirotrichonymphella*** Grassi, 1917

Spirotrichonymphids with a spindle-shaped body (40–100 µm) and 4 flagellar rows reaching the posterior end. Microtubules of the axostyle run into the columella and do not extend beyond the nucleus; no axostylar trunk described. A parabasal lamina underlies the flagellar rows. One species, *S. pudibunda,* occurs in *Porotermes adamsoni* from Australia (Grassi, 1917) and another, *S. psammotermitidis,* symbiotic in *Psammotermes hybostoma,* has been described by light and electron microscopy by Hollande and Carruette-Valentin (1971).
TYPE SPECIES: *Spirotrichonymphella pudibonda* Grassi, 1917
TYPE HOST: *Porotermes adamsoni* (Froggatt)

Genus ***Spironympha*** Koidzumi, 1921
(incertae sedis)

Flagellates with a spindle-shaped body and spiraled flagellar rows which do not reach the posterior end. Tubular axostyle posterior to the nucleus. Koidzumi (1921) created another genus, *Microspironympha,* which is a synonym of *Spironympha,* for similar flagellates occurring in *Reticulitermes speratus* from Japan (Grassé, 1952b). These flagellates are difficult to distinguish from young *Spirotrichonympha* present in the same host and need more investigation.

Fig. 56. *Micromastigotes psammotermitidis* from *Postelectrotermes* (formerly *Kalotermes*) *praecox*, different stages of development with an axostyle and several flagellar rows, (after Hollande and Carruette-Valentin, 1971). Bar=10 µm

Genus ***Micromastigotes***
Hollande & Carruette-Valentin, 1971
(*incertae sedis*)

The species *M. grassei* lives in *Postelectrotermes* (formerly *Kalotermes) praecox* of the island of Madeira. Flagellates vary in size (10–30 µm long) and may have a pseudo-rostrum. They have

1 to 6 spiraled flagellar rows which do not reach the posterior end and large subjacent dictyosomes (Fig. 56). Flagella are associated with the plasma membrane in their proximal portion (hemi-desmosome). It has similar features to other spirotrichonymphines: *Spirotrichonympha, Holomastigotoides,* and particularly to *Holomastigotes,* but has a stout axostylar trunk projecting at the posterior. Ultrastructural study shows that the pelta-axostyle microtubular rows form an anterior cape; the flagellar rows are organized as in *Holomastigotoides* except for the absence of the microfibrillar bundle lining each row of basal bodies. Axostylar ribbons originating anteriorly enclose the nucleus to form the axostylar trunk. These features could correspond to *Spironympha* (=*Microspironympha* (Koidzumi, 1921). Light and EM by Hollande and Carruette-Valentin (1971).

TYPE SPECIES: *Micromastigotes grassei* Hollande & Carruette-Valentin, 1971

TYPE HOST: *Postelectrotermes* (formerly *Kalotermes*) *praecox*

FAMILY HOLOMASTIGOTOIDIDAE

Pseudo-rostrum absent, or when present, lacks a columella; no anterior cap; however, the genus *Rostronympha* develops a proboscis. All the flagellar rows do not arise from the apex and do not reach the posterior end. Basal bodies of a row are interconnected by a thin fibrillar bundle. A parabasal striated fiber and an axostylar fiber underlie each flagellar row, and dictyosomes are subjacent. At division, flagellar rows are equally separated. Chromosomes apparent during the interphase. Xylophagous, occurring in Psammotermitidae, Hodotermitidae, and Rhinotermitidae (Grassé, 1952b; Cleveland, 1949b; Hollande and Carruette-Valentin, 1971).

Genus ***Holomastigotoides*** Grassi & Foà, 1911

Large (80–180 µm) spirotrichonymphids with an anterior pole generally spiraled and bare, from which originate the helical flagellar rows which may reach to the posterior end. Basal bodies interconnected by fibrillar desmoses and by a lateral bundle of microfibrils. A parabasal lamina, a lane of amorphous material, and axostylar ribbons underlie each flagellar row. Spot-shaped dictyosomes situated between the rows (about 1 dictyosome per 7-8 basal bodies). Among the 5 species *H. hemigymnum* is a symbiont of *Coptotermes sjöstedti*; *H. tusitala* and *H. diversa* occur in *Prorhinotermes.* Light microscopic study in Grassi (1917), Mackinnon (1926), Grassé and Hollande (1945), and Grassé (1952b); division in Cleveland (1949b) and Grassé and Hollande (1963); EM in Hollande and Carruette-Valentin (1971), Lingle and Salisbury (1995).

TYPE SPECIES: *Holomastigotoides hertwigi* Hartmann, 1910

TYPE HOST: *Coptotermes hartmanni* Holmgren

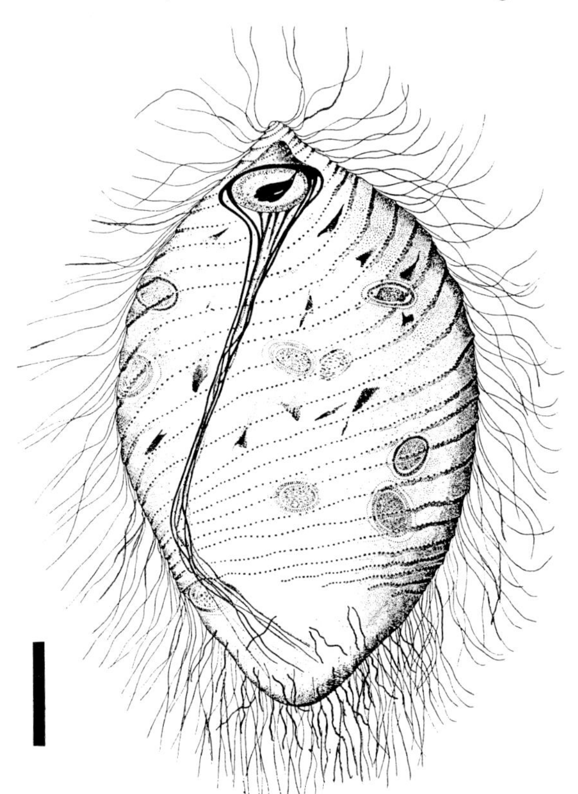

Fig. 57. *Holomastigotoides hemigymnun* of *Copotermes sjostedti* (from Grassé and Hollande, 1945, in Grassé, 1952). Bar=20 µm.

Genus ***Rostronympha***
Duboscq, Grassé & Rose, 1937

Large (135–180 µm) spirotrichonymphid very similar in structure to *Holomastigotoides* except for the retractile proboscis, which attaches the

flagellate to the host's gut wall (Fig. 58). The flagellar rows do not reach the posterior end, which is covered by spirochetes. Monospecific. Light microscopic study by Duboscq and Grassé (1943), Grassé (1952b), and Grassé and Hollande (1963). No EM study.
TYPE SPECIES: *Rostronympha magna* Duboscq, Grassé, & Rose, 1937
TYPE HOST: *Anacanthotermes*

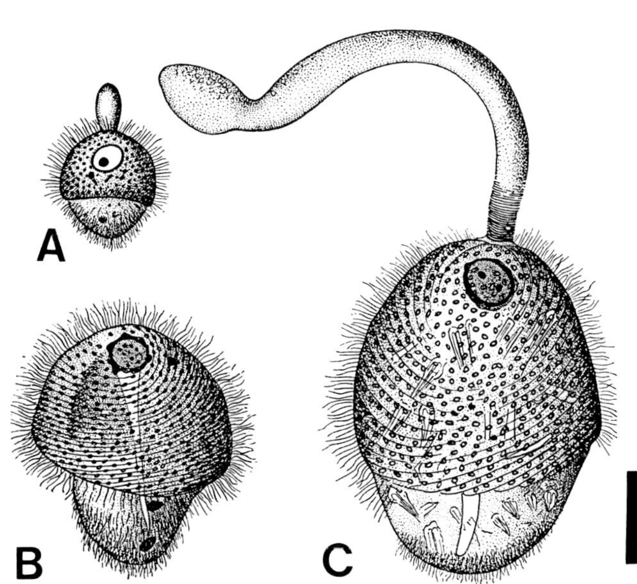

Fig. 58. *Rostronympha magna* of *Anacanthotermes ochraceus*. A. Young stage. B. Advanced stage without proboscis. C. Large organism with extended proboscis (from Duboscq and Grassé, 1943, in Grassé, 1952b). Bar=20 µm.

FAMILY HOLOMASTIGOTIDAE

Small variable-sized spirotrichonymphids with flagellar rows originating from the apex and progressing helically in a counterclockwise direction. No rostrum; no axostyle described; nucleus anterior; dictyosomes under or between the flagellar rows. One genus *Holomastigotes.*

Genus ***Holomastigotes*** Grassi, 1892

Variable in size; adults are 30 µm long, with a flat leaf-like body and 4–5 prominent flagellar rows, which always reach the posterior end. Dictyosomes scattered under or between the flagellar rows. Osmotrophic, living in the rectal pocket of *Reticulitermes, Anacanthotermes,* and *Macrohodotermes*, e.g. *H. elongatum.* Light microscopy by Grassi (1917), Koidzumi (1921), Duboscq and Grassé (1933), and Grassé (1952b). No EM study.
TYPE SPECIES: *Holomastigotes elongatum* Grassi, 1892
TYPE HOST: *Reticulitermes lucifugus* (Rossi)

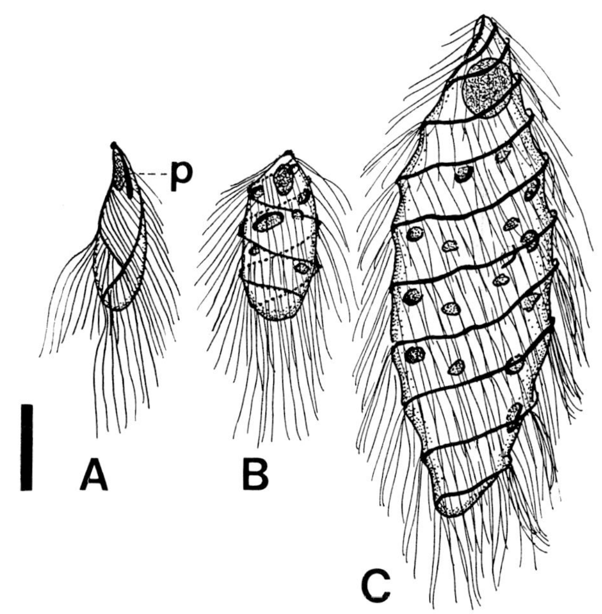

Fig. 59. *Holomastigotes elongatum* from *Reticulitermes lucifugus.* A. Stage with two flagellar rows and one parabasal (p). B. Stage with two flagella rows and seven parabasals; C. Adult stage (after Duboscq and Grassé, 1925, in Grassé, 1952b). Bar=10 µm.

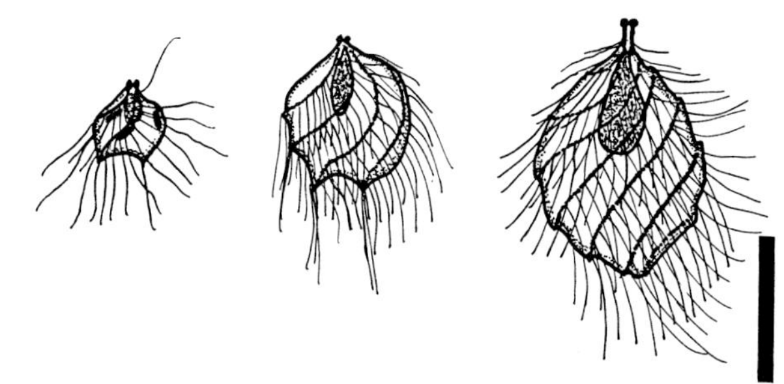

Fig. 60. *Spiromastigotes rosei* from *Anacanthotermes ochraceus* at different stages of development with three, five, and six flagella (after Duboscq and Grassé, 1943, in Grassé, 1952b). Bar=10 µm.

Genus *Spiromastigotes* Duboscq & Grassé, 1943 (*Incertae sedis*)

Flagellates vary in size (10–20 µm) with flagellar rows spiraled (1 turn) and originating from an apical disc or from 2 parallel small rods. Nucleus attached to the apex; no axostyle. Light microscopy by Duboscq and Grassé (1943) and reported in Grassé (1952b). No EM study.
TYPE SPECIES: *Spiromastigotes rosei* Duboscq & Grassé, 1943
TYPE HOST: *Anacanthotermes ochraceus* (Burmeister)

Literature Cited

Amos, W. B., Grimstone, A. V. & Rothschild, L. J. 1979. Structure, protein composition and birefringence of the costa: a motile flagellar root fiber in the flagellate *Trichomonas*. *J. Cell Sci.*, 35:139-164.

Alexeieff, A. G. 1910. Sur les flagellés intestinaux des poissons marins. *Arch. Zool. Exp. Gén.*, Paris, VI: série 5, I-XX.

Beams, H. W. & Sekhon, S. S. 1969. Further studies on the fine structure of *Lophomonas blattarum* with special reference to the so-called calyx, axial filament, and parabasal body. *J. Ultrastruct. Res.*, 26:296–315.

Bêlar, K., (ed.) 1926. *Der Formwechsel der Protistenkerne*. G. Fischer, *Jena*. pp 39–40

Bishop, A. 1939. A note upon the systematic position of "*Trichomonas*" *keilini* (Bishop, 1935). *Parasitology*, 31:469-472.

Boykin, M. S., Stockert, L., Buhse Jr., H. E. & Smith-Sommerville, H. E. 1986. *Trichomitus trypanoides* (Trichomonadida) from the termite *Reticulitermes flavipes*. II. Fine structure and identification of the cloned flagellate. *Trans. Am. Microsc. Soc.*, 105:223-238.

Brown E. 1930. Hypermastigote flagellates from the termite *Reticulitermes*: *Torquenympha octoplus* gen. nov., sp. nov., and two new species of *Microjoenia*. *Univ. Calif. Publ. Zool.*, 36:67–80.

Brugerolle, G. 1971. Ultrastructure du genre *Trichomitus* Swezy 1915, Zooflagellata Trichomonadida. *Protistologica*, 7:171-176.

Brugerolle, G. 1973. Sur l'existence de vrais kystes chez les trichomonadines intestinales. Ultrastructure des kystes de *Trichomitus batrachorum* Perty 1852, *Trichomitus sanguisugae* Alexeieff 1911, et *Monocercomnas tipulae* Mackinnon 1910. *C. R. Acad. Sci.*, Paris, 277:2193-2196.

Brugerolle, G. 1975. Etude de la cryptopleuromitose et de la morphogénèse de division chez *Trichomonas vaginalis* et chez plusieurs genres de trichomonadines primitives. *Protistologica*, 11:457-458.

Brugerolle, G. 1975-76. Cytologie ultrastructurale, systématique et évolution des Trichomonadida. *Ann. Stat. Biol. de Besse-en-Chandesse*, 10:1-90.

Brugerolle, G. 1980. Etude ultrastructurale du flagellé *Protrichomonas legeri* (Léger 1905) parasite de l'estomac des Bogues (*Box boops*). *Protistologica*, 16:253-258.

Brugerolle, G. 1981. L'ultrastructure de la membrane ondulante différencie 2 espèces de *Tritrichomonas* des rongeurs correspondant à *T. muris* (Grassi) et à *T. minuta* (Wenrich, 1924). *Protistologica*, 17:431-438.

Brugerolle, G. 1986a. Séparation des genres *Trimitus* (Diplomonadida) et *Tricercomitus* (Trichomonadida) d'après leur ultrastructure. *Protistologica*, 22:31-37.

Brugerolle, G. 1986b. Structural diversity of trichomonads as the basis for systematic and evolutionary considerations. *Acta Univ. Carol. Biol.*, Prague, 30:199-210.

Brugerolle, G. 1991a. Flagellar and cytoskeletal systems in amitochondrial flagellates: Archamoebae, Metamonada and Parabasala. *Protoplasma*, 164:70-90.

Brugerolle, G. 1991b. Cell organization in free-living amitochondriate heterotrophic flagellates. *In:* Patterson, D. J. & Larsen, J. (eds.). *The Biology of Free-living Heterotrophic Flagellates*. The Systematics Association, Clarendon Press, Oxford, 45:133-158.

Brugerolle, G., Gobert, J. G. & Savel, J. 1974. Etude ultrastructurale des lésions viscérales provoquées par l'injection intrapéritonéale de *Trichomonas vaginalis* chez la souris. *Ann. Parasit. Hum. Comp.*, 49:301-318.

Brugerolle, G., Breunig, A. & König, H. 1994. Ultrastructural study of *Pentatrichomonoides*, a trichomonad flagellate from *Mastotermes darwiniensis*. *Eur. J. Protistol.*, 30:372-378.

Buttrey, B. W. 1954. Morphological variations in *Tritrichomonas augusta* (Alexeieff) from Amphibia. *J. Morphol.*, 94:125-164.

Camp, R. R., Mattern, C. F. T. & Honigberg, B. M. 1974. A study of *Dientamoeba fragilis* Jepps & Dobell. I. Electron microscopic observation of binucleate stages. II. Taxonomic position and revision of the genus. *J. Protozool.*, 21:69-82.

Cleveland, L. R. 1928. *Tritrichomonas fecalis* nov. sp. of man; its ability to grow and multiply indefinitely in faeces diluted with tapwater and in frogs and tadpoles. *Am. J. Hyg.,* 8:232-255.

Cleveland, L. R. 1938a. Morphology and mitosis of *Teranympha. Arch. Protistenkd.,* 91:441–451.

Cleveland, L. R. 1938b. Longitudinal and transverse division in two closely related flagellates. *Biol. Bull.,* 74:1–40.

Cleveland, L. R. 1949a. Hormone-induced sexual cycles of flagellates. I. Gametogenesis, fertilization, and meiosis in *Trichonympha. J. Morphol.,* 85:197–295.

Cleveland, L. R. 1949b. The whole cell cycle of chromosomes and their coiling systems. *Trans. Am. Philos. Soc.,* 39:1–100.

Cleveland, L. R. 1950. Hormone-induced sexual cycles of flagellates. V. Fertilization in: *Eucomonympha. J. Morphol.,* 87:349–368.

Cleveland, L. R. 1951a. Hormone-induced sexual cycles of flagellates. VII. One-division meiosis and autogamy without cell division in *Urinympha. J. Morphol.,* 88:365–440.

Cleveland, L. R. 1951b. Hormone-induced sexual cycles of flagellates. VI. Gametogenesis, fertilization, meiosis, oöcysts, and gametocysts in *Leptospironympha. J. Morphol.,* 88:199–244.

Cleveland, L. R. 1952. Hormone-induced sexual cycles of flagellates. VIII. Meiosis in *Rhynchonympha* in one cytoplasmic and two nuclear divisions followed by autogamy. *J. Morphol.,* 91:269–324.

Cleveland, L. R. 1953. Hormone-induced sexual cycles of Flagellates. IX. Haploid gametogenesis and fertilization in *Barbulanympha. J. Morphol.,* 93:371–403.

Cleveland, L. R. 1956a. Brief accounts of sexual cycles of the flagellates of *Cryptocercus. J. Protozool.,* 3:161–180.

Cleveland, L. R. 1956b. Hormone-induced sexual cycles of flagellates. XIV. gametic meiosis and fertilization in *Macrospironympha. Arch. Protistenkd.,* 101:99–169.

Cleveland, L. R. 1957. Correlation between the molting period of *Cryptocercus* and sexuality in its Protozoa. *J. Protozool.,* 4:168–175.

Cleveland, L. R. 1958a. A factual analysis of chromosomal movement in *Barbulanympha. J. Protozool.,* 5:47–62.

Cleveland, L. R. 1958b. Movement of chromosomes in *Spirotrichonympha* to centrioles instead of the ends of central spindle. *J. Protozool.,* 5:63–68.

Cleveland, L. R. 1960. The centrioles of *Trichonympha* from termites and their functions in reproduction. *J. Protozool.,* 7:326–341.

Cleveland, L. R. 1961. The centrioles of *Trichomonas* and their functions in cell reproduction. *Arch. Protistenkd.,* 105:149-162.

Cleveland, L. R. 1966a. General features of the flagellate and amoeboïd stages of *Deltotrichonympha operculata* and *Deltotrichonympha nana,* sp. nov. *Arch. Protistenkd.,* 109:1–7.

Cleveland, L. R. 1966b. Reproduction in *Deltotrichonympha. Arch. Protistenkd.,* 109:8–14.

Cleveland, L. R. 1966c. Reproduction by binary and multiple fission in *Gigantomonas. J. Protozool.,* 13:573-485.

Cleveland, L. R. 1966d. General features and reproduction in *Koruga bonita,* gen. et sp. nov. *Arch. Protistenkd.,* 109:18–23.

Cleveland, L. R. & Day, M. 1958. Spirotrichonymphidae of *Stolotermes. Arch. Protistenkd.,* 103:1–53.

Cleveland, L. R. & Grimstone, A. V. 1964. The fine structure of the flagellate *Mixotricha paradoxa* and its associated microorganisms. *Proc. Roy. Soc. B.,* 159:668-686.

Cleveland, L. R., Hall, S. R., Sanders, E. P. & Collier, J. 1934. The wood-feeding roach *Cryptocercus,* its protozoa, and the symbiosis between protozoa and roach. *Mem. Am. Acad. Arts* and *Sci.* (N. S.), 17:185–342.

Culberson, D. E., Pindak, F. F., Gardner, W. A. & Honigberg, B. M. 1986. *Tritrichomonas mobilensis* n. sp. (Zoomastigophora: Trichomonadida) from the bolivian squirrel monkey *Saimiri boliviensis boliviensis. J. Protozool.,* 33:301-304.

Cutler, D. W. 1920. Protozoa parasitic in termites. II. *Joenopsis polytricha* n. g. , n. sp. with brief notes on two new species, *J. cephalotricha* and *Microjoenia axostylis. Q. J. Microsc. Sci.,* 64:383–411.

Daniel, W. A., Mattern, C. F. T. & Honigberg, B. M. 1971. Fine structure of the mastigont system in *Tritrichomonas muris* (Grassi). *J. Protozool.,* 18:575-586.

Dobell, C. 1940. Researches on the intestinal protozoa of monkeys and man. X. The life-history of *Dientamoeba fragilis:* observations, experiments, and speculations. *Parasitology,* 32: 417-461.

Duboscq, O. & Grassé, P.-P. 1925. L'appareil parabasal et son évolution chez *Holomastigotes* elongatum Grassi. *C. R. Soc. Biol., Paris,* 92:154–156.

Duboscq, O. & Grassé, P.-P. 1928. Notes sur les protistes parasites de France. V. Les *Spirotrichonympha* et leur évolution. *Arch. Zool. Exp. Gén. Notes Rev.,* 67:159–178.

Duboscq, O. & Grassé, P.-P. 1933. L'appareil parabasal des Flagellés. *Arch. Zool. Exp. Gén.,* 73:381–621.

Duboscq, O. & Grassé, P.-P. 1943. Les flagellés de l'*Anacanthotermes ochraceus* Burm. *Arch. Zool. Exp. Gén.,* 82:401–438.

Farmer, M. 1993. Ultrastructure of *Ditrichomonas honigbergii* n. g., n. sp. (Parabasalia) and its relationship to amitochondrial protists. *J. Euk. Microbiol.,* 40:619-626.

Gabel, J. R. 1954. The morphology and taxonomy of the intestinal Protozoa of the American woodchuck, *Marmota monax* Linnaeus. *J. Morphol.,* 94:473-549.

Grassé, P.-P. 1952a. Ordre des trichomonadines. *In:* Grassé, P.-P. (ed.), *Traité de Zoologie.* Vol.1. Masson et Cie, Paris. pp 704-779

Grassé, P.-P. 1952b. Ordre de Joeniides. Ordre des Lophomonadines. Ordre des Trichonymphines. Ordre des Spirotrichonymphines. Flagellés termiticoles *Incertae sedis.* La symbiose flagellés-termites. *In:* P.-P. Grassé (ed.), *Traité de Zoologie, I. Flagellés.* Masson Cie, Paris. pp 836–962

Grassé, P.-P. & Hollande, A. 1945. Les flagellés du *Coptotermes sjöstedti* Silvestri. *Ann. Sc. Nat., Zool.,11° série,* VI:91–96.

Grassé, P.-P. & Hollande, A. 1947. La structure d'une hypermastigine complexe, *Staurojoenina caulleryi. Ann. Sc. Nat., Biol. Ani. 11° série,* VII:147–158.

Grassé, P.-P. & Hollande, A. 1950. Recherches sur les flagellés termiticoles. Les sous-familles des Devescovininae Kirby et des Macrotermitinae nov. *Ann. Sci. Nat. Zool.,* Paris,11° série, 12:25-64.

Grassé, P.-P. & Hollande, A. 1951a. Recherches sur les symbiotes des termites Hodotermitidae nord-africains. I. Le cycle évolutif du genre *Kirbynia.* II. Les Rhizomastigidae fam. nov. III. *Polymastigotoides,* nouveau genre de Trichomonadidae. *Ann. Sci. Nat., Zool.* Paris, 11° série, 13:1-32.

Grassé, P.-P. & Hollande, A. 1951b. Cytologie et mitose de *Pseudotrichonympha. Ann. Sci. Nat. , Biol. Ani., 11° série,* XIII:237–248.

Grassé, P.-P. & Hollande, A. 1963. Les flagellés des genres *Holomastigotoides* et *Rostronympha.* Structure et cycle de spiralisation des chromosomes chez *Holomastigotoides psammotermitidis. Ann. Sci. Nat. Zool., 12° série,* V:750–792.

Grassi, B. 1917. Flagellati viventi nei termiti. *Mem. R. Accad. Lincei.* (Ser. 5), 12:331–394.

Grimstone, A. V. & Gibbons I. R. 1966. The fine structure of the centriolar apparatus and associated structures in the complex flagellates *Trichonympha* and *Pseudotrichonympha. Phil. Trans. R. Soc. Lond. B. Biol.Sci.,* 250:215–242.

Helson, G. H. A. 1935. *Spirotrichosoma magna* n. sp. from a New-Zealand termite. *Trans. R. Soc. N. Z.,* 64:251–256.

Hibber, C. P., Hammond, D. M., Caskey, F. H., Johnson, A. E. & Fitzgerald P. R. 1960. The morphology and incidence of the trichomonads of swine, *Tritrichomonas suis* (Gruby and Delafond), *Tritrichomonas rotunda* n. sp. and *Trichomonas buttreyi* n. sp.. *J. Protozool.,* 7:159-171.

Hirschler, J. 1927. Studien über die sich mit Osmium schwärzenden Plasmakomponenten (Golgi-Apparat, Mitochondria) einiger Protozoenarten. *Z. Zellforsch. Mikrosk. Anat.,* 5:704–786.

Hollande, A. 1979. Complexes parabasaux et symétrie bilatérale du rostre chez *Joenia annectens. Protistologica,* 15:407–426.

Hollande, A. 1986a. Morphologie, ultrastructure et affinités de *Deltotrichonympha operculata* Sutherland (1933) symbionte du termite australien *Mastotermes darwiniensis* Froggatt. *Protistologica,* 22:415–439.

Hollande, A. 1986b. Les hypermastigines du genre *Staurojoenina.* Précisions sur la structure de leur rostre et description d'espèces nouvelles. *Protistologica,* 22:331–338.

Hollande, A. & Carruette-Valentin, J. 1971. Les atractophores, l'induction du fuseau et la division cellulaire chez les Hypermastigines. Etude infrastructurale et révision systématique des Trichonymphines et des Spirotrichonymphines. *Protistologica,* 7:5–100.

Hollande, A. & Carruette-Valentin, J. 1972. Le problème du centrosome et la crypt-pleuromitose atractophorienne chez *Lophomonas striata. Protistologica,* 8:267–278.

Hollande, A. & Valentin, J. 1967. Morphologie et infrastructure du genre *Barbulanympha,* hypermastigine symbiotique de *Cryptocercus punctulatus* Scudder. *Protistologica,* 3:257–267.

Hollande, A. & Valentin, J. 1968. Morphologie infrastructurale de *Trichomonas (Trichomitopsis* Kofoid & Swezy, 1919) *termopsidis,* parasite intestinal de *Termopsis angusticollis* Walk.

Critique de la notion de centrosome chez les polymastigines. *Protistologica*, IV:127-139.

Hollande, A. & Valentin, J. 1969a. La cinétide et ses dépendances dans le genre *Macrotrichomonas* Grassi. Considérations générales sur la sous-famille des Macrotrichomonadinae. *Protistologica*, 5:335-343.

Hollande, A. & Valentin, J. 1969b. Appareil de Golgi, pinocytose, lysosomes, mitochondries, bactéries symbiotiques, atractophores et pleuromitose chez les Hypermastigides du genre *Joenia*. Affinités entre Joeniides et Trichomonadines. *Protistologica,* 5:39–86.

Honigberg, B. M. 1951. Structure and morphogenesis of *Trichomonas prowazeki* Alexeieff and *Trichomonas brumpti* Alexeieff. *Univ. Calif. Pub. Zool.,* 55:337-394.

Honigberg, B. M. 1953. Structure, taxonomic status and host list of *Tritrichomonas batrachorum* (Perty). *J. Parasitol.,* 39:191-208.

Honigberg, B. M. 1963. Evolutionary and systematic relationships in the flagellate order Trichomonadida Kirby. *J. Protozool.*, 10:20-63.

Honigberg, B. M. 1978. Trichomonads of veterinary importance. *In*: Kreier, J. P. (ed.), *Parasitic Protozoa*. Vol. 2. Academic Press, New York. pp 163-173

Honigberg, B. M. 1970. Protozoa associated with termites and their role in digestion. *In*: Krishna, K. & Weesner, F. M., (eds.), *Biology of Termites*. Vol. 2. Acad. Press, New York. pp 1–36

Honigberg, B. M. & Christian, H. H. 1954. Characteristics of *Hexamastix batrachorum* Alexeieff. *J. Parasitol.*, 40:508-514.

Honigberg, B. M. & Bennett, C. J. 1971. Light microscopic observations on structure and division of *Histomonas meleagridis* (Smith). *J. Protozool.,* 18:687-697.

Honigberg, B. M. & Kuldova, J. 1969. Structure of a nonpathogenic histomonad from the cecum of galliform birds and revision of the trichomonad family Monocercomonadidae Kirby. *J. Protozool.,* 16:526-535.

Honigberg, B. M. & Lee, J. J. 1959. Structure and division of *Trichomonas tenax* (O. F. Muller). *Am. J. Hyg.,* 69:177-201.

Honigberg, B. M., Mattern, C. F. T. & Daniel, W. A. 1968. Structure of *Pentatrichomonas hominis* (Davaine) as revealed by electron microscopy. *J. Protozool.,* 15:419-430.

Honigberg, B. M., Mattern, C. F. T. & Daniel, W. A. 1971. Fine structure of the mastigont system in *Tritrichomonas foetus* (Riedmüller). *J. Protozool.,* 18:183-198.

Honigberg, B. M., Mattern, C. F. T. & Daniel, W. A. 1972. Fine structure of *Trichomitus batrachorum* (Perty). *J. Protozool.,* 19:446-453.

Honigberg, B. M. & Brugerolle, G. 1990. Structure. *In:* Honigberg, B. M. (ed.), *Trichomonads Parasitic in Humans*. Springer Verlag, Berlin, Heidelberg, New York, Tokyo. pp 2-35

Janicki, C. 1910. Untersuchungen an parasitischen Flagellaten. I. Teil. *Lophomonas blattarum* Stein. *L. striata* Bütsch. *Z. Wiss. Zool.,* 95:243–315.

Janicki, C. 1915. Untersuchungen an parasitischen Flagellaten. II. Teil: Die Gattungen *Devescovina, Parajoenia, Stephanonympha, Calonympha. Z. Wiss. Zool.,* 112:573-691.

Joyon, L., Mignot, J. P., Kattar, M. R. & Brugerolle, G. 1969. Compléments à l'étude des Trichomonadida et plus particulièrement de leur cinétide. *Protistologica,* V:309-326.

Keeling, P. J., Poulsen, N. & McFadden, G. I. 1998. Phylogenetic diversity of parabasalian symbionts from termites, including the phylogenetic position of *Pseudotrypanosoma* and *Trichonympha*. *J. Euk. Microbiol.*, 45:643-650

Kimura, G. G. 1934. *Cochlosoma rostratum* sp. nov., an intestinal flagellate of domestic ducks. *Trans. Am. Microsc. Soc.,* 53:102-115.

Kirby, H. 1926a. On *Staurojoenina assimilis* sp. nov., an intestinal flagellate from the termite *Kalotermes minor* Hagen. *Univ. Calif. Publ. Zool.,* 29:25–102.

Kirby, H. 1926b. The intestinal flagellates of the termite, *Cryptotermes hermsi* Kirby. *Univ. Calif. Publ. Zool.,* 29:103-120.

Kirby, H. 1929. *Snyderella* and *Coronympha*, two new genera of multinucleate flagellates from termites. *Univ. Calif. Publ. Zool.,* 31:417-432.

Kirby. H. 1930. Trichomonad flagellates from termites. I. *Tricercomitus* gen. nov., and *Hexamastix* Alexeieff. *Univ. Calif. Publ. Zool.,* 33:393-444.

Kirby, H. 1931. Trichomonad flagellates from termites. II. *Eutrichomastix,* and the subfamily Trichomonadidae. *Univ. Calif. Publ. Zool.,* 36:171-262.

Kirby, H. 1932. Flagellates of the genus *Trichonympha* in termites. *Univ. Calif. Publ. Zool.,* 37:349–476.

Kirby, H. 1939. Two new flagellates from termites in the genera *Coronympha* Kirby and *Metacoronympha* Kirby new genus. *Proc. Calif. Acad. Sci.,* 22:207-220.

Kirby, H. 1941. Devescovinid flagellates of termites. II. The genus *Devescovina*. *Univ. Calif. Publ. Zool.,* 45:1-92.

Kirby, H. 1942a. Devescovinid flagellates of termites. II. The genera *Caduceia* and *Macrotrichomonas. Univ. Calif. Publ. Zool.,* 45:93-166.

Kirby, H. 1942b. Devescovinid flagellates of termites. III. The genera *Foaina* and *Parajoenia. Univ. Calif. Publ. Zool.,* 45:167-246.

Kirby, H. 1944. Some observations on cytology and morphogenesis in flagellated Protozoa. *J. Morphol.,* 75:361-421.

Kirby, H. 1945a. The structure of the common intestinal trichomonad of man. *J. Parasitol.,* 31:163-175.

Kirby, H. 1945b. Devescovinid flagellates of termites. IV. The genera *Metadevescovina* and *Pseudodevescovina. Univ. Calif. Publ. Zool.,* 45:247-318.

Kirby, H. 1946. *Gigantomonas herculea* Dogiel a polymastigote flagellate with flagellated and amoeboid phases of development. *Univ. Calif. Publ. Zool.,* 53:163-226.

Kirby, H. 1947. Flagellates and host relationships of trichomonad flagellates. *J. Parasitol.,* 33:214-228.

Kirby, H. 1949a. Systematic differentiation and evolution of flagellates in termites. *Rev. Soc. Mex. Hist. Nat.,* X:57–79.

Kirby, H. 1949b. Devescovinid flagellates of termites. V. The genus *Hyperdevescovina,* the genus *Bullanympha,* and undescribed or unrecorded species. *Univ. Calif. Publ. Zool.,* 45:319-422.

Kirby, H. 1951. Observations on the trichomonad flagellate of the reproductive organs of cattle. *J. Parasitol.,* 37:445-459.

Kirby, H. & Honigberg, B. M. 1950. Intestinal flagellates from a wallaroo, *Macropus robustus* Gould. *Univ. Calif. Publ. Zool.,* 55:35-66.

Koidzumi, M. 1921. Studies on the intestinal protozoa found in the termites of Japan. *Parasitology,* 13:235–309.

Kotlàn, A. 1923. Zur Kenntnis der Darmflagellaten aus der Hausente und anderen Wasservögeln. *Centrbl. Bakt. Parasit.,* Abb. I. Orig., 90:24-28.

Kozloff, E. N. 1945. The morphology of *Trichomonas limacis* Dujardin. *J. Morphol.,* 77: 53-61.

Kudo, R. 1926. Observation on *Lophomonas blattarum*, a flagellate inhabiting the colon of the cockroach *Blatta orientalis. Arch. Protistenkd.,* 53:158–214.

Kulda, J. & Nohynkovà, E. 1978. Flagellates of the human intestine and of intestines of other species. *In:* Kreier J. P. (ed.), *Parasitic Protozoa.* Vol. 2. Academic press, New York. pp 2-138

Kulda, J., Nohynkovà, E. & Ludvik, J. 1986. Basic structure and function of the trichomonad cell. *In:* Kulda, J. & Cerkasov; J. (eds.). *Proceedings of the International Symposium on trichomonads & Trichomoniasis,* Prague. *Acta Univ. Carol. Biol.,* 30:181-198.

Lavette, A. 1970. Sur le genre *Projoenia* et les affinités des Joeniidae (Zooflagellés Metamonadina). *C. R. Acad. Sci.* Paris, série D, 270:1695-1698.

Lavette, A. 1971a. Sur la structure de *Foainia brasiliensis* n. sp. , un Devescovinidae symbiote d'un termite brésilien du genre *Neotermes. C. R. Acad. Sci.* Paris, D, 272:1394-1397.

Lavette, A. 1971b. *Stephanonympha chagasi* n. sp., un Flagellé trichomonadidé et ses bactéries symbiotiques. *C. R. Acad. Sci.* Paris, D, 272: 1785-1788.

Lee, J. J. 1960. *Hypotrichomonas acosta* (Moskowitz) gen. nov. from reptiles. I. Structure and division. *J. Protozool.,* 7:393-401.

Light, S. F. 1926. On *Hoplonympha natator,* gen. nov., sp. nov. A non-xylophagous hypermastigote, from the termite, *Kalotermes simplicicornis* Banks, characterized by biradial symmetry and a highly developed pellicle. *Univ. Calif. Publ. Zool.,* 29:123–139.

Light, S. F. 1927. *Kofoidia*, a new flagellate, from a California termite. *Univ. Calif. Publ. Zool.,* 29:467–492.

Lingle, W. L. & Salisbury, J. L. 1995. Ultrastructure of the parabasalid protist *Holomastigotoides. J. Euk. Microbiol.,* 42:490-505.

Lund, E. E. 1963. *Histomonas wenrichi* n. sp. (Mastigophora: Mastigamoebidae), a non-pathogenic parasite of galliform birds. *J. Protozool.,* 10:401-404.

Lund, E. E. & Chute, A. M. 1973. Means of acquisition of *Histomonas meleagridis* by eggs of *Heterakis gallinarum. Parasitology,* 66:335-342.

Ludwig, W. F. 1946. Studies on the protozoan fauna of the larvae of crane-fly, *Tipula abdominalis.* Flagellates, amoebae, and grégarines. *Trans. Am. Microsc. Soc.* 75:189-214.

Mc Dougald, L. R. & Reid, W. M. 1978. *Histomonas meleagridis* and relatives. *In:* Kreier, J. P. (ed.), *Parasitic Protozoa.* Vol. 2. Academic Press, New York. pp 139-161

McDowell Jr., S. 1953. A morphological and taxonomic study of the caecal Protozoa of the common fowl *Gallus gallus* L. *J. Morphol.,* 92:337-400.

Mackinnon, D. L. 1913. Studies of parasitic protozoa II. *Tetratrichomastix parisii* n. subgen., n. sp. *Q. J. Microsc. Sci.,* 59:459-470.

Mackinnon, D. L. 1926. Obsrvations on trichonymphids. I. The nucleus and axostyle of *Holomastigotoides hemigymnum* Grassi. *Q. J. Microsc. Sci.,* 70:173–191.

Mattern, C. F. T., Daniel W. A. & Honigberg, B. M. 1969. Structure of *Hypotrichomonas acosta* (Moskowitz) (Monocercomonadidae, Trichomondida) as revealed by electron microscopy. *J. Protozool.*, 16:668-685.

Mattern, C. F. T., Honigberg, B. M. & Daniel, W. A. 1867. The mastigont system of *Trichomonas gallinae* (Rivolta) as revealed by electron microscopy. *J. Protozool.,* 14:320-339.

Mattern, C. F. T., Honigberg, B. M. & Daniel, W. A. 1972. Structure of *Monocercomonas sp.* As revealed by electron microscopy. *J. Protozool.,* 20:265-274.

Mattern, C. F. T., Honigberg, B. M. & Daniel, W. A. 1973. Fine structure changes associated with pseudocyst formation in *Trichomitus batrachorum. J. Protozool.,* 19:222-229.

Mignot, J. P., Joyon, L. & Kattar, M. R. 1969. Sur la structure de la cinétide et sur les affinités systématiques de *Devescovina striata* Foà, protozoaire flagellé. *C. R. Acad. Sci.* Paris, D., 268:1738-1741.

Moskowitz, N. 1951. Observations on some intestinal flagellates from reptilian hosts (Squamata). *J. Morphol.,* 89:257-321.

Müller, M. 1988. Energy metabolism of protozoa without mitochondria. *Ann. Rev. Microbiol.,* 42:465-488.

Nie, D. 1950. Morphology and taxonomy of the intestinal protozoa of the guinea-pig *Cavia porcella. J. Morphol.,* 86:381-493.

Nielsen, M. H. 1972. Electron microscopy of *Trichomonas vaginalis* Donné. Negative staining of the mastigont. *J. Microsc. (Paris),* 15:121-134.

Nielsen, M. H., Ludvik, J. & Nielsen, R. 1966. On the ultrastructure of *Trichomonas vaginalis. J. Microsc. (Paris),* 5:229-250.

Nurse, F. M. 1945. Protozoa of New-Zealand termites. *Trans. R. Soc. N. Z.,* 74:305–314.

Pecka, Z., Nohynkovà, E. & Kulda, J. 1996. Ultrastructure of *Cochlosoma anatis* Kotlan, 1923 and taxonomic position of the family Cochlosomatidae (Parabasalidae: Trichomonadida). *Europ. J. Protistol.*, 32:190-201.

Radek, R. 1997. *Spirotrichonympha minor*, n. sp., a new hypermastigote termite flagellate. *Europ. J. Protistol.*, 33:360-374.

Radek, R. & Hausmann, K. 1994. *Placojoenia sinaica* n. g., n. sp. A symbiotic flagellate from the termite *Kalotermes sinaicus. Europ. J. Protistol.*, 30:25-37.

Rybicka, K., Honigberg, B. M. & Holt, S. C. 1972. Fine structure of the mastigont system in culture forms of *Histomonas meleagridis* (Smith). *Protistologica*, 8:107-120.

Saleudin, A. S. M. 1971. Fine structure of *Tetratrichomonas limacis* (Dujardin). *Canad. J. Zool.,* 50:695-701.

Samuels, R. 1957. Studies of *Tritrichomonas batrachorum.* I. The trophic organism. *J. Protozool.,* 4:110-118.

Stabler, E. E. 1947. *Trichomonas gallinae,* pathogenic trichomonad of birds. *J. Parasitol.,* 33:207-213.

Schuster, F. L. 1968. Ultrastructure of *Histomonas meleagridis* (Smith) Tyzzer, a parasitic amebo-flagellate. *J. Parasitol.,* 54:725-737.

Sutherland, J. L. 1933. Protozoa from Australian termites. *Q. J. Microsc. Soc.,* 76:145–153.

Tamm, S. L. 1972. Free kinetosomes in Australian flagellates. I. Types and spatial arrangement. *J. Cell Biol.,* 54:39-55.

Tamm, S. L. 1978. Laser microbeam study of a rotatory motor in termite flagellates. *J. Cell Biol.,* 78:76-92.

Tamm, S. & Tamm S. L. 1973a. The fine structure of the centriolar apparatus and associated structures in the flagellates *Deltotrichonympha* and *Koruga.* I. Interphase. *J. Protozool.,* 20:230–245.

Tamm, S. & Tamm S. L. 1973b. The fine structure of the centriolar apparatus and associated structures in the flagellates *Deltotrichonympha* and *Koruga*. II. Division. *J. Protozool.,* 20:245–252.

Travis, B. V. 1938. A synopsis of the flagellate genus *Cochlosoma* Kotlàn, with the description of two new species. *J. Parasitol.,* 24:343-351.

Tyzzer, E. E. 1920. The flagellate character and reclassification of the parasite producing "blackhead" in turkeys-*Histomonas* (gen. nov.) *meleagridis* (Smith). *J. Parasitol.,* 6:124-131.

Tyzzer, E. E. 1930. Flagellates from the Ruffed Grouse. *Am. J. Hyg.,* 11:56-73.

Vickerman, K. 1982. Mastigophora. *In*: Parker, S. P. (ed.), *Synopsis and Classification of Living Organisms.* Vol. I. McGraw-Hill Book Company, New York. pp 496–508

Viscogliosi, E., Philippe, H., Baroin, A., Perasso, R. & Brugerolle, G. 1993. Phylogeny of trichomonads based on partial sequences of large subunit rRNA and on cladistic analysis of morphological data. *J. Euk. Microbiol.,* 40:411-421.

Warton, A. & Honigberg, B. M. 1979. Structure of trichomonads as revealed by scanning electron microscopy. *J. Protozool.,* 26:56-62.

Watkins, R. A. , O'Dell, W. D. & Pinter A. J. 1989. Redescription of flagellar arrangement in the duck intestinal flagellate, *Cochlosoma anatis,* and description of a new species, *Cochlosoma soricis* n. sp. from shrews. *J. Protozool.,* 36: 527-531.

Wenrich, D. H. 1921. The structure and division of *Trichomonas muris* (Hartmann). *J. Morphol.,* 36:119-155.

Wenrich, D. H. 1943. Observations on the morphology of *Histomonas* (Protozoa, Mastigophora) from pheasants and chickens. *J. Morphol.,* 72:279-303.

Wenrich, D. H. 1944a. Studies on *Dientamoeba fragilis* (Protozoa). IV. Further observations, with an outline of present-day knowledge of this species. *J. Parasitol.,* 30:322-338.

Wenrich, D. H. 1944b. Comparative morphology of the trichomonad flagellates of man. *Am. J. Trop. Med. Hyg.,* 24:39-51.

Wenrich, D. H. 1944c. Morphology of the intestinal trichomonad flagellates in man and of similar forms on monkeys, cats, dogs and rats. *J. Morphol.,* 74:189-211.

Wenrich, D. H. & Emerson, M. A. 1933. Studies on the morphology of *Tritrichomonas foetus* (Riedmüller) (Protozoa, Flagellata) from American cows. *J. Morphol.,* 55:193-205.

Wenrich, D. H. & Nie, D. 1949. The morphology of *Trichomonas wenyoni* (Protozoa, Mastigophora). *J. Morphol.,* 85:519-531.

Yamin, M. 1979a. Flagellates of the orders Trichomonadida Kirby, Oxymonadida Grassé, and Hypermastigida Grassi & Foà reported from lower termites (Isoptera families Mastotermitidae, Kalotermitidae, Hodotermitidae, Termopsidae, Rhinotermitidae, and Serritermitidae) and from the wood-feeding *Cryptocercus* (Dictyoptera: Cryptocercidae). *Sociobiology,* 4:3-117.

Yamin, M. 1979b. Scanning electron microscopy of some symbiotic flagellates from the termite *Zootermopsis*. *Trans. Am. Microsc. Soc.,* 98:276-279.

ORDER RETORTAMONADIDA, GRASSÉ

by

GUY BRUGEROLLE and JOHN J. LEE

Retortamonads are small (5-20 µm) bacterivorous flagellates with two or four flagella; one of them is turned backward and associated with a conspicuous antero-ventral cytostomal aperture (Fig. 1a). Basal bodies of the flagella are closely associated with the anterior nucleus and with the two fibrils lining the cytostome. The right lip of the cytostome is bordered by a striated or paracrystalline fiber which terminates by a posterior hook (Fig. 1d,e). The cell surface is underlain by a corset of microtubules (Fig. 1c) except in the bottom of the cytostome where endocytosis takes place. The recurrent flagellum has typically two to three fin-like lamellae along the axoneme (Fig. 1d,e). There are no mitochondria, Golgi apparatus, hydrogenosomes, or axostyles. They reproduce by binary fission and the mitotic spindle is intranuclear. They are transmitted from host to host by resistant cysts. Most of them are parasitic in the digestive tract of animals; however, free-living species have been reported (Bernard et al., 1997). The group has been described in Brugerolle (1991); Brugerolle and Mignot (1990); Vickerman (1982). Only two genera have been described: *Retortamonas* and *Chilomastix.*

FAMILY RETORTAMONADIDAE

Genus ***Retortamonas*** Grassi

Retortamonas (syn=*Embadomonas, Plagiomonas, Waskia*) are small pyriform or fusiform flagellates with only 2 flagella. The recurrent flagellum is associated with the cytostomal pocket but extends posteriorly (Fig. 2). Electron microscopy has shown that 1 basal body in each pair has no flagellum (barren basal body) and a striated fiber connecting the nucleus to 1 pair of basal bodies. They form ovoid cysts. Among the 18 species described, several live in the lower gut of insect larvae of Trichoptera, Coleoptera, Diptera, such as *R. alexeieffi, R. agilis* (Mackinnon 1916) from the crane fly larvae *Tipula*, Orthoptera such as *R. gryllotalpae* from the mole cricket (Wenrich, 1932), also in water bugs, roaches, termites; list of invertebrate species in Grassé (1952). Other species live in intestine of vertebrates such as *R. intestinalis* from humans and monkeys (Wenyon,

1926), *R. caviae* from the guinea pig caecum (Nie, 1950), *R. mitrula* from *Marmotta* (Kirby and Honigberg, 1950). Still other species have been reported in the caecum of mammals (monkeys, sloth, sheep) and in the rectum of lizards, snakes, tortoises, anurans, and urodeles (Bishop, 1931). A complete list of vertebrate species is provided by Kulda and Nohynkovà (1978). Electron microscopic study by Brugerolle (1977).

Genus ***Chilomastix*** Alexeieff

Chilomastix (syn=*Macrostoma, Fanapetia*) have a pyriform and twisted cell with 3 anteriorly directed flagella and 1 recurrent flagellum beating in the ventral cytostomal pocket (Figs. 1,3). The right fibril bordering the cytostome is thicker and forms a hook at its posterior extremity where the

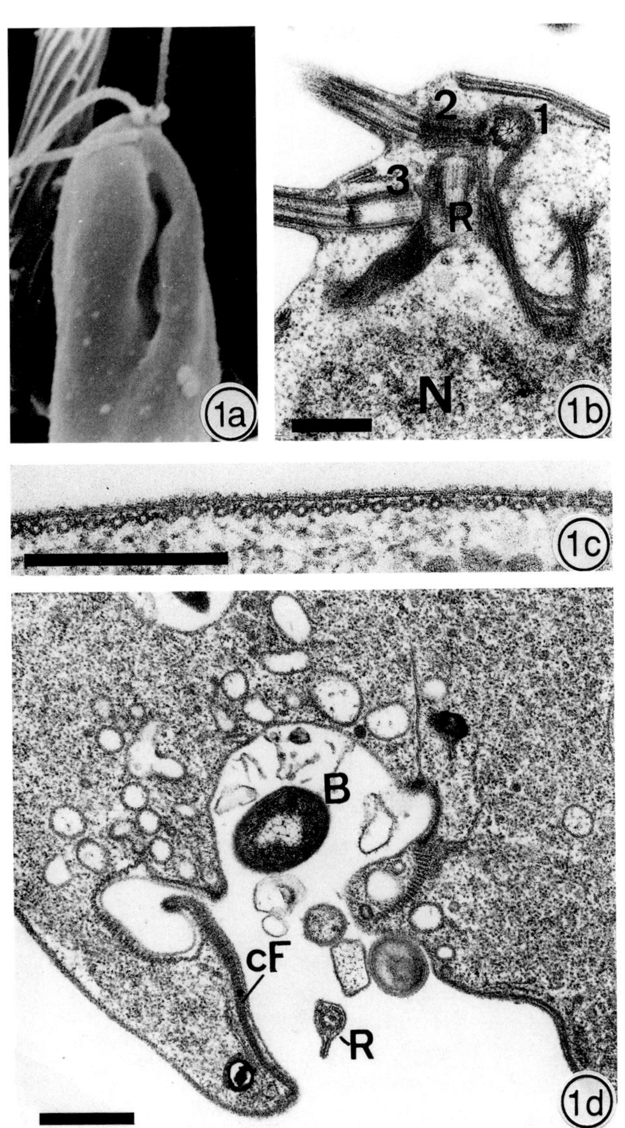

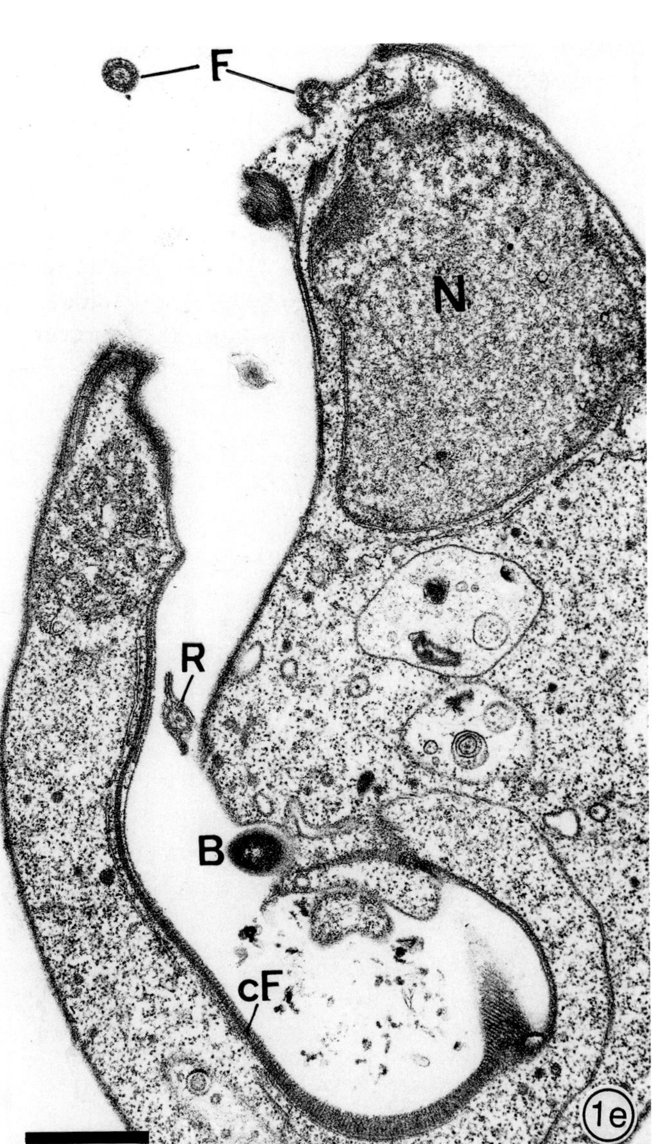

Fig. 1a,b,c,d,e. Scanning and transmission electron microscopic views of *Chilomastix*. Fig. 1a. Scanning view of *C. bettencourti* from the caecum of mouse, showing the three apical flagella inserted at the top of the elongated cytostomal opening. Fig. 1b. Basal bodies 1,2,3 of the anteriorly directed flagella, arranged around the basal body of the recurrent intra-cytostomal flagellum (R). Fig. 1c. Corset of interlinked microtubules under the plasma-membrane. Figs. 1d,e. Transverse and longitudinal sections of the cytostome/cytopharynx containing the recurrent flagellum (R) and bordered by a paracrystalline right fiber (cF). Bacteria in course of phagocytosis at the bottom of the cytopharynx; nucleus (N). Bar=10 µm. (Photographs from G. Brugerolle)

cytopharynx takes place. Pyriform cysts have a single nucleus and internal cytostomal fibers (Fig. 3). Among the 29 species described, several live in invertebrates hosts (see list in Grassé, 1952): termites, sea urchins, and leeches such as *C. aulastomi* (Bêlâr, 1921). Among the species living in vertebrates, several live in mammals (Geiman, 1935), *C. mesnili* from man (Fig. 3)(Kofoid and Swezy, 1920; Wenyon, 1926) is a common parasite which multiplies in diarrheic stools; *C. bettencourti* (Fig. 1a) is common in the caecum of rodents. Other species occur in primates, rodents, lagomorphs, horse, goat, ducks, galliform birds, lizards, amphibians, and marine fishes; *C. gallinarum* (Boeck and Tanabe, 1926) from the caecum of poultry is a potential pathogen. A complete list of *Chilomastix* species of vertebrates is provided by Kulda and Nohynkovà (1978). Electron microscopic study by Brugerolle (1973).

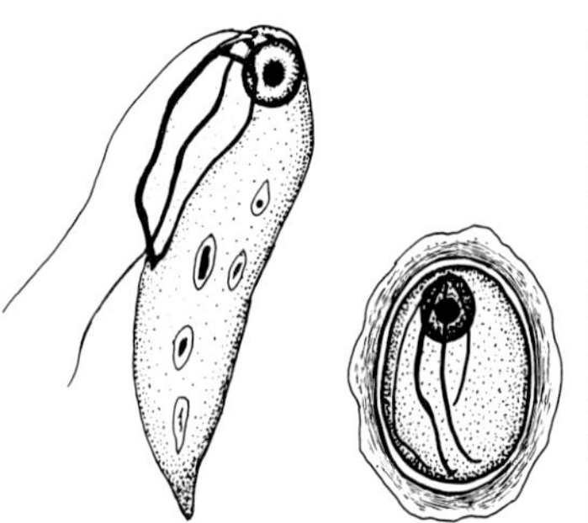

Figs. 2. *Retortamonas agilis* from *Tipula* larvae (after Mackinnon, 1916); flagellate and cyst. Bar=1 µm.

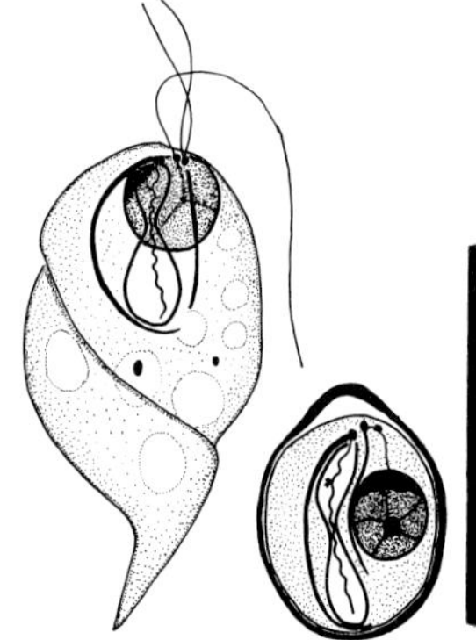

Fig. 3. *Chilomastix mesnili* from man (after Kofoid and Swezy, 1920); flagellate and cyst. Bar=10 µm.

LITERATURE CITED

Bêlar, K. 1921. Protozoenstudien III *Chilomastix aulastomi. Arch. Protistenkd.*, 43:439-446.

Bernard, C., Simpson, A. G. & Patterson, D. J. 1997. An ultrastructural study of a free-living retortamonad, *Chilomastix cuspidata* (Larsen et Patterson, 1990) n. comb. (Retortamonadida, Protista). *Europ. J. Protistol.*, 33:254-240.

Bishop, A. 1931. A description od *Embadomonas* n. sp. from *Blatta orientalis, Rana temporaria, Bufo vulgaris, Salamandra maculosa*, with a note upon the "cyst" of *Trichomonas. Parasitology,* 23:286-300

Boeck, W. C. & Tanabe, M. 1926. *Chilomastix gallinarum*, morphology, division and cultivation. *Am. J. Hyg.,* 6:319-336.

Brugerolle, G. 1973. Etude ultrastructurale du trophozoite et du kyste chez le genre *Chilomastix* Alexeieff, 1910 (Zoomastigophorea, Retortamonadida Grassé, 1952). *J. Protozool.*, 20:574-585.

Brugerolle, G. 1977. Ultrastructure du genre *Retortamonas* Grassi 1879 (Zoomastigophorea, Retortamonadida Wenrich 1931). *Protistologica,* 13:233-240.

Brugerolle, G. 1991. Flagellar and cytoskeletal systems in amitochondrial flagellates: Archamoebae, Metamonada and Parabasala. *Protoplasma,* 164:70-90.

Brugerolle, G. & Mignot, J. P. 1990. Phylum Zoomastigina, class Retortamonadida. *In*: Margulis, L., Corliss, J. O., Melkonian, M. & Chapmann, D. J. (eds.), *Handbook of Protoctista*. Jones & Bartlett Publishers Boston. pp 259-265

Geiman, Q. M. 1935. Cytological studies of the *Chilomastix* of man and other mammals. *J. Morphol.,* 57:430-453.

Grassé, P.- P. 1952. Ordre des Retortamonadines. *In:* Grassé, P.-P. (ed.), *Traité de Zoologie*. Vol. I. Masson et Cie, Paris. pp 824-831

Kirby, H. & Honigberg, B. M. 1950. Intestinal flagellates from a wallaroo, *Macropus robustus. Univ. Calif. Publ. Zool.,* 55:35-66.

Kofoid, C. A. & Swezy, O. 1920. On the morphology and mitosis of *Chilomastix mesnili* (Wenyon), a common flagellate of the human intestine. *Univ. Calif. Publ. Zool.*, 20:177-244.

Kulda, J. & Nohynkovà, E. 1978. Flagellates of the human intestine and of the intestines of other species. *In*: Kreier, J. P. (ed.), *Parasitic Protozoa*. Vol. 2. Academic Press, New York. pp 2-138

Mackinnon, D. L. 1916. Studies on parasitic protozoa. III. (a) Notes on the flagellate *Embadomonas.* (b) The

multiplication cysts of Trichomastigine. *Q. J. Microsc. Sci.*, 61:105-118.
Nie, D. 1950. Morphology and taxonomy of the intestinal protozoa of the guinea-pig, *Cavia porcella*. *J. Morphol.*, 86:381-493.
Vickerman, K. 1982. Mastigophora. *In*: Parker S. P. (ed.), *Synopsis and Classification of Living Organisms.* Vol. I. McGraw-Hill Book Company, New York. pp 496-508
Wenrich, D. H. 1932. The relation of the protozoan flagellates, *Retortamonas gryllotalpae* (Grassi, 1879) Stiles, 1902 to the species of the genus *Embadomonas* Mackinnon, 1911. *Trans. Am. Microsc. Soc.*, 51:225-238.
Wenyon, C. M. 1926. *Protozoology*. Ballière, Tindall & Cox, London.

CLASS PRASINOPHYCEAE

T. Christensen ex P.C. Silva, 1980

By Øjvind Moestrup

The Prasinophyceae (synonyms=Micromonadophyceae, Prasinomonadida) comprises a small number of unicellular algae (probably 100-200 species), mainly flagellates measuring 2-30 µm in length. Much larger stages occur in the life cycle of some species, e.g. up to nearly 1 mm in *Halosphaera,* and an increasing number of ex tremely small species are presently being found in the marine picoplankton. The latter appear to be reduced rather than primitive forms. Members of the Prasinophyceae are considered to represent some of the most primitive green flagellates, and numerous fossils resembling extant prasinophyceans have been found in precambrian deposits (e.g., *Tasmanites*; 60 genera are mentioned by Tappan (1980) but not included below). The extant genera are morphologically very diverse, probably reflecting the long evolutionary history of the group. More advanced green algae are thought to derive from prasinophyceans, and attempts have been made to classify some of the genera in other classes of green algae with which these genera share common ancestors. However, the genera of the Prasinophyceae share many features indicating genuine phylogenetic relationship and specialists classify prasinophyceans into a single class (Moestrup and Throndsen, 1988; Melkonian, 1990; Sym and Pienaar, 1993). Classification into orders and families is based mainly on ultrastructural features and gene studies have provided no clear answers (Daugbjerg et al., 1995). A stable classification at the order and family level must await studies of additional genes.

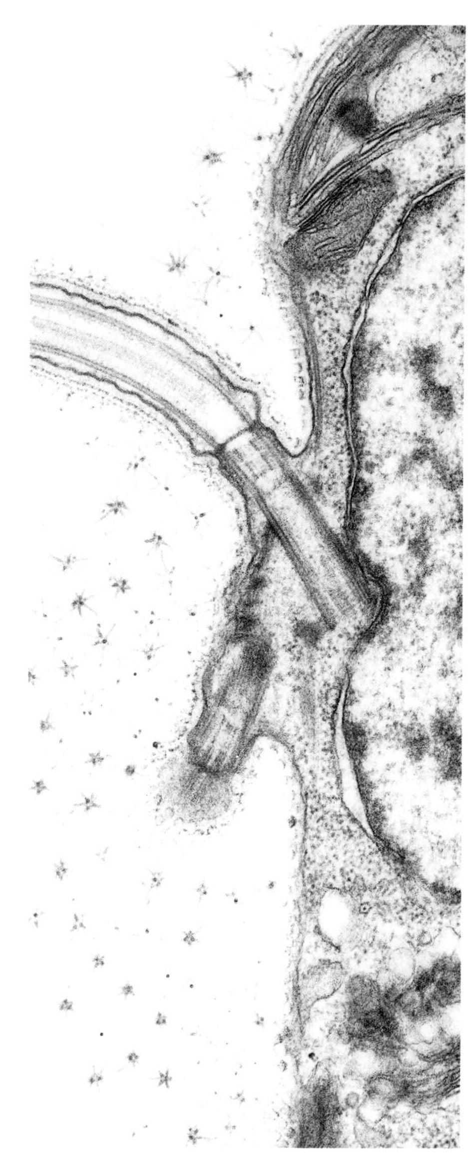

Fig. 1. *Nephroselmis rotunda,* showing periplast scales on cell body and flagella and the long basal bodies characteristic of prasinophyceans. x30,000. (From Moestrup and Ettl, 1979)

Key Characters

1. Flagella and cell body covered with sub-microscopic organic scales manufactured in the cisternae of the Golgi apparatus (Figs. 1-5). The flagella may carry up to three layers of scales, of three different types, in addition to hair-shaped ones, which occur in two opposite rows (Figs. 2, 3A). The same flagellum may sometimes carry more types of hair scales, and other hair-like scales may be present on the cell surface near the insertion of the flagella. *Mesostigma* lacks hair-shaped scales. In the most complex species (e.g. *Pyramimonas* or *Tetraselmis*) the underlayer scales on the flagella are very small, almost square and arranged in 24 longitudinal rows (Fig. 2). Two nearly opposite pairs of these are slightly different and the hair scales insert in these rows (Fig. 2). In some genera the underlayer scales are covered by larger scales in nine longitudinal rows, in others again by rod-like scales in 24 double rows. Body scales occur in up to four layers, each layer usually containing only a single type of scale (Fig. 3B). A few genera have naked cells or the scales fuse immediately after release into a wall-like periplast (formerly known as a 'theca'. Cells usually possess 1, 2, or 4 flagella. Some lack flagella, but a few species of *Pyramimonas* have 8 or in one species 16 flagella (the only known 16-flagellated photosynthetic flagellate). The flagella on a cell are morphologically identical (isokont) in all genera except in a few genera, e.g. *Nephroselmis* and *Pseudo-scourfieldia,* which have anisokont flagella, i.e. the flagella differ in length but are otherwise identical (Figs. 19,20). In *Mantoniella one* of the two flagella is very short and knob-like (Fig. 13B), and in *Micromonas, the* smallest known flagellate, only the proximal part of the (single) flagellum contains a typical axoneme (Fig. 15). The flagella are inserted apically (e.g. *Pyramimonas*, *Tetraselmis*), laterally (e.g. *Nephroselmis*) or posteriorly (*Pseudoscourfieldia*). In most genera the flagellar basal bodies are parallel and notably long compared to other green algae (Fig. 1). Multilayered structures resembling those of some other protists and charophycean green algae are associated with one or two microtubular flagellar roots in some members of the Pyramimonadales

2. The scales contain 2-keto-sugar acids, a compound not found in any other algal flagellates.

3. Cells contain a single (in a few genera two), green or yellow-green chloroplast, generally with one, rarely more pyrenoids. A few species lack pyrenoids. An eyespot is usually present, occasionally two. The chloroplasts contain chlorophylls *a* and *b*, prasinoxanthin (in some genera), Mg 2,4 D (in some genera), and other pigments such as lutein, carotenes, zeaxanthin, violaxanthin, neoxanthin, uriolide, and siphonaxanthin.

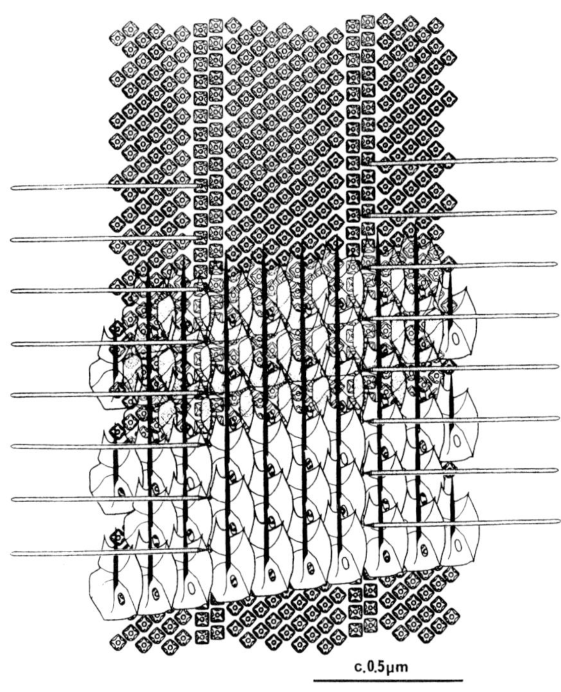

Fig. 2. The scaly periplast on the flagella of the prasinophycean *Pyramimonas mitra* unfolded to show the 24 rows of minute square scales, two opposite pairs of which are arranged in exact longitudinal rows. The small scales are overlain by nine rows of limuloid scales. Hair-shaped scales emanate from the interspace between every third scale in the longitudinal rows of scales, but are not exactly opposite. x42,000. (From Moestrup and Hill, 1993).

4. All species are uninuclear

5. Some species form ejectile organelles or mucilaginous bodies.

6. Freshwater species possess a contractile vacuole system.

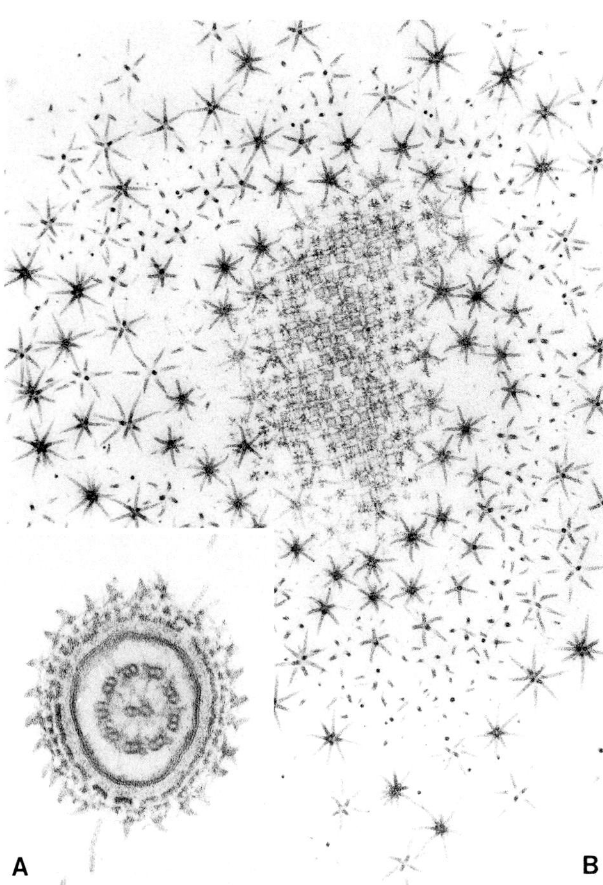

Fig. 3. *Nephroselmis olivacea.* A. Transverse section through flagellum, showing three layers of scales, in addition to two almost opposite rows of hair-shaped scales. x60,000. B. Tangential section of the cell surface showing the scaly periplast. Scales of the innermost layer are very small and square and overlain by scales shaped somewhat like a Maltese cross. The outermost layer comprises larger, stellate scales. x36,000. (Both figures from Moestrup and Ettl, 1979)

7. All known species are photosynthetic, and one, *Cymbomonas,* here considered the most primitive prasinophycean, has been found with food vacuoles containing other eukaryotes (Throndsen, 1988) (Fig. 5). This indicates mixotrophy, presently the only case of mixotrophy found in any green alga. The mixotrophic habit is an indication that the ancestors of green algae were heterotrophic flagellates.
8. The cells store starch in the chloroplast(s).
9. Reproduction is by longitudinal fission. Isogamic sexual reproduction has been described in *Nephroselmis* (Suda et al., 1989). Many species form resting cysts.
10. Most prasinophyceans are members of the marine plankton, and a few are benthic or epiphytic. A small number are confined to fresh water. Many are very widely distributed.
11. Cell shape, number of flagella and scale morphology are diagnostic features. Several species cannot be identified without the aid of transmission electron microscopy of the scales.

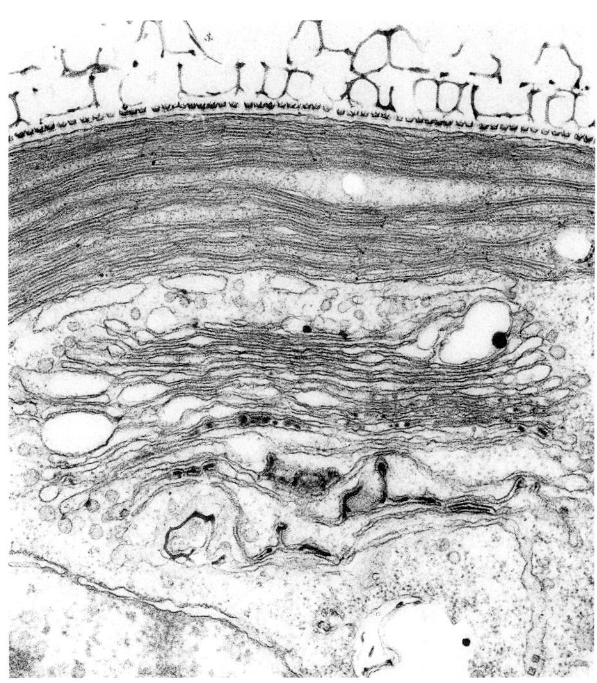

Fig. 4. *Pyramimonas tetrarhyncus.* The cell surface is covered by three layers of scales, all manufactured in the cisternae of the Golgi apparatus. X24,000. (Walne and Moestrup, 1990 Original)

KEY TO THE ORDERS

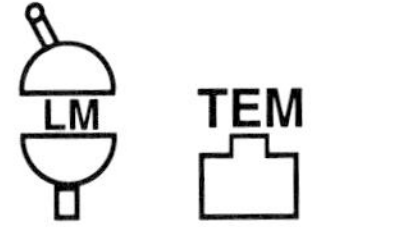

1. Flagella with square or nearly square underlayer scales often overlain by other scale types 2
1´. Flagella lack square underlayer scales. Only one type of flagellar scale present in addition to hair scales .... **Mamiellales**

2. Underlayer scales covered with rod-shaped scales in 24 double rows ...... **Chlorodendrales**
2´. Underlayer scales covered with limuloid scales or without additional layers of scales................ .... **Pyramimonadales**

ORDER PYRAMIMONADALES

Flagella with underlayer of small square scales in 24 longitudinal rows, in one family overlain by 9 rows of limuloid scales.

KEY TO THE FAMILIES

1. Cells more or less pyramidal or rounded, quadri-flagellate.................. **Halosphaeraceae**
2. Cells strongly flattened, biflagellate .... **Mesostigmataceae**

FAMILY HALOSPHAERACEAE

Flagella with small square scales in 24 longitudinal rows, overlain by 9 rows of limuloid ones.

KEY TO THE GENERA

1. Flagella absent (*Phycoma* stage)...................... 2
1´. Flagella present 3

2. Cells up to 230 µm in diameter, with one or more wings .... ***Pterosperma***
2´. Cells without wings, up to 800 µm in diameter .... ***Halosphaera***

3. Cells radially symmetrical................................ 4
3´. Cells bilaterally symmetrical........................... 5

4 With one pyrenoid posteriorly (2 in a 16-flagellate species)............... ***Pyramimonas***
4´. With two or four pyrenoids halfway along the cell. .... ***Halosphaera***

5. With four posterior flagella, eyespot absent. .... ***Pterosperma***
5´. With four anterior flagella, more rapid swimming with the flagella posteriorly may also occur eyespot present.................. ***Cymbomonas***

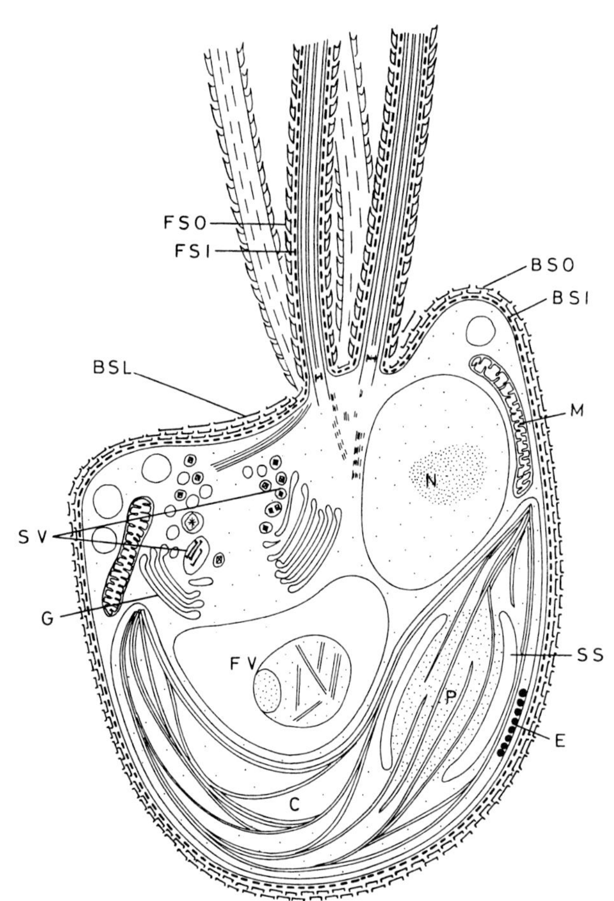

Fig. 5. *Cymbornonas tetramitiformis*. BSI, BSL, BSO, body scales; FSI, FSO, flagellar scales; C, chloroplast, E, eyespot; FV, food vacuole; G, Golgi body; M, mitochondrion; N, nucleus; P, pyrenoid; SS, starch grains; SV, scale vesicles. Not to scale. (From Thondsen, 1988).

Genus ***Cymbomonas*** Schiller, 1913

Bilaterally symmetrical cells with an apical depression surrounded by a horseshoe-shaped ridge. Four flagella emerge from the depression near the inner part of the horseshoe. With parietal chloroplast containing a pyrenoid in the posterior part. An eyespot is located on the pyrenoid surface. The cell body covered with a continuous layer of very small square scales overlain by 2 layers of regularly arranged larger ones. Flagella covered with square underlayer scales overlain by limuloid scales. Two almost opposite rows of hair-shaped scales are also present. *Cymbomonas* is unusual in its swimming mode which includes rapid swimming with the flagella held behind the cell and more slow swimming with the flagella anterior but directed backwards along the cell. Three species described, but only one is well known. See Throndsen (1988).

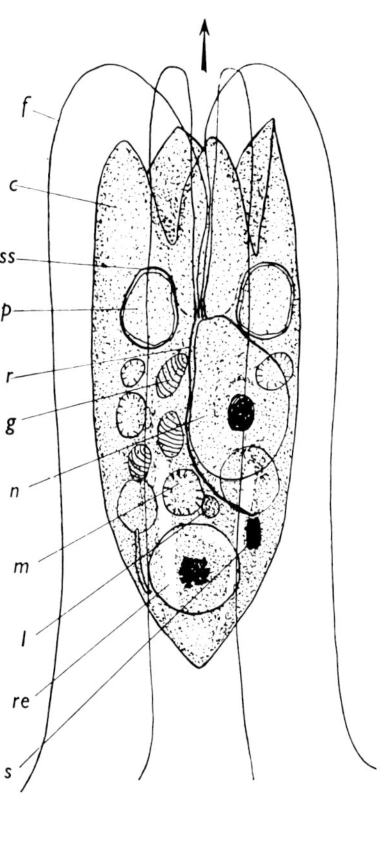

Fig. 6. Pyramimonas-stage of *Halosphaera viridis*. c, chloroplast; f, flagellum; g, Golgi body; l, lipid body; m, mitochondrion; n, nucleus; p, pyrenoid; r, rhizoplast; re, reservoir; s, eyespot; ss, starch sheath. x2500. (From Parke and den Hartog-Adams, 1965)

Cymbomonas. tetramitiformis Schiller, 1913 (Fig. 5). Cells almost isodiametric, 12-16 µm in diameter. Large body scales square. Two additional scale types present on the cell near the flagellar bases, one type small and elongate, the other complex plates (Throndsen, 1988; Moestrup, Inouye and Hori, unpubl.). Throndsen (1988) illustrates food vacuoles in this species, indicating mixotrophy. Very widely distributed marine flagellate.

Genus ***Halosphaera*** Schmitz 1879

Unicellular algae occurring in 2 stages, a coccoid so-called "Phycoma" stage and a motile flagellated stage. The Phycoma stage is spherical, up to 800 µm in diameter and surrounded by a two-layered wall. It contains numerous chloroplasts, usually with pyrenoids, and is initially uninucleate. At maturity it divides into a large number of somewhat *Pyramimonas*-like flagellates, each with 4 flagella emanating from an anterior pit. The single parietal chloroplast of the motile cell has 2 - 4 pyrenoids located along the lateral sides of the cell and a single eyespot in the posterior part of the cell. Cell surfaces with organic scales in several layers, on the body up to 4 types, including an underlayer of very small square scales and 2 layers of larger scales. Flagella with an underlayer of small square scales overlain by limuloid scales. Hair-shaped scales are located in two opposite rows.

The switch between the Phycoma stage and the motile cells appears to be related to the lunar cycle (Boalch & Parke 1971). The Phycoma stage contains large amounts of lipid and cells may accumulate at the sea surface in calm weather. The motile stage divides by fission. *Halosphaera* differs from the closely related *Pachysphaera* (see Parke 1966) by the cytoplasmic contents of the Phycoma stage being distributed peripherally in *Halosphaera* but filling the Phycoma in *Pachysphaera.* Motile cells of *Pachysphaera* are very similar to those of *Pterosperma*. *Pachysphaera* is sometimes merged with *Tasmanites*, a genus of fossil species.

Halosphaera is widely distributed in the marine plankton.

Halosphaera viridis Schmitz, 1879 (Fig. 6). *Phycoma* stage 400-800 µm in diameter when mature. Outer wall smooth. Motile cells 20-28 µm long, cell body covered with 4 types of scales, one type restricted to the flagellar pit.

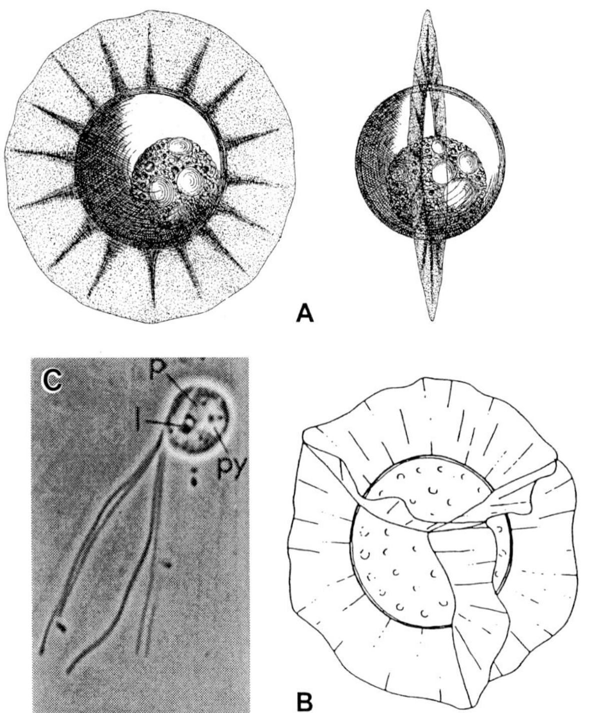

Fig. 7. *Pterosperma*. A. B, the Phycoma stage of *Pterosperma moebii* and *P. vanhoeffenii.* C, the motile stage. l, lipid globule; p, chloroplast; py, pyrenoid. A=360x. (From Cleve, 1900) and Meunier, 1910). B=400x. (From Throndsen, 1993). C=1000x. (From Parke et al., 1978).

Genus ***Pterosperma*** Pouchet, 1893

Unicellular algae occurring in 2 very different stages. The main, coccoid, Phycoma stage is spherical, 14-230 µm in diameter and surrounded by a double wall. The surface of the wall is smooth, papillose, or poroid and bears 1 or more wings protruding from the surface. The mature Phycoma is initially uninuclear and contains numerous chloroplasts, each with a pyrenoid. Cell divisions give rise to formation of a large number of motile cells, which are released through a slit in the outer wall. Each motile cell possesses 4 flagella, held together as a single, thick 'compound' flagellum behind the swimming cell. The motile cell contains a chloroplast with a pyrenoid but lacks an eyespot. Extrusomes are also present. All surfaces covered with organic scales, the flagella with an underlayer of small square scales overlain by limuloid ones, in addition to 2 opposite rows of hair-shaped scales. The cell body has a three-layered scale cover: an inner layer of small square scales and 2 layers of larger, octagonal or nonagonal scales (3-4 types). The change between Phycoma stage and motile cells is apparently determined by the lunar cycle, release of the motile cells occurring 2-3 days on either side of the new or full moon (Parke et al., 1978). The motile stage divides by fission. The Phycoma stage of *Pterosperma* may be misidentified as a pollen grain, resting spore, etc. It is widely distributed in the marine plankton. The genus probably comprises c. 20 species.

Pterosperma moebii (Jørgensen) Ostenfeld (Fig. 7). Phycoma with single, equatorial, undulate wing, 120-200 µm in total diameter. Width of the wing c. half the body diameter. The cell body subglobose to ellipsoidal with flattened poles. Motile cells 5-8 x 4-6 x 3-4 µm; flagella 4-7 times the body length. Cells covered with 4 types of large scales, in addition to small underlayer scales. This species, known from both the North Atlantic and the North Pacific, was studied in detail by Parke et al., (1978).

Genus ***Pyramimonas*** Schmarda 1850

More or less radially symmetrical cells with 4, 8 (rarely), or 16 flagella (one species only) emerging from an anterior depression in the cell, which is more or less inversely pyramidal. A single cup-shaped chloroplast and a basal pyrenoid. One or 2 eyespots. Freshwater species with a contractile vacuole system near the flagellar bases. Some species with tubular ejectile organelles. The 16-flagellate species differs in having 2 chloroplasts, 2 pyrenoids, and 4 eyespots. The flagella with square underlayer scales, limuloid scales and two opposite rows of hair-shaped scales on the flagella. Cell body covered with scales in several layers, underlayer scales on the body are overlain by 1 or 2 layers of larger, usually crown or box-like scales. Scales accumulate in a scale reservoir before release through a canal to the cell surfaces. Mainly distributed in the marine phytoplankton, a few from fresh water. Some species epiphytic or episammic. Probably around

50 species, but most freshwater species have not been adequately examined and some have been shown to belong to the genus *Hafniomonas* (Volvocales), which lacks scales. Marine species of *Pyramimonas* occur in all seas. *Prasinochloris* is a monotypic genus apparently related to *Pyramimonas*. Cells are unicellular epiphytes surrounded by a thick wall. They multiply by quadriflagellate zoospores resembling *Pyramimonas* (Belcher, 1966).

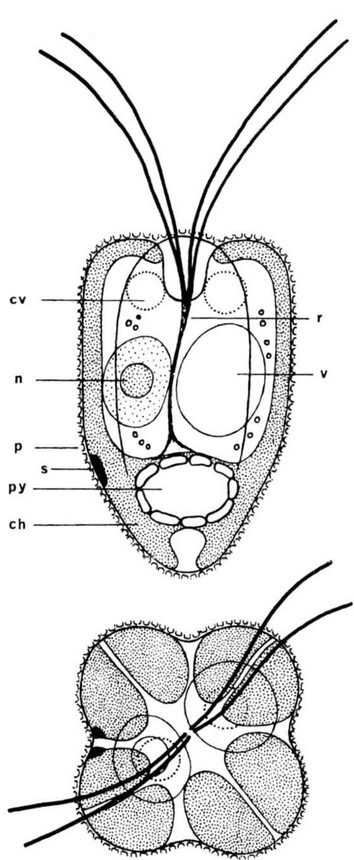

Fig. 8. *Pyramimonas tetrarhynchus* in side and top view. ch, chloroplast; cv, contractile vacuole; n, nucleus; p, scaly periplast (only one layer of scales indicated); py, pyrenoid; r, rhizoplast; v, vacuole. In top view the eight chloroplast-lobes are visible, two containing an eyespot. Not to scale. (Ettl and Moestrup, 1980).

Pyramimonas tetrarhynchus Schmarda, 1850 (Fig. 8). Cells 20-28 μm long, 12-20 μm wide near the front end. Chloroplast with 8 anterior lobes and 1 basal pyrenoid. Two eyespots in the posterior part of the cell. Ejectile organelles absent. Two contractile vacuoles. Underlayer body scales covered with 2 layers of larger scales, box-shaped scales overlain by crown-like ones. Asexual cysts known. Widely distributed fresh-water species. Type species of the genus.

FAMILY MESOSTIGMATACEAE
Fott ex Moestrup and Throndsen

Flagella with a single layer of small scales. Flagellar hair-scales not known. With a single genus.

Genus ***Mesostigma*** Lauterborn, 1894

Strongly flattened cells with 2 flagella arising subapically from a depression in 1 of the flat sides. Cells oval or rounded when seen from the flat sides. From the narrow sides concave-convex. The flagella emerge from the convex side. Chloroplast a curved plate, covering most of the cell periphery. One species with 2 pyrenoids, the other without pyrenoids. Eyespot near the middle of the concave side. A system of contractile vacuoles near the flagellar bases. Cells covered with several types of scales, the largest visible in the light microscope, giving the cell a punctate appearance. Two known species, both from fresh water.

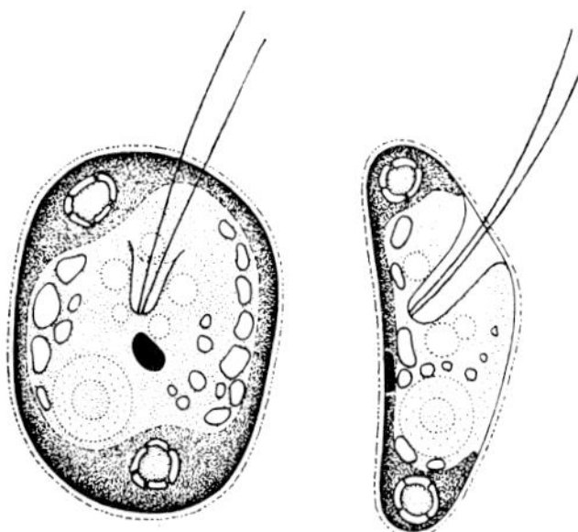

Fig. 9. *Mesostigma viride*. Upper view (left), side view (right). x1500. (From Manton and Ettl, 1965).

Mesostigma viride Lauterborn 1894 (Figs. 9-11). Cells 15-18 μm long, 11-13 μm wide, and 5-7 μm thick. Cells covered with scales of 3 types: small subrhomboid maple-like underlayer scales, usually confined to the upper cell surface, several intermediate layers of naviculoid scales, and an outer layer of larger basket-like ones. The flagella covered with the rhomboid type arranged in a single layer. Known from fresh water in Europe and North America.

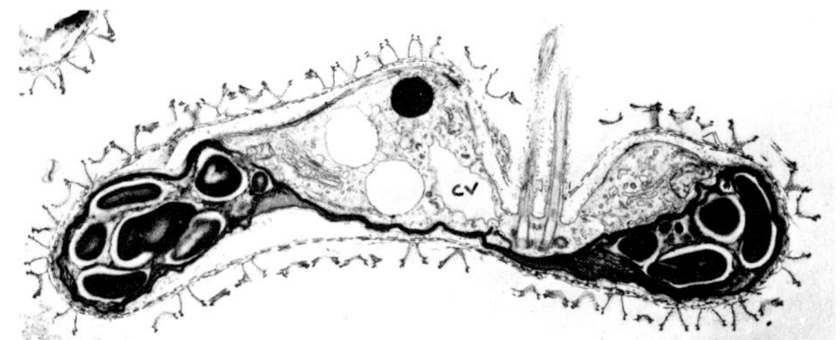

Fig. 10. *Mesostigma viride*. Longitudinal section of cel showing chloroplast swollen with starch at periphery and very flat in center, the two flagella in longitudinal section inserted in the conical pit, contractile vacuole (CV), and scale layers on body and flagella. x5000. (From Manton and Ettl, 1965)

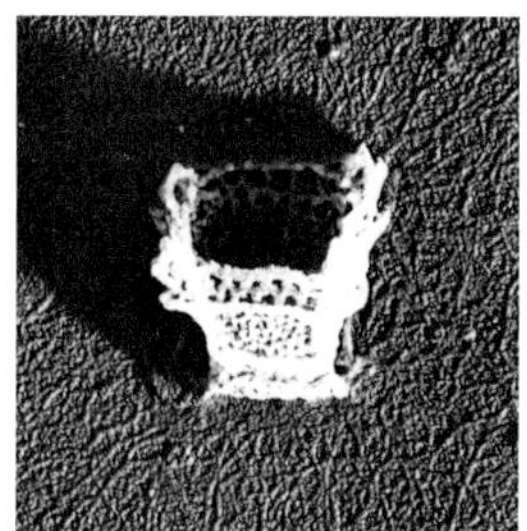

Fig. 11. *Mesostigma viride*. Single basket scale from body. Shadowcast, electron micrograph. x25,000. (Manton and Ettl, 1965).

Order Mamiellales Moestrup, 1984

Cell and flagella covered with spiderweb-like scales of several types. Some species lack scales but possess pigments similar to those of the scale-bearing species.

KEY TO THE FAMILIES

TEM

1.Cells covered with spiderweb-like scales.. **Mamiellaceae**
1´. Cells naked .... **Micromonadaceae**

FAMILY MAMIELLACEAE Moestrup, 1984

Cell and flagella covered with spiderweb-like scales.

KEY TO GENERA

1. Cells flagellate... 2
1´. Cells without flagella............... ***Bathycoccus***

2. Cells with two long flagella.......... ... ***Mamiella***
2´. Cells with one long and one stubby flagellum .... ***Mantoniella***

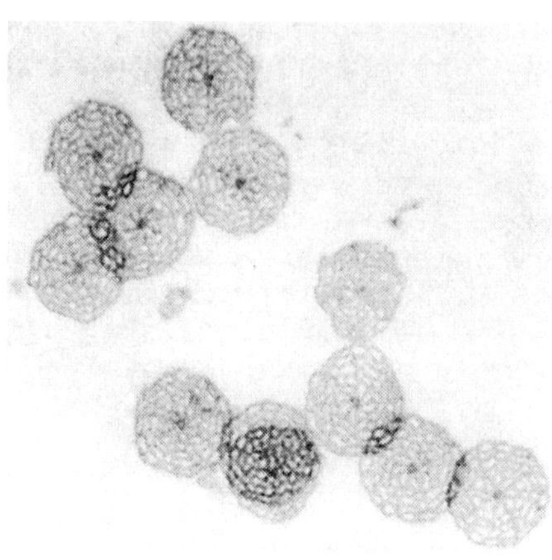

A

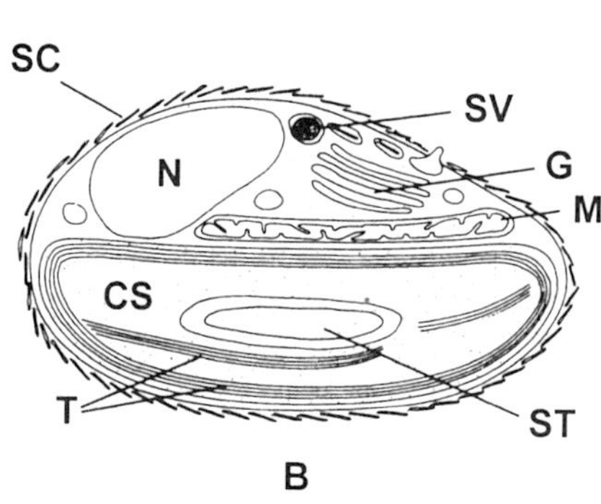

B

Fig. 12. *Bathycoccus prasinos*. A. Scales. B. The cell, including CS, chloroplast; G, Golgi body; M, mitochondrion; SC, scaly periplast; ST, starch; SV, scale vacuole; T, chloroplast thylakoids. A=60,000x. B=26,000x. (From Eikrem and Throndsen, 1990).

Genus *Bathycoccus*
Eikrem and Throndsen, 1990

Picoplanktonic marine unicell. Cells covered with spiderweb-like scales of a single type. Cells contain a parietal chloroplast but lack a pyrenoid.

Bathycoccus prasinos Eikrem and Throndsen, 1990 (Fig. 12). Cells 1.5-2.5 µm long, 1-2 µm wide. Described from 100-m depth in the Gulf of Naples (Eikrem and Throndsen 1990) and subsequently found in the north-eastern Atlantic and Norway. Probably overlooked in most areas.

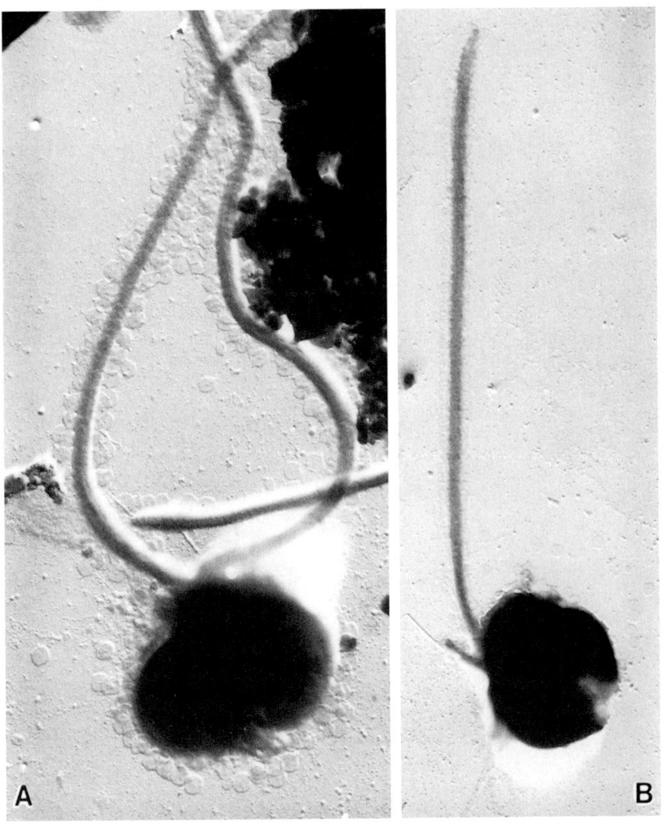

Fig. 13. A. *Mamiella gilva*. B. *Mantoniella squamata*. The scaly periplast on flagella and cell body are visible also at this low magnification. Shadowcast whole mounts. A=10,500x. (From Moestrup, 1984). B=6,300x. (From Moestrup, 1990).

Genus *Mamiella* Moestrup, 1984

Bean-shaped flagellates with 2 lateral flagella of nearly equal length inserted in the cavity. Cells contain a single parietal chloroplast with a pyrenoid. An eyespot located on the pyrenoid surface. With numerous muciferous bodies. Cell covered with 2 types of spineless spiderweb-like scales, the scales on the flagella with a central spine, resembling the limuloid flagellar scales of *Pyramimonas*. One described species; see Moestrup (1984). *Dolichomastix* was described by Manton (1977) based on shadowcast preparations. The 4 described species differ from *Mamiella* in scale structure, notably the lack of a spine on the flagellar scales, but the genus appears to be in need of critical revision (see also Throndsen and Zingone, 1997).

Mamiella gilva (Parke and Rayns) Moestrup, 1984 (Figs. 13A, 14). Cells 4-6.5 x 3-4.5 x 3-4 µm. Marine widely distributed flagellate.

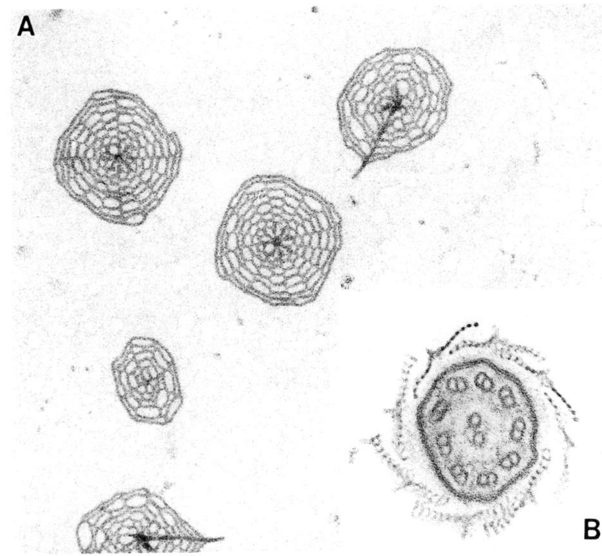

Fig. 14. *Mamiella gilva*. A. the three scale types, the spine-scales are from the flagella. B. flagellum in transverse section showing the layer of imbricated scales covering the surface. A=73,000x. Stained whole mount (orig.). B=75,000x. (From Moestrup, 1984).

Genus *Mantoniella* Desikachary

With 1 long and 1 very short flagellum which is readily overlooked. The long flagellum directed posteriorly during swimming. Cells contain a single parietal chloroplast with a pyrenoid. An eyespot is situated on the surface of the pyrenoid. With numerous muciferous bodies. Cells covered with spiderweb-like scales on cell body and flagella.

Scales on the flagella lack the spine characterizing *Mamiella.* Very widely distributed marine flagellate, comes often up in enrichment cultures. Two described species. See Barlow and Cattolico (1980); Marchant et al. (1989); Moestrup (1990)

Mantoniella squamata (Manton and Parke) Desikachary (Fig. 13B). Cells 2.4-4.0 µm in length. The long flagellum 2.5-4 times the cell length. The short stubby flagellum up to 1 µm long.

FAMILY MICROMONADACEAE nom. illeg.

Cells without scales.

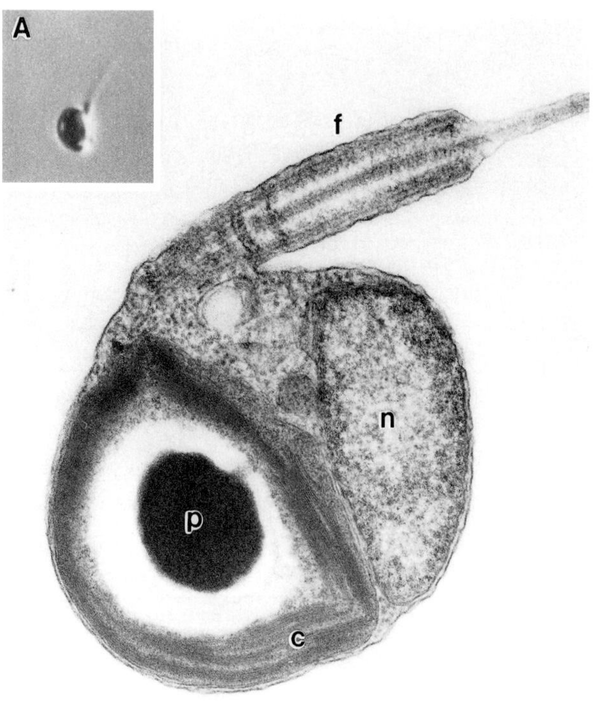

Fig. 15. *Micromonas pusilla*. A. Light micrograph, the cell was fixed in iodine to show the thin whip-like part of the flagellum. B. Transmission electron micrograph. c, chloroplast; f, flagellum; n, nucleus; p, pyrenoid. A=3000x, (From Moestrup, 1991). B=40,000x. (Courtesey of M. Vesk, orig.).

Genus ***Micromonas*** Manton and Park, 1960

Extremely small flagellate with single emergent flagellum, single parietal chloroplast and single pyrenoid. Eyespot absent. Cells naked. The axoneme-containing part of the flagellum very short but the central pair of microtubules extends into a long thin whip. This is usually visible only in the electron microscope. Healthy cells swim quickly with the short stub-like part of the flagellum directed backwards. One species. See Manton and Parke (1960).

Micromonas pusilla (Butcher) Manton and Parke, 1960 (Fig. 15). One of the smallest known flagellates, 1-3 µm long. Distributed in all seas and often very numerous. May readily be mistaken for a bacterium but the stubby flagellum is characteristic in moving cells. Non-moving. cells can hardly be identified in the light microscope.

ORDER CHLORODENDRALES

The flagella with an underlayer of small square scales in 24 rows overlain by 24 double rows of rod-shaped scales, in addition to hair-shaped scales in two opposite rows on the flagella. Some recently discovered marine picoplanktonic unicells without flagella also belong in this order. They resemble *Pseudoscourfieldia* in pyrenoid characters and in pigments.

KEY TO THE FAMILIES

1. Cells flagellate 2
1´. Cells without flagella......... **Pycnococcaceae**

2. Cells quadriflagellate, with distinct cell wall (periplast of fused scales)..**Chlorodendraceae**
2´. Cells biflagellate, covered with scales in two or more layers .... **Nephroselmidaceae**

FAMILY CHLORODENDRACEAE Oltmanns, 1904

Cells surrounded by a wall-like periplast of fused organic scales. The main constituent of the wall in *Tetraselmis* and *Scheffelia* are three 2-keto-sugar-acids as in scales of other prasinophytes (Becker et al., 1989, 1991).

KEY TO THE GENERA

1. Cells with basal pyrenoid. Cells more or less compressed, often slightly curved.. *Tetraselmis*

1′. Cells lacking pyrenoid. Cells markedly compressed and often twisted ...... ***Scherffelia***

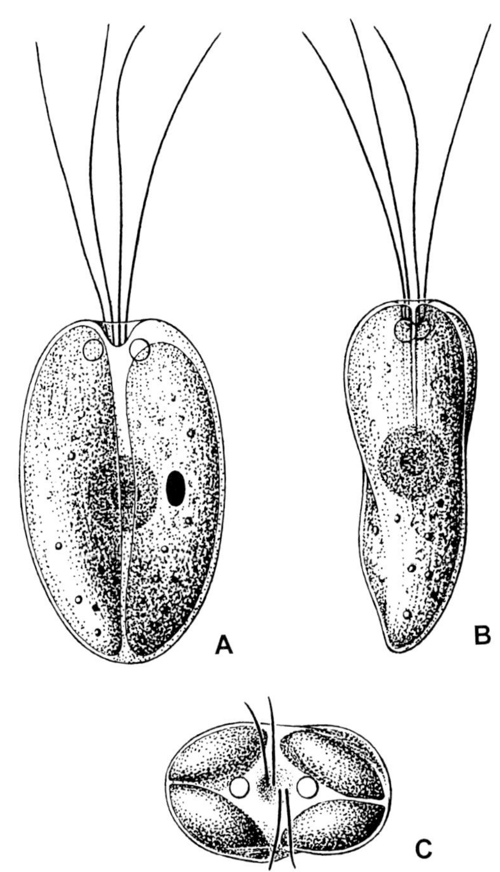

Fig. 16. *Scherffelia dubia,* in side views (A, B) and top view (C). Two lobed chloroplasts are present in each cell. X3150. (From Melkonian and Preisig, 1986)

Genus ***Scherffelia*** Pascher, 1911

Cells markedly compressed and often twisted, in face view more or less elliptical, oval, or heart-shaped. In lateral view narrower, rhombic, elliptic, or wedge-shaped. The 4 flagella emerge from an anterior pit, as 2 opposite pairs. Two lateral chloroplasts without pyrenoids. A single eyespot is usually present. Freshwater species contain 2 contractile vacuoles. Flagella covered with small square scales overlain by rod-shaped nes. Two opposite rows of hair-shaped scales. Asexual reproduction by longitudinal fission into 2 or 4 cells within the parental wall. Daughter cells are usually inverted with respect to each other. Rarely sighted flagellates, in fresh water and brackish water. Six species. See Melkonian and Preisig (1986).

Scherffelia dubia (Perty) Pascher, 1911 (Fig. 16). Cells 10-18 μm long, 7-12 μm wide (*forma maxima* 24-26 x 20-21 μm), in side view obovate to wedge-shaped, often with a short posterior protuberance. This species (the type species of the genus) was examined in detail by Melkonian and Preisig (1986), using TEM. Other species remain incompletely known.

Genus ***Tetraselmis*** Stein 1878

Cells more or less compressed, often slightly curved but never twisted. Cells cordiform, elliptic, or almost spherical. Similar to and very closely related to *Scherffelia* , but differing in cell shape and in the presence of a single chloroplast (very rarely 2), a pyrenoid, and an eyespot. Asexual reproduction as in *Scherffelia*. Thick-walled cysts known in several species, but sexuality not reported. Many species described, but the taxonomy requires revision. Marine species of *Carteria* probably also belong to *Tetraselmis*. Total number of species probably less than 50, distributed in both freshwater and marine biota, some species in plankton, others benthic in green sand, or in symbiosis with metazoa, e.g. the turbellarian *Convoluta roscoffensis* (Provasoli et al. 1968). See Hori et al. (1982, 1983, 1986).

Tetraselmis cordiformis (Carter) Stein, 1878 (Figs. 17, 18). Cells 14-20 μm long, 16-23 μm wide, 9-13 μm thick, flattened, elliptic, or nearly round when seen in side view. Flagellar length equal to or shorter than cell length. With cup-shaped chloroplast and a basal pyrenoid. Eyespot near the mid-region of the cell, 2 contractile vacuoles anteriorly. Cell wall distinct, sometimes slightly brownish or yellowish (Melkonian, 1979). Geographically widespread, in many types of freshwater.

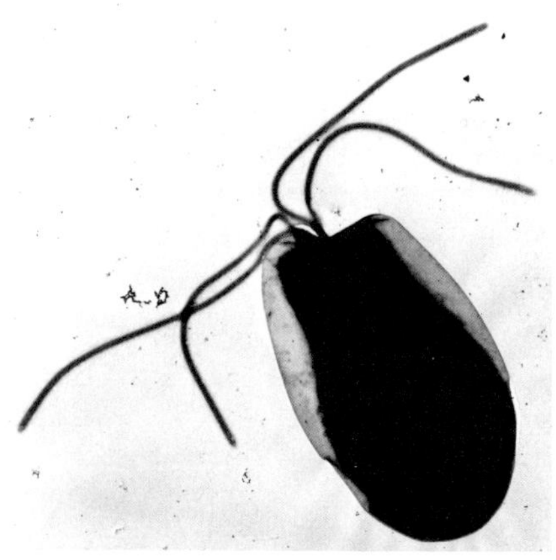

Fig. 17. *Tetraselmis marina*. Motile phase. Shadowcast whole cell. Electron micrograph, x2500. (From Parke and Manton, 1965).

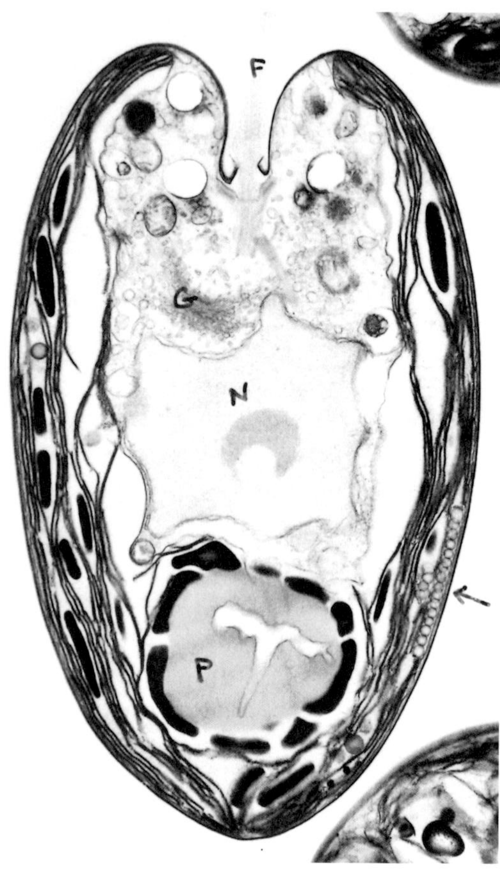

Fig. 18. *Tetraselmis cordiformis*. Longitudinal section of cell showing chloroplast around periphery (with stroma starch), pyrenoid (P) with cytoplasmic intrusions and starch shell, eyespot (arrow), nucleus (N), one of the Golgi bodies (G), one of the flagella (F) with scales, basal body and root to the nucleus, and the closely adpressed theca. x10,000. (From Manton and Parke, 1965).

FAMILY NEPHROSELMIDACEAE Skuja, 1948 ex P.C. Silva, 1980

The cell body enclosed in several layers of scales.

Key to the genera

1. Flagella laterally inserted. Shorter flagellum directed anteriorly during swimming, longer flagellum trailing. The flagella inserted at a wide angle to each other.. ***Nephroselmis***
1. Flagella inserted in the cell axis, both directed backwards during swimming, almost parallel .... ***Pseudoscourfieldia***

Fig. 19. *Nephroselmis olivacea*., a_1, a_2, starch grains around the pyrenoid; b, unknown body; ch, chloroplast; cv, contractile vacuole; g, Golgi body; n, nucleus; p, pyrenoid; s, eyespot. x2100. (From Moestrup and Ettl, 1979).

Genus **Nephroselmis** Stein, 1878

More or less bean-shaped cells with 2 unequal flagella inserted in the lateral invagination. The shorter flagellum directed anteriorly during swimming, the longer trailing. Cells with a single plate-like chloroplast and a pyrenoid in a lateral position opposite the flagellar insertion. An eyespot is present in most species, located near the base of the short flagellum. Freshwater species with a contractile vacuole system near the flagellar bases. Cells covered with up to 7 types of scales: the flagella with an underlayer of square scales overlain by pairs of rod-shaped scales and in some species by small stellate ones. Two opposite rows of hair-shaped scales. The cell body with underlayer of small square scales overlain by 1 - 3 layers of other types. These are usually stellate and belong to several morphological types. The flagellar pit sometimes with a special type of hair-shaped 'pit-scale.' Sexual reproduction by hologamy reported in one species (Suda et al., 1989). Geographically widespread in marine and freshwater plankton; less than 10 species known.

Nephroselmis olivacea Stein, 1878 (Fig. 19). Cells flattened, 6-10 µm long, 7-15 µm wide, and 3-6 µm thick. Short flagellum 1.5-2 times cell length, long flagellum 2-3 times cell length. Pyrenoid surrounded by 2 distinct starch grains. Flagella covered with 3 layers of scales in addition to hair-shaped scales. Cell body with scales in 4 layers, the three outer layers comprising stellate scales of 3 morphological types.

Widespread organism in fresh water. See Moestrup and Ettl (1979).

Genus ***Pseudoscourfieldia*** Manton 1975

Oblong-truncate cells with 2 distinctly anisokont flagella projecting from the longitudinal axis of the cell. The flagella are held directly backwards during swimming. The cell contains a disc-shaped chloroplast with a basal pyrenoid in the region directed anteriorly during swimming. Eyespot absent. Cell covered with 2 types of small body scales in 2 layers: an inner layer of square scales and an outer layer in which each scale apparently comprises a pair of rod-shaped scales with opposite polarity. Widespread but rarely reported marine nanoplankton organism. One species.

Pseudoscourfieldia marina (Throndsen) Manton, 1975 (Fig. 20). Cells flattened, 3.2-5 µm long, 2.4-3.2 µm wide, and 1.5-1.8 µm thick. Other characters above.

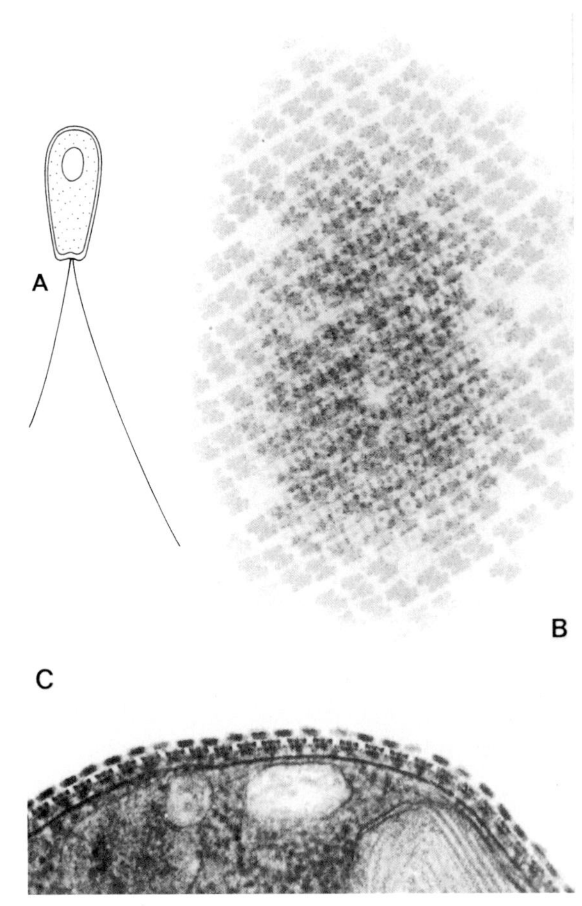

Fig. 20. *Pseudoscourfieldia marina*, an anisokont prasinophycean flagellate. A. Whole cell. B,C. Scaly periplast in tangential (B) and transverse section (C), showing underlayer of tiny square scales overlain by 'double-scales'. A=5400x. B=90,000x. C=75,000x. (After Moestrup and Throndsen, 1988).

FAMILY PYCNOCOCCACEAE Guillard, 1991

Members of this family (presently 3, including *Prasinoderma coloniale* Hasegawa and Chihara, a recently described colony-forming species from the Western Pacific (Hasegawa et al., 1996) all belong in the marine picoplankton. Species are unicellular (rarely colonial), surrounded by a wall, and lack flagella. A channel in the pyrenoid contains a protrusion from the mitochondrion, a very unusual character shared with the flagellated prasinophyte *Pseudoscourfieldia*. Studies on the Rubisco-gene have confirmed phylogenetic relationship between these coccoid species and *Pseudoscourfieldia* (Daugbjerg et al., 1995), but a relationship to the Mamieliales, as originally perceived, has not been confirmed. Like members of the Mamiellales, species of the Pycnococcaceae are probably secondarily reduced rather than primitive forms. The recently described *Ostreococcus tauri* Courties and Chrétiennot-Dinet (Chrétiennot-Dinet et al., 1995) differs in pyrenoid structure and its phylogenetic position is presently unknown.

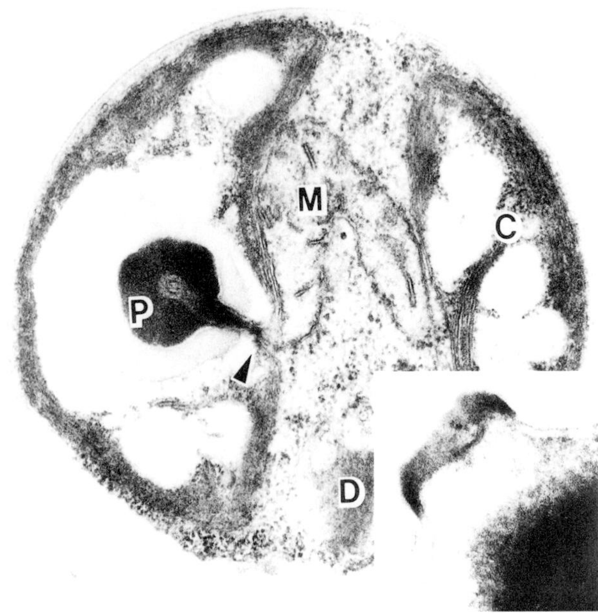

Fig. 21. *Pycnococcus provasolii*, a picoplanktonic member of the Prasinophyceae that lacks scales. The cell Fig. 21A) is surrounded by a wall, which at one point possesses an operculum-like structure (enlarged as Fig. 21. B). C, chloroplast; D,. dictyosome M, mitochondrion; P, pyrenoid. Fig. 21 A=28,000x. Fig. 21 B=48,000x. (From Guillard et al, 1991).

KEY TO THE GENERA

1. Cells with circular ring projecting slightly from one side, surrounded by circle of holes (visible only in EM) .... ***Prasinococcus***
1. Cells with an operculum............ ***Pycnococcus***

Genus ***Pycnococcus*** Guillard, 1991

Picoplanktonic unicells with single parietal chloroplast and 1-2 pyrenoids. Cell surrounded by wall-like structure. Flagella and eyespot lacking. A flagellated stage has been seen but not examined in detail. It possesses a single flagellum, which arises anteriorly, folds back along the cell, and is directed posteriorly during swimming. The cell has an operculum-like structure in the wall.

Pycnococccus provasolii Guillard, 1991 (Fig. 21). Cells spherical, 1.5-4 µm in diameter. This picoplanktonic coccoid cell is known from the North Atlantic and from the Gulf of Mexico (Guillard et al., 1991). It is probably generally overlooked because of its small size.

Genus ***Prasinococcus*** Miyashita and Chihara 1993

Unicellular coccoid cells lacking flagella and eyespot. Each cell contains a single parietal Chloroplast with a pyrenoid. A protrusion from the mitochondrion enters and fills a channel in the pyrenoid (see also *Pycnococcus*). Cells surrounded by a thick wall, comprising a well- defined thin inner layer and a very thick mucilaginous outer layer. A ring-shaped collar is present in one area of the wall, surrounded by large holes in the wall. Marine picoplanktonic cells recently described from the western Pacific (Miyashita et al., 1993). One species.

Prasinococcus capsulatus Miyashita and Chihara, 1993 (Fig. 22). Cells spherical, 3.5-5.5 µm in diameter. Found commonly from the surface to 200 m depth, in both temperate and tropical waters of the western Pacific.

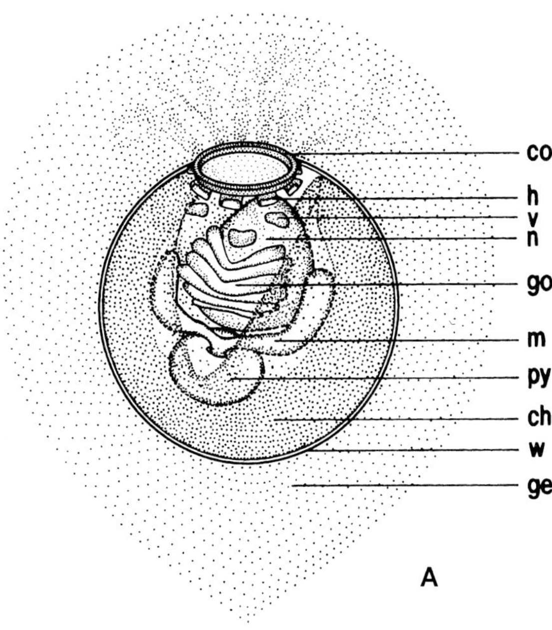

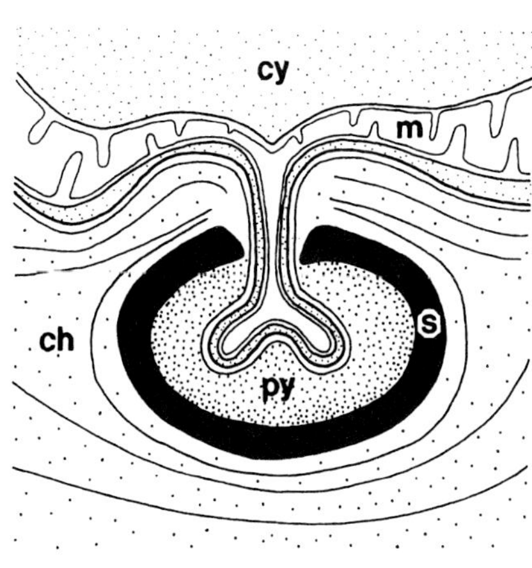

Fig. 22. *Prasinococcus capsulatus*. A. Whole cell. B. Detail of pyrenoid. ch, chloroplast; co, operculum surrounded by a ring of holes (h); cy, cytoplasm; ge, gelatinous envelope around the cell; go, Golgi body, m, mitochondrion; n, nucleus; py, pyrenoid, containing projection of the mitochondrion; s, starch grain; w, cell wall. (From Miyashita, 1993).

LITERATURE CITED

Barlow, S. B. & Cattolico, R. A. 1980. Fine structure of the scale-covered green flagellate *Mantoniella squamata* (Manton & Parke) Desikachary. *Br. Phycol. J.*, 15:321-333.

Becker, B., Becker, D., Kamerling, J. P. & Melkonian, M. 1991. 2-keto-sugar acids in green flagellates: a chemical marker for prasinophycean scales. *J. Phycol.*, 27:498-504.

Becker, B., Becker, D., Kamerling, J. P., Melkonian, M. & Vliegenthart, F. G. 1989. Identification of 3-deoxymanno-2-octulosonic acid, 3-deoxy-5-0-methyl-manno-2-octulosonic acid and 3-deoxy -lyxo-2-heptulosaric acid in the cell wall (theca) of the green alga *Tetraselmis striata* Butcher (Prasinophyceae). *Eur. J. Biochem.*, 182:153-160.

Belcher, J. H. 1966. *Prasinochloris sessilis* gen. et sp. nov., a coccoid member of the Prasinophyceae, with some remarks upon cyst formation in *Pyramimonas*. *Br. Phycol. Bull.*, 3:43-51.

Boalch, G. T. & Parke, M. 1971. The prasinophycean genera (Chlorophyta) possibly related to fossil genera, in particular the genus *Tasmanites*, *in* Farinacci, A. (ed.), *Proc. II. Planctonic Conf., Rome 1970*, Edizioni Tecnoscienza, Rome, pp. 99-15.

Chrétiennot-Dinet, M-J., Courties, C., Vaquer, A., Neveux, J., Claustre, H., Lautier, J. & Machada, M. C. 1995. A new marine picoeucaryote: *Osterococcus tauri* gen. et sp. nov. (Chlorophyta, Prasinophyceae). *Phycologia*, 34:285-292.

Daugbjerg, N., Moestrup, Ø. & Arctander, P. 1995. Phylogeny of genera of Prasinophyceae and Pedinophyceae (Chlorophyta) deduced from molecular analysis of the rbcL gene. *Phycological Research*, 43:203-213.

Eikrem, W. & Throndsen, J. 1990. The ultrastructure of *Bathycoccus* gen. nov. and *B. prasinos* sp. nov., a non-motile picoplanktonic alga (Chlorophyta, Prasinophyceae) from the Mediterranean and Atlantic. *Phycologia*, 29:344-350.

Guillard, R. L., Keller, M. D., O'Kelly, C. J. & Floyd, G. F. 1991. *Pycnococcus provasolii* gen. et sp. nov. a coccoid prasinoxanthin-containing phytoplankter from the western North Atlantic and Gulf of Mexico. *J. Phycol.*, 27:39-47.

Hasegawa, T., Miyashita, H., Kawachi, M., Ikemoto, H., Kurano, N., Miyachi, S. & Chihara, M. 1996. *Prasinoderma coloniale* gen. et sp. nov., a new pelagic coccoid prasinophyte from the Western Pacific. *Phycologia*, 35:170-176.

Hori, T. 1983. Studies on the ultrastructure and (Prasinophyceae). II. Subgenus *Prasinocladia*. *Bot. Mag. Tokyo*, 96:385-392.

Hori, T. 1986. Studies on the ultrastructure and taxonomy of the genus *Tetraselmis* (Prasinophyceae). III. Subgenus *Parviselmis*. *Bot. Mag. Tokyo*, 99:123-135.

Hori T, Norris, R. E. & Chihara, M. 1982. Studies on the ultrastructure and taxonomy of the genus

Tetraselmis (Prasinophyceae). I. Subgenus *Tetraselmis*. *Bot. Mag. Tokyo*, 95:49-61.

Manton, I. 1977. *Dolichomastix* (Prasinophyceae) from Arctic Canada, Alaska and South Africa: a new genus of flagellates with scaly flagella. *Phycologia*, 16:427-438.

Manton, I. & Parke, M. 1960. Further observations on small green flagellates with special reference to possible relatives of *Chromulina pusilla* Butcher. *J. Mar. Biol. Ass. U.K.*, 39:275-298.

Marchant, H. J., Buck, K. R., Garrison, D. L. & Thomsen, H. A. *1989*. *Mantoniella* in antarctic waters including the description of *M. antarctica* sp. nov. (Prasinophyceae). *J. Phycol.*, 25:167-174.

Melkonian, M. 1979. An ultrastructural study of the flagellate *Tetraselmis cordiformis* Stein (Chlorophyceae) with emphasis on the flagellar apparatus. *Protoplasma*, 98:139-151.

Melkonian, M. 1990. Phylum Chlorophyta. Class Prasinophyceae. *In*: Margulis L., Corliss, J. O., Melkonian, M. M. & Chapman, D. J., (eds.) *Handbook of Protoctista*. Jones and Bartlett, Boston. pp 600-607.

Melkonian, M. & Preisig, H. 1986. *A* light and electron microscopic study of *Scherffelia dubia,* a new member of the scaly green flagellates (Prasinophyceae). *Nord. J. Bot.*, 6:235-256.

Miyashita, H., Ikemoto, H, Kurano, N., Miyachi, S. & Chihara, M. 1993. *Prasinococcus capsulatus* gen. et sp. nov., a new marine coccoid prasinophyte. *J. Gen. Appl. Microbiol.*, 39:571-582.

Moestrup, Ø. 1984. Further studies on *Nephroselmis* and its allies (Prasinophyceae). II. *Mamiella gilva* gen. nov., Marmellaceae fam. nov., Mamiellales ord. nov. *Nord. J. Bot*, 4:109-121.

Moestrup, Ø. 1990. Scale structure in *Mantoniella squamata,* with some comments on the phylogeny of the Prasinophyceae (Chlorophyta). *Phycologia*, 29: 437-442.

Moestrup, Ø. & Ettl, H. 1979. *A* light and electron microscopical study of *Nephroselmis olivacea* (Prasinophyceae). *Opera Bot.*, 49:1-39.

Moestrup, Ø. & Throndsen, J. 1988. Light and electron microscopical studies on *Pseudoscourfieldia marina* a primitive scaly green flagellate (Prasinophyceae) with posterior flagella. *Can. J. Bot.*, 66: 1415-1434.

Parke, M. 1966. The genus *Pachysphaera* (Prasinophyceae), *In*: Barnes, H (ed.) *Some Contemp. Studies in Mar. Science,* Allen & Unwin Ltd., London, pp 555-563.

Parke M, Boalch GT, Jowett R. & Harbour IDS. 1978. The genus *Pterosperma* (Prasinophyceae): species with a single equatorial ala. *J. Mar. Biol. Ass. U.K.* 58: 239-276.

Provasoli, L., Yamasu, T. & Manton, I. 1968. Experiments on the resynthesis of symbiosis in *Convoluta roscoffensis* with different flagellate cultures. *J. Mar. Biol. Ass. U.K.* 48:465-479.

Suda, S., Watanabe, M. M. & Inouye, I. 1989. Evidence for sexual reproduction in the primitive green alga *Nephroselmis olivacea* (Prasinophyceae). *J. Phycol.*, 25:596-600.

Sym, S. D. & Pienaar, R. N. 1993. The Class Prasinophyceae, *In*: Round R. E. & Chapman D. J. (eds.) *Prog. Phycol. Res.*, 9:281-376. Biopress Ltd., Bristol,

Tappan, H. 1980. *The Paleobiology of Plant Protists.* W.H. Freeman and Co., San Francisco.

Throndsen, J. 1988. *Cymbomonas* Schiller (Prasinophyceae) reinvestigated by light and electron microscopy. *Arch. Protistenk.,* 136:327-336.

Throndsen, J. & Zingone, A. 1997. *Dolichomastix tenuilepis* sp. nov., a first insight into the microanatomy of the genus *Dolichomastix* (Maliellales, Prasinophyceae, Chlorophyta): *Phycologia*, 36:244-254.

ORDER PRYMNESIIDA

J. C. GREEN and R. W. JORDAN

The Prymnesiida is considered here to be the equivalent of the algal division Haptophyta (Hibberd, 1972), class Prymnesiophyceae (Hibberd, 1976). It comprises a group of organisms, many of them flagellates, that were for many years retained as a group within the Chrysomonadida (algal class Chrysophyceae). However, pioneering fine-structural studies on species of *Chrysochromulina* demonstrated a number of unusual features, including the presence of an organelle unique to the group, the **haptonema** (Parke et al., 1955), one or more layers of unmineralized body-scales, and, usually, two equal or sub-equal smooth flagella (Parke et al., 1955, 1956, 1958, 1959). At the same time, von Stosch noted the presence of an abbreviated haptonema in a flagellated coccolithophorid (Stosch, 1958), thus linking the unmineralized forms with the better known calcified 'chrysomonads'. This led to the erection of a new algal class, the Haptophyceae (Christensen, 1962), subsequently renamed by Hibberd (1976) using the typified form Prymnesiophyceae. Recently, it has been accepted by most workers that the haptophytes should be separated from other protists at the level of division (phylum), possibly as part of a separate kingdom, Chromista (Cavalier-Smith, 1989, 1993). Within this phylum, some recent schemes include either one class and two sub-classes (Cavalier-Smith, 1989; Jordan and Green, 1994) or two classes (Cavalier-Smith, 1993, 1994), separating *Pavlova* and related genera, whose representatives lack plate-scales and have **anisokont flagella**, from other haptophytes (see below). Here the two subgroups are referred to, using zoological terminology, as the suborders Pavlovina and Prymnesiina. For a recent review of taxonomic systems within the algal division Haptophyta, see Green and Jordan (1994).

The most familiar members of the order Prymnesiida are probably the coccolithophorids, whose cells characteristically carry one or more layers of calcified scale-like structures, the **coccoliths**. In those coccolithophorids that have been studied in detail, it has been observed that the coccoliths often overlie a layer of unmineralized organic plate-scales, similar to those found in other members of the Prymnesiida (suborder Prymnesiina) and, in a number of species, the coccoliths are known to develop by the deposition of calcite on a base-plate scale or other organic matrix (for a review, see Pienaar, 1994).

The basic scale-like form of the coccolith is often obscured by extreme elaboration to produce a diversity of complex structures characteristic of the species; indeed, coccolithophorid taxonomy depends heavily on coccolith morphology as relatively few species have been described from living material. However, this can be misleading as it is now known (Parke and Adams, 1960; Thomsen et al., 1991) that at least some coccolithophorids have complex life-cycles involving alternation of calcified and uncalcified forms or forms with different types of coccolith. In some cases, the different forms involved in the life-cycle have been described as independent species.

The systematics of the unmineralized forms also depends heavily on scale and, as in the coccolithophorids, the basic plate-scale form can be strongly modified to form spines, tubs, cylinders, and other types of scale(e.g. Parke et al., 1955, 1956, 1958, 1959; Manton and Parke, 1962; Leadbeater, 1972; Manton and Leadbeater, 1974; Manton, 1978a,b; Estep et al., 1984, *inter alia*). A number of prymnesiomonad species, however, are now available in culture, and there is information on, for example, cell morphology, biochemical characteristics (fatty acids, sterols, alkenones, alkenoates, alkenes; for a review, see Conte et al., 1994), pigment characteristics (see reviews by Jeffrey, 1989; Bjørnland and Liaaen-Jensen, 1989; Jeffrey and Wright, 1994), cell division (review by Hori and Green, 1994), flagellar root systems (reviews by Inouye, 1993; Green and Hori, 1994) and there are also recent data from molecular genetic studies (Barker et al., 1994; Fujiwara et al., 1994; Medlin et al., 1994a,b, 1996), the work of Fujiwara et al. (1994) supporting the separation of the Pavlovina from the remainder of the Prymnesiida and confirming that relationships in the rest of the division are in urgent need of reassessment. It is still difficult, therefore, to deduce relationships between genera, and Christensen (1980) and Green and Jordan (1994) and Jordan and Green (1994) in their phycological classifications have included all members of the sub-class Prymnesiophycidae in one order, the Prymnesiales, and all members of the sub-class Pavlovophycidae in one order, the Pavlovales. This is certainly a temporary arrangement pending the acquisition of further information; however, it does reconcile, for the present, the grouping of some coccolithophorids with principally non-coccolithophorid genera on the grounds of other shared characters, and vice-versa.

The Prymnesiida as a group have a fossil record extending back to the late Triassic/Early Jurassic (c. 200 Ma) where there are abundant coccolith remnants and other cryptic calcareous nanofossils (Perch-Nielsen, 1985; Green et al., 1990; Young et al., 1994) and the formation of extensive chalk deposits is a measure of their success during the

Cretaceous (c. 125 Ma). Coccolith synthesis, however, is linked to processes involving the Golgi body, unmineralized scales, and photosynthesis (Pienaar, 1994) and it is clear, therefore, that the history of prymnesiomonads as a whole must extend further back than the Triassic, though this cannot be traced in view of the lack of fossil material. Some species, e.g. *Emiliania huxleyi* (Lohmann) Hay & Mohler and *Gephyrocapsa oceanica* Kamptner, have been been shown to produce unusual lipids (alkenones etc.) that are conserved in sediments and provide useful markers in dating sediments and in palaeoclimatological studies (Conte et al., 1994, 1995; Volkmann et al., 1995).

Modern members of the Prymnesiida form an important component of the phytoplankton, especially in the sea (Thomsen et al., 1994; Marchant and Thomsen, 1994) sometimes occurring in 'bloom' concentrations. For example, the coccolithophorids *Coccolithus pelagicus* (Wallich) J. Schiller and *Emiliania huxleyi* (Lohmann) Hay & Mohler frequently form large blooms in the North Atlantic and blooms of *E. huxleyi* occur regularly in the North Sea and Norwegian coastal waters with concentrations up to 115 x 10^6 cells l^{-1} (Berge, 1962) having been recorded. Such blooms can be detected using remote sensing techniques as the discarded coccoliths are highly reflective, and images of recent blooms in the North Atlantic have shown that they may extend over wide areas (thousands of km^2; Holligan et al., 1983, 1993). These massive blooms of coccolithophorids have an impact on the transfer of carbon from the photic zone to the ocean depths through coccolith sedimentation and also on ocean/atmosphere CO_2 equilibria. Blooms of prymnesiomonads, including coccolithophorids and unmineralized forms, such as species of *Phaeocystis,* have also been implicated in the production of atmospheric sulphur compounds. These are derived from the photo-oxidation of labile DMS (dimethyl-sulphide), itself derived from intracellular DMSP (dimethyl-sulphoniopropionate) (Malin et al., 1994). Blooms of species of *Phaeocystis* also induce changes in food-webs, affect fisheries, and cause massive foaming on beaches (see reviews by Davidson and Marchant, 1992; Lancelot and Rousseau, 1994; Moestrup, 1994).

At least two species of *Chrysochromulina*, *C. polylepis* Manton & Parke (Dahl et al., 1989) and *C. leadbeateri* Estep et al. (Rey, 1991; Aune et al., 1992) are toxic to a variety of organisms, and old cultures of *C. brevifilum* Parke & Manton and *C. kappa* Parke & Manton have been shown to be toxic to the bryozoan *Electra pilosa* (Jebram, 1980). There is also circumstantial evidence for toxicity in *C. parva* Lackey (Hansen et al., 1994). For further literature on the toxic blooms of *Chrysochromulina*, see Granéli et al. (1990), Rey (1991), Smayda and Shimizu (1993), and Moestrup (1994). Icthyotoxins are also produced by species of *Prymnesium*, including *P. parvum* N. Carter and *P. patelliferum* Green et al. Fish-kills attributed to species of *Prymnesium* have been reported from N. Europe to China (Chen and Tseng, 1986; Moestrup, 1994). The toxic component comprises a family of similar substances, with cytotoxic, haemolytic, and neurotoxic activity which particularly affect the membranes of the gills (Shilo, 1981). The composition of the toxin has been discussed by Ulitzur and Shilo (1970), Shilo (1971), Paster (1973) and Kozakai et al. (1982).

Some of the main features of members of the Prymnesiida are summarized below, but for detailed discussions of prymnesiomonad biology and ecology, see Hibberd (1980); Green et al. (1990), Winter and Siesser (1994), and Green and Leadbeater (1994).

Key Characters

1. The **haptonema**: The haptonema is inserted close to the flagella so that the two flagellar bases and the haptonematal base form a triangle, with the latter situated dorsally (terminology of Beech and Wetherbee, 1988) with respect to the axis joining the flagellar bases. The haptonematal base is linked to the flagellar bases and to other organelles by a number of fibers (Inouye, 1993; Green and Hori, 1994).

Information on the haptonema has been summarized by Inouye and Kawachi (1994). In its most developed form it can be very long (up to approx. 120 μm in some species of

Chrysochromulina), and it can actively coil and uncoil. However, it may also be short and non-coiling (e.g. species of *Prymnesium* or the motile cell of *Phaeocystis* and species of *Pavlova*) or it may be rudimentary (e.g. *Isochrysis galbana* Parke) or absent (e.g. the motile cell of *Emiliania huxleyi*). Although associated with the flagella, it differs in structure. Typically, in the emergent part, there is an axoneme of seven microtubules (occasionally six, eight or, rarely, 14) surrounded by a fenestrated cylindrical cisterna, confluent with a peripheral cisterna lying just beneath the plasma membrane and interpreted as an extension of the ER. There is often electron-dense material situated in the central core of the haptonema linking the microtubules to each other, and ribosomes may also be present. Immediately distal to the cell, the microtubules become arranged to form a C-shaped group with the open side of the C facing the left flagellum. At the same level, the cylinder of ER forms a finger-like projection of the inner membrane which extends into the arc of microtubules. At this level, the cytoplasmic core is electron-dense, giving the appearance of a membranous partition across the haptonema when viewed in LS. Within the cell, the microtubules become rearranged to form at first two superposed rows and, subsequently, following the addition of further microtubules, a hexagonally close-packed group with two additional microtubules peripherally.

In the rudimentary haptonemata of species of *Chrysotila* and in *Isochrysis galbana* there are only three to five microtubules, respectively, with no increase in number in the haptonematal base. In species of *Pavlova*, the number of microtubules is rapidly attenuated in the free part, though it increases proximally to give the normal basal structure. Some coccolithophorids also have a modified bulbous free part with six microtubules surrounded by fragmented profiles of ER.

The functional behavior of the hapotonema has recently been studied by Kawachi et al. (1991) and Kawachi and Inouye (1995). In *Chrysochromulina hirta* Manton particulate material is aggregated at the haptonematal base, apparently by dynamic activity of the haptonematal surface, and subsequently translocated back to the tip before transfer, by haptonematal bending, to the antapical part of the cell where it is ingested (Kawachi et al., 1991). In *C. ericina* Parke et al., particulate material is aggregated on the spine-scales before being swept up by the haptonema and transferred to the cell surface before ingestion (Kawachi and Inouye, 1995). Haptonematal coiling, which can be a very rapid process (faster than 5.0 milliseconds) in contrast to uncoiling which is always slow (seconds), is accompanied by changes in flagellar activity and appears to be an avoidance reaction. Gregson et al. (1993b) and Inouye and Kawachi (1994) have demonstrated that coiling and uncoiling is calcium-mediated, but the mechanism by which the mechanical forces are transmitted are unknown.

2. *Flagella*: The arrangement of the flagella and basal bodies in the Prymnesiida has been reviewed recently by Inouye (1993) and Green and Hori (1994). Members of the Prymnesiida typically have two flagella (occasionally more) associated with the haptonema, the flagellum whose basal body faces the open side of the C-shaped arc of haptonematal microtubules being referred to as the left flagellum. Ontogenetic studies (Beech et al., 1988) have also demonstrated that this is the mature or No. 1 flagellum. Members of the Prymnesiina typically have equal, sub-equal, or occasionally unequal flagella, sometimes with a fine **tomentum** (Day et al., 1986; Birkhead and Pienaar, 1994), and often with attenuated tips to form short or long hair-points. The flagella are inserted either apically (many prymnesiomonads), sub-apically (e.g. species of *Prymnesium)* or laterally (e.g. flattened species of *Chrysochromulina*). They may be inserted in a depression or a groove (e.g. *Corymbellus aureus* Green) or may emerge from a **papilla** (e.g. *Imantonia rotunda* Reynolds). Flagellar action is variable, being either **homodynamic** or **heterodynamic** and may vary within one species. Within the cell, the basal body is associated with a number of microtubular roots, typically four, though not all may be present and there may be additional roots in some species. Root 1 is associated with the left basal body and the haptonema, passing from the haptonema around the left side of the basal body. Root 2 has its origin close to a distal fiber connecting the basal bodies and then passes ventrally into the cell, whilst roots

3 and 4 arise each side of the right basal body and follow a course along the dorsal side of the cell, sometimes combining to form a composite root. The microtubular roots may consist of few microtubules (e.g. *Imantonia rotunda, Chrysochromulina apheles* Moestrup & Thomsen) or large bundles of microtubules (e.g. compound roots 1 and 2 of some coccolithophorids). Within the cell, the basal bodies are connected to each other, to the haptonema, and to other cell organelles by a number of fibers.

Members of the Pavlovina have two unequal flagella inserted subapically or laterally. The longer flagellum carries an array of small dense bodies interpreted either as scales (Leadbeater, 1994) or modified hairs (Cavalier-Smith, 1994), and non-tubular hairs. The shorter flagellum is often distally attenuated, does not have scales and may be reduced to a vestigial structure. The flagellar action is markedly heterodynamic. There are two microtubular roots associated with the basal body of the shorter flagellum, which represents the left or mature flagellum.

3. *Chloroplasts*: Cells may have one, two, or four chloroplasts usually situated parietally. Lamellae consist of three thylakoids except those penetrating the pyrenoid which are composed of two only. There are no girdle lamellae. The chloroplasts are enclosed in periplastid ER that is confluent with the nuclear envelope. Pyrenoids are often immersed within the chloroplasts and in some species may be enclosed by a fine single layered membrane. In some species, however, the pyrenoid may be only semi-immersed forming a bulge on the inner face of the chloroplast, and in a few species, it may be stalked, connected to the chloroplast by a narrow neck. Such pyrenoids are, nevertheless, part of the chloroplast being surrounded by the chloroplast membrane and the periplastid ER. They may also be capped by a shell of poorly staining material, probably reserve metabolite.

4. *Eyespots*: Eyespots (stigmata) have been recorded only in some species of *Pavlova* and related genera. In *Pavlova pinguis* Green, for example, the eyespot is a group of osmiophilic globules at the inner periphery of the chloroplast, clustered round the inner end of a pit or canal penetrating the cell close to the flagellar insertion. In *Diacronema vlkianum* Prauser, the eyespot lies at the outer periphery of the chloroplast about half-way along the cell, beneath the short flagellum which carries a unique swelling on the side towards the cell surface.

5: *Pigments*: In bloom conditions or in culture, prymnesiomonads may vary from golden-brown to greenish-gold, but individual cells often appear as pale yellow-green. With respect to pigment content, the prymnesiomonads fall into four groups (Jeffrey and Wright, 1994). All prymnesiomonads examined to date have chlorophylls *a*, c_1 + c_2, ββ-carotene, diadinoxanthin, and diatoxanthin. Type 1 prymnesiomonads have, in addition, fucoxanthin; type 2 have chlorophyll c_3 and fucoxanthin; type 3 have chlorophyll c_3 and 19'-hexanoyloxy-fucoxanthin, fucoxanthin, and a trace of 19'-butanoyloxyfucoxanthin, and type 4 have chlorophyll c_3, 19'-butanoyloxyfucoxanthin, together with variable quantities of fucoxanthin and 19'-hexanoyloxy-fucoxanthin (Jeffrey and Wright, 1994). However, information is needed from yet more species before the taxonomic value of such information can be fully appreciated, since at present not all prymnesiomonads can be fitted into particular pigment types.

6. *Reserve metabolites*: The presence of a β-D-glucan with a (1-6) linked backbone and short (1-6) side chains has been confirmed in *Emiliania huxleyi* (Vårum et al., 1986). Leucosin (chrysolaminarin), a β1-3 glucan, is also said to occur, but this has not been confirmed. Reports of paramylon in species of *Pavlova* have now been shown to be incorrect (Kiss and Triemer, 1988).

7. *Golgi body*: The prymnesiomonad Golgi body has a number of characteristic features. It is situated close to the flagellar apparatus, with the cisternae bunched together beneath the basal bodies, so that in a section passing through the dictyosome from forming to mature face and transverse with respect to the cisternae, the Golgi body has a fan-shaped profile. The Golgi body is actively involved in scale synthesis which takes place from the forming face, subtended by an ER vesicle, to the mature face, the completed scales being released in vesicles and transported to the cell membrane. In the central part of the Golgi body, the cisternae may be swollen with densely staining material and this seems to be unique to the

Prymnesiida ('peculiar Golgi'; Parke et al., 1959; Manton, 1966, 1967).

8. *Cell covering*: The basic unit of cell covering in the Prymnesiina is a two-layered scale composed of microfibrils, those on the proximal face being arranged radially in a pattern of four quadrants. On the distal face, the microfibrils are often spirally wound though the pattern on this face may be obscured by elaboration, such as the formation of spines, cups, or tubes, or by suppression of the distal face completely. Cells of a particular species may carry several different types of scale, and these may be arranged in different layers with the more elaborate scales outermost. Certain scales may be confined to particular parts of the cell, e.g. spine-scales at the poles, or small plate-scales covering reduced haptonemata. The chemical nature of scales is complex. The principal components of the organic scales of *Pleurochrysis carterae* (Braarud & Fagerland) Christensen are (1) the sulphated polysaccharide of the proximal radial fibrils, (2) a cellulose-like polysaccharide and protein forming the distal spiral microfibrils, (3) a glycopeptide covering the spiral microfibrils, and (4) an amorphous matrix of acidic polysaccharide. Scales are formed in the Golgi body (see above) in the sequence radial fibrils, spiral microfibrils, then amorphous material.

Calcified scales (coccoliths) may occur as a layer overlying unmineralized scales, but in some coccolithophorids, the unmineralized scales are apparently absent. In *Pleurochrysis carterae* and some other coccolithophorids, calcification takes place on a base-plate scale, though such a scale is absent in, for example, *Emiliania huxleyi*. In heterococcolithophorids, coccolith formation is intracellular, either mediated by the Golgi body or by a special coccolith vesicle, but there is evidence that in holococcolithophorids calcification may take place extracellularly, beneath an enveloping 'skin' that encloses the cell and scale and coccolith layers (Rowson et al., 1986). For detailed discussions of scale and coccolith formation, see Leadbeater and Green (1993), Leadbeater (1994) and Pienaar (1994).

The filamentous benthic stages of *Pleurochrysis carterae* have been shown to have thick walls composed of many layers of closely adpressed scales that presumably act as a protection against desiccation. Similar protection is afforded by the thick wall of lamellated mucilage in non-motile cells of *Chrysotila*.

Plate-scales have not been recorded in the Pavlovina. Knob-like structures occur, principally on the longer flagellum and these, since they are formed in the Golgi body, have been interpreted as scales (Green, 1980). However, Cavalier-Smith (1994) has suggested that they may be modified hairs.

9. *Cell shape*: Many species are planktonic unicellular flagellates of variable form, the cells of some species being more or less spherical whilst others are elongate, flattened, or saddle-shaped. Many coccolithophorids, though planktonic, are known only as non-motile cells. In other prymnesiomonad species, the cells may be arranged in filaments, palmelloid masses, or colonies, possibly alternating with other stages (see below). For discussions on prymnesiomonad cell form and its terminology, see Throndsen (1993), Heimdal (1993), Jordan et al. (1995).

10. *Locomotion*: Locomotion varies between and within species. Many species swim with the haptonema directed forwards and the flagella, beating homodynamically or heterodynamically, directed outwards or backwards. However, in a number of species both flagella and haptonema may be directed backwards during swimming, though this may change rapidly if the cell encounters an obstacle. In *Pavlova*, the long flagellum is always directed forwards, beating with an S-form wave, whilst the short flagellum is directed laterally and beats stiffly.

11. *Nutrition*: Most species are photosynthetic and many are mixotrophic, supplementing photosynthesis by osmotrophy or phagotrophy (for reviews, see Green, 1991; Jones et al., 1994). Recent evidence, however, suggests that some coccolithophorids from polar waters may be non-photosynthetic, obligately heterotrophic organisms, possibly as a response to the low light, low temperature, low salinity environment (Marchant and Thomsen, 1994). Cells identified as prymnesiomonads have been identified in presumed symbiotic relationships (Gaarder and Hasle, 1962; Okada and McIntyre, 1977; Reid, 1980; Febvre and Febvre-Chevalier, 1979).

12. *Mitosis*: Mitosis is preceded by replication of the basal bodies and chloroplasts, though the exact timing of chloroplast replication varies from soon after the completion of mitosis to immediately preceding the next mitosis. The extent to which the nuclear envelope breaks down is variable, ranging from almost complete disruption to retention of integrity apart from some fenestration, particularly towards the poles. The nuclear envelope-periplastid ER link is usually broken by prophase, though its restoration may begin as early as metaphase when the primordia of each daughter nuclear envelope appear. Microtubule organizing centers (MTOCs) have been reported only in *Pavlova* and *Phaeocystis.* In the former, a fibrous flagellar root at each spindle pole acts as an MTOC (Green and Hori, 1986), whilst in the motile cells of *Phaeocystis globosa* Scherffel, MTOCs occur as osmiophilic material on the surface of mitochondrial profiles at each pole (Hori and Green, 1994). For a comprehensive review of mitosis in the Prymnesiida, see Hori and Green (1994).

13. *Life-cycles*: Alternate stages are known for many prymnesiomonads. For example, palmelloid or filamentous stages alternate with motile cells in *Chrysotila*, and some species of *Pavlova* are known both as motile cells and as cells embedded in masses of mucilage. Evidence is accumulating that suggests that life-cycles involving a sexual process may be common in the Prymnesiida. In *Phaeocystis*, recent flow cytometric studies have shown that the colonial stage is diploid whilst the microzoospores are haploid (Rousseau et al., 1994; Vaulot et al., 1994). Paasche et al. (1990) observed cells with different types of scale occurring in clonal cultures of *Chrysochromulina polylepis* indicating the occurrence of at least two motile phases in the life-cycle of this species, probably stages in a seasonal life-cycle (Simon et al., 1997), later work using flow cytometry has confirmed that scale-bearing motile cells of the coccolithophorid *Emiliania huxleyi* are haploid whilst the coccolith-bearing cell is diploid (Green et al., 1996). In coastal genera of coccolithophorids, diploid coccolith-bearing cells alternate with haploid scale-covered forms, and it has been demonstrated that the organic scales of each phase are differently patterned (Billard, 1994). In addition, there are increasing numbers of reports of coccolithophorid life-cycles involving alternating heterococcolithophorid and holococcolithophorid phases (Thomsen et al., 1991; Billard, 1994).

ARTIFICIAL KEY TO SELECTED PRYMNESIOMONAD GENERA*

1. Cells with an investiture of coccoliths, sometimes with underlayers of unmineralized scales.......12
1'. Cells lacking coccoliths, but often with unmineralized scales or a mucilaginous investment....2

2. Cells motile.. 3
2'. Cells non-motile.. 10

3. Cells forming ring-shaped colonies..................... .... ***Corymbellus***
3'. Cells solitary... 4

4. Cells with markedly unequal, strongly heterodynamic flagella and a short haptonema. Haptonema length less than that of cell. Shorter flagellum sometimes rudimentary.................. 5
4'. Cells with two equal or subequal smooth flagella. Haptonema present or absent; if present, sometimes many times longer than the flagella.........6

5. Appendages usually inserted sub-apically, longer flagellum with an investment of small knob-like bodies (scales), shorter flagellum sometimes rudimentary. Eyespot sometimes present near flagellar insertion, two refringent reserve metabolite bodies, plastid single, sometimes with bulging posterior pyrenoid............***Pavlova***
5'. Appendages inserted more or less centrally on the concave face of a saucer-like cell; longer flagellum lacking knob-like scales; eye-spot pale, on ventral face of cell and lying beneath the shorter flagellum........................... ***Diacronema***

6. Cells spherical or elongate; if elongate, appendages inserted sub-apically. Haptonema present or absent, but if present reduced and inconspicuous and detectable only with the EM. Body-scales present, but very small and detectable only with EM, occasionally absent.....7

6'. Cells variously shaped (spherical, elongate, or saddle-shaped); haptonema obvious and sometimes very long. Body-scales always present, sometimes large plates or spines detectable with the optical microscope............. 8

7. Cells more or less spherical; chloroplasts two (four); haptonema absent................. ***Imantonia*** ....(if scales about, ***Dicrateria***)

7'. Cells usually elongate; appendages inserted sub-apically; chloroplasts one (two); haptonema present, but very reduced............ ***Isochrysis*** or motile cells................................. ***Chrysotila***

8. Haptonema longer than cell length (up to 16x); cells sub-spherical, elongate with appendages inserted apically, or saddle-shaped with appendages inserted in the concave face of the cell................................... ***Chrysochromulina***

8'. Haptonema shorter than the cell length............ 9

9. Cell usually elongate with appendages inserted sub-apically................................ ***Prymnesium***

9'. Cells irregularly spherical, appendages inserted in a groove or furrow. Colonial.. ***Corymbellus*** or colonial and single motile cells. ***Phaeocystis***

10. Cells with thick, sometimes lamellate, gelatinous walls; occasionally forming short filaments .... ***Chrysotila***

10'. Cells in a mucilaginous mass........................ 11

11. Cells arranged, often peripherally, in a colorless bladder.................................. ***Phaeocystis***

11'. Cells randomly disposed in an unorganized mass of mucilage; cells, though inactive, bearing strongly unequal flagella..................... ***Pavlova***

12. Cells bearing holococcoliths...........................13

12'. Cells bearing heterococcoliths.......................16

13. Microcrystals arranged as hollow pyramids partially covering an organic scale.. ***Balaniger***

13'. Microcrystals covering an organic scale...... 14

14. Microcrystals arranged in one layer on a plate-scale, with a second layer around the periphery ...... ***Coccolithus*** (***Crystallolithus*** phase)

14'. Microcrystals arranged in more than one layer to form elaborate structures........................... 15

15. Microcrystals arranged to form a tubular structure.. ....***Papposphaera*** (***Turrisphaera*** phase)

15'. Microcrystals arranged to form a cap-like structure.............................. ***Calyptrosphaera***

16. Organic scales only partially calcified; calcified parts arranged in the form of a wigwam on an organic plate .... ***Wigwamma***

16'. Scales more heavily calcified......................... 17

17. Cells partially covered by a rosette of pentagonal or quadrangular flattened plates arranged in layers around one hemisphere of the cell ....***Florisphaera***

17'. Cells entirely covered with coccoliths.......... 18

18. Outline of coccoliths rhombic... ***Calciosolenia***

18'. Outline of coccoliths circular or elliptic...... 19

19. Coccoliths bearing distinct distal and proximal shields 20

19'. Coccoliths lacking shields............................. 24

20. Distal shield composed of T-shaped elements..... .... ***Emiliania***

20'. Distal shield solid and composed of wedge-shaped elements.. 21

21. Proximal part of central area of coccolith partially covered by variously shaped structures 22

21'. Central area structure absent....................... 23

22. Central area structure cruciform.................... .... ***Cruciplacolithus***

22'. Central area structure a bar.......................... .... (in part)***Coccolithus***

23. Coccoliths small, generally less than 3.0 μm, with narrow shields............. ***Pleurochrysis*****

23'. Coccoliths large, usually between 5.0 μm and 13.0 μm, with wide shields................................ ....(in part)***Coccolithus***

24. Coccoliths with single-layered, crenate walls. 25

24'. Coccoliths with single to multi-layered, flat-topped walls, sometimes associated with an outer layer of flattened or dome-shaped coccoliths .... ***Syracosphaera***

25. Coccoliths bearing a central area structure...... .... ***Papposphaera***

25'. Central area structure absent..***Hymenomonas***

* For other recent guides to the identification of extant genera and species, see Chrétiennot-Dinet (1990), Moestrup and Larsen (1992), Heimdal (1993), and Throndsen (1993).

** Except *P. placolithoides* Fresnel & Billard with coccoliths 2.0 µm-4.0 µm

NOTE: Members of the genus *Derepyxis*, previously included in the Prymnesiida (e.g. Hibberd and Leedale, 1985), are now considered to be loricate rhizopodial chrysomonads (Meyer, 1986).

SELECTED GENERA

SUBORDER PAVLOVINA Cavalier-Smith

Genus ***Diacronema*** Prauser

Cells free-swimming, compressed, asymmetrical. Flagella 2, unequal, with conspicuous hair-points, inserted laterally; flagella action heterodynamic, the longer directed anteriorly, but forming a loop position in the settled cell, the shorter directed posteriorly; shorter flagellum swollen on side adjacent to cell, the swelling enclosing a series of densely staining projections attached to the axonemal doublet nearest the cell. Haptonema short, distally tapered, non-coiling; haptonematal microtubules terminating intracellularly at a densely staining plaque from which fibrillar roots radiate. Chloroplast single, parietal, yellow-green to olive-green; eyespot pale, situated posteriorly underlying the short flagellum; pyrenoid absent. Cell penetrated by a pit or canal close to the anterior flagellum. One euryhaline species recorded.

Diacronema vlkianum Prauser (Figs. 1-4). Cell cordate to orbiculate in dorsoventral aspect, reniform in lateral aspect, 3.5-7.5 µm x 4.0-5.0 µm x 1.5-3.0 µm. Flagella and haptonema inserted centrally in concave face; long flagellum 7.0-10.0 µm, shortly tapered, short flagellum 6.0-9.0 µm, swollen proximally (4.0-5.5 µm), attenuated distally (2.0-3.5 µm); haptonema approx. 1.0 µm, distally attenuated. Chloroplast sometimes markedly 2-lobed; eyespot composed of numerous droplets, peripherally in the chloroplast, beneath a groove in the plasma membrane in which runs the short flagellum. Lipid droplets several; contractile vacuole in freshwater isolates discharging into the pit. Two large reserve bodies often present. Swimming action an irregular stuttering motion superimposed on a slow side to side rocking at slow speeds. See Prauser (1958), Fournier (1969), Green and Hibberd (1977).

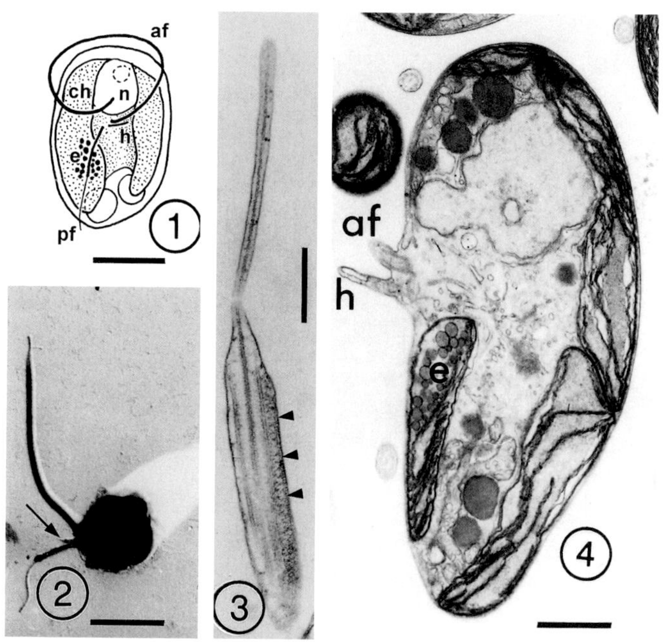

Figs. 1-4. *Diacronema vlkianum* Prauser. Fig. 1. Diagrammatic representation of a cell showing the disposition of the principal organelles: af, the anterior flagellum; ch, the parietal chloroplast; e, eyespot; h, haptonema; n, nucleus; pf, posterior flagellum. Scale bar=2.5 µm Fig. 2. Shadowed whole-mounted cell with the long anterior flagellum, short flagellum with a distal attenuation, and, between them, the short haptonema (arrow). Scale bar=2.5 µm. Fig. 3. longitudinal section of a short flagellum with densely staining 'teeth' attached to one pair of axonemal MTs (arrowheads). The central pair of MTs extends into the attenuated flagellar tip. Scale bar=0.5 µm. Fig. 4. Longitudinal section of a cell; for abbreviations see Fig. 1 (the posterior flagellar base is not included in this plane). Scale bar=1.0 µm. Figs. 2-4 from Green and Hibberd (1977).

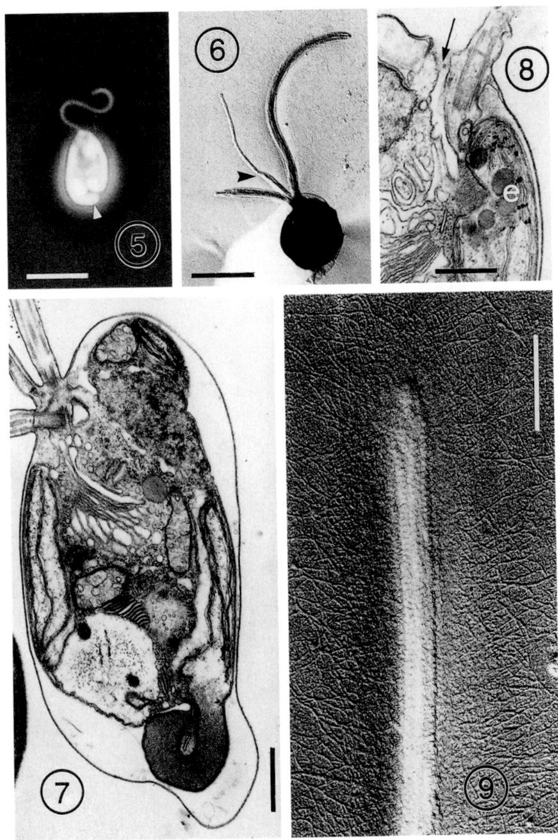

Figs. 5-8. *Pavlova pinguis* Green. Fig. 5. Live cell showing the characteristic flagellar action and the bulging pyrenoid (arrowhead). Negative phase contrast. Scale bar=5.0 µm. Fig. 6. Shadow-cast whole mounted cell; the haptonema (arrowhead) has a long distal attenuation. Scale bar=2.5 µm. Fig. 7. Longitudinal section of a cell. Note the subapical insertion of the two flagella and the haptonema, the anterior nucleus, and the posterior bulging pyrenoid. Scale bar=0.6 µm. Fig. 8. Section through the eyespot (e) close to the base of the anterior flagellum and the termination of the pit (arrow). Scale bar=0.4µm. Fig. 9. *Pavlova gyrans* Butcher. Shadow-cast whole mounted anterior (long) flagellum with its tomentum of fine hairs and covering of regularly arranged dense knob-scales. Scale bar=1.0 µm. Figs. 5-8 from Green (1980); Fig. 9 from Green and Manton (1970).

Genus *Pavlova* Butcher

Cells of variable form, sometimes strongly metabolic, with 2 markedly unequal flagella and a short haptonema inserted subapically to laterally. Long flagellum shortly tapering distally, usually with a dense covering of fibrillar hairs and small dense bodies (knob-scales), thus appearing thick relative to the short flagellum in the optical microscope and beating with a characteristic S-shaped wave pattern; short flagellum sometimes with fibrillar hairs, but no knob-scales, extending outwards and beating with a stiff jerky action; occasionally rudimentary. Haptonema short, sometimes distally attenuated, non-coiling. Chloroplast usually single parietal, yellow-green, strongly bilobed; pyrenoid present or absent; if present, usually posterior and bulging; eyespot, if present, situated on the internal face of a chloroplast lobe, close to the basal bodies. A pit or canal penetrating the cell close to the flagellar apparatus and terminating close to the inner face of the eyespot. Two large translucent reserve metabolite bodies of unknown nature present together with lipid vacuoles and other inclusions. A conspicuous cavity beneath the plasma membrane, the latter adorned with various tomenta and scattered knob-scales. Reproduction by longitudinal fission, flagellar fibrous roots acting as microtubule organizing centers during mitosis. Marine, brackish, or freshwater. See Butcher (1952) for original description and Green (1980) for a review and literature until 1980, Green and Hori (1994) for a summary of features of the flagellar apparatus, and Green and Hori (1988) and Hori and Green (1994) for details of mitosis. Some species, notably *P. lutheri* (Droop) Green, have proved to be valuable food organisms in aquaculture (Jeffrey et al., 1994) and their biochemical composition is well documented (Jeffrey et al., 1994; Conte et al., 1994). The dense bodies on the long flagellum (Fig. 9) were termed knob-scales as they originate in the cisternae of the Golgi body and were thought to be homologous with the plate-scales of the Prymnesiina, but Cavalier-Smith (1994) suggests that they may be modified hairs.

Pavlova pinguis Green (Figs. 5-8). Cells 5.0-8.0 µm x 3.0-4.0 µm, ovate to obovate, not or only slightly compressed. Long flagellum 10.0-12.0 µm with fibrous hairs and superficial knob-scales, the latter 0.05 µm x 0.03 µm with a median constriction; hairs and knob-scales replaced proximally by a garniture of small ± spherical dense bodies carried on fine hairs. Short flagellum approximately 4.0 µm. Haptonema approximately 2.0 µm with a distal attenuation of variable length;

number of microtubules reduced to 2 or 3 in the free part with an incomplete cylinder of ER, but basal region normal. Chloroplast single with 2 large lateral lobes; eyespot obvious; pyrenoid conspicuous, basal, bulging. Reserve metabolite bodies, 2. Non-motile cells irregularly spherical, 6.0-8.0 µm diameter; flagella present, but inactive. Marine.

SUBORDER PRYMNESIINA Cavalier-Smith

NON-COCCOLITHOPHORID GENERA

Genus *Chrysochromulina* Lackey

Cells variously shaped, sub-spherical, cylindrical, ovoid, pyriform, or saddle-shaped, sometimes metabolic, 5.0-10.0 µm in longest dimension. Two (rarely 4) equal or sub-equal flagella and a haptonema inserted apically, subapically, or in the center of the concave face of saddle-shaped cells. Flagella 1.5-3 times the cell length, homodynamic or heterodynamic. Haptonema variable in length, but often long or very long (many times longer than the diameter of the cell) and often coiling, usually clavate distally and swollen proximally; haptonematal microtubules usually 7, but some species with 6 only. Chloroplasts up to 8 in number, parietal, variously lobed or incised. Pyrenoids immersed or bulging, the latter sometimes with a cap of transparent material that may be reserve metabolite. Eyespot absent. One or more types of bodyscale present in 1 to several layers; the outermost scales often elaborated as saucers, spines, tubs, cylinders, baskets, and other forms; under-layers of scales often circular or elliptical plates with a reflexed rim; the proximal surface of all scales with radially arranged fibrils, the distal surface patternless or with spiral or roughly concentrically wound fibrils, though pattern sometimes obscured by morphological transformation. Reproduction generally by longitudinal fission, but recent evidence indicates the possibility of sexual stages; amoeboid phase recorded in older cultures of some species; thick-walled resting stages also recorded but not confirmed. Swimming action variable, either with apical pole and flagella directed backwards, the haptonema coiled and the cell rotating or with the apical pole forwards, and the flagella extended laterally or backwards, and the haptonema extended forwards, the cell exhibiting a 'gliding' motion. Cell attaching to substrates by haptonema either at the tip or along its length. Contractile vacuole in freshwater species. Many species mixotrophic, particle ingestion being effected by the extrusion of pseudopodia, or by particle aggregation by the haptonema followed by transfer to the ingestion site, or by aggregation on spine-scales followed by haptonematal transfer to ingestion site. Most recorded species marine.

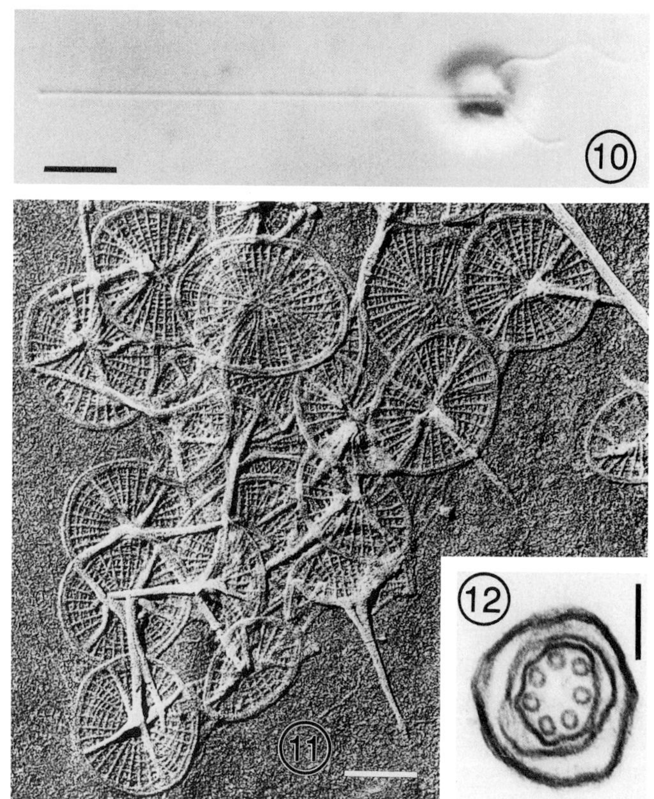

Figs. 10-12. *Chrysochromulina acantha* Leadbeater and Manton. Fig. 10. Living cell with two posteriorly directed flagella and a long extended haptonema. Interference contrast. Scale bar=0.6 µm. Fig. 11. Shadow-cast whole-mounted body scales (plate scales and spined scales). Scale bar=0.25 µm. Fig. 12. Transverse section of the free part of the haptonema showing the cylinder of ER enclosing a ring of seven MTs. Scale bar=0.1 µm. Fig. 10 from Gregson et al. (1993a); Figs. 11,12 from Leadbeater and Manton (1971).

Chrysochromulina is one of the larger genera of the Prymnesiida with approximately 50 species recorded (Jordan and Green, 1994). However, many species are known only from electron microscopic examination of material collected in the field, and information on the live cells is lacking. Nevertheless, the diversity of structure exhibited suggests that *Chrysochromulina* is an artificial polyphyletic assemblage of organisms. This has been confirmed by molecular systematic studies which also show that, as suggested by earlier workers, some species of *Chrysochromulina* have close affinities with species of *Prymnesium* (Simon et al., 1997). With regard to the range of structure, the type species, *C. parva* Lackey and a number of other species are saddle-shaped with long, coiling haptonemata, and relatively simple scales whilst many others are sub-spherical or elongate, haptonematal length and the ability to coil are variable, and their scales may be very complex. Haptonematal structure is variable. Most species for which information is available have haptonemata with 7 axial microtubules, but a few species, e.g. *C. strobilus* Parke & Manton, *C. cymbium* Leadbeater & Manton, *C. camella* Leadbeater & Manton, *C. bergenensis* Leadbeater & Manton and *C. apheles* Moestrup & Thomsen have 6 only (Leadbeater and Manton, 1969; Manton and Leadbeater, 1974; Moestrup and Thomsen, 1986). Flagellar root systems vary from simple to complex, *C. apheles*, for example, having a simple system with few microtubules (Moestrup and Thomsen, 1986) whilst *C. acantha* Leadbeater & Manton, *C. brevifilum* Parke & Manton, and an unnamed species of *Chrysochromulina* have more complex systems (Gregson et al., 1993a; Birkhead and Pienaar, 1994, 1995). The possibility that sexual reproduction occurs in species of *Chrysochromulina* is suggested by the detection of polymorphism within cultures of *C. polylepis*. Two types of cell with different scale patterns have been regularly observed (Paasche et al., 1990; Edvardsen and Paasche, 1992) and the distribution of haploid and diploid DNA indicates that they are part of a sexual life-cycle (Edvardsen and Vaulot, 1996). There is an extensive bibliography on other aspects of *Chrysochromulina*. For references to original descriptions of individual species, see Jordan and Green (1994); see Leadbeater and Green (1993) and Leadbeater (1994) for discussions and literature on cell coverings and scale formation; Inouye (1993), Green and Hori (1994), Birkhead and Pienaar (1994, 1995) for details of flagellar ultrastructure; Kawachi et al. (1991), Gregson et al. (1993b), Inouye and Kawachi (1994), Kawachi and Inouye (1995) for discussions of haptonematal function and physiology, and feeding mechanisms; Jones et al. (1994) for a discussion of mixotrophy in *Chrysochromulina*.

Chrysochromulina acantha Leadbeater & Manton (Figs. 10-12). Cells ephippiomorphic, concavo-convex, 6.0-10.0 µm in length and breadth. Chloroplasts 2, pale golden-brown with immersed pyrenoids. Two flagella and a haptonema arising from the concave surface near the posterior end, insertion closer to one chloroplast than to the other. Flagella approx. 2-2.5 times cell length, haptonema up to c. 40.0 µm when fully extended, extended or coiled during swimming. Cells with an investment of 2 layers of scales; proximal scales plate-like, 0.5 x 0.6 µm with a pattern of approx. 30 radial fibers extending to a thickened margin, and fine spiral fibers; distal scales similar in patterning, but circular 0.4 - 0.5 µm in diameter, each with a single central spine, c. 0.5 µm long, supported by 4 decurrent struts reaching the scale margin and raised above it centrally. Marine, planktonic. See Leadbeater and Manton (1971), Gregson et al. (1993a,b).

Genus **Chrysotila** Anand, 1937

Vegetative cells non-motile, young cells forming cubic masses, older cells with a thick stratified mucilage sheath, the latter often forming colorless unbranched stalks with 1 or more cells situated distally. Calcified bodies formed extracellularly. Cells with 1 chloroplast with immersed pyrenoids, chrysolaminarin and other vesicles present together with a carotenoid body. Asexual reproduction by vegetative cell division and production of motile zoospores. Zoospores with 2 subequal, sub-apically inserted flagella, homo- or heterodynamic; haptonema rudimentary with few (c. 2 or 3) microtubules and reduced endoplasmic reticulum. Chloroplast single with immersed pyrenoid close to inner face; eyespot absent. Species of *Chrysotila* were first reported from the 'Chrysophyceae belt' of the chalk cliffs of Kent, but the limited records suggest that they are ubiquitous on damp basic substrates. They are of particular interest in that they produce extracellular calcified bodies corresponding in form to some Mesozoic nannofossils. See Anand (1937); Parke (1971), Green and Parke (1975), Green and Course (1983), Green (1986).

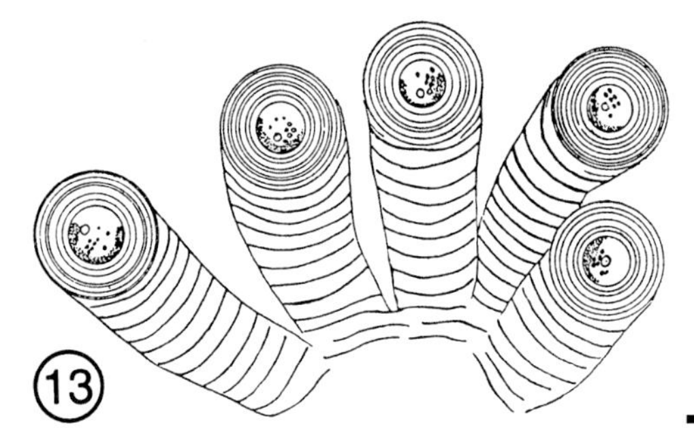

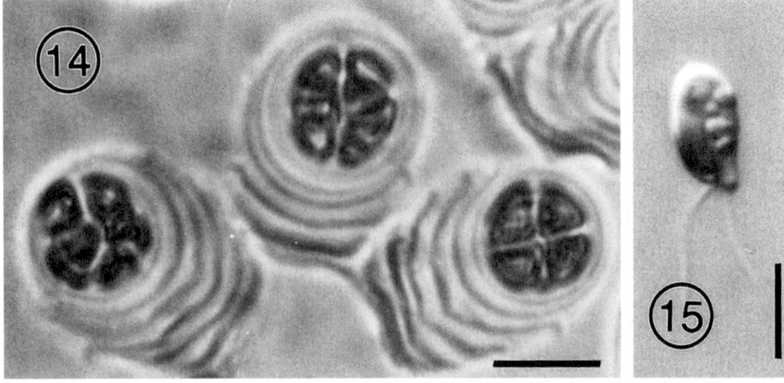

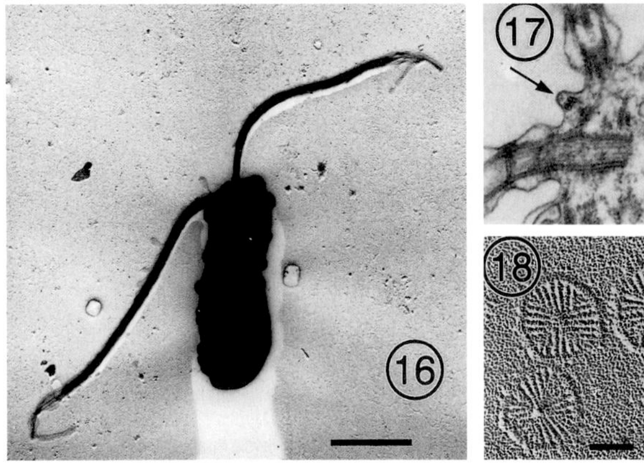

Figs. 13-18. *Chrysotila lamellosa* Anand. Fig. 13. Lamellate mucilaginous stalks, each with a single cell terminally. Scale bar=20.0 µm. Fig. 14. Mucilaginous stalk with the terminal cells dividing to form zoosporangia. Scale bar=10.0 µm. Fig. 15. Zoospore. Interference contrast. Scale bar=5.0 µm. Fig. 16. Shadow-cast whole-mounted zoospore with two more or less equal flagella and apparently no haptonema. Scale bar=1.8 µm. Fig. 17. Section through the apical pole of a zoospore showing the rudimentary haptonema (arrow) between the flagella. Scale bar=0.3 µm. Fig. 18. Shadowed whole-mounted body-scales from a zoospore. Scale bar=0.1µm. Fig. 13 from Anand (1937); Figs. 14,16,17 from Green and Parke (1974); Figs. 15,18 from Green and Course (1983).

Chrysotila lamellosa Anand (Figs. 13-18). Benthic colonies at first globose, later becoming laminar. Young cells spherical or hemispherical, 4.5-11.0 µm, surrounded by lamellate mucilage sheaths, the latter developing asymmetrically to form short stalks (up to 20.0 µm) with 1 or 2 cells distally. Chloroplast single, parietal yellow-green to golden-brown; carotenoid body often present. Extracellular mineralized bodies deposited initially as rods or cruciform bodies, later with striated conical protuberances, the whole up to 50.0 µm across. Zoospores elongate, pyriform, 4.5-8.0 µm x 3.0-6.0 µm, compressed, metabolic. Flagella 2, unequal, 8.0-10.0 µm and 6-8 µm inserted sub-apically; haptonema rudimentary with approximately 2 microtubules. Chloroplast single, parietal, carotenoid body present. Body scales in 1 layer, approx. 0.2 µm diameter. Peripheral muciferous bodies numerous. Habitats: coastal in the splash zone of calcareous cliffs, but also inland on damp basic substrates. See Anand (1937), Green and Parke (1975), Green and Course (1983).

Genus ***Corymbellus*** Green

Yellow-green cells aggregated to form spherical or annuloid motile colonies. Each cell with 2 sub-equal smooth flagella and a haptonema arising from an apical depression. Unmineralized body-scales present. Plastids 2, each with a pyrenoid.

Corymbellus aureus Green (Figs. 19-25). Colonies up to 200.0 µm diameter. Cells irregularly rectangular or cordate in dorso-ventral view, 8.0-11.0 µm x 7.5-10.0 µm. Flagella 2, sub-equal, heterodynamic in colonial cells, hetero- or homodynamic in monads; haptonema short (c. 3.0 µm), non-coiling. Appendages arising from a furrow between 2 anterior lobes of the cell, the lobes sometimes more or less acute, the cell thus appearing irregularly pyriform in lateral aspect. Larger body scales oval, 0.30-0.35 µm x 0.21-0.25 µm, with raised concavo-convex rims, the convex face of the rim facing inwards, and a distal surface bearing a pattern of radiating ridges and a short 4-strutted spine bridging a central pore. Small oval spineless scales (0.2-0.24 µm x 0.10-0.15 µm) associated with the apical groove. Plastids 2, parietal, lobed, yellow-green, each with an immersed fusiform pyrenoid.

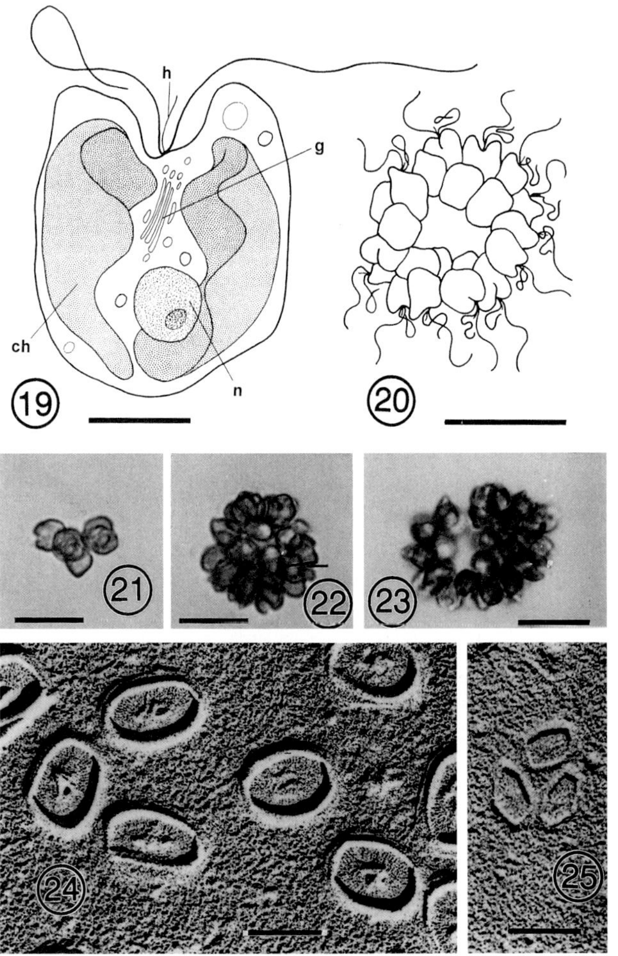

Figs. 19-25. *Corymbellus aureus* Green. Fig. 19. Diagrammatic representation of a single cell: ch, chloroplast; g, Golgi body (dictyosome); h, haptonema; n, nucleus. Scale bar=4.0 µm. Fig. 20. A ring-shaped colony. Scale bar=15.0 µm. Figs 21-23. Stages in colony formation. Scale bars=20.0 µm. Fig. 24. Large body-scales, some with a short 4-strutted spine. Scale bar=0.3 µm. Fig. 25. Scales from the region of the apical groove. Scale bar=0.3 µm. Figs. 19-25 from Green (1976).

Corymbellus aureus has been recorded from the English Channel (Green, 1976) and also from the North Sea where it may occur in substantial concentrations (Gieskes and Kraay, 1986). It occurs also in the N. Atlantic where it may be occasionally abundant (D. Harbour, pers. comm.) suggesting that *C. aureus* may, therefore, be more widespread than the records indicate.

Genus ***Imantonia*** Reynolds

Cells motile, not metabolic with 2 equal to sub-equal flagella; haptonema vestigial. Chloroplasts parietal, with immersed pyrenoids. Cell with unmineralized body-scales.

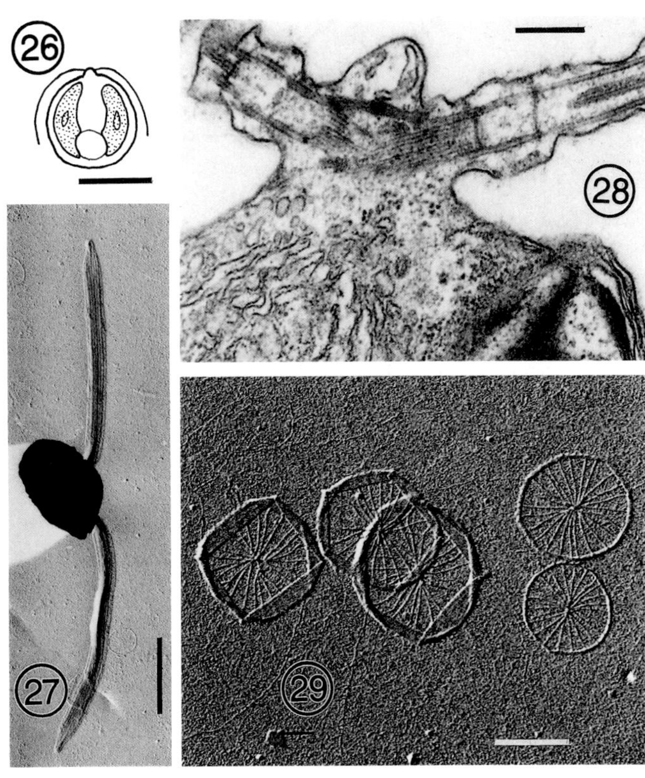

Figs. 26-29. *Imantonia rotunda* Reynolds. Fig. 26. Representation of a cell at rest. Scale bar=2.5 µm. Fig. 27. Shadow-cast whole-mounted cell. Scale bar=2.0 µm (approx.). Fig. 28. Section through the apical pole showing the rudimentary haptonema between the flagella. Scale bar=0.3 µm. Fig. 29. Shadowed, whole-mounted, rimmed and unrimmed body scales. Scale bar=0.5 µm. Fig. 26 Redrawn from Throndsen (1993); Figs. 27-29 from Green and Pienaar (1977).

Only one species recorded, *Imantonia rotunda* Reynolds (Figs. 26-29). Cells motile, more or less spherical, 2.0-4.0 µm diameter, not metabolic. Flagella 2, smooth, distally tapered, 4.5-7.0 µm, homodynamic, arising from a papilla. Haptonema rudimentary consisting of a short proboscis (c. 0.2 µm) containing profiles of endoplasmic reticulum. Body scales of 2 types, the first 0.45-0.68 µm diameter with 16-23 superficial radiating ridges, the second 0.72-0.8 µm diameter with similar superficial pattern of ridges, but also with upturned rim. Chloroplasts 2 or 4, parietal, with immersed pyrenoids. Marine, planktonic.

Imantonia rotunda is difficult to identify with certainty using the light microscope only, being easily confused with the morphologically similar but scale-less species *Dicrateria inornata* Parke (Parke, 1949; Green and Pienaar, 1977). The distribution of the latter species is not known, but electron microscopical examination of plankton samples indicate that *I. rotunda* is cosmopolitan (Reynolds, 1974; Marchant and Thomsen, 1994; Thomsen et al., 1994). Mitosis in *I. rotunda* and the flagellar root system have been studied by Hori and Green (1985a) and Green and Hori (1988), respectively.

Genus ***Isochrysis*** Parke

Cells generally elongate, ellipsoid, dorso-ventrally flattened, but metabolic and occasionally subspherical. Flagella 2 inserted subapically, directed anteriorly or posteriorly during swimming, heterodynamic. Haptonema rudimentary, very short, barely detectable with the optical microscope. Chloroplast 1, parietal, with pyrenoid bulging on inner face. Small oval body-scales present; very small scales also covering the haptonema. Marine, planktonic.

Isochrysis galbana Parke (Figs. 30-35). Cells 5.0-6.0 µm x 2.0-4.0 µm wide x 2.5 -3.0 µm thick. Ellipsoid, anteriorly truncate, posteriorly rounded, but overall morphology variable. Flagella approx. 7.0 µm long, equal. Haptonema short, 0.3 -0.4 µm with 5 microtubules in the free part and in the base. Body scales 0.3-0.4 µm x 0.2-0.3 µm, each with a superficial pattern of approx. 40 radial ridges and a raised central swelling, present in several layers. Small oval scales, 0.08-0.10 µm x 0.05-0.08 µm with approximately 12 radial ridges and a central swelling, covering the haptonema. Chloroplast usually single, parietal, yellow-brown with an immersed fusiform pyrenoid; eyespot absent, but the cell often containing a conspicuous carotenoid body.

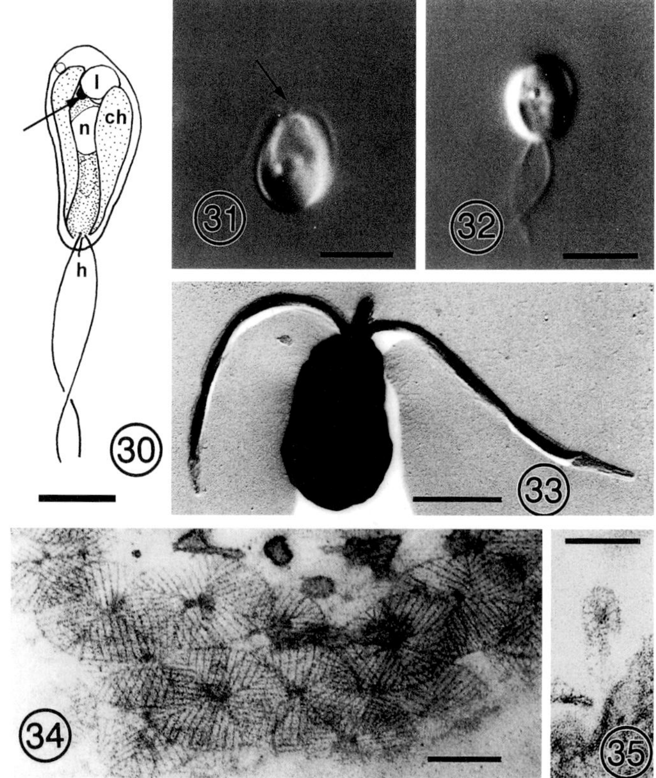

Figs. 30-35. *Isochrysis galbana* Parke. Fig. 30. Cell showing the disposition of the principal organelles: ch, chloroplast; h, abbreviated haptonema; l, chrysolaminarin (leucosin) body; n, nucleus. The arrow indicates the small carotenoid body. Scale bar=1.8 µm. Fig. 31. Living cell at rest showing clearly the small haptonema (arrow). Interference contrast. Scale bar=5.0 µm. Fig. 32. Swimming cell. Interference contrast. Scale bar=5.0 µm. Fig. 33. Shadow-cast whole-mounted cell showing the two flagella and the short haptonema. Scale bar=2.0 µm. Fig. 34. Glancing section of a cell showing the body scales in several layers. Scale bar=0.22 µm. Fig. 35. Haptonematal scale. Scale bar=0.13 µm. Figs. 33-35 from Green and Pienaar (1977).

Isochrysis galbana is best known for its value as a food organism in the aquaculture industry, and it has been the subject of numerous biochemical analyses (see Jeffrey et al., 1994). Little is known about its distribution in the wild, but *Isochrysis*-like motile stages are released by benthic algae of the genus *Chrysotila* Anand known from basic substrates in the United Kingdom and France (Billard and Gayral, 1972; Green and Parke, 1975). One species with a non-motile stage consisting of mucilage covered cells, *I. litoralis* Billard & Gayral, is at present retained in *Isochrysis*, but is in need of further investigation. Mitosis in *I. galbana* has been studied by Hori and Green (1985b) and the flagellar apparatus, which has some interesting features comparable with some coccolithophorids, by Hori and Green (1991).

Genus ***Phaeocystis*** Lagerheim

Cells spherical to ovoid, 4.5-8.0 μm arranged peripherally in large (up to 9.0 mm or more) gelatinous colonies of variable shape, gelatinous investment multilayered. Cells with 2 or 4 chloroplasts and chrysolaminarin vesicles; no flagellar or haptonematal bases detected; body scales absent. Motile cells of several types, including vegetative swarmers and gametes, with 2 equal or sub-equal flagella, a short haptonema, plate-like body scales, and, in some cases, releasing thread-like material, the threads arranged to form stars or other characteristic patterns. Colonial cells diploid, motile cells haploid or diploid. Reproduction by vegetative division of non-motile cells and fragmentation of colonies, vegetative division of motile cells, or by fusion of gametes. Species of *Phaeocystis* occur in all the world's oceans, though there is some doubt about the distribution of individual species in view of the problems of identification. The systematics of *Phaeocystis* are at present under active review. Baumann et al. (1994) distinguish 4 species based on colony morphology and motile cell morphology, and variation in temperature and light requirements. These are *P. globosa* Scherffel, *P. pouchetii* (Hariot) Lagerheim, an unnamed Antarctic form, and *P. scrobiculata* Moestrup, known from the motile cell only. These conclusions are supported by genetic analysis of the 3 colony-forming species, Medlin et al. (1994b) referring to the Antarctic form as *P. antarctica* Karsten.

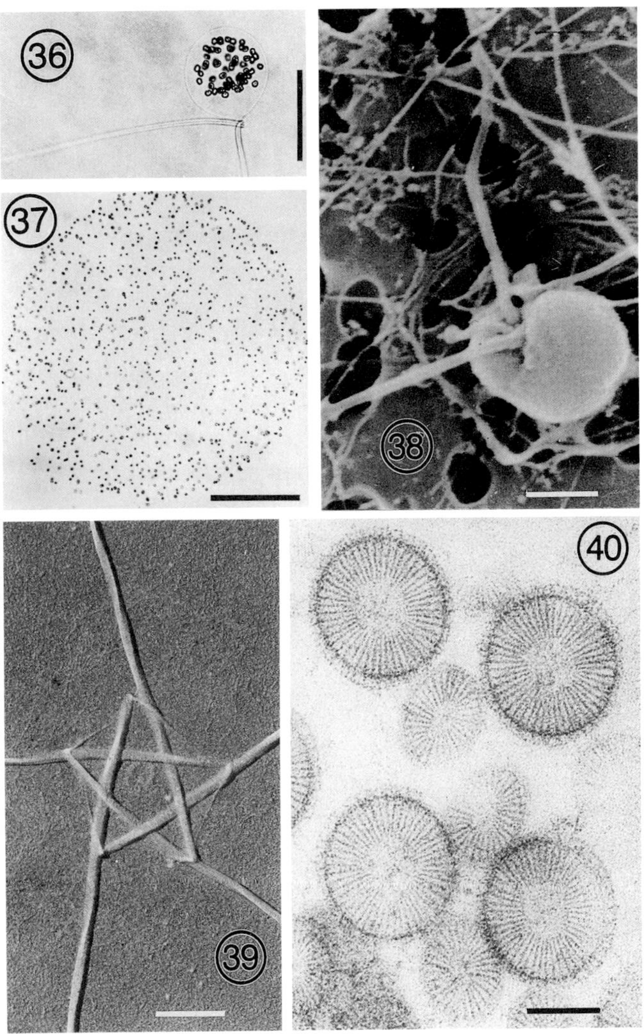

Figs. 36-40. *Phaeocystis* aff. *globosa* Scherffel. Fig. 36. Young colony developing on a spine of *Chaetoceros* sp. Scale bar=100.0 μm. Fig. 37. Mature spherical colony. Scale bar=200.0 μm. Fig. 38. Motile cell showing the two flagella and the short haptonema arising from an apical depression, and masses of thread-like material in the background. Scale bar=1.0 μm. Fig. 39. Shadowed whole-mounted thread-like material from a motile cell showing the pentagonal stellate arrangement. Scale bar=0.5 μm. Fig. 40. Glancing section of body scales. Scale bar=0.1 μm. Fig. 38 from Vaulot et al. (1994); Figs 39,40 from Parke et al. (1971).

Other species are recorded in the literature (Sournia, 1988; Jordan and Green, 1994), but they need detailed re-investigation. Recently, Vaulot et al. (1994) found evidence for a variety of possibly new species from studies of cell morphology, ploidy level, cell size, pigment composition, and genome size. See also Kornmann (1987), Billard (1994), Hori and Green (1994), Lancelot and Rousseau (1994), and Rousseau et al. (1994).

Phaeocystis globosa Scherffel (Figs. 36-40). Colonies 8.0-9.0 mm, spherical with numerous derived forms, cells distributed evenly at the periphery, mucilage solid. Colonial cells 4.5-8.0 µm with 2 to 4 chloroplasts, flagella, haptonema, and body-scales lacking. Free-living non-motile cells with the characters of the above. Swarmers short-lived, 4.5-8.0 µm, diploid, with 2 equal flagella and a short haptonema, details of body-scales not known. Microzoospores haploid, 3.0-6.0 µm, more or less spherical, chloroplasts 2 with immersed fusiform pyrenoids, each with a delicate bounding layer. Two flagella and a haptonema arising from a depression; flagella equal, 5.0-7.5 µm, heterodynamic, 1 flagellum directed forwards with a wave-like beat, the other extended backwards and sideways beating stiffly; haptonema approximately 2.5 µm, non-coiling, with a terminal bulge. Cell covering with 2 types of body-scale: small oval plates, 0.1-0.13 µm proximally, and circular plates, 0.2 µm diameter with vertical rims, distally. Some microzoospores releasing thread-like material from peripheral vesicles, the threads up to 20.0 µm in length, apparently tube-like, c. 0.05 µm wide when flattened, distally tapered; arranged in groups of 5, the proximal ends arranged to form a 5-pointed star. Macrozoospores arising within colonies and forming new colonies either within the mother colony or on release; morphology and cytology not known. The life-cycle of *P. globosa* is complex, but poorly understood; see Kornmann (1955), Kayser (1970), Billard (1994), Rousseau et al. (1994).

Genus ***Prymnesium*** Massart

Cells sub-spherical to elongate, sometimes somewhat compressed with t2 equal or sub-equal flagella and a flexible non-coiling haptonema, sometimes with a covering of scales, arising sub-apically from the obliquely truncate apical pole. Flagella 1 - 1.5 x cell length, haptonema approx. 1/3 to 1/5 the length of the flagella. Flagellar action heterodynamic, 1 flagellum often held

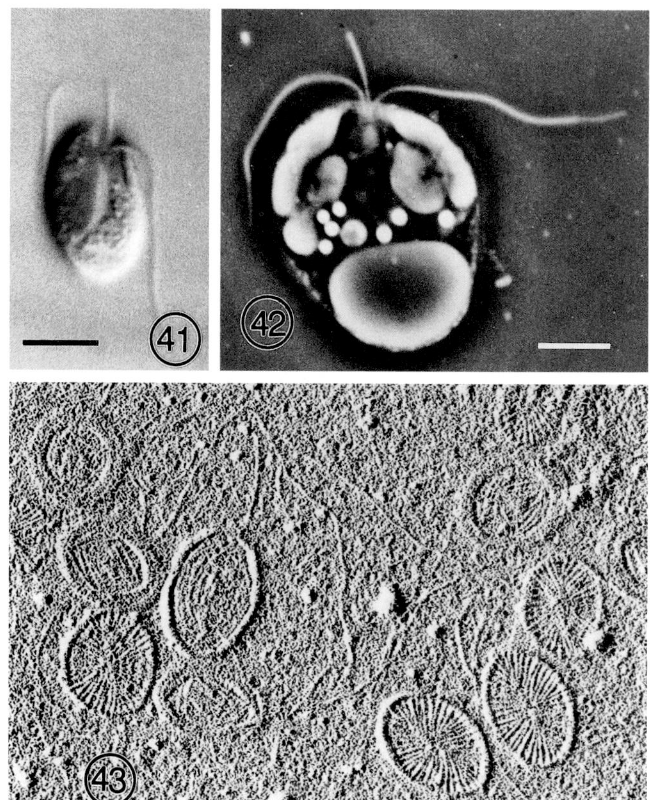

Figs. 41-43. *Prymnesium parvum* N. Carter. Fig. 41. Swimming cell. Interference contrast. Scale bar=5.0 µm. Fig. 42. Compressed living cell showing the two flagella and the haptonema, parabasal Golgi body, two parietal chloroplasts, central nucleus, lipid globules, and large posterior chrysolaminarin vesicle. Scale bar=5.0 µm. Fig. 43. Shadowed, whole-mounted body scales. Scale bar=0.25 µm. Figs. 41,43 from Green et al. (1982); Fig. 42 from Manton and Leedale (1963a).

closely to the cell body, the other extended away from the cell; haptonema directed forwards during swimming. Chloroplasts usually 2 with immersed or bulging pyrenoids; eyespot absent. A contractile vacuole sometimes present anteriorly; muciferous bodies often situated peripherally. Body scales present, with 2 or more morphotypes arranged in 2 or more layers. Some species toxic to fish and other gill-breathing animals (see above and review by Moestrup, 1994). Both *P. parvum* N. Carter and *P. patelliferum* Green et al. have been recorded as heterotrophic; the latter species has been observed to undertake communal heterotrophy whereby a number of individuals can jointly attack a larger prey organism which becomes invested by the confluent cytoplasm of the *P. patelliferum* cells (U. Tillmann, pers. comm; 1995). There are a number of species in the literature (Jordan and Green, 1994), but several of the older species have not been re-isolated or re-examined for details of scales and other structures detectable only with the electron microscope. See Conrad (1926, 1941), Carter (1937), Green et al. (1982), Billard (1983), Chang and Ryan (1985), Green and Hori (1990), Pienaar and Birkhead (1994).

Prymnesium parvum N. Carter (Figs. 41-43). Cells 8.0-15.0 µm long x 4.0-10.0 µm broad; flagella equal or sub-equal, 12.0-20.0 µm long; haptonema 3-5 µm long, non-coiling, with 7 microtubules in TS. Cell covered with 2 types of oval plate-scale in 2 layers, those of the outer layer 0.3-0.4 µm x 0.23-0.3 µm with narrow inflexed rim, those of the inner layer 0.3-0.36 µm x 0.26-0.32 µm with a wider rim, strongly inflexed over the distal face. Cysts ovoid, wall composed of layers of scales with siliceous material deposited on their distal faces. Planktonic in marine and brackish water world-wide. See Manton and Leedale (1963a), Manton (1964a, b, 1966, 1968), Green et al. (1982). *Prymnesium parvum* is strongly toxic to fish and other marine organisms (see above). Species of *Platychrysis* Geitler are distinguished from species of *Prymnesium* by the dominance of the non-motile stage in *Platychrysis* and the pronounced metaboly of the motile cells (Chrétiennot, 1973; Gayral and Fresnel, 1983a).

COCCOLITHOPHORID GENERA

Heterococcolithophorid genera (sometimes with holococcolithophorid phases known in life-cycle)

Genus *Calciosolenia* Gran

Coccolithophorids, possibly biflagellate, possessing cylindrical coccospheres terminating at each end with a number of pole spines. Monomorphic body coccoliths, rhomboliths (syn.=scapholiths), rhombic in shape. Marine, planktonic, generally in warmer waters. The shape and total length (up to 90 µm) of coccosphere may allow the cell to maintain a position transverse to the direction of incident light, thus maximizing the photosynthetic efficiency of the lateral chloroplasts (Manton, 1986).

Calciosolenia murrayi Gran (Fig. 44). Cells reported to have 2 flagella. Coccospheres cylindrical, 20.0-50.0 µm x 2-5 µm, with 1-4 pole spines, about 20.0 µm long, at each end. Rhomboliths covering the entire coccosphere, about 2.5-4.0 µm x 1.0-1.6 µm, with 2 unequal sets of parallel sides; polar rhomboliths narrower. Rim defined by upright wall elements, 0.2 µm high; from the proximal part of the rim approx. 40 laths converging into the central area, sometimes overlapping to form a low central ridge. Pole-spine structure not well known. See Halldal and Markali (1955), Gaarder and Hasle (1971), Manton and Oates (1985).

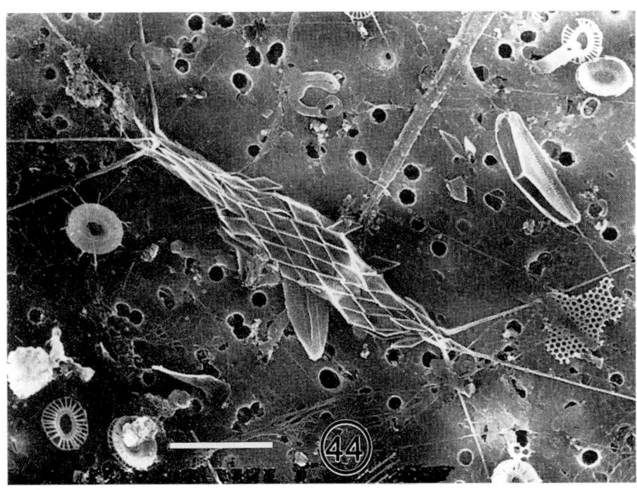

Fig. 44. *Calciosolenia murrayi* Gran. Cylindrical coccosphere composed mainly of abutting rhomboliths, but with specialized pole-spines at each end. Note the narrow circumpolar rhomboliths. Scale bar=5.0 µm. (From Kleijne in Winter and Siesser, 1994)

Genus *Coccolithus* Schwarz

Coccolithophorids with non-motile hetero-coccolith-bearing and motile holococcolith-bearing phases. Non-motile phase bearing placoliths, with both shields composed of overlap-ping wedge-shaped elements and the central area wide and open. Motile phase with 3 flagella and a rudimentary bulbous haptonema, and bearing crystalloliths. Marine, planktonic, commonly found in boreal waters and less frequently along the coastlines of most continents (including Australasia and Africa).

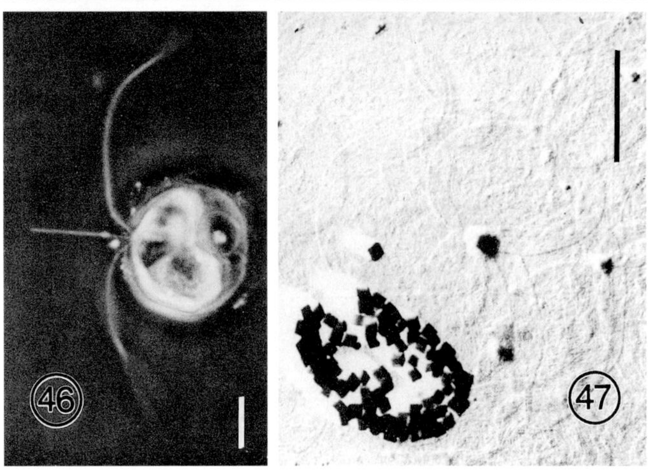

Figs. 45-47. *Coccolithus pelagicus* (Wallich) J. Schiller. Fig. 45. Coccosphere of interlocking coccoliths. Scale bar=5.0 µm. Fig. 46. '*Crystallolithus*' phase; living cell. Negative phase contrast. Scale bar=5.0 µm. Fig. 47. '*Crystallolithus*' phase; crystallolith and organic scales. Scale bar=1.0 µm. Fig. 45 from Faber and Preisig (1994); Figs. 46, 47 from Manton and Leedale (1963b).

Coccolithus pelagicus (Wallich) Schiller (Figs 45-47). Heterococcolith-bearing cells, ovoid to spherical, 10.0-35.0 µm in diameter, non-motile, covered proximally by small, circular to oval, organic scales, about 2.0 µm in diameter, and distally by large, oval, organic base-plate scales, 7.5 µm x 5.0 µm, supporting the placoliths. Placoliths elliptical, 4.4-13.0 µm x 5.3-9.0 µm, with 2 shields of elements surrounding a wide open central funnel, the funnel sometimes with a bar across the short axis. The distal shield monocyclic with straight sutures, the proximal shield bicyclic. Holo-coccolith-bearing cells, globose to subglobose, 6.0-20.0 µm in diameter, with 2 equal or subequal flagella, about 20.0 µm long, a short, coiling haptonema, 2.0-10.0 µm long; chloroplasts 2-4. Organic body-scales and crystalloliths in several layers with an outermost envelope, the "skin", more than one system of scales, crystalloliths and "skin" sometimes present. Organic body-scales 2.0 x 1.3 µm; smaller circular to elliptic organic scales, 0.8 µm x 0.5 µm, present around the flagellar area. Organic base-plate scales 2.9 µm x 1.8 µm; crystalloliths 2.0-3.0 µm x 1.3-1.85 µm. Crystalloliths composed of a single layer of rhombohedral crystals, with a double layer around the base plate periphery.

Coccolithus pelagicus has 2 crystallolith-bearing cell types (previously referred to as *Crystallolithus hyalinus* and *C. braarudii*). In the *Cr. hyalinus* -form there are small gaps between neighboring crystals, whilst in the *Cr. braarudii*-form the gaps are large and the crystals are arranged in an almost radial pattern. Billard (1994) has suggested that the hetero-coccolith/holococcolith phases are part of a sexual life-cycle. See also Gaarder and Markali (1956), Parke and Adams (1960), Gaarder (1962), Manton and Leedale (1963b, 1969), Rowson et al. (1986).

Genus *Cruciplacolithus* Hay & Mohler

Cells either biflagellate or non-motile, surrounded proximally by organic scales and distally by a coccosphere of interlocking placoliths. Life cycle unknown. Marine, coastal. The genus was described for fossil forms presumed to have become extinct during the Miocene. The coccoliths of *C. neohelis* (McIntyre & Bé) Reinhardt, the only recorded extant species, strongly resemble the coccoliths of fossil species in this genus. Thus, the above generic description is restricted to *C. neohelis.*

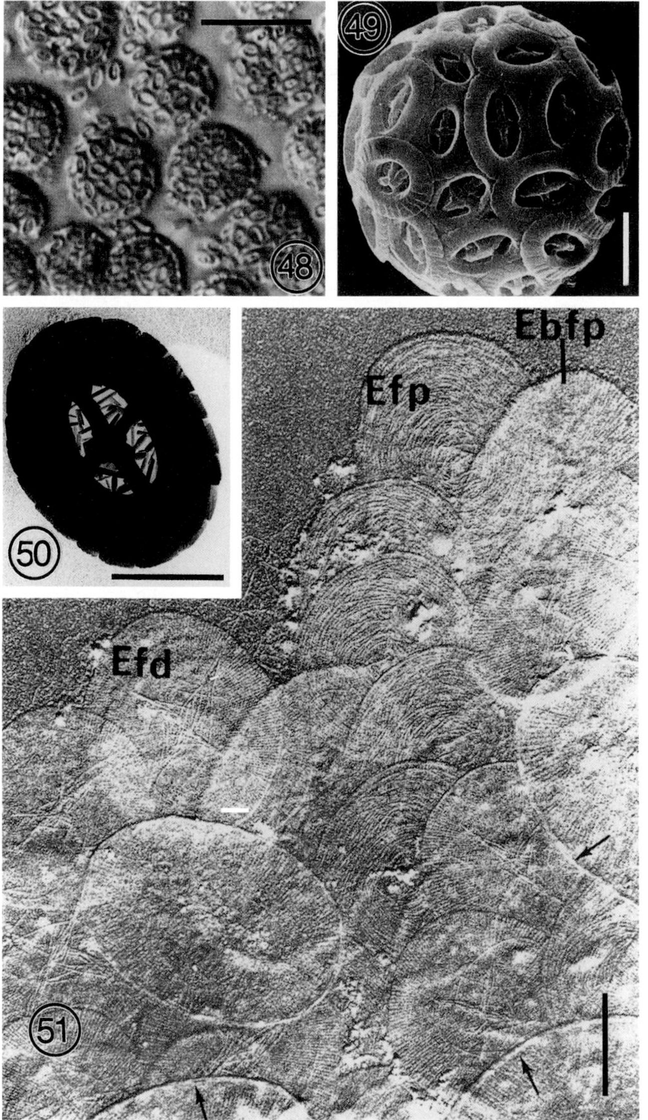

Figs. 48-51. *Cruciplacolithus neohelis* (McIntyre & Bé) Reinhardt. Fig. 48. Coccospheres. Interference contrast. Scale bar=10.0 μm. Fig. 49. Coccosphere with interlocking placoliths. Scale bar=2.0 μm. Fig. 50. Placolith in proximal view showing cruciform bridge and rhombohedric crystals in the central area. Scale bar=1.0 μm. Fig. 51. Distal (Efd) and proximal (Efp) views of unmineralized organic scales and proximal view of a base-plate scale (Ebfp); the arrows indicate the raised rims of the base-plate scales. Scale bar=0.5 μm. Figs. 48-51 from Fresnel (1986).

Cruciplacolithus neohelis (McIntyre & Bé) Reinhardt (Figs. 48-51). Motile cells, 6.0-10.0 μm, with 2 subequal flagella about 7.0 μm long; haptonema non-emergent but haptonematal base present consisting of 3-5 microtubules. Subspherical nucleus, 1.5-3.0 μm in diameter, located between the inner faces of the 2 chloroplasts, each chloroplast containing an immersed lenticular pyrenoid. Cell surrounded by several layers of circular organic scales about 1 μm in diameter, one face with a radial fibrillar pattern, the other with a spiral pattern. Elliptical organic base plates, 1.3-1.6 μm x 1.0 μm, with a pattern of radial fibrils on each face and a slightly thickened rim, supporting elliptical coccoliths (placoliths). Placoliths with distal shield 2.2-3.2 μm x 1.8-2.6 μm, composed of 19-28 dextrally imbricate elements with straight sutures; proximal shield bicyclic; central area with a cruciform bridge spanning the long and short axes of the coccolith, and a number of rhombohedric crystals apparently originating from the proximal part of the coccolith tube. Non-motile cells, 6.0-7.0 μm, identical to motile cells, although with no emergent flagella. Coccosphere of non-motile cells, 7.0-11.0 μm in diameter.

As in *Emiliania huxleyi* , cells in old cultures may possess many layers of coccoliths. Cell division involves invagination of coccosphere and separation of daughter cells. See McIntyre and Bé (1967), West (1969), Fresnel (1986).

Genus ***Emiliania*** Hay & Mohler

Coccolithophorids with non-motile hetero-coccolith-bearing and motile naked phases. Motile phase with 2 flagella, but lacking a haptonema, and covered by a single layer of organic scales. Non-motile phase covered by 1 or more layers of interlocking placoliths (composed of 2 shields of subhorizontal elements connected by a short tube), but lacking underlying organic scales or base-plate scales. Distal shield elements T-shaped, those of the proximal shield similar or flattened into blades. Marine, planktonic.

Emiliania huxleyi (Lohmann) Hay & Mohler in Hay et al. (Figs. 52-54). Non-motile cells, 5.0-7.0 µm in diameter, with a parietal chloroplast with immersed pyrenoid. Placoliths elliptical, approx. 3.5 µm x 2.9 µm in 1 or more layers (multiple layers in older cultures or at the end of a bloom). Placoliths composed of either 2 shields of T-shaped elements connected by a short tube, or with the elements of the proximal shield flattened. Central area wide and partially covered proximally by a reticulum/grill. Naked non-motile cells also recorded. Motile cells, 4.0-6.0 µm in diameter, with 2 equal or unequal flagella, but lacking a haptonema. Cells covered by a single layer of oval organic body scales, 0.3-0.5 µm in diameter, with radial fibrillar patterning in distinct quadrants. Marine, planktonic.

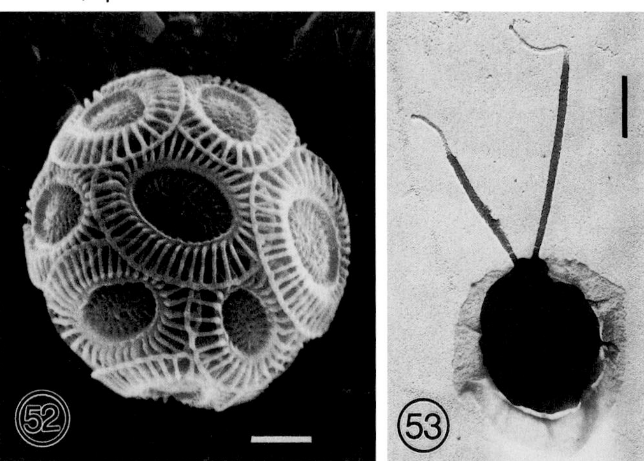

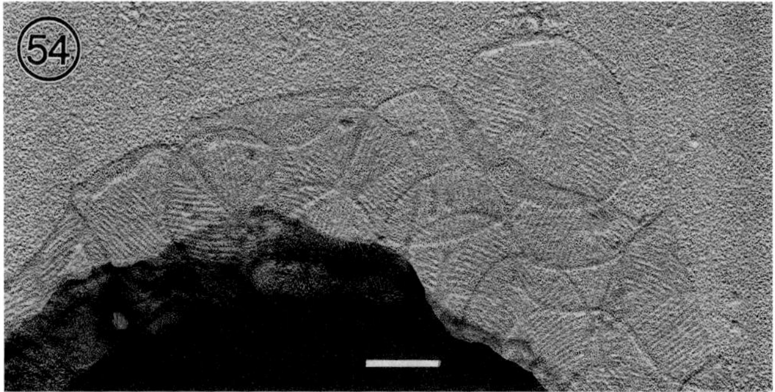

Figs. 52-54. *Emiliania huxleyi* (Lohmann) Hay & Mohler. Fig. 52. Coccolith-bearing cell (C-cell). Scale bar=1.0 µm. Fig. 53. Motile scale-bearing cell (S-cell). Scale bar=1.0 µm. Fig. 54. Unmineralized body-scales from an S-cell. Scale bar=0.2 µm. Figs. 53, 54 from Green et al. (1996).

Emiliania huxleyi is cosmopolitan and arguably the commonest extant prymnesiomonad. It forms massive blooms covering wide areas of ocean (see above) and has been the subject of intense recent

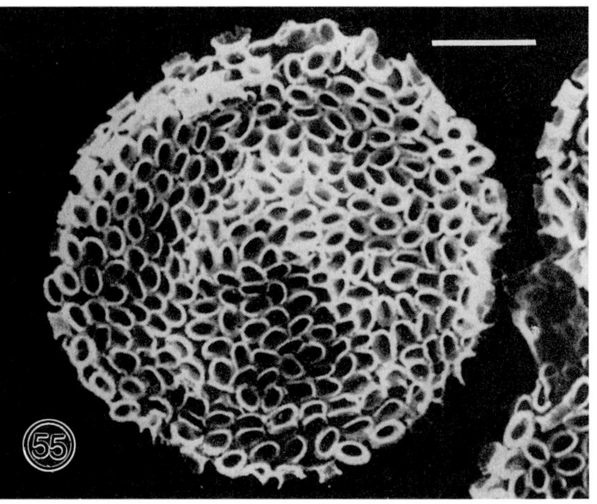

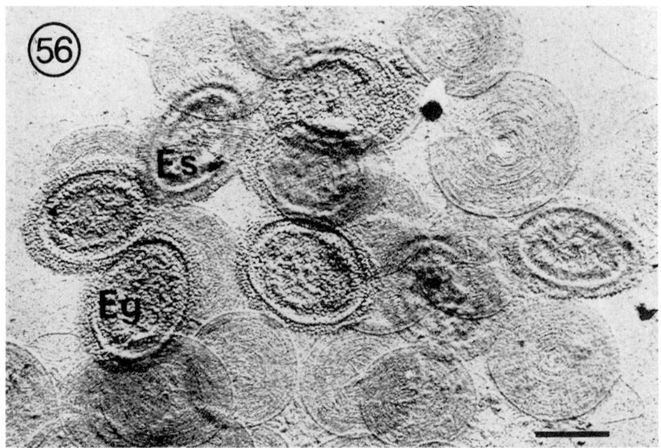

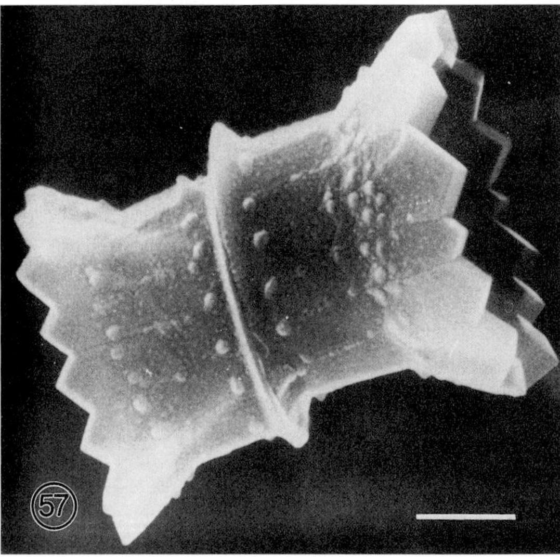

Figs. 55,56. *Hymenomonas globosa* (Magne) Gayral & Fresnel. Fig. 55. Coccosphere composed of abutting tremaliths. Scale bar=5.0 µm. Fig. 56. Distal (Eg) and proximal (Es) faces of coccolith base-plates together with circular body-scales. Scale bar=0.5 µm. Fig. 57. *Hymenomonas lacuna* Pienaar. Two tremaliths (abutting base-base) in lateral view. Scale bar=0.5 µm. Figs. 55,56 from Gayral and Fresnel (1976); Fig. 57 from Gayral and Fresnel (1979).

research in both laboratory and the field (Holligan et al., 1983, 1993; Westbroek et al., 1993, 1994; Barker et al., 1994; Heimdal et al., 1994; Medlin et al., 1994a; Conte et al., 1995; Harris, 1996). The coccoliths are variable in form and those exhibiting a distal projection from the tube wall are assigned to the variety *E. huxleyi* var. *corona* (Okada & McIntyre, 1977). Recent morphological and genetic investigations (Young and Westbroek, 1991; Medlin et al., 1996) have led to the erection of other varieties based on both morphological studies and genetic analysis. There is evidence suggesting that the life-cycle includes a sexual process (Green et al., 1996; Medlin et al., 1996).

Genus ***Hymenomonas*** Stein

Coccolithophorids generally non-motile, but when motile with 2 subequal flagella and a non-emergent haptonema (i.e. internal trace) or with a short, bulbous haptonema. Both cell types bearing hundreds of monomorphic crown-shaped coccoliths, tremaliths, overlying layers of circular organic scales, sometimes also with small elliptical scales. Coastal, brackish, and freshwater. See Gayral and Fresnel (1979).

Hymenomonas globosa (Magne) Gayral & Fresnel (Figs. 55, 56). Cells spherical, approx. 15.0 µm diameter, with 2 smooth sub-equal flagella, longer than the diameter of the cell. Emergent haptonema absent. Chloroplasts 2, each with a globular bulging pyrenoid, traversed by thylakoids, the latter not continuous with those of the chloroplast. Coccoliths (tremaliths) formed from 14-16 wall elements united to form a cylindrical tube proximally, but splayed outwards distally, each element pointed distally, the point sometimes slightly asymmetrical; the whole tremalith bearing protuberances on the external face (as in *H. lacuna* Pienaar; see Fig. 57). Coccolith base-plate scale elliptical, 1.1 µm x 0.8 µm, with radial striations on 1 face, the other face granular. Body scales circular, 0.7 µm-1.0 µm, ornamented with concentric fibrils. See Gayral and Fresnel (1976, 1979).

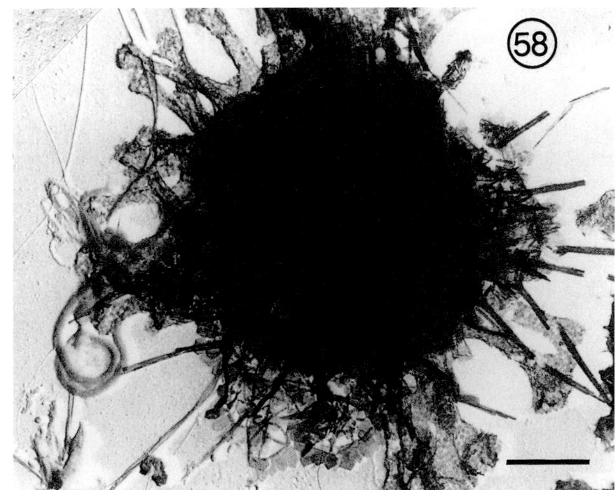

Figs. 58, 59. *Papposphaera arctica* (Manton et al.) Thomsen et al. Fig. 58. Combination cell showing both '*Papposphaera*' and '*Turrisphaera*' coccoliths. Scale bar=2.0 µm. Fig. 59. Pappoliths showing the arrangement of the wall and process. Scale bar=1.0 µm. Fig. 58 from Thomsen et al. (1991); Fig. 59 from Thomsen (1981).

Genus ***Papposphaera*** Tangen

Coccolithophorids having heterococcolith- and holococcolith-bearing phases (see below). Heterococcolith-bearing cells in some species seen with 2 flagella and a coiling haptonema. Cells surrounded by monomorphic process-bearing pappoliths. Unmineralized organic scales apparently absent. Holococcolith-bearing cells biflagellate with a short coiling haptonema, and surrounded by small, oval rimmed organic scales and an outer layer of goblet-shaped coccoliths composed of hexagonal crystallites. Circum-flagellar coccoliths morphologically similar but slightly larger than the body coccoliths. At present there are 7 recognized species (Jordan and Green,

1994) although only 2 are known to form combination cells; i.e. cells bearing 2 complete sets of morphologically different coccoliths. These combination cells are considered to be transitions between 2 phases in the same life cycle, and have resulted in the transfer of *Turrisphaera* species (holococcolithophorids) into *Papposphaera* (Thomsen et al., 1991). Marine, planktonic, mostly recorded from polar regions, but also present in subtropical areas (Hoepffner and Haas, 1990).

Papposphaera arctica (Manton et al.) Thomsen et al. (Figs. 58,59). Heterococcolithophorid cells spherical, about 7.0 µm in diameter, apparently non-motile and lacking a haptonema, and bearing monomorphic pappoliths. Pappoliths consisting of an oval base-plate, 1.2 µm x 1.7 µm, supporting a rim of 2 alternating types of element; 1 rod-shaped and confined to the proximal part of the rim, the other pentagonal, 0.5-0.8 µm high and vertically expanded; 4 struts spanning the base plate, meeting in the center from which arises a central process, 1.9-3.6 µm long, consisting of 4 vertical rows of elements, and ending distally in 4 long elements, 1.0 µm long. Holococcolithophorid cells slightly oval, about 6.0-8.0 µm in diameter, biflagellate with a short coiling haptonema. Unmineralized organic scales not reported. Holococcoliths, up to 2.5 µm in length, with the larger coccoliths near the flagellar pole. Coccoliths tubular, supported by an oval organic base plate; tubes consisting of a wide base and narrow stem, widening slightly distally. Tube composed of small hexagonal crystallites, each about 0.1 µm in diameter with a central perforation. Calcification sometimes absent. See Manton et al. (1976a,b), Thomsen (1981), Thomsen et al. (1991).

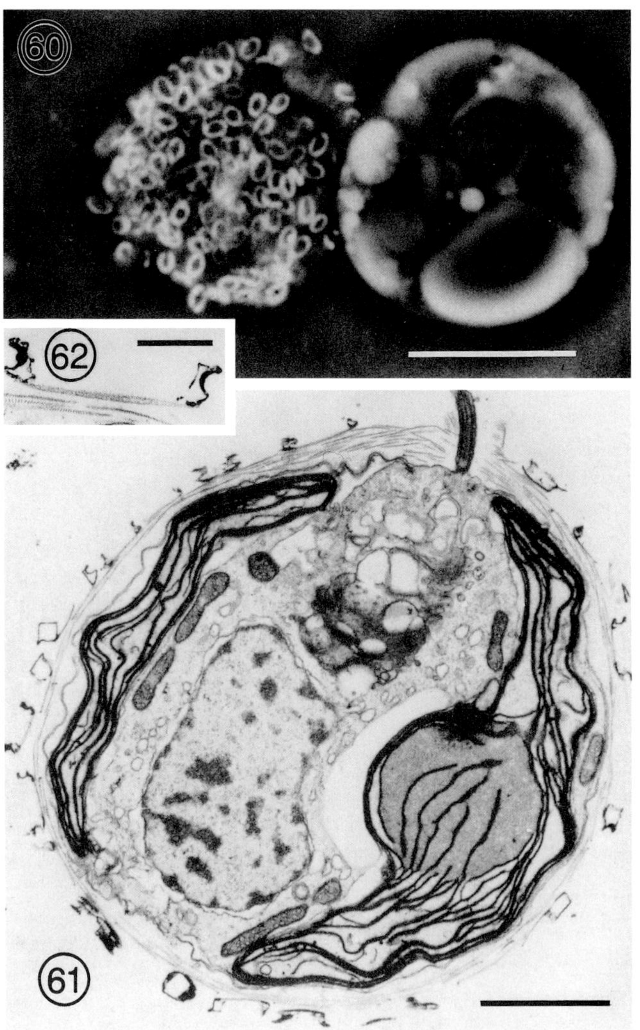

Figs. 60-62. *Pleurochrysis carterae* (Braarud & Fagerland) Christensen. Fig. 60. Living cell with the protoplast squashed out of the coccosphere. Negative phase contrast. Scale bar=10.0 µm. Fig. 61. Longitudinal section through the motile cell showing the peripheral chloroplasts with large pyrenoid, the central nucleus, parabasal Golgi complex, and the surface covering of scales and coccoliths. Scale bar=2.0 µm. Fig. 62. Section showing a coccolith with underlying organic scales. Scale bar=0.4 µm. Figs. 60, 62 from Manton and Leedale (1969); Fig. 61 from Manton and Peterfi (1969).

Genus *Pleurochrysis* E. G. Pringsheim

Cells either motile or non-motile. Diploid cells with elliptical organic base plate scales associated with monomorphic coccoliths (cricoliths) composed of 2 narrow shields of elements. Naked meiospores released from coccolith-bearing cells settling and forming non-motile pseudofilaments (benthic phase); pseudo-filaments surrounded by many layers of rimless elliptical organic scales. Motile cells, with 2 slightly to strongly unequal flagella, a short bulbous haptonema (in some species the haptonema either absent or reduced to an internal trace), and 2 chloroplasts, each with a bulging pyrenoid. Cells surrounded by several layers of circular organic scales and smaller elliptical scales restricted to the flagellar area. Both types of scale generally rimmed, although rimless in gametes and zoospores. Coastal, planktonic-benthic, probably widespread distribution. Species of *Pleurochrysis* have been intensively studied with particular attention having been given to elucidation of the life-cycle. See Beuffe (1978), Gayral and Fresnel (1983b); Fresnel and Billard (1991); Billard (1994).

Pleurochrysis carterae (Braarud & Fagerland) Christensen (Figs. 60-62); (=*Hymenomonas carterae* [Braarud & Fagerland] Braarud). Pseudofilamentous cells (haploid), spherical to elongate, about 20.0 µm long, containing 1 or more chloroplasts, each with a bulging pyrenoid. Cells surrounded by many layers of rimless elliptical scales, approx. 1.0 µm-1.5 µm x 0.8 µm x 1.7 µm with a distal pattern of spiral fibrils and a proximal pattern of radial fibrils in close-packed arrangement. Motile cells (haploid) surrounded by 1-2 layers of rimless elliptical organic body scales similar to above; smaller scales associated with the haptonema; 2 subequal or equal flagella, haptonema short or non-emergent. Coccolith-bearing cells (diploid), pear-shaped to spherical, 3.9-7.5 µm, motile, with 2 slightly unequal flagella and a short haptonema. Cells surrounded by 1 or more layers of rimmed circular organic scales, 0.80-0.89 µm in diameter, and distally by a layer of cricoliths, each supported by a large elliptical organic base plate, 1.01-1.90 µm x 0.67-0.90 µm. Smaller, elliptical to rectangular organic scales, 0.36-0.5 µm x 0.23-0.34 µm, associated with the haptonema. Cricoliths oval, 1.41-2.20 µm x 0.7-1.0 µm, composed of 2 narrow shields of elements surrounding a large open central area. See Manton and Leedale, 1969, Manton and Peterfi (1969), Pienaar (1969), Leadbeater (1970).

Genus *Syracosphaera* Lohmann

Coccolithophorids bearing monomorphic or dimorphic caneoliths, sometimes with an exothecal layer of monomorphic cyrtoliths. Circum-flagellar caneoliths often bearing a central process. Cells often with 2 flagella and a haptonema. Marine, planktonic, worldwide distribution.

Syracosphaera is at present the largest coccolithophorid genus, containing 21 species (Jordan and Green, 1994), although many undescribed species still remain (Heimdal and Gaarder, 1981; Kleijne, 1993). See Gaarder and Heimdal (1977) for a review. Recently, some species assigned to *Caneosphaera* Gaarder have been returned to *Syracosphaera* (Jordan and Young, 1990).

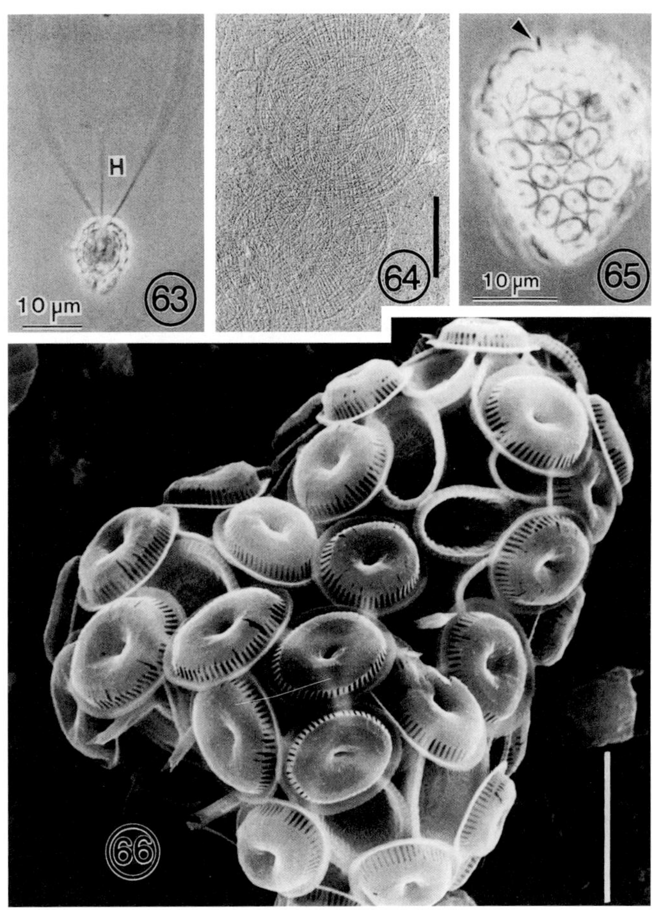

Figs 63-66. *Syracosphaera pulchra* Lohmann. Fig. 63. Living cell with two flagella and extended haptonema (H). Scale bar = 10.0 µm. Fig. 64. Organic scale showing both radial and spiral fibrillar patterning. Scale bar = 4.0 µm. Fig. 65. Coccosphere composed of caneoliths with some overlying cyrtoliths. The arrow head indicates a circum-flagellar caneolith. Scale bar = 10.0 µm . Fig. 66. Coccosphere with distal layer of cyrtoliths and proximal layer of caneoliths. Note the circum-flagellar caneoliths (bottom left of picture) with central processes. Scale bar = 5.0 µm. Figs 63-65 from Inouye & Pienaar (1988).

Syracosphaera pulchra Lohmann (Figs. 63-66). Cells teardrop-shaped, obpyriform, or spherical, 12.0-39.0 µm x 12.0-18.0 µm, with 2 nearly equal flagella, 34.0-51.0 µm long, and a coiling haptonema, 15.0-28.0 µm long. Cells usually covered by several layers of rimless, circular to slightly elliptical organic scales, 0.8-1.1 µm in diameter, underlying 2 layers of heterococcoliths. Proximal coccolith layer (endotheca) composed of dimorphic caneoliths; basket-shaped coccoliths, 4.5-7.0 µm long, 1.2-1.4 µm high, surmounting a base plate scale, 3.5-3.8 µm x 4.8-5.6 µm. Caneolith with proximal, midwall, and distal flanges, and with a central area composed of radiating laths. Circum-flagellar caneoliths bearing a central helatoform process. Outer layer (exotheca) composed of monomorphic cyrtoliths; coccoliths (without a base plate scale) with a peripheral rim, 3.4-7.2 µm in diameter, supporting a highly vaulted roof of laths, 0.9-1.5 µm high. Central depression associated with a tube directed proximally. See Halldal and Markali (1954, 1955), Gaarder and Heimdal (1977), Inouye and Pienaar (1988).

HOLOCOCCOLITHOPHORIDS

Genus ***Calyptrosphaera*** Lohmann

Coccolithophorids with coccospheres composed of monomorphic cap-shaped holococcoliths, calyptroliths. At present the genus contains 6 species (Jordan and Green, 1994). Marine, planktonic, worldwide.

Calyptrosphaera sphaeroidea Schiller (Figs. 67-69). Cells spherical or slightly pyriform, about 8.0-12.0 µm in diameter, with 2 equal flagella, twice the length of the cell and a conspicuous coiling haptonema, 5.0-8.0 µm long when extended, with an intermediate or distal swelling. Chloroplast single, usually located diametrically opposite the flagellar insertion, 2 chloroplasts, if present, situated laterally. Pyrenoid fusiform, completely immersed in the chloroplast, difficult to see in living material. Nucleus spherical, situated slightly eccentrically. Cell surrounded proximally by several layers of rimless organic scales, 0.9-1.3 µm x 0.8-1.0 µm, the proximal face with a pattern of radial fibrils, the distal face with concentric or spiral fibrils, and some smaller scales, 0.3-0.6 µm x 0.3-0.4 µm, around the flagellar area; distal investment of small, fragile, monomorphic calyptroliths, 1.1-1.3 µm x 0.9-1.0 µm and 0.6-1.1 µm high, and an organic covering, termed the skin. The calyptroliths are extremely fragile as the comparatively large microcrystals are arranged in only one layer. See Gaarder (1962), Klaveness (1973).

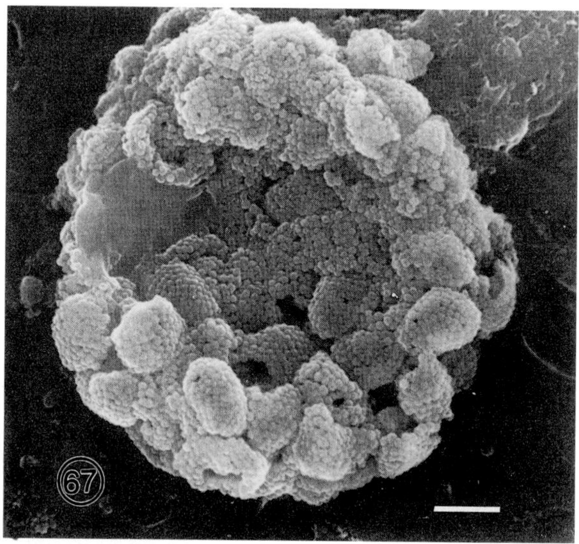

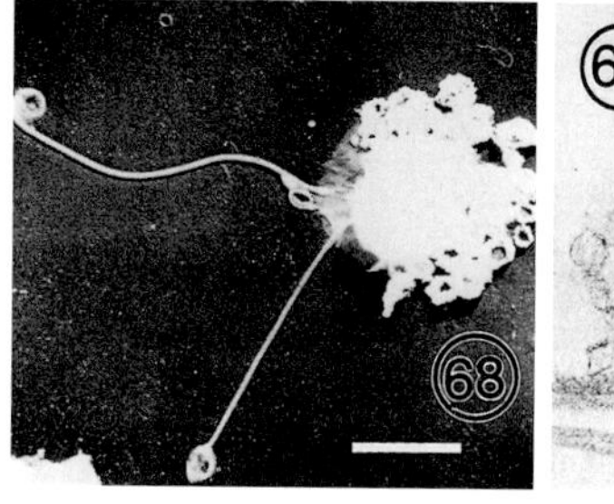

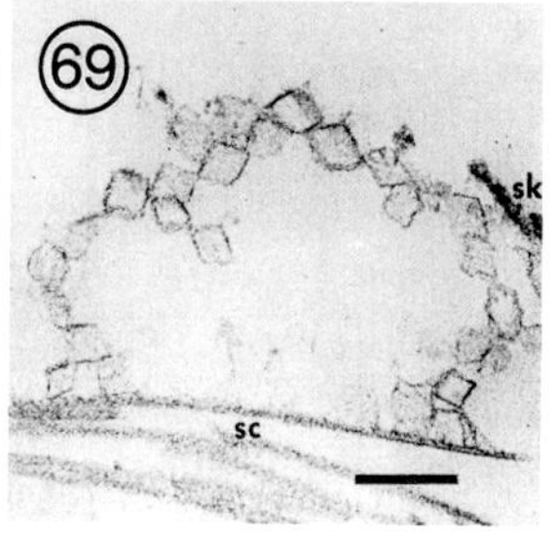

Figs. 67-69. *Calyptrosphaera sphaeroidea* Schiller. Fig. 67. Coccosphere composed of holococcoliths (calyptroliths). Scale bar=1.0 µm. Fig. 68. Flagellate cell bearing haptonema and collapsed coccoliths. Scale-bar=2.5 µm. Fig. 69. Section through a demineralized coccolith, scales (sc) and 'skin' (sk). Scale bar=0.2 µm. Fig. 67 from Kleijne in Winter and Siesser (1994); Figs. 68, 69 from Klaveness (1973).

GENERA *INCERTAE SEDIS*

Genus ***Balaniger*** Thomsen & Oates, 1978

Coccolithophorids with saddle-shaped cells bearing 2 flagella, a haptonema, and partially calcified organic scales. Brackish-marine, planktonic, from the Baltic Sea and West Greenland.

Balaniger balticus Thomsen & Oates (Figs. 70-72). Cell saddle-shaped, 3.1-4.7 µm in length, 3.8-6.0 µm in width, with 2 flagella, 12.5-26 µm long, and a haptonema, 7.0-8.8 µm long. Cell surrounded by oval (slightly angular), rimmed organic scales, 0.4-0.5 µm x 0.3-0.35 µm, covered on the distal face by a single layer of small, calcified, pyramid-shaped hollow crystals, 0.10-0.15 µm high and with a triangular base, 0.10-0.15 µm.

See Thomsen and Oates (1978), Thomsen (1979, 1981).

Genus ***Florisphaera*** Okada & Honjo

Coccolithophorids with hemi-ellipsoidal coccospheres composed of 30-100 polygonal plates arranged in up to 10 or more layers, and with coccoliths oriented in the same direction. One end of the coccosphere covered by overlapping coccoliths, the other open. The open side is, presumably, the site of the disintegrated cell; cell ultrastructure and coccosphere-to-cell orientation are still unknown. One species, with 2 varieties, recorded.

Figs. 70-72. *Balaniger balticus* Thomsen & Oates. Fig. 70. Whole-mounted, shadowed cell bearing two flagella and a haptonema. Scale bar=3.0 µm. Fig. 71. The cell covering of pyramidal coccoliths. Scale bar=1.0 µm. Fig. 72. Individual coccoliths. Scale bar=0.2 µm. Figs 70,72 from Thomsen and Oates (1978); Fig. 71 from Thomsen (1979).

See Okada and Honjo (1973), Okada (1983). *Florisphaera profunda* inhabits the lower photic zone (approx. 100-220 m water depth) of subtropical and tropical open ocean waters, where it is often the dominant species. The upper limit of its vertical distribution is apparently governed by the position of the nutricline (Molfino and McIntyre, 1990). Its presence in temperate regions is restricted to the lower photic zone during the summer/autumn months, whilst it is absent from high latitudes (Okada and McIntyre, 1979). In addition, it is not usually abundant in coastal or upwelling environments, as it appears to be affected by the water column depth (Okada, 1983) and turbidity of the surrounding water (Ahagon et al., 1993).

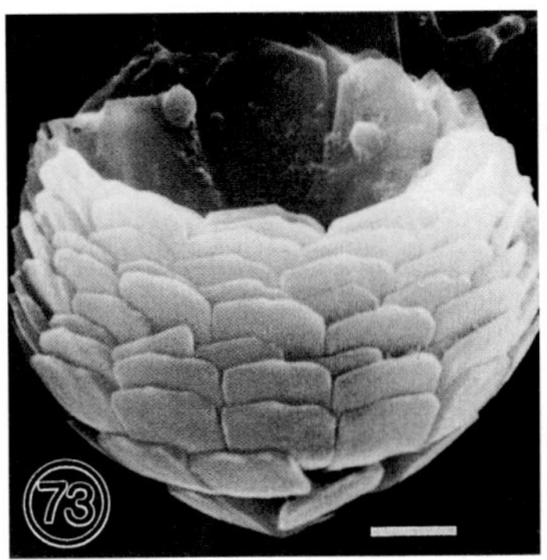

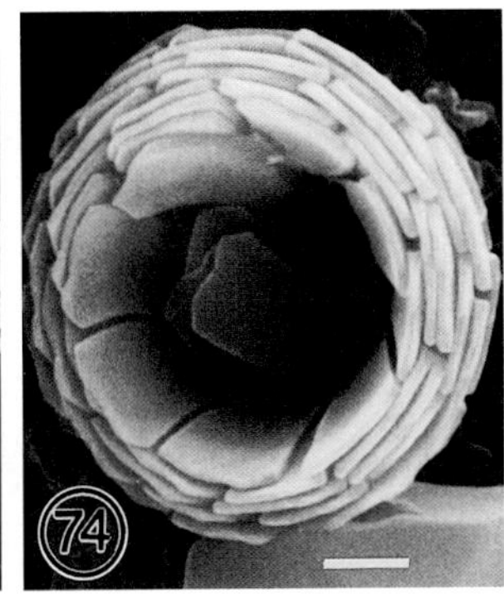

Figs. 73,74. *Florisphaera profunda* Okada & Honjo. Fig. 73. Hemi-ellipsoidal coccosphere in side view showing the coccoliths overlapping in the same orientation. Scale bar=1.0 µm. Fig. 74. The open end of the coccosphere showing the multilayered plate-like coccoliths. Scale bar=1.0 µm.

Genus ***Wigwamma*** Manton et al.

Coccolithophorids with 2 equal flagella, a short coiling haptonema, organic scales, and partially calcified coccoliths. Coccoliths with calcified rims of 2 vertical rows of elements supporting a wigwam-like superstructure (except in *W. scenozonion* Thomsen). Marine, planktonic, largely restricted to polar regions.

One species, *W. aff annulifera* Manton et al. has been seen in a combination cell with *Calciarcus aff. alaskensis* Manton et al. (Thomsen et al., 1991), suggesting that other *Wigwamma* species may have coccolith-bearing alternate phases.

Wigwamma arctica Manton et al. (Figs. 75,76). Cells globose, 4.0-6.0 µm, with 2 equal flagella up to 24.0 µm long and a much shorter haptonema (approx. 5.0 µm). Cells surrounded by rimless organic scales, approx. 0.1 µm diameter with an open pattern of radial and spiral fibrils, underlying monomorphic coccoliths. Each coccolith comprising a circular to anisodiametric organic base plate, 1.1-2.0 µm with a form similar to the scales of the underlayer, and a calcified rim of rod-shaped elements, 2 vertical rows high, on the distal surface of the base plate. Rim elements identical except (in Northern Hemisphere specimens only) for 4 specialized elements of the uppermost row which each support a crystallite, 1.5-2.5 µm long, these crystallites meeting distally to form a conical superstructure. See Manton et al. (1977), Thomsen (1981), Thomsen et al. (1988).

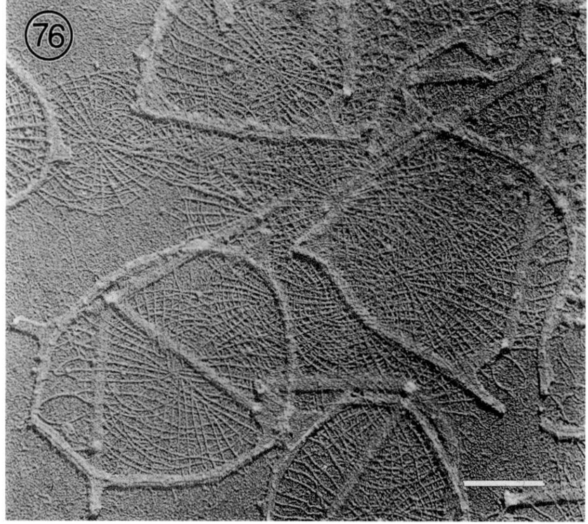

Figs. 75, 76. *Wigwamma arctica* Manton et al. Fig. 75. Complete protoplast covered by calcified wigwam-like superstructures. Scale bar=2.0 µm. Fig. 76. Uncalcified coccoliths showing base-plate patterning, together with a rimless scale (top left) of the unmineralized scale layer. Scale bar=0.5 µm. Fig. 76 from Thomsen et al. (1988); Fig. 77 from Manton et al. (1977).

Literature Cited

Ahagon, N., Tanaka, Y. & Ujiié, H. 1993. *Florisphaera profunda*, a possible nannoplankton indicator of late Quaternary changes in sea-water turbidity at the northwestern margin of the Pacific. *Mar. Micropaleontol.*, 22:255-273.

Anand, P. L. 1937. A taxonomic study of the algae of British chalk-cliffs. *J. Bot., Lond.*, 75, suppl. II:1-51.

Aune, T., Skulberg, O. M. & Underdal, B. 1992. A toxic phytoflagellate bloom of *Chrysochromulina* cf. *leadbeateri* in coastal waters in the North of Norway, May-June 1991. *Ambio*, 21:471-474.

Barker, G. L. A., Green, J. C., Hayes, P. K. & Medlin, L. K. 1994. Preliminary analyses using the RAPD analysis to screen bloom populations of *Emiliania huxleyi* (Haptophyta). *Sarsia*, 79:301-306.

Baumann, M. E. M., Lancelot, C., Brandini, F. P., Sakshaug, E. & John, D. M. 1994. The taxonomic identity of the cosmopolitan prymnesiophyte *Phaeocystis*: a morphological and ecophysiological approach. *J. Mar. Syst.*, **5**:5-22.

Beech, P. L. & Wetherbee, R. 1988. Observations on the flagellar apparatus and peripheral endoplasmic reticulum of the coccolithophorid, *Pleurochrysis carterae* (Prymnesiophyceae). *Phycologia*, 27:142-158.

Beech, P. L., Wetherbee, R. & Pickett-Heaps, J. D. 1988. Transformation of the flagella and associated flagellar components during cell division in the coccolithophorid *Pleurochrysis carterae*. *Protoplasma*, 145:37-46.

Berge, G. 1962. Discoloration of the sea due to *Coccolithus huxleyi* 'bloom'. *Sarsia*, 6:27-40.

Beuffe, H. 1978. Une Coccolithophoracée marine nouvelle: *Cricosphaera gayraliae* nov. sp. *Protistologica*, 14:451-458.

Billard, C. 1983. *Prymnesium zebrinum* sp. nov. et *P. annuliferum* sp. nov., deux nouvelles espèces apparentées à *P. parvum* Carter (Prymnesiophyceae). *Phycologia*, 22:141-151.

Billard, C. 1994. Life cycles. *In*: Green, J. C. & Leadbeater, B. S. C. (eds.), *The Haptophyte Algae*. Systematics Association Special Volume No. 51. Clarendon Press, Oxford. pp 167-186

Billard, C. & Gayral, P. 1972. Two new species of *Isochrysis* with remarks on the genus *Ruttnera*. *Br. Phycol. J.*, 7:289-297.

Birkhead, M. & Pienaar, R. N. 1994. The ultrastructure of *Chrysochromulina brevifilum* (Prymnesiophyceae). *Eur. J. Phycol.*, 29:267-280.

Birkhead, M. & Pienaar, R. N. 1995. The flagellar apparatus of *Chrysochromulina* sp. (Prymnesiophyceae). *J. Phycol.*, 31:96-108

Bjørnland, T. & Liaaen-Jensen, S. 1989. Distribution patterns of carotenoids in relation to chromophyte phylogeny and systematics. *In*: Green, J. C., Leadbeater, B. S. C. & Diver, W. L. (eds.), *The Chromophyte Algae*, Systematics Association Special Volume No. 38. Clarendon Press, Oxford. pp 37-60

Butcher, R. W. 1952. Contributions to our knowledge of the smaller marine algae. *J. Mar. Biol. Assoc. U.K.*, 31:175-191.

Carter, N. 1937. New or interesting algae from brackish waters. *Arch. Protistenkd.*, 90:1-68.

Cavalier-Smith, T. 1989. The Kingdom Chromista. *In*: Green, J. C., Leadbeater, B. S. C. & Diver, W. L. (eds.), *The Chromophyte Algae*. Systematics Association Special Volume No. 38. Clarendon Press, Oxford. pp 381-407

Cavalier-Smith, T. 1993. Kingdom Protozoa and its 18 phyla. *Microbiol. Rev.*, 57:953-994.

Cavalier-Smith, T. 1994. Origin and relationships of Haptophyta. *In*: Green, J. C. & Leadbeater, B. S. C. (eds.), *The Haptophyte Algae*. Systematics Association Special Volume No. 51. Clarendon Press, Oxford. pp 413-435

Chang, F. H. 1984. The ultrastructure of *Phaeocystis pouchetii* (Prymnesiophyceae) vegetative colonies with special reference to the production of new mucilaginous envelope. *N. Z. J. Mar. Freshwater Res.*, 18:303-308.

Chang, F. H. & Ryan, K. G. 1985. *Prymnesium calathiferum* sp. nov. (Prymnesiophyceae), a new species isolated from Northland, New Zealand. *Phycologia*, 24:191-198.

Chen, J. F & Tseng, C. K. 1986. Two species of *Prymnesium* from the North China. *Oceanol. Limnol. Sin.*, 17:394-399. (Chinese with English abstract)

Chrétiennot, M.-J. 1973. The fine structure and taxonomy of *Platychrysis pigra* Geitler (Haptophyceae). *J. Mar. Biol. Assoc. U.K.*, 53:905-914.

Chrétiennot-Dinet, M.-J. 1990. *Atlas du Phytoplancton Marin.* Vol. 3. Éditions du CNRS, Paris.

Christensen, T. 1962. Alger. *In*: Böcher, T. W., Lange, M. & Sørensen, T. (eds.), *Botanik*, Bd. 2, *Systematisk Botanik*, Nr. 2. Munksgaard, Copenhagen. pp 1-178

Christensen, T. 1980. *Algae.* AiO Tryk As, Odense.

Conrad, W. 1926. Recherches sur les flagellates de nos eaux sumâtres. 2e Partie: Chrysomonadines. *Arch. Protistenkd.*, 56:167-231.

Conrad, W. 1941. Notes protistologiques. XXI. Sur les Chrysomonadines à trois fouets. Aperçu synoptique. *Bull. Mus. R. Hist. Nat. Belg.*, 17:1-16.

Conte, M. H., Thompson, A., Eglinton, G. & Green, J. C. 1995. Lipid biomarker diversity in the coccolithophorid *Emiliania huxleyi* (Prymnesiophyceae) and the related species *Gephyrocapsa oceanica.* *J. Phycol.*, 31:272-282.

Conte, M. H., Volkman, J. K. & Eglinton, G. 1994. Lipid biomarkers of the Haptophyta. *In*: Green, J. C. & Leadbeater, B. S. C. (eds.), *The Haptophyte Algae*. Systematics Association Special Volume No. 51. Clarendon Press, Oxford. pp 351-377

Dahl, E., Lindahl, O., Paasche, E. & Throndsen, J. 1989. The *Chrysochromulina polylepis* bloom in Scandinavian waters during Spring, 1988. *In*: Cosper, E. M. Bricelj, V. M.& Carpenter, E. J. (eds.), *Coastal and Estuarine Studies.* Vol. 35, *Novel Phytoplankton Blooms.* Springer-Verlag, Berlin. pp 383-405

Davidson, A. T. & Marchant, H. J. 1992. The biology and ecology of *Phaeocystis* (Prymnesiophyceae). *Progr. Phycol. Res.*, 8:1-45.

Day, A. W., Gardiner, R. B. & Brown, L. M. 1986. Extracellular protein fibrils in *Chrysochromulina breviturrita* (Prymnesiophyceae) and their serological relationship to fungal fimbriae. *Arch. Mikrobiol.*, 146:207-213.

Edvardsen, B. & Paasche, E. 1992. Two motile stages of *Chrysochromulina polylepis* (Prymnesiophyceae): morphology, growth and toxicity. *J. Phycol.*, 28:104-114.

Edvardsen, B. & Vaulot, D. 1996. Ploidy analysis of the two motile forms of *Chrysochromulina polylepis* (Prymnesiophyceae). *J. Phycol.*, 32:94-102.

Estep., K. W., Davis, P. G., Hargraves, P. E. & Sieburth, J. McN. 1984. Chloroplast containing microflagellates in natural populations of North Atlantic nanoplankton, their identification and distribution; including a description of five new species of *Chrysochromulina* (Prymnesiophyceae). *Protistologica*, 20:613-634.

Faber, W. W & Preisig, H. R. 1994. Calcified structures and calcification in protists. *Protoplasma*, 181:78-105.

Febvre, J. & Febvre-Chevalier, C. 1979. Ultrastructural study of zooxanthellae of three species of Acantharia (Protozoa: Actinopoda) with details of their taxonomic position in the

Prymnesiales (Prymnesiophyceae, Hibberd, 1976). *J. Mar. Biol. Assoc., U.K.*, 59:215-226.

Fournier, R. O. 1969. Observations on the flagellate *Diacronema vlkianum* Prauser (Haptophyceae). *Br. Phycol. J.*, 4:185-190.

Fresnel, J. 1986. Nouvelles observations sur une Coccolithacée rare: *Cruciplacolithus neohelis* (McIntyre et Bé) Reinhardt (Prymnesiophyceae). *Protistologica,* 22:193-204.

Fresnel, J. & Billard, C. 1991. *Pleurochrysis placolithoides* sp. nov. (Prymnesiophyceae), a new marine coccolithophorid with remarks on the status of cricolith-bearing species. *Br. Phycol. J.,* 26:67-80.

Fujiwara, S., Sawada, M., Someya, J., Minaka, N., Kawachi, M. & Inouye, I. 1994. Molecular phylogenetic analysis of *rbc*L in the Prymnesiophyta. *J. Phycol.*, 30:863-871.

Gaarder, K. R. 1962. Electron microscope studies on holococcolithophorids. *Nytt Mag. Bot.*, 10:35-51.

Gaarder, K. R. & Hasle, G. R. 1962. On the assumed symbiosis between diatoms and coccolithophorids in *Brenneckella. Nytt Mag. Bot.*, 9:145-148.

Gaarder, K. R. & Hasle, G. R. 1971. Coccolithophorids of the Gulf of Mexico. *Bull. Mar. Sci.*, 21:519-544.

Gaarder, K. R. & Heimdal, B. R. 1977. A revision of the genus *Syracosphaera* Lohmann (Coccolithineae). *"Meteor" Forsch.-Ergebnisse*, Serie D, 13:54-71.

Gaarder, K. R. & Markali, J. 1956. On the coccolithophorid *Crystallolithus hyalinus* n. gen., n. sp. *Nytt Mag. Bot.*, 5:1-5.

Gayral, P. & Fresnel, J. 1976. Nouvelles observations sur deux Coccolithophoracées marines: *Cricosphaera roscoffensis* (P. Dangeard) comb. nov. et *Hymenomonas globosa* (F. Magne) comb. nov. *Phycologia*, 15:339-355.

Gayral, P. & Fresnel, J. 1979. Révision du genre *Hymenomonas* Stein. A propos de l'étude comparative de deux Coccolithacées: *Hymenomonas globosa* (Magne) Gayral et Fresnel et *Hymenomonas lacuna* Pienaar. *Rev. Algol., N.S.*, 14:117-125.

Gayral, P. & Fresnel, J. 1983a. *Platychrysis pienarii* sp. nov. et *P. simplex* sp. nov. (Prymnesiophyceae): description et ultrastructure. *Phycologia*, 22:29-45.

Gayral, P. & Fresnel, J. 1983b. Description, sexualité et cycle de développement d'une nouvelle Coccolithophoracée (Prymnesiophyceae): *Pleurochrysis pseudoroscoffensis* sp. nov. *Protistologica*, 19:245-261.

Gieskes, W. W. & Kraay, G. W. 1986. Analysis of phytoplankton pigments by HPLC before, during and after a mass occurrence of the microflagellate *Corymbellus aureus* during the spring bloom in the open North sea in 1983. *Mar. Biol.*, 92:45-52.

Granéli, E., Sundström, B., Edler, L. & Anderson, D. M. (eds.) 1990. *Toxic Marine Phytoplankton.* Elsevier, New York.

Green, J. C. 1976. *Corymbellus aureus* gen. et sp. nov., a new colonial member of the Haptophyceae. *J. Mar. Biol. Assoc. U.K.*, 56:31-38.

Green, J. C. 1980. The fine structure of *Pavlova pinguis* Green and a preliminary survey of the order Pavlovales (Prymnesiophyceae). *Br. Phycol. J.*, 15:151-191.

Green, J. C. 1986. Biomineralization in the algal class Prymnesiophyceae. *In*: Leadbeater, B. S. C. & Riding, R. (eds.), *Biomineralization in Lower Plants and Animals*. Systematics Association Special Volume No. 30. Clarendon Press, Oxford. pp 173-188

Green, J. C. 1991. Phagotrophy in prymnesiophyte flagellates. *In*: Patterson, D. J. & Larsen, J. (eds.), *The Biology of Free-living Heterotrophic Flagellates.* Systematics Association Special Volume 45. Clarendon Press, Oxford. pp 401-414

Green, J. C. & Course, P. A. 1983. Extracellular calcification in *Chrysotila lamellosa* (Prymnesiophyceae). *Br. Phycol. J.*, 18:367-382.

Green, J. C., Course, P. A. & Tarran, G. A. 1996. The life-cycle of *Emiliania huxleyi*: a brief review and a study of relative ploidy levels analysed by flow cytometry. *J. Mar. Syst.*, 9:33-44.

Green, J. C. & Hibberd, D. J. 1977. The ultrastructure and taxonomy of *Diacronema vlkianum* (Prymnesiophyceae) with special reference to the haptonema and flagellar apparatus. *J. Mar. Biol. Assoc. U.K.*, 57:1125-1136.

Green, J. C., Hibberd, D. J. & Pienaar, R. N. 1982. The taxonomy of *Prymnesium* (Prymnesiophyceae) including a description of a new cosmopolitan species, *P. patellifera* sp. nov., and further observations on *P. parvum* N. Carter. *Br. Phycol. J.*, 17:363-382.

Green, J. C. & Hori, T. 1986. The ultrastructure of the flagellar root system of *Imantonia rotunda* (Prymnesiophyceae). *Br. Phycol. J.*, 21:5-18.

Green, J. C. & Hori, T. 1988. The fine structure of mitosis in *Pavlova* (Prymnesiophyceae). *Can. J. Bot.*, 66:1497-1509.

Green, J. C. & Hori, T. 1990. The architecture of the flagellar apparatus of *Prymnesium patellifera* (Prymnesiophyta). *Bot. Mag.*, Tokyo, 103:191-207.

Green, J. C. & Hori, T. 1994. Flagella and flagellar roots. *In*: Green, J. C. & Leadbeater, B. S. C. (eds.), *The Haptophyte Algae.* Systematics Association

Special Volume No. 51. Clarendon Press, Oxford. pp 47-71

Green, J. C. & Jordan, R. W. 1994. Systematic history and taxonomy. *In*: Green, J. C. & Leadbeater, B. S. C. (eds.), *The Haptophyte Algae*. Systematics Association Special Volume No. 51. Clarendon Press, Oxford. pp 1-21

Green, J. C. & Leadbeater, B. S. C. (eds.) 1994. *The Haptophyte Algae*. Systematics Association Special Volume No. 51. Clarendon Press, Oxford.

Green, J. C. & Manton, I. 1970. Studies in the fine structure and taxonomy of flagellates in the genus *Pavlova*. I. A revision of *Pavlova gyrans* , the type species. *J. Mar. Biol. Assoc. U.K.*, 50:1113-1130.

Green, J. C. & Parke, M. 1974. A reinvestigation by light and electron microscopy of *Ruttnera spectabilis* Geitler (Haptophyceae), with special reference to the fine structure of the zoids. *J. Mar. Biol. Assoc. U.K.*, 54:539-550.

Green, J. C. & Parke, M. 1975. New observations upon members of the genus *Chrysotila* Anand, with remarks upon their relationships within the Haptophyceae. *J. Mar. Biol. Assoc. U.K.*, 55:109-121.

Green, J. C., Perch-Nielsen, K. & Westbroek, P. 1990. Phylum Prymnesiophyta. *In*: Margulis, L., Corliss, J. O., Melkonian, M. & Chapman, D. J. (eds.), *Handbook of Prototictista*. Jones and Bartlett Publishers, Boston (USA). pp 293-317

Green, J. C. & Pienaar, R. N. 1977. The taxonomy of the order Isochrysidales (Prymnesiophyceae) with special reference to the genera *Isochrysis* Parke, *Dicrateria* Parke and *Imantonia* Reynolds. *J. Mar. Biol. Assoc. U.K.*, 57:7-17.

Gregson, A., Green, J. C. & Leadbeater, B. S. C. 1993a. Structure and physiology of the haptonema in *Chrysochromulina*. I. Fine structure of the flagellar/haptonematal root system in *C. acantha* and *C. simplex*. *J. Phycol.*, 29:674-686.

Gregson, A., Green, J. C. & Leadbeater, B. S. C. 1993b. Structure and physiology of the haptonema in *Chrysochromulina*. II. Mechanisms of haptonematal coiling and the regeneration process. *J. Phycol.*, 29:686-700.

Halldal, P. & Markali, J. 1954. Observations on coccoliths of *Syracosphaera mediterranea* Lohm., *S. pulchra* Lohm., and *S. molischi* Schill. in the electron microscope. *Cons. Int. Expl. Mer J.*, 19:329-336.

Halldal, P. & Markali, J. 1955. Electron microscope studies on coccolithophorids from the Norwegian Sea, the Gulf Stream and the Mediterranean. *Avh. norske Vidensk.-Akad. Oslo. I. Mat.-Naturv. Klasse,* 1955 (1):1-30.

Hansen, L. R., Kristiansen, J. & Rasmussen, J. V. 1994. Potential toxicity of the freshwater *Chrysochromulina* species, *C. parva* (Prymnesiophyceae). *Hydrobiologia*, 287:157-159.

Harris, R. P. 1996. Coccolithophorid dynamics: The European *Emiliania huxleyi* Programme (EHUX). *J. Mar. Syst.*, 9:1-11..

Heimdal, B. R. 1993. Modern coccolithophorids. *In*: Tomas, C. R. (ed.), *Marine Phytoplankton: a Guide to Naked Flagellates and Coccolithophorids.* Academic Press, London. pp 147-249

Heimdal, B. R. & Gaarder, K. R. 1981. Coccolithophorids from the northern part of the eastern central Atlantic. II. Heterococcolithophorids. *"Meteor" Forsch.-Ergebnisse*, Serie D, 33:37-69.

Heimdal, B. R., Egge, J. K., Veldhuis, M. J. W. & Westbroek, P. 1994. The 1992 Norwegian *Emiliania huxleyi* experiment. An overview. *Sarsia*, 79:285-290.

Hibberd, D. J. 1972. Chrysophyta: definition and interpretation. *Br. Phycol. J.*, 7:281.

Hibberd, D. J. 1976. The ultrastructure and taxonomy of the Chrysophyceae and Prymnesiophyceae (Haptophyceae): a survey with some new observations on the ultrastructure of the Chrysophyceae. *Bot. J. Linn. Soc.*, 72:55-80.

Hibberd, D. J. 1980. Prymnesiophytes (= Haptophytes). *In*: Cox, E. R. (ed.), *Phytoflagellates.* Elsevier/North Holland, New York. pp 273-317

Hibberd, D. J. & Leedale, G. F. 1985. Prymnesiida. *In*: Lee, J. J., Hutner, S. H. & Bovee, E. C. (eds.), *Illustrated Guide to the Protozoa*. Society of Protozoologists, Lawrence, Kansas, 74-88.

Hoepffner, N. & Haas, L.W. 1990. Electron microscopy of nanoplankton from the North Pacific Central Gyre. *J. Phycol.*, 26:421-439.

Holligan, P. M., Fernández, E., Aiken, J., Balch, W. M., Boyd, P., Burkill, P. H., Finch, M., Groom, S. B., Malin, G., Muller, K., Purdie, D. A., Robinson, C., Trees, C. S., Turner, S. M. & Wal, P. van der, 1993. A biochemical study of the coccolithophore, *Emiliania huxleyi*, in the North Atlantic. *Global Biogeochemical Cycles*, 7:879-900.

Holligan, P. M., Viollier, M., Harbour, D. S., Camus, P. & Champagne-Philippe, M. 1983. Satellite and ship studies of coccolithophore production along the shelf edge. *Nature*, 304:339-342.

Hori, T. & Green, J. C. 1985a. The ultrastructural changes during mitosis in *Imantonia rotunda* Reynolds (Prymnesiophyceae). *Bot. Mar.*, 27:67-78.

Hori, T. & Green, J. C. 1985b. The ultrastructure of mitosis in *Isochrysis galbana* Parke (Prymnesiophyceae). *Protoplasma*, 125:140-151.

Hori, T. & Green, J. C. 1991. The ultrastructure of the flagellar root system of *Isochrysis galbana* (Prymnesiophyta). *J. Mar. Biol. Assoc. U.K.*, 71:137-152.

Hori, T. & Green, J. C. 1994. Mitosis and cell division. *In*: Green, J. C. & Leadbeater, B. S. C. (eds.), *The Haptophyte Algae*. Systematics Association Special Volume No. 51. Clarendon Press, Oxford. pp 91-109

Inouye, I. 1993. Flagella and flagellar apparatuses of algae. *In*: Berner, T. (ed.), *Ultrastructure of Microalgae*. CRC Press, Boca Raton pp 99-133

Inouye, I. & Kawachi, M. 1994. The haptonema. *In*: Green, J. C. & Leadbeater, B. S. C. (eds.), *The Haptophyte Algae*. Systematics Association Special Volume No. 51. Clarendon Press, Oxford. pp 73-89

Inouye, I. & Pienaar, R. N. 1988. Light and electron microscope observations of the type species of *Syracosphaera, S. pulchra* (Prymnesiophyceae). *Br. Phycol. J.,* 23:205-217.

Jahnke, J. & Baumann, M. E. M. 1987. Differentiation between *Phaeocystis pouchetii* (Har.) Lagerheim and *Phaeocystis globosa* Scherffel. I. Colony shapes and temperature tolerances. *Hydrobiol. Bull.*, 21:141-147.

Jebram, D. 1980. Prospection for sufficient nutrition for the cosmopolitic marine bryozoan *Electra pilosa* (Linnaeus). *Zool. Jb. (Syst.)*, 107:368-390.

Jeffrey, S. W. 1989. Chlorophyll *c* pigments and their distribution in the chromophyte algae. *In*: Green, J. C., Leadbeater, B. S. C. & Diver, W. L. (eds.), *The Chromophyte Algae*. Systematics Association Special Volume No. 51. Clarendon Press, Oxford. pp 13-36.

Jeffrey, S. W., Brown, M. R. & Volkman, J. K. 1994. Haptophytes as feedstocks in mariculture. *In*: Green, J. C. & Leadbeater, B. S. C. (eds.), *The Haptophyte Algae*. Systematics Association Special Volume No. 51. Clarendon Press, Oxford. pp 287-302

Jeffrey, S. W. & Wright, S. W. 1994. Photosynthetic pigments in the Haptophyta. *In*: Green, J. C. & Leadbeater, B. S. C. (eds.), *The Haptophyte Algae*. Systematics Association Special Volume No. 51. Clarendon Press, Oxford pp 111-132

Jones, H. L. J., Leadbeater, B. S. C. & Green, J. C. 1994. Mixotrophy in haptophytes. *In*: Green, J. C. & Leadbeater, B. S. C. (eds.), *The Haptophyte Algae*. Systematics Association Special Volume No. 51. Clarendon Press, Oxford. pp 247-263

Jordan, R. W. & Green, J. C. 1994. A check-list of the extant Haptophyta of the world. *J. Mar. Biol. Assoc., U.K.*, 74:149-174.

Jordan, R. W., Kleijne, A., Heimdal, B. R. & Green, J. C. 1995. A glossary of the extant Haptophyta of the world. *J. Mar. Biol. Assoc. U.K.,* 75:769-814.

Jordan, R. W. & Young, J. R. 1990. Proposed changes to the classification system of living coccolithophorids. *I.N.A. Newsletter*, 12:15-18.

Kawachi, M. & Inouye, I. 1995. Functional roles of the haptonema and spine scales in the feeding process of *Chrysochromulina spinifera* (Fournier) Pienaar et Norris (Haptophyta = Prymnesiophyta). *Phycologia*, 34:193-200.

Kawachi, M., Inouye, I., Maeda, O. & Chihara, M. 1991. The haptonema as a food-capturing device: observations on *Chrysochromulina hirta* (Prymnesiophyceae). *Phycologia*, 30:563-573.

Kayser, H. 1970. Experimental-ecological investigations on *Phaeocystis poucheti* (Haptophyceae): cultivation and waste water test. *Helgoländer Wiss. Meeresunters.*, 20:195-212.

Kiss, J. Z. & Triemer, R. E. 1988. A comparative study of the storage carbohydrate granules from *Euglena* (Euglenida) and *Pavlova* (Prymnesiida). *J. Protozool.*, 35:237-241.

Klaveness, D. 1973. The microanatomy of *Calyptrosphaera sphaeroidea*, with some supplementary observations on the motile stage of *Coccolithus pelagicus*. *Norw. J. Bot.*, 20:151-162.

Kleijne, A. 1993. *Morphology, taxonomy and distribution of extant coccolithophorids (calcareous nannoplankton).* Enschede: FEBO. [Ph.D. Thesis, Free University of Amsterdam]

Kornmann, P. 1955. Beobachtungen an *Phaeocystis*-Kulturen. *Helgol. Wiss. Meeresunters.*, 5:218-233.

Kozakai, H., Oshima, Y. & Yasumoto, T. 1982. Isolation and structural elucidation of hemolysin from the phytoflagellate *Prymnesium parvum*. *Agric. Biol. Chem.*, 46:233-236.

Lancelot, C. & Rousseau, V. 1994. Ecology of *Phaeocystis*: the key role of the colony form. *In*: Green, J. C. & Leadbeater, B. S. C. (eds.), *The Haptophyte Algae*. Systematics Association Special Volume No. 51. Clarendon Press, Oxford pp 229-245

Leadbeater, B. S. C. 1970. Preliminary observations on differences of scale morphology at various stages in the life-cycle of '*Apistonema-Syracosphaera*' *sensu* von Stosch. *Br. Phycol. J.*, 5:57-69.

Leadbeater, B. S. C. 1972. Fine structural observations on six new species of

Chrysochromulina (Haptophyceae) from Norway with preliminary observations on scale production in *C. microcylindra* sp. nov. *Sarsia*, 49:65-80.

Leadbeater, B. S. C. 1994. Cell coverings *In*: Green, J. C. & Leadbeater, B. S. C. (eds.), *The Haptophyte Algae*. Systematics Association Special Volume No. 51. Clarendon Press, Oxford. pp 23-46

Leadbeater, B. S. C. & Green, J. C. 1993. Cell coverings of microalgae. *In*: Berner, T. (ed.), *Ultrastructure of Microalgae*. CRC Press, Boca Raton. pp 71-98

Leadbeater, B. S. C. & Manton, I. 1969. *Chrysochromulina camella*, sp. nov., and *C. cymbium* sp. nov., two new relatives of *C. strobilus* Parke & Manton. *Arch. Mikrobiol.*, 68:116-132.

Leadbeater, B. S. C. & Manton, I. 1971. Fine structure and light microscopy of a new species of *Chrysochromulina (C. acantha)*. *Arch. Mikrobiol.*, 78:58-69.

Malin, G., Liss, P. S. & Turner, S. M. 1994. Dimethyl sulfide: production and atmospheric consequences. *In*: Green, J. C. & Leadbeater, B. S. C. (eds.), *The Haptophyte Algae*. Systematics Association Special Volume No. 51. Clarendon Press, Oxford, pp 303-320

Manton, I. 1964a. Observations with the electron microscope on the division cycle in the flagellate *Prymnesium parvum* Carter. *J. Roy. Microsc. Soc.*, 83:317-325.

Manton, I. 1964b. Further observations on the fine structure of the haptonema in *Prymnesium parvum*. *Arch. Mikrobiol.*, 49:315-330.

Manton, I. 1966. Observations on scale production in *Prymnesium parvum*. *J. Cell Sci.*, 1:375-380.

Manton, I. 1967. Further observations on the fine structure of *Chrysochromulina chiton* with special reference to the haptonema, 'peculiar' Golgi structure and scale production. *J. Cell Sci.*, 2:265-272.

Manton, I. 1968. Further observations on the microanatomy of the haptonema in *Chrysochromulina chiton* and *Prymnesium parvum*. *Protoplasma*, 66:35-53.

Manton, I. 1978a. *Chrysochromulina hirta* sp. nov., a widely distributed species with unusual spines. *Br. Phycol. J.*, 13:3-14.

Manton, I. 1978b. *Chrysochromulina tenuispina* sp. nov. from Arctic Canada. *Br. Phycol. J.*, 13:227-234.

Manton, I. 1986. Functional parallels between calcified and uncalcified periplasts. *In*: Leadbeater, B. S. C. & Riding, R. (eds.), *Biomineralization in Lower Plants and Animals*. Systematics Association Special Volume 30. Clarendon Press, Oxford. pp 157-172

Manton, I. & Leadbeater, B. S. C. 1974. Fine-structural observations on six species of *Chrysochromulina* from wild Danish marine nanoplankton, including a description of *C. campanulifera* sp. nov. and a preliminary summary of the nanoplankton as a whole. *K. Danske Vidensk. Salsk. Biol. Skr.*, 2:1-26

Manton, I. & Leedale, G. F. 1963a. Observations on the fine structure of *Prymnesium parvum* Carter. *Arch. Mikrobiol.*, 45:285-303.

Manton, I. & Leedale, G. F. 1963b. Observations on the micro-anatomy of *Crystallolithus hyalinus* Gaarder and Markali. *Arch. Mikrobiol.*, 47:115-136.

Manton, I. & Leedale, G. F. 1969. Observations on the microanatomy of *Coccolithus pelagicus* and *Cricosphaera carterae*, with special reference to the origin and nature of coccoliths and scales. *J. Mar. Biol. Assoc. U.K.*, 49:1-16.

Manton, I. & Oates, K. 1985. Calciosoleniaceae (coccolithophorids) from the Galapagos Islands: unmineralized components and coccolith morphology in *Anoplosolenia* and *Calciosolenia*, with a comparative analysis of equivalents in the unmineralized genus *Navisolenia* (Haptophyceae = Prymnesiophyceae). *Phil. Trans. R. Soc. Lond.*, B, 309:461-477.

Manton, I. & Parke, M. 1962. Preliminary observations on scales and their mode of origin in *Chrysochromulina polylepis* sp. nov. *J. Mar. Biol. Assoc., U.K.*, 42:565-578.

Manton, I. & Peterfi, L. S. 1969. Observations on the fine structure of coccoliths, scales and the protoplast of a freshwater coccolithophorid, *Hymenomonas roseola* Stein, with supplementary observations on the protoplast of *Cricosphaera carterae*. *Proc. Roy. Soc*, B, 172:1-15.

Manton, I., Sutherland, J. & McCully, M. 1976a. Fine structural observations on coccolithophorids from South Alaska in the genera *Papposphaera* Tangen and *Pappomonas* Manton and Oates. *Br. Phycol. J.*, 11:225-238.

Manton, I., Sutherland, J. & Oates, K. 1976b. Arctic coccolithophorids: two species of *Turrisphaera* gen. nov. from West Greenland, Alaska and the Northwest Passage. *Proc. Roy. Soc.*, B, 194:179-194.

Manton, I., Sutherland, J. & Oates, K. 1977. Arctic coccolithophorids: *Wigwamma arctica* gen. et sp. nov. from Greenland and Arctic Canada, *W. annulifera* from S. Africa and S. Alaska and *Calciarcus alaskensis* gen. et sp. nov. from S. Alaska. *Proc. Roy. Soc.*, B, 197:145-168.

Marchant, H. J. & Thomsen, H. A. 1994. Haptophytes in polar waters. *In*: Green, J. C. & Leadbeater, B. S. C. (eds.), *The Haptophyte Algae*. Systematics Association Special Volume No. 51. Clarendon Press, Oxford. pp 209-228

McIntyre, A. & Bé, A. W. H. 1967. *Coccolithus neohelis* sp. n., a coccolith fossil type in contemporary seas. *Deep-Sea Res.*, 14:369-371.

Medlin, L. K., Barker, G. L. A., Baumann, M., Hayes, P. K. & Lange, M. 1994a. Molecular biology and systematics. *In*: Green, J. C. & Leadbeater, B. S. C. (eds.), *The Haptophyte Algae*. Systematics Association Special Volume No. 51. Clarendon Press, Oxford. pp 393-411

Medlin, L. K., Lange, M. & Baumann, M. E. M. 1994b. Genetic differentiation among three colony-forming species of *Phaeocystis*: further evidence for the phylogeny of the Prymnesiophyta. *Phycologia*, 33: 199-212.

Medlin, L., Barker, G. L. A., Campbell, L., Green, J. C., Hayes, P. K., Marie, D., Wrieden, S. & Vaulot, D. 1996. Genetic characterisation of *Emiliania huxleyi* (Haptophyta). *J. Mar. Syst.*, 9:13-31.

Meyer, R. L. 1986. A proposed phylogenetic sequence for the loricate rhizopodial Chrysophyceae. *In*: Kristiansen, J. & Andersen, R. A. (eds.), *Chrysophytes: Aspects and Problems*. Cambridge University Press, Cambridge. pp 75-85

Moestrup, Ø. 1979. Identification by electron microscopy of marine nanoplankton from New Zealand, including the description of four new species. *N.Z. J. Bot.*, 17:61-95.

Moestrup, Ø. 1994. Economic aspects: 'blooms', nuisance species, and toxins. *In*: Green, J. C. & Leadbeater, B. S. C. (eds.), *The Haptophyte Algae*. Systematics Association Special Volume No. 51. Clarendon Press, Oxford. pp 265-285.

Moestrup, Ø. & Larsen, J. 1992. Potentially toxic phytoplankton. 1. Haptophyceae. *In*: Lindley, J. A. (ed.), *ICES Identification Leaflets for Plankton*, Leaflet No. 179. International Council for the Exploration of the Sea, Copenhagen.

Moestrup, Ø. & Thomsen, H. A. 1986. Ultrastructure and reconstruction of the flagellar apparatus in *Chrysochromulina apheles* sp. nov. (Prymnesiophyceae = Haptophyceae). *Can. J. Bot.*, 64:593-610.

Molfino, B. & McIntyre, A. 1990. Precessional forcing of nutricline dynamics in the equatorial Atlantic. *Science*, 249:766-769.

Okada, H. 1983. Modern nannofossil assemblages in sediments of coastal and marginal seas along the western Pacific Ocean. *In*: Meulenkamp, J. E. (ed.), *Reconstruction of Marine Paleoenvironments. Utrecht Micropal. Bull.*, 30:171-187.

Okada, H. & Honjo, S. 1973. The distribution of oceanic coccolithophorids in the Pacific. *Deep-Sea Res.*, 20:355-374.

Okada, H. & McIntyre, A. 1977. Modern coccolithophores of the Pacific and North Atlantic Oceans. *Micropaleontology*, 23:1-55.

Okada, H. & McIntyre, A. 1979. Seasonal distribution of modern coccolithophores in the western North Atlantic Ocean. *Mar. Biol.*, 54:319-328.

Paasche, E., Edvardsen, B. & Eikrem, W. 1990. A possible alternate stage in the life cycle of *Chrysochromulina polylepis* Manton et Parke (Prymnesiophyceae). *Beih. Nova Hedwigia*, 100:91-99.

Parke, M. 1949. Studies on marine flagellates. *J. Mar. Biol. Assoc. U.K.*, 28:255-286.

Parke, M. 1971. The production of calcareous elements by benthic algae belonging to the class Haptophyceae (Chrysophyta). *In*: Farinacci, A. (ed.), *Proc. II Planktonic Conf., Roma, 1970*. Vol. II. Edizioni Tecnoscienza, Rome, 929-937.

Parke, M. & Adams, I. 1960. The motile (*Crystallolithus hyalinus* Gaarder & Markali) and non-motile phases in the life-history of *Coccolithus pelagicus* (Wallich) Schiller. *J. Mar. Biol. Assoc. U.K.*, 39:263-274.

Parke, M., Green, J. C. & Manton, I. 1971. Observations on the fine structure of zoids of the genus *Phaeocystis* (Haptophyceae). *J. Mar. Biol. Assoc. U.K.*, 51:927-941.

Parke, M., Manton, I. & Clarke, B. 1955. Studies on marine flagellates. II. Three new species of *Chrysochromulina*. *J. Mar. Biol. Assoc. U.K.*, 34:579-604.

Parke, M., Manton, I. & Clarke, B. 1956. Studies on marine flagellates. III. Three further species of *Chrysochromulina*. *J. Mar. Biol. Assoc. U.K.*, 35:387-414.

Parke, M., Manton, I. & Clarke, B. 1958. Studies on marine flagellates. IV. Morphology and microanatomy of a new species of *Chrysochromulina*. *J. Mar. Biol. Assoc. U.K.*, 37:209-228.

Parke, M., Manton, I. & Clarke, B. 1959. Studies on marine flagellates. V. Morphology and microanatomy of *Chrysochromulina strobilus* sp. nov. *J. Mar. Biol. Assoc. U.K.*, 38:169-188.

Paster, Z. 1973. Pharmacognosy and mode of action of *Prymnesium*. *In*: Martin, D. F. & Padilla, G. M. (eds.), *Marine Pharmacognosy*. Academic Press, New York. pp 241-263

Perch-Nielsen, K. 1985. Mesozoic calcareous nannofossils. *In*: Bolli, H. M., Saunders, J. B. & Perch-Nielsen, K. (eds.), *Plankton Stratigraphy*. Cambridge University Press, Cambridge. pp 329-426

Pienaar, R. N. 1969. The fine structure of *Cricosphaera carterae*. I. External morphology. *J. Cell Sci.*, 4:561-567.

Pienaar, R. N. 1994. Ultrastructure and calcification of coccolithophores. *In*: Winter A. & Siesser, W. G. (eds.), *Coccolithophores.* Cambridge University Press, Cambridge. pp 13-37

Pienaar, R. N. & Birkhead, M. 1994. Ultrastructure of *Prymnesium nemamethecum* sp. nov. (Prymnesiophyceae). *J. Phycol.*, 30:291-300.

Prauser, H. 1958. *Diacronema vlkianum*, eine neue Chrysomonade. *Arch. Protistenkd.*, 103:117-128.

Rey, F. (ed.) 1991. The *Chrysochromulina leadbeateri* bloom in Vestfjorden, North Norway, May-June, 1991. *Fisken og Havet*, No. 3. Institute of Marine Research, Bergen. [In Norwegian]

Reid, F. M. H. 1980. Coccolithophorids of the North Pacific Central Gyre with notes on their vertical and seasonal distribution. *Micropalaeontology*, 26:151-176.

Reynolds, N. 1974. *Imantonia rotunda* gen. et sp. nov., a new member of the Haptophyceae. *Br. Phycol. J.*, 9:429-434.

Rousseau, V., Vaulot, D., Casotti, R., Cariou, V., Lenz, J., Gunkel, J. & Baumann, M. 1994. The life cycle of *Phaeocystis* (Prymnesiophyceae): evidence and hypotheses. *J. Mar. Syst.*, 5:23-39.

Rowson, J. D., Leadbeater, B. S. C. & Green, J. C. 1986. Calcium carbonate deposition in the motile (*Crystallolithus*) phase of *Coccolithus pelagicus. Br. Phycol. J.*, 21:359-370.

Shilo, M. 1971. Toxins of the Chrysophyceae. *In*: Kadis, S., Ciegler, A. & Ajl, S. J. (eds.), *Microbial Toxins*. Vol. 7. Academic Press, New York. pp 67-103

Shilo, M. 1981. The toxic principles of *Prymnesium parvum*. *In*: Carmichael, W, W. (ed.), *The Water Environment. Algal Toxins and Health.* Plenum Press, New York, 37-47.

Simon, N., Brenner, J., Edwardsen, B. & Medlin, L. K. 1997. The identification of *Chrysochromulina* and *Prymnesium* species (Haptophyta, Prymnesiophyceae) using fluorescent or chemiluminescent oligonucleotide probes: a means for improving studies on toxic algae. *Eur. J. Phycol.*, 32:393-401.

Smayda, T. J. & Shimizu, Y. (eds.) 1993. *Toxic Phytoplankton Blooms in the Sea.* Elsevier, Amsterdam.

Sournia, A. 1988. *Phaeocystis* (Prymnesiophyceae): How many species? *Nova Hedwigia*, 47:211-217.

Stosch, H. A. von 1958. Die Geißelapparat einer Coccolithophoride. *Naturwissenschafte*, 45:140-141.

Thomsen, H. A. 1979. Electron microscopical observations on brackish-water nannoplankton from the Tvärminne area, SW coast of Finland. *Acta Bot. Fennica*, 110:11-37.

Thomsen, H. A. 1981. Identification by electron microscopy of nanoplanktonic coccolithophorids (Prymnesiophyceae) from West Greenland, including the description of *Papposphaera sarion* sp. nov. *Br. Phycol. J.*, 16:77-94.

Thomsen, H. A., Buck, K. R. & Chavez, F. P. 1994. Haptophytes as components of the marine phytoplankton. *In*: Green, J. C. & Leadbeater, B. S. C. (eds.), *The Haptophyte Algae.* Systematics Association Special Volume No. 51. Clarendon Press, Oxford. pp 187-208

Thomsen, H. A., Buck, K. R., Coale, S. L., Garrison, D. L. & Gowing, M. M. 1988. Nanoplanktonic coccolithophorids (Prymnesiophyceae, Haptophyceae), from the Weddell Sea, Antarctica. *Nord. J. Bot.*, 8:419-436.

Thomsen, H. A. & Oates, K. 1978. *Balaniger balticus* gen. et sp. nov. (Prymnesiophyceae) from Danish coastal waters. *J. Mar. Biol. Assoc. U.K.,* 58:773-779.

Thomsen, H. A., Østergaard, J. B. & Hansen, L. E. 1991. Heteromorphic life histories in Arctic coccolithophorids (Prymnesiophyceae). *J. Phycol.*, 27:634-642.

Throndsen, J. 1993. The planktonic marine flagellates. *In*: Tomas, C. R. (ed.), *Marine Phytoplankton: a Guide to Naked Flagellates and Coccolithophorids.* Academic Press, London. pp 1-145

Ulitzer, S. & Shilo, M. 1970. Procedure for purification and separation of *Prymnesium parvum* toxins. *Biochim. Biophys. Acta*, 201:350-363.

Vårum, K. M., Kvam, B. J., Myklestad, S. & Paulsen, B. S. 1986. Structure of a food-reserve β-D-glucan produced by the prymnesiophyte alga *Emiliania huxleyi* (Lohmann) Hay and Mohler. *Carbohydr. Res.*, 152:243-248.

Vaulot, D., Birrien, J.-L., Marie, D., Casotti, R., Veldhuis, M. J. W., Kraay, G. W. & Chrétiennot-Dinet, M.-J. 1994. Morphology, ploidy, pigment composition, and genome size of cultures strains of *Phaeocystis* (Prymnesiophyceae). *J. Phycol.*, 30:1022-1035.

Volkman, J. K., Barrett, S. M., Blackburn, S. I. & Sikes, E. L. 1995. Alkenones in *Gephyrocapsa oceanica*: implications for studies of paleo-climate. *Geochim. Cosmochim. Acta*, 59:513-520.

West, J. A. 1969. Observations on four rare marine microalgae from Hawaii. *Phycologia*, 8:187-192.

Westbroek, P., Brown, C. W., Bleijswijk, J. van, Brownlee, C., Brummer, G. J., Conte, M., Egge, J., Fernández, E., Jordan, R., Knappertsbusch, M., Stefels, J., Veldhuis, M., Wal, P. van der & Young, J. 1993. A model system approach to biological climate forcing. The example of *Emiliania huxleyi*. *Global. Planet. Change*, 8:27-46.

Westbroek, P., Hinte, J. E. van, Brummer, G.-J., Veldhuis, M., Brownlee, C., Green, J. C., Harris, R. & Heimdal, B. R. 1994. *Emiliania huxleyi* as a key to biosphere-geosphere interactions. *In*: Green, J. C. & Leadbeater, B. S. C. (eds.), *The Haptophyte Algae*. Systematics Association Special Volume No. 51. Clarendon Press, Oxford. pp 321-334

Winter, A. & Siesser, W. G. (eds.) 1994. *Coccolithophores*. Cambridge University Press, Cambridge.

Young, J. R., Bown, P. R. & Burnett, J. A. 1994. Palaeontological perspectives. *In*: Green, J. C. & Leadbeater, B. S. C. (eds.), *The Haptophyte Algae*. Systematics Association Special Volume No. 51. Clarendon Press, Oxford. pp 379-392

Young, J. R. & Westbroek, P. 1991. Genotypic variation in the coccolithophorid species *Emiliania huxleyi*. *Mar. Micropaleontol.*, 18:5-23.

RESIDUAL FREE-LIVING AND PREDATORY HETEROTROPHIC FLAGELLATES

DAVID J. PATTERSON, NAJA VØRS, ALASTAIR G. B. SIMPSON & CHARLES O'KELLY

The heterotrophic flagellates—previously referred to as zooflagellates (or zoomastigina or zoomastigotes)—are an adaptive group which includes organisms drawn many different branches of the eukaryotic tree of life. The heterotrophic flagellates do not constitute a phylogenetically coherent group (Patterson, 1993). The territory embraces primitively amitochondriate organisms from the base of the evolutionary tree of eukaryotes, as well as organisms which have close affinities with lineages which appeared much later. Some heterotrophic flagellates are clearly members of groups which contain autotrophs (euglenids, dinoflagellates, stramenopiles, and cryptomonads). Other heterotrophic flagellates belong to well circumscribed groups (such as the kinetoplastids and choanoflagellates), but the nature and taxonomy of many genera remain understudied, or groupings are innovative. These genera are covered in this chapter. We expect that as identities become clear, so new or resurrected suprageneric groupings of flagellates will become accepted, and other generic names will be rendered into synonymy.

This grouping is arbitrary. Some genera not covered in detail are listed at the end of the accounts of genera. The free-living heterotrophic flagellates resemble each other only in using one or a few flagella for motion or for feeding. The free-living flagellates have recently been reviewed (Patterson and Larsen, 1991). These taxa have also been placed in an overall phylogenetic scheme (Patterson, 1994). Readers are referred to Larsen and Patterson (1990), Patterson (1993), Patterson and Larsen (1991), or Patterson and Zölffel (1991) for information on most genera, additional references are given below to publications not included in these general articles.

KEY CHARACTERS

Many flagellates are under 20 micrometers in size. The majority of generic descriptions are based on the light microscopical appearance of cells. Such descriptions refer to little more than size, shape, and presence and location of major organelles. It is on these characteristics that most of the following generic descriptions are based; however, ultrastructural studies are often required to establish the affinities of a genus unambiguously (Patterson, 1994). The resulting "ultrastructural identity" allows the flagellates to be assigned to robust taxa and includes complements of organelles, cytoskeletal structures, structure of flagellar apparatus, etc. For practical purposes, descriptions here are restricted to those features that can be used by light microscopists in the identification of these organisms.

KEY TO GENERA

This key refers to a polyphyletic and paraphyletic subset of free-living flagellates (those taxa that have not yet been assigned to familiar higher taxa). Readers seeking to identify free-living heterotrophic flagellates should also refer to the chapters on the heterolobosea, kinetoplastids, spironemids, colpodellids, and pelobionts. See also step 106 for cells which lack flagella but behave as if they are flagellates (i.e. glide or otherwise move without the use of pseudopodia).

1\. Cells actively moving by swimming or gliding 27
1'. Cells attached to substrate, most activity limited to flagellar beating 2

2\. Cells forming colonies............................ 3
2'. Cells solitary or in irregular clusters...........7

3\. Branching colonies of triangular cells, two flagella near apex of cell.. ***Pseudodendromonas***
3'. Cells embedded in mucoid material................. 4

4\. Cells with a single flagellum..***Phalansterium***
4'. Cells with two flagella.................................. 5

5\. Colony branching, one cell per terminating branch.................................. ***Cladomonas***
5'. Colony constituted of mucus globules; if branching, typically more than one cell per terminating branch.. 6

6\. Colony flattened.................... ***Rhipidodendron***
6'. Colony formed of bulbous branches............... .... ***Spongomonas***

7\. Cells within lorica.. 8
7'. Cells not within lorica.................................. 17

8\. Lorica comprised of mucus globules............... .... ***Spongomonas***
8'. Lorica a discrete shell.................................. 9

9\. With fine branching pseudopodia......................... .... ***Microcometes***
9'. No fine branching pseudopodia......................... 10

10\. With one flagellum.. 11
10'. With two flagella.. 14

11\. .Lorica double.............................. ***Diplocalium***
11'. Lorica single.. 12

12\. Lorica lying, narrowed ends....... ***Platytheca***
12'. Lorica with open end directed away from substrate.. 13

13\. Cell with antero-ventral furrow... ***Histiona***
13'. Cell rounded, flagellum apical..... ***Codonoeca***

14\. Both flagella free and projecting away from the cell.. 15
14'. One flagellum recurrent.............................. 16

15\. Flagella arise from one corner of the cell .... ***Cyathobodo***
15'. Flagella arise from center of rounded cell .... ***Diploselmis***

16\. With angle between stalk and remainder of lorica; cell lying with one projecting flagellum and one in a groove ....***Reclinomonas***
16'. No angle between stalk and remainder of lorica, with hood around projecting flagellum...................... ***Histiona/Stenocodon***

17\. No arms emanating from body........................ 18
17'. With arms radiating from body...................... 23

18\. Attaching by flagellum (a)............................ 19
18'. Not attaching by flagellum, but by the body, or part of it.. 20

19\. Attaching by a hooked flagellum that appears to be posterior .... ***Metromonas***
19'. Attaches by hook of anterior flagellum, with second unattached flagellum in a ventral groove.. ***Jakoba***

20. Posterior drawn out as a stalk.................... 21
20'. Cell may be embedded in mucus, single long flagellum .... ***Rhizomonas***

21. One or two flagella, with hood around projecting flagellum; body attached directly to substrate................................. ***Stomatochone***
21'. No hood 22

22. With two flagella..................... ***Amphimonas***
22'. With one flagellum arising centrally from a depression.................................. ***Cyathomonas***

23. Arms of equal length, not branching.......... 24
23'. Arms varying in length, branching............ 26

24. Arms from all areas of body, may slowly move .... ***Artodiscus***
24'. Arms restricted to periflagellar region of body 25

25. Two flagella, arms stiff................ ***Dimorpha***
25'. Four flagella, arms stiff......***Tetradimorpha***

26. With two flagella curved over unattached face of cell .... ***Massisteria***
26'. Flagella reduced to one or two short stubs (may not be visible with light microscope) .... ***Gymnophrys***

27. Gliding... 28
27'. Swimming... 54

28. With one flagellum.................................. 22
28'. With two flagella..................................... 43

29. Flagellum directed to rear.......................... 30
29'. Flagellum directed antero-laterally...........37

30. Recurrent flagellum short, anterior of cell; with snout .... ***Amastigomonas***
30'. Recurrent flagellum easily visible behind cell 31

31. Cell twisted............................... ***Phyllomonas***
31'. Cell not twisted....................................... 32

32. Flagellum arises near rear margin of moving cell... 33
32'. Flagellum inserts subapically or laterally34

33. Body flattened and triangular in profile .... ***Metromona***
33'. Body rounded................................ ***Kiitoksia***

34. Flagellum emerges from the middle of a depressed side of the cells.............. ***Discocelis***
34'. Flagellum emerges subapically, and trails parallel to the axis of the cell................... 35

35. With ventral or lateral depression of body 36
35'. Without any visible depression 39

36. Body ventrally grooved............. ***Clautriavia***
36'. Groove/indentation is lateral...................... 37

37. Small (about 10 μm or less) 38
37'. Medium to large flagellate....... ***Fromentella***

38. With bulbous antero-lateral swelling of body, flagellum appearing subapically from one side and trailing, may also be biflagellated .... ***Metopion***
38'. With thin anterior flagellum inserted in apical depression and groove cut into ventral and ventro-lateral part of cell. May also have second flagellum in middle of anterior margin of cell...................................... ***Ancyromonas***

39. Body without evident depression; body rigid .... ***Allantion***
39'. Cells may become amoeboid......... ***Sainouron***

40. Flagellum emerges from tip of thickened projection of the cell.................................. 41
40'. No thickened projection of the cell, flagellum directed anteriorly .... ***Rigidomastix***

40. Flagellum arising from the end of a long mastigophore .... ***Apusomonas***
41'. Flagellum from end of short lateral extension of body 42

42. Emergent part of flagella short.............. .... ***Amastigomonas***
42'. Emergent part of flagellum long............ .... ***Cruzella***

43. Cell body can be amoeboid......................... 44

43'. Cell body not amoeboid, but pseudopodia may be produced.. 45

44. Posterior flagellum trails under cell to which it is attached; cells usually very pliable almost amoeboid...................... *Cercomonas*
44'. Cell body occasionally amoeboid, usually uni-flagellated.................................... *Sainouron*

45. Cells under 10 µm 46
45'. Cells longer than 10 µm.............................. 50

46. Able to produce anterior feeding pseudopodia.. *Heteromita*
46'. Anterior feeding pseudopodia not reported.47

47. Flagella emerging subapically and both trailing.. *Metopion* (see also ***Bodomorpha*** and ***Sainouron***)
47'. One flagellum or both flagella directed anteriorly.. 48

48. Anterior flagellum extremely short..... *Allas*
48'. Anterior flagellum not extremely short.... 49

49. Emerging flagellum arising from middle of broad anterior margin of cell..*Ancyromonas*
49'. Emerging flagellum arising from center of narrowed anterior margin of cell........ ....*Caecitellus*

50. Both flagella long and adhering to the substrate.. *Glissandra*
50'. Typically with only one flagellum adhering to the substrate.. 51

51. Laterally flattened.................... *Hyaloselene*
51'. Dorso-ventrally flattened or not flattened 52

52. With scales and spicules visible by light microscopy.......................... *Thaumatomastix*
52'. Cells naked.. 53

53. With flagellar pocket......... *Thaumatomonas*
53'. No flagellar pocket, flagella at best inserting in a slight depression or groove.......*Protaspis*

54. With one flagellum.. 55
54'. With more than one flagellum.................... 60

55. With scales.. 56
55'. Without scales.. 57

56. Scales flattened.............................. *Petasaria*
56'. Spine scales................................ *Trichonema*

57. Body rounded, anteriorly depressed at site of flagellar emergence .... *Paramonas*
57'. Body not rounded.. 58

58. Body flattened, flagellum inserting centrally .... *Peltomonas*
58'. Body sculpted or irregular.......................... 59

59. .Twisted irregular body............. *Cyclomonas*
59'. Body with broad equatorial groove.... *Errera*

60. With two flagella.. 61
60'. With more than two flagella........................ 94

61. Internal siliceous skeleton.......................... 62
61'. No internal siliceous skeleton.................... 63

62. Skeleton with three branches............... *Ebria*
62'. Skeleton with four branches....*Hermesinum*

63. With radiating arms...................................... 64
63'. Without radiating arms................................ 66

64. Spherical unattached body......... *Acinetactis*
64'. Body may adhere to substrate.................... 65

65. Adhering cells naked................ *Gymnophrys*
65'. Adhering cells develop lorica.*Microcometes*

66. With adhering bacteria............... *Postgaardi*
66'. Without adhering bacteria.......................... 67

67. Kills marine diatoms by attaching and inserting a pseudopodium.................. *Pirsonia*
67'. Feed by other means.................................. 68

68. Body amoeboid.. 69
68'. Body not amoeboid...................................... 70

69. Body with narrow aspect............ *Dinomonas*
69'. Body with broad aspect, may produce pseudopodia .... *Toshiba*

70. One flagellum directed anteriorly, one directed posteriorly or otherwise functionally anisokont 71
70'. Both flagella directed anteriorly, laterally, or posteriorly.. 84

71. Flagella thick, subapical insertion, cells with extrusomes 72
71'. Flagella not noticeably thickened.............. 74

72. Dorso-ventrally flattened.***Platychilomonas***
72'. Not flattened.. 73

73. Marine.................................. ***Leucocryptos***
73'. Freshwater.......................... ***Kathablepharis***

74. Cells flattened.. 75
74'. Cells rounded in cross-section.................. 76

75. With flagella inserting at one antero-lateral margin ....***Adriamonas***
75'. Flagella inserting in center of anterior margin or subapically...................... ***Parabodo***

76. With recurrent flagellum adhering to the body..77
76'. Recurrent flagellum not adhering to the body..78

77. Recurrent flagellum lies along length of narrow elongate body and extends from posterior end, metabolic.............. ***Phanerobia***
77'. Recurrent flagellum does not project from behind posterior end of cell.. ***Proleptomonas***

78. Flagella inserting into two opposed grooves .... ***Palustrimonas***
78'. With single ventral or ventro-lateral groove 79

79. Cells compressed, with single groove located along the ventral face; flagellum located in this groove................................ ***Carpediemonas***
79'. Ventral gutter which may form pseudopodia .... ***Bodopsis***

80. Food ingested by a mouth at the anterior end of the cell 81
80'. Ingestion of food occurs ventrally or laterally 83

81. Body grooved.. 82
81'. Body not grooved; recurrent flagellum free from body; fast swimming cells, flagella inserting subapically; with an anterior mouth that it presses against attached bacteria in order to ingest them.................. ***Bordnamonas***

82. With posterior flagellum usually following a spiral path around the body, body can be metabolic.................................... ***Cryptaulax***
82'. Groove a deep cut setting of anterior end of cell.................... ***Entomosigma/Hemistasia***

83. Medium sized cells, with ventral groove in which lies the recurrent flagellum ....***Colponema***
83'. Anterior flagellum with a hook, posterior flagellum in a groove, filter feeding..... ***Jakoba***

84. Body with radiating cytoplasmic strands...... .... ***Heliobodo***
84'. Not with radiating strands.......................... 85

85. Flagella adhering..................... ***Phyllomitus***
85'. Flagella not adhering.................................. 86

86. Flagella directed anteriorly or posteriorly in motile cells.. 87
86'. Flagella directed laterally............... ***Bilorum***

87. Cell keeled.............................. ***Streptomonas***
87'. Cell not keeled.. 88

88. Flagella arise from beside the mouth, but are directed posteriorly when the cell is swimming; small and predatory...... ***Telonema***
88'. Flagella long, cells usually 10 μm or longer.. 89

89. With pouch or groove................................. 90
89'. No pouch or groove.................................... 92

90. .Broad groove extending along length of body .... ***Diphylleia***
90'. Groove/pouch located subapically............. 91

91. With anterior pouch................ ***Phyllomitus***
91'. With subapical groove......... ***Pleurostomum***

92. Small cells.................................... ***Kamera***
92'. Medium cells.. 93

93. .Cells with anterior papilla between flagella, pseudopodia produced laterally, with theca .... ***Cryothecomonas***
93'. No papilla .... ***Gyromitus***

94. With three flagella.................................. 95
94'. With more than three flagella.................... 96

95. Anterior flagellum short.............. ***Macapella***
95'. Anterior flagellum long............ ***Dallingeria***

96. .With four flagella.................................... 97
96'. With more than four flagella................... 100

97. Flagella largely similar in behavior and orientation, inserting apically..................... 98
97'. Flagella directed in dissimilar directions 99

98. With broad ventral groove...... ***Collodictyon***
98'. Without groove, may produce pseudopodia .... ***Quadricilia***

99. Pyramidal cell with flagella at each of four corners..................................... ***Tetracilia***
99'. With broad ventro-lateral groove, one anterior flagellum, two lateral flagella, and one posterior flagellum .... ***Trimastix***

100. Flagella restricted to two rows.............. 101
100'. Flagella not in two rows.......................... 103

101. Body not pliable......................... ***Hemimastix*** (see chapter: **Hemimastigophora**)
101'. Body pliable.. 102

102. Body contractile........................... ***Spironema*** (see chapter: **Hemimastigophora**)
102'. Body not contractile................ ***Stereonema*** (see chapter: **Hemimastigophora**)

103. Flagella in sub-apical wreath............. .... ***Paramastix***
103'. Flagella arise from all over body.......... 104

104. Flagella beat very slowly, cell moves very little... ***Artodiscus***
104'. Cell moves at speeds typical for flagellates or ciliates.. 105

105. Flattened with anterior mouth surrounded by thickened lips .... ***Stephanopogon***
105'. Cell rounded.............................. ***Multicilia***

Not amoebae, without flagella orcilia, but believed to be relatedto flagellates

106. Cells mostly attached, not moving......... 107
106'. Cells gliding 110

107. Cell with stalk.................... ***Salpingorhiza***
107'. Cell not stalked...................................... 108

108. .Without lorica....................... ***Gymnophrys***
108'. With lorica.. 109

109. With arms.......................... ***Microcometes***
109'. Without arms........................... ***Platytheca***

110. Anterior end tapered, no visible source of motive power.................................. ***Anehmia***
110'. Gliding, but with pseudopodial strands.... .... ***Rhizaspis***

Figure legends: All scale bars represent 10μm.

Genus ***Acinetactis*** Stokes, 1886

Biflagellated cells, rounded bodies measuring about 10 μm, giving rise to fine stiff pseudopodia with extrusome-like structures. The flagella arise from the same face of the cell; 2 contractile vacuoles (Fig.1). Few species, all from freshwater. Arms may be withdrawn from the anterior part or the whole cell when swimming.
TYPE SPECIES: *Acinetactis mirabilis* Stokes, 1886

Genus ***Allantion*** Sandon, 1924

Flagellate (7-14 μm) with smooth rapid gliding, rigid, with single trailing flagellum and refractile inclusions. From soils and seawater, forms cysts. Distribution widespread and common. Monospecific (Fig. 2). Ref. Vørs, 1992.
TYPE SPECIES: *Allantion tachypoon* Sandon, 1924

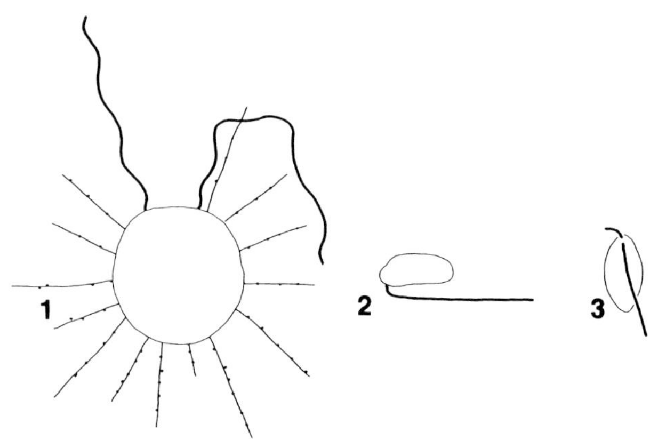

Fig. 1. *Actinectactis mirabilis* (after Patterson and Zölffel, 1991). Fig. 2. *Allantion tachyploon* (after Patterson and Zölffel, 1991). Fig. 3. *Allas diplophysa* (after Patterson and Zölffel, 1991).

Genus ***Allas*** Sandon, 1927

Flattened oblong gliding cell, plastic body, short anterior flagellum and a trailing flagellum projecting behind the cell (Fig. 3). Monospecific, from soils. Identity uncertain, but may be a thaumatomastigid (C. O'Kelly, unpubl. obs.).
TYPE SPECIES: *Allas diplophysa* Sandon, 1927

Genus ***Alphamonas*** see *Colpodella.*

Genus ***Amphimonas*** Dujardin, 1841

Biflagellated cells, plastic bodies, flagella equal or nearly so, freely swimming or attached, with or without stalk, many species (Fig. 4). None of the species are well studied, and this genus is in need of urgent attention. Type species not determined.

Genus ***Ancyromonas*** Kent, 1880

Small (up to 8 µm) gliding cells, one flagellum inserting subapically but directed to the posterior of the cell, usually with groove from point of flagellar insertion and along the lateral margin of the cell. Second flagellum is absent or is delicate and projects forwards. (Fig. 5) Widespread and common, several described species.
TYPE SPECIES:*Ancyromonas sigmoides* Kent, 1880

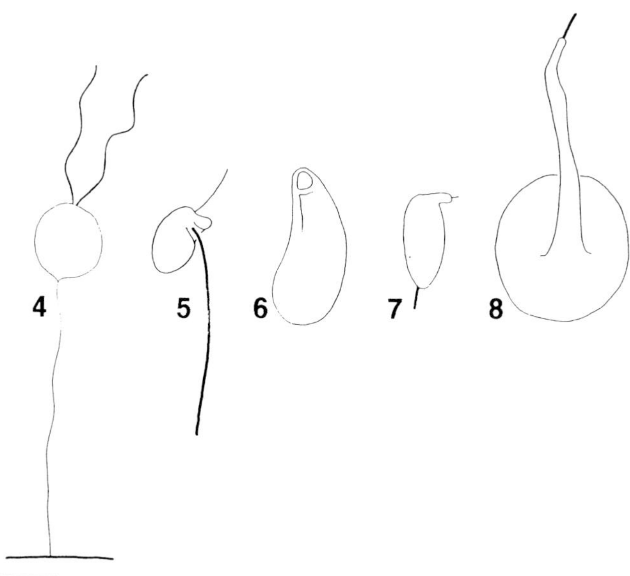

Fig. 4. *Amphimonas globosa* (after Patterson and Zölffel, 1991) (scale on left for this Fig.). Fig. 5. *Ancyromonas sigmoides* (original). Fig. 6. *Anehmia exotica* (original). Fig. 7. *Amastigomonas debruynei* (original). Fig. 8. *Apusomonas proboscidea* (after Patterson and Zölffel, 1991).

Genus ***Anehmia***
Ekebom, Patterson & Vørs, 1996

Gliding organism without flagella, but with anterior rostral region with mouth and with plastic sausage-shaped body. Marine. Although without flagella and not metabolic, the body shape and gliding behavior is reminiscent of *Diplonema*, *Rhynchopus,* and other flagellates believed to be related to the Euglenozoa. Monospecific. (Fig. 6) Ref. Ekebom et al., 1996
TYPE SPECIES: *Anehmia exotica* Ekebom, Patterson & Vørs, 1996

FAMILY APUSOMONADIDAE
Karpov & Mylnikov, 1989

Biflagellated, gliding flagellates with flexible organic outer sheath (theca) visible by electron microscopy. Tubulocristate. Two genera.

Genus ***Amastigomonas*** de Saedeleer, 1931

Biflagellate, gliding protists; flagella insert subapically and to one side; dorsal surface of the body is covered with a thin organic theca, and the ventral surface produces pseudopodia. Anterior flagellum is enclosed basally or completely by the theca. Posterior flagellum trails under the body; both flagella may be very difficult to see. From marine and freshwater sites. Several species (Fig. 7).
TYPE SPECIES: *Amastigomonas debruynei* de Saedeleer, 1931

Genus ***Apusomonas*** Aléxéieff, 1924

Flattened, gliding flagellate, with anterior mastigophore; 2 flagella insert near the distal end of the mastigophore. One flagellum extends forwards and a thin filament projects beyond the tip of the mastigophore; the other runs backwards to lie under the cell. Mastigophore beats slow during motion. Body: 8 - 14 µm. Bacterivorous; food taken ventrally. Temporary cysts formed, cryptobiotic (can dry out). Common in soil, also from fresh-waters. Monospecific. (Fig.8)
TYPE SPECIES: *Apusomonas proboscidea* Aléxéieff, 1924

Genus ***Artodiscus*** Penard, 1890

Polyhedral body with 5-10 slowly moving flagella/projections moves adjacent to substrate; body with coat or similar investment. One rarely reported species from freshwater (Fig. 9).
TYPE SPECIES: *Artodiscus saltans* Penard, 1890

Genus ***Bilorum*** Calaway, 1960

Biflagellated colorless cells isolated from probably anoxic regions of warm mineral springs in Florida (USA). One species which varies in length from 14 - 90 µm, plastic, occasionally amoeboid (retractile) posteriorly. Two flagella equal in length, inserting anteriorly but directed laterally; basally the flagella are sometimes surrounded by a sheath. Cells may glide, and may produce small anterior pseudopodia. Gliding motion may suggest affinities with other gliding flagellates such as cercomonads, apusomonads, thaumatomonads, diplonemids, or euglenids, etc. Monospecific (Fig. 10).
TYPE SPECIES: *Bilorum multicorme* Calaway, 1960

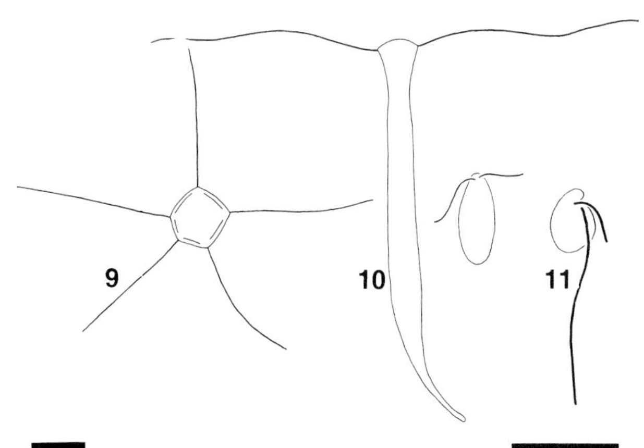

Fig. 9. *Artodiscus saltans* (original) Fig. 10. *Bilorum multicorne* (after Patterson and Zölffel, 1991). Fig. 11. *Bodomorpha minimus* (after Patterson and Zölffel, 1991). Right hand scale is for Fig. 11 only.

Genus ***Bodomorpha*** Hollande, 1952

Kidney-shaped flagellates, 2 flagella inserting in a shallow subapical depression, but lacking a kinetoplast. Non-metabolic. Phagotrophic. Nature and identity uncertain. With similarities to *Sainouron*, *Metopion*, and *Heteromita*. (Fig. 11).
TYPE SPECIES: *Bodomorpha minima* (Hollande, 1942) Hollande, 1952

Genus ***Bodopsis*** Lemmermann, 1913

Ovate swimming or gliding cells with 2 flagella, one beating anteriorly, one trailing, often curving around the body. Flagella insert in depression at anterior of cell from which pseudopodia emerge (Fig. 12). This genus may now contain unrelated taxa, in need of further work.
TYPE SPECIES: *Bodopsis alternans* (Klebs, 1893) Lemmermann, 1913

Genus ***Bordnamonas*** Larsen & Patterson, 1990

Swimming cell, 2 thick flagella arising in a subapical crease, one held in an arc in front of the cell, the other trailing behind. Coming to rest to

feed, at which time an apical ingestion apparatus is used to detach particles of food which are attached to the substrate. Affinities suggested to lie with stramenopiles (because of swimming behavior) or kinetoplastids (because of anterior mouth and the thick flagella). Marine. Monspecific. (Fig. 13). Reference: Larsen and Patterson, 1990.
TYPE SPECIES: *Bordnamonas tropicana* Larson & Patterson, 1990

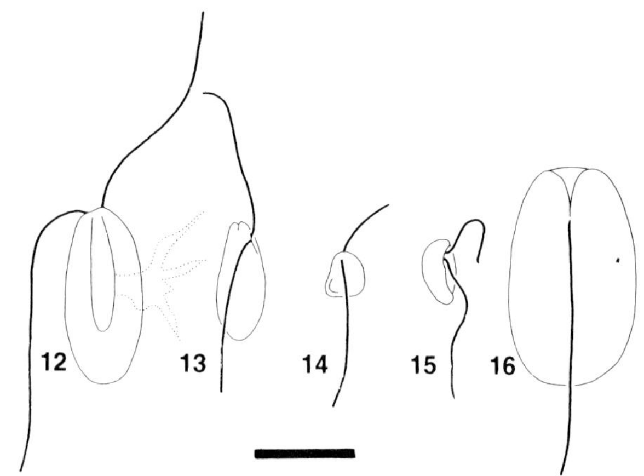

Fig. 12. *Bodopsis alternans* (after Patterson and Zölffel, 1991). Fig. 13. *Bordnamonas tropicana* (after Patterson and Zölffel, 1991). Fig. 14. *Caecitellus parvulus* (after Patterson and Simpson, 1997). Fig. 15. *Carpediemonas membranifera* (after Ekebom *et al.*, 1995/6). Fig. 16. *Clautriavia mobilis* (after Patterson and Zölffel, 1991).

Genus ***Caecitellus***
Patterson, Nygard, Steinberg & Turley, 1993

Small gliding flagellates from marine sediments and detritus,. one trailing flagellum, one stiffly and slowly moving anterior flagellum, flattened and often triangular in profile; with discrete ingestion organelle visible at one posterior lateral margin by light-microscopy. (Fig. 14). Affinities with stramenopiles suggested. Monspecific. Refs. O'Kelly and Nerad, 1998; Patterson et al., 1993.

Genus ***Carpediemonas***
Ekebom, Patterson & Vørs, 1996

Protists with 2 unequal flagella, the anterior projects antero-laterally when beating, the posterior one is longer and lies in a ventral groove where it beats actively. The flagellates usually move by skidding, but may attach to substrate by tip of posterior flagellum. Pseudopodia or cytoplasmic strands may also be produced. Usually encountered under conditions of diminished oxygen. Marine. Monospecific (Fig. 15). Ref. Ekebom et al., 1996.
TYPE SPECIES: *Carpediemmonas membranifera* Ekebom, Patterson & Vørs, 1996

Genus ***Clautriavia*** Massart, 1900

Gliding flagellate with a single trailing flagellum and a mid-ventral groove. Able to ingest large particles. Body not plastic. With similarity to some thaumatomastigids, e.g. *Protaspis*. Two species (Fig. 16).
TYPE SPECIES: *Clautriavia mobilis* Massart, 1900

Cercomonads

With two naked flagella inserting apically or subapically, gliding or swimming, mitochondria with irregular tubular cristae, cells with homogeneous paranuclear body. Some species with small concentric capped extrusomes. Individuals in some species may fuse, cells usually with a facility to produce pseudopodia. Mostly bacterivorous using pseudopodial engulfment, not with a discrete mouth; marine, freshwater, and soil; common and widespread. Three genera.

1. Genus ***Cercomonas*** Dujardin, 1841

Gliding, with flexible body usually able to produce pseudopodia more often from posterior end; anterior flagellum varies in length from one species to another, it beats slowly and is flexible; posterior flagellum trails adhering in part to the ventral surface, usually lying in a groove. Some of the species (e.g. *C. agilis*) may actively swim.
TYPE SPECIES: *Cercomonas longicauda* Dujardin, 1841

2. Genus ***Heteromita*** Dujardin, 1841

Biflagellated cells, subapical anterior insertion of flagella. In normal form of locomotion, gliding, the posterior flagellum trails behind the cells. The anterior flagellum beats stiffly, and is often directed posteriorly. Cells are phagotrophic, ingesting bacteria using pseudopodia mostly from the front end. *Heteromita* is distinguished from *Cercomonas* because pseudopodial formation in *Heteromita* is mostly anterior, and the body of Cercomonas is more plastic, and in *Cercomonas* the recurrent flagellum adheres to the body surface. (Fig. 18) Type species not specified.

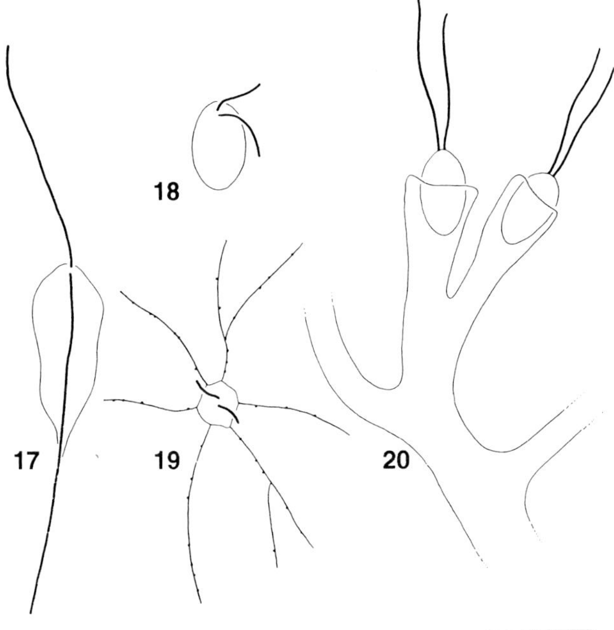

Fig. 17. *Cercomonas parva*. Fig. 18. *Heteromita globosa*. Fig. 19. *Massisteria marina* Fig. 20. *Cladomonas fruticulosa*. (Figs. 17-20 after Patterson and Zölffel, 1991).

3. Genus ***Massisteria*** Patterson & Fenchel, 1990

Small body, usually under 5 µm, which in trophic organisms emits curving, sometimes branched fine pseudopodia with extrusomes, and have the flagella lying inactive over body surface. Under some circumstances, the arms may be resorbed and the organism will swim actively. Trophont does not move actively, feeds usually on bacteria. Multicellular forms are encountered. To date, reported only from marine sites. (Fig. 19)
TYPE SPECIES: *Massisteria marina* Patterson & Fenchel, 1990

Genus ***Cladomonas*** Stein, 1878

Round-bodied flagellates with 2 equal flagella, forming arborescent colonies, cells located at termini of tubular mucus stalks. Affinities unclear. Monspecific (Fig. 20).
TYPE SPECIES: *Cladomonas fruticulosa* Stein, 1878

Genus ***Codonoeca*** Kent, 1880

Uniflagellate cell in stalked lorica. Cell small, flagellum apical, freshwater. Affinities uncertain. Two species (Fig. 21).
TYPE SPECIES: *Codonoeca costata* Kent, 1880

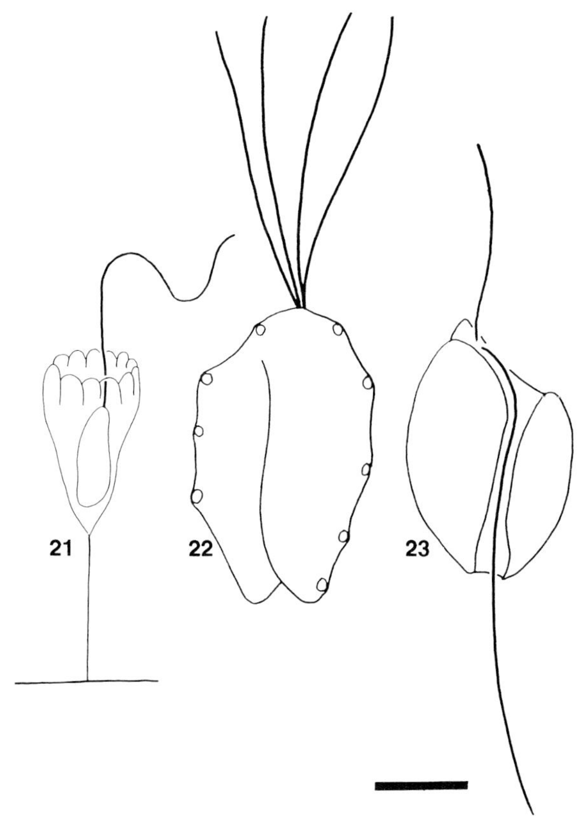

Fig. 21. *Codonoeca costata*. Fig. 22. *Collodictyon triciliatum*. Fig. 23. *Colponema loxodes*. (Figs. 21-23 after Patterson and Zölffel, 1991)

Genus ***Collodictyon*** Carter, 1865

Medium to large swimming cell with ventral gutter and 4 flagella; phagotrophic via gutter. Several species, freshwater. Initially described with 3 flagella. Medium sized, 20-40 µm. Eating

protists. Form and feeding behavior very like *Diphylleia*. (Fig. 22).
TYPE SPECIES: *Collodictyon triciliatum* Carter, 1865

Genus ***Colponema*** Stein, 1878

Phagotrophic flagellate, medium size; 2 flagella insert subapically in ventral gutter, one directed anteriorly. With tubulocristate mitochondria, extrusomes, and subpellicular membranous sacs (alveoli?). (Fig. 23).
TYPE SPECIES: *Colponema loxodes* Stein, 1878

Genus ***Cruzella*** de Faria et al., 1922

Biflagellated organism, anterior flagellum emerging from the tip of a short anterior rostrum, a feature which may suggest similarities to kinetoplastids such as *Dimastigella*. The second flagellum recurrent. Metabolic, marine. Monospecific (Fig. 24).
TYPE SPECIES: *Cruzella marina* de Faria et al., 1922

Genus ***Cryothecomonas*** Thomsen et al., 1991

Medium-sized cells, 2 flagella, enclosed in a 2 - layered theca, flagella emerge anteriorly through channels in theca. With lateral cytostome, ingesting particulate food using pseudopodia. Mitochondria with tubular cristae. Three species from Antarctic waters (Fig. 25).
TYPE SPECIES: *Cryothecomonas armigera* Thomsen et al., 1991

Genus ***Cryptaulax*** Skuja, 1948

Elongate cell; subapical flagella insert at the head of a groove that spirals around the body, one flagellum projecting forwards and one is recurrent often associated with the groove. With anterior rostrum. Often metabolic, several species, some or all of which may be related to the euglenozoan territory which includes *Rhynchobodo* (Vørs, 1992). (Fig. 26).
TYPE SPECIES: *Cryptaulax akopos* (Skuja, 1939) Skuja, 1948

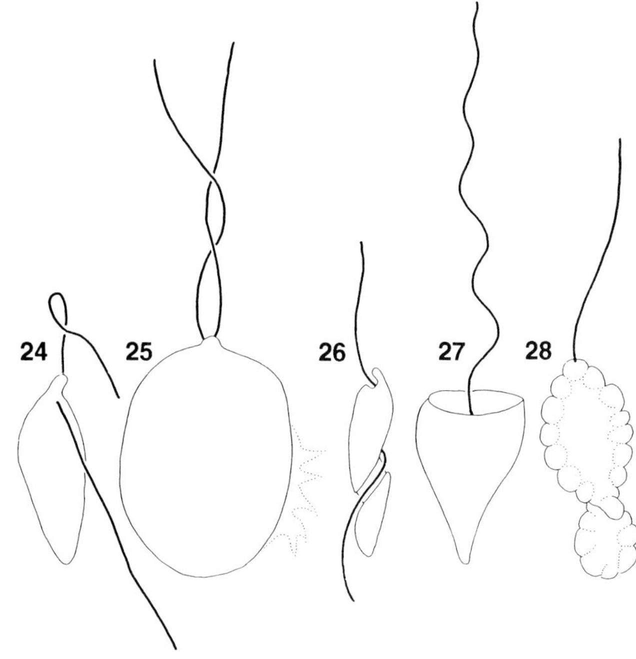

Fig. 24. *Cruzella marina*. Fig. 25. *Cryothecomonas* sp. Fig. 26. *Cryptaulax akopos*. Fig. 27. *Cyathomonas turbinata*. Fig. 28. *Cyclomonas distortum*. (Figs. 24-28 after Patterson and Zölffel, 1991).

Genus ***Cyathomonas*** Fromentel, 1874

Attached cell with anterior depression from which emerges a single flagellum. At one time, this genus housed species currently assigned to the heterotrophic cryptomonad, *Goniomonas*. Now monospecific (*C. turbinata*, Fig. 27, above), the identity of which is uncertain. Reference: Novarino et al., 1994.

Genus ***Cyclomonas*** Fromentel, 1874

Flattened cell, twisted, 14 - 25 μm long, with single apical flagellum. Cytoplasm with globular consistency. Monospecific (Fig. 28).
TYPE SPECIES: *Cyclomonas distortum* Fromental, 1874

Genus ***Dallingeria*** Kent, 1880

Small flagellate with 3 flagella, 1 apical and 2 projecting laterally from the center of the elongate elliptical body. Nature and identity uncertain.
TYPE SPECIES: *Dallingeria drysdali* Kent, 1880

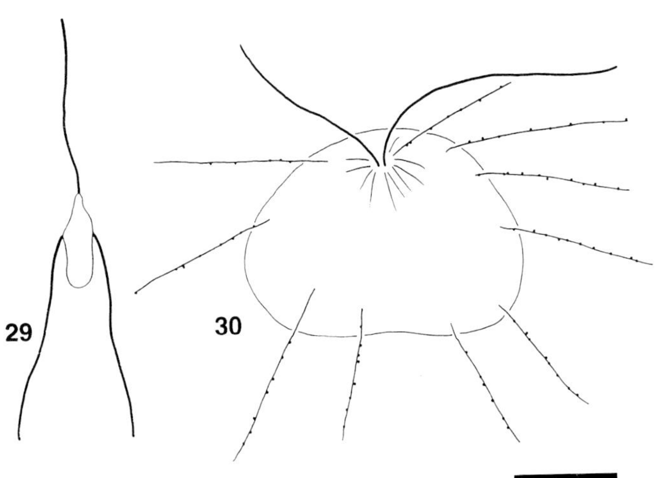

Fig. 29. *Dallingeria drysdali.* Fig. 30. *Dimorpha mutans.* (Figs. 29,30 after Patterson and Zölffel, 1991).

Dimorphids

Flagellates with axopodia emerging from most regions of the body. Axopodial microtubules end on a body lying between the nucleus and the anterior pole of the cell, axopodia may be withdrawn to produce a motile flagellated stage. With concentric extrusomes, elongate kinetosomes, and tubulocristate mitochondria. Superficial similarity with pedinellid helioflagellates but dimorphids may be distinguished ultrastructurally by the arrangement of microtubules, the axonemes, the axonemal termination site, and extrusome form, etc. With two genera.

1. Genus ***Dimorpha*** Gruber, 1881

Rounded body, with 2 flagella, axonemes of retractible axopodia insert on a body lying between the nucleus and the anterior pole of the cell (Fig. 30).
TYPE SPECIES: *Dimorpha mutans* Gruber, 1881

2. Genus ***Tetradimorpha*** Hsiung, 1927

Dimorphid with 4 long flagella inserting adjacent to each other. Tetradimorphids are much larger than *Dimorpha* (Fig. 31).

TYPE SPECIES: *Tetradimorpha tetramastix* Hsiung, 1927

Genus ***Dingensia***: See *Colpodellidae*

Genus ***Dinomonas*** Kent, 1880

Irregularly shaped body measuring about 10 µm, granular cytoplasm, not metabolic; 2 equal flagella inserting apically. Swimming cell, posterior contractile vacuole, eats flagellates and other large particles in the region of the flagella; freshwater and marine. Monospecific (Fig. 32); a second species being a colpodellid (see Colpodellidae).
TYPE SPECIES: *Dinomonas tuberculata* Kent, 1880

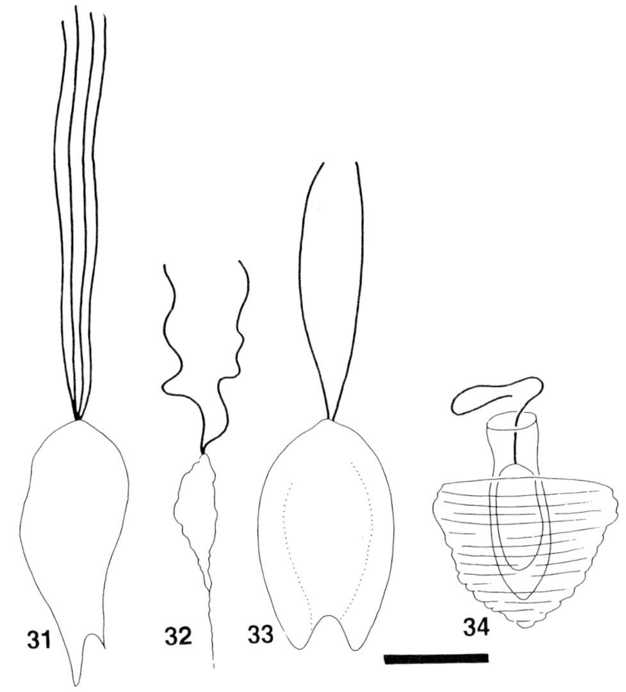

Fig. 31. *Tetradimorpha tetramastix* (swimming). Fig. 32. *Dinomonas tuberculata.* Fig. 33. *Diphylleia hyalina.* Fig. 34. *Diplocalium inopinatum.* (Figs. 31-34 after Patterson & Zölffel, 1991).

Genus ***Diphylleia*** Massart, 1920

Swimming flagellate, with 2 long apical and equal flagella inserting near the anterior end at the top of a ventral groove formed by the curving lateral margins of the cell. Phagotrophic, probably all descriptions refer to one species (Fig. 33).

TYPE SPECIES: *Diphylleia rotans* Massart, 1920

Genus ***Diplocalium*** Grassé & Deflandre, 1952

Uniflagellated cell, living in a cylindrical lorica itself located in a bowl-shaped lorica, flagellum apical. Monospecific (Fig. 34).
TYPE SPECIES: *Diplocalium inopinatum* (Middlehoek, 1950) Grassé & Deflandre, 1952

Genus ***Diploselmis*** Kent, 1881

Cell attached by a strand to the base of a stalked brownish lorica. Said to have an eye-spot. Two similar long flagella emerging from the open end of the lorica; the flagella beat actively. Occurring in aggregates but not colonies. Monospecific from freshwater (Fig. 35).
TYPE SPECIES: *Diploselmis socialis* (Kent, 1880) Kent, 1881

Genus ***Discocelis*** Vørs, 1988

Dorsoventrally flattened gliding protists, 2 flagella arising from anterior or lateral depression, one trailing under and behind the cell. Anterior or lateral velum next to flagella and peripheral extrusomes. Marine, several species (Fig. 36).
TYPE SPECIES: *Discocelis saleuta* Vørs, 1988

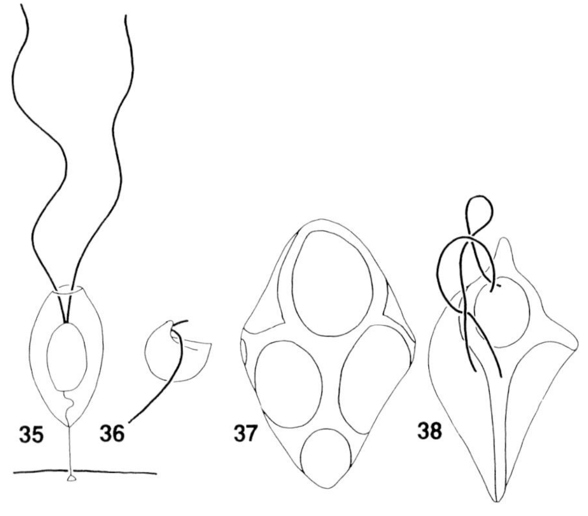

Fig. 35. *Diploselmis socialis*. Fig. 36. *Discocelis saleuta*. Fig. 37. *Ebria tripartita*. Fig. 38. *Hermesinum adriaticum*. (Figs. 35-38 after Patterson and Zölffel, 1991)

FAMILY EBRIIDAE

Medium to large marine taxon with two subapically inserting flagella. With an internal siliceous sheleton. With an extensive fossil record, but only two genera extant. At times in the past this group was considered related to dinoflagellates; TEM studies showed that the nucleus is not a dinokaryon.

1. Genus ***Ebria*** Borgert, 1891

Internal siliceous element with 3 branches. Flagella hard to see. Sometimes with multiple nuclei. Widespread in coastal temperate, tropical, and boreal waters, less common in oceanic areas; eurythermal and apparently somewhat euryhaline. Possibly only one species (Fig. 37). Nutrition herbivorous, primarily nanoplanktonic diatoms, occasionally dinoflagellates.
E. tripartita (Shumann, 1867) Lemmermann, 1899

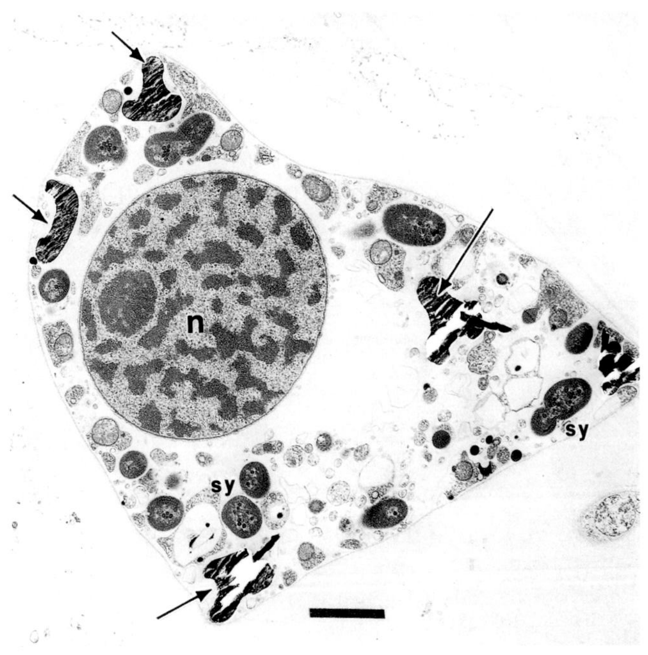

Fig. 38A. *Hermesium adriaticum* Zacharias. TEM showing siliceous endoskeletal elements (arrows); nucleus (N); and endosymbiotic *Synechococcus-like* cyanobacteria (Sy), some of which are dividing within their host. Photograph by P. Hargraves, Univ. Rhode Island. Scale = 2 μm.

2. Genus ***Hermesinum*** Zacharias, 1906

Cells approximately rhomboid surrounding an internal siliceous skeleton composed of solid elements which partly enclose the nucleus. Skeleton tetraxial in arrangement (triaene), cytoplasm pale yellow, bluegreen, or pink in color. Blue-green cells have endosymbiotic *Synechococcus*-like cyanobacteria (Fig. 38A) (Hargraves and Miller, 1974; Hargraves, in progress). Nutrition is primarily bactivorous. Widespread in coastal temperate and tropical waters (>20°C and 15-30 ppt salinity). (Figs. 38, 38a).
TYPE SPECIES: *Hermesinum adriaticum* Zacharias, 1906

Genus ***Entomosigma*** Schiller, 1925

Ovoid cell, with longitudinal anterior groove setting off a rostral region, 2 flagella emerging from the groove. Two species. Originally and erroneously described with a chloroplast. Probably the same as *Hemistasia* and related to the diplonemids (Fig. 39).
TYPE SPECIES: *Entomosigma peridinoides* Schiller, 1925

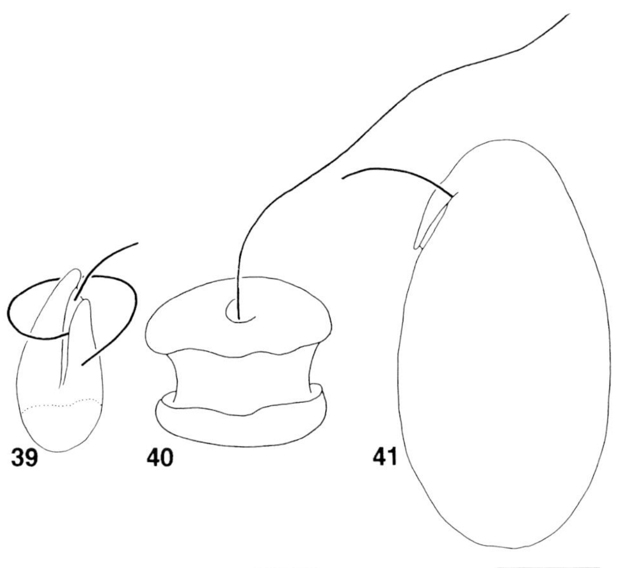

Fig. 39. *Entomosigma peridinioides*. Fig. 40. *Errera mirabilis.* Fig. 41. *Fromentella granulosa.* (Figs. 39-41 after Patterson and Zölffel, 1991, right hand scale refers to Figs. 39 and 41)

Genus ***Errera*** Schouteden, 1907

Protist with single apical flagellum emerging from an anterior depression, anterior and posterior borders extended and recurved. Affinities uncertain. Monospecific (Fig. 40).
TYPE SPECIES: *Errera mirabilis* Schouteden, 1927

Genus ***Fromentella*** Patterson & Zölffel, 1991

Medium to large cell with a single flagellum inserting in a subapical groove and directed posteriorly. Monospecific (Fig. 41).
TYPE SPECIES: *Fromentella granulosa* (Fromentel, 1874) Patterson & Zölffel, 1991

Genus ***Glissandra*** Patterson & Simpson, 1996

With 2 flagella inserting laterally into an anterior ventral groove and extending posteriorly and anteriorly. Glides with most of both flagella held against the substrate. Marine, monospecific (Fig. 42). Ref. Patterson and Simpson, 1996.
TYPE SPECIES: *Glissandra innuerende* Patterson & Simpson, 1996

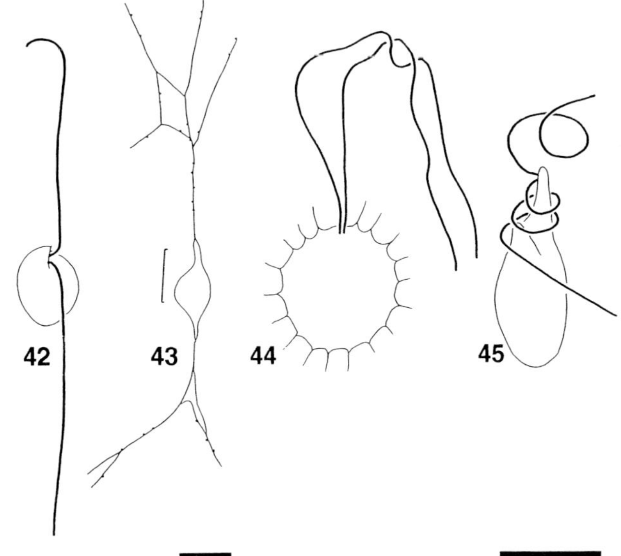

Fig. 42. *Glissandra innuerende* (after Patterson and Simpson, 1997). Fig. 43. *Gymnophrys cometa*. Fig. 44. *Heliobodo radians*. Fig. 45. *Hemistasia klebsii*. (Figs. 43-45 after Patterson and Zölffel, 1991, left hand scale for Fig. 43 only)

Genus ***Gymnophrys*** Cienkowski, 1876

Amoeboid body form, small, usually under 10 µm, giving rise to a small number of branching thin pseudopodia that bear extrusomes. Short flagella evident by electron microscopy. The boundaries of this genus and *Penardia* appear confused. From soils and freshwater, probably monotypic (Fig. 43). Ref. Mikrjukov and Mylnikov, 1995
TYPE SPECIES: *Gymnophrys cometa* Cienkowski, 1876

Genus ***Heliobodo*** Valkanov , 1928

Rounded body giving rise to 2 long equal flagella and numerous fine pseudopodia. Monospecific (Fig. 44).
TYPE SPECIES: *Heliobodo radians* Valkanov, 1928

Genus ***Hemistasia*** Griessmann, 1913

Cell with 2 equal flagella, inserting at anterior pointed apex, posterior end broader and rounded. Usually with 2 grooves setting off narrowed anterior part. Body and snout contractile. Crawls or swims; when swimming, with one flagellum projecting forwards, the other trailing. Phagotrophic using snout at which time trailing flagellum is curved around the cell. Contractile, marine, forms cysts. Probably the same as *Entomosigma.* One species is a kinetoplastid (Elbrächter et al., 1996) with ultrastructural similarities to diplonemids said to be synonymous with the type species (Fig. 45).
TYPE SPECIES:*Hemistasia klebsii* Griessmann 1913

Jakobids

Biflagellated organisms grouped on the basis of ultrastructural characters. Flagella without hairs or scales, one held anteriorly, and the other recurrent and lying within a ventral groove. In those taxa in which studies have been carried out, the recurrent flagellum has a **vane**. Mitochondria with tubular or flat cristae. With two families, the naked Jakobidae and the loricate Histionidae. The boundaries and composition of this area are unclear (Flavin and Nerad, 1993; O'Kelly, 1993), resolution of which awaits studies of the type species of *Histiona*. The phylogenetic position is in dispute. Structurally, the jakobids are most similar to the retortamonads, which lack mitochondria. Ref. O'Kelly (1993).

A. FAMILY JAKOBIDAE Patterson, 1990

Naked jakobids, trophic cells uninucleate, ventral groove extending the length of the cell; a prominent cytoplasmic lip defining the right side of the ventral groove. With one or two mitochondria with flattened cristae; single golgi body; structured extrusomes present; asexual reproduction by binary fission; cysts and sexual reproduction unknown. With one genus and species, widespread in marine habitats. Reference: O'Kelly (1993).

Genus ***Jakoba*** Patterson, 1990

Anterior flagellum held crook-like, the second lying in the groove, suspension feeding. Marine. Monospecific (Fig. 46).
TYPE SPECIES:*Jakoba libera* (Ruinen) Patterson, 1990

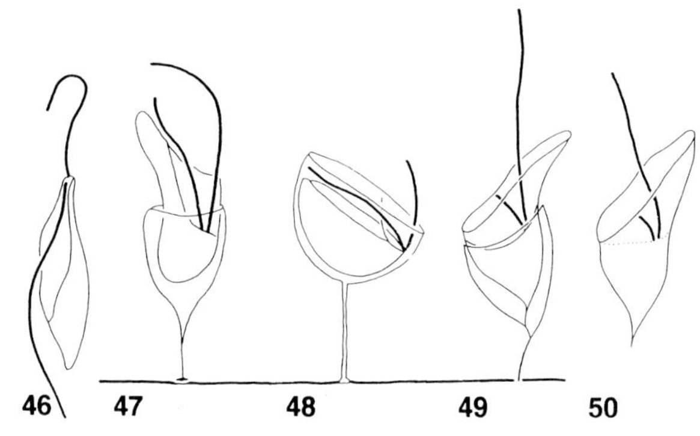

Fig. 46. *Jakoba libera*. Fig. 47. *Histiona aroides* (after Mylnikov, 1994). Fig. 48. *Reclinomonas americana* (after O'Kelly, 1993). Fig. 49. *Stenocodon* sp. Fig. 50. *Stomatochone infundibuliformis*. (Figs. 46, 49, 50 after Patterson and Zölffel, 1991)

B. FAMILY HISTIONIDAE Flavin & Nerad, 1993

Loricate jakobids, the lorica is constructed of organic materials and not easily visible with light microscopy; dorsal surface of trophic cells faces the base of the lorica; ventral groove extending the length of the cell; a prominent cytoplasmic lip defining the right side of the ventral groove. Asexual reproduction by binary fission; cyst spherical, plugged, retained in parent lorica, wall unmineralized, apparently formed asexually from a single cell; sexual reproduction unknown. With two genera and four species, all from freshwater habitats.

1. Genus ***Histiona*** Voigt, 1902

Lorica upright, originally described with one flagellum, second flagellum revealed by electron-microscopy. With 3 nominal species, (Fig. 47). Found in Europe, Siberia, and North America. Ref. Mylnikov, 1984, 1989.
TYPE SPECIES: *Histiona velifera* Voigt, 1902

2. Genus ***Reclinomonas*** Flavin & Nerad, 1993

Lorica stalked, laterally compressed, the exterior studded with nail-shaped scales; portion of left ventral microtubular root in trophic cells terminating blindly in an emergent structure, the epipodium. Found in freshwater habitats in North America and New Zealand. Here considered monotypic (Fig. 48) although Flavin and Nerad (1993) transferred *Histiona campanula* to this genus. Ref. Flavin and Nerad, 1993.
TYPE SPECIES: *Reclinomonas americana* Flavin & Nerad, 1993

3. Genus ***Stenocodon*** Pascher,1942 *incertae sedis*

Sessile flagellates, 2 flagella unequal in length, usually with hood around the projecting flagellum in which form the cells closely resemble *Histiona*. Unhooded cells resemble *Ochromonas* and similar stramenopiles with which affinities may subsequently be shown to lie. The status of this genus is uncertain. One species (Fig. 49).

4. Genus ***Stomatochone*** Pascher, 1942 *incertae sedis*

Sessile flagellates attaching directly to substrate by means of a cytoplasmic attachment, species described with 1 or 2 (unequal) flagella and with cytoplasmic hood around the long flagellum. Superficial similarity with cell of *Histiona*, but awaiting ultrastructural or other studies to confirm affinities with this taxon or with the stramenopiles. (Fig. 50).TYPE SPECIES: *Stomatochone infundibuliformis* Pascher, 1942

Genus ***Kamera*** (Woodcock, 1916) Patterson & Zölffel, 1991

Biflagellated cell with ovoid body, and subapically inserting flagella. The description is based on an imprecise account of the form of locomotion. Probably not recognizable. Monospecific (Fig. 51).
TYPE SPECIES: *Kamera lens* (Woodcock, 1916) Patterson & Zölffel, 1991

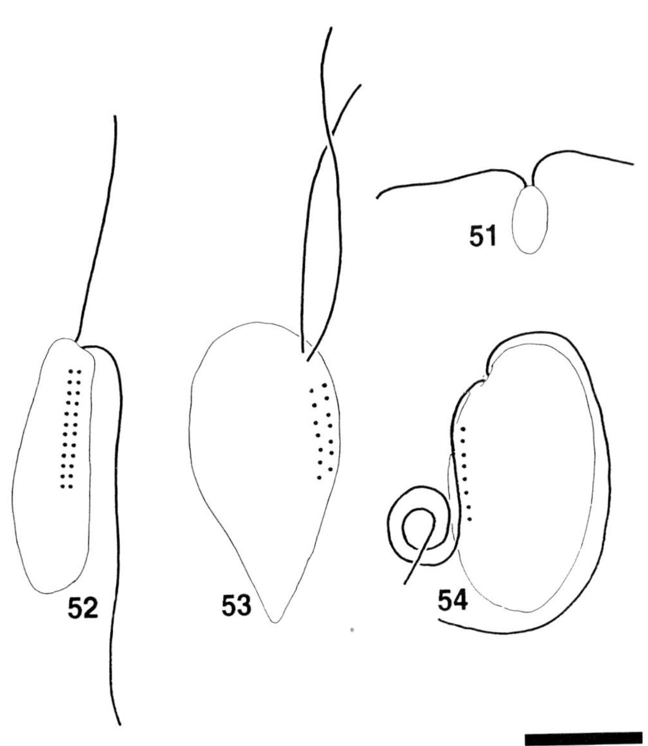

Fig. 51. *Kamera lens*. Fig. 52. *Kathablepharis phoenikoston* Fig. 53. *Leucocryptos marina*. Fig. 54. *Platychilomonas psammobia*. (Figs. 51-54 after Patterson and Zölffel, 1991)

FAMILY KATHABLEPHARIDAE Vørs, 1992

Medium sized free-swimming flagellates with 2 subapically inserting flagella, flagella long and thick. Cells with 1 to 3 types of peripheral ejectisome-like extrusomes. Tubulocristate mitochondria. Ingest algae and other protists via anterior end. In marine and freshwater habitats. Three genera, but there is little to distinguish *Leucocryptos* and *Kathablepharis*.

1. Genus ***Kathablepharis*** Skuja,1939

Flagella inserting subapically in an anterior longitudinal depression. Small peripheral extrusomes and several rows of extrusomes in a longitudinal band along ventral side of cell (Fig. 52). Reported from freshwater sites. The genus name may be spelled *Katablepharis* in the botanical literature.
TYPE SPECIES: *Kathablepharis phoenikoston* Skuja, 1939

2. Genus ***Leucocryptos*** Butcher, 1967

Common member of the marine plankton, flagella inserting subapically, usually at the anterior end of a furrow with 1 or 2 longitudinal, parallel rows of few or several extrusomes (Fig. 53). Two more types of scattered peripheral extrusomes. Body circular in cross-section. With a microtubule-lined ingestion apparatus opening at the apex of the cell. Two species.
TYPE SPECIES: *Leucocryptos marina* (Braarud, 1935) Butcher, 1967

3. Genus ***Platychilomonas*** Larsen & Patterson, 1990

Flattened cells, observed only after coming to rest against the substrate, with extrusomes and thickened flagella suggestive of affinities with kathablepharids, but awaiting ultrastructural examination. (Fig. 54). Monospecific.
TYPE SPECIES: *Platychilomonas psammobia* Larsen & Patterson, 1990

Genus ***Kiitoksia*** Vørs, 1992

Small spherical cell (Fig. 55) with 1 or 2 short, smooth, ventrally inserted trailing flagellum/a. Cells glide; body measuring 1 - 3 um in diameter. Ultrastructure unknown. Two free-living marine species, *Kiitoksia ystava* (the type from Finland) and *K. kaloista* often found in association with immersed surfaces.
TYPE SPECIES: *Kiitoksia ystava* Vørs, 1992

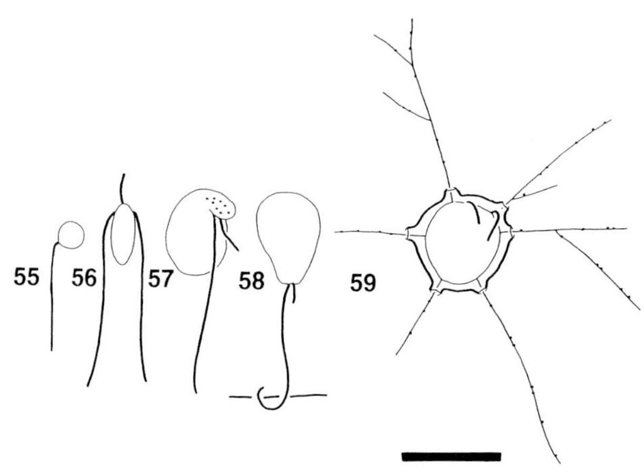

Fig. 55. *Kiitoksia ystava* (after Vørs, 1992). Fig. 56. *Macapella lapsa*. Fig. 57. *Metopion fluens*. Fig. 58. *Metromonas simplex*. (Figs. 56-58 after Patterson and Zölffel, 1991). Fig. 59. *Microcometes paludosa* (original).

Genus ***Macapella*** Patterson & Zölffel, 1991

Ovoid to elliptical cells with 3 apically inserting flagella, 1 is directed anteriorly, the other 2 trail. Similar to *Dallingeria*. From freshwater. Monospecific (Fig. 56).
TYPE SPECIES: *Macapella lapsa* (Stokes, 1890) Patterson & Zölffel, 1991

Genus ***Metopion*** Larsen & Patterson, 1990

Flattened gliding cells, with 1 or 2 flagella emerging from an anterior ventro-lateral groove, one flagellum trailing behind the cell, the other, if present, is short and also trailing (Fig. 57). Resembles *Ancyromonas* but distinguished by the nature and orientation of the second flagellum.
TYPE SPECIES: *Metopion fluens*, Larsen & Patterson, 1990

Genus ***Metromonas*** Larsen & Patterson ,1990

Flattened, with 1 or 2 flagella (second, if present, is very short), normally attached to substrate by posterior end of long flagellum, moves with nodding motion, predatory on other flagellates, may glide with flagellum trailing behind (Fig. 58). Two species, both marine.
TYPE SPECIES: *Metromonas simplex* (Griessmann, 1913) Larsen & Patterson, 1990

Genus ***Microcometes*** Cienkowski, 1876

Small, 6-14 μm diameter, polygonal, flexible test with a small number of apertures with or without short necks and from which thin, branching and granular pseudopodia emerge, (Fig. 59). De Saedeleer's observations of flagella have generally been overlooked; normally with 2 short flagella (Patterson and Simpson, 1999). De Saedeleer (1934) holds that *Patavia* is a free-swimming stage of the same organism. Bacterivorous, and from freshwater.
TYPE SPECIES: *Microcometes paludosa* Cienkowski, 1876

Genus ***Multicilia*** Cienkowsky, 1881

Rounded bodies from which emanate numerous flagella not in rows. Cells may have short proboscis (Fig. 60). Tubulocristate mitochondria. Several species, freshwater and marine. Ref. Mikrjukov and Mylnikov (1996)
TYPE SPECIES: *Multicilia marina* Cienkowsky, 1881

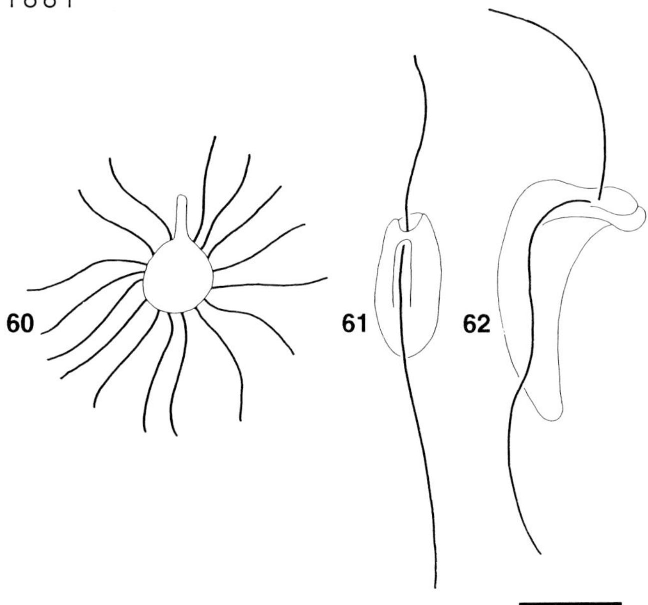

Fig. 60. *Multicilia lacustris*. Fig. 61. *Palustrimonas yorkeensis* (after Patterson and Simpson, 1996). Fig. 62. *Parabodo sacculifera*. (Figs. 60,62 after Patterson and Zölffel, 1991)

Genus **Nephromonas** see *Colpodellidae*

Genus ***Palustrimonas*** Patterson & Simpson, 1996

Heterotrophic, free-living, swimming protist with 2 flagella inserting subapically into separate grooves or pockets, one directed anteriorly, one posteriorly (Fig. 61). Cells capable of consuming large items of food. From high saline content habitats. Monospecific. Ref. Patterson and Simpson, 1996.
TYPE SPECIES: *Palustrimonas yorkeensis* (Ruinen, 1938) Patterson & Simpson, 1996

Genus ***Parabodo*** Skuja, 1939

Laterally compressed flagellates, 2 flagella with an apical or subapical insertion, plastic (Fig. 62). Identity and affinities unclear, possibly referable to *Colpodella* or kinetoplastids (Zhukov, 1991). One species assigned to this genus has been shown to have bodonid affinities (Mylnikov, 1986)
TYPE SPECIES: *Parabodo sacculiferus* Skuja, 1939

Genus ***Paramastix*** Skuja, 1948

Ovoid cells, about 20 μm long, with anterior papilla, surrounded by a wreath of 12-24 flagella (Fig. 63). Two species.
TYPE SPECIES: *Paramastix truncata* Skuja, 1948

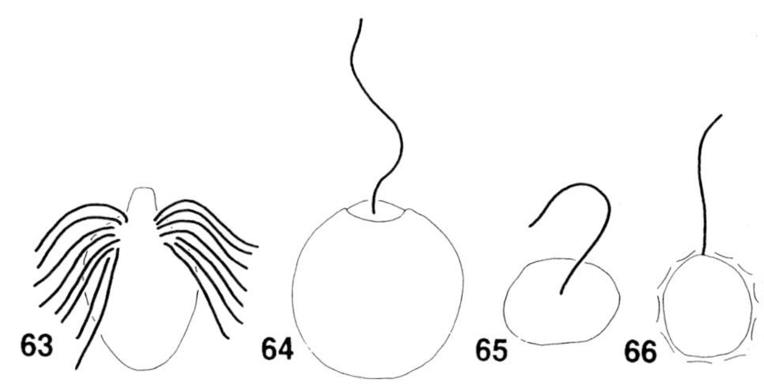

Fig. 63. *Paramastix conifera*. Fig. 64. *Paramonas globosa*. Fig. 65. *Peltomonas volitans*. Fig. 66. *Petasaria heterolepis*. (Figs. 63-66 after Patterson and Zölffel, 1991)

Genus **Paramonas** Kent, 1880

Kent created this genus for globular taxa described by Fromentel (1874) as having a single flagellum arising from an apical depression (Fig. 64).
TYPE SPECIES: *Paramonas globosa* (Fromentel, 1874) Kent, 1880

Genus **Peltomonas** Vlk, 1942

Dorso-ventrally flattened, with one anteriorly projecting flagellum inserting ventrally in the front part of the cell (Fig. 65). Freshwater, identity unclear.
TYPE SPECIES: *Peltomonas volitans* Vlk, 1942

Genus **Petasaria** Moestrup, 1979

Uniflagellated cell, rounded body 4-6 µm diameter, with siliceous scales (Fig. 66). Marine and widespread in distribution, known only from whole mount preparations.
TYPE SPECIES: *Petasaria heterolepis* Moestrup, 1979

Genus **Phalansterium** Stein, 1878

Uniflagellated cells with single flagellum surrounded by narrow collar of cytoplasm (Fig. 67). Tubulocristate mitochondria. Freshwater, forms cysts; several species, the most common of which is colonial and lives within a gelatinous matrix.
TYPE SPECIES: *Phalasterium digitatum* Stein, 1878

Genus **Phanerobia** Skuja, 1948

Elongate cell, 2 flagella, one projecting forwards, the other recurrent and attached to cell surface (Fig. 68), metabolic, one species, freshwater. Affinities are unclear.
TYPE SPECIES: *Phanerobia pelophila* Skuja, 1948

Genus **Phyllomitus** Stein, 1878

Described originally for organisms with 2 adhering flagella at the head of a depression or groove (Fig. 69). Several other kinds of organism appear to have been assigned to this genus subsequently.
TYPE SPECIES: *Phyllomitus undulans* Stein, 1878

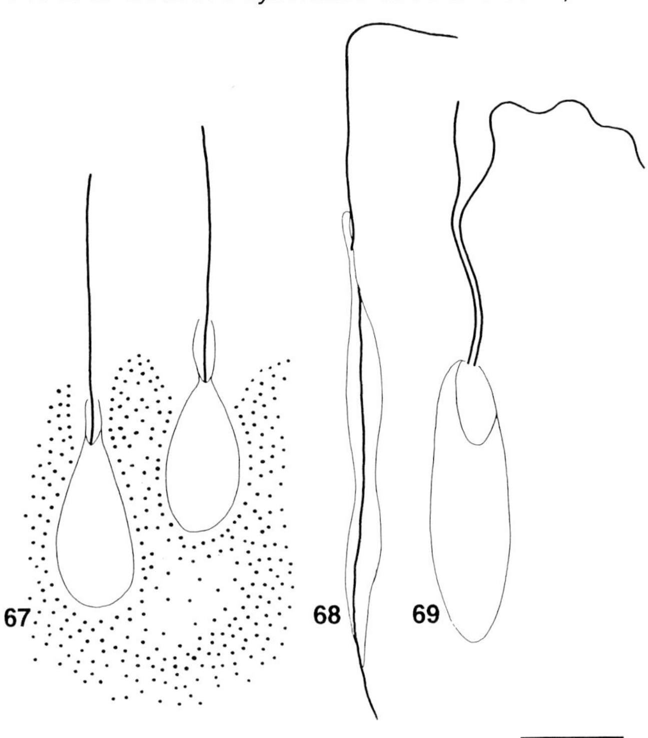

Fig. 67. *Phalansterium digitatum*. Fig. 68. *Phanerobia pelophila*. Fig. 69. *Phyllomitus undulans*. (Figs. 67-69 after Patterson and Zölffel, 1991)

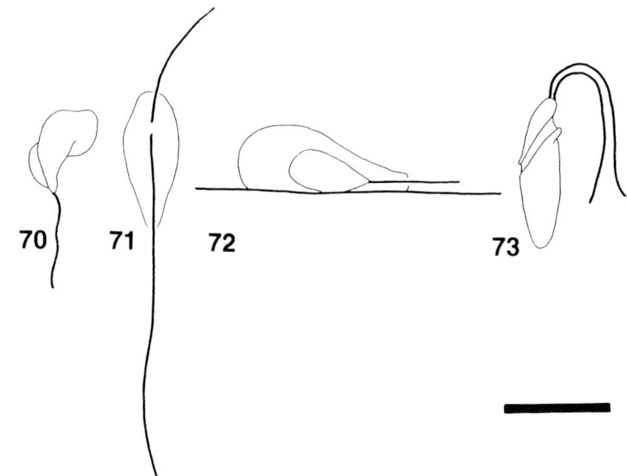

Fig. 70. *Phyllomonas contorta* (after Patterson and Zölffel, 1991). Fig. 71. *Pirsonia mucosa* (after Kühn et al., 1996). Fig. 72. *Platytheca microspora* (after Patterson and Zölffel, 1991). Fig. 73. *Pleurostomum flabellatum* (after Patterson and Simpson, 1997).

Genus ***Phyllomonas*** Klebs, 1892

Leaf-like flagellate 6-7 µm long, the edges of the cell curled up (Fig. 70). Single posteriorly directed flagellum which does not beat; the cell moves with agitated trembling and rotates. Identity not clear. One species assigned to *Metromonas*.
TYPE SPECIES: *Phyllomonas contorta* Klebs, 1892

Genus ***Pirsonia*** Schnepf et al., 1990

Marine predatory flagellates which feed on diatoms by attaching to the outside and intruding a pseudopodium through the frustule. Attacking algae (diatoms). With 2 flagella, one anterior, one posterior (Fig. 71); the flagella may or may not be lost during feeding. Exclusively reported from Europe, several species. Ref. Kühn et al., 1996.
TYPE SPECIES: *Pirsonia guinardiae* Schnepf et al., 1990

Genus ***Platytheca*** Stein, 1878

Uniflagellated pear-shaped cells living in flattened brownish lorica that attached to the substrate along one long face. The single flagellum emerges through an apical aperture (Fig. 72). Freshwater.
TYPE SPECIES: *Platytheca micropora* Stein, 1878

Genus ***Pleurostomum*** Namyslovsky, 1913

Two long equal flagella inserting apically, beating together; more or less metabolic body with subapical groove or depression (Fig. 73). Some similarity with *Cryptaulax* with its spiral groove and one flagellum usually located within the groove. With 5 species, typically from hypersaline habitats. Type species not specified. Ref. Patterson and Simpson, 1996.

Genus ***Postgaardi***
Fenchel, Bernard, Estaban, Finlay, Hansen & Iversen, 1995

Swimming biflagellated cell measuring about 20 µm, with 2 long flagella inserting subapically in a pocket, and with cell surface covered with long bacteria (Fig. 74). Reported from anoxic brackish habitat in Europe, but since identified from Antarctica. Recent work indicates this organism is a euglenozoon but neither a euglenid nor a kinetoplastid. Monotypic. References: Fenchel et al., 1995; Simpson et al., 1997.
TYPE SPECIES: *Postgaardi mariagerensis* Fenchel, Bernard, Estaban, Finlay, Hansen & Iversen, 1995

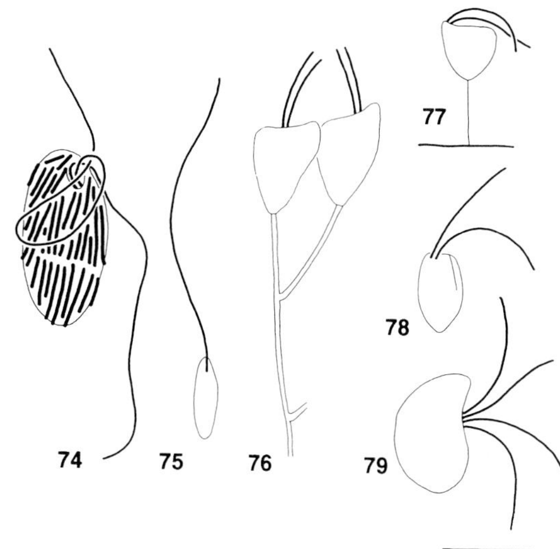

Fig. 74. *Postgaardi mariagerensis* (after Fenchel et al., 1995). Fig. 75. *Proleptomonas faecicola* (after Patterson and Zölffel, 1991). Fig. 76. *Pseudodendromonas vlkii* (after Patterson and Zölffel, 1991). Fig. 77. *Cyathobodo stipitatus* (after Patterson and Zölffel, 1991). Fig. 78. *Adriamonas peritocrescens* (after Verhagen et al., 1994). Fig. 79. *Quadricilia rotundata* (after Vørs, 1992)

Genus ***Proleptomonas*** Woodcock, 1914

Elongate ellipsoidal body, one anterior flagellum emerging from the apex but anchored near the nucleus, second recurrent and adhering flagellum (Fig. 75). Darting motion. Soils and goat dung. Monospecific.
TYPE SPECIES: *Proleptomonas faecicola* Woodcock, 1914

ORDER PSEUDODENDROMONADIDA
Hibberd, 1985

Biflagellated protists, triangular profile, 2 unadorned flagella arising side by side, of similar length but one with finer tip; tubulocristate mitochondria, phagotrophic with microtubule-supported ingestion area, mostly with surface unmineralized scales. Three genera—one of attached colonies, one of attached solitary cells, one of unattached solitary cells.

1. Genus ***Pseudodendromonas*** Bourrelly, 1953

Forming dichotomously branching arborescent colonies, stalk substantial (Fig. 76).
TYPE SPECIES: *Pseudodendromonas vlkii* Bourrelly, 1953

2. Genus ***Cyathobodo*** Petersen & Hansen, 1961

Cells solitary either swimming or attached to substrate via delicate stalks (Fig. 77). Species divisions principally by surface scales.
TYPE SPECIES: *Cyathobodo stipitatus* Petersen & Hansen, 1961

3. Genus ***Adriamonas*** Verhagen, Zölffel, Brugerolle & Patterson, 1994

Unattached pseudodendromonad, one species, from soils. With triangular to elongate profile characteristic of the family; with 2 flagella at one anterior antero-lateral corner, and an ingestion apparatus at the other (Fig. 78). No scales. Reported from soil, but subsequently observed in freshwater (unpubl. observ.). Monospecific. Ref. Verhagen et al., 1994.
TYPE SPECIES: *Adriamonas peritocrescens* Verhagen, Zölffel, Brugerolle & Patterson, 1994

Genus ***Quadricilia*** Vørs 1992

Spherical cell with 4 smooth, slightly unequal and acronematic flagella, which are almost twice as long as the cell (Fig. 79). Pseudopodia may be formed and adopt amoeboid motion. Ultrastructure unknown. Free-living, marine, swimming. Often found in association with diatom blooms. Ref. Vørs, 1992.
TYPE SPECIES: *Quadricilia rotundata* (Skuja, 1948) Vørs, 1992

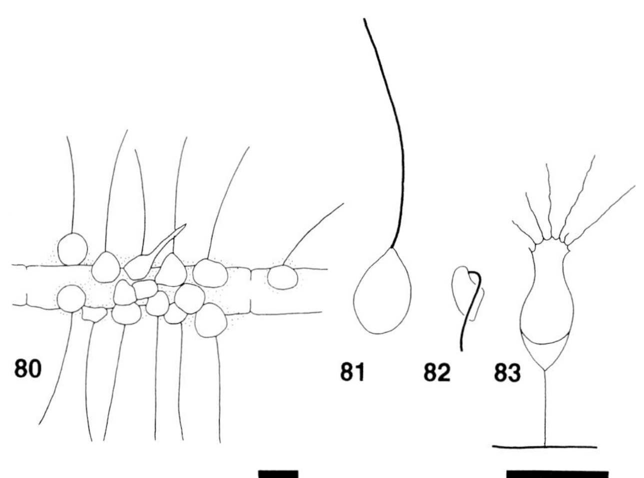

Fig. 80. *Rhizomonas setigera* (after Patterson et al., 1993). Fig. 81. *Rigidomastix coprocola*. Fig. 82. *Sainouron microtekton*. Fig. 83. *Salpingorhiza pascheriana*. (Figs. 81-83 after Patterson and Zölffel, 1991). Left hand scale is for Fig. 80 only.

Genus ***Rhizomonas*** Kent,1880

Uniflagellated, cell body adhering by its flagellar tip, body may be embedded in mucus (Fig. 80). Now embracing taxa previously assigned to *Solenicola*. Ref. Patterson et al., 1993.
TYPE SPECIES: *Rhizomonas verrucosa* Kent, 1880

Genus ***Rigidomastix*** Aléxéieff, 1929

Small cells (under 10 µm), rounded body, one anterior projection (flagellum?) (Fig. 81), eats bacteria. Coprozoite.
TYPE SPECIES: *Rigidomastix coprocola* Aléxéieff, 1929

Genus ***Sainouron*** Sandon, 1924

Gliding protist, usually with one trailing flagellum visible, but a short anterior flagellum also present (Fig. 82). Recurrent flagellum not attached to body; occasionally amoeboid; with cysts and possibly conjugation; from soils. May be related to *Heteromita* and the other cercomonads.
TYPE SPECIES: *Sainouron mikrotekteron* Sandon, 1924

Genus ***Salpingorhiza*** Klug, 1936

One species, loricate, with apical filose pseudopodia and without flagella in trophic form (Fig. 83), but produces uniflagellated stage without arms and with a rounded body. During transformation back to trophic form the apical flagellum is surrounded by a wreath of short filamentous pseudopodia. Similarities at this stage with *Stephanomonas*.
TYPE SPECIES: *Salpingorhiza pascheriana* Klug, 1936

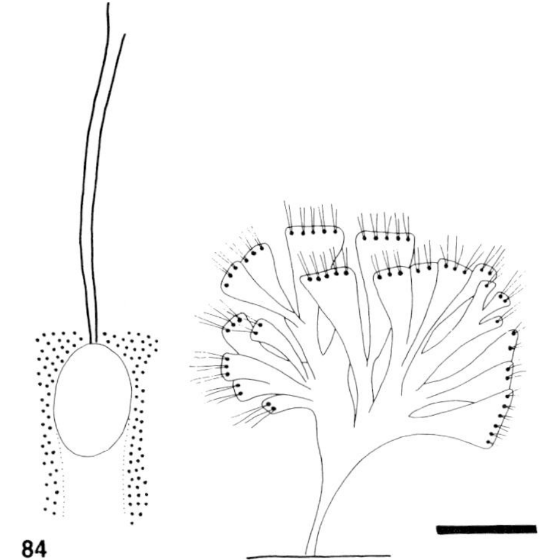

Fig. 84. *Rhipidodendron splendidum* (after Patterson and Zölffel, 1991)

Genus ***Schewiakoffia*** Corliss, 1960

Treated here as a ciliate.

FAMILY SPONGOMONADIDAE Hibberd, 1983

Mostly colonial flagellates, two long similar unadorned flagella, phagotrophic, without discrete mouth but with a cytoplasmic extension near the flagella. Tubulocristate mitochondria, cells embedded in granular mucus. Two genera. Phylogenetic relationships obscure.

1. Genus ***Rhipidodendron*** Stein, 1878

Colony of spongomonads flattened and fan-shaped (Fig. 84).
TYPE SPECIES: *Rhipidodendron splendidum* Stein, 1878

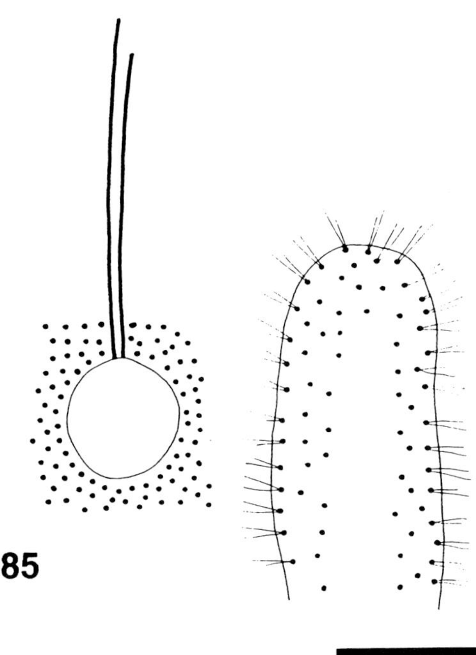

Fig. 85. *Spongomonas intestinum* (after Patterson and Zölffel, 1991)

2. Genus ***Spongomonas*** Stein, 1878

Colonies of spongomonads globular or cylindrical(Fig. 85); one species solitary.
TYPE SPECIES: *Spongomonas intestinum* (Cienkowsky, 1870) Stein, 1878

Genus ***Stephanomonas*** Kent, 1881

Rounded cell, single apical flagellum surrounded by a wreath of cilia/pseudopodia/bacteria? (Fig. 86). Monospecific.
TYPE SPECIES: *Stephanomonas locellus* (Fromentel, 1876) Kent, 1881

Genus ***Stephanopogon*** Entz, 1884

Protist with multiple cilia/flagella arising in rows (kineties), dorso-ventrally flattened, anterior end with mouth with thickened lips and with or without barbs; many similar nuclei and with capacity to produce cysts; marine phagotroph (Fig. 87). Previously assigned to the ciliates and later to the euglenozoa, though lacks the distinguishing features of both taxa. With a few species.
TYPE SPECIES: *Stephanopogon colpoda* Entz, 1884

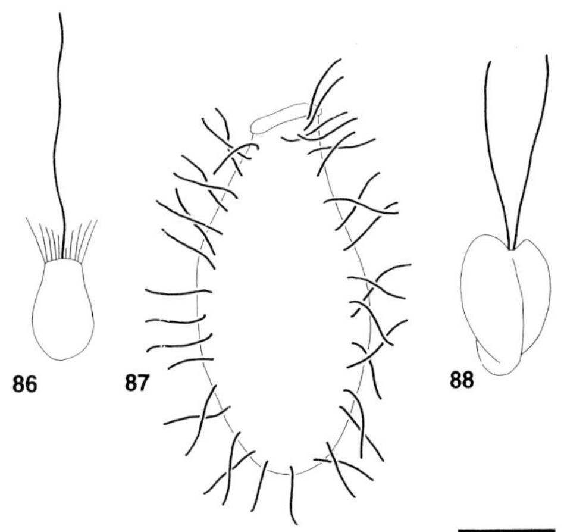

Fig. 86. *Stephanomonas* sp. Fig. 87. *Stephanopogon apogon* (original). Fig. 88. *Streptomonas locellu.* (Figs. 86, 88 after Patterson and Zölffel, 1991)

Genus ***Streptomonas*** Klebs, 1892

Rigid cell, heart-shaped, 2 equal flagella inserting into depressed anterior end, posterior end tapering, with dorsal and ventral flanges (Fig. 88).
TYPE SPECIES: *Streptomonas cordata* (Perty, 1852) Klebs, 1892

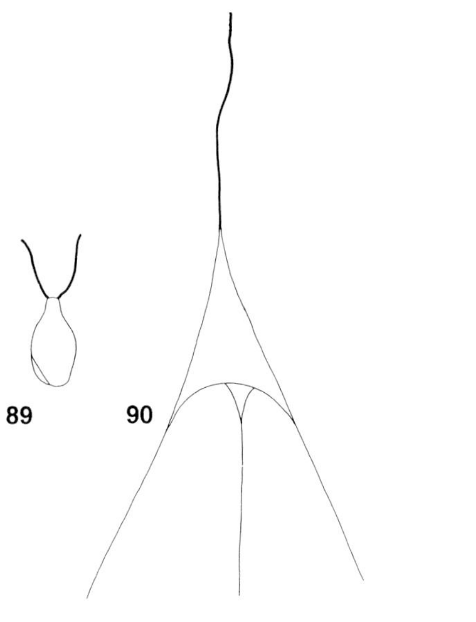

Fig. 89. *Telonema subtilis*. Fig. 90. *Tetracilia paradoxa*. (Figs. 89,90 after Patterson and Zölffel, 1991)

Genus ***Telonema*** Griessmann, 1913

Ovoid cell, 6-8 μm long, anterior end drawn out; 2 similar flagella insert subapically; anterior region may produce pseudopodia for feeding (Fig. 89); swimming cells move 'backwards' with flagella pointing away from the direction of movement. Phagotrophic, eating detritus or protists. Tubulocristate mitochondria, but otherwise affinities unclear.
TYPE SPECIES: *Telonema subtilis* Griessmann, 1913

Genus ***Tetracilia*** Valkanov, 1970

Elongate tetrahedral cells with 1 flagellum at each of 4 corners (Fig. 90). Described from brackish water. Monospecific.
TYPE SPECIES: *Tetracilia paradoxa* Valkanov, 1970

FAMILY THAUMATOMASTIGIDAE
Patterson & Zölffel, 1991

Typically biflagellated swimming or gliding cells capable of producing thin pseudopodia from one (ventral) face of the body which may be grooved. Mostly medium-sized (15-50 μm). All species studied to date by electron microscopy have surface scales that form in association with the tubulocristate mitochondria; dictyosomes adpressed to anterior face of nucleus, and concentric extrusomes. Six genera are brought together here, but the distinctions between genera are not substantial and further work is desirable. This family was identified as the Thaumatomastigaceae by Patterson & Zölffel 1991) and provided the 'zoological' spelling by Vørs, 1992) with the same authority. The taxon was incorrectly referred to as the Thaumatomonadidae, again with the same authority, (Ekebom et al., 1996).

1. Genus ***Gyromitus*** Skuja, 1939

Thaumastomastigid with 2 flagella of equal length (Fig. 91), typically swimming but may settle, produce pseudopodia, and become amoeboid. With

surface scales visible only by electron-microscopy. Ref. Swale and Belcher, 1974, 1975.
TYPE SPECIES: *Gyromitus disomatus* Skuja, 1939

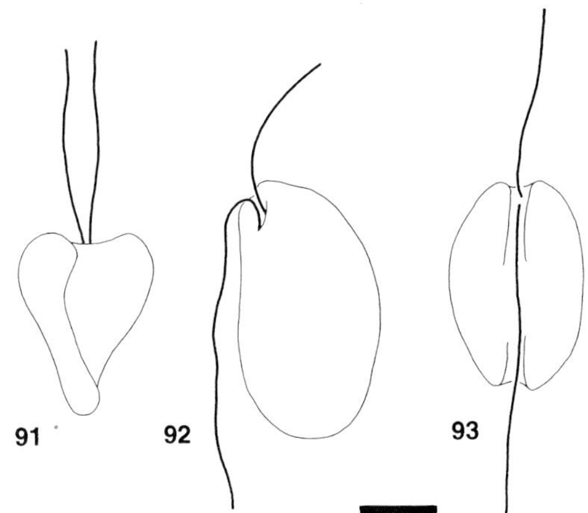

Fig. 91. *Gyromitus disomatus*. Fig. 92. *Hyaloselene compressa*. Fig. 93. *Protaspis glans*. (Figs. 91-93 after Patterson and Zölffel, 1991)

2. Genus **Hyaloselene** Skuja, 1956

Thaumatomastigid with slightly unequal flagella that insert anteriorly into a pocket, one projecting forwards, one trailing (Fig. 92). Capable of producing pseudopodia. Naked by light microscopy, with extrusomes. Laterally compressed. We assign here the genus *Synoikomonas* Skuja, 1964, which differs primarily by its inability to produce pseudopodia and by the presence of endosymbionts. As pseudopodial formation may be observed only rarely in this group, we do not believe that there are sufficient grounds for segregation at the genus level.
TYPE SPECIES: *Hyaloselene compressa* Skuja, 1956

3. Genus **Protaspis** Skuja,1939

Thaumatomastigid with flagella inserting subapically on ventral side, no flagellar pocket, flagella equal or subequal (Fig. 93). Ventral surface usually grooved, and nuclei often with nuclear caps (=dictyosomes?). May be dorso-ventrally flattened. Appear naked by light microscopy; no ultrastructural studies have been carried out.
TYPE SPECIES: *Protaspis glans* Skuja, 1939

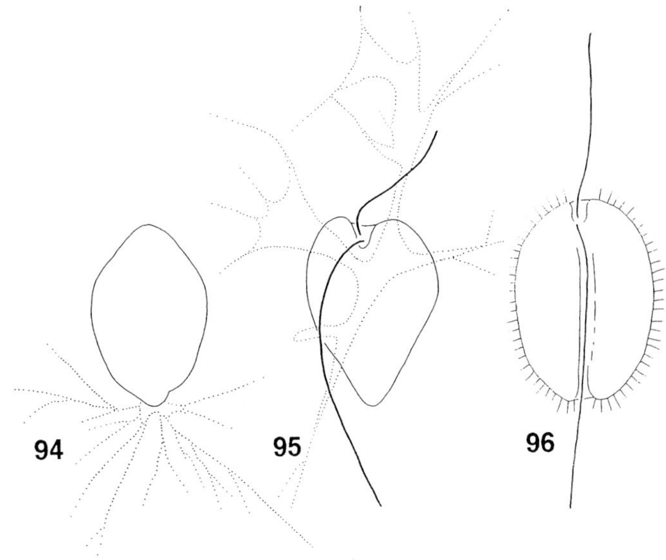

Fig. 94. *Rhizaspis granulatus*. Fig. 95. *Thaumatomonas lauterborni*. Fig. 96. *Thaumatomastix setifera*. (Figs. 94-96 after Patterson and Zölffel, 1991) Right scale for Fig. 96 only.

4. Genus **Rhizaspis** Skuja,1948

This is an aflagellate protist (Fig. 94), but body form and capacity to form pseudopodia suggest affinities with *Protaspis* and the other thaumatomastigids.
TYPE SPECIES: *Rhizaspis granulata* Skuja, 1948

5. Genus **Thaumatomonas** de Saedeleer, 1931

Thaumatomastigid with 2 unequal flagella inserting into a flagellar pocket, form pseudopodia (Fig. 95), and may form aggregates. With scales visible only by electron microscopy.
TYPE SPECIES: *Thaumatomonas lauterborni* de Saedeleer, 1931

6. Genus **Thaumatomastix** Lauterborn, 1899

Thaumatomastigid with flagella inserting subapically and ventrally, with spines and scales visible by light microscopy (Fig. 96).
TYPE SPECIES: *Thaumatomastix setifera* (Lauterborn, 1896) Lauterborn, 1899

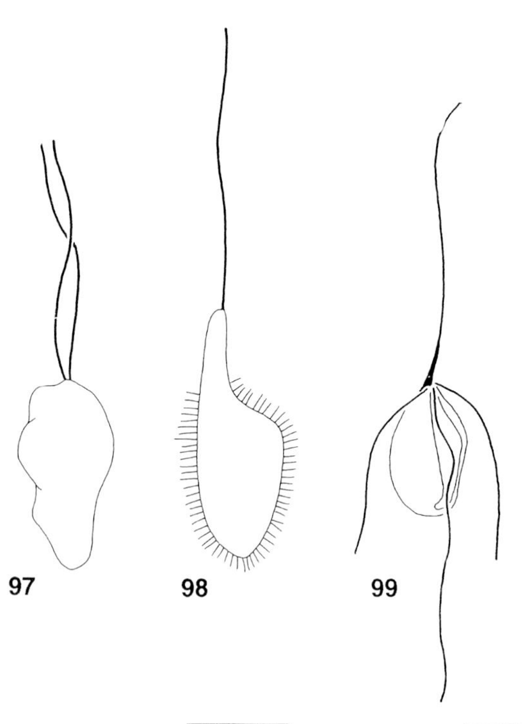

Fig. 97. *Toshiba vorax* (after Patterson and Zölffel, 1991). Fig. 98. *Trichonema hirsuta* (after Patterson and Zölffel, 1991). Fig. 99. *Trimastix marina* (original). Right scale for Fig. 99 only.

Genus ***Toshiba*** Patterson & Zölffel, 1991

Gliding or swimming cell, 2 equal flagella inserting apically, producing pseudopodia when swimming or gliding (Fig. 97). The taxon was first described under the name *Dimastigamoeba vorax*. Identity uncertain.
TYPE SPECIES: *Toshiba vorax*, (Massart, 1920) Patterson & Zölffel, 1991

Genus ***Trichonema*** Fromentel, 1872

Uniflagellate, metabolic, flagellum apical, body completely or partly with short filamentous projections (Fig. 98), freshwater and marine. Two species.
TYPE SPECIES: *Trichonema hirsuta* Fromentel, 1872

Genus ***Trimastix*** Kent, 1880

With 3 or 4 flagella inserting subapically, and with one recurrent flagellum and associated with a groove (Fig. 99). Typically from anaerobic habitats. Amitochondriate (Brugerolle and Patterson, 1997). Two species, type species *T. marina*, Kent, 1880, which has 4 flagella.

A number of genera of uncertain nature have not been included above. They are: *Dimastigamoeba* (of Blochmann), *Dinoasteromonas, Endostelium, Euglenacapsa, Hermisenella, Ligniera, Pansporella, Perkinsiella, Phloxamoeba*, and *Spongospora*. Several genera (*Peliainia, Strombilomonas*, and *Cyanomastix*) with "blue-green algal" inclusions have been described. It is possible that these are heterotrophic flagellates which have consumed cyanobacteria—but the nature of the 'host' is not clear. The 'proteomyxid' taxa with flagellated stages in the life cycle have been excluded. Skvortzkov has described a number of genera which lack a clear identity. Those which are not assignable to other higher taxa are: *Akaiyamamonas*, *Brasilobia*, *Colemania*, *Foliomonas*, *Gordymonas*, *Jolya*, *Kisselivia*, *Kuzminia*, *Liouamonas*, *Loukashkinia*, *Meyerella*, *Nodamastix*, *Pavloviamonas*, *Proctormonas*, *Prowsemonas*, *Serpentomonas*, *Skvortzoviella*, *Spiromonas*, and *Stigmobodo*. These are excluded from detailed coverage for the reasons given elsewhere (Patterson, 1994). The genus *Harbinia* (Zhou et al., 1993) has also been described without sufficient details to give it a clear identity. The status of *Ministeria* (Patterson et al., 1993) remains unclear as vibratile motility has been identified in one species. We have been unable to obtain papers by Dumas in which several genera are named (Foissner, 1995).

Acknowledgements:

Support from the Australian Research Council, the Australian Biological Resources Study, and the University of Sydney is gratefully recognized.

References

Brugerolle, G. & Patterson, D. J. 1997. Ultrastructure of *Trimastix convexa* Hollande, an amitochondriate anaerobic flagellate with a previously undescribed cellular organization. *Eur. J. Protistol.*, 33:121-130.

Ekebom, J., Patterson, D. J. & Vørs, N. 1996. Heterotrophic flagellates from coral reef sediments (Great Barrier Reef, Australia). *Arch. Protistenkd.,* 146:251-272.

Elbrächter, M., Schnepf, E. & Balzer, I. 1996. *Hemistasia phaeocysticola* (Scherffel) comb. nov., redescription of a free-living, marine, phagotrophic kinetoplastid flagellate. *Arch. Protistenkd.,* 147:125-136.

Fenchel, T., Bernard, C., Esteban, G., Finlay, B. J., Hansen, P. J. & Iversen, N. ,1995. Microbial diversity and activity in a Danish fjord with anoxic deep water. *Ophelia*, 43:45-100.

Flavin, M. & Nerad, T. A. 1993. *Reclinomonas americana* n. g., n. sp., a new freshwater heterotrophic flagellates. *J. Euk. Microbiol.*, 40:172-179.

Foissner, W. 1995. Forgotten protist species:The monographs by Abbé E. Dumas. *Eur. J. Protistol.*, 31:124-126.

Hargraves, P. E. & Miller, B. T. 1974. The ebridian flagellate *Hermesinum adriaticum* Zacharias. *Arch. Protistenkd.*, 116:280-284.

Kühn, S. F., Drebes, G. & Schnepf, E. 1996. Five new species of the nanoflagellate *Pirsonia* in the German Bight, North Sea, feeding on planktonic diatoms. *Helgol. Meeresunter.*, 50:205-222.

Larsen, J. & Patterson, D. J. 1990. Some flagellates (Protista) from tropical sandy beaches. *J. Nat. Hist.,* 24:801-937.

Mikrjukov, K. A. & Mylnikov, A. P. 1995. Fine structure of an unusual rhizopod, *Penardia cometa*, containing extrusomes and kinetosomes. *Europ. J. Protistol.*, 31:90-96.

Mikrjukov, K. A. & Mylnikov, A. P. 1996. Protist *Multicilia marina* Cienk. Flagellate or a heliozoon? *Doklady Akad. Nauk.*, 346:136-139.

Mylnikov, A. P. 1984. The morphology and life cycle of *Histiona aroides* Pascher (Chrysophyta). *Information Bulletin Biologiya Vnutrennikh Vod*, 62:16-19. (in Russian)

Mylnikov, A. P. 1986. The ultrastructure of the flagellate *Parabodo nitrophilus* (Bodonina) *Tsitologiya*, 28:1056-1059 (in Russian)

Mylnikov, A. P. 1989. The fine structure and systematic position of *Histiona aroides* (Bicoecales). *Bot. Zh.*, 74:184-189 (in Russian).

Novarino, G., Warren, A. Kinner, N. E. & Harvey, R. W. 1994. Protists from a sewage-contaminated aquifer on Cape Cod, Massachusetts. *Geomicrobiol. J.*, 12:23-36.

O'Kelly, C. J. 1993. The jakobid flagellates: structural features of *Jakoba*, *Reclinomonas* and *Histiona* and implications for the early diversification of eukaryotes. *J. Euk. Microbiol.*, 40: 627-636.

O'Kelly, C. J. & Nerad, T. 1998. Kinetid architecture and bicosoecid affinities of the marine heterotrophic nanoflagelate *Caecitellees parvulus* (Griessmann, 1913) Patterson *et al.,* 1993. *Eur. J. Protistol.*, 34:369-375.

Patterson, D. J. 1993. The current status of the free-living heterotrophic flagellates. *J. Euk. Microbiol.*, 40:606-609.

Patterson, D. J. 1994. Protozoa, Evolution and Classification. *Prog. Protozool., Proc. IX Int. Cong. Protozool.* Fischer Verlag, Stuttgart. pp 1 - 14

Patterson, D. J., Nygaard, K., Steinberg, G. & Turley, C. 1993. Heterotrophic flagellates and other protists associated with oceanic detritus throughout the water column in the mid North Atlantic. *J. Mar. Biol. Ass. UK.*, 73:67-95.

Patterson, D. J. & Larsen, J. (eds.) 1991. *The Biology of Free-Living Heterotrophic Flagellates*, Clarendon Press, Oxford.

Patterson, D. J. & Simpson, A. G. B. 1996. Heterotrophic flagellates from coastal marine and hypersaline sediments in Western Australia. *Eur. J. Protistol.*, 32:423-448.

Patterson, D. J. & Simpson, A. G. B. 1999. Heterotrophic flagellates from freshwater sites in Tasmania (Australia) In prep.

Patterson, D. J. & Zölffel, M. 1991. Heterotrophic flagellates of uncertain taxonomic position. *In*: Patterson, D. J. & Larsen, J. (eds.), *The Biology of Free-Living Heterotrophic Flagellates*, Clarendon Press, Oxford. pp 427-475

Simpson, A. G. B., van den Hoff, J., Bernard, C., Burton, H. & Patterson, D. J. 1997 The ultrastructure and affinities of *Postgaardi mariagerensis*, Fenchel et al., an unusual free-living euglenozoon. *Arch. Protistenkd.*, 147:213-225.

Swale, E. M. F. & Belcher, J. H. 1974. *Gyromitus disomatus* Skuja—a free-living colourless amoebo-flagellate. *Arch. Protistenkd.*, 116:211-220.

Swale, E. M. F. & Belcher, J. H. 1975. *Gyromitus limax* nov. sp.—a free-living colourless amoebo-flagellate. *Arch. Protistenkd.* 117:20-26

Verhagen, F. J. M., Zölffel, M., Brugerolle, G. & Patterson, D. J. 1994. *Adriamonas peritocrescens* gen. nov. sp. nov., a new free-living soil flagellate (Protista, Pseudodendromonadidae Incertae sedis). *Eur. J. Protistol.*, 30:295-308.

Vørs, N. 1992. Heterotrophic amoebae, flagellates and heliozoa from the Tvärminne area, Gulf of Finland, in 1988-1990. *Ophelia*, 36:1-109

Zhou, Y., Yang, G-T. & Mu, L-q. 1993. New flagellatae from Harbin, China. *Bull. Bot. Res.*, 13:113-117.

Zkukov, B. F. 1991. The diversity of bodonids. *In*: Patterson, D. J. & Larsen, J. (eds.), *The Biology of Free-Living Heterotrophic Flagellates.* Clarendon Press, Oxford. pp 177-184

PHYLUM HAPLOSPORIDIA
Caullery & Mesnil, 1899

By FRANK O. PERKINS

The phylum Haplosporidia Caullery & Mesnil, 1899 (syns. (H)Aplosporidia Caullery & Mesnil, 1899; Acetospora Sprague, 1979; Balanosporida Sprague, 1979) includes a group of parasites which cause diseases in fresh water and marine invertebrates. They are apparently world-wide in distribution having been described from Australia, both sides of the Atlantic Ocean, Mediterranean Sea, Europe, and North America. There are 31 named species and three genera with the hosts being an ascidian, annelids, crustaceans, molluscs, a crinoid and a nemertean as well as trematodes and nematodes which are, in turn, parasitic in marine molluscs and decapod crustaceans. Due to their significance in causing extensive mortalities in American oysters along the east coast of the United States, from Massachusetts to North Carolina, most of the literature concerns two species, *Haplosporidium nelsoni* and *H. costale.*

None of the life cycles have been elucidated, all of the species descriptions involving accounts of plasmodial structure and sporulation sequences in a single host species. When spores or infected host tissues have been used in attempts to transmit infections to other individuals of the same host species or other potential host species, the attempts have failed except for possibly one study on fresh water snails (Barrow, 1965).

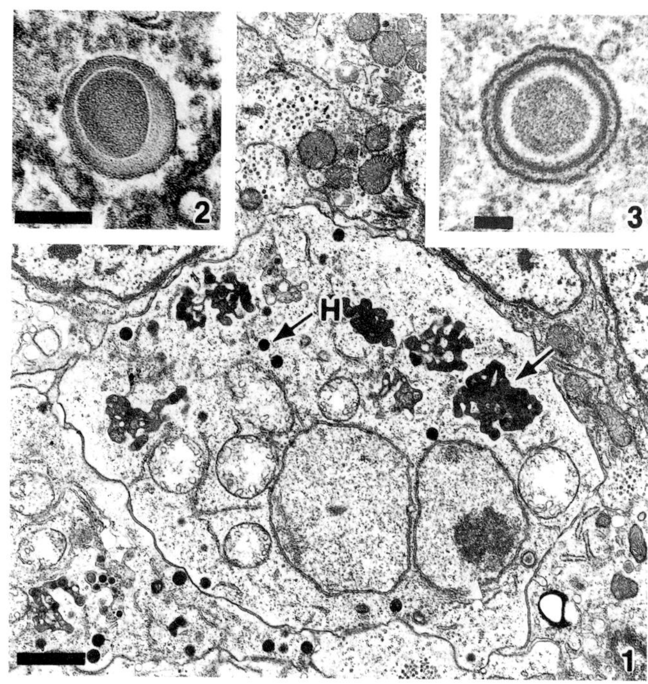

Figs. 1-3. Plasmodium of *Haplosporidium nelsoni* in connective tissue of the oyster, *Crassostrea virginica.* Note close juxtapositioning of nuclear pair. Haplosporosome formative region (arrow); haplosporosome (H). Bar=1 μm. Fig. 2. Haplosporosome in plasmodium of *H. nelsoni.* Note delimiting unit membrane and membrane (thin electron-light zone) separating cortex from medulla of the organelle. Bar=0.1 μm. Fig. 3. Haplosporosome-like body formed from Golgi-derived cisternae. The core of the body is believed to contain material derived from the nucleus. Bar=0.1 μm.

Development within the host probably begins with a naked, uninucleated cell which undergoes karyokinesis to form larger naked cells or **plasmodia** (Fig. 1) with two or more nuclei. The earliest cells seen in *de novo* infections are plasmodia with 2-4 nuclei. Nuclear multiplication and increase in cell mass occurs to yield plasmodia with as many as 50 or more nuclei found most often in pairs in which the nucleii are closely pressed against each other. The

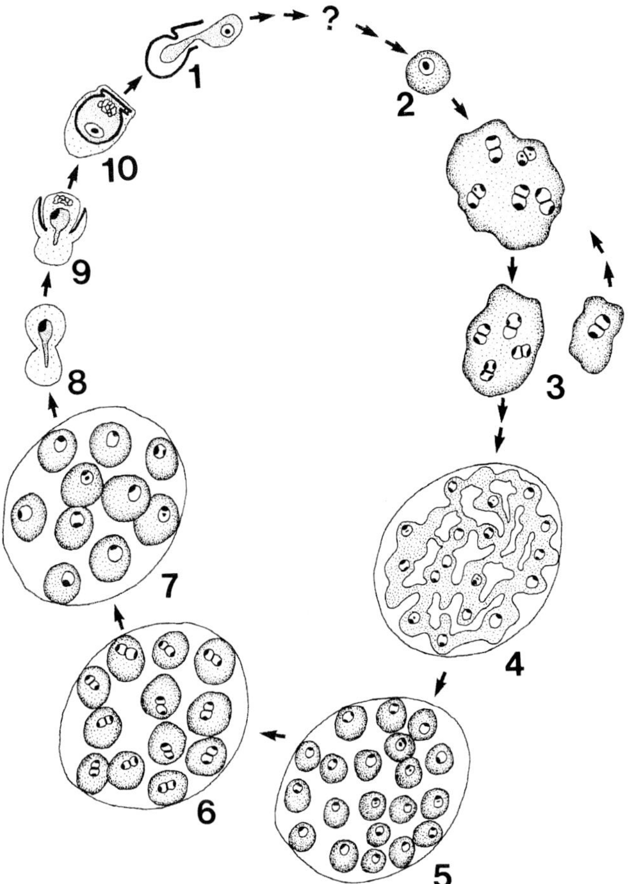

Fig. 4. The life cycle of Haplosporidia as typified by *Haplosporidium* spp. or *Minchinia* spp. The spore lid lifts and the sporoplasm moves out as an ameboid cell (1). The next stage is unknown. The cell type which infects the host is believed to be an uninucleate, naked cell (2) which undergoes repeated karyokineses and enlarges to form plasmodia with paired nuclei in the host connective tissue or epithelium. Multiple, irregular fission of the plasmodia results in spread of the infection (3). Sporulation is initiated in a plasmodium which undergoes three successive nuclear divisions, involving meiosis to yield a large sporont in which cytokinesis occurs (4) to yield uninucleate, haploid sporoblasts(5). Pairs of sporoblasts fuse (6) followed by karyogamy to form zygotes (7) each of which forms a spore by pinching into an hour-glass cellular form (8). The anucleate posterior half grows around the anterior half (9) forming the epispore cytoplasm configuration in which the spore cell wall is formed (i.e., flask with a hinged lid) (10). The epispore cytoplasm is lost as the spore matures; however, depending on the species, tubules, ribbons, and fibers, which are attached to and/or wrapped around the spore wall, are left behind. The sporulation sequence is as proposed by Desportes and Nashed (1983).

plasmodia then undergo irregular multiple fission (plasmotomy) to yield daughter cells with varying numbers of nuclei. The cycle is repeated indefinitely, resulting in spread of the cells throughout the connective tissue of the host and between epithelial cells (Fig. 4). Generally the plasmodia are found extracellularly, but the smaller ones may be phagocytized by hemocytes.

The ultrastructure of the plasmodia includes tubulo-vesicular mitochondria, Golgi bodies, smooth and rough endoplasmic reticulum, and an unusual membrane-bounded organelle which is tripartite, having an electron-dense cortex and medulla separated by an internal membrane (Fig. 2). The latter are called haplosporosomes and are also found in the Paramyxa and Myxosporidia (where they are termed sporoplasmosomes; Lom et al., 1986). The function of **haplosporosomes** is unknown, but there is some evidence to indicate that the organelle contains lytic components which cause damage to host cells (Perkins, 1979; Scro and Ford, 1990). Another possibility is that they are virus-like particles in that Perkins (1968) noted that large haplosporosome-formative regions in the plasmodial cytoplasm of *H. nelsoni* are Feulgen positive. In at least some species, haplosporosome-like bodies are also formed from Golgi-derived cisternae and nuclear material (Fig. 3). The details of structure and morphogenesis have been described in detail by Hine and Wesney (1992) for a problematic haplosporidian, *Bonamia* sp. (see section on *Incertae Sedis*). They suggested that the structures are virus-like and may be related to the virulence of the parasite.

Mitosis occurs without breakdown of the nuclear envelope and the mitotic spindle terminates at two spindle pole bodies (microtubular organizing centers). Centrioles are not formed (Perkins, 1975a).

Since the main criterion for generic and species identity is the spore structure, sporulation has been described in all known species. A number of other possible species have been noted, but they are described with only a generic designation since no spores were observed (La Haye et al., 1984; Dyková et al., 1988). At the beginning of sporulation the plasmodia acquire a thin cell wall signaling the formation of the **sporont** stage (Perkins, 1975b). Among the various species there may be variations in the details of spore formation yet to be described; however, in general the best known sequence, described by Desportes and Nashed (1983) from *Minchinia dentali,* may typify species in the families Haplosporidiidae and Urosporidiidae (Fig. 4). Three successive nuclear divisions, involving meiosis, appear to occur in the sporont resulting in a marked increase in cell size. This is followed by internal cleavage whereby the cytoplasm appears to condense around individual nuclei to yield uninucleate **sporoblasts** within the sporont wall. Pairs of sporoblasts fuse to yield binucleate cells in which the two nuclei fuse. Thus it is believed that there is a brief haplophase following meiosis in the sporont then a restoration of diplophase in a zygote. The observations of 1) nuclear structures believed to be synaptonemal complexes in sporont nuclei (Perkins, 1975b) and 2) binucleate sporoblasts, observed also by La Haye et al. (1984), lend support to this suggestion. Haplosporosomes usually disappear from the sporont during sporoblast formation. Since none of the Haplosporidia have been established in culture, the sporulation sequences thus far described have all been deduced from fixed and sectioned cells.

The proposed developmental sequence described above must be considered to be tentative, since I have found evidence in *Haplosporidium* sp. (probably *H. louisiana*; see below) that there are two size-classes of nuclei in the sporont leading to the suggestion that karyogamy occurs between the paired nuclei observed in plasmodia followed by meiosis then subdivision into sporoblasts (Perkins, 1975b). I found no evidence of sporoblast fusion and other workers (La Haye et al., 1984) found no evidence of an increase in nuclear size following nuclear fusion in the binucleate sporoblasts. Thus instead of the life cycle being dominated by diplophase in which there is only a brief haplophase after meiosis followed by sporoblast fusion then karyogamy, the life cycle may be dominated by haplophase and diplophase occurs only when paired nuclei in plasmodia fuse followed immediately by meiosis. In the latter version, there would be no sporoblast fusion, followed by zygote formation. In the Apicomplexa haplophase dominates (Barta, 1989) but differs from the above latter version in that meiosis follows zygote formation. Obviously the life cycles of the Haplosporidia need to be studied further, utilizing techniques such as those of Cornelissen et al. (1984). There is also some uncertainty concerning the formation of spores from the zygotes or sporoblasts, but most workers believe that a constriction forms around the mid-region of the cell, dividing it into a nucleated and an anucleated half with a slender connecting isthmus (Ormières et al., 1973; Desportes and Nashed, 1983). Thus the sporoblast assumes an hour-glass form. The anucleated half then almost completely surrounds the other half and the isthmus breaks, resulting in an uninucleated sporoplasm primordium which is not in continuity with the surrounding cytoplasm (epispore cytoplasm) (Fig. 4). The latter lacks a nucleus but appears to have nuclear components which are contributed from the sporoplasm nucleus by extrusion of such material into the epispore cytoplasm before the cytoplasmic isthmus breaks (Desportes and Nashed, 1983). Thus the spores are unicellular, not multicellular as in the Paramyxa because the sporoplasm is unicellular and the epispore cytoplasm does not have an ultrastructurally visible nucleus.

At the anterior end of the forming spore where the anucleate, epispore cytoplasm converges on itself, the resulting orifice may be organized in one of two ways. In the genera *Haplosporidium* and *Minchinia* one side of the convergence zone extends over the other side to form a lid of cytoplasm resting on a flange of cytoplasm formed by a flaring of that side. In *Urosporidium* spp. a flap of cytoplasm is formed on one side which extends under the leading edge of the other side. Thus there is a orifice covered by extensions of cytoplasm.

As the epispore cytoplasm extends around the

sporoplasm, wall synthesis commences in the epispore cytoplasm against its inner cell membrane to form a wall which conforms to the shape of the epispore cytoplasm (Figs. 4,5,10). The first signs of wall deposition are around the posterior end of the sporoplasm as envelopment occurs (Fig. 4). In the epispore cytoplasm there are also formed a variety of tubular, filamentous or ribbon-like strands of material (ornaments) which generally persist after spore maturation when the epispore cytoplasm disintegrates. They may be organized into bundles to form two to four prominent extensions from the spore which are visible in the light microscope (*Minchinia* spp.) (Figs. 6,9) and which 1) may be fused in part to the wall (Azevedo, 1984; Perkins and van Banning, 1981; Ormières, 1980) or 2) may not be attached to the wall (McGovern and Burreson, 1990). The strands may also be wrapped around the spore without forming extensions visible in the light microscope (*Haplosporidium* spp.) but may be fused to the spore wall at various sites (Perkins, 1968; Perkins, 1975b) (Figs. 7,8). The fine structure of the strands is believed to be species specific (Perkins, 1979).

In the sporoplasm, haplosporosomes are reformed and a Golgi body-like, complex membranous structure, called a spherule or **spherulosome** (Figs. 5,10) (Azevedo and Corral, 1987) is synthesized in the anterior end of the cell. This organelle is suspected to be involved in spore excystment, but emergence of the sporoplasm has not been induced in vitro except by crushing the cell, and the next stage of the life cycle has not been observed (Fig. 4). Azevedo et al. (1985) and Desportes and Nashed (1983) have suggested that spore emergence may occur within the host after sporulation occurs, followed by further internal spreading of the infection. The characteristics of the phylum are based on the spore structure since the plasmodia have no structures found exclusively in the phylum. However, of some taxonomic value is the fact that haplosporosomes are found in the cytoplasm and the nuclei are often found in closely apposed pairs. In addition the mitotic apparatus functions within an intact nuclear envelope, and there are no centrioles, but rather the spindle is attached to microtubular organizing centers, which are not visibly attached to the nuclear envelope.

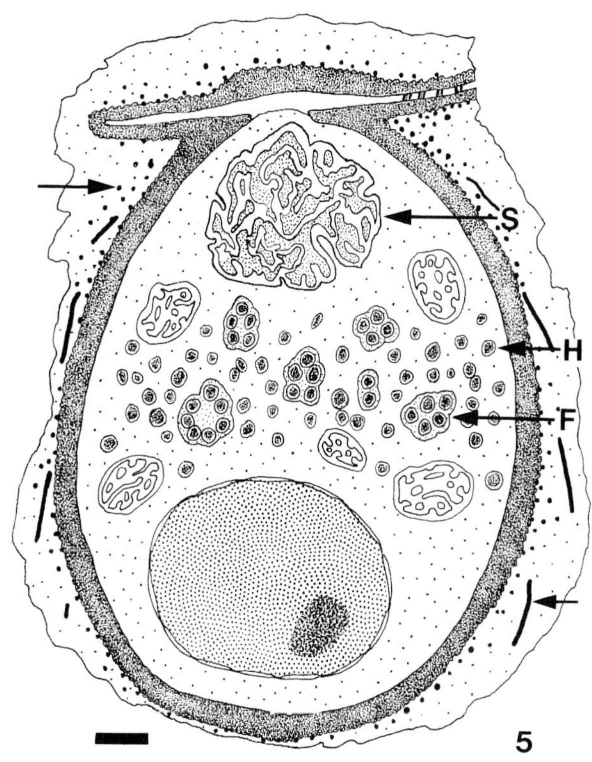

Fig. 5. Nearly mature haplosporidian spore typified by *H. louisiana.* Note hinged lid of spore wall covering orifice and fastened with strands of electron-dense material at the non-hinged side. Spherulosome (S); haplosporosome (H); formative regions for haplosporosomes (F); fibers formed in epispore cytoplasm which persist around the spore after the spore matures (arrows). Bar=1 µm.

Definitive structural features are found mainly by using transmission electron microscopy; however, preliminary screening can be accomplished using light microscopy. If spores are present, one should look for a lid of wall material covering an orifice (*Haplosporidium* or *Minchinia*) (Figs. 5-9) or the orifice may appear to be occluded by wall material within a ring of raised wall material (*Minchinia*) (Fig. 10). By staining with a variety of cytological and histological stains such as hematoxylin and eosin, the nucleus can be seen in the sporoplasm as well as the anteriorly located spherulosome. The layer of cytoplasm containing ornaments will most often be found around the spore wall except in fully mature spores where the exospore cytoplasm disperses, leaving only the ornaments which may also disperse if not attached to the spore wall (McGovern and Burreson, 1990). If the strands are organized into bundles, they may be visible as one to four spikes extending from the wall in species-specific extensions or arrays.

At the ultrastructural level the organelles described above should be found if the microbe is one of the Haplosporidia. Preparations for fine structural observations can be accomplished using a variety of glutaraldehyde and osmium tetroxide fixations. Procedures outlined by Paulin (1992) and Perkins et al. (1975) are recommended.

Phylogenetic affinities of the phylum are not clear. Throughout the last 100 years since the Haplosporidia were described as a separate taxon, non-photosynthetic, non-rhizopodean, spore-forming protozoa which do not synthesize a polar filament, polar capsules, or apical complex, and do not exhibit internal cleavage as in the Paramyxa have been placed in the Haplosporidia. Thus the taxon has served as a systematist's "trash bin". If the affinities of a spore-forming protist were not known, then the organism was allied with the Haplosporidia. Those genera which have been described as Haplosporidia, but do not meet the criteria for inclusion (particularly the lack of an orifice in the cell wall) as described above, are treated in the *Incertae Sedis* section of this chapter with the recognition that the characters of the phylum may be too restrictive.

Although one would expect that the rather heavily walled spores of Haplosporidia would be preserved in the fossil record, thus far none have been found. Whether they are primitive organisms is subject to debate. The presence of microtubular organizing centers in the mitotic apparatus, nuclei in the dikaryon configuration, the occurrence of meiosis in the sporont nuclei, and the expression of sexuality as only nuclear processes (possibly diploidy followed by haploidy then diploidy) (Desportes and Nashed, 1983) could be evidence for affinities with the Microspora, which are among the most primitive living eukaryotes as shown by small subunit ribosomal RNA analyses (Sogin, 1989); however, the Ascomycetes of the Eumycota, which evolved much later than the Microspora and which even share a common ancestor with the Metazoa (Wainright et al., 1993), also possess such structures and exhibit similar nuclear cycles. The phylogeny of the Haplosporidia will not be elucidated solely on the basis of morphology. Molecular analyses such as those reported by Sogin (1989) are necessary, if further insights into its evolutionary relationships are to be obtained.

Other descriptions of the phylum can be found in Sprague (1979, 1982), Corliss (1984) and Perkins (1990, 1991).

KEY CHARACTERS OF THE PHYLUM

1) Ovoid, walled spores with an orifice covered externally by a hinged lid or internally by a flap of wall material. One to four projections visible in the light microscope may be found extending from the spore wall and consist of either extensions of the cell wall or bundles of filaments, tubules or ribbons (ornaments) formed in the epispore cytoplasm, attached or not attached to the spore wall. In spores lacking projections—tubules, filaments, or ribbons are also found around the cell wall but are not organized into bundles which support projections. Anucleate cytoplasm covers the spore during its formation and is the site of wall and ornament formation.

2) Sporoplasm contains a) haplosporosomes which are spheroidal, pyriform, or cuneiform, membrane-bounded organelles with an electron-dense cortex and medulla separated by an unit membrane, b) an anteriorly situated organelle (spherulosome) consisting of anastomosing, membrane-bounded cisternae, c) tubulo-vesicular mitochrondria and d) one, or rarely two, nuclei.

3) Sporulation occurs within sporonts which are derived from plasmodia. In the process of sporont

formation, plasmodia synthesize a delimiting, thin cell wall and haplosporosomes are lost. Several cycles of nuclear multiplication then occur during which meiosis is suspected to be involved, resulting in a marked increase in cell size. Multiple fission results in uninucleate sporoblasts contained within the wall. It has been suggested that pairs of sporoblasts then fuse, followed by fusion of the paired nuclei, thus resulting in a return to diplophase. Each sporoblast pinches almost in half creating a nucleated half and an anucleated half. The anucleated half grows around the other half leaving an opening at the anterior end where a hinged lid or flap of cytoplasm occludes the orifice. The sporoplasm then becomes a separate entity when the cytoplasmic connection between the two halves breaks. The spore wall, including the lid or flap, is formed in the surrounding anucleate cytoplasm as are the spore wall ornaments.

4) Plasmodia contain haplosporosomes. Mitosis occurs within an intact nuclear envelope and involves a mitotic spindle attached to microtubular organizing centers. The spindle is persistent through interphase in some species.

KEY TO TAXA

CLASS HAPLOSPOREA Caullery,1953

With characters of the phylum.

ORDER HAPLOSPORIDA
Caullery & Mesnil,1899

With characters of the phylum.

Spores with a hinged operculum externally covering an anterior orifice in the spore wall .... **Family Haplosporidiidae**.....1
Spores with a flap of wall material which internally covers the anterior spore wall orifice ....**Family Urosporidiidae** 2

1. Spores with tubules, filaments, or ribbons of material around the external surface of the spore wall and not organized into pronounced extensions visible in the light microscope........................ ***Haplosporidium***

1'. Spores with filaments or tubules supporting two to four prominent extensions visible in the light microscope.................... ***Minchinia***

2. With characters of the Family.***Urosporidium***

In an earlier review of the phylum (Perkins, 1990), I did not recognize the family Urosporidiidae, including *Urosporidium* spp. in the family Haplosporidiidae. This may have been an error since 1) generic distinctions are being made between *Haplosporidium* and *Minchinia* on the basis of the rather tenuous difference as to whether the epispore cytoplasm exhibits prominent extensions, whereas 2) the wall structure of *Urosporidium* is fundamentally different from species of the other two genera. Once knowledge of nucleic acid base sequences, as for example, small subunit ribosomal RNA sequences, are made available, the choice as to whether to accept a second family may become less arbitrary.

There has been much debate in the literature concerning the distinctions between the genera *Haplosporidium* and *Minchinia* as well as whether the genus *Minchinia* should be recognized (Sprague, 1963, 1970, 1978, 1982; Perkins, 1990; McGovern and Burreson, 1990; Lauckner, 1983; Ormières, 1980). Much of the debate is due to the lack of ultrastructural observations and inadequacies with the original type species descriptions, all resulting in a taxonomic "nightmare". Even though 12 different species have been examined in the electron microscope, there continues to be uncertainty. Thus the taxonomic scheme presented herein will undoubtedly be subject to continued debate.

The original description of the type species for *Minchinia, M. chitonis* (Lankester, 1885) Labbé, 1896, presents complications in that the description by Labbé (1896) involved observations of the spores of a haplosporidian and other stages of a coccidian. This followed the original brief notation of Lankester (1885) in which he drew a picture of a haplosporidian spore with two prominent extensions and briefly wrote that it was a new sporozoan species which he called *Klossia chitonis.* He stated that it was from a chiton. This problem of two species being involved was noted by Debaisieux (1920), who renamed the protist with haplosporidian spores, *Haplosporidium chitonis*, and the coccidian protist, *Pseudoklossia chitonis.* This name change held until Sprague (1963) re-established the genus *Minchinia* and species *M. chitonis*, on the basis of characters of the spore lid which have since been shown by electron microscopy to be invalid (Sprague, 1970). Despite Debaisieux's (1920) contention that *Minchinia* is not a valid genus, followed by Lauckner's (1983) excellent review of the history of the genus, in which he supported Debaisieux, *Minchinia* has continued to be used as a genus. The net result is an array of species with the generic designation of *Minchinia,* some of which have projections from the spore and some which do not have projections. This confusion also exists for *Haplosporidium* spp. For example, *M. nelsoni* became *H. nelsoni* and *M. louisiana* became *H. louisiana* because of the lack of projections. Some species such as *H. costale* have even experienced shifts from *Haplosporidium* to *Minchinia* and back again to *Haplosporidium* (Sprague, 1978).

Despite the recognition that *Minchinia* was described from two species of protists, I propose to utilize the genus for species of Haplosporidia which have spores with an external hinged lid and which form projections from the spore which are visible in the light microscope. My reasons are: 1) the genus is well established in the literature and has been used recently (Hillman et al., 1990); 2) it was originally described in part from spores of a haplosporidian with very distinctive structures (i.e., prominent extensions of a spore with an orifice and lid covering the spore wall orifice); 3) the genus has not been used for other species; and 4) it would result in less confusion in the literature if another generic name were not introduced. A formal, new description of the genus will be necessary in the future. It was agreed that no new generic descriptions would be included in this volume.

Another component of the controversy over *Minchinia* vs. *Haplosporidium* centers around whether or not extensions of the spores can be seen at the light microscope level, and, if so, the significance of the fine structure of the extensions. The problem started with the description of the type species, *H. scolopli* (Caullery and Mesnil, 1899) Lühe 1900 of the genus *Haplosporidium*, the question being whether there are extensions from the posterior end of the spore. I agree with Sprague (1963) that *H. scolopli* does not have such extensions as shown in Caullery and Mesnil's 1905a paper in Fig. 1 and Pl XI, Figs. 17a,b and their 1905b paper in Fig. 2. Ormières (1980) disagrees, stating that the drawing in Fig. 17c of Caullery and Mesnil (1905a) contains a spore with extensions. Two delicate lines were drawn, trailing from the posterior end of the spore, but I believe that they probably represent membranes from remnants of the epispore cytoplasm and not extensions supported by internal filaments as in other species (see below). This interpretation is supported by the Caullery and Mesnil's (1905a) statement: "Il y a en outre une membrane externe beaucoup plus délicate et qui n'est souvent reconnaissable qu'à quelques débris (fig. 17b et c)". Thus, I accept the type species, *H. scolopli,* as being a species without prominent extensions.

With the acceptance of the genus *Minchinia* and the assumption that the original description of *H. scolopli* did not include extensions visible in the light microscope, I have proposed the above key to the genera *Haplosporidium* and *Minchinia.* Thus the following species needs to be transferred from genus *Minchinia* to *Haplosporidium: M. cadomensis* Marchand & Sprague, 1979. *Haplosporidium parisi* Ormières, 1980; *H. lusitanicum* Azevedo, 1984; *H. heterocirri* Caullery & Mesnil, 1899; and *H. potamillae* Caullery & Mesnil, 1905 need to be designated as species of *Minchinia.* The following species remain as currently described: *M. teredinis* Hillman, Ford & Haskin, 1990; *H. dentali* Arvy,

1949; *M. armoricana* van Banning, 1977; *M. chitonis* (Lankester, 1885) Labbé, 1896; *H. nelsoni* (Haskin, Stauber & Mackin, 1966) Sprague, 1978; *H. costale* Wood & Andrews, 1962; *H. louisiana* (Sprague, 1963) Sprague, 1978; *H. ascidiarum* Duboscq-Harant, 1923; *H. cernosvitori* Jirovec, 1936; *H. nemertis* Debaisieux, 1920; *H. caulleryi* Mercier & Poisson, 1922; *H. scolopli* Caullery & Mesnil, 1899; *H. vejdovskii* Caullery & Mesnil, 1905; *H. marchouxi* Caullery and Mesnil,1905; *H. pickfordi* Barrow, 1961; *H. tapetis* Vilela, 1951; and *H. limnodrili* Granata, 1913.

As noted by La Haye et al. (1984), *H. comatulae* La Haye, Holland & McLean, 1984 appears to be very similar to *H. parisi*, particularly in terms of the location and ultrastructure of the epispore filaments. However, they did not demonstrate whether the filaments were attached to the posterior end of the spore and whether they were visible as prominent extensions visible in the light microscope. Because of the marked similarity I propose to tentatively consider *H. comatulae* also to be a species of the genus *Minchinia*. *Haplosporidium tumefacientis* Taylor, 1966 appears to be very similar to *H. comatulae* and *H. parisi,* therefore it should also be placed in the genus *Minchinia*.

Haplosporidium aselli and *H. gammari* have been removed from the genus and have become *Claustrosporidium asellii* (Pflugfelder, 1948) Larsson, 1987 and *C. gammari* (Ryckeghem, 1930) Larsson, 1987 (see below). I do not consider *H. aulodrili* and *H. mytilovum* to be members of the Haplosporidia because they lack spores with orifices. Thus, with the above proposed changes, there would be 9 species of *Minchinia* and 15 species of *Haplosporidium.*

Genus ***Haplosporidium***
(Caullery & Mesnil, 1899) Lühe, 1900

Spores are oval with the anterior end having an orifice and appearing to be slightly flattened in the light microscope (Figs. 5,7,8). The orifice is covered by a hinged lid. Spores range in size from 2.8-12 X 4.3-15 μm. Tubules or solid filaments (42-80 nm diameter) or ribbons (ca. 80-140 nm wide), formed in the epispore cytoplasm, are wrapped around the spore wall with at least some of them being fused in part to the spore wall. These wall ornaments are not organized into prominent extensions visible in the light microscope. There are 15 named species (as revised above), found as parasites of marine molluscs, decapod Crustacea, an ascidian urochordate, a crinoid echinoderm, polychetes, and a nemertean worm as well as freshwater oligochetes and snails.
TYPE SPECIES: *Haplosporidium scolopi* (Caullery & Mesnil, 1899) Lühe, 1900

Genus ***Minchinia***
(Lankester, 1895) Labbé, 1896

Spores have the same general shape and structure as species of *Haplosporidium.* They range in size from 2.1-9 X 3.0-13 μm. Solid filaments (20-83 nm diameter), cylindrical filaments (32-135 nm diameter) comprised of a coiled ribbon or microtubule-like units (ca. 290 nm diameter) are formed in the epispore cytoplasm (Figs. 6,9).

Genus ***Urosporidium***
Caullery & Mesnil, 1905

Spores are spherical or oval and have an anterior orifice closed by a flap of wall material (Fig. 10). The epispore cytoplasm extends posteriorly into a tapering extension except for *U. pelseneeri* (Caullery & Chappellier, 1906) Dollfus, 1925. The shape of the extension is maintained by bands or ribbons of material curved in cross-sectional view and composed of folded subunits which lie mainly beneath the cell membrane of the epispore cytoplasm giving the cytoplasm a corrugated appearance in surface view. The spores, minus the epispore cytoplasm, range from 3.0-5.9 X 5 - 13.3 μm, and the posterior extensions range from 10 to 43 μm long. The species are mostly hyperparasites of trematode or nematode sporocysts, metacercariae or juveniles; however, one species, *U. fulginosum* Caullery & Mesnil, 1905 is found in the coelom of a polychete. When spores are mature and found in large numbers, their presence causes the host to be black. Seven named species. TYPE SPECIES: *U. fulginosum* Caullery & Mesnil, 1905. The genus *Anurosporidium* Guyenot, 1943 is considered to be synonymous with *Urosporidium* (Ormières et al., 1973).

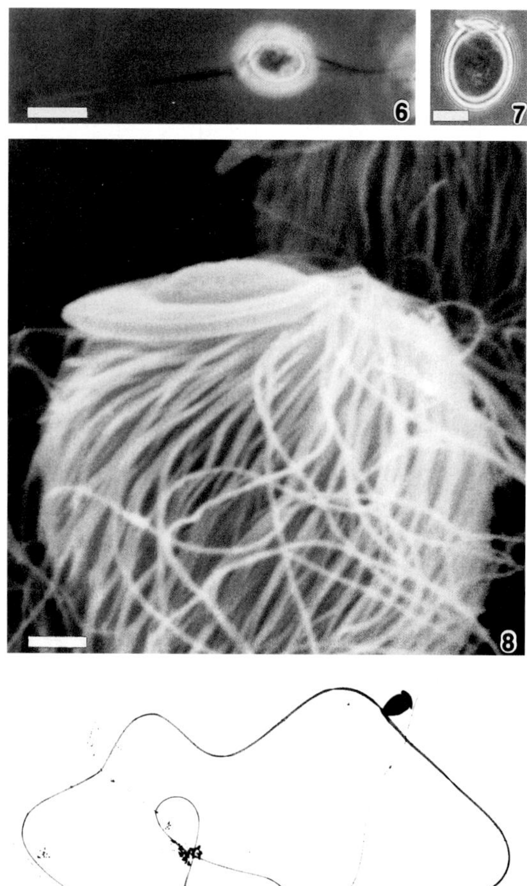

Figs. 6-9. Phase contrast micrograph of *Minchinia armoricana* spore. A prominent, anteriorly directed extension (left) originates from beneath the hinge of the lid and a similar, posteriorly directed extension (right) arises from the posterior of the spore. Bar=5 µm. Fig. 7. Phase contrast micrograph of mature spore of *H. louisiana.* The spore lacks prominent episporal extensions and the fibers around the wall are not visible. Bar=5 µm. Fig. 8. Scanning electron micrograph of *H. louisiana* spore as in Fig. 7. Fibers are clearly visible. Bar=1 µm. Fig. 9. Transmission electron micrograph of a whole mount of a *M. lusitanicum.* Note the two long extensions arising as bifurcations from the posterior end of the spore and the spore lid (From Azevedo, 1984). Bar=5 µm.

These units, singly or in groups, form 2 - 4 prominent extensions of the spore which are visible in the light microscope. The extensions are 10-112 µm long. There are 9 named species (as revised above), found as parasites of marine polychetes and molluscs. TYPE SPECIES: *Minchinia chitonis* (Lankester, 1895) Labbé, 1896

INCERTAE SEDIS

Larsson (1987) has erected a third family Claustrosporidiidae, for the order Haplosporida with a single genus *Claustrosporidium* Larsson, 1987 and two species, *C. gammari* (Ryckeghem, 1930) Larsson, 1987 (syn.=*Haplosporidium gammari* Ryckeghem, 1930); and *C. aselli* (Pflugfelder, 1948) Larsson, 1987 (syn.=*Haplosporidium aselli* Pflugfelder, 1948). The uninucleate spores of both species lack orifices. However, in the only species, *C. gammari,* examined by electron microscopy, the cytoplasm contains haplosporosomes in plasmodia, sporoblasts, and spores. There is no spherulosome in the sporoplasm.

Claustrosporidium gammari is found in the adipose tissue of the marine amphipod, *Rivulogammarus pulex* (Larsson, 1987). Prior to sporulation the protist exists as plasmodia with small (1.7 to 2.0 µm diameter) nuclei which form pairs then appear to fuse, creating large (2.3 to 3.0 µm diameter) nuclei followed by division to form small (ca. 2.0 µm) nuclei again. Subdivision of the plasmodium then yields uninucleate sporoblasts, each of which acquire a cell wall. The wall and protoplast become the spore wall and sporoplasm, thus one spore is formed per sporoblast. The mitotic apparatus is the same as in the Haplosporidiidae and Urosporidiidae and appears to be persistent into interphase as in *Haplosporidium nelsoni* (Perkins, 1975a).

The main reason I have not yet accepted Larsson's placement of the Claustrosporidiidae in the Haplosporidia is that the spores have no orifice and the spore wall is formed on the plasmalemma of the sporoblast, not in the epispore cytoplasm

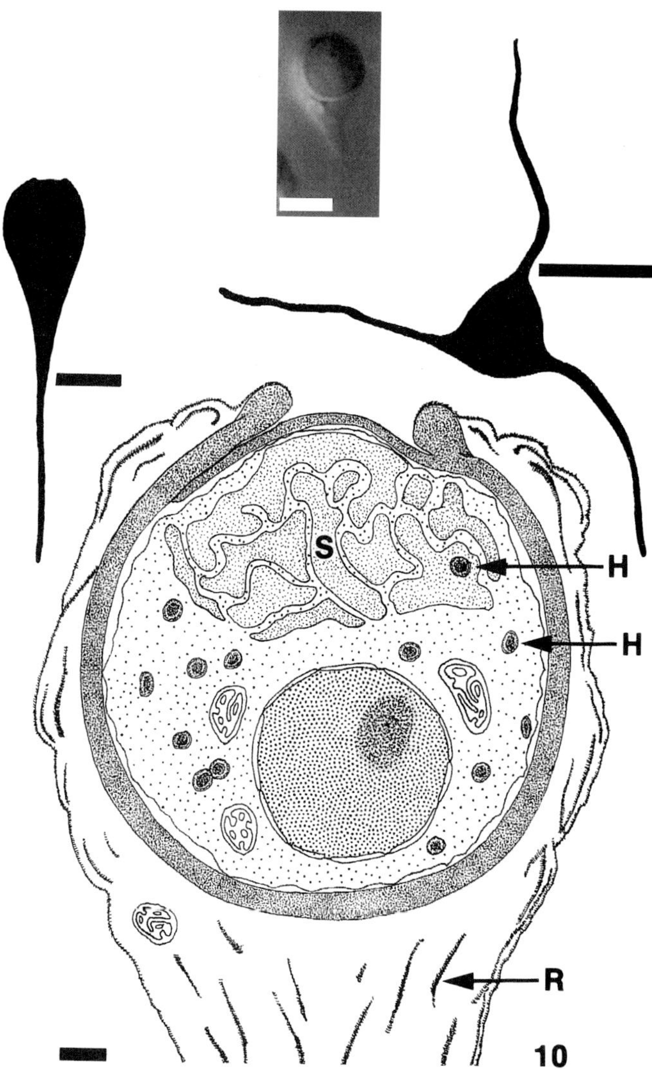

Fig. 10. Spores of *Urosporidium* spp. Upper left silhouette is typical of most species where the epispore cytoplasm persists around the spore wall and is drawn into a posteriorly directed extension as represented in the interference contrast micrograph of *U. crescens*. The upper right silhouette is representative of *U. spisuli* where three episporal extensions are formed. The drawing of the internal detail typifies *U. crescens.* Spherulosome (S) in which haplosporosomes (H) form; ribbon (R) of episporal material which provides support to the epispore cytoplasm. Bar=10 µm for top three figures and 1 µm for drawing of spore internal detail.

and on the membrane which faces the plasmalemma of the sporoplasm as occurs in the genera *Haplosporidium*, *Minchinia*, and *Urosporidium.* However, there are a number of similarities which make Larsson's classification attractive. The presence of haplosporosomes in a spore-forming protist, which lacks polar filaments as in the Microspora, internal cleavage as in the Paramyxa, and polar capsules as in the Myxospora, strengthens the argument that it is one of the Haplosporidia. As noted above haplosporosomes appear to be present in the Myxospora and Paramyxa, but not in the Microspora. The presence of a mitotic apparatus which consists of a bundle of microtubules attached to spindle pole bodies in the nucleoplasm, not centrioles, also strengthens the argument. In attempts to determine the taxonomic affinities of the Claustrosporidiidae, it will be useful to examine the ultrastructure of other non-photosynthetic, non-rhizopodean, spore-forming protists which do not form an apical complex, polar capsules. or polar filaments in their life

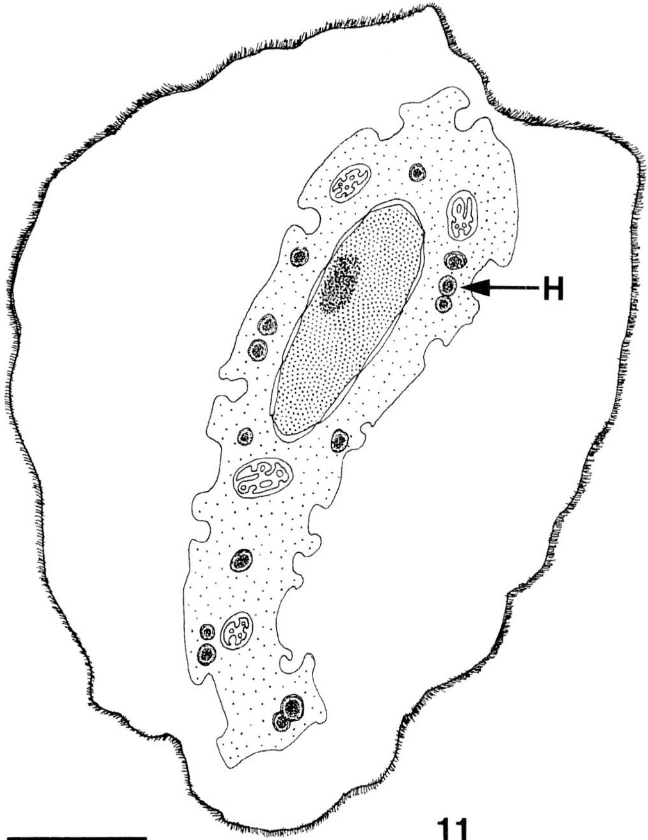

Fig. 11. *Claustrosporidium gammari* spore drawn from Larsson (1987). Spore wall loosely enrobes the sporoplasm and has ridges on the outer surface. Haplosporosomes (H). Bar=1 µm.

cycle, nor form multicellular spores. Species of the genera *Coelospora* and *Peltomyces* will be of special interest. Lange (1993) has examined the ultrastructure of *Nephridiophaga periplanetae* Lutz & Splendore, 1903 and did not find haplosporosomes: Neither were they found in *Coelosporidium chydoricola* (Manier et al., 1976), *Oryctospora alata* (Purrini and Weiser, 1990), *N. tangae* (Purrini et al., 1988), *N. blattellae* (Woolever, 1966), nor *N. ormieresi* (Toguebaye et al., 1986). Lange (1993) included the above 5 genera in the family Nephridiophagidae Sprague, 1970, because species in the 5 genera form similar spores and are all found in the Malpighian tubules of arthropods. Since haplosporosomes appear to be absent (at least in species of three of the genera) and orifices are not present in the spores, he did not consider the Nephridiophagidae to be haplosporidians and further suggested that they may need to be placed in a separate phylum, placing them in *incertae sedis* status for the present.

If the Claustrosporidiidae can be accepted as haplosporidians, then the question arises as to whether species like those in the Nephridiophagidae can be considered to also be haplosporidians. It might be reasonable to accept a gradation from species of accepted Haplosporidia (i.e., unicellular spores with orifices) to those species with no orifice, but with haplosporosomes (the Claustrosporidiidae) to species with no orifice and no haplosporosomes (the Nephridiophagidae). All of this would be predicated on the assumption that none of the species of uncertain affinities formed spores with polar filaments or polar capsules and spores were not multicellular nor were there apical complexes formed in any cellular stages in the life cycle. It should be noted that species of *Urosporidium*, *Haplosporidium*, and *Minchinia* are not considered to be multicellular as found in the Paramyxa (see above).

FAMILY CLAUSTROSPORIDIIDAE Larsson, 1987

Spores have one uninucleate sporoplasm with haplosporosomes and no spherulosome. Spore wall has no orifice. One named genus and two species.

Genus ***Claustrosporidium*** Larsson, 1987

Spores are inoperculate and have one uninucleate sporoplasm without a spherulosome but with haplosporosomes (Fig. 11). Spore wall is loosely fitted around the sporoplasm and is covered with fine ridges on the external surface of the wall. Plasmotomy yields uninucleate sporoblasts, each of which becomes a spore with the addition of a wall synthesized on the sporoblast plasmalemma. Found in a fresh water isopod and a marine amphipod. Two named species.
TYPE SPECIES: *Claustrosporidium gammari* (Ryckeghem, 1930) Larsson, 1987

OTHER SPECIES

Species of the genera *Bonamia* and *Mikrocytos* are oyster parasites which have not yet been assigned to a family but are suspected to be Haplosporidia because they possess haplosporosomes. However, they have not been observed to form spores, and, therefore, can not be definitively allied with any established group. Species of both genera are small (2 to 6 µm) parasites which multiply in oyster hemocytes and branchial epithelia (Montes et al., 1994) where they are found in phagosomes or parasitophorous vacuoles. In contrast and where descriptions are of sufficient detail to make a judgement, the established species of Haplosporidia multiply extracellularly and only the smaller cells are found in hemocytes where multiplication has not been observed (Scro and Ford, 1990). The cells are simple, containing usually one nucleus or sometimes two nuclei. Rarely, plasmodia with four or five nuclei can be found (Brehelin et al, 1982). The mitotic apparatus is the same as in the established species of Haplosporidia (Pichot, et al., 1980). Haplosporosomes are found in the cytoplasm and, at least in *Bonamia* sp., are derived in part from multivesicular bodies (Hine and Wesney, 1992) as in *Haplosporidium nelsoni* (Perkins, 1979).

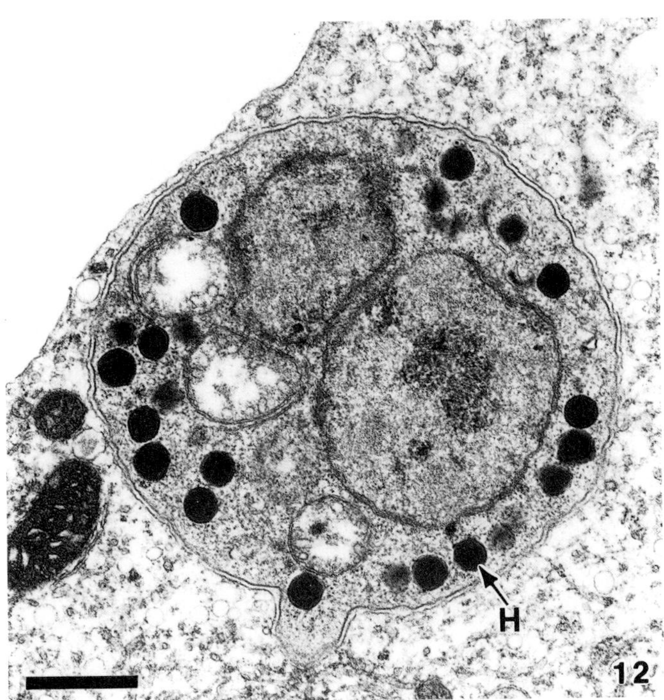

Fig. 12. Transmission electron micrograph of *Bonamia* sp. from the oyster *Tiostrea chilensis*, with paired nuclei and located in hemocyte. Haplosporosome (H). Provided by P.M. Hine. Bar=1 µm.

Genus ***Bonamia***
Pichot, Comps, Tige, Grizel & Rabouin, 1980

Spores have not been observed. Small (2-7 µm), rounded, uninculeate or binucleate cells or, rarely, plasmodia with 4-5 nuclei are found in the cytoplasm of granular hemocytes and gill epithelia (Fig. 12). The ultrastructure includes tubulovesicular mitochondria, haplosporosomes, multi-vesicular bodies, and Golgi bodies where haplosporosomes are formed. Nuclei divide within an intact nuclear membrane, utilizing a mitotic apparatus which is attached to 2 spindle pole bodies (microtubule organizing centers). There is one named species found in oysters (*Ostrea edulis* Linné, 1758) from Europe and North America. A member of the genus, either *B. ostreae* or a new species yet to be described, has been reported from New Zealand oysters (*Tiostrea chilensis* Phillipi, 1845; syn.=*T. lutaria* Hutton, 1873) by Dinamani et al. (1987). One named species.
TYPE SPECIES: *Bonamia ostreae* Pichot, Comps, Tige, Grizel & Rabouin, 1980.

Genus ***Mikrocytos*** Farley, Wolf & Elston, 1988

No spores have been observed. Small (1-4 µm) cells with essentially the same structure as *B. ostreae* except plasmodia have not been reported. Membrane-bounded, 5- and/or 6-sided dense bodies of 40-45 nm are found in the cytoplasm of one species (*M. mackini*). Always associated with abscess-type focal inflammatory lesions in the gill, connective, and gonadal tissues. Found in the Pacific oyster, *Crassostrea gigas* Thunberg, 1793, from Denman Island, British Columbia and from the Sydney rock oyster, *Saccostrea commercialis.* Two named species.
TYPE SPECIES: *Mikrocytos mackini* Farley, Wolf & Elston, 1988.

LITERATURE CITED

Azevedo, C. 1984. Ultrastructure of the spore of *Haplosporidium lusitanicum* sp. n. (Haplosporida, Haplosporidiidae), parasite of a marine mollusc. *J. Parasitol.,* 70:358-371.

Azevedo, C. & Corral, L. 1987. Fine structure, development and cytochemistry of the spherulosome of *Haplosporidium lusitanicum (Haplosporida). Europ. J. Protistol.,* 23:89-94.

Azevedo, C., Corral, L. & Perkins, F. O. 1985. Ultrastructural observations of spore excystment, plasmodial development and sporoblast formation in *Haplosporidium lusitanicum* (Haplosporida, Haplosporidiidae). *Z. Parasitenkd.,* 71:715-726.

Barrow, J. H. 1965. Observations on *Minchinia pickfordae* (Barrow 1961) found in snails of the Great Lakes region. *Trans. Am. Microsc. Soc.,* 84:587-593.

Barta, J. R. 1989. Phylogenetic analysis of the class Sporozoea (phylum Apicomplexa Levine, 1970): evidence for the independent evolution of heteroxenous life cycles. *J. Parasitol.*, 75:195-206.

Brehelin, M., Bonami, J.-R., Cousserans, F. & Vivares, C. P. 1982. Existence de formes plasmodiales vraies chez *Bonamia ostreae* parasite de l'Huître plate *Ostrea edulis. C. R. Acad. Sci. Paris,* 295:45-48.

Caullery, M. 1953. Appendice aux Sporozoaires: Classe des haplosporidies (Haplosporidia Caullery et Mesnil 1899). *Traite de Zoologie,* 1:922-934.

Caullery, M. & Mesnil, F. 1899. Sur les

Aplosporidies, order nouveau de la classe des Sporozoaires. *C. R. Acad. Sci. Paris,* 129:616-619.

Caullery, M. & Mesnil, F. 1905a. Recherches sur les haplosporidies. *Arch. Zool. Exp. Gen., Ser.IV,* 4:101-181.

Caullery, M. & Mesnil, F. 1905b. Sur quelques nouvelles Haplosporidies d'Annélide. *C. R. Soc. Biol.,* 58:580-583.

Corliss, J. O. 1984. The kingdom Protista and its 45 phyla. *Biosystems,* 17:87-126.

Cornelissen, A. W. C. A., Overdulve, J. P. & Van Der Ploeg, M. 1984. Determination of nuclear DNA of five Eucoccidian parasites, *Isospora* (*Toxoplasma*) *gondii, Sarcocystis cruzi, Eimeria tenella, E. acervulina* and *Plasmodium berghei*, with special reference to gamontogenesis and meiosis in *I.* (*T.*) *gondii. Parasitology,* 88:531-553.

Debaisieux, P. 1920. *Haplosporidium (Minchinia) chitonis* Lank., *Haplosporidium nemertis,* nov. sp., et le groupe des haplosporidies. *Cellule,* 30:291-313.

Desportes, I. & Nashed, N. N. 1983. Ultrastructure of sporulation in *Minchinia dentali* (Arvy), an haplosporean parasite of *Dentalium entale* (Scaphopoda, Mollusca); taxonomic implications. *Protistologica,* 19:435-460.

Dinamani, P., Hine, P. M. & Jones, J. B. 1987. Occurrence and characteristics of the haemocyte parasite *Bonamia* sp. in the New Zealand dredge oyster *Tiostrea lutaria*. *Dis. Aquat. Org.*, 3:37-44.

Dyková, I., Lom, J. & Fajer, E. 1988. A new haplosporean infecting the hepatopancreas in the penaeid shrimp, *Penaeus vannamei. J. Fish Dis.,* 11:15-22.

Hillman, R. E., Ford, S. E. & Haskin, H. H. 1990. *Minchinia teredinis* n. sp. (Balanosporida, Haplosporidiidae), a parasite of teredinid shipworms. *J. Protozool.,* 37:364-368.

Hine, P. M. & Wesney, B. 1992. Interrelationships of cytoplasmic structures in *Bonamia* sp. (Haplosporidia) infecting oysters *Tiostrea chilensis*: an interpretation. *Dis. Aquat. Org.,* 14:59-68.

Labbé, A. 1896. Recherches zoologiques, cytologiques et biologiques sur les coccidies. *Arch. Zool. Exp. Gén. (Ser. 3),* 4:517-654.

La Haye, C. A., Holland, N. D. & McLean, N. 1984. Electron microscope study of *Haplosporidium comatulae* n.sp. (Phylum Acetospora: Class Stellatosporea), a haplosporidian endoparasite of an Australian crinoid, *Oligometra serripinna* (Phylum Echinodermata). *Protistologica,* 20:507-515.

Lange, C. 1993. Unclassified protists of arthropods: The ultrastructure of *Nephridiophaga periplanetae* (Lutz & Splendore, 1903) n. comb., and the affinities of the Nephridiophagidae to other protists. *J. Protozool.*, 40:689-700.

Lankester, E. R. 1885. Protozoa. *In*: *Encyclopaedia Britannica.* 9th ed. 19:830-866.

Larsson, R. 1987. On *Haplosporidium gammari*, a parasite of the amphipod *Rivulogammarus pulex*, and its relationships with the phylum Ascetospora. *J. Invertebr. Pathol.,* 49:159-169.

Lauckner, G. 1983. Diseases of Mollusca: Amphineura. *In*: Kinne, O. (ed.), *Diseases of Marine Animals.* II. Biologische Anstalt Helgoland, Hamburg, Germany. pp 963-975

Lom, J., Molnár, K. & Dyková, I. 1986. *Hoferellus gilsoni* (Debaisieux, 1925) comb. n. (Myxozoa, Myxosporea): redescription and mode of attachment to the epithelium of the urinary bladder of its host, the European eel. *Protistologica*, 22:405-413.

Manier, J.-F., Akbarieh, M. & Bouix, G. 1976. *Coelosporidium chydoricola* Mesnil et Marchoux, 1897: observations ultrastructurales, données nouvelles sur le cycle et la position systématique. *Protistologica*, 12:599-612.

McGovern, E. R. & Burreson, E. M. 1990. Ultrastructure of *Minchinia* sp. spores from shipworms (*Teredo* spp.) in the western north Atlantic, with a discussion of taxonomy of the Haplosporidiidae. *J. Protozool.*, 37:212-218.

Montes, J., Anadón, R. & Azevedo, C. 1994. A possible life cycle for *Bonamia ostreae* on the basis of electron microscopy studies. *J. Invertebr. Pathol.*, 63:1-6.

Ormières, R. 1980. *Haplosporidium parisi* n. sp., Haplosporidie parasite de *Serpula vermicularis* L. étude ultrastructurale de la spore. *Protistologica,* 16:467-474.

Ormières, R., Sprague, V. & Bartoli, P. 1973. Light and electron microscope study of a new species of *Urosporidium* (Haplosporida), hyperparasite of trematode sporocysts in the clam *Abra ovata. J. Invertebr. Pathol.,* 21:71-86.

Paulin, J. J. 1992. General comments on fixation of protozoa for transmission electron microscopy. *In:* Lee, J. J. & Soldo, A. T. (eds.), *Protocols in Protozoology*. Publ. by Society of Protozool. C-16.1.

Perkins, F. O. 1968. Fine structure of the oyster pathogen *Minchinia nelsoni* (Haplosporida,

Haplosporidiidae). *J. Invertebr. Pathol.*, 10:287-305.

Perkins, F. O. 1971. Sporulation in the trematode hyperparasite *Urosporidium crescens* De Turk, 1940 (Haplosporida: Haplosporidiidae)—An electron microscope study. *J. Parasitol.*, 57:9-23.

Perkins, F. O. 1975a. Fine structure of the haplosporidan *Kernstab*, a persistent, intranuclear mitotic apparatus. *J. Cell Sci.*, 18:327-346.

Perkins, F. O. 1975b. Fine structure of *Minchinia* sp. (Haplosporida) sporulation in the mud crab, *Panopeus herbstii*. *Mar. Fish. Rev.*, 37:46-60.

Perkins, F. O. 1979. Cell structure of shellfish pathogens and hyperparasites in the genera *Minchinia, Urosporidium, Haplosporidium* and *Marteilia*-taxonomic implications. *Mar. Fish. Rev.*, 41:25-37.

Perkins, F. O. 1990. Phylum Haplosporidia. *In*: Margulis, L., Corliss, J. O., Melkonian, M. & Chapman, D. J. (eds.), *Handbook of Protoctista*. Jones & Bartlett Publ. Boston, Massachusetts. pp 19-29

Perkins, F. O. 1991. Sporozoa. *In*: Harrison, F. W. & Corliss, J. O. (eds.), *Microscopic Anatomy of Invertebrates*. Wiley-Liss, New York, New York, 1:261-331.

Perkins, F. O. & van Banning, P. 1981. Surface ultrastructure of spores in three genera of Balanosporida, particularly in *Minchinia armoricana* van Banning, 1977—The taxonomic significance of spore wall ornamentation in the Balanosporida. *J. Parasitol.*, 67:866-874.

Perkins, F. O., Zwerner, D. E. & Dias, R. K. 1975. The hyperparasite, *Urosporidium spisuli* sp.n. (Haplosporea), and its effects on the surf clam industry. *J. Parasitol.*, 61:944-949.

Pichot, Y., Comps, M., Tige, G., Grizel, H. & Rabouin, M.-A. 1980. Recherches sur *Bonamia ostreae* gen. n., sp. n., parasite nouveau de l'huitre plate *Ostrea edulis* L. *Rev. Trav. Inst. Pêches Marit.*, 43:131-140.

Purrini, K. & Weiser, J. 1990. Light and electron microscopy studies on a protozoan, *Oryctospora alata* n. gen., n. sp., (Protista: Coelosporidiidae) parasitizing a natural population of the Rhinoceros beetle *Oryctes monoceros* Oliv. (Coleoptera: Scarabaeidae). *Zool. Beitr.*, 33:209-220.

Purrini, K., Weiser, J. & Kohring, G.-W. 1988. *Coelosporidium tangae* n. sp. (Protista), a new protist parasitizing a natural population of a field cockroach, *Blatta* sp. (Blattaria). *Arch. Protistenkd.*, 136:273-281.

Scro, R. A. & Ford, S. E. 1990. An electron microscopic study of disease progression in the oyster, *Crassostrea virginica*, infected with the protozoan parasite, *Haplosporidium nelsoni* (MSX). *In*: Perkins, F. O. & Cheng, T. C. (eds.), *Pathology in Marine Science*. Academic Press, San Diego, California. pp 229-254

Sogin, M. L. 1989. Evolution of eukaryotic microorganisms and their small subunit ribosomal RNA's. *Amer. Zool.*, 29:487-499.

Sprague, V. 1963. Revision of genus *Haplosporidium* and restoration of genus *Minchinia* (Haplosporidia, Haplosporidiidae). *J. Protozool.*, 10:263-266.

Sprague, V. 1970. Recent problems of taxonomy and morphology of Haplosporidia. *J. Parasitol.*, 56:(Sec. II, Part I):327-328.

Sprague, V. 1978. Comments on trends in research on parasitic diseases of shellfish and fish. *Mar. Fish. Rev.*, 40:26-30.

Sprague, V. 1979. Classification of the Haplosporidia. *Mar. Fish. Rev.*, 41:40-44.

Sprague, V. 1982. Acetospora. *In*: Parker, S. P. (ed.), *Synopsis and Classification of Living Organisms*. McGraw-Hill Book Co., New York, New York. 1:599-601.

Toguebaye, B. S., Manier,J.-F., Bouix, G. & Marchand, B. 1986. *Nephridiophaga ormieresi* n. sp., protiste parasite d'*Aspidomorpha cincta* Fabricius, 1781 (Insecte Coléoptère: Chrysomelidae). Étude ultrastructural. *Protistologica*, 22:317-325.

Wainright, P. O., Hinkle, G., Sogin, M. L. & Stickel, S. K. 1993. Monophyletic origins of the Metazoa: an evolutionary link with fungi. *Science*, 260:340-342.

Woolever, P. 1966. Life history and electron microscopy of a Haplosporidium, *Nephridiophaga blattellae* (Crawley) n. comb., in the Malpighian tubules of the German cockroach, *Blattella germanica* (L.). *J. Protozool.*, 13:622-642.

PHYLUM PLASMODIOPHORA
(PLASMODIOPHOROMYCOTA)

JAMES P. BRASELTON

The plasmodiophorids include 10 genera in a single family. They are treated as Protozoa or as fungi, and have been classified up to the rank of Phylum (Phylum Plasmodiophora = Plasmodiophoromycota, with one class, Plasmodiophorea (Plasmodiophoromycetes), one order, Plasmodiophorida (Plasmodiophorales); with the family Plasmodiophoridae (Plasmodiophoraceae).

The relationship of Phylum Plasmodiophora to other taxa is not known, and in past classifications the group has been included in the protoctists (Margulis et al., 1989, Olive, 1975) or fungi (Sparrow, 1960, Waterhouse, 1972). The informal term "plasmodiophorids" often is used for the group since there are a variety of formal names for phylum, class, order, and family, depending on whether the group is considered in the protozoa or fungi. The group was reviewed by Cook (1933), Karling (1968), and Dylewski (1989).

Regardless of where the plasmodiophorids are classified, they are a discrete taxonomic unit, and may be considered a monophyletic group: all members share the derived character state **cruciform nuclear division**, a type of mitotic division in which a persistent nucleolus is elongated perpendicularly to the metaphase plate of chromatin, centriolar pairs occur at each pole, and the nuclear envelope remains intact through metaphase (Fig. 1). Other shared character states for the group include a) **zoospores** with two, anterior **whiplash flagella** (Fig. 2); b) multi-nucleated protoplasts (generally referred to as **plasmodia**); c) obligate, intracellular parasitism; and d) environmentally resistant resting spores (cysts) (Figs. 3,4). Several plasmodiophorids cause hypertrophy and (or) hyperplasia of host tissues, producing either galls or distortion of the infected organ.

Economically significant members include *Plasmodiophora brassicae* Woronin, the causative agent of club root of cabbage and other brassicaceous (cruciferous) crops; *Spongospora subterranea* (Wallroth) Lagerheim f. sp. *subterranea* Tomlinson, the causative agent of powdery scab of potato; *S. subterranea* (Wallroth) Lagerheim f. sp. *nasturii* Tomlinson, the causative agent of crook root of watercress; and *Polymyxa betae* Keskin, which, along with a virus, is associated with rhizomania of sugar beet. Also, *Polymyxa graminis* Ledingham, *P. betae*, and *S. subterranea* serve as vectors for pathogenic viruses of crops including barley, wheat, oats, peanuts, potatoes, and watercress (Cooper and Asher, 1988).

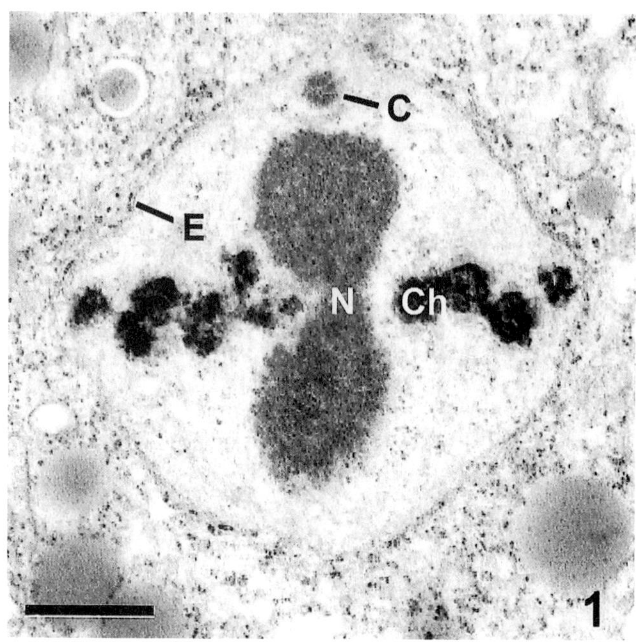

Fig. 1. Transmission electron micrograph of *Tetramyxa parasitica.* Cruciform nuclear division has intact nuclear envelope (E), centrioles (C) at the poles, and a persistent nucleolus (N) elongated perpendicularly to the metaphase plate of chromatin (Ch). Scale=1 µm.

The terminology for structures and stages in life cycles has been varied and confusing; terminology used here is based on that advocated by Karling (1981), with common synonyms included within parentheses. There are two major phases in the life cycles of plasmodiophorids: **sporogenic** (cystogenous or secondary) and **sporangial** (sporangiogenous or primary). Each phase is initiated when a single zoospore encysts on a host cell and injects the zoospore contents through the host cell wall and plasma membrane into host cytoplasm. Each phase has plasmodia as growth forms and cruciform nuclear divisions during growth of plasmodia.

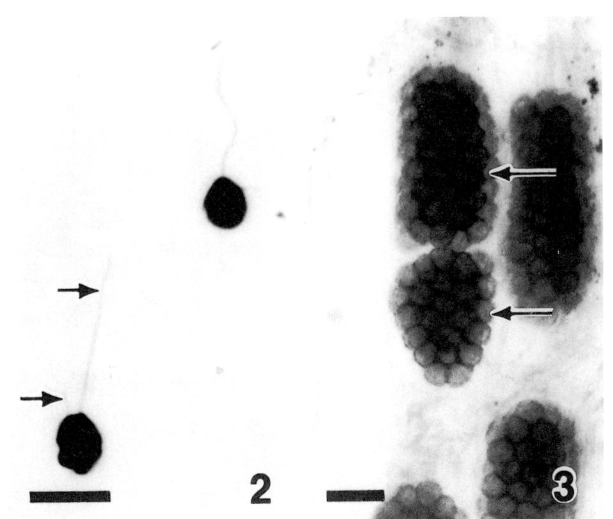

Fig. 2. Optical micrograph of *Spongospora subterranea*. Secondary zoospores have been air-dried and stained with methylene blue to show two, heterokont, anterior flagella (arrows). Line scale=5 µm. Fig. 3. Optical micrograph of *Ligniera verrucosa* in root of *Veronica* sp. Sporosori (arrows) are stained with cotton blue. Scale=10 µm.

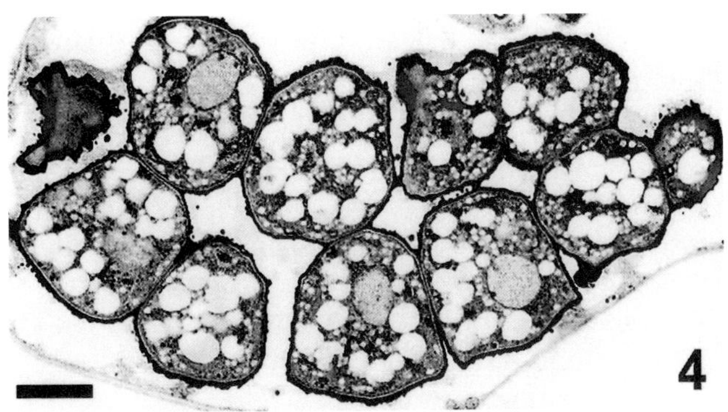

Fig. 4. Transmission electron micrograph of resting spores of *Polymyxa graminis* in root of *Triticum monococcum*. Scale=2 µm.

Non-cruciform divisions in sporogenic plasmodia are considered to be meiosis since synaptonemal complexes are observed in prophases (Fig. 5), hence resting spores are believed to be haploid. Haploid chromosome numbers of members of most genera have been determined though transmission electron microscopic analysis of serial sections of synaptonemal complexes (Braselton, 1992). Resting spores may be grouped into **sporosori** (**cystosori**), the morphologies of which are the major taxonomic characters for the genera (see Key to the Genera). During germination each resting spore releases one primary zoospore, which upon infection of a host cell develops into a sporangial plasmodium.

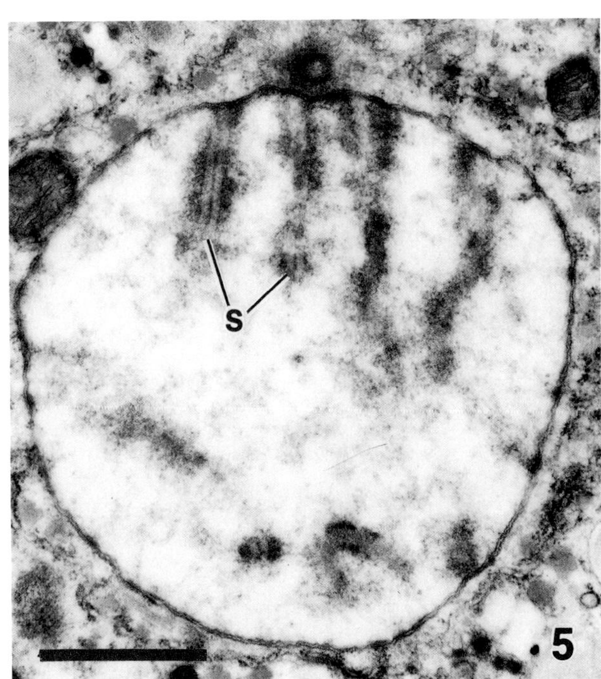

Fig. 5. Transmission electron micrograph of *Polymyxa betae*. During cleavage of sporogenic plasmodia into resting spores, nuclei contain synaptonemal complexes (S), indicators of prophase I of meiosis. Scale=1 µm.

Identification of plasmodiophorids in hosts often can be accomplished through compound optical microscopy of free-hand sections of gall material mounted in water. Similarly, small scrapings from herbarium sheets can be used to identify genera based on the groupings of resting spores. In some cases pieces of roots may be mounted in water or in lacto-phenol with a stain such as cotton blue for bright-field optical microscopy to

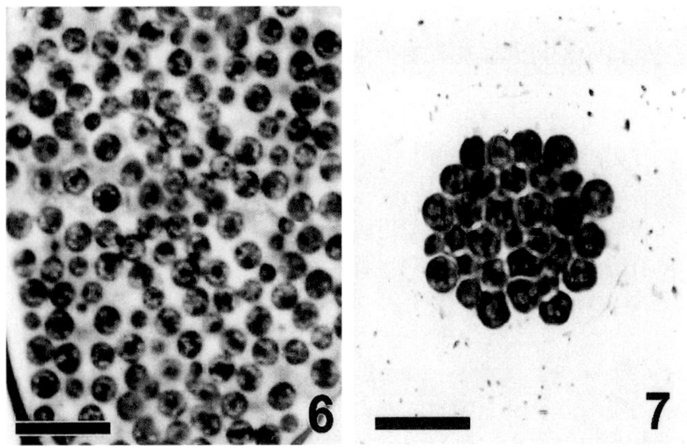

Fig. 6. Optical micrograph of *Plasmodiophora brassicae* resting spores in root of Chinese cabbage. Line scale=10 µm. Fig. 7. Optical micrograph of sporosorus of *Woronina pythii* in cell of *Pythium* sp. Scale=10 µm.

observe plasmodia, sporangia, and (or) **sporosori** (Fig. 3). **Sporangial plasmodia** produce thin-walled **zoosporangia** which contain secondary **zoospores**. After cruciform divisions have ceased, non-cruciform nuclear divisions occur in secondary plasmodia during cleavage of the plasmodia into sporangial lobes and subsequently into incipient, secondary zoospores. Non-cruciform divisions in sporangial plasmodia are not meiotic as are those in **sporogenic plasmodia**. Secondary zoospores produced by sporangial plasmodia are similar to primary zoospores. Upon infection of host cells, secondary zoospores may develop into either sporangial or sporogenic plasmodia. Although meiosis is documented to occur during cleavage of sporogenic plasmodia into resting spores, convincing evidence for location of karyogamy in the life cycle is lacking. Material for optical microscopy should be fixed and embedded in plastics following procedures for transmission electron microscopy, sectioned 0.5-1 mm thick, and stained with toluidine blue (Figs. 6-11).

KEY TO THE GENERA

1. Resting spores occurring singly or in loose masses................................ ***Plasmodiophora***
1'. Resting spores in sporosori.......................... 2

2. Sporosori definite in size and shape............... 3
2'. Sporosori indefinite in shape and size........... 8

3. Sporosori with four or eight resting spores... 4
3'. Sporosori with more than eight resting spores 5
4. Four resting spores per sporosorus.................. .... ***Tetramyxa***
4'. Eight resting spores per sporosorus ***Octomyxa***

5. Sporosori ellipsoidal to spherical, spongy or hollow within.. 6
5'. Sporosori disk-shaped.................................... 7

6. Sporosori hollow sphere............ ***Sorosphaera***
6'. Sporosori spongy, ellipsoidal to spherical........ .... ***Spongospora***

7. Sporosori two-layered disks........ ***Sorodiscus***
7'. Sporosori single-layered disks......................... .... ***Membranosorus***

8. Infect protists................................. **Woronina**
8'. Infect roots of vascular plants....................... 9

9. Walls between lobes of zoosporangia partially or completely dissolve, resulting in a single, multilobed zoosporangium from a single thallus .... ***Polymyxa***
9'. Walls between lobes remain intact, delineating each lobe as an indivual zoosporangium within a zoosporangiosorus.......................... ***Ligniera***

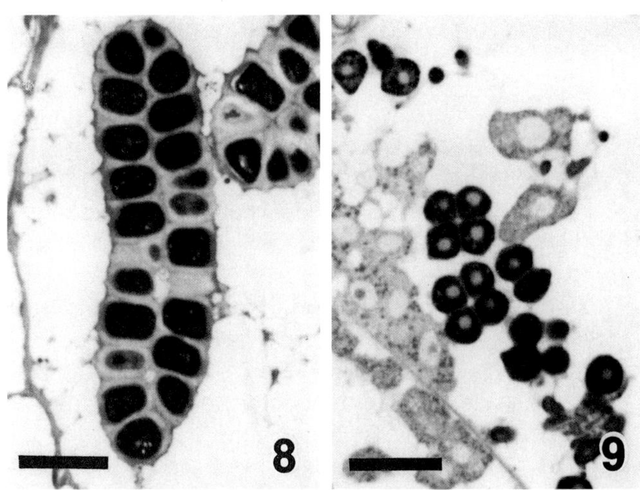

Fig. 8. Optical micrograph of disk-shaped sporosorus of *Sorodiscus callitrichis* in stem cell of *Callitriche* sp. Section sliced through disk showing two layers of resting spores. Line scale=10 µm. Fig. 9. Optical micrograph of four-celled sporosori of *Tetramyxa parasitica* in stem cell of *Ruppia maritima*. Line scale=10 µm.

LITERATURE CITED

Braselton, J. P. 1992. Ultrastructural karyology of *Spongospora subterranea* (Plasmodiophoromycetes). *Can. J. Bot.,* 70:1228-1233.

Cook, W. R. I. 1933. A Monograph of the Plasmodiophorales. *Arch. Protistenkd.,* 80:179-254.

Cooper, J. I. & Asher, M. J. C. (eds.). 1988. *Viruses with Fungal Vectors*. The Association of Applied Biologists, Wellesbourne, Warwick, United Kingdom.

Dylewski, D. P. 1989. Phylum Plasmodiophoromycota. *In*: Margulis, L., Corliss, J. O., Melkonian, M. & Chapman, D. J. (eds.), *Handbook of Protoctista*. Jones and Bartlett Publishers, Boston. pp 399-416

Karling, J. S. 1968. *The Plasmodiophorales*. 2nd ed. Hafner Publishing Co., New York.

Karling, J. S. 1981. *Woronina leptolegniae* n. sp., a plasmodiophorid parasite of *Leptolegnia. Nova Hedwigia,* 35:17-24.

Margulis, L., Corliss, J. O., Melkonian, M. & Chapman, D. J. (eds.). 1989. *Handbook of Protoctista.* Jones and Bartlett Publishers, Boston.

Olive, L. S. 1975. *The Mycetozoans*. Academic Press, New York and London.

Sparrow, F. K. 1960. *Aquatic Phycomycetes*. 2nd ed. University of Michigan Press, Ann Arbor.

Waterhouse, G. 1972. Plasmodiophoromycetes. *In*: Ainsworth, G. C., Sparrow, F. K. & Sussman, A. S. (eds.), *The Fungi.* Vol. 4B. Academic Press, New York and London. pp 75-82.

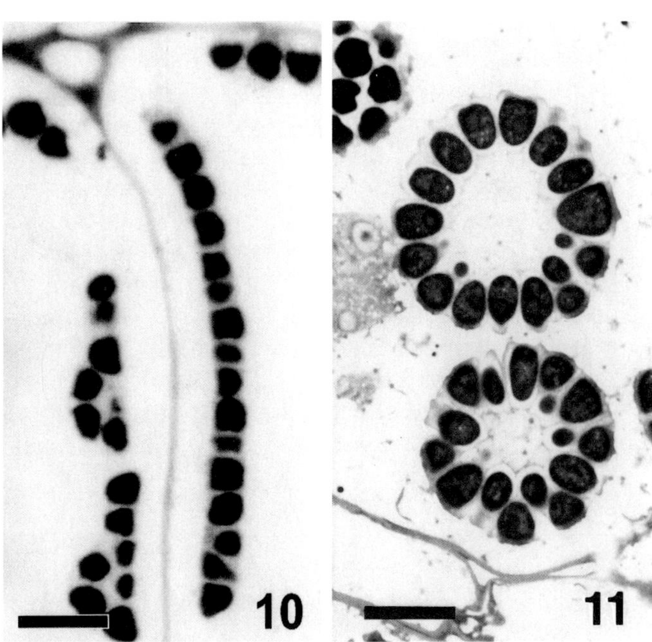

Fig. 10. Optical micrograph of thin sheets of resting spores of *Membranosorus heterantherae* in root cell of *Heteranthera dubia*. Line scale=10 µm Fig. 11. Optical micrograph of sections through spherical sporosori of *Sorosphaera veronicae* in stem cell of *Veronica* sp. Line scale=10 µm.

AN ANNOTATED GLOSSARY OF PROTOZOOLOGICAL TERMS

Updated from the first edition by JOHN J. LEE with the aid of many of the contributers to this edition

First Edition Version by John O. Corliss and Jiri Lom

Because of space restriction, this glossary, regretfully, is limited in scope. Unlike the glossary written for the first edition, which was written out of context with the rest of the contents of the book, this glossary is derived from terms in each chapter. These terms were placed in bold and are defined here, or in context in the chapter where they are found. Many words considered to be of a very general biological nature (e.g., centriole, chromosome, cytoplasm, endoplasmic reticulum, karyoplasm, meiosis, mitosis, nucleus), or of minor importance, or rare usage were purposely omitted. Little known or long discarded synonyms are usually left out, and popular or persisting (even if superseded) synonyms are generally given by cross reference. With rare exception, concepts or hypotheses are excluded. Also, lack of space precludes inclusion of detailed chemical terminology.

The authors of the first edition's glossary freely consulted earlier glossaries published by experts in subfields of protozoology (and of phycology and mycology) as well as relying heavily on research papers for an understanding of many entries.

I wish to express deep and sincere thanks to various colleagues and students—and to authors of the chapters for their helpfulness, along the way, in providing critical remarks concerning many of the descriptions presented on the following pages. In the last analysis, however, the author must assume responsibility for the overall coverage (i.e., for the particular terms finally chosen from the chapters for inclusion) and for the definitions themselves. It is truly amazing how terminology changes or requires refinement with increase in knowledge over even a relatively short period of time. Comments from users of this glossary are most warmly invited and will be deeply appreciated.

Abdomen: See **Apophyses**

Aboral, Aboral Pole: Surface, side, or pole away (or opposite) from the oral area; often, but not necessarily, this means the posterior or antapical end of the organism.

Acephaline: See **Aseptate,** when the word is used in reference to certain gregarines.

Actinopodia: A type of pseudopodia which are thin and usually radiate outward from the central pores of polycystines. Found also in phaeodarians, acantharians, and heliozoans. Actinopods are often supported by skeletal spines or in the case of heliozoans, by microtubular arrays. They serve to capture food and often form anastomosing webs around prey.

Acronematic Flagellum: Flagellum with a single, terminal mastigoneme or flagellar hair.

Actinophore: Structure bearing several or a bundle (or fascicle) of suctorial tentacles; characteristic of certain suctorian ciliates.

Adhesive Disc: Cup-shaped organelle at the aboral pole of mobiline peritrichs and some other ciliates that is used for attachment to the substrate (usually the surface of another organism serving as host; also see **Sucking Disc).**

Adhesive Organelle: Term often restricted to a secretory structure or system that is responsible for or structurally involved in the production of a substance allowing the organism possessing it to adhere or attach to the substrate; also see **Holdfast Organelle.**

Adoral Zone of Membranelles: Orderly arrangement of three or more membranelles *sensu lato* that are serially arranged along the left side of the oral area of a ciliate; the AZM is typically found in a buccal cavity or peristome; a synonym for the AZM of some ciliates is the adoral ciliary spiral.

Aethalium (plural, aethalia): Relatively large, sessile, round or mound-shaped fruiting body of plasmodial mycetozoans. It is formed from one plasmodium (e.g. *Lycogala* and *Fuligo*). This largest myxogastric fruiting body sometimes exceeds 20 cm across (*Fuligo septica*). The world's record specimen measured 70 X 5 4 cm growing on sawdust in Gotland, Sweden.

Afferent Canal: Cytoplasmic channel (usually several in number) active in ciliates in transporting fluid secreted by the spongiome or nephridioplasm to a contractile vacuole; it is also known as a pulsating, nephridial, collecting, or radial canal.

Agglutinated Shells (tests): Test of a foraminiferan, ciliates, testate amoebae, and other protozoa which is composed of small sedimentary particles which have been gathered from the environment and cemented onto an organic basal envelope. The cement may be mineralized itself.

Aggregation Center: Site of convergent streams of amoebae or amoebulae in the life cycle of certain cellular slime molds (e.g., the acrasian *Dictyostelium).* The aggregation is caused by a chemical attractant (e.g., cAMP) that is secreted by the amoebae themselves.

Aggregated pedicel: In choanoflagellates the aggregated pedicel is constructed from linearly arranged and closely packed costal strips. Costal strips diverging from the posterior end of the lorica, like a tripod, constitue a **compound pedicel.**

Akaryomastigont: Condition of a flagellate with its mastigont(s) lacking (an association with) a nucleus; this is characteristic of certain polymastigont trichomonads and hypermastigotes (and the opalinids as well); contrast with **Karyomastigont.**

Akontobolocyst: Extrusome found in certain dinoflagellates that is considered by some to be a kind of fusiform trichocyst; the term is also spelled acontobolocyst; see **Trichocyst.**

Alveolus, Pellicular (alveolar sacs): Flattened vesicle or sac, bound by a unit membrane, lying just beneath the surface of the plasma membrane of ciliated protozoa, with a layer of epiplasm often below it; such alveoli are also present in dinoflagellates, enclosing the thecal plates in "armored" forms.

Amastigote stage: Developmental stage or morphological form of a trypanosomatid flagellate in which the flagellar base is anterior to the nucleus but there is no emergent flagellum; this ovoid form is the intracellular stage in the vertebrate host in the life cycle of species of the genus *Leishmania,* and it has been known classically as the **Leishmanial Stage.**

Amoebula: Small amoeboid stage following reproduction in certain rhizopod amoebae; it is also a common name for the form found prior to aggregation in cellular slime molds and on germination of the spore in plasmodial slime molds; the term has also been used in the past for the sporoplasm of micro- and myxosporidians.

Amphiesma: Special term for the cellular covering or pellicle of a dinoflagellate (in its so-called "vegetative" stage), comprised of only membranes in the "naked" species or thecal plates enclosed in membranes in "armored" species (amphiesmal [thecal] plates).

Amphosome: Widely used synonym of Paranucleus.

Ampulla: "Glandular" organelle or organellar complex in some hypostome ciliates that produces a thigmotactic substance or **structure (see Adhesive Organelle),** but the term is also used for a collecting canal (or its enlarged distal end) feeding into the contractile vacuole of certain (other) ciliates.

Anarchic Field: Unorganized array of barren (nonciliferous) kinetosomes that gives rise to the infraciliary bases of major oral organelles during early stages of stomatogenesis in some ciliates.

Anisogamous gametes: Referring to gametes (or gametic cells) that consistently differ in size, the larger designated "female" and the smaller "male"; contrast with **Isogamous.**

Anisokont flagella: Organism with flagella of unequal length. Reference is usually to biflagellate species. This term is preferable to **Heterokont,** which now is given a quite *different* meaning.

Annulus: General term for a girdle or an equatorial belt, band, or groove; its meaning is sometimes modifed for specific uses by specialists on the various major protozoan taxa: for example, it is the transverse or helical groove or constriction (covered by a plate called the **Cingulum** in thecate species) characteristic of many dinoflagellates.

Antapex: The posterior extremity or area of a dinoflagellate when oriented in the direction of swimming.

Antapical series. A latitudinal series of plates in dinoflagellates at antapex of the cell.

Anterior intercalary plate(s): Plate(s) in a dinoflagellate theca which lies between the apical and precingular series of plates.

Anterior sulcal plate: The sulcal plate lying closest to the apex and usually aligned with the proximal end of the cingulum of a dinoflagellate.

Aperture: General term for an opening that is much larger than that of a pore; it is specifcally used for a major opening of a foraminiferan test (there can be primary, secondary, umbilical, sutural, terminal, and trematophore apertures) through which the granular reticulopodia emerge.

Apertural lip: Projection of chamber wall above and along an interiomarginal aperture.

Apertural tooth (denticulate aperture): Projection of wall into the test aperture. It may be simple or complex, single or multiple.

Apex: The anterior extremity or area of a dinoflagellate when oriented in the direction of swimming.

Aphanoplasmodium: Mycetozoan plasmodial type characterized in its early stages of development by a network of flattened, thread-like, transparent strands which are almost invisible and lack polarity and directional movement. At this time there is an apparent affinity for growing submerged in water or under very wet conditions. The young strands lack a distinct ectoplasm and endoplasmic region and the protoplasm is not coarsely granular. Members of the order Stemonitida all have this type of plasmodium.

Apical Complex: Anteriorly located ultrastructures or organelles (e.g. polar rings, conoid, rhoptries, micronemes) characteristically found in some stage of the life cycle of all sporozoan (apicomplexan) species.

Apical pore: A large pore or series of pores at the apex of the episome of peridiniphycideans (dinoflagellates).

Apical series: A latitudinal series of plates in dinoflagellates which is in contact with the apical pore or apical pore complex.

Apical Funnel: Distally drawn-out anterior end of many chonotrich ciliates, sometimes lined with ciliature and leading posteriorly to the cytostome.

Apokinetal: Type of stomatogenesis in ciliates in which the involved kinetosomes have no apparent preassociation with either somatic kineties or the parental oral apparatus, as in entodiniomorphids and many spirotrichs; the origin of the kinetosomal anlage appears to be of a *de novo* nature.

Aplanospore: A non-motile dispersal cell.

Apophyses: Sometimes forked or branched rods, apophyses extend outward from the center of the skeleton of radiolarian. **Cephalis**: In Nassellaria, the spicule is a complex structure, basically tripodal in design, and often forming the major scaffolding of the skeleton, either as the main skeletal element or as a central structure to which other structures are attached or enclosed (hidden) within a porous (simple or segmented) shell. The shell may consist of only a basal tripod; the tripod and associated ring(s), a single chamber known as the cephalis; or two or more chambers connected serially, designated cephalis, thorax, abdomen, and postabdominal segments.

Aragonite: Orthorhombic crystal form of calcium carbonate.

Arboroid Colony: Individual cells or zooids are disposed in a branching or tree-like manner, typically interconnected either by stalks or by their loricae; found among various flagellate and ciliate groups, most strikingly exhibited by species of several major genera of peritrich ciliates; synonyms, dendroid and **Dendritic Colony.**

Areal aperture: Aperture in face of final chamber.

Areolate surface: Term used to describe the surface of tests of some amoebae (Microchlamyiidae, Arcellidae) constructed of regularly arranged hollow building units.

Argentophilia: "Silver-loving," taking silver stains positively; many such organelles and structures are thus made visible by **Silver Impregnation Techniques; see Argyrome.**

Armored dinoflagellates: Dinoflagellates with a theca or thecal plates in their amphiesmal vesicles.

Argyrome: Whole system of pellicular or cortical structures, organelles, or markings that are argentophilic and thus revealable by application of techniques of silver impregnation; it is well known in ciliates and revealed in some dinoflagellate and sporozoan species; synonym, **Silverline System;**

Aseptate: Condition of there being no partitions, no septa dividing a body into segments; this is commonly the case among many protozoan groups, but the term is mostly used by protozoologists to distinguish certain gregarine sporozoa that exhibit such an "acephaline" condition (whereas certain other species of gregarines *do* have septa); contrast with **Septate.**

Asexual Reproduction: Reproduction by means of fission or budding, binary or multiple (e.g., schizogony), of a single parental organism or by production of motile or nonmotile spores (from uni- or multicellular sporangia). Such methods of multiplication do *not* directly involve meiosis, gametes, syngamy (zygote-formation), etc. (therefore, there is no change in ploidy) which are associated with sexual reproduction. Asexual reproduction is widespread among protozoa, with many entire taxonomic groups seemingly reproducing *only* that way.

Astomy: Condition of being mouthless, without a true cytostome or oral feeding apparatus of any kind; the species of some entire taxonomic groups of the protozoa are astomatous.

Astropyle, astropylum: Main opening of the central capsule in phaeodarian actinopods; it is usually accompanied by two or more secondary orifices or accessory tubular apertures, the **Parapyles.**

Atractophore: Prominent, if transient, fibrillar, rod-like structure, arising from the basal body complex, that is or serves the role of a centriole or centrosome in the formation of the mitotic spindle fibers (microtubules) in certain trichomonad flagellates and in some foraminiferans and radiolarians; synonym, **Batacchio.**

Attachment Knob: Enlarged distal end of a suctorial tentacle of a suctorian or a sucking tube of rhynchodid hypostome ciliates; it adheres to or embeds itself in the body of the prey or host cell; when the knob is particularly prominent, the tentacle bearing it is said to be capitate; in suctorians, it contains haptocysts.

Attachment Organelle: Nonspecific name for all sorts of stalks, filaments, tentacles, knobs, hooks, crochets, spines, mucrons, suckers, discs, or thigmotactic fields used in a temporary or permanent manner to attach an individual (or colony) to some living or inanimate substrate; see **Holdfast Organelle.**

Autogamy, autogamic reproduction: In ciliates and foraminifera, a self-fertilization type of sexual phenomenon involving only a single individual and thus to be contrasted with **Conjugation;** for some of the other protozoan groups, it refers to a kind of sexual reproduction in which a zygote is formed by the fusion of two haploid nuclei from one individual.

Auxokineties: Perioral somatic kineties of densely arranged kinetids, to the right of the oral cavity in ciliates such as loxocephalids and frontoniids.

Axoneme: See **Cilium** or **Flagellum.**

Axonemes: The **axopodia** of heliozoans are supported internally by geometric arrays of microtubules (axonemes).

Axoplast: See **Central Granule.**

Axopodium (plural Axopodia): Straight, slender pseudopodium extending radially from the surface of the body. Some possess an axial rod (sometimes called an axoneme) composed of microtubules. Others are supported by spines. Axopodia are characteristic of the actinopod sarcodines (heliozoa, radiolarians, etc.), where they are used primarily in feeding rather than in locomotion. Synonym, actinopodium.

Axostylar Capitulum: Anterior, flattened part of the axostyle characteristic of certain trichomonads.

Axostylar Granules: Granules aligned along the axostyle in trichomonads which are stained by hemotoxylin. Superseded by the term **Hydrogenosomes.**

Axostylar Trunk: Postnuclear part of the axostyle, which may form a tube by a recurving of the microtubular sheet on itself.

Axostyle: Rod-like, often prominent, supporting organelle composed of microtubules arranged in sheets or ribbons, present as part of the mastigont system of various flagellates. In some species it runs through the body from the apical end toward, even piercing, the posterior pole of the organism.

AZM: See **Adoral Zone of Membranelles.**

Barren Kinetosome: Basal body not associated (never or for only a given period of time in the life cycle) with a cilium or flagellum; at such a time, it is said to exhibit a nonciliferous (or aciliferous) or naked state; for certain kinetosomes, this condition is particularly common during some stages of stomatogenesis in quite a number of ciliates.

Basal Body: Kinetosome, blepharoplast (of flagellates), centriole, etc.; a popular usage of the term is as a synonym of the kinetosome of ciliated protozoa; a long time superseded (and inaccurate) synonym is basal granule.

Basal Disc: See **Adhesive Disc.**

Basal Fibers: See **Basal Microtubules.**

Basal Microtubules: Set, group, ribbon, or bundle (though very few in number) of microtubules extending along the left side of the somatic kinetics at the level of the base (proximal end) of the kinetosomes of a kinety; they are found in a number of oligohymenophoran ciliates.

Basal Plate: See **Kinetosome.**

Batachio: See **Atractophore.**

Biconvex: Both sides of a spiral foraminiferal test are convex or inflated.

Bilamellar: Chamber wall of hyaline calcareous foraminiferan test consisting of two primarily formed layers, deposited on each side of organic layer, the outer layer extending over the previously formed test, resulting in secondary lamellae.

Biloculine: Test of a foraminiferan with only two chambers visible externally.

Biumbonate: Enrolled test of a foraminiferan with central elevated umbilical boss on both sides.

Biserial: Foraminiferan test formed by alternating two rows of chambers from side to side.

Blepharoplast: Older term for basal body of flagellates; typically associated with a flagellum and sometimes additional structures; the term is replaced by **Basal Body** or **Kinetosome.**

Border Membrane: Finely striated circumferential band, with fibers, proteinaceous radial pins, etc., associated with and reinforcing the **Adhesive Disc** of mobiline peritrich ciliates.

Bradyzoite: Slowly developing merozoite; the term is especially used for some stages in the life cycles of species of certain genera (e.g., *Sarcocystis, Toxoplasma)* of coccidian sporozoa.

Bristles: Common name for generally single lengthy rather stiff cilia of several kinds and functions (sensory, tactile, thigmotactic, locomotor, attachment, etc.) occurring on one or more parts of the body of various ciliates.

Brosse (brosse ciliature, brosse kineties): Distinctive "brush" of cilia arising from some specialized short kineties or kinetal segments, often oriented obliquely to the body axis, which is present on the anterodorsal surface of nondividing individuals belonging to certain genera of ciliates (e.g., *Prorodon).*

Brosse cleft: Groove in the somatic cortex of some nassophorean ciliates in which the brosse resides.

Buccal Apparatus: Whole complement of compound ciliary organelles, the bases of which are located in or associated with the oral area (specifically, the buccal cavity or peristome) in oligohymenophoran and polyhymenophoran ciliates; it includes the paroral membrane(s) and membranelles *sensu lato* (plus homologues and possible nonhomologues of these structures) and their infraciliary bases (=buccal infraciliature); the whole apparatus functions primarily in food-getting, sometimes in locomotion.

Buccal Area: Region around the mouth in those ciliates that possess a **Buccal Apparatus;** therefore, strictly speaking, this designation is not a synonym of the much broader and more generalized term **Oral Area.**

Buccal Cavity: Pouch or depression at or near the apical end of the body and/or on the ventral surface that contains compound ciliary organelles and leads inwardly to the organism's cytostome-cytopharyngeal complex; typically quite deep though sometimes secondarily flattened out or everted, it may have a specialized, still deeper funnel-like portion known as the **Infundibulum;** it is especially characteristic, under the name "buccal cavity," of many oligohymenophoran ciliates, but it is considered to be the structural equivalent of the peristome (a kind of everted buccal cavity) of peritrichs and polyhymenophorans.

Buccal Ciliature and Infraciliature: See **Buccal Apparatus.**

Buccal Membranelles: See **Membranelle.**

Buccokinetal: Type of stomatogenesis in ciliates in which (at least some of) the fields of kinetosomes involved have an apparent origin from organelles of the parental buccal apparatus *sensu lato,* as in many hymenostomes, peritrichs, and spirotrichs; superseded synonyms, autonomous and semi-autonomous.

Bud, Budding: In ciliates and some monothalmic foraminifera, the filial product of single or multiple fission that is characteristically much smaller than the parental form and typically quite unlike it in both form and function. In ciliates it is generally ciliated and plays a migratory role in the organism's life cycle. A number of methods of budding exist among the various major protozoan taxa. In ciliates, the phenomenon occurs universally among suctorians and chonotrichs and is not uncommonly found among rhynchodid hypostomes, peritrichs, and apostomes.

Calyculus: The lower portion of the sporangium remaining as a persistent cup-like or saucer-like base in some species of mycetozoans (e.g.*Cribraria* and *Arcyria*).

Calcareous Bodies: deposits of calcium carbonate ($CaCO_3$) characteristically found on the peridium, stalk, columella, or in the capillitium of many species in the Physarida. In *Physarum* expansions in the capillitial threads contain granules of calcium carbonate (lime knots) interconnected by noncalcareous threads. The use of the word lime (calcium oxide, CaO) in myxogastrid literature is incorrect when it is substituted for calcium carbonate. In *Badhamia* the capillitium is a network of calcareous tubules.

Calymma: Vacuolated region of the cytoplasm around the central capsule of radiolarians, functioning to keep the organism afloat.

Calyx: Cup-shaped structure surrounding the single nucleus in hypermastigote flagellates of the genus *Lophomonas;* it is composed of microtubular lamellae.

Canal: Anterior portion of the apical invagination in euglenoid flagellates; it is followed by the reservoir. Many rotallid foraminifera have canal systems (e.g. elphidiids, calcarinids, nummulitids)

Capitate Tentacles: A tentacle having a swollen tip.

Capitulum: Term with two distinctly different meanings 1) Amorphous material capping the proximal ends of the nematodesmata in some cyrtophorid hypostome ciliates; when prominent in appearance, these capitula are sometimes called maxillae or teeth; 2) the expanded anterior end of the axostyle which, in some parabasalians, cups the organism's nucleus.

Capillitium: a system of sterile threads formed within the spore mass of mycetozoans (e.g. *Stemonitis*).

Capsulogenic Cell: Cell (sometimes more than one) within a myxosporidian or actinomyxidian sporoblast producing the **Polar Capsule.**

Carina: Peripheral keel of the test of an imperforate foraminiferan test.

Case: See **periplast**.

Catenulation, Catenoid Colony: Temporary line or chain of individuals brought about by repeated (and generally anisotomic) binary fissions without separation of the resulting filial products; the most common examples known among the protozoa occur in certain ciliate and dinoflagellate species.

Caudal Appendage: Almost any posterior extension of an organism's body; but the term is particularly used in protozoology to indicate the posterior ends of the shell valves in myxosporidian spores and the posterior extension of the spore wall in certain microsporidians.

Caudal Cilium: Distinctly longer somatic cilium (occasionally more than one) at or near the posterior or antapical pole of a ciliate; it is sometimes used for temporary attachment to the substrate.

Caudiform Division: See **Nudiform Division**

Central Capsule: Nucleus-containing central mass of cytoplasm which is surrounded by a spheroidal central capsular wall sharply delimiting the ectoplasm from the endoplasm in radiolarians. Often with numerous apertures or **Fusules,** uniformly distributed or clustered at one pole, for outward passage of the axial rods of the axopodia.

Central Granule: Structure in the center of the body of some heliozoa and radiolarians from which the axial rods of the axopodia radiate; synonyms: **axoplast** and **centroplast**.

Centroplast: See **Central Granule.**

Cephaline Gregarine: See **Septate Gregarine.**

Cephalis: Innermost framework or first division (nearest the center) of the segmented lattice shell of many nassellarid species of the polycystine radiolarians; also see **Skeletons, Internal.**

Cephalized, Cirri Transverse: Cirri oriented transversely to the long axis of the ciliate.

Chamber: Test cavity of foraminifera formed at one instar and always connected to other cavities or to exterior by pores, intercameral foramina, or stolons.

Chamberlet: Subdivision of primary chamber of foraminifera by axial or transverse septula or other partitions.

Chromidium: Posterior denser zone of arcrelinids which stains with basic dyes.

Choanomastigote: Developmental stage or morphological form of a trypanosomatid flagellate in which the flagellum arises just anterior to the nucleus from a broad flagellar pocket. This short bodied form is found *only* in the (revised) genus *Crithidia.*

Chondriome: Total mitochondrial complex of a cell.

Chromatin Granules: Small darkly staining masses, usually displaced peripherally under the nuclear membrane, which are particularly typical of the nucleus in many rhizopod amoebae.

Chromatoid Body: Accumulation of a dense mass of RNA in the cytoplasm of some naked and testaceous amoebae; it appears as a refractile rod-shaped body (may be several in number, all staining deeply with hematoxylin) in cysts of organisms such as species of *Entamoeba.*

Ciliary Girdle: See **Locomotor Fringe;** in a general way, the term is restricted to peritrich ciliates, yet it may be used for any encircling band of somatic ciliature; synonym, ciliary wreath.

Ciliary Meridian: See **Kinety**.

Ciliary Row: Ordinarily a more or less longitudinal line of somatic cilia, characteristically found in the great majority of the ciliated protozoa; see **Kinety.**

Cingulum: Strictly speaking, the thecal plate covering the annulus of "armored" dingflagellates; it is also sometimes used as a synonym of the **Annulus** itself.

Circumoral Ciliature, Circumoral kinetids: Line, circle, or band of essentially simple somatic cilia encircling (periorally) all or part of the apical end (including the cytostome) of the body of a number of gymnostome ciliates; the cilia are basically organized in pairs (but not dyads) of those kinetosomes (only one of which is typically ciliferous) that also represent the anterior extremities of the more or less regularly arranged somatic kinetics.

Cirrus (cirri, plural**)**: Composite tuft of cilia, few to more than a hundred in number, functioning as a single unit, although with no special or additional enveloping membrane, and tapering or fimbriate distally; its infraciliary kinetosomes, arranged in a hexagonal pattern, are interlinked and joined to other cirral bases by connecting fibers or tracts of microtubules. This kind of compound somatic ciliature, which is typical of hypotrich and stichotrich ciliates (although it is not exclusively found there) is principally a locomotor organelle, but it is also used in feeding; cirri occur in lines or groups in definite patterns on the ventral surface of hypotrichs, with subtypes identifiable by location: buccal, frontal, frontoventral, midventral, transverse (anal), caudal, and marginal (right and left).

Clavate Cilia, Clavate Kineties: Short immobile cilia that lack the central pair of microtubules in their axoneme; often occurring in a special field or area on the ciliate's body, they are allegedly sensory in function (e.g. in *Didinium).*

Clustered spores (**Spore Balls**): Mycetozoan spores (e.g. *Badhamia*) that adhere together in clusters of 2-40. Spore clustering varies both in the number of spores in a cluster, the degree of their adherence, and the presence or absence of an interior cavity (hollow) formed by the interface of the inner surfaces of the individual spores.

Cnidocyst: See **Nematocyst**.

Coccoliths: Calcified areas of the often disc-shaped body scales that cover much of the surface of most species of haptomonads. Special terminology has been developed for various subtypes of this kind of body scale.

Coenobium: Colony in which the total number of cells or zooids is fixed before its release from the parent colony; it is best exemplified by the spherical *Volvox,* which has its interconnected cells embedded in a common gelatinous matrix.

Coenocytic: Presence of multiple nuclei in a cytoplasm that is not partitioned by cross-walls, cell membranes, or complete septa of any other sort; this nonseptate condition is also known by the term **Syncytial,** the synonym preferred by protozoologists (while coenocytic is mainly used by phycologists and mycologists).

Collar: Several meanings. 1) In choanoflagellates, a corona composed of slender, very closely appressed cytoplasmic projections or filopodia; 2) in ciliates, the term has been used: for a portion of loricae, and any contracted neck region of the body (e.g., in some chonotrichs).

Collarette: An oval, antero-ventral ring, with thin, clear, anterior scalloped edge found in some filose testate amoebae.

Colony: Permanent or semi-permanent physically very close association of a group of adult individuals in some specific morphological arrangement with, further, an exhibition of various degrees of physiological integration; for further definitions, see the following widely recognized types **Arboroid** (synonyms, dendritic and dendroid), **Catenoid, Coenobial** (see **Coenobium**), **Discoid, Crenuloid** and **Spherical** (including "hemispherical"); still others have been identified but are not defined in this glossary.

Columella: central sterile structure within the fruiting body of mycetozoans which represents the continuation of the stipe inside the sporangium, or, in sessile fruiting bodies, a dome-shaped, spherical, or elongated structure arising from the base. It may be of various shapes, sizes, and textures and may serve as a supporting structure for the capillitium. This structure is absent in all species of the Trichlida and Liceida.

Compound Ciliature: General term for all ciliature of ciliates that is *not* composed of single, separate, or isolated individual cilium or of simple pairs of somatic cilia; there are various kinds, of both somatic and oral (or buccal) origin, throughout the ciliate taxa— outstanding examples are the cirri of the hypotrichs and the membranelles of the spirotrichs.

Compound Pedicel: See **Aggregated Pedicel.**

Concrement Vacuole: Quite complex subpellicular cytoplasmic inclusion of certain ciliates, one to an organism, containing refractile grains (calcium carbonate crystals?), having no opening, and strengthened by surrounding microtubules; this "vacuole" is characteristic of certain gymnostome and trichostome endocommensal species; its function is unknown, but if it is a kind of statocyst with statoliths, then it might serve as a "balancing" organelle; see also **Muller's Vesicle** and **Statocyst**.

Cone, Funnel-shaped cone: See **Proboscis**.

Conjugation: In ciliates, a reciprocal-fertilization type of sexual phenomenon that occurs only between members of different mating types (e.g., as those well known in the hymenostomes *Paramecium* and *Tetrahymena);* iso- or macro- and microconjugants may be involved, with temporary (most widespread) or total (as in all peritrichs, chonotrichs, and many suctorians) fusion of the members of the pair; the process in ciliates is always followed by the occurrence of binary fission in the separated exconjugants (or single exconjugant when total fusion has taken place); genetically, conjugation produces the same results that are obtainable via **Syngamy,** the true sexual reproduction so commonly found among many nonciliate taxa of protozoa; for some workers, a synonym of conjugation is **Gamontogamy.**

Conoid: Electron-dense, hollow, truncated conical structure found inside the polar ring at the anterior pole of sporozoites, merozoites, and/or certain other stages of most sporozoa; it is composed of spirally coiled microtubules.

Conopodium: A seldom used term for the conical, clear round-tipped lobopodium formed from the body surface and/or advancing hyaline margin of many naked amoebae and some testate amoebae.

Contractile Vacuole: Liquid-filled organelle (sometimes multiple), serving as an osmoregulator in the cytoplasm of many species of protozoa. Though predominantly limited to fresh-water forms, the CV generally pulsates with a regular frequency, expands (diastole) to a certain size, and then "contracts" (systole), typically emptying its contents (dissolved waste materials) to the exterior via one or more pores.

Contractile Vacuole Pore (CVP): In ciliates, it is a minute permanent opening (reinforced by microtubular fibers and with an argentophilic rim) in the cortex and pellicle through which the contents of the contractile vacuole are expelled to the outside; it is a corticotypic structure (see **Corticotype)** that is stable in number and location and thus of value in both taxonomic and morphogenetic studies; it is also known as an expulsion vesicle pore.

Cornuspirine: Foraminiferan test which has undivided tubular chambers and is planispirally enrolled.

Corona: Apical, cytostome-bearing extremity of the body of certain haptorid gymnostome ciliates; it is often set off by longer cilia (=coronal ciliature) from the extensible posteriorly adjacent neck region of the body; a collar is sometimes thought of as a corona.

Cortex: In the broadest sense, the outer portion or layer of the body, sometimes termed the cell envelope; in ciliates, it includes the pellicle and the infraciliature *sensu lato* and it bears the cilia; its various openings, ridges, alveoli, ciliary kinetosomes, their fibrous and microtubular associates, etc. essentially comprise the corticotype of a ciliate. In mycetozoans it is the thick outer calcareous covering of an aethalium (e.g. *Fuligo septica*)

Cortical Shell: When there is more than one shell in polycystines, the outer one is called the **cortical shell** and the innermost is the **medullary shell.**

Cortical Vesicle: See **Alveolus, Pellicular.**

Costa (plural, **Costae**): Slender, rib-like structure subtending the undulating membrane of certain trichomonad flagellates. Essentially a modified flagellar rootlet with striations showing ~400Å periodicity; often contractile, sometimes used in motility of entire organism. **A totally different meaning is given to the structures found in the Acanthoecidae** (loricate choanoflagellates). They have a unique basket-like lorica made of siliceous **costae** (ribs) each made up of **costal strips** of approximately equal length. Usually costae are either parallel or perpendicular to the long axis of the cell. In the former case they are referred to as **longitudinal costae** and in the latter as **transverse costae**.

"Couronne ciliaire": see **Corona**.

Cresta: Fibrillar, noncontractile structure located below the basal portion of the trailing flagellum of devescovinid flagellates, its length (and thus prominence) is highly variable in different genera, but its shape in all cases is basically subtriangular.

Cribrate Aperture: An aperture of a foraminiferan test which is perforated with round holes, sievelike.

Cruciform Nuclear Division: A type of mitotic division characteristic of plasmidiophorids in which a persistent nucleolus is elongated perpendicularly to the metaphase plate of chromatin, centriolar pairs occur at each pole, and the nuclear envelope remains intact through metaphase.

Cryptozoite: Primary exoerythrocytic meront of certain members of the haemosporidian sporozoa.

Crystalline Calcareous Bodies: loose crystals scattered on the surface of the peridium or compacted to form a smooth eggshell-like outer crust of a mycetozoan as *Didymium*.

Cyanellae: Species of cyanobacteria ("blue-green algae") that are intimately associated with some protozoan species. Essentially, such photosynthetic endosymbionts are often the functional equivalent of chloroplasts or plastids in their "host" cytoplasm.

Cyrtos: Tubular cytopharyngeal apparatus in ciliates, often curved with walls strengthened by longitudinally arranged nematodesmata derived from apically located kinetosomes and lined with extensions of postciliary microtubules; the nematodesmata may be interconnected and/or wrapped circumferentially by annular sheaths of diffuse fibrous material, an amorphous electron-dense substance that may form capitula proximally; it contains no toxicysts; the cyrtos is a highly characteristic structure of hypostome ciliates; synonyms, **Nasse** and **Pharyngeal** or **Cytopharyngeal Basket.**

Cyst: Nonmotile, dehydrated, resistant, inactive, dormant stage in the life cycle of many protozoa that is generally considered to serve an important role in either protection or dispersal of the species. The organism is typically rounded up, surrounded by one or more layers of secreted cystic membranes, envelopes, or walls. In some cases the walls may be quite thick, sculptured on the outside, and with or without an emergence pore.

Cystozoite: Stage found in certain sporozoan (e.g., *Sarcocystis)* life cycles.

Cystozygote: General term for a zygote that has encysted immediately after fusion of gametes in syngamy; this type of zygote is found in various groups of flagellates and sporozoa.

Cytokalymma: Skeletal matter of polycystines is siliceous and is secreted within a thin cytoplasmic sheath or **cytokalymma.**

Cytomere: Separated multinucleate portion within a grown schizont (meront) in some sporozoa; it develops into numerous merozoites.

Cytopharyngeal Apparatus: For ciliates, a term generally reserved for reference to either the **Cyrtos** or the **Rhabdos;** also see **Cytopharynx.**

Cytopharyngeal Basket: See **Cyrtos.**

Cytopharyngeal Pouch: Reservoir-like enlargement or receiving vacuole of the cytopharynx of a few ciliates (e.g., the hymenostome *Tetrahymena vorax),* present in the carnivorous macrostome stage of the life cycle and permitting ingestion of large prey; synonym, receiving vacuole.

Cytopharyngeal Rod: See **Nematodesma,** especially as it relates to the cytopharynx of certain ciliates.

Cytopharynx: In ciliates, a nonciliated tubular passageway, sometimes of different lengths in species of different higher taxa, leading from the cytostome proper into the inner cytoplasm of the organism; typically, food vacuoles are formed at its inner or distal end; when its walls are particularly strengthened in certain ciliates, it then becomes a cytopharyngeal apparatus with such specialized names as **Cyrtos or Rhabdos;** a cytopharynx is also present in a few flagellate groups, generally limited to species that can exhibit holozoic feeding habits.

Cytoproct: Cell anus of ciliates; generally a permanent (if present at all), slit-like opening through the pellicle, near the posterior end of the body, through which egesta may be discharged; its argentophilic sides or edges, usually closely apposed, are reinforced with microtubules; synonym **Cytopyge.**

Cytoskeleton: Generalized term referring to any secreted inorganic or organic material, or various other microtubular or microfibrillar organelles, found in, on, or below the surface of a protozoon, covering or involving all or some specific part of the body and often lending considerable rigidity to the shape of the organism; synonym, Skeletogenous Structure.

Cytostome: Cell mouth, the "true" mouth or oral opening; strictly speaking, it is simply a two-dimensional aperture, most commonly permanently open (if present at all), through which food materials pass into the endoplasm of the organism (via a more or less distinct cytopharynx, which is immediately behind the cytostome). In ciliates, it may open directly to the exterior or be sunken into a depression or cavity of some kind (e.g. an atrium, vestibulum, or buccal, or peristomial cavity); the ciliate cytostome is also definable as the level at which the organism's pellicular alveolar sacs are no longer present (end of the ribbed wall); it may occur or appear as an angled or tipped elliptical opening with a long axis of considerable length.

Dactylopodia: Unbranched, probably sensory pseudopodia which extend through multiple apertures in the tests of trichosidae. They are probably tactile in function. Movement is by means of **lobopodia.**

Dargyrome: Silverline pattern on the dorsal surface of hypotrich ciliates that is visible under light microscopy following use of some silver impregnation technique.

Definitive Host: Host organism in which the sexual or final stages in the life cycle of a parasite develop.

Dehiscence: Lines or patterns marking openings in the peridial wall of mycetozoans.

Dendritic Colony: See **Arboroid Colony.**

Denticle: Tooth-like structure. Two common uses among protozoa: 1) proteinaceous subpellicular skeletal element of a supporting ring that underlies the adhesive disc of mobiline peritrich ciliates; one of many identical interlocked structures, each is typically a hollow cone, fitting into the one next to it; associated with the conical centrum of the denticle are usually an inwardly directed spine or ray and an outer blade; 2) Tooth-like structure in the fossae of elphidiid foraminifera and other foraminiferan groups which form comb-like sieves to collect prey from reticulopodia.

Denticulated lip: An extended part of a foraminiferan aperture which bears denticles or teeth.

Denticulate Ring: Circular skeletal organelle, made up of denticles, found in mobiline peritrich ciliates; for further description, see **Denticle.**

Deutomerite: Nucleus-containing, posteriormost segment of a septate gregarine.

Dictydine granules (plasmodic granules): Minute (0.5 to 3.0 ~m in diameter) strongly refractive granules that are conspicuous on the peridium of certain mycetozoa (e.g. *Cribraria, and Lindbladia*).

Dikaryon: Pair of associated nuclei in a single cell, usually with each being of different parentage or derivation.

Digenetic life cycle: A life cycle that involves two different host species.

Dikinetid: Organellar complex in ciliates composed of two more or less adjacent basal bodies or kinetosomes plus their associated cilia and infraciliary organelles; this is a general term for all kinds of "paired" basal body situations, from the side-by-side arrangement underlying the "coronne ciliaire" of some gymnostomes to the zigzag pairing at the base of paroral membranes and to the condition (paired kinetosomes) so frequently appearing in all or part of a somatic kinety on the body of many ciliates; under the word dikinetid can be included such specialized terms as **Dyad** and Stichodyad**.**

Dinokaryon: Term that has sometimes been used for the rather unique nucleus of dinoflagellates, which is characterized by "herring bone" chromosomes at TEM level of examination, and DNA with little or no associated basic protein (histones).

Dinospore: Free-swimming, infective stage in the life cycle of some parasitic dinoflagellates, thus a **Swarmer** or **Zoospore** in a broad sense. They are morphologically quite similar to members of the free-living genus *Gymnodinium;* synonym, gymnospore.

Dinosporin: A highly resistant organic compound forming the enclosing wall of fossilizable dinoflagellate cysts.

Diplokaryon Stage: Binucleate cell, the terminal product of schizogony in certain microsporidians (e.g., *Nosema*), in which the nuclei—the membranes of which have been adhering to each other over a large area—allegedly undergo autogamy prior to sporogony.

Ectolobopodia: A term used to describe some of the pseudopods of Arcellinida, which anastomose outside the test (**reticulolobopodia).**

Elaters: free, usually unbranched, elastic capillitial threads marked with spiral bands (e.g.*Oligonema* and *Trichia*). Other genera (e.g *Arcyria, Hemitrichia, Metatrichia,* and *Perichaena*) also have elastic capillitial threads (elaters) that may have varying degrees of branching and anastomosing. These elastic threads are hygroscopic and undergo twisting movements, disseminating spores in response to changes in humidity in the surrounding environment.

Endolobopodia: A term used to describe some of the pseudopods of Arcellinida, which are granular or completely hyaline.

Endomitotically: Replication of chromosomes with out mitosis, often resulting in polytene chromosomes.

Endosome: The nucleus of pelobionts has a large, central endosome, and mitosis includes division of the endosome**Endospore**: See **Exospore**

Endosprits: Term for the very short tentacles of the suctorian *Cyathodinium*.

Endosymbiosis: A type of symbiosis in which a smaller symbiont (endosymbiont) lives within a larger one (host).

Entosolenian tube: Internal tubular extension from the aperture of a foraminiferan.

Epicone: That part of an athecate dinoflagellate anterior to the cingulum.

The anteriormost segment or organelle of a gregarine. It contains no nucleus and attaches the gregarine to its host cell. It generally breaks off when the gregarine becomes detached from the host cell.

Epipellicular scales: Scale-like structures secreted on to the cell pellicle of some ciliates.

Epispore cytoplasm: The cytoplasm which surrounds and is not in continuity with the sporoplasm of a developing haplosporidian spore.The spore wall and ornaments of the wall are formed within the epispore cytoplasm. The latter is anucleate with the possible exception of a fragment of the sporoplasm nucleus, contributed before the epispore cytoplasm becomes separated from the sproplasm during development. The epispore cytoplasm degenerates upon maturity of the spore.

Equatorial aperture: Symmetrical opening in the plane of coiling of an enrolled foraminiferal test.

Euglenoid movement: The term used for the rapid changes of body shape when swimming ceases in some euglenids. Other euglenids are completely rigid.

Euryxenous: Having a broad host range.

Evanescent: used in reference to the peridium of mycetozoans that often disappears early in development (e.g. *Arcyria, Comatricha,* and *Stemonitis*).

Evaginative Budding: Type of fission in ciliates involving formation of a temporary brood pouch but one (in contrast to the situation in

Endogenous Budding) in which the larval form is not freed within the parental body; in emergence, the entire wall of the pouch evaginates and cytokinesis takes place outside the parental body; it is the characteristic mode of reproduction of members of an entire suborder of suctorians; it may also be called evaginogemmy.

Evolute: A type of foraminiferan test, having each whorl not embracing any earlier whorls, and all whorls are visible; contrast with **Involute.**

Exoerythrocytic Stage: Phase in the life cycle of haemosporidians occurring in host tissue cells other than erythrocytes; synonym, EE-stage.

Exogenous Budding: Type of single or multiple fission in ciliates taking place essentially on the surface of the parental body; the larvae are pinched off singly or multiply—in the latter case, synchronously or consecutively; it is the characteristic mode of reproduction of certain chonotrichs and in one large group of suctorians; the phenomenon is also known as exogemmy.

Exogenous Cycle: That part of the life cycle of an organism (e.g., in parasitic sporozoa) taking place outside the host.

Exospore: A term used in the description of microsporidian spores. The spore wall consists of an outer electron-dense **exospore** (sometimes seen as a layered structure) and an electron-lucent **endospore**, which is usually much thicker. Occasionally the wall is very thin, without differentiation as layers. The exospore layers and some profound modifications of the spore wall structure are of diagnostic value for some genera.

Explosive Trichocyst: See **Trichocyst.**

Expulsion Vesicle: See **Contractile Vacuole;** but the present term is preferred by certain workers today (e.g., some ciliate physiologists).

External Budding: See **Exogenous Budding.**

External Tube: Coiled tube, a part of the capsular primordium in myxosporidian sporoblasts.

Extracapsulum, Extracapsular Zone: The actinopods radiating from the pores in the central capsular wall of radiolarians produce a frothy or web-like extracapsular cytoplasm known as the extracapsulum. Algal symbionts, when present, are enclosed within perialgal vacuoles usually in the extracapsulum.

Extramacronuclear microtubules: Microtubules arranged outside the macronuclear envelope of ciliates during cell division.

Extrusome: Membrane-bounded extrusible body, subpellicularly located, found in some flagellates and ciliates. It is a generalized term useful in referring to various types of often, admittedly, probably nonhomologous structures (e.g discobolocysts, ejectisomes, fibrocysts, haptocysts, muciferous bodies, mucocysts, nematocysts, rhabdocysts, toxicysts, trichocysts, clathrocysts, conocysts, crystallocysts, and kinetocysts (in certain heliozoan species). Extrusion typically occurs under conditions of some appropriate chemical or mechanical stimulation, but the function of many extrusomes (more than 20 have been described in the literature) is still largely unknown.

Eyespot: Reddish, carotenoid-containing body in the pigmented "phytoflagellates" *sensu lato* and in some of their colorless relatives; it shields the photoreceptor or flagellar swelling in euglenoids; while a part of the chloroplast in most groups, it is an independent organelle in euglenoids; it probably functions in the light-sensitivity of most organisms possessing one, influencing their phototactic or phototropic responses; synonyms, **Ocellus** (of certain dinoflagellates) and **Stigma** (plural: stigmata).

Falx: Specialized area of the cortex along the front edge of the body of opalinid flagellates; the falx is a region of kinetidal proliferation that results in the increased length of the kineties; it is usually bisected during the symmetrogenic fission of the organism.

Fibrocyst: Fusiform, explosive trichocyst (an extrusome) with a conspicuous parachute- or umbrella-like quadripartite tip on a long paracrystalline filament after discharge, characteristic of microthoracine ciliates, it has been called a **Fibrous Trichocyst** and a **compound trichocyst** and has also been included under the term **Spindle Trichocyst** by many authors.

Filamentous Annulus: Elastic, expansible binding of fine fibrils surrounding (and considered part of) the cytopharyngeal rhabdos near its proximal (outer) end; it allows the great expansion required by certain carnivorous gymnostome ciliates, which possess a **Rhabdos,** when they are feeding.

Filamentous Reticulum: Three-dimensional lattice of kinetosome-associated microfibrils present in the wall of the buccal cavity or infundibulum of certain ciliates; often the fibrils are united at so-called condensation nodes, giving a striking hexagonal pattern at the ultrastructural level (e.g. as found in some peritrichs).

Filopodium: Long, filamentous, but relatively simple pseudopodium (no central axial rod, seldom branched, etc.) without anastomoses; it may function in both food-capture and locomotion, and it is characteristic of various rhizopod sarcodinid taxa.

Fission: Cell division. This is the sole mode of reproduction in ciliates and is widespread among all other protozoan (and protist) phyla as their main method of asexual reproduction; there are many kinds or types: iso- or anisotomic (filial products of equal or unequal size) and mono- or palintomic (two or multiple filial products); strobilation (products temporarily held together, chainlike); and budding (exogenous, endogenous, evaginative; mono- or polygemmic; synchronous or consecutive); binary fission (two resulting cells) is the most common kind of fission; a cystic stage is sometimes regularly involved. In the usual binary fission of ciliates, the anterior filial product is called the proter, the posterior the opisthe; ciliates are said to exhibit homothetogenic (and usually perkinetal or transverse) fission, while flagellates show a symmetrogenic (mirror-image, interkinetal in opalinids, and often longitudinal) type; see also **Plasmotomy** and **Schizogony.**

Flabelliform, Flabellulate: Descriptive term for foraminifera with fanlike tests.

Flagellar Base-Kinetoplast Complex: In trypanosomatids, the mitochondrial DNA (as represented by the kinetoplast) and the basal bodies associated with the typically emergent flagellum are always in very close juxtaposition, although an actual physical connection has not been demonstrated ultrastructurally; this association of the flagellar base and the kinetoplast persists throughout the morphogenetic changes involved in the various complex life cycles of these organisms: for example, the entire complex shifts *as a unit* between prenuclear and postnuclear positions in all of the developmental stages defined in this glossary (a-, choano-, epi-, opistho-, pro-, sphaero-, and trypomastigote).

Flagellar Hairs: Structures or fine filamentous appendages associated with or coating the flagella of many flagellates. Typically appearing at right angles to the flagellar shaft and arranged in one or more rows; sometimes they appear in groups (=thick hairs). Simple hairs consist of a single file of similar subunits, are very common, thin, delicate, and nontubular; whereas tubular hairs are composed of two or more distinct regions, one thick and tubular. Partial synonyms are the older terms **Flimmer** and **Mastigonemes,** which are now considered by a number of workers as inappropriate and confusing to use because of the different meanings given to them by different authors (but mastigoneme, in particular, is still very popular as a generalized term); see also **Flagellum** for the types of flagella based on kinds and arrangements of their hairs.

Flagellar Pocket: Preferred term for the **Reservoir** of kinetoplastid flagellates; this flask-shaped intucking of the body surface may be shifted posteriorly in development, and one or two flagella may emerge from it.

Flagellar Pore: Opening in the theca of dinoflagellates through which a flagellum passes outward.

Flagellar Rootlet: Fibrillar structure extending from the basal body into the (deeper) cytoplasm of the cell, typically cross-banded (striated), often establishing contact or near-contact with other organelles, and sometimes running posteriorly all the way to the antapical pole of the organism; such a flagellar root system, most commonly found among "phytoflagellate" taxa, may likely play a role as an anchoring device or else have a skeletal function; some may exhibit elasticity or contractility; the present somewhat restricted definition excludes *microtubular* roots (or rootlets), which are relatively rare in flagellates; **Rhizoplast** (sometimes **Rhizostyle),** essentially a synonym, is the term that has most often been used in the literature for striated flagellar rootlets, but there is no evidence that the diverse structures described under either name are all homologous; kinetorhiza, an outmoded term, is a partial synonym; for ciliates, ciliary rootlet may be used but such organelles—if/when non-microtubular—are rare among ciliophoran species.

Flagellar Rows: Rows or lines of flagella, often spiraled, covering all or parts of the body of such flagellates as the hypermastigotes; they are considered as kineties by some workers, but ultrastrueturally they do not resemble the kineties of ciliates at subpellicular levels.

Flagellar Swelling: See **Photoreceptor**; but the present morphologieally oriented term is becoming preferred by many workers because it is more "neutral" and implies no specific function.

Flagellar Transition Region: Region between the proximal end of the axoneme of an emergent flagellum and its associated basal body which lies within the cell; it contains some distinct (though little understood) structures, particularly among "phytoflagellate" groups, that are considered useful indicators of phylogenetic relationships; most striking is the so-called "transitional helix" charaeteristic of most heterokont groups (e.g., chrysomonads, xanthophytes, and eustigmatophytes).

Flagellipodium: A long, tapered, clear, highly vibratile pseudopodium formed from the ectoplasm of a few species of naked amoebae; it is *not* a flagellum, but may be mistaken for one.

Flap: Extension of chamber wall of a foraminiferan test covering an interiomarginal aperture.

Flexostyle: Enrolled tube immediately following the proloculus of some foraminifera.

Floor: Bottom wall of a foraminiferan test chamber.

Flimmer: See **Flagellar Hairs.**

Foramen: Opening in the septum between two chambers of a foraminiferan test; it may represent a previous aperture or be formed as a fresh one.

Fossettes: Short deep grooves between ponticuli of the tests of some rotaliid foraminifera.

Frange: Band of perioral ciliature that varies in composition from an extensive line of specialized ciliature winding helically around much of the anterior end of the organism to a short linear group of as few as three **Pseudomembranelles** (or **paves**) adjacent to the cytostome proper; characteristic of certain hypostome eiliates, it is sometimes called an adoral ciliary fringe but more often the **Hypostomial Frange.**

Fruiting Body: A spore-bearing structure of a mycetozoan. Also called a sporophore, a sporocyst, spore case, or fructification. Four types are recognized (see **sporangium, plasmodiocarp, aethalium,** and **pseudo-aethalium**).

Fugacious: disappearing soon after development, as in species of *Cribraria*, where portions of the peridium disappear at maturity to form a net-like upper half of the spore case and a cup-like base. The interstices of the peridial network portion of the spore case appear to be represented by such a thin and ephemeral membrane.

Fundital plate: One of the plates posterior to the postcingular series and external to the sulcus of armored dinoflagellates.

Furrow/gullet complex: The external architecture of the cryptomonads is asymmetrical due to a furrow/gullet complex. All cells, whether they are oval, compressed, lunate, caudate, acute, elongate, sigmoid, or contorted, possess an anterior, outwardly facing depression called a **vestibulum**, from which the flagella arise on the right side. In addition, there may be a **furrow**, a **gullet,** or combination of both. Furrows are ventral grooves originating in the vestibulum extending posteriorly, but terminating in the anterior half of the cell. A gullet is a tubular invagination which extends posteriorly from the vestibulum or the end of the furrow.

Fusule: See mention under central capsule.

Galea: The top of the cephalis of a nassellarian can be augmented by a galea (cupola) and may have a cylindrical or three-bladed horn.

Gametocyst: Cyst formed by the union of two gregarine gamonts, with secretion of a wall around them; within it, the gametes are formed that will then fuse to form the oocysts in which the sporozoites develop; synonym, gamontocyst.

Gametocyte: See **Gamont.**

Gametogamy: See **Syngamy.**

Gametogony: Formation of gamete cells; among the sporozoa, this often occurs by schizogony (in the case of microgametes); a possible synonym is **Gamogony,** although this latter term implies inclusion also of the formation of the gamonts (which, in turn, will produce the gametes).

Gamogony: See mention under **Gametogony.**

Gamont: Typically, the cell or stage in the life cycle that will produce one or more gametes.

Gamontocyst: A term used to describe the periphery of the area between two gamonts of foraminifera which are undergoing gamontogamy. The gamonts deposit residua from food in a ring to enclose the space between the umbilicus of both gamonts.

Gamontogamy (Plastogamy): Union or pairing of two (or more) gamonts, with subsequent production of gametes. In foraminifera this type of sexual reproduction (fertilization) takes place inside the space formed by gamonts attached by their umbilical faces.

Gemmation: See **Budding.**

Generative Cell: One of the cells within the myxosporidian plasmodium; they join in pairs, giving rise to sporoblasts; see **Pericyte** and **Sporogonic Cell.**

Generative (Germ) Nuclei: Nuclei which will be passed on to the next generation. In heterokaryotic foraminifera they are usually retained in the earliest formed chambers in contrast to the somatic nuclei, which are found in outer whorls. Generative nuclei are homologous to micronuclei of ciliates.

Germinal Row: Line of nonciliferous kinetosomes associated with the terminal portion of the infraciliary base of the paroral membrane (or haplokinety) of peritrichs; it plays a productive role (as an anlage) in stomatogenesis in such ciliates, and it may be homologous with the **Scutico-vestige** of the scuticociliates.

Girdle: See **Cingulum.**

Gleocystic Stage: Nonmotile trophozoites within a mucilaginous capsule of a characteristic structure; this is a permanent or transitory stage in the life cycle of certain groups of flagellates.

Glycocalyx: Finely hairy secreted layer covering the outer cell membrane of many of the "naked" rhizopod amoebae. This can be thin and barely detectable in some genera or well developed as in the thick amorphous cuticle of *Mayorella* or the complex cuticle of *Paradermamoeba*.

Granellare System: The protoplasm of xenophyophores is organized as a branched, multinucleate plasmodium surrounded by a tubular, transparent organic sheath. These two elements together constitute the **granellare system,** which occupies an estimated 0.1-5% of the test volume. Rose Bengal staining can be used to make the granellare system more easily visible.

Granular Calcareous Bodies: Granular calcium carbonate ($CaC0_3$) found in some mycetozoans (e.g. Physaridae).

Gullet: Unnecessary term for an oral cavity of various sorts in the protozoa; it is often employed, improperly, for the buccal cavity of such ciliates as *Paramecium,* and is also a frequently used misleading term for the anterior invagination (canal plus reservoir) of euglenoid flagellates.

Gymnospore: Rosette-like formation of gamonts in the gregarine family Porosporidae; in the second host, these gamonts develop into gametes; synonym, heliospore.

Haplokinety: Typically, a double row of kinetosomes (paired tangentially, as stichodyads) associated with the oral ciliature and exhibiting a zigzag pattern; generally only the outermost basal bodies bear cilia; haplokinety was a once-popular term for the infraciliary base of a generalized paroral membrane, especially as found in hymenostome and peritrich ciliates.

Haplomonad stage: Developmental form of certain trypanosomatid flagellates, a stage that is attached to the walls of organs in the body of the vector or intermediate host.

Haplosporosome: An electron-dense, membrane-bound organelle found in the cytoplasm of plasmodia and sporoplasms of Haplosporidia Paramyxa, and Myxozoa. The term sporoplasmosome has been proposed for the organelles found in the Myxozoa. The organelle is generally spherical but may be cuneiform, pyriform, vermiform or oblate. The distinguishing structure is an internal unit membrane which is free of the delimiting one and found in various configurations. The function is unknown but there is evidence to suggest that when haplosporosomes are released from plasmodia into the surrounding host cells, lysis of the latter is induced.

Haptocyst: Minute complex extrusive organelle (an extrusome) in the suctorial tentacles of all kinds of suctorian ciliates; it is presumed to contain lytic enzymes useful in the capture of prey organisms; the haptocyst is still sometimes referred to as a microtrichocyst, a missile-like body, or a phialocyst, all of which are now superseded names, and some workers classify it as a subtype of **Toxicyst.**

Haptonema: Typically, a filamentous appendage of considerable length arising close to the site of flagellar insertions in haptomonad flagellates. Superficially, the haptonema resembles a flagellum, but it lacks a true basal body, is enclosed within three membranes, and has only a few singlet microtubules in its axoneme; some shorter ones move by coiling and uncoiling; the organelle possibly functions as an anchoring device.

Haptotrichocysts: A rod-like extrusome found in rhynchodine ciliates.

Heterogenetic: Used to describe parasitic protozoa whose life cycles manifest an alternation of (sexual and asexual) generations; a great many sporozoa, especially, fit this description.

Heterokaryotic: Possessing more than one kind of nucleus (i.e., exhibiting nuclear dualism); this nuclear condition is especially characteristic of ciliates, with their micro- and macronuclei, but it is also true for agamonts of some foraminifera in which the two types of nuclei are called somatic and generative nuclei.

Heterokont Flagellate: Biflagellate organism, with one smooth flagellum and one hairy one; the older meaning—possession of two flagella of unequal length—is now widely being discarded; see **Anisokont.**

Heteromerous (Heteromeric) Macronucleus: Macronucleus that is partitioned into two moieties or karyomeres (orthomere and paramere) with strikingly different DNA and RNA contents and, therefore, differential staining capacities; such a nucleus is found especially in cyrtophorid and chonotrich hypostome ciliates; contrast with **Homomerous Macronucleus.**

Different sporogonies may take place within a single host or in hosts as different as insects and crustacea. These different spore types, arising from fundamentally different sporogonic sequences are not to be confused with "early" and "late" spore types arising from the same type of sequence of sporogony.

Heteroxenous: Having two (or more) hosts in the life cycle; this is typical of many sporozoan species, most trypanosomatids, some apostome ciliates, and scattered other small groups; contrast with **Homoxenous.**

Histophagous, Histozoic: Literally, tissue-eating; this is the feeding habit of several kinds of protozoa.

Holdfast Organelle: When broadly used (see **Attachment Organelle**. The term refers to any structure(s) by which a given organism can afffix or attach, temporarily or permanently, to some living or inanimate substrate (e.g. by use of cilia, hooks, uncini, crochets, tails, loricae, mucous filaments, spines, stalks, epimerites, suckers, tentacles, etc.). In the usual, more restricted sense, it refers to specific, specialized organelles: for example, in ciliates, to stalks of various kinds, the adhesive disc of mobiline peritrichs, the sucker of some astomes or clevelandelline heterotrichs, and the localized thigmotactic ciliature of many thigmotrichine scuticociliates.

Holotrichous: Evenly or completely ciliated; typically refers to somatic ciliature.

Homokaryotic: Possessing only one kind of nucleus; this nuclear condition is characteristic of all protists except the ciliates (see **Heterokaryotic)** and a few foraminiferans.

Homomerous Macronucleus: Macronucleus that shows no differentiation into zones containing differing DNA and RNA contents; an essentially uniform staining capacity is exhibited; this is the type of macronucleus found in the great majority of ciliates; contrast with **Heteromerous Macronucleus.**

Homothetogenic Fission: Type of division (generally transverse or perkinetal) of a parental organism in such a manner that there is a point-to-point correspondence between structures or landmarks in both filial products, proter and opisthe; typical of ciliates, it is to be contrasted with the **Symmetrogenic** or mirror-image **Fission** of flagellates, opalinids (interkinetal division), and possibly other nonciliate groups.

Homoxenous: Having only one host in the life cycle; overall, this is typical of most symbiotic species from the various protozoan phyla; contrast with **Heteroxenous.**

Host: Independent or so-called dominant member of a symbiotic or parasitic pair. The smaller dependent partner lives in or on the host.

Hyaline: Glassy, clear, transparent.

Hyaline Cap: The clear, ectoplasmic, more or less rounded tip at the advancing end of a lobopodium, especially of *Amoeba proteus* and related amoebae that form endolobopodia.

Hyaline Margin: The clear, ectoplasmic material that surrounds and/or precedes the granular body mass of certain amoebae during locomotion.

Hyalosome: Hyaline part of the ocellus in certain dinoflagellate species; it serves as a lens, presumably focusing or concentrating light on the pigment-containing **Melanosome.**

Hydrogenosomes: Single-membrane-bounded, enzyme-rich cytoplasmic organelles or "microbodies" characterized by the presence of hydrogenases, and functionally substituting, in effect, for mitochondria in some anaerobic protozoa (e.g. trichomonads and entodiniomorphid ciliates); the term supersedes axostylar, paracostal, and paraxostylar granules.

Hypothallus: A term used to describe the layer at the base of the fruiting bodies of mycetozoa deposited by the plasmodium during fruiting. It ranges from a membranous and transparent structure, not always evident, to a tough or spongy conspicuous layer.

Hypnozygote: Very thick-walled immobile zygote of certain dinoflagellates, often with spiny to warty ornamentation; following its excystment, the hypnozygote may release a single, biflagellate swarmer. Also a thick-walled resting stage zygote found in volvocids. Germination involves meiosis and formation of flagellated **meiospores**

Hypocone, Hyposome: Part of the body posterior to the annulus in dinoflagellates; it is covered by the **Hypotheca** in armored forms.

Hypostomial Frange: See **Frange.**

Hypotheca: Strictly speaking, the part of the theca in "armored" dinoflagellates that is posterior to the annulus, covering the hypocone; the term is sometimes used to mean the whole posterior half of the body itself.

Hystrichospheres: Cysts or resting stage of certain fossil and some recent dinoflagellates that show characteristic projections and markings and often an apparent excystment aperture called the archeopyle.

Idiosomes: Particles, usually siliceous or calcareous, made in the cytoplasm the protozoan and extruded to cover the surface of the body or of a shell or test that surrounds the body, e.g. scales on the shells of filose testate amoebae..

Infraciliature: In ciliates, the total assembly of all kinetosomes and their associated subpellicularly located microfibrillar (or microfiamentous) and microtubular structures, both somatic and oral in location; its taxonomic and phylogenetic significance is summed up in Corliss' dictum, (Once a ciliate, always at least a bit of an infraciliature)—in all or in some part of the organism's full life cycle.

Infundibulum: Lower or inner or posterior part or section of the buccal cavity in certain ciliates, particularly peritrichs; it is often funnel-shaped or in the form of a long tube or canal; it may contain some buccal ciliature and infraciliature, but it does not dedifferentiate or regress during division.

Intercapsular Appendix: Thickened rim of the shell valves between the discharging canals of the polar capsules at the anterior end of spores in the myxosporidian genus *Myxobolus.*

Interiomarginal aperture: Foraminiferan test where the basal opening is at the margin of final chamber.

Intermediate Host: Typically, the host in which the juvenile, intermediate, or asexual stages of a parasite develop.

Intracytoplasmic Pouch: Temporary depression, cavity, or vacuole, indenting or below the surface of the organism (although opening to the outside), which is present in the opisthe during fission (especially in entodiniomorphid, oligotrich, and some hypotrich ciliates) and in which the anlagen of (new) oral ciliature appear—or come to be located—during the overall process of stomatogenesis.

Involute: A type of enrolled foraminiferan test in which each whorl completely overlaps earlier whorls; only the final whorl is visible; contrast with **Evolute.**

Isogametes, Isogamous: Referring to gametes (or gametic cells) that are morphologically indistinguishable from each other; contrast with **Anisogamous.**

Isokont Flagellate: Organism with two or more flagella of equal length; the term is especially usable when referring to biflagellate; contrast with **Anisokont,** *not* **Heterokont.**

Karyogamy: Fusion of nuclei most often leading to zygote formation.

Karyomastigont: Mastigont system plus its own associated nucleus. The term is typically used, in the case of some of the hypermastigids to designate the unit of mastigont-plus-nucleus; see **Mastigont System,** and contrast with **Akaryomastigont.**

Karyophore: Strands or sheets of specialized and generally conspicuous fibers emanating from subpellicular locations in the cell and surrounding and suspending the (macro)nucleus; a karyophore is a characteristic structure of certain clevelandellid ciliates, and the one known from some flagellates is very likely a nonhomologous organelle.

Keel: Peripheral thickening of a foraminiferan shell.

Kinetal Suture System: See **Suture Lines,** although the present term is probably more appropriate.

Kinetid: Elementary repeating organizational unit of the typical ciliate cortex. It consists basically of a kinetosome (or pair or occasionally more) and certain intimately associated structures or organelles; commonly, the latter include cilia, unit membranes in the area, pellicular alveoli, kinetodesmata, and various ribbons, bands, or bundles of microtubules (including some nematodesmata); sometimes also included are microfibrils, myonemes, parasomal sacs, such extrusomes as mucocysts or trichocysts, and even additional organelles, also see **Dikinetid, Monokinetid,** and **Polykinetid.**

Kinetodesma, Kinetodesmal Fibrils (Kd): Periodically striated, longitudinally oriented, subpellicular fiber (of component fibrils) arising close to the base of a somatic kinetosome (posterior one, if paired) of ciliates, near its microtubular triplets 5-8 (Grain convention), and extending anteriad toward or parallel to the organism's body surface and on the right side of the kinety involved (**Rule of Desmodexy**); when it is of a length greater than the interkinetosomal distance along its kinety, the kinetodesmata overlap, shingle-fashion, producing a bundle of fibers; kinetodesmata are well developed in apostomes, hymenostomes, scuticociliates, and peritrichs, and are present as large and heavy bundles (part of the endoskeletal system) in certain astomes; but in some other groups, it may be reduced to a fibrous spur extending upwards in a pellicular crest.

Kinetofragments: Segments, patches, or short files of basically somatic kinetics in the general vicinity of the more or less apical (or subapical and ventral) cytostomal or oral area in certain ciliates (many belonging to the class Kinetofragminophora); these kinetal segments are sometimes only partially ciliferous; the **Hypostomial Frange** may be considered composed of kinetofragments, so complex in some species that "pseudomembranelle" or the French term "pave" is used to describe some of them..

Kinetome: Assembly of all (generally longitudinal rows of) kinetids (i.e., the kineties) covering the body of a given ciliate; the kinetome may be considered as the total mosaic of an organism's repeating kinetosomal territories; see **Kinetid.**

Kinetoplast: Specialized DNA-containing moiety or "nucleoid" of the single mitochondrion of nonpigmented flagellates belonging to the Kinetoplastida; its DNA is typically localized in a single large stainable (Feulgen positive) mass close to the base of the locomotory apparatus (see **Flagellar Base-Kinetoplast Complex)**: in some bodonid species, the "kinetoplast-mitochondrion" is a long, ribbon-like organelle with multiple DNA containing nodules dispersed throughout the organelle.

Kinetosome: Subpellicularly located tubular cylinder of nine longitudinally oriented (at right angles to the cell's surface), equally spaced, skewed, peripheral structures each composed of three microtubules, its typical size, in ciliates, is ~1.2 x 0.25 μm; when ciliferous, this argentophilic embedded basal body produces a cilium or flagellum at its distal end; when viewed from deeper in the cytoplasm of the organism looking outward, the nine triplets of microtubules (numbered by the Grain convention) are skewed inwardly, clockwise; distally, they are continuous with the paired peripheral microtubules of the cilium or flagellum; the generally electron-lucent central core may contain an electron dense basal plate (often continuous laterally with the epiplasm) distally; proximally, the core contains a cartwheel-like arrangement of fine filaments; synonyms, **Basal Body, Blepharoplast,** and, by accepted homology, **centriole**; "basal granule" is now a completely discarded earlier synonym.

Kinety: Single structurally and functionally integrated (somatic) row, typically oriented longitudinally, of single or paired (or, in special cases, possibly a greater number of) kinetosomes (not all necessarily ciliferous) plus their cilia and other associated cortical organelles and structures, viz., a line of **Kinetids**; these ciliary and infraciliary meridians are basically bipolar (although some may be interrupted, fragmented, intercalated, partial, shortened, etc.), with an asymmetry allowing recognition of anterior and posterior poles of the organism itself; the term is commonly restricted to use with the ciliated protozoa (plus the opalinid flagellates), but some workers consider that the "flagellar rows" of certain other flagellates also represent a kind of kinety); a synonym, at least superficially, is **Ciliary Meridian.**

Kinety n: The last somatic kinety; see **Kinety Number 1** (below).

Kinety Number 1: Meridian or ciliary row which, in a number of hymenostome ciliates, has three unique features or properties: it is the stomatogenic kinety (or stomatogenous meridian), functioning in new mouth-formation during fission; it bears, or is topologically associated with, the cytoproct at its extreme posterior end, and it is terminated anteriorly by the posterior margin of the buccal overture as the rightmost postoral meridian.

Km Fiber: As used for a structure in certain heterotrich ciliates, the term has now been superseded by **Postciliodesma.**

Laminin, Laminin Receptor: Recognition of a host cell by *Toxoplasma gondii* involves at least three molecular interactions. There is a parasite **laminin** and host-cell **laminin receptor**; there are parasite surface lectin-like molecules involved and a specific parasite-surface protein known as SAG-1(P30) which binds to a glycosylated host cell receptor.

Left sulcal plate: The sulcal plate located to the left side of the sulcus, anterior to the posterior sulcal plate of armored dinoflagellates.

Leucoplast: Pigmentless plastid, often serving as a starch storage organelle.

Light-Spored: spore color in mass of brighter colors such as hyaline, gray, red, and yellow characterizes orders such as the Echinostellida and Trichlida.

Linellae: Xenophyophores are divided into two groups according to whether long fibers, linellae, composed of a silk-like protein are present (Order Stannomida) or absent (Order Psamminida).

Lip: Border of the aperture of a foraminiferan test.

Lithosome: Vesicular, membrane-bounded cytoplasmic inclusion or smal1 body, found in a number of ciliates, that is composed of some inorganic material laid down in concentric layers.

Lobopodium (plural, **Lobopodia)**: Lobular, more or less broad and rounded, or cylindroid, type of pseudopodium; it often has a hyaline cap at its advancing tip; used both in locomotion and feeding.

Longitudinal Flagellum: Term used for the flagellum which lies partly in the sulcus of dinoflagellates and beats posteriorly. It may be cylindrical or flattened and has one, or only a few waves. It is used mainly for steering the cell.

Lorica, Loricate: Test, envelope, case, basket, or shell (or, occasionally, theca) secreted and/or assembled (using various materials) by protozoa from a number of different major taxa, generally fitting (the body) loosely, opening at one (anterior) end (or occasionally both ends), and being either attached to the substrate (common) or carried about by the freely moving organism (e.g. in testaceous amoebae and tintinnine ciliates); loricae may occur in a multiple (arboroid) arrangement; they may be occupied only temporarily; in composition, a lorica may be calcareous, composed of some proteinaceous or mucopolysaccharide secretion (including chitin, pseudochitin, or tectin), or be made up of foreign matter (sand grains, diatom frustules, coccoliths, debris) cemented together in species-specific patterns. The distinctive siliceous basket-like periplast of members of the Acanthoecidae (loricate choanoflagellates) also is known as the **lorica** and is composed of **costae** made up of **costal strips**. Usually costae are either parallel or perpendicular to the long axis of the cell. In the former case they are referred to as **longitudinal costae** and in the latter as **transverse costae**.

L-type sulcus: A term used to describe the longitudinal nature of certain dinoflagellates. The sulcus is more or less straight. The sixth precingular plate is bounded to left and right by almost parallel sutures and anteriorly by a gabled margin due to its contact with both the fourth and first apical plate or their homologues. The fourth plate is or its homologue is typically not long and narrow and does not contact the anterior sulcal plate. There is commonly a broad contact between the last precingular plate and the first apical plate or its homologue.

Macrogamete: Larger one of the pair of anisogametes (thus often a relatively large-sized cell), considered female, and generally nonflagellated; synonym, **oogamete**; contrast with **Microgamete**.

Macrogametocyte: See **Macrogamont**.

Macrogamont: Gamont that will form or become a female gamete (macrogamete); it already has the haploid (N) number of chromosomes, so some workers consider the term to be, in effect, superseded by the word "macrogamete" itself; synonym, **Macrogametocyte**.

Macronucleus: Transcriptionally active nucleus of ciliates, responsible for the organism's phenotype; it may be multiple, but even then is typically much larger than the micronucleus. It is polyploid (diploid in karyorelictids) with respect to its genomic content, which is the result of selective amplification of 10-90% of the micronuclear sequences; It may be **Homomerous** or **Heteromerous** (see those terms). Its presence in ciliates is an important character, the **Heterokaryotic** condition, contrasted with the **Homokaryotic** nuclear condition typical of almost all other protozoan (or protist) taxa.

Macrostome: Stage in the lifecycle of some ciliates in which the oral cavity enlarges considerably from the typical microstome form; enables ingestion of large prey, including conspecifics.

Marginal Cilia: Circumferential band of long, stout cilia located above the aboral locomotor fringe of many mobiline peritrich ciliates; they are sometimes called cirri because of their stoutness.

Marginal cirri: Cirral files extending along the edges of the cell body of hypotrich and stichotrich ciliates.

Marsupium: See **brood pouch.**

Mastigonemes: See **Flagellar Hairs**; although the present rather popular term is well entrenched in the literature, it has been used in different senses in recent years, making its continued usage a source of confusion.

Mastigont System in **Chrysomonads**: Flagellated cells have 2 obliquely inserted heterodynamic flagella of unequal length; the longer (immature) flagellum is directed anteriorly, beating in a sine wave; the shorter (mature) flagellum is directed laterally when emergent (in several genera it is vestigial). The long flagellum bears tripartite tubular hairs ("mastigonemes"; the short flagellum is smooth and often has a photoreceptor-related swelling at the base.The basal body of the short flagellum usually lies at an angle of approximately 90° with respect to the long flagellum, but narrower or wider angles also occur (e.g. ca. 30° in *Chrysonebula* [Hydrurales], 160° in *Hibberdia* [Hibberdiales]). The basal bodies are interconnected by fibrous bands. A typically cross-striated fibrous root (rhizoplast) is normally present.

Mastigont System in **Parabasalids**: Complex of flagella-associated organelles defined by its major components (basal bodies, flagella, undulating membrane, costa, or cresta, parabasal apparatus, axostyle, etc.); most strikingly seen in many of the trichomonad and hypermastigote flagellates. The basic unit is four flagella, and their basal bodies are typically arranged with the basal body of the recurrent flagellum (R) perpendicular to the basal bodies of the anteriorly directed flagella. These basal bodies bear special appendages which make them recognizable: basal bodies 1 and 3 bear a hooked lamina, F1 and F3, respectively, and basal body 2, (F2) bears **sigmoid fibers.** This typical arrangement is found from the small monocercomonads to the hypermastigids, such as *Joenia* and *Lophomonas*, and is partially conserved in other hypermastigids. These basal bodies are designated as privileged basal bodies.

Median Body: Paired organelle (actually masses of microtubules) located in the area of the striated disc and of the two ventral flagella in such diplomonad parasitic flagellates as *Giardia*; the two organelles were formerly called parabasal bodies, terminology now unacceptable for them.

Medullary shell: When there is more than one shell in polycystines, the outer one is called the **cortical shell** and the innermost is the **medullary shell.**

Megalospheric Form: Stage in the life cycle (gamont) of some foraminiferans characterized by a large proloculum and a haploid (N) uninucleate condition; contrast with **Microspheric Form.**

Membranelle: One of the several serially arranged compound ciliary organelles (together known as the AZM) found on the left side of the buccal cavity or peristomial field in oligo- and polyhymenophoran ciliates; its cilia may be used in food-getting or locomotion; its generally rectangular infraciliary base is commonly composed of three rows of densely set kinetosomes which may be associated with parasomal sacs and/or microtubular or microfibrillar structures in specific patterns that may vary in different taxonomic groups. The term includes organelles very likely not homologous. For some specific types, see **Peniculus**, **Polykinety**, and **Quadrulus**; and contrast with **Paroral** or **Undulating Membrane**.

Merocyst: Large meront of some haemosporidian sporozoa that breaks up into separate multinucleate fragments (cytomeres) before producng merozoites.

Mergony: Phenomenon or process of the formation or production of merozoites from an antecedent cell (a schizont of the meront type).

Meront: Asexual stage in the life cycle of certain sporozoa that undergoes merogony, forming merozoites.

Merotelokinetal: Development of new oral structures from the ends of several parental kineties (e.g. colpodean ciliates).

Merozoite: Stage of an intracellular sporozoan that is produeed by merogony. In turn, it develops into either a new meront itself or, at an appropriate time and place, a gamont (i.e., a microgamont or a macrogamont, the latter, in effect, being the macrogamete); such development takes place in the vertebrate bloodstream in the case of malarial species.

Mesoxenous: The host specificity or host restriction of most Apicomplexa. Some are mesoxenous, occurring in hosts of a single order, while a few are euryxenous, occurring in hosts of more than one order or class.

Metacryptozoite: Merozoite produced by a meront which itself has been produced by a cryptozoite; in the life cycles of certain haemosporidian sporozoa, the metacryptozoite may develop into another (exoerythrocytic) cryptozoite or enter into an erythrocytic stage.

Metacyclic Form: Stage or form in the life cycle of parasitic kinetoplastid flagellates that represents the termination of the developmental phase in the invertebrate (vector) host and that has (now) become infective for the vertebrate host.

Metamitosis: The "typical" mitotic division with a spindle, typical centrioles, asters at the ends, with dissolution of the nuclear membrane; characteristic of many eukaryotic cells.

Methanogen: An archaebacterium which produces methane as a metabolic endproduct. Often found as endosymbiotic terminal respiration acceptors inside anaerobic protozoa.

Metrocyte: Large, round proliferative "mother" cell found in the last generation meronts of the coccidian genera *Sarcocystis* and *Frenkelia*. It divides repeatedly by to form fresh metrocytes, which become progressively more elongate, each eventually becoming a **Cystozoite,** which resembles (or, in effect, becomes) a typical merozoite.

Microconjugant: Smaller member of a conjugating pair of ciliated protozoa, when there is any pronounced size distinction. Its body is typically completely absorbed by the macroconjugant (e.g., chonotrichs, peritrichs, and many suctorians).

Microcysts: Under unfavorable environmental

conditions for growth, myxogastrid amoebae may encyst to form round, tiny (4-7 μm in diameter), thin-walled, unornamented, uninucleate, protoplasts that serve as dormant resting stages.

Microfibrils Generalized term, usually referring to the composition of some protozoan structure or organelle: if it is "microfibrillar" in its nature, it is composed of nonhollow fibers of very small diameter (4-10 nm); contrast with microtubular.

Microgamete: Smaller one of the pair of anisogametes, considered male and often flagellated; contrast with **Macrogamete**.

Microgametocyte: see **Microgamont**.

Microgamont: Gamont that will produce a number of microgametes by binary or multiple fission; synonym, **Microgametocyte**.

Micronemes: Numerous, convoluted and elongate, electron-dense organelles, of circular cross-section, extending antapically from the conoid region through the anterior part of the body of sporozoites (and other stages) of most sporozoa; often they are attached to and/or give rise to the rhoptries.

Micronucleus: Transcriptionally inert nucleus of ciliates, serving as a repository of the unaltered germ line; it may be multiple, but is typically much smaller than the macronucleus. It is generally spherical or ovoid in shape, diploid (2N) and without nucleoli. The micronucleus divides mitotically in association with cell division (reproduction) in ciliates. Periodically it undergoes meiosis, prior to sexual phenomena.

Micropore: Minute opening in the cell surface of certain stages in the life cycles of sporozoa through which particulate food or other material can be taken into the body; the micropore is sometimes called a cytostome, a cuticular pore, or—erroneously—a **Micropyle.**

Micropyle: Opening (or predetermined locus of the break-through) in the wall of a sporozoan oocyst; it is not to be confused with **Micropore.**

Microstome: Stage in the life cycle of some ciliates in which the oral cavity is relatively small, adapted for feeding on bacteria. See **Macrostome**.

Microspheres: See mention at end of **Skeletons, Internal.**

Microspheric Form: Stage in the life cycle of some foraminiferans that is characterized by a small proloculum and a multinucleate diploid (2N) condition; contrast with **Megalospheric Form.**

Microspore: Unusually small spore in polysporous microsporidians. It is a product of aberrant sporont division, with the sporont undergoing more than the usual number of cell fissions and so producing spores half the size of the normal ones. The functional significance of this phenomenon is unknown.

Microtubule, Microtubular Organizing Center (MTOC), Microtubular Rootlets: Hollow, cylindrical structure of an indeterminate length, ~20-25 nm in diameter, composed of α and β subunits of tubulin; MT's are rigid, often cross-linked with others to form a ribbon, band, or bundle; many—but not all—of the microtubules in the cytoplasm of ciliates are associated with, if not derived from, kinetosomes; they comprise the "9 + 2" axonemal structure of all cilia and flagella, and they also occur in many locations in protozoan cells, from pellicular positions to deep in the cytoplasm, often with different functions, some MT's occur in the nucleoplasm (with no kinetosomal association there); in all eukaryotic cells, the spindle apparatus is composed of microtubular fibers, so MT's play a prominent role in karyokinesis. MT's represent a conspicuous part of the sporozoan apical complex; the cytoskeletal axostyle of trichomonads is microtubular in composition, and in certain sarcodinids the axopodial rod is composed of them. The most striking microtubular configurations in ciliates include the ribbons of transverse and postciliary microtubules, the bundles of nematodesmata, and the microtubular arrays in suctorial tentacles. In synurophytes the rhizoplast acts as a microtubular organizing center (MTOC) for the spindle microtubules. The nuclear envelope remains partially intact during mitosis in *Synura* and *Chrysodidymus* .

Monocystid Gregarine: Term that is still in use, but generally considered today as having been superseded by **Aseptate** (gregarine).

Monokinetid, Monokinetidal Rows: Organellar complex in ciliates composed of one basal body or kinetosome plus its cilium and the associated infraciliary organelles; monokinetids are organized in rows. This is a common condition of the somatic kinetid in ciliates.

Monosporous: See mention under **Polysporous.**

Monothalamic: Foraminiferan with a test or shell composed of only one chamber.

Monoxenous, Monozoic Form: Having but a single taxonomic species of host (i.e., the protozoan parasite is species-specific with respect to its "choice" of host); this is a highly restricted example of the **Homoxenous** condition.

Muciferous Body: Sac-like vesicle (an extrusome) located subpellicularly in euglenoid and some other flagellates that secretes or extrudes an amorphous mucoid substance to the outside. It is considered to be a special kind of **Mucocyst** by some workers. The secreted mucin is composed largely of carbohydrates and may function in locomotion of the organism; synonym, mucigenic body. Species with muciferous bodies may have mucocysts at the same time. In Chrysophytomonads many species have peripheral muciferous bodies that can be extruded as long threads; in some the muciferous bodies are large complex discobolocysts, explosively releasing discoid projectiles.

Mucocyst: Subpellicularly located, membrane-bounded, saccular or rod-shaped organelle (an extrusome) of paracrystalline structure, typically dischargeable through an opening in the pellicle as an expanded mucus-like mass, although retaining its polyhedral paracrystalline form; mucocysts are probably involved in cyst formation in many species of flagellates and ciliates (the presence of a proteinaceous or mucopolysaccharide substance, sometimes called "tectin," supports this). In ciliates, they often appear in regular, longitudinal, interkinetal rows. Synonyms include the following (generally superseded) terms: protrichocyst, mucous trichocyst, mucigenic body, and tektin rods or granules. Some workers consider several additional kinds of extrusomes to be "subtypes" of the mucocyst: for example, the ampullocyst, clathrocyst, conocyst, crystallocyst, pigmento-cyst, and even the kinetocyst (at one time considered a kind of haptocyst) found in the axopodia of certain heliozoan sarcodinids.

Mucous Envelope: Thick covering of mucus on the surface of many micro- and myxosporidian spores; it apparently enhances their flotation capacity.

Mucron: Anterior attachment organelle of many aseptate gregarines that becomes embedded in the host cell; the mucron is similar to an epimerite, but is not set off from the rest of the gregarine cell by a septum; analogous structures are also found in some symbiotic flagellates and ciliates.

Müller's Law: Grouping of the 20 radial skeletal spines of acantharians in a highly specific pattern. The spines and spicules emerge from the center of the cell along five equatorial circles.

Müller's (Müllerian) Vesicle: Small vacuole containing mineral concretions, its function unknown; described from various karyorelictid ciliates, it is possibly homologous with the more complex concrement vacuole of certain commensal prostomatid and trichostomatid ciliate groups. An organelle considered to be a kind of statocyst with statoliths (thus a "balancing" organelle?) by some workers; see **Concrement Vacuole** and **Statocyst.**

Multilocular: Refers to a foraminiferan test shell that is composed of many chambers.

Myoneme: Fibrillar protozoan organelle with a known or presumed contractile function and thus the unicellular equivalent of a muscle. Its fibrils, sometimes running deep in the cytoplasm, may be (inter)connected to one another, the pellicle, and/or certain basal bodies. In the broadest sense, The term very likely does not always refer to homologous structures: It is used to include such organized organelles as the spasmoneme found in the stalk of many peritrich ciliates, the M-band and M-fibers coursing beneath the ciliary rows in the bodies of certain contractile heterotrichs, the retractors and sphincters in various other ciliate groups, and still additional (micro)filamentous strands, bands, sheets, or bundles active in contraction or retraction of all or part of the body in species of various ciliate and nonciliate protozoan groups.

Myophrisk: Special type of myoneme in certain radiolarians that attaches the ectoplasmic layer of the cell to the skeletal rods or spines.

Myxamoeba: Term found in the myxogastrid literature for a non-flagellated, amoeboid, usually haploid cell which ultimately functions as a gamete.

Nematocyst: Spindle-shaped organelle, capable of extruding a filament coiled within it, that is found in the peripheral cytoplasm of certain dinoflagellates (e.g. *Polykrikos).* This extrusome strongly resembles the typical cnidocyst of cnidarians and is also reminiscent of the myxosporidian polar capsule and an organelle described from karyorelictid ciliates. Synonym, **Cnidocyst.**

Nematodesma (plural **Nematodesmata)**: Birefringent bundle of parallel microtubules, often showing a hexagonal, paracrystalline arrangement in cross- section; typically, the nematodesmata of ciliates appear to arise from, or at least be associated with, kinetosomes. They plunge into the cytoplasm at right angles to the pellicle, forming a major reinforcement of the walls of the cytopharyngeal apparatus (rhabdos and cyrtos) of gymnostome and hypostome ciliates, respectively, but they are found in other ciliates and a few flagellates. They comprise the armature of the oral area that was formerly identified by light microscopy as trichites, cytopharyngeal rods, or a cytopharyngeal basket.

Nucleomorph: In cryptomonads nucleomorphs are located in the periplastidial compartment. They represent a vestigial nucleus which remains from an ancestral red algal endosymbiont. The nucleomorph is small, is limited by a double membrane, and contains DNA. In addition, the nucleomorph contains a fibrillar-granular region and dense bodies

Nudiform Asexual Reproduction, **Nudiform Division**: A descriptive term for one type of division in choanoflagellates. Species with siliceous loricae may divide longitudinally and produce a naked swarmer (nudiform division), or a juvenile covered with costal strips that is pushed backwards out of the parent's lorica (**tectiform division**). **Caudiform division**, which is probably a variation of tectiform division, is observed in species of *Bicosta*, *Calliacantha,* and *Saroeca.* Shorter costal strips are accumulated at the top of the collar whilst the longer strips are produced towards the end of interphase and project within the rear of the protoplast.

Ocellus: Complex light-sensitive organelle in certain dinoflagellates, consisting of a hyaline lens (the **Hyalosome)** and a **Melanosome;** also see **Eyespot.**

Ogival field: Field of thigmotactic cilia that appears during tomite formation of an apostome ciliate.

Operculum: Lid or covering flap; used variously in different groups of protozoa: 1) as the cover of the emergence pore of some protozoan cysts; 2) lid found in some species of Licea where its apical position on sporangia aids in the exposure and release of spores; and two quite different structures in species of sessiline peritrich ciliates: 3) the stalked **Epistomial Disc** present in many opercularids; and 4) an organelle attached to the anterior end of the body—as a stalked "cap" at an oblique angle to the epistomial disc—which may wholly or partially cover the opening of the lorica (on retraction of the organism into its case) in some loricate vaginicolids.

Ophryokineties: See **Autokineties**.

Opisthe: Posterior filial product (daughter) of a regular binary fission in ciliates; contrast with **Proter**.

Opisthomastigote: Developmental stage or morphological form of a trypanosomatid flagellate in which the flagellum arises well behind the nucleus with a very long flagellar pocket running all the way to the anterior end of the body (from which the flagellum projects for a considerable distance); this elongate form, with no undulating membrane, is one of the characteristic stages in the life cyle of species of the genus *Herpetomonas,* and a superseded synonym is **Herpetomonad Stage**.

Oral Area: General term for that part of the protozoan body that bears the mouth or cytostome; in ciliates, so many of which are mouthed forms, it may be contrasted with the **Somatic Area.**

Oral Ciliature: General term for cilia, simple or compound, that are directly associated with the oral apparatus of ciliated protozoa; the subpellicular *bases* of all such structures represent the oral infraciliature (as opposed to kinds of *somatic* infraciliature).

Oral Groove: Generalized term for the depression leading to the buccal cavity or cytostome; it has been widely used in the past for the ciliate *Paramecium* to indicate what was later called a vestibulum and which *is now* considered to be a kind of prebuccal area in that organism **(Vestibulum** is thus reserved for a different usage).

Oral Cavity: Pouch or depression in oral area. See **buccal cavity**.

Oralized Somatic Kineties: Somatic kineties that function in feeding.

Oral Polykinetids, Oral Polykinetid 1 or **Membranelle 1**; **Oral Polykinetid 2** or **Membranelle 2**; and **Oral Polykinetid 3** or **Membranelle 3**, **OPk's, OPk1, OPk2, OPk3**: See **oral area**, **polykinetid**.

"Organelle of Lieberkuhn": See **Watchglass Organelle**.

Oral Ribs: Pellicular crests of a so-called non-naked ribbed wall in the buccal cavity of certain ciliates; argentophilic structures, they are associated with the infraciliature of the right-hand paroral membrane, as has now been verified ultrastructurally for many oligohymeno-phoran species.

Orthomitosis: A term used to describe the closed nuclear division of filose testate amoebae which have intranuclear MTOC's.

Ovular Nuclei: One of the two principal types of nuclei that are found in testate (arcellinids) amoebae; it has several to many small nucleoli. The other type is a **Vesicular Nucleus.**

Palintomy: Rapid sequence of binary fissions, typically within a cyst and without intervening growth; it results in the production of numerous, small-sized filial products or tomites; the phenomenon is characteristic of various parasitic ciliates, and/but it is also used to describe microgamete production in *Volvox.*

Palmella Stage: Nonmotile trophozoite within a simple, unstructured mucilaginous envelope. If/when division has occurred, the term palmelloid colony may be used; it is a characteristic permanent or transitory stage in the life cycle of certain flagellate species.

Papilla (Plural **papillae**): A raised inverted cone-tike structure from which the flagella may arise in volvocids. They can be characterized as a cell wall papilla or a plasmatic papilla.

Pansporoblast: Multinucleate body or multicellular complex that gives rise to one or more sporoblasts (later, spores) contained within a single common two-celled envelope; it is found in stages of sporogony in the life cycle of myxosporidians.

Parabasal Apparatus: Complex consisting of a **Parabasal Body** and a single (or pair of) parabasal filament(s), often connected anteriorly to a basal body; it is present in trichomonad and hypermastigote flagellates.

Parabasal Body: Organelle of various shapes (e.g oval, disc, bean-shaped) and sizes, sometimes wound about the axostyle. It is a modified Golgi apparatus characteristic of the Parabasalia.

Parabasal Filaments: Microfibrillar, periodically striated organelles, often in pairs, intimately associated with the **Parabasal Body** and arising from a basal body-complex.

Paracostal Granules: Term superseded by **Hydrogenosomes.**

Paracrystalline Fiber: The right lip of the cytostome of retortamonads is bordered by a striated or paracrystalline fiber which terminates by a posterior hook.

Paradiploid Macronuclei: Macronucleus that contains approximately the diploid amount of DNA.

Paraflagellar Body: see **Photoreceptor;** also a synonym, **Flagellar Swelling**.

Paraflagellar Rod: See **Intraflagellar Structure,** although the present term is more popular with many workers.

Parapyles: **See Astropyle**.

Parasitophorous Vacuole: Fluid-filled vacuole or "space" around the developing intracellular coccidian microgamonts and macrogametes (also around microsporidian sporonts) formed by confluence of spaces or vesicles arising in the host's cytoplasm and bounded by a unit membrane of the host cell; a disused synonym is "periparasitic vacuole."

Parasomal Sac: Small, membrane-lined, pit-like invagination in the pellicle of numerous ciliates, characteristically to the right of a ciliferous kinetosome and a regular component of the kinetid; a site of Pinocytosis.

Paratene: An alignment of kinetids from different kineties and typically at right angles to the longitudinal axis of a ciliate's body.

Paraxial Rod: See **Paraflagellar Rod;** both are popular synonyms.

Paraxostylar Granules: See **Hydrogenosomes**.

Paroral Membrane: Preferred term, used broadly, for the ciliary organelle (sometimes multiple) lying along the right side or border of the buccal cavity of ciliates with such an oral structure its kinetosomes are in a haplokinetal (zigzag) arrangement, with the inner ones barren and outer ciliferous; major synonyms, **Endoral Membrane** and **Undulating Membrane.**

Paroral Dikinitid: Dikinetid usually in a file, along the right side of the oral area of ciliates. It has *a*, *b*, and *c* segments

Pectinelle: One of a circumferential band of short rows of closely apposed cilia oriented at an oblique angle to the long axis of the body; the term pectinelle is used by some workers to describe the composition of both the **Locomotor Fringe** (or telotrochal band) of the peritrichs and the **Ciliary Girdle** of the didiniid gymnostomes.

Peduncle, Pedicel: General name for a holdfast organelle, usually long and slender and located at the posterior or aboral pole of the organism, found among diverse groups of free-living protozoa; it may be a secreted structure (as in ciliate stalks) or a part of the body proper of the organism; it is attached more or less securely to a substrate; also see **Stalk.**

Pellicle, pellicular strips: Outermost "living" layer of a protozoon, lying beneath any nonliving secreted materials; the pellicle sometimes exhibits ridges or folds or distinct crests, and it contains the typical cell or plasma membrane plus the pellicular alveoli (in ciliates, dinoflagellates, a few others) and sometimes an underlying epiplasm or other membranes; the term is loosely used by many workers as a synonym of **Cortex,** but cortical structures in ciliates (e.g., the infraciliary organelles of the kinetid) may be considered primarily subpellicular in location, as may the "independent" sheets of microtubules characteristic of trypanosomes, euglenoids, and other protozoan groups. The euglenoid pellicle consists of inter-locking proteinaceous strips that spiral along the cell and intussuscept at both ends or fuse along their entire lengths; the strips lie within the cell membrane and are flexible and elastic in some species, rigid in others; the cell is naked except for a thin layer of mucilage secreted from subpellicular muciferous bodies; some species eject large lumps of mucilage on irritation; in others, copious secretion of mucilage results in cysts or palmellae.

Pellicular Alveolus: See **Alveolus**.

Pellicular (Cuticular Pores) Pores: Numerous minute pores or openings through the pellicle that are known especially in sessiline peritrich ciliates; they are presumably used for extrusion of various secretions of a mucoid nature.

Pellicular Striae: Superficial ridges or markings in or on the pellicle; the term includes the argentophilic circumferential annuli on the "bell" of the body of many sessiline peritrich ciliates, markings possibly of taxonomic importance.

Pelta, Pelta-Axostyle Complex: Crescent-shaped structure, by light microscopy, composed of microtubules and associated with the axostylar capitulum (it actually may be an extension of that organelle that is wrapped around the blepharoplast-complex); the pelta is a characteristic structure of many trichomonads.

Peniculus: Kind of ciliary membranelle or special type of oral polykinetid found running along the left wall in the buccal cavity of certain hymenostome ciliates (e.g., *Paramecium),* its infraciliary base is usually three to seven kinetosomes in width, tapering at each end; the cilia of the two (one dorsal, one ventral) peniculi in *Paramecium* seem to move as if fused.

Pericyte: In myxosporidians, the outer of two generative cells uniting pairwise within the plasmodium to give rise to a **Pansporoblast**; it envelopes the inner one, the sporogonic cell, and becomes the pansporoblast envelope.

Peridium: Acellular structural wall that surrounds the spore mass of some mycetozoans. At maturity the peridium may be persistent as in *Perichaena*, either single as in *P. microspore*, double as in *P. depressa*, or triple as in *Physarum bogoriense*. It may have a predetermined lid as in *Perichaena corticalis* or be evanescent as in *Stemonitis*.

Periflagellar Area, Platelets: In prorocentraean dinoflagellates, the area around, and including two large anterior pores, the flagellar pore and the accessory pore, and several plates (called periflagellar platelets).

Periplasts: These envelopes are characteristic of cryptomonads. Periplasts consist of the plasma membrane and inner and surface components, and both can be variable in their composition. The inner periplast component (IPC) is protein and may be composed of many variously shaped plates or a sheet. The surface periplast component (SPC) may consist of plates, heptagonal scales, mucilage, or combinations of these. The plates of the inner periplast component are connected to the cell membrane by intramembrane particles or proteins. The arrangement of these IMP domains conform to the plate shapes which have been used to characterize genera. The periplast has pores through which the ejectisomes dock with the cell membrane.

Peristome: Term used to refer to the entire expansive oral area or peristomal field with its encircling AZM of highly evolved groups of ciliates. Synonym of buccal cavity.

Peristomal Disc: Disc-like peristomal field.

Phaeodium: Prominent pigmented (brownish-blackish) mass of (mostly) waste products around the astropyle of the central capsule of, especially, species of phaeodarian radiolarians.

Phaneroplasmodium: Members of the order Physarida have the best examples of this type of plasmodium which are the largest and often most colorful and frequently seen plasmodial type in the field. It frequently gives rise to many fruiting bodies that may cover an extensive area. In some species, this large plasmodium heaps up to form a single, massive fruiting body. During initial phases of development, it is a microscopic puddle of granular protoplasm similar to the proto-plasmodium. It soon grows larger and is visible to the naked eye, even in early development. At maturity it exhibits polarity and directional movement, terminating anteriorly in an advancing, fan-shaped feeding edge, and posteriorly in a trailing network of veins. The veins deposit excreted matter along their margins, frequently appearing as plasmodial tracks on the substratum. Veins typically have two distinct regions: an inner, fluid sol layer, which undergoes rapid rhythmic, reversible streaming surrounded by a thickened, nonstreaming gel layer. The entire plasmodium has a raised, three-dimensional aspect with definite margins. This type of plasmodium grows best under drier conditions where free water is absent.

Phanerozoite: Sporozoan merozoite that has been formed in a host red blood cell and then has entered another kind of cell to develop exoerythrocytically (e.g., in species of bird malaria).

Pharyngeal Basket: See **Cyrtos.**

Pharynx: See **Cytopharynx.**

Phoront: Stage in the polymorphic life cycle of some apostome ciliates during which the organism is carried about in an encysted state ("phoretic cyst") on the carapace of the crustacean hosts. In the full cycle, this form is preceded by a **tomite** stage and followed by a **trophont**.

Photoreceptor: Swelling near the base of the main (emergent) flagellum (and within its membrane) of the euglenoids. It is always in a position associated with or directly opposite the eyespot, near the transition region from reservoir to canal; presumably it functions, in conjunction with the eyespot, in bringing about phototactic responses of the organism; the relationship of this electron-dense crystalline structure, which contains flavins, to the rather similar appearing (though of much smaller diameter) **Intraflagellar Structure** or **Paraflagellar Rod,** which is also present within the flagellar membrane running along the length of the axoneme is not clear if there is any at all; synonyms include **Flagellar Swelling** (widely used and preferred over photoreceptor by many workers) and **Paraflagellar Body.**

Phycoplast: An assembly of microtubules oriented in the plane of cytokinesis which develops in the volvocids.

Phyllae: Radially-arrayed ribbon of microtubules lining the cytopharynx of phyllopharyngean ciliates.

Pigmentocysts: See pigment granules.

Planispiral: A foraminiferan test which is coiled in a single plane.

Planospore: An asexual reproductive cell which is motile.

Planozygote: Motile zygote of certain dinoflagellates; it may form a resting cyst from which two swarmers will be released.

Plasmodiocarp: Sessile, elongated, simple, wormlike, branched, ring-shaped, or netted fruiting body of some mycetozoans. It is formed by the plasmodium concentrating and remaining intact in the main veins during fruiting. *Hemitrichia serpula* is one example of a true plasmodiocarpous fruiting body.

Plasmodesma: Protoplasmic connection between any two adjoining "cells" of a colony or a coenobium; the plasmodesmata of *Volvox,* for example, render that organism nonmulticellular, in a strict sense.

Plasmodium (plural: Plasmodia): Multinucleate mass of protoplasm enclosed by a plasma membrane; it is a nonhomologous trophic stage in the life cycle of members of various protozoan taxa, serving a role appropriate to the group: a plasmodium may move and feed in amoeboid fashion, may undergo plasmotomy, schizogony, or encyst, etc.; thus, the word is a general term with different meanings in different situations. In the Haplosporidia the nuclei are often found in pairs

Plasmogamy: Fusion of two protoplasts, but without fusion of their nuclei.

Plasmotomy: Form of binary or occasionally mutiple fission of a multinucleate single-celled organism: the term is typically used in reference to certain "giant amoebae" (but also for opalinids and some other forms) in which the more or less evenly distributed nuclei may exhibit mitosis following, rather than during or immediately preceding, the process of somatic fission or may undergo nuclear divisions totally asynchronously, so that some mitoses may be found at any time in the two (or more) separable multinucleate masses.

Pleurotelokinetal: Development of new oral structures in ciliates within several right lateral parental kineties (e.g. colpodean ciliates).

Podite: Conical foot-like appendage projecting from the ventral surface, near the posterior pole, of certain hypostome ciliates; a mucus-like filament, used for temporary attachment to the substrate, may be secreted through it; a podite is also known as a styles (or stylus or style), and it may be homologous to the stalk of chonotrich ciliates.

Polar Cap: Chromophilic (PAS-positive) body beneath the spore wall at the anterior pole of a microsporidian spore, contained within a so-called polar sac.

Polar Capsule: Thick-walled vesicle, one to six in number, in myxosporidian spores, containing the polar (anchoring) filament; it is found in no other protozoan or even protist group.

Polar Filament: Extrusible organelle of the myxosporidians, it is a hollow elastic tube spirally wound within the individual polar capsules and everting on discharge; it anchors the hatching spore to the intestinal wall of the host's digestive tract; the nonhomologous organelle of the microsporidians is better termed a **Polar Tube**; strictly speaking, a polar filament or tube is not an **Extrusome** because they are *not* membrane-bound.

Polar Granule: Structure found in the oocyst of some coccidian sporozoa, formed at the time of the first division of the zygote in the oocyst.

Polar masses: Found in schizopyrenids in which the nucleolus forms two **polar masses** during mitosis (**promitosis**).

Polar Ring: Thickened electron-dense ring at the anterior pole of sporozoites, merozoites, and certain other stages in the life cycles of sporozoa; synonym, apical ring.

Polar Tube: Extrusible organelle of the microsporidians; a long, fine, hollow elastic tube spirally coiled within the whole spore; everting on discharge, it injects the sporoplasm into a cell of an appropriate host tissue; this "polar injecting filament," as it is often called, is neither structurally nor functionally homologous with the "polar anchoring filament" of the myxosporidians.

Polyenergid: State of having either multiple nuclei and/or multiple ploidy (see **Polyploid)** in a nucleus or nuclei within a single cell or protist body; many examples of this are to be found among diverse protozoan groups.

Polygemmic: Pertaining to production of multiple buds, synchronously or consecutively; this is a mode of fission known primarily in certain chonotrich and suctorian ciliates.

Polygenomic: See mention under **Polyploid.**

Polykinetid: Organellar complex in ciliates composed of two (=**Dikinetid)** or more kinetosomes, plus associated cilia and infraciliary organelles, in a set; they may be found in a somatic location (e.g., the ventral cirri of hypotrichs) or, more often, as parts of a compound oral (buccal) apparatus (e.g. of the AZM, or its homologue, which is present in many groups of "higher" ciliates).

Polykinety: Group or series or file of sets of polykinetids in a close arrangement of some sort (e.g., membranelle number 1 of an AZM may be considered to represent a single oral polykinety).

polymorphic Life Cycle: Life cycle consisting of different morphological forms.

Polyploid: Several or multiple sets of chromosomes in a single nucleus; this is a typical condition in the nuclei of many radiolarians (e.g., up to 3000 N), in some somatic nuclei (e.g., 30 N) of certain foraminiferans. In the usual macronuclei (16-13,000 N, according to the literature) of the ciliates; maybe "poiyploid" should often be replaced by the term "polygenomic," implying gene amplification of some kind rather than multiple sets of whole chromosomes; contrast with Diploid and Haploid.

Polysporous, Polysporoblastic: Condition in micro- and myxosporidian species in which several to many spores have been produced within a single pansporoblast; the term is to be contrasted with the **Monosporous** or **Disporous** condition (one or two spores, respectively).

Polythalamic: A foraminiferan test consisting of multiple chambers.

Polytomic: Pertaining to the division (generally rapid) of a single individual into numerous filial products, as in **Palintomy.**

Porcelanous: Wall of a calcareous foraminiferal test, characteristic of the order Miliolida which is white, shiny sometimes translucent in reflected light and commonly imperforate.

Postabdominal segments: See **Cephalis**.

Postciliary Fibers: Long-used but now a term superseded by **Postciliary Microtubules**.

Postciliary (Pc) Microtubules, Pc Ribbon: Ribbon or band of microtubules occurring widely throughout the ciliate taxa (known as the **Postciliodesma** when organized into a conspicuous stack or sheet of ribbons), arising very close to triplet number 9 of a somatic kinetosome; the ribbon first extends diagonally, to the right, upward into a pellicular crest. If well developed, it continues posteriorly parallel to and midway between the kinety containing its kinetosomes of origin and the next kinety to the right, with the sheet perpendicular to the ciliate's pellicle. They are part of the ribbed wall of hymenostomes, the cyrtos of hypostomes, etc. The term supersedes such earlier names as **Postciliary Fibers**, postciliary fibrils, and posterior microtubules.

Postciliodesma (plural **Postciliodesmata**): Specialized but now popular term for a conspicuous, posteriorly directed "fiber" composed of stacks or sheets of **Postciliary Microtubules.** It is found in karyorelictid and heterotrich ciliates, and the members of these taxonomic groups which possess postciliodesmata seldom simultaneously have kinetodesmata (which represent another, totally different, principal organelle of a kinetid): thus the structure may be of considerable phylogenetic value; a superseded synonym for this organelle in the heterotrichs is **Km Fiber**.

Postcingular Plates: A latitudinal series of plates posterior to, and contiguous with, the cingulum of dinoflagellates.

Posterior Intercalary (p) Plates: A latitudinal series of plates which lie between the postcingular series and the antapical series of plates of dinoflagellates.

Posterior Sulcal Plate: The most posterior plate of the sulcus of dinoflagellates.

Postoral Meridian: Ventral kinety of ciliates that terminates anteriorly at the posterior border of the oral area; in tetrahymenine hymenostomes, "POM #1" (rightmost if more than one) is the kinety that is active in stomatogenesis and that also bears the cytoproct at its posterior extremity (see **Kinety Number 1).**

Preaxostylar Fibers: The microtubules in parabasalids which arise from an organizing center which is close to the sigmoid fibers F2 are called **preaxostylar fibers**. This **axostyle** differs by its organization and origin from the crystalline axostyle of the oxymonads.

Prebuccal Ciliature (Area): Replacement term for the depression or oral groove leading to the inner oral area (buccal cavity) in ciliates like *Paramecium;* it is lined with somatic or slightly modified somatic ciliature—formerly called vestibular ciliature—that may now be referred to as **Prebuccal Ciliature.**

Precingular Plate Series: A latitudinal series of plates in dinoflagellates which are anterior to, and contiguous to the cingulum.

Precyst: A characteristic cell type found in arcellinids which is enclosed by a relatively thin membrane and with a diaphragm, which closes the aperture. Often this diaphragm is externally enforced by a detrital plug. Some Microcoryciidae have external cysts.

Precytostomal depression: A depression on the cell surface of ciliates preceding the cytostome.

Prehensile Tentacle: Specialized suctorian tentacle used only in the capture of prey, not in sucking out their cytoplasm; such a tentacle is particularly well known in members of the genus *Ephelota.*

Privileged Basal Bodies: See **Mastigont System** in parabasalids.

Proboscis: Trunk-like extension of the anterior end of the body of certain ciliates (e.g. *Dileptus)* with the organism's oral area at its base (not at its distal extremity); a true proboscis is not particularly shortenable; the nonhomologous "proboscis" of other ciliates, like *Didinium,* is an everted cytopharyngeal apparatus (thus a feeding organelle); also see **Rostellum.**

Procyclic Form, Procyclic Stage: Stage in the life cycle of parasitic kinetoplastid flagellates that represents the start of the developmental phase in an invertebrate host; in the *Trypanosoma brucei* group, the procyclic form is usually identified with midgut stages.

Proloculum: Initial embryonic chamber, of a foraminiferan test; an alternative spelling is **proloculus**.

Projection: See **Spine**.

Promastigote: Development stage or morphological form of a trypanosomatid flagellate in which the flagellum arises very near the anterior end of the body, well in front of the nucleus, projecting forward for some distance out of the flagellar pocket; it is the predominant form found in plants and in the insect host in species of several genera (e.g. *Leptomonas* and *Phytomonas);* superseded synonym, **Leptomonad Stage.**

Promitosis, Promitotic Nuclear Division: In heteroloboseans nuclear division has persistent nucleoli and an endonuclear division spindle.

Proter: Anterior filial product of regular binary fission of cilliates.

Protoplasmodium: This is the smallest of the plasmodial types of mycetozoans and remains microscopic throughout its existence (e.g. *Echinostelium*). The highly granular and homogeneous protoplasm has a plate-like shape that fails to develop the vein-like strands, reticulum, and advancing fans seen in other mycetozoan plasmodial types. Its protoplasm streams slowly and irregularly. Each plasmodium gives rise to a single, tiny sporangium..

Protomerite: Anucleate portion or segment of the trophont stage of a septate (cephaline) gregarine that is positioned between the **epimerite** and the **deutomerite**.

Protomite: Relatively rare stage in the polymorphic life cycle of a few ciliates (e.g., certain apostomes) that may occur between the **Tomont** and the tomont's usual products, the **Tomites.**

Protomont: Relatively rare stage in the polymorphic life cycle of a few ciliates (e.g., certain apostomes) that may occur (and be recognizable) between the feeding **Trophont** and the true, often encysted, **Tomont** stage.

Pseudoaethalioid, **Pseudoaethalium**: A term used to describe mycetozoans in which many sporangia are packed so closely together as to suggest an aethalium (e.g. *Dictydiaethalium*), or only partially fused (e.g. *Tubifera*), where the individual sporangia are clearly distinguished at maturity.

Pseudocapillitium: This structure is limited to the order Liceida (mycetozoa). Information is lacking on the mode of development. In *Dictydiaethalium* it takes the form of slender threads which are remnants of angles between adjoining peridial walls. In species of *Lycogala* (Fig. 19) it consists of irregular, non-calcareous tubules and is usually over 6 μm in diameter. In *Enteridium* it occurs as perforated plates or membranes and in *Tubifera* as bristles.

Pseudocolumella: Calcareous mass freely suspended in the center of the fruiting body and resembling a columella, but disconnected from the tip of the stalk and at the base (e.g. *Graterium leucocephalum*, *Physarella oblonga*, and *Physarum stellatum*).

Pseudocyst: Cyst-like stage or structure, often containing numerous cells of a protozoan parasite, the wall or outer membrane(s) of which has (have) been formed by the host cell, *not* by the host's intracellular parasite; the word is also a superseded term for the residuum produced in the gametocysts of some gregarines.

Pseudoplasmodium: Sausage-shaped amoeboid structure that consists of an aggregation of many myxamoebae and behaves as a unit; this is a stage that occurs in the life cycle of the cellular slime molds (e.g. *Dictyostelium)* and gives rise to one or more fruiting bodies or sorocarps; synonyms, grex and slug.

Pseudopodial types: Cylindrical amoebae can be **polypodial** or **monopodial**, the latter often being referred to as **limax**. An important dichotomy is whether locomotion is steady or eruptive; amoebae with markedly eruptive pseudopodia are frequently in the class Heterolobosea. The form of the pseuodopodia is important. Most amoebae advance by a single pseudopodium or by a succession of progressing pseudopodia. From the anterior edge, extending subpseudopodia are common. These can be short, long, slender, or thick and may be distinctive in shape. Some are **digitiform** or **mamilliform** in shape, **furcate** (branched) or **non-furcate**. Subpseudopodia that branch usually do so close to their base. The anterior edge of leading pseudopodia often have a clear zone, called a hyaline cap. The extent of this zone is important for identification as it can be thin, broad, or even absent.

Pseudopodium: "False-foot," a generally temporary, retractile cytoplasmic protrusion typical of protozoan cells devoid of a rigid pellicle; thus, the transient organelle is most characteristic of "sarcodinid" *sensu lato* protozoa, although it occurs in other taxonomic groups as well. Pseudopodia (or pseudopods) function in feeding and/or locomotion. See further description under these four main types: **Axopodium, Filopodium, Lobopodium**, and **Reticulopodium.**

Pseudostome: The aperture of the test or shell of a free-living testate amoeba. Pseudopodia protrude from it during locomotion and/or feeding. The pseudostome of arcellids can be protected by teeth or located at the end of a **tubus** (tubular section of test leading inwards from aperture).

Pusule: Distinct vacuole near the annulus (girdle) of both fresh-water and marine dinoflagellates, opening to the outside by a tubular canal near the base of the transverse flagellum; the pusule sometimes forms an extensive system of (smaller) vacuoles and additional canals in a specialized region of the cytoplasm, its exact function is unknown, but is generally presumed to be osmoregulatory.

Pylome: A tube-like extension from the surface of the spumellarian skeleton (either entire, latticed, and/or ornamented with spines), or a funnel-like groove in the margin of the spongy disk, is known as a pylome.

Quadrulus: Compound ciliary organelle in the buccal cavity of the hymenostome ciliate *Paramecium,* on the left wall not far from the location of the two peniculi; it possesses long cilia arising from an infraciliary base that is four kinetosomes in width and many in length; few other ciliates possess this particular buccal organelle, a specialized oral polykinetid.

Quinqueloculine: Test form of some miliolid foraminifera whose chambers are one-half coil in length. Successive chambers are added 144° apart, resulting in chambers in five planes 72° apart, as in *Quinqueloculina*.

Radial Transverse Ribbons: A ribbon of transverse microtubules oriented radially to the kinetosome's perimeter.

Radial Wall: Test of a foraminiferan that is composed of calcite or aragonite crystals with the c-axis perpendicular to the surface. It is seen between crossed Nicols as a black cross with concentric rings of color mimicking the negative uniaxial figure.

Rectilinear: A test of a foraminifera which is formed from succesive chambers which have been added in a straight line. Terms refers to uniserial or biserial forms.

Recurrent Flagellum: Backwardly directed flagellum, running posteriorly over the body of the flagellate in a loose or attached state; when attached, it often becomes part of an undulating membrane, one edge of which is, in effect, continuous with the surface of the body while the other—the outer—contains the axoneme of the flagellum; a recurrent flagellum is common (e.g. trichomonads).

Reorganization Band: Lightly staining cross-band of a ciliate macronucleus that typically moves through one-half the length of the nucleus (to meet a similar band sweeping through the nucleoplasm from the other end) preceding division. The area is involved in DNA replication and histone synthesis. These bands are commonly found in hypotrichs, but in some species they start midway and move out to the ends. In chonotrichs and certain cyrtophorid hypostomes there is only one band, which moves across the orthomere to its free end; a newer and more appropriate name is **Replication Band**.

Reserve Plates: Found in testate filose amoebae (Euglyphida). They are stacked at the margins of the posterior part of the parental cell prior to construction of daughter cells. Golgi complexes (dictyosomes) participate in the production of siliceous building elements and in the organic cement. In well nourished cells there are large electron-dense vesicles situated anterior to this zone.

Reservoir: Specifically used to identify an ampule-shaped invagination (or its posteriorly enlarged portion) of the cell surface from which one or two flagella may emerge and which is found in species belonging to the Euglenozoa.

Residuum: Literally, remaining material; in coccidian biology, the term is preceded by an *adjectival noun* descriptive of the stage in the life cycle involved: viz., the material remaining in a gametocyst after formation of gametes and zygotes is the *gametocyst* residuum; in an oocyst after formation of sporocysts, the *oocyst* residuum; and in a sporocyst after formation of sporozoites, the *sporocyst* residuum.

Reticulate: The textured shell surface of a foraminiferan marked by a network of raised, but not necessarily regular, ornamental ridges.

Reticulopodium (Granuloreticulopodium): Very slender, repeatedly branching and anastomosing, thread-like granule containing pseudopodium. Throughout the network that is formed, a two-way flow of cytoplasm and food particles is detectable.

Retractor Fibers: Generalized term for bundles of myonemes, found in various ciliates, used to draw back (retract) a protruding oral area or some other extended part of the body.

Retral Processes: Proximally projecting extensions of a foraminiferan chamber cavity located beneath the external ridges on the chamber wall and ending blindly at the face of the previous chamber.

Rhabdocyst: Short, rod-like organelle (an extrusome) found subpellicularly in the cytoplasm of certain karyorelictid ciliates.

Rhabdos: Tubular cytopharyngeal apparatus in ciliates, noncurved, with its walls strengthened by bundles of nematodesmata and its interior surface often lined with extensions of transverse microtubules derived from circumoral kinetosomes; it may contain toxicysts, and it may be bound, near its distal (outer) end, by an expansible filamentous annulus; although, like the Cyrtos, the rhabdos is found among members of the less advanced ciliate groups, it is considered evolutionarily the more primitive organelle; synonym, **Pseudonassc.**

Rhizoplast: See **Flagellar Rootlet**.

Rhizostyle: A rhizostyle is an intergral component of the flagellar apparatus in most cryptomonads, and it consists of microtubules that originate near the basal bodies and then extend posteriorly into the cell .

Rhoptry, (plural Rhoptries): Elongate electron-dense tubular, saccular, or club-shaped organelle, two to eight in number, often enlarged at its posterior end, extending back from the anterior pole of a sporozoan sporozoite or merozoite through the conoid toward the equatorial level of the cell (although terminating considerably more anteriorly in many species); only a single pair occurs in sporozoites of *Plasmodium* and in merozoites of *Eimeria;* this term supersedes various earlier words (generally formed from some taxonomic group-name), including the, at first seemingly quite appropriate term **Paired Organelle.**

Ribbed Wall: Nonciliated surface of the right side of the buccal cavity of many oligohymenophoran ciliates; it appears "ribbed' because of the regular presence of crests of the pellicular alveoli underpinned by numerous postciliary microtubules (from kinetosomes of the nearby paroral membrane) that are arranged perpendicular to the outer pellicular surface at orderly spaced intervals.

Right Accessory Sulcal Plate, Right Sulcal Plate: In gonyaulacalean dinoflagellates the right sulcal plate is located on the right side of the sulcus and is bounded anteriorly by the anterior sulcal plate or the right accessory sulcal plate and bounded posteriorly by the posterior sulcal plate.

Ring Form: Young haemosporidian trophozoite produced from a merozoite that has just invaded an erythrocyte; it is named from its appearance (that of a signet ring) following Romanowsky staining; synonym, ring stage.

Rootlet: In a very general sense, any fibrillar or microtubular structure originating from, or near, a basal body or cluster of basal bodies (in flagellates or ciliates) that courses into the cytoplasm away from the pellicle while remaining more or less perpendicular to it; today, the term—in a more restricted sense—is most often used for a fibrillar, cross-striated flagellar rootlet or root (or rhizoplast), of which there are several kinds (not all homologous), and there may be *no* homologous structure in ciliates; see **Flagellar Rootlet**.

Rosette, Rosette Organelle: Group of cells in a circle-formation (i.e., forming a rosette) loosely attached, in the case of trypanosomatid flagellates, by their flagella (e.g., as exhibited by certain epimastigotes in insect gut) or by the posterior poles of their bodies in an incomplete multiple fission stage (e.g., as shown by certain trypomastigotes in rat blood); occasionally, members of other protozoan groups (e.g., some piroplasmids; and see **Gymnospore)** may temporarily exhibit rosette formation in their life cycles; in addition, the term is explicitly given to a unique septate structure near the cytostome of many apostome ciliates, an ultrastructurally complex organelle with function unclear.

Rostellum: Trunk-like extension of the anterior end of certain parasitic flagellates (e.g., oxymonads) that serves as an attachment organelle; it contains a part of the axostyle and may be enlarged at its own anterior tip to form a sucker-like cup; also see **Proboscis**.

Rostral Tube: Cone-shaped tube in the apical region of certain hypermastigotes. It consists primarily of densely packed Golgi bodies.

Rostrum: The apical end or tip of a protozoan body, especially of ciliates and flagellates, when its shape is that of a beak or when there is some other sort of distinctive protuberance in that area of the body. It may bear the cytostome in some ciliates, such as *Chaenea,* or a unique **Sucking Tube,** as in all members of the order Rhynchodida; a rostrum is generally much less extensive and less conspicuous than either a **Rostellum** or a **Proboscis**.

Rotallid Wall: A foraminiferan test in which each chamber is composed of a single layer that continues as a thin lamella over the entire previously formed exterior surface.

Sagittal Suture: The serrated suture which separates the theca of two major groups of dinoflagellates (dinophysialean and prorocentralean) and the hyposome of nannoceratopsialeans into two lateral halves.

Saltatorial Cilia: Long, and often stiff or thick-looking, cilia typically distributed sparsely on or around the body of certain ciliates (e.g. the oligotrich *Halteria)* and used in a quick, jerky sort of locomotion readily recognizable by its unusualness.

Sarcocyst: Last-generation meront of the sporozoan *Sarcocystis* in the muscles of the intermediate host.

Schizogony: Term restricted primarily to mean formation of daughter cells among sporozoan groups (e.g., in the haemosporidians) and foraminifera, by multiple fission; In the former group the products are merozoites, the process can be (sub)termed **Merogony**; if gametes, **Gametogony**; if sporozoites, **Sporogony**.

Schizont: Stage or form in the life cycle that undergoes **Schizogony** (multiple fission); often reserved or substituted for the term Meront; also see Tomont, which is primarily a ciliate term; schizont is essentially a word used with reference to sporozoa and foraminifera.

Sclerotium (plural **Sclerotia**): Dormant, resting structure of mycetozoans formed from the plasmodium under unfavorable environmental conditions. Lower temperatures, decreased moisture, depleted nutrient conditions, and aging are probably responsible for sclerotium formation. This stage consists of multinucleate, walled, plasmodial segments of various sizes sometimes called "spherules". It may remain dormant for long periods of time but upon return of favorable environmental conditions may revive quickly giving rise to the plasmodial stage. Many species of corticolous myxogastrids on living trees and vines form piasmodia that undergo encystment and become sclerotized on the bark surface. In this case the sclerotium dries as a thin sheet of horny material quite unlike that of many species of

Physaridae, (e.g. *Badhamia, Badhamiopsis,* and *Physarum*) which may form small kernel-like masses under decaying logs. The formation of sclerotia on the bark of living trees accounts in part for the rapid re-emergence of plasmodia after brief rains and the rather short periods of time (24-48 h) before sporulation.

Scopula, Scopular Disc: Compound organelle, structure, or area, often cup-shaped with a thickened peripheral border or lip, at the aboral pole of sessiline peritrich ciliates that is composed of a field of kinetosomes bearing very short and immobile cilia (lacking the central pair of microtubules); it may function directly as a holdfast organelle or, more commonly, as the source of a secreted peduncle or stalk of considerable complexity; a peritrich scopula is possibly homologous with the **Thigmotactic Ciliature** area of some scuticociliates, but not at all with the **Scopuloid** of suctorians.

Scopuloid: Organelle or area found at the posterior pole of the body of most suctorian ciliates; it is analogous, but *not* homologous, with the **Scopula** of peritrichs in that it also functions in production of a stalk, but the kind of stalk produced is characteristic only of suctorians (differing in a number of significant ways from those of both peritrichs and chonotrichs, the latter ciliates with still a third type of stalk).

Scutica: Transient multi-kinetosomal structure found in each presumptive filial organism during the process of stomatogenesis (and cell division) in a major taxonomic group (Scuticociliates) of ciliates. Typically, it manifests a hook-like or whiplash configuration at the last ontogenetic stage of fission in which it appears as a primordium, and it is without cilia at that time.

Scutico-field: See **Scutica**, but the present term is generally used to refer to earlier stages of the stomatogenic anlage in scuticociliates, at a time when it includes more kinetosomes.

Segmenter: Late meront or schizont (e.g. haemosporidian sporozoa) which is or has been producing merozoites.

Sensory Bristles: Name rather widely applied to many so-called bristles or setae in ciliated protozoa although the function implied has seldom been demonstrated experimentally; it has very often been used, for example, with respect to the several short rows of clavate cilia of *Didinium* and the very short cilia found in pits on the dorsal surface of *Euplotes* and other such hypotrichs.

Sensory Organelle: Generalized term widely used, especially in studies on ciliates, for a variety of very likely nonhomologous structures or organelles considered to play a role in the cell's reactions to environmental stimuli, but, to date, seldom has there been solid experimental evidence of the "sensory" functioning of such structures.

Septate, Septum, Septa: Condition of having partitions that separate parts of the body into segments (as in certain gregarine sporozoa) or that divide a test into several chambers (as in certain foraminiferans); contrast with **Aseptate.**

Shell: See **Test** and **Lorica**. But, separate from its use as direct synonyms of those two terms, the word is also employed in specific reference to the allegedly chitinous spore membrane characteristic of myxosporidians. The myxosporidian shell consists of two to six valves or sections with suture planes between them and often with sculpturing, markings, or processes on the outside surfaces.

Sigmoid Fibers: See **Mastigont system** in parabasalids.

Silica Deposition Vesicles (SDVs): In Chrysomonads scales are formed endogenously in silica deposition vesicles (SDVs) near the periphery of the cell, and silica is then deposited in the vesicles. The SDVs ultimately fuse with the plasma membrane, and the mature scales are extruded onto the cell surface where they are adhered in some unknown way.

Siliceous Tests: Produced by arcellinids; composed of endogenous rods, nails or rectangular plates

Silverline System: See **Argyrome**; but the present term is generally the more popular one.

Silver Impregnation Techniques: Cytological methods allowing cellular structures or organelles (or their sites) of the morphology of certain flagellates and of many ciliates with affinity for silver ions to be revealed on stained microscopic slides. See details in *Protocols in Protozoology.*

Simple Pedicel: In choanoflagellates the term simple pedicel refers to a single costal strip or a line of costal strips attached end-to-end.

Skeletal Plates: Usable in a general way for any plate-shaped structures lending strength or thickness to a wall, pellicle, or cortical region of a protozoan body, also see **Thecal Plates**. In protozoology, the term is usually reserved for the often conspicuous intracytoplasmic structures composed of amylopectin embedded in a fibrillar lattice that are characteristic of many of the rumen-dwelling entodiniomorphid (ophryoscolecid) ciliates.

Skeletons, External: Few protozoan structures fit this category specifically: The usual "external supporting structures" are much more widely represented by the numerous loricae, tests, shells, thecae, scales, walls, and/or pellicular thickenings of various sorts—though even some of these are not always strictly "external" in position and certainly are not all homologous—found throughout the major taxonomic groups. The silicoflagellates, however, are said to have a true external skeleton composed of quite delicate siliceous tubular elements, often fused together at junction points, plus projecting spines.

Skeletons, Internal: "True" supporting skeletons, composed of either crystalline strontium sulfate or hydrated amorphous silica (sometimes plus organic substances of unknown nature), are characteristic solely of certain major groups within the actinopods. They are composed of rods, spines, needles, or spicules—often supported by a fused grid or framework of various shapes—typically forming an overall lattice-like structure of high intricacy, diverse symmetry, and exquisite beauty; skeletons of the taxa lumped together by some workers as radiolarians are basically of silicon, those of the acantharians are of strontium sulfate; the skeletal spines of the latter group are arranged in a definite geometrical pattern known as **Muller's Law**. A few heliozoa also possess siliceous internal skeletons, generally delicate and of little complexity, as well as surface scales or spines. Fossil ebriids have internal skeletons, in some actinopod groups (e.g. the spumellarid polycystine radiolarians), the lattice-like skeleton is commonly termed a **Shell**, and, in cases of its multiple concentric arrangement, the outermost one(s) is(are) called **Cortical Shell(s)** and the innermost, **Medullary Shell(s)**. The spheres so represented are sometimes called **Macrospheres** and **Microspheres,** respectively.

Slug: See **Pseudoplasmodium**.

Somatic Area: General term for all parts of the body surface of the ciliated protozoa that are not intimately associated with the oral structures; it may have its own subdivisions; contrast with **Oral Area**.

Somatic Ciliature: All-inclusive term for cilia or compound ciliary organelles found anywhere on the surface of the body other than in the oral area; associated with it would be the bases of all such structures, viz., the **Somatic Infraciliature**; contrast with **Oral Ciliature**.

Somatonemes: Tubular hairs, ultrastructurally resembling the thick **Flagellar Hairs** or mastigonemes of chrysomonads, found externally covering the posterior part of the body of certain proteromonad flagellates; associated with subpellicular microtubules. They are presumably synthesized in Golgi bodies and transported to the cell surface via secretory vacuoles.

Sorocarp: Fruiting body of acrasids and dictyostelids (the "cellular slime molds" of the mycetozoa *sensu lato),* developing atop an aerial stalk after earlier aggregation of amoeboid cells that have differentiated into a complex compound structure (sometimes called a sorogen); spores mature in the sorocarp (now called a **Sorus** in dictyostelids), either in chains or in sort, and are eventually released into the air, becoming the source of new populations of the amoeboid stage again. One remarkable ciliated protozoon *(Sorogena)* exhibits sorogenesis, producing a sorocarp with sorocysts (spores), etc., but the phenomenon is characteristic only of the two groups mentioned above. The term is not to be confused with **Sporocarp** (see below), although both structures are part of a process for the ultimate production of multiple spores, a means for propagation and dispersal of the species.

Sorogenesis: Process leading to the production of a sorocarp; see **Sorocarp**.

Sorus: Mature sorocarp, with developing spores, in dictyostelids; see **Sorocarp**.

Spasmoneme: Term limited to a myoneme found in the contractile stalks of peritrich ciliates; see **Myoneme**.

Spherulosome (**Spherule**): Membranous organelle found at the anterior end of the haplosporidian sporoplasm. It consists of convoluted membrane-bound cisternae and may serve as a site for formation of haplosporosomes. In addition, it appears to be involved in excystment.

Spindle Trichocyst: See **Trichocyst**; but the present term is the preferred synonym in the view of many workers.

Spine: In loricate choanoflagellates longitudinal costae may project above the anterior transverse costa. Two terms are applied to describe these projections, namely **spine**, being a continuation of the longitudinal costa above the transverse costa for less than the length of one costal strip and **projection**, being a continuation of a longitudinal costa above the transverse costa that is equal or longer than one costal strip.

Spiral side: Descriptive term for the side of trochospiral foraminiferal test in which all whorls are visible.

Spongiome: See **Spongioplasm.**

Spongioplasm: Specialized secretory cytoplasm, of spongy appearance, found in the vicinity of the contractile vacuoles in ciliates and some flagellates; synonym, **Spongiome.**

Sporadin: Older term for the gamont of a gregarine that has lost its epimerite or mucron and is moving about freely in the lumen of its host's gut; a sporadin is often considered to be a mature trophozoite, although it will soon engage in syzygy leading to the production of numerous gametes.

Sporangium (plural **sporangia**): Small, stipitate (stalked) or non-stipitate (sessile) structure of mycetozoans, generally less than 1 mm in diameter, of definite shape and restricted size, often roughly spherical in shape, formed by the clumping of the plasmodium. In certain minute species a plasmodium may form only one sporangium but under ideal conditions for phaneroplasmodial species (Physaridae) hundreds may be present.

Spore Case: The spore-filled portion of the fruiting body of mycetozoans without the stalk.

Spore: Very general term in protozoology widely used for a variety of infective, propagative, or resistant stages in the life cycles of species belonging to different major taxa. Unfortunately, the stages so named or described are seldom homologous, sometimes even *within* a single high-level group (e.g. the coccidians). Universally recognized and accepted, however, are the propagative spores developing from the fruiting bodies (or sporangia) of free-living mycetozoan groups and the highly resistant spores of the micro- and myxosporidian groups. Controversial is the use of "spore" for the oocysts of gregarines or the sporocysts of coccidians, even though these stages are both resistant and infective (i.e. they carry the sporozoites that invade fresh hosts—and thus propagate the species—under appropriate conditions). In mycetozoans spore color, shape, size, and ornamentation are among the more constant characteristics used in species descriptions. Spores generally are globose in shape, free as single units, or aggregated into loose or tight clusters; size ranges from 4 μm to 22 μm in diameter, and wall surfaces may be smooth, spiny, warty, or reticulate (partially, evenly or banded), or variations of these basic wall markings. In mycetozoans mature spores are generally uninucleate and typically haploid. Spore walls serve to protect this resting stage during airborne dissemination until arrival at a site with adequate free water for germination.

Sporoblast: Cell formed by division of the sporont (essentially the zygote) in the life cycle of coccidians, thus an immature coccidian sporocyst. On the other hand, in myxosporidians, it is the cell group within a pansporoblast envelope that will be transformed into the multicellular spore typical for that group. In the Haplosporidia the term is used for one of numerous uninucleate cells formed in a sporont and from which spores are formed.

Sporocarp: Fruiting body of the myxogastrians (Mycetozoa), developing atop an aerial stalk that has arisen from an often large multinucleate plasmodial stage and containing one or more sporangia (each with its developing spores). It releases mature spores under appropriate stimulation, individuals that germinate and grow as amoeboid or flagellated cells to eventually produce, first by fusion, the plasmodial stage, by **Sporogenesis**, mature sporangia are again produced. In the protostelid mycetozoa, the process is much simpler: the mature sporocarp, again stalked, is often composed of only one (maximum number, four) spore(s); see **Aethalium** and **Sorocarp.**

Sporocyst: Cyst formed within the oocyst in the life cycle of most coccidians, it will contain the sporozoites. The term, erroneously is also often applied to the oocysts of gregarines, which have no sporocyst stage; for the microsporidians and haplosporidians, the sporocyst is an envelope or cyst containing spores.

Sporoducts: Long tubes radiating out from the spherical gametocyst of certain gregarines, used for extrusion of the oocysts ("spores") to the outside; often the minute oocysts—numbering in the thousands—pass out in a single file in lengthy chains; synonym, **Ooducts**.

Sporogenesis: Term used by mycologists and "mycetozoologists" to describe the development of sporocarps; see **Sporocarp.**

Sporogonic Cell: In myxosporidians, the inner of two generative cells that unite pairwise within the plasmodium to give rise to a **Pansporoblast**; it produces the cells constituting the spores.

Sporogony: Formation of sporocysts (if present in the life cycle) and sporozoites, in coccidians and other sporozoa, by (repeated) division of a zygote. It is a type of schizogony *sensu lato;* nuclear one-step meiosis takes place in the first division, but only mitosis occurs in the subsequent fissions.

Sporokinete: Seldom-used term for the motile division product of an oocyst in a few coccidians (e.g. *Karyolysus);* the vermiform organism apparently represents a homologue of the sporoblast in other coccidians.

Sporont: Stage in the life cycle of coccidians that will form sporocysts; in other words, it is the zygote within the oocyst wall. In the haplosporidians the sporont is a multinucleate cell which is enclosed in a thin wall and which is derived from a plasmodium. It is the first cellular type in sporulation and becomes a sporocyst in which spores are formed.

Sporoplasm: Amoeboid "germ" cell, or amoebula, within the spores of micro- and myxosporidians; it is the infective stage on emergence (or extrusion) from such spores.

Sporoplasmosome: Term used for haplosporosomes in the Myxozoa (see haplosporosome)

Sporozoite: Motile infective stage in gregarine and coccidian life cycles that is produced by sporogony; this usually takes place within an envelope or shell (although not so for the malarial organisms).

Sporulation: Broadly, the process of spore production; for the sporozoa, it may be considered the process of sporozoite production (especially in the shed coccidian oocyst), that is, **Sporogony.**

Stalk: General term, broadly used, for various sorts of pillar- or threadlike supporting structures for the body, usually of a single protozoan; of varying lengths, it is typically attached to the posterior pole of the body, at one of its own ends, and, at its other end, to one of a great number of possible substrates. In the case of the unusual stalks of the mycetozoa, each one supports (fruiting bodies containing) multiple organisms (ultimately, as spores). Stalks may be of varying sizes and composition, may be living or nonliving, may be secreted or produced some other way, may or may not be contractile, etc.. Specific kinds ean be diagnostic for species of an entire major group (e.g., chonotrich, peritrich, and suctorian ciliates).

Statocyst: See **Muller's Vesicle.**

Statospore: Resting stage or resistant cyst of certain flagellates (e.g., chrysomonads, Synurophyceae), generally spherical with a silicified (and sometimes iron-impregnated) outer surface that is smooth or adorned with warts, spines, etc. and a distinctive pore with a collar closed by a plug. This unique cyst is formed from the vegetative cell by a complex process. On germination, the plug is dissolved and the organism (protoplast) emerges by amoeboid movement, forming flagella by the time it is entirely freed from the statospore. Synonym: **Aplanospore**.

Stenoxenous: A parasite having a narrow host range, occurring in species of hosts that all belong to the same taxonomic group at the familial level. If the parasite is further restricted to only a single species of host, the proper term is Monoxenous.

Stercomata: Fecal pellets produced by some granuloreticulopods.

Stichodyad: See **Haplokinety**.

Stieda Body: Knob-like thickening at one end of the sporocyst wall in coccidians.

Stigma: See **Eyespot.**

Stomatocysts or **Statospores**: In Chrysomonada and Synurophyceae these cysts are common, usually spherical with either smooth or variously ornamented surfaces. They are composed of silica and have a narrow neck that is closed at maturity by a plug. At germination the plug is lost and the amoeboid protoplast emerges through the neck. The stomatocyst is a fundamental and diagnostic character of the Chrysomonada and Synurophyceae. Undifferentiated cells may act as gametes, fuse apically, and produce a short-lived, binucleate, quadriflagellate planozygote that subsequently undergoes encystment to become a binucleate hypnozygote or sexual stomatocyst.

Stomatogenesis: Literally, (new) mouth-formation; a dynamic phenomenon that may become quite complex in, for example, the so-called "more highly evolved" ciliates, since it embraces neoformation or replacement of all oral structures and infrastructures, plus associated openings, cavities, and the like, in both the proter and the opisthe during binary fission of the ciliate proper.

Stomatogenic Field: See **Stomatogenesis**.

Streptospiral: Foraminiferal test coiled like a ball of string, in continually changing planes.

Strobilation: Kind of rapid multiple fission in which successive tomites or buds (or daughter organisms) are typically—at first—only partially separated or pinched off from one another, resulting in a (temporary) linear chain of small individuals. The outcome is thus a kind of **Catenoid Colony**.

Style, Stylet, or **Stylus**: See **Podite**.

Subkinetal Microtubules: One or more ribbons of microtubules coursing along under, and very near, the proximal ends of the kinetosomes of a kinety in certain kinetofragminophoran ciliates; they are sometimes confused with **Basal Microtubules**.

Subpellicular Microtubules: Ribbons or sheet of microtubules lying just under the pellicle and coursing along parallel to the outside surface of the organism's body; they are found in species belonging to various major taxonomie groups; origins of these particular microtubules and whether or not such ribbons are homologous in all eases remain largely unresolved questions.

Sucker: Variously used, but generally for any cup-shaped concavity or structure used as an adhesive or holdfast organelle; it may involve most of the body of an organism or only some specialized part or appendage; it may bear **Thigmotactic Ciliature**. Suckers in species of different protozoan taxa are usually very likely nonhomologous structures.

Sucking Disc: Concave area of the anterior part of the body in the diplomonad flagellate *Giardia by* which the organism adheres to the host's epithelium; also see **Sucker**.

Sucking Tube: Apically located (rostrum-borne), complex septate structure composed of microtubules and serving as the ingestatory apparatus of rhynchodid hypostome ciliates; it is presumably homologous with the **Cyrtos**.

Suctorial Tentacle, Suctorial Tube: Tubular extension of the body of suctorian ciliates, extensible and retractable, containing a complex array of longitudinally arranged microtubules and equipped with (mature) haptocysts at its capitate tip and elsewhere, and vesicles functioning in the formation of food vacuolar membrane. The organelle, typically several in number and distributed variously (sometimes in specific patterns or grouped in fascicles) over the body of the organism, serves both for prey capture and as a holozoic ingestatory apparatus (thus, suctorians are "polystomic").

Sulcus: Longitudinal furrow or groove on the body of dinoflagellates, generally coursing from the posterior end anteriorly to an equatorial level where it becomes confluent with the transverse annulus (girdle, cingulum). It contains the insertion of (and often the proximal part of) the longitudinal flagellum, which runs posteriorly beyond the edge of the body.

Surface Net: The system of branching capillitial threads found in the mycetozoan *Stemonitis* which arise from the columella and eventually fuse at the surface into a net.

Suture, Suture Lines: Interfaces, planes, or lines of union between different parts of the body (surface) or of its coverings found quite widespread among protozoan groups (e.g line of union between adjacent chambers [intercameral suture] of a foraminiferal test or between two whorls [spiral suture] of evolutely coiled tests, between plates of a pellicle in dinoflagellates, between the contiguous shell valves of the body in myxosporidian spores, etc.); in ciliates, they are associated with what French workers call the **Systeme Secant**: here they represent various lines of convergence of fields of kinetics over the surface of the body, not limited to the long known pre- and postoral sutures but also including the boundaries of large areas-particularly striking in various taxonomic groups (e.g., in thigmotrichs, astomes, and heterotrichs)—that do not have regular rows of simply bipolar kineties.

Swarmer, Swarm Cell: Dispersive form in the life cycle of numerous protozoa; the term is used for the biflagellated zoospores of marine actinopods, and the motile asexual propagative cell in the life cycle of many pigmented flagellates (e.g. the dinoflagellate, *Noctiluca*). In the case of ciliates, the word is typically employed to denote the detached, free-swimming stages arising from sessile or sedentary adult forms (peritrichs, suctorians, etc.); In mycetozoans swarm cells are comma-shaped, usually unequally biflagellate, uninucleate, haploid amoebo-flagellate cells capable of functioning as gametes.

Synaptonemal Complexes: In microsporidia sporogony starts with the formation of plasmodia. Meiosis is indicated by synaptonemal complexes in the nuclei.

Syncilium: Group of closely packed cilia forming a special tuft exhibiting considerable internal coherence and arising from a packet of kinetosomes interconnected subpellicularly with like packets in the vicinity; the compound ciliary structure is especially characteristic of many of the entodiniomorphid (ophryoscolecid) ciliates living in ruminants.

Syncytial: See **Coenocytic,** although the term syncytial is the synonym preferred by most protozoologists.

Syngamy: Fusion of haploid gametes or gametic cells in **Sexual Reproduction** to form a diploid zygote; the phenomenon is exhibited by various members of all major protozoan taxa except the ciliates.

Syzygy: Association side-by-side or end-to-end of gamonts (especially of gregarines) prior to formation of gametocysts and gametes.

Tabulation: Refers to the arrangement of thecal plates on the surface of dinoflagellates. There are several systems (Kofoid, Taylor-Evitt) for naming and comparing the plates.

Tachyzoite: Fast-developing merozoite; the term is especially used in reference to merozoite production in aggregates of the sporozoan *Toxoplasma* during the acute stage of the infection.

Tangential Transverse Ribbon: A ribbon of transverse microtubules oriented tangentially to the kinetosome's perimeter.

Tectiform Asexual Reproduction, Tectiform Division: See **Nudiform Division**.

Tectum: A flexible, secreted covering over the body surface of certain amoebae.

Teeth, Tooth: A projection from the lip of the aperture, or larger projections in the umbilicus of foraminiferal tests which act as sieves for food as the reticulopodia are drawn through and past them. Also the term is used for spine-like protrusions called "teeth" on the rim of the terminal chamber of polycystines, either the thorax or abdomen, may bear teeth.

Tela Corticalis: A microfilamentous layer separating the cortex from the rest of the cytoplasm in some ciliates (e.g. litostomes).

Telokinetal: Type of stomatogenesis in ciliates in which formation of the new oral area(s) occurs principally by direct involvement of kinetosomes at the anterior extremities of all or some of the somatic kinetics. The resulting oral areas (cytostomes, circlets of cilia,. etc.) are essentially apical, or subapical, in position.

Telomerozoites: Merozoites of the last schizogonic generation preceding gamogony in the coccidian life cycle.

Telotroch: Motile migratory stage in the life cycle of sessiline peritrich ciliates; also see **Larval Form**.

Telotroch Band: Ring of cilia at the posterior end of a telotroch.

Tentacle: Term broadly usable for any "tentacle-like" projection of or from a protozoan body (e.g. see **Proboscis, Rostellum,** and **Sucking Tube**), but generally reserved for three quite different organelles: the conspicuous, mobile, contractile cytoplasmic projection arising in the area of the sulcus in certain dinoflagellates (notably *Noctiluca)* and projecting posteriorly; the nonsuctorial but highly extensible and retractable prey-capturing organelle of such haptorid gymnostomes as *Actinobolina,* composed of microtubular arrays often enclosing a prominent toxicyst and found in abundance on the body in association with clusters of somatic cilia; and the **Prehensile** and **Suctorial Tentacles** of suctorian ciliates, the second of the last-mentioned types (in the Suctoria) is the most commonly known, widely distributed kind of tentacle among all protozoa. Tentacles (microvilli) are also found at the anterior end of Choanoflagellates. The single flagellum is surrounded by a funnel-shaped collar of tentacles or microvilli. The number of tentacles varies between 20-50 and is approximately consistent for an individual species.

Tentaculoid: Small, contractile, finger-like extensions of the cytoplasm found among the oral membranelles of some tintinnine spirotrich ciliates.

Test: General term, essentially synonymous with **Lorica** and **Shell**, for an agglutinated or secreted outer covering, or "house" in which an organism may live (or leave, via a sizable aperture). Thus, unlike a theca, it is not very closely fitting, although its dweller may attach itself more or less temporarily to it by protoplasmic strands. In the case of the perforate calcareous test of foraminiferans, there is actually a layer of cytoplasm (often with reticulopodia) *outside* the shell. The term is applied to the testate amoebae with the shape, composition, origin, fate during fission, etc. different for different groups. Tests may be gelatinous, tectinous, pseudochitinous, siliceous, calcareous, membranous, or arenaceous, The exterior of tests may be smooth or rough, bear spines or other decorations or attached granules of various kinds.

Tetrahymenal Buccal Apparatus: See **Buccal apparatus**; but the present term refers to the "four-membraned" nature of the oral ciliature so characteristic of many oligohymenophoran ciliates; that is, the basic pattern consists of a paroral or undulating membrane on the right of the buccal cavity and three membranelles, forming a "primitive" AZM, on the left.

Theca: Closely fitting envelope or covering secreted by the organism itself, composed principally of cellulose or some cellulose-like polysaccharide, with small openings or pores for emerging flagella. In some cases, the theca is impregnated with mineral salts, giving it additional rigidity; in armored dinoflagellates, the cellulose **Thecal Plates** are actually within the membrane-lined pellicular alveoli, underlain by sheets of microtubules. Numbers, shapes, and arrangements of these plates are of taxonomic importance. In certain dinoflagellates, plates are shed periodically by a process referred to as ecdysis by specialists on the group.

Thecal Plates: See **Theca**.

Theront: "Hunter" stage in the polymorphic life cycle of a number of histophagous and parasitic ciliates (e.g., ophryoglenine hymenostomes); a theront is essentially a developing, free-swimming **Tomite** searching for a new host or for a fresh source of food (which will permit further transformation into the **Trophont** stage).

Thigmotactic Ciliature, Thigmotactic Ventral Cilia, Thigmotactic Field: More or less specialized somatic cilia, generally in a patch, area, tuft, field, or zone, that have been modified to serve a tactile or, more commonly, an adhering function (see **Holdfast Organelle**); the area is often localized (e.g. on the anterodorsal surface of many thigmotrich ciliates); in some astomes and a few other ciliates, the surface of the body covered by these cilia may be concave and even called a **Sucker**; single, scattered cilia also may sometimes be classified as thigmotactic in nature (see **Bristles).**

Thorax: See **Cephalis**.

Toothplate: Extension of chamber wall of a foraminiferan test within the chamber lumen from the aperture to the margin of the preceding foramen.

Tomite: Fission product of a **Tomont**, a stage in the polymorphic life cycle of a number of parasitic (e.g., apostome) or histophagous (e.g., some hymenostome) ciliates; the organism is small, free-swimming, and nonfeeding; it has usually emerged with numerous congeners from a cyst (e.g., ophryoglenine hymenostomes) in which the tomont has rapidly divided; surviving tomites will become **Theronts, Trophonts**, or even **Phoronts,** depending on their taxonomic group.

Tomont: Prefission and dividing stage in the polymorphic life cycle of a number of parasitic or histophagous ciliates (e.g., apostomes and ophryoglenine hymenostomes). The tomont is a relatively large, nonfeeding form, typically encysting before starting to undergo (multiple) fission, which it usually carries out very rapidly to yield numerous **Tomites** (by tomitogenesis); a seldom used synonym is **Schizont**.

Toxicyst: Slender tubular organelle (an extrusome), rich in acid phosphatase, located subpellicularly in the body of many kinetofragminophoran ciliates, especially the predaceous, carnivorous haptorids (e.g. *Didinium* and *Dileptus*). Typically, this structurally complicated (inverted tube inside a capsule), toxic organelle is concentrated in great numbers at or near the apical end of the organism and in the oral area. On discharge, the toxicysts evert and penetrate, immobilize, and commence to cytolyze (kill) the prey with which it has made successful contact. Several other extrusomes are considered subtypes of the toxicyst by some authors: the curved cyrtocyst and the possibly nontoxic pexicyst (both in *Didinium,* along with toxicysts proper) and even the **Haptocyst** of suctorians (treated as an entirely separate organelle in this glossary); toxicysts have been reported from one flagellate (*Colponema*).

Trabecula: An unbranched column or pillar bridging a space as in the vertically aligned calcareous columns in some mycetozoans (e.g. *Badhamiopsis ainoae*).

Transverse Flagellum: The ribbon-like flagellum of dinoflagellates which winds to the cell's left and encircles the cell, usually in the cingulum. It has multiple waves, an axoneme running along its outer edge, and an inner fiberous contractile striated strand. It is the main propulsive organelle of the cell.

Transition Region: See **Flagellar Transition Region**.

Transitional Helix: See **Flagellar Transition Region**.

Transverse Microtubular Ribbon: Ribbon or band of microtubules occurring widely throughout the major ciliate taxa, always in association with a kinetosome, arising tangentially to that organelle at its left anterior side close to triplets number 3, 4, and sometimes 5; the ribbon, usually composed of four to six cross-linked microtubules, first extends upward toward the pellicle and then continues across to the left, parallel to the nearby surface of the body and at right angles to the kineties, stopping just short of the next line of kinetosomes on the left; extensions of transverse microtubules are involved in the composition of the cytopharyngeal apparatus known as the **Rhabdos**.

Triactinomyxon Actinospore: In Myxozoa an actinosporean stage, a triactinomyxon actinospore, that is infective to fish is in the oligochaete worm, *Tubifex tubifex* (Tubificidae).

Trichite: Word with two unrelated protozoological uses,: it is a superseded—but, unfortunately, still employed—term for the **Nematodesma**, a microtubular structure so prominent especially in the ciliate cytopharyngeal apparatus known as the **rhabdos**; and it is the name used for each of the unique skeletal structures, hollow, rod-like, and of a proteinaceous nature, arranged radially just below the pellicle in the posterior hemisphere of the body of certain oligotrich ciliates.

Trichocyst: Term limited here to include only the rather prominent spindle-shaped, nontoxic, paracrystalline, proteinaceous, explosive organelle (an extrusome) located subpellicularly in predictable positions within the body of quite a number of protozoa (hymenostome ciliates *Frontonia, Paramecium,* etc., dinoflagellates *Gymnodinium, Oxyrrhis, Peridinium,* etc.). In the mature resting stage, the 3-4 µm long organelle consists (e.g., in *Paramecium)* of an apical inverted, golf tee-shaped tip and a long fusiform, fibrous shaft with periodic cross-striations. On ejection, following an appropriate stimulus, it appears thread-like, with a length of 25-35 nm, and the characteristic periodic structure, increased by a factor of 8, shows division into subperiods. Spindle trichocysts are known to originate in the endoplasmic reticulum. The dinoflagellate organelle is sometimes separated out as an **Akontobolocyst**.

Triloculine: Miliolid foraminiferan with chambers one-half coil in length, added successively in three planes.

Triserial: Foraminiferan test with a high trochospiral, with three chambers in each whorl, that results in three columns of chambers.

Trochal Band: See **Locomotor Fringe;** synonym, telotrochal band.

Trochospiral: Foraminiferan test with chambers that are spirally coiled, evolute on one side and involute on the opposite side; both low and high trochospiraling occur.

Trophont: Mature, vegetative, feeding, adult stage in the life cycle. It is an interfissional form, a growing or trophic stage. The term is mostly used in reference to ciliated protozoa, for those types that have a preceding **Tomite** or **Theront** form and a succeeding **Tomont** stage, as in the polymorphic life cycles of various parasitic or histophagous apostome and hymenostome species. A synonym is **Trophozoite** (see below).

Trophozoite: Term with exactly the same meaning as that given above for **Trophont** (first two lines); but it is used by protozoologists primarily with reference to parasitic species of nonciliate taxa, especially to forms of direct biomedical importance; students of totally free-living groups of protozoa have essentially shunned both terms, using some entirely different general word, such as feeding stage, vegetative stage, mature form, or nondividing form.

True Capillitium: Thread-like structures present in the fruiting body of Myxogastria. Development occurs inside a system of preformed vacuoles which coalesce into channels, giving rise to threads that are of uniform diameter, often less than 6.0 µm. Examples include many taxa belonging to the endosporous orders Echinostellida, Physarida, Stemonitida, and Trichiida.

Trypomastigote: Developmental stage or morphological form of a trypanosomatid flagellate in which the flagellum arises posterior to the nucleus and emerges laterally to form a long undulating membrane as it runs along most of the length of the body to the anterior end; this is the classical and typical vertebrate bloodstream form—and thus the predominant form or stage—of all members of the genus *Trypanosoma* itself; superseded synonyms, trypanosomal and **Trypaniform Stage.**

Umbilical Side: Commonly the involute side of trochospirally coiled foraminiferan test, with chambers of the final whorl visible around the umbilicus.

Umbilicus: Foraminiferan term for the center (possibly depressed) of involute planispiral forms, or the **Umbilical** (involute) Side of trochospiral forms where all chambers of the last-formed whorl meet; it may be open or closed.

Undulating Membrane: Term with two meanings related to completely nonhomologous structures or organelles in the protozoa. 1) The undulating membrane in a number of flagellates is an extension or fold of the body surface membrane coalesced with the flagellar membrane so that the (axoneme of the flagellum is supported and attached to the organism's body by a thin fold (although sometimes for only a part of the length of the entire locomotory organelle), whether the flagellum itself is anteriorly directed (as in trypanosomes) or lying in a recurrent or posteriorly directed position (as in trichomonads). 2) For the many ciliates that possess a compound oral ciliary apparatus, the undulating membrane is the organelle typically lying on the right side of the buccal cavity—see **Paroral Membrane,** the preferred synonym today, for further description of the structure as it appears in ciliates.

Unilocular: Single-chambered foraminiferan test, not one divided by septa into separate chambers; the term is also used to describe tubular forms that arise from an initial proloculum; contrast with **Multilocular.**

Uniserial: Foraminiferan test growth pattern in which all chambers are in a single row or column.

Uroid: Posteriormost part of a moving "naked" lobose amoeba, often trailing sticky projections; by the presence of this structure, polarity of the organism can be determined; the area is active in pinocytosis in some amoebae and also may represent an area of defecation and expulsion of water from the water-expelling vesicle (contractile vacuole). When morphologically differentiated the shape of the uroid (e.g. bulbous, globular, morulate, villous, papillate, is a useful diagnostic feature as is the presence or absence of trailing uroidal filaments.

Valves: Major sections, usually halves, of a theca, test, or shell that are more or less recognizable as separate structures; for example, the **Shell Valves** of myxosporidian spores.

Valvogenic Cells: In myxosporidians, the cells in the sporoblast that ultimately produce the shell valves of the mature spore; see **Valves**.

Vector: Living carrier or transporter of a parasite) from a definitive host back to another individual of the same host species; if the parasite goes through a developmental stage in the carrier, the latter is known as a metacyclic or biological vector; if there is no change (including no multiplication) of the parasite in the carrier, it is called a mechanical or contamination vector (an outstanding example is the housefly); the metacyclic vector has also been considered an essential **Intermediate Host**; principal metacyclic vectors of medically important parasitic protozoa are (also) insects: mosquitoes, tsetse flies, sandflies, and triatomid bugs.

Vegetative Nuclei: The term used for the transcriptionally active nucleus (i) of heterokaryotic foraminifera. It usually is larger than the generative nuclei and found in young chambers. Little is known of the genetic organization of these nuclei, but they seem to be the functional equivalents of ciliate macronuclei.

Velum: Two meanings. 1) The curtain-like paroral cilia of scuticociliates. 2) The wing-like extension found in the leptodiscacean dinoflagellates.

Vesicular Nuclei: One of the two principal types of nuclei that are found in testate (arcellinid) amoebae. It has one, often central nucleolus, sometimes with a few additional very small nucleoli. The other type is an **Ovular Nucleus.**

Ventrolateral Flange: Motile cytoplasmic lip surrounding the ventral adhesive sucking disc of such diplomonad flagellates as *Giardia.*

Vestibulum, Vestibular Groove, Vestibular Cavity, Vestibular Ligule: Depression or invaginated area of the body of some ciliates, that is covered by extensions of somatic kineties (thus vestibular ciliature) and that leads directly into the oral cavity (e.g. *Paramecium*).

Visor: A structure found in arcellids, especially in drier habitats such as mosses and soil. The apertural apparatus has evolved to a slit-like structure often covered by the anterior lip (*Plagiopyxis*) or a **visor** (*Planhoogenraadia*).

Watchglass Organelle: Lenticular refractile structure, intensely Feulgen-positive, found in ophryoglenine hymenostome ciliates. It lies beneath the pellicle close to the left side, or in the left wall of, the buccal cavity of these histophagous organisms. Its function and origin are enigmatic.

Whorl: Single turn or volition of a coiled foraminiferan test through 360°.

***x*, *y*, and *z* kineties**: Oral polykineties of apostome ciliates.

X-group: Oral polykinetid anterior to oral polykinetid 2 in glaucomid ciliates.

Xenodemes: Natural populations of a particular taxonomic species appearing in different hosts; sometimes they may represent incipient subspecies or even species; other demes useful to parasitologists, especially those interested in ecology and epidemiology, include the following (which are generally self-explanatory): monodemes, nosodemes, serodemes, and topodeme.

Xenophyae: Most xenophyophores possess an agglutinated test composed of foreign particles, **xenophyae,** held together with an organic cement.

Xenosomes: Literally, "alien bodies," referring to minute bodies or structures being found in the cytoplasm and nucleus of protozoa; in the broadest sense, these are either endosymbiotic entities as the bacteria (Kappa and Omicron particles) of *Paramecium, Euplotes,* etc. and zoochlorellae, zooxanthellae, and cyanellae. The term **Xenosome** is also used for those foreign particles (sand granules, diatom frustules, etc.) agglutinated to the test of some testate amoebae, or similar adherent materials attached to loricae or thecae of other protozoa.

Xylophagous: Organisms which feed on plant material.

Zig-zag Kinetosomes: Kinetosomes, typically dikinetids, arranged so that they appear to zig-zag.

Zoochlorellae: Endosymbiotic green algae found widely in the cytoplasm of various protozoa, especially ciliates, amoebae, and foraminifera.

Zoospore: Motile asexual reproductive cell; a motile spore; a swarm cell or swarmer. This stage is found in some pigmented flagellates, mycetozoan groups (and their nonprotozoan protist relatives), and the actinopod sarcodinids. Its mode of locomotion may be by either flagella or pseudopodia; see **Swarmer**.

Zooxanthellae: Endosymbiotic brown or golden brown (dinoflagellates, diatoms, cryptomonads, or chrysomonads), found in the cytoplasm of various protozoa (radiolarians, foraminifera, and ciliates).

TAXONOMIC INDEX

A

B

C

D

E

F

G

H

I

J

K

L

M

N

O

P

Q

R

S

T

U

V

W

SUBJECT INDEX

C

D

E

L

M

N

O

P

R

S

T

U

V

W

X

Z

GROUPS OF PROTOZOA

CHANGING VIEWS OF PROTISTAN SYSTEMATICS (see page 2)

	Related Groups	Level	Common Name or Well Known Genera
Related to Multicellular Groups	Choanoflagellata	Class	Collared Flagellates
	Microsporida	Phylum	*Nosema, Abelspora*
	Myxozoa	Phylum	*Myxobolus*
	Volvocida	Order	*Volvox, Chlamydomonas*
Alveolates	Apicomplexa	Phylum	*Malaria, Toxoplasma gragarines*
	Colpodellidae	Family	*Copodella*
	Ciliophora	Phylum	*Paramecium, Tetrahymena*
	Dinozoa	Phylum	*Gymnodinium, Peridinium*
Stramenophiles	Bicoecids	Order	*Bicosoeca, Cafeteria*
	Chrysomonada	Class	*Ochromonas, Dinobrion*
	Pelagiophyceae	Class	*Pelagomonas, Aureococcus*
	Raphidomonadida	Order	*Heterosigma, Olisthodicus*
	residual heterotrophic stramenophiles	"Group"	*Bicosoeca, Cafeteria*
	Slopalinida	Order	*Opalina, Zelleriella*
	Synurophyceae	Class	*Synura, Mallomonas*
	Silicoflagellata	Class	Silicoflagellates, pedinellids
Amoeboid Protists	Acantharia	Class	*Acanthometra, Staurocantha*
	amoebae of uncertain affinities	"Group"	*Malamoeba, Endamoeba, Entamoeba*
	Arcellinida	Order	*Arcella, Difflugia*
	Heliozoa	Phylum	"Sun animaccules"
	Granuloreticulosea	Phylum	Foraminifera
	Mycetozoa	Phylum	Slime molds
	Phaeodaria	Class	*Aulacantha, Challengeria*
	Polycystina	Phylum	*Thalassicolla, Collozoum*
	ramicristate amoebae	"Group"	*Amoeba, Mayorella, Vexillifera*
	Schizocladidae	Family	*Schizocladus*
	testate amoebae with filopodia	"Group"	*Euglypha, Amphitrema*
	Trichosidae	Family	*Trichosphaerium*
	Xenophyophorea	Class	Xenophyophores
Amoeboid & Flagellated Protists	Pelobiontida	Order	*Pelomyxa, Mastigella*
	Heterolobosea	Class	*Vahlkampfia, Tetramitus*
Flagellated Protists	Cryptomonadida	Order	*Cryptomonas, Rhodomonas*
	Diplomonadida	Order	*Giardia, Hexamita*
	Euglenozoa	Phylum	*Euglena, Trypanosoma*
	Hemimastigophora	Phylum	*Spironema, Hemimastix*
	Oxymonads	Order	*Oxymonas, Pyrsonympha*
	Parabasalia	Phylum	*Trichomonas, Trichonympha*
	Pedinophyceae	Class	*Pedinomonas*
	Prasinophyceae	Class	*Tetraselmis, Mesostigma*
	Prymnesiida	Order	*Prymnesium, Coccolithus*
	residual heterotrophic flagellates	"Group"	*Reclinomonas, Jakoba*
	Retortamonadida	Order	*Retortamonas, Chilomastix*
Other	Haplospora	Phylum	*Haplosporidium, Minchinia*
	Plasmodiophora	Phylum	*Tetramyxa, Plasmodiophora*